35th Polymer Processing Society Annual Meeting 2019 (PPS-35)

Cesme-Izmir, Turkey
26 - 30 May 2019

ISBN: 978-1-7138-0200-6

Printed from e-media with permission by:

Curran Associates, Inc.
57 Morehouse Lane
Red Hook, NY 12571

Some format issues inherent in the e-media version may also appear in this print version.

Printed with permission by Curran Associates, Inc. (2020)

For permission requests, please contact Polymer Processing Society at the address below.

Polymer Processing Society
810 E. 10th St.
Lawrence, KS 66044
USA

Phone: +1-800-627-0326
+1-785-865-9403
Fax: +1-785-843-6153

pps@allenpress.com

Additional copies of this publication are available from:

Curran Associates, Inc.
57 Morehouse Lane
Red Hook, NY 12571 USA
Phone: 845-758-0400
Fax: 845-758-2633
Email: curran@proceedings.com
Web: www.proceedings.com

TABLE OF CONTENTS

INJECTION MOLDING

MIXING AND COMPOUNDING

MORPHOLOGY AND STRUCTURAL DEVELOPMENT

POLYMER BLENDS AND ALLOYS

POLYMER COMPOSITES

POLYMER FOAMS

POLYMER NANOCOMPOSITES AND NANOTECHNOLOGY

POLYMERIZATION AND SYNTHESIS

PROCESS MODELING AND SIMULATIONS

PROCESS MONITORING, CONTROL AND SENSORS

REACTIVE PROCESSING

RHEOLOGY AND RHEOMETRY

RUBBER AND ELASTOMERS

SPECIAL SYMPOSIUM: ADDITIVE MANUFACTURING (3D PRINTING)

SPECIAL SYMPOSIUM: HIGH PERFORMANCE POLYMERS

SPECIAL SYMPOSIUM: BIOPOLYMERS, ECO-FRIENDLY POLYMERS AND RECYCLING

SPECIAL SYMPOSIUM: SMART MATERIALS AND POLYMERS FOR RENEWABLE ENERGY

Ageing Behavior of Thermally Joined Hybrids of Laser Pre-treated Metal and Thermoplastic Polymers

Niclas Emrich (presenting author)[a], Stefan P. Meyer (reference author) [b]

[a] *Institute for Product Engineering, University Duisburg-Essen, Lotharstr. 1, 47057 Duisburg Continue Here*

[b] *Institute for Machine Tools and Industrial Management (iwb), Technical University Munich, Bolzmannstr. 15, 85748 Garching*

Abstract. Cost-efficient, load bearing and chemically resistant hybrid joints are necessary for future applications within the automotive industry. Thermal joining can be applied to create a direct connection between thermoplastic polymers and metal components and is thus seen as a promising prospect. Yet, the ageing resistance and seal tightness of joints remains uncertain. Within this work, a pulsed laser system (λ=1064 nm, f=20 kHz) is used to create microscopic random structures on aluminum and copper surfaces. The surface topology is analyzed using optical focus variation. The metal is then connected to different thermoplastic polymers. Polymers are selected with a particular focus on chemical resistance to electrolytes used in modern automotive battery systems. Shear strength and seal tightness for helium gas are measured before and after exposing the joints to humidity and heat. The cause of failure is investigated using REM- and EDX-analysis and by comparing observations from different ageing tests. It is demonstrated that thermoplastic elastomers, such as TPC and TPU, enable long-term resistant hybrid joints (6 N/mm^2), with no tendency of a decreasing shear strength.

Keywords: Polymer, Copper, Aluminum, TPE, Hybrid Joint
PACS: Replace this text with PACS numbers; choose from this list: http://www.aip.org/pacs/index.html

MOTIVATION

E-mobility is a major trend within the automotive industry. Following 100 years of research in drive trains powered by fossil fuels, we currently observe a massive shift towards electric drive trains. The electric drivetrain is not reliant on endless fossil fuels, it generates no emissions and can be integrated into urban mobility concepts and decentralized power grids.

The political framework in Germany is set. Average CO_2 emissions of 95 g/100 km by 2020 in the intensified WLTP test cycle, will only be achieved by an increased number of electrified vehicles. Currently the German target of 1 million electrified vehicles by 2020 is far out of reach. Only 0,054 million vehicles are registrated at the Federal Motor Transport Authority. From the customer's point of view, high acquisition cost are an inhibitor in comparison to conventional fossil fueled drive trains (Lienkamp 2012). When it comes to battery electric vehicles (BEV), the battery system is the dominant cost driver. With a share of 77 % of the cost for an electric drive train, a major impact on the vehicles performance, safety and range the battery system is considered as the key component in the technological shift towards electric drive trains.

Consequently, cost-down approaches are a major effort for engineers in the automotive industry. Innovative joining technologies, such as thermal joining are thus seen as a promising prospect. The process of thermal joining was researched and patented within a publicly funded Project NEXHOS (Next Generation Battery System). Thermal joining can be applied to create a direct connection between thermoplastic polymers and metal components. The direct connection enables construction with a minimum amount of parts, while achieving high load bearing stabilities up to 30 MPa (Heckert und Zaeh 2014). The decay of stability over time, was not the focus of research in the field of thermal joining so far. When it comes to thermal joints in the use case of battery systems for the automotive industry, very specific requirements are set. These include a chemical resistance to electrolytes, seal tightness for helium and preservation of the joints stability, when being exposed to humidity and heat.

This paper examines whether thermal joints are feasible for applications in the field battery systems for the automotive industry.

STATE OF THE ART

Laser-Pre-Treatment for thermal joining

Laser pretreatment is being used to improve the properties of thermally joint hybrids. Laser pretreatment is categorized as a physical surface pretreatment and aims to create a defined surface (Habenicht 2009). Compared to chemical or electrochemical surface pre-treatments laser structuring does not rely on toxic acids and is more cost efficient. (Rechner 2012), (Kurtovic 2014) and (Hose 2008) compared laser pretreatment to several other pertreatment methods, finding that laser pretreatment leads to superior mechanical strength in bonding. In the field of thermal joining the surface structuring via laser is currently the dominant method.

Within the field of laser pretreatment there are several ways to modify the surface. A categorization can be done by the size of a surface modification. Taking into account the proposition of (Heckert und Zaeh 2014) three categories for laser treatment with the purpose of thermal joining are defined.

TABLE 1. Categorization of laser surface modifications by its size.

Category	Size of structure	Description
1	< 100 nm	almost no modification in of the surface topology
2	100 nm – 200 µm	order of magnitude used in this paper
3	< 200 µm	larger structures, requiring shorter cycle times

Category 1 treatment is commonly used for surface cleaning. Native oxide layers, dirt or other remains are removed from the surface to promote the bonding properties. This correlation was observed in studies done by (Rechner 2012). In particular the native oxide layers are seen as a weak link between the metal substrate and polymer. Category 2 and 3 are most commonly used for thermally joined connections. Mechanical interlocking with the surface topology increases the bonding strength. Modifications with a size of 200 µm and more (category 3) can be recognized by eye. In this paper laser structures of category 2 are evaluated.

Beside the modification size, the type of modification determines the properties of a hybrid connection. As shown by (Demir 2013) there are numerous types of surface structures. These include line- or grid-shaped structures, dimpling, pin and random structures. Depending on the type of structure thermal joints show different properties. While deep line shaped structures promote the shear strength of thermal joints, they also tend to lower the media-barrier-properties of the hybrid connection. As shown by (Heckert et al. 2016) in particular thin line shaped modifications lead to the formation of air pockets at the bottom of a groove and act as an escape pat for helium gas. Structures for thermal joining that require gas-seal-tightness are therefore limited in its depth and shape.

Thermal Joining Process and qualification of thermal Joints

Thermal joining is a post mold joining process to create a direct connection between a laser-pretreated metal substrate and a thermoplastic polymer. The metal substrate is being heated to a temperature above the melting temperature of the thermoplastic polymer and then squeezed upon that polymer. The heat conduction within the boundary layer leads to a local melting of the polymer, which allows a deformation and adaption to the metal surface. The heating is interrupted and the polymer is quickly cooled to a temperature below its melting point. The polymer is interlocked with the surface topology.

There are several methods for heat generation within the metal substrate: heating elements, ultrasonic, induction or laser radiation (Ageorges und Ye op. 2002). In this paper ceramic heating elements are used to heat the metal substrate via heat conduction. For this particular experimental setup there are several parameters, such as joining pressure, cooling rate and joining temperature, influencing the joining strength. An analysis and optimization of these parameters was done by (Kornas 2017) and adopted in this paper.

The major evaluation-method used for a joints quality is shear lap testing. Other vital quality criteria, such as ageing behavior, gas seal tightens, or peel strength are not investigated in detail. In particular the ageing behavior is vital for most applications in the automotive industry. Polymer-wise, the research is focused on several variants of PA and some versions of PBT. These polymers are known for their positive adhesion properties, but they are limited when it comes to chemical resistance or cost-efficiency and are not investigated for joints with media-barrier properties.

The research done by (Heckert et al. 2016) is one of few, regarding seal tightness and ageing behavior. (Heckert et al. 2016) shows that there are limitations in the structure depth for joints created with the polymer PP. Yet, other polymers are not considered. The laser-nano-structure (category 2) recommended for seal thigh applications is adopted in this paper. Based on this treatment-method the properties of joints with several additional polymers is evaluated.

EXPERIMENTAL SETUP

Laser Pre Treatment

For the laser pre-treatment a pulsed single mode Yb-fibre laser was used. The pulse is generated with a frequency of 20 kHz, at a mean power output of 20 W. Each pulse is focused with a diameter of 50 um onto the metal substrate, while surrounded by ambient atmosphere. The pulses are arranged in a rectangular grid pattern as shown in FIGURE 1

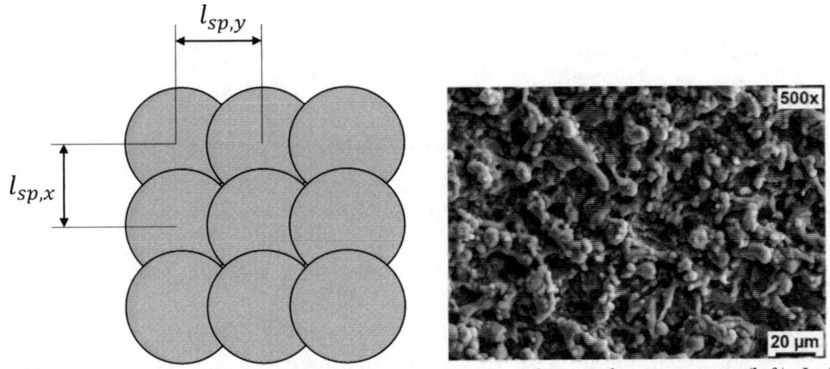

FIGURE 1. Pattern of laser pulse spots on a metal surface to create a random surface structure (left). Indeterministic surface structure after application of laser spot pattern (right).

The distance between to pulse spots $l_{sp,x}$ and $l_{sp,y}$ is varied with equal proportions to x- and y-direction. The random structures are then evaluated following the approach of (Heckert et al. 2016) by comparing structures in regard to their cumulative laser Energy E_{cum} that is applied upon the metal surface. The cumulated energy is defined by the spot distance $l_{sp,x}$ and $l_{sp,y}$ and the number of repetition in which the defined spot pattern is irradiated upon the metal surface.

$$E_{cum} = \frac{4 \cdot E_p}{\pi \cdot l_{sp,x} \cdot l_{sp,y}} \cdot n_{rep} \tag{1}$$

In this paper the following parameter set was used to create laser nano structures:

TABLE 2. Laser parameters used for surface treatment.

Wave length of laser	Spot distance in x- and y-direction $l_{sp,x}$ and $l_{sp,y}$	Cumulated energy input E_{cum}	Mean power output	Pulse frequency	Number of repetitions n_{rep}
1064 nm	20 µm	1000 J/cm^2	20 W	20 kHz	4

Thermal Joining Process and Polymers

The influence of different polymer types on the ageing resistance and seal tightness of a connection is evaluated. Each polymer is selected regarding its suitability for applications in automotive battery systems. The polymers are also selected with particular “property sets” that might lead to an improvement in ageing resistance and seal tightness.

TABLE 3. Polymers used for thermal joining with copper and aluminum.

Polymer		PP-H	PA 66 GF 30	TPC	TPU	PVD F	PA-PP Blend	POM
Melting Temperature	°C	170	254	212	ca 200	171	222	166
lin. Thermal expansion	$\cdot 10^{-6}\ /K$	160	50	220	-	160	-	130
Shore D	-	-	-	45	48	-	-	-
Young´s Modulus	MPa	1400	5500	95	-	2200	1800	2800
Density	g/cm^3	0.9	1.34	1.15	1.15	1.78	1.01	1.41

For each polymer, shear lap joints to copper and aluminum are produced. To get the optimal strength in each polymer-metal combination a parameter study was done, evaluating the influence of the metal surface temperature on the joint stability and the loss in strength due to ageing. It is shown, that there is no further increase in joint strength once the surface temperature reaches a certain limit. This temperature limit is recognizable by a rapid increase in the melting path. The melting path can be measured as the distance between the solid upper part of the polymer and the metal surface. For every polymer metal combination, the velocity of the melting path increase was measured. The joining temperature was adapted accordingly, so that each combination was produced with a similar velocity in of the melting path.

Test Specimens, Ageing Test and Helium Leak Test

The dimensions of the shear lap testing specimen (FIGURE 2) were adapted, originating from DIN EN 1465. The polymer dimensions are 100 mm x 28 mm x 4-6mm. The dimension of the metal component was enlarged in its length and width to 140 mm x 30 mm x 2 mm. Both modifications were done to adapt the specimen to the thermal joining facility and for comparison to previous work done by (Kornas 2017).

The sealing test specimen was manufactured with the most promising polymer, resulting from the spear lap test. The specimen represents a combined sealing test for thermally joined connections to copper and to aluminum. Both are needed simultaneously in applications for automotive battery systems. The joint dimensions are according to FIGURE 2. The reservoir is filled with helium while the specimen is positioned in a vacuum chamber. Helium that is able to escape through the joining layer between metal and polymer, is measured by a mass spectrometer, which is connected to the vacuum chamber. Any helium leakage test specimen, with a leakage rate of below $q_l < 10^{-5}\ mbar * l/s$ is considered as helium seal thigh.

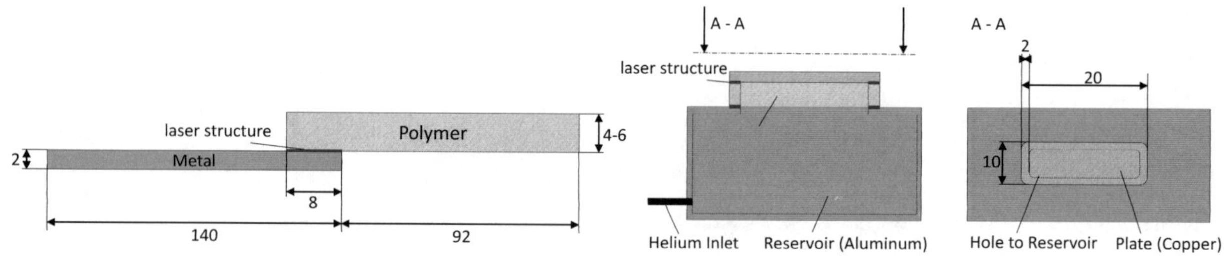

FIGURE 2. Shear lap joint (left) and helium sealing test specimen (right).

To simulate an ageing behavior of 10 years, a standard ageing test for bonding applications in the automotive sector was performed. The ageing test consists of temperature and humidity variation under ambient atmosphere conditions. The temperature range is from -30 °C to 90 °C and the humidity from 0 % to 80 %. The cycle is shown in (FIGURE 3) and is repeated for 20 times, leading to a total duration of 240 h.

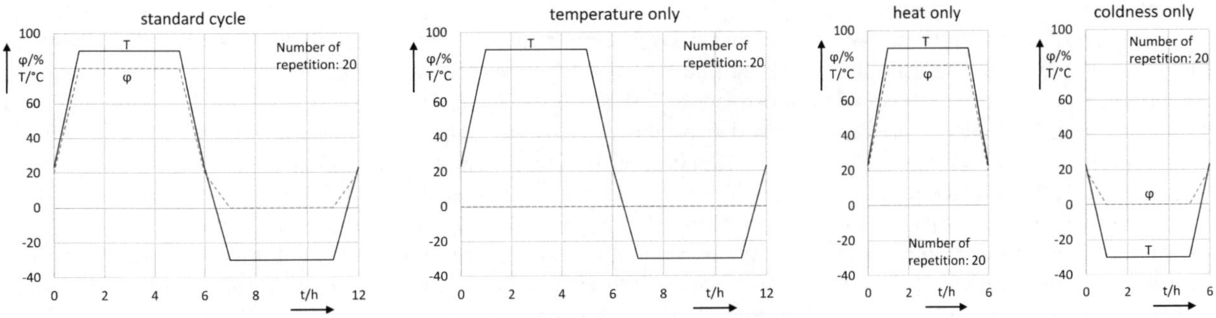

FIGURE 3. Standard ageing test and modified ageing tests for bonding applications in the automotive industry.

To examine the influence of the different components (heat, coldness, humidity) of the ageing test, three modified ageing tests were performed, each consisting of heat only, coldness only and one test without humidity. The modified ageing tests were repeated with the same number of repetitions as the standard cycle.

For surface analysis, the fracture pattern was analyzed optically and by using REM and EDX analysis. To evaluate the influence of oxide layers on nanostructures, the surface of the pure metal was analyzed by REM. Specimen without nanostructures were aged, to evaluate the influence of the native metal ageing condition upon the laser structure. In addition, metal specimens with laser structures were aged to evaluate the ageing behavior of metal surfaces after laser treatment. Changes in the surface topology due to ageing were observed by REM. The metal specimen with ageing before laser treatment and with ageing after treatment where then thermally joined. Their shear strength was compared to the strength obtained from joints with a metal surface that had not been exposed to the ageing test.

RESULTS AND DISCUSSION

The shear strength was analyzed for each combination of polymer and metal, before and after being exposed to the standard ageing test. The results are demonstrated in FIGURE 4. Connections with PA66 GF30 serve as a reference to other research, confirming the effectiveness of the thermal joining process. Different thickness and elastic modulus of the polymer lead to varied stress conditions in the boundary layer. Comparison between joints with different polymers is therefore only partially possible.

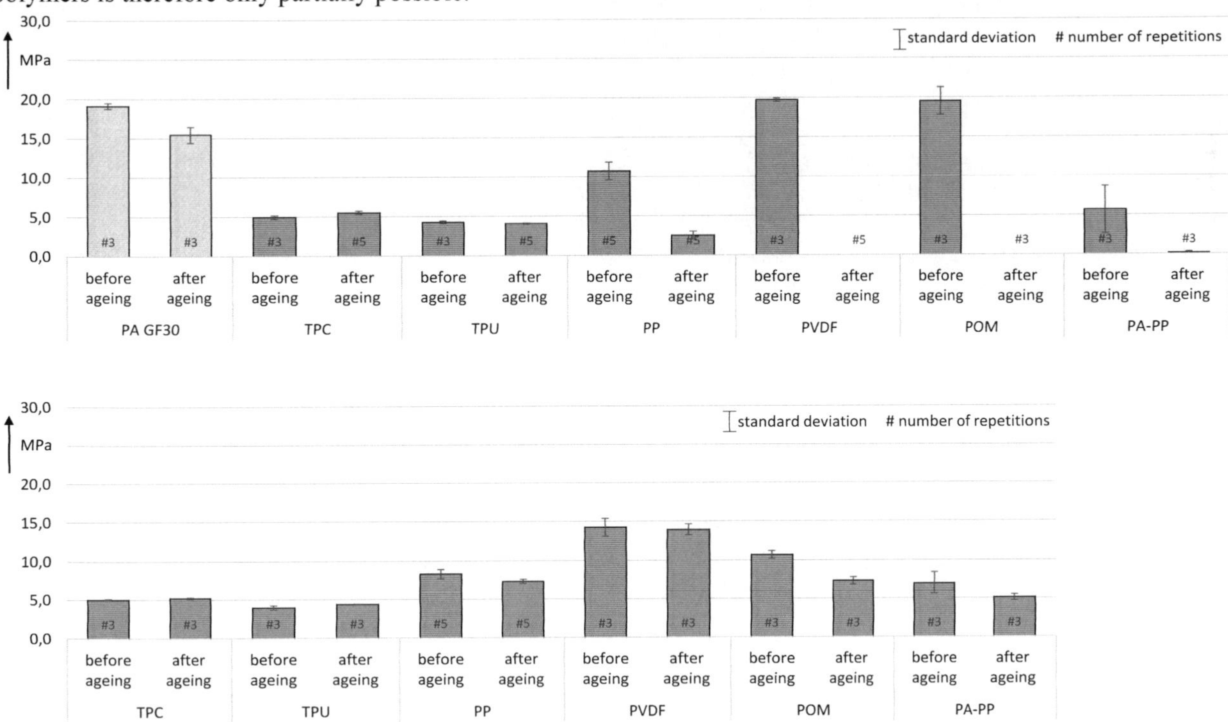

FIGURE 4. Shear lap testing of varied polymer-metal-connections before and after ageing

Most connections show a reduction in strength after ageing. While joints with aluminum show no or few rates of loss in strength, copper joints are usually decreasing rapidly in strength. For the connections with PVDF, POM and PA-PP, this will lead to a total failure, meaning that the strength is reduced to 0 MPa. Joints with PA GF 30 and PP show a loss in strength of 20 % and 75 % respectively. If this behavior is extrapolated, none of the joints with these polymers are suitable for long term use. Using thermoplastic elastomer polymers, joints show no decrease in strength. The absolute amount of strength is lower. While the other polymers maintain their shape, thermoplastic elastomers show an elongation behavior up to +100 % in the shear test. This also leads to a deformation of the polymer in the joining area, which causes different stress conditions. A comparison between the absolute amounts of shear strength of joints with thermoplastic elastomers and the other polymers, is therefore not possible. Joints with aluminum show far less loss in joint strength. Joints with PP drop about 12 % and joints with PVDF only 2 %. Joints with thermoplastic elastomers remain uncritical, same as for copper. A decrease is only observed for POM and PA-PP.

The surface topology of the metal substrate was analyzed using REM and EDX analysis (FIGURE 5). Copper and aluminum surfaces where treated with the same amount of cumulated energy. Both surface structures show a random pattern of solidified melt droplets. On top of the surface topology, a fuzzy oxide layer is recognizable. The fuzzy appearance can be traced back to a porosity of the oxide layer, which is recognizable by cross section analysis. Some local gaps are recognizable in the oxide layer. While aluminum specimens (1) show bigger and vertically protruding drops, the copper surface (2) shows less surface roughness (S_a of 1.4 µm compared to 4.2 µm) with smaller droplets. This might be an explanation for the different absolute strength values obtained (e.g. joint with PP or PVDF). Less interlinkage of the polymer on copper surface may be the reason for lower ageing resistance too. Especially for copper surfaces it is noticeable that the fuzzy oxide layer is reducing its thickness and gains in compactness after ageing. Aluminum surfaces show a similar behavior, but less intense. Joints consisting of the aged surface and PP show a 21 % decay in shear strength for aluminum and 45 % loss for copper. The compaction of the oxide layer is reducing the joint strength. This behavior is intensified for copper surfaces. A similar correlation was observed by (Rechner 2012) for joints with aluminum and epoxy bonding.

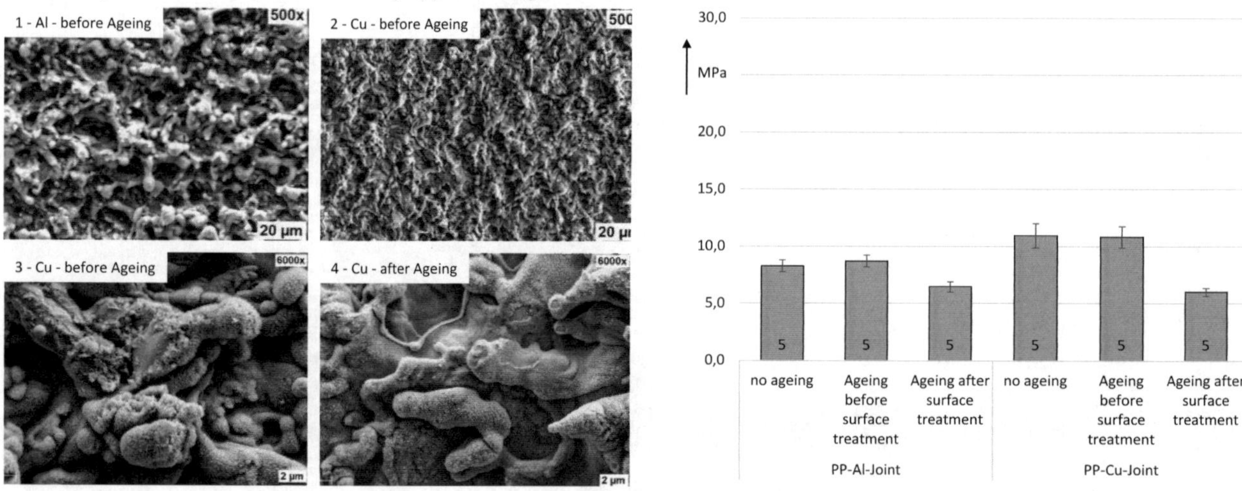

FIGURE 5. REM surface analysis of non-joined metal surfaces after being treated by laser (left). Shear strength obtained with different ageing conditions of the metal surface (right).

By using REM and EDX Mapping on the surface of fracture areas it is also recognizable that the major proportion oxide layer remains on the metal substrate. The interface between oxide layer and metal base material is therefore not the weakest link.

To analyze the impact of cold, hot, and dry phases of the ageing test, modified test cycles where performed on joints between copper and PP, PVDF and POM. The impact of each modified cycle on joints with PVDF and POM is so high, that a total loss of strength is observed, enabling no further insights. Joints with PP are at the tipping point, so that the influence of different ageing cycles is visible.

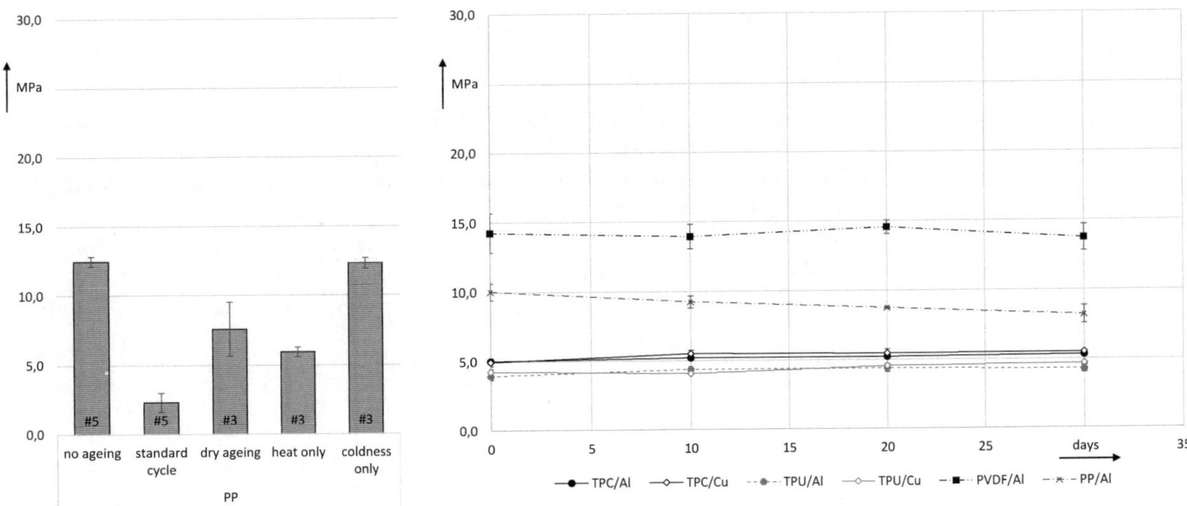

FIGURE 6. Shear strength after ageing with modified ageing tests (left) and long term ageing behavior (right).

It can be observed that the cold part of the cycle hast no effect. This is only valid for PP-connections, as other polymer-copper connections respond critically. Both, the cycle with no humidity and the cycle with heat only lead to a reduction of shear strength of around 45 %. It is shown that coldness-only shows no impact, but when combined with heat-only (i.e. the standard cycle) it leads to a drop in strength of 80 % in strength. This draws to the conclusion that the humidity applied during the heat-cycle has an effect on the cold cycle which follows after the heat cycle. To influence the ageing behavior, the humidity must intrude (at least partially) into the boundary layer (Emrich 2003). This is also supported by the REM and EDX analysis of different fracture patterns.

EDX mappings of fracture areas on joints with TPC (representative for an ageing-resistant connection) show that about 50 % of the metal-surface is covered with remaining polymer. On the other hand, fracture areas of joints with PVDF (representative for non-ageing-resistant joints) show few to no remains of polymer on the metal surface. Connections between copper and PP show a combination of these two observations, locally separated in two areas, one with polymer remains and one without. The impairment of the joint seem to be initiated from the border towards the center of the joining area. The influence of humidity seems plausible to explain this behavior.

Furthermore, the long term behavior was analyzed for the two polymers with the most promising ageing behavior (TPC and TPU), as well as the two most promising polymers for joints with aluminum. All connections with thermoplastic elastomers show no loss in strength for the applied conditions of the standard test cycle. This is valid for three repetitions of the standard test procedure (total amount of ageing time = 720 h), simulating 30 years. Previous research done by (Kornas 2017) showed that connections to copper are dropping in shear strength under similar ageing conditions. While the polymers PVDF, POM, PA-PP and PA66 GF30 maintain similar a behavior, joints with thermoplastic elastomers seem to be unaffected those ageing conditions. The most obvious assumption to explain this observation, is the thermal expansion behavior as root cause of failure on copper surfaces. All joints that show low or no loss in strength, possess either a rate of thermal expansion that is close to the one of copper (PA GF30) or possess an elastic behavior, so that displacements induced by thermal expansion can be compensated.

To test the suitability of thermally joined hybrids for application with sealing function, a number of 15 specimens were tested on helium leakage. None of the specimens showed a helium leakage value of above $1 \cdot 10^{-6}\, mbar \cdot l/s$. All specimens can be considered as seal thigh (leakage rate < $10^{-5}\, mbar \cdot l/s$). In addition, 5 specimens were tested after an ageing duration of 30 days (similar to the long term ageing), showing equal results.

CONCLUSIONS

- The oxide layer, covering the surface topology of nanostructures is decreasing in thickness and increasing in compactness, when being exposed to heat and humidity. This behavior is more intense for copper than for aluminum. The deformation leads to a loss in bonding strength.
- The polymers TPC and TPU enable connections to copper and aluminum. These joints are not influenced by standard ageing conditions for automotive applications.
- The positive ageing behavior is most likely due to the elastic deformation behavior (in perspective to thermoplastic materials) of thermoplastic elastomers.
- TPC enables long term helium seal thigh joint that can be used for applications with sealing function.

REFERENCES

Ageorges, Christophe; Ye, Lin (op. 2002): Fusion bonding of polymer composites. [from basic mechanisms to process optimisation]. London, New York: Springer (Engineering materials and processes).

Demir, et al (2013): Fibre laser texturing for surface functionalization. In: *Physics Procedia* (41), S. 759–768.

Emrich (2003): Untersuchungen zum Einfluss der Oberflächenchemie und -morphologie auf die Langzeitbeständigkeit geklebter Aluminiumverbunde.

Habenicht (2009): Kleben. Berlin, Heidelberg: Springer Berlin Heidelberg.

Heckert, André; Singer, Christian; Zaeh, Michael F.; Daub, Ruediger; Zeilinger, Tobias (2016): Gas-tight Thermally Joined Metal-thermoplastic Connections by Pulsed Laser Surface Pre-treatment. In: *Physics Procedia* 83, S. 1083–1093. DOI: 10.1016/j.phpro.2016.08.114.

Heckert, André; Zaeh, Michael F. (2014): Laser Surface Pre-treatment of Aluminium for Hybrid Joints with Glass Fibre Reinforced Thermoplastics. In: *Physics Procedia* 56, S. 1171–1181. DOI: 10.1016/j.phpro.2014.08.032.

Hose, Ralf (2008): Laseroberflächenvorbehandlung zur Verbesserung der Adhäsion und Alterungsbeständigkeit von Aluminiumklebungen. Dissertation.

Kornas (2017): Wärmeleitungsfügen von Thermoplast-Metall-Verbindungen für prismatische Batteriezellen. Abschlussarbeit.

Kurtovic (2014): Laserinduzierte Nanostrukurierung von Titanoberflächen für das strukturelle Kleben. Universität Paderborn.

Lienkamp, Markus (2012): Elektromobilität. Berlin, Heidelberg: Springer Berlin Heidelberg.

Rechner, Rico (2012): Laseroberflächenvorbehandlung von Aluminium zur Optimierung der Oxideigenschaften für das strukturelle Kleben. Dissertation. TU Dresden, Dresden. Fakultät Maschinenwesen.

Investigations On The Influence Of The Extrusion Process On The Foam Structure Of Mechanical Supporting Phenolic Sheets

Daniel Methe and Michael Gehde

University of Technology Chemnitz, Institute of Materials Handling, Conveying and Plastics Engineering, Professorship of Plastics Engineering, Reichenhainer Straße 70, 09126 Chemnitz, Germany

Abstract. The foam extrusion of thermoplastic resin is a well known and established manufacturing process for insulation materials. On the other hand, the continuous manufacturing of insulation materials made of thermosets is limited to liquid thermoset systems, which exhibit very good insulation properties but poor mechanical characteristics. Because of the rising demand to lower CO_2 emissions, the prospect of new insulation material with mechanical supporting characteristics to reduce thermal bridges in construction leads to new approaches in material, process and part development. After the underlying extrusion process development to unsure stable processing conditions over longer periods in regards to the temperature and time induced curing of the phenolic resin was developed, several variations of processing conditions were executed. For this research, we investigated the processing influences of the extrusion on the properties of the manufactured parts. Especially the overall size of the pores and the formation of foam gradient over the part thickness were examined and compared. The results show that pore structures and in return part properties such as density and mechanical properties are vastly influenced by the processing conditions.

Keywords: Foam Extrusion, Phenolic Resin, Thermoset, Free Flowing, chemical blowing agent, supporting insulation.

INTRODUCTION

Compact components made of free flowing phenolic molding compounds show very good static mechanical behavior in applications in the automotive sector with high thermal load [1]. Furthermore, insulating phenolic foams possess with 0,018-0,035 W/(m*K) good insulation properties [2]. Those attributes can be combined to create supporting insulation parts to open up potential new markets like the civil engineering sector. The development of a continuous manufacturing process such as the extrusion is very challenging because of the complex flow-cure behavior of the phenolic molding compound [3,4]. Especially a potential clogging of the extrusion die can disrupt and prevent stable processing windows. With a new approach to the die design, a stable processing and foaming of the phenolic molding compound is possible. While exiting the die, the molding compound foams to a semi stable extrudate that can be further processed to cured foamed sheets. The processing to foamed sheets can be done semi-continuously via a stationary calibration press or full-continuously via a belt puller. The processing conditions such as shearing rate, dwell time and melt temperature have a big influence on the degree of cure of the phenolic molding compound at the exit of the die. Consequently, the degree of cure influences the foam formation because of its dependency on the extensional viscosity. By varying the process parameters such as screw design, screw speed and throughput, the process conditions can be regulated/influenced and therefore the degree of cure of the molding compound while leaving the die. The aim of this study is to investigate the influence of the screw speed on the foam formation, density and the resulting bending properties of the foamed sheets for the semi-continuous molding via stationary calibration press.

EXPERIMENTAL SETUP

For the process development, we used the experimental free flowing phenolic molding compound "FoamSet 18135" provided by Hexion GmbH. The resin-hardener-system consists of novolac (phenolic resin) and hexamethylenetetramine (hardener). Curing starts with the beginning of the hydrolysis of hexamethylenetetramine between 110 and 120°C to formaldehyde and amino methylated products [5,6]. The molding compound also consists of endothermic chemical blowing agents and glass fibers as fillers. The chemical agent was chosen by its decomposition temperature to complement the curing window of the molding compound. As processing setup the co-rotating twin screw extruder ZE 25x32D from KraussMaffei Berstorff with a circular orifice was used. To manufacture the foamed sheets with defined thickness a stationary calibration press with heatable plates was constructed that allowed to generate sheets up to 450 mm long with variable width and thickness. Figure 1 shows the process steps. In difference to the extrusion of foamed thermoplastics sheets, the chemical blowing agent decomposes only partially while the thermoset melt strand leaves the die. Further decomposition of the blowing agent occurs while shaping and curing at 180°C in the calibration press. In Table 1 the chosen processing parameters are presented. To investigate the influence of the extrusion process on the foam structure the screw speed was varied.

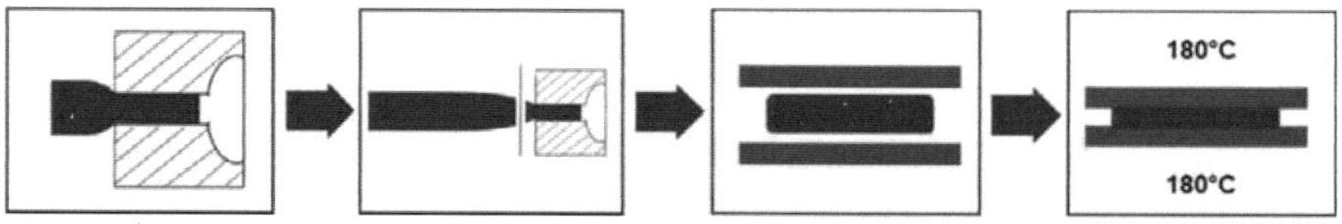

FIGURE 1. Manufacturing steps of foamed phenolic sheets: a) Extrusion of foamed semi stable strand, b) portioning of the extruded strand, c) insertion into calibration press, d) Calibrating to predefined thickness and curing

TABLE 1. Processing parameters

Extruder				Unit
Material throughput		10		kg/h
Screw speed	100	150	200	rpm
Haul-Off				
Haul-Off speed		65		mm/min
Average cooling time at room temperature		330		s
Calibration press				
Press temperature		180		°C
Curing time		450		s

For this study foamed sheets with a thickness of 25 mm were manufactured. For the analysis, test specimen with the dimensions of 140x25x25 mm were prepared. The thickness and width possess a tolerance range of ±0,5 mm. Before the specimen were mechanical tested, their density were determined according to DIN EN 1602. In order to investigate if the specimen exhibit anisotropic behavior, bending tests in calibration direction and perpendicular were conducted. The specimens were tested with the universal testing device Zwick/Roell Z250 with a bearing distance of 125 mm, a shoulder diameter of 30 mm and a testing speed of 10 mm/min. The samples for the optical analysis were prepared out of center between the bending specimens to get a representation of the foam structure at the point of maximal stress in perpendicular direction. The images along the sheet thickness were taken with the reflected-light microscope Olympus BX51. Afterwards the pore structure analyses were carried out with the image processing tool ImageJ.

RESULTS AND DISCUSSION

The results of the density analysis reveal that an increase in screw speed from 100 up to 200 rpm causes also an increase in density of the foamed sheets. Figure 2 shows the density over screw speed. Whereas at 100 rpm the density lies around 0,79 g/cm³, it rises gradually to 0,92 g/cm³. Compared to the compact material the foamed specimen exhibit an overall weight reduction ranging from 46 (200 rpm) to 53,5 % (100 rpm). The increase in density can be explained with temperature and time depending curing of the phenolic resin. Through the increase in screw speed the shear rate and in turn the melt temperature increases well above 120°C. Therefore, the blowing agent and the hardener decompose simultaneously and the phenolic resin begins to chemical bond to polymeric networks, which exhibit higher viscosity and melt stability. As a result, the foaming is more constricted when the melt strand is consolidated inside the calibration press. Also more decomposition gases might escape the melt strand and decreases the foaming. In strong contrast to thermoplastic foam extrusion, thermoset foam extrusion adds the dimension of time depending molecular size and in turn viscosity. Therefore, different temperature and dwell time regiments lead to different foaming behavior in the calibration press, which are further examined in the pore structure analysis later on.

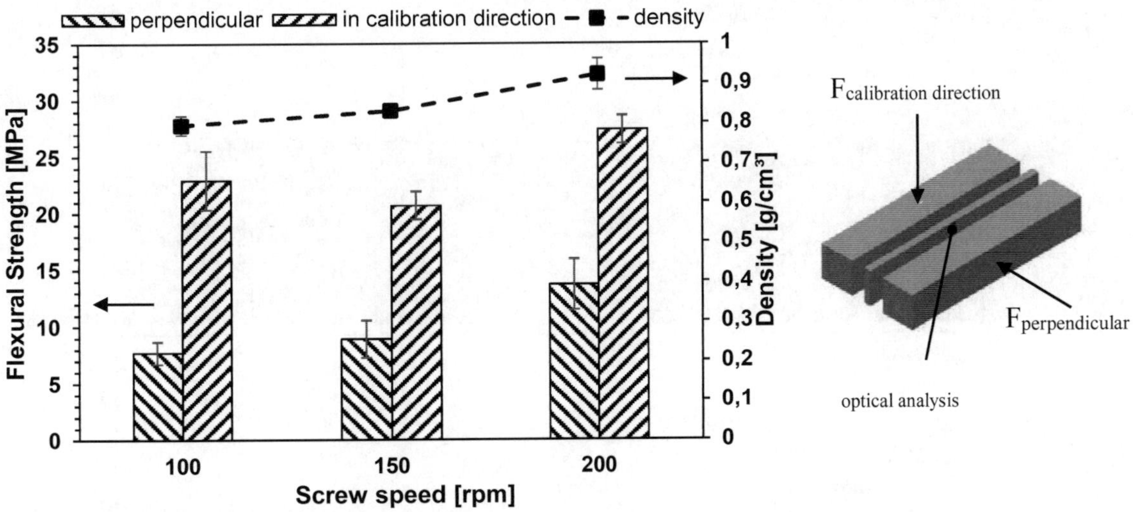

FIGURE 2. Density and flexural strength in calibration direction and perpendicular over screw speed (right); specimen preparation (right)

Consequently, the bending properties are also influenced by the screw speed. However, contrary to the gradually increase of density with increasing screw speed the flexural strength in calibration direction decreases slightly with a screw speed of 150 rpm and increases significantly with a screw speed of 200 rpm. The achieved flexural strengths range from 20,5 MPa up to 27,5 MPa. Figure 2 depicts the flexural strength over screw speed in calibration direction and perpendicular. In contrast to the flexural strength in calibration direction, the flexural strength in perpendicular direction lies significantly lower. Therefore, anisotropic behavior of the foamed parts is shown, which can be expected to occur at other mechanical properties such as compression and shear strength. Two reasons may be responsible for the anisotropic behavior. Through the mechanical preparation such as sawing and grinding, the surfaces in perpendicular testing direction possess an open pore structure, which are structural notches. The other reason for the reduced flexural strength lies in directional foaming vertical to the calibration direction, which is caused by pressing the extruded strand into the desired dimension. Furthermore, the calibration press acts like a geometric resistance. Therefore, a geometric expansion is only possible perpendicular. The flexural moduli follow the trend of flexural strengths and they are ranging from 1300 MPa (100 rpm, perpendicular) up to 2050 MPa (200 rpm, in calibration direction). When comparing the anisotropic factors

$$\frac{\sigma_{fM}}{\sigma_{fM,perpendicular}} \tag{1}$$

for different screw speeds it shows that the increase in screw speed leads to a decrease of the anisotropic flexural behavior. Whereas the anisotropic factor is 3,3 at 100 rpm, it sinks to 2,4 at 150 rpm and to 2,2 at 200 rpm.

To interpret the results of the flexural test further, it is necessary to analyze the foam structure via microscopic imagery. Figure 3 shows the foam structures and the pore ratio is represented by the mean grey value at different screw speeds. The phenolic specimens exhibit continuous foaming without a skin(compact)-core(foamed) structure over the sheet thickness. [7,8] have shown that thermoplastic foams made by injection molding exhibit integral foam structures where the degree of foam is lowest at the surface and highest in the center. Consequently, the injection molded foamed thermoplastics exhibit good flexural properties because the highest stress and elongation occurs at the compact surface while testing [9]. The surface of all shown phenolic sheets with maximal outer fiber load (left side figure 3) consists of closed cells with diameters up to 800 μm. When comparing the influence of the screw speed on the foam structure, it is observable that 150 rpm screw speed produces a more macropore structures then 100 rpm and 200 rpm. This explains the decrease in flexural strength at 150 rpm seen in figure 2. The coarsely pore structure increases the notch effect at the outer fiber. The formation of pores at the surface originates from the process sequence. Opposite to the thermoplastic foam extrusion, thermoset foams are usually being plasticized at temperatures around 100 °C

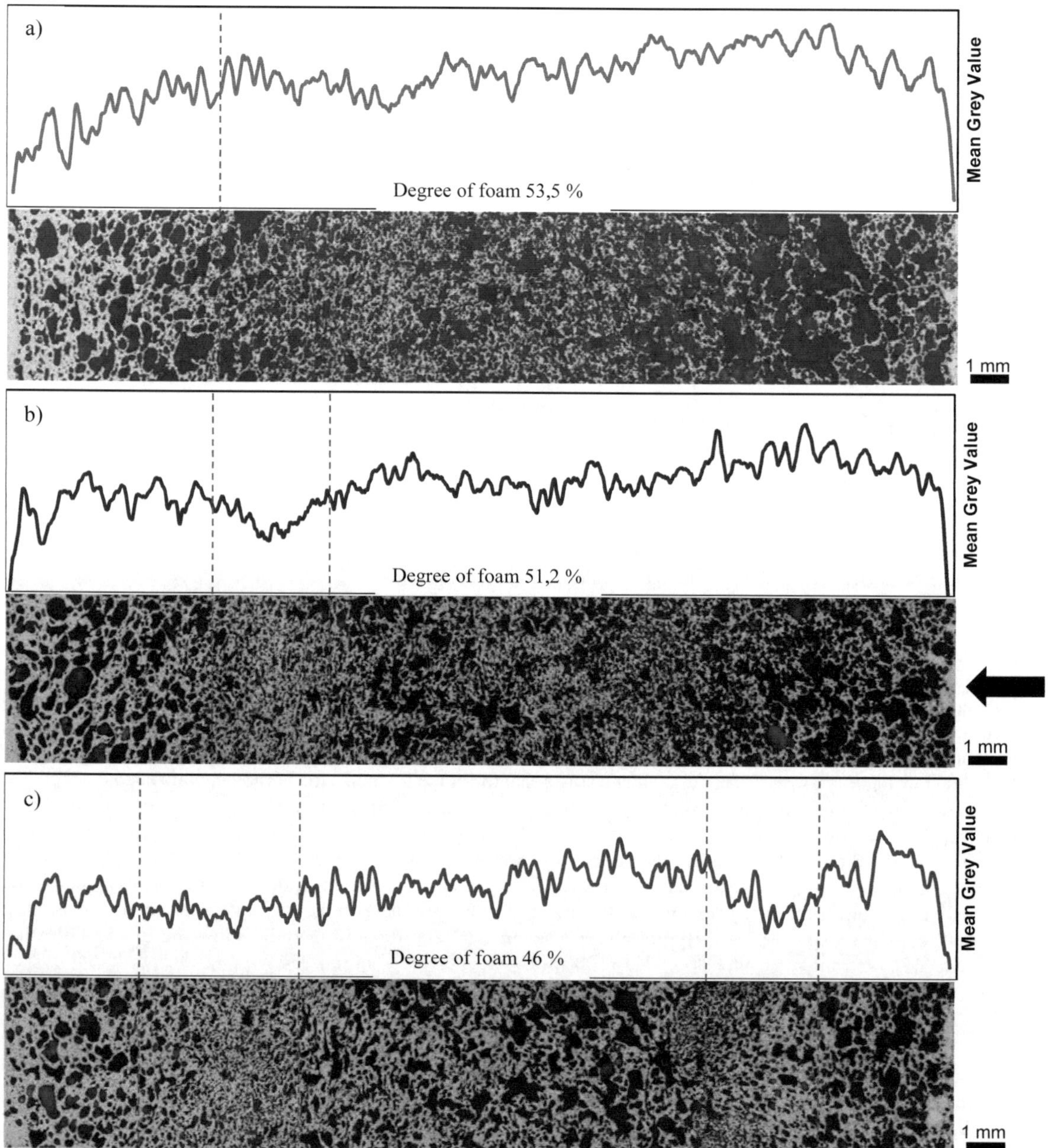

FIGURE 3. Microscopy and mean grey value over sheet thickness for a) 100 rpm, b) 150 rpm, c) 200 rpm; the arrow represents the testing direction in calibration direction; the dashed red lines symbolize distinguishable boundaries between different foam structures

and afterwards molded and cured at temperatures over 160 °C. As a consequence, the melt at the surface experiences a decrease in melt viscosity before the time-delayed crosslinking increases the viscosity again. At the same time, the rate of decomposition of the blowing agent increases rapidly. Therefore, coarsely pore structures are expected in the skin area. To counteract the formation of coarsely pore structure the degree of pre crosslinking could be increased before the final molding to reduce the drop in viscosity and increase melt stability while being heated up. The comparison of the overall foam structures support that. With increasing screw speed the foam structure changes from an open cell structure more and more to a close cell structure. Furthermore, the generated foam structure at 200 rpm screw speed shows five distinguishable areas where the pore sizes and the degree of foam vary. How the area form is also strongly influenced by the self-insulating effect that causes a different foam and cure behavior over the sheet sickness while being molded. In figure 3c) two areas with a fine pore structure can be identified, which are favorable for the mechanical properties. The position of those areas also explain the decrease in the anisotropic factor of the flexural tests. They serve as a reinforcement for the outer fiber in perpendicular direction whereas the more coarsely pore structures at the surfaces in calibration direction determine the flexural properties in that direction.

CONCLUSION

With the variation of the screw speed, the influence of the extrusion process on the final part properties could be shown. With increasing screw speed the density rises steadily, whereas flexural strength does not increases steadily. The best flexural properties could be achieved with a screw speed of 200 rpm. The perpendicular flexural properties lie well below the flexural properties in calibration direction. The foam structure itself is also influenced by the screw speed. At lower screw speeds the pore structure shows a higher ratio of open-cell formations. With increasing screw speed the ratio of open-cell foam structure sinks and distinct foam layers form. The combining results lead to the conclusion that the pre crosslinking is detrimental to the formation of pores. Because of the shear sensitivity of the phenolic molding compound, small differences in screw speed and design lead to a different pre crosslinking progress and in turn to different extensional viscosity. However, with increasing shearing and crosslinking inside the extruder and die, the process stability is a limiting factor.

Further studies encompass the quantification of other influences on the formation of pore structures and optimization of processing conditions. The aim lies in generating an integral foam with optimal mechanical to insulation ratio. The gained process comprehension will be transferred to a full-continuously extrusion process via a belt puller. Additionally, the die design will be further improved to expand the processing window to higher processing temperatures.

ACKNOWLEDGMENTS

The authors acknowledge the financial support by the Federal Ministry for Economic Affairs of Germany in the project FoamSet (project number 03ET1424A) supervised by the project executing organization Juelich. The responsibility with regard to contents lies with the authors.

REFERENCES

1. K. Koizumi, T. Charles and H. De Keyser, Phenolic Molding Compounds, In: L. Pilato, Phenolic Resins: A Century of Progress, Berlin, Springer-Verlag, 2010.
2. A. Gardziella, L. A. Pilato and A. Knob, Phenolic Resins: Chemistry, Applications, Standardization, Safety and Ecology, Berlin, Springer-Verlag, 2000.
3. M. R. Kamal and S. Sourour, Kinetics and Thermal Characterization of Thermoset Cure, POLYMER ENGINEERING AND SCIENCE, January, Vol. 13, No. 1, 59-64 (1973).
4. M. B. Roller, Rheology of Curing Thermosets: A Review, POLYMER ENGINEERING AND SCIENCE, March, Vol. 26, No. 8, 432-440 (1986).
5. G. W. Becker and D. Braun, Kunststoff Handbuch Band 10 – Duroplaste, München, Carl Hanser Verlag, 1988.
6. J. Patel, ANALYSES OF DENSELY CROSSLINKED PHENOLIC SYSTEMS USING LOW FIELD NMR, Dissertation, 2017.
7. A. Spörrer and V. Altstädt, Controlling Morphology of Injection Molded Structural Foams by Mold Design and Processing Parameters, Journal of Cellular Plastics, July/September, Vol. 43, 313-330, 2007.
8. S.-C. Chen, P.-H. Leeand J.-S. Huang, Study on the Molding Characteristics and Mechanical Properties of Injection-molded Foaming Polypropylene Parts, Journal of Reinforced Plastics and Composites, Vol. 23, No.4/2004, 429-444
9. V. Altstädt and A. Mantey, Thermoplast-Schaumspritzgießen, München, Carl Hanser Verlag, 2011

Plant Design, Stress-Strain Characteristics in Compression and Lifecycle Description of Coiled Filament Mats Based on Production According to Fitzer

Elmar Moritzer, Tobias Westhues

Kunststofftechnik Paderborn, Paderborn University, Warburger Str. 100, 33098 Paderborn, Germany
elmar.moritzer@ktp.upb.de; tobias.westhues@ktp.upb.de

Abstract. Coiled filament mats (CFM) are manufactured within an extrusion process and special downstream peripherals. A large number of endless polymer strands are joined to form a spatial mat structure. The individual, randomly arranged melt strands are connected at the contact points, thus enables the structure to be held together. Possible fields of application can be found wherever foam mats are currently used. CFM products offer a number of advantages, such as good air circulation, suitability for allergy sufferers and good cleaning possibilities. In order to achieve defined and reproducible product characteristics as a foam mat substitute, it is necessary to determine the physical properties using a suitable test setup. In this publication a description of the laboratory plant design for the production of CFM is given [1]. The effects of process variables like mass throughput, take-off velocity and distance between die to water surface on the achievable compression hardness and bulk density of CFMs are described. Occurring mat defects are presented and the process limits are explained. The test method of the mats is introduced and the results of the investigation are presented. A life cycle analysis shows the decrease in strength over the service life of the product. The investigation of loop formation during strand deposition of the single strand allows conclusions about optimum process settings. Tensile tests allow an insight into the failure mechanisms of the mat structure under mechanical load.

Keywords: Coiled filament mats; strand deposition, compression hardness, bulk density.

POTENTIAL APPLICATION FOR CFMS

Depending on the field of application, CFMs are also known as spaghetti mat [2], filament loop mat [3], coiled filament mat [4] or coil mattress [5], these terms all refer to a similar product whose definition is based on [6]: A spatial mat structure produced in an extrusion process from a multitude of endless, randomly arranged intertwined polymer strands or filaments connected to each other at the contact points.

The open structure of the random mat offers several advantages, such as good air circulation, suitability for allergy sufferers and good cleaning possibilities. Possible areas of application can be found wherever foam mats are currently used. In order to achieve defined and reproducible product properties as a foam mat substitute, it is necessary to determine the physical properties using a suitable test setup. Foams can mainly be characterized by three criteria: raw density or density [kg/m³], compression hardness [kPa] (DIN EN 3386-1) and the actual material. Typical values for foams are between 17 - 60 kg/m³ for the bulk density and 1.5 - 6.0 kPa for the compression stress-strain properties. PUR (Polyurethane), cold or viscose foams are used.

CFM PRODUCTION PLANT DESIGN

Description of the plant structures within patents for the production of CFMs go back to the 60s of the last century. In these patents different setups can be found. The construction range reaches from no calibration [US3687759] to fixed calibration ramps [US4952265], rotating calibration rollers [US4859516] to combinations thereof [US3691004]. In the case of multi-layer mats, the use of several extruders [US5972463] or the lateral feeding of semi-finished products [US4913757] is also required.

In the underlying investigations, the plant design according to Fitzer [US4227350] is used. FIGURE 1 shows an overview of the system and FIGURE 2 the adjustable take-off unit.

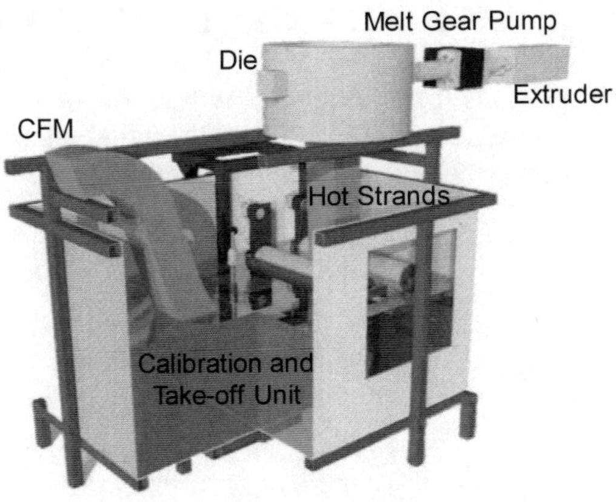

FIGURE 1. Laboratory Plant Design for the Production of CFMs [1]

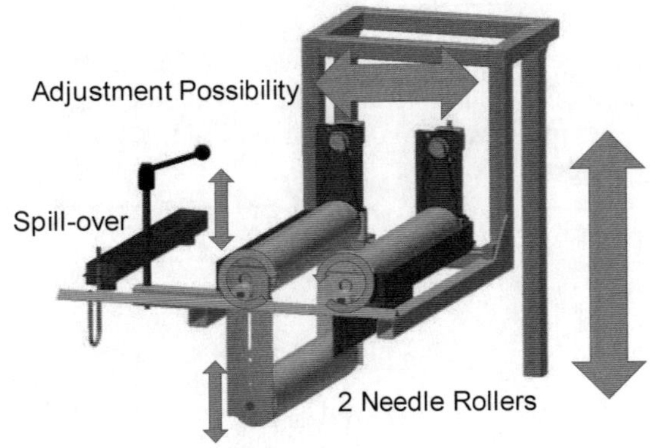

FIGURE 2. Calibration and Take-off Unit [1]

A 45 mm Battenfeld extruder was used to provide a constant melt throughput. Via a heated melt flow-way tube, the mass flow reaches a Maag melt gear pump. The melt pump generates a maximum pressure increase of 160 bar. The melt gear pump is connected to the die. Within the die the melt flow is deflected by 90° and exits through a perforated plate. The number of holes in the system is 232. The melt strands are deposited over a cooling bath in the gap between the two needle rollers. The difference in speed between the extrusion and rotation speeds of the needle rollers causes the hot strands to accumulate. The strands are deposited proportionally on the needle rollers and in the roll gap on the cooling water surface. The Queo 0230 material from Borealis with a density of 0.902 g/cm³ floats on the cooling water. If the hot strands are superimposed, they connect at the contact points (autohesion). The continuous rotation of the needle rollers in combination with a downstream take-off causes the strands to sink and solidify vertically in the cooling bath. The strands deposited on the outer surface of the needle rollers form a compact surface layer. The strands deposited in the roll gap on the water surface form the core of the coiled filament mat structure. The system design of the calibration and take-off unit allows the distance of the needle rolls to each other, the distance of the rolls to the underside of the tool and the water level to the center of the roll to be set. This makes it possible to place the filaments both with (directly onto the rolls) and without calibration (only into the water bath).

TEST METHOD OF THE MATS

A test specification has been developed to determine the mat properties based on the standards [DIN EN 1957, DIN EN ISO 2439, DIN EN ISO 178]. After cutting the specimen, this test specification specifies the specimen conditioning in a standard climate (≥ 24 h; 23 °C & 50 % rel. humidity). In the beginning of the test routine, two quasi-static pressure tests (from 0 % compression to 50 % compression at 5 mm/s test speed) are stipulated. After that 60,000 cycles of dynamic pressure testing (0 % compression to 50 % compression at 1 Hz) and then, two quasi-static pressure tests (from 0 % compression to 50 % compression at 5 mm/s test speed) must be conducted.

The following figure shows the test setup with the specimen inserted (left) and the state between the compression states 0 % and 50 % (right). The CFM specimens are 75 mm high. The dynamic test system Electropuls E 10000 from Instron is used for the test.

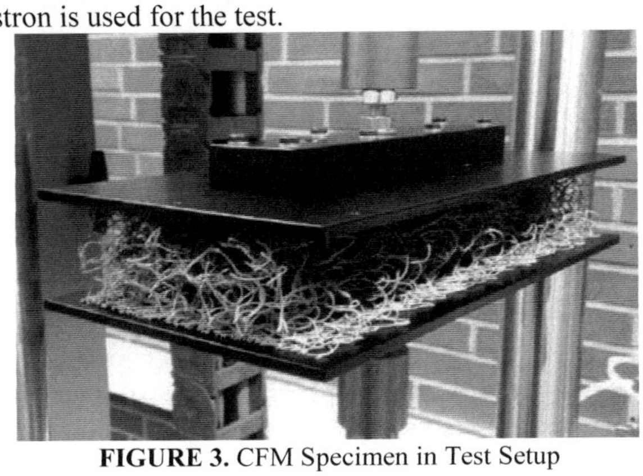

FIGURE 3. CFM Specimen in Test Setup

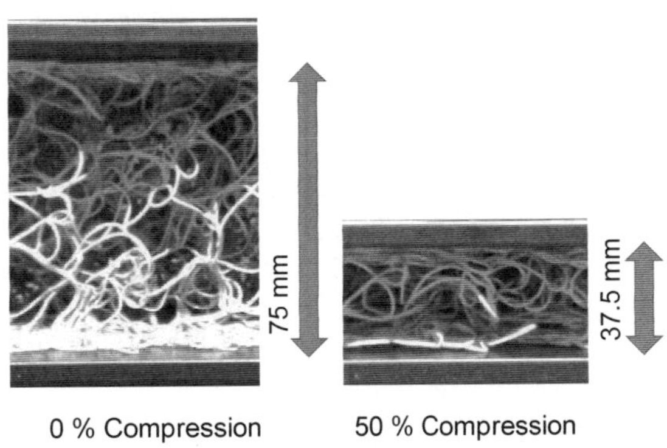

FIGURE 4. CFM Specimen Between 0 % to 50 % Compression

STRENGTH DECREASE DURING THE SERVICE LIFE (LIFE CYCLE ANALYSIS)

A decrease in strength occurs with increasing dynamic test cycles. This reduction in strength can be described by means of a decay curve. A number of 60,000 cycles corresponds to a standard [DIN EN 1957] lifetime of 10 years. The following figure FIGURE 5 shows on the y-axis the course of the standardized compression force and on the x-axis the number of cycles.

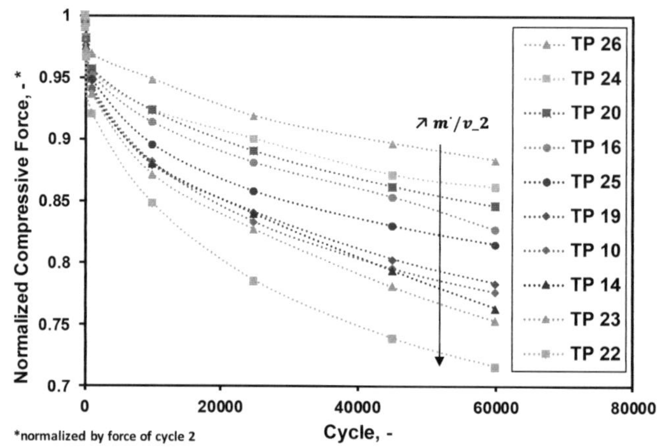

FIGURE 5. Normalized Compressive Force Versus the Number of Cycles

EFFECTS OF PROCESS VARIABLES ON CFMS COMPRESSION HARDNESS AND GROSS DENSITY

The following figures show the influence of the process variables (mass throughput, circumferential velocity of the needle rollers) on the achievable CFM gross density and the compression hardness of the CFMs. FIGURE 6 shows the gross density [kg/m³] determined per sample versus the ratio of mass throughput to circumferential velocity of the rolls in [kg/m]. The range of the abscissa goes from 0.125 to 0.5 and results from throughputs of 15, 22.5 and 30 kg/h as well as the circumferential velocity of the rolls of 1, 1.5 and 2 m/min. The diagram shows the direct influence of the process variables on the gross density and thus on the strength of the mat. The gross density describes the ratio of sample mass to the volume of the mat structure including the air surrounded by filament loops. With increasing gross density, the number of loops per unit of space also increases.

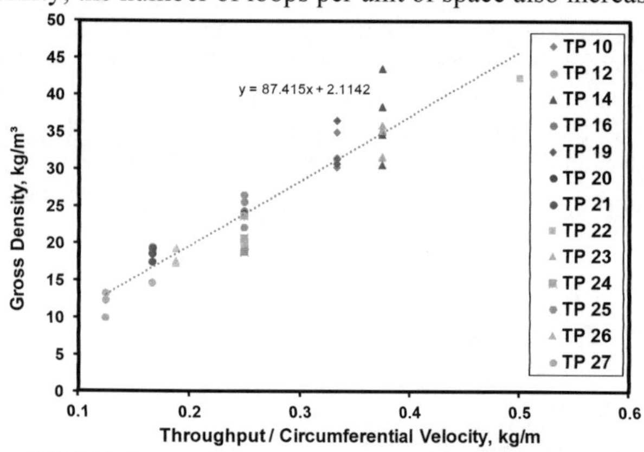

FIGURE 6. Gross Density Versus Ratio of Mass Throughput to Circumferential Velocity of the Rolls

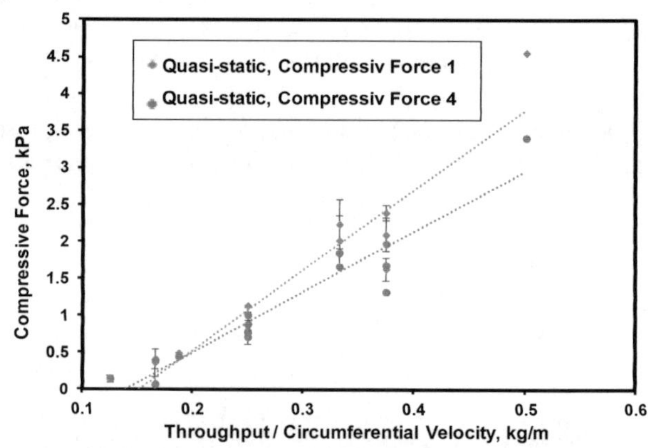

FIGURE 7. Compression Hardness Versus Ratio of Mass Throughput to Circumferential Velocity of the Rolls

The measured compression hardness [kPa] of the quasi-static pressure test plotted over the throughput to take-off speed ratio is shown in FIGURE 7. It can be seen that the process parameters have a direct influence on the compression hardness. In good approximation, the relationship between these two parameters is linear.

Thus it is possible to adjust the desired strength of the CFMs precisely. In addition, a direct correlation between bulk density and compression hardness is revealed.

MAT DEFECTS AND PROCESS LIMITS

During the production process, unsuitable process parameter settings can result in different error patterns within the mats. In addition to an inhomogeneous mat take-off speed, which results, for example, in a too small number of nodes (FIGURE 7 (a)), material agglomerations in the finished product can occur as filament nubs. These defects appear when minor soilings adhere to the tool , with results in an impure filament outlet. Individual extruded strands adhere to the surface of the tool and form an accumulation of many loops (adherent dross formation). If the mass of the accumulation is large enough, it dissolves and the filament bundle can later be found in the mat as a defect. FIGURE 7 (b) shows the structure of a filament knob and an accumulation in the finished mat. Clean tool surfaces and release agents reduce the adhesion tendency.

In addition, two states can be defined which describe the upper and lower process limits. The lower process limit is characterized by a filament length that is too short. Due to the insufficient free filament length when filaments immerse in the cooling water, no nodal points are formed, compare FIGURE 7 (c). The bulk density of the random mats is too low and no load-bearing structure can be formed. The upper process limit is determined by the cooling capacity in the gap between the two needle rollers. As the accumulated filament length increases, a larger amount of heat is introduced per time unit. If the cooling capacity is not sufficient to solidify the melt, the mat structure collapses in the further process, see FIGURE 7 (d). This can be remedied by lower melt throughput or increasing cooling capacity.

FIGURE 7. (a) Error Caused by Inhomogeneous Take-off Speed (b) Adhesion at Nozzle Outlet (c) Not Enough Contact Points (d) Insufficient Heat Dissipation

SUMMARY AND OUTLOOK

The investigations have shown that reproducible mat properties can be achieved with the test setup. There is a linear correlation between the considered process variables and the mat strength.

In further investigations, the strand deposition behavior of the single strand at different melting temperatures, mass flow rates, strand diameters, distances (nozzles to water) and different materials is examined. The aim is to create a better understanding of the process in order to increase the bond strength and to be able to carry out a tool adjustment derived from the findings. FIGURE 8 (a) shows an image of the test setup with a high-speed camera.

Within the dynamic tests on CFMs, a higher decrease in strength occurs with increasing initial strength over the cyclic loading. In order to examine the failure mechanisms, individual nodes, individual filaments and CFMs as a whole are tested in tensile tests. This procedure makes it possible to determine the failure mechanisms as a function of the manufacturing parameters. FIGURE 8 (b) shows a filament node clamped in the jaws of a tensile testing machine.

As reference measurements, different foams with varying densities and compressive strengths are taken to the same test cycle as the CFMs. The performance of the different products can be compared. FIGURE 8 (c) shows a PUR foam sample with a density of 20 kg/m^3 and a compressive strength of 2.0 kPa within the test setup according to DIN EN ISO 3386-1.

FIGURE 8. (a) Strand Deposition Behavior of the Single Strand (b) Contact Point Strength (c) Reference Measurements of Different Foams

ACKNOWLEDGMENTS

The cooperation project was funded by the AiF within the "Zentrales Innovationsprogramm Mittelstand (ZIM)". The used material was a donation by Borealis. We are grateful for the financial support and the donation.

TABLE 1. Test Point Parameters.

Test Point (TP)	**Throughput** kg/h	**Circumferential Velocity** m/min	**Throughput / Circumferential Velocity** kg/m	**Temperature** °C
10	30.0	1.5	0.3333	150
12	15.0	1.5	0.1667	150
14	22.5	1.0	0.3750	150
16	30.0	2.0	0.2500	150
19	30.0	1.5	0.3333	180
20	22.5	1.5	0.2500	180
21	15.0	1.5	0.1667	180
22	30.0	1.0	0.5000	180
23	22.5	1.0	0.3750	180
24	15.0	1.0	0.2500	180
25	30.0	2.0	0.2500	180
26	22.5	2.0	0.1875	180
27	15.0	2.0	0.1250	180

REFERENCES

1. A. Ostapez; M. Dann; S. Richter; D. Amirov; *Entwurf einer variablen Abzieh- und Kalibriereinrichtung für den Extrusionsprozess von Wirrmattenstrukturen;* Paderborn, Paderborn University, 2017,
2. BOB MATS B.V, 2015. Spaghetti Matte [online]. Cuijk-Vianen: Bob Mats B.V [Access at: 09-27-2018]. Available at: http://www.sauberlaufmattenshop.de/aussen/spaghetti-matte
3. KABUSHIKI KAISHA RISURON, 1989. *Verfahren zum Herstellen einer Matte bestehend aus Filamentgeschlinge*. Inventors: Minoru YAMANAKA and Tetsuo AMAWA. 08-24-1989. Filed: 09-27-1988. DE, Patent Document DE3832792A1
3. MINNESOTA MINING AND MANUFACTURING COMPANY, 1972. *Method of making coiled filament mat*. Erfinder: Warren R. MANNER und Richard J. KINDERMAN. 08-22-1972. Filed: 07-03-1969. U.S., Patent Document US3686049
5. DALIAN PLASTICS RESEARCH INSTITUTE CO., LTD., *Plastic coil matress production line* [online]. Dalian Plastics Research Institute Co., Ltd. [Access at: 09-27-2018]. Available at: http://en.plasticdl.com/products_detail/productId=50.html
6. MINNESOTA MINING AND MANUFACTURING COMPANY, 1980. *Low-density abrasive product and method of making the same*. Inventor: Robert C. FITZER. 10-14-1980. Filed: 11-02-1977. U.S., Patent Document US4227350

Energy Consumption of High Speed Multiscrew Extruders

Mansour M. Albareeki[a], Javier Vera Sorroche[a], and Carol Forance Barry[a]

[a]*Department of Plastics Engineering, University of Massachusetts Lowell, Lowell, MA, USA*

Abstract. For extruders, energy consumption depends on the energy required by the drive motor and for heating and cooling of the extruder barrel. This work investigated the energy consumption of high speed twin and quad extruders. Both extruders had 15-mm-diameter screws with length-to-diameter ratios of 45:1 and the same 22-kW motors, but the different number of screws changed the clearance between the screws. Energy consumption was evaluated as a function of screw speed, feed rate, and polyethylene type and molecular weight. On-line measurements of viscosity were correlated with the energy consumed by the extruders. Results are discussed with respect to material structure and viscosity as well as extruder design.

Keywords: Extrusion, Twin screw extrusion, Quad screw extrusion.

INTRODUCTION

There is considerable interest in the power consumption required for plastics processing, including single screw extrusion and twin screw extrusion. Prior work with a traditional twin screw extruder[1] showed that drive power (drive torque) decreased with increasing screw speed, increased with feed rate, and was greater when polymer exhibited less shear thinning (i.e., LLDPE > LDPE). With LDPE and LLDPE, the specific energy consumption (in J/g) increased significantly with higher feed rates, but it increased only slightly with increasing screw speed and less shear thinning of the polymers[1]. In traditional co-rotating twin screw extruders, screw speeds can reach 1200 rpm with issues[2-4], whereas when ultra-high speed twin screw extruders, which can achieve screw speeds as high as 4500 rpm, have produced degradation of even relatively stable polymers[5-8]. Although these ultra-high speed twin screw extruders exhibit decreased drive torque with increasing screw speed, the drive torque plateaus at high screw speeds, due to the lower viscosity associated with polymer degradation[9]. Quad screw extruders with four co-rotating, parallel, fully intermeshing screws have a greater free volume, but a greater number of intermeshing regions ("bite points") which produce high shear stresses compared to twin screw extruders. Albareeki et al.[9] reported relatively constant drive torque for screw speeds of 0-2000 rpm, but a greater decrease in melt viscosity compared to ultra-high speed twin screw extruders. With the quad screw extruder, drive torque increased with increasing feed rate[10].

This work compared the energy consumption for ultra-high speed twin and quad screw extruders. Energy consumption was measured directly from the extruders and correlated with drive torque, melt temperature, and viscosity obtained from in-line rheology. The objective was to determine the effects of extruder design (i.e., the number of screws), screw speed, and feed rate on energy consumption.

METHODOLOGY

Three commercially-available polymers, two low density polyethylenes (LDPE) and one linear low density polyethylene (LLDPE), were used for this work. The complex viscosity of the neat resins was measured at temperatures of 190, 205, and 220°C using parallel plate rheometry. Table 1 lists the reported densities and melt flow indices (MFI) as well as the calculated melt indices (n) and activation energies for flow (E_a). LDPE B exhibited lower viscosity and less shear thinning than LDPE A. LLDPE was a specialty material that had similar viscosities to LDPE A at low shear rates, but it showed significantly less shear thinning.

TABLE 1. Materials Properties

Material	Density (kg/m^3)	MFI (g/ 10 min)	n	E_a (kDa)
LDPE A	920	0.18	0.42	15.9
LDPE B	920	1.85	0.59	39.9
LLDPE	921	0.60	0.68	14.8

These polymers were processed in two extruders: 1) an ultra high speed co-rotating 15-mm-diameter twin-screw-extruder with a length-to-diameter (L/D) ratio of 45:1 (Technovel, model: KZW15TW-45/60MG-NH (-4400)) and (2) an ultra high speed co-rotating 15-mm-diameter quad-screw extruder with four parallele screws and an L/D ratio of 45:1 (Technovel, model: WDR15QD-45MG-NH (-2200)). As shown in Figure 1b, the screws for the both extruders contained three mixing sections with respective lengths of 15, 25, and 75 mm (i.e., about 17% of the total screw length); Therefore, both the twin and quad screw extruders had the same number and type of mixing elements. Melt exiting the extruders passed through an in-line slit die rheometer – i.e., a slit die with three pressure transducers and a flush-mounted thermocouple; the die for the quad screw extruder is shown in Figure 1a. The extrudate strands was cooled in a water bath. For extrusion, the six barrel and two die temperatures were 190, 195, 200, 200, 205, 205, 205, and 205°C; the barrel was not cooled. The polymers were fed through the first feed port at a constant feed rate of 2 kg/h; this feed rate was limited by the capacity of the twin screw extruder and the measured error was ± 5%. The quad screw extruder also was fed at a rate of 4 kg/h. The screw speeds were 500, 1000, 1500, and 2000 rpm.

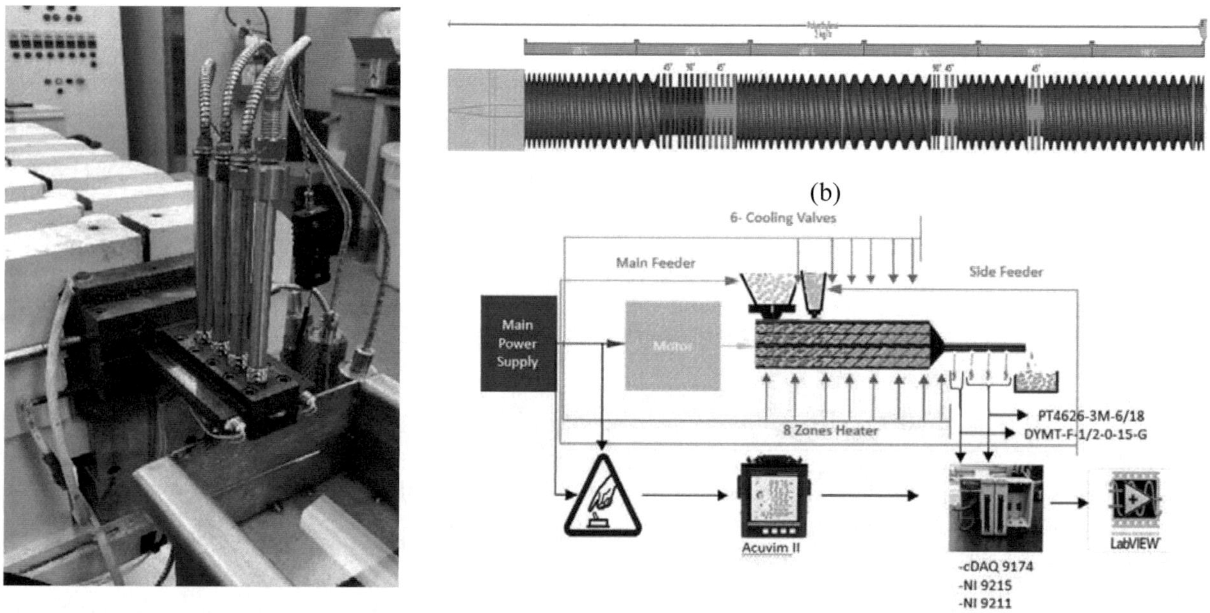

FIGURE 1. For extrusion, (a) melt viscosity was measured using a slit die system, (b) the screw design was the same for the TSE and QSE, and c) power consumption were measured from the extruders.

During each extrusion trial, actual barrel temperatures and drive torque were recorded from extruder's readouts. The melt temperature and pressure were determined by using 'K' type thermocouple and Dynisco pressure transducer, respectively, located in the die adaptor. As shown in the Figure 1c, power consumption (from the motor) was measured by using a three-phase power meter (Acuvim II) connected to the control panel and cDAQ 9174 to convert to LabVIEW. Melt viscosity was determined from the in-line slit die rheometers using the pressure drop between the first and third transducers in the slit dies; melt temperature at the die was measured with a flush-mounted thermocouple (Figure 1a).

RESULTS AND DISCUSSION

Figure 2 presents the ratio of the measured barrel temperature to the set barrel temperature along the length of the twin and quad screw extruders (i.e., from feed zone 1 to the die); the feed rate was 2 kg/h. Changing the feed for the quad screw extruder to 4 kg/h did not significantly affect barrel temperatures (not shown). With the twin screw extruder, the barrel temperatures did not increase when the polymer flowed through the first set of kneading blocks (at zone 2), but increased sharply when the polymer reached the major set of kneading blocks (in zone 5); the barrel temperatures then decreased gradually as the melt flowed through the end of the extruder and into the die. In the quad screw extruder, however, the barrels increased slightly in zone 3 and again in in zone 4. The relative temperature rise in the quad screw extruder (~ 1.8) was significantly greater than the temperature rise of 1.6 for the twin screw extruder. Since the screw designs in both extruders were similar, this increase in temperature was attributed to the greater number of intermeshing regions the quad screw extruder. The temperature ratio generally increased with screw speed, indicating the higher shear rates had a greater impact on barrel temperature than the decreased barrel residence time. In both extruders, LDPE A and LLDPE provided greater increases in barrel temperature than LDPE B. This effect was not unexpected since the rise in melt temperature increases with melt viscosity.

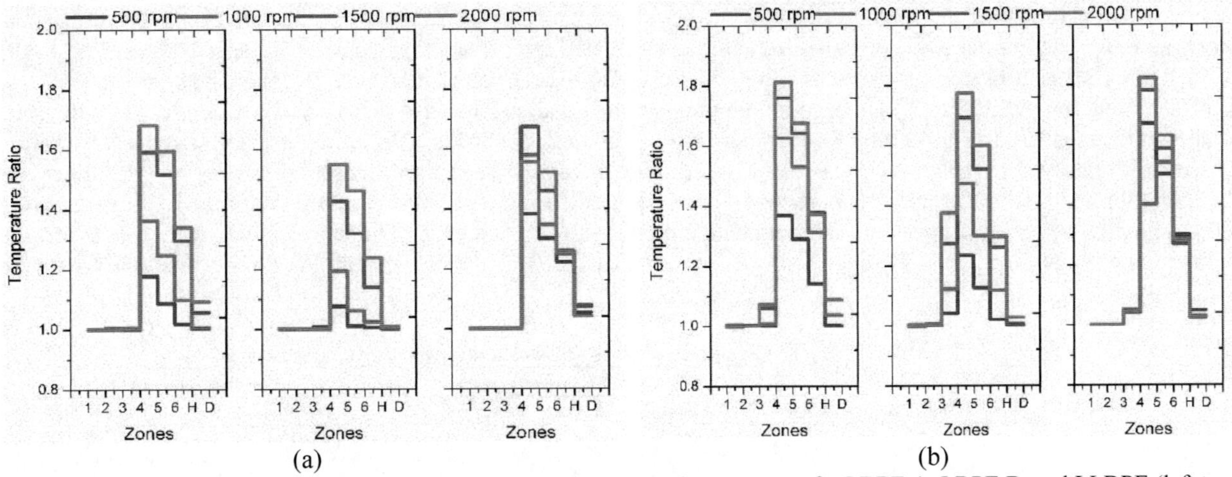

FIGURE 2. Ratio of the measured barrel temperature to the set barrel temperature for LDPE A, LDPE B, and LLDPE (left to right) processed in the (a) twin and (b) quad screw extruders.

In contrast to the barrel temperatures, the melt temperatures measured in the die adaptor were significantly greater for the twin screw extruder than for the quad screw extruder (Figure 3a). This difference was attributed to the greater free volume in the quad screw extruder; (the calculated free volumes for the twin and quad screw extruders were 114.5 cm^3 and 273 cm^3, respectively). Increasing the feed rate to the quad screw extruder produce little change in the melt temperature of the LDPEs, but it increased the overall temperature of the LLDPE melt. For all three conditions, the melt temperature of LDPE B increased linearly with screw speed. With LDPE A, the melt temperature increased linearly with screw speed in the twin screw extruder, but they leveled off at high speeds in the quad screw extruder. The LLDPE, however, showed little change in melt temperature with increases in screw speed; this behavior was consistent with the limited shear thinning of the LLDPE melt (i.e., n = 0.68). Consistent with their melt rheology, the low viscosity LDPE B exhibited lower melt temperatures compared to the more viscous LDPE A. At the original feed rate, LLDPE had somewhat lower melt temperatures than LDPE A, but the temperatures were similar when the feed rate was increased for the quad screw extruder.

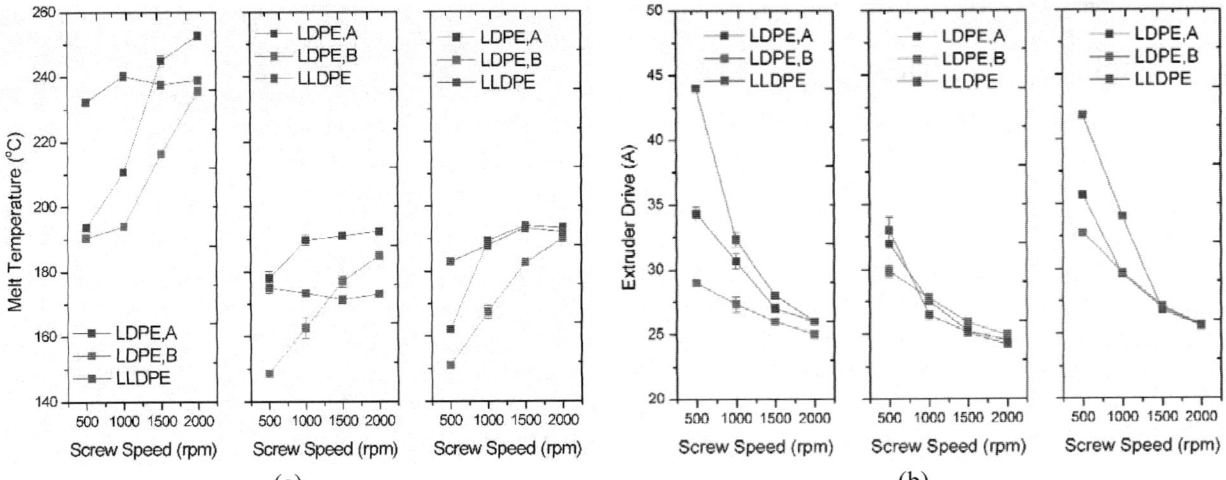

FIGURE 3. The (a) melt temperature measured in the die adaptor and (b) measured drive torque for (left to right) the twin screw extruder, the quad screw extruder, and the quad screw extruder with the higher feed rate

As shown in Figure 3b, the drive power (extruder drive torque) decreased with screw speed, probably due to increases in melt temperature, and the corresponding decreases in melt viscosity. For the same feed rate, the drive torque was generally greater in the twin screw extruder than in the quad screw extruder. This difference was attributed to the greater free volume in the quad screw extruder. Increasing the feed rate for the quad screw extruder, however, significantly increased the drive torque. With the lower feed rate, the drive torques were similar for all three materials; this behavior may reflect the limited filling of the screw channels. When the screw channels in the twin and quad screw extruder had greater fill levels, the drive torques reflected the expected viscosities of the polymers – i.e. LDPE B < LDPE A < LLDPE. Similar patterns were observed with the measured head pressure (not shown).

The melt temperatures (T_m) measured in the slit die (Table 2) were significantly different than the melt temperatures measured at the end of the extruder (Figure 3a). Causes for this difference are still under investigation. These melt temperatures did not vary as much and did not always increase with increasing screw speed. In general, the lower viscosity LDPE B had the lowest melt temperature; the temperatures of LDPE A and LLDPE were similar although LLDPE's melt temperature did not change much with screw speed. The temperatures were highest for the twin screw extruder, lowest for the quad screw extruder fed at 2 kg/h, and somewhat higher for the quad screw extruder fed at 4 kg/h.

TABLE 2. Melt Temperature Measured in Die and Power Consumption

	TSE at 2 kg/h			**QSE at 2 kg/h**			**QSE at 4 kg/h**		
	LDPE A	**LDPE B**	**LLDPE**	**LDPE A**	**LDPE B**	**LLDPE**	**LDPE A**	**LDPE B**	**LLDPE**
T_m (°C)	216-240	215-224	228-236	196-211	190-194	195-206	197-221	192-218	217-223
P (kW)	1.02-1.69	0.81-1.58	1.34-1.71	1.75-2.88	1.56-3.08	1.91-2.80	2.15-3.58	1.85-3.64	2.67-3.56
$\Delta P/\Delta N$ (W/rpm)	0.4	0.5	0.2	0.7	1.0	0.6	0.9	1.2	0.6

The power (motor energy) consumption and viscosity measured for the four screw speeds in twin and quad screw extruders are presented in Figure 4. As expected, the power increased and the melt viscosity decreased with increasing screw speed. Graphs of power vs. time were used to determine the power consumption when the extruder stabilized. Similar graphs of viscosity vs. time were used to determine the melt viscosity for stabilized extrusion.

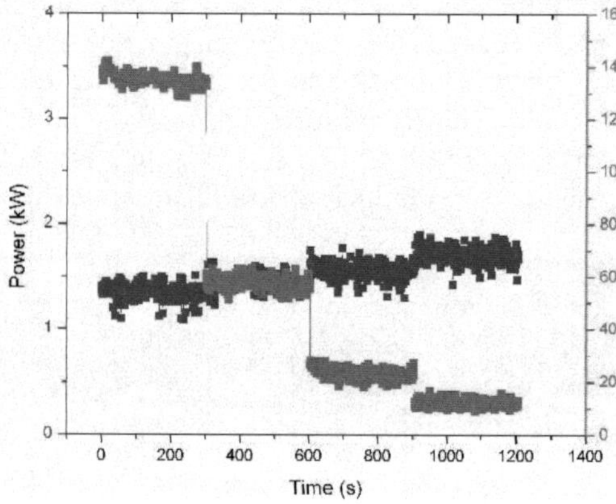

(a)

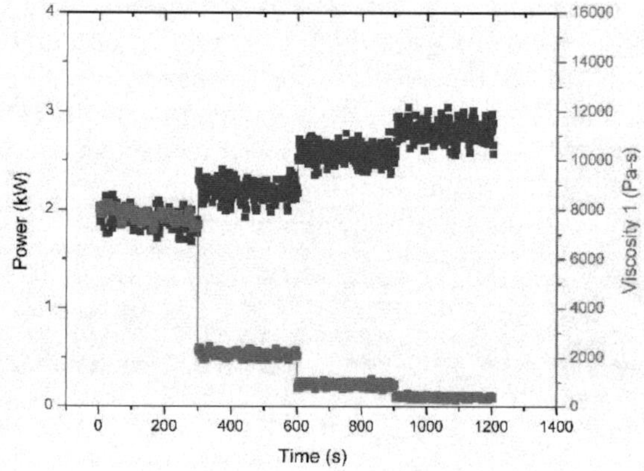

(b)

FIGURE 4. The measured power and viscosity of LLDPE for the (a) twin screw extruder and (b) quad screw extruder.

As illustrated in Figure 5a, the measured viscosities of each polymer were similar for the twin and quad screw extruders with a feed rate of 2 kg/h, but were significantly lower when the quad screw extruder was fed at 4 kg/h. In general, LDPE B (MFI = 1.85 g/10 min. and n = 0.59) had the lowest viscosity and that viscosity decreased less with increasing screw speed. LDPE A (MFI = 0.18 g/10 min. and n = 0.42) had a higher viscosity and greater decrease in viscosity with increasing screw speed. This behavior was consistent with their rheological values measured using parallel plate rheology. In contrast, LLDPE exhibited the highest viscosities at low screw speeds and the greatest decrease in viscosity with increasing screw speed. These viscosities were not consistent with the rheological values (MFI = 0.60 g/10 min. and n = 0.68) and may reflect the unchanging melt temperatures shown in Figure 3a.

For the twin screw extruder, the measured power (motor energy) increased linearly with screw speed from about 1.02 kW to 1.71 kW (Figure 5b). With the quad screw, the power also increased relatively linearly with screw speed. This power increased from 1.56 kW to 3.08 kW with the lower feed rate and 1.85 kW to 3.64 kW with the higher feed rate (Figure 5b). The increase in power consumption with increasing screw speed ($\Delta P/\Delta N$) was 0.2-0.5 W/rpm for the twin screw extruder, 0.6-1.0 W/rpm for the quad screw extruder fed at 2 kg/h, and 0.6-1.2 W/rpm for the quad screw extruder fed at 4 kg/h (Table 2). As expected, power consumption increased with the number of screws and with the feed rate. For all three conditions, the low viscosity LDPE B produced the lowest power consumption, but the greatest increase in power consumption with increasing screw speed - $\Delta P/\Delta N$ = 0.5, 1.0, and 1.2 W/rpm. The latter behavior has attributed to greater backflow with the LDPE B. Extrusion of LDPE A required greater power, but it produced lower $\Delta P/\Delta N$ values (0.4, 0.7, and 0.9 W/rpm). Powder consumption increased further with the highest viscosity resin (LLDPE).

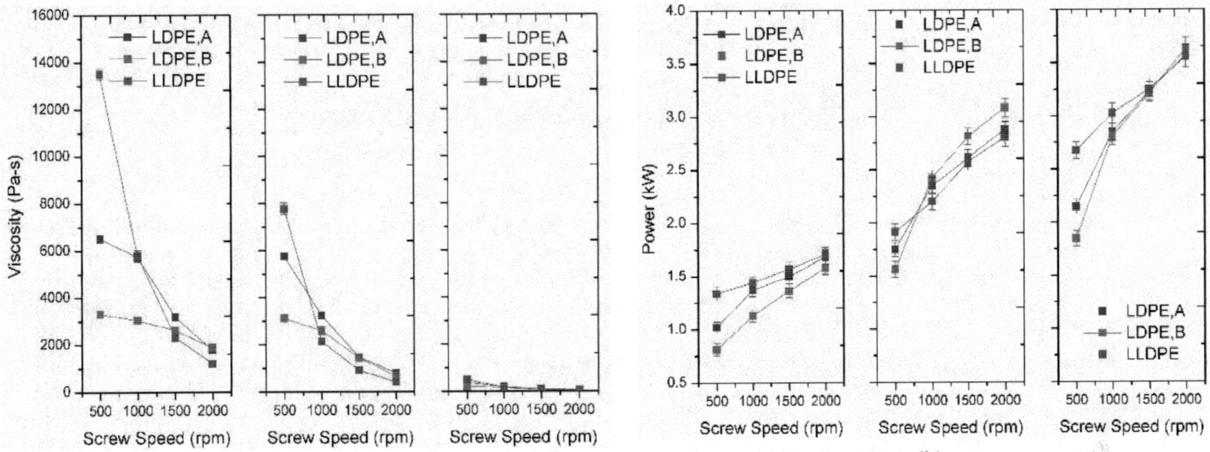

FIGURE 5. The (a) calculated melt viscosity and (b) measured power consumption for (left to right) the twin screw extruder, the quad screw extruder, and the quad screw extruder with the higher feed rate.

CONCLUSIONS

In comparison a twin screw extruder, a quad screw extruder exhibited a greater increase in barrel temperatures and a lower increase in melt temperature as the screw speed was increased from 500 to 2000 rpm; increasing the feed rate had little impact on these temperatures. This behavior produced decreases in drive torque in the quad screw extruder that increased again with higher feed rates. The power (motor energy) consumption, however, increased significantly with screw speed and was greater for the quad screw extruder. Feed rate and polymer viscosity had less impact on the power consumption in these extruders, although lower viscosity polymers showed greater increases in power consumption with increasing screw speed.

ACKNOWLEDGMENTS

The authors would like to thank Technovel Corporation (Osaka, Japan) for the use of their extruders and Emile Barry for his assistance with the power measurement equipment.

REFERENCES

1. J. Vera Sorroche, A. Kelly, E. Brown, and P. Coates, unpublished data.
2. K. Kohlgrüber, *Co-Rotating Twin-Screw Extruders: Fundamentals, Technology, and Applications*, Cincinnati, OH: Hanser Publications, 2012.
3. P. Peltola, E. Välipakka, J. Vuorinen, S. Syrjälä, and K. Hanhi, *Polym. Eng. Sci*, **46**, 995-1000 (2006).
4. T. Villmow, B. Kretzschmar, and P. Pötschke, *Compos. Sci. Technol.*, **70**, 2045-2055 (2010).
5. J. J. Boone, G. Pigeon, C. Thellen, J. A. Ratto, and R. Whitehouse, *Proc. Ann. Tech. Conf. Soc. Plast. Eng.*, **71**, 1591699 (2013).
6. A. Farahanchi, R. Malloy, and M. J. Sobkowicz Kline, *Polym. Eng. Sci.*, **56**, 743-751 (2016).
7. A. Farahanchi, E. Boehm, N. Orbey, and R. Malloy, *Polym. Eng. Sci.*, **57**, 60-68 (2017)
8. A. Farahanchi and M. J. Sobkowicz Kline, *Polymer Degradation and Stability* **138**, 40-46 (2017).
9. M. Albareeki, S. B. Driscoll, and C. F, Barry, *Proc. Ann. Tech. Conf. Soc. Plast. Eng.*, **75**, (2017).
10. M. Albareeki, S. B. Driscoll, and C. F, Barry, unpublished data.

Studies on Single-Screw Extrusion of Wood-Polymer Composites with Yield Stress and Slip Effects

K. Wilczyński (presenting author, reference author), K. Buziak, A. Lewandowski, K.J. Wilczyński and A. Nastaj

, *Warsaw University of Technology, Institute of Manufacturing Technologies, Polymer Processing Department 02-524 Warsaw, Narbutta 85, Poland*

Presenting author's email: k.wilczynski@wip.pw.edu.pl
Reference author's email: k.wilczynski@wip.pw.edu.pl

Abstract. Studies on rheology and processing of wood-polymer composites are very limited in the literature [1]. These are pseudoplastic and viscoelastic materials that exhibit yield stress and wall slipping. It is important to understand the processing of these materials. Slip effects and flow with yield stress have been investigated numerically for single screw extrusion using ANSYS-Polyflow software. Fully three-dimensional non-Newtonian analysis has been performed both in the extruder (slip on the screw/barrel surfaces) and in the die. The power-law model of Ostwald-de Waele has been used for flow modeling, the generalized law of Navier has been applied for a slip analysis, and Bingham law has been used for an analysis of the flow with yield stress. An effect of the parameters of these models on the flow in the extruder/die system has been studied. Screw pumping characteristics and die characteristics (i.e. flow rate vs. pressure gradient) have been computed and implemented into the recently developed [2] (at the Warsaw University of Technology) global model of the single screw extrusion process of wood-polymer composites. This model allows to predict various process parameters, e.g. flow rate, pressure and temperature profiles, melting profile (the rheological model of Klein is used for flow modeling). The effect of slipping and yield stress on the single screw extruder operation, i.e. location of the operating point on the screw/die characteristics has been discussed.

Keywords: Extrusion, Wood-Polymer Composites, Modeling.
PACS: 47.85.md, 81.20. Hy.

INTRODUCTION

Wood polymer composites are used in many branches of economy to replace wood. The main advantage of these over wood is their higher resistance to decay. The most commonly used are the PP-, HDPE-, and PVC-composites. Due to extensive use of WPC in the form of profiles, the basic technology for their processing is extrusion.

Studies on rheology and extrusion of wood-polymer composites are very limited [1]. It is known that these composites are pseudoplastic and viscoelastic, and exhibit yield stress and wall slipping.

When modeling extrusion process, it is generally assumed that flowing material, in the screw and in the die, adheres to the wall, and several models of extrusion process have been developed using this approach. Extensive reviews on modeling of extrusion process have been presented in the literature [3-7].

The phenomenon of wall slippage was studied for the first time by Mooney [8]. A number approaches to explain this have been later presented, and a number of fundamental papers appeared, e.g. [9-12]. Now, it is generally accepted that wall slippage can occur above a certain wall shear stress. However, much less research has been devoted to wall slippage in extrusion process. And, only a few papers have been published to answer the question how to consider wall slippage when designing the extruder, e.g. [13-20], which have been recently reviewed [21].

When modeling extrusion, it is assumed that flowing material has no yield stress (critical value of stress below which a material does not flow). However, many materials have a yield stress (viscoplastics), among them WPC. Bingham [22] was the first to describe this type of materials. A number of works related to viscoplastics appeared later which were reviewed by Bird et al. [23], and later by other researchers, e.g. [24,25]. However, much less research has been devoted to viscoplastic flows in extrusion process [7,14-16].

Summarizing, the flow of filled polymers, e.g. WPC, that usually exhibit yield stress and wall slippage, is still poorly understood in extrusion. Thus, the global extrusion models are generally limited to viscous flows. In this paper, the effect of both wall slippage and yield stress on the extruder operation has been studied (both for a plasticating unit and for a die).

MATERIAL & RHEOLOGICAL CHARACTERISTICS

The wood-polymer composite of 50% wood flour content based on the polypropylene matrix (WPC 50/50 PP copo inj 4, Beologic) was used in the study. Polypropylene has density 0.9 g/cm^3, mass flow rate index MFR = 25 g/10 min (230 ^oC, 2.16 kg), and a melting point of 160 ^oC. The WPC composite has density 0.95 g/cm^3 (melt), and mass flow rate index MFR = 4 g/10 min (190 ^oC, 5 kg).

Viscous flow properties of the material were determined at different temperatures using a high-pressure capillary rheometer. The measurements were performed in the temperature range 170-200 ^oC and the shear rate range 5-2000 s^{-1}. The corrections of Rabinowitsch, Bagley, and Mooney have been applied. It has been observed that the material viscosity decreased with an increase of the shear rate (pseudoplastic), and with an increase of the temperature.

FEM MODELING & FLOW SIMULATION

The ANSYS Polyflow v.17.0 CFD program [26] has been used for simulations*. The modeling and meshing procedures have been described elsewhere [27].

The power-law model has been used for flow modeling

$$\eta = m\dot{\gamma}^{n-1} \tag{1}$$

where η is the viscosity, $\dot{\gamma}$ is the shear rate, m is the consistency factor and n is the power-law exponent.

The parameters of the model were as follows: the consistency factor m=30111 Pa·s^n, and the flow index n=0.25.

The generalized Navier law [26] has been used for wall slippage modeling

$$f_s = F_{slip}(v_{wall} - v_s)|v_s - v_{wall}|^{e_{slip}-1} \tag{2}$$

where f_s is the shear stress, v_s is the tangential velocity of the fluid, v_{wall} is the tangential velocity of the wall (it is assumed to be zero by default), F_{slip} and e_{slip} are the model parameters (the full slip is obtained for F_{slip}=0, the law is linear when e_{slip}=1)

The Bingham model [26] has been used for viscoplastic flow modeling

$$\eta = \begin{cases} \eta_0 + \frac{\tau_0}{\dot{\gamma}}, \dot{\gamma} \geq \dot{\gamma}_c \\ \eta_0 + \tau_0 \frac{\left(2-\frac{\dot{\gamma}}{\dot{\gamma}_c}\right)}{\dot{\gamma}_c}, \dot{\gamma} < \dot{\gamma}_c \end{cases} \tag{3}$$

where η is the viscosity, η_0 is the zero viscosity, τ_0 is the yield stress, $\dot{\gamma}$ is the shear rate, $\dot{\gamma}_c$ is the critical shear rate,

The parameters of the model were as follows: the zero viscosity η_0=51 Pa·s, the yield stress τ_0=104132 Pa, and the critical shear rate $\dot{\gamma}_c$ =29,1 s^{-1}.

An effect of wall slippage (F_{slip}=0,002, e_{slip}=1) and yield stress ($\tau_0/2$, τ_0, $2\tau_0$) on the melt flow in the extrusion die has been depicted in Fig.1. The flow through the rod cylindrical die has been simulated (3-sectional die of D=45/5 mm in diameter, and L=180/30/60 mm in length, Q=0,0022 kg/s). The flow geometry and the boundary conditions have been presented elsewhere [21].

It is clearly seen that velocity profile drastically changes and pressure substantially drops when slipping appears (red and blue lines, Fig.1a). Pressure/velocity distributions for power-law model and slip/no slip conditions are shown in Fig.1b. If yield stress is allowed, velocity profile also substantially changes, and pressure drop increases when yield stress increases (green, purple and navy lines, Fig.1a). Pressure/velocity distributions for Bingham model at various yield stress are shown in Fig.1c.

An effect of wall slippage and yield stress (F_{slip}=0,01, e_{sli}=1, Q=0,0022 kg/s) and yield stress (0, τ_0, $4\tau_0$, Q=0,0066 kg/s) on the melt flow in the screw has been depicted in Figs.2-3. The flow through the conventional screw has been studied (screw element of D=45 mm in diameter, H=3 mm in channel depth, and L=180 mm in length). The screw flow geometry and the boundary conditions have been presented elsewhere [21]. .

It is clearly seen that velocity profile drastically changes and pressure substantially drops when slipping appears, both at the screw and barrel surfaces (blue line to compare with red line, Fig.2). Velocity at the barrel wall increases while at the screw root decreases. Pressure/velocity distributions for power-law model and slip/no slip conditions are shown in Fig.2.

If yield stress is allowed, velocity and pressure profiles also substantially change (red, blue and green, lines, Fig.3). Pressure/velocity distributions for Bingham model at various yield stress are shown in Fig.3.

PROCESS OPERATION

Extrusion process is usually described by extruder operating characteristics which is defined by the screw/die characteristics (Fig.4). Actual operating conditions are determined by the extruder operating point which is the point of intersection of the screw/die characteristics. This point defines the extrusion output and the extrusion pressure.

An example of the effect of wall slippage and yield stress on the extruder operating point has been depicted in Fig.4. The starting point (1) has been obtained for no slip/no yield stress conditions (brown lines). When slip in the die/screw system is allowed (green lines), the operating point moves to lower flow rate and lower pressure, i.e. to the point (2). When yield stress is allowed and it increases (blue lines), the operating point moves to higher flow rate and higher pressure, i.e. to the point (3). It should be noted that these relations can change in dependence on the material parameters and process parameters.

CONCLUSIONS

It has been concluded that slip in the screw/die system as well as material yield stress have substantial effect on the extruder operation defined by the operating point, i.e. is on the extrusion output and extrusion pressure. When slip and yield stress appear, the extruder operation can drastically change. Therefore, slip effects and yield stress should be taken into account when global modeling the extrusion process, especially for filled polymers, e.g. WPC, and for elastomers, polymers like PVC and HDPE, polymer suspensions, as well as ceramic materials and some foodstuffs which may exhibit wall slip and yield stress effects under certain conditions. This may be valid not only for the melt flow but also for melting of polymers. For example, slipping may change the classical Tadmor melting mechanism which is observed, e.g. for PVC. FEM modeling of melting using CFD software allows to study this problem.

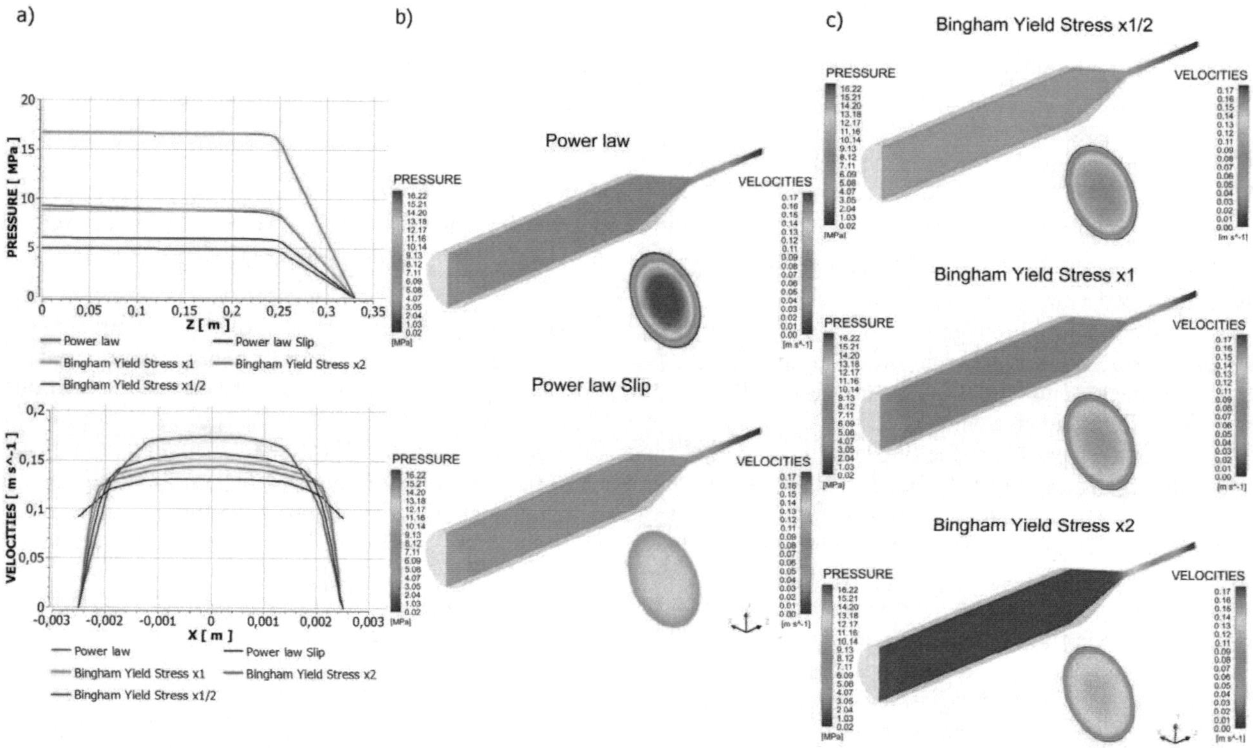

FIGURE 1. Die flow simulations: a) pressure/velocity profiles for power-law model and slip/no slip conditions, and Bingham model at various yield stress ($\tau_0/2$, τ_0, $2\tau_0$), b) pressure/velocity distributions for power-law model and slip/no slip conditions, c) pressure/velocity distributions for Bingham model at various yield stress ($\tau_0/2$, τ_0, $2\tau_0$).

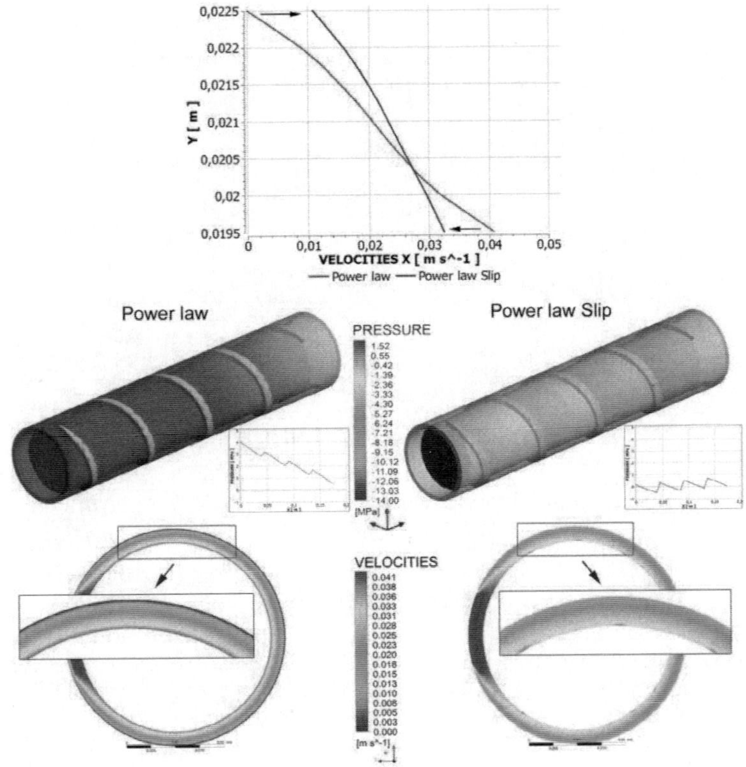

FIGURE 2. Screw flow simulations: pressure/velocity distributions for power law model at slip/no slip conditions

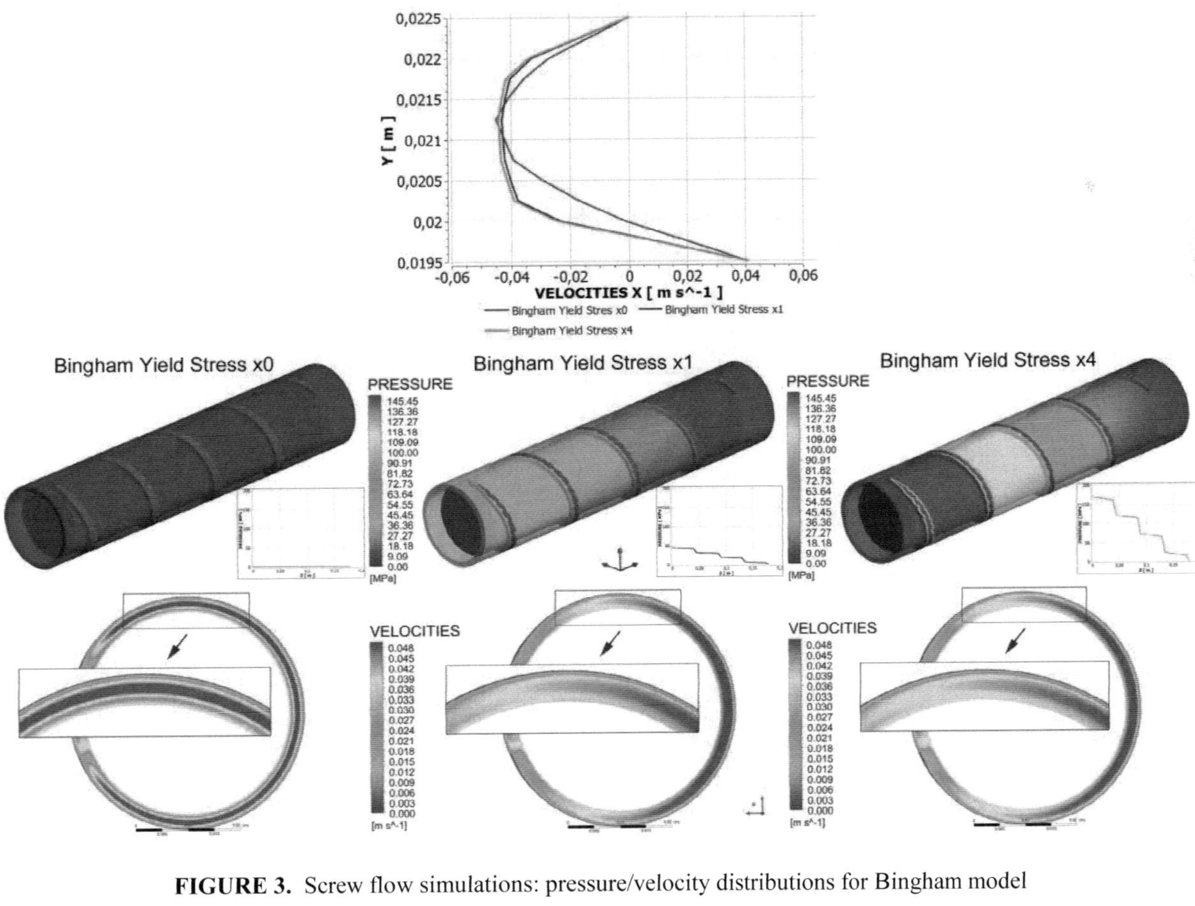

FIGURE 3. Screw flow simulations: pressure/velocity distributions for Bingham model

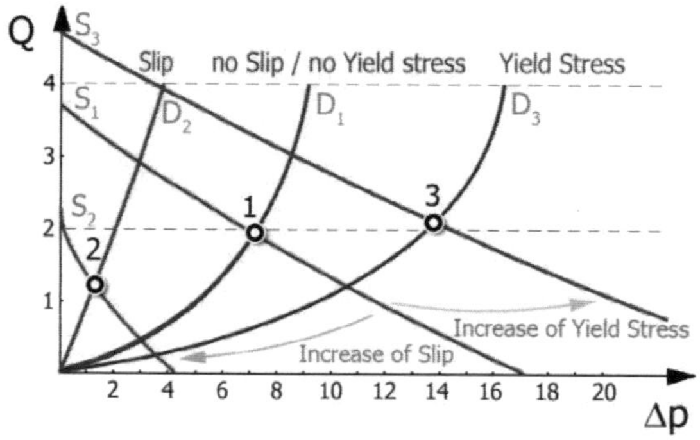

FIGURE 5. Extruder operation: an effect of slip and yield stress conditions

ACKNOWLEDGMENTS

Acknowledgments: The authors would like to acknowledge support from National Science Center, Poland (DEC-2015/19/B/ST8/00948).

REFERENCES

1. K. Wilczyński, A. Nastaj, A. Lewandowski, and K.J. Wilczyński, K. Buziak, *Int. Polym. Proc.*, **30**, 114-120 (2015)
2. K. Wilczyński, K. Buziak, K.J. Wilczyński, A. Lewandowski, and A. Nastaj, *Polymers*, **10**, 295. doi:10.3390/polym10030295 (2018)
3. A. Ariffin and M.S.B. Ahmad, *Polym. Plast. Technol. Eng.*, **50**, 395-403 (2011)
4. A. Altinkaynak, M. Gupta, M. Spalding, and S.I. Crabtree, *Int. Polym. Proc.*, **26**, 182-196 (2011)
5. C. Teixeira, A. Gaspar-Cunha, and J.A. Covas, *Polym. Plast. Technol. Eng.*, **51**, 1567-1577 (2012)
6. K. Wilczyński, A. Nastaj, A. Lewandowski, and K.J. Wilczyński, *Polym. Plast. Technol. Eng.*, **51**, 626-633 (2012)
7. M. Malik, D.M. Kalyon, and J.C. Golba Jr., *Int. Polymer Proc.*, **29**, 51-62 (2014)
8. M. Mooney, *J. Rheol.* **2**, 210–222 (1931)
9. S.G. Hatzikiriakos and J.M. Dealy, *J. Rheol.* **36,** 703–741 (1992)
10. S.G. Hatzikiriakos, C.W. Stewart, and J.M. Dealy, *Int. Polym. Process.*, **8**, 36-43 (1993)
11. S.G. Hatzikiriakos, *Polym. Eng. Sci.*, **34**, 1441-1449 (1994)
12. S.G. Hatzikiriakos, *Progress Polym. Sci.*, **37**, 624-643 (2012)
13. H.F.H. Meijer and C.P.J.M. Verbraak, *Polym. Eng. Sci.*, **28**, 758-771 (1988)
14. A. Lawal, D. Kalyon, and U. Yilmazer, *Chem. Eng. Comm.*, **122**, 127-150 (1993)
15. A. Lawal, and D. Kalyon, *Polym. Eng. Sci.*, **34**, 1471-1479 (1994)
16. D. Kalyon, A. Lawal, A., R. Yazici, P. Yaras, and S. Railkar, *Polym. Eng. Sci.*, **39**, 1139-1151 (1999)
17. H. Potente and H. Ridder, *Int. Polym. Process.*, **17**, 102-107 (2002)
18. H. Potente, M. Bornemann, and M. Kurte-Jardin, *Int. Polym. Proc.*, **24**, 23-30 (2009)
19. R.C. Santi, E. Hage Jr., J. Vlachopoulos, and C.A. Correa, *Int. Polym. Proc.*, **24**, 346-353 (2009)
20. S.G. Hatzikiriakos and E. Mitsoulis, *Polym. Eng. Sci.*, **49**, 1960-1969 (2009)
21. A. Lewandowski and K. Wilczyński, *Int. Polym. Proc.*, **34**, (2019)
22. E.C. Bingham, "*Fluidity and plasticity*", McGraw-Hill, New York (1922)
23. R.B. Bird, G.C. Dai, and B.J. Yarusso,*Rev. Chem. Eng.*, **1**, 1-70 (1983)
24. E. Mitsoulis, "Flows of Viscoplastic Materials: Models and Computations" in *Rheology Reviews*, 135-178, The British Society of Rheology (2007)
25. P. Coussot, *J. Non-Newt. Fluid Mech*,.**211**, 31-49 (2014)
26. ANSYS Polyflow v.17, www.ansys.com (2017)
27. K. Wilczyński and A. Lewandowski, *Int. Polym. Proc.*, **29**, 649-659 (2014)

*The simulations have been performed at the Centre for Computational Science (Interdisciplinary Centre for Mathematical Modeling, ICM) at the University of Warsaw, and were supported by PL-Grid Infrastructure (HP/Supermicro Hydra computer of x86 architecture, with eight 2.8 GHz processors, 12288–40960 MB of RAM).

PVC Gelation Inside A Single Screw Extruder And Its Effect On Product Properties

A. Demirci[a, b], I. Teke[b], A. Goger[c], E. Canbaz[a], J. Vlachopoulos[d*]

[a]*Mikrosan Inc., Gebze, Kocaeli, Turkey*
[b]*Department of Mechanical Engineering, Yildiz Technical University, Besiktas, Istanbul, Turkey*
[c]*Department of Chemical Engineering, Karabuk University, Karabuk, Turkey*
[d]*Department of Chemical Engineering, McMaster University, Hamilton, Ontario, Canada*

Abstract. Two different PVC formulations were extruded in a single screw extruder with the same screw speed and different processing temperatures. The gelation level of different extrudates were analyzed by differential scanning calorimetry, capillary rheometry and solvent absorption techniques. The methods were compared qualitatively and quantitatively. The differential scanning calorimetry method appears the most accurate method for determining the PVC gelation degree with very small sample usage. The capillary rheometry has accurate results as well but could be inaccurate because of the temperature rise during the test. The temperature rise changes the viscosity of the material non-linearly and the gelation curve could have inaccurate results. The solvent absorption method is simply applied but only gives qualitative results when compared with the other two methods. Ultimate tensile strength, elongation at break, solid density and color change were also measured and compared for different samples processed at different temperatures. The extruded specimens have satisfactory properties for both PVC resins between 60%-90% gelation degree.

Keywords: PVC extrusion, gelation degree, capillary rheometer, DSC, solvent absorption
PACS: 61.25.hk, 61.41.+e, 81.20.Hy, 81.05.Lg, 82.35.Lr, 83.80.Sg, 83.80.Tc

INTRODUCTION

Poly (vinyl chloride) (PVC) is one of the most used commodity polymers in different extrusion applications. Processing conditions and additives in the formulation play a dominant role on the end-product properties of PVC [1]. Small crystallites fuse together with heating and creating a partially gelled state during PVC extrusion. The quality of extruded PVC is determined by measuring the gelation degree (GD) with different methods such as differential scanning calorimetry (DSC) [2–12], capillary rheometry (CR) [2, 5, 8], hardness analysis [13, 14], solvent absorption (SA) [3, 9, 15], laser surface scattering [3] and microscopy analysis [3]. The most frequently used technique for the determination of GD is DSC. The DSC technique is based on two endothermic enthalpy peaks of PVC which were identified and explained for the first time by Gilbert and Vyvoda [8]. Potente et al. [7] quantified this method by using the ratio of one of the enthalpy peaks against the sum of both. Fillot et al. [4] optimized the DSC protocol for a wide range of GD (0% to 100%) with high resolution DSC measurements. Chen et al. [16] and Gramann [11] indicated the DSC benefits in obtaining reproducible results with very small amount of sample. Hinrichsen and Thornsteinsen [17] discussed the fact that the DSC technique does not give any information on the mechanical shearing history. A less frequently used method for the determination of the GD is based on entrance pressure drop over a zero length die in a capillary rheometer (CR) [2,5,8,16,18]. Chen et al. [16] indicated that the main advantage of capillary rheometry is the use of large enough sample to eliminate local inhomogeneity effects on GD. Moghri et al [6] and Benjamin [19] found that the optimum GD is approximately between 60%-80% for best mechanical properties of PVC extrudates. Covas and Gilbert [20] studied the behaviour of unplasticized PVC (uPVC) inside a single screw extruder for barrel temperatures between 170 °C to 200 °C. They concluded that post-yield parameters reach maximum values for GD in the range of 70%-80%. Thomas and Harvey [21] investigated the effect of processing conditions on Charpy impact strength, color, gloss and GD of PVC window profiles and made analysis to determine the optimal processing conditions.

In the present study, differential scanning calorimetry (DSC), capillary rheometry (CR) and solvent absorption (SA) methods are used for Gelation Degree (GD) determination of extrudates. The advantages and disadvantages of the two quantitative methods of DSC and CR were compared. In addition, important end-product properties like tensile strength, color change and solid density are correlated with the determined GD of the extrudates.

EXPERIMENTAL

Two different uPVC powder grade formulations (Table 1) used in this study. Apparent viscosity values without any corrections measured at 100 s^{-1} and 200 s^{-1} at 190 °C are shown in Table 1. Each formulation was mixed in a Zeppelin Reimelt Henschel FML10 / KM 23 hot-cold mixer (Kassel, Germany). Extrusion trials were made using a 25 L/D, 30 mm diameter Mikrosan MTV 30/25D single screw extruder (SSE) (Gebze, Turkey). The extruder was operated at a constant screw speed of 10 rpm with nine different melt temperatures (MT). The samples were cut in different lengths and cooled immediately in a cold water bath for to avoid continuation of gelation after the die exit.

TABLE (1). The formulations of resin A and resin B used in this study

	Formulation	
Material	Resin A (phr)	Resin B (phr)
PVC, *K* value: 67	100	100
Calcium carbonate ($CaCO_3$)	20	9
Stabilizer one-pack (Ca/Zn)	4	4.15
Titanium dioxide (TiO_2)	3.7	5
Impact modifier	–	5.5
	Apparent viscosity	
100 s^{-1} at 190°C	1,911 Pa.s	2,459 Pa.s
200 s^{-1} at 190°C	1,211 Pa.s	1,670 Pa.s

A differential scanning calorimeter Netzsch 200F3 Maia DSC (Selb, Germany) was used to determine the gelation degree (GD) of the extruded PVC samples. The device was operated under ramp mode from room temperature to 260 °C with a heating rate of 20 °C/min. The quantification of the gelation degree (GD) is based on the work of Gilbert and Vyvoda [8] and Potente et al. [7].

A Ceast Smart Rheo 2000 model capillary rheometer (Torino, Italy) was used as a second method for GD determination. The extruded samples were cut into pellets and the capillary rheometer bore was filled completely. The determination of GD is based on the entrance pressure drop measurement, apparently first proposed by Gonze [22] and subsequently used by other investigators, including Summers et al. [18], Krzewki and Collins [23], Fujiyama and Kondou [5], Fillot et al. [3]. All measurements were made at 145 °C bore temperature and nominal shear rate of 100 s^{-1} with a zero length capillary. Based on the work of Summers et al. [18], Krzewki and Collins [23], Fillot et al. [3] and our own DSC measurements, the entrance pressure drop (P) for the resin A and resin B samples extruded at 143 °C was assumed as 0% GD. The entrance pressure drop for the resin A and resin B samples extruded at 216 °C was assumed as the 100% GD.

A third method was used to assess the gelation level of our samples by immersing them into dichloromethane (DCMT) for 30 minutes at 15 °C, according to ISO 9852. Inadequately gelled PVC samples are highly damaged, while the adequately gelled samples have high resistance to chemical attack by DCMT and swelling. This method does not provide quantitative results.

The color changes (ΔE) of processed PVC samples were quantified by a TBT Scietech HP-200 precise color reader (Nanjing, China) according to ASTM D6290. After color calibration was done for a reference white PVC sample, the individual color measurements were recorded. ΔE is a single number that represents the total color difference between two colors.

Mechanical properties of the extruded PVC samples were determined by Instron 5969 Dual Column Tensile Test Machine (Massachusetts, USA). Crosshead speed of 5 mm/min was applied to samples at room temperature to determine the ultimate tensile strength (UTS) values. Test samples were prepared according to the rod tension tensile specimen in ASTM D638 standard method.

Solid density values were measured using the standard test method ASTM D1505. An A&D GR-200 semi-micro analytical balance (Tokyo, Japan) was used for precisely monitoring the weight of the PVC samples in air and ethanol. At least, three specimens were measured for each set of conditions and the average solid density values were calculated.

RESULTS AND DISCUSSION

In this study, two different PVC formulations were extruded in a single screw extruder at different melt temperatures to obtain extrudate samples at various gelation degrees (GD). The GD of extrudates were determined with differential scanning calorimetry (DSC), capillary rheometry (CR) and solvent absorption (SA) methods, which were described in the previous section. The effect of GD on product properties such as tensile strength, solid density as well as color change were investigated and the variation of these properties with processing temperature was established.

The two most important parameters during thermal analysis with DSC are the heating rate and the final test temperature. Final temperatures of DSC analysis were always set to 260 °C which is high enough to detect the primary crystallites endothermic peak "Area B" from the DSC curve. Also, Fillot et al. [4] measured the weight loss of unprocessed PVC window formulation from 50 °C to 300 °C and obtained dramatic weight loss after 245 °C. GD results for heating rate of 20 °C/min and 30 °C/min are about the same, while it is somewhat less at heating rate of 40 °C/min and 50 °C/min. The GD is significantly higher at 10 °C/min heating rate, which shows that the GD of PVC samples is affected by the long residence time in the DSC during the test. Therefore, it was decided to use 20 °C /min as the most suitable heating rate for the tests.

We used a zero-length capillary as explained earlier to determine GD. The choice of temperature of 143 °C for 0% gelation degree (GD) and 216 °C for 100% GD is based on our DSC results (approximately) and is consistent with other investigations [3,6,29,30].

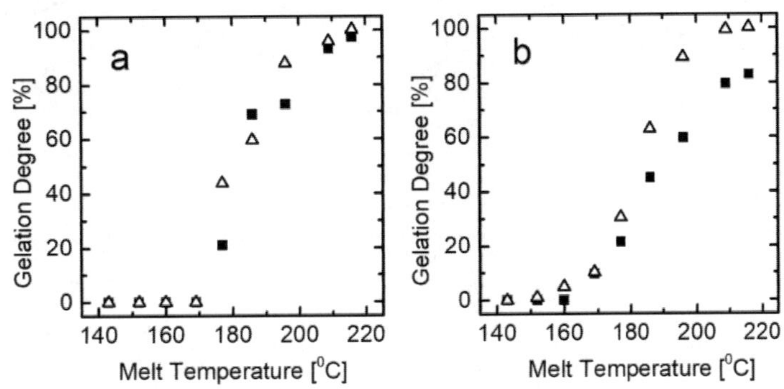

FIGURE 1. Comparison of GD determined with thermal and rheological methods for resin A and resin B processed at different melt temperatures. (a) GD for resin A samples (b) GD for resin B samples. ■ DSC, Δ CR

Figure 1a compares the calculated GD, based on DSC to that on CR for different melt temperatures (MT) of processed resin A samples. It shows that GD stays roughly at 0% from 143 °C to about 169 °C as determined both by DSC and CR. Then, GD starts to increase reaching 97% with DSC. The differences in the two methods are within the expected range. In Figure 1b, a similar comparison between DSC and CR is made for GD calculation of processed resin B samples. It shows that GD stays roughly at 0% from 143 °C to about 160 °C and starts to increase with melt temperature reaching a maximum value of about 82% with DSC. For the resin B the differences are larger between DSC and CR measurements than for resin A. We believe that the differences are due to temperature, which is difficult to measure in the entry region of the rheometer. For resin A the difference between max. and min. pressure is roughly 3 MPa while for resin B is about 14 MPa during the capillary rheometer tests, Higher pressure implies higher amount of mechanical energy converted into heat during the extrusion through our zero length capillary. If we assume an adiabatic process then the average temperature rise will be:

$$\Delta T = \frac{\Delta P}{\rho * c_p} \tag{1}$$

where ΔT is the temperature rise, ΔP is the pressure drop, ρ is the melt density and c_P is the thermal capacity. So, for a pressure difference of 14 MPa we would have a temperature rise of about 7 °C for resin B. The assumption of linear relationship between GD and entrance pressure drop is an oversimplification, especially for resin B. On the other hand, DSC is calculated directly from the thermogram curve on very small samples, but it is very important to select test specimens which have similar processing histories. In both Figures 1a and 1b, the GD versus temperature curve is defined by a characteristic S-shape in accordance with other investigations available in the open literature [3,6,16,29–32]. The bottom part of the S-shape curve implies that the transformation of primary crystallites cannot be started until the temperature rises to a certain level and slows down when certain higher level is reached. From our results it appears that primary crystallites are formed mainly between 170 °C and 195 °C, and the crystallite formation becomes slower between 195 °C and 210 °C for both resin A and resin B.

The solvent absorption (SA) method is used frequently in industry due to its low cost and simplicity. The extrudates processed at different melt temperatures were categorised as poor, good and very good based on examination for signs of attack. Resin A has poor result for the temperatures between 143 – 163 °C while Resin B has between 142 – 177 °C. Resin A have good results for 177 °C and 186 °C MT while Resin B has good results for 186 °C and 196 °C MT. Both Resin A and B have poor result at 216 °C MT due to degradation of PVC material.

Tensile strength, color and solid density are very important PVC product properties. Figure 2a shows tensile strength for resin A and resin B plotted against the GD obtained from the DSC measurements. The horizontal dash-line corresponds to the minimum tensile strength value of 34.5 MPa recommended in ASTM D1784 as standard specification of PVC compounds. Figure 2b shows the tensile strength for resin A and resin B versus GD obtained from the CR measurements. From Figures 2a and 2b it appears that max. values were attained for gelation degree (GD) roughly between 50% and 90% for both resins. The error bars are indicative of the max. and min. values obtained in our averaging procedure for each set of measurements. The extremely low value of the tensile strength for resin A at GD=100% is probably due to degradation. Covas and Gilbert [20] and Covas et al. [33] found that stress at break reaches max. level when the GD varied between 40% and 85% for their PVC extrudates.

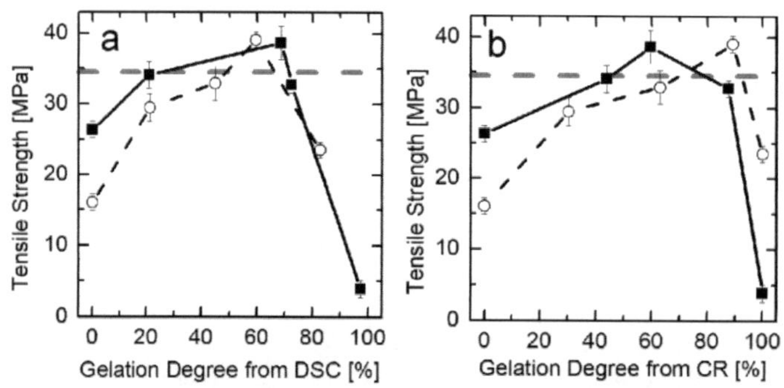

FIGURE 2. Tensile strength for resin A and resin B. (a) Gelation degree calculated from DSC. (b) Gelation degree calculated from CR. ━■━ resin A, ━ ○━ resin B. *Horizontal dash-lines indicate the minimum required tensile strength for PVC compound based on ASTM D1784

It is expected that structural changes during the extrusion of PVC compounds do not only affect mechanical properties, but also influence other product properties such as density and color. The density of the material is a very important parameter for PVC profile producers. Figures 3a and 3b show solid density, color and tensile strength as a function of melt temperature for resin A and resin B. The density tends to stay at a plateau until around 190 °C, then it starts to decrease with the lowest value at 216 °C. The decreasing density, which corresponds to increasing GD, especially after 190 °C, is apparently due to destruction of primary crystallites [20]. Figures 4a and 4b also show that the color of the extruded samples started to change towards brown color after 196 °C MT for resin A and 209 °C MT for resin B. Figure 3 also implies that for MT range between 175-185 °C we obtain over the minimum standard mechanical strength value with density of 1.52 g/cm^3 without color changes for resin A, while resin B can be extruded at MT range between 185-195 °C with over minimum standard mechanical strength value with density of 1.47 g/cm^3 without color changes.

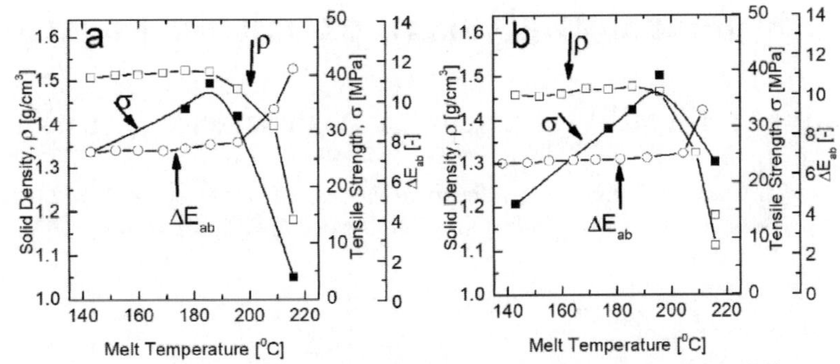

FIGURE 3. Solid density, color change and tensile strength of extrudates processed at different melt temperatures. (a) resin A, (b) resin B —■— Tensile strength, —□— Solid density, —○— Color difference ΔE_{ab}

ACKNOWLEDGEMENTS

The authors wish to thank Mikrosan Inc. for provision of resins and access their polymer processing laboratory. Furthermore, we wish to express our thanks to Instron Turkey for diligence with countless number of mechanical tests.

REFERENCES

1. M. Schiller, *PVC Additives: Performance, Chemistry, Developments, and Sustainability* (Carl Hanser Verlag GmbH & Co. KG, München, 2015).
2. C. H. Chen, R. D. Wesson, J. R. Collier, and Y. W. Lo, J. Appl. Polym. Sci. **58**, 1101 (1995).
3. L.-A. Fillot, P. Hajji, C. Gauthier, and K. Masenelli-Varlot, J. Vinyl Addit. Technol. **12**, 98 (2006).
4. L.-A. Fillot, P. Hajji, and C. Gauthier, J. Vinyl Addit. Technol. 108 (2006).
5. M. Fujiyama and M. Kondou, J. Appl. Polym. Sci. **90**, 1808 (2003).
6. M. Moghri, H. Garmabi, and M. Akbarian, J. Vinyl Addit. Technol. **9**, 81 (2003).
7. H. Potente, Schultheis S.M., and Gollner M., Kunststoffe **78**, (1987).
8. M. Gilbert and J. C. Vyvoda, Polymer **22**, 1134 (1981).
9. K. Piszczek, T. Sterzynski, and E. Łukaszewicz, Polym. Test. **22**, 115 (2003).
10. K. Piszczek and J. Tomaszewska, J. Vinyl Addit. Technol. **18**, 147 (2012).
11. P. Gramann, J. Cruz, and B. Ralston, SPE ANTEC (2009).
12. J. Tomaszewska, T. Sterzynski, and K. Piszczek, J. Appl. Polym. Sci. **93**, 966 (2004).
13. J. A. Covas, Plast. Rubber Compos. Process. **9**, 91 (1988).
14. E. Drenth, T. C. Bor, R. Visser, and M. Wolters, (2012).
15. B. Terselius, J. F. Jansson, and J. Bystedt, J. Macromol. Sci. Part B **20**, 403 (1981).
16. C. H. Chen, R. D. Wesson, J. R. Collier, and Y. W. Lo, J. Appl. Polym. Sci. **58**, 1093 (1995).
17. E. L. Hinrichsen and P. Thorsteinsen, J. Vinyl Addit. Technol. **2**, 18 (1996).
18. J. W. Summers, E. B. Rabinovitch, and P. C. Booth, Vinyl Addit. Technol. **8**, (1986).
19. P. Benjamin, in *PVC Process.* (1978).
20. J. A. Covas and M. Gilbert, Polym. Eng. Sci. **32**, 743 (1992).
21. N. L. Thomas and R. J. Harvey, Plast. Rubber Compos. **28**, 157 (1999).
22. A. Gonze, Chim Ind. - Genie Chim **104**, 422 (1971).
23. R. J. Krzewki and E. A. Collins, J. Macromol. Sci. Part B **20**, 443 (1981).
24. F. N. Cogswell, Polym. Eng. Sci. **12**, 64 (1972).
25. C. Kwag and J. Vlachopoulos, Polym. Eng. Sci. **31**, 1015 (1991).
26. M. Padmanabhan, C. W. Macosko, and M. Padmanabhan, Rheol. Acta **36**, 144 (1997).
27. V. Hristov and J. Vlachopoulos, Appl. Rheol. **17**, 57191 (2007).
28. M. E. Mackay and G. Astarita, J. Non-Newton. Fluid Mech. **70**, 219 (1997).
29. J. W. Summers and E. B. Rabinovitch, J. Vinyl Addit. Technol. **13**, 54 (1991).
30. M. Fujiyama and M. Kondou, J. Appl. Polym. Sci. **92**, 1915 (2004).
31. J. W. Summers, J. Vinyl Addit. Technol. **14**, 105 (2008).
32. S. Zajchowski, Polimery **11**, (2005).
33. J. A. Covas, M. Gilbert, and D. E. Marshall, Plast. Rubber Process. Appl. **9**, 107 (1988).

Mechanical properties of additively manufactured die with numerical analysis in extrusion process

A. Davoudinejad[a*], M. Bayat[a], A. Larsen[b], D. B. Pedersen[a], J. H. Hattel[a], G. Tosello[a]

[a]Department of Mechanical Engineering, Technical University of Denmark, Kgs. Lyngby, Denmark
[b]Inter Primo, Højbro Plads 6, Copenhagen K, Denmark

*alidav@mek.dtu.dk

Keywords: Extrusion; Additive manufacturing; soft tooling;; Finite element modelling.

Abstract

Extrusion process has been used for many decades as a transformation and shaping technology in the plastic processing industry. In this study the die deformation, critical for the extrudate product quality, was investigated in the plastic extrusion process by numerical modelling and compared with the experimental results. Dies with various thickness were fabricated by stereolithography (SLA) additive manufacturing (AM). COMSOL Multiphysics was applied as a modelling tool to simulate the experiments. Hence in this work a Fluid-Structure Interaction (FSI) numerical model based on the finite element method has been developed. The FSI model solves full Navier-Stokes (N-S) equation for the insert flow, while its output serves as an applied distributed load, e.g. as a boundary condition, for the structural analysis of the extruder. Eventually, this model can enable to identify the potential areas in the extrusion die where structural failure might occur.

Keywords: Additive manufacturing; extrusion; polymer; soft tooling.

INTRODUCTION

Extrusion represents one of the most established transformation and shaping processes. Different methods are applied to process materials, yielding production of finished or semi-finished products, commonly known as extrudates [1]. Various materials are formed through an extrusion process such as metals, polymer, clays, ceramics, foodstuffs, etc. However, polymer extrusion in particular is a process which covers a very large portion of the plastics processing industry to produce profiles with different cross-section geometry profiles. In the extrusion process die design and the corresponding extrudate are dependent on numerous issues such as how to make the flow distribution more uniform and how to anticipate post-extrusion effects [2]. The single screw extruders have been extensively used for thermoplastic materials since the thirties of the past century until today, and there have been numerus studies focusing on the plasticising process since that time such as Starve feeding and flood feeding [1] and a screw pulling-out technique to study polymer behaviour along the screw [3]. In terms of extrusion, dies manufacturing by conventional processes are time and cost consuming in comparison to additive manufacturing processes which have the capability of producing complex geometries, both external and internal to the component (such as conformal cooling channels without additional post-processing steps), and to fabricate a monolithic die design [4]. Die manufacturing was investigated recently by fabricating metal dies by AM processes for profile extrusion with Selective Laser Melting (SLM) method. One of the main objective is to reach a specific surface quality of the extruded plastics profiles due to the high roughness of the directly manufactured die [5]. A promising approach to minimise the necessary amount of design and manufacture iterations consists in the usage of numerical flow simulations in the design process before the die is manufactured [6]. In this study different experimental tests were carried out with the additively manufactured dies made of photopolymer material. The failure of the die was investigated in several processing conditions, and numerical analysis were applied in order to predict the weak point of the die for further consideration during the die design phase. The model will help the improvement of the AM design for optimized die manufacturing.

PRINCIPLE OF THE STEREOLITHOGRAPHY (SLA)

The AM technology used for fabrication of the die is the stereolithography (SLA) technology. In a Vat Photopolymerization method, an ultraviolet (UV) light cures, i.e. hardens, the liquid photopolymer resin placed in a vat, and the features are produced by slices one on top of another by tracing 2D contours of a CAD model using UV light. In this specific case, the SLA technology uses a laser beam to radiate the UV light, solidifying the resin in the vat in a point-by-point fashion [7]. The Somos® PerFORM material was used for this study, which is a suitable material for the current application due to its high stiffness and high temperature resistance. The platform based on SLA technology, is able to provide a relatively cost-effective method of rapidly creating 3D parts. Figure 1 and Figure 2 show the drawing and the part fabricated with SLA method for a U form profile respectively. The die was fabricated in two different thicknesses of 12 and 15 mm.

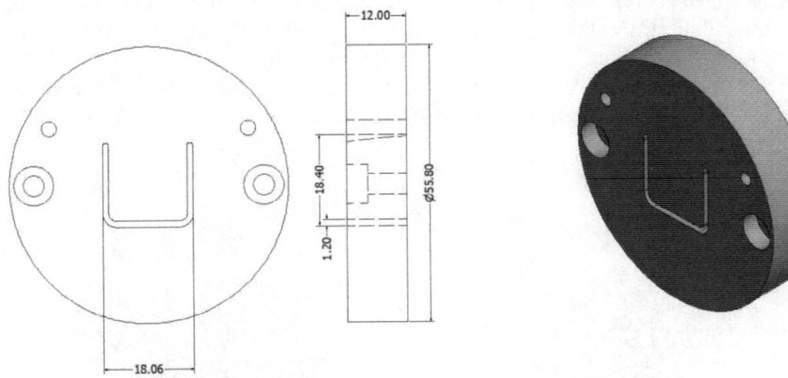

FIGURE 1. The detail drawing of AM fabricated die.

EXTRUSION PROCEDURE

Experimental tests were carried out with a single screw extrusion machine. The extruder used for all the test runs was a Cincinnati Proton 60 equipped with a 60 mm screw diameter characterized by a ratio L/D (Length/Diameter) equal to 25. The extrusion process material was selected as TPE Thermolast ® K TP6AFA. Table 1 presents the plastic material properties used during the extrusion experiments.

Prior to the die setup on the extrusion machine the die plate was pre-heated at 90 °C for 35 minutes warm up phase and 20 minutes on the extruder to reach to the proper temperature range. The initial screw speed was 1 rpm, the initial temperature of the tool was 125 °C. As the melt reached the tool and the pressure started to increase, a pressure of 5 bar was recorded when the first melt came out of the die. Table 2 shows the selected extrusion process parameters for the experiments.

TABLE 1. TP6AFA Mechanical properties.

Hardness	Density	Tensile strength	Elongation at Break	Tear Resistance
54 ShoreA	1,162 g/cm3	4,5 MPa	650 %	20 N/mm
DIN ISO 7619-DIN EN ISO 1183		DIN 53504/ISO 37	DIN 53504/ISO 37	ISO 34–1

TABLE 2. Test run process parameters details.

Temperature		Die	Melt Temperature	Melt Pressure	Screw Speed
Inlet	Next to die				
138 °C	149 °C	125 °C	142 °C	5 bar	1 rpm

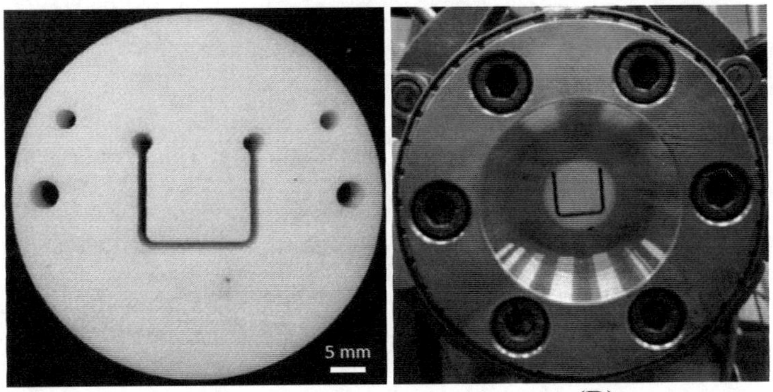

(A) (B)

FIGURE 2. (A) Additively manufactured die and (B) Experimental setup.

NUMERICAL CALCULATION PROCEDURE

To investigate the flow pattern during the extrusion process of the melt material and to analyze its impact on the extruder's mechanical response, a finite element model (FEM) is developed in COMSOL Multiphysics 5.3a software package. The model implements fluid-solid interaction (FSI) where first the fluid flow equations for a non-Newtonian fluid are solved for the melt and then the Hooke's linear elasticity along with force equilibrium are solved for the extruder die to find the resultant stress-strain field. In this study, the model consists of a computational solid mechanics (CSM) sub-model for extruder's mechanical analysis. According to Figure 3(A), the CFD model only holds for the domain in which the melt flows, while the CSM model applies to the die(see Figure 3(B)), which is the complement of the CFD model shown on the left. The melt starts flowing from the circular region on the left with an inlet pressure of 5 bar. The outlet region on the right side has a fixed atmospheric pressure. The internal domain belonging to the CFD model is densely meshed with quadratic tetrahedral elements to have high resolution of the fluid flow to capture the velocity gradients of the melt properly.

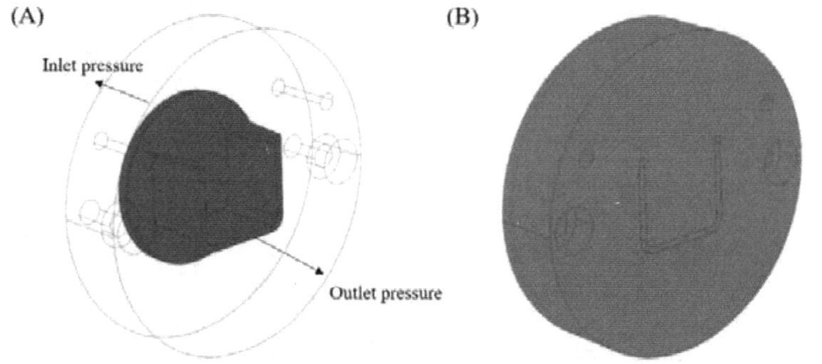

FIGURE 3. Model domain for: (A) CFD (melt) and (B) CSM (extruder) models.

COMPUTATIONAL SOLID MECHANICS

To find the corresponding stress/strain field induced by the fluid motion, it is necessary to solve the equilibrium partial differential equation in three dimensions.

$$\sigma_{ij,j} + F_{Vi} = 0 \tag{1}$$

where F_{Vi} (N/m^3) is the volumetric force field which is absent in the current model. The extruder material is assumed to follow Hooke's general linear elasticity.

$$\sigma_{ij} = E_{ijkl}\varepsilon_{kl}^{el} \tag{2}$$

$$E_{ijkl} = \frac{E}{1+v}\left[\frac{1}{2}\left(\delta_{ik}\delta_{jl} + \delta_{il}\delta_{jk}\right) + \frac{v}{1-2v}\delta_{ij}\delta_{kl}\right]\varepsilon_{kl}^{el} \tag{3}$$

Where E (Pa) and υ (-) are Young modulus and Poisson's ratio in equation (3), respectively. The finite element representation of equation (4) in a matrix form and derived from virtual work principle (VWP) can be expressed as follows.

$$[K]_{neqn \times neqn}\{D\}_{neqn\times 1} = \{P\}_{neqn\times 1} \tag{4}$$

where $\{D\}$ and $\{P\}$ stand for nodal displacements and nodal external forces of all degrees of freedom (d.o.f) respectively. $[K]$ is the global stiffness matrix of the domain which is calculated as follows [8].

$$[K] = \sum_{e=1}^{ne}\left(\iiint_{V_e}[B]^T[C][B]\cdot dV\right) \tag{5}$$

where $[C]$ and $[B]$ are element stiffness and their corresponding differentiated shape function matrixes.

RESULTS AND DISCUSSION

After proper tool setup and machine warm up for stable process condition, the extrusion started with low screw speed that was then progressively increased. The failure of the die occurred after a short run of approximately 5 minutes when the die pressure reached a peak of 6 bar with the process parameters listed in Table 2. The extrudate product obtained was acceptable only for a small production since the die failed shortly after running. Figure 4 illustrates the extrudate by AM die and the failure, with the critical regions of the die where the cracks happened. Figure 4(A) shows the cracks occurred by elevating the die temperature in both top corners of the U shape. The cracks and the failure of the die affect the extrudate mainly in terms of the thickness of the profile, which was not uniform after the defect appeared. The numerical analysis carried out in the die with 12 mm thickness performed better in the experiments than 15 mm thickness.

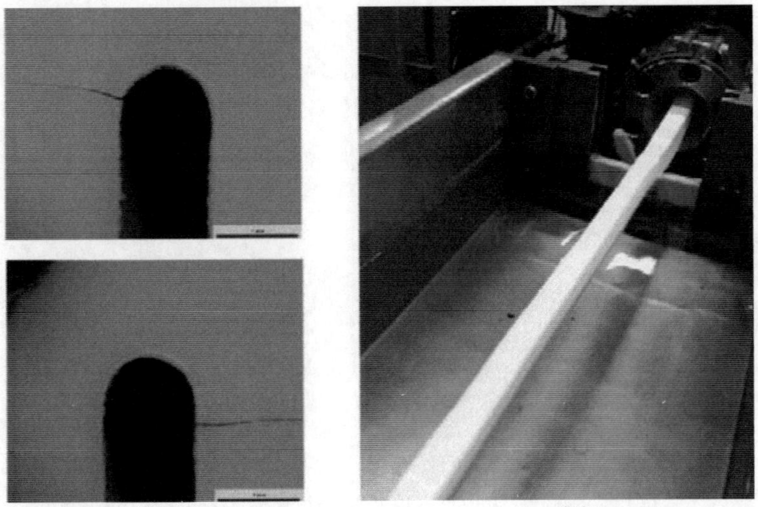

(A) (B)

FIGURE 4. (A) Cracks area on the dies and (B) extrusion process with AM die. Scale bars indicate 1 mm.

Based on Figure 5(A), maximum amount of z-direction deflection occurs on the lower side of the insert's wing and its maximum value is found to be around 0.014 mm. The corresponding Von Mises stress contour is shown in Figure 5(B). Two zones with high stress concentration are formed right on the upper corners of the insert gate. These two critical zones are found out to be the main cause of crack formation in the die. To give a further insight, the principal stresses along with directions are shown with colored arrows in Figure 6. According to Figure 6 the most critical zone prone to crack propagation or formation would occur on the upper corners of the insert gate. Based on this analysis, the direction of principal stresses are inclined towards 30°, measured from the horizon. This means that the crack will most probably occur along this direction where it can release the highest amount of energy. This prediction of the model is aligned with the experiments, as shown in the crack propagation in the die top corners presented in Figure 4(A).

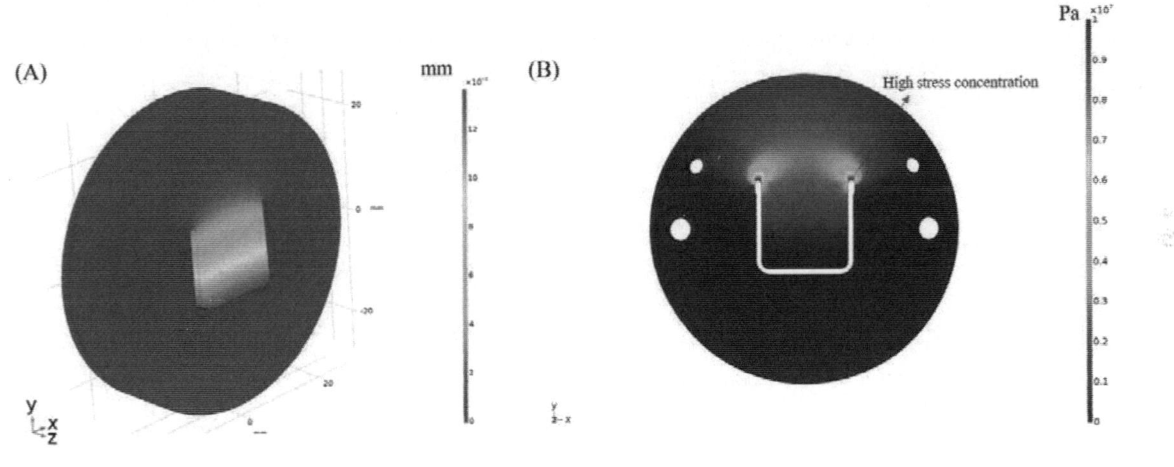

FIGURE 5. (A) z-direction displacement and (B) the Von Mises stresses contours for the insert.

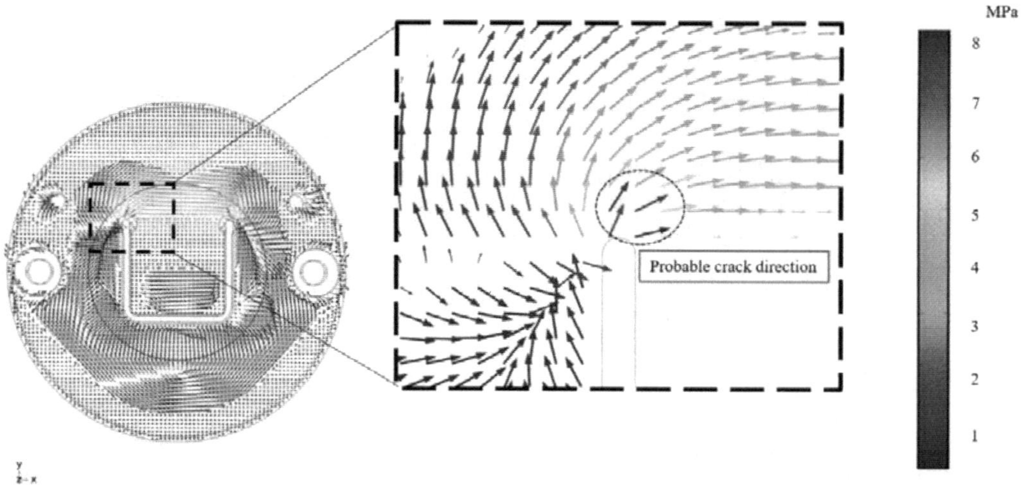

FIGURE 6. Vectors of first principal stresses. The colors show the magnitude of the principal stresses.

CONCLUSIONS

In this work the SLA AM process was used for fabrication of the die with photopolymer material in a plastic extrusion process. The failure of the die was investigated in different experiments and with a numerical model applied in order to investigate the thermos-mechanical properties of the die during processing conditions. An acceptable shape of the extrudate profile was achieved with AM die however, after a short production time, cracks defects occurred and breakage of the U shape geometry compromised the reliability of the extrusion process. The model accurately predicted the cracks area and the stress distribution on the die region of interest.

ACKNOWLEDGMENTS

The research leading to these results has received funding from the People Programme (Marie Curie Actions) of the European Union's Seventh Framework Programme (FP7/2007-2013) under REA grant agreement no. 609405 (COFUNDPostdocDTU).

REFERENCES

[1] C. Rauwendaal, *Polymer extrusion*. Carl Hanser Verlag GmbH Co KG, 2014.

[2] J. M. Nóbrega, O. S. Carneiro, P. J. Oliveira, and F. T. Pinho, "Flow Balancing in Extrusion Dies for Thermoplastic Profiles," *Int. Polym. Process.*, 2003.

[3] K. Wilczyński, A. Lewandowski, and K. J. Wilczyński, "Experimental study for starve-fed single screw extrusion of thermoplastics," *Polym. Eng. Sci.*, 2012.

[4] C. Hopmann and W. Michaeli, "Extrusion Dies for Plastics and Rubber," in *Extrusion Dies for Plastics and Rubber (Fourth Edition)*, Fourth Edi., C. Hopmann and W. Michaeli, Eds. Hanser, 2016, pp. I–XVIII.

[5] N. Yesildag, C. Hopmann, C. Windeck, S. Bremen, K. Wissenbach, and S. Merkt, "Opportunities and challenges of profile extrusion dies produced by additive manufacturing processes," *AIP Conf. Proc.*, vol. 1914, no. 1, p. 40002, 2017.

[6] C. Brecher *et al.*, "Integrative production technology for high-wage countries," in *Integrative production technology for high-wage countries*, Springer, 2012, pp. 17–76.

[7] A. Davoudinejad *et al.*, "Geometrical and feature of size design effect on direct stereolithography micro additively manufactured components," *Procedia Struct. Integr.*, vol. 13, pp. 1250–1255, Jan. 2018.

[8] R. D. Cook, *Concepts and applications of finite element analysis.* John Wiley & Sons, 2007.

Analysis of Fiber Structure and Properties of Drawn Fibers prepared by Cold Drawing in Ethanol and subsequent Hot Drawing

Go Dongwoo[1], Takarada Wataru[1, a)] Aneja Arun[2] and Kikutani Takeshi[1]

[1]*Department of Materials Science and Engineering, Tokyo Institute of Technology*
2-12-1-S8-32, O-okayama, Meguro-ku, Tokyo 152-8550 Japan
[2]*Department of Engineering, East Carolina University, Greenville, NC 27858, USA*

[a)]takarada.w.aa@m.titech.ac.jp

Abstract. In order to improve the mechanical properties of the fiber produced by the continuous drawing process using ethanol, drawn fibers were produced under various drawing conditions including multi-stage drawing. It was confirmed that the fiber prepared through the two-stage drawing and the annealing process had a tensile strength of up to 1.2 GPa. By comparison with the fibers made by conventional multi-stage drawing process, it has similar strength and higher elongation at break. Crystallinity or crystalline orientation of these fibers were quite similar, but results of microscopic Raman spectroscopy suggest that the molecular orientation and residual stress distribution of ethanol drawn fiber was lower than that of conventionally drawn fibers especially at the center of the fiber.

INTRODUCTION

It has been reported that in the tensile drawing of amorphous low oriented PET fibers in an organic solvent [1-4], multiple-necking phenomenon occurs because of the reduction of surface free energy from polymer-air interface to polymer-solvent interface. This drawing method is called multiple-neck drawing. In addition, infusion of an organic solvent, i.e. sucking in of organic solvent with the aid of tensile drawing, was demonstrated. The infusion phenomena occurs when a fiber is drawn in an organic solvent [1-4] without requiring long time intervals or high temperature which are typically involved in conventional diffusion methods. Crystallization of molecules by organic solvents infused into fibers has also been reported [1-4]. The mechanism of infusion in the drawing process is that when the tensile strain is applied to the polymer, since the poisson's ratio is less than 0.5, the volume increases with the tensile strain. On the other hand, since the volumetric elastic modulus of the polymer is not sufficiently high, negative pressure is generated, thereby sucking the medium surrounding the polymer [2]. At this moment, it has been reported that the surrounding media penetrating into the polymer matrix is not only the solvent but also accompany the solute dissolved in solvent [3-4]. For instance, in case of using a solution in which a dye is dissolved as a solute, it is possible to dye the fiber by simultaneous infusion behavior of solvent and dye.

These phenomena occur when the organic solvent or solution is used as the surrounding media, simultaneously manifest its effect in a single step drawing process. The solvent-induced crystallization of the polymer, and the assignment of characteristics by the incoming solute, i.e. dyeing of fibers using dyes, are possible simultaneously. Thus it is expected that the number of steps in the manufacturing process can be reduced innovatively if such a process is applied in industrial.

Our earlier studies focused on the fiber formation behavior upon continuous drawing of as-spun amorphous poly(ethylene terephthalate) (PET) fibers through an ethanol bath at room temperature. Formation of multiple-neck, and infusion of ethanol into filament along with dye were confirmed. Intermittent initiation of a group of necks and fluctuation of drawing tension were effectively suppressed with introduce of the snubber pin, and also efficiency of infusion was remarkably increased. Furthermore, it succeeded in producing an ethanol-drawn fiber having a tensile strength comparable to that of a fiber produced by a conventional drawing method. Based on previous studies, stable continuous ethanol-drawing technique has been developed [5-6]. In this study, subsequent drawing was tried on the fiber produced by the continuous cold drawing process using ethanol to improve mechanical properties. Fiber structure of resultant fibers was also investigated.

EXPERIMENT

Materials

As spun amorphous poly(ethylene terephthalate) (PET) fiber was prepared by high-viscosity PET polymer (IV= 1.0, manufactured by Unitika Co., Ltd) for the purpose of obtain further high-strength fibers through multi-stage continuous drawing process and conventional thermal drawing process. The PET polymer was extruded at 300 °C from the spinneret with a single circular capillary of 1 mm diameter and taken-up using a winder placed at 350 cm below the spinneret. The through-put rate and take-up speed were maintained to 5 g/min and 400 m/min. Diameter of the as-spun PET fiber was about 105 μm, and tensile strength and elongation at break of that filament were 135 MPa and 450%, respectively. The natural draw ratio of that fiber in continuous ethanol drawing process was DR 3.6 and DR 3.65 at the feed speed of 0.3 m/min and 3.2 m/min, respectively.

Cold drawing with ethanol and subsequent hot drawing

The equipment for the continuous ethanol drawing of the as-spun amorphous PET filaments consisted of a feed roller and a take-up roller. Ethanol was continuously flowed into a glass channel of 300 mm long, which served as a drawing bath installed between the feed and take-up rollers. A snubber pin was placed in the glass channel at 10 mm from the filament entrance to stabilize drawing process. The drawn PET fibers were prepared through the single-stage continuous cold drawing with ethanol at the feed speed of 3.1 m/min.

For the purpose of producing further high-strength PET fibers, in some experiments, the multi-stage drawing was performed including the ethanol drawing step as described above. The first-stage drawing carried out in an ethanol drawing process with a drawing condition of DR 3.65, which correspond to natural draw ratio of PET fibers After first-stage drawing, the ethanol contained in the filament was sufficiently dried over one week at room temperature under atmospheric pressure. Then, second-stage drawing of the fiber was conducted to DR 1.5 at 80 °C, and then annealed at 200 °C.

The PET fibers also prepared by conventional thermal drawing for comparison. The amorphous fibers spun from high viscosity PET polymer were used in the thermal drawing process. The drawing process was carried out using continuous drawing equipment capable of heating. The fibers firstly drawn to DR 4 at the temperature of 75 °C, then second drawing was conducted to DR 1.25 at the temperature of 130 °C, and then annealed at 200 °C.

Analysis of fibers

The stress versus strain behavior of the drawn filaments were measured using a tensile testing machine (AG-I, Shimadzu Inc., Japan) at the strain rate of 50 %/min. Initial Young's modulus, yield stress, tensile strength and elongation at break were obtained through the analysis of the stress-strain curve.

Wide angle X-ray diffraction and small angle X-ray scattering patterns of the drawn filaments were obtained using an X-ray diffractometer (RMT-18HFVE, Rigaku, Japan). The birefringence was calculated from images taken using an interference microscope (JenaVert, Carl-Zeiss Jena GmbH, Germany) and CCD camera (Retiga R1, Qimaging Corp. Canada).

Drawn PET filament was analyzed utilizing a laser Raman spectrometer (NRS-5100, Jasco Inc., Japan) in order to evaluate the molecular structure of drawn fibers. As the laser source, a polarized green laser with a single wavelength of 532.26 nm was used for laser Raman measurements. The polarization direction of the laser was set parallel to the fiber axis. Cross-sectional distribution of fiber structure was also investigated by using confocal optical system.

RESULTS AND DISCUSSIONS

Tensile properties

Stress-strain curves of drawn fibers were shown in Figure 1. Single-stage ethanol drawn fibers showed 80% elongation at break, it was nearly double as drawing without ethanol fibers. From the previous studies, ethanol drawn fibers were already crystallized. However, this stress-strain curve suggests that the further drawability of ethanol-drawn fibers.

The mechanical properties of the multi-stage ethanol drawn fiber exhibited a tensile strength of ca. 1200 MPa at the elongation at break of about 22%. Meanwhile, the tensile strength of the fiber produced by the conventional thermal drawing process was ca. 1200 MPa at the elongation at break of about 17%. This can be evaluated that the tensile strength of multi-stage ethanol drawn fiber is comparable to the tensile strength of a fiber produced through a conventional thermal drawing process and slightly higher toughness because of higher elongation until break.

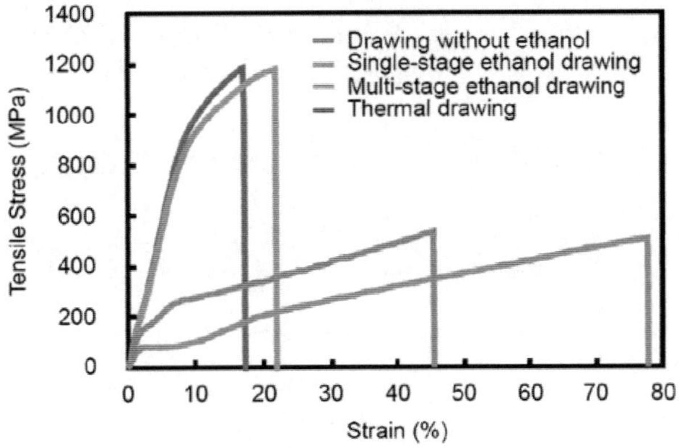

FIGURE 1. Stress-strain curves of PET fibers drawn with various drawing stages and drawing conditions.

Structure analysis

Highly oriented triclinic crystals of PET was also observed by 2D-WAXD in the fibers produced by the ethanol multi-stage continuous drawing process. Calculated crystallinity and crystalline orientation of PET were shown in Table 1. Birefringence was also listed in Table 1. Multi-stage ethanol drawn fibers had similar crystallinity and slightly lower orientation.

TABLE 1. Results of orientation analysis

1st Stage drawing method	Crystallinity (%)	Crystalline orientation	Birefringence
Ethanol	45.11	0.950	0.1783
Conventional	45.07	0.972	0.1850

Results of confocal Raman spectroscopy were showed in Figure 2 and Figure 3. Peak area ratio in Figure 2 indicating the degree of orientation of the PET molecules. There was little changes in the thermal drawn fiber samples in the degree of orientation depending on the depth from the surface to the inner center of the fiber. However, in both fiber sample drawn using ethanol showed decrease of the degree of orientation in the inner part compared to the surface of the filament. In the comparison between the ethanol multi-stage drawn fiber and the conventional thermal drawn fiber, the ethanol multi-stage drawn fiber sample exhibited relatively low degree of orientation as WAXD measurement and birefringence measurement. Full width at half maximum (FWHM) of 1616 cm^{-1} peak indicates residual stress distribution on each molecules [7]. Smaller FWHM means uniform residual stress. Ethanol drawn fiber showed smaller FWHM than thermal drawn fibers except for fiber surface. It may cause of larger elongation at break of ethanol drawn fibers.

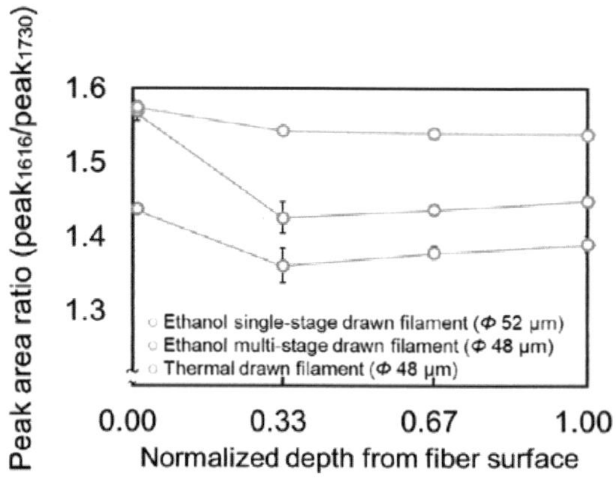

FIGURE 2. Raman analysis results of the peak area ratio of 1616 cm^{-1} peak and 1730 cm^{-1} peak across the fiber cross-section of various drawn fibers.

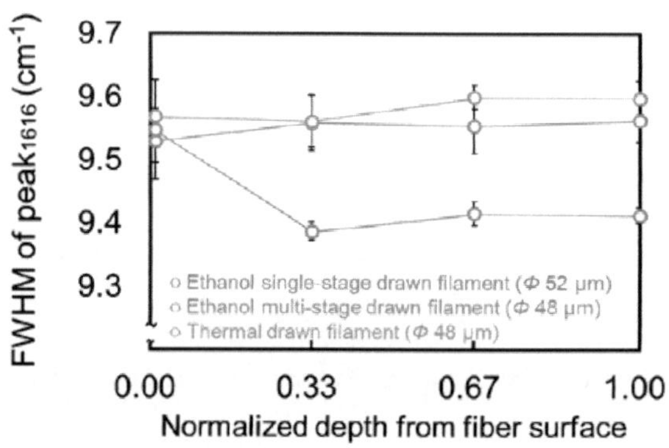

FIGURE 3. Raman analysis results of the full width at half maximum (FWHM) of 1616 cm^{-1} peak across the fiber cross-section of various drawn fibers.

CONCLUSION

For the purpose of producing high-strength PET fiber, a multi-stage continuous drawing using ethanol was carried out, and it was able to produce PET fiber having the same level of tensile strength compared to the high-strength PET fiber produced by conventional thermal drawing process. Both the PET fiber produced by the ethanol multi-stage drawing and the fiber produced by the thermal drawing showed the same level of crystallinity, and the thermal drawn PET fiber showed slightly higher degree of orientation than that of the ethanol multi-stage drawn PET fiber. Furthermore, in the case of thermal drawn fibers, there was little difference in the molecular structure between the surface part of the fiber and the inner part of the fiber, but the ethanol drawn fiber shows the degree of orientation has been decreased in the inner part of the fiber as compared with the surface part of the fiber.

As a conclusion, it has been found that it is possible to produce a PET fiber having the outstanding level of crystallinity and tensile strength even the process includes the continuous ethanol drawing process. Furthermore, the PET fiber produced in this study by ethanol multi-stage drawing process has the same level of tensile strength as conventional thermal drawn PET fiber, but there is still margin for improvement in the degree of orientation and the elongation at break. Thus it can be evaluated that the ethanol drawn fiber having a potential for further improvement of the mechanical properties

REFERENCES

1. O. V. Arzhakova, A. A. Dolgova, L. M. Yarysheva, A. L. Volynskii, N. F. Bakeev, Polymer, **56**, 256-262 (2015)
2. R. Khanum, W. Takarada, A. Aneja and T. Kikutani, Polymer, 59, 26-34 (2015)
3. R. Khanum, D. Go, W. Takarada, A. Aneja and T. Kikutani, Sen'i Gakkaishi, **71** (9), 273-283 (2015)
4. R. Khanum, D. Go, W. Takarada, A. Aneja and T. Kikutani, Polymer, **70**, 19-29 (2015)
5. D. Go, W. Takarada, A. Aneja, T. Kikutani, Materials Today Comm., **19**, 98-105 (2019)
6. D. Go, W. Takarada, A. Aneja, T. Kikutani, AIP Conf. Proc. **2055,** 060010 (2019)
7. D. Go, W. Takarada, T. Kikutani, IOP Conf. Series: Materials Sci. Eng. **254,** 032003 (2017)

The effect of different process parameters on poly amide 66 nano fiber by forcespinning method

Vahid Bazrafshan[a], Ardeshir Saeidi[a], Afshin Mousavi[a](S03-217)

[a]*Islamic Azad University, Science and Research Branch Tehran, Iran*

Abstract:

Nowadays, polymer Nano fibers have been extensively used in the industrial and research fields. At the present, products such as filters, wound healing, vessels, batteries, and many other types of products which are manufactured from polymeric Nano fibers are find in the markets. There are various methods for the polymeric nanofibers production. In this research, poly amide 66 nano fibers were produced by forcespinning®. In this way, the effect of different parameters such as concentrations of polymer solution, chamber temperature and rotation speed (RPM) were investigated. In this research we find out the average diameter PA66 nanofiber it was within nano scale. Also the production time was during few second and the speed flow rate was 6.5 meter over second.

Introduction:

Nowadays, nanofibers are highly researched by the researchers. There are several methods for producing Nano Fibers, One of these methods is ForceSpinning Method. The Force Spinning Method with the help of the inherent properties of the material, produces polymer fibers in the shortest possible time. Sarkar et al., in 2010 at the University of Texas introduced a project called Force spinning, This method has the capability to produce a very large volume of nano-fibers in little time, Dependent on the dielectric solution and the concentration used is very high and because of the high speed of production, the nozzle is not blocked [1]. G.L. Dotto, et al, have been developed with the Forcespinning technology of chitosan / polyamide nanofibres. The results showed that in the experimental conditions, chitosan / polyamide nanofibers with a diameter of 100 to 500 nm were obtained using the forcepination technique at a rate of production of 3 gr per hour[2]. Polyamide 66 has many usage in the industry, It was the same thing to use this material in this research. The solvent of this formic acid polymer is presented in Table 1, combining the percentage of PA66 and formic acid. This polymer is made of 20, 25 and 30 Weight percent in formic acid.

Table 1: Weight percent of polyamide solutions 66

	Formic acid(gr)	PA66(gr)	Magnet stirrer temperature (°C)
a	100	20	95
b	100	25	95
c	100	30	95

Figure 1 shows the mean diameter of the fiber with the variable solution, In this system, the RPM, temperature and nozzle are constant And the concentration of the solution is variable .In solutions with concentrations of 20, 25 and 30 Weight percent, The average fiber diameter is 664, 74 and 59 nanometers, Therefore, by increasing the concentration of the polymer, the mean diameter of the fiber increases. In Figure 2, the system temperature, nozzle and constant solution and the only variable is the RPM. It is observed in this material that by increasing RPM, we can see nano fibers thicker. Respectively, at 6000, 9000 and 12000 RPM, the fiber diameter is 73, 114, and 135 nanometers. In Figure 3, Parameters such as nozzle, rpm and concentration are considered constant and only the temperature varies. Respectively, The temperatures at 70, 90 and 110 ° C were investigated, with a mean diameter of 148, 118 and 114 nm. We see that the diameter of the fiber decreases with increasing temperature. Images 1 and 2 show SEM images of nano-fibers PA66. Figure 4 is captured with a magnification of 100 nm. The minimum diameter observed in this fiber is 18 nm, with a maximum diameter of 61 nano fibers and an average fiber diameter of 36 nm. Figure 5 is also shown with a magnification of 400 nm, with a minimum diameter of 55 nm and a maximum diameter of 231 nanometers, with an average fiber diameter of 114 nm, Using the Image j software, all of these fibers are measured. Figure 6 shows the rheometer test, which is carried out at a temperature range of 84 to 95 ° C, which increases the viscosity of formic acid evaporation from the polymer solution.

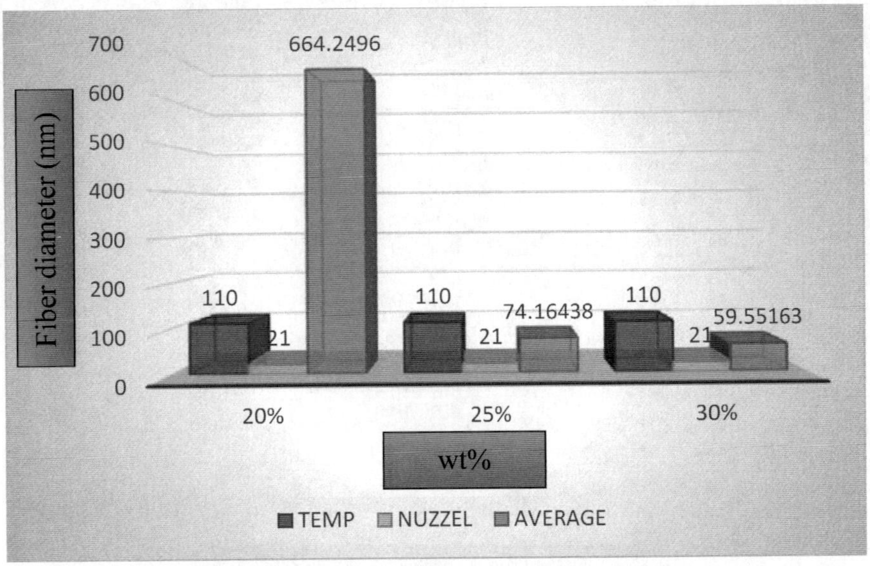

Figure 1: Average fiber diameter with variable PA66 solution

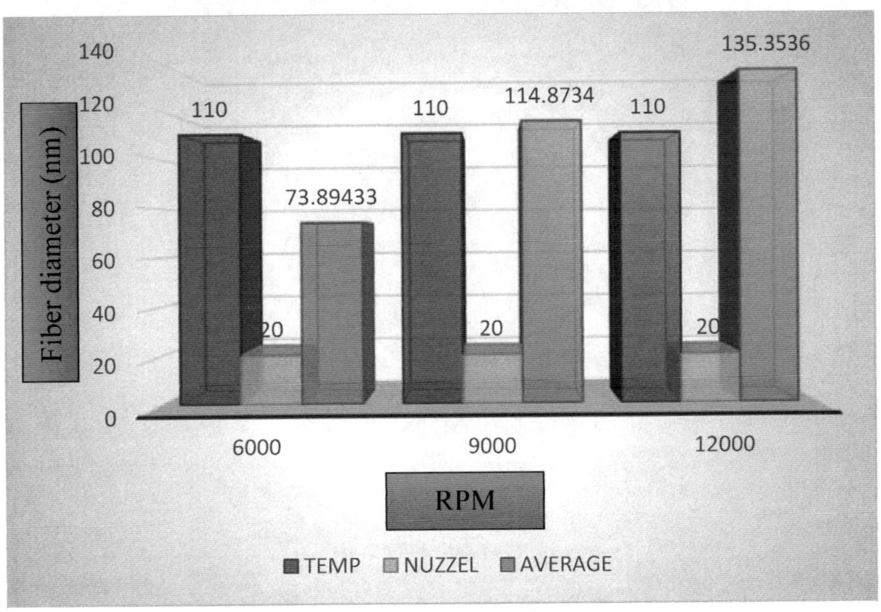

Figure 2: Average fiber diameter with varying variable speed PA66

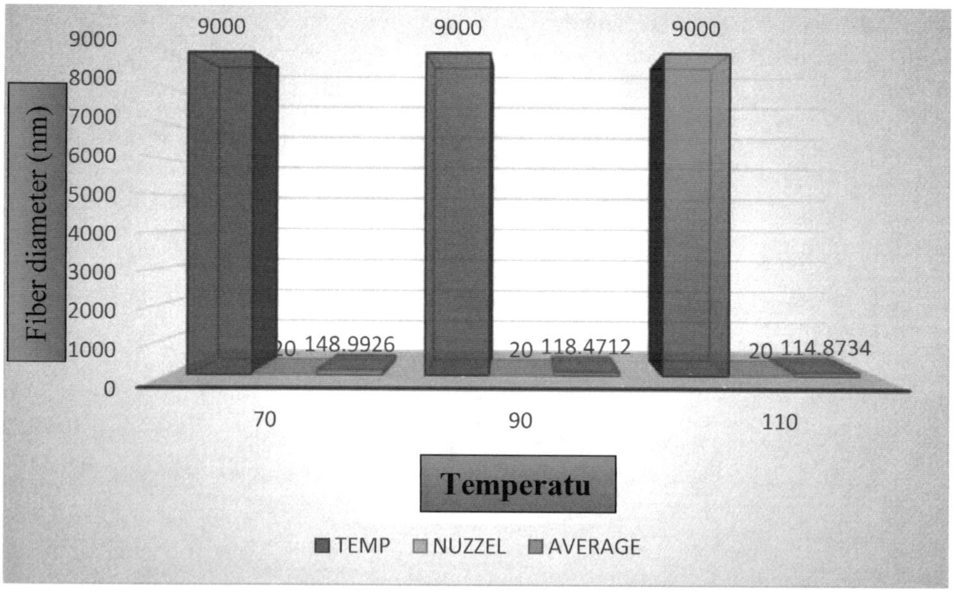

Figure 3: Average fiber diameter with variable temperature PA 66

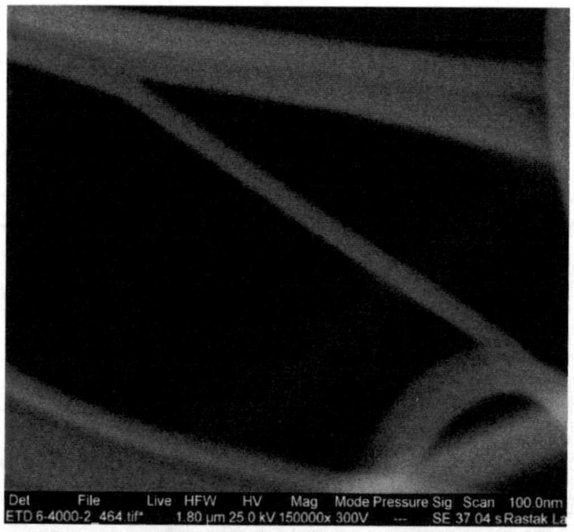

Figure 4: SEM image - PA66

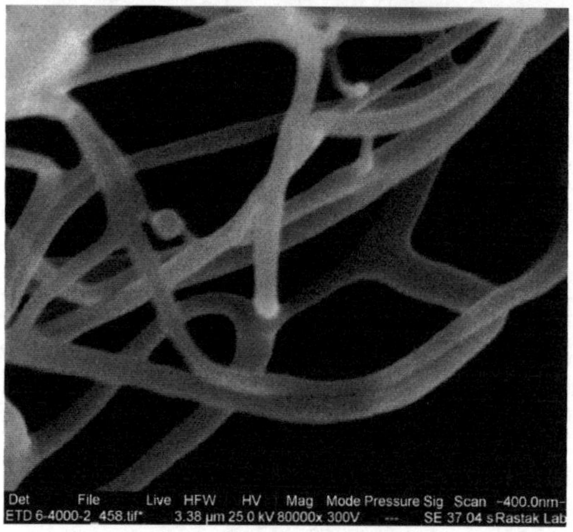

Figure 5: SEM image - PA66

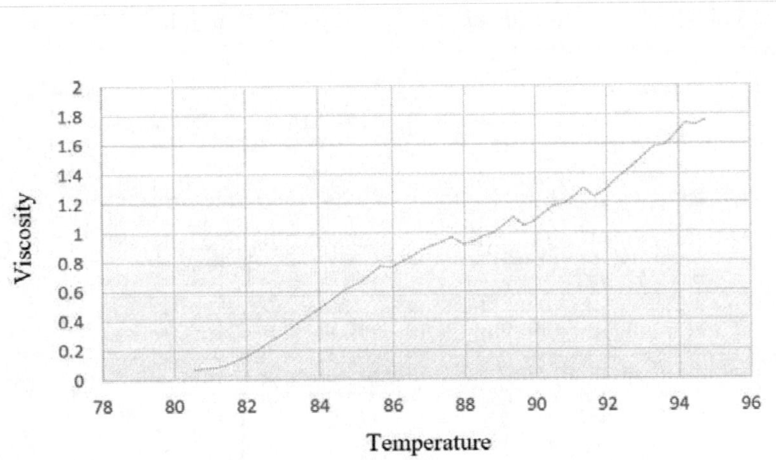

Figure 6: Rheometer Test 66 PA

In this study, considering the viscosity and the various factors involved in the fiber production process, we examined the rate of discharge of the solution from the spintern and the volumetric flow rate of each of the materials, and we arrive at the following results. First, using the following formula for each material, the speed of the solution of the spinet is calculated and, given the velocity obtained, we obtain the volume of discharge[3]:

$$U = -\frac{8L\mu}{\rho r^2} + \frac{1}{2}\sqrt{256(\frac{L\mu}{\rho r^2})^2 + 4\Omega^2[D^2 + 2L(C - \frac{L}{2})} \quad (1)$$

$L=$ 3/8 cm

$\mu=$ 1/764 Pa.s

$\rho=$ 1/14 g/ml

$r=$ 0/03 cm

$\Omega=$ 9000 RPM

$D=$ 4 cm

$C=$ 7/8 cm

U = 650/03 cm/s

or

U = 6/5 m/s

Here, taking into account the production conditions based on volumetric flow rates in two processes of electrospinning and forcing spinings, we will have a comparison with the production process of these two methods. In an electrospinning method, different values of 0.8 ml of polymer solution are obtained per hour, which can be obtained according to the following formula:

$$Q_V = \frac{V}{t} \quad (2)$$

$$Q_{V(ES)} = \frac{0.8\ cm^3}{3600s} = 0.0002\ \frac{cm^3}{s}$$

and in spurting forwards, according to the achieved fluxes, the slowest production time was observed to be about 30 seconds. Due to the volume of 25 ml of the reservoir, we can obtain the volumetric flow as follows:

$$Q_{V(FS)} = \frac{25\ cm^3}{30s} = 0.8333\frac{cm^3}{s}$$

With respect to the values obtained, we can compare the rate of production of the spin-forming force system with electrospinning as follows:

$$\rho_{Fluid} = \frac{Q_{V(FS)}}{Q_{V(ES)}} \approx 4000$$

As can be seen, the speed of spinning fibers production is at least 4000 times greater than that of electrospinning.

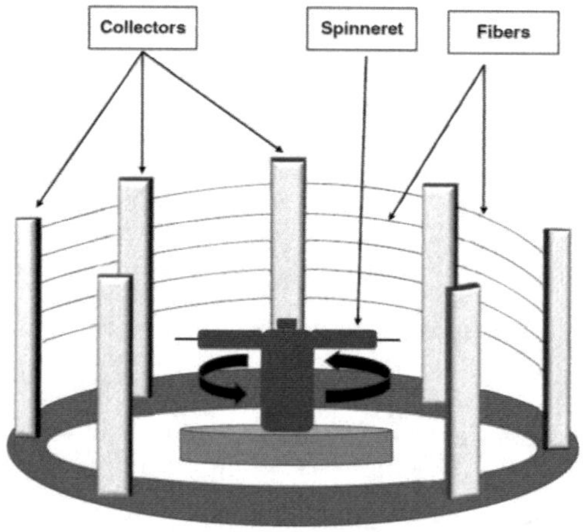

Figure 7: force spinning schematic [4]

ACKNOWLEDGMENTS:

Thanks My dear parents, Dr. Majid Berehi, Dr. arezoo Bazrafshan and my dear professors,Dear friends Mr. AmirHossein Najafi Mr. Hamed Mohammad Gholiha, Mr. Hossein Aghatabar. Appreciate the Science and Technology Park of Tarbiat Modares University for supporting us and baspar faravaresh Iranian company for the production of Polymer Nano-Fibers by force spinning Machine.

Conclusion:

The Force spinning machine (Figure 7) has the ability to produce polymer nanofibers, The speed of fiber production in this process is at least 4000 times greater than that of electrospinning, which is a factor in reducing production costs. In addition, the Force spinning system has the ability to produce nanoscale nanoparticles at the nanoscale. The device also has the ability to receive at different temperatures different fibers of different diameter, which indicates the effect of temperature on the concentration of polymer solution and evaporation of solvent from the fiber at different times. Also, the nozzle type changes the conditions for the fiber that can be obtained with different diameters. The high speed of this system has the ability to produce more fiber at a low cost, which can be used to prepare and use non-woven fabrics for sale to the market. The minimum diameter received was 18 nm, and the production speed was calculated to be 6.5 m / s.

REFERENCES

[1] Sarkar kamal et al; Electrospinning to Forcespinning ;Materialstady, 2010, volume 13 namber 11, page12-14

[2] G.L. Dotto, et al.; Chitosan/polyamide nanofibers prepared by Forcespinning technology: A new adsorbent to remove anionic dyes from aqueous solutions, Journal of Cleaner Production,2017,144, 120e129

[3] Simon Padron, et al.; Experimental study of nanofiber production through forcespinning; J. Appl. Phys. 2013; 113, 024318

[4]David De la Garza, et al.; Fabrication and characterization of centrifugally spun poly(acrylic acid) nanofibers, J. APPL. POLYM. SCI. 2019, DOI: 10.1002/APP.47480

Pilot-scale Melt-spinning of Poly (ε-caprolactone): Solid and Liquid-core Filaments

Figen Selli[1,2], Ümit Halis Erdoğan[1], Edith Perret[2] and Rudolf Hufenus[2]

[1]*Dokuz Eylül University, Department of Textile Engineering, Buca-İzmir, Turkey*

[2]*Empa, Laboratory of Advanced Fibers, St.Gallen, Switzerland*

Abstract. Being a member of biodegradable polyesters, Poly (ε-caprolactone) (PCL) has attracted significant interest in biomedical applications such as sutures, implants and tissue scaffolds. The low melting point of PCL makes it a good candidate for processing with additives that are thermally sensitive. Several research studies have investigated the production of PCL fibres via melt spinning, which is the most cost-effective fibre production method. However, the production parameters and fibre properties so far are not compatible with large-scale production. Therefore, in this study, we tried to establish different conditions and parameters for melt spinning of pure PCL filaments with improved properties. Furthermore, we utilized the knowledge of PCL melt spinning and low melting temperature of PCL to spin liquid-core PCL fibres by an economic process to expand the field of application.

PCL fibres were produced on a pilot scale melt-spinning plant. As a semi-crystalline polymer with extension-thinning behaviour, PCL shows necking (ductile failure) during drawing in the spinline. This abrupt and drastic decrease in the cross-sectional area produces oscillations in the filament tension (draw resonance) which impede melt-spinning. To overcome the draw instability of this biodegradable polyester, drawing setup was modified. The modified drawing system, in which the PCL filament was running over the second godet in reverse direction, prevented the draw instability during fibre production. WAXD patterns show distinct diffraction peaks for the drawn fibres which indicate highly oriented crystallites. The low melting temperature of polymer allowed processing below 100 °C and water-filled PCL fibres were also successfully spun in a continuous process. Tensile test results suggest that PCL can be melt-spun in a large scale with decent mechanical properties

Keywords: Biodegradable fibres, poly (ε-caprolactone), melt spinning, liquid-core fibres

Optimization of the UV Stabilization of a Plasticized PVC Film for Exterior Automotive Applications

Hélène Delinière[a, b], Jacques Lefranc[a], Fabienne Deuzé[a], Laurent Darcheville[a], Jérémie Soulestin[b], Sébastien Charlon[b], Marie-France Lacrampe[b]

[a]*DOURDIN SAS, 25 rue Emile Vandamme, F-59530 Saint-André-lez-Lille, France*
[b]*IMT Lille Douai, Institut Mines Télécom, Polymers and Composites Technology & Mechanical Engineering Department (TPCIM), 941 rue Charles Bourseul, F-59508 Douai, France*

Abstract. Poly(vinyl chloride) (PVC) formulation for outdoor applications should include the best suitable UV stabilization system. Two main families of UV stabilizers for polymer matrices exist: ultraviolet absorbers (UVA) and hindered amine light stabilizers (HALS). The effects of one UVA, one HALS and the combination of both were investigated in a PVC plastisol by comparing the colorimetric profiles, mechanical and viscoelastic properties of different films after artificial photo-irradiation. The UVA acts immediately by absorbing the major part of UV radiation which limits film degradation. Conversely, the HALS has no effect on the UV stabilization of PVC at the early stage, indeed the combination of both UVA and HALS behaves like single UVA, but at long term the presence of HALS accentuates PVC degradation. The tested HALS, being a strong base, is probably deactivated by acid-base reaction with HCl and is not able anymore to inhibit the PVC oxidation; thus, it should be replaced by a non-basic HALS (NOR HALS).

Keywords: poly(vinyl chloride), plastisol, ultraviolet stabilization, exterior automotive application.

INTRODUCTION

Formulation of poly(vinyl chloride) (PVC) plastisol offers the possibility to create a wide range of soft films with different colors and surface aspects. This is of great interest for decorative automotive solutions. Nowadays the specifications required by automotive constructors for outdoor applications demand an ever stronger ultraviolet resistance. However, PVC is degraded in presence of any source of energy. Due to structure defects implying labile chlorine atoms on the PVC chain, a dehydrochlorination occurs by creating carbon-carbon double bonds, i.e. formation of polyene sequences, and eliminating hydrogen chloride (HCl). Formed polyenes react by producing free radicals which undergoes an oxidative chain reaction. These mechanisms increase yellowing of PVC films and a degradation of mechanical properties [1]. Therefore, PVC formulation should include the best suitable ultraviolet (UV) stabilization system.

Two main families of UV stabilizers for polymer matrices exist: ultraviolet absorbers (UVA) and hindered amine light stabilizers (HALS). UVA absorb ultraviolet radiation, while HALS react with free radicals and thus inhibit the oxidative chain reaction [2]. In this study, the role of one widely used stabilizer of each family was investigated in a plasticized PVC. The studied UVA (Figure 1) belongs to the hydroxyphenyl benzotriazole class. This UVA exhibits strong absorbance in the 300 – 400 nm region and changes to an isomer structure when absorbing the UV radiation.

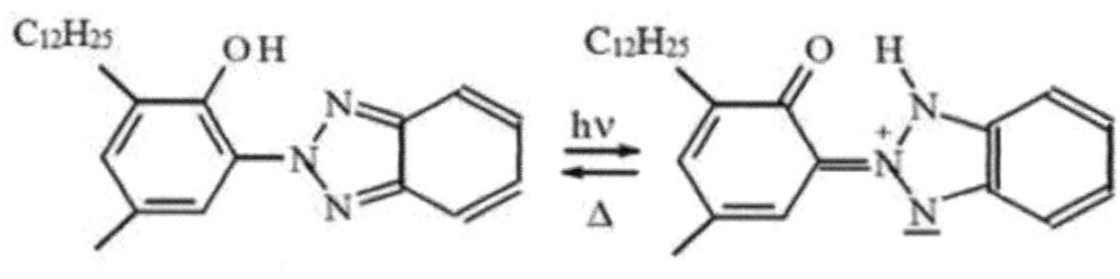

FIGURE 1. Mechanism of action of the studied UVA.

The examined HALS deactivates free radicals during a mechanism called Denisov cycle [3], where it changes to very stable nitroxyl radicals (Figure 2).

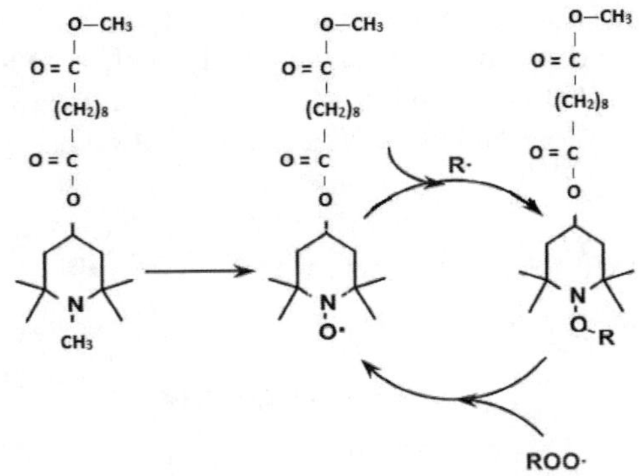

FIGURE 2. Mechanism of action of the studied HALS, R˙= free radical.

Their role is investigated by determining their single or combined effects on colorimetric profile, mechanical and viscoelastic properties of a plasticized PVC film after artificial photo-irradiation.

MATERIALS AND METHODS

Preparation of PVC Films

The PVC films are prepared by processing plastisols. PVC resin is dispersed in a polymeric plasticizer in presence of a stabilization system. Then the plastisol is poured on a non-stick paper support and heated at 110 to 200°C to obtain PVC films with a thickness of 300 microns.

The basic formulation of PVC films without the UV stabilizer is as follows: PVC resin 100 parts by weight per hundred parts of resin (phr), polymeric plasticizer 38 phr and mix of organotins as thermal stabilizer 2 phr. The UV stabilizers added in the different studied films are described in Table 1.

TABLE 1. UV stabilization system of the studied films.

Film Identification	HALS introduced (phr)	UVA introduced (phr)
UVA + HALS	0.6	0.6
HALS	0.6	-
UVA	-	0.6

Artificial photo-irradiation

The samples of the different films are photo-irradiated in a Weather-Ometer ATLAS Ci4000. They are exposed to the filtered radiation of a Xenon lamp at 340 nm with an intensity of 0.55 $W.m^{-2}$. A watering system is integrated to simulate rain phases. The temperature is (70 ± 2) °C with (50 ± 5) % of relative humidity outside watering periods. Watering last 18 min and stop during 102 min. Samples of each film are taken out after 166 h, 336 h, 522 h and 686 h.

Characterization of the films

The color evolution of the films during aging is measured with a Dataflash 100 colorimeter. The colorimetric profile is based on three parameters, DL which represents the lightness, Da the redness and Db the yellowing in comparison with a standard chosen here as the virgin UVA + HALS film. These three data define the global change of color DE as follow:

$$DE = \sqrt{(DL^2 + Da^2 + Db^2)} \tag{1}$$

The tensile properties of the virgin and aged films are evaluated with an Adamel Lhomargy DY30 dynamometer at the rate of 300 mm.min^{-1} (EN ISO 527-3).

The samples are also characterized by dynamic mechanical analysis (DMA) under shear stress, with a Metravib DMA +150. The measurement is realized at 10 Hz in linear viscoelastic region, the maximal force is configured at 2 N. The properties are examined during a second temperature elevation between -40 and 110 °C at the rate of 5 °C.min^{-1}.

RESULTS AND DISCUSSION

Colorimetric profile

First the color evolution of the different films during photo-irradiation is observed with naked eyes (Table 2). Yellowing of the HALS film occurs from the first aging week and continues to increase quickly and significantly. The UVA and the UVA + HALS films yellow very slightly between 0 and 686 h of irradiation. The yellowing is associated to polyenes formation during dehydrochlorination of PVC. Therefore, the single use of the HALS cannot prevent the dehydrochlorination initiation, whereas the presence of UVA, and thus the absorption of UV irradiation, limits considerably the plasticized PVC photo-degradation.

Samples of the UVA and UVA + HALS films were also observed after 3354 h irradiation to differentiate the effects of the single UVA and of the combination with the HALS. The UVA + HALS film is more degraded than the UVA film, thus the presence of the HALS promotes the dehydrochlorination of the PVC after a long exposition to UV radiation. The inefficiency of HALS to protect PVC could result from its deactivation by acid-base reaction between HCl and its basic amine [4].

TABLE 2. Coloration of the different films during photo-irradiation.

Film	0 h UV	166 h UV	336h UV	522 h UV	686 h UV	3354 h UV
UVA + HALS						
HALS						-
UVA						

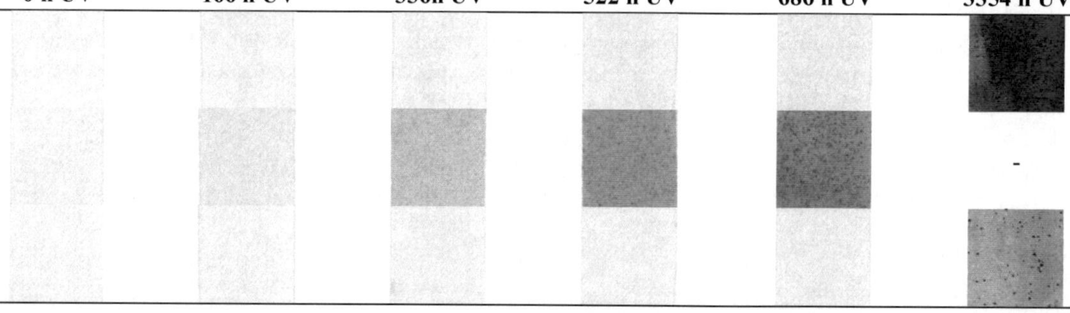

Only the evolution of DE and Db are presented in Figure 3 because the global change of color is mainly due to the yellowing. The results indicate that the effect of the studied UVA is similar as the effect of the combination of this UVA with the studied HALS; the color of these films is quite stable between 0 and 686 h of irradiation, showing no clear sign of PVC degradation.

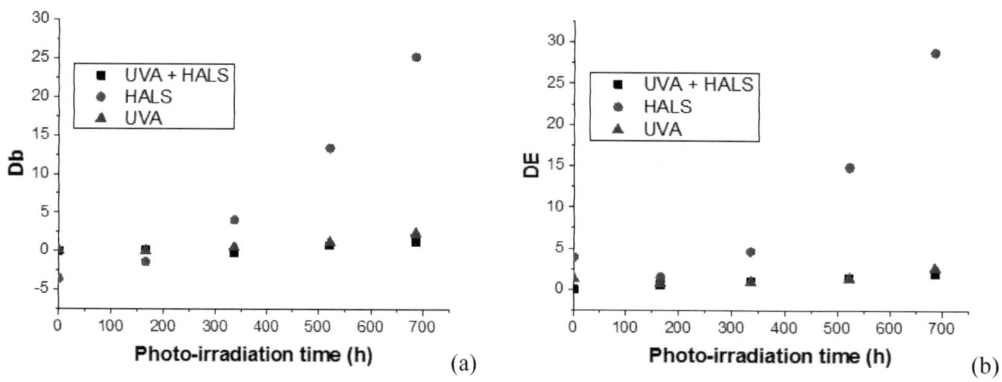

FIGURE 3. Evolution of (a) the yellowness (Db) and (b) the global change of color (DE) during photo-irradiation.

Mechanical properties

The mechanical properties of the HALS film degrade rapidly during the photo-irradiation: between 166 h and 686 h of aging, the tensile strength decreases almost linearly by 53 %, the Young modulus rises by 65 % and the elongation at break drops until being around 0 %. The presence of the HALS does not protect the plasticized PVC from a substantial flexibility loss during irradiation. The mechanical properties of PVC with UVA and PVC with UVA + HALS degrade progressively with notably a respective increase of the Young modulus of 20 and 30 % and a loss of elongation at break of 39 and 38 %, between 166 h and 686 h. Therefore, the degradation of PVC is limited but not inhibited in presence of the UVA, implying a stiffness increase.

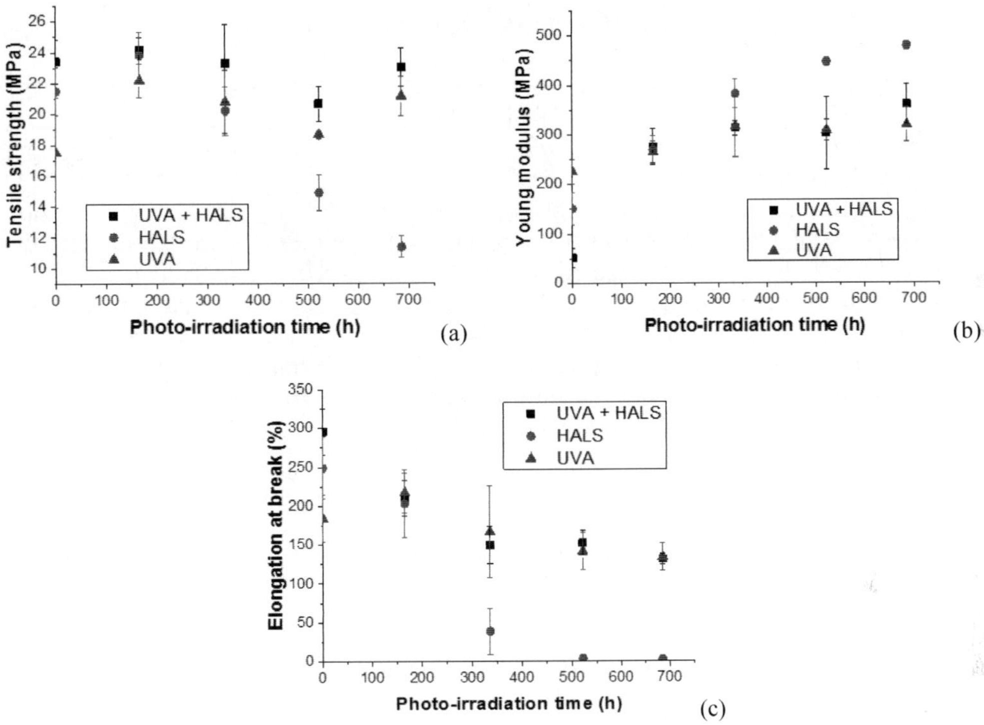

FIGURE 4. Evolution of (a) the tensile strength, (b) the Young modulus and (c) the elongation at break during photo-irradiation.

Dynamic mechanical analysis

The figure 5 shows the evolution of the curve tan(δ) = f(T) during the photo-irradiation for each studied formulation. The HALS film is the most impacted by the irradiation. The temperature at which the loss factor tan(δ) is maximal represents the α relaxation temperature associated with the glass transition temperature. Thus, the glass transition temperature of the HALS film increases from 51.3 to 64.5 °C between 0 and 686 h. The glass transition temperatures increase of the UVA and the UVA + HALS films is more limited during irradiation, respectively from 51.5 to 57.9 °C and from 51.3 to 57.8 °C. These increases of glass transition temperatures, and thus the increase of films stiffness could result from two combined phenomena: the scission of PVC chains implying the plasticizer migration and the reticulation of PVC chains, both occurring during the photo-oxidation of the PVC [4], [5].

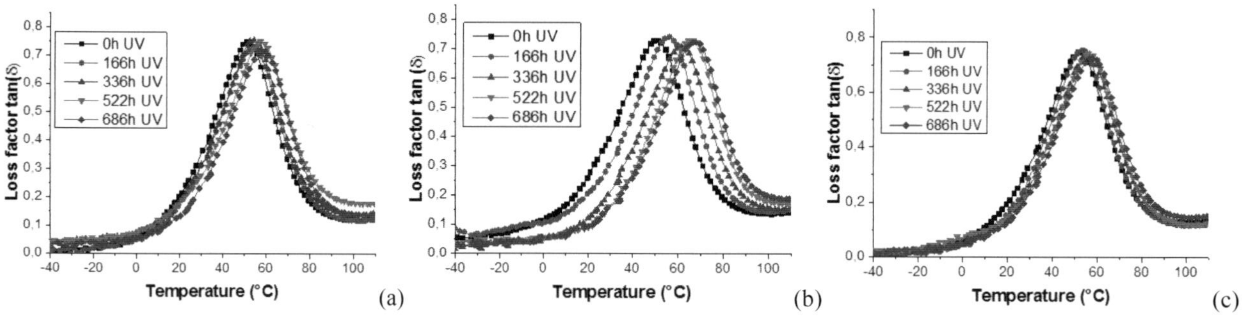

FIGURE 5. Loss factor evolution during irradiation of (a) UVA + HALS, (b) HALS, and (c) UVA films measured by DMA.

CONCLUSIONS

The role of a UVA and a HALS, used singly or in combination, as UV stabilizers of a plasticized PVC was investigated. The studied UVA, a benzotriazole, has an immediate effect on the plasticized PVC protection from irradiation by absorbing UV irradiation. The degradation is slowed down but not inhibited, occurring by an increase of yellowing and stiffness of the film.

The studied HALS does not stop the photo-oxidation of the film, the degradation begins in the first hours of irradiation and expands rapidly. The HALS, which is a strong base, could react with HCl released during the PVC dehydrochlorination, and thus be deactivated, thereby preventing it from scavenging the free radicals.

In the film stabilized by the combination of the UVA and the HALS, the effect of the UVA dominates at middle term: the degradation is limited as good as in the PVC stabilized singly by the UVA. However, the presence of the HALS accentuates the photo-degradation after a long time of irradiation. Therefore, the use of a non-basic HALS with the UVA is currently being studied to combine the absorption of UV rays with the deactivation of free radicals and thus to inhibit the propagation of the PVC photo-oxidation [6].

REFERENCES

1. G. Wypych, "Principles of UV degradation", in PVC Degradation and Stabilization, Third Edition., ChemTec Publishing, 2015, p. 167-203.
2. G. Wypych, "Principles of stabilization", in PVC Degradation and Stabilization, Third Edition., ChemTec Publishing, 2015, p. 287-412.
3. P. Gijsman, J. Hennekens, et D. Tummers, *Polym. Degrad. Stab.*, **39**, 225–233 (1993).
4. R. Chai, S. Chen, et J. Zhang, *J. Appl. Polym. Sci.*, **125**, 3376-3384 (2012).
5. M. Ito et K. Nagai, *Polym. Degrad. Stab.*, **92**, 260-270 (2007).
6. G. Capocci et M. Hubbard, *J. Vinyl Addit. Technol.*, **11**, 91-94 (2005).

Preparation of Nanofibers by Force Spinning, Based Polyvinylpyrrolidone and its Application in Wounds Healing

(Amirhossein Najafi)[a], Ardeshir Saeidi[b] and Afshin Mousavi[b] (S03-216)
[a] Islamic Azad University, Science and Research Branch Tehran, Tehran,Iran - tehran - Iran

[b] Islamic Azad University, Science and Research Branch Tehran, Tehran,Iran - tehran - Iran

Abstract:
nowadays, polymer nanofibres are used in various industries on a daily basis. Important industries such as military, electronics, filtration, sensors and medical and pharmaceutical industries. Worldwide, various programs are used as scars. The wound is usually made of fiber used polymer. Researchers at nanofibers have recently loaded drugs on fiber wounds to release drugs at a specified time. There are two types of chemicals and herbs in this study. We use several types of polymers to produce nano-fibers. Also, the Forcespinning system for fiber production is a new method in the production of high-speed fibers. Also, polymers used to produce polyvinylpyrrolidine fibers, which were also used in some compounds of chitosan

Keywords: Nano fiber-Forcespinning-Wound healing-polyvinylpyrrolidine

Introduction:
Lozano et al;To overcome present limitations encountered in typical electrospinning processes, a novel process was developed and tested. This new method called ForcespinningTM uses centrifugal forces, rather thanelectricalforces as in electrospinning[1].
Suganya et al;Herbal drug incorporated PCL/PVP nanofibers were prepared by electrospinning technique. The nanofibers were bactericidal to the testing microorganisms due to the strong antibacterial ability of bark extract from T.undulata. The produced electrospun nanofiber has great potential in biomedical applications[2].
Nowadays, polymer nanofiber has many applications in the field of medical and wound. The Forcespinning technique is one of the newest methods of production Nanofibre is high speed process, nanofibers have a great effect on wound healing.

Method:
This method has the ability to produce the volume in a short time, so that the minimum production speed of this device is 1000 times the Electrospinning method.The Forcespinning method also converts to a fiber for uniformity in the system due to its high speed, or to better explain the percentage of solution and drug composition during the whole process.However, the Electrospinning due to the production long time.The high speed fiber production makes it possible to use several drugs simultaneously in a polymeric material.

EXPERIMENTAL:

The materials used are PVP K30(Mv = 30,000) polymer purchased from chemie HSH Germany

As you can see in Table, we produced 2 solutions A and B with an engine speed RPM 9000 at 120 ° C. The fibers are shown uniformly in these two samples, indicating the repeatability of the Forcespinning, Sample A contains the Aloe Vera, chitosan and PVP, as shown in Figure 1, and the diameter of the fiber is 796 nm. Figure 2 also belongs to material B, as the change in the image shows, the fiber diameter is 115 nm.

Sample	Water(gr)	Ethanol(gr)	PVP(gr)	Alco Vera(gr)	chitosan(gr)	Vitamin C(gr)	Phenytoin(gr)	Zinc(gr)	Atorvastatin(gr)
A	50	50	65	0/5	0/1	0	0	0	0
B	50	50	65	0	0	0/2	1	2	2

Table 1

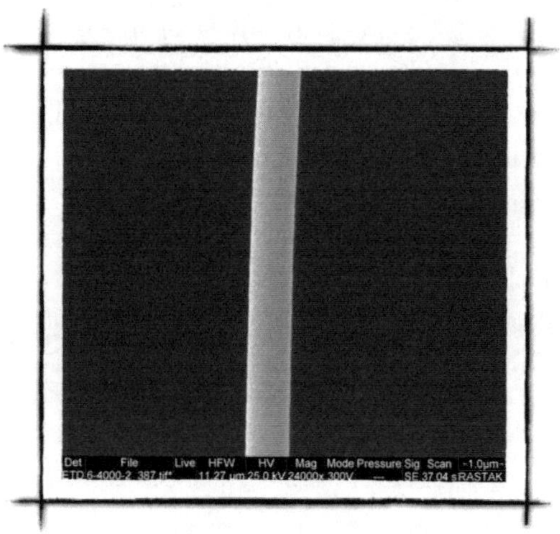

Figure 1:A

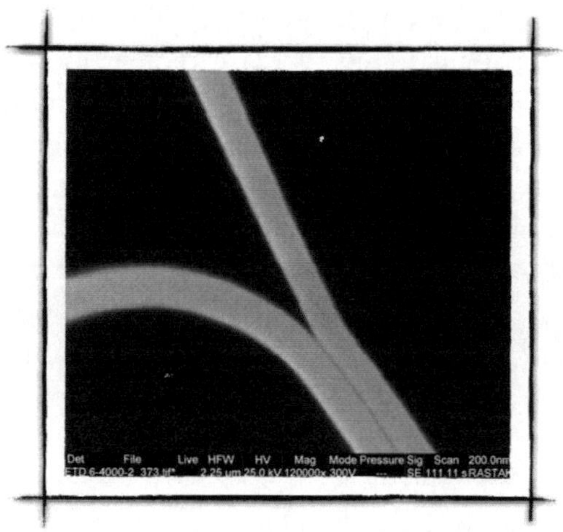

Figure 2:B

In this sample showed in Figure3, we see the release of a drug carrying chemical drugs that has risen in the first 6 hours, which is the rule of its release, followed by mildly slippery hours.

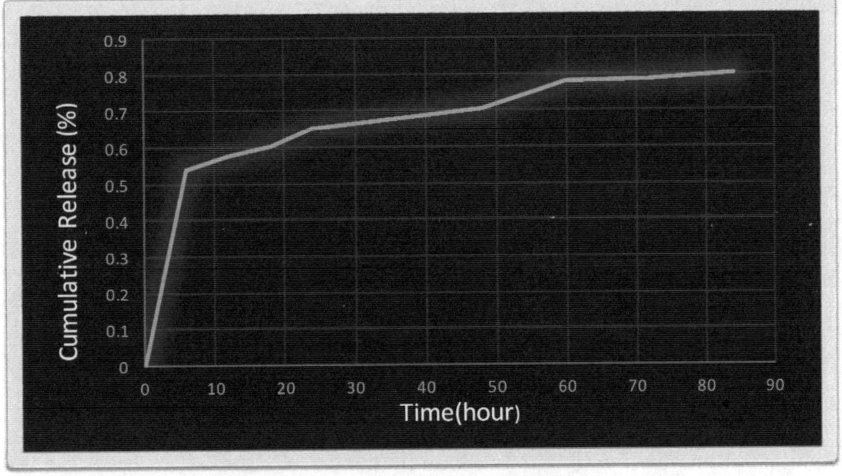

Figure 3: Chemical drug release test

The chitosan drug release test for the herbal drug is shown in the following diagram, which indicates that it was released early in the early hours and, furthermore, due to the presence of antibacterial properties of chitosan polymeric material, its release rate compared to the aloe Vera Pure Slower.

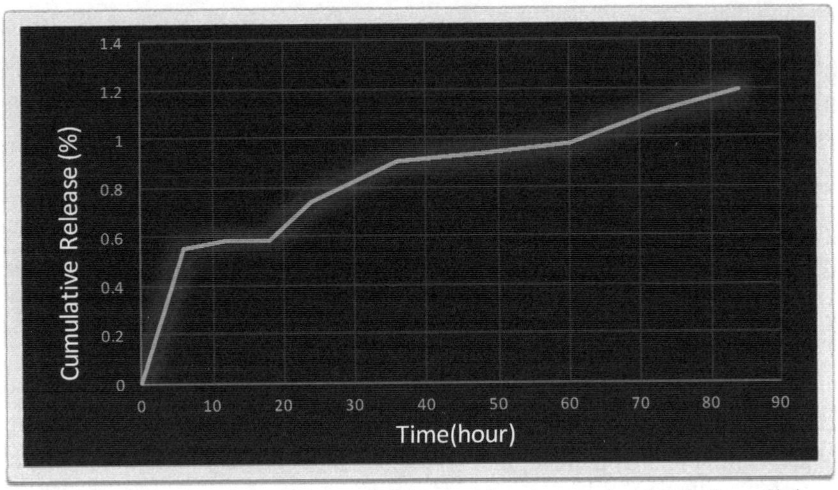

Figure 4: Test for the release of chitosan drug carrying herbal medicine

Figure 5 shows the TGA test, which we can see at about 395 degrees of degradation. Also, in the temperature range of 80 ° C, we see a degradation courier that indicates water evaporation.

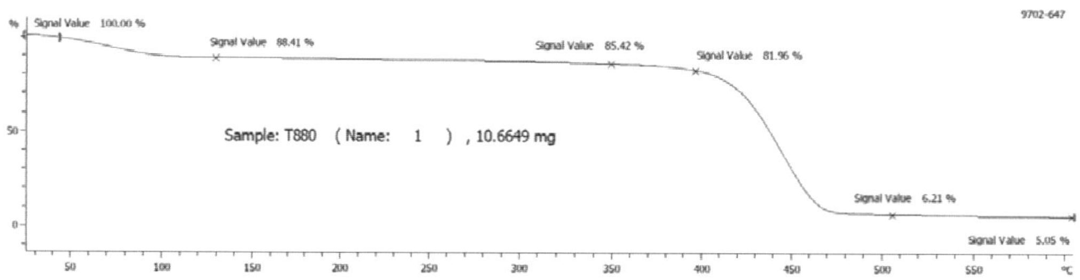

Figure 5:TGA test for Chemical drug release test

Figure 6 shows the TGA test, which we can see at the 410 temperature range at the 410 temperature degradation. Also, in the temperature range of 70 ° C, we see a degradation courier that indicates water evaporation.

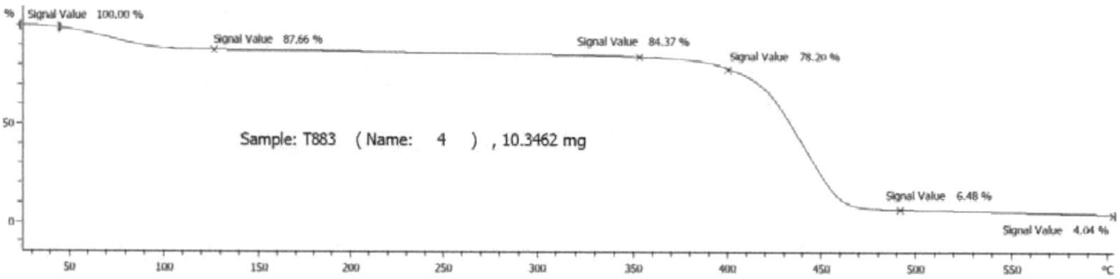

Figure 6:TGA test for chitosan drug carrying herbal medicine

Acknowledgments:

Thanks My dear parents, ,Dear friends Mr. Vahid Bazrafshan . Appreciate the Science and Technology Park of Tarbiat Modares University for supporting us and baspar faravaresh Iranian company for the production of Polymer Nano-Fibers by force spinning Machine.

Conclusion:

Forcespinning machine has a high power in producing nano fiber polymer, and polyvinylpyrrolidone is also highly used in wound healing due to its inherent adhesion. Chitosan increases the thickness of nanofibers to the extent that the nano fibers are about We observed 800 nm. However, fibers in which chitosan was not used did produce very low-diameter fibers. In other samples without chitosan, we produced fibers in the range of 100 to 200 nm. In the forcing spinning method, one-handedness and repeatability of the system and fibers Production is observed.

REFERENCES

1-Lozano et al. ; (2010); Electrospinning to Forcespinning; materialstoday13,11,12-14

2-Suganya et al ;(2011); Herbal Drug Incorporated Antibacterial Nanofibrous Mat Fabricated by Electrospinning:

An Excellent Matrix For Wound Dressings; Journal ofAppliedPolymer Science,Vol. 121, 2893–2899
3- Semnani et al;2018;Electrospun polyvinylidene pyrolidone/gelatin membrane impregnated with silver sulfadiazine as wound dressing for burn treatment
4- Miguel Manzano and Marı´a Vallet-Regı´;2010; New developments in ordered mesoporous materials for drug delivery; Journal of Materials Chemistry
5- Arezoo Mashak,Azam Rahimi;2012;Polymeric Drug Delivery systems;Tehran;Iran Polymer Society
6- Park, S.J., Carbon Fibers. 2015, Inha University, Incheon, Korea: Springer

An Experimental Study for Determination of Convective Heat Transfer Coefficients on Yarn Production Line

Mert Patkavak[a], PhD. Candidate[a], Prof. Dr. Hakan Demir[b]

[a]*Global Engineering Project Engineer, Kordsa Teknik Tekstil A.S., Alikahya/Izmit, Kocaeli, 41310, Turkey*
[b]*Heat and Thermodynamics Division, Department of Mechanical Engineering, Yildiz Technical University (YTU), Besiktas, Istanbul 34349, Turkey*

Abstract. In this study, the convective heat transfer coefficient around small diameter circular wires is determined in an effort to obtain main parameters affecting quenching in melt fiber spinning. The different diameter wires and air velocities were employed for detailed research in parallel and cross flow by using a wind tunnel. These correlations were based on conditions used in monofilament yarn production line. The wire diameters which are used in the experiments were varied from 0,2 mm to 1 mm and air velocities were varied from 0,5 m/s to 5m/s. These wires were heated with power supply and cooled with specified air flows for the determination of convective heat transfer coefficients. A multimeter was used to measure the voltage and current values; and the resulting wire resistances were calculated for each wire surface temperature. The convective heat transfer coefficients were obtained from total electrical power consumption and formulas of convection heat coefficient and radiative heat transfer. The experiments were carried out with different Reynolds numbers. Finally, the resulting Nusselt correlations were compared to the correlations reported by Ohkoshi,, Glicksman, Copley, Kase and Matsuo, Churcill and Bernstein, Hilpert, Fand and Kesvani, and, Dito. The present results are in good agreement with literature and they are in the same range for these Reynolds numbers.

Keywords: Convection heat transfer, Nusselt number, Reynolds number, Wind tunnel, Melt Spinning, Quenching
PACS: Replace this text with PACS numbers; choose from this list: http://www.aip.org/pacs/index.html

INTRODUCTION

Heat transfer (or heat) is thermal energy in transit due to a spatial temperature difference. The transfer of energy as heat is always from the higher-temperature medium to the lower-temperature one, and heat transfer stops when the two mediums reach the same temperature. Heat can be transferred in three different modes: conduction, convection, and radiation [1].

The heat transfer mechanisms used in this study are convection and radiation heat transfer mechanisms. These mechanisms are formed on the wires in the wind tunnel environment. The heat transfers with the radiation of the environment in which the wire is located is a size that cannot be negligible.

The regional convective heat flux of a fluid over a surface and total heat transfer are expressed as:

$$Q = h.\mathrm{A}.(\mathrm{Ts} - \mathrm{T}\infty) \qquad (1)$$

where A is cross sectional area of heat surface, Ts and T∞ are the temperature of surface and environment fluid, respectively.

Heat transfer is a function of Reynolds Number,

$$\mathrm{Re} = \rho.v.\mathrm{L} / \mu \qquad (2)$$

where ρ is the fluid density, v is a bulk velocity, μ is dynamic viscosity. The Nusselt number (Nu) is the ratio of convective to conductive heat transfer across (normal to) the boundary [1].

$$Nu = h.\mathrm{L} / \mathrm{k} \qquad (3)$$

Heat transfer coefficient h can be derived from (3) by placing the Nusselt number into the equation in which k is the fluid conductivity, and L is a characteristic length, such as a diameter,

In a study by Ravi and others in 2014, air flow and heat transfer in a two-dimensional circular cylinder unsteady flow regime were investigated. The effects of the Reynolds number on the drift and lift coefficients, the number of Strouhal, and the effect of a long circular cylinder on the heat transfer characteristic were investigated in the crossflow of Reynolds numbers from 50 to 180. At the same time CFD (Computational Fluid Dynamics) solver was also simulated using the Fluent program. The results obtained were consistent with the experimental / numerical data available for the circular cylinder in the literature. Nusselt number increased with increasing Reynolds number. At the same time, as the number of Reynolds increased, the drift coefficient increased. As a result, the relative drift coefficient for the specified operating range and the Nusselt number were obtained as a function of the Reynolds number [2].

In Manohar and Ramroop's experiment in 2010, the flow was made through two steel pipes of 34 mm and 49 mm in diameter along the air flow of 1.1 m/s and 2.5 m/s, with the horizontal axis at 30 and 60 degrees and heat transfer characteristics. These experimental studies are compared with literature. 30 ° slope and Nu values at an air velocity of 1.1 m / s were observed to agree well with the literature, however, considerable differences were observed at 60 degrees slope and air velocity of 2.5 m/s. The experimentally determined Nusselt number for 30 degrees of inclination varied from 2% to 18% in the air velocity of 1.1 m/s and from 12% to 41% in the air velocity of 2.5 m/s. The experimentally determined Nusselt number for 60 degrees of inclination varied from 19% to 45% in the air velocity of 1.1 m / s and from 29% to 65% in the air velocity of 2.5 m/s [3].

Experiments by Yuge in 1960 to find convection heat transfer for the sphere were carried out in the range of 1.44x105 to 3.5 x 105 Reynolds number and 1 to 105 Grashof number. Empirical formulas for forced, natural and combined convection heat transfer have been obtained and compared with other experiments in the literature. The results obtained are consistent with literature studies [4].

EXPERIMENTAL STUDY

A schematic figure of Wind Tunnel Test Apparatus is shown in **FIGURE 1**. The experimental test apparatus comprises tunnel, fan, power supply, digital multimeter, flow meter, frequency converter and wires with diameters between 0.2 mm to 1.0 mm. These are used to represent monofilament yarns. The model of the wires is Resistohm 30 and their chemical compositions is Ni-Cr-Fe alloy which is in conformity with ASTM B 344.

FIGURE 1. Wind Tunnel Test Apparatus

Air passes over around the wires and it is cooled. Air flow rate is adjusted with a frequency converter and measured with Testo 435 model flow-meter with a full scale accuracy of 1.0%. Wires holder is used to keep the wires stable during air flow. The cylinder position for cross flow is 90 degrees as shown in **FIGURE 2**.

FIGURE 2. Wire and Wire Holder for Cross Flow

Before testing, uncertainty of measurement test was realized. Repeatability and reproducibility of testing system was analyzed. Test system can recognize differences stemming from changes in different parameters on test. Also it gives repetitive results for the same inputs. Test system provides significant results according to uncertainty of measurement analysis. Tests are carried out at air velocities from 0,5 to 5 m/s and with wires having diameters varying from 0,2 to 1,0 mm. All parameters were selected to simulate monofilament yarn production line.

RESULTS

Wind Tunnel Test Apparatus was used to determine Nusselt and Reynolds numbers correlations to calculate convective heat transfer coefficients around circular wires with small diameters. Parallel and cross flow type air was used for these experiments to represent quenching process of monofilament yarn. Analysis of the results and drawing of the graphics were done using Minitab Software.

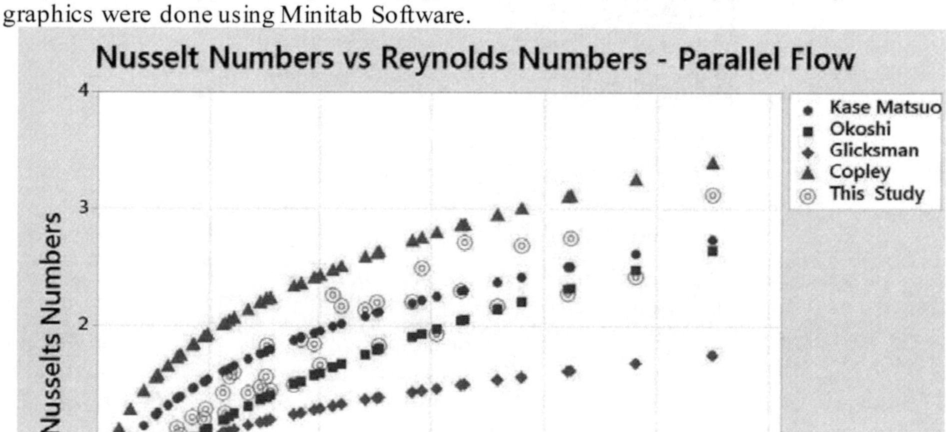

FIGURE 3. Comparison of Nusselt and Reynolds Number for Parallel Flow with Kase Matsuo, Ohkoski, Glicksman, Copley

FIGURE 3 shows that Nusselt and Reynolds numbers changes which they were calculated from experimental results for parallel flow. The experiments were carried out in the range of Reynolds numbers from 9 to 275, with an r^2 value of 0,93. The convective heat transfer coefficients were found from 24 to 160 $W/m^2.K$. The results were compared with the ones reported by Kase and Matsuo, Ohkoshi, Glicksman and, Copley [5-8].

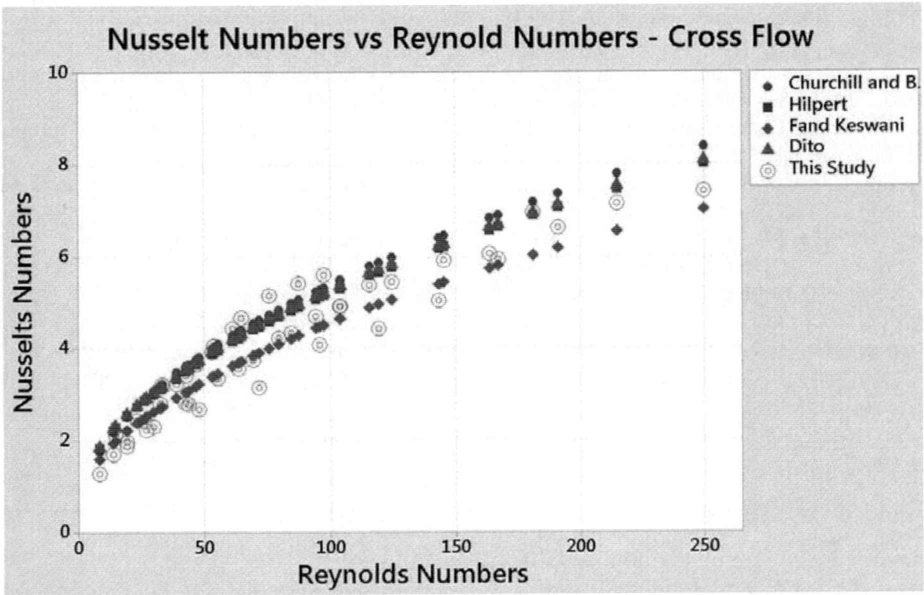

FIGURE 4. Comparison of Nusselt and Reynolds Number for Cross Flow with Churchill and B., Hilpert, Fand and Keswani, Dito

The cross flow experiments were carried out in the Reynolds numbers range from 8 to 250, with an r^2 value of 0,93. The Nusselt correlations which are determined from the experiments were compared with the ones reported by Churcill and Bernstein, Hilpert, Fand and Kesvani, and, Dito on **FIGURE 4** [9-12].

CONCLUSIONS

In this paper, wires with different diameters and air velocities were used by using a wind tunnel. These parameters play an important role on convective heat transfer coefficient. The results showed that the higher diameter value of wire provides a greater forced convection heat transfer for the same air velocity. For the same experimental conditions and wires, the convective heat transfer coefficient increases with air velocity.

For parallel flow, the experiments were carried out in the Reynolds numbers ranging from 9 to 275. The correlations were compared with the ones reported by Golzar, Ohkoshi, Zieminski, Glicksman, Copley, Brünig and, Kase and Matsuo. For cross flow, the experiments were carried in the Reynolds numbers ranging from 8 to 250. The Nusselt correlations which are determined from experiments were compared with the correlations of Churcill and Bernstein, Hilpert, Fand and Kesvani and Dito. The present results are in good agreement with literature and they are in same range for these Reynolds numbers. These present correlations give the Nusselt numbers for various flow conditions and the heat transfer coefficient *h* is found.

ACKNOWLEDGMENTS

The authors acknowledge financial and materials support from Yildiz Technical University and TUBITAK. Special thanks to Murat Gokten and Yasin Sen for the support and feedback.

REFERENCES

1. T. L. Bergman, A.S. Lavine, F.P. Incropera, D.P. Dewitt, Fundamentals of Heat And Mass Transfer [7th Edition], 2011, pp. 2-128.
2. G. Ravi, A. K. Dhiman, "Fluid flow and heat transfer across a circular cylinder in the unsteady flow regime", *The International Journal of Engineering and Science IJES*, 2319:1805 (2014).
3. M. Krishpersad M., R. Kimberly,"A Comparison of Correlations for Heat Transfer from Inclined Pipes", *International Journal of Engineering (IJE)*, Volume: 4, 269:278 (2010).
4. T. Yuge, "Experiments on Heat Transfer from Spheres Including Combined Natural and Forced Convection", Transactions of the ASME, pp. 214:220 (1960).

5. S. Kase, T. Matsuo, "Studies on melt spinning. I. Fundamental equations on the dynamics on melt spinning", *J. Polym. Sci A3.*, pp 2541–2554 (1695).
6. Y. Ohkoshi, T. Kikutani, A. Konda, and J. Shimizu, "Melt spinning of polyether-ether-ketone (PEEK) – cooling, thinning, and structure development on spin-line", Sen-i Gakkaishi 49, pp. 211–219 (1993).
7. L. R. Glicksman, 1968 "The dynamics of a heated free jet of variable viscosity at lowReynold numbers", *J. Amer. Soc. Mechan. Eng. 90D*, pp. 343–354 (1968).
8. M. Copley, N. H. Chamberlain, "Filament attenuation in melt spinning and its effect on axial temperature gradient", *Appl. Polym. Symp.,* pp. 27–50 (1967).
9. Churchill S.W., Bernstein M., 1977, "A Correlating Equation for Forced Convection from Gases and Liquids to a Circular Cylinder in Cross Flow", J. Heat Transfer, 99, 300-306.
10. R. Hilpert, "Wärmeabgabe von geheizten Drähten und Rohren im Luftstrom", *Forschung 4,* 1933.
11. R. M. Fand, K. K. Keswani, "A Continuous Correlation Equation for Heat Transfer from Cylinders to Air in Crossflow for Reynold's Numbers from 102 to 2x105", International Journal of Heat and Mass Transfer, Volume 15:559-562 (1972).
12. H. Brünig, R. Beyreuther, "Dynamics of Fibre Formation and Processing", Modelling and Application in Fibre and Textile Industry, pp. 55, (2007).

Influence of Mw weight average mass of the polymer support on the evolution of performances relating to extraction membrane processes of Cr(VI) ions.

Imane. MOURTAH[1*], R. OUCHEN [1,2], Y. CHAOUQI [1,2], T. ELJADDI[1], N. SEFIANI [1], M. HLAIBI [1]

[1] *Equipe Interaction Matière – Matière et Procédés Membranaires (**I3MP**) laboratoire GeMEV, Université Hassan II, Faculté des Sciences Aïn Chock, B.P. 5366, Maârif, Casablanca, Maroc.*

[2] *Laboratoire de Recherche sur les Matériaux Textiles (REMTEX), ESITH Casablanca, Maroc.*

(Projects: PPR2 and ERANETMED3-166)

Abstract. Currently, the use of polymers is a growth necessity in several areas, particularly, medicine, chemical, and food industries, because of their characteristics such as their low cost, light weight and excellent chemical stability..

Polysulfone (PSU) is one of the technical polymer classes with high thermal, oxidative and hydrolytic stability; in addition, it is inert and resistant to chemical degradation. For all these excellent properties it is adopted as a support for the development of polymer membranes, used to treat polluted industrial discharges.

In the present work, we have studied the effect of the average mass (**M_w**) of the PSU polymer support on the evolution of Cr(VI) ions extraction processes. Indeed, to develop our membranes, we adopted the same polysulfone polymer (**PSU**) with the following three average masses **M_w**: 35.000, 26.000 and 22.000 g/mole and the same **extractive agent** Ethylene diamine tetra acetic **(EDTA)**. Elaborated membranes were characterized, using **TFIR** spectroscopic technique which confirms the integration of **EDTA** agent into the polymer matrix, whereas **SEM** technique made it possible to characterize the morphology and the porosity of these membranes.

Obtained membranes were adopted to carry out several oriented processes relating to extraction and elimination of Cr(VI) ions, in order to study the influence of several factors (**M_w** polymer support average mass, **C_0** initial concentration of Cr(VI) ions substrate, and temperature factor) on the evolution of macroscopic (initial flux J_0, permeability P) and microscopic (association constant K_{ass} and apparent coefficient diffusion D^*) parameters, as well as the determination of activation parameters (Energy E_a, entropy $\Delta S^{\#}$ and enthalpy $\Delta H^{\#}$). The analysis of all results indicates a clear influence of the size PSU polymer strings on the characteristics of synthesized membranes and the evolution of all parameters. When the **M_w** weight average mass of the **PSU** polymer decreases the membrane performance becomes better.

Keywords: grafted polymer membrane, oriented processes, flux; Permeability, apparent coefficient diffusion, association constant, activation parameters.

PACS : pacs-reg60

1. Introduction

The protection of the environment is now a key priority to protect our planet, the harmful evolution of environmental pollution is mainly due to the pollution of air, soil and water, due to human activities and industrial development. One of the most serious pollution is the water pollution by metal ions, especially toxic heavy metal ions, including ions of chromium element.

Chromium can exist in several chemical forms with oxidation degrees ranging from 0 to +VI, the most common and stable forms in the environment are species: Cr (VI) and Cr (III) [1]. Chromium (+VI) ions are classified as carcinogenic to humans and their high toxicity is associated with a high oxidizing power in the body. In fact, dissolved Cr(VI) can be disseminated through cell membranes [2]. High doses of Cr(VI) can stimulate many systemic effects: dizziness, thirst, abdominal pain, hemorrhagic diarrhea, lung cancer, severe cases, irregular coma and death [3].

Various methods have been carried out for the removal and separation of chromium Cr(VI) as a preferential pollutant from wastewater, multiple industrial applications such as: metallurgy, paints, textile industry, wood, etc.). Among these methods are liquid-liquid extraction, which requires large quantities of volatile organic solvents often considered as pollutants for the environment [4], microbial treatment [5], coagulation [6], membrane separation [7], desalination by freezing [8], photo catalytic reduction [9], and ion exchange [10].

Today, membrane processes are part of the techniques increasingly adopted for the treatment of problems related to the environment, because of their multiple advantages, simplicity, efficiency and low energetic.

In this study we are interested in the extraction of Cr(VI) ions through a graft polymer membrane (GPM) based on polysufone as a polymeric support belonging to the class of amorphous thermoplastic polymers, characterized by a high thermal, mechanical and chemical stability.

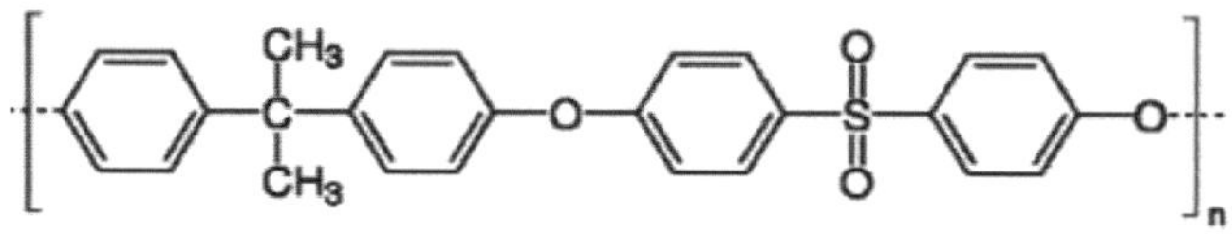

FIGURE 1: Repeated monomer formula of the polysulfone polymer (PSU).

2. Result and discussion

2-1) preparation of the GPM membrane

We used the phase inversion procedure [11] for the elaboration of three membranes by changing only Mw weight average mass of the polymer support PSU: 3 g of Polysulfone are dissolved in the DMF solvent (13 ml) at 30° C, for 6 hours until the total solubilization of the polysulfone, then 0.6 g of polyvinylpyrrolidone (PVP) is added. After total solubilization of this mixture (PSU-PVP-DMF), a weighed mass of the extractive agent Ethylene diamine tetra acetic disodium salt is added. The resulting paste is cast on a plate glass, spread with a ruler and immersed in a bath containing distilled water. DMF solvent leaves the matrix of the pulp spread and a rigid membrane is obtained in the form of a sheet (phase inversion method), then the membrane is stored in water during 48 hours to eliminate the entire DMF phase.

2-2) Extraction cell

To carry out the experiments related to the facilitated extraction processes of the Cr (VI) ions, we used a cell composed of two glass compartments: feed (S) and receiving (R) separated by the membrane (M). The system is immersed in thermostatic bath (TB) containing water in order to keep a constant temperature along the experiment. Homogeneity is ensured by a multi-station magnetic stirrer

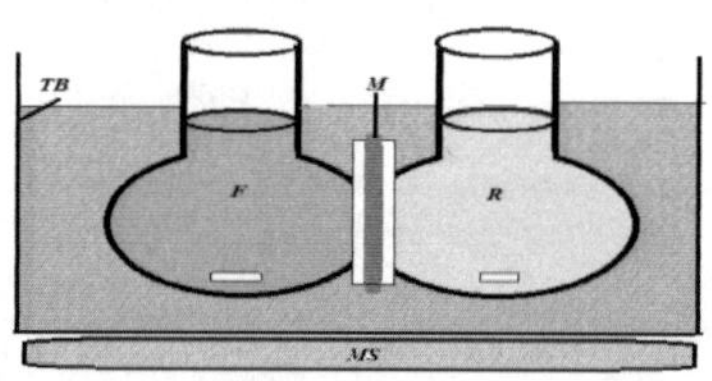

FIGURE 2. Scheme of an extraction cell:

2-3) Modeling and theoretical calculations:

We have used the Kinetic and Thermodynamic Models developed and adopted in our laboratory for several extraction and separation processes [12,13,14,15] to calculate macroscopic parameters, permeability (***P***) and initial flux (J_0) as well as microscopic parameters, apparent coefficient diffusion (D^*) and association constant (K_{ass}) relative to the movement of Cr(VI) ions across the phases of three elaborated membranes. The influence of the Cr(VI) ions initial concentration C_0 and the medium acidity (**pH**) factors on the evolution of these different parameters (J_0, P, K_{ass} and D^*) was studied. On the other hand, the examination of the temperature factor influence on the evolution of these parameters was carried out and the values of the activation parameters (Energy E_a, entropy $\Delta S^{\#}$ and enthalpy $\Delta H^{\#}$) were determined.

2-4. The influence Study of the polymer weight average mass on the facilitated extraction of Cr(VI) ions:

We carried out studies on the processes of the facile extraction of Cr(VI) ions through each of the three prepared membranes at fixed temperature and pH (pH = 3 and T = 298 K). Every 30 min, a negligible quantity is taken from the receiving phase using a micropipette in order to follow the evolution of the absorbance of the Cr(VI) ions in this phase as a function of time, that allowed us to calculate the concentration $\mathbf{C_R} = \mathbf{C_0} - \mathbf{C_S}$ as a function of time and therefore we have been able to study the evolution of the kinetic function $-\ln(\mathbf{C_0}-2\mathbf{C_R})$ as a function of time, which made it possible to verify the model, and to determine the permeability ***P*** and the initial flux $\boldsymbol{J_0}$ through each of the elaborated membranes:

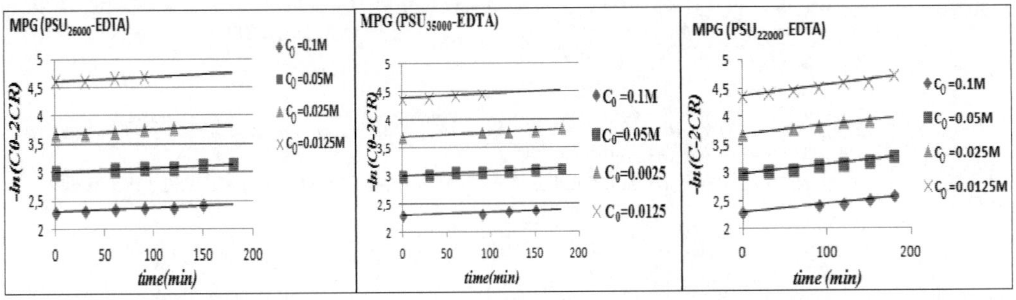

FIGURE2. Presentation of the kinetic function $-\ln(\mathbf{C_0}-2\mathbf{C_R}) = f(t)$ for facilitated extraction processes of Cr (VI) ions through the three elaborated membranes.

The representations of the function $-\ln(\mathbf{C_0}-2\mathbf{C_R}) = f(\mathbf{t})$ for each of the membranes are linear, which shows that the results of the studied processes clearly verify the proposed kinetic model. From the slopes of these lines we have calculated the macroscopic parameters ***P*** and $\boldsymbol{J_0}$ according to the equations provided by the kinetic model and the results are grouped in the following table:

TABLE(1). Evolution of the parameters ***P*** and $\boldsymbol{J_0}$ for the facilitated extraction processes of Cr (VI) ions through the three elaborated membranes

Mmembranes	C_0 mole/L	$P*10^7$ $cm^2.s^{-1}$	J_0 $*10^5$ mmol /s.cm^2	$1/C_0$ L /mol	$1/J_0$ s.cm^2/mmol
PSU_{35000}-EDTA	0.1	7.12	0.50	10	1.97
	0.05	8.18	0.29	20	3.44
	0.025	9.94	0.17	40	5.67
	0.0125	10.07	0.089	80	11.20
PSU_{26000}-EDTA	0.1	8.60	0.64	10	1.54
	0.05	9.14	0.34	20	2.90
	0.025	10.09	0.18	40	5.27
	0.0125	10.56	0.099	80	10.06
PSU_{22000} –EDTA	0.1	15.13	1.33	10	0.74
	0.05	16.14	0.71	20	1.40
	0.025	17.15	0.37	40	2.63
	0.0125	18.16	0.20	80	4.97

The obtained results (Table 1) clearly indicate a reverse evolution for the two macroscopic parameters (***P*** and $\boldsymbol{J_0}$).as a function of the initial concentration factor $\mathbf{C_0}$ of the Cr (VI) ions substrate. Indeed, the initial flux parameter $\boldsymbol{J_0}$ varies in the same direction as the substrate initial concentration $\mathbf{C_0}$, whereas the permeability parameter ***P*** varies in the opposite direction of this factor $\mathbf{C_0}$. Similarly, we note that there is a clear influence of the weight average mass Mw (chain length of the support polymer (PSU)) on the evolution of the values of ***P*** and $\boldsymbol{J_0}$ parameters. Indeed, the values of these two parameters are inversely proportional to the length of the polymeric chains (Mw of the PSU), and when the Mw of the PSU polymer decreases the values of these two parameters become larger and membrane performances improves for all initial concentrations of the extracted substrate This can be explained by an easy interaction and association of the Cr(VI) ions with extractive agent EDTA, grafted onto short chains of the polymeric support, less entangled. From the experimental results summarized in Table (1) we can plot Lineweaver-Burk presentation $1/\boldsymbol{J_0} = f(1/\mathbf{C_0})$ lines for the studied processes through three elaborated membranes.

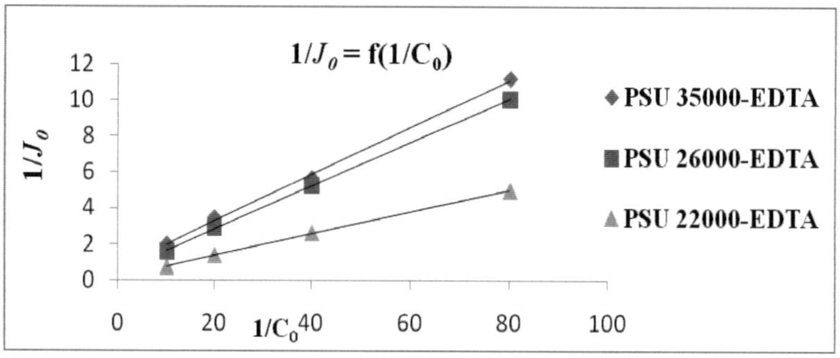

FIGURE 3 1. Lineweaver-Burk representation for facilitated extraction processes of Cr (VI) ions through the studied membranes.

L-B representation ($1/J_0$ = f($1/C_0$)) relating to the obtained results for the studied processes through elaborated membranes, indicates linear evolutions. Slopes and intercepts of the three obtained segments, make it possible to calculate the values of D^* and K_{ass} parameters, regrouped in the following table:

TABLE (2): Influence of the weight average mass Mw of the polymer support on the evolution of D^* and K_{ass} parameters relating to the extraction processes of Cr(VI) ions.

Membrane	D^*10^5 (cm^2/s)	K_{ass} ($L.\ mol^{-1}$)
PSU_{35000}-EDTA	1.11	5.15
PSU_{26000}-EDTA	2.49	3.,41
PSU_{22000}-EDTA	5.70	2.99

According to the results grouped in Table (2), the apparent diffusion coefficient D^* increases with a decrease in the length of the polymer chains (high value of D^* for PSU22000), while the association constant K_{ass} decreases. This evolution explains perfectly the increase of the P and J_0 parameter values and thus the best performance of the membrane PSU22000-EDTA.

To explain these results and to elucidate the mechanism of studied processes, we studied at different temperatures (T = 298, 303 and 308 K), facilitated extraction oriented processes of the Cr(VI) ion substrate through PSU35000-EDTA and PSU22000-EDTA membranes. The values of the parameter J_0 processed by the Arrhenius relation, makes it possible to calculate values of activation parameters E_a, $\Delta H^{\neq}$ and $\Delta S^{\neq}$ regrouped in following figure.

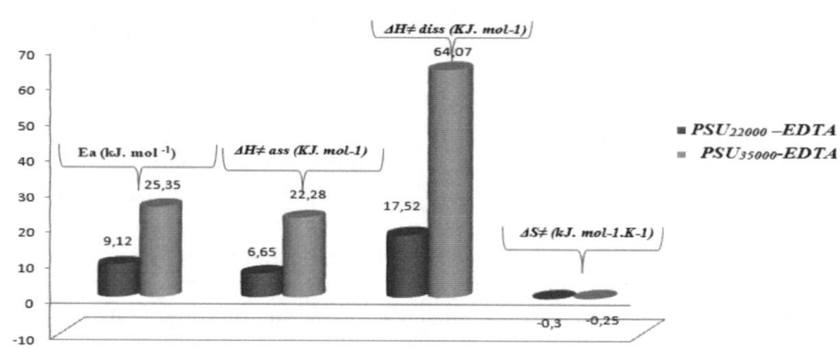

FIGURE 4: 2. Evolution of Activation Parameters for facilitated extraction processes of Cr (VI) ions across (PSU35000-EDTA) and (PSU22000-EDTA) membranes.

3. Conclusions:

- All obtained results indicate a better performance for facilitated extraction oriented processes of the Cr (VI) ion substrate through the membrane elaborated from the PSU polymer consisting of short polymer chains (low Mw), whatever the concentration of the extracted substrate. Moreover, these results are confirmed by the evolution of E_a, $\Delta H^{\neq}$ and $\Delta S^{\neq}$ parameters, with a better diffusion of Cr (VI) ions through short chains of the polymer support, less entangled and well spaced, according to a mechanism by successive jumps of Cr (VI) ions on fixed sites of the extractive agent EDTA.
- This result is very **original** and shows the importance of the **lengths** (sizes) of the polymer chains which constitute the polymer support, on the performance evolution of affinity membranes adopted for the oriented processes of extraction, separation or elimination.

References

1. A.D. Bokare, W. ChoiAdvanced oxidation process based on the Cr(III)/Cr(VI) redox cycle Environ. Sci. Technol., 45 (2011), pp. 9332-9338.

2. W. X. Wang, S. B. Griscom, N. S. Fisher, Bioavailability of Cr(III) and Cr(VI) to Marine Mussels from Solute and Particulate Pathways. Environmental Science & Technology, 31, 1997, p. 603-611.

3.E. Rifkin, P. Gwinn, E. Bouwer, Chromium and Sediment Toxicity. Environmental Science &Technology, 15, 2004, p. 267A-271A.

4.Ana M., Eliceche S.M., Corval M., Rom F.S., Ortiz I. Minimum membrane area of an emulsion pertraction process for Cr(VI) removal and recovery, Comput. Chem. Eng. 2005; 29: 456-472.

5.Singh R., Kumar A., Kirrolia A., Kumar R., Yadav N., Bishnoi N.R., Lohchab R.K. Removal of sulphate, COD and Cr(VI) in simulated and real wastewater by sulphate reducing bacteria enrichment in small bioreactor and FTIR study, Bioresour. Technol. 2011; 102: 677-682.

6.Lu J., Wang Z.-R., Liu Y.-L., Tang Q. Removal of Cr ions from aqueous solution using batch electro coagulation: Cr removal mechanism and utilization rate of in situ generated metal ions, Process. Saf. Environ. Prot. 2016; 104: 436-443.

7.Korus I., Loska K. Removal of Cr(III) and Cr(VI) ions from aqueous solutions by means of polyelectrolyte-enhanced ultra filtration. Desalination 2009; 390–395.

8.Melak F., Laing G.D., Ambelu A., Alemayehu E. Application of freeze desalination for chromium (VI) removal from water. Desalination 2016; 377: 23-27.

9.Liu W., Ni J., Yin X. Synergy of photocatalysis and adsorption for simultaneous removal of Cr(VI) and Cr(III) with TiO_2 and titanate nanotubes, Water Res. 2014; 53: 12-25.

10.Cao Dang Z., Yi. X-Y., Yang C., Lu G-N. Removal of chromium (VI) from electroplating wastewater using an anion exchanger derived from rice straw, Environ. Technol. 2013; 34: 7-14.

11.J.F. Blanco, J. Sublet, Q. T. Nguyen and P. Schaetzel, J. Membr. Sci. 283 (2006), 27–37.

12.N. Tbeur, T. Rhlalou, M. Hlaïbi, D. Langevin, M. Métayer, J.-F. Verchère, Carbohydr. Res. 329 (2000) 409–422.

13. Abdelkhalek Benjjar, Tarik Eljaddi, Oussama Kamal, Khalifa Touaj, Laurent Lebrun, Miloudi Hlaibi, Journal of Environmental Chemical Engineering **2014**, 2, 503–509

14. T. ELJADDI, O.KAMAL, El-H. EL ATMANI, I. TOUARSSI, L. LEBRUN, M.Hlaibi, The Canadian Journal of Chemical Engineering, 93 (3), **(2015)** 613–621.

15. I. Mourtah, I. Touarssi, Y. Chaouqi, N. Sefiani, L. Lebrun, M. Hlaïbi, *Materials Today Proceeding*: **Accepté** XX (2018) XXX - XXX

Processing of Ultra-High Molecular Weight Polyethylene (UHMWPE) with Regular and Special Injection Molding Technologies

Galip Yilmaz[a,b] and Lih-Sheng Turng[a,b*]

[a]*Polymer Engineering Center*
Department of Mechanical Engineering
University of Wisconsin–Madison, Madison, WI 53706, USA
[b]*Wisconsin Institute for Discovery*
University of Wisconsin–Madison, Madison, WI 53715, USA

Corresponding author: turng@engr.wisc.edu

Abstract. Ultra-high molecular weight polyethylene (UHMWPE) in pellet form was injection molded into box-shaped parts using a specially designed mold by both conventional and special, high-pressure microcellular injection molding (MIM). Super critical nitrogen (scN_2) was utilized as the physical blowing agent or, more precisely, plasticizer, in MIM. It was found that persistent short shot issues associated with conventionally injection molded samples could be solved and complete parts can be injection molded when scN_2 was used. Furthermore, not only did scN_2 effectively promote processability of UHMWPE, it also reduced the very high injection pressure requirement and high part shrinkage issues.

Keywords: Ultra-high molecular weight polyethylene (UHMWPE), injection molding, microcellular injection molding, super critical nitrogen

INTRODUCTION

As a linear homopolymer polyethylene, ultra-high molecular weight polyethylene (UHMWPE) holds a special place among other PE grades. It has an extremely high molecular weight, typically more than 1 million g/mol, and possesses many favorable properties for industrial and medical applications. However, its high molecular weight adversely affects its melt processability because of its extremely high viscosity and highly entangled molecular structure [1].

Injection molding is preferred over other types of polymer processing methods for complex part designs and highly repeatable and automated manufacturing [2]. So far, most of the research that has investigated the molding behavior and resulting part quality of UHMWPE has been limited to simple geometries such as tensile bars and rectangular parts [3–5]. However, complex part designs, which are more representative of potential and targeted UHMWPE applications, will yield more useful information and in-depth understanding of UHMWPE's molding behavior.

Conventional injection molding of UHMWPE has been hampered by the lack of fluidity, short shots, and other molding difficulties resulting from the ultra-high molecular weight and molecular entanglement of UHMWPE. It has been reported that supercritical fluids (SCFs) could improve the moldability of UHMWPE [5]. In this study, scN_2 was chosen because it was found that scN_2 and $scCO_2$ had a similar effect on the processing of UHMWPE in our early study [5]. Moreover, compared to CO_2, using N_2 in microcellular injection molding (MIM) requires 50% less back pressure, which is beneficial when a large dosage volume is required [6].

This study aims to utilize SCF to improve the injection molding of solid UHMWPE parts using more complicated molds and part designs. This effort will allow for the development of processing guidelines and engineering know–how for the injection molding of this unique material. For this purpose, experiments of molding solid UHMWPE with and without scN_2 into a box-shaped mold were conducted. Herein, the images of injection molded parts, part weights, injection pressures, and percentage of linear shrinkage measurements will be presented.

MATERIAL AND EQUIPMENT

UHMWPE resin (GUR® 5129) in pellet form was provided by the Celanese Corporation. It had an average molecular weight of 4.7 million g/mol and a density of 0.93 g/cm^3. An Arburg Allrounder 320S equipped with MIM technology was used to produce regular solid parts without scN_2 and special solid parts by first introducing scN_2 into the injection barrel to be dissolved into the melt and then by packing the melt in the cavity after the cavity is completely filled to suppress the foaming and force the emerging N_2 to dissolve back into the melt again.

Figure 1 shows a 3D model of the box-shaped part with overall dimensions of 100 × 75 × 45 mm. Intended to produce molded parts with severe shrinkage and warpage for research purposes, the design of the box-shaped part was intentionally flawed, with a decreasing thickness (from 3.5 mm to 2.0 mm in steps of 0.5 mm) on the shorter wall. The wider wall had a very small thickness of 1.25 mm. The thickness of the top wall of the box was 2.25 mm, as shown in the figure.

Sample types produced in this study are tabulated in Table 1, together with major processing differences in scN_2 content, injection speed, and packing pressure. For the scN_2-laden MIM samples, the nitrogen content was fixed at 0.5 wt%, which is a commonly used SCF dosage in the MIM process. However, packing pressure was applied to the scN_2-laden samples so that the gas emerging from the polymer–gas mixture during filling would be forced to dissolve back into the polymer melt until the material was fully solidified. Eventually, the dissolved gas will diffuse out from the solidified parts. With this approach, scN_2-laden solid samples were produced.

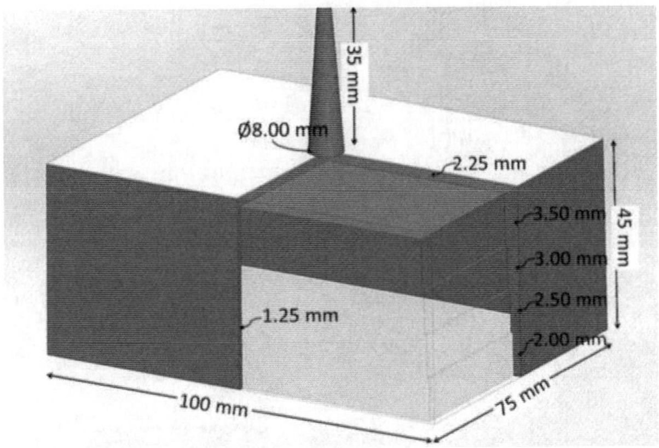

FIGURE 1. 3D model of the box-shaped part. A quarter of the part is cut away to show the variable wall thickness.

TABLE (1). Sample types and difference in relevant processing conditions.

Sample Name	Condition	scN_2 Loading Weight %	Injection Speed cm^3/s	Packing Pressure MPa
Regular	Regular injection molding as a reference sample	0	40	130
High-Q	Regular molding with a doubled flow speed (Q)	0	80	130
High-P	Regular molding with an increased packing pressure (P)	0	40	150
scN_2-laden	Similar conditions to the regular sample except processed with scN_2	0.5	40	130

A moderate injection speed (40 cm^3/s) and a high packing pressure (130 MPa) typical of UHMWPE molding were used for the regular injection molded samples (cf. Table 1). A relatively high injection speed (80 cm^3/s) was set for the High-Q samples. The packing pressure was increased from 130 to 150 MPa (very close to the injection pressure of the sample) for High-P samples. scN_2-laden samples had similar conditions to the regular samples, but with the addition of a 0.5 wt% gas dosage. Table 2 shows the rest of the processing parameters for the box-shaped part used in this study.

TABLE (2). Processing parameters used in injection molding.

Processing Parameters	Units	Value
Injection volume	cm^3	53.5
Cooling time	s	25
Packing time	s	7
Nozzle temp.	°C	255
Mold temp.	°C	80
Barrel temp. profile	°C	255-255-240-180-135-18

For all cases studied, at least five molded samples were measured and averaged. The sizes of the shrinkage samples were measured with a caliper. The height of the samples was measured from the thick wall side, due to the short shot problem seen on the thin walls for some samples.

RESULTS

Mold Filling

Figure 2 shows the various injection molded samples. The average weight %, which was normalized based on the average part weight of the scN_2-laden samples (37.01 g), can be seen in Figure 3. The edges of the thin walls of all samples, except the scN_2-laden sample, had short shot issues to various degrees. This is an expected problem of UHMWPE molding with thinner wall thicknesses (1.25 mm).

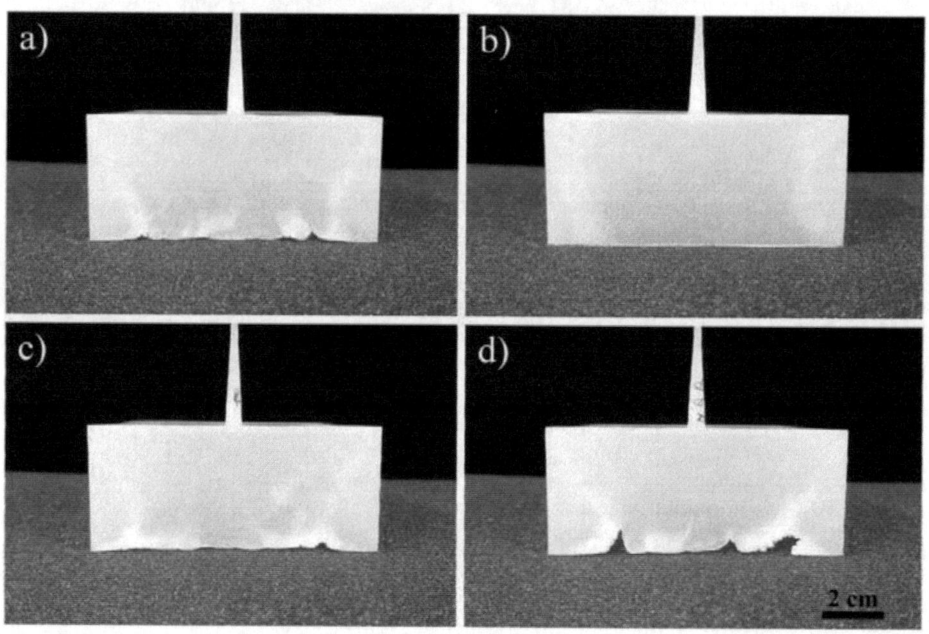

FIGURE 2. Sample images: (a) Regular molded sample, (b) scN2-laden sample, (c) High-P molded sample, and (d) High-Q molded sample.

The regular injection molded samples weighed considerably less, about 4% less than the scN_2-laden samples. Increasing the packing pressure from 130 MPa to 150 MPa (High-P) seemed to help the filling process by injecting more material under a higher packing pressure. However, the parts were still 3.7% lighter than the scN_2-laden samples. Therefore, High-P molded samples were unable to completely fill the mold, as can be seen in Figure 2 (c). The High-Q molded samples (one of them can be seen in Figure 2 (d)) interestingly weighed 2.0% less than the regular molded samples, and 6.0% less than scN_2-laden samples, despite having two-times faster filling. This result was counterintuitive and against the conventional wisdom and intuition. One possible reason for this could have been due to polymer leakage at the nozzle and/or clearance between other functional parts like the check ring (non-return valve) caused by the extreme pressure [7,8]. For this box-shaped mold, the average injection pressure for commodity polymers typically ranges from 55–70 MPa. However, the average injection pressures for regular and High-Q-molded UHMWPE samples were 153.4 MPa and 184.0 MPa, respectively. Therefore, controlling the injection pressure is very important for successfully molding UHMWPE, as well as the precise control of its part weight. However, the leakage and pressure spike alone do not explain the whole story when one compares the weights of all molded samples.

Another reason for the High-Q samples' weight shortage could be an increased frictional force due to the high pressure and high flow speed of the sliding UHMWPE melt during the filling cycle. The frictional force between the melt and mold interface is simply a function of the coefficient of friction and the normal force. However, the complicated tribological behavior of the UHMWPE melt on cold metal, especially for the injection molding process, is not well understood and has not been reported upon in the literature.

Among the sample types, only scN_2-laden samples were able to fill the cavity completely. When the thin-wall-edge scN_2-laden sample was examined in Figure 2 (b), it could be seen that its edge was flat on the bottom and it was also the heaviest sample, as shown in Figure 3 (a). This not only indicates that the emerging gas was either vented or dissolved into the polymer completely, but it also shows that there was possibly a lower surface friction due to the escaping scN_2 that created a cushion between the melt and mold surface and hence caused a lubricating effect on the melt during the filling cycle. The lowest injection pressure in Figure 3 (b) proved that the escaping N_2 reduced the frictional force between the melt and the mold surface.

Furthermore, it is well-known that the plasticizing effect of scN_2 lowers viscosity due to induced chain mobility. This effect may also contribute to the lower injection pressure [9]. In our early study, we found that SCFs yielded lower thermal degradation. By assuming that a majority of the thermal degradation occurred during the dosage and injection stages, the plasticizing effect of scN_2 will have had a considerable impact on the degradation behavior of UHMWPE and could have led to a lower injection pressure.

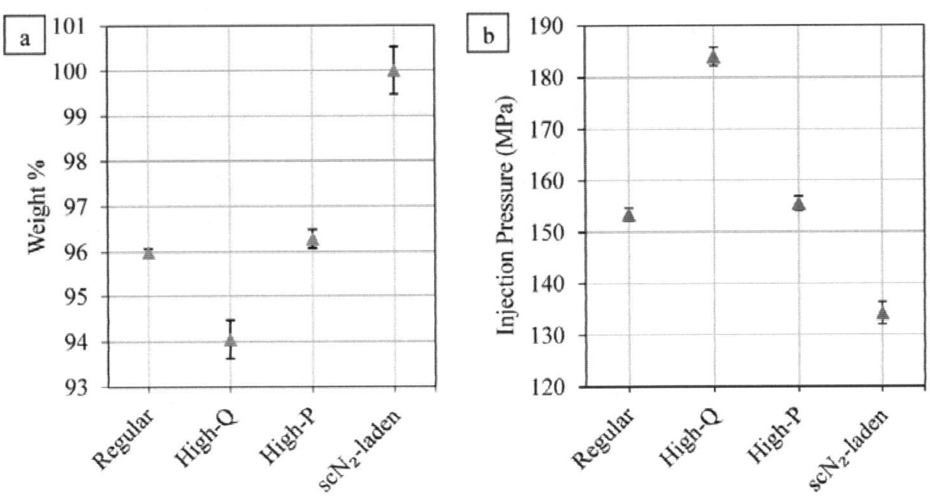

FIGURE 3. (a) Average weight % and (b) injection pressure.

Shrinkage Discussion

Shrinkage of the polymer is inevitable during cooling. To the best of our knowledge, there is no study that focuses on the shrinkage of injection molded UHMWPE. However, as a semi-crystalline polymer, UHMWPE tends to shrink a considerable amount. Figure 4 shows the linear shrinkage % based on the dimensions of the nominal box-shaped mold. All samples showed the highest shrinkage in the direction of the box's height. All samples, except scN_2-laden samples, showed approximately 6% shrinkage, which was considerably high for dimensional accuracy. The length and width shrunk relatively less than the height of the box. This was due to the mold core restricting shrinkage in the width and length directions during cooling, while the height dimension had more freedom to shrink due to the lack of mold constraints. Compared to other samples, the scN_2-laden samples yielded less shrinkage in all directions. This was due to more material having been packed into the mold with the aid of the scN_2.

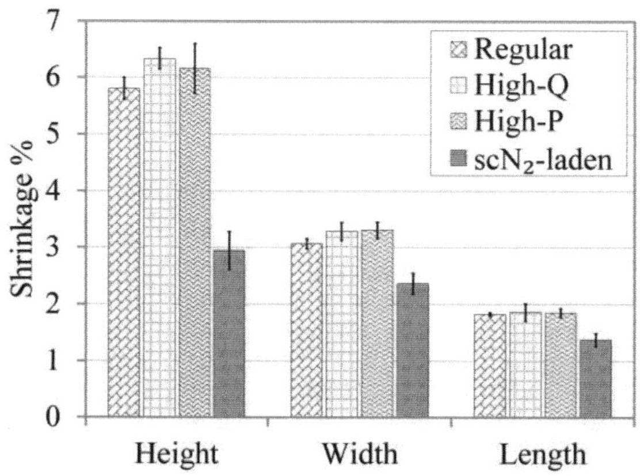

FIGURE 4. Shrinkage % of the samples.

CONCLUSIONS

As the preliminary results show, the injection molding of UHMWPE with scN_2 is a promising method. It helps solve the persistent and challenging problem of short shots when injection molding UHMWPE. It is also beneficial for improving the dimensional accuracy of the final part. Despite some interesting behaviors revealed by this preliminary study, it is hard to draw a complete conclusion on the relationships among filling speed, injection pressure, and sliding frictional force. Therefore, more research is needed to explore the optimum filling speed, mold temperature, and melt temperature settings to facilitate the proper mold filling of UHMWPE in injection molding.

ACKNOWLEDGMENTS

The authors acknowledge the support of the Wisconsin Institute for Discovery (WID) and the Kuo K. and Cindy F. Wang Professorship at the University of Wisconsin–Madison. The authors would also like to thank the support of the SPE Milwaukee Education Foundation and Celanese Corporation for the material donations.

REFERENCES

1. S. M. Kurtz, *The UHMWPE Biomaterials Handbook: Ultra-High Molecular Weight Polyethylene in Total Joint Replacement and Medical Devices*, 2nd ed. ,Academic Press, Burlington, MA, USA, (2009).
2. T. A. Osswald, L.-S. Turng, and P. J. Gramann, *Injection Molding Handbook*, 2nd ed. ,Hanser, München, Germany, (2008).
3. H.-C. Kuo and M.-C. Jeng, *Materials & Design*, **31**, 2 (2010).
4. G. Yilmaz, T. Ellingham, and L.-S. Turng, *Polymer Engineering & Science*, (2018).
5. G. Yilmaz, T. Ellingham, and L.-S. Turng, *Polymers*, **10**, 1 (2017).
6. K. T. Okamoto, *Microcellular Processing* ,Hanser, München, Germany, (2003).
7. B. D. M. Bryce, *Plastic Injection Molding Vol I: Manufacturing Process Fundamentals* ,Society of Manufacturing Engineers, Dearborn, MI, USA, (1996).
8. T. A. Osswald, E. Baur, S. Brinkmann, K. Oberbach, and E. Schmachtenberg, *International Plastics Handbook* ,Hanser, München, Germany, (2006).
9. I. Kikic, *Journal of Supercritical Fluids*, **47**, 3 (2009).

Development of The Monitoring Methods of the Vent Clogging in the Injection Mold

Bongju Kim[a], Jinsu Gim[a], Joohyeong Jeon, Taejoon Jeong, Eunsu Han and Byungohk Rhee[a*]

[a]*Ajou university (Woncheon-dong, Yeongtong-gu, Suwon-si, Gyeonggi-do, South Korea)*

Abstract. The vent of the injection mold is an essential part for mold operation. It makes it possible for the incoming gas during filling to be ventilated out of mold preventing molding defects such as burnt mark or short shot. Many additives such as flame retardants, UV stabilizers or lubricants are compounded to resins to enhance products quality and processing capability. The resins are likely to produce a large amount of gas during processing. The gas includes organic substances that can easily condense and be accumulated on the surface of the vent, which reduces the vent to make it difficult for the gas to be ventilated easily. The mold with the clogged vent should be cleaned regularly. If the cleaning time could be predicted by monitoring, it can make mold operators to set up the working schedule with proper preparation ahead. In this work, how to measure the temperature and pressure changes of the gas in the cavity during filling is examined to make the vent clogging to be monitored efficiently. The gas in the cavity during filling is compressed adiabatically in a moment due to the fast filling speed. As the vent gets clogged, the adiabatic compression would become severe. When the adiabatic compression occurs, the temperature and pressure of the gas in the cavity will increase sharply. We developed temperature and pressure sensors to detect the adiabatic compression in the cavity. A sensitivity of the sensors was tested in a compression chamber. We developed and tested methods to measure the temperature and pressure changes in the chamber in this work. We discussed the problems to affect the sensitivity and the long-term reliability of the sensors.

Keywords: Injection mold, The vent, Vent clogging, Monitoring

INTRODUCTION

The vent is a passage between cavity and ambient atmosphere in the injection mold. It generally has a very small depth of 20 to 70 μm. In the filling process of the injection molding, polymer melts are injected from the barrel into the mold cavity and the gas in the cavity should be properly ventilated from the mold. If the gas is not ventilated properly, polymer melts cannot fully fill the cavity and will result in a short shot. It may cause burnt mark defect due to a high temperature at the end of the filling [1]. Or tiny air bubbles could remain in the injection molded product [2].

Nowadays, many additives such as plasticizers, UV stabilizers or flame retardants are compounded in resins to improve the properties of the product or the moldability. The additives are easily vaporized because of its light molecular weight. Therefore, not only the air but also the vaporized gas need to be ventilated out of the mold. The vent management is becoming more important because the gas from additive includes organic substances that can easily condense and be accumulated on the surface of the vent.

The vent monitoring is a way to improve the efficiency of the mass production plant by reducing the machine down time with respect to the process schedule management. The vent cleaning often needs to be done suddenly after a defect was detected. The irregular vent cleaning could deteriorate the productivity. If the vent clogging could be monitored and the vent cleaning time can be predicted without deteriorating the productivity.

Research on the vent of the injection mold has focused primarily on developing ways to improve a clogging problem. Bowen has developed a method of discharging the gas in the cavity by porous metal [3]. Ken mold service in Japan has developed a degassing device operated by melt pressure. However, the methods have a disadvantage of additional maintenance required for the equipment. In the case of porous molds, if the micro-pore is clogged by organic substances, it must be cleaned or replaced. The degassing apparatus also requires periodic maintenance. Also, an installation may be restricted depending on the cavity. Buja attempted to measure a cavity trapped gas temperature by a temperature sensor on the surface of the vent [4]. He asserts that this method can sense a burnt mark, melt flow in the cavity, clamping force change and so on. However, it is not practical to install such a fragile sensor on the surface of the vent.

The gas in the mold is almost adiabatically compressed by the melt that flows in rapidly during the filling stage of the injection molding process. Then the gas temperature and pressure are elevated sharply. The flow rate of the gas through the vent changes by the depth of the vent, and the gas temperature elevation due to the adiabatic compression varies. Shin used this phenomenon to monitor the vent by measuring the gas temperature elevation [6]. The structure of the cavity gas temperature sensor developed is shown in Figure 1(a). Holes of a 30 μm diameter are drilled on the top surface of the sensor so that gas can flow into the sensor. This cavity gas temperature sensor was installed at the end of the cavity.

The experimental results are shown in Figure 1(b). The temperature elevation of the gas due to the adiabatic compression is regarded as an indicator of the vent clogging. Temperature elevation is the difference between the peak temperature and the initial temperature. In the experiment with ABS, the temperature elevations measured as different vent depths changed. At a constant injection rate of 147cc/sec and the vent depth of 50 μm, the temperature elevation was

10.3 °C, while the temperature elevation was 31.6 °C at the vent depth of 0 μm. The change in vent depth can be monitored successfully by the cavity gas temperature sensor.

However, the sensor developed by Shin did not always show a high enough sensitivity with various materials. The experiment with HIPS showed only 8 °C of the temperature elevation in the condition of the vent depth of 0 μm and the injection speed of 147 cc/sec. There was little temperature elevation at injection speed lower than 147 cc/sec. The low temperature elevation could be explained by the small amount of gas vaporized from HIPS resin. When the gas vaporizes more, the temperature elevation will be larger at the same compression ratio. Therefore, it is desirable to improve the sensitivity of the cavity gas temperature sensor to monitor the vent clogging effectively in the realistic molding conditions. The first goal of this study is to find out the causes of the low sensitivity of the sensor and to improve the sensor. The second goal is investigating another venting hole monitoring method with a higher sensitivity and reliability.

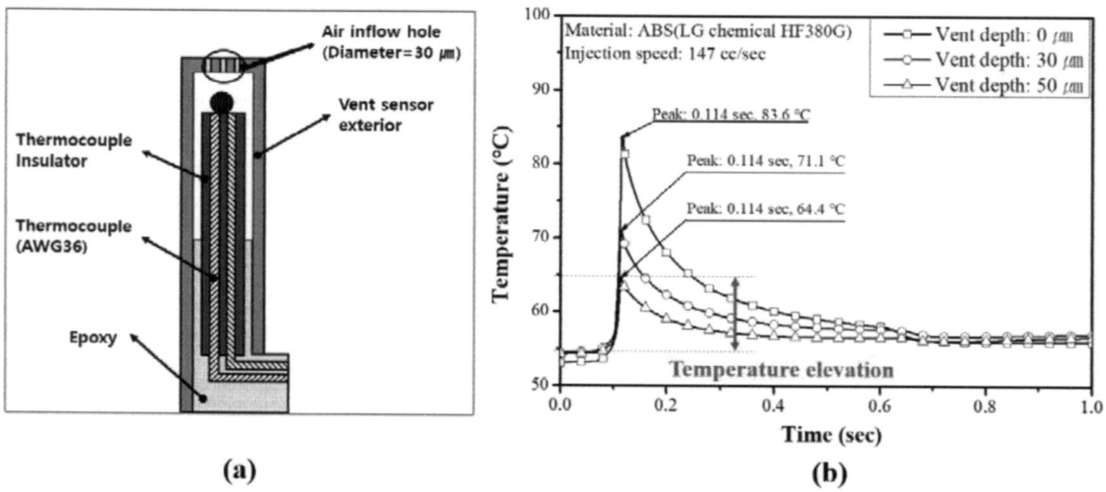

FIGURE 1. (a) Structure of the cavity gas temperature sensor in the previous research, (b) Temperature graph measured by the cavity gas temperature sensor installed in the mold

EXMPERIMENTS

Apparatus for Experiment

In this study, an equipment was constructed and tested to examine the vent clogging monitoring method and the sensor. A schematic view of the equipment is shown in Figure 2. The air compressor is connected to the cylinder, and the cylinder is connected to a chamber. When the air compressor is activated, the air in both the cylinder and the chamber is compressed. This simulated the situation that gas was compressed in the cavity by the polymer melt flow. The installed needle valve was used to simulate the vent state. Pressure and temperature in the chamber can be measured by a pressure sensor, a bare thermocouple and a new gas temperature sensor. The volume compression ratio calculated is 9.56.

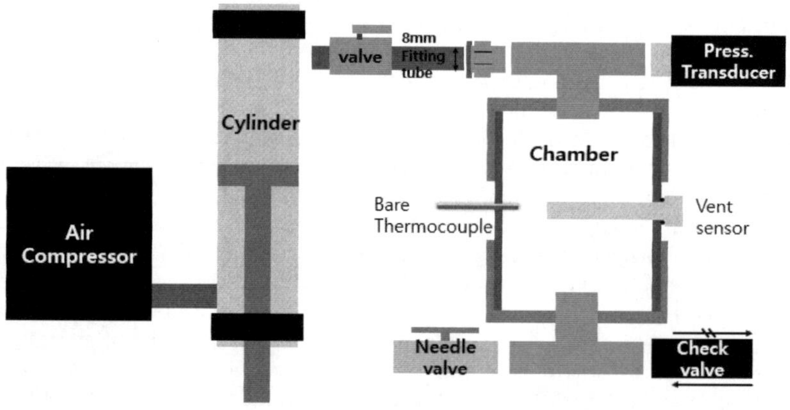

FIGURE 2. Schematic of simulated chamber with needle valve as a simulated vent

Possible Cause of Low Temperature Elevation

The temperature in the reversible adiabatic process can be calculated by Equation (1). Assuming that the initial temperature of the air is 20 °C and the compression ratio is 9.5. The calculated temperature of the air after adiabatic compression is about 448 °C. Where y is the specific heat ratio of the air.

$$T_{compressed} = T_{initial}\left(V_{initial}/V_{compressed}\right)^{y-1} \tag{1}$$

Since the above calculation result is based on the assumption of adiabatic compression, it is natural that there is a great difference between a calculation and a measured temperature by the sensor at the compression experiment. The temperature elevation of the sensor should be at least equal to the one of the bare thermocouple, but it was not. This phenomenon was attributed to the fact that heat of the air increased by the compression was transferred to the metal case of the sensor. The following experiment was conducted to verify the heat transfer. First, a metallic nut was installed near the bare thermocouple in the chamber to see the effect. The experimental results are shown in Figure 3. When the nut was installed around the bare thermocouple, the temperature elevation was reduced by more than half. Thus, heat transfer to the surrounding material of the sensor significantly decreases the amount of temperature elevation, which indicated a possible reason for the low sensitivity of the sensor.

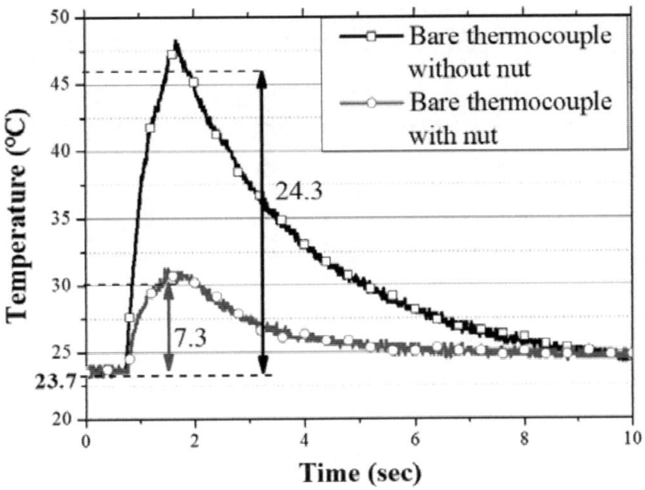

FIGURE 3. Temperature elevation of the bare thermocouple with and without the presence of a metallic nut

Experiment of Improved Sensor

A new sensor was designed to extend the distance of the thermocouple tip from the case wall. The inner diameter of the case was increased from 1 mm to 3 mm to reduce the heat transferred to the case. In addition, the flow resistance was reduced for a smooth air inflow into the sensor. The cross-sectional area of the inlet was increased by about 290 times from 2,120 μm^2 to 620,464 μm^2 by removing the holes in the head and reducing the diameter of the head of the pin by about 50 μm instead. The shape of the new sensor is shown in Figure 4 (a). The position of the thermocouple tip in the sensor is shown in Figure 4 (b).

In order to investigate the performance of the improved sensor, an experiment was conducted in the simulated apparatus. The air was compressed at a volume compression ratio of 5.6 for 0.5 seconds. The needle valve was adjusted to simulate the clogging of the vent. The experimental results are shown in Figure 5(a). After the temperature increased with the rapid compression of the air, the temperature decreased by a heat loss and a leakage through the needle valve. A bare thermocouple showed a temperature elevation about 20 °C, and the improved sensor showed 5.5 degrees. This is due to the heat transfer to the sensor case, as mentioned above. There was little difference in temperature elevation depending on the opening level of the needle valve. The low temperature elevation of the improved sensor in this experiment is due to the lower compression ratio and longer compression time than the realistic conditions of the injection molding process. The compression time at the injection ratio of 147 cc/sec is about 0.04 seconds that is much smaller than the experimental condition. And an additional verification experiment will be carried out with an injection mold and the improved sensor.

Unlike cavity gas temperature sensors with a low sensitivity due to the heat loss, cavity gas pressure sensors have shown more sensitivity to the compression conditions. The result shown in Figure 5(b) indicates the feasibility as an effective way of the vent monitoring. Even if the needle valve was opened a little, the difference of the reduction in the pressure elevation appeared. Additional validation experiments were conducted to confirm the availability of the cavity gas pressure sensor at the end of the fill of the cavity.

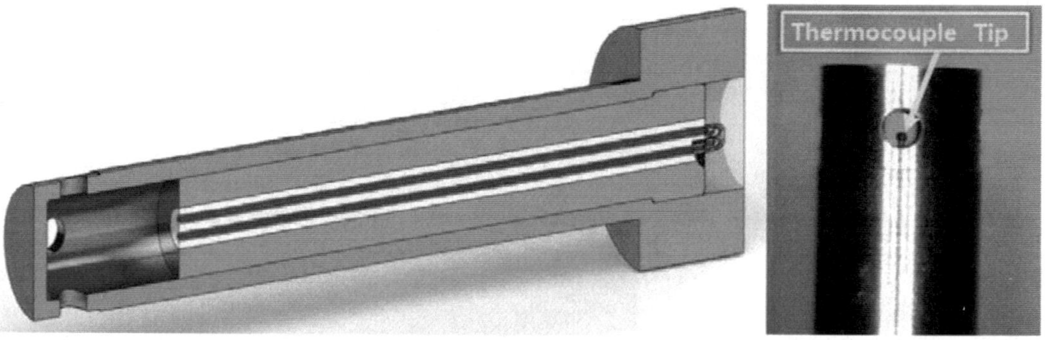

(a) (b)

FIGURE 4. **(a)** Cross-section of the improved sensor, **(b)** Position of the thermocouple tip

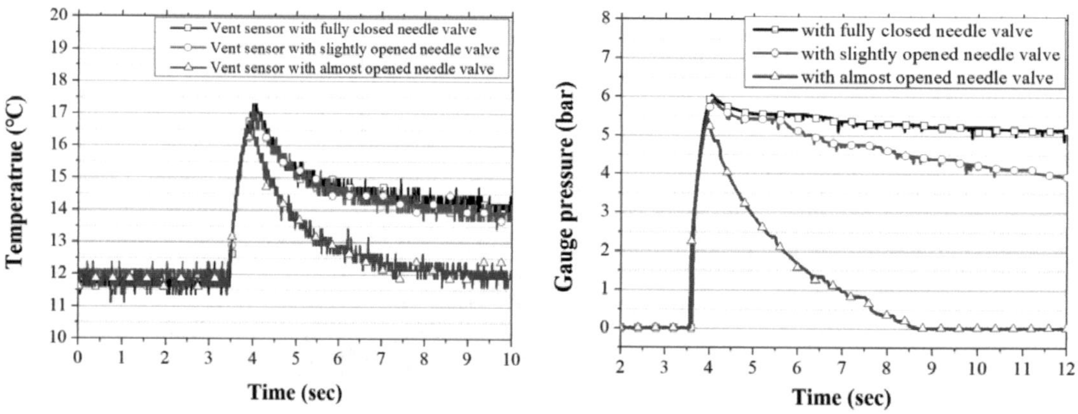

(a) (b)

FIGURE 5. **(a)** Results of the improved sensor verification experiment, and **(b)** pressure data measured by a pressure sensor

Experiment of Pressure Sensor

When a pneumatic sensor was installed at the end of the fill in the cavity, the gas pressure should be measured through a small gap that the polymer melt cannot flow into. Therefore, we examined whether there was a problem in the sensitivity of the measurement even if the gas pressure in the cavity was measured through a small gap. An experiment was conducted with a pneumatic pressure sensor with a hole of diameter 0.9 mm and depth of 1 mm. As shown in Figure 6, the small hole does not affect the sensitivity of the pressure sensor.

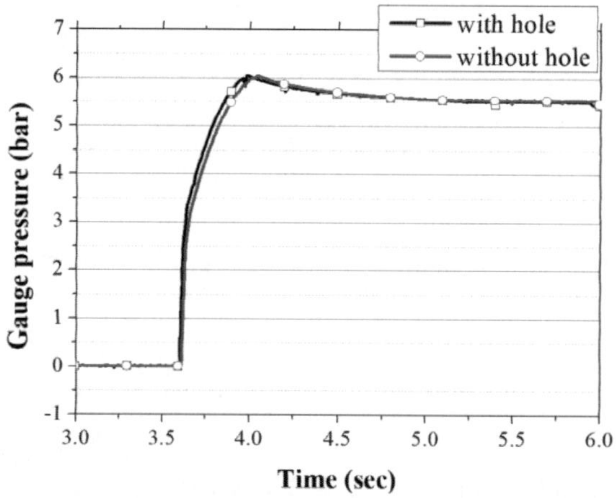

FIGURE 6. Results of the pressure sensor experiment with and without a small hole

CONCLUSION

The cavity gas temperature sensor to monitor the vent clogging developed in the previous research had a problem of a low sensitivity. This is because the heat of the compressed gas is transferred to the metallic sensor case. In order to improve this, a sensor with an expanded inner space was newly made. The new cavity gas temperature sensor showed an improvement compared with the previous sensor in the simulated chamber with a needle valve as a simulated vent. However, the experimental condition in the simulated chamber has a lower compression ratio and longer compression time than the realistic condition of the injection molding process. An additional experiment of the cavity gas temperature sensor is necessary in the actual injection mold. As an alternative to the gas temperature sensor, the pneumatic pressure sensor through a small hole showed a feasible result in the simulated chamber experiment with the needle valve. It gives us chances to explore the possibility of the pneumatic pressure sensor as well as the cavity gas temperature sensor to monitor the vent clogging effectively.

REFERENCES

1. P.C. Chang et al., Journal of material processing technology, volume 184, issues1-3, pp. 163-172 (2007)
2. B.V.Hooreweder, et al., Polymer testing, volume 32, issue 5, pp 972-981 (2013)
3. R. Bowen, ANTEC1999 proceedings, International mold steel (1999)
4. F. Buja, ANTEC 2006 proceedings, FjB PlasTechnology, Rochester, NY, 2006, pp. 1414-1417
5. F.J. Buja, U.S. Patent No. 7,585,166 (Sep. 8, 2009)
6. D.G. Shin, "A Study on the monitoring technique for the venting defect in injection molding", Master's Thesis, Ajou University, 2017

Development of a screw-mounted melt temperature measuring system during plasticizing of the injection molding

Joohyeong Jeon[a], Eunsu Han[a], Jinsu Gim[a], Bongju Kim[a], Taejoon Jeong[a]

and Byungohk Rhee[a*]

[a] *Ajou Univ, Woncheon-dong, Yeongtong-gu, Suwon-si, Gyeonggi-do, South Korea*

jeon13806@gmail.com
rhex@ajou.ac.kr

Abstract. The melt temperature variation during plasticizing of the injection molding has been reported for a long time, but it has not been considered as an important factor in the field because it is very difficult to measure it during processing. In the previous research, we examined the melt temperature variation according to the screw rotation speed, metering stroke, the dwelling time and barrel heater profile by a newly developed wire-type sensor with a higher sensitivity than the conventional plug-type sensors mounted in the nozzle adaptor. The wire-type sensor was in the middle of the melt stream of the nozzle adaptor and measured the melt temperature passing by the sensor during slow purging. In this work, we developed a temperature sensor system to measure the melt temperature at the screw tip. A thermocouple is installed at the screw tip so that the melt temperature passing through the check ring could be measured. In this type of sensing system, the data acquisition has a difficulty to be sent out of the rotating screw. Conventionally, the signal is transmitted through a slip ring or an analog telemetric device to external data logger. We developed a digital data acquisition system using a Raspberry Pi computer with a Wi-Fi communication. The temperature signal line is connected through the screw center hole to an amplifier and an analog-to-digital converter (ADC) installed at the back of the screw. The converted digital data is acquired and stored by the computer. The signal conditioning system and the computer are battery powered. The data stored in the computer is uploaded to an external server after one molding cycle via Wi-Fi communication. The sensitivity of the screw-tip sensor and the performance of the system were evaluated. The same experiments as in the previous research were carried out to compare the temperature measured by the new system with the previous one measured by the sensor installed at the nozzle adaptor.

Keywords: Injection molding, Melt temperature, Temperature measurement, Plasticizing, Digital telemetric system

INTRODUCTION

In the injection molding process, the melt temperature is one of the important parameters. The melt temperature affect features such as the viscosity, cycle time, degree of crystallinity and product quality. Inadequate melt temperature results in various defects and consequent productivity degradation. The melt temperature is commonly assumed to be similar to barrel heater temperature. According to the assumption, a single value is input to commercial injection molding analysis software as the melt temperature. In injection molding machines, only the heater temperature can be set, not the melt temperature since they are assumed to be similar to each other. Contrary to the assumption, many experimental results were reported on the melt temperature variation in the barrel of injection molding machine [1]. Amano et al. reported that the melt temperature has a relationship with some major operating conditions such as cycle time, screw rotation speed, barrel temperature, etc.[1-3] the authors measured the melt temperature by a sheathed thermocouple probe in the nozzle of injection molding machine. In addition to the experimental evidence, the melt temperature variation was reported in numerical studies. K. Straka et al. developed a numerical model in order to analyze the plasticizing process in an injection molding machine [4]. On the basis of the calculated pressure, the temperature profile of the polymer melt were examined and it was found that the melt temperature varied along the screw axis. According to previous studies, the melt temperature measured at the nozzle outlet is significantly different from the barrel heater temperature. In other words, resin with different temperature from the intention of the operator or engineer is been injected into the mold.

The melt temperature has been usually measured by sheathed thermocouples. Generally the melt has a high viscosity, and the sensors in it undergo high stress. In this respect, the rigid sheathed thermocouple is considered appropriate. However, sheathed thermocouples with large bodies have a large heat capacity and generate shear heating in the flow of melt. Furthermore, the sensor is not thermally insulated from the nozzle adaptor or the barrel. The condition causes a retardation in the measuring time and a low sensitivity. In this regard, accurate measurement of the melt temperature is required to examine the melt temperature variation in the barrel. It was necessary to use a sensor with a less thermal capacity and insulation from barrel.

Peischl and Bruker studied the influence of the screw geometry [5]. A ring-bar device was introduced to measure the radial distribution of the melt temperature in the barrel as a representation of the melt homogeneity. Dontula et al. used an IR temperature sensor at the nozzle to measure the melt temperature. Since IR sensors have a high sensitivity, which have an advantage in measuring fluctuated temperature [6]. Sombatsompop et al. measured the radial distribution of melt temperature in the nozzle using a thermocouple mesh originally designed by Wood et al. [7, 8]. One of the advantages of the sensor was minimization of both flow disturbance and thereby generation of shear heating caused by an intrusive temperature sensor. Praher et al. introduced an ultrasound-based system for the temperature measurement [9]. The method ensures the measuring stability in a harsh condition since the system is not invasive.

In the previous studies, a nozzle-mounted wire-type temperature sensor with a less thermal capacity and a smaller size was developed for a faster and more accurate temperature measurement [10]. The schematic of the wire-type sensor mounted in the nozzle adaptor is shown in Figure 1. The effect of temperature fluctuation of melt temperature on various process variables was investigated and several estimations of the causes of temperature fluctuations are also presented. In this study, a screw-mounted temperature sensing system was developed to accurately analyze how each individual factor affects melt temperature.

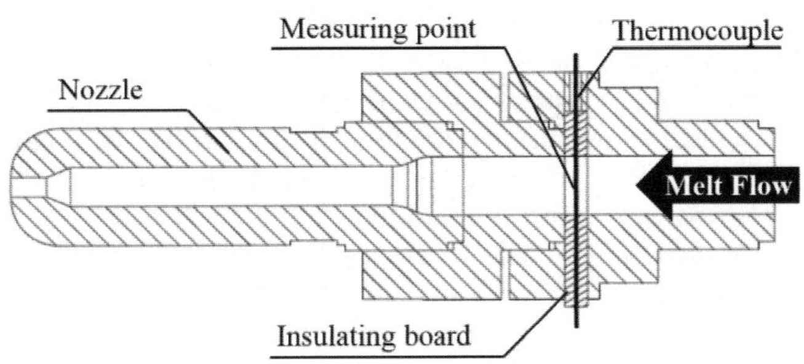

FIGURE 1. Schematic diagram of nozzle-mounted melt temperature sensor.

MELT TEMPERATURE MEASURING SYSTEM

During plasticizing, resin introduced through the hopper undergoes a very complex process in the barrel. Resin is melted by frictional and shear heating due to the screw rotation as well as conduction heating from the barrel heater. Unlike extrusion process that is a continuous process, the screw retreats backward during plasticizing in an injection molding machine, which causes the resins to experience a non-uniform thermal history. In addition, several factors influence the melt temperature. The result of complex thermal history in plasticizing was revealed by the experiments of measuring the melt temperature through the temperature sensor mounted on flow channel of the nozzle.

In order to analyze exactly how each individual factor affects the melt temperature, it is necessary to measure the melt temperature right after the check ring in real time. This requires a temperature sensor mounted on the screw tip. Then it can measure the melt temperature during the backward movement of the screw. While the melt temperature measured by the sensor mounted on the nozzle may include influences from any unidentified sources, a screw-mounted temperature sensor can measure the melt temperature right after the melt is made in the screw without unidentified influences.

The screw-mounted temperature measurement system with a thermocouple in the screw tip is shown in Figure 2. Since the screw rotates in one direction, it is needed to find a way of transmitting the signal from the thermocouple mounted at the screw-tip to the external device. We made the central hole along the screw so that the thermocouple wires were inserted through the hole to the back of the screw. Instrument including amplifiers, ADC and small computer (Raspberry Pi) were installed in the screw adapter (A part for fixing the screw to the injection molding machine). Raspberry Pi with Linux is connected to the server through a wireless router and uploads measured data to the server. The embedded devices mounted to the screw adaptor are shown in Figure 3.

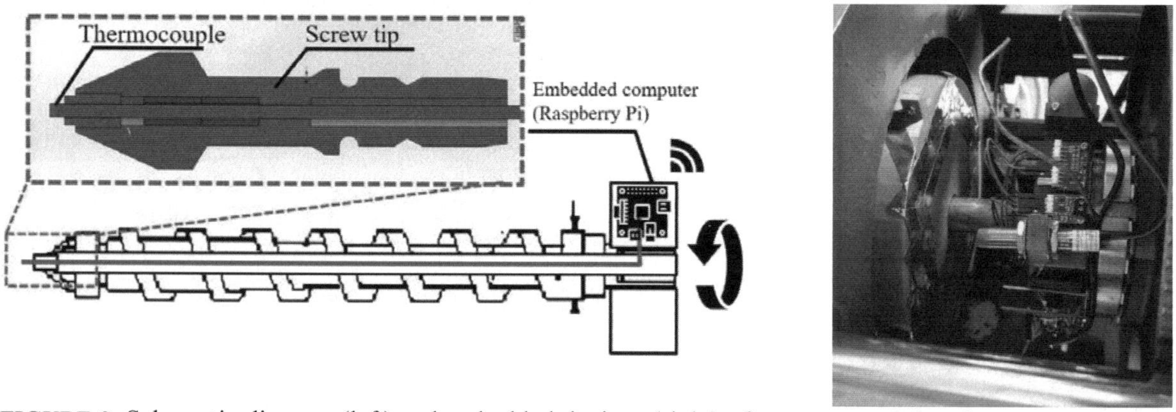

FIGURE 2. Schematic diagram (left) and embedded devices (right) of screw-mounted melt temperature measuring system.

POSITION OF TEMPERATURE MEASURING POINT

In the initial design, the temperature measuring position was at the end of the screw tip. However, no temperature variation was measured in the experiment using this structure. The measuring position turned out to be wrong position to measure the melt temperature variation. The specimen in Figure 4 was made by a screw quenching experiment to visualize the flow in the screw channel and the chamber in front of the screw. It shows a flow pattern in the chamber well. As shown in Figure 3, the fresh melt flows along the wall of the chamber and is accumulated toward the center of the chamber. The previously accumulated melt pushes the screw backward by the pressure in the chamber. The fresh melt does not through the center the screw tip where the original measuring position is located. It did not measure the melt temperature, but the temperature of the screw tip that is much lower than the melt due to a higher conduction through the metallic screw body. To measure the melt temperature flowing into the chamber beyond the check ring, the measuring point must locate in the passage in the screw after the check ring as shown in Figure 4.

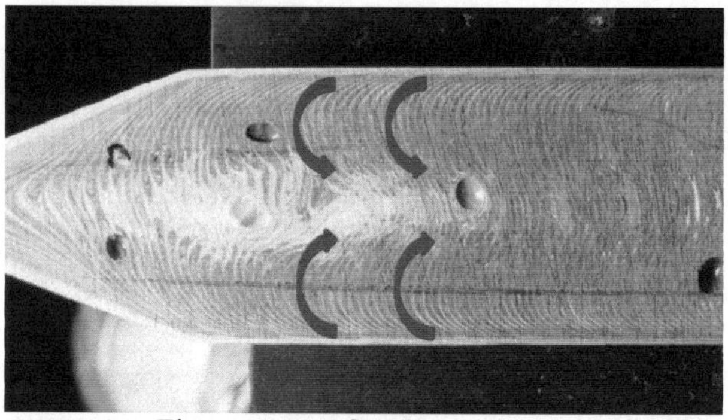

FIGURE 3. Flow pattern of molten resin accumulated in the nozzle chamber during plasticizing.

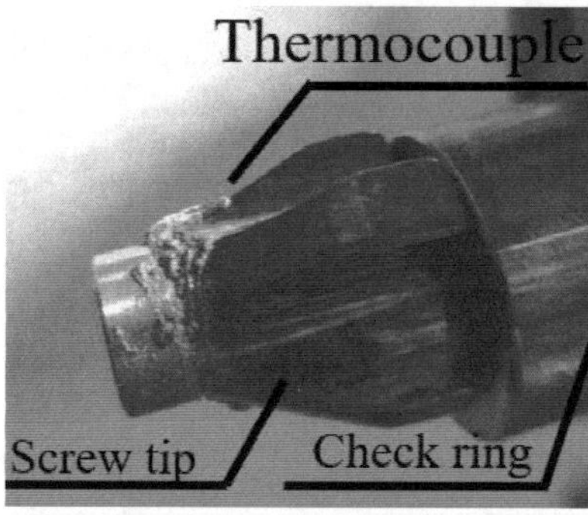

FIGURE 4. Thermocouple mounted to screw tip.

EXPERIMENT

The melt temperature was measured using a screw-mounted temperature sensor. The experimental conditions are shown in Table 1. The melt temperature was measured by changing the screw rotation speed that was identified as the most significant influencing factor to the melt temperature variation in the previous research. The LSMtron 150-

ton high speed electric injection molding machine was used. After the plasticizing was completed, the screw was advanced slowly to inject the melt into the mold, the nozzle-mounted wire-type sensor measured the melt temperature. The measured melt temperatures by the two sensors were compared.

Conditions	*Value*	*Conditions*	*Value*
Material	LDPE	Injection rate (mm/s)	10
Screw rotation speed (rpm)	50, 300	Back pressure (bar)	10
Heater temperature (°C)	190	Holding pressure	Not applied

TABLE 1. Process conditions for experiment.

RESULT AND DISCUSSION

Figure 5(left) shows temperature data measured by the nozzle-mounted sensor during the injection stage. The temperature as reported in the previous research showed a large amount of temperature variation at high screw rotational speeds. The melt temperature dropped less than 5 °C at 50 rpm of the screw rotation speed after it reached the peak value, but the melt temperature dropped more than 20 °C at 300 rpm.

Figure 5(right) shows temperature data measured by the screw-mounted sensor during the plasticizing stage. In the case of 50 rpm of the screw rotation speed, almost a constant temperature was measured. In the case of 300 rpm, a temperature variation of about 1 °C was measured. The average melt temperatures measured by the screw-mounted sensor were lower than the ones by the nozzle-mounted sensor. The variation trends looked similar in both cases. Relatively smaller amount of temperature variation was measured at the screw-mounted sensor compared with the temperature measured at nozzle-mounted sensor. The higher temperature variation cannot be made in the chamber from the melt that has a lower temperature variation after the check ring. As a possible cause of the smaller amount of the melt temperature variation, the screw-mounted sensor has a lower sensitivity than the wire-type sensor mounted in the nozzle adaptor. Both sensors have a diameter of 0.8 mm (AWG20), but in the case of a nozzle-mounted sensor, L/D ratio from the nozzle wall to the measuring point is 15. However, in the case of a screw-mounted sensor the L/D ratio is about 8 from the mounting point. The closer distance between the thermocouple tip and the metallic body in the screw-mounted sensor may have a higher heat transfer to the screw body to make the sensitivity lower. From the result, a way to increase the sensitivity of the screw-mounted sensor should be made to make the sensor feasible enough to measure the melt temperature accurately.

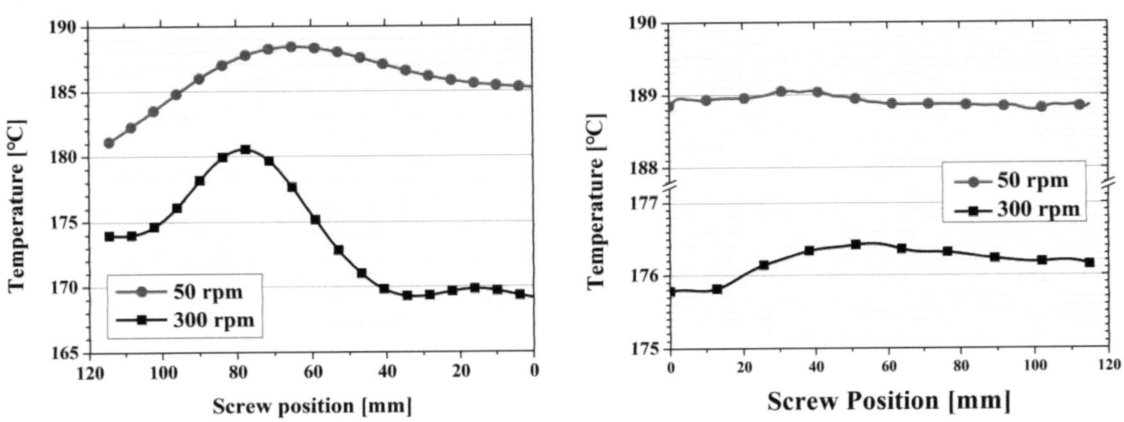

FIGURE 5. Melt temperature profile measured by nozzle-mounted temperature sensor (left) and screw-mounted temperature sensor (right) for the different screw rotation speed.

CONCLUSION

A screw-mounted temperature sensor has been introduced to more accurately analyze the effects of influencing factors to the melt temperature during plasticizing. Since the nozzle-mounted wire-type sensor was capable of measuring the melt temperature during the injection stage, it has a drawback of installation in the production type nozzle adaptor due to the structure and the low durability. Neither the nozzle-mounted sensor gives us the melt temperature right after the melt is made in the screw. We developed a new sensor mounted at the screw tip to measure the melt temperature during plasticizing. A digital telemetric system was developed by battery-powered computer of Raspberry pi with amplifier and ADC. The original measuring position at the center the screw tip could not measure the melt temperature due to the unique melt flow pattern in the chamber. The measuring position was changed to the middle of the melt passage of the screw tip. The melt temperature could be successfully measured, but it turned out to have an insufficient sensitivity to measure the melt temperature variation. In the future, it is necessary to find a way to increase the sensitivity of the screw-mounted sensor to analyze the effects of influencing factors to the melt temperature. Subsequently, it is also necessary to study to minimize the melt temperature variation and predict the melt temperature.

REFERENCES

1. O. Amano and S. Utsugi, Polym. Eng. Sci., 28, 1565, (1988)
2. O. Amano and S. Utsugi, Polym. Eng. Sci., 29, 171, (1989)
3. O. Amano and S. Utsugi, Polym. Eng. Sci., 30, 385, (1990).
4. K. Straka, B. Praher and G. Steinbichler, AIP Conference Proceedings, 1558, 94, (2013)
5. G.C. Peischl and I. Bruker, Polym. Eng. Sci., 29, 202, (1989).
6. N. Dontula, P.C. Sukanek, H. Devanathan, and G.A. Campbell, Polym. Eng. Sci., 31, 1674, (1991).
7. N. Sombatsompop and W. Chaiwattanpipat, Polym. Test., 19, 713, (2000)
8. Wood, Y. Judeh, M. Yue, UK Patent Application, GB 2291197, (1996)
9. B. Praher; K. Straka; G. Steinbichler Meas. Sci. Technol., 24, 084004, (2013)
10. J. Jeon, J. Gim and B. Rhee, Proceedings of SPE ANTEC 2016, 1156, (2016).

Numerical Simulation for Predicting Sink Marks on Injection Molding and Injection Compression Molding Process

Yen-Chih Chen, Chih-Chung Hsu, and Chia-Hsiang Hsu

CoreTech System (Moldex3D) Co., Ltd., Chupei City, Hsinchu Country, 302, Taiwan

Abstract. Sink marks are small depressions on the surface of molded parts due to the localized uneven shrinkage and varying thicknesses. Although the sink mark depth is so shallow that it has no influence on the strength of the product, sink marks would be visible through the reflection of lights in different directions. Therefore, sink marks are often regarded as quality defects in the molding process. This study presents a numerical method to predict sink marks on different molding processes. The molded part composed of a plate and four ribs with different widths is used to validate sink marks from the numerical simulation and the experiment. The result shows good agreement between the simulation and the experiment. Moreover, the proposed method can be applied to the injection compression molding process as well. The sink mark displacements decrease due to the uniform compression effect in the injection compression molding process.

Keywords: Sink Marks, Sink Mark Displacements, Injection Molding, Injection Compression Molding

INTRODUCTION

With the development of the I.T. industry, numerous products are asked for stylish appearance, strength structure, and light weight. Thin-wall sections are typically applied to reduce weights in the injection molded parts. To enhance mechanical properties of thin-wall parts, reinforcing ribs, bosses and similar structures are frequently attached and hided in invisible locations. However, the small depression named as the sink mark may often occur on the surface opposite to a rib or a boss. Due to varying thicknesses at the intersection of the part surface and the rib, the heat of the polymer melt tends to accumulate at the base of the rib during the packing stage. In general, the base of the rib with a large quantity of heat is the last area of the part to solidify. When the material in this area cools, the shrinking material leads to the inward deformation of the outer frozen layer and forms the sink mark on the surface.

To reduce sink marks, modifying the packing condition or the part geometry is commonly used. In addition, the application of advanced molding process can also improve sink marks effectively. In the molding experiment, Chen et al.[1] has demonstrated the improvement of sink marks by using the external gas-assisted injection molding (EGAIM) compared with the conventional injection molding (CIM). Due to the uniform packing pressure provided by the gas in the EGAIM, the shrinkage of the part can be compensated. Therefore, the depth of the sink mark only ranges from 1.24 to 3.54 micrometers for the EGAIM while the sink mark ranges from 5.07 to 26.88 micrometers for the CIM. With the similar concept to reduce sink marks, this study applies the injection compression molding (ICM) compared with the injection molding. Firstly, a numerical approach on evaluating the sink marks would be verified with the aforementioned experiment in the injection molding. Finally, the proposed method can be extended to the ICM. On the basis of the uniform pressure applied by the clamping unit, the result shows less sink mark displacements with the ICM.

The main purpose of this work is to develop a numerical method for predicting sink marks with a given part geometry, material, process condition, and molding type. This method employed by Moldex3D software can serve as a design tool to optimize the formation of sink marks without expensive molded tryouts.

MATERIALS AND METHODS

For a general resin molding process, the polymer melt is filled into a cavity and is packed after the filling is finished. The dynamic model of the melt flow in filling and packing stages can be described by following governing-conservation equations:

Equation of continuity:

$$\frac{\partial}{\partial t}\rho + \nabla \cdot \rho \mathbf{u} = 0 \tag{1}$$

Equation of motion:

$$\frac{\partial}{\partial t}(\rho \mathbf{u}) + \nabla \cdot (\rho \mathbf{u}\mathbf{u} + \boldsymbol{\tau}) = -\nabla p + \rho \mathbf{g} \tag{2}$$

Equation of energy:

$$C_p \frac{\partial}{\partial t}(\rho T) + \nabla \cdot (\rho \mathbf{u} T - k \nabla T) = \eta \dot{\gamma}^2 \tag{3}$$

where ρ is density, $\mathbf{u}$ is velocity vector, t is time, $\boldsymbol{\tau}$ is stress tensor, $\mathbf{g}$ is acceleration vector of gravity, p is pressure, C_p is specific heat, T is temperature, k is thermal conductivity, η is viscosity, and $\dot{\gamma}$ is shear rate. In present study, the polymer melt is considered to be the Generalized Newtonian Fluid (GNF), so the term of the stress tensor in Eq. (2) can be rewritten as:

$$\tau = \eta(\nabla \mathbf{u} + \nabla \mathbf{u}^T) \tag{4}$$

Moreover, the viscosity of the melt flow involves temperatures and shear rates yielding to the Cross Willam-Landel-Ferry (Cross-WLF) model expressed as:

$$\eta(T, \dot{\gamma}) = \frac{\eta_0}{1 + (\eta_0 \dot{\gamma}/\tau^*)^{1-n}}, \eta_0 = D_1 \exp\left[\frac{-A_1(T - T_c)}{A_2 + (T - T_c)}\right] \tag{5}$$

with

$$T_c = D_2 + D_3 P, A_2 = \widetilde{A}_2 + D_3 P$$

The finite volume method (FVM) is used to solve aforementioned equations numerically. In the FVM, the flow field is separated into numerous cells also known as mesh. Properties of the polymer melt, such as the velocity, the temperature, and the pressure, in each cell can be computed during filling and packing stages. The specific volume can be further determined by substituting the pressure and the temperature into the p-V-T diagram of the polymer material. Thus, the volumetric shrinkage can be estimated by:

$$S_V = \frac{v_{EOP} - v_{1\,\mathrm{atm}, 25\,^\circ\mathrm{C}}}{v_{EOP}} \tag{6}$$

where v_{EOP} is the specific volume at the end of packing status and $v_{1\,\mathrm{atm}, 25\,^\circ\mathrm{C}}$ is the specific volume at conditions of the room temperature and the ambient pressure. Since sink marks are caused by linear shrinkages along the direction of the part thickness, the linear shrinkage can be approximated by assuming isotropic shrinkage of the polymer material. For the three dimensional simulation, the linear shrinkage is computed by:

$$S_L = 1 - (1 - S_V)^{1/3} \tag{7}$$

Therefore, depths of depressions on the surface of the part, named as sink mark displacements, are evaluated by the product of the linear shrinkage and the part thickness.

RESULTS AND DISCUSSIONS

To verify sink marks estimated by the proposed method in this study, the molded part is established and illustrated in **FIGURE 1**. The dimension of the main flat plate of the part is $100\,\text{mm} \times 50\,\text{mm} \times 1.5\,\text{mm}$. Four ribs with different widths connect to the main plate. The height of ribs is 4.5 mm and aspect ratios (rib width to part thickness) of ribs are designated as 0.6T (0.9 mm width), 0.8T (1.2 mm width), 1.0T (1.5 mm width), and 1.2T (1.8 mm width) respectively. The material used in the molding process is ABS CHIMEI PA757. **FIGURE 2** shows the viscosity and p-V-T curves of the plastic material. The process conditions of the injection molding are listed in **TABLE 1**.

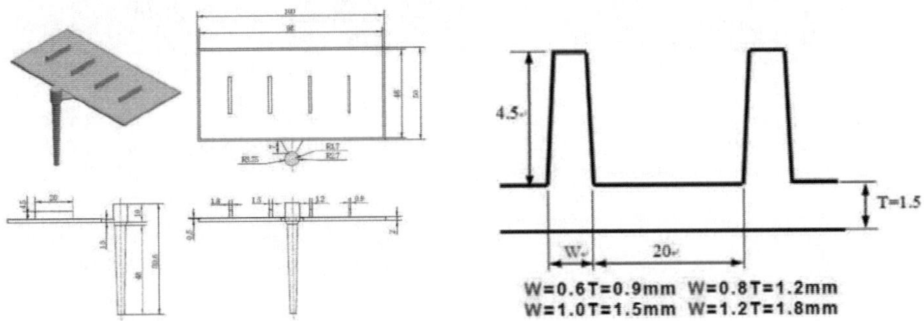

(a) (b)

FIGURE 1. (a) The dimension of the molded part and the runner system. (b) The dimension of ribs. [3]

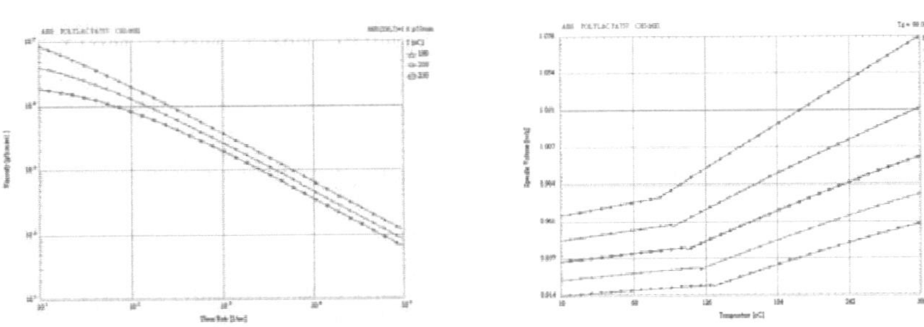

(a) (b)

FIGURE 2. Properties of ABS material involved (a) viscosity and (b) p-V-T.

TABLE 1. The process conditions of injection molding.

Injection molding process	
Machine	CLF-80T
Material	ABS CHIMEI PA757
Injection velocity	80 mm/s
Packing pressure	40 MPa
Packing time	3 seconds
Melt temperature	210 °C
Mold temperature	55 °C

The experimental results of the injection molding process are presented by Chen et al.[1] The comparison of sink mark results between the experiment and the simulation are shown in **FIGURE 3** and **FIGURE 4**. As the results indicated, the sink mark displacement increases while the rib width increases correspondingly. The wider rib tends to accumulate more heat in the core at the end of packing. As **FIGURE 5 (a)** indicates, the polymer has high temperature and volumetric shrinkage in the base of rib due to the unremovable heat. Therefore, the highly uneven distribution of volumetric shrinkages leads to the deep sink mark on the surface of wide rib. Furthermore, the comparison of sink mark displacements of the injection molding between the experiment and the simulation demonstrates that proposed numerical approach can be applied to predict sink marks and the numerical results is relevant to the experiment.

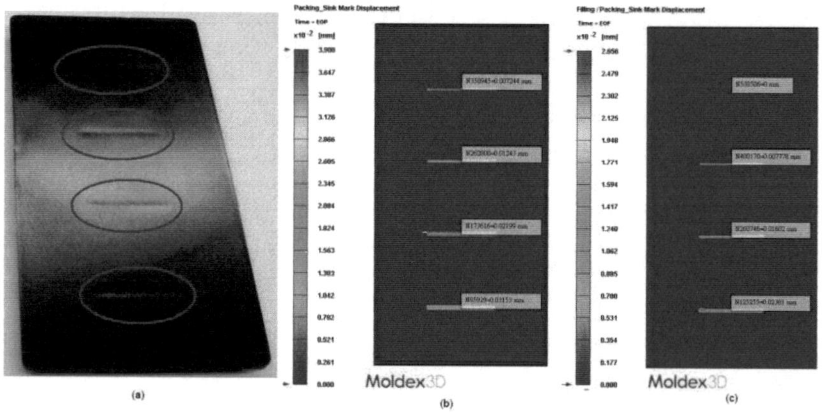

FIGURE 3. Comparisons of sink mark appearance between the experiment and simulations: (a) The experimental result of injection molding, (b) The simulated result of injection molding, (c) The simulated result of injection compression molding.

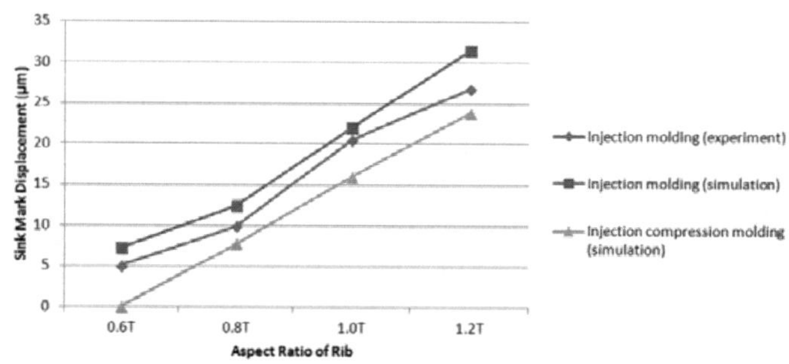

FIGURE 4. Comparisons of sink mark displacements between the experiment and simulations on various widths of ribs.

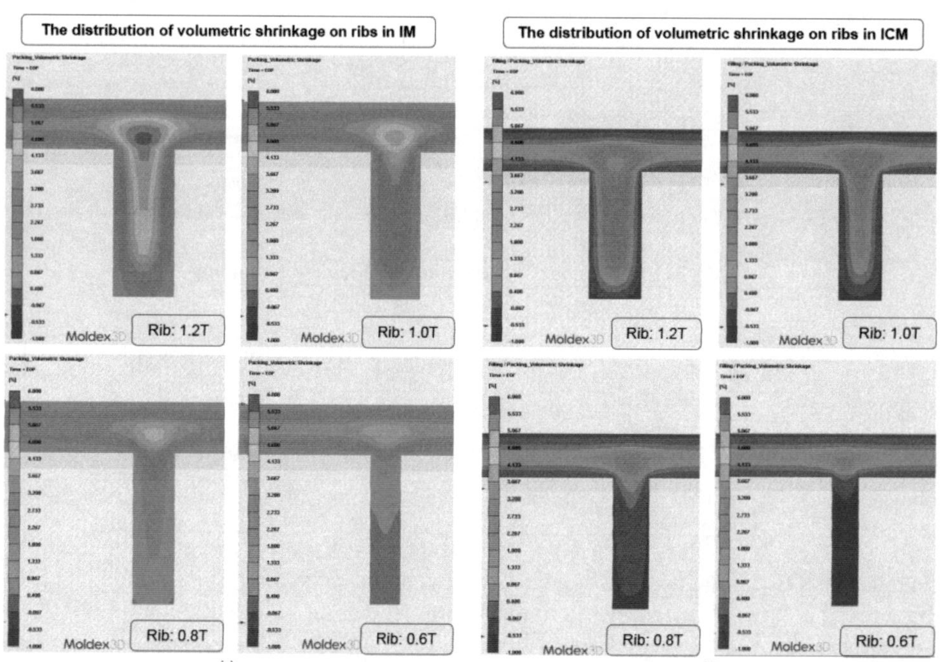

FIGURE 5. Comparisons of the distribution of volumetric shrinkage between (a) the injection molding and (b) the injection compression molding on various widths of ribs.

To reduce sink mark displacements, the injection compression molding (ICM) process is employed. The injection process of the ICM is identical to that of the aforementioned injection molding. In the ICM, when the part is filled over 70 percent in the injection stage, the screw is holding and the clamping unit starts compressing the melt on the vertical direction of the melt flow. In present study, the compression time and the compression gap are set to be 0.5 second and 1 mm respectively. The different flow patterns between the injection molding and the ICM are illustrated by **FIGURE 6**. The polymer material at the flow front is pushed forward and fountains to the cavity surface in the injection molding while it is pushed along the compression direction in the compression stage of the ICM. Furthermore, the simulated sink marks of the ICM are shown in **FIGURE 3 (c)** and the green line of **FIGURE 4**. Due to the uniform holding pressure applied by the clamping unit, the shrinkages of the polymer can be effectively compensated at packing stage. As a result, volumetric shrinkages of the ICM shown in **FIGURE 5 (b)** are more uniform than that of the injection molding and sink mark displacements of the ICM are less than that of injection molding correspondingly. Moreover, the sink mark of the narrowest rib is nearly eliminated in the ICM process.

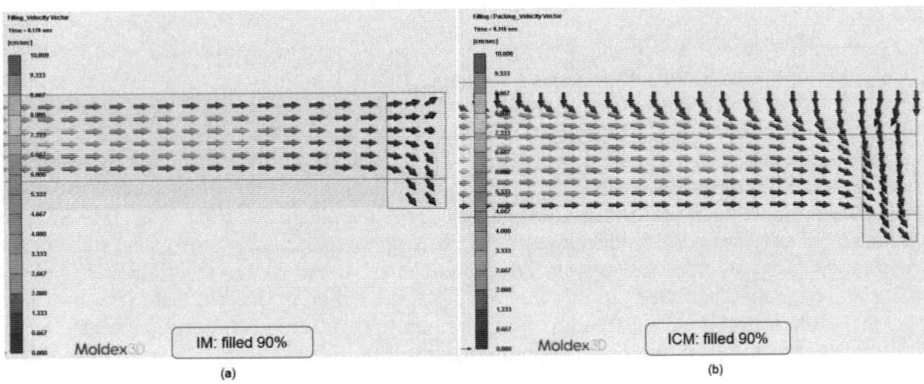

FIGURE 6. Comparisons of velocity vectors between (a) the injection molding and (b) the injection compression molding in the 90 percent of filling volume.

CONCLUSIONS

The numerical approach proposed in this study is applied to evaluate sink marks on different molding processes. To verify simulated sink marks, the molded part composed of a plate and four ribs with different widths is used in the injection molding process. The results indicate that the wide rib leads to deep sink mark on the surface of the upper plate due to the unremovable heat accumulation in the base of rib. Both simulated results and experimental results show the similar tendency of sink mark displacements. Furthermore, the injection compression molding process is employed to reduce sink marks. On the base of the uniform pressure applied on the molded part, the shrinkages of the polymer are effectively compensated. Hence, sink mark displacements of the ICM are less than that of the injection molding and the surface quality of the part molded by the ICM is better than that molded by the injection molding correspondingly. In summary, Moldex3D software provides a simulation tool based on the proposed method to evaluate sink marks in various molding process. It is a great benefit to find out potential sink marks and improve the outward appearance of products for the molding industrial application.

REFERENCES

1. S.-C. Chen, Y.-C. Lin, and S.-W. Huang, *Polym. Eng. Sci.*, **50**, 2085-2092 (2010).
2. C.-C. Su, "Study on the Part Surface Quality of Rib with External Gas-Assisted Injection Molding", Master Thesis, Chung Yuan University, 2007.
3. S.-C. Chen, S.-W Huang, C.-Y Chang, H.-W. Hung and W.-H. Yang, *ANTEC*, 2008, pp. 1504-1507
4. C. A. Hieber and H. H. Chiang, *Polym. Eng. Sci.*, **32**, 931-938 (1992)
5. R.-Y Chang and W.-H. Yang, *Int. J. Numer. Meth. Fluids*, **37,** 125-148 (2001)

Effects of rapid cavity temperature variations on the crystallinity of PLA

Annarita De Meo[a*], Felice De Santis[b] and Roberto Pantani[a]

[a]*Department of Industrial Engineering, University of Salerno, Fisciano - SA – Italy*
[b]*Borealis Polyolefine GmbH - Linz - Austria*

Abstract. The extraordinary growing demand for environmentally friendly materials leads to an increasing attention on the use of biopolymers. The Poly-lactid acid is one of the most promising biopolymers, with very interesting physical and mechanical properties. However, the low rate of crystallization of the PLA compared to other semi-crystalline polymers limits its use in technical applications. In this study, the effect of the application of a system able to rapidly change the temperature of the cavity surface on the final crystallinity of Poly-lactid acid micro-parts was studied. The technique adopted is based on the use of very thin heating elements, thermocouples and insulation layers that can be attached near the surface of the cavity allowing a precise and local control of the temperature during each stage of the molding cycle. The system was adopted to increase the temperature of the cavity up to 160°C during the filling stage to permit the filling of a cavity with a thickness of 0.2 mm and an aspect ratio equal to 50. After the filling, the system allowed the application of rapid temperature variations locally along the sample. Different thermal histories were applied to tune the morphology of the parts. Calorimetric analyses permitted to understand the crystallization behavior of the processed material and determine the final crystallinity in the produced samples.

Keywords: Micro-injection molding, Crystallization, Polymer processing
PACS: 83.50.Uv, 44.10.+i, 07.05.Tp, 47.85.md

INTRODUCTION

Poly(lactic acid) is a biodegradable polymer obtained by synthesis of lactic acid that can be produced from renewable resources such as corn or sugarcane. PLA is an enantiomeric polyester including poly(L-lactic acid) and poly(D-lactic acid). High content of poly(D-lactic acid) permits to obtain fully amorphous materials whereas highly crystalline material is obtained when the D content is low [1]. Obtaining a highly crystalline injection-molded PLA part, however, remains difficult due to PLA's slow crystallization rate [2]. The polymer's crystallinity plays a significant role in the performance of the material. Obtaining a highly crystalline part of PLA remains difficult due to the low crystallization rate. The minimum crystallization halftime, $t_{1/2}$, of a pure sample of PLA is of the order of several minutes, which is an extremely long time if compared to the cooling times experienced by the polymer during extrusion and injection molding, which is of the order of 1s. Many researchers have spent significant efforts to study the crystallization kinetics of PLA and the possibility of selecting suitable nucleating agents to improve the crystallization rates [3]. However, significant crystal microstructure is sometimes found in injection molded parts [4], thus suggesting that flow can significantly influence the phenomenon of crystallization. The flow generates a local alignment and a stretch of chain segments that can induce the formation of oriented precursors. The flow induced precursors can persist at high temperature and long times, even above the melting temperature. The precursors accelerate the crystallization kinetics and can cause the generation of anisotropic flow-induced crystalline structures. The final physical properties are also influenced by the oriented morphology [5].

In general, an increase in overall crystallinity leads to improvements in stiffness, strength, heat deflection temperature and chemical resistance [6, 7]. To maximize property enhancements with a high degree of crystallinity in the injection-molded part, the cooling rate has to be slow, or alternatively, annealing of post injection-molded part is needed [8, 9].

In this work, a special surface-conditioning concept described in detail in the work of De Meo et al [10] is adopted to rapidly change the temperature of the cavity surface and affect the morphology of a semicrystalline polymer.

MATERIALS AND METHODS

Raw Material

Poly(lactic acid) 3251D, hereafter called PLA 3251D, supplied by NatureWorks LLC, is a grade designed for injection molding applications, having a number average molecular weight M_n of 83 kDa, dispersity of M_w/M_n=1.6, and D-isomer content of 1.4%. As recommended by the supplier, before any test or processing, the material was dried for 8 h under vacuum at a temperature of 60°C to prevent degradation during the subsequent processing.

The Rapid Temperature Control

In this work, the injection-molding machine HAAKE Minijet II by Thermo Scientific is used. The temperature control is based on the use of very thin resistances as heating source and thermocouples that can be attached near the surface of the cavity that has a rectangular shape (11 mm long and 4 mm wide) and a thickness of 200 μm. The electrical resistances are grids of constantan foil sealed in a polyimide film with a nominal resistance of 120 Ω. Each heating element is 3.8 mm wide, 5.7 mm long and 50 μm thick. A thin wire thermocouple, (Type T, about 50 μm in diameter supplied by Omega Engineering Ltd.) between each resistance and the cavity surface is attached. The insulation layer is a PolyEtherSulfone (PES) film. The insulation has the fundamental role of lowering the heat loss towards the rest of the mold and permits a necessary conduction to allow a reasonable cooling rate. A data acquisition board (DAQ NI-USB 6210 supplied by the National Instruments) allows transferring the temperature data to a pc where a dedicated software controls the values of each zone at 10 kHz. This system, shown in Figure 1, allows regulating the evolution of the local mold temperature switching the linked relays and thus imposing a given heating and cooling rate (in the limit of the maximum values determined by the heating power and the ballistic cooling without power supply). A Labview code was written to control differently two zones of the cavity. Each experiment in this work is carried out by using a power supply equal to 12 V. In Table 1 the main physical properties and the thickness of the layers in the system are reported.

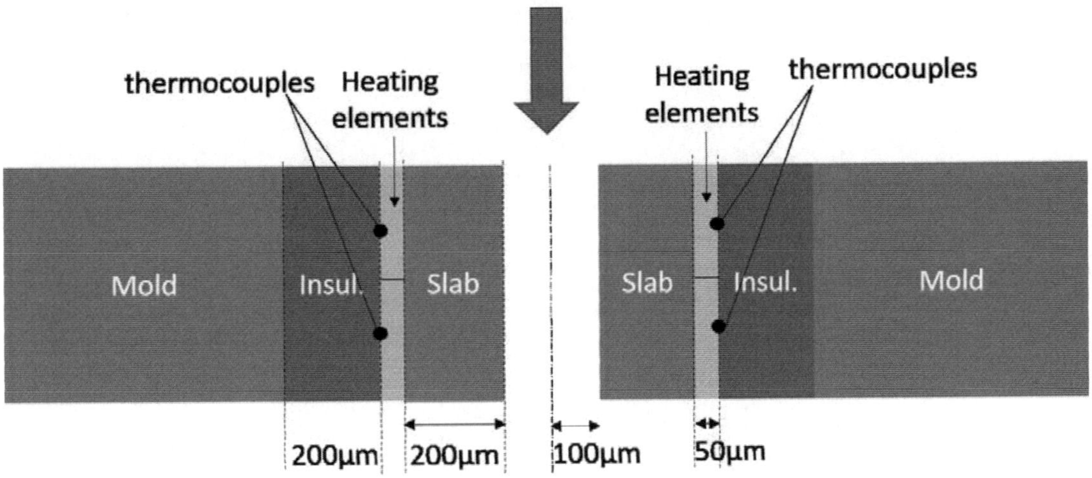

FIGURE 1. The mold and heating system section.

TABLE 1. Layers properties and thickness

Layer	Material	density [g/cm3]	Specific heat [J/kg K]	Thermal conductivity [W/m K]	Thickness [mm]
Insert	Steel P20	7.8	460	29	0.200
Thermocouple	Polymide	2.138	1090	2.74	0.155
Heater	Polymide	2.138	1090	2.74	0.140
Mold	Steel	7.8	460	29	10

Micro-injection Molding Tests

The micro-injection molded bars were obtained by applying the process parameters showed in Table 1 and the experimental protocol is represented in Figure 2. For each experiment, two steps were considered. In the first one, the cavity surfaces were heated up to 160°C to permit the complete filling of the microcavity. When the temperature was reached, the polymer is injected applying a pressure of 100 bar for a time of 20 s. After that, the power supply are switched off and because of the ballistic cooling the polymer injected solidifies. In the second step, the temperature control is used to heat only one the zone after the gate (called zone 1). Half sample experiences an isotherm at 105°C for 500 s. Previous investigations pointed out that the narrowest crystallization peak of the PLA 3251D is at 105 °C, i.e. it requires the shortest time at this temperature to complete the crystallization. The semi-crystallization time, i.e. the time necessary for half crystallization of the material $t_{1/2}$, at the temperature of 105 °C, is about 1000 [10].

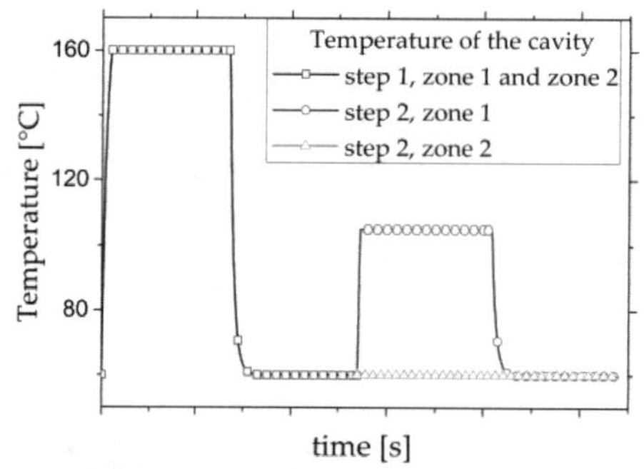

FIGURE 2. Experimental protocol

Table 1 Set Parameters for injection molding

Melting Temperature [°C]	Mold Temperature [°C]	Injection Pressure [bar]	Cavity Temperature [°C]
180	60	100	160

Samples Characterization

Polarized light microscopy was used to examine the morphology of microinjection-molded PLA. For this purpose, 15 μm sections were obtained in the flow directions across the sample, using a Leica RM2265 microtome. The microtomed sections were observed under transmitted polarized light by using an Olympus BX51 optical microscope system.

Differential scanning calorimetry was carried out by a Mettler DSC822 in flowing nitrogen atmosphere. The analysis was used to assess the crystallinity developed locally. The samples were initially heated at a rate 10 °C/min from 0°C to 200 °C and then cooled to 0°C.

RESULTS AND DISCUSSION

The acquisitions of the mold surface temperature, showed in the Figure 3, demonstrate that by using the technique for the rapid temperature control less than 10 s is required for the temperature to rise from 60°C to 160°C, and the same time is required to cool from 160°C to 60°C. In addition, the system is able to tune in different manner the temperature in two zones of the cavity surface therefore the sample after the injection and the cooling remains inside the mold and undergoes different thermal histories in different positions.

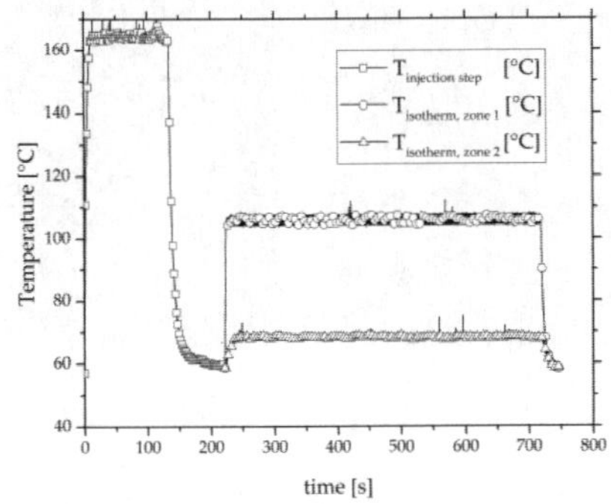

FIGURE 3. Temperature evolution during the test.

The calorimetric analysis carried out on the injected samples confirmed that for an annealing of 500 s the amount of crystalline degree obtained in the opaque zone is about 19 % whereas in the transparent zone a negligible crystallinity is found.

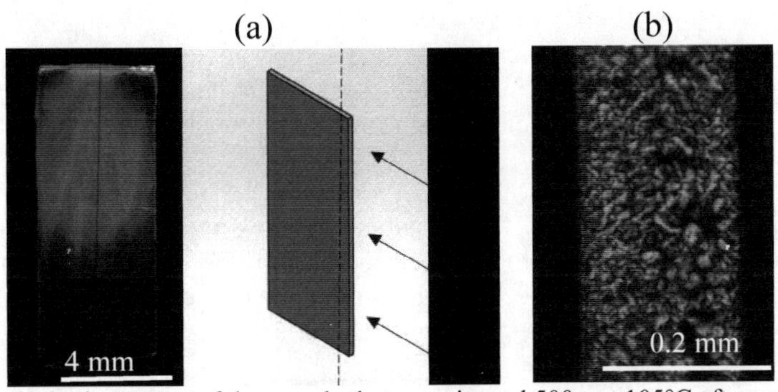

FIGURE 4. Polarized optical microscopy of the sample that experienced 500 s at 105°C after a permanence at 160°C for 5 s: (a) the whole sample before cutting, (b) the morphology of the opaque zone along the thickness of the sample.

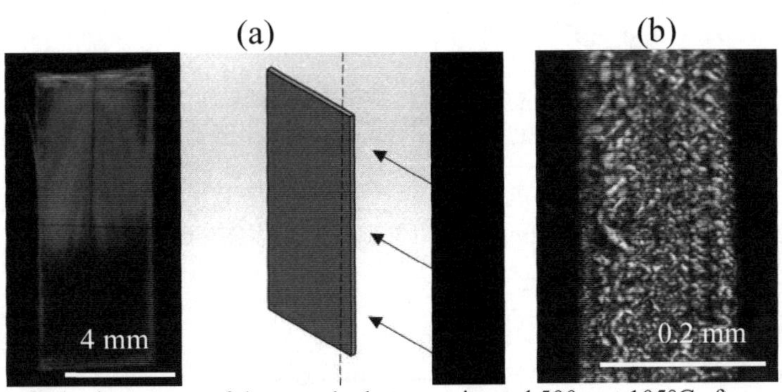

FIGURE 5. Polarized optical microscopy of the sample that experienced 500 s at 105°C after a permanence at 160°C for 20 s: (a) the whole sample before cutting, (b) the morphology of the opaque zone along the thickness of the sample.

The Figures 4 and 5 show the morphology of the produced parts. The zone that experienced the annealing appears opaque whereas the other zone is completely transparent. In the Figures 4a and 5a the samples before the microtoming. In the Figures 4b and 5b slices from the samples are observed along the thickness with core region in the center. Contrary to what is present in literature, the microinjected parts do not present different layers with different morphologies. In fact, the spherulitic structures are predominant with a small spherulitic skin region. The result is that the sample appears completely homogenous in both cases.

(a) (b)

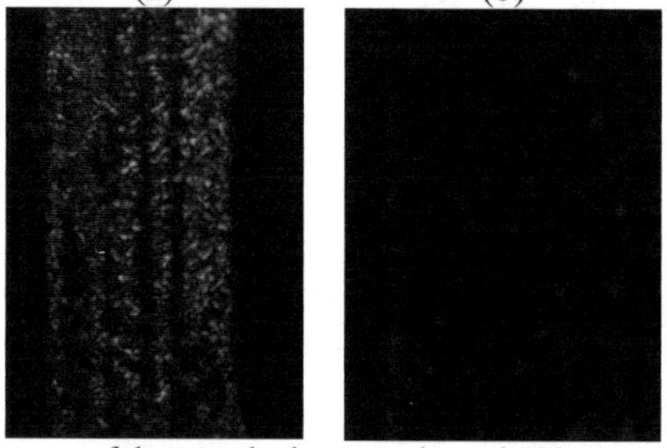

FIGURE 6. Polarized optical microscopy of the sample that experienced 500 s at 105°C after a permanence at 160°C for 5 s (a) in the center between the opaque zone and the transparent, (b) at the tip.

In the center of the sample, as showed in Figure 6a, a lower number of spherulites is visible, they are smaller in diameter and most of them seem to be arranged in sequence. This part is in the middle of the sample, between the area heated during the annealing and the subsequent that is at a lower temperature where no crystalline structure is visible (Figure 6b) because the sample appears completely transparent.

CONCLUSIONS

The investigations performed on the PLA micro-injection molded parts show that the system for the temperature control of the cavity is an excellent strategy to tune the final properties of the parts by controlling the amount of crystallinity. The parts were injected by applying the same pressure when the cavity surfaces were at 160°C. The polymer, after the filling, remained at 160°C for different time up from 5 s to 20 s After that, only one zone of the sample is kept at 105°C for 500 s.

The calorimetric and optical analyses revealed that the time of permanence at 160°C considered for the experiments, after the injection in the cavity, does not affect the crystallization process: the same amount of crystallinity was measured and similar morphology was observed along the section of each sample. The longest time applied, 20 s, is not sufficient to cancel the mechanical history. The process yields enhanced point-like nuclei and the morphology remains isotropic.

REFERENCES

1. B. Gupta, N. Revagade, and J. Hilborn, Prog. Polym. Sci., 2007, 32, 4.
2. Li H, Huneault MA. Polymer 2007; 48(23):6855-66.
3. Tang et al, Journal of Applied Polymer Science, Vol. 125, 2012, 1108–1115.
4. V. Volpe, M. De Filitto, V. Klofacova, F De Santis, R. Pantani, Polym Eng Sci, 2017, 58: . doi: 10.1002/pen.24730
5. X. Monnier et al. / Polymer 126, 2017, 141-151.
6. T.A. Osswald, International Plastics Handbook: The Resource for Plastics Engineers, Hanser Verlag, Munich, 2006.
7. Y. Srithep, A. Javadi, S. Pilla, L.S. Turng, S. Gong, C. Clemons, and J. Peng, Polym. Eng. Sci., 2011, 51, 6.
8. A.M. Harris and E.C. Lee, J. Appl. Polym. Sci., 2008, 107, 4.
9. K. Goto, S. Nakano, and T. Kuriyama, Strength. Fract. Complex. 2006, 4, 3.
10. A. De Meo, F. De Santis, R. Pantani, Polym. Eng. Sci., 2018, 58:586–591, https://doi.org/10.1002/pen.24784.

Simulation of Interface Strength for a Two-Component Injection Moulded Wheel Product of HDPE and EPDM Rubber

W. Six[a], G.J. Bex[b], J. De Keyzer[c], A. Van Bael[b], F. Desplentere[a]

[a]*Materials Technology TC, KU Leuven Bruges Campus*
Spoorwegstraat 12, 8200 Bruges, Belgium
[b]*Materials Technology TC, KU Leuven Technology Campus Diepenbeek,*
Wetenschapspark 27, 3590 Diepenbeek, Belgium
[c]*Sustainable Chemical Process Technology TC, KU Leuven Technology Campus Diepenbeek,*
Wetenschapspark 27, 3590 Diepenbeek, Belgium
Corresponding author: wim.six@kuleuven.be

Abstract. As polymer products are being used in ever more demanding applications e.g. automotive and aerospace, the demand for high performance polymers is rising. For a hard – soft material combination in an overmolding situation, TPE's or TPV's are often used. For high demanding applications, these materials can be replaced by thermoset rubbers e.g. EPDM, NBR, ... However, until recently, only little was known about the quality of adhesion between these two components in an overmolded product, being an important quality parameter for two-component injection molding. In previous research performed by the researchers, a mold for test plates has been developed for the thermoset rubber, thermoplastic material combination. With this mold, knowledge of the adhesion in this process has been acquired. By using this knowledge, various samples could be produced and tested for their interfacial strength. Combining these test results with numerical simulations led to a method to simulate and to predict the interface strength for the given material combination. Furthermore, the simulated product properties acquired by Autodesk Moldflow can be mapped on a structural mesh, to perform numerical simulations on the failure behavior of the product and the interface. This procedure has been used to optimize the interfacial strength for a new product: a HDPE ring overmolded with a thick rubber layer of EPDM, performing as a wheel. Experimental tests and simulations have been performed on the product by peeling the rubber from the HDPE core. The experimental work, being the peel tests, shows good correlation with the simulations

Keywords: Two-component injection molding, thermoset rubbers, thermoplastics, simulation, process optimization.
PACS: 68.35 Gy, 81.70-q 81.050 Lg

INTRODUCTION

The use of numerical simulation to predict the flow behavior in the processing phase of injection molding is very common, even for thermoset rubbers. It is often done for complex product shapes to detect possible problems in the production phase of the part. More recently, the simulations predicting the production phase are being combined with "performance" simulations. These include structural, thermal, fatigue or failure behavior. The combination of production and performance, results in "as made" simulations. In the case of two-component injection molding, the strength of the interface can be considered as one of the most important product parameters, which should be taken into account in these kind of simulations. For two-component products, consisting of thermoplastics and thermoplastic rubbers, various attempts have been made for the prediction of this interfacial strength. Interfacial strength can be governed by three different ways: chemical bonding, chain entanglement and mechanical interlocking [1]. The latter is mainly governed by product design and will not be taken into account in this work. Chain entanglement can only happen if the local temperature is sufficiently high and Brownian motion of the polymer chains is allowed to happen. If the polymer of the first component can chemically react with the material of the second component, and form hydrogen bonds for example, the resulting interface strength will be very high [2]. In most injection molding situations, the latter will not happen as commercial available polymeric material tend to be chemically stable. For the thermoplastic-thermoplastic material combinations, prediction of adhesion is mainly done by the reptation theory [3]. This theory describes the mobility of polymer chains in a melt where the movement is restricted by neighboring and entangled polymer chains. In the given theory, polymer chains are allowed to move back and forth within an idealized tube. This theory, based on probability of chains moving across the interface has been used to predict the inter-diffusion of polymeric chains across polymer-polymer interfaces [3]. As soon as polymer chains cross the interface, the interfacial strength σ increases, reaching a maximum σ_∞ if inter-diffusion comes to an equilibrium. The characteristic time in which this happens is defined as the reptation time t_r

$$D_h = \frac{\sigma}{\sigma_\infty} \propto \left(\frac{t}{t_r}\right)^{\frac{1}{4}} \tag{1}$$

Wool et al.[4] showed that the reptation theory is mainly valid for low molecular weight polymers where M_c<M<$8M_c$, where M_c is the critical entanglement molecular weight. For engineering polymers where the molecular weight is higher than $8M_c$ the interfacial strength reaches its maximum before interfusion comes to an end. In the given case, Yang et al. [5] defined the degree of healing, D_h by equation 2 defining a welding time (t_w) instead of a reptation time. This welding time is stated to be very short (e.g. 0.11 s for PEEK [5])

$$D_h(t) = \frac{\sigma(t)}{\sigma_\infty} = \left[\int_0^t \frac{1}{t_w(T(t))} dt\right]^{1/4} \tag{2}$$

Six et al [6]. showed that this theory is also valid to estimate the adhesive strength for the presented material combination, being Sabic HDPE M80064SE and Hercorub EPDM 005K, a sulfur curing rubber. Next to the adhesive strength, cohesive strength is of key importance in the given material combination, as the stiffness and strength of the vulcanizing rubber develops in function of curing time. They also state that by combining the strength of the rubber σ_α, thus the cohesive strength and the adhesive strength σ_{Dm}, due to interface healing, determined by the local interface melting the total strength of the interface σ_I can be estimated, shown by equation 3

$$\sigma_I = \begin{cases} \sigma_\alpha, & \sigma_\alpha < \sigma_{Dm} \\ \sigma_{Dm}, & \sigma_\alpha \geq \sigma_{Dm} \end{cases} \tag{3}$$

SIMULATION AND PRODUCTION METHOD

Production Method for a Wheel Product

To test the theory earlier developed by Six et al [6], a new mold and product was designed for the presented work which has the shape of a wheel. The product consists of a thermoplastic core, a thickness of 3 mm which is overmolded with the thermoset rubber having a thickness of 8 mm. The product and the cross section of the mold are shown in fig. 1. This mold is built according to the specifications of Bex et al. [7]. The most important factor is the thermal separation of the "cool" parts, used to control the temperature of the thermoplastic material and the "hot" parts, used to control the temperature of the thermoset rubber. The location of the cooling channels was optimized in order to achieve a uniform mold surface temperature. The molding sequence is also similar as described in the work of Bex et al. In a first phase, several thermoplastic components are molded, while a metal insert is placed into the mold to fill the thermoset cavity. Then, the metal insert is removed and a thermoplastic component is placed into the mold to be overmolded with the thermoset rubber. Between insertion of the thermoplastic part and the overmolding, a waiting time of 120 s is applied, to allow thermal stabilization of the thermoplastic insert. The process parameters used to mold the thermoplastic and thermoset component are shown in table 1. The production of the given parts was performed on an ENGEL 100t ES330H 80V 80HL-F injection molding machine having a thermoplastic injection

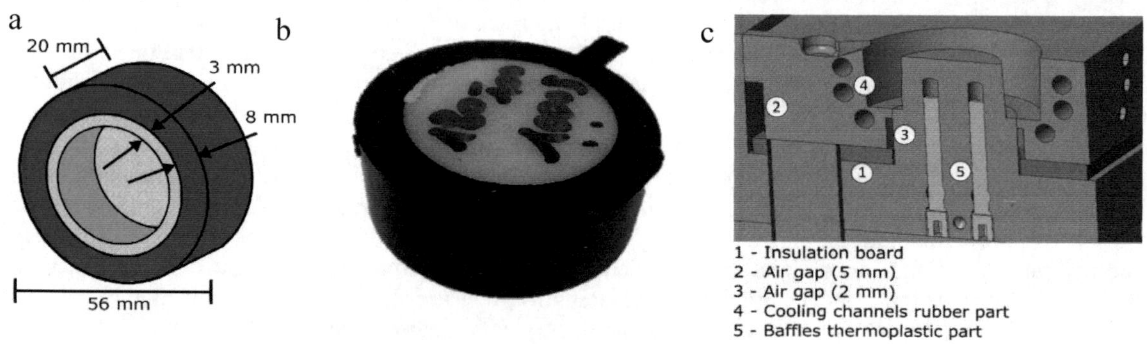

FIGURE 1. Wheel product, drawing and real product (a and b), cross section of the mold showing the thermal separation between the mold surface for the thermoplastic material and thermoset material.

screw of ø 35 mm with an length over diameter ratio (L/D) of 20 and a thermoset injection screw of ø 25 mm with an L/D of 16.

TABLE 1. Process parameters for the production of thermoset and thermoplastic component of the wheel product.

Process parameters	HDPE	EPDM
Injection temperature [°C]	230	80
Mold temperature [°C]	20	180
Injection flow rate [cm³/s]	40	60
Holding pressure [bar]	790	200
Holding time [s]	15	200

To investigate the influence of the curing time, the "hot" mold surface temperature of 180 °C, achieved by water under a pressure of 12 bars, can rapidly by transformed to a "cold" mold surface of 80 °C by a variotherm apparatus. Due to this drastic temperature drop, the curing phase stops. This makes it possible to produce samples with a specific curing status.

Simulation Method for Interface Strength of the Wheel Product

The structural simulation of the adhesion between the thermoplastic part and the thermoset product, is performed by combining the processing simulation software and the structural simulation software. For the processing simulation, the complete mold and product are meshed with linear tetrahedral elements. The cooling channels are meshed with beam elements. To simulate the presented process, the reactive solver is used to model the mold temperature, the filling, packing and curing phase of the thermoset rubber. The thermoplastic component is inserted as a 'part insert' into the simulation software, having the exact polymer properties. During the curing phase, a curing time of 3000 s an output result is requested every 50 seconds. to have a clear understanding of the curing at the interface of the rubber component near the thermoplastic insert. After performing the simulation, the insert temperature and the curing of the thermoset rubber as a function of time are exported. By the use of scripting, the two given result files were combined to calculate the strength of the interface. The cohesive strength is determined by using the local degree of cure, resulting in a local stress strain curve for the rubber material. The adhesive strength is determined by using the local interface temperature. This temperature is utilized to calculate the local degree of melting, which results in a local degree of healing [6]. This local degree of healing is a measure for the local adhesive strength, which was experimentally defined. The result of this scripting is a plot for local interface temperature, and local interface cure followed by the plot which shows the local adhesive and cohesive strength. By comparing these two strength values, the type of failure can be estimated. The adhesive and cohesive strength values also can be mapped on to a structural mesh. The complete procedure is schematically shown in fig. 2. The structural model will be used in a later phase to perform the structural simulation.

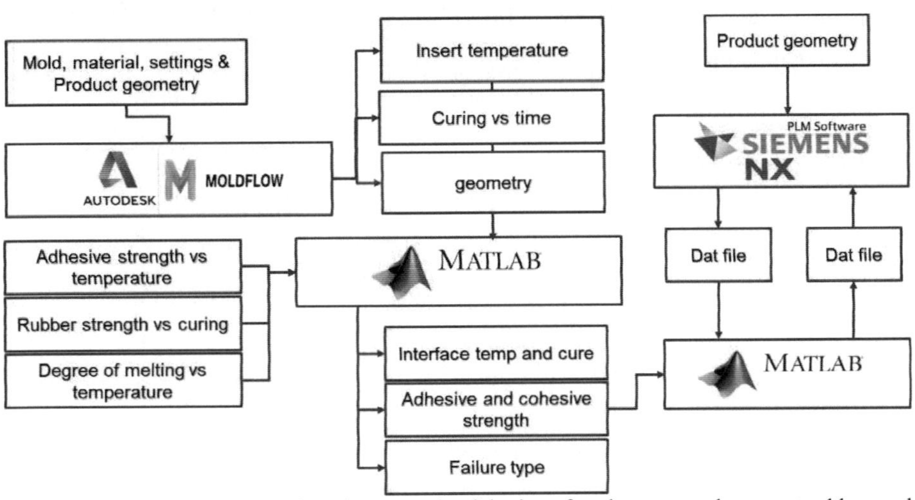

FIGURE 2. Simulation workflow to simulate the strength of the interface between a thermoset rubber and thermoplastic polymer in an overmolding application

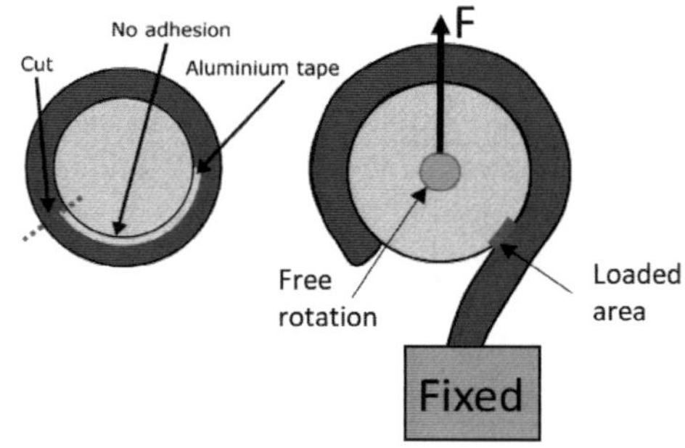

FIGURE 3. Peel test method for wheel product, real test method

RESULTS

The peel tests have been performed on samples with two temperature settings, being 180-100 and 180-90. This means, a rubber curing temperature of 180 °C, combined with a thermoset cavity temperature of 100 and 90 °C respectively. Both temperature settings have been tested at two curing times being 1200 s and 1500 s for 180-90, 1000s and 1350s for 180-100. The script created for this project, shows three results, shown in fig. 4. The first plot shows the expected failure mode, being cohesive or adhesive (fig 4 a). The second plot shows the local interface strength σ_i,defined by equation 3. The third plot (fig. 4 c) shows the expected average interfacial strength and the distribution of the interface strength..

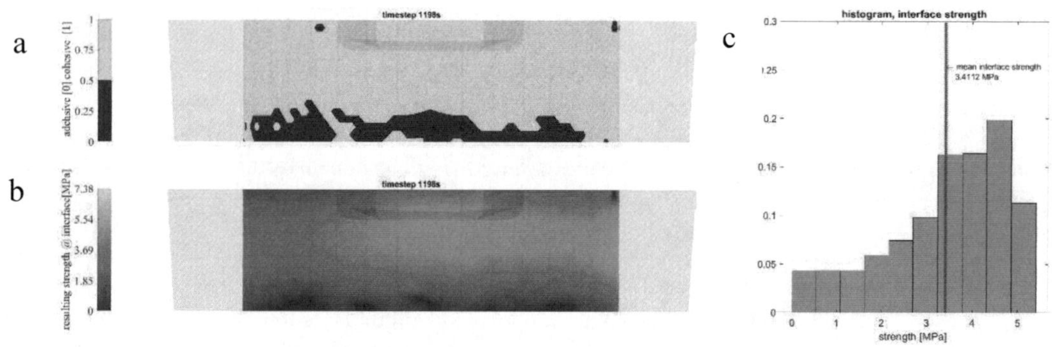

FIGURE 4. Expected failure mode (a), local interface strength (b) and distribution and average interface strength (c) for the wheel product for setting 180-90-1200s

The maximum peel force, by conducting a test according to the setup shown in fig. 3 for setting 180-90-1200s is shown in fig. 5. An average force of 210N was found. The results for the three other settings are shown in table 2. This table also shows the average expected interface strength for the given process settings, being the result as shown in fig. 4 c. A similar trend can be found between the expected interface strength and the measured peel force. If the peel force is divided by the expected interface strength, a loaded area of ~55 mm² can be found for all settings. This area results in a loaded strip where the peel forces act on the product of 20 by 2.75 mm , shown by the red strip in fig. 3. This is a realistic value for the tested product, and thus shows confidence in the simulated strength values

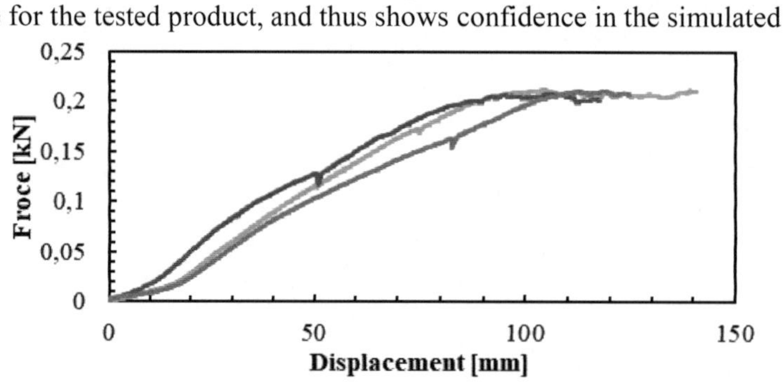

FIGURE 5. Experimental result of peel test for the 180-90-1200s setting.

TABLE 2. Peel results for test 4 wheel samples with different temperature and curing, combined with the expected, average interface strength.

Setting	Peel force [N]	Std. [N]	Expected interface strength [MPa]
180-90-1200s	210	2.2	3.41
180-90-1500s	237	11.2	4.38
180-100-1000s	188	8.8	3.15
180-100-1350s	245	12	4.71

CONCLUSIONS

By using the presented approach, a new set of results can be created estimating the interfacial strength between components for two-component injection molding. These results can be used to optimize the process settings in order to increase the resulting interface strength. Optimization can be done by increasing the rubber curing and hence increasing the cohesive strength. However, this is only useful if the adhesive strength, determined by the interfacial temperature is higher than the expected cohesive strength. The results shown in this project show good agreement with the measured results. The next step includes the structural simulations for the given product geometry. Increasing accuracy of the results can be done by optimizing material data and in-depth material characterization of the stress strain curve of the partially cured rubber.

ACKNOWLEDGMENTS

The authors acknowledge the KU Leuven Impulsfonds for the financial support through project IMP/14/043 and the Company Hercorub NV for providing the uncured rubber material. G.J. Bex acknowledges the Flanders Research Foundation (FWO) for his PhD grant strategic basic research.

REFERENCES

[1] R. M. Gouker, S. K. Gupta, H. A. Bruck, and T. Holzschuh, “Manufacturing of multi-material compliant mechanisms using multi-material molding,” *Int. J. Adv. Manuf. Technol.*, vol. 30, no. 11–12, pp. 1049–1075, 2006.

[2] H. R. Brown, “The Adhesion Between Polymers,” *Annu. Rev. Mater. Sci.*, vol. 21, no. 1, pp. 463–489, 1991.

[3] S. Prager and M. Tirrell, “The healing process at polymer-polymer interfaces,” *J. Chem. Phys.*, vol. 75, no. 10, pp. 5194–5198, 1981.

[4] R. P. Wool, “Polymer Entanglements,” *Macromolecules*, vol. 26, no. 7, pp. 1564–1569, 1993.

[5] F. Yang and R. Pitchumani, “A kinetics model for interphase formation in thermosetting matrix composites,” *J. Appl. Polym. Sci.*, vol. 89, no. 12, pp. 3220–3236, 2003.

[6] W. Six and F. Desplentere, “Two-component injection moulding of thermoset rubber and thermoplastics : Absolute interface strength prediction,” in *Polymer & mould innovations*, 2018.

[7] G. Bex, F. Desplentere, J. De Keyzer, and A. Van Bael, “Two-component injection moulding of thermoset rubber in combination with thermoplastics by thermally separated mould cavities and rapid heat cycling,” *Int. J. Adv. Manuf. Technol.*, 2017.

A comparison between additively and conventionally manufactured injection moulding inserts

Thomas Mitterlehner[a] and Georg Steinbichler[a]

[a]*Johannes Kepler University Linz, Altenberger Straße 69, A4040 Linz, Austria*

Abstract. The cooling time in injection moulding can constitute up to 70% of the cycle time. With the additive manufacturing technology there is a new method for designing and optimizing cooling channels available. With this method it is possible to manufacture conformal cooling channels randomly orientated inside injection moulds. This enables the possibility to design more efficient cooling and more even heat dissipations, which results in a reduction of thermal stresses and higher accuracies of injection moulding parts. Against this background, the central question that motivates this paper is: What are the main differences between additive manufactured conformal cooling channels and conventional manufactured cooling channels? To directly compare both manufacturing strategies a test injection mould was manufactured which was designed in a way that both a conventional and an additive manufactured insert can be operated simultaneously. In this context one section of the injection mould was manufactured by using the selective laser melting technology. The second section of the mould was manufactured conventionally by milling. This opens the possibility to compare both manufacturing technologies within the same injection moulding cycle. After the additive and the conventional inserts of the mould were manufactured several injection moulding experiments were accomplished. For evaluating the heat distribution on the injection moulding part thermographic measurement methods were used. Furthermore the effects of mould temperature, coolant volume flow and cooling time on the part surface temperature were examined.

Keywords: Additive Manufacturing, Injection Moulding, Conformal Cooling

INTRODUCTION

Injection moulding parts play a major role in daily life and in industrial sectors as well. Shorter life cycle spans and increasing numbers of plastic parts lead to great demands in today's plastics industry. The injection moulding process is already a highly productive and automated process but there are still some potentials to increase productivity and quality. One of these potentials is the additive manufacturing mould components. This enables the possibility to manufacture free shaped cooling channels whereby the designing of cooling channel layouts is no longer limited to straight drilled holes or milled contours in combination with hard soldering.

This additional degree of freedom allows the manufacturing of cooling channels which follow the part surface conform to its shape. These additive manufactured channels are commonly called conformal cooling channels and are a persistent theme when cycle time reduction should be achieved. An injection moulding process can be separated in six steps: plasticization, injection, holding pressure, cooling phase, mould movement and ejection. The cooling phase is typical 2/3 of the whole cycle time which implies a high potential for reducing production time and costs. These factors are directly linked because an efficient cooling typically reduces the cooling time which leads to an increase in productivity. Nevertheless the cooling shouldn't be reduced on its rapidity because it could affect the quality in a negative way and can cause for example warpage, residual stresses and inaccurate products. Therefore a careful management of the surface temperatures and heat transfer rates are essential.

If we have a look in the literature there are several examples for appropriate implementations of conformal cooling channels using different additive manufacturing technologies ([1]–[4]). Ismet Ilyas [1] reported that they achieved a 16% reduction in cycle time by using conformal cooling channels. Another example is given by Park and Dang [5] where they showed a geometry with several hotspots. By using additive manufactured conformal cooling channels they achieved a reduction in cycle time of 23%. The aim of this work is to get a closer look on the differences between additive and conventional manufactured cooling channels. Therefore a test injection mould was designed and manufactured which includes both an additively and a conventionally manufactured mould insert. To identify the differences between the two cooling concepts a factorial design of experiments was set up to examine the effects of mould temperature, coolant volume flow and cooling time on the part surface temperature.

CAD DESIGN

For our intention to compare both cooling channel layouts at the same time we had to design and manufacture a test injection mould at first. Therefore a shell-shaped geometry was chosen. The injection moulding part had a total length of 168 mm, a width of 44 and a height of 35 mm (without runner). All drawings were made with PTC Creo Parametric 3.0, using the injection moulding extension. After the injection moulding part was designed the mould and the cooling channels were construed from these drawings. The geometrical drawings of the test geometry are shown in Figure 1.

Figure 1. Shell-shaped test geometry, 2D drawing with dimensions (left), 3D geometry of the injection moulding part (middle), injection moulding part with additive and conventional cooling channel concepts (right)

Based on these designs, the injection mould and the inserts were construed. The mould contains a carrier mould were the inserts are mounted. In order to directly compare additive manufactured conformal cooling designs and conventional drilled cooling designs two pairs of inserts were manufactured. The first one was additive manufactured by selective laser melting and includes conformal cooling channels. The second one was conventional manufactured and includes a conventional cooling layout out of steel 1.2083. Both of them were mounted in the carrier mould and were then processed under the same conditions. The manufacturing of the mould was done by conventional milling and drilling. The inserts with the conformal cooling channels were additive manufacturing by selective laser sinter process on a EOS M280 machine which was equipped with a 200 watt laser. The inserts were made out of steel 1.2709 (MS1) and were post processed by milling and polishing afterwards.

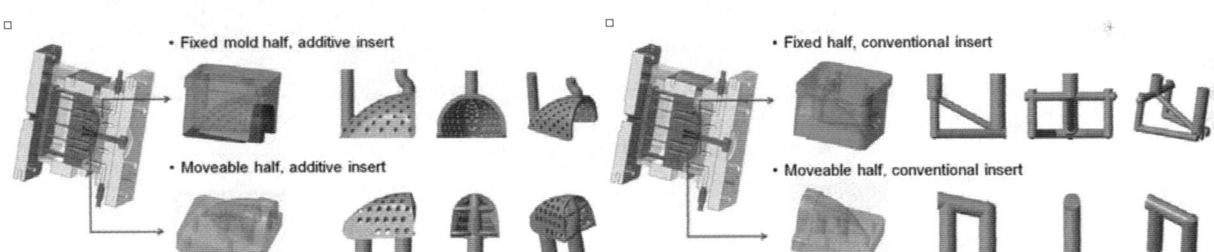

Figure 2. Overview of the injection mould and the inserts; additive manufactured insert and the conformal cooling channel geometry (left), conventional manufactured insert and the drilled cooling channel geometry (right)

In Figure 2 the manufactured inserts and the corresponding cooling channel geometries are shown. The dimensions of the fixed mould half are 80x70x60 mm and those of the moveable half are 80x70x50 mm.

EXPERIMENTS

As mentioned before, the objective was to determine the effects of the mould temperature, coolant temperature and the cooling time on the part surface temperature. Therefore a factorial design of experiments (DOE) was set up. The mould temperature was varied from 20 to 40°C. The volume flow was set to 3, 6 or 12 l/min and the cooling time was set to 2, 4 or 6 seconds. An overview of the resulting factorial design of experiments is shown in Table 1.

Table 1. Factorial design of experiments

Level	Factor		
	Mould temperature [°C]	**Volume flow [l/min]**	**Cooling time [s]**
-1	20	3	2
0	-	6	4
1	40	12	6

This results in 18 different injection moulding parameter settings. The injection moulding was carried out with a fully electric machine with a maximum clamping force of 110 tons. Because of the easy use and the wide range of applications Polypropylene HE125MO manufactured by Borealis was chosen for the experiments. Since thermographic measurements were carried out 1% of black master batch were added. The shot volume was 22 cm^3 and the injection speed 10 cm^3/s. The melt was metered with a screw speed of 0,6 m/s. The temperature of the melt was 200°C. The holding pressure was $500 \cdot 10^5$ Pa and acted 5 seconds. The cooling time was varied according the DOE definitions.

At the beginning of each experimental setup at least 20 injection moulding cycles were proceeded in order to get constant conditions. Afterwards the surface temperature of the following 30 injection moulding parts was measured with a thermographic measurement system. Therefore the thermographic measurement system was placed inside the injection moulding machine upon the fixed clamping plate. For the measurements the thermographic system Variocam Hr was used which has a resolution of 1280 x 960 pixels and an accuracy of ±1%. To provide accurate and cyclic measurements the camera was connected to the injection moulding machine and was trigged by the mould open signal. Thus a thermographic measurement was carried out every time the mould was fully opened. This experimental setup makes it possible to measure the part surface temperature at the exact same time of any injection moulding cycle. In Figure 3 a schematic drawing of the experimental setup and a thermographic picture which results from the measurement is shown.

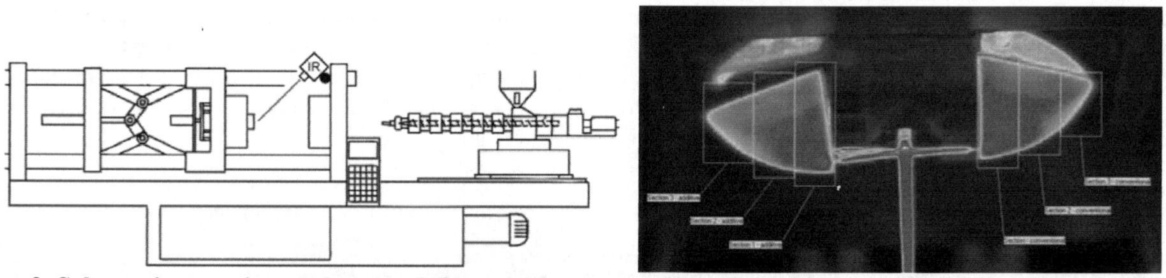

Figure 3. Schematic experimental set up (left), resulting thermographic measurement picture with the defined result sections (right)

After the thermographic measurement was completed the data were analyzed with the software IRBIS3. As shown in Figure 3 three measurement sections were defined on each part surface. For each of these sections and every relevant cycle the maximum part surface temperature was determined.

RESULTS

At least 30 thermographic measurements were accomplished for any experimental setup. For the evaluation of the part surface temperature main effect diagrams were created and divided into manufacturing method and measurement section. In Figure 4 the main effect diagram for the maximum part surface temperature of section 1 is shown.

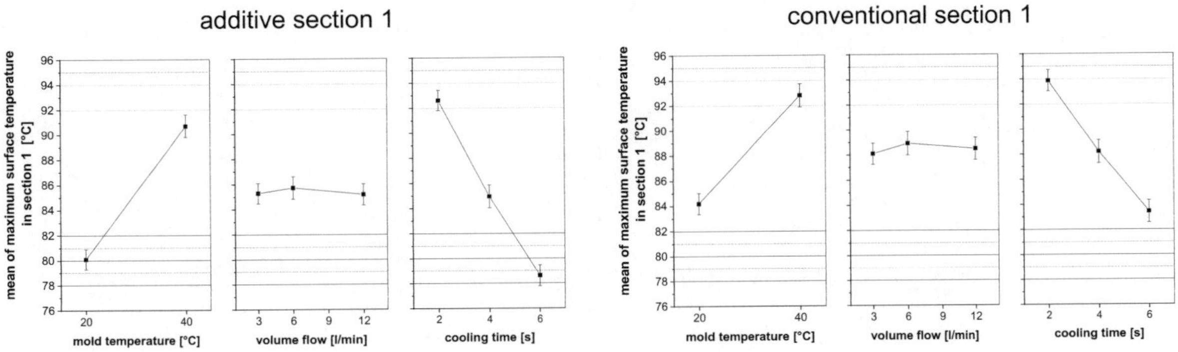

Figure 4. Main effect diagram for the maximum part surface temperature of section 1, additive (left), conventional (right)

The obtained data indicates that in every experimental setup the additive manufactured insert with the conformal cooling concept has a lower part surface temperature compared to the conventional one. A closer look reveals that at a mould temperature of 20°C the difference of the maximum part surface temperature between additive and conventional is 4,1°C and at a mould temperature of 40°C the difference is 2,1°C. The temperature difference at the volume flow is 3,3°C at 3 l/min, 3,2°C at 6 l/min and 12 l/min. The temperature difference at the cooling time is 1,2°C at 2 seconds, 3,3°C at 4 seconds and 4,8°C at 6 seconds. It can be seen from these values that there is a mean difference of the maximum surface temperature in section 1 of 3,2 °C across all experimental setups. Beside the difference in surface temperature there is no statistically, significant effect in section 1 between the additive and conventional cooling concepts because both show similar curve gradients.

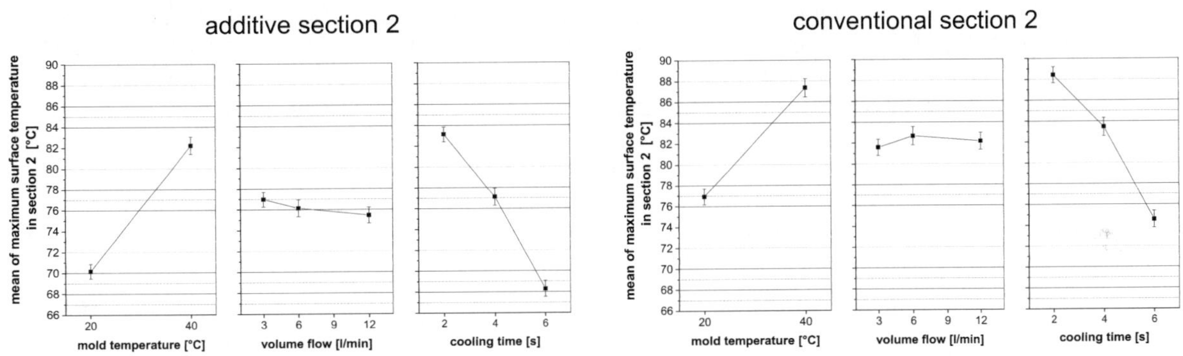

Figure 5. Main effect diagram for the maximum part surface temperature of section 2, additive (left), conventional (right)

Figure 6 shows the main effect diagrams for the maximum part surface temperature of section 2. It can be seen that there is a similar behavior of the part surface temperature of section 2 in respect to section 1. In each experimental setup the part surface temperature of the additively manufactured concept is lower than the conventional one. The biggest difference occur at a mould temperature of 20°C and amounts to 6,5°C. The mean difference of the maximum surface temperature in section 2 across all experimental setups is 6°C. Regarding the effects there is a noticeable behavior in the volume flow of additive section 2. It shows a slightly higher part surface temperature at a volume flow of 3 l/min compared to 12 l/min. That behavior can't be seen in the graphs of conventional section 2.

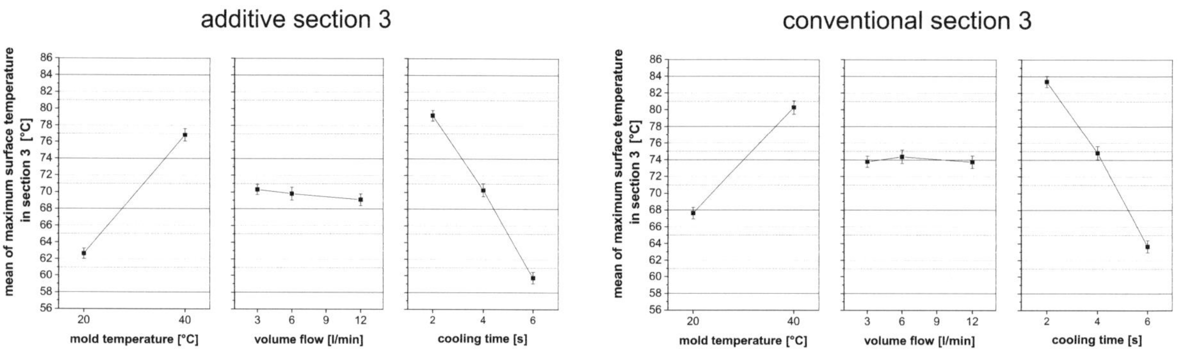

Figure 6. Main effect diagram for the maximum part surface temperature of section 3, additive (left), conventional (right)

The shapes of the graphs in Figure 6 are showing similar patterns as seen in the figures before. Again the maximum part surface temperature of the conformal cooling layout is significantly lower than the conventional layout. In this particular case the mean difference of the maximum surface temperature in section 3 across all experimental setups is 4,2°C.

CONCLUSION

Taking into account the statistical data, we can summarize that in every experimental setup the conformal cooling layout has a significantly lower part surface temperature compared to the conventional cooling layout. The mean difference in section 1 was 3,2°C, in section 2 6°C and in section 3 4,2°C. The effects of the mould temperature, volume flow and cooling time on the maximum part surface temperature can be assumed as equal. Due to the measurements of the effects of the coolant volume flow it was found that in almost all cases an increase of the volume flow has no effect on the maximum surface part temperature neither at the conformal nor the conventional cooling channel layout. Only at the additive section 2 was a slight, significant difference existent.

ACKNOWLEDGMENTS

The authors gratefully acknowledge the financial and technical support for the work presented in this paper by Austrian Research Funding Association (FFG), Horitschoner Werkzeugbau GmbH, Oberhumer Klaus + Partner GmbH, Schöfer GmbH and ZKW Group GmbH.

REFERENCES

[1] K. Dalgarno, C. Taylor, J. Gosden, and I. Ilyas, "Design and manufacture of injection mould tool inserts produced using indirect SLS and machining processes," *Rapid Prototyp. J.*, vol. 16, no. 6, pp. 429–440, Oct. 2010.

[2] C. Gustafson, L. Rännar, and A. Glad, "Efficient cooling with tool inserts manufactured by electron beam melting," *Rapid Prototyp. J.*, vol. 13, no. 3, pp. 128–135, Jun. 2007.

[3] Nagahanumaiah, K. Subburaj, and B. Ravi, "Computer aided rapid tooling process selection and manufacturability evaluation for injection mold development," *Comput. Ind.*, vol. 59, no. 2, pp. 262–276, Mar. 2008.

[4] S. Kitayama, H. Miyakawa, M. Takano, and S. Aiba, "Multi-objective optimization of injection molding process parameters for short cycle time and warpage reduction using conformal cooling channel," *Int. J. Adv. Manuf. Technol.*, pp. 1–10, May 2016.

[5] H.-S. Park and X.-P. Dang, "Development of a Smart Plastic Injection Mold with Conformal Cooling Channels," *Procedia Manuf.*, vol. 10, pp. 48–59, Jan. 2017.

Influences of micro-ridges orientation and position on the replication of micro-structured surfaces by Injection Molding

Francesco Regi[a], Alberto Basso[a], Martin Kain[a], Dario Loaldi [a], Dongya Li[a], Yang Zhang[a] and Guido Tosello[a]

[a]*Department of Mechanical Engineering, Technical University of Denmark, Produktionstorvet, Building 427A, 2800 Kgs. Lyngby, Denmark*

Abstract. In this work, directional optical properties were achieved with a microstructure composed of a close array of ridges, defined by their width and by the angle with respect to the main surface. Under constrained lighting, the reflectance was maximized from a certain viewing angle and direction, and minimized from its horizontally orthogonal position. The purpose was generating quick response codes (i.e. Data matrix) that could be easily scanned by means of commercially available software. The ridges were 1.2 mm long, 0.2 mm wide and with a 10° slope angle, for a depth of about 0.03 mm. Each code was defined by 10x10 modules, each composed of 6 ridges. The code was machined on a tool insert and replicated by means of Injection Moulding. This study investigated the effect of the ridge orientation with respect to flow front, of the distance from the gate and the effect of the surrounding modules, i.e. isolated modules vs modules surrounded by the full geometry, on the replication while varying the processing conditions. To this scope, 4 inserts have been milled. One insert presented the full structures (100 modules), whilst only 9 modules were machined on each of the remaining inserts. The modules were selected to maintain the same topography of the original surface at 3 different distances from the gate. The inserts have then been replicated using Acrylonitrile butadiene styrene (ABS). Inserts and replicates were characterized using a Olympus Lext OLS4100 Laser Confocal microscope. Values of height, slope and roughness of the milled ridges and their replicates have been measured. The results showed that ridges parallel to the flow on the replicates had better replication accuracy. Furthermore, the isolated modules had worse replication if compared with the corresponding surrounded module for all the involved positions.

Keywords: Functional Surface, Micro-Structured Surface, Injection Molding , Thermoplastic.
PACS: 83.50.Uv.

INTRODUCTION

In this study, directional optical properties were achieved with a microstructure composed of a close array of ridges (Figure 1), as described by Li in [1]. These structures are similar to those obtained by Levin in [2] but use durable micro-milled structures. The chosen geometry was designed to maximize the reflectance from a certain point of view and minimize it from its orthogonal position. Using this principle, two adjacent orthogonally oriented arrays of ridges produce a black and white contrast effect. The end purpose was generating quick response codes (i.e. Data matrix) that could be easily scanned by means of commercially available software.

This functionality requires precise and homogenous replication of the micro-structures over the entire code surface: a few faulty modules may in fact impede the proper decoding of the information.

Extensive research exists in literature regarding the best practice for micro-structures replications: many researchers report that the mold temperature is the most important process parameter, as it has the highest impact on the replication quality of micro- and nanostructured structured surfaces. Rytka [3] has for example shown that the mold temperature has the biggest influence on the replication ratio (RR), and that there exists a profound relation between the mold temperature and the no-flow temperature. Nevertheless, high mold temperature may also have negative side effects: Matschuk [4] osbserved sticking of the polymer on the mold surface when the temperature was too high, even if an elevated temperature was needed to replicate nanostructures. Several studies documented also the positive effect of high injection speed, depending of the used polymer [5]. Tofteberg [6] achieved almost perfect replication of a surface grating with amplitude 0.5 μ, and wavelength 3 μm using high mold temperature and high injection velocity for four different materials. It was also observed that increasing the holding pressure had a positive effect on replication. Anyhow, the same author [7] noted that increasing the holding pressure could have

either a positive or a negative effect on the replication of an optical surface. In fact, the definition of the feature was improved increasing the pressure when the microfeatures were not fully filled.
Another interesting aspect of the replication of microstructures on surfaces is the effect of the directionality of the structures in respect to the melt flow direction. Lucyshyn [8] found that the micro-structures orientated in melt flow direction were filled better than those orientated perpendicular to melt flow direction. Similarly, the experiments from Zhang [9] revealed that the quality of micro/nano features is very sensitive to the flow direction. The ridges that were perpendicular to the flow direction had round edges, while the ridges that were aligned along the flow direction had sharp edges and better replicated heights.
Unfortunately, common denominators to these researches was the low number and the low density of the structures, i.e. detached and far from each other.
Furthermore, other authors have highlighted the influence of geometry of the substrate and the distance from the gate. Han [10] has shown that a smaller wall thickness of the macro part is favorable for the filling of μ-v-grooves and Lucchetta [11] obtained improved replication for micro features closer to the gate due to the higher pressure in those positions.
In this paper, optimized parameters were used and two process parameters were varied (mold temperature and packing pressure) to evaluate their effect on the replication of a highly dense of structures surface. The effects of directionality of the ridges, geometry and distance from the gate while varying parameters were investigated.

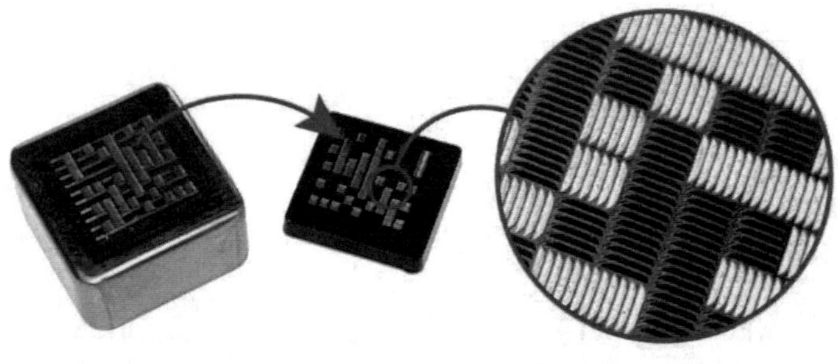

FIGURE 1. Insert and the correspondent plastic part. The structures are highlighted.

EXPERIMENTAL

Materials and Equipment

The substrate part is shown in Figure 2a, it is 15,47 mm on the top, about 2,7 mm in height and with constant thickness of 1,11 mm.

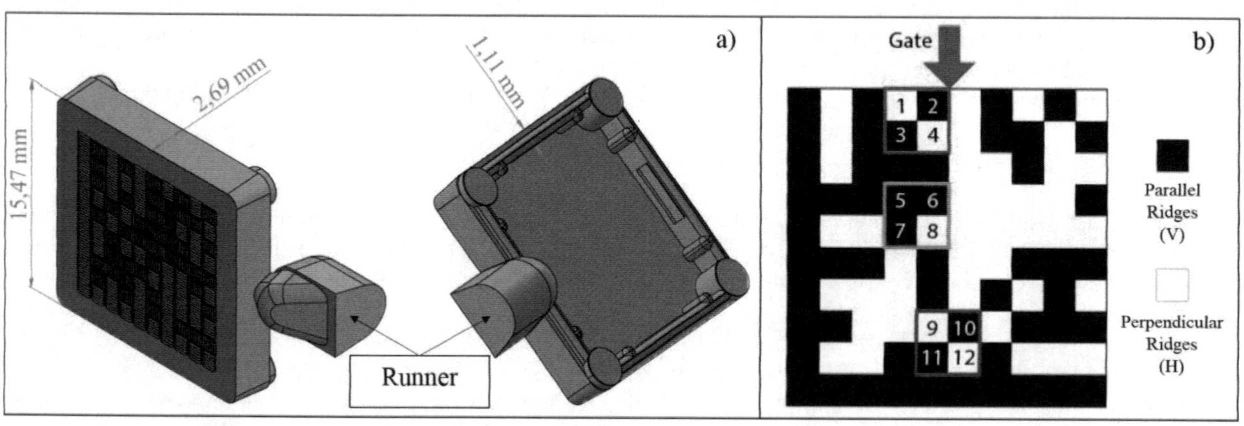

FIGURE 2. a) Schematic of the substrate. **b)** Code orientation and measured modules.

The ridges were 1.2 mm long, 0.2 mm wide and with a 10° slope angle, for a depth of about 0.03 mm (Figure 2). Each full code was defined by 10x10 modules and each module was composed by 6 ridges.
The structures have been milled on the inserts by a precision five-axis milling machine. The milling parameters have been chosen experimentally and optimized accordingly to minimize burrs and defects and enhance the functionality [1]. The resulting code is visible in Figure 1 and Figure 2b, it contains the letters "DTU".
The molding machine used to produce the parts was an Arburg Allrounder 370A, with a 18 mm diameter reciprocating screw. The maximum clamping force was 600 kN.
Screening experiments were conducted beforehand to evaluate the optimized processing parameters. Then, for each insert, a small 2^2 design of experiment (DOE) has been performed varying mold temperature and packing pressure to evaluate their effect on the replication. ABS was injected at 100 mm/s at a melt temperature of 237° C. The mold temperature had two levels, 40° and 60 °, whilst the packing pressure varied from 50 MPa to 100 MPa and was applied for 4 seconds. The part was cooled for 6 seconds before being ejected. Injection velocity is also usually regarded as an influential factor in the replication of high-ratio microstructures, however in this case the aspect ratio was approximately 0.15, sufficiently low to be uninfluential at high injection velocities.
The effects of the orientation and of the distance from the gate were then assessed at the two combinations of high pressure-high temperature (HH) and low pressure-low temperature (LL) to reduce the required measuring time.
The position of the gate and the geometry of the part is shown in Figure 2.

Measurements

The measurements were carried out using the analytical software for microscopy SPIP®, where they were imported and investigated.
Several parameters might be used to characterize the part during the analysis: height, width, slope of the ridges and roughness. Since width and roughness did not present significant differences when varying the processing conditions, the results in this work focused on height and slope of the ridges. A small deviation in these two parameters may in fact affect the required functionality. The selected measurement points are shown in Figure 2b.
The average section profile, perpendicular to the ridges in a user-selected zone, was extracted (Figure 3) and each ridge on the insert has been characterized and compared to the correspondent ridge on the replicates.
The relative deviation in height (ΔH) and in slope (ΔS) was then calculated and the average deviation over module was used for the analysis.
The uncertainty U of the measurements was evaluated following the standard ISO 15530-3[12].

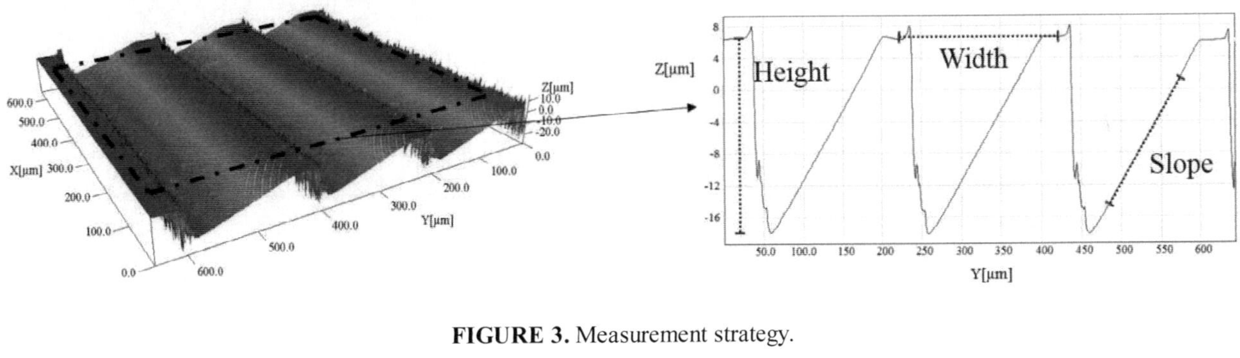

FIGURE 3. Measurement strategy.

ANALYSIS AND RESULTS

The data collected was analyzed using the statistic software Minitab ®. The results are reported in figures 4 and 5.
As mentioned, the relative difference between the replicated parts and the insert was considered. ΔH corresponds to the replication ratio of the structures (ΔH = 100% - RR). However, for the optical functionality, it was also necessary to investigate the slope of the structures and their morphology. Small variations may in fact change the desired reflection. The relative difference ΔS between the insert and the replicates was therefore evaluated.
Figure 4a shows the influence of the process parameters on the replication ratio: as expected, high mold temperature was required to minimize ΔH. The difference was 4.1 % when low packing pressure was applied, and 5.1 % at high packing pressure. On the opposite, ΔH was 10.5 % and 7.5 % respectively at low mold temperature.

Mold temperature has clearly a larger impact on the replication than the packing pressure. In fact, in the filling phase a frozen layer is formed as soon as the melt gets in contact with the surface. At this point applying higher packing pressure did not improve dramatically the fill of the structures. Packing pressure effect was also larger at lower mold temperature, while at 60 °C the difference between high and low mold temperature was within the uncertainty of the measurement.

High mold temperature was also necessary to achieve the desired slope (Figure 4b). In this case, higher packing pressure corresponded also to more accurate morphological replication.

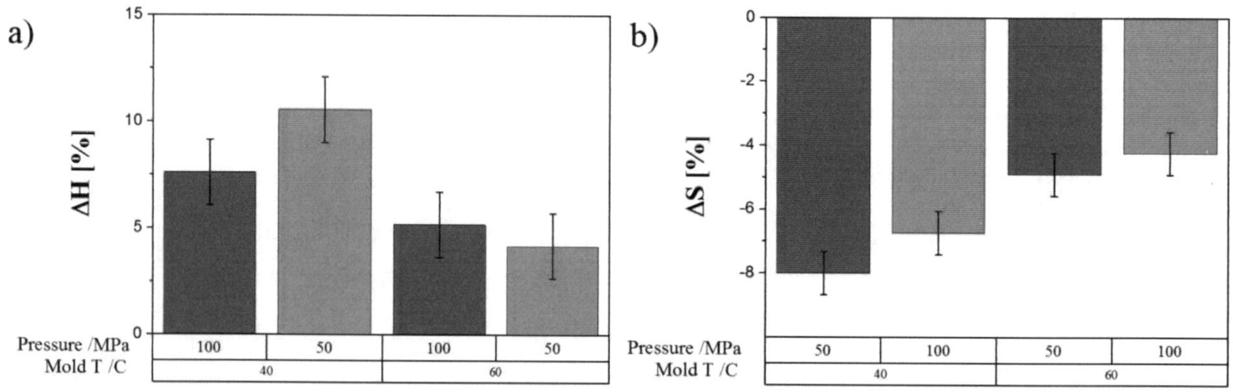

FIGURE 4. a) Interval plot of the parameters effect on RR. b) Interval plot of the parameters effect on the slope of the structures. The error bars show the uncertainty associated with the measurements.

The analysis of ridge orientation and gate distance influence on the replication produced some interesting results. Figure 5 shows the mean value and the associated confidence interval for ΔH and ΔS respectively.

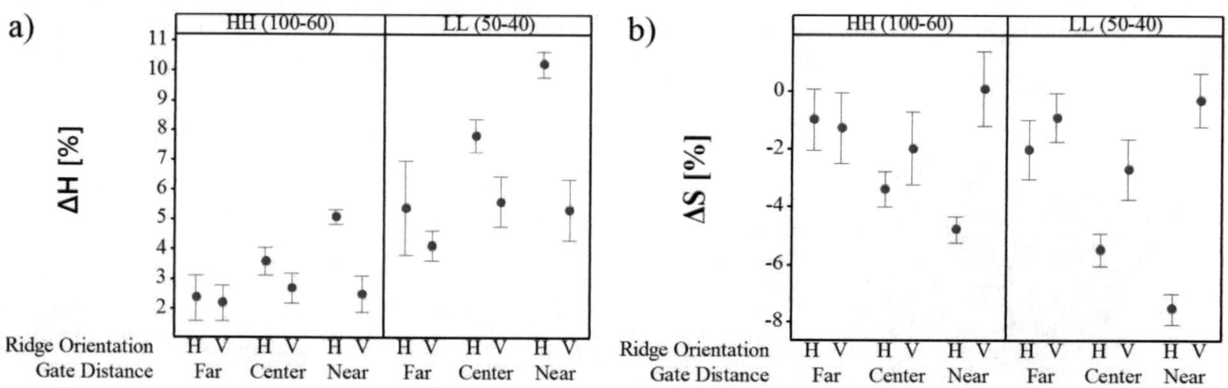

FIGURE 5. a) Effect of ridge orientation and distance from the gate on RR. b) Effect of ridge orientation and distance from the gate on the slope replication. The error bars indicate the 95 % confidence intervals for the mean.

As many authors have reported, structures parallel (V) to the melt flow front were systematically better replicated, both for ΔH and ΔS. The effect was reduced when using high mold temperature and high packing pressure, but it was evident for the worse molding conditions.

Unexpectedly, the position in the cavity, i.e. distance from the gate, was particularly important for the replication. Ridges far from the gate had lower deviations from the mold geometry both in height and in slope. Closer to the gate, the ridge replication was very affected by the directionality of the structures, so that the ridges perpendicular (H) to the flow were substantially worse.

Pressure in the cavity is expected to decrease linearly from the gate, so that microstructures close to the gate have generally better replication accuracy [11].

In this case, the structures are probably too small to require high pressure for proper replication and the geometry of the part is preponderant.

As the melt enters the cavity, it accelerates while being pushed through the gate in the thin-wall sections. When it reaches the first ridges, the melt is too fast to fill them completely and flows over them, possibly preventing proper venting. The phenomenon is consequently accentuated for the perpendicular structures.
Advancing, the melt slows while gaining heat due to the shear heating effect, achieving an improved filling of both parallel and perpendicular ridges.

CONCLUSION

This paper reported the results of an investigation on the influence of geometry and molding conditions on the replication by injection molding of a micro-structured surface with optical functionality. A datamatrix code composed of ridges 1.2 mm long, 0.2 mm wide and with a 10° slope angle, for a depth of about 0.03 mm was replicated on a small thin-wall part. 10x10 modules, each composed of 6 ridges, defined each code.
Molding parameters were varied while investigating the replication in terms of height and angle of the structures.
High mold temperature was necessary to minimize the deviation from the mold geometry and while the packing pressure effect was not preponderant, higher packing pressure was advisable.
Furthermore, the ridges parallel to the flow had better replication accuracy. However, the ridges far from the gate showed less sensitivity regarding to their orientation and were generally more accurately replicated. In fact, position in the cavity of the part deeply affected the replication of the structures that was substantially worse in the region close to the gate.

REFERENCES

1. D. Li et al., “Process chain for fabrication of anisotropic optical functional surfaces on polymer components” in *Euspen's 17 th Int. Conf. Proc.*, 1994, pp 4–5
2. A. Levin et al., “Fabricating BRDFs at high spatial resolution using wave optics” in *ACM Trans. Graph. 32.*, 2013, pp. 144:1-14
3. C. Rytka, “Replication of Functional Polymeric Micro- and Nanostructures.” *PhD thesis*, 2016, pp. 145
4. M. Matschuk et al., “Injection molding of high aspect ratio sub-100 nm nanostructures.” in *Journal of Micromechanics and Microengineering 23.2*, 2013, pp 25
5. X. Guojun et al., “Experimental and numerical studies of injection molding with microfeatures.” in *Polymer Engineering & Science 45.6*, 2005, pp. 866–875
6. T. Tofteberg and E. Andreassen, “Injection moulding of microfeatured parts.” in *Proceedings of the Polymer Processing Society 24*, 2008
7. T. Tofteberg et al., “Effects of Injection Molding Holding Pressure on the Replication of Surface Microfeatures.” in *International Polymer Processing 25*, 2010, pp. 236-41
8. T. Lucyshyn et al., “Filling behavior of micro-ribs on injection molded parts depending on micro-structure orientation and processing parameters” in *AIP Conference Proceedings 1779*, 2016
9. N. Zhang et al. “Replication of micro/nano-scale features by micro injection molding with a bulk metallic glass mold insert” in Journal of Micromechanics and Microengineering 22.6, 2012, pp 65
10. X. Han et al. “Effects of cavity conditions on transcription molding of microscale prism patterns using ultra-high-speed injection molding” in *International Polymer Processing 21*, 2006, pp. 473-9
11. G. Lucchetta et al., “Investigating the technological limits of micro-injection molding in replicating high aspect ratio micro-structured surfaces” in *CIRP Annals – Manufacturing Technology 63*, 2014, pp. 521-4
12. Evaluation of measurement data — Guide to the expression of uncertainty in measurement. JCGM 100:2008

Intelligent Parameter Optimization of Injection Molding Process

S.C Nian [a*], M.S. Huang [b], and Y.C. Fang [c]

[a]*Department of Power Mechanical Engineering, National Taitung College, Taitung, ROC*
[b,c]*Department of Mechanical and Automation Engineering, National Kaohsiung University of Science and Technology, Kaohsiung, ROC*

Abstract. Injection molding is an established technology with many decades of use; however, factors such as the processed raw material, the mold and machine, and the processing parameters result in significant variations in product qualities. Traditionally, researchers have attempted to improve the quality of injection-molded components by controlling screw position, injection and holding pressures, and mold and barrel temperatures. However, the geometrical dimensions of molded components still tend to vary from shot to shot even when such controls are applied. In other words, a further work is required to properly understand the factors affecting melt quality in each cycle such that more effective control strategies can be put in place. Conventionally injection molding is a 'black box' and thus process parameters setting is often time consuming and has limited effect basing on statistically experimental methods, computer aided simulations, or operator's experiences. With the advances of sensing technology, understanding of injection molding process is transformed into a 'grey box' that reveals the physical information about molten resin's flowing behaviors in the cavity. Thereby, this study develops an intelligent optimal parameters tuning approach that analyzes the profiles of injection speed, injection pressure, cavity pressure, screw position of injection molding, and the process parameters setting data provided from the controller of injection molding machines. Further, the major parameters at each stage are determined individually including injection speed/pressure in the mold filling stage, velocity-to-pressure switchover point, holding pressure, and packing time. Through injection molding of an IC tray as an example for experimental verification, this study presents the initial results of developing an intelligent computing approach that determines the above-mentioned process parameter optimization. In addition, the molding curves can be used for process monitoring.

Keywords: intelligent parameter optimization; intelligent mold testing; scientific molding; injection molding
PACS: 81.20.HY

1. INTRODUCTION

The injection molding processes of screw plasticizing, cavity filling, and melt cooling are at a state of invisible "black box". Therefore, the traditional molding factory accustomed to control the molding quality by method of try and error that depends on the engineers' experience to adjust machine parameters. In fact, the effect factors of molding quality are not only the motion control of injection machine, but also the variations of materials, mold, peripheral equipment, and environment. Therefore, even for the parts molded under precision machine control also difficult to avoid the occurrence of molding defect parts.

The scientific molding emphasizes a concept that injection molding must be changed from the "machine-focused" to the "plastic's point of view" [1] [2]. In particular, control of injection molding must be evolved from traditional machine parameters control to melt status control.

The first step in achieving intelligent injection molding is to install the appropriate sensors at key positions to obtain critical molding data during the molding process. The sensors installation can be divided into three categories according to the installation position [3] : (1) Nozzle sensor: at the nozzle position to measure the melt statue inside the nozzle [4]; (2) In-mold sensor: inside the mold to measure the state of the melt in the cavity [5-7]; (3)Tie-bar sensor: strain gages on the tie-bar surface to monitor the tie-bar [8].

This study develops an intelligent mold testing approach that analyzes the profiles of injection speed, injection pressure, cavity pressure, screw position of injection molding, and the process parameters setting data provided from the controller of injection molding machines.

2. EXPERIMENTAL METHOD AND SETUP

This study is based on concept of scientific molding and injects 4.0 to develop an intelligent mold testing technology. One nozzle pressure sensor and three cavity pressure sensors (SSB01KN08x06, Futaba co., Japan) are used to detect the melt pressure variation during the molding process. The detected pressure curves of sensors will be used to achieve the following four purposes: (1) to formulate rules for determining V/P switchover time, (2) to establish gate freezing time rules, (3) to improve component shrinkage, and (4) to obtain robust molding parameters through PVT-based packing pressure adjustment, which can be used as standard profiles of molds running in different machines and quality monitoring in mass production.

In this study, an IC-tray (figure 1) with length 322.6 mm and width 135.9 mm was molded by AH200 injection molding machine (Fu Chun Shin, Taiwan). The employed plastic material was 35% glass fiber reinforced PPE.

3. RESULTS AND DISCUSSION

3.1 Experiment of V/P Switchover

The V/P switching position will seriously affect the quality and stability of the molded parts since it determines the melt initial pressure state of the packing stage. If the switching position is too late, the melt in the cavity will be over-extruded and the cavity pressure will rise sharply, causing the flash of molded parts. On the contrary, if the switch over was too early the molded parts is easily to cause short shot.

This study developed a single-shot identification technique in which a cavity pressure curve is used to determine the screw position of cavity fully filled. Figure 2 (a) shows the pressure curve of the screw stroke acting on a fully filled cavity. The pressure curve shows that when the cavity is completely filled, the behavior of the melt inside cavity changes from flow to compression, which causes a sudden rise in cavity pressure. The point on the sudden elevation of pressure curve indicates the time at which the cavity is completely filled, and the screw position of the cavity completely filled can be determined at the same time point (see Fig. 2(b)). Figure 2(b) also shows the corresponding tie-bar elongation during the molding process, the cavity pressure rise will cause the corresponding tie tie-bar elongation.

Figure 3 shows the pressure curve, corresponding tie-bar elongation, and molded short-shot plastic parts when the V/P position is set at 95% screw stroke (no-packing). The results show that when the cavity is not full filled, the P1 pressure curve does not have a rapidly increasing value, and the elongation of the tie-bar can be greatly reduced. Figure 4 shows the pressure variation when the packing pressure is applied. The pressure profiles during the molding process are smoothly, which is beneficial to the stability of the molding quality. And the corresponding tie-bar force profile is also smooth which is beneficial to reduce the occurrence of fatigue fracture of tie-bar.

3.2 Gate frozen time experiment

The current packing stages of injection molding are controlled by time segmenting method. Therefore, before we set the time parameters of packing stages, it is necessary to find out the gate frozen time. The traditional mold testing determines the gate frozen time by measuring molded parts weight change with various packaging time setting. This traditional method requires repeated operations when the mold temperature, melt temperature, and packing pressure are changed. This study developed a quickly determining method of gate frozen time by observing the molding pressure profiles, which can saves mold testing time and improve experiment accuracy.

Figure 5 (a) shows the pressure variation of near gate sensor P1 under different packing time. The results show that when the packing time is lower than the gate frozen time (less than 4 seconds), the P1 pressure curve will sharply drop with the end of the packing time. However, when the holding time is higher than the gate frozen time (4 seconds), the cavity pressure does not fluctuate with the release of the packing pressure. It can be conformed that, when the near gate pressure without dropping down with the end of the packing time is reached, the packing time is greater than the gate frozen time and the increase of packing time will not affect to the pressure profiles.

Since the gate is frozen, the melt cannot enter the cavity and the melt pressure decrease with the melt cooling shrinks. Therefore, it is possible to use a long packing time shot to get the pressure curve peak of near the gate as the feature of the gate frozen time (Figure 5 (b)).

3.3 Experiment of packing pressure setting

The packing pressure profile is the dominate parameter of parts shrinkage and warpage. Figure 6 (a) shows the pressure curves of molding with 70 bar single-stage packing. When the P3 pressure dropped to zero, the pressure of P1 was still over 60 MPa. According to the PVT theory [9], the curves of P1, P2 and P3 should be drawn closer to decrease the shrinkage difference.

Figure 6(b) shows the pressure profiles of three-stage packaging molding. The pressures of P1 and P2 are first raised at the high pressure of the first stage packing, then the pressure of P1 is released through the second stage packing, and finally in the third packing stage, the injection pressure is maintained at a system pressure of 20 bar. Through the pressure adjustment of the three-stage packing process, the pressure difference between P1 and P3 can be successfully reduced to within 10 MPa.

Table 1 shows the measurement results of 10 molded parts' (sampled after 40 shots) average warpage and surface appearance. In the single-stage packing, the 70 bar system pressure has the lowest warpage, but molded IC-Tray at the flowing end has sink marks. Although the average warpage of the three-stages packing is approximately the same as that 70 bar single-stage packing, but on the parts' surface without sink mark. In addition, the standard deviation and range of the three-stages packing are very small. It means that the three-stages packing with excellent molding stability, and therefore could greatly improve the mass production yield.

3.4 Experiment of robust molding curve

It is well known that the molding pressure peak dominates the geometry quality of the molded part. In this experiment, we disturb the P1 pressure peak of the mold testing curve (Fig. 6(b)), which was obtained in the previous section, to establish an operating window depending on the molding curve, and from the window we are able to find out the robust molding curve. In this section, the warpage requirement was increased to within 0.4 mm. By adjusting the injection speed and first stage packing, the pressure peak of P1 can be disturbed in the horizontal and vertical directions.

Figure 7(a) shows that the warpage quality window depends on various peak positions of the P1 profile. The pressure peak of the robust curve must be at the center of the moldable area, thus having a maximum range of motion under the quality window to allow for greater molding variations. In this study, the peak position of the robust curve was at pressure 70 MPa and injection time 2.8 seconds (machine operating parameters: injection speed of 70 mm / s, pressure of 80 bar in the first packaging stage).

Figure 7(b) shows the corresponding machine operation window with various injection speeds and packaging pressures. Robust operation parameters are located at the center of the moldable area, approximately at injection speed 140 mm / s and packing pressure 80 bar, which is different from the molding curve quality window.

4. CONCLUSIONS

This research successfully developed an intelligent parameter optimization technique based on the scientific molding. We established important decision methods for V/P switchover timing, gate frozen time, and robust molding curve. We believe that the proposed method is helpful for increasing the yield rate in injection molding.

Table 1 molding quality with different packing processes

	Single-staged packing			**3-staged**
	70 bar	**80 bar**	**90 bar**	**packing**
Max. warpage	0.50	0.61	0.67	0.42
Averaged warpage	0.42	0.53	0.58	0.40
Standard deviation	0.03	0.07	0.06	0.01
Range	0.12	0.23	0.22	0.03
Surface appearance	sink-mark	good	good	good

unit: mm

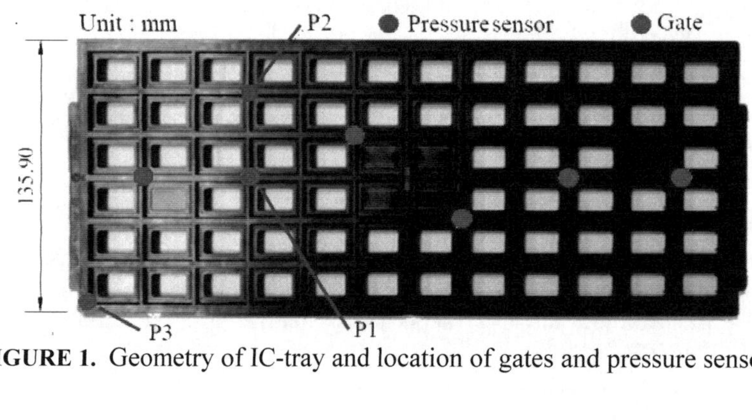

FIGURE 1. Geometry of IC-tray and location of gates and pressure sensors

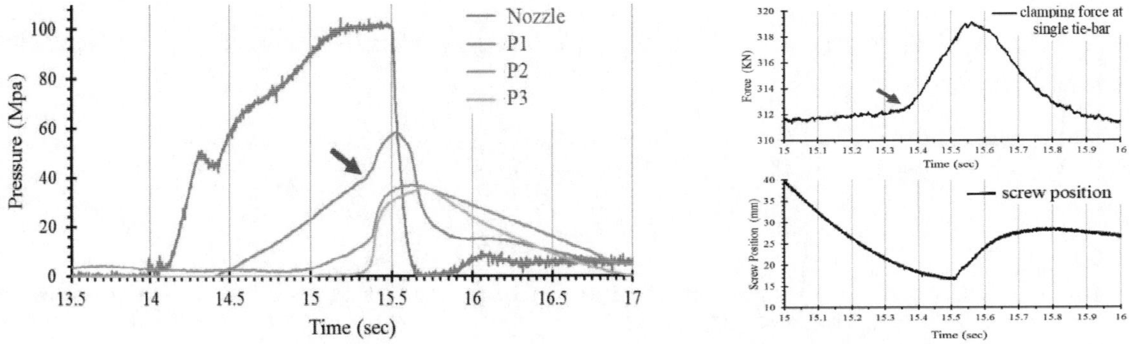

(a) Pressure profiles (b) Tie-bar tension force and screw position

FIGURE 2. Pressure profiles with over-stroke (free of packing)

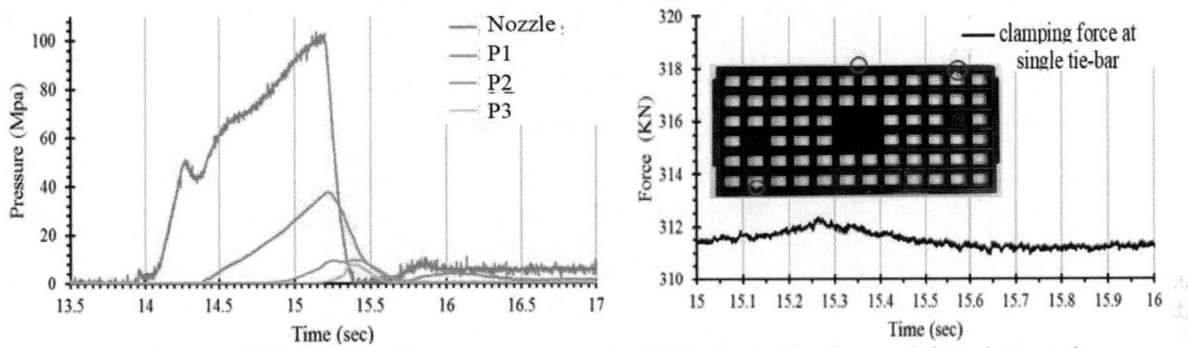

(a) Pressure profiles (b) Tie-bar tension force and short shot sample

FIGURE 3. Profiles of pressure and screw position with adequate V/P position (free of packing)

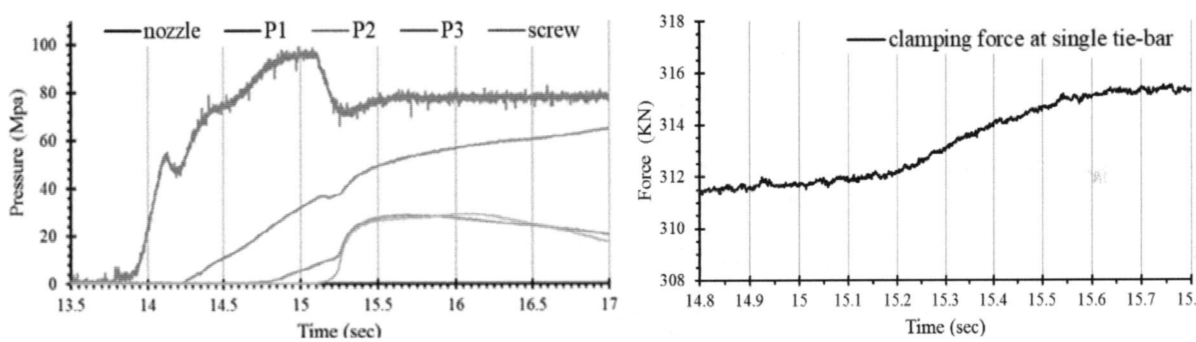

(a) Pressure profiles (b) Corresponding Tie-bar tension force

FIGURE 4. Profiles of pressures and screw position with adequate V/P switchover point (with packing)

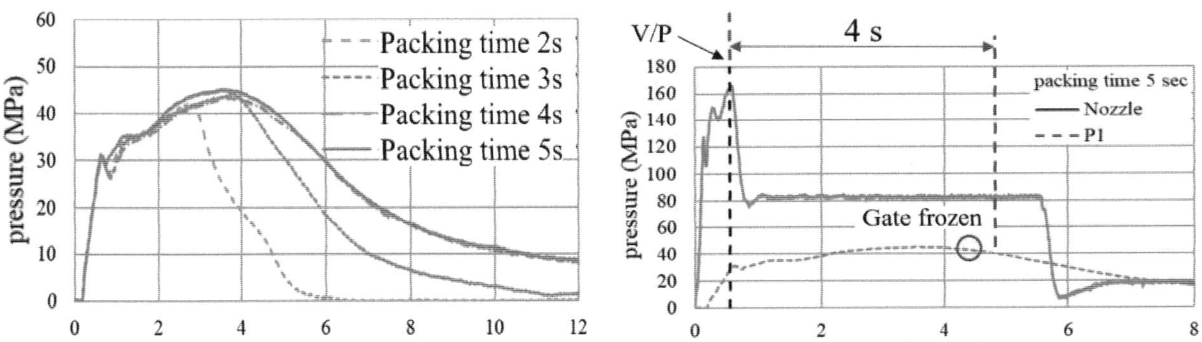

(a) Pressure profiles w.r.t. various packing time (b) Pressure profiles and gate frozen time

FIGURE 5. Pressure profiles of gate frozen time experiment

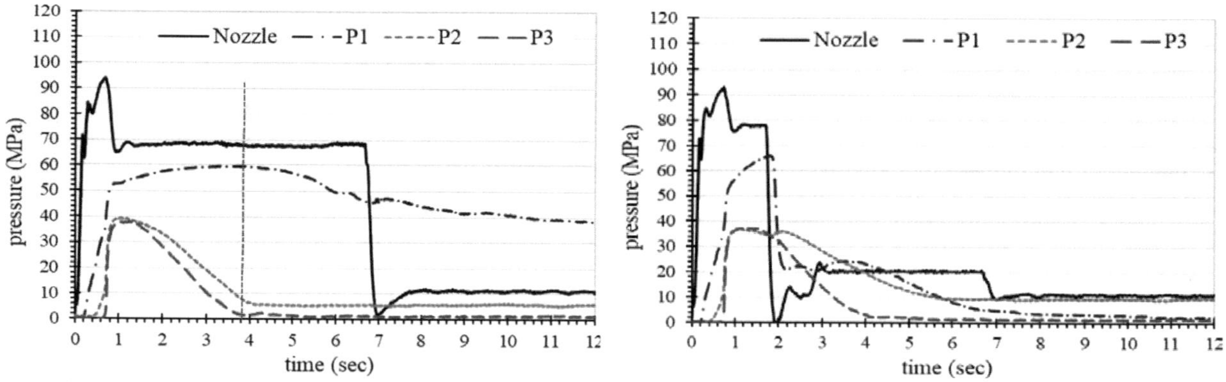

(a) Pressure profiles with one stage packing pressure (b) Pressure profiles with one stage packing pressure

FIGURE 6. Pressure profiles with various packing pressure

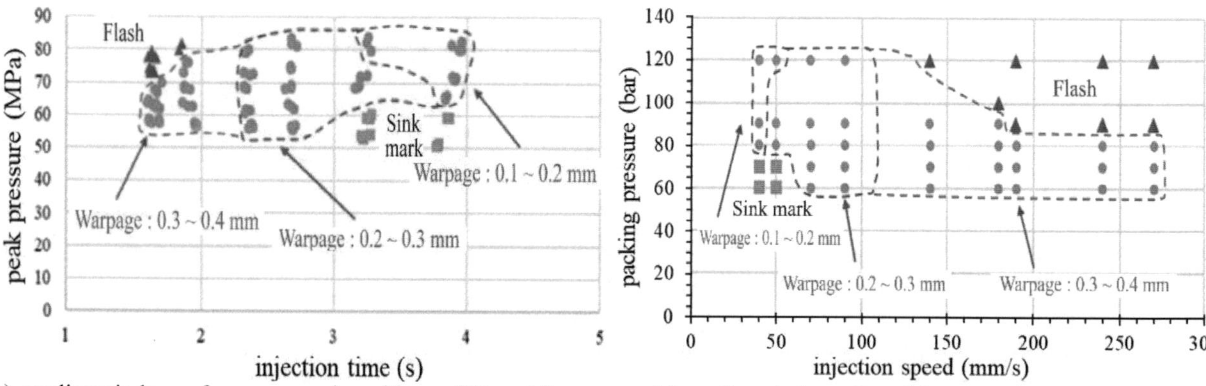

(a) quality window of pressure peak positions of P1 molding curve (b) quality window of machine operation parameters

FIGURE7. The warpage quality window of molded parts with various injection speed and packing pressure

ACKNOWLEDGMENTS

This work was financially supported by the Frontier Mold and Die Research and Development Center from The Featured Areas Research Center Program within the framework of the Higher Education Sprout Project by the Ministry of Education (MOE), Taiwan, and the research project supported from the Ministry of Science and Technology (Project number: 106-2218-E-327 -002 -MY2), Taiwan.

REFERENCES

1. S. Kulkarni, *Robust proces*s development and scientific molding: theory and practice, Carl Hanser Verlag: Munchen, Germany, 2014
2. www.scientificmolding.com
3. Z. Chen, L.S. Turng, *Adv Polym Tech* **24**, 165–182 (2005)
4. C. Gornik, *Mater Sci Forum* **591–593**, 174–178 (2008)
5. Gordon, et al., *Int J Adv Manuf Technol* **78**, 1381–1391 (2015)
6. N. Zhang, M.D., *Polym Test* **31**, 748–758 (2012)
7. V. Speranza, U. Vietri, and R. Pantani, *Macromolecular Research* **19**, 542-554 (2011)
8. M.S. Huang, et al., Precision Engineering 51, 647-658 (2018)
9. J. Wang, Q. Mao, *Adv Polym Tech* **32(S1)**, E474-E485 (2013)

TECHNICAL PROGRESS IN FILLER COMPOUNDING USING A TWIN SCREW EXTRUDER

Natinee Lopattananon[1)], Tadamoto Sakai*[2)]

[1] *Faculty of Science and Technology, Prince of Songkla University, Pattani, 94000 Thailand,*
[2] *Tokyo Office, Shizuoka University, 3-3-6 Shibaura, Minato-ku, Tokyo, 108-0023 Japan,*

Abstract: A twin-screw extruder to conduct compounding or blending of nano-fillers and polymer alloys is used for the innovative creation of functionalization and reinforcement of various polymers. In this study recent progresses in compounding technology were reviewed concerning newly developed compounding techniques such as the separate feeding for optimizing mixing of each component, the water assist mixing for superior dispersion and the dewatering/devolatilization for process rationalization by use of a twin-screw extruder. As examples, two kinds of the hybridized compounding technology combined with polymer blending and inorganic filler addition were introduced to obtain higher physical and chemical performance of polymer composites mainly based on natural rubber. Furthermore, innovative compounding systems using sophisticated twin screw extrusion technology were proposed for polymer composites based on natural rubber to apply commercial production.

Keywords: Twin screw extruder, Compounding, Natural rubber, Polymer blending, Composite, Reinforcement,
PACS: 06.60.Vz, 81.07.-b, 81.20.-n, 81.70.-q,

1. Introduction

A twin-screw extruder to conduct compounding or blending such as inorganic fillers/fibers and polymer alloys is used for the creation of the functionalization and the reinforcement of various polymers as shown below.

1. Compounding of fillers, fibers, and additives.
2. Morphology control of polymer blends.
3. Chemical reactions of polymers, monomers and additives.
4. Devolatilization for solvent and by-products
5. Dewatering of polymers after polymerization.

The main advantages of a twin screw extruder are to be able to obtain higher additional-value-products and higher productivity, when other mixing devices are compared. Table 1 shows the comparison between a continuous twin screw extruder and a batch type mixer such as Banbury mixer. It is clear that number of operational parameters of a twin screw extruder are much superior to those of a batch type mixer. This means the flexibility and the multiplicity in the creation of new polymer properties can be obtainable when a twin screw extruder is efficiently used.

Table 1 Operational parameters for each mixer

Operational parameters for batch compounding	Operational parameters for TSE compounding
1) Material mixing time	1) Screw speed
2) Material feed rate	2) Screw geometry
3) Barrel temperature	3) Screw length
4) Material feed order	4) Driving power
5) Screw geometry	5) Barrel geometry
	6) Barrel length
	7) Barrel temperature
	8) Barrel cooling
	9) Material feed rate
	10) Material feed location
	11) Material shape
	12) Material feed temp.
	13) Devolatilization
	14) Die pressure
	15) Process aid(water, SCF)

In this study two different compounding techniques using a twin screw extruder were introduced from our research results. Natural rubber (NR), one of the most important bio-based sustainable polymers, is applied as research targets. The first research was carried out concerning flame-retardant NR compounds[1] and the other was NR composites with self-reinforcement[2]. The fundamental research works to establish practical compounding processes were carried out at each process step by step separately by using batch type mixing procedures. Based on these research results, two kinds of new compounding scheme were proposed how to apply to practical production using a twin screw extruder.

▪2. Compounding study (1)

Flame-retardant thermoplastic vulcanizates (TPVs) of NR/polypropylene (PP) blends filled with alumina trihydrate (ATH) were prepared with an internal mixer. To improve the properties of flame-retardant NR/PP TPV, a stepwise ATH mixing with NR and PP separately, that is, "stepwise masterbatch mixing (SMB)" method, was developed. A direct ATH mixing together with NR and PP, "one-step mixing (CV)" method, was also evaluated for comparison.

2.1 Preparation of flame-retardant NR/PP/ATH TPVs

ATH filled NR/PP TPVs were prepared by melt mixing with Brabender plastograph. The NR/PP composition was selected at 60/40 %wt. The TPVs with various ATH loading were prepared with CV mixing and SMB mixing processes, respectively.

2.2 Results and Discussions

(1) **Morphology difference between CV and SMB mixing procedures.**

Figure 1 shows TEM images of TPV-ATH60 fabricated with two different mixing methods. It is clearly observed that two samples showed significant difference in the homogeneity of spatial distribution of ATH in the blends. In the CV-TPV- ATH sample, the ATH preferentially located in the dispersed NR phase, and there was almost no ATH imbedding in the PP phase. On the other hand, in the SMB-TPV-ATH sample the ATH was homogeneously distributed in each polymer matrix (both of NR and PP phases).

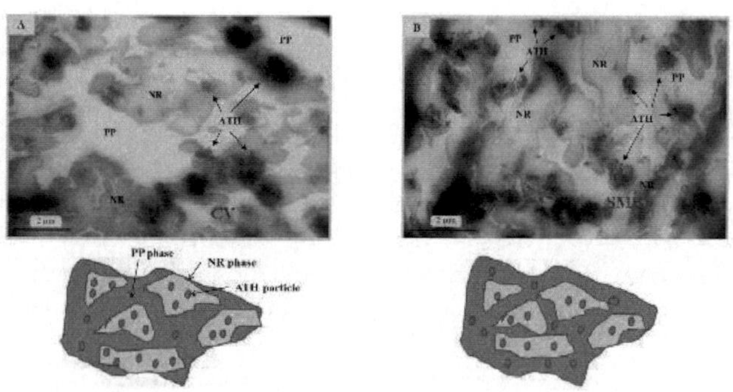

Figure 1. TEM photomicrographs of (A) CV-TPV-ATH60 and (B) SMB-TPV-ATH60

(2) Fire retardant properties of NR/PP/ATH TPVs

Figure 2 shows the variation of LOI as a function of ATH loading and each mixing method. LOI values were increased depending on the mixing procedure and ATH loading. Fig. 2 clearly shows that the flame-retardancy of the SMB-TPV-ATH was higher, when the CV-TPV-ATH was compared. Cone-calorimetry tests supported the same burning behavior as LOI analysis, too.

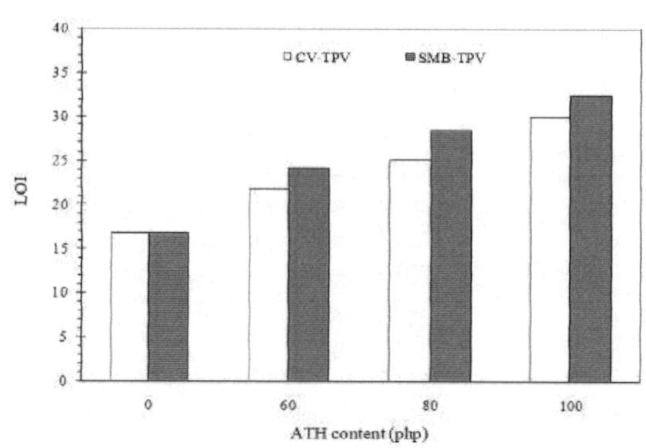

Fig. 2. Comparison of LOI for CV-TPV-ATH and SMB-TPV-ATH at different ATH loadings.

(2) **Pyrolysis-GC-MS analysis on thermal decomposition of NR/PP/ATH TPVs**

The thermal decomposition behavior by Pyrolysis GC-MS was applied to investigate the effect of each mixing method on the flammability of the NR/PP/ATH compounds analyzed from LOI and cone-calorimetry. The pyrolyzed products for both samples are given in Table 2. The thermally decomposed products from both NR/PP/ATH blends were similar flammable fragments but the thermal breakdown of PP phase in the SMB-TPV-ATH60 was clearly less than that of the CV-TPV-ATH60 due to higher concentration of ATH in the PP.

Table 2. Pyrolyzed products from CV-TPV-ATH60 and SMB-TPV-ATH60

Polymer	Pyrolysis Component (%)	Sample	
		CV-TPV-ATH60	SMB-TPV-ATH60
PP Phase	Propene	17.2	11.9
	Methyl propene	15.3	12.0
	Dimethyl heptene	3.7	7.1
NR Phase	Pentadiene	47.6	53.0
	Methyl cyclopentane	4.5	3.9
	Toluene	5.6	5.4
	Xylene	6.3	6.7

From these analyses, it is concluded that the fire-retardancy behavior of NR/PP blends were governed not only by the concentration of ATH, but also by their morphology (ATH dispersion).

2.3 Conclusions

The results showed that the homogeneity of the distribution of ATH in the NR/PP TPVs by SMB mixing was led to be practically beneficial to the improvement in mechanical strength, flame retardancy and thermal stability. Based on these fundamental research results, a practical compounding process using a tandem type twin screw extruder was proposed as shown in Fig. 3.

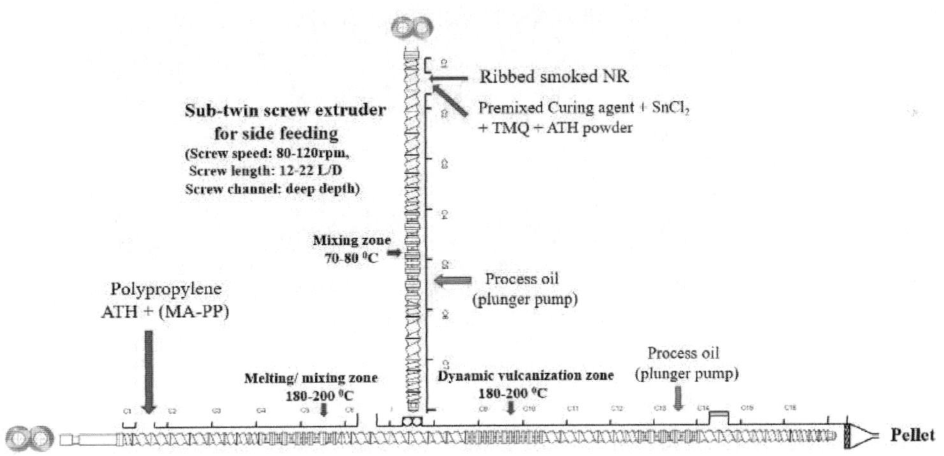

Figure 3. A flame retardant NR compounding process using a twin screw extruder.

▪3. Compounding study(2)

Blending of NR together with thermoplastics such as PP is always an attractive method for developing new polymer materials, showing synergistic properties of NR and plastic materials. Interestingly, an addition of nano-clay into the immiscible mixtures of NR and various plastics has been found to be a key benefits of functional property improvement [3]. In this study, NR/PP blend nanocomposites with a fixed content of NR/PP of 80/20%wt were prepared. The nano-clay was used as a filler. The blend morphology and mechanical properties of NR/PP/Clay nanocomposites were characterized. The results were also compared with those of the NR/PP blends without nano-clay filling.

3.1 Preparation of NR/PP/Clay nanocomposites

Suspension of nano-clay (2wt%) was added into NR latex. The NR and clay were mixed thoroughly under vigorous stirring at ambient temperature to obtain NR/clay masterbatch. The content of clay was 5 phr. Later, the NR/nano-clay masterbatch was mixed with PP in a batch mixing device at 170°C. The NR/PP blend nanocomposite was mixed with phenolic resin (PhOH) and stannous chloride at 120°C as curative agents. Finally, the nanocomposite compounds were statically vulcanized with a hot-press machine at 180°C and 14.0 MPa to obtain the NR/PP/Clay nanocomposite (NR/PP/NC) sheet. The NR/PP binary blend without clay filling was also prepared by using the same procedure for comparison. The content of NR/PP blends was fixed with 80/20 wt%.

3.2 Results and Discussions

(1) NR/PP/clay blend morphology

Figure 4 shows TEM images of the NR/PP and the NR/PP/NC, respectively. The NR/PP sample showed a typical dispersed PP phase within NR matrix. However, in the NR/PP/NC filled with nano-clay, the PP domain was smaller when compared with those of NR/PP and the large proportions of the PP phase tended to turn into a fiber-like structure. This morphological change suggested that the nano-clay addition and mechanical shearing brought about the formation of fibrillar structure, as well as the refinement of blend morphology. The mechanism of the fibrillar structure formation was discussed using a proposed model schematically as shown in Fig. 5.

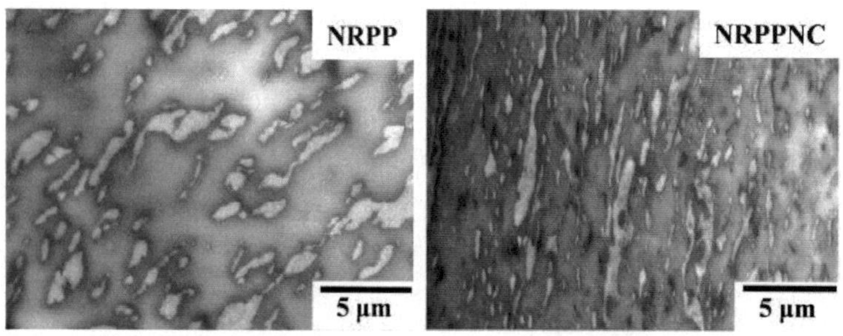

Fig. 4 TEM images of NR/PP and NR/PP/NC

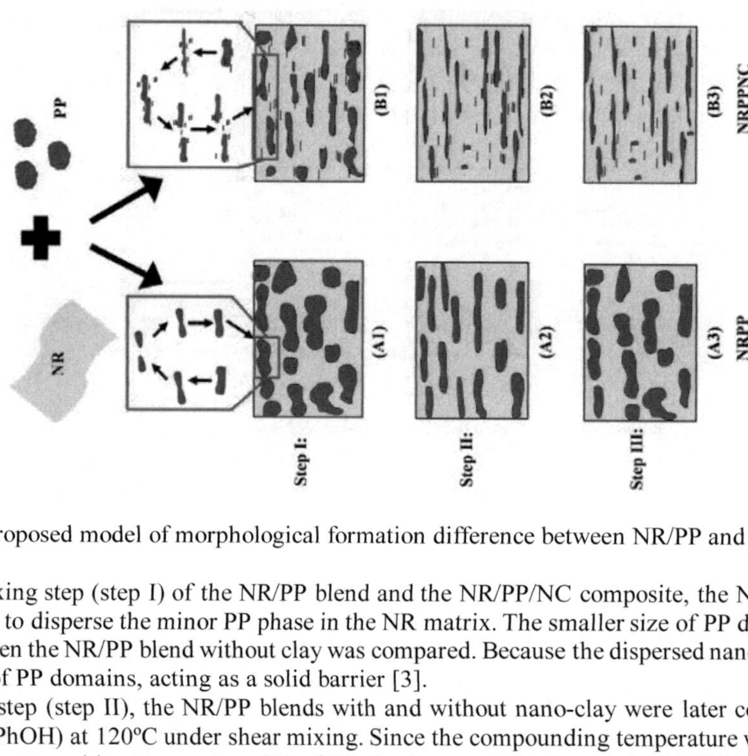

Fig. 5 Proposed model of morphological formation difference between NR/PP and NR/PP/NC.

In the first mixing step (step I) of the NR/PP blend and the NR/PP/NC composite, the NR was melt-blended with PP at 170°C to disperse the minor PP phase in the NR matrix. The smaller size of PP domain in NR/PP/NC was obtained, when the NR/PP blend without clay was compared. Because the dispersed nano-clay would prevent the coalescence of PP domains, acting as a solid barrier [3].

In the second step (step II), the NR/PP blends with and without nano-clay were later compounded with the curative agents (PhOH) at 120°C under shear mixing. Since the compounding temperature was between melting point (T_m) and glass transition temperature (T_g) of PP, the dispersed PP phase in both NR/PP and NR/PP/NC was deformed into the fiber-like structure by shearing force.

Finally, in the third step (step III), NR in the blend was vulcanized by PhOH at high temperature of 180°C in the mold. Since the vulcanization temperature was above the melting of PP, the PP in the NR/PP blends without clay was melted, and the fiber-like domain was shrunk-back to the droplets again. On the other hand, in case of NR/PP/NC with nano-clay, the shrunk-back phenomenon of the fiber-like domain was strongly suppressed due to the local existence of clay at each interface layer.

(2) Tensile properties

Figure 6 shows stress-strain curves of the NR/PP and NR/PP/NC. It is obvious that the tensile stress level was greatly increased due to the morphological improvement of PP fibril structure.

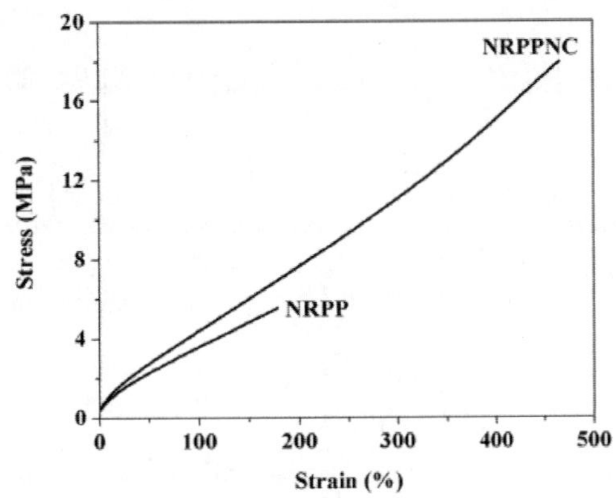

Fig.6 Comparison of mechanical properties of normal NR/PP and newly developed NR/PP/Clay

3.3 Conclusions

NR/PP blends (80/20) were fabricated by a melt-mixing method. The addition of nano-clay into NR/PP blends and mechanical shear mixing led to the generation of dispersed PP fibers within the NR matrix, which highly improved the tensile properties of the NR/PP/NC nanocomposites. Based on these fundamental research results, a practical compounding process using a tandem type twin screw extruder were proposed as shown in Fig. 7.

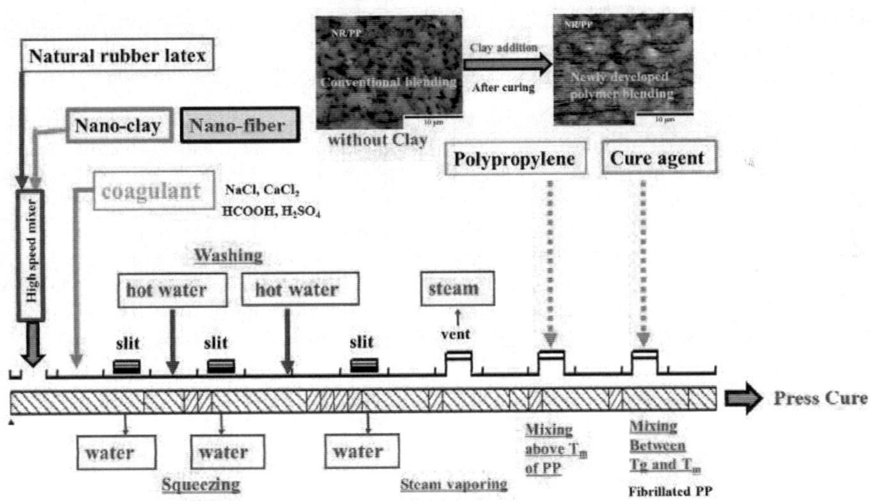

Figure 7. A compounding process using a twin screw extruder for producing NR/PP/Clay nano-composites with high mechanical properties by self-reinforcement

Acknowledgment

Financial supports from Thailand Research Fund of Royal Golden Jubilee Ph.D. Program and Prince of Songkla University are greatly acknowledged.

References

1. N. Lopattananon, A.Walong, T. Sakai, *J .Appl. Polym. Sci.* DOI.10.1002/APP46231 (2018)
2. A.Masa, S.Iimori, R.Saito, H.Saito, T.Sakai, A.Kaesaman, N.Lopattananon, *J. App. Polym. Sci.* **134** (10), p.44574 (2017)
3. N.Lopattananon, S.Tanglakwaraskul, A. Kaesaman, M. Seadan, T. Sakai, *Int. Polym. Proc.*, **29**, p332 (2014)

Influence of the Degree of Crystallinity and the Surface Free Energy on the Adhesion Properties of Different PLA/PBS Blends in Multicomponent Injection Molding

Kübra Güzel[1, a)], Marco Klute[2, b)], Naci Kurgan[3, c)] and Hans-Peter Heim [4, d)]

[1]*Master student, Ondokuz Mayıs University (Turkey) and exchange student, Kassel University (Germany)*
[2]*Scientific assistant, Kassel University (Germany)*
[3]*Scientific researcher, Ondokuz Mayıs University (Turkey)*
[4]*Director of the polymer engineering department, Kassel University (Germany)*

[a)]kubra.guzel@yahoo.com.tr
[b)]marco.klute@uni-kassel.de
[c)]naci.kurgan@omu.edu.tr
[e)]heim@uni-kassel.de

Abstract. The climatic consequential losses of the utilization of raw resources such as oil, gas, coal influence the producers to use more renewable, environmental resources. Poly (lactic acid) (PLA) and poly (butylene succinate) (PBS) are such biodegradable polymers that are getting ever-increasing attention due to their noteworthy mechanical and thermal properties.

A series of PLA/PBS blends was compounded with the weight ratios of 50/50, 60/40, 70/30, 80/20, 90/10 and fabricated to tensile test specimen. As the second component in the two-component injection molding process, thermoplastic polyurethane (TPU) was injected onto the PLA/PBS blends. The change in the adhesive bonding strength between the two components was measured regarding the increase of PBS content in the PLA/PBS blends under the influence of time and temperature (storage at room temperature and 100 °C in the course of 1, 3, 7 and 14 day(s)).

The adhesion properties of the PLA/PBS blends were examined with a contact angle measurement device and FTIR/ATR measurements. A decrement in the surface free energy was observed by the reason of the increase in the PBS content. This decrement led to a decrease in the interfacial tension between the blends and the TPU soft component which correlates to better bonding abilities. Moreover, differential scanning calorimetry (DSC) thermograms of the blends displayed the effect of the increase in PBS content and the storage (at 100 °C in the course of 0, 3 and 14 days) on the degree of crystallization. The resistance to de-linking between the blends and TPU was measured by a universal testing machine. According to the results, the maximal load (552,69 N) was reached on the PLA/PBS blend with a weight ratio of 60/40 w% after the storage of 7 days at 100 °C. This result was associated with the low interfacial tension (0,54 mN/m) between the two surfaces and compared to the results of crystallinity (DSC) and ATR measurements.

INTRODUCTION

In recent years, there is a growing interest to increase the sustainability of products by using bio-based polymers. According to the study of the Nova-Institute GmbH, the bio-based polymers have a share of 1,5 % in the field of construction polymers. This share represented a production of 3.5 million annual tons of the bio-based polymers in 2011 and it will be tripled to nearly 12 million tons by 2020 [1]. PLA is an aliphatic polyester that is obtained from renewable plant sources such as potatoes, sugar from beet and starch from corn. Because of the complex processing standards, the cost and the brittleness of the PLA, it is reasonable to blend PLA with other bio-based polymers.

According to studies on PBS, films as well as molded objects show a significant biodegradation after a couple of months in soil, water with activated sludge or in the ocean [2]. Bhatia et al. investigated the possibility to replace commodity polymers with PLA/PBS blends for the use in packaging applications. It could be shown, that the tensile strength and the elastic modulus both decrease with increasing amount of PBS [3]. Ojijo et al. concluded that samples, that were tempered at 80 °C show a reduction in the yield strength for PLA-dominated blends since the crystallization of the PLA component during the annealing process causes a phase separation [4]. When it comes to the thermal characteristic of the PLA/PBS blends, the addition of PBS enhances the crystallization of the PLA and the DSC diagrams display two melting peaks over the entire composition range indicating that both polymers are semi-crystalline. [5].

EXPERIMENTAL

The PLA used in this study is a commercial grade (PLA 2003D), that was purchased from Natureworks LLC. (Nebraska, USA), with an organic content of 100 %. On the other hand, PBS (bio PBS FZ71 PM), was obtained from Mitsubishi Chemical Co. (Tokyo, Japan) with an organic content of 50 %. As a soft component for the hard- soft- combinations Elastollan N 95A 12 000 (TPU95) was used. It has an organic content of 43 % and was produced by BASF Polyurethanes GmbH (Lemförde, Deutschland).

Since both, the PLA as well as the PBS have hydrophilic characteristics, they were dried for 4 hours at 60 °C prior to processing. The PLA/PBS blends were compounded in a twin-screw extruder (ZSE18 HPe, Leistritz Extrusionstechnik GmbH, Nürnberg, Germany) at different weight-ratios (50/50, 60/40, 70/30, 80/20, 90/10). The temperature (180 °C), throughout rate (6 kg/h) and the screw speed (200 rpm) were fixed for all experimental runs.

The one- and two-component test specimens were produced by using an injection molding machine (Allrounder 320 C Golden Edition, Arburg GmbH + Co KG, Loßburg, Germany). The sample geometry was chosen according to A1 tensile test specimen (ISO 527-1). The temperature of the barrel to produce the one-component specimens (composed of PLA/PBS blends) was set to 190, 185, 180, 175 and 170 °C from nozzle to hopper. The mold temperature was kept at 30 °C. While the holding time was set to 13 s, the cooling time and the injection velocity were adjusted to the amount of PBS in the blends (25-60 s and 30-40 cm^3/s). To produce two-component test specimen, the one-component tensile test pieces of the PLA/PBS-blends were cut perpendicular to the melt flow direction in a distance of 80 mm from the sprue distant side. The halved test specimens were placed as inserts in the cavity of the injection mold, so that the second component (TPU 95) could be injected onto the cut surface to create a two- component tensile test specimen. For the injection molding of the TPU, barrel temperatures of 220, 215, 210, 205 and 200 °C were used from nozzle to hopper. The mold temperature of 30 °C, the holding time of 2 s, the injection velocity of 10 cm^3/s and the cooling time of 20 s were selected.

To determine the influence of raised temperatures on the mechanical and thermal behavior as well as the crystallinity, surface free energy and bonding strength of the blends, storage tests were carried out. The storage types consisted of the combination among 2 different temperatures (room temperature and 100 °C) and 5 different storage lengths (0, 1, 3, 7 and 14 day(s)). For the storage at 100 °C, the samples were placed in an oven right after the production. The room temperature storage was carried out at a standardized climate of 23 °C and 50 % relative humidity (ISO 527-1).

On the purpose of measuring the maximal load that is needed to separate the two components from each other, tensile tests were carried out by using a universal testing machine (Zwick Z010, Ulm, Germany). According to ISO 527-1, the specimens that were stored at 100 °C were kept at 23 °C and 50 % relative humidity for at least 16 hours prior to the tensile test. The tests were conducted with a constant speed of 10 mm/min and 5 samples were tested for each combination of the storage length and the temperature.

Thermal analysis was performed by using the differential scanning calorimetry (DSC-Module Q 1000, TA Instruments, New Castle, DE, USA) under a nitrogen atmosphere. Preliminarily, the samples (~10 mg) that were cut out of the tensile test specimen, were sealed in an aluminum pan before placing into the chamber. Thereafter, the samples were heated to 200 °C at a rate of 10 K/min, cooled down to 0 °C at the same rate and then reheated to 200 °C. Data was analyzed using the TA universal analysis software package. Each combination of the storage type and composition ratio was tested. The degree of the crystallinity (X_c) of the samples was calculated by the following relationship:

$$X_C(\%) = \frac{\Delta H_f}{W_{PLA/PBS} \times \Delta H_f^0} \times 100 \tag{1}$$

Where ΔH_f is the heat of fusion of the PLA or PBS in the blend, ΔH_f^0 is the heat of the fusion for 100 % crystalline PLA (93 J g^{-1}) or PBS (110,3 J g^{-1}) [6] and W_{PLA}, W_{PBS} are the weight fractions of the PLA and PBS in the blend.

A static contact angle measurement was performed with the "ADVENCE" software and a contact angle measuring device (both Krüss GmbH, Hamburg, Germany) that measured the left and right tangent angle of the fluid droplet on the material surface and averaged the contact angle value. As part of the two component model based on Fowkes [7] and OWRK [8] (Owens, Wendt , Rabel and Kaelble), the interfacial surface energy (γ_{LS}) was obtained by the sum of the surface free energies of the individual phases divided into their dispersive (D) and polar (P) interaction between phases. The equation is given as following:

$$\gamma_{LS} = \sigma_L + \sigma_S - 2\left(\sqrt{\sigma_L^D \times \sigma_S^D} + \sqrt{\sigma_L^P \times \sigma_S^P}\right) \quad (2)$$

σ_L is the surface tension of the liquid and σ_S indicates the surface free energy of the solid. To calculate the dispersive and polar phase of the surface free energy of the samples, two liquids of known surface tensions had to be used. Almost purely dispersive diiodo-methane (surface tension 50,8 mN/m, polar component 2,3 mN/M, dispersive component 48,5 mN/m [7]) and highly polar water (surface tension 72,8 mN/m, polar component 51,0 mN/M, dispersive component 21,8 mN/m [9]) were chosen as the test liquids.

Fourier transform infrared spectroscopy (FTIR spectroscopy) with an FTIR spectrometer (Shimadzu, IRAffinity - 1S, Kyoto, Japan) was used for further investigation. The specimens were examined using the attenuated total reflection (ATR) without prior preparation of the specimen.

RESULTS AND DISCUSSION

Mechanical Properties of Blends

The averaged tensile strength results are shown in Fig.1. (a). It can be seen that with increasing content of PBS up to 40 wt%, a reduction in the tensile strength eventuates. This is indicative of the influence of the weaker PBS component on the strength of the material. The highest tensile strength of 67 MPa was obtained after 14 days storage of the PLA/PBS (90/10) at 100 °C. The longer storage time ensures a gradual drop in the strength with increasing amount of PBS. Fig. 1. (b) shows the results for the elongation at break as a function of the PLA content. According to the results, the highest elongation of 25 % occurred for the PLA/PBS (60/40) that was stored for 3 days at 100 °C. Analogous to the decrease in tensile strength, the increasing content of PBS leads to an increase in elongation at break up to a weight ratio of 40 wt% PBS. Moreover, it must be noticed that the highest standard deviation was measured at both plots in fig. 1. when the PLA content was 70 wt%.

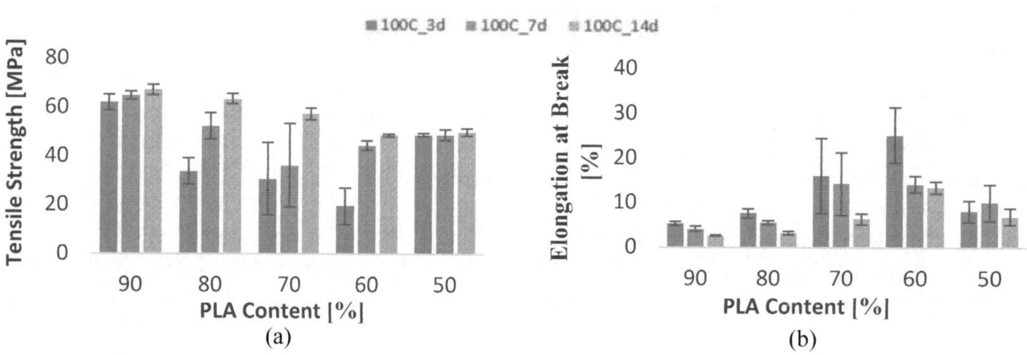

FIGURE 1. Tensile strength (a) and elongation at break (b) of the PLA/PBS blends

Crystallinity and Thermal Analyses

Figure 2 (a) shows the results of the DSC measurements from the first heating scans. The first heating scan was used, to show the effects of the storage conditions. The none-stored pure PLA (not shown in Figure 2. (a)) exhibited a glass transition at 61,25 °C and the crystalline melting at 148,09 °C. However, PLA, and PLA/PBS blends did not show a noticeable glass transition after the storage at 100 °C. In addition, the melting peaks of PBS in the composition range of 30-50 % comprise a shoulder, verifying that the melting is overlaid by a recrystallization [10]. Referring to Fig. 2 (b), the longer storage length increased the crystallinity of the PLA in the blend. The fact, that the crystallinity of the PLA increases with weight ratios of 20-30 wt% PBS after 3 days storage at 100 °C, shows that the PBS works as a plasticizer for the PLA [11].

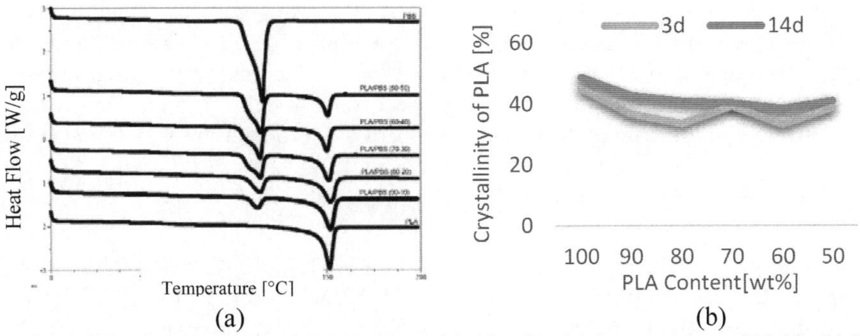

FIGURE 2. (a) DSC traces of PBS, PLA and PBS/PLA blends after 14 days storage at 100 °C, (b) The crystallinity in the PLA after the storage 3- and 14-days at 100 °C as a function of PLA weight fraction.

Surface Free Energy and Interfacial Tension

The contact angle measurement was used to calculate the surface free energies of the PLA/PBS blends and the TPU95 under the influence of storage temperature and length. To reach a good bonding strength between the two components, the surface energies should have similar phase (dispersive or polar) interaction which eventuate in small interfacial tension [12] [13]. By comparing Fig. 3 and Table 3, it can be observed that the calculated interfacial tension between PLA/PBS (60/40) and TPU95 correlates with the required load to disengage these two components. The highest interfacial tension (9,03 mN/m) was obtained after 1 day of storage at room temperature. This storage condition also leads to the lowest load (411,85 N). Moreover, Table 3 and Fig. 3 revealed that a longer thermal storage at 100 °C contributes to an increase in the adhesive bonding strength. Since the decrease in the polar phase of the soft component as well as hard component causes the attenuation in interfacial tension.

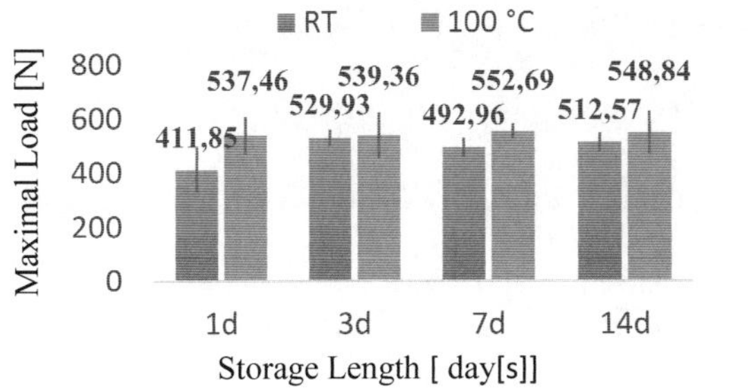

FIGURE 3. Load that was needed to separate the two-component test specimen (PLA/PBS (60/40) and TPU 95)

TABLE (3). Interfacial tension between the PLA/PBS (60/40) blend and TPU95, regarding storage temperature and length

Temperature \ Length	Interfacial Tension [mN/m] 1 day	3 days	7 days	14 days
RT	9,03	0,08	2,05	0,64
100 °C	0,60	0,63	0,54	0,46

Phase Morphology

Figure 4 illustrates the absorption peak shifts in specific region of pure PLA, pure PBS and the PLA/PBS (70/30) blend after the storage at 100 °C (3 and 14 days), measured by the ATR infrared spectroscopy to determine the intermolecular interactions.

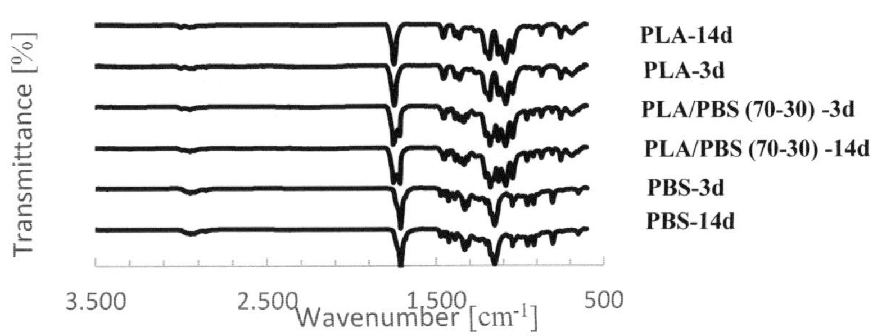

FIGURE 4. Infrared spectra in the 500-3,500 cm^{-1} region obtain from pure PBS, pure PLA and PLA/PBS (70-30) stored 3 and 14 days at 100 °C

The spectrum shows the C=O stretching for all samples with a single strong and sharp peak in the region of 1755-1707 cm^{-1}. With increasing amount of PBS content, the stretching vibration of the C-O, C- C, and C-H groups signals appeared less but sharper in the region of 1100-1300 cm^{-1} indicating the increase in the polar phase of the PLA/PBS blend which results in the enhancement of the surface energy. According to calculation of the surface free energy and the tensile test results (not shown in the article), this increase causes a decrement in the bonding strength.

CONCLUSIONS

PBS was added to the PLA to improve the crystallinity and the adhesive bonding strength in two component injection molding. The calculated interfacial tension between the PLA/PBS blends and the TPU estimates that the storage at 100 °C increases the bonding strength. These results also correlate with the increment in crystallinity of the PLA shown by the DSC measurement. The measurement of the reached maximal load needed to disengage the two-component test specimen complies with the calculated interfacial tension. The highest interfacial tension (9,03 mN/m) was obtained after 1 day of storage at room temperature. This storage condition also leads to the lowest load (411,85 N). The crystallization density and spherules' size of the PLA was affected by the addition of PBS regarding DSC measurement. In terms of mechanical properties, it was found that the tensile strength decreased with increased concentration of the PBS. The longer storage time ensures a less gradual drop in the strength with increasing amount of PBS. In contrast, the elongation at break shows an increase up to 40 wt% of PBS.

ACKNOWLEDGMENT

The author is grateful for the funding of this work from the polymer engineering department, University of Kassel.

REFERENCES

1. Mirabal A.S., Schalz L., Carus M., "Bio-based Polymers in the World Capacities, Production and Applications: Status Quo and Trends towards 2020", pp.6-7, (2013).
2. Nishioka, M., T. Tuzuki, Y. Wanajyo, H. Oonami and T. Horiuchi, Japan Stud. Polym. Sci. 12, pp. 584-590, (1994).
3. Deng, Y. and Thomas, N.L., "Blending poly (butylene succinate) with poly (lactic acid): ductility and phase inversion effects", European Polymer Journal, 71, pp. 534 - 546. (2015).
4. Bhatia A., Gupta R. K., Bhattacharya S. N., and Choi H. J., "Compatibility of Biodegradable Poly (lactic acid) and Poly (butylene succinate) Blends for Packaging Application," Korea-Australia Rheol., vol. 19, no. 3, pp. 125–131, (2007)
5. Park J. W. and Im S. S., "Phase Behavior and Morphology in Blends of Poly (L-Lactic Acid) and Poly (Butylene Succinate)," Journal of Applied Polymer Science, Vol. 86, No. 3, pp. 647-655, (2002)
6. Hassan, E., Wei, Y., Jiao, H., & Yu, M., "Dynamic Mechanical properties and Thermal stability of Poly (lactic acid) and poly (butylene succinate) blends composites", Journal of Fiber Bioengineering and Informatics, 6(1), 85-94. (2013).
7. Fowkes F. M., "Attractive Forces at Interfaces. In: Industrial and Engineering Chemistry" 56,12, pp. 40-52, (1964)
8. Kaelble D. H., "Dispersion-Polar Surface Tension Properties of Organic Solids". In: J. Adhesion 2), pp. 66-81, (1970)
9. Janczuk, B.; Bialopiotrowicz, T., "Surface Free-Energy Components of Liquids and Low Energy Solids and Contact Angles" in Journal of Colloid and Interface Science, 127, pp. 189-204 (1989).
10. Mustapa, I. R., Shanks, R. A., & Kong, I., "Multiple melting behavior of poly (lactic acid)-hemp-silica composites using modulated-temperature differential scanning calorimetry", Journal of Polymer Engineering, 34(9), (2014).
11. Gachter, R.; Muller, H. Plastics Additives Handbook, 3rd ed.; Hanser: New York, 1990; Chapter 5.
12. Kleeschulte, R. "Modellierung der Verbundfestigkeit von Hart-Weich-Kombinationen polymerer Werkstoffe", PhD diss., Paderborn University, (2011).
13. Kaiser, E., "Scale-Up - Methoden zur Prozessführung von 2K Spritzgussteilen im Hinblick auf ihre Verbundhaftung", PhD diss., Paderborn University (2012).

SCTRACTH RESISTANCE IMPROVEMENT OF THE PLASTIC MATERIALS

Seda Köksal Yeğin[a], Muhammet Tekin[a], Koray Gonce[a], Halit Gocuklu[a]

[a]*Farel Plastic Research and Development Center, Çerkezköy, Tekirdağ, Turkey*

Abstract. Scratch resistance of plastic materials has become important for many applications. In this study, the effects of boron compounds, fumed silica and commercial scratch resistance increasing additives have been investigated scratch resistance and surface tension on polycarbonate (PC), polymetylmethacrylate (PMMA).

Keywords: Transparent polymers, scratch resistance, anhydrous borax, boron oxide, clearity, transparency, PMMA, PC.
PACS: S05-263

INTRODUCTION

Transparent plastics have been used in many sectors such as electronics and automotive. Improving their features will increase usage areas. It is possible to increase the usage areas instead of glass. Polymethyl methacrylate and polycarbonate are the most widely known transparent thermoplastics. Improving of their sctracth resistance have been widely discussed.[1,2] Unfortunately, most polymers have low scratch resistance[3] Blending entered the plastics industry half a century ago.[3] Blends of PMMA ve PC thermoplastics were investigated and their scratch resistance, transparency, gloss, hardness of shore D, surface tension were evaluated.

EXPERIMENTAL

PMMA [PLEXIGLAS® Resist zk6BR, Evonik Performance Materials GmbH, density= 1.16 g/cm3, melt flow rate = 1.6 cm3/10 min] was injection molded with 1 wt % of Aerosil 200 fumed silica (Evonik), 1 wt % TEGOMER® H-Si 6441 P (Evonik Nutrition & Care GmbH, melting point = approx. 54 °C, water content= < 0.1 %), 1 wt % IncroMax™ 100 (Croda Polymer Additives), 1 wt % Anhydrous Borax ($Na_2B_4O_7$, Etibor-68, CAS No: 1330-43-4, Disodium Tetraborate Anhydrous, Etimaden, Turkey), 1 wt % Boron oxide (B_2O_3 , CAS No: 1303-86-2, Etimaden, Turkey).

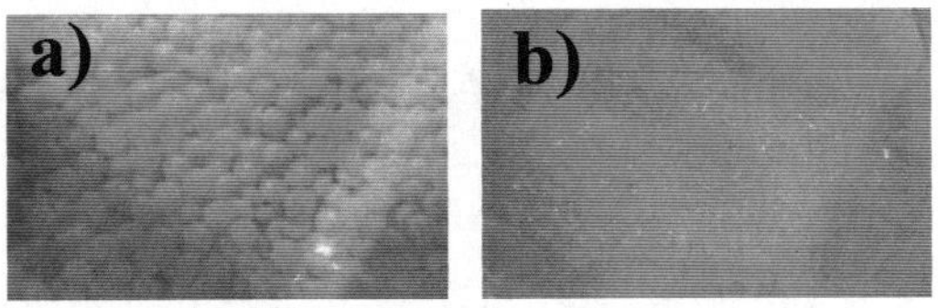

FIGURE 1. a) Granule of TEGOMER® H-Si 6441 P, b) Anhydrous Borax

PC [HopeLex PC-1220 U, Lottee Chemical, density = 1.20 g/cm^3 , melt flow index= 22 g/10 min] was injection molded with 1wt % Aerosil 200 fumed silica (Evonik), 1 wt % Boron oxide (B_2O_3, CAS No: 1303-86-2, Etimaden, Turkey), 1 wt % IncroMax™ 100 (Croda Polymer Additives). Samples were prepared by 250 tones injection machine. After the additives were added to the two polymers at 1wt% rate, the samples were prepared. The twin screw extruder was not used to mix the raw materials with additives. Polycarbonate polymer and its blends injection process parameters are seen TABLE 1.

TABLE 1. Polycarbonate Injection Process Conditions

Drying Condition		
Drying Temperature		120 ºC
Drying Time		4 hr
Maximum moisture content		0.02 %
Injection molding		
Melt Temperature		290-310 º C
Nozzle Temperature		280-300 ºC
Barrel	Rear zone	209-310º C
	Middle zone	280-300º C
	Front zone	270-290º C
Hopper Temperature		60-80 ºC
Mold Temperature		60-90 º C

RESULTS AND DISCUSSION

In some results the range is specified because the mold which was used for preparation of the samples was prone. As different results were obtained in different regions of the samples, the highest values were showed.

Appearance-Transparency

The highest transparency was obtained at composition of %1 Anhydrous Borax & PMMA. Addition of Aerosil 200 created a dusty image. The IncroMax™ 100 and boron oxide were similar to transparency of original PMMA.

TABLE 2. Appearance Comprasion of the Blends of PMMA with the Additives

PMMA	
Composition	**Appearance-Transparency**
Original PMMA	Good
1 % Aerosil 200 & PMMA	Dusty Image
1 % Tegomer & PMMA	Good but lower than original PMMA
1 % IncroMax™ 100 & PMMA	Good (similar to Original PMMA)
1 % Anhydrous Borax & PMMA	Higher transparency than Original PMMA
1 % Boron oxide & PMMA	Lower than original PMMA

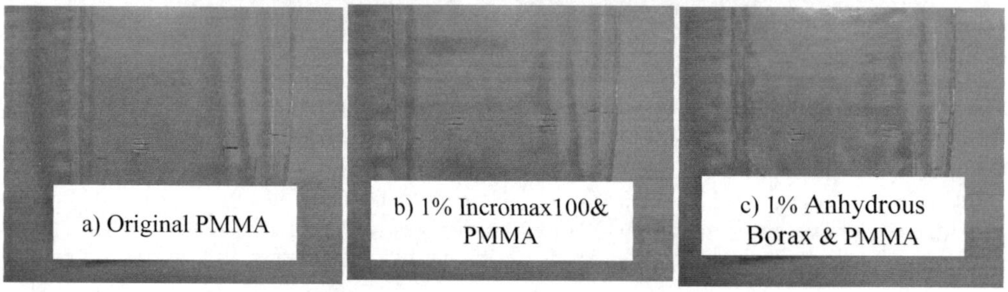

FIGURE 2. Image of PMMA's Sample

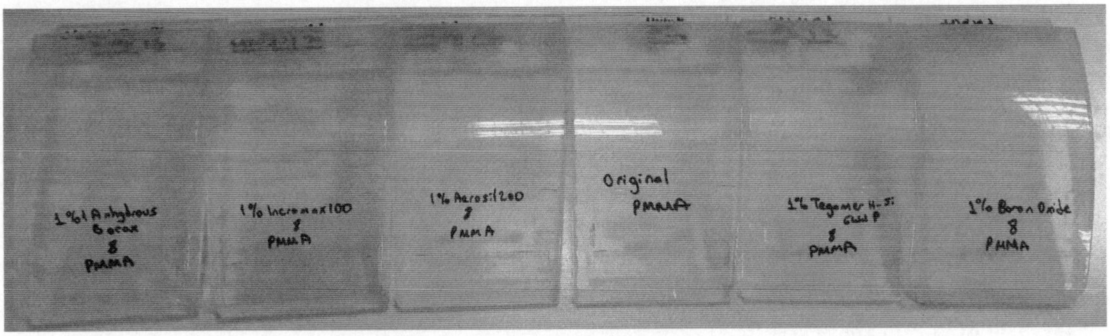

FIGURE 3. Comprasion of Original PMMA and Blends of PMMA with the Additives

As the full distribution of silica particles could not be achieved, visible silica clusters formed in the transparent PMMA polymer.

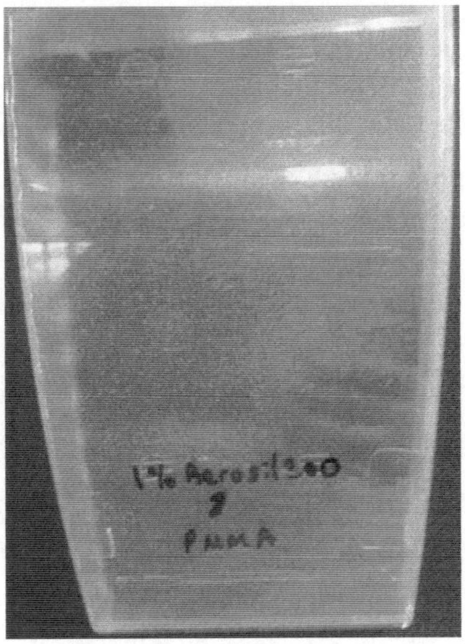

FIGURE 4. 1% Aerosil 200 & PMMA Blend. Silica particles are visible.

Adding 1% boroxide to the polycarbonate polymer turned the color to green.

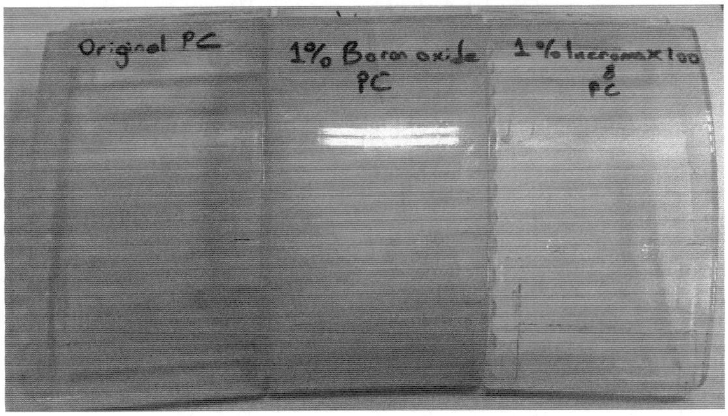

FIGURE 5. PC and PC's Blends

TABLE 3. Appearance Comprasion of PC and PC's Blends

PC	
Composition	**Appearance-Transparency**
Original PC	Good
1% Boron oxide&PC	Boron oxide was not dissolved in the PC.
1% Aerosil 200 & PC	Dusty Image
1% IncroMax™ 100	Good (similar to Original PMMA)

Gloss Measurement

Gloss value of the PMMA polymer was measured as 90 with TQC Glossmeter 20º/60º/85º gloss gauge. The highest gloss value was measured as 90.3 on 1% IncroMax 100 & PMMA sample. The lowest gloss value was measured as 89.4 on 1% Aerosil 200 & PMMA sample. %1 Anhydrous Borax & PMMA and %1 Boron oxide & PMMA. The gloss values of borax and boron oxide were measured at the same gloss value as the original PMMA. %1 Tegomer & PMMA was 89.6.

TABLE 4. Gloss Comprasion of PMMA and PMMA's Blends

PMMA	
Composition	**Gloss**
Original PMMA	90
%1 Aerosil 200 & PMMA	89.4
%1 Tegomer & PMMA	89.6
%1 IncroMax™ 100 & PMMA	90.3
%1 Anhydrous Borax & PMMA	90
%1 Boron oxide & PMMA	90

The addition of boroxide and silica in the PC polymer reduced the gloss value of the PC. Incromax 100 has increased the gloss value of the PC.

TABLE 5. Gloss Comprasion of PC and PC's Blends

PC	
Composition	**Gloss**
Original PC	95
1% Boron oxide&PC	75
1% Aerosil 200 & PC	77
1% IncroMax™ 100	104

Hardness Measurement

Hardness measurements was made according to Shore Hardness ASTM D2240 with Insize Shore Durometer Code ISH-SD Original PMMA was measured 80 Shore D. %1 Tegomer & PMMA composition had the lowest Shore D value; 77. %1 IncroMax™ 100 &PMMA, %1 Aerosil 200&PMMA, %1 Anhydrous Borax&PMMA and %1 Boron oxide&PMMA compositions were measured the same hardness as the original PMMA.

TABLE 6. Shore D Hardness Comprasion of PMMA and PMMA's Blends

PMMA	
Composition	**Shore D**
Original PMMA	75-80
%1 Aerosil 200 & PMMA	80
%1 Tegomer & PMMA	77
%1 IncroMax™ 100 & PMMA	80
%1 Anhydrous Borax & PMMA	80
%1 Boron oxide & PMMA	80

Shore D hardness on PC increased by addition Incromax 100. the other additives did not provide any changes.

TABLE 7. Shore D Hardness Comprasion of PC and PC's Blends

PC	
Composition	**Shore D**
Original	80
1% Boron oxide&PC	80
1% Aerosil 200 & PC	80
1% IncroMax™ 100	84

Scratch Resistance

Scratch Resistance of the samples was measured with Erichsen Hardness Test Pencil Model 318S. Scratch resistance of original PMMA was 15N. %1 Tegomer&PMMA, %1 IncroMax™100&PMMA, %1 Anhydrous Borax&PMMA and %1 Boron oxide&PMMA were measured above 20 N. Aerosil 200 did not have an effect on scratch resistance. tegomer, incromax, anhydrous borax and boron oxide increased the scratch resistance of PMMA. Erichsen Hardness Test Pencil Model 318S only can measure up to 20 N. Therefore, the upper limit of the increase in scratch resistance was not measured.

TABLE 8. Scratch Resistance Comprasion of PMMA and PMMA's Blends

PMMA	
Composition	**Scratch**
Original PMMA	15 N
%1 Aerosil 200 & PMMA	15 N
%1 Tegomer & PMMA	20 N ≤
%1 IncroMax™ 100 & PMMA	20 N ≤
%1 Anhydrous Borax & PMMA	20N ≤
%1 Boron oxide & PMMA	20 N≤

Boron oxide and Aerosil 200 added into the PC polymer did not affect scratch resistance. However, the incromax100 additive increased the scratch resistance of the polymer. Since the Erichson Hardness Test Pencil could measure up to 20N, the upper limit could not determine.

TABLE 9. Scratch Resistance Comprasion of PC and PC's Blends

PC	
Composition	**Scratch**
Original	15 N
1% Boron oxide&PC	15 N
1% Aerosil 200 & PC	15 N
1% IncroMax™ 100	20 N ≤

Surface Tension

Surface Tension of the samples was measured with Arcotest surface tension test pencils. Original PMMA and 1% Aerosil 200 and 1% anhydrous borax added PMMA samples were 36. Aerosil 200 and anhydrous borax did not affect surface tension of PMMA. Tegomer, IncroMax™ 100 and boron oxide decreased surface tension of PMMA.

TABLE 9. Surface Tensions of PMMA and PMMA's Blends

PMMA	
Composition	**Surface Tension**
Original PMMA	36
%1 Aerosil 200 & PMMA	36
%1 Tegomer & PMMA	34
%1 IncroMax™ 100 & PMMA	34
%1 Anhydrous Borax & PMMA	36
%1 Boron oxide & PMMA	34

The contribution of the additives to the PC polymer in general did not cause a significant change in surface tension.

TABLE 10. Surface Tensions of PC and PC's Blends

PC	
Composition	**Surface Tension**
Original PC	34
%1 Aerosil 200 & PC	34
%1 IncroMax™ 100 & PMMA	34
%1 Boron oxide & PMMA	34

CONCLUSIONS

Anhydrous Borax addtion to PMMA increased clearity and sctratch resistance. In order to increase the use of PMMA instead of glass, anhydrous borax looks promising. The development of commercial products can be achieved by increasing the studies. The obtained data showed that The IncroMax™ 100 had a positive effect for improving scratch resistance of PMMA and PC polymers. Boron oxide increased the scratch resistance of PMMA. However, transparency was keeped. In PC polymer, it was observed that boron oxide did not affect scratch resistance. The addition of boron oxide changed the appearance to opaque green. The twin screw extruder was not used to mix the raw materials and additives. It is expected that more homogeneous samples will be prepared by the injection process, since it will provide a more homogenous distribution of additives in the raw material by the use of twin screw extruders.

ACKNOWLEDGMENTS

I would like to thank Haluk Akçalı (Farel Plastic's General Manager) and Ömer Burhanoğlu (Farel Plastic's CEO) for all their support. Moreover, I thanks to both of Farel Plastic injection and laboratory test teams for their assistance in the preparation of samples and making the tests.

REFERENCES

1. M. Muhammad, "Improving the Scratch Resistance of PMMA", Ph.D. Thesis, Strasbourg University, 2011.
2. V. H. Carl, "Organoclay Imparts Scratch Resistance", *Kunststoffe International 1-2/2017*.
3. J. Magdalena, L. Antoni and O-S. Martin "Friction and Scratch Resistance of Polyamide 6 Modified wit Ionomeric Etyhlene/Methacrylic Acid Copolymer," journal of Applied Polymer Science, Vol. 91, 3866-3870, 2004, Wiley Periodicals, Inc.

High Performance Silane Coupled Basalt Fiber – Polypropylene Composites Prepared by Melt Compounding

Orkun Kaymakçı and Nurseli Uyanık[c]

[a] *Arçelik Central R&D, Tuzla - Istanbul - Turkey*
[b] *Istanbul Technical University, Polymer Science & Technology Prog. - Istanbul - Turkey*
[c] *Istanbul Technical University, Faculty of Science and Letter, Department of Chemistry - Istanbul - Turkey*

Abstract. Despite the popularity of fiber reinforced polypropylene (PP) composites both in academia and industry, the research on basalt fiber (BF) reinforced thermoplastic PP composites is quite limited. Due to their superior properties, BFs have a great commercial potential for industrial applications. In this study, six different silane coupled BF/PP composites were prepared with fiber contents up to 15wt% using a twin-screw extruder. Maleic anhydride grafted PP (MA-g-PP) coupling agent content was fixed at 10wt%. Mechanical properties were investigated by tensile, flexural, and Izod impact tests. Thermal, viscoelastic, and morphological properties were studied by differential scanning calorimetry (DSC), dynamic mechanical analysis (DMA), and scanning electron microscopy (SEM), respectively. The results showed that BFs have substantial effect on the mechanical and viscoelastic properties of the composites. The considerable increase in the tensile strength, flexural strength, and Izod impact strength was attributed to the higher interfacial adhesion between the BF and PP matrix in the presence of MA-g-PP coupling agent. The SEM images proved the increased interfacial adhesion between the fibers and the PP matrix. The improved interfacial adhesion was also affected the crystallization behavior of PP.

Keywords: Basalt fiber, mechanical properties, compounding, polypropylene
PACS: 81.05.Lg

INTRODUCTION

Thermoplastic polypropylene (PP) composites have been widely used in engineering applications due to their comprehensive mechanical performance and low cost. These composites have been selected especially in automotive, home appliances, construction, furniture and sporting goods industries due to their balanced toughness and stiffness, good chemical resistance, low moisture absorption, and recyclability [1-3]. In addition to these properties, the processability of these composites are relatively easier than most of the other polymer composites such as composites based on polyesters and polyamides.

In industry, different types of additives and fillers are used to tailor the mechanical properties of PP composites. Inorganic fillers such as calcium carbonate, talcum, mica, glass beads, glass fibers, carbon fibers, and organic fillers such as wood powder, jute fibers, kenaf fibers, and hemp fibers are the most common reinforcers used in PP [4-5]. Due to its non-polar nature, PP has poor interfacial adhesion with most of the fibers and particulate fillers. To improve the interfacial adhesion between the polymer matrix and the fillers, different types of coupling agents are used. MA-g-PP is one of the most commonly used coupling agents in the industry due to high reactivity of the maleic anhydride groups.

Depending on its application area, mechanical properties such as stiffness, elongation, and toughness are the most important criteria for selection of PP composites. PP composites that are reinforced with stiff fibers such as glass fibers and carbon fibers possess high mechanical properties and they are widely used in industry. However, considering the industry's demand on PP composites with high mechanical properties, the variety of industrial stiff fibers are very limited. Even though basalt is a well-known and extremely abundant material in nature, its application as a polymer reinforcing fiber in industry is limited. The chemical structure of BFs is similar to glass fibers (GFs,) however, BFs are stiffer than GFs. In addition, they possess better thermal properties, less water absorption and better chemical resistance. Therefore, for industrial applications, there is a great potential for BFs as reinforcing fibers for PP composites.

Farsani et al. studied BF reinforced PP/nanoclay hybrid composites that were prepared through melt-mixing method. They found that increasing nano-clay content and BF content increases the tensile modulus and stiffness of the composites and decreases materials' elongation at break and toughness [6]. Prajapati et al. investigated BF reinforced Nylon-6 polymer composites with MA-g-PE coupling agent and showed that the length of the BFs were reduced due to the shear effect during compounding. In addition, MA-g-PE increased the stiffness and the impact strength of the composites [7]. Botev et al. also investigated the effect of untreated BFs on the mechanical properties of PP composites. Even though there was some increase in the mechanical properties with increased loading levels of BF, at 10% BF and 10% coupling agent loading levels, the improvements in tensile strength and Izod impact strength were only about 5% and 10% respectively [8].

In this work, the effect of MA-g-PP coupling agent and the effect of BF content on the properties of the PP/BF composites were investigated. The composites produced by melt compounding were characterized by tensile tests, flexural tests, Izod impact tests, differential scanning calorimetry (DSC), dynamic mechanical analysis (DMA), and scanning electron microscopy (SEM).

EXPERIMENTAL

Materials

Borealis HE125MO homopolymer PP with 12 g/10min melt flow index (230°C/2,16 kg) was supplied by Borealis AG, Austria. BFs were kindly supplied by Polotsk-Steklovolokno Joint Stock Company from Belarus. Nominal diameter of the basalt fibers was 16 microns and fiber length was 6mm. The fibers had silane chemical coupling. Maleic anhydride grafted PP coupling agent was kindly supplied by BYK Additives, Germany. BYK 9212 MAH-g-PP was a solid phase grafted coupling agent with 1.8wt% chemically bound maleic anhydride content.

Composite Preparation

BF reinforced PP compounds were prepared using PRISM TSE24 IIC, with the screw diameter 24 mm and L/D ratio 28:1 at a screw speed of 200 rpm. Temperature profile of the compounder was set to 180°C/190°C/200°C/220°C/225°C/225°C/210°C. Formulations that were produced in the compound extruder are listed in Table 1. Test specimens for tensile testing, Izod Impact testing and DMA are produced using an Arburg Allrounder 3200C injection machine.

TABLE 1: Sample Codes and Compositions of PP/BF Composites

Sample Code	PP (%)	BF (%)	MA-g-PP (%)
PP100	100	0	0
PP95BF5C0	95	5	0
PP90BF10C0	90	10	0
PP85BF15C0	85	15	0
PP85BF5C10	85	5	10
PP80BF10C10	80	10	10
PP75BF15C10	75	15	10

Mechanical Characterization

Tensile tests were performed using Zwick Z020 universal tensile testing machine according to ISO 527 standards at room temperatures. Cross-head speed was kept constant at 50 mm/min. Flexural tests were performed in three-point bending mode using Instron 4505 universal testing machine according to ISO 78 standards at room temperature. Tests were performed at 5 mm/min cross-head speed rate. Izod Impact tests were performed using Zwick HIT Series Impact Tester according to ISO 180 standards at room temperature.

Differential Scanning Calorimetry (DSC)

DSC measurements were performed using a TA Instruments DSC Q-2000. About 10 mg of compound sample was sealed in aluminum pans and an empty aluminum pan and a lid were used as reference. Two heating/cooling scans at 10°C/min heating were performed in flowing nitrogen at 50 mL min^{-1}.

Dynamic Mechanical Analysis (DMA)

DMA analyses were performed using TA Instruments DMA Q800. 50x10x2 mm samples were tested in three-point bending mode. Frequency ramp tests were performed between 0.1 Hz to 100 Hz at constant temperature of 30°C. The DMA was used in strain-controlled mode and the strain rate was set to 0.05%.

Scanning Electron Microscopy (SEM) / Energy Dispersive X-Ray (EDS) Analysis

SEM Analyses were performed by a Zeiss SEM at an acceleration voltage of 20kV. The specimens were fractured under liquid nitrogen and gold-sputtered before observation.

RESULTS AND DISCUSSION

Mechanical properties of PP/BF composites were investigated by tensile tests. Figure 1 shows the effect of BF content and the effect of coupling agent on tensile modulus (a), yield stress (b), and elongation at yield (c) properties of the composites. BFs showed significant effect on the tensile modulus of the composites due to the reinforcing effect of the fibers. 15 wt% BF loading increased the modulus of polypropylene by about 120%. However, the effect of coupling agent on the tensile modulus was quite limited. Since the modulus depends on only the first 0.25% strain [10], the interfacial adhesion between the BFs and the PP matrix did not significantly affect the modulus values [11,12]. However, the effect of interfacial adhesion between the polymer matrix and the BFs on yield stress and elongation at yield was highly distinguishable. Without the presence of the coupling agent, the addition of BF resulted a slight decrease in yield stress values of the composites due to the poor interfacial adhesion between the matrix and fibers. On the contrary, due to the hydrogen bonding interactions between the silane coating of the BF and anhydride groups of the grafted PP, yield strength of the samples with coupling agent showed a very prominent increase with the increasing BF content. Yield strength values increased by about 30% and 50% at BF loading levels of 10 wt% and 15 wt% respectively. In a similar study, Botev et. al has investigated the yield strength of untreated BF reinforced PP composites [8]. At 10 wt% BF loading with 10 wt% MA-g-PP coupling agent, they only observed about 5% increase in the yield strength, in contrast to 30% increase in this study. That difference possibly resulted from the promoted adhesion between the silane coating of the BFs and the anhydride groups [13]. The effect of high interfacial adhesion was also evident in tensile strain results. Normally, fiber type reinforcers decrease the elongation at yield values. Due to the high interfacial adhesion, BFs restricted the elongation of the polymer matrix. Therefore, at 10 wt% BF loadings, elongation at yield value decreased about 50%.

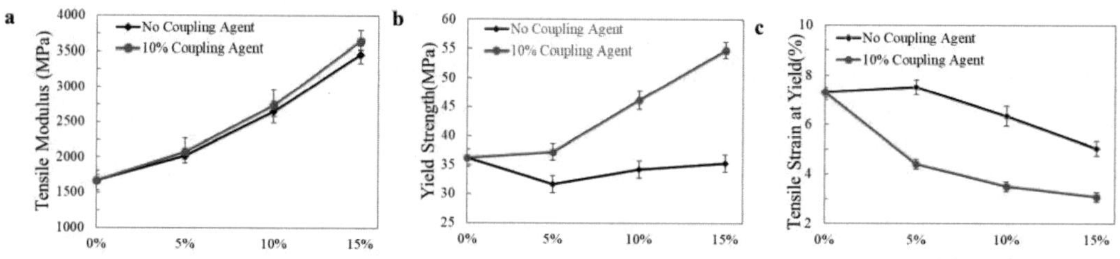

FIGURE 1: Tensile modulus (a), Yield Strength (b) and Elongation at Yield (c) values of PP/BF Composites

As shown in Figure 2, the effect of BF content and the effect of coupling agent on the flexural mechanical properties were similar to the effect on tensile properties. However, the effect of BF loading was less prominent on flexural modulus than tensile modulus possibly due to the orientation of the BFs on the flow direction during injection molding. As in the case of tensile modulus, coupling agent did not affect the flexural modulus. In the presence of the coupling agent, increasing BF content significantly improved the flexural strength of the composites. Without the coupling agent, at 5wt% and 10wt% BF loading levels, BFs showed negative impact on the flexural modulus due to the poor interfacial adhesion between the matrix and the fibers. As expected, Izod impact strength values increased with the increasing BF loading levels due to the reinforcing effect of the BFs. However, in the presence of the coupling agent, due to the high interfacial adhesion between the fiber and the matrix, the effect of BF loading on the impact strength was more prominent. 15wt% BF loading increased the Izod impact strength of the composites by about 85%, whereas the increase was only about 40% when there was no coupling agent in the system.

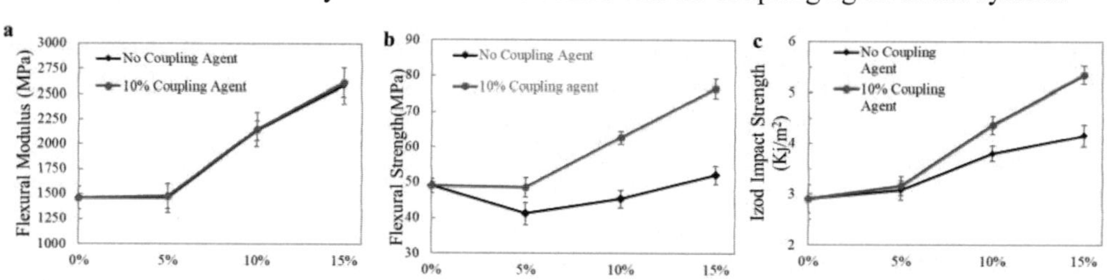

FIGURE 2: Flexural Modulus (a), Flexural Strength (b) and Izod Impact Strength (c) of PP/BF Composites

Table 2 shows melting temperature (T_m), melting enthalpy (ΔH_m), crystallization temperature (T_c), crystallization enthalpy (ΔH_c) and percent crystallinity (X_c) of the PP/BF composites. The apparent increase in the T_c values of the composites was related with the high interfacial interaction between the PP matrix and the BF. In these composites, BFs acted as effective nucleation centers increasing PP crystallization growth rate [14,15]. As a result, percent crystallization increased with the increasing BF content. These results compromised with the mechanical properties of the composites. BF content d the coupling agent did not show a significant effect on the T_m of the compounds.

TABLE 2: Thermal properties of PP/BF Composites

Sample Code	ΔH_m	T_m (°C)	T_c(°C)	ΔH_c	X_c
PP100	97.8	164.3	117.2	99.5	46.8
PP95BF5C0	87.4	161.7	116.1	92.35	44.0
PP90BF10C0	87.3	160.7	117.6	90.9	46.4
PP85BF15C0	82.1	161.0	116.3	83.3	46.2
PP85BF5C10	97.2	160.7	117.4	94.9	48.9
PP80BF10C10	95.8	160.8	117.9	90.7	50.9
PP75BF15C10	96.9	160.8	119.2	93.6	54.6

Dynamic mechanical properties of the PP/BF composites are demonstrated on Figure 3. Similar to the changes in static modulus values, storage moduli of the PP composites were significantly higher at higher BF loadings. Introduction of rigid BFs into the polymer matrix substantially decreased the polymer chain mobility. Decreased chain mobility resulted higher storage moduli for the composites. The effect of the reduced chain mobility with increasing BF content was also prominent on the tan delta values. Samples with lower BF contents showed higher dissipation behavior due to their relatively higher chain mobility. Samples with higher BF contents showed less viscous dissipation.

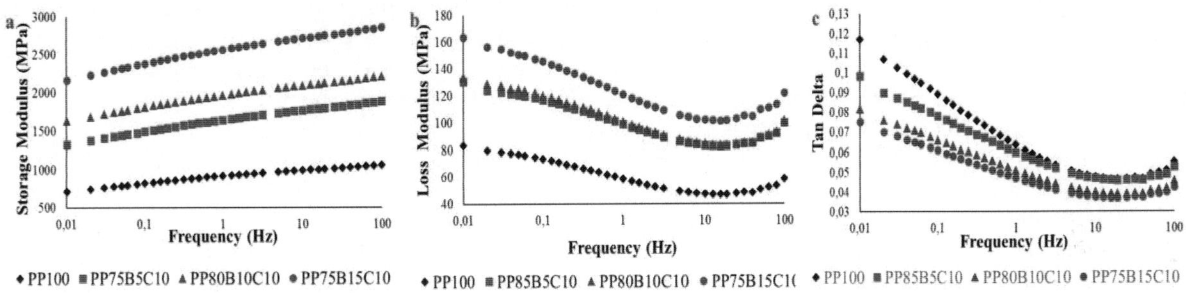

FIGURE 3: The effect of BF content on storage modulus (a), loss modulus (b) and loss tangent (tanδ) (c)

Fractured surfaces of PP/BF composites were investigated by SEM. SEM images in Figure 5 show that BFs were mostly oriented in the flow direction. Because of the fiber orientation, BFs showed stronger effect on the tensile strength than flexural strength of the composites. Composite without coupling agent (Figure 6 a) shows that surfaces of the BFs and fiber-matrix contact points in the composites were very smooth and clear, indicating poor interfacial adhesion between the matrix and the fibers. In the contrary, the surfaces of the fibers in the SEM image of the sample with coupling agent (Figure 6 c) were coated with PP polymer. In addition, the contact points between the fiber and the matrix indicates strong adhesion between the fibers and the matrix due to the hydrogen bonding between the silane coating of the fibers and the anhydride groups of MA-g-PP. It was also observed that there were more fiber pull-outs on the fracture surface of the composites without coupling agent due to poorer fiber matrix adhesion.

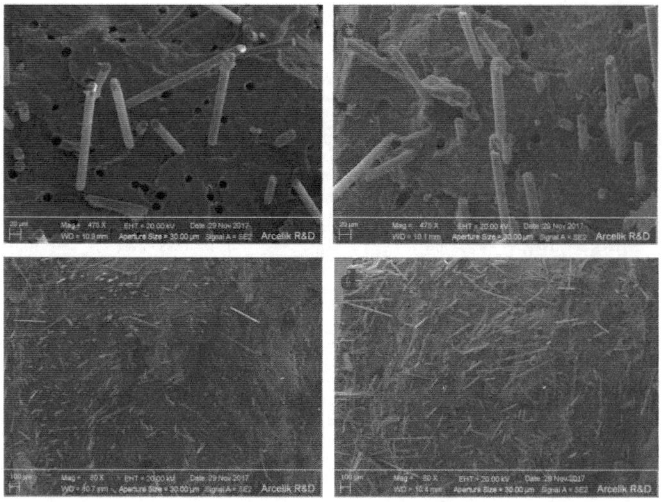

FIGURE 4: SEM Images of PP85B15C0 - 475x (a), PP85B15C0 - 80x (b), PP75B15C10 - 475x (c), PP75B15C10 - 80x (d)

CONCLUSIONS

In this study, the effect of BF loading level and the effect of coupling agent on the mechanical, thermal, and morphological properties of silane coupled PP/BF composites were investigated.

Tensile tests indicated that BF content had a significant effect on the tensile modulus of the composites. It was also found that the presence of coupling agent tremendously affected the yield strength and the yield strain values of the PP/BF composites as it provided interfacial adhesion between the fibers and the matrix. Similar results were also obtained in flexural test results. However, the effect of BF loading was less effective on the flexural modulus possibly due to the orientation of the fibers on the flow direction. In addition, Izod impact test results showed that BF content increased the impact strength significantly and the was more prominent in the presence of coupling agent. Thermal characterization showed that improved interaction between the BFs and the PP matrix increased crystallization temperatures and the percent crystallization. Composites with higher BF contents demonstrated higher dynamic storage modulus and lower viscoelastic energy dissipation. Higher interfacial adhesion between the BF and the PP matrix was demonstrated by SEM images.

The results of this study showed that silane coated BF significantly improves, mechanical, thermal, and viscoelastic properties of PP composites in the presence of MA-g-PP coupling agent.

REFERENCES

1. S. B. Cordeiro, J. N. Lunz and M. F. V. Marques, Polymer-Plastics Technology and Engineering **53**, 189 (2014).
2. J.-Z. Liang, Q. Du, G. C.-P. Tsui and C.-Y. Tang, Composites Part B: Engineering **95**, 166 (2016).
3. H. Karian, Handbook of polypropylene and polypropylene composites, revised and expanded, CRC press, New York, 2009.
4. J. Andrzejewski, N. Tutak and M. Szostak, Journal of Applied Polymer Science **133**, (2015).
5. S. Y. Fu, G. Xu and Y. W. Mai, Composites Part B: Engineering **33**, 291 (2002).
6. R. Eslami-Farsani, S. M. Reza Khalili, Z. Hedayatnasab and N. Soleimani, Materials & Design **53**, 540 (2014).
7. R. S. Prajapati, S. Jain and S. C. Shit, Polymer Composites **38**, 2798 (2015).
8. M. Botev, H. Betchev, D. Bikiaris and C. Panayiotou, Journal of Applied Polymer Science **74**, 523 (1999).
9. M. Arroyo, M. A. Lopez-Manchado and F. Avalos, Polymer **38**, 5587 (1997).
10. ISO, Plastics - Determination of tensile properties, (2012).
11. M. Sain, P. Suhara, S. Lawand and A. Bouilloux, Journal of Reinforced Plastics and Composites **24**, 121 (2005).
12. S. M. Zabihzadeh, G. Ebrahimi and A. A. Enayati, Journal of Thermoplastic Composite Materials 24, 221 (2010).
13. D. Bikiaris, P. Matzinos, A. Larena, V. Flaris and C. Panayiotou, Journal of Applied Polymer Science **81**, 701 (2001).
14. P. Wambua, J. Ivens and I. Verpoest, Composites Science and Technology **63**, 1259 (2003).
15. M. A. López-Manchado and M. Arroyo, Polymer **41**, 7761 (2000).
16. M. J. John, R. Tlili, R. D. Anandjiwala, A. Boudenne and L. Ibos, Composites Part B: Engineering **43**, 526 (2012).

Low Density, High Modulus Polypropylene Wood Composites Prepared by using Thermo Kinetic Mixer

Çağla Girişken[1], Ceyda Yanmaz[1], Buse Nur Can[2], Yakup Ülçer[2], Yusuf Ziya Menceloğlu[13],

1 Department of Material Science and Nano Engineering, Sabanci University, Istanbul 34956, Turkey
E-mail: cgirisken@sabanciuniv.edu, ceydayanmaz@sabanciuniv.edu

2 Ravago Petrochemicals Production Inc., Akse Mahallesi, Fevzi Çakmak Cd. No:130, 41420 Çayırova/Kocaeli

3 Sabanci University Integrated Manufacturing Technologies Research and Application Center & Composite Technologies Center of Excellence, Teknopark İstanbul, 34906 Pendik, İstanbul

Abstract. The increasing usage of polymers in automotive industry requires low density and high modulus polymers without losing its tensile strength. Therefore, the main challenge is to replace high density talc-based additives with low density and sustainable filler for density reduction. At this context, wood fiber is remarkable in terms of its low density. The wood fiber filler incorporated into polymer by using ultrahigh speed thermo kinetic mixing method. A series of experiment was performed by using totally 100 grams of sample in different loads (30% or 40%) of organic filler. In addition, 2% compatibilizer was used for comparison. After compounding, test samples were prepared by using mini injection molding and then their mechanical properties as well as thermal properties and homogeneity of the mixture were characterized by UTM, TGA, DSC and SEM. The test results showed that 40% wood fiber addition increased the tensile strength of the PP 4 times more (from 8.6 MPa to 36 MPa) than the talc added PP. Meanwhile the tensile modulus did not undergo a change but managed to kept same as with the talc added PP. Also, density measurements proved that wood fiber addition decreased density by 14%. SEM images revealed that addition of compatibilizer increased the homogeneity and results confirm tensile strength increase. Likewise talc added PP, crystallinity increase with wood fiber addition also proves that wood fiber can replace talc.

INTRODUCTION

The need of high performance plastics in automotive industry is increasing constantly. High performance plastics being low weight but high strength, provides more fuel efficient vehicles. For every 10% reduction of vehicle weight brings in 5-7% reduction in fuel consumption [2,8]. So, the environmental and economic concerns encourage the automotive industry to produce more fuel efficient vehicles by using high performance plastics. The use of high performance plastics also provides minimal corrosion which allows longer vehicle life, flexibility in integrating components, safety and recyclability [7,8]. Commonly used high performance plastics in automotive industry which constitutes the 66% of the total high performance plastics used in a car are polypropylene (32%), polyurethane (17%) and PVC (16%) [8].

Being the most commonly used high performance plastics according to this ratio, the studied polymer in this project is polypropylene (PP) which is a thermoplastic polymer and used in various components of vehicles such as automotive bumpers. Due to the low weight high strength demand of automotive industry, this project aims to obtain low density and high modulus polymers without losing the tensile strength. Thus the main task of the project is to replace the inorganic material talc that causes density increase in PP with low density and sustainable organic fillers. Wood fiber offers some significant advantages like low density, high specific properties, non-abrasive to processing equipment, low cost and most importantly biodegradability. Even it causes poor mechanical properties due to its poor interfacial adhesion between polar-hydrophilic wood–fibers and nonpolar-hydrophobic plastics (PP), this problem can be improved by using compatibilizers or coupling agents [1]. So being the most appropriate organic filler for this task, the wood fiber (WF) is replaced with the talc by introducing it into PP by using ultrahigh speed thermo kinetic mixing method. This method provides a compound of polymer and the organic filler by mixing them at high temperatures subjected to high turbulence as a result from impact against the chamber wall, mixing blades and themselves [3].

EXPERIMENTAL PROCESS

In the scope of the project a series of experiments are performed by incorporating wood fiber into polymer in different loads reinforced with compatibilizer. Wood fibers are named Kastamonu wood fiber which is provided by Kastamonu Entegre Company. After addition process their mechanical tests besides the density measurements are carried out. Also TGA, DSC and SEM tests are studied in order to examine thermal behavior and homogeneity of the mixture. The obtained results are compared with the pure PP.

During the experimental process, a series of test samples are prepared by using totally 100 grams of samples. Wood fiber is incorporated into PP in different loads such as 30% and 40%. Also, by adding 2% compatibilizer totally 4 sets of experiments are prepared. Details of test samples are given in Table 1.

Table 1. Number of experiment set with experiment contents

Number of experiment sets	Content of the experiment set		
	Amount PP (%)	Amount Wood Fiber (%)	Amount Compatibilizer (%)
Set 1	70	30	-
Set 2	60	40	-
Set 3	68	30	2
Set 4	58	40	2

Prepared set mixtures are fed into the thermo kinetic mixing machine (Gelimat, Dusatec) and the materials are compounded with reached temperature around 200° C, at 5600 rpm and during around 100 seconds. After the mixing process the compounded samples are obtained.

These compounds are left to cool down and then are copped into little pieces. The copped material is treated in injection molding machines in order to have mechanical test bars. Finally the mechanical and thermal tests besides the density measurement of the samples are done and compared with the pure PP.

RESULTS & DISCUSSION

The mechanical tests and density measurements are done for the 4 sets of experiment samples. For the mechanical tests (UTM, Instron, ISO 527-2), tensile strength at break and modulus values are examined and compared with 30% and 40% talc added PP.

Graph 1. Tensile Modulus values of compounds compared with 30% and 40% talc added PP (b)

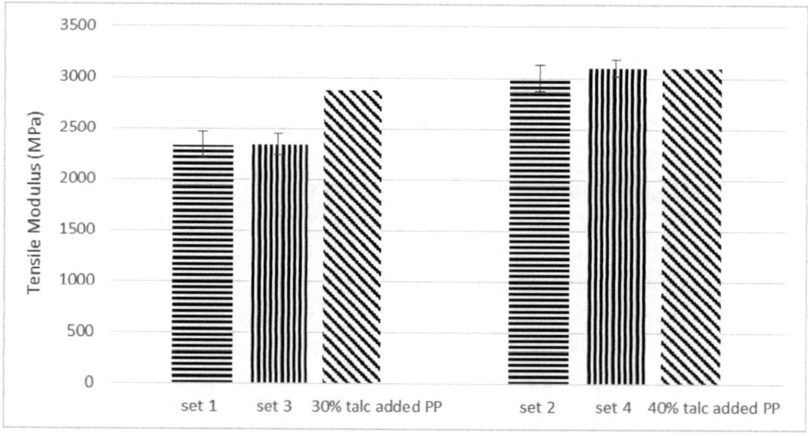

Graph 2. Tensile strength at break values of compounds compared with 30% and 40% talc added PP

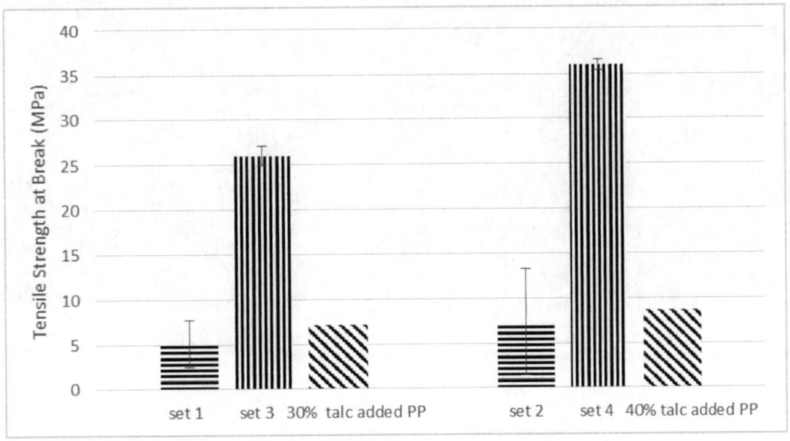

Comparing the 30% and 40% talc added PP mechanical test results with wood fiber added PP mechanical test results, one can observe that the highest increase in tensile strength at break is obtained in set 4 which is composed of 58% PP + 40% Kastamonu wood fiber + 2% compatibilizer mixture. Increasing from 8.6 MPa to 36 MPa, set 4 has increased 4 times of the tensile strength at break value of 40% talc added PP. Meanwhile there is no change in tensile modulus value and for set 4 managed to reach the tensile modulus of 40% talc added PP.

Also density measurements are done in order to observe the density decrease between talc added PP and wood fiber added.

Graph 3. Density values of compounds compared with 30% and 40% talc added PP

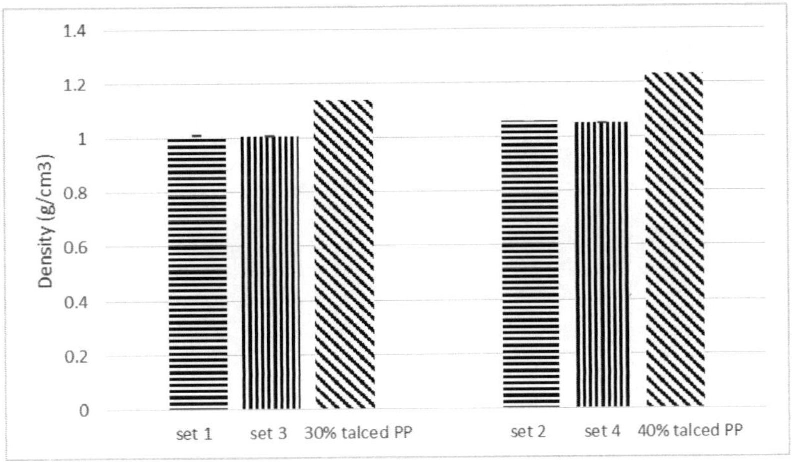

According to graph 3, one can observe that for experiment sets 1 and 3 which contains 30% wood fiber have approximately 11% density decrease while experiment sets 2 and 4 which contains 40% wood fiber have approximately 14% density reduction.

Additionally SEM (Zeiss LEO Supra 35VP SEMFEG) is studied for wood fiber added PP with and without compatibilizer addition, in order to see the effect of compatibilizer on the homogeneity of the compound. When the breaking surface of the mechanical test bars are studied under SEM, it is observed that PP with compatibilizer addition shows a more homogeneous structure.

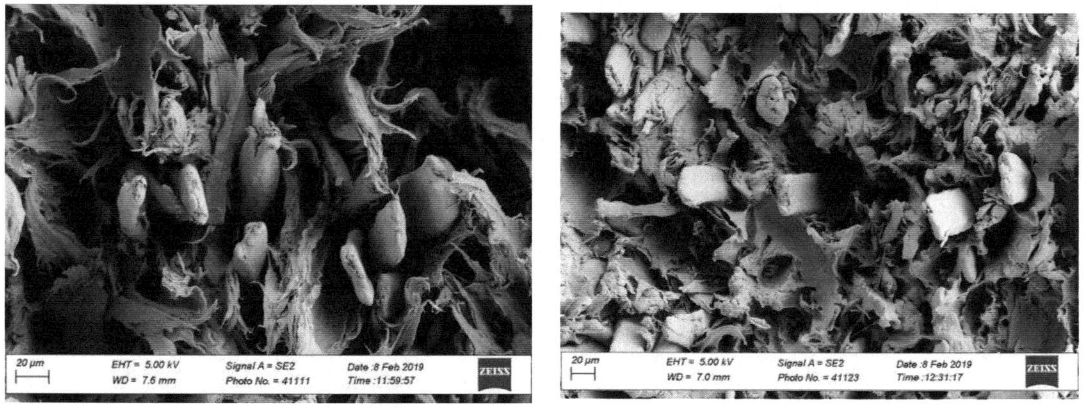

(a) (b)

Figure 2. Comparison of compatibilizer effect between PP with and without compatibilizer. (a) 40% Kastamonu wood fiber added PP without compatibilizer with 1 kx magnification (b) 40% Kastamonu wood fiber added PP with compatibilizer wtih 700x magnification

Thermal properties of wood fiber was studied by thermo gravimetric analysis instrument (TGA; DTG60H, Schimadzu) up to 800° C with heating rate 10 K/min. Due to the test results of Kastamonu wood fiber, shown in fig. 3, one can observe a mass decrease around 100° C which corresponds to evaporation of the water inside the fiber. DTG curve shows two peaks at 300° C and 320° C which might give information about decomposition temperatures of two different wood type in the filler [4]. Since there is not a significant mass loss between 148° – 226° C decomposition of wood fiber cannot take place therefore, temperature range can be considered as working temperature for thermokinetic mixing method.

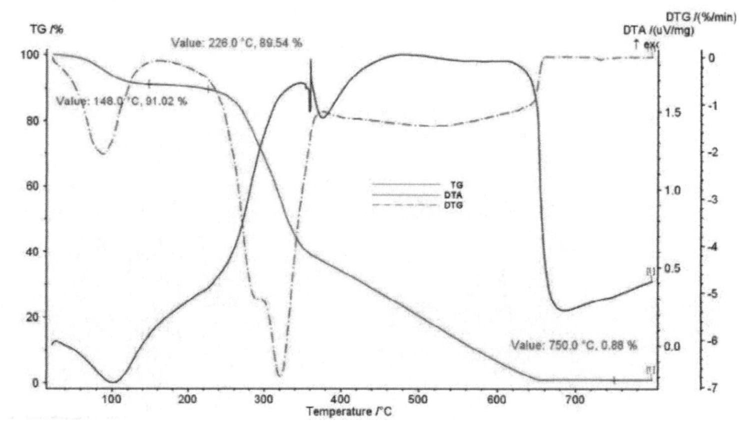

Figure 3. TGA results of Kastamonu wood fiber

Also thermal properties of wood fiber added PP are studied by Differential Scanning Calorimetry (DSC; TA2000, TA instruments). One can observe in table 2 that the increase in wood fiber addition between set 1 (30% WF) and set 2 (40% WF) increases the crystallinity of compounds by 11%. Also increase in crystallinity by 8% due to addition of compatibilizer can be observed by comparing set 1 and set 3 (30% + 2% compatibilizer). Although these increases are not very significant, it can indicate that fibers are acting as nucleating agent [5]. The increase in crystallinity has a contribution in increase of the strength of the PP [6]. One can observe the increasing crystallinity from set 1 to set 2 and also increasing tensile strength from graph 2. The crystallinity increase is observed both for talc added and wood fiber added PP.

Table 2. Crystallisation of wood fiber added PP compounds compared to pure PP

	T_c (°C)	ΔH_c (J/g)	$\Delta H_{normalized}$	X_c (% increased)
Pure PP	171. 32	65.16	-	-
Set 1	168.71	56.07	81	24
Set 2	168.23	44.44	88	35
Set 3	167.28	51.19	86	32
Set 4	167.76	44.58	85	30
30% talc added PP	168.97	52.50	87	33
40% talc added PP	169.85	39.30	99	52

CONCLUSION

In this project, the incorporation of wood fiber into PP is carried out and aimed to decrease the density of compound while increasing the tensile modulus without losing mechanical strength. It is observed that 40% wood fiber addition can be considered as a replaceable filler for inorganic talc material in PP. It increased the tensile strength value up to 36 MPa from 8.6 MPa. Also density measurements supported that wood fiber addition decreased density by 14% while the polymer keeps its original modulus. The increased homogeneity effect of compatibilizer is observed from the SEM images and also proved by the tensile strength increase. Furthermore the increasing crystallinity trend of both talc added PP and wood fiber added PP proves that wood fiber is an organic filler that can replace inorganic talc.

REFERENCES

1. A. Karmarkar, S. S. Chauhan, J. M. Modak, M. Chanda, "Mechanical properties of wood–fiber reinforced polypropylene composites: Effect of a novel compatibilizer with isocyanate functional group", (2006), pp.228-233, available at https://www.sciencedirect.com/science/article/pii/S1359835X06001722
2. "Automotive Industry Seeks Light-Weight and High-Performance Materials", (2012), available at https://www.icis.com/explore/resources/news/2012/07/27/9581541/automotive-industry-seeks-light-weight-and-high-performance-materials/
3. Gelimat™ Technology Ultrahigh-Speed Thermokinetic Mixing,Compounding & Fluxing, available at https://www.academia.edu/34219550/Gelimat_Technology_Ultrahigh-Speed_Thermokinetic_Mixing_Compounding_and_Fluxing
4. Poletto M., Zattera A. J., Forte M. M. C., Santana R. M. C., Thermal decomposition of wood: Influence of wood components and cellulose crystallite size (2012), pp. 149-151, available at https://www.sciencedirect.com/science/article/pii/S0960852411017330
5. Reixach, R., Puig, J., Méndez, J. A., Gironès, J., Espinach, F. X., Arbat, G., Mutjé, P., "Orange Wood Fiber Reinforced Polypropylene Composites: Thermal Properties", (2015), pp. 2156-2163 available at https://www.researchgate.net/publication/276426380_Orange_Wood_Fiber_Reinforced_Polypropylene_Composites_Thermal_Properties
6. Sharma, R., Maiti, S.N., "Effects of crystallinity of polypropylene (PP) on the mechanical properties of PP/styrene-ethylenebutylene-styrene-g-maleic anhydride (SEBS-g-MA)/teak wood flour (TWF) composites", (2015), pp. 631
7. Szeteiová, K., "AUTOMOTIVE MATERIALS PLASTICS IN AUTOMOTIVE MARKETS TODAY", available at https://pdfs.semanticscholar.org/e2d3/16ca62ec296bfc66ef3f2f5a4daf974bd65c.pdf
8. 13 high performance plastics used in the automotive industry, available at http://www.craftechind.com/13-high-performance-plastics-used-in-the-automotive-industry/

The Role of Nanoparticles Localization on Morphology and its Influence on Fracture Toughness Enhancement of Polypropylene/Ethylene-Octene-Copolymer Blend

Fatemeh Morshedi[a] and Hossein Nazockdast[b]

[a]Polymer Engineering Department, Amirkabir University of Techollogy, Tehran, Iran
[*) Corresponding Auther:nazdast@aut.ac.ir

Abstract. Toughening enhancement has always been an important target of polymer blending. This work was to explore the localization effect of carbon nanotubes (CNTs) on morphology and mechanical properties of PP/EOC blends. The melt viscosity measurement were used to study the effect of the mixing sequence on CNTs localization. The results of frequency sweep experiments showed a pronounce enhancement in storage modulus together with viscosity upturn for the nanocomposites, which were prepared by PP based master batch feeding method. Whereas, no enhancement in storage modulus (G') was observed for the nanocomposites, prepared by direct or sequential feeding. In addition, for the nanocomposites prepared through master batch feeding, CNTs located in PP matrix while for the samples prepared by later two methods, the nanoparticles were accumulated in EOC phase or interface. The results of mechanical test showed the greater improvement in toughness for the samples in which nanotubes were located in the PP matrix as a result of CNTs induced reduction in EOC particle size as evidenced by scanning electron microscope (SEM) micrographs.

Keywords: Polypropylene, Toughening, Carbon nanotube, master batch.

INTRODUCTION

Introducing elastomer like ethylene-octene copolymer (EOC) into Polypropylene has demonstrated improvement in impact strength especially in low temperature and/or high strain rates [1]. In recent years, Carbon Nanotubes (CNTs) has been widely applied in polymer blends to improve mechanical and electrical properties [2-4] due to their outstanding properties, including high aspect ratio and high electrical conductivity[5]. Toughening effect of CNTs in these immiscible systems are greatly influenced by the localization of CNTs, their dispersion and distribution [6]. CNTs induced toughening mechanisms such as bridge effect [7], forming percolation network of CNTs [8], interfacial adhesion enhancement [4] and increasing degree of crystallinity [9] have been proposed which depends on localization of nanoparticle are different.

In this work, localization of the CNTs by sequence mixing including master batch, direct and sequential feeding as kinetics factor are studied. Then different contents of CNTs are introduced in immiscible blends in adequate mixing method and its effect is studied. The effects of localization of CNTs and different contents of CNTs are investigated on morphological features, fracture toughness and toughening mechanisms.

EXPERIMENT

Materials

Polypropylene (HP552 R with melt flow 25 gr/10 min) was purchased from Jam Petrochemical. EOC (trade name POE LC670) was obtained from Korea synthetic rubber. Multiwalled CNTs (with a trade name Nanoclay NC7000) were purchased from Nanocyl. CNTs had average diameter and average length of about 9.5 and 1.5 micrometer, respectively. In order to reach better dispersion of CNTs in PP matrix, Polypropylene grafted maleic anhydride which was purchased from Honam Petrochemical was used.

Characterization

In order to investigate dispersion of CNTs and rheological response of blend nanocomposites, linear viscoelastic measurements were performed using a rotational and oscillatory rheometer (Paar Physica USD 200).The melt temperature of sample was set at 190℃ and the test was conducted within the linear viscoelastic strain range under the measuring frequency range of 0.01-1000 rad/s.

Notched Izod impact strength was measured using a U-F-impact tester (Ueshima seisakusho Ltd, Japan). Samples were prepared by compression molding at 220℃ and with dimensions of 64*13*3 mm.

SEM was used to characterize sample morphology and impact-fractured surface morphologies. The samples were first cryogenically fractured in liquid nitrogen, and etched in n-Heptane at 50℃ for 5 h and then were dried at 70℃ for 3h.

RESULTS AND DISCUSSIONS

Selective localization of CNTs

The preferential location of Carbon Nanotubes in the immiscible polymer blends is estimated by thermodynamic and kinetic factors. Wetting parameter, ω_a is calculated according to the following equation:

$$w_a = \cos\theta = \frac{\gamma_{PP/CNT} - \gamma_{EOC/CNT}}{\gamma_{PP/EOC}} \tag{1}$$

Where $\gamma_{PP/CNT}$, $\gamma_{EOC/CNT}$ and $\gamma_{PP/EOC}$ are the interfacial energies between CNT and PP, CNT and EOC, and PP and EOC. If $w_a > 1$, CNTs preferentially locate in EOC and if $w_a < -1$, CNTs preferentially locate in PP else if $-1 < w_a < 1$, CNTs locate at the interface between the two phases. The interfacial energy can be calculated according to the Harmonic-mean equation and Geometric-mean equation:
Harmonic- mean equation:

$$\gamma_{12} = \gamma_1 + \gamma_2 - 4\left(\frac{\gamma_1^d \gamma_2^d}{\gamma_1^d + \gamma_2^d} + \frac{\gamma_1^p \gamma_2^p}{\gamma_1^p + \gamma_2^p}\right) \tag{2}$$

And Geometric-mean equation:

$$\gamma_{12} = \gamma_1 + \gamma_2 - 2\left(\sqrt{\gamma_1^d \gamma_2^d} + \sqrt{\gamma_1^p \gamma_2^p}\right) \tag{3}$$

Where γ_1 and γ_2 are the surface free energy of the blend component 1 and 2 respectively, γ_1^d and γ_2^d are the dispersion parts of the surface energies of blend component 1 and 2 respectively and γ_1^p and γ_2^p are the polar parts of the surface energies of blend component 1 and 2 respectively. The calculated values of surface free energy of blend components and CNT at 190℃ are given in **TABLE i**.

TABLE i). the surface free energy and polarity of blend components and CNT at 190℃

Component	γ	γ^d	γ^p
PP	21.8	0.5	21.2
EOC	17.5	1.9	15.6
CNT	27.8	10.2	17.6

By calculating wetting parameter according to Harmonic-mean equation and Geometric-mean equation, the values of ω_a are shown in **TABLE *ii***.

TABLE ii. wetting parameter as calculated using Harmonic-mean and Geometric-mean equations

Calculation method	$\gamma_{PP/CNT}$	$\gamma_{EOC/CNT}$	$\gamma_{APP/EOC}$	**Wetting parameter (ω_a)**
Harmonic-mean	5.814	9.227	1.769	1.929
Geometric-mean	3.3357	6.4507	0.979	3.161

The value of wetting parameter obtained from both methods is more than 1 which indicates CNT tends to localize in EOC phase. On the other hand, viscosity of EOC is more than PP whereas its melting point is less than PP. Therefore these two factors are in competition with each other in carbon nanotubes localization. So in this work nano localization is investigated by using different feeding methods as another kinetic factors.

Rheological properties

The influence of different feeding methods on dispersion of nanoparticles was studied. The complex viscosity (μ^*) and storage modulus (G') of samples with different feeding methods in comparison with samples without CNTs are shown in **FIGURE 1**. As can be seen in low frequencies, the nanocomposite with master batch feeding has a plateau and upturn in storage modulus and complex viscosity respectively. However, in sequential or direct methods no apparent enhancement are observed. This indicates that in sample prepared by PP based master batch feeding method a percolated network structure of CNTs in the matrix of PP are formed. Whereas based on the thermodynamic factor in other feeding methods, more CNTs are presented in EOC phase.

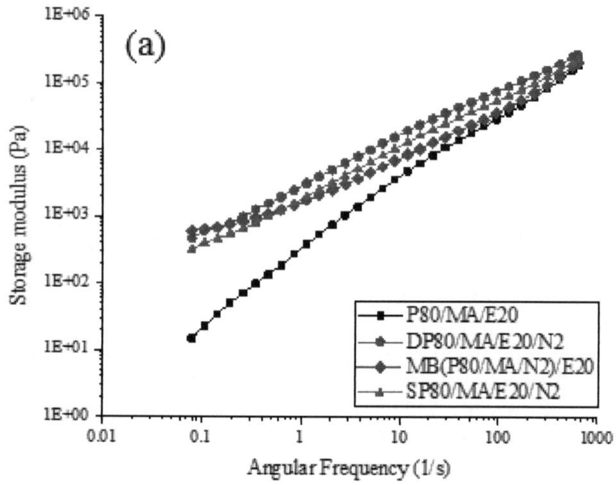

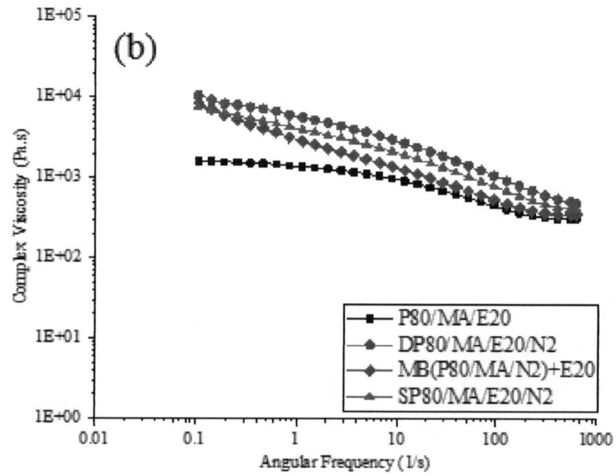

FIGURE 1. Rheological properties of nanocomposite-blends of different methods of feeding. (a) storage modulus, (b) complex viscosity

Mechanical properties

The toughening effect of CNTs and different feeding methods in the blend nanocomposites specimens are measured by notched Izod impact test at room temperature (23°C). As shown in **FIGURE** *2*, addition of CNTs achieve the impact strength of PP/EOC blend. However, in the E20+MB (P80/MA/N2) sample, the degree of the enhancement in impact strength is more than other methods. It can be due to the formation of CNTs network structure. The presence of most nanoparticles in PP matrix and less of them in the interface facilitates the stress transfer between components under loading condition. Also existing results in the morphological examination in next section show that the selective localization of CNTs in the PP component decreases the size of EOC particles.

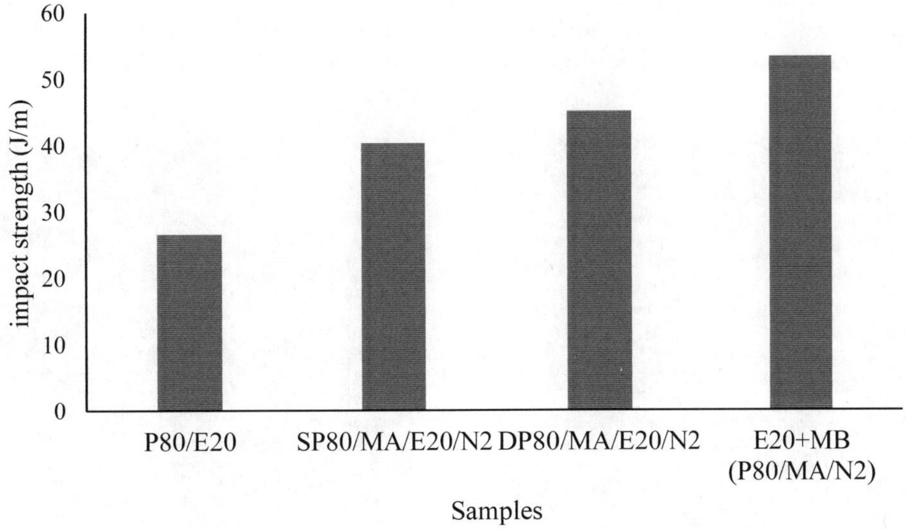

FIGURE 2. Comparison of the Izod impact strength of the blend and samples with different methods of feeding at room temperature

Blend nanocomposites morphology and dispersion of CNTs

In order to investigate morphology and better understanding toughening mechanisms of blend nanocomposite prepared by master batch feeding compared with PP/EOC blend, samples were characterized using SEM. In **FIGURE 3**, the black holes represent the EOC particles which were removed by n-Heptane. As can be seen the diameter of EOC particles in samples prepared by master batch feeding (**FIGURE 3-b**) are decreased.

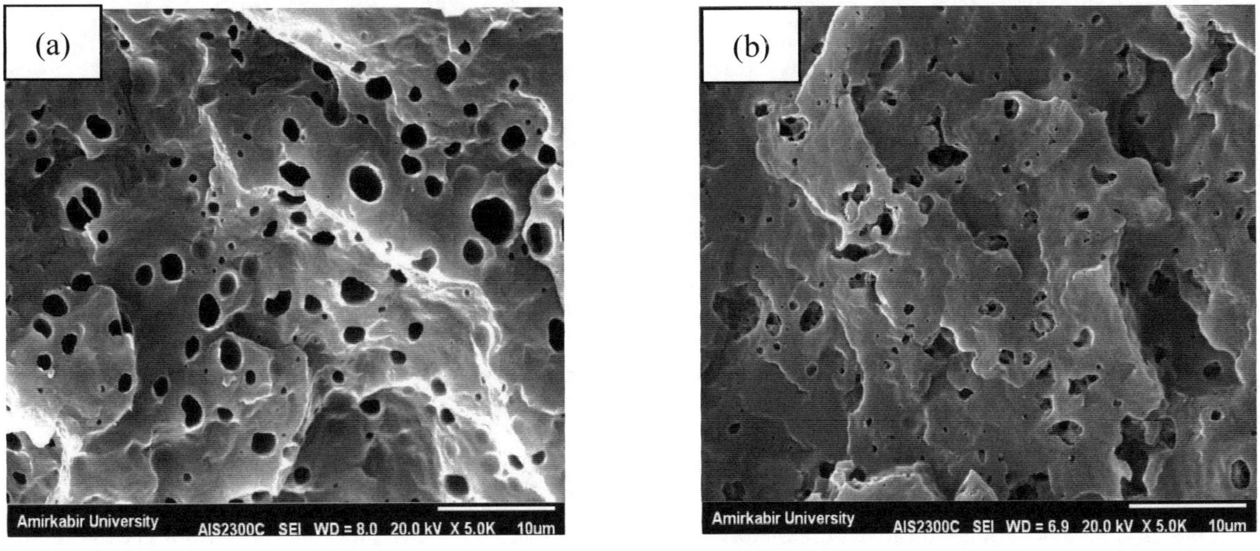

FIGURE 3. SEM images showing the morphologies of (a) P80/E20 blend, (b) MB (P80/MA/N2) + E20 blend nanocomposite

According to **FIGURE 4-a**, there is no obvious deformation of PP matrix and major mechanism of toughening is considered cavitation of rubber particles. But adding CNTs to the blend and their localization in the PP matrix decrease the diameter of EOC particles and the matrix ligament thickness. CNTs can exhibit the bridge effect and lead to stress transfer between components. Also CNTs facilitate the superposition of the stress field around EOC particles under the impact loading condition, especially when CNTs form the network structure. Therefore apparent plastic deformation of PP matrix around the rubber particles is observed for the MB (P80/MA/N2) + E20 sample (shown by arrows).

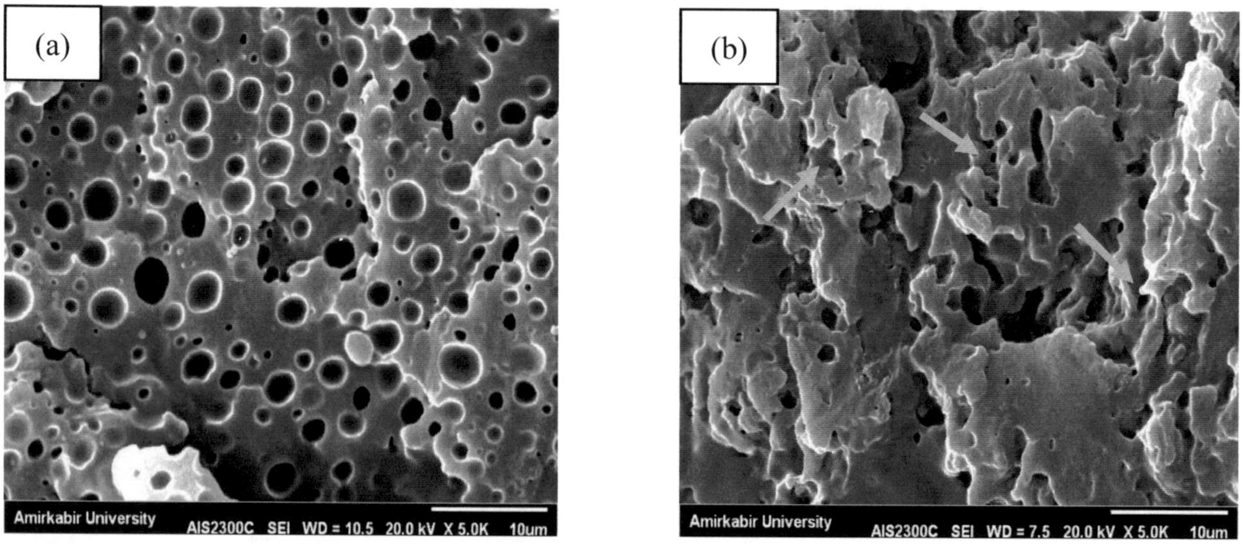

FIGURE 4. SEM images of impact fracture surface morphologies of (a) P80/E20 blend (b) MB (P80/MA/N2) + E20 blend nanocomposite

To understand the toughening mechanism better, more visualized schematic representations for sample obtained from master batch feeding are shown in **FIGURE 5a-b**.

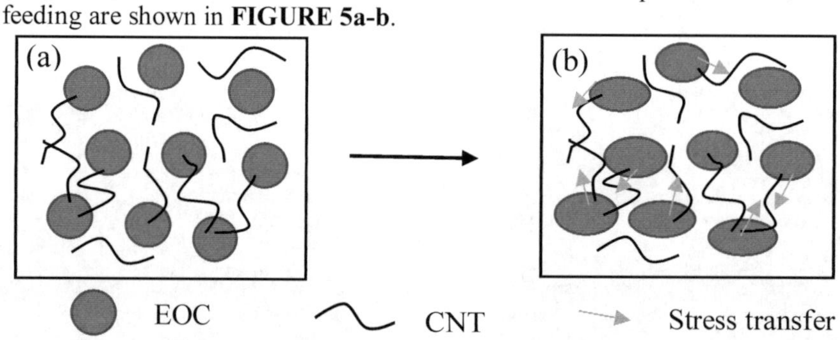

FIGURE 5.schematic representation showing the changes of morphologies under the impact load (a) the dispersion of EOC and CNT (b) plastic deformation of EOC component and stress transfer by CNTs (specified by arrows).

It represents the dispersion of EOC particles and CNTs (**FIGURE 5-a**) and stress transfer by CNTs (**FIGURE 5-b**).

CONCLUSION

In this study, the effect of different feeding methods on CNT localization and mechanical properties was explored by rheological and impact measurements. Frequency sweep test result revealed that in the master batch feeding method CNTs localized in PP matrix and formed a solid-like network structure leading to more enhancement in impact strength. SEM images showed that presence of CNTs in matrix and interface decreased the particles diameter of EOC and the ligament matrix thickness and leading to stress transfer between components.

REFERENCES

1. M. P. Brown and K. Austin, *The New Physique*, Publisher City: Publisher Name, 2005, pp. 25-30.
2. M. P. Brown and K. Austin, *Appl. Phys. Letters* **85**, 2503-2504 (2004).
3. R. T. Wang, “Title of Chapter,” in *Classic Physiques,* edited by R. B. Hamil, Publisher City: Publisher Name, 1999, pp. 212-213.
4. C. D. Smith and E. F. Jones, “Load-Cycling in Cubic Press” in *Shock Compression of Condensed Matter-2001*, edited by M. D. Furnish et al., AIP Conference Proceedings 620, American Institute of Physics, Melville, NY, 2002, pp. 651-654.
5. B. R. Jackson and T. Pitman, U.S. Patent No. 6,345,224 (8 July 2004)
6. D. L. Davids, "Recovery Effects in Binary Aluminum Alloys", Ph.D. Thesis, Harvard University, 1998.
7. R. C. Mikkelson (private communication).

The in situ micro/nano-fibrillation concept as an efficient route for the enhancement of the processability and mechanical properties of bio-based polymeric blown films

T. Dadouche, M. Yousfi, C. Samuel, J. Soulestin, M-F. Lacrampe

IMT Lille Douai, Institut Mines-Télecom, Department of Polymer and Composite Technology & Mechanical Engineering (TPCIM), 941 rue Charles Bourseul, CS 10838, F-59508 Douai, France

Abstract. Multiphase biosourced nano-fibrillar polymer/polymer composites based on poly(butylene succinate-co-adipate)/poly(amide-11) blends (PBSA/PA11, 85/15 w/w) were successfully produced via a fast and simple melt extrusion process followed by a tubular film blowing process. The effect of the take-up ratio on the morphological and rheological properties of resulted nano-fibrillar blown films was highlighted. The formation of stable nano-fibrillar structures is deeply discussed via rheological investigation to highlight the contribution of viscosity ratio, elasticity ratio and interfacial tension. SEM analyses indicated that increasing the take-up ratio during the film blowing leads to stretch the microfibrils and to generate fine nano-fibrils domains. Uniaxial extensional viscosity measurements showed a significant increase in the strain hardening intensity of the blends exhibiting nano-fibrillar structure compared to blends with nodular morphology. Maintaining the fibril morphology seems to stabilize the bubble during film blowing process without meanwhile sacrificing the mechanical properties of thin films obtained. The concept of in situ micro/nano-fibrillation is an efficient, cost-effective, and environmentally friendly technique for the enhancement of the processability and tensile properties of bio-based polymeric blown films and could find interesting applications in the design of novel bio-based polymer composites thin films for packaging industry.

Keywords: Blow molding, morphology development.
PACS: Replace this text with PACS numbers; choose from this list: http://www.aip.org/pacs/index.html

INTRODUCTION

Film blowing is one of the most important industrial processes used to produce thin films dedicated for packaging industry [1]. Blending polymers is an effective route to improve the bulk physical characteristics of bio-based polymers. But most of polymers used are not miscible, and in these blends, physical properties can be optimized by controlling the quality of the dispersion, which in turn depends on the rheological properties as well as on the mode of deformation involved during the different processing steps. In particular, generating fibrillar morphology from droplets can be done in very short time and depending on which some specific rheological properties are satisfied or not [2]. According to several literature reports [3, 4], in the particular case of fiberlike morphology, the droplet-fibril transition could be controlled by the viscosity ratio (k) between the dispersed phase and the matrix (a more viscous matrix promotes the breakdown of droplets [5, 6]), the elasticity ratio (k') between the dispersed phase and the matrix (a more elastic matrix promotes deformation and extension of the nodules into fibrils [6, 7]) and a compromise between deformation (shear or elongation) and relaxation phenomena should be achieved to insure a stable filamentous morphology [6, 8]. This work presents relationships between the morphology, rheological behavior in extensional flow of bio-based immiscible polymer blends and their stability during film blowing process. Presented here are binary blends of PBSA and PA11. PBSA is a promising biodegradable aliphatic polyester which was chosen to form the matrix phase. A strategy was adopted in the present study consisting in the in-situ generation, during the manufacturing process, of well-aligned flexible PA11 fibers having strong interfacial affinity with the PBSA matrix. Correlations between the stability of PBSA/PA11 blends during the extrusion film blowing, morphological evolution, rheological properties of blend components in extensional flow are discussed through this report.

EXPERIMENTAL SECTION AND MAIN RESULTS

The binary blends used in this study were prepared from poly(butylene succinate-co-adipate) (PBSA Bionolle 3001 Showa Denko, Japan), and three grades of biobased polyamide 11 with different viscosities were chosen for the dispersed phase.

TABLE 1: Characteristics of the polymers used in this study.

Name	Supplier	Grade	T_m (°C)	MVR* (cm^3/10min)
PBSA	Showa Denko	Bionolle 3001	96	-
PA11-LV	Arkema	LMFO	189	39
PA11-MV	Arkema	BMNO	189	36
PA11-HV	Arkema	BESVO	189	6

***:** *Melt volume-flow rate according to ISO 1133 (235 °C, 2.16 Kg)*

FILMS PREPARATION

The films were prepared in two extrusion stages. In a first step, the PBSA/PA11 (85/15 w/w) blends were made by melt extrusion using a co-rotating twin screw extruder (ZSK 26, Coperion, Germany) with a screw diameter of 25mm and a length to diameter ratio of 40:1. Then, blowing was achieved using a single screw extruder (diameter: 19 mm, L/D ratio: 25) (Rheomex 254, Haake Buchler, Germany) coupled to an annular 25 mm diameter blow die regulated by a Rheocord System 40 device (Haake Buchler, Germany). The ratio of the take-up speed to the extrudate polymer velocity is called the take-up ratio (TUR), and the ratio of the final bubble diameter to die diameter is called the blow-up ratio (BUR). For all blends, the blow-up ratio is maintained constant and different take-up ratios are fixed (table 2). The extrusion parameters were: screw speed = 30 rpm and the thermal regulation was made over 4 zones with a temperature profile of 190, 230, 230 and 210 °C in the die. The height of the freeze line is kept constant around 40 mm. The mass flow rate « Q_m » of the melt was determined by collecting the extrudate for a predetermined time interval. Then the volumetric flow rate « Q_v » was deduced using a melt density of 0.90 g cm^{-3} for PA11 [6] and 0.95 g cm^{-3} for PBSA and PBSA/PA11 blends [9]. Having measured the bubble radii « r » at the axial distance « z », film thickness « h », and the angle « θ » as function of machine direction « z », the elongation rate acting in the direction of flow (i.e., machine direction « z ») was determined at different values of « z » using equation of Park and Han [10].

$$\dot{\varepsilon} = \frac{Q_v \cos\theta}{\pi\, r\, h^2}\,\frac{dh}{dz}$$

Note that « θ » is the angle between the bubble surface and the vertical direction « z » (i.e., tan θ = dr/dz).

TABLE 2: Parameters used to prepare PBSA/PA11 blown films.

Blend	Film	BUR	TUR	Take-up speed ($cm.min^{-1}$)	Film thickness (μm)
PBSA/PA11-LV	Film 1	3.6	4.6	200	50
PBSA/PA11-LV	Film 2	3.6	6.9	300	30
PBSA/PA11-LV	Film 3	3.6	9.2	400	15
PBSA/PA11-MV	**Film 1**	**3.6**	**4.6**	**200**	**50**
PBSA/PA11-MV	**Film 2**	**3.6**	**6.9**	**300**	**30**
PBSA/PA11-MV	**Film 3**	**3.6**	**9.2**	**400**	**15**
PBSA/PA11-HV	Film 1	3.6	4.6	200	50
Neat PBSA	**Film 1**	**3.6**	**4.6**	**200**	**50**
Neat PBSA	**Film 2**	**3.6**	**6.9**	**300**	**30**
Neat PBSA	**Film 3**	**3.6**	**9.2**	**400**	**15**

MORPHOLOGYCAL ANALYSES

Figure 1 shows SEM images of the morphology of PBSA/PA11-LV blends at the exit of the annular die and in the blown films for different take-up ratios. The extrudate morphology (before stretching and inflation) appears nodular (Figure 1(a)) while stretched blown films show a fine fibrillar morphology (Figures 1(b,c,d))). It's observed that the diameter of the fibrils in the films decreases when the take-up ratio increases (averaged diameter of 0.5 μm for

PBSA/PA11-LV films prepared with the highest take-up ratio). The same tendency is observed for PBSA/PA11-MV blend, but the transition from nodular (Figure 1 a (extrudate) and b) to fibrillary (Figure 1 c and d) morphology requires imposing higher take-up ratios. Her again, increasing the take-up ratio induces a decrease in the dispersed phase size. This fine morphology of PA11 in PBSA matrix can be explained by their lowest interfacial tension as reported by Yousfi et al [6].

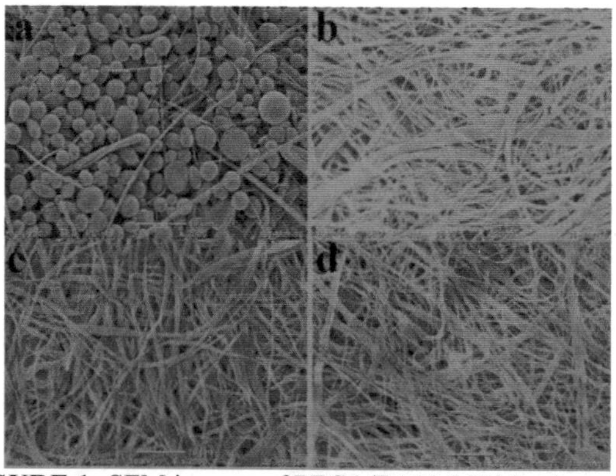

FIGURE 1: SEM images of PBSA/PA11-LV blends (a) The not inflated extrudate, b) Film 1 (thickness = 50 μm), c) Film 2 (thickness = 30 μm), d) Film 3 (thickness = 15 μm).

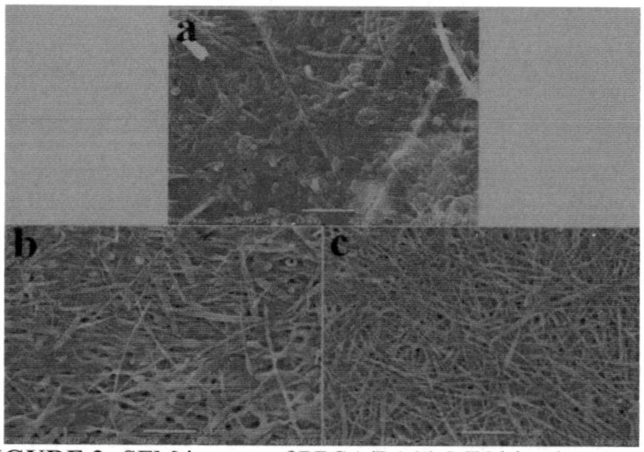

FIGURE 2: SEM images of PBSA/PA11-MV blends a) Film 1 (thickness = 50 μm), b) Film 2 (thickness = 30 μm), c) Film 3 (thickness = 15 μm).

It was not possible to prepare films from PBSA/PA11-HV blend with high take-up ratio, due to the bubble instability (Figure 4). This one is mainly helical with a slight draw resonance (the diameter of the bubble is not stable along the bubble). The only film prepared has a thickness of 50 μm (TUR = 4.6, BUR = 3.6).

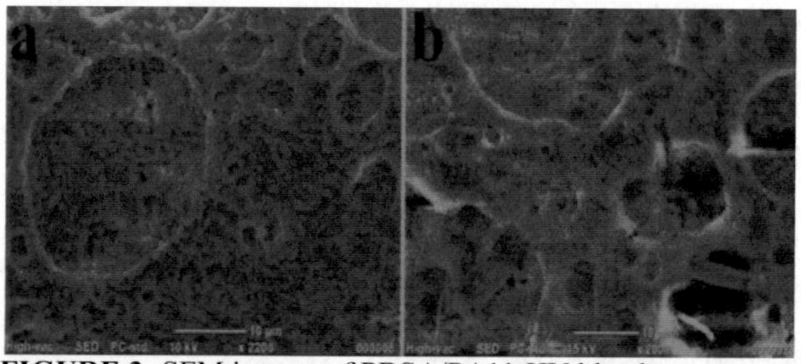

FIGURE 3: SEM images of PBSA/PA11-HV blends, a) the not inflated extrudate, b) Film 1 (thickness = 50 μm).

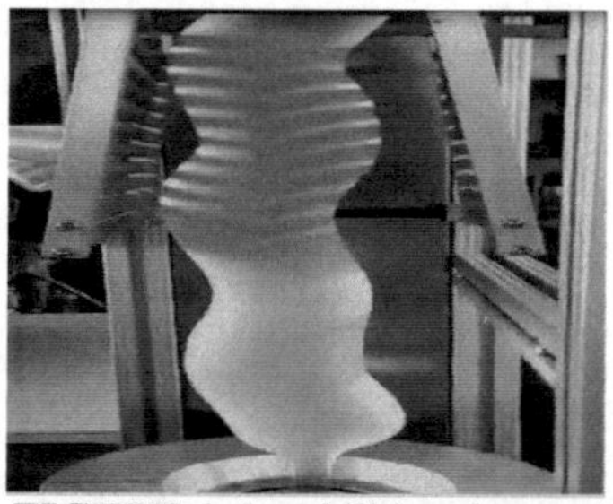

FIGURE 4: Instabilities observed in high take-up ratios during film blowing of PBSA/PA11-HV.

The morphology of the not elongated extrudate is nodular for all the PBSA/PA11 blends prepared in this study, and the extensional flow applied at the exit of the annular die seems to be crucial for the morphology development of the blown films. For this purpose, it was decided to study the morphology evolution as a function of the elongation rate. In fact, the elongation rate changes along the bubble (from the exit of the annular die to the freeze line). Here, the elongation rate as a function of the distance from the die was calculated using Han and Park model [10] as mentioned above. It is well known that the elongation rate in the transverse direction is negligible compared the one in the machine direction one, thus, using Han and Park model, one can infer the elongation rate from the evolution of the thickness of the film h and the radius of the bubble r as function of the distance from the die exit z. The calculated elongation rate profile in the machine direction as a function of the distance from the die exit for PBSA/PA11-LV blown films using the highest TUR value (TUR = 9.2 and BUR = 3.6) is presented in Figure 5. Thus, the maximum of the strain rate corresponds to a Z distance from the die of 20 mm which matches with the half of the freeze line position (40 mm).

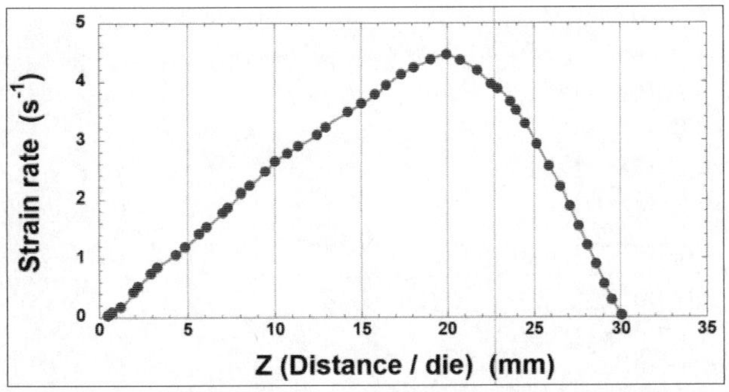

FIGURE 5: Strain rate profile in the machine direction for PBSA/PA11-LV blown film 1 (BUR = 3.6 and TUR = 9.2).

Figure 6 presents SEM pictures in different distances from the die exit for PBSA/PA11-LV bubble which allows us to determine the morphology development as a function of the applied elongation rate. It would be noted that in the case of PBSA/PA11-LV blends, the droplets can be deformed into fibrils by applying a low elongation rate (an elongation rate of 1.2 s^{-1} was enough to observe fibrils in SEM images). Increasing the elongation rate allows to deform PA11 (PA11-LV) droplets into nano-fibrils. However, the same extensional flow was applied in the case of PBSA/PA11-HV blends and the PA11-HV droplets could not be deformed, this indicates that the rheological properties of PA11 dispersed phase and PBSA matrix should be suitable in the flow conditions used in order to generate fibril morphology. Correlations between the rheological properties and the morphology development are detailed below.

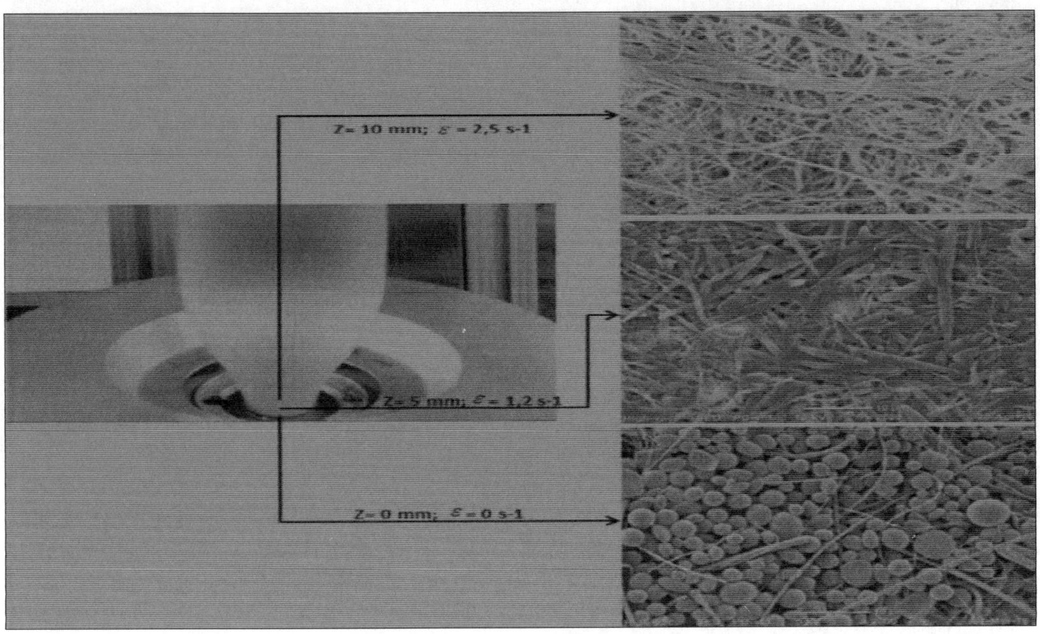

FIGURE 6: SEM images in different distances/die for PBSA/PA11-LV blown film 1 (BUR = 3.6 and TUR = 9.2).

RHEOLOGICAL STUDY

Figure 7 displays the transient extensional viscosity of neat PBSA and PBSA/PA11 blends. It is worth mentioning that it was too difficult to characterize the polyamide 11 elongational viscosity. Indeed, because of its fast creep behaviour, PA11 flows before reaching a stable temperature (190 °C in our case as a measuring temperature). A pronounced strain hardening behavior can be noted for neat PBSA, PBSA/PA11-LV and PBSA/PA11-MV blends against a very weak strain hardening for PBSA/PA11-HV blend. In fact, strain hardening behavior is crucial to promote the stability of a polymer during film blowing process. Therefore, despite the extensional viscosity of PBSA/PA11-HV blend is higher than the other PBSA/PA11 blends ones, PBSA/PA11-HV blend is less stable during film blowing because of its weak strain hardening.

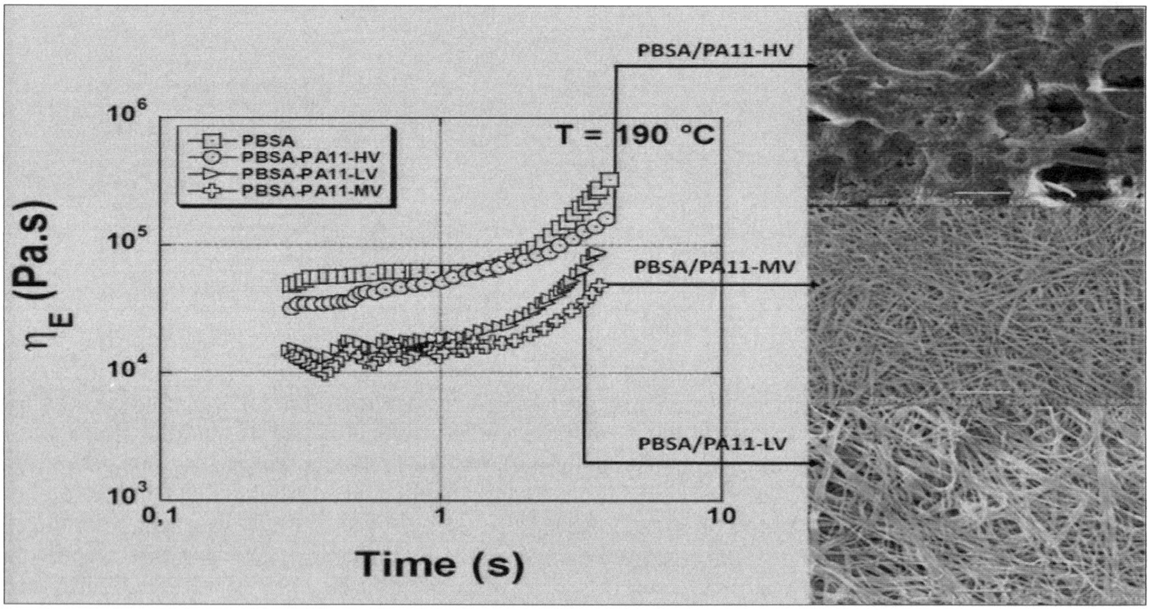

Figure 7: Transient extensional viscosity at 190 °C, $\dot{\varepsilon} = 1\ s^{-1}$ corresponding to generated morphologies during elongation of PBSA/PA11 blends.

The resulted rheological parameters are reported in Table 3. One can observe that all the calculated rheological parameters were suitable to generate fibrils under extensional flow, except the PBSA/PA11-HV blend which the capillary number ratio is less than 4 in the case. That could explain the totally nodular morphology found for PBSA/PA11-HV blown film, and the stable nano-fibrillar morphology observed for the two other blends.

TABLE 3: Calculated rheological parameters of PBSA/PA11 blends corresponding to the extensional flow at the exit of the die.

Blend	K	Ca	Ca_{crit}	Ca/Ca_{crit}
PBSA/PA11-LV	0.2	3.2	0.5	6.4
PBSA/PA11-MV	0.2	3.5	0.5	7.0
PBSA/PA11-HV	1.9	1.8	0.5	3.6

The formation of nano-fibrils generates a pronounced strain hardening behavior and improves the stability of the blend under extensional flow. However, the droplets of PA11-HV grade remains undeformable under the same extensional flow conditions, these droplets are less resistant to melt stretching than fibrils and generate a very weak strain hardening for PBSA/PA11-HV blend resulting in an unstable bubble. These results explain relationships between the morphology development of polymer blend and their stability during film blowing process.

REFERENCES

1. Y. Ikada and H. Tsuji, Macromolecular Rapid Communications 21:117-132 (2000).
2. D. Chomat, J. Soulestin, M. F. Lacrampe, M. Sclavons and P. Krawczak, Journal of Applied Polymer Science 132:(2015).
3. S. Torza, R. Cox and S. Mason, Journal of Colloid and Interface Science 38:395–411 (1972).
4. F. Rumscheidt and S. Mason, Journal of Colloid Science 16:238-261 (1961).
5. L. Levitt, C. W. Macosko and S. D. Pearson, Polymer Engineering and Science 36:1647-1655 (1996).
6. M. Yousfi, T. Dadouche, D. Chomat, C. Samuel, J. Soulestin, M.-F. Lacrampe and P. Krawczak, Rsc Advances 8:22023-22041 (2018).
7. W. Lerdwijitjarud, A. Sirivat and R. G. Larson, Polymer Engineering and Science 42:798-809 (2002).
8. S. Tomotika, Proceedings of the Royal Society A 153:302-318 (1936).
9. Y. Sato, K. Inohara, S. Takishima, H. Yamamoto and M. Takasugi, Polymer Engineering and Science 40:2602-2609 (2000).
10. C. D. Han and J. Y. Park, Journal of Applied Polymer Science 19:3257-3276 (1975).

Flexible Electrospun PET/TiO_2 Nanofibrous Structures For Dye-Sensitized Solar Cell (DSSC) Photoanodes

Hajer Gallah[a,b], Frej Mighri[a,b*], Abdellah Ajji[a,c], and Jayita Bandyopadhyay[d]

[a]*Research center for high performance polymer and composite systems, CREPEC*
[b]*Dep. of Chem. Eng., Laval University, QC, Canada, G1A 0A6*
[c]*Dep. of Chem. Eng., Ecole Polytechnique of Montreal, QC, Canada, H3T 1J4*
[d]*Nat. Centre for Nanostructured Mat., SCIR Materials Sciences and Manufacturing, Pretoria 0001, South Africa*

Abstract. In this study, a solvothermal method was used to synthesize Titanium dioxide (TiO_2) nanoparticles in the presence of oleic acid (OA) and oleylamine (OM) as morphology-directing agents. Functional nanocomposite fibers composed of poly(ethylene terephtalate) (PET) mixed with OA-OM capped TiO_2 nanoparticles were developed by electrospinning process. The different variables that affect the stability of the process were optimized to produce uniform PET/TiO_2 nanofibrous mats without beads. The morphology and thermal stability of PET/TiO_2 nanofibers were investigated by Scanning Electron Microscopy (SEM), Transmission Electron Microscopy (TEM) and Thermo-gravimetric analysis (TGA).

Keywords: Electrospinning, Nancomposites, TiO_2, Nanofibers, Flexible DSSC photoanode

INTRODUCTION

In recent years, there has been a growing interest for the development of flexible Dye-sensitized solar cells (DSSCs) because of their potential application as cost effective, highly flexible, lightweight and scalable photovoltaic devices [1]. The fabrication of the photo-electrode, a key component of the DSSC, includes the sintering step that involves high temperatures (450-500 °C) in order to remove the organic additives and ensure a good electrical connection between the TiO_2 nanoparticles [2]. However, the thermal instability of plastic substrates at high temperature limits the photo-electrode manufacturing temperature to less than 150 °C [3]. Numerous studies have been conducted to develop flexible photo-electrodes at low temperature with major focuses on exploring alternatives to the sintering process and the rigid glass used in the conventional DSSC [4, 5]. However, there are still many challenges to be overcome in terms of DSSC performance and its long-term stability [6]. Approaches using polymer-inorganic composites could offer promising solutions to produce flexible photoelectrodes [7]. These composites are quite attractive because of their interesting properties that combine those of the polymer, such as its lightweight, flexibility and good processability, with those of the inorganic materials, such as their thermal stability and chemical resistance [7]. Among the available methods for synthesizing various composite nanostructures, the electrospinning technique has been proven to be simple, versatile and effective for manufacturing ultrafine fibers from a variety of materials, such as polymers, inorganic and hybrid systems (polymer-inorganic) [8]. Briefly, the mechanism of fiber formation consists of applying a high voltage to create an electric field between a polymeric solution droplet at the tip of a needle and the metallic collector. A fiber jet is ejected as the electric field strength exceeds the surface tension of the solution, leading to the formation of solid fibers [9]. This process is known for its ability to produce porous flexible structures with high surface areas suitable for DSSC's photoelectrodes [10]. Over the last decades, there have been many studies on the preparation of polymer/TiO_2 nanocomposites using the electrospinning process. Some of them report the preparation of these nanocomposite materials using a combination of sol-gel and electrospinning techniques [11], or using the electrospinning technique to develop the polymeric fibers followed by electrospraying the TiO_2 nanoparticles on the surface of theses electrospun fibers [12]. However, the direct incorporation of TiO_2 nanoparticles during fiber electrospinning is more recent. Reitzenstein's group reported the development of electrospun PVP/TiO_2 and PS/TiO_2 nanocomposite mats with up to 5 wt.% of TiO_2[13]. It was reported that the incorporation of TiO_2 tends to increase the surface roughness and structural defects, like beads and branching compared to neat PVP and PS fibers. An increase of surface roughness means less contact area for the electrospun mat with the dye, which can result in reduced energy conversion efficiency. One of the major challenges during the preparation of TiO_2-Polymer nanocomposites is the adequate dispersion of the hydrophilic nanoparticles in the polymer matrix before and during the electrospinning process. Herein, we developed for the first time, by using the electrospinning process, flexible electrospun PET/TiO_2 nanofibrous mats composed of well-dispersed hydrophobic surfactant capped TiO_2 inside the PET matrix.

MATERIALS AND METHODS

The following chemicals were used as received without further purification. Titanium (IV) butoxide (TB, 97%), oleic acid (OA, 90%), oleylamine (OM, 70%), absolute ethanol and toluene (analytical grade) were purchased from Sigma Aldrich. Polyethylene terephthalate (PET) pellets (Laser +C9921) used in this work were purchased from DAK Americas. Electrospinning solvents, trifluoroaceticacid (TFA) and dichloromethane (DCM), which were all of analytical grade, were respectively supplied by Alfa Aesar and Acros Organic.

Synthesis of TiO_2 nanoparticles: The synthesis of TiO_2 nanoparticles was performed using a solvothermal method in the presence of OA and OM as capping agents. 1 mmol of TB was mixed with 5 mmol of OA and 5 mmol of OM. The obtained mixture was stirred for 15 min prior the addition of 1ml of absolute ethanol then stirred again for another 15 min and placed into a Teflon cup before being transferred into a Teflon-lined stainless steel autoclave containing 4 ml of ethanol 95%. The autoclave was then heated at 180 °C inside an oven for 18h. After cooling to room temperature, the resulting TiO_2 nanoparticles (white solid precipitate) were collected by vaccum filtration, washed for at least four times with ethanol and toluene and dried at room temperature for 24h.

Preparation of spinnable solutions: 15 wt.% PET were first dissolved in a (70/30 v/v) mixture of TFA and DCM and mixed overnight. Following total polymer dissolution, the as-synthesised TiO_2 nanoparticles were added into the solution at various weight concentrations (5, 10, 20 and 30 wt.% with respect to PET). Commercial P25 TiO_2 was also used for comparison. The mixture was vigorously stirred for 18 h at room temperature, then sonicated for 2 h to obtain a homogenous solution with a good dispersion of the TiO_2 nanoparticles. Before the electrospinning process, each suspension was further sonicated for 5 minutes to prevent agglomeration.

Electrospinning process: Electrospinning was conducted using a lab-scale setup composed of a programmable syringe pump (Harvard Instrument), a 10 mL glass syringe, a high voltage power supply and a grounded aluminum foil coated collector. The following electrospinning conditions, which were maintained unchanged during the process, were the result of an initial optimisation to obtain a stable jet and uniform nanofiber mats. Each of the prepared spinnable solutions was loaded into the glass syringe, which was connected to a stainless steel needle with inner diameter (22 G). Solution flow rate was 1mL/h, the applied voltage was 22 kV, and tip/collector distance was 15 cm. Electrospun fibers were collected randomly on the static collector and dried at room temperature under a chemical fume hood for 24 h to ensure complete solvent evaporation.

Morphological characterization: In order to investigate the morphology of the electrospun PET/TiO_2 nanocomposite fibers, SEM was performed with a JOEL JSM-840A microscope operating at 10 kV. All specimens were coated with a thin layer of gold/palladium then characterized at different magnifications. The fiber average diameter and the fiber diameter distribution of the prepared mats were analyzed using image J software. The reported values are based on the analysis of 100 fibers taken from 3 different samples. The detailed morphology of the synthesised TiO_2 nanoparticles and electrospun PET/TiO_2 nanocomposite fibers was studied with high-resolution transmission electron microscope HRTEM (JEOL JEM-2100) operating at 200 kV. TEM images of TiO_2 nanoparticles were obtained by dispersing a small fraction of TiO_2 powder in toluene. Ultrasonic dispersion for 5 min was performed before mounting them on carbon-coated girds. TEM images for PET/TiO_2 mats were obtained by dispersing the electrospun fibers in ethanol then depositing them onto copper grids.

Thermogravimetric (TGA) characterization: TGA was used to study the effect of the TiO_2 nanoparticles incorporation on the thermal stability of PET/TiO_2 mats and to estimate the real amount of TiO_2 inside the PET nanofibers. The analysis was performed using a Q-5000 thermogravimetric analyzer (TA Instruments). Samples were heated up to 700 °C at a heating rate of 10°C/min under air environment.

RESULTS AND DISCUSSION

Morphology characterization of TiO_2 nanoparticles and PET/ TiO_2 nanofiber mats: Figure 1(a) shows a TEM image of the surface morphology of TiO_2 nanoparticles prepared by the solvothermal method. Uniform TiO_2 nanoparticles of nanorhombic shape with an average size of 7x14 nm were obtained by using a 5:5 molar ratio of OA/OM and a synthesis temperature of 180°C. The TGA curve of OA-OM capped TiO_2 (Fig. 1(b)) shows an initial weight loss (starting from 50°C) corresponding to the water absorbed on the surface of TiO_2 nanoparticles. The major weight loss occurs between 200°C and 490°C, which is attributed to the presence of OA and OM on the surface of

TiO_2 nanorhombic, as already observed in literature [14]. The weight loss corresponding to surfactants calculated from the curve is about 31wt.%. This relatively high amount of the organic part could be explained by the presence of an excess of free surfactants.

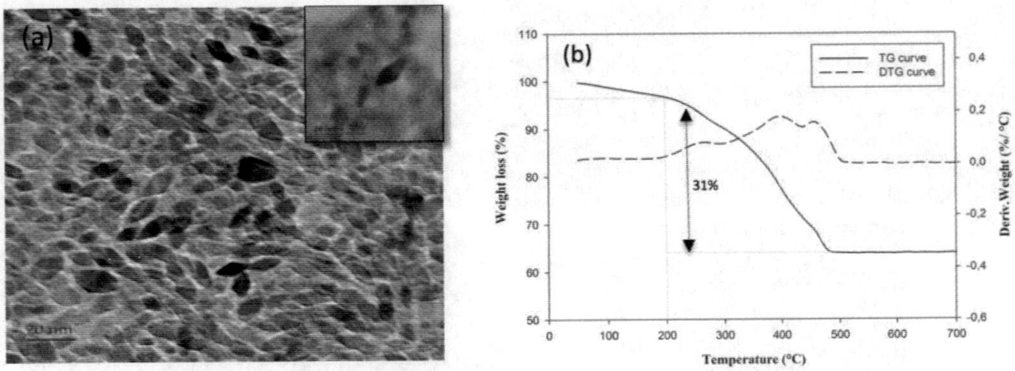

FIGURE 1. TEM image of nanorhombic TiO_2 (a), TGA and DTG curves of TiO_2 (b)

Figure 2 shows SEM pictures of PET/TiO_2 electrospun fibrous mats with different TiO_2 concentrations. SEM was performed to investigate the stability of the process and the effect of the incorporation of TiO_2 on fibers morphology. As shown, the addition of TiO_2 did not affect the electrospinning process. The process was stable and continuous and bead-free nanofibers with a normal fiber's diameter distribution were obtained for the four TiO_2 concentrations studied. For the electrospinning process, the applied voltage is the crucial factor. To produce bead-free uniform fibers, the ejecting jet has to be stable throughout the jet trajectory until it reaches the collector. Figure 2 shows that the addition of TiO_2 nanoparticles did not affect the surface morphology of the PET nanofibers since no TiO_2 clusters were observed on their surface. This indicates that the nanoparticles were dispersed inside the PET nanofibers. Figure 2 insets show the distribution of PET/TiO_2 fibers diameter (from which the average fibers diameter was calculated) for the different TiO_2 concentrations. The average nanofibers diameter decreased significantly from 453 nm (neat PET fiber) to 386 nm (PET fibers with 5 wt.% of TiO_2) and reached 357 nm with 10 wt.% of TiO_2. This decreasing trend in fibers diameter by the incorporation of metal oxide particles was also reported in literature [15]. It was explained by the modification of the electrical conductivity of the electrospinning solutions. Higher suspension conductivity is achieved by the addition of TiO_2 nanoparticles, leading to higher stretchability of the polymeric solution between the syringe and the collector and consequently the formation of thinner fibers. Furthermore, high TiO_2 concentration into the PET solution (higher than 10 wt. %) results in an increase in the average fiber diameter. This increase is be attributed to the increase of the viscosity of the spinnable solution with high TiO_2 concentration.
Figure 3 shows TEM pictures of two PET/TiO_2 nanofibers containing 10 wt.% of TiO_2: the first one (Fig. 3(a)) was prepared with 10 wt.% of nanorhombic 7 x 7 nm (TB/OA/OM= 1:5:5, 180 °C), and the second one with 10 wt.% of commercial TiO_2 P25 (Fig. 3(b)). Light areas in both pictures correspond to the PET matrix and dark areas to TiO_2 nanoparticles. As shown, both commercial and synthesized TiO_2 nanoparticles were successfully incorporated into the PET nanofibers. However, the commercial TiO_2 P25 nanoparticles tend to agglomerate within the PET nanofiber leading to increased surface roughness. This agglomeration is due to the inhomogeneous dispersion of the TiO_2 P25 nanoparticles in the original spinnable solution. On the other hand, the nanofibers prepared with our synthesized TiO_2 nanoparticles show a smooth surface and better TiO_2 dispersion. This is probably due to the surfactants used as capping agents. These surfactants reduce TiO_2 surface tension and ensure a better stability of the prepared spinnable solution.

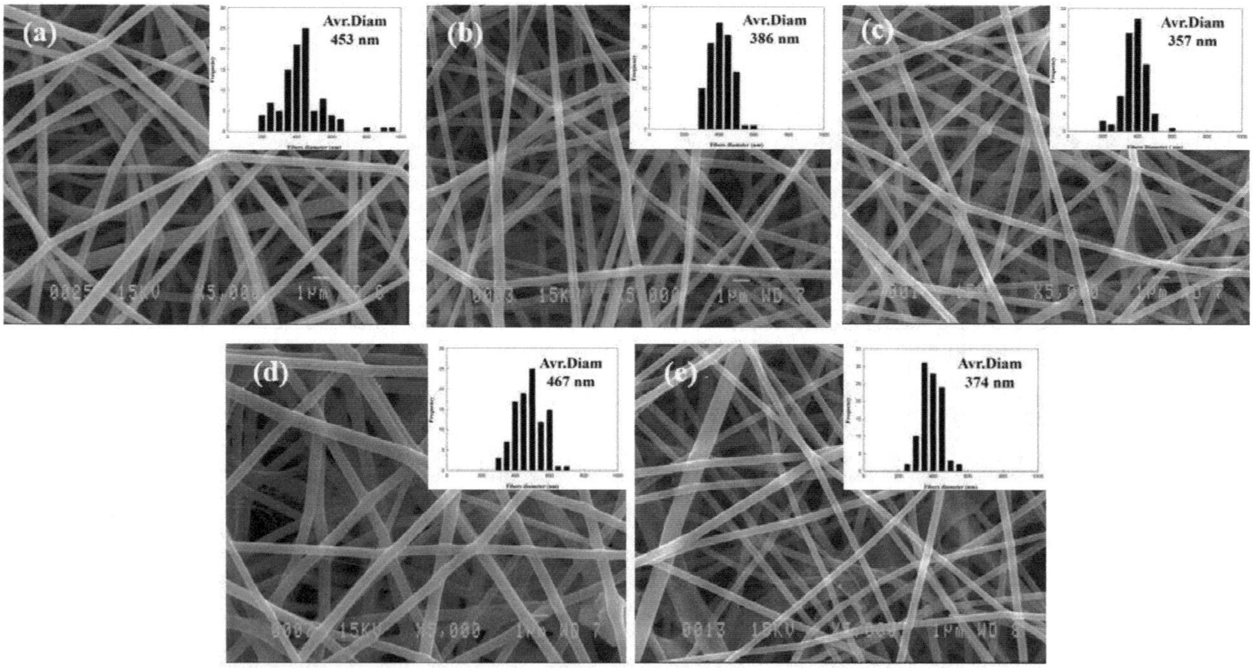

FIGURE 2. SEM of PET nanofibers (a) neat PET, (b) 5 wt.% TiO_2, (c) 10 wt.% TiO_2, (d) 20 wt.% TiO_2, (e) 30 wt.% TiO_2

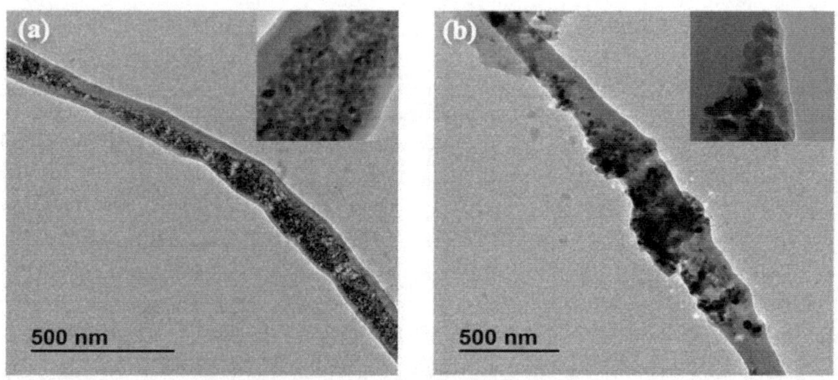

FIGURE 3. TEM images of PET/TiO_2 (10 wt.%) nanofibers containing (a) synthetized Nanorhombic TiO_2, and (b) Commercial TiO_2 P25

TGA characterization: The TGA and DTG curves are shown in Figure 4. Two weight loss stages are observed for both neat PET and PET/TiO_2 nanocomposite fibers. The first stage with a large slope occurs between 300 and 435 °C and is due to the decomposition of PET matrix. The second one appears in the range between 435 and 525 °C and is attributed to the consumption of char under the air atmosphere [16]. It can be noticed from the TGA curves (Fig. 4(a)) that thermal degradation of PET/TiO_2 samples starts at lower temperatures compared to the neat PET. This is mainly due to the lower temperature degradation of OA and OM surfactants. However, with the addition of up to 20 wt.% of TiO_2 nanoparticles into the PET matrix, the temperature of the first major weight loss ($Tmax_1$) for PET/TiO_2 samples was higher than that of neat PET, as shown in Figure 4(b). This is an indication that, up to 20 wt.% of TiO_2, the thermal stability of PET/TiO_2 nanocomposites was enhanced compared to the neat PET.

As shown in Figure 4(b), $Tmax_1$ for the samples with 5 wt.% and 10 wt.% of TiO_2 are respectively 396 and 402°C, showing a respective increase by 4 °C and 10 °C compared to the neat PET. This increase reflects the interaction between the TiO_2 nanoparticles and the PET matrix. The presence of the TiO_2 nanoparticles increases the rigidity of PET chains and consequently the energy required to break them. However, when TiO_2 concentration is increased up to 20 wt.%, $Tmax_1$ of the sample decreased to around the same value as of samples with 5 wt.% TiO_2. A further increase of TiO_2 concentration up to 30 wt.% led to $Tmax_1$ of 388 °C, approximately 4°C lower than that of neat PET sample. The thermal degradation mechanism of PET/TiO_2 nanocomposites seems to be influenced in a complex way by the degree of agglomeration of TiO_2 particles inside the fiber. Furthermore, the value of the residual amount at 700 °C also increased for all PET/TiO_2 nanocomposite fibers compared to the neat PET fiber, which confirms the presence of inorganic materials in the nanocomposite. However, there is no correlation between the amount of TiO_2 in the samples and the residual amount at 700°C. Previous studies have also shown a slightly increase in thermal stability of polymer matrix when metal oxide nanoparticles are incorporated [17].

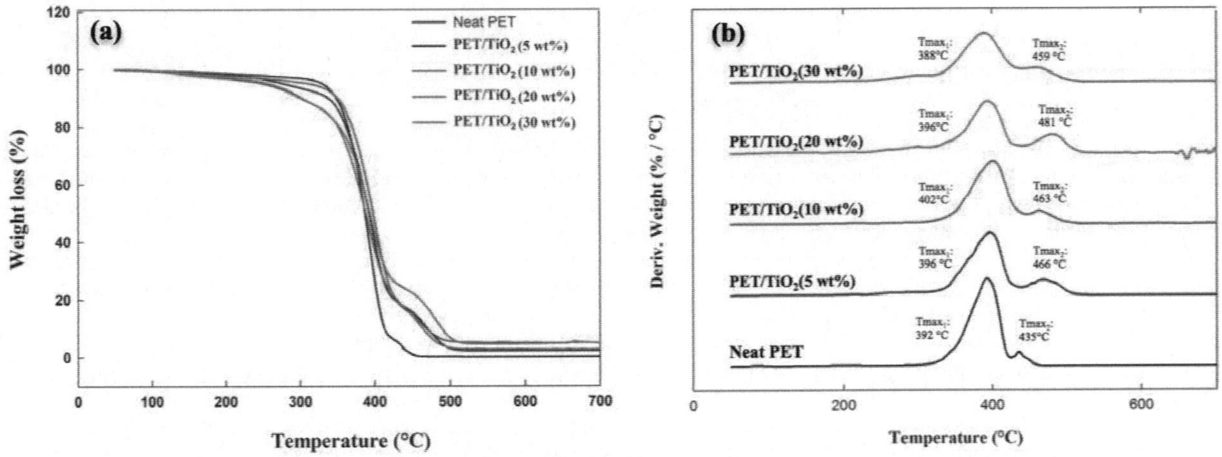

FIGURE 4. TGA (a) and DTG (b) curves of neat PET and PET/TiO_2 nanocomposites

CONCLUSIONS

In the present work, PET/TiO_2 nanocomposite fibrous mats were developed by a simple and low cost electrospining process. The different process parameters were carefully adjusted in order to find an optimum fibrous structure that could be advantageously used to develop DSSC photoanodes. Morphological and thermal properties characterization of the developed PET/TiO_2 nanofibers using our synthesized TiO_2 nanoparticles shows promising results with a better dispersion, a decrease of surface roughness and enhanced thermal stability. Future research work is underway (optical and mechanical properties characterization) in order to study the potential use of the developed PET/TiO_2 nanofibers as flexible photoanodes for DSSCs.

REFERENCES

1. L. Song, X. Yin, X. Xie, P. Du, J. Xiong and F.Ko, *Electrochimica 255*, 2017, 256-265.
2. Y. Oh, S. N. Lee, H. K. Kim and J. Kim, *J. of The Electroch. Soc. 159 (10),* 2012, H777-H781.
3. X. Li, H. Lin, J. Li, N. Wang, C. Lin and L. Zhang, *J. of Photochem. and Photobiol. A 195, 2008,* 247-253.
4. D. S. Zhang, T. Yoshida and H. Minoura, *Adv. Materials 15, 2003,* 814-817.
5. T. Yamaguchi, N. Tobe, D. Matsumoto and H. Arakawa, *Chem. Communication 45,* 2007, 4767-4769.
6. K. Fan, R. Li, J. Chen, W. Shi and T. Peng, *Sci. of Adv. Materials 5(11),* 2013, 1596-1626.
7. Y. Li, K. Yoo, D. K. Lee, J. H. Kim, N. G. Park, K. Kim and M. J. Ko, *Current Appl. Phys.* 10, 2010, pp. 171-175
8. J. Xue, J. Xie, W. Liu and Y. Xia, *Account of chem.l Resesearch* 50 (8), 2017, 1976-1987.
9. K. Garg and G. L. Bowlin, *Biomicrofluidics* 5, 2011, 1-19.
10. T. Wan, S. Ramakrishna and Y. Liu, *J. of Appl. Polym. Sc.*, 2018, 45649-45659.
11. X. Meng, N. Luo, S. Cao, S. Zhang, M. Yang and X. Hu, *Materials Letters* 63, 2009, 1401-1403.
12. Y. Li, D. K. Lee, J. Y. Kim, N. G. Park, K. Kim, B. S. Kim, J. H. Shin, I. S. Choi and M. J. Ko, *Energy and Env. Sc.* 5(10), 2012, 8950-8957.
13. N. H. V. Reitzenstein, X. Bi, Y. Yang, K. Hristovski and P. Westerhoff, *J. of Applied Polym. Sci.*, 2016, 43811-43819.
14. T. D. Vu, F. Mighri, T. O. Do and A. Ajji, *J. of Nanosc. and Nanotech.* 12 (3), 2012, 2815-2824.
15. A. Karimi, H. Izadan, A. Khoddami, Z. M. Sebdani and S. A. H. Ravandi, *The J. of the Text. Inst.* 109 (5), 2018, 569-576.
16. G. Gorrasi, C. Milone, E. Piperopoulos, R. Pantani, *Composites Part B 81*, 2015, 44-52.
17. E. A. S. González, D. Olmos, M. A. Lorente, I. Vélaz and J. González-Benito, *Polymers* 10, 2018, 1365.

The Influence of Pressure on Crystallization Kinetics of Isotactic Polypropylene

Felice De Santis[a,b*] and Roberto Pantani[b]

[a]*Borealis Polyolefine GmbH - Linz - Austria*
[b]*Department of Industrial Engineering, University of Salerno, Fisciano - SA - Italy*

Abstract. Isotactic polypropylene (iPP) presents a complex polymorphism, influenced by the processing conditions. Flow fields, cooling rate, and pressure influence the crystallization rate and crystalline structure of polymers. As an example, in the complex polymorphism of isotactic polypropylene, the main way to obtain γ structure for a high molecular weight homopolymer is to perform crystallization under high pressure. The effect of pressure on the processing and, in particular, on the crystallization kinetics of a polymeric material sometimes is neglected, because of the difficulties involved in the experimental measurements. Consequently, crystallization behaviors were rarely described by using a model that takes into account the effect of pressure on the crystallization. On the other hand, the crystallization kinetics under pressure can be investigated by dilatometry. In this study, experimental data on the dependence of the specific volume on temperature and pressure of isotactic polypropylene using a high-pressure dilatometer up to 150 MPa are reported and discussed. Isothermal crystallization experiments at high pressures confirmed a relevant fraction of γ-form, as revealed by wide-angle X-ray scattering. From the subsequent heating step after the isothermal crystallization, the melting temperatures at the same pressures have been determined to assess the pressure effect on the equilibrium melting temperature by the Hoffman-Weeks analysis. The crystallization times of isothermal experiments have been analyzed according to the nucleation and growth models. In particular, the influence of pressure on the equilibrium melting temperature and the diffusion process is introduced in the modeling.

INTRODUCTION

Isotactic polypropylene (i-PP) is one of the most investigated polymers because of its industrial importance. Depending on thermal treatments, molecular mass, and degree of tacticity, when crystallized from the melt, the i-PP chains can organize in several spatial arrangements giving rise to different crystalline polymorphs (α-monoclinic, β-hexagonal, and γ-triclinic) and a mesomorphic phase. Furthermore, the effect of the flow, often relevant in the processing, also affects crystallization both for the final phases (microstructure) and macrostructure. Consequently, the crystallization behavior of i-PP is very complex. Numerous groups have aimed their research on a better understanding of the different crystal modifications.

The existence of the γ crystalline phase of isotactic polypropylene was reported soon after the elucidation of the more common α phase. The γ-phase is usually associated with the α-phase, although individual single crystals of the γ-phase have also been observed. The γ-modification may be formed in degraded, low molecular weight isotactic polypropylene and in samples crystallized under high pressures. Specifically, about 50 years ago, it was established that the ratio between the γ -iPP and the α -iPP polymorphs grows with increasing pressure.

In polymer processing, the specific volume plays a major role. It is influenced by processing conditions like temperature, pressure, and flow history and it determines shrinkage, which expresses itself by dimensional (in)stability.

EXPERIMENTAL

Material

The resin adopted for the experiments performed in this work is a commercial Ziegler-Natta isotactic polypropylene homopolymer[1-4] (T30G, M_w = 376 kg mol^{-1}, polydispersity index M_w/M_n of 6.7) with melt flow rate (at 2.16 kg/230°C) of 3.6 g/10 min.

Dilatometry

The experiments were performed with a PVT dilatometer, Gnomix-PVT apparatus (GNOMIX Inc., Boulder, CO, USA), developed by Zoller et al.[5] using the confining-fluid technique. The sample is enclosed in a rigid sample chamber, and it is submerged into a fluid (mercury). The cell is closed using a flexible wall or bellows. The bellows are used to apply hydrostatic pressure to the fluid and polymer by reducing the sample chamber volume, and sensing the cumulative volume change of fluid and polymer. The absolute specific volume of the polymer can be obtained by correcting the relative volume difference with the specific volume of the confining fluid. Both pressure and temperature can be varied. The advantages of this technique are:

- pressure is purely hydrostatic as the sample is surrounded by the confining fluid in both a melted and solid state;
- there is an absence of leakage and friction.

The disadvantage is:

- the volumetric changes measured are not that of the polymeric sample only.

The PVT data, most of them obtained on Gnomix equipment, are currently used in the simulation of injection molding, i.e. by Moldflow. The PVT instrument has some advantages compared to a capillary rheometer if accurate numbers are desired (0.0001 cm^3/g)[6].

The Gnomix manufacturer suggests three "modes" to perform experimental tests:

- in the isobaric mode (IBA) the sample will be heated or cooled at a defined rate and a defined pressure.
- in the isothermal standard (ITS) mode, after reaching thermal equilibrium, the pressure is raised from 10 to 200 MPa and all data are recorded in steps of 10 MPa
- in a special (data acquisition, DAQ) mode it is possible to follow changes in the specific volume at constant pressure and constant temperature, which may be caused by a phase transition.

In particular, the special DAQ mode is adopted in this work as summarized in Table 1.

Isobaric isothermal crystallization

To erase the previous thermomechanical history the sample in the PVT apparatus was heated to 240°C and held at this temperature for 10 minutes. In order to study melt isothermal crystallization, the sample was cooled from 240°C directly to isothermal test temperature (crystallization temperature), with a cooling rate of 3 K/min. This experimental protocol is just the same modus operandi usually adopted in calorimetry, in which thermal effects linked to crystallization are detected. Obviously in dilatometry, during the isothermal step, the volumetric changes related to crystallization, i.e. specific volume decrease, are measured under pressure.

The cooling rate of 3 K/min before the isothermal step is substantially restricted by the total heat capacity of the system and the sample size, as in calorimetry or nanocalorimetry[2] it could be of the order of 10 K/min or 1000 K/s, respectively. As consequence, this cooling rate limits the minimum reachable temperature of the isothermal step, without evident previous crystallization during the cooling step. The temperatures of the isothermal step, using the special DAQ mode at various pressure, are summarized in Table 1.

TABLE (1). Crystallization temperatures of experimental PVT isotherms.

P [MPa]	10	100	150
Temperature [°C]	127.1	152.0	162.5
	131.9	154.8	164.2
	133.6	155.5	167.8
	136.6	157.6	169.5
	138.6	159.6	
	141.3	162.5	
		167.3	
		169.3	

Furthermore, as known by the previous literature[7], the pressure increase has the effect to accelerate crystallization rate of iPP and thus the higher is the pressure the greater is the minimum reachable temperature of the isothermal step in Table 1, without evident previous crystallization during the cooling step.

An example of an isothermal experimental test is reported in Figure 1: during the cooling, the specific volume change is almost linear with the temperature until the isothermal step temperature is reached.

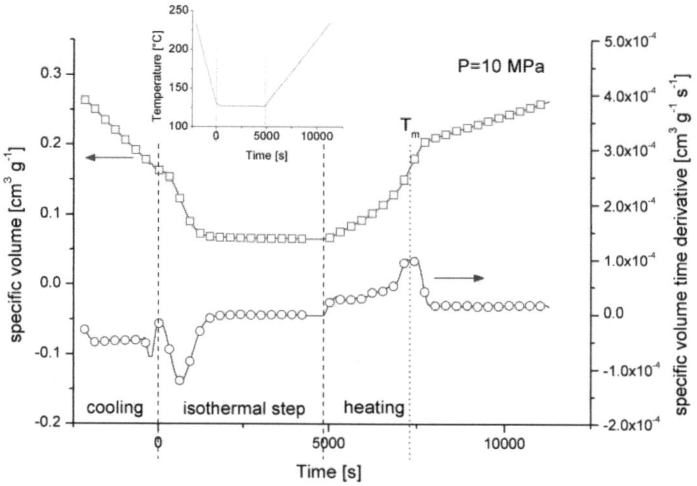

FIGURE 1. Experimental data of specific volume, *v*, during an isobaric crystallization test (on the left axis), and resultant time derivative (on the right axis) to evaluate melting temperature. In the inset the experimental protocol: temperature evolution during the test.

During the isothermal step, the specific volume shows an inverse sigmoid curve related to the crystallization. After the isothermal step, the sample is heated again to 240 °C, at a heating rate of 1 K/min, to explore the corresponding melting behavior. The peak of the time derivative of the specific volume permits to evaluate the melting temperature during the heating. After the complete melting, the specific volume change is again linear with the temperature during the heating to 240 °C.

The temperature experimental protocol is shown in the inset in Figure 1, for the isothermal test at 127.1 °C.

RESULTS AND DISCUSSION

From the analysis of the specific volume evolution during the isothermal step is possible to evaluate crystallization half-time for each test, performed changing the temperature and pressure of the PvT experiments.

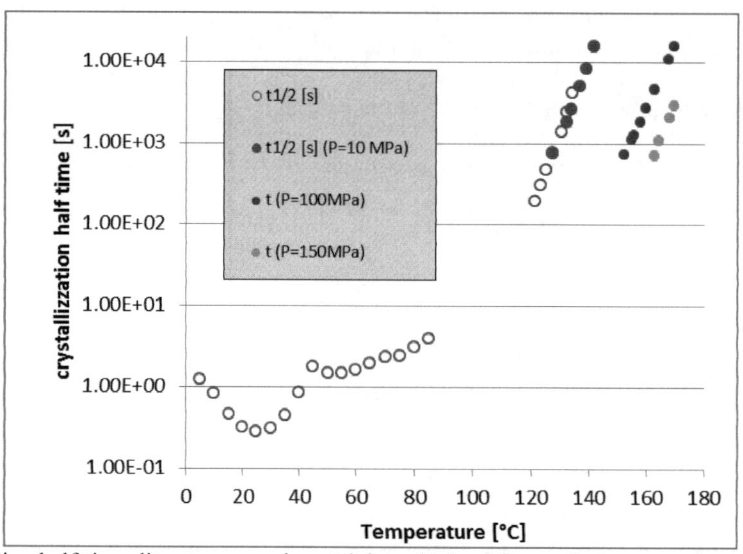

FIGURE 2. Crystallization half-time: literature experimental data obtained from calorimetry and nanocalorimetry[2] (open symbol) and data from the analysis of isobaric isothermal crystallization PvT tests (full symbol) with different pressures

The crystallization half-times are reported in Fig. 2: the data obtained from low pressure PvT experiments, at 10 MPa, compares well with isothermal calorimetry results taken from literature.

Furthermore, for the tests at higher pressures, i.e. at 100 and 150 MPa, a clear effect on crystallization is reported and crystallization rate is improved by the effect of pressure. The analysis of these data is still in progress, in order to take into account the influence of pressure in a crystallization kinetics model.

From the subsequent heating step after the isothermal crystallization, the melting temperatures at the same pressures have been determined to assess the pressure effect on the equilibrium melting temperature.

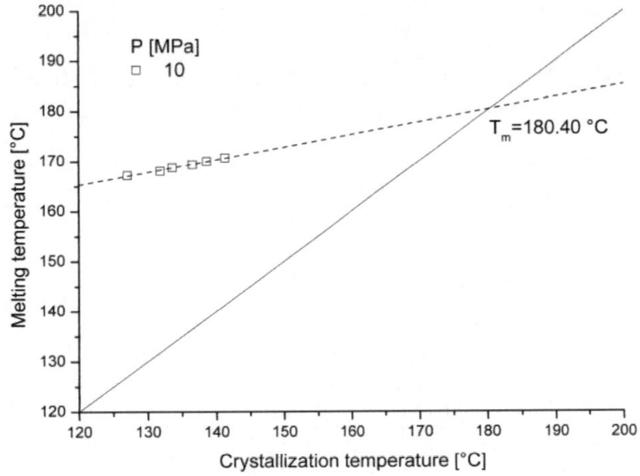

FIGURE 3. Hoffman-Weeks plot for the pressure of 10 MPa

The effect of the pressure on the equilibrium melting temperature was analyzed by means of Hoffman-Weeks plots of crystallization and melting. The results are summarized in Figure 3, for the pressure of 10 MPa, evaluating the corresponding equilibrium melting temperature, with a linear extrapolation, of 180.4°C. Actually, this is value is lower than equilibrium melting temperature of 194°C, reported for this resin at ambient pressure but estimated through the non-linear Hoffman—Weeks analysis[4, 8]. On the other hand a higher equilibrium melting temperature, if compared to the one at ambient pressure, could be expected at higher pressure[7].

CONCLUSIONS

The crystallization was affected by temperature, pressure, and shear stress. In particular, the change of temperature and pressure is an important factor governing the crystallization progress during injection molding.

Isobaric isothermal crystallization test were performed in a PvT apparatus, following the evolution of specific volume during the crystallization and subsequent melting.

The investigation of these experiments allowed to evaluate the effect of pressure on crystallization kinetics. A further analysis should descript the crystallization kinetics model taking into account the effect of pressure.

REFERENCES

1. F. De Santis, R. Pantani and G. Titomanlio, Polymer **90**, 102-110 (2016).
2. F. De Santis, S. Adamovsky, G. Titomanlio and C. Schick, Macromolecules **40** (25), 9026-9031 (2007).
3. F. De Santis, S. Adamovsky, G. Titomanlio and C. Schick, Macromolecules **39** (7), 2562-2567 (2006).
4. I. Coccorullo, R. Pantani and G. Titomanlio, International Polymer Processing **20** (2), 186-190 (2005).
5. P. Zoller, P. Bolli, V. Pahud and H. Ackermann, Review of Scientific Instruments **47** (8), 948-952 (1976).
6. P. Zoller and Y. A. Fakhreddine, Thermochimica acta **238**, 397-415 (1994).
7. C. Angelloz, R. Fulchiron, A. Douillard, B. Chabert, R. Fillit, A. Vautrin and L. David, Macromolecules **33** (11), 4138-4145 (2000).
8. P. Juhász, J. Varga, K. Belina and H. Marand, Journal of thermal analysis and calorimetry **69** (2), 561-574 (2002).

Shear-Induced Nucleation and Growth in Poly(Lactic Acid)

Fabiana Foglia[a], Valentina Volpe[a], Felice De Santis[b] and Roberto Pantani[a*]

[a]*Department of Industrial Engineering, University of Salerno via Giovanni Paolo II, 132, 84084 Fisciano (SA), Italy*
[b]*Borealis Polyolefine GmbH - Linz - Austria*

Abstract. Poly(lactic acid) (PLA) is biodegradable thermoplastic polyester produced from renewable sources that can be easily processed to obtain articles to be used in many applications, from commodities to electronics. The large influence of the crystalline structure and morphology on the biodegradation rate, as well as on the other material properties, led to an ever-increasing interest in the study of the crystallization kinetics of PLA along the years. Several investigations on the crystallization kinetics of PLA under quiescent condition were carried out. However, the description of the evolution of morphology under shear conditions is still challenging and even the basic principles of the phenomenon are not well understood yet, because of the experimental difficulties in obtaining direct measurements of nucleation and growth rates during flow. In this work, experiments of nucleation density and growth rate of spherulites were carried out in quiescent condition and under shear flow. In particular, in order to investigate the nucleation and growth rate at different shear rates, step shear tests were conducted by applying the shear flow when the crystalline structures have already reached dimensions easily detectable by an optical microscope. On the basis of the experimental results, the effect of the shear rate on the nucleation phenomenon of PLA was assessed.

Keywords: Crystallization; Poly (Lactic Acid); Nucleation and growth rate.

INTRODUCTION

In the last decades biobased polymers have gained increasing importance, with applications ranging from the biomedical to the agricultural and packaging. Among the bio-based polymers, the most promising alternatives to petrochemical based polymers is the poly(lactic acid) (PLA), which is produced from renewable resources such as corn and exhibits special properties including compostability or biocompatibility [1-3]. Nowadays, thanks to the large variety of grades provided by suppliers, it is adopted in various processes, such as extrusion or injection molding. However, one of the drawbacks still exhibited by PLA is the low thermal stability under processing conditions, which leads to degradation due to a decrease in molecular weight. Another problem of the PLA is its slow crystallization kinetics, which determinates an almost amorphous final part with consequent low mechanical performances [4]. Therefore, the degree of crystallinity is of fundamental importance as it influences the final properties of the product and the degradation process [5]. Several investigations on the crystallization kinetics of PLA under quiescent condition were carried out. In 2003 Di Lorenzo et al. presents data of spherulite growth rate obtained in quiescent conditions with combined isothermal and non-isothermal methods. Il 2012 Saeidlou et al. [6] presented an overview of the fundamentals of PLA in quiescent conditions in order to provide comprehensive relations between crystallization kinetics and the main molecular structures characteristics of PLA. Although several studies on conventional polymers (iPP and PE) demonstrated that the shear flow accelerates nuclei formation and growth rate [7,8], flow-induced crystallization of PLA still received very limited attention. In particular, the experimental difficulties in obtaining direct measurements of growth rates during flow has led to focus the attention of the researchers only on the effect of the flow on the nucleation rate. Li et al. [9] studied the crystalline morphology in terms of nucleation of PLLA under steady shear flow at different crystallization temperature. Refaa et al. [4] focused their work on the crystallization kinetics of PLA under quiescent conditions as well as flow conditions. They observed an enhancement of the nucleation mechanism due to the effect of the applied flow. In this work, experiments of nucleation density and growth rate of spherulites were carried out in quiescent condition and under shear flow. In particular, in order to investigate the nucleation and growth rate at different shear rates, step shear tests were conducted by applying the shear flow when the crystalline structures have already reached dimensions easily detectable by an optical microscope. On the basis of the experimental results, the effect of the shear rate on the nucleation phenomenon of PLA was assessed.

MATERIALS AND METHODS

The material adopted in this work is a commercial grade PLA produced by NatureWorks with the trade name of 4032D with a D-isomer content of approximately 2% and with a maximum degree of crystallinity of about 45% [5]. PLA 4032D has a molecular weight distribution characterized by Mn = 119×10^3 g/mol and Mw = 207×10^3 g/mol. Before any test or processing, the material was dried for 8 h under vacuum at a temperature of 60°C.

A Linkam CSS-450 high-temperature optical shear stage equipped with video camera coupled to a polarized optical microscope, Olympus BX51 model, was employed for the observation of the crystallization process in quiescent condition and under a shear field. The Linkam shear stage has a plate-plate geometry with two quartz windows: the top window is fixed while the bottom one can rotate at a constant velocity to produce a shear rate, which is constant in the observation field located at a fixed radial position (7.5 mm).

The thermal protocol adopted both in quiescent condition and under shear, and the mechanical protocol adopted for the crystallization experiments under shear is described in Figure 1.

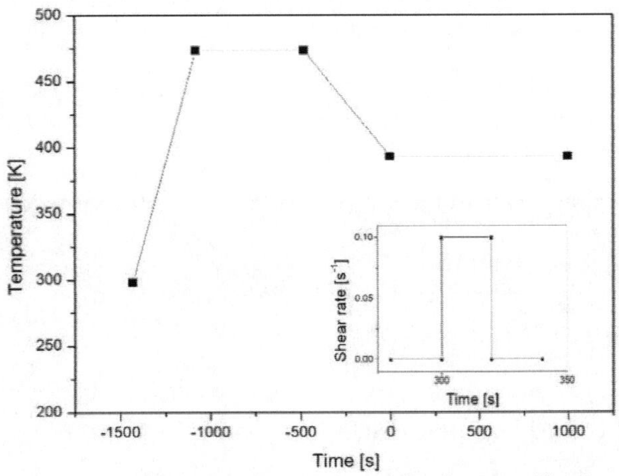

FIGURE 1. Thermo-mechanical protocol adopted for the crystallization experiments.

The samples were heated at the rate of 30 °C/min from the room temperature to 200 °C then were kept at this temperature for 10 min to cancel their thermal history. At this temperature the gap between the two windows was reduced to 350 μm by slowly lowering the upper glass plate of the Linkam device. Then the temperature was lowered at the rate of 10 °C/min until it reached the desired crystallization temperature. Step shear tests were conducted by applying the shear flow for 40 s after 5 min from reaching the crystallization temperature, when the crystalline structures have already reached dimensions easily detectable by an optical microscope. In order to avoid degradation during the experiment, in each experiment a hydrogen flow was injected in the system.

Nucleation density was determined by counting the number of nuclei in the sample and growth rate was determined by monitoring the radius of the spherulites during time.

RESULTS

Quiescent crystallization

The crystallization experiments under quiescent conditions were performed in the temperature range of 110-135°C at intervals of 5 °C. During the experiments performed under quiescent conditions, the nucleation density and the growth rate of the spherulites were monitored. Experimental determinations showed that a fixed number of nuclei appeared at the test temperature, and their number did not change in time. The nucleation density, evaluated as the number of nuclei in the observed volume, decreases with increasing crystallization temperature. Furthermore, spherulite radial growth rate was found constant with time. The growth rate increases with the crystallization temperature until it reaches a maximum at 125°C. In fact, spherulite growth rate of PLA is low at high temperatures, then gradually increases with the increase of the undercooling, namely the difference between the melting temperature and the test temperature, to reach a maximum at about 125°C [10]. Results of nucleation density and growth rate in quiescent condition are reported in Figure 2.

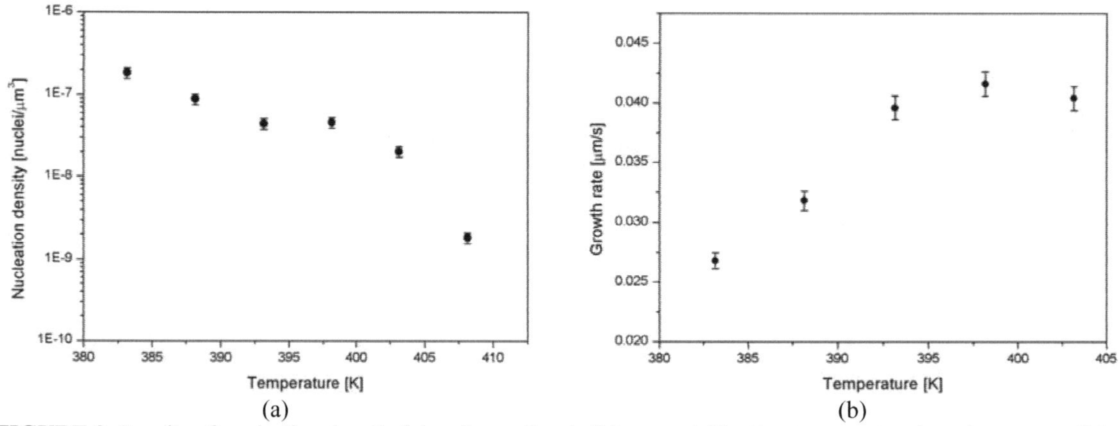

FIGURE 2. Results of nucleation density (a) and growth rate (b) vs crystallization temperature in quiescent condition.

In Figure 3, in which micrographs collected by a Polarized Optical Microscope during a quiescent test at 410 s and 950 s from the time at which temperature reached the isothermal crystallization temperature are represented, a constant number of nuclei that increase in dimensions with the time can be observed.

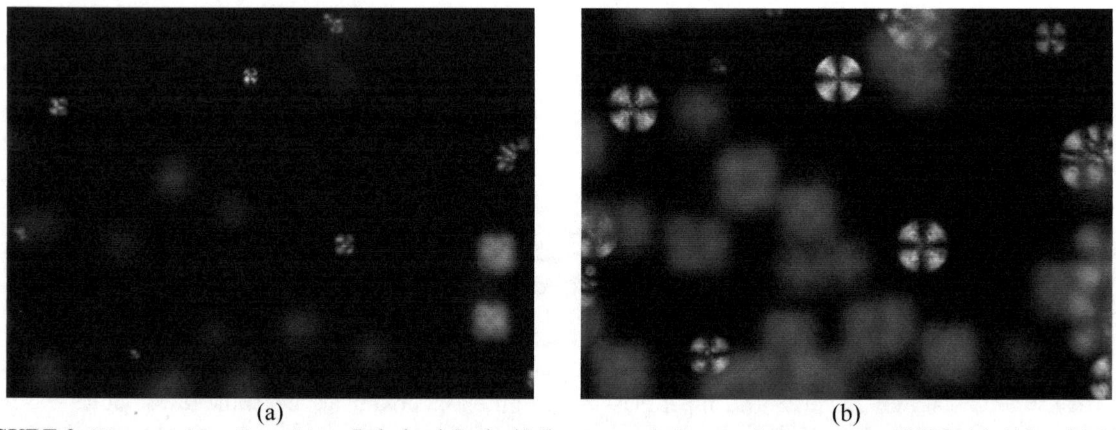

FIGURE 3. Micrographs collected by a Polarized Optical Microscope during a quiescent test at 120°C. (a) 410 s; (b) 950 s.

Crystallization under Shear Flow

The crystallization experiments under shear flow were performed at 120°C in a shear rate ranging from 0.1 to 1.5 s^{-1}. Under shear flow the nucleation density increases with time whereas in quiescent conditions it remained constant and it was function only of temperature. The nucleation rate, namely the slope of the evolution of the nucleation density during the crystallization time, increases with increasing shear rate (Figure 4 a). Values of growth rate, obtained by a linear fitting of the evolution of spherulites dimensions for all shear rates applied in this work, were reported as a function of the shear rate in Figure 4 b. From the graph it is possible to observe an initial increase until it reaches a maximum at shear rate 0.15 s^{-1} and a subsequent decrease of growth rate, phenomenon that is subject of further analysis.

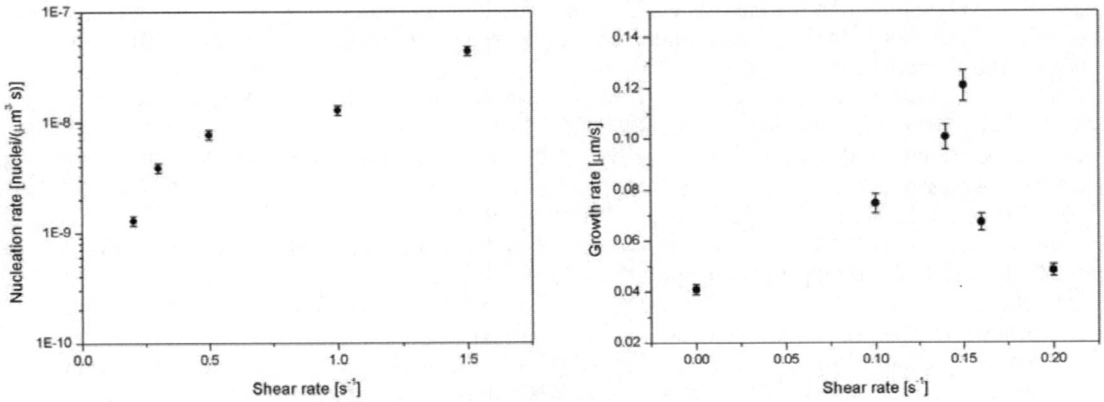

(a) (b)

FIGURE 4. Nucleation rate (a) and growth rate (b) vs shear rate.

In Figure 5 it is possible to observe micrographs collected by a Polarized Optical Microscope during a test at a shear rate of 0.5 s^{-1} before and after the application of the shear step. From the micrographs it is evident how the number of nuclei increases considerably after the application of a shear step.

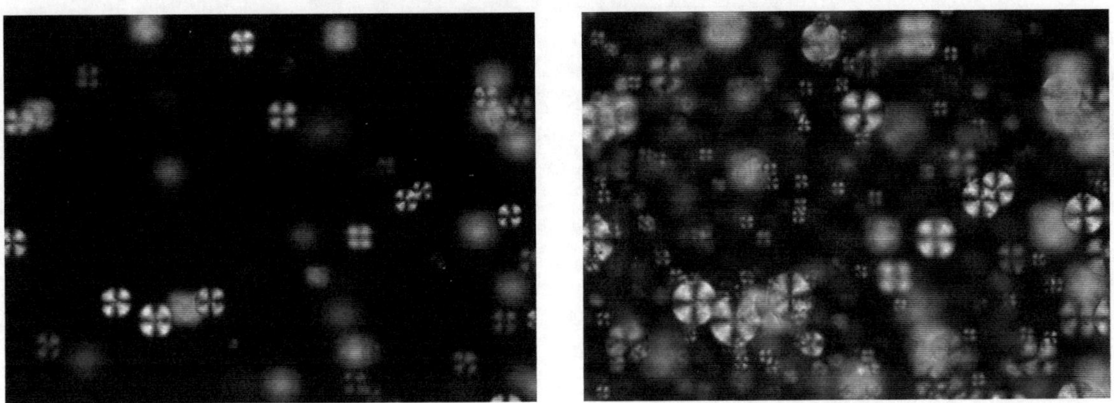

(a) (b)

FIGURE 5. Micrographs collected by a Polarized Optical Microscope during a test at 120°C with a shear rate of 0.5 s^{-1}. (a) before the shear step; (b) after the shear step.

CONCLUSIONS

In this work, experiments of nucleation density and growth rate of spherulites were carried out in quiescent condition and under shear flow, in order to investigate the crystallization of the PLA 4032D. In quiescent conditions a fixed number of nuclei appeared at the test temperature, and their number did not change in time. The nucleation density decreases with increasing crystallization temperature. Furthermore, the growth rate increases with the crystallization temperature until it reaches a maximum at 125°C. Under shear flow the nucleation density increases with time whereas in quiescent conditions it remained constant and it was function only of temperature. The nucleation rate increases with increasing shear rate. The growth rate initially increases with the shear rate until it reaches a maximum at shear rate 0.15 s^{-1}. After this value of shear rate, a decrease of growth rate can be observed.

REFERENCES

1. Androsch, R.; Di Lorenzo, M.L. Kinetics of crystal nucleation of poly(L-lactic acid). *Polymer* **2013**, *54*, 6882-6885, doi:10.1016/j.polymer.2013.10.056.
2. Pantani, R.; Sorrentino, A.; Volpe, V.; Titomanlio, G. Foam Injection Molding of Poly(Lactic Acid) with Physical Blowing Agents. *Aip Conf Proc* **2014**, *1593*, 397-400, doi:10.1063/1.4873808.

3. Volpe, V.; De Filitto, M.; Klofacova, V.; De Santis, F.; Pantani, R. Effect of mold opening on the properties of PLA samples obtained by foam injection molding. *Polym Eng Sci* **2018**, *58*, 475-484, doi:10.1002/pen.24730.
4. Refaa, Z.; Boutaous, M.; Siginer, D.A. PLA Crystallization Kinetics and Morphology Development. *Int Polym Proc* **2018**, *33*, 336-344, doi:10.3139/217.3525.
5. Gorrasi, G.; Pantani, R. Effect of PLA grades and morphologies on hydrolytic degradation at composting temperature: Assessment of structural modification and kinetic parameters. *Polym Degrad Stabil* **2013**, *98*, 1006-1014, doi:10.1016/j.polymdegradstab.2013.02.005.
6. Saeidlou, S.; Huneault, M.A.; Li, H.B.; Park, C.B. Poly(lactic acid) crystallization. *Prog Polym Sci* **2012**, *37*, 1657-1677, doi:10.1016/j.progpolymsci.2012.07.005.
7. De Santis, F.; Pantani, R.; Titomanlio, G. Effect of shear flow on spherulitic growth and nucleation rates of polypropylene. *Polymer* **2016**, *90*, 102-110, doi:10.1016/j.polymer.2016.02.059.
8. Pantani, R.; Coccorullo, I.; Volpe, V.; Titomanlio, G. Shear-Induced Nucleation and Growth in Isotactic Polypropylene. *Macromolecules* **2010**, *43*, 9030-9038, doi:10.1021/ma101775h.
9. Li, X.J.; Li, Z.M.; Zhong, G.J.; Li, L.B. Steady-shear-induced isothermal crystallization of poly(L-lactide) (PLLA). *J Macromol Sci B* **2008**, *47*, 511-522, doi:10.1080/00222340801955313.
10. Di Lorenzo, M.L. Determination of spherulite growth rates of poly(L-lactic acid) using combined isothermal and non-isothermal procedures. *Polymer* **2001**, *42*, 9441-9446, doi:Doi 10.1016/S0032-3861(01)00499-2.

Orientation in LLDPE Cast Films Manufactured under Different Processing Parameters

Volker Schöppner[a], Hoda Bayazian[a], Ali Yadegari[b*], Hoda Haddad[a]

[a]*Kunststofftechnik Paderborn (KTP), Paderborn University, Paderborn, Germany*
[b]*Faculty of Petroleum, Gas and Petrochemical Engineering, Persian Gulf University (PGU), Bushehr, Iran*

hoda.bayazian@ktp.upb.de
* yadegari@pgu.ac.ir

Abstract. Linear low density polyethylene (LLDPE) stretch films under different processing conditions in terms of line speed, air gap and film thickness are manufactured using a cast film line. The influence of processing parameters on crystal orientation in the cast LLDPE films is evaluated by wide angle X-ray scattering (WAXD). Diffraction patterns reveal no preferred crystal orientation in the cast films. To evaluate the amorphous orientation, shrinkage measurements are carried out in the both machine direction (MD) and transverse direction (TD). It is found that higher line speeds lead to a higher MD shrinkage in the films, and some TD shrinkages are also noticed in the films prepared at high line speeds. However, no significant differences are observed in the shrinkage behavior of films as air gap and film thickness change. The shrinkage detected in the cast films of high line speeds is taken as a measure of chain orientation in the amorphous phase.

Keywords: Linear low density polyethylene (LLDPE), Stretch film, Orientation, Cast film extrusion
PACS: 81.05.Lg, 61.05.C, 81.10.Aj

INTRODUCTION

Packaging applications dominate production of polyolefin films which are being continuously developed as an important section of the polymer industry. For example, stretch films are widely used to wrap the goods for protecting them during transportation. Linear low density polyethylene (LLDPE) is the main material used to manufacture stretch films due to its good tensile strength, high stretchability and great puncture and tear resistance. Such properties are required to keep an appropriate article bundling [1-4]. Cast film extrusion and blown film process are the most common techniques for preparation of polymer films. Each process has its own advantages regarding ease of processing and final film properties. However, in the cast film process a rapid cooling is manageable, in which film morphology can be more optimized to obtain desired properties [1, 5, 6]. Altering the fabrication factors of cast film process such as line speed, air gap and film thickness changes the molecular orientation and the final film properties. Mechanical performance of films including tear resistance and tensile properties are primarily determined by molecular orientation [1, 7]. Therefore, understanding the micro-structural features, including orientation in crystal or amorphous phases, in LLDPE cast films prepared under various processing conditions which contribute to the mechanical properties of cast films is of great interest.

In this study, the effects of processing conditions on the crystal orientation in the LLDPE stretch films are investigated. Accordingly, using an SML cast film line, the influence of line speed, air gap and film thickness on the orientation state of LLDPE films is evaluated. In this regard, wide angle X-ray scattering (WAXS) and also shrinkage measurements are utilized to study the orientation state of crystalline and amorphous phases in the LLDPE films.

EXPERIMENTAL

Material

Four different grades of linear low density polyethylene supplied by Braskem Petrochemical company, were selected. The properties of the LLDPE resins used are listed in Table 1. In addition, a processing aid masterbatch obtained from A. Schulman was also used.

TABLE 1. Materials properties.

LLDPE Resin	**MFI (g/10 min)** ASTM D1238 (190 °C/2.16 kg)	**Density** (g/cm^3)
LL1	3.5	0.918
LL2	2.7	0.919
LL3	4.5	0.912
LL Cling	4.3	0.906

Film Preparation

All films in this study were prepared using an SML cast film line (SML Maschinengesellschaft mbH, Austria). The line consisted of four extruders and a feedblock to produce five-layered films. The polymer extrudate exited the die vertically and contacted a chill roll having a diameter of 1600 mm. To keep the film in contact with the chill roll a vacuum box web pinning and an electrostatic edge pinning were used. The films were fabricated at nominal thicknesses of 12 and 23 μm. The air gap was set to 21 mm and 81 mm. Different mechanical line speeds of 200, 400, 600, 800 and 1000 m/min were adjusted for film preparation.

Film Characterization

Wide Angle X-Ray Scattering (WAXS)

The WAXS patterns were obtained utilizing a D/Max Rapid II diffractometer with a Cu anode. Diffraction patterns were corrected and afterwards converted into the corresponding diffractogram by integration. The meridian and equatorial area was separately integrated.

Shrinkage Experiments

Films were cut as rectangles, 10×1.5 cm^2, with longer side parallel to the machine direction (MD) and transverse direction (TD), respectively. Afterwards, a piece of film was inserted into an oven at 90 °C for 20 seconds. Shrinkage measurements were carried out for both machine and transverse directions after removing the films from oven.

RESULTS AND DISCUSSION

Characterizing the molecular orientation as an important structural feature provides the opportunity to describe final properties of polyolefin films. Therefore, WAXS measurements were performed to evaluate the effect of air gap, film thickness and line speed on crystal orientation of the LLDPE stretch films. The representative 2D WAXS diffraction patterns of LLDPE cast films are presented in Figure 1. The reflections corresponding to the (110) and (200) crystal planes are obvious in the patterns. However, various processing conditions were considered to fabricate LLDPE cat films, the reflections are practically uniform which no preferred crystal orientation in the cast films is indicated.

It is observed that, moreover, the reflections of the film prepared at line speed of 1000 (Figure 1d) are very similar to those of the films fabricated at lower line speeds. However, it is expected that using higher line speeds leads to higher molecular orientations in the films regarding lower time scales for chain relaxations.

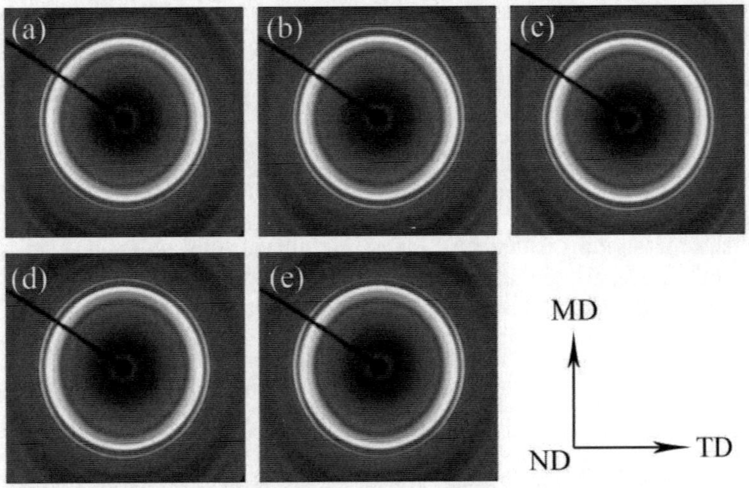

FIGURE 1. 2D WAXS patterns for representative LLDPE cast films prepared under different processing conditions: (a) 21 mm air gap, 23 μm thickness and 400 m/min line speed; (b) 21 mm air gap, 12 μm thickness and 400 m/min line speed; (c) 21 mm air gap, 12 μm thickness and 200 m/min line speed; (d) 21 mm air gap, 12 μm thickness and 1000 m/min line speed; (e) 81 mm air gap, 12 μm thickness and 400 m/min line speed.

The orientation measurements obtained from WAXS patterns are primarily due to the crystalline phase, therefore, using only WAXS measurements no information about the orientation state of the amorphous part can be extracted [8]. Therefore, this might be implied that applying higher line speeds could result in orientation in the amorphous phase. For this reason, thermal shrinkage measurements were also carried out to describe orientation in the amorphous component.

It was observed that shrinkage of samples occurred immediately after insertion in the oven. Therefore, 10 seconds was considered for shrinkage time of all the samples. The shrinkage values of LLDPE films manufactured under various processing conditions were obtained and the results are summarized in Table 2.

TABLE 2. Shrinkage values of LLDPE films prepared under different processing parameters.

Sample	Air gap (mm)	Thickness (μm)	Line speed (m/min)	MD shrinkage (%)	TD shrinkage (%)
1	21	23	400	2	1
2	21	12	400	3	1
3	21	12	200	1	0
4	21	12	1000	5	1
5	81	12	400	3	1

The results listed in Table 2 show that there is no significant differences in the TD shrinkage values of different films, however, it is clear that a decrease in the film thickness (comparing samples 1 and 2) leads to a slightly higher shrinkage value along the MD. The observed shrinkage is mainly due to relaxation of extended chains in the amorphous phase. Reducing the film thickness requires an increase in take-up speed of the chill roll, therefore, higher elongation stresses are applied to the molten film and leads to higher degrees of orientation in the film. Although, the orientation is observed in the amorphous phase.

Increasing the line speed, while keeping the other processing conditions constant, results in higher shrinkage values (comparing samples 2, 3 and 4 in Table 2), which imply higher orientation in the amorphous phase as well. An increase in the line speed provides lower time scales for molten film to relax before contacting the chill roll and solidification. Therefore, polymer chains retain their extended or oriented conformation and higher orientation is noticed. On the other hand, changing the air gap has no effects on shrinkage values of films (comparing samples 2 and 5), neither along MD nor TD.

CONCLUSIONS

LLDPE cast films were manufactured under different processing factors to study the processing effects on the film microstructure in term of crystal or amorphous orientation. To evaluate the crystal orientation of LLDPE cast films, WAXS experiments were utilized which the corresponding 2D WAXS patterns revealed no preferred crystal orientation in the cast films. However, the shrinkage behavior of LLDPE cast films was influenced by line speed and to a lesser degree by film thickness in the cast film process. The observed shrinkage was due to frozen chain orientation in the amorphous phase during solidifying the molten film before contacting the chill roll.

ACKNOWLEDGMENTS

The authors would like to gratefully thank the SML Maschinengesellschaft mbh in Lenzing, Austria and Mr. Robert Preuner (Head of R & D Department at SML) for their support and also providing the opportunity to use an MSL cast film line for preparing the LLDPE stretch films.

REFERENCES

1. C.M. Small, G.M. McNally, G. Garrett, and W.R. Murphy, "The Characteristics of Polyethylene Film for Stretch and Cling Film Applications", *Dev. Chem. Eng. Mineral Process.* **12**, 5-20 (2004).
2. C. A. Cateto, M. J. Vinek, A. J. Van Loon, B. Lauwers, U.S. Patent No. 2018/0257286 Al (13 September 2018).
3. S. E. Pirtle, U.S. Patent No. 2015/0148492 Al (28 May 2015).
4. J. Park, L. Horvath, M. White, P. Araman, R. J. Bush, "The Influence of Stretch Wrap Containment Force on Load Bridging in Unit Loads", *Packag Technol Sci.* **31**, 701-708 (2018).
5. C. M. Small, G. M. McNally, A. Marks, and W. R. Murphy, "The Effect of Extrusion Processing Conditions and Polyisobutylene Concentration on the Properties of Polyethylene for Stretch and Cling Film Applications", *Journal of Plastic Film & Sheeting*, **18**, 245-258 (2002).
6. J. A. Degroot, A. 1. Doughty, K. B. Stewart, and R. M. Patel, "Effects of Cast Film Fabrication Variables on Structure Development and Key Stretch Film Properties", *Journal of Applied Polymer Science*, **52**, 365-376 (1994).
7. R.K. Krishnaswamy, A.M. Sukhadia, "Orientation Characteristics of LLDPE Blown Films and Their Implications on Elmendorf Tear Performance", *Polymer* **41**, 9205-9217 (2000).
8. S. H. Tabatabaei, P. J. Carreau, and A. Ajji "Effect of Processing on the Crystalline Orientation, Morphology, and Mechanical Properties of Polypropylene Cast Films and Microporous Membrane Formation", *Polymer* **50**, 4228-4240 (2009).

Time-dependent rheological and electrical properties of LLDPE/EVA blends containing graphene

Emna Helal[a], Rafael S. Kurusu[a], Eric David[a], Nima Moghimian[b] and Nicole R. Demarquette[a]

[a]*École de Technologie Supérieure, Département de Génie Mécanique, Montreal, Canada*
[b]*NanoXplore Inc., Montreal, Canada*

Abstract. In this study, the dynamic viscoelastic moduli and electrical conductivity of blends of linear low density polyethylene (LLDPE) and ethyl vinyl acetate (EVA) presenting a co-continuous morphology, filled with commercial graphene (GN) nanoplatelets, have been assessed as function of thermal annealing time and correlated with changes in the morphologies of the blends. It was observed that both the storage modulus and electrical conductivity of the filled blends increased during the early stage of annealing and then stabilized over time, These evolutions were attributed to a coarsening of blend's morphology which was observed by scanning electron microscopy.

Keywords: dynamic percolation threshold, thermal annealing, co-continuous blends.
PACS: 83.80.Tc (Polymer blends)

INTRODUCTION

Among several methods investigated to reduce the percolation threshold of conductive polymer composites, blending has been proven to be an efficient tool, especially in the case of co-continuous blends [1-3]. However, when annealed at temperatures above their softening or melting points, the morphologies of the as-obtained blends usually evolve over time. Besides, their rheological and electrical percolation thresholds are often affected as a consequence of the diffusion and migration of particles and subsequent reconstruction of the networks [4-8]. This concept is known as dynamic percolation [6-7]. As these changes occur simultaneously, assessing the evolution of the electrical and rheological properties as well as the evolution of the morphology over time might reveal strong correlation between electrical and rheological percolations despite the different natures of the responsible elastic and conductive networks.

In a previous work [8], co-continuous linear low density polyethylene/ethyl vinyl acetate/graphene (LLDPE/EVA/GN) blends, have been prepared by melt mixing and characterized in terms of microstructure, rheological properties and electrical properties. In particular, it was found that blending reduced remarkably the electrical percolation threshold and that the presence of GN nanoplatelets induced an important refinement of the blend's morphology, compared to the unfilled blend. Moreover, it was observed that upon thermal annealing for 2 hours, the electrical conductivity of the blends in the near-percolation region increased by several orders of magnitude. Simultaneously, their viscoelastic moduli increased progressively during thermal annealing, especially during the first 30 minutes. In this study, the evolutions of the electrical conductivity and blend's morphology were also monitored as function of annealing time and compared to the observed time-dependant viscoelastic behavior in order to evaluate the potential correlations. To emphasize further the effect of thermal annealing on the properties, the rheological and electrical percolation thresholds were estimated quantitatively, before and after annealing, using a scaling Power law. In particular, corresponding threshold concentrations and critical exponents were discussed in light of the observed changes of the properties over time.

EXPERIMENTAL

Materials and Processing

The polymers used in this study are LLDPE and EVA. LLDPE (grade NOVAPOL® PI-2024-A) was purchased from Nova Chemicals and EVA (grade Repsol, Alcudia® PA-540) was supplied from Entec Polymers. Graphene nanoplatelets (GN) (heXo-G V20) were provided by NanoXplore, with nanoplatelets mean thickness of 20 nm and average flake size of 50 μm. This grade is produced by mechano-chemical exfoliation of graphite and is functionalized with hydroxyl and carboxyl groups [8-9].

LLDPE/EVA/GN composites were fabricated by extrusion using a twin screw extruder Haake Rheomix OS PTW16 operated at a temperature equal to 160 °C in all zones. Two masterbatches have been used in order to tailor the processing sequence in the blends: an EVA masterbatch containing 40 wt% GN and an LLDPE masterbatch containing 25 wt% GN. The two masterbatches are referred to as EVAMB and LLDPEMB, respectively. The weight ratio LLDPE/EVA was equal to 50/50 in all the materials to obtain composites with co-continuous morphology. Selected GN concentrations up to 10 wt% were prepared. In the rest of the paper, LLDPE/EVA/XGN (LLDPEMB) and LLDPE/EVA/XGN (EVAMB) will stand, respectively, for composites prepared from LLDPEMB and EVAMB masterbatches, with X the wt% of graphene. After extrusion, samples were hot pressed at 160 °C in order to get films for electrical and rheological characterization. Besides, some samples were annealed inside the press for different periods up to 2 hours at 160 °C. More details regarding the processing and preparation of composites specimens can be found in our previous study [8].

Characterization

Scanning electron microscopy (SEM) was performed, using S3600N microscope (Hitachi) operated at 5kV, to characterize the morphologies of the blends and their evolution during thermal annealing. Before observation, the samples were fractured in liquid nitrogen and the EVA phase was extracted with toluene.

The complex permittivity, and subsequently the complex conductivity, of the composites were evaluated by broadband dielectric spectroscopy in the frequency range from 10^{-2} Hz to 300 kHz. The conductivity values presented in this work refer to the real part of the complex conductivity (σ') at the frequency 10^{-2} Hz, which can be assumed equal to DC conductivity at these conditions. All samples were coated with a thin layer of platinum before measurements.

The rheological measurements were performed using a Rotational Rheometer (MCR 501, Anton Paar) equipped with 25 mm-diameter parallel plate geometry. In particular, time Sweep tests were conducted, at a strain in the linear viscoelastic region, to assess the evolution of the composites morphology and viscoelastic behavior as function of time. The tests were carried out at 160 °C for 2 hours at 10^{-1} rad.s^{-1}.

More details regarding the characterization can be found in our previous study as well [8].

RESULTS

In our previous study [8], it was observed that the viscoelastic moduli of the different LLDPE/EVA/GN composites were increasing when subjected to thermal annealing during time sweep tests. Moreover, it was observed that the electrical conductivity increased and the morphology of the blends coarsened after 2 hours of thermal annealing. In order to elucidate better this correlation, the electrical conductivity and the characteristic domain size ζ were evaluated at different annealing times ranging from 0 to 2 hours for different LLDPE/EVA/GN composites. In particular, the characteristic domain size of each blend composite was estimated from the SEM images using the equation (1) described below.

$$\zeta = \frac{A_{SEM}}{L_{int}} \qquad (2)$$

Where:
A_{SEM}: total area of the SEM image.
L_{int}: interface length estimated by the software imageJ

The results are summarized in Figure 1 for LLDPE/EVA/5GN (LLDPEMB) nanocomposite as a representative material. Similar trend was observed for the rest of the tested composites as well. From Figure 1(a) and data reported in previous study [8], it can be seen that the storage modulus G' of LLDPE/EVA/5GN (LLDPEMB) nanocomposite increases especially during the first 30 minutes of annealing then stabilizes progressively. The electrical conductivity follows the same trend with an initial increase of more than 1 order of magnitude at 28 minutes of annealing, followed by a more stable behavior. As far as the morphology is concerned, the characteristic domain size increases at 30 and 70 minutes of annealing indicating a coarsening of the morphology, then ζ dropped but remained higher than its initial value. Overall, it can be concluded from these results that the substantial increase of both rheological and electrical quantities occurs during the first 30 minutes of annealing. This increase is accompanied by a coarsening of the blend's morphology. As mentioned earlier, this behavior is attributed to diffusion of GN particles and reconstruction of elastic and conductive networks. More precisely, transmission electron microscopy (TEM) images performed on the annealed materials confirm the migration of GN nanoplatelets to the blend's interface during the first 30 minutes of annealing. The lateral size of the diffusing GN nanoplatelets was around 1 µm and less. These TEM results will be discussed more in details in a separate study.

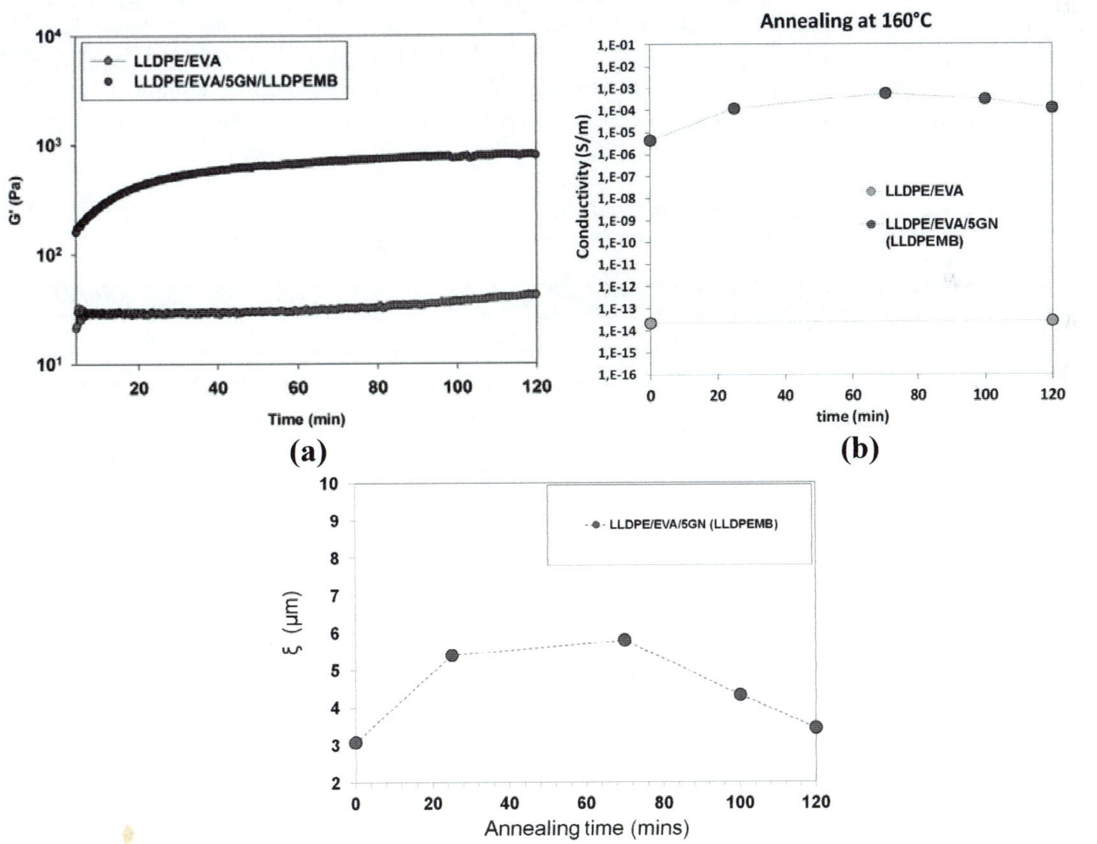

(c)

Figure 1: Evolution of (a) storage modulus G', (b) electrical conductivity and (c) characteristic domain size ζ of LLDPE/EVA/5GN (LLDPEMB) nanocomposite as function of annealing time

To quantify the effect of annealing on the percolation thresholds and highlight further the observed correlations, the electrical and rheological percolation threshold concentrations Φ_c were estimated using a scaling Power law (equation (2)). In addition to the prediction of Φ_c, the analysis of the critical exponent gives an insight about the network structure [10]. Before using this simplified approach, originally designed to model the percolation threshold of single-phase systems, on the LLDPE/EVA/GN blend composites featuring complex morphology, the storage modulus of these materials was successfully modeled using a two-phase model indicating that the viscoelastic response of these composites can be described irrespective of the complexity of their morphologies. The modeling of the viscoelastic behavior of the LLDPE/EVA/GN composites will be discussed further in a different study.

$$\sigma'_0, G'_0 = A(\varphi - \varphi_c)^t \tag{2}$$

Where

- G'_0: storage modulus above percolation
- σ'_0: electrical conductivity above percolation
- A: constant reflecting the strength of the network
- φ_c : percolation threshold concentration
- t: critical exponent depending on the aspect ratio of the filler and describing the spatial distribution of the network

The fitting parameters related to rheological percolation are reported in Table 1 while the fitting parameters related to electrical percolation are reported in Table 2. In each case, the best fitting has been determined by selecting Φ_c value corresponding to the highest regression coefficient. It can be seen from Table1 that LLDPE/EVA/GN (LLDPEMB) composites have higher rheological percolation threshold concentration compared to LLDPE/EVA/GN (EVAMB): Φ_c = 1 vol% (2.4 wt%) vs. Φ_c = 0.5 vol% (1.2 wt%). Besides, the values of the critical exponent t are similar for both sets of materials: 1.53 vs. 1.48 for LLDPE/EVA/GN (LLDPEMB) vs. LLDPE/EVA/GN (EVAMB). In the context of rheological percolation, it has been reported that values of t lower than 2.1 correspond to particle networks mediated by polymer chains [11].

Table 1: Power law parameters for estimation of rheological percolation threshold of LLDPE/EVA/GN (LLDPEMB) and LLDPE/EVA/GN (EVAMB) nanocomposites

	Φ_c	t	R^2
LLDPE/EVA/GN (EVAMB)	**0.5**	1.48	0.9704
LLDPE/EVA/GN (LLDPEMB)	**1**	1.53	0.9909

In terms of electrical percolation, it can be seen from Table 2 that before annealing, Φc values were estimated to be 1.7 vol% (3.9 wt%) and 2.5 vol% (5.7 wt%) for LLDPE/EVA/GN (LLDPEMB) and LLDPE/EVA/GN (EVAMB) sets, respectively. However, after annealing, Φc values were reduced to 1.21 vol% (2.8 wt%) and 0.5 vol% (1.2 wt%) for LLDPE/EVA/GN (LLDPEMB) and LLDPE/EVA/GN (EVAMB) sets respectively, both very close to those estimated for rheological percolation. Before annealing, the critical exponent values were estimated to be 2.59 and 4.05 for LLDPE/EVA/GN (LLDPEMB) and LLDPE/EVA/GN (EVAMB) sets, respectively. After annealing, they increased to 3.88 and 5.28, respectively, indicating more dependence of the electrical conductivity above the percolation threshold on the filler loading and the network spatial distribution. High values of the critical exponent t, in the same range, have been reported in the literature for different GN nanocomposites and attributed to the high aspect ratio of the plate-like fillers [12-13].

The similarity in electrical and rheological percolation thresholds after annealing agrees well with the observed strong correlation in the time-dependent increase of both electrical conductivity and viscoelastic modulié This similarity has been reported in the literature in some cases [14] while in other studies the rheological percolation threshold was observed earlier than electrical percolation threshold [12]. In the latter case, the discrepancy was attributed to the different nature of elastic and electrical networks.

Table 2: Power law parameters for estimation of electrical percolation threshold (before and after thermal annealing) of LLDPE/EVA/GN (LLDPEMB) and LLDPE/EVA/GN (EVAMB) nanocomposites

	Φ_c	t	R^2
LLDPE/EVA/GN (EVAMB)	**2.5**	4.05	0.9989
LLDPE/EVA/GN (LLDPEMB)	**1.7**	2.59	0.9912
LLDPE/EVA/GN (EVAMB) (annealed)	**0.5**	5.28	0.9348
LLDPE/EVA/GN (LLDPEMB) (annealed)	**1.21**	3.88	0.9813

CONCLUSION

In this study, the correlation between the time-dependent electrical conductivity and time-dependent viscoelastic storage modulus of LLDPE/EVA/GN co-continuous blend nanocomposites, during thermal annealing, has been established. An increase at a similar time scale has been observed for both quantities and attributed to diffusion and migration of particles to the blend's interface leading to the reconstruction of electrical and elastic networks. In support of this correlation, the rheological and electrical percolation threshold concentrations, estimated using a scaling Power law, were reduced and almost equal after thermal annealing; the lowest value is equal to 0.5 vol% and corresponds to LLDPE/EVA/GN (EVAMB) set of composites.

ACKNOWELDGEMENTS

Mitacs, Natural Sciences and Engineering Research Council of Canada (NSERC) and École de Technologie Supérieure (ÉTS) are acknowledged for financial support.

REFERENCES

1. L. Bai, S. He, J. W. Fruehwirth, A. Stein, C. W. Macosko and X. Cheng, J. Rheol., 2017, 61, 575–587.
2. R. Altobelli, M. Salzano de Luna and G. Filippone, Interfacial crowding of nanoplatelets in co-continuous polymer blends: assembly, elasticity and structure of the interfacial nanoparticle network, *Soft Matter*, 2017, 13, 6465-6473.
3. L. Bai, R. Sharma, X. Cheng, C.W. Macosko, Kinetic Control of Graphene Localization in Co-continuous Polymer Blends via Melt Compounding, Langmuir, vol 34(3), pp. 1073-1083, 2018.
4. Ingo Alig, Petra Pötschke, Dirk Lellinger, Tetyana Skipa, Sven Pegel, Gaurav R. Kasaliwal, Tobias Villmow, Establishment, morphology and properties of carbon nanotube networks in polymer melts, Polymer, Volume 53, Issue 1, 5 January 2012, Pages 4–28.
5. A. Göldel, G.R. Kasaliwal, P. Pötschke, G. Heinrich, "The kinetics of CNT transfer between immiscible blend phases during melt mixing", Polymer, vol. 53, pp. 411-421, 2012.
6. W. Li, Y. Zhang, J. Yang, Jun Zhang, Y. Niu, and Z. Wang, Thermal Annealing Induced Enhancements of Electrical Conductivities and Mechanism for Multiwalled Carbon Nanotubes Filled Poly(Ethylene-co-Hexene) Composites, ACS Appl. Mater. Interfaces, 2012, Volume 4 (12), pp 6468–6478.
7. C. Zhang, P. Wang, C. Ma, G. Wu, M. Sumita,Temperature and time dependence of conductive network formation: Dynamic percolation and percolation time, Polymer, Volume 47, Issue 1, 2006, Pages 466-473.
8. R. S. Kurusu, E. Helal, N. Moghimian, E. David, N. Demarquette, The role of selectively located commercial graphene nanoplatelets in stabilizing the morphology and electrical properties of EVA/LLDPE blends, manuscript submitted to Macromolecular Materials and Engineering (March 2018).
9. N. Moghimian, S. Saeidlou, H. Lentzakis, G.F. Rosi, N. Song, E. David, "Electrical conductivity of commercial graphene polyethylene nanocomposites", 2017 IEEE 17th International Conference on Nanotechnology (IEEE-NANO), pp. 757-761, 2017.
10. Altobelli, R., M. Salzano de Luna, and G. Filippone, *Interfacial crowding of nanoplatelets in co-continuous polymer blends: assembly, elasticity and structure of the interfacial nanoparticle network*. Soft Matter, 2017. **13**(37): p. 6465-6473.
11. Filippone, G., et al., Elasticity and structure of weak graphite nanoplatelet (GNP) networks in polymer matrices through viscoelastic analyses. Polymer, 2012. 53(13): p. 2699-2704.
12. Gao, C., et al., Non-contact percolation of unstable graphene networks in poly(styrene-co-acrylonitrile) nanocomposites: Electrical and rheological properties. Composites Science and Technology, 2018. 155: p. 41-49.
13. Gao, C., et al., Graphene Networks with Low Percolation Threshold in ABS Nanocomposites: Selective Localization and Electrical and Rheological Properties. ACS Applied Materials & Interfaces, 2014. 6(15): p. 12252-12260.
14. Kim, H., et al., Graphene/polyethylene nanocomposites: Effect of polyethylene functionalization and blending methods. Polymer, 2011. 52(8): p. 1837-1846.

Influence of Coupling Agents on Blend Properties

Johannes Heyn[a] and Christian Bonten

IKT, Institut für Kunststofftechnik, University of Stuttgart, Pfaffenwaldring 32, 70569 Stuttgart, Germany
[a)]Corresponding author: Johannes.Heyn@ikt.uni-stuttgart.de

Abstract. According to Gibbs, the free mixing energy is decisive for the thermodynamic compatibility of two substances, e. g. polymers. If the free mixing energy is positive, the two mixing partners are separated. For such molecularly immiscible polymers, a finely distributed disperse phase within the matrix can only occur if drop breakage happens during processing. This depends on the strain and shear rates and the capillary number, which must exceed a critical value. If all preconditions are met, coupling agents can be used to improve the phase adhesion of two given polymers and to stabilize the morphology. This is due the reduction of the interfacial tension, thus facilitating drop breakage and simultaneously increasing interfacial tension, which suppresses coalescence. In this work, the effect of three different coupling agents is investigated to evaluate their influences on the property changes in polyamide 6/ethylene-octene-copolymers-blends. It can be shown that the rheological behavior as well as the blend morphology strongly depends to the type and the amount of the used coupling agent. A direct correlation between the rheological behavior and the blend morphology can be assumed but not fully confirmed. For this purpose, the particle size of the soft phase within the matrix is analyzed optically by a scanning electron microscope. Fracture mechanical investigations are carried out to evaluate the blend modifications to the mechanical properties. The results obtained using the method of essential work of fracture shows a significant influence due to the type of coupling agent, but less due to morphology. Therefore, in order to better understand the structural property relationships, the selection of coupling agent is as important as the blend morphology.

Keywords: fracture mechanics, morphology, polyamide, polymer blend, rheology, thermoplastic elastomers
PACS: 64.75.Va, 81.05.Lg, 81.40.Np, 83.80.Tc, 83.80.Va

INDRODUCTION

There are many ways to optimize or create materials with new or improved properties. A common and cost-effective method for polymers is a mixture of two or more components into a so-called blend. Basically, there are two types of blends, homogeneous and heterogeneous blends. The blending result in heterogeneous blends can be significantly improved by using various modifiers, such as coupling agents, radical formers, cross-linking agents and so on. By the selection and the mixing ratio of the basic components, the substantial blend properties can be estimated relatively well in advance. However, the final blend properties and the achievement of the entire blend potential require a closer look inside the material. The blend morphology, which can be varied over a broad range using modifiers and processing parameters, is the focus of interest. Only those who understand the morphology and the relationship to the blend properties can achieve synergistic effects from the blend components and achieve the best properties for the blend.

STATE OF THE ART

The chemical structures as well as the structure of the blend partners are crucial for the development of the blend morphology. According to Gibbs, heterogeneous polymer blends have a positive free mixing energy. The thermodynamic incompatibility increases with increasing chain length and the individual components therefore tend to separate. These segregated components can form different morphologies. In many cases, a spherical morphology is desired. One component dominates and forms the matrix, the other component is finely distributed within the matrix. Double-spherical, fibrillary, platelet- and layer-shaped or a co-continuous morphology are also possible. The morphology is influenced by the material properties and the processing conditions. For processing, the blend and viscosity ratio of the components are important factors. Therefore, a critical capillary number of about 3.8 should not be exceeded, so that shear and extensional flow together result in drop breakage [1]. If additional modifiers such as coupling agents and radical formers are used to influence the morphology, the activation energy, the chemical conversion rate and quantity as well as the residence time of all components must be taken into account. Coupling agents reduce the interfacial tension of the grafted phase and increase the interfacial adhesion to the matrix. This facilitates droplet breakage and suppresses coalescence, thus producing a stabilizing fine spherical morphology. In this respect, controlling the specific morphology is of interest, as many blend properties such as impact strength and crack resistance are closely related to morphology. According to Bucknall [2] there is an optimal particle size at which the impact strength is maximized, according to own investigations [3] the fracture toughness increases with decreasing particle sizes. Many other blend properties, such as strength, follow the linear mixing rule.

In this work, the influence of three coupling agents in different amounts on various blend properties of a polyamide 6/thermoplastic elastomer blend will be investigated. A relationship between rheological, morphological and fracture mechanical properties will be developed in order to be able to predict blend properties from structure-property relationships.

EXPERIMENTAL

Materials

The materials used are polyamide 6, Typ Ultramid® B40 (PA6) provided by BASF SE, Germany, and thermoplastic ethylene-octene-copolymer ENGAGE™ 8100 (EOR) provided by The Dow Chemical Company, Spain. Maleic anhydride (MAH), itaconic acid (ITA) and glycidyl methacrylate (GMA) are used as coupling agents and dicumyl peroxide (P) as radical former. PA6/EOR-blends with a fixed amount of 20 wt.-% EOR and varying amounts of the coupling agents (1.0, 2.0 and 4.0 wt.-%) and peroxide (0.1 and 1.0 wt.-%) are produced on a co-rotating, intermeshing twin screw extruder, type ZSK 26 from Coperion, Germany, with a l/d ratio of 40. A vacuum furnace at 80 °C is used to dry the PA6 for 4 h. The blends are produced at a throughput of 8 kg/h and a screw speed of 180 rpm. After the compounding process plates (60 mm x 40 mm x 2 mm) are prepared on an injection molding machine, type Allrounder 520S from Arburg, Germany. The mold temperature is set to 80 °C and the temperature profile is varied from 180 °C to 290 °C. All plates were annealed for 12 h at 180 °C in a vacuum furnace.

In the following, the blends are named according to their composition. The first number describes the soft phase amount and the second number the amount of the coupling agent (MAH, ITA, GMA) related to the EOR amount multiplied by a factor of 100. The amount of peroxide (P) is specified analog to the amount of coupling agent. For example, a blend with 20 wt.-% EOR, 1 wt.-% coupling agent and 0.1 wt.-% P is called 20EOR100M10P.

CHARACTERIZATION

Thermal Investigations

The used modifier could shift characteristic temperatures or transitions of the blend. So, it is possible to deduce their chemical modes of action. For this purpose, the crystallite melting temperature, the melting enthalpy and the extrapolated onset crystallization temperature are determined for each blend. The analyses of the thermal properties are carried out on a calorimeter type DSC 2/700 from Mettler Toledo GmbH, Germany, with a heating and cooling rate of 10 K/min in a nitrogen atmosphere. The first heating cycle starts from -100 °C to 280 °C, the subsequent cooling cycle from 280 °C to 100 °C, and the second heating cycle from 100 °C to 300 °C.

Morphological Investigation

The determination of the particle size and distribution of the EOR particles within the PA6 matrix is a fundamental part of the determination of the mode of action of the modifiers. In addition, particle size is an important parameter in the evaluation of the blend morphology and is necessary for correlation with fracture mechanical properties. To determine the particle size, the injection-molded plates are divided in the middle and cut with a microtome. Then the interfaces are immersed in boiling xylene for 4 h to remove the EOR particles from the surface of the PA6 matrix. After a subsequent drying process of 2 h at 80 °C in a vacuum furnace, the samples are vaporized with a fine gold layer. Subsequently, images of the prepared surface are taken under a scanning electron microscope (SEM) type JSM 6300 F from Jeol, Germany. Using the image analysis software Leica Qwin from Leica, Germany, the mean particle sizes are determined.

Rheological Investigations

Rheological investigations are performed to analyze the thermal stability of each blend as well as to investigate the mode of action of the modifier. The rheological properties of the produced blends are determined for the injection-molded plates by means of a plate-plate rheometer type R HR-2 from TA Instruments, USA. All measurements are done in an oscillation mode under nitrogen atmosphere in the linear viscoelastic region. The determination of the storage and loss moduli, the complex viscosities and the viscosity ratios is carried out in frequency sweeps. The analyzed angular frequency range is 0.1 rad/s to 100 rad/s, whereby the measurement starts at 100 rad/s. The set measuring gap is uniformly 2 mm. In addition, time sweeps are performed in the linear viscoelastic region for 30 min at 240 °C and at a constant angular frequency of 1 rad/s to evaluate the thermal stability of the blends.

Fracture Mechanical Investigation

Fracture mechanical investigations by means of the method of essential work of fracture (EWF) are performed to analyze the blend behavior and to evaluate the mode of action of the modifier. Furthermore, a dependence of the fracture mechanical properties on the blend morphology will be shown. For this purpose, the injection-molded plates are prepared with a deep notch in the middle on both sides, with a 3 mm long razor blade cut at the notch base. Double edge notched tension (DENT) specimens with different ligament lengths are investigated.

RESULTS AND DISCUSSION

Thermal Properties

Fig. 1 left shows the specific melting enthalpy as well as the melting temperature of the each blend depending on different coupling agents and amounts. The two dashed red lines show the values of the unmodified PA6/EOR blends. For molecularly non-miscible blends, the melting temperature and the onset crystallization temperature (not shown here) should not be changed without adding modifiers [4, 5]. However, the coupling agents and amounts as well as the radical formers used here have only a minor influence on the onset crystallization temperature. All values lie with small deviations ($\Delta\vartheta \pm 1$ K) around the initial value of the unmodified blend. When considering the specific melting enthalpy, the influence of the modifiers becomes more obvious. As soon as EOR is covalently bonded to PA6, this should have a crystallization inhibiting effect on PA6 [6]. This can be confirmed for the coupling agents MAH and GMA, but not for ITA. A high amount of coupling agent even increases the specific melting enthalpy. This result indicates that ITA cannot produce covalent bonds between the two blend partners. Although the amount of P used influences the thermal properties in combination with all coupling agents, the results cannot show a trend.

Rheological Properties

Fig. 1 right shows the viscosity ratio of the two blend partners over a temperature range from 240 °C to 280 °C, which corresponds to the processing temperatures. With a maximum viscosity ratio of 1.0 at a test temperature of 280 °C, which is well below the critical viscosity ratio for shear flow of 3.8, the optimum conditions for droplet breakdown by means of shear and extension stress during the processing process are given.

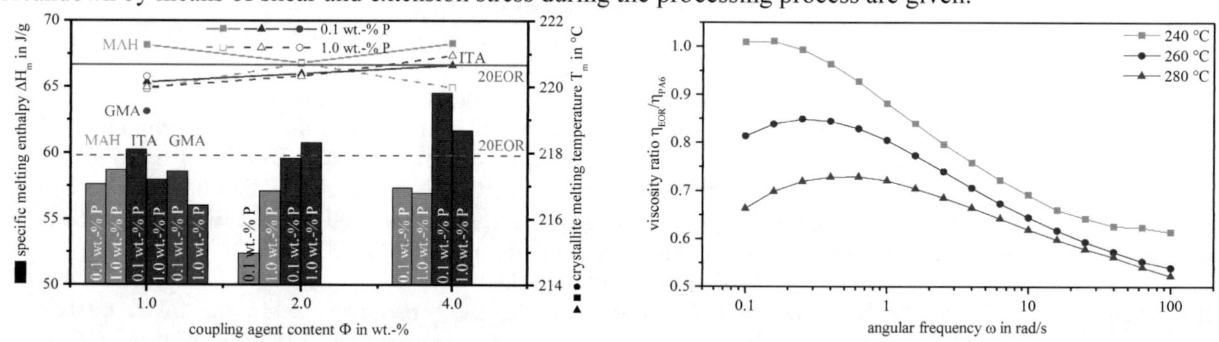

FIGURE 1. Left: thermal properties of the blends; right: viscosity ratio of the basic blend partners

Fig. 2 left shows exemplary time sweeps of selected blends and the two blend partners at a measurement time of up to 30 min and a specified angular frequency of 1 rad/s. If the temporal course of the complex viscosity for the blends is considered, they are thermally stable. If ITA is used, a slight degradation can be observed; if GMA is used, a slight build-up can be observed. The typical post-condensation of PA6 as well as the typical degradation of EOR under thermomechanical stress compensate each other by blending. Based on the different levels of the time sweeps, a different mode of action of the modifiers can be assumed. The frequency sweeps in Fig. 2 right provide a detailed analysis. Fig. 2 right shows the complex viscosity plotted over the angular frequency to illustrate the influence of the modification. The unmodified blend has a similar flow behavior to the two basic materials. Thermomechanical grafting can never be excluded in PA6/EOR blend production, which explains the low increase in complex viscosity at low angular frequencies [7]. The complex viscosity changes drastically only after the modification. First, the complex viscosity increases with increasing MAH content until it decreases again at the maximum concentration of MAH. A higher P amount at the same MAH amount leads to a lower increase in viscosity than with a lower amount. Only for the maximum concentration of MAH, where the complex viscosity decreases again, an increase can be achieved by a higher P content. Obviously, there is a maximum amount of MAH added to the coupling agent, above no further improvement can be achieved. This should also be apparent in fracture mechanics examination. The increase in viscosity is attributed to successful grafting and coupling of the EOR to the PA6. Fig. 3 right shows a comparison between the effects of different coupling agents on the flow behavior of selected blends. For all coupling agents, the low amount of P always leads to a lower complex viscosity. The coupling agent MAH leads to the most significant increase in complex viscosity, followed by GMA. ITA, on the other hand, shows only a minor influence on the flow behavior compared to the unmodified blend. Thus, the findings from the thermal investigations in Fig. 1 left can be confirmed for ITA. The different flow behavior of the blends, which is caused by chemical and physical interactions between the two blend partners, also affected the blend morphology.

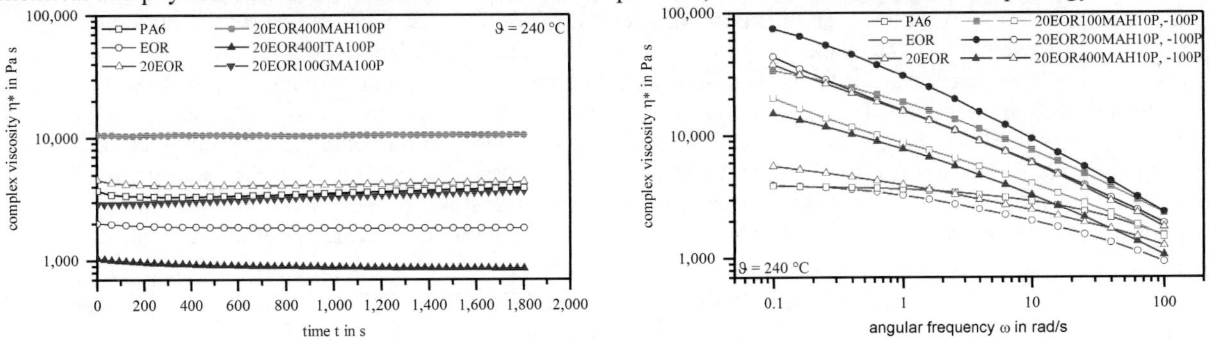

FIGURE 2. Left: time sweeps of different modified blends; right: frequencies sweeps of MAH modified blends

Morphological Properties

Fig. 3 right shows SEM images of modified blends. The 20EOR100GMA100P blend has an irregular morphology and is taken out of consideration. The remaining blends show an explicit dependency of the resulting particle sizes on the complex viscosity. The higher the complex viscosity of the blend, the smaller and more homogeneous the particles are. If the particle sizes or the complex viscosity do not vary significantly, this statement cannot be confirmed in detail. Tab. 1 shows an overview of the particle sizes of all modified blends.

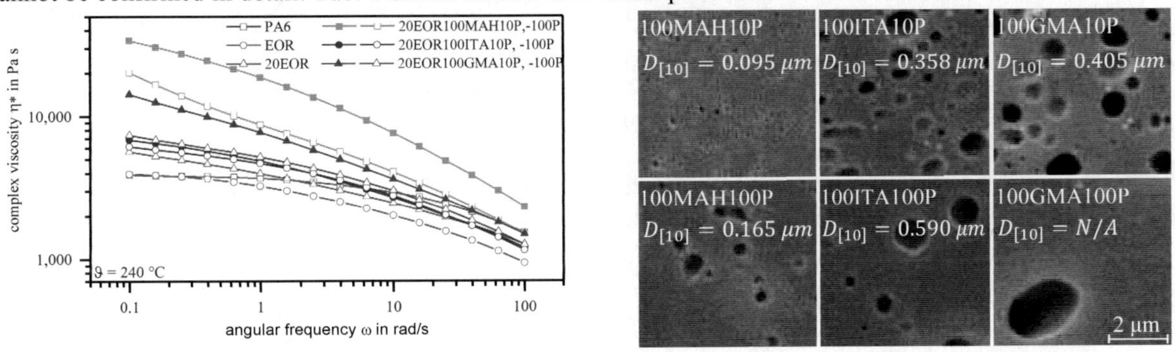

FIGURE 3. Left: frequencies sweeps of different modified blends; right: SEM-images of different modified blends

TABLE 1. Overview of the particle size of all blends

blend recipe		particle size d in µm		
Peroxide	**Coupling agent**	**100**	**200**	**400**
10 P	MAH	0.095	0.079	0.198
100 P		0.169	0.179	0.146
10 P	ITA	0.358	0.574	0.632
100 P		0.590	0.511	0.662
10 P	GMA	0.405	N/A	
100 P		N/A		

The blend morphologies of the GMA modified blends show a poor dispersive as well as distributive distribution of the particles. The blend morphology differs considerably from the blend morphology of the unmodified blend. Thus, no reliable evaluation via particle size can be performed for these blends. For the remaining blends a good agreement with the complex viscosities of Fig. 3 left can be found again, especially if the shear thinning region is considered at high angular frequencies. But here, as well, blurring occurs, especially when the particle sizes and the complex viscosities of the blends are close together. The results for the ITA modified blends are conspicuous again because the variation of the achievable particle sizes is only minimal.

Fracture Mechanical Properties

The consideration of the specific EWF (Fig. 5 left) as a function of the blend modifications and the rheological properties confirm the dependency of the blend morphology (Tab. 1) on the complex viscosity (Fig. 2 right, Fig. 3 left). A comparison of the same coupling agent at the same amount, but with varying P amount, always leads to a higher specific EWF for smaller particle sizes and larger complex viscosities than for larger particle sizes or smaller complex viscosities. A general derivation that the particle size can be used as the exclusive criterion can be clearly disproven. Since the coupling agents influence both the particle size and the phase adhesion of the blend partners, and this in different ways, depending on the coupling agent, it is not possible to formulate a statement covering all coupling agents. An analysis of the results of the specific n-EWF of the each blend modifications in Fig. 5 right does not provide any significant further information. A lower P amount always leads to better values than a higher P amount in a MAH modification. For an ITA addition, the opposite is true. In general, the initial value of the unmodified blend can almost be exceeded. There is no relationship between particle size, or rheological properties, and the specific EWF. This confirms statements by Bárány et al. [8] that the specific n-EWF is not sensitive enough for changes at the molecular level and that the consideration of the specific EWF for such questions should be put in the foreground. This is supported by the fact that the specific n-EWF does not provide material information, but also depends on the specimen geometry.

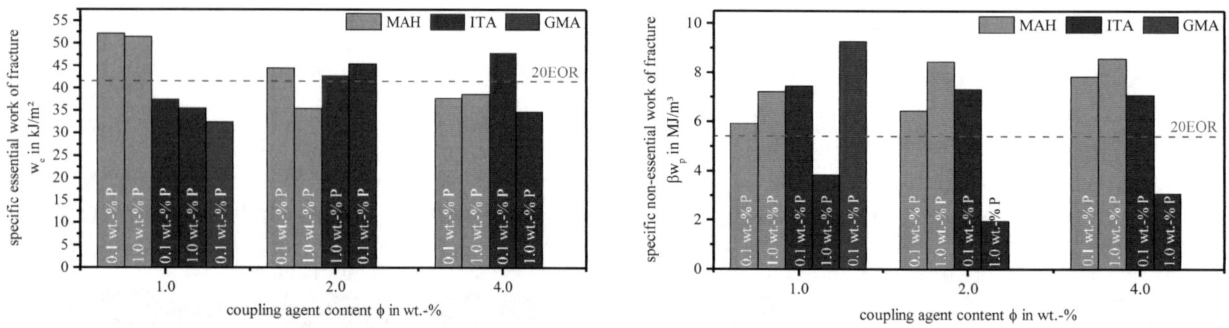

FIGURE 4. Left: results of the specific n-EWF; right: results of the specific EWF

CONCLUSION

The investigation of the influence of three different coupling agents on PA6/EOR blends showed that they have a strong influence on the blend morphology, the flow behavior and the fracture mechanical properties. A shift of the the complex viscosity to higher values is a clear indication of a strong chemical and / or physical interaction. The surface-volume ratio between the particles and the matrix is also strongly influenced by the modification, which is also expressed in the changed flow behavior. Thermal analyses can only provide indications of this behavior, if at all. A pure consideration of the particle sizes within the matrix in order to deduce fracture mechanical properties is not suitable, since the phase adhesion is not fully recorded. Thus, the mode of action of different coupling agents cannot be assessed on the basis of the particle size alone. If the individual coupling agents are considered isolated from each other, smaller particle sizes usually lead to improved fracture toughness. The method of EWF and especially the consideration of the specific EWF can be reliably applied here.

ACKNOWLEDGMENTS

This research project was funded by the German Research Foundation (DFG) under the code BO 1600/30-1.

REFERENCES

1. H.P. Grace, *Chemical Engineering Communications*, **14**, 225 (1982).
2. C.B. Bucknall and D.R. Paul, *Polymer*, **54**, 320 (2013).
3. J. Heyn and C. Bonten, *AIP Conference Proceedings*, **2055**, 30002 (2019).
4. K. Premphet-Sirisinha and S. Chalearmthitipa, *Polym. Eng. Sci.*, **43**, 317 (2003).
5. R.J.M. Borggreve, R.J. Gaymans, J. Schuijer et al. *Polymer*, 1489 (1987).
6. C. Marco, G. Ellis, M.A. Gómez, et al. *Journal of Applied Polymer Science*, **65**, 2665 (1997).
7. G. Illig, *Die Angewandte Makromolekulare Chemie*, **95**, 83 (1981).
8. T. Bárány, T. Czigány, and J. Karger-Kocsis, *Progress in Polymer Science*, **35**, 1257 (2010).

Influence of Reactive Melt Mixing on Mechanical Behavior of Chain Extended PLLA/Thermoplastic Polyurethane Blends

Mohammad Moin Garmabi[a], Elnaz Erfanian[a], Hamid Garmabi[a] and Omid Yousefzade [a*]

[a]*Department of Polymer Engineering and Color Technology, Amirkabir University of Technology, Iran*

Abstract. Control of thermal degradation and melt strength of Poly (lactic acid), PLA, during melt processing while improving the toughness of the obtained samples is still a challenge for the polymer industry. Poly (L-lactide acide) (PLLA)/Thermoplastic polyurethane (TPU) blends were prepared in compositions of 70/30% and 50/50% in presence of multifunctional epoxy chain extender. A laboratory internal mixer was used to prepare the samples via reactive melt mixing. The effect of TPU on tensile modulus and elongation at break of chain extended PLLA were investigated. The tensile properties of the obtained samples show significant improvement in ductility of the PLA/TPU blends whereas chain extender contributed positive effect on the elastic modulus and strain at break of prepared blends. SEM micrographs of the samples show improved adhesion at the interface for samples incorporating multifunctional chain extender.

Keywords: Poly (lactic acid), Thermoplastic polyurethane (TPU), Toughness, Chain extender

PACS: 87.15.Nn

INTRODUCTION

Because of the excessive use of non-degradable petro-chemical based polymers in the past decades, the world's environment has been harmed severely since a large volume of these materials has been discarded in the environment after their usage. In addition, the incineration of these polymers lead to a dramatic increase in concentration of the carbon dioxide in atmosphere, and the global warming is a consequence of this phenomenon; as a result, many attempts have been made in order to find friendly-environmental replacements for these materials. Therefore, great attention has been paid to biodegradable and biocompatible polymers, which are derived from renewable sources [1]. Among all of these renewable bio-based polymers, poly (lactic acid) (PLA), has received significant attention because of its unique properties. In general, PLA has high strength; however, some difficult issues act as obstacles and restricts its application as a general-purpose plastic. PLA suffers from poor thermal stability, insignificant ductility and low melt strength [2]. In order to overcome these drawbacks, considerable efforts have been made to improve the toughness of PLA including alloying with other polymers [3]. Among all of these methods, blending PLA with flexible polymers such as Thermoplastic polyurethane (TPU) has been proven to possess significant influence in improving PLA toughness [4]. TPU elastomer are used widely in diverse areas including medical application because of their unique properties like high strength, high modulus, flexibility and biocompatibility. Although blending PLA with elastomers have been proven to have significant effects in improving the toughness of PLA, adding elastomers to PLA obviously decrease its modulus and tensile strength. It is clear that in order to reach a practical system, a delicate balance between different properties such as modulus and toughness has to be achieved. In order to achieve this goal, chain extenders have been used to improve thermal stability, melt strength as well as to work as compatible agents at interfaces between different phases in polymer blends. It is believed that thermal degradation of PLA is related to terminal degradation reactions within PLA chains such as backbiting. Using chain extenders neutralize the effects of these reactions by relinking the polymer chains that are broken due to any kind of degradation reactions and thereby, increase the molecular weight of PLA chains [5]. Branched chains can be obtained with respect to the functionality and concentration of chain extender in the blend. Multifunctional epoxy chain extender (Joncryl) has been proven to have dramatic effect in improving mechanical properties of PLA blends due to its epoxy groups.

EXPERIMENTAL

Materials

Poly (L-lactide) (PLLA), grade 4032D, was purchased from Natureworks LLC, USA. It is a semi-crystaline material containing 2% of D-lactide units. A thermoplastic polyurethane (TPU, grade RAVATHANE® R120A65) was supplied by Ravago Manufacturing Americas. Commercial multifunctional styrene-acrylic oligomers (BASF, Joncryl®ADR-4386) were used as a reactive agent.

Sample Preparation

PLLA was prepared in the presence or absence of the selected chain modifier (i.e., Joncryl ADR) for reactive and nonreactive melt mixing. A lab-scale internal mixer (Brabender mixer, PL2200) operating at 190C and 60 r/min was used. Chain extender reactions were considered completed when a plateau in torque curves was achieved. This time was 15 min for the longest case and therefore was applied for all studied samples, including those prepared without Joncryl. The samples prepared via internal mixer were put in the oven for 10 hours in order to achieve certainty about absence of water in the samples and then, sheets were obtained using hot press machine under the condition of 200°C for 10 minutes. The amount of chain modifier was 0.5 wt. % for samples prepared via reactive mixing. The studied samples are named by the abbreviation xPLA/JC, which "x" shows PLA weight fraction in the samples and obviously 100-x shows weight fraction of TPU, and presence or absence of JC attributed to the presence and absence of Joncryl chain extender in the blends respectively.

All epoxy groups can be consumed during the chain extension reaction and the risk of having crosslinking reactions minimized. Thus, it can be hypothesized that the default epoxy groups of Joncryl can react in the melt state with the carboxylic groups of polymer chains, giving rise to a certain ratio of this hybrid structure.

Measurements

Mechanical properties of samples were evaluated by tensile test using a Galdabini Su 2500 testing machine according to the ISO 527 and the tensile speed was 5 mm/min; in addition, three samples were tested under the ambient temperature (20-25 °C) for each specimen, and the average values was reported. Elongation at break was reported as the maximum elongation observed for each sample before they were broken.

The surface morphology of tensile fractured samples of PLA/TPU blends was observed via MIRA3 model scanning electron microscope designed by TESCAN company (Czech Republic) after the samples were sputter-coated with a thin gold layer for better conductivity and at last, the final pictures were prepared under 10 Kv voltage.

RESULTS AND DISCUSSION

Tensile Properties of Chain Extended PLA/TPU Blends

Toughness and strength of materials are considered to present two completely different behavior of materials when combined together, represent general mechanical behavior of the material. The area underneath the stress-stain curve represents the term tensile toughness which is the energy of mechanical deformation per unit volume prior to fracture [6]. The tensile stress-strain curves of neat PLA and chain extended PLA are shown in Figure 1. As shown in Figure 1, PLA is a rigid material and the area underneath the stress-strain curve is much less compared to the curve for chain extended PLA, where it presents much lower elongation at break compared to other samples. In addition, no obvious yield occurred for the neat PLA, which presented a poor toughness; nevertheless, an obvious yield occurred in the chain extended PLA curve and much higher elongation at break detected in comparison with neat PLA which present better ductile behavior for chain extended PLA compared to neat PLA.

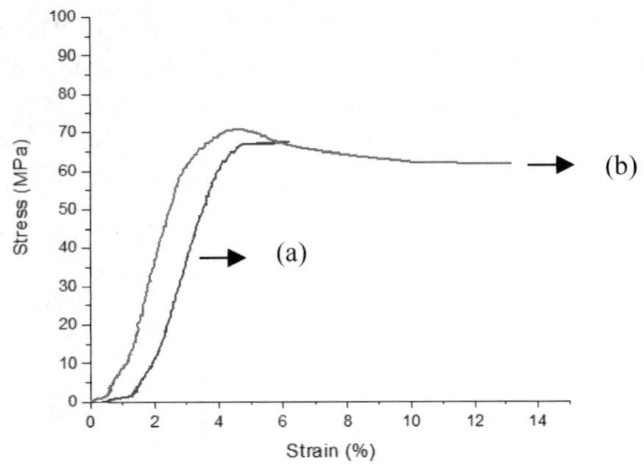

FIGURE 1. Tensile stress-strain curves of the PLA (a) and 100PLA/JC (b).

In order to evaluate the effect of increasing TPU weight fraction on mechanical behavior of samples, the tensile stress-strain curves for 70PLA/JC and 50PLA/JC blends are shown in Figure 2. It is observed that both samples present tough fracture behavior with an obvious yield; however, with increasing TPU content from 30% to 50%, a considerable increase occurred for elongation at break in the 50PLA/JC sample and a significant drop in the stress at yield observed for 50PLA/JC sample. It should be noted that the 50PLA/JC sample is relatively tougher compared to 70PLA/JC since the area within stress-strain curves is larger for 50PLA/JC specimen.

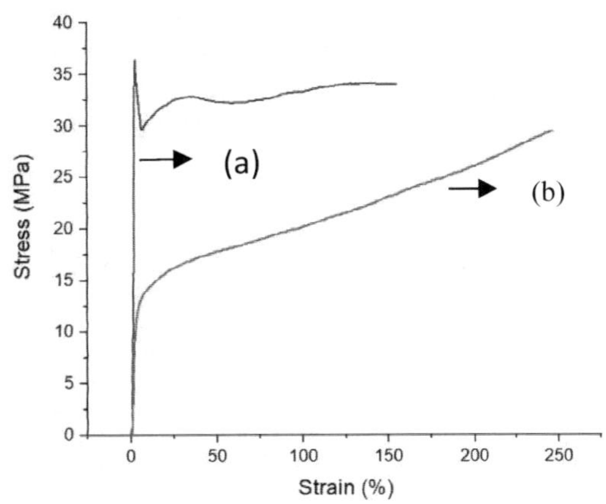

FIGURE 2. Tensile stress-strain curves of the 70PLA/JC (a) and 50PLA/JC (b).

Mechanical properties such as tensile modulus, elongation at break and stress at yield were shown in Table 1 in order to evaluate the effects of chain extender and TPU content on mechanical behavior of samples. It is observed that with adding chain extender to the neat PLA, the modulus and stress at yield increased from 2825.45 MPa to 3464 MPa and from 68MPa to 72Mpa, which indicates a significant improvement in strength of the sample with adding chain extender. Moreover, an increase in elongation at break observed from 6.5% to 13% with adding chain extender to the neat PLA.

With adding TPU in the 70PLA/JC sample to 30 wt%, a decrease in the modulus from 3464 MPa for the PLA/JC sample to 2110.16 MPa for 70PLA/JC sample, and another reduction in stress at yield from 72 MPa for PLA/JC to 37MPa for 70PLA/JC observed. This reduction in modulus and stress at yield attributed to the increasing content of TPU in the sample since TPU is considered as an elastomer with high ductile behavior, which results into reduction in modulus and stress at yield within the samples.

With increasing content of TPU in samples from 30 wt% in 70PLA/JC to 50 wt% in 50PLA/JC dramatic decreases was shown in modulus and stress at yield within 50PLA/JC sample; in fact, the modulus and stress at yield decreased from 2110.16 MPa for 70PLA/JC to 683.59 MPa and from 37 MPa to 12.5 MPa, respectively. In addition, an increase in elongation at break from 160% for 70PLA/JC sample to 250% for 50PLA/JC sample was observed which indicates a more tough fracture behavior.

TABLE1. Modulus, elongation at break and stress at yield for PLA, PLA/JC, 70PLA/JC AND 50PLA/JC samples

Sample Name	Modulus	Elongation at Break	Stress at Yield
PLA	2825.45 MPa	6.5%	68 MPa
PLA/JC	3464.00 MPa	13%	72 MPa
70PLA/JC	2110.16 MPa	160%	37 MPa
50PLA/JC	683.59 MPa	250%	12.5 MPa

Morphology and Toughening Mechanism of the Chain Extended PLA/TPU Blends

It is believed that controlling morphology of materials is a vital factor in controlling its mechanical behavior since mechanical properties of materials depend heavily on their morphology. SEM images of the tensile test fractured surfaces of neat PLA and PLA/JC are shown in Figure 3 in order to evaluate the effect of Joncryl chain extender on neat PLA. It is believed that although the results of tensile test indicate tougher behavior for 100PLA/JC sample compared to the neat PLA, the results of SEM shows minor difference between these samples. The surface of 100PLA/JC seems smoother compared to neat PLA indicating negligible reduction in toughness for 100PLA/JC sample compared to the neat PLA. A possible explanation for these result may be related to increase in molecular weight of 100PLA/JC due to presence of Joncryl chain extender [7]; however, further studies is required to explain this behavior more clearly.

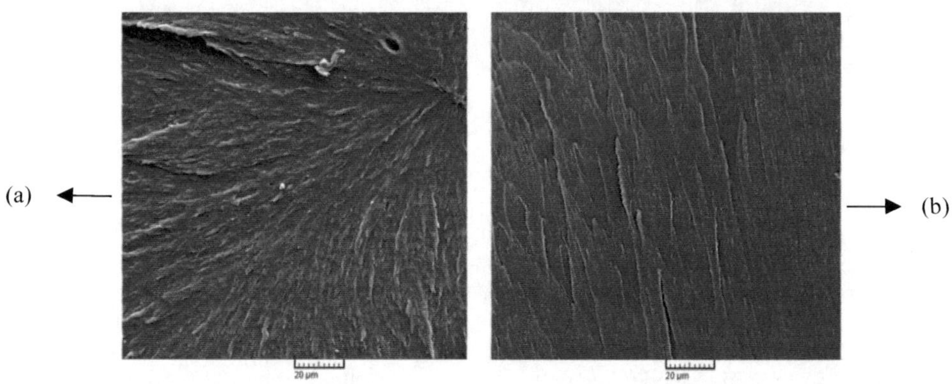

FIGURE 3. SEM images of the Tensile test fractured surfaces of neat PLA (a) and 100PLA/JC. (magnification: 1000)

CONCLUSIONS

To improve the toughness of neat PLA, different contents of TPU were added to the neat PLA, and to prevent significant drop in strength of the samples, addition of Joncryl chain extender to the blends were considered, and samples' properties were studied in terms of mechanical and morphological properties. Its observed that with using Joncryl chain extender in the neat PLA, improved toughness and higher modulus could be achieved. With increasing TPU content in the blends, sample 70PLA/JC showed much higher ductility than neat PLA and 100PLA/JC with a slight decrease in its strength compared to the other samples. As a result, a desirable strength-toughness balance could be achieved in the 70PLA/JC sample. In the 50PLA/JC sample, however, a significant drop in modulus was detected compared to neat PLA and 70PLA/JC while no significant improvement in the toughness was observed. As a result, samples 70PLA/JC considered to have the best desired and balanced strength-toughness behavior between all the samples.

REFRENCES

1. R. Zullo, S. Iannace, *Carbohydrate Polymers* **77 (2),** 376-383 (2009).
2. Alexandre M, Dubois P, *Mater Sci Eng* **28**,1-63 (2000).
3. P. Zhao, W. Liu, Q. Wu, J. Ren, *Journal of Nanomaterials* **2010**, 1-8 (2010).
4. Fei Feng, Lin Ye, *J. Appl. Poly*. Sci **119**, 2778-2783 (2011).
5. H. Li, M.A. Huneault, *J. Appl. Poly. Sci* **122**, 134-141 (2011).
6. N. Bitinis, A. Sanz, A. Nogales, R. Verdejo, M.A. Lopez-Manchado, T.A. Ezquerra, *Soft Matter* **8**, 8990-8997 (2012).
7. Liliane Cardoso Arruda, Marina Magaton, Rosario Elida Suman Bretas, Marcelo Massayoshi Ueki, *Polymer Testing* **43**, 27-37 (2015).

Effect of Compatibilizer EMA-GMA on Rheological Behavior and Mechanical Property of PPC/PBAT Blends

Jiang Guo, Wang Feng

Lab for Micro Molding and Polymer Rheology, South China University of Technology, 381 Rd, Wushan, Tianhe, Guangzhou, 510640, China

Abstract: In this work, Poly(propylene carbonate) (PPC)/poly(butylenes adipate-co-terephthalate) (PBAT) blends was prepared at the weight ratio of 50:50 via incorporating ethylene-methyl acrylate-glycidyl methacrylate copolymer (EMA-GMA) as a reactive compatibilizer using a twin screw extruder. Effect of compatibilizer content on rheological and mechanical properties of PPC/PBAT blends was investigated. Results of rheological measurement for EMA-GMA compatibilized PPC/PBAT blends showed a higher storage modulus and a decreased phase angle (δ) in the van Gurp-Palmen plot. This indicated that there exhibited an improved melt strength and miscibility for PPC and PBAT, owing to the reaction between epoxide groups of EMA-GMA and end groups (-OH, -COOH) of PPC and PBAT. The incorporation of EMA-GMA to the binary PPC/PBAT blends was helpful to the significant improvement for tensile strength, elongation at break and impact strength. When the content of EMA-GMA was 5 wt%, the tensile strength increased from 5.5MPa to 8.5MPa and elongation at break increased from 24.1% to 159.1%, compared with that of PPC/PBAT blends. Impact strength was also improved from 3.70KJ/m^2 to 6.28KJ/m^2, increased by 69.7% when the content of EMA-GMA was increased from 0 to 10wt%. The enhanced mechanical properties of the EMA-GMA compatibilized PPC/PBAT blends indicated that the interfacial interaction between PPC and PBAT was improved by reactive compatibilization. This will make a very potential degradable material for food packaging application.

Keywords: Poly(propylene carbonate), Poly(butylenes adipate-co-terephthalate), Reactive compatibilization, Rheological properties, Mechanical properties.
PACS: 83.80.Tc

INTRODUCTION

Development of plastic products from biodegradable and renewable materials has received considerable research interests in recent years with a primary aim to reduce environmental pollution and waste of petroleum resources. Poly(propylene carbonate) (PPC) is a biodegradable aliphatic polycarbonate synthesized by the alternating copolymerization of propylene oxide and carbon dioxide. It has been paid much attention for the consumption of CO_2 in the production process and its biodegradability, biocompatibility as well as excellent oxygen barrier performance. However, the amorphous nature and low glass-transition temperature (T_g) of PPC result in poor mechanical properties and inferior thermal stability, which severely limits its wide application. For this reason, many efforts have been devoted to enlarge its application [1-3]. Blending of PPC with other polymers is seen as a simple, economical and effective process to enhance its comprehensive performance and potential application [4].

Poly(butylene adipate-co-terephthalate) (PBAT) is an aliphatic aromatic copolyester manufactured from petroleum-based resources 1,4-butanediol, terephthalic acid and adipic acid consisting of a rigid butylene terephthalate segment and a flexible butylene adipate section. It has been studied more extensively due to its excellent physical properties, such as high toughness, good heat resistance, and high impact performance. PBAT exhibits good thermo-mechanical properties and extrusion blown film forming properties. The obtained film possesses good transparency, high toughness, impact resistance, and resistance to water and humidity [5-7]. It is possible to obtain a material with excellent comprehensive performance by blending PPC with PBAT [8]. Nevertheless, these polymers are thermodynamically immiscible and require compatibilizers to achieve stable morphologies and superior mechanical properties. Reactive compatibilization is often used to improve the miscibility of polymer blends. Ethylene-methyl acrylate-glycidyl methacrylate copolymer (EMA-GMA) is one of the chain extenders with reactive multiple epoxy functional groups, widely used in two component systems. It can highly react with carboxyl and hydroxyl groups under heat and high shear stress in polymer processing. Thus EMA-GMA may be useful for interfacial compatibility improvement [9-10].

In this work, the biodegradable blends of PPC/PBAT were prepared in the presence of EMA-GMA as a compatibilizer using a twin screw extrusion. Then, the rheological and mechanical properties of the blends were investigated. The purpose is to prepare a biodegradable material with high melt strength, good compatibility and outstanding mechanical properties.

EXPERIMENTAL

Materials

PPC used in this work was provided by the Henan Tianguan Group Co., Ltd., China. PBAT was supplied by Xinjiang Blue Ridge Tunhe Polyester Co., Ltd. Ethylene-methyl acrylate-glycidyl methacrylate (EMA-GMA) terpolymer containing 8% of glycidyl methacrylate was a product of the Arkema Company with a trade name Lotader AX 8900, obtained from Quadra Chemicals, Canada.

Sample Preparation

Prior to blending, PBAT and PPC were dried overnight in the oven at 60°C and 40°C, respectively. All samples were prepared on a co-rotating twin-screw extruder with a temperature profile of 110/115/125/130/135/140/140/140°C from hopper to die. The extrudates were quenched in a cold-water bath and pelletized. Then all the samples were compression molded into sheets with thicknesses of 1.0 mm at 150°C. The formulation and nomenclatures of the PPC/PBAT/EMA-GMA blends were shown in table 1.

TABLE 1 Formulation and nomenclatures of PPC/PBAT/EMA-GMA blends.

PPC/PBAT/EMA-GMA(wt%/wt%/wt%)	Designation
50/50/0	PPC/50PBAT
50/50/1	PPC/50PBAT-1
50/50/3	PPC/50PBAT-3
50/50/5	PPC/50PBAT-5
50/50/7	PPC/50PBAT-7
50/50/10	PPC/50PBAT-10

Rheological Measurements

The rheological behavior was measured on a rotational rheometer (Bohlin Gemini 200, UK) equipped with a 25 mm (diameter) parallel plate geometry. The disks samples with a size of 25 × 1 mm were tested at 150 °C under an oscillatory mode with a frequency from 0.01 to 100 s^{-1}.

Mechanical Test

The tensile properties of samples were evaluated as performed on a universal testing machine (104B, Shenzhen Wance Testing Machine Co., Ltd., China) with a crosshead speed of 50 mm/min according to ISO 1184-1983. Prior to the testing, all samples were cut into dumbbell-shaped specimens (75 × 4 × 1 mm^3). Notched Izod impact tests were performed according to standard GB/T1843-2008 on an impact tester 501B-2 (Shenzhen Wance Test equipment Co., Ltd., China). The samples with dimensions 80.0 × 10.0 × 4.0 mm^3 were obtained from compression-molded specimens.

FTIR Spectroscopy Analysis

The infrared (IR) spectra of samples were analyzed using a Vertex 70 spectrometer (Bruker Corporation, Germany) in an ATR module at room temperature from 4000 to 560 cm^{-1} at a resolution of 4 cm^{-1}. The blends were extracted with N,N-dimethylformamide(DMF) and cyclohexane to remove PPC and EMA-GMA, respectively.

RESULTS AND DISCUSSION

Rheological behavior plays an important role in polymer processing [11]. Figure 1(a) shows the storage modulus (G′) versus oscillatory frequency for blends with different EMA-GMA contents. G′ gradually increases with increasing of EMA-GMA content and also increases with increasing frequency. This indicates that the addition of EMA-GMA changes the rheological properties of the PPC/PBAT blends. In addition, at low frequency, the slopes of compatibilized blends show a slight leveling off as compared to the uncompatibilized blends. However, at high frequency, the G′ of both uncompatibilized and compatibilized blends display similar values. The higher G′ of compatibilized blends indicates the formation of entanglement structures in PPC/PBAT melts, which is mainly attributed to epoxy functional groups of EMA-GMA react with terminal carboxyl groups (terminal hydroxyl groups) of PPC and PBAT. This reaction generates a polymer with branched structure and enhances the molecular interaction of the blends system [12-14].

Figure 1(b) presents the plots of phase angle (δ) versus the absolute value of the complex modulus (|G*|) for the PPC/PBAT/EMA-GMA blends. In the literature, the plots were known as the van Gurp-Palmen plot, which was used to detect the rheological percolation of the filled polymeric composites [15]. The δ–|G*| curves for linear polymers almost overlap to form one 'universal' curve and δ values are nearly 90° at the low values of |G*| due to the narrow molecular mass distribution of linear polymers. Any deviations from the 'universal' curve can be referred to the influence of long chain branches or broadening of the molecular mass distribution [16]. It can be seen from Figure 1(b), The low frequency δ decreases with increasing EMA-GMA content indicating the transition from viscous fluid-like to an elastic solid-like behavior [17]. Such changes in rheological response also reveals an enhanced elastic behavior, which would be benefit for blends during the practical production process. Furthermore, the shape of the vGP curve also implies that the blends are a mixture of linear and randomly branched polymers [18].

A plot of G' versus G'' (Han plots) is applied as criteria for the evaluation of the miscibility and interfacial interaction between the polymer matrix in a composite. It generates a composition independent correlation for miscible blends, whereas the correlation is composition dependent for immiscible blends [19]. As shown in Figure 1(c), the curves of blends clearly present EMA-GMA loading dependence, confirming the presence of strong interaction between EMA-GMA and PPC/PBAT matrix. In addition, the reduced slope with the addition of EMA-GMA indicates that the composites become more heterogeneous, which may be that an interface phase at the interface of the blends was formed through the compatibilizer [20-21]. It is worth mentioning that the inflection point where the slope is changed shifts to a higher frequency with increasing EMA-GMA loadings. This indicates that much energy is necessary to change the degree of heterogeneity due to the increased physical association. The enhanced interfacial interaction would restrict chain mobility inevitably [22].

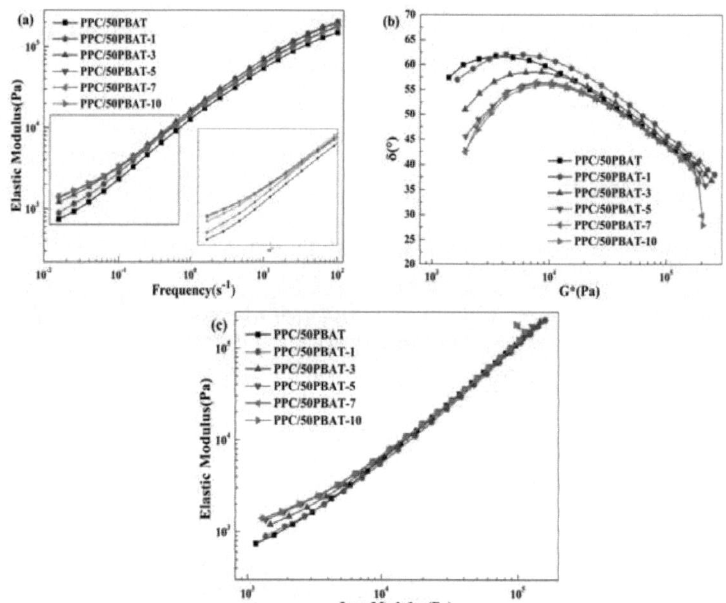

FIGURE 1. (a) Elastic modulus (G') of PPC/PBAT/EMA-GMA blends at 150 °C. (b) Phase angle (δ) of PPC/PBAT/EMA-GMA blends at 150 °C. (c) Han plots of G' versus G'' of PPC/PBAT/EMA-GMA blends at 150 °C.

Figure 2 summarizes some tensile properties and notched Izod impact strength of the blends. For the binary blend, the tensile strength and Yong's modulus are only 5.5MPa and 123.1MPa due to the limited compatibility of the components. With the incorporation of EMA-GMA content, the tensile properties are obviously improved. A tremendous increase in the tensile strength and Yong's modulus is achieved with values of 8.5MPa and 168.2MPa, increased by 54.5% and 36.6% after addition of 5wt% EMA-GMA. With further increasing the compatibilizer content, the tensile strength and Yong's modulus of the blends are decreased due to the excessive crosslinking reaction [23-24]. The uncompatibilized blend present poor ductility and toughness. Upon the addition of EMA-GMA as a compatibilizer, the elongation at break and impact strength display a significant improvement, compared with the binary blend. When the content of EMA-GMA is 5wt%, the elongation at break increases by 5.6 times from 24.1% to 159.1%. The impact strength is also improved by 69.7% from 3.70KJ/m^2 to 6.28KJ/m^2 when the content of EMA-GMA increases from 0 to 10wt%, suggesting a very effective toughening behavior of this blend. This suggests that interfacial modification is necessary for obtaining enhanced mechanical properties. The improvement in mechanical properties may be attributed to the reactive groups of EMA-GMA reacting with the terminal groups (hydroxyl and carboxyl groups) of PPC/PBAT blends and forming grafted copolymers at interface. This compatibilization reaction increases the molecular weight and enhances the miscibility as well as interfacial adhesion [25-26].

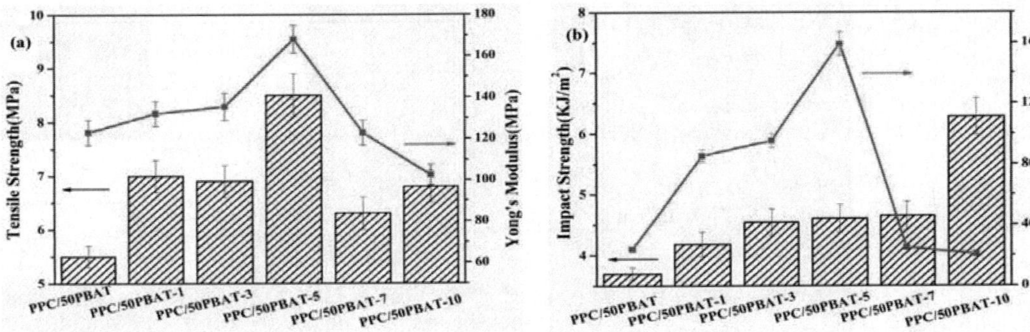

FIGURE 2. Mechanical properties of the blends:(a) The tensile strength and Yong's modulus. (b) The impact strength and elongation at break.

The residue after extraction from blends was characterized by FTIR spectroscopy. The results are shown in Figure 3. Compared to the uncompatibilized blends, the absorbance peak at 2850cm^{-1} is assigned to the stretching of methylene groups(-CH_2-) of EMA-GMA, indicating that the grafting reaction between PBAT and EMA-GMA can take place. For compatibilized blends, the characteristic carbonate group vibration attributed to PPC at 1733 appears, whereas it is absent in the uncompatibilized blends.This suggests that PPC is grafted onto PBAT via EMA-GMA [27].

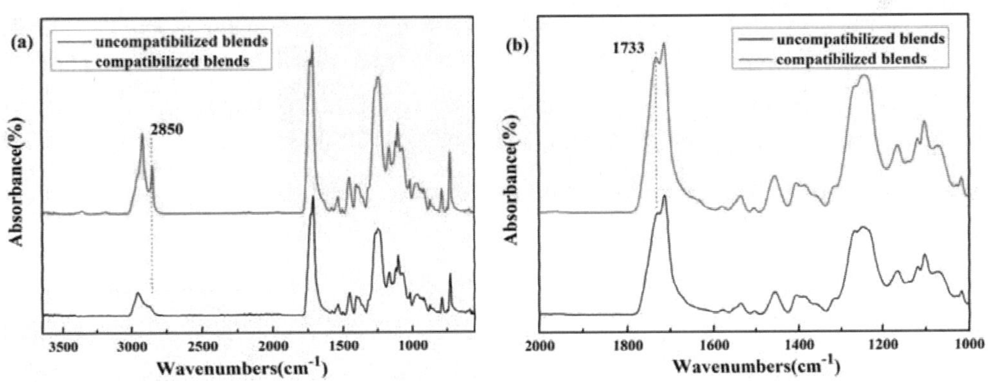

FIGURE 3. ATR-FTIR spectra of uncompatibilized and compatibilized blends.

CONCLUSIONS

PPC/PBAT/EMA-GMA blends were successfully prepared through melt blending. Results show that the incorporation of EMA-GMA improves the melt strength and miscibility. The addition of EMA-GMA to the binary PPC/PBAT blend shows a significant improvement in tensile strength, elongation at break, Yong's modulus and impact strength. When the content of EMA-GMA is 5wt%, the tensile strength increases from 5.5MPa to 8.5MPa by 54.5%, the elongation at break changes from 24.1% to 159.1%, increased by 5.6 times and the Yong's modulus increases from 123.1MPa to 168.2MPa, increased by 36.6%. The impact strength increases from 3.70KJ/m^2 to

$6.28 KJ/m^2$ by 69.7% at EMA-GMA loading of 10wt%. In summary, the adding of EMA-GMA plays a role of compatibilizer, remarkably improving rheological behavior and enhancing the mechanical properties. This will make the biocomposites as a very potential degradable material for food packaging application.

ACKNOWLEDGMENTS

This work was supported by the National Natural Science Foundation of China (51573062) and Guangzhou Scientific Research Project.

REFERENCES

1. J. J. Kong, Z. L. Li, Z. W. Cao, C. Y. Han, L. S. Dong, *Polym. Bull.* **74**, 5065-5082 (2017).
2. D. M. Han, Z. Guo, S. Chen, M. Xiao, X. H. Peng, S. J. Wang, Y. Z. Meng, *Polymers* **10**, 771 (2018).
3. D. M. Han, G. Chen, M. Xiao, S. J. Wang, S. Chen, X. H. Peng, Y. Z. Meng, *International Journal of Molecular Sciences* **19**, 2032 (2018).
4. R. Muthuraj, T. Mekonnen, *Polymer* **145**, 348-373 (2018).
5. S. Kashi, R. K. Gupta, N. Kao, S. A. Hadigheh, S. N. Bhattacharya, *Journal of Materials Science & Technology* **34**, 1026-1034 (2018).
6. L. B. Tavares, N.M. Ito, M.C. Salvadori, D. J. dos Santos, D. S. Rosa, *Polymer Testing* **67**, 169-176 (2018).
7. J. B. Olivato, M. V. E. Grossmann, A. P. Bilck, F. Yamashita, *Carbohydate Polymers* **90**, 159-164 (2012).
8. H. W. Pan, Y. P. Hao, Y. Zhao, X. Z. Lang, Y. Zhang, Z. Wang, H. L. Zhang, L.S. Dong, *Korean J. Chem. Eng.* **34**, 1294-1304 (2017).
9. Y. W. Xu, J. Loi, P. Delgado, V. Topolkaraev, R. J. McEneany, C. W. Macosko, M. A. Hillmyer, *Ind. Eng. Chem. Res.* **54**, 6108-6114 (2015).
10. G. F. Brito, P. Agrawal, E. M. Araujo, T. J. A. Melo, *Polymer Engineering and Science* **54**, 1922-1930 (2014).
11. H. W. Pan, D. D. Ju, Y. Zhao, Z. Wang, H. L. Yang, H. L. Zhang, L. S. Dong, *Fibers and Polymers* **17**, 1540-1549 (2016).
12. R. Alitry, K. Lamnawar, A. Maazouz, *Polymer Degradation and Stability* **97**, 1898-1914 (2012).
13. G. F. Brito, P. Agrawal, E. M. Araujo, T. J. A. Melo, *POLYM. ENG. SCI.* **54**, 1922-1930 (2014).
14. Y. P. Hao, H. L Yang, G. B. Zhang, H. L. Zhang, G. Gao, L. S. Dong, *Chinese Journal of Polymer Science* **33**, 1702-1712 (2015).
15. S. Trinkle, P. Walter, C. Friedrich, *Rheol. Acta.* **41**, 103-113 (2002).
16. H. G. Fang, Y. Q. Zhang, J. Bai, Z. K. Wang, Z. G. Wang, *RSC Adv.* **3**, 8783-8795 (2013).
17. D. Das, B. K. Satapathy, *Materials Chemistry and Physics* **147**, 127-140 (2014).
18. R. Alitry, K. Lamnawar, A. Maazouz, *European Polymer Journal* **58**, 90-102 (2014).
19. L. Bazli, A. Khavandi, M. A. Boutorabi, M. Karrabi, *Polymer* **113**, 156-166 (2017).
20. H. T. Wang, X. Yang, Z. A. Fu, X. W. Zhao, Y. J. Li, J. Y. Li, *Macromolecules* **50**, 9494-9506 (2017).
21. J. K. Kim, K. H. Yoon, D. S. Bang, Y. B. Park, H. U. Kim, Y. H. Bang, *Journal of Applied Polymer Science* **107**, 272-279 (2008).
22. D. F. Wu, L. A. Wu, M. Zhang, Y. L. Zhao, *Polymer Degradation and Stability* **93**, 1577-1584 (2008).
23. N. W. Zhang, Q. F. Wang, J. Ren, L. Wang, *J. Mater. Sci.* **44**, 250-256 (2009).
24. X. Li, X. Y. Yan, J. Yang, H. W. Pan, G. H. Gao, H. L. Zhang, L. S. Dong, *POLYM. ENG. SCI.* **58**, 1868-1878 (2018).
25. A. M. Zolali, B. D. Favis, *Macromolecules* **51**, 3572-3581 (2018).
26. Y. W. Xu, J. Loi, P. Delgado, V. Topolkaraev, R. J. McEneany, C. W. Macosko, M. A. Hillmyer, *Ind. Eng. Chem. Res.* **54**, 6108-6114 (2015).
27. J. Li, C. R. Sun, X. Q. Zhang, *POLYM. COMPOS.* **33**, 1737-1749 (2012).

Microstructure and Mechanical Properties of PA66/PP Blends Compatibilized by Electron Beam Irradiation

D. Guo[a*] and Z. Q. Liao[b]

[a] *Chengdu Technological University*
No.1, Section 2, Zhongxin Avenue, Pidu, Chengdu, Sichuan 611730, P.R.China
[b] *Sichuan Institute of Atomic Energy*
No. 4128, Yudu West Road, Longquanyi, Chengdu, Sichuan 610101, P.R.China

Abstract. In order to obtain a good PA66/PP alloy, and resolve the incompatibility of polar PA66 and nonpolar PP, electron beam irradiation was used to compatibilize PA66/PP blends in this paper. The effects of irradiation on the microstructure and mechanical properties of PA66/PP blends were investigated. And the results show that the interface compatibility between PA66 and PP is improved through irradiation; the tensile strength, flexural strength and impact strength of the blends can be also improved at appropriate irradiation dose range. Therefore, electron beam irradiation compatibilize PA66/PP solves the problem that PA66 is highly incompatible with nonpolar PP, and is not ideal for composite performance. It is a new approach based on radiation processing for the preparation and structural performance regulation of high performance PA66/PP blends, and with a wide range of applications.

Keywords: PA66/PP Blend; Irradiation Compatibilization; Microstructure; Mechanical properties
PACS: 81.20.−n

INTRODUCTION

Polymer blending or alloying is one of the important routes to realize its high performance and functionalization. PA66/PP blends maintain the excellent performance of PA66 and PP, and also improve the performance of their own deficiencies. However, the polar PA66 is incompatible with the non-polar PP. It is necessary to compatibilize the blends to obtain good performance alloy materials. At present, different monomers grafted PP are used as compatibilizer of PA66/PP blend. Although the graft copolymer can improve the interfacial compatibility of the blends, the graft ratio is a key factor affecting the comprehensive properties of the polymer blend, thus restricting the preparation of high performance polymer blends. In various compatibilized strategies of polymer blends, it is important to study the grafting or cross-linking reaction induced by electron beam irradiation between continuous phase and dispersed phase. In this paper, In order to obtain a good PA66/PP alloy, and resolve the incompatibility of polar PA66 and nonpolar PP, electron beam irradiation was used to compatibilize PA66/PP blends. The effects of irradiation on the microstructure and mechanical properties of PA66/PP blends were investigated. And the results show that the interface compatibility between PA66 and PP is improved through irradiation; the tensile strength, flexural strength and impact strength of the blends can be also improved at appropriate irradiation dose range. Therefore, electron beam irradiation compatibilize PA66/PP solves the problem that PA66 is highly incompatible with nonpolar PP, and is not ideal for composite performance. It is a new approach based on radiation processing for the preparation and structural performance regulation of high performance PA66/PP blends, and with a wide range of applications.

EXPERIMENTAL

Materials

PA66: DuPont 101F; PP: Maoming Petrochemical T30S; Antioxidant: BASF (Ciba) 1089.

Preparation and Radiation Processing of PA66/PP Blends

PA66 was dried in an oven at 105 °C for 12 h, stored in a desiccator for use. Equal mass of pre-dried PA66 and PP, and antioxidant 1089 0.25wt% was mixed in a high-mixer, and the uniformly mixed material was extruded through screw extruder to obtain PA66/PP blend pellets, and the PA66/PP blend pellets were injection molded through injection molding machine to obtain test samples, the injection temperature was 280 °C, and the injection pressure was 100 MPa. The injection speed is 50 mm/s and the mold temperature is 40 °C.

The samples were irradiated at different dose using a GJ-2 high frequency high pressure electron accelerator. The preparation and irradiation modification process is shown in Figure 1.

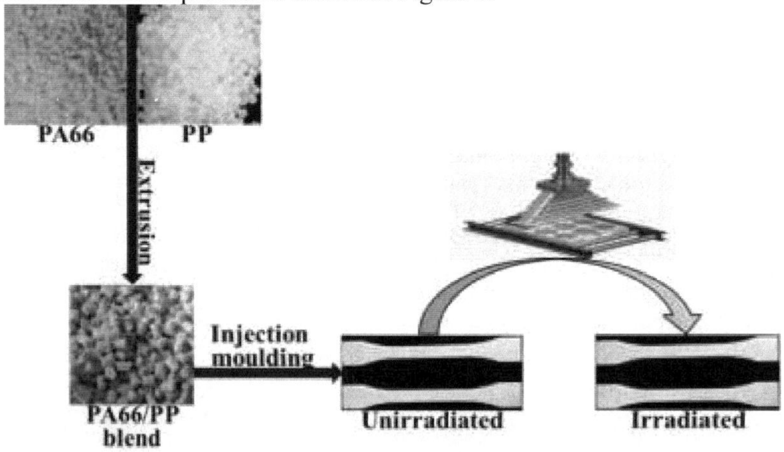

FIGURE 1. Process of preparation and radiation processing.

Characterizations

The sample was cryofractured in liquid nitrogen, and the vacuum was sprayed for 90 s. The morphology of the cryofractured section of the sample was observed by scanning electron microscopy (SEM). The accelerating voltage was 10 kV and the magnification was 2000 times.

All test samples were pre-conditioned for 48 h in a standard environment. The mechanical properties were tested in accordance with GB/T 1040.1-2006, GB/T 9341-2008, GB/T 1843-2008.

RESULTS AND DISCUSSION

Microstructure of Irradiation Modified PA66/PP Blends

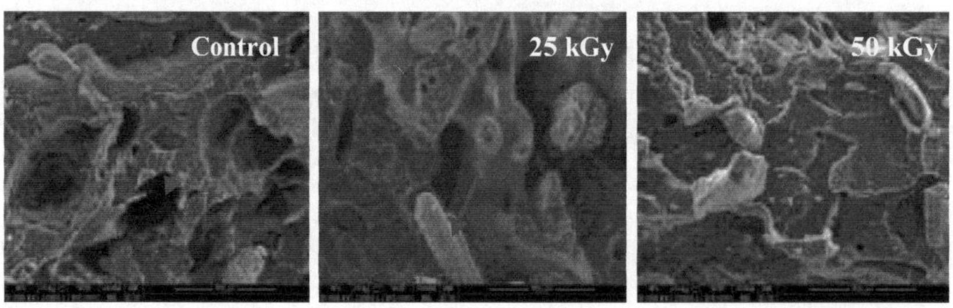

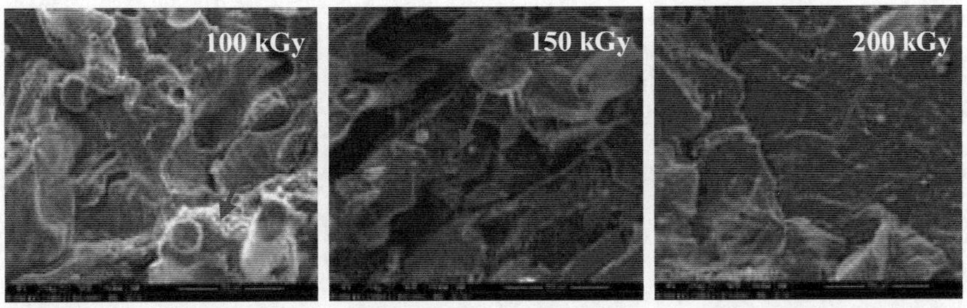

FIGURE 2. SEM micrographs of cryofractured surfaces of PA66/PP blend irradiated at 25, 50, 100, 150 and 200 kGy.

It can be seen from the SEM micrographs that there are many defects in the cryofractured section of the unirradiated samples, with the increase of irradiation dose, the defects gradually decrease, and when irradiated at 200 kGy, the interface between the fiber phase and the matrix phase disappears. The polar PA66 fibrils are incompatible with the non-polar PP phase in PA66/PP blends, the phase separation are serious, and the PA66 fibrils are easily pulled out from the PP phase. When the sample is irradiated by electron beam, the free radicals from the irradiated PA66/PP blends rearranged to form many branching and cross-linking structures, and at the same time, the free radicals generated from irradiated PP can form strong hydrogen bonds with PA66. The presence of cross-linking network and hydrogen bonds improved the interface compatibility between PA66 fibrils and PP phase.

Mechanical Properties of Irradiation Modified PA66/PP Blends

TABLE 1. Mechanical properties of PA66/PP blends irradiated at 25, 50, 100, 150 and 200 kGy.

Irradiation dose (kGy)	Tensile strength (MPa)	Flexural strength (MPa)	Impact strength (kJ/m^2)
Control	22.9	57.8	10.7
25	26.2	60.2	11.2
50	26.7	60.7	9.7
100	23.5	59.9	8
150	19.3	56.8	6.2
200	17.9	55.6	4

The mechanical properties test results of electron beam irradiated PA66/PP blends show that electron beam irradiation simultaneously enhanced and toughened PA66/PP blends. This can be attributed to radiation-induced grafting, cross-linking and degradation of PA66/PP blends. At low irradiation dose, radiation grafting toughen the blends, and radiation cross-linking reinforce the blends, and the mechanical properties of PA66/PP blends improve with the increase of irradiation dose. When the irradiation dose exceeds the optimum dose, excessive radiation crosslinking and radiation degradation reaction lead to a sharp drop in the mechanical properties of the blends. This can be also confirmed from the microstructure of the blends, when the irradiation dose exceeds 100 kGy, no obvious fiber pull-out appears, especially from the quenched section of the sample irradiated at 200 kGy, It can be seen that the material has shown obvious brittle fracture characteristics.

CONCLUSIONS

1) Electron beam irradiation modification can improve the interfacial compatibility of PA66/PP blends, and promote the close combination of PA66 and PP, thus improving the overall performance of the material.

2) Electron beam irradiation modification can simultaneously improve the mechanical strength and toughness of PA66/PP blends. When the irradiation dose is 50 kGy, the comprehensive mechanical properties of the materials are optimal.

3) Electron beam irradiation modified PA66/PP blends solves the problem that the compatibility between polar PA66 and non-polar PP, and it is a good new technology to prepare blends and regulate the structural and properties.

ACKNOWLEDGMENTS

Financial support from the Applied Basic Research Programs of Science and Technology Commission Foundation of Sichuan Province (No. 2018JY0123) is gratefully acknowledged.

REFERENCES

1. C. Koning, M. V. Duin, C. Pagnoulle and R. Jerome, *Prog. Polym. Sci.* **23**, 707-757 (1998).
2. V. Mittal, Functional Polymer Blends: Synthesis, Properties and performance. CRC Press. 2012.
3. K. Makuuchi and C. Song, Radiation processing of polymer materials and its industrial applications. John Wiley & Sons, Inc., Hoboken, New Jersey, 2012.
4. M. Porubská, D. Babić, I. Janigová, M. Šlouf, K. Jomová and I. Chodák, *Polym. Bull.* **73**, 1775-1794 (2016).

Study of Morphology and Melt Rheological Behavior of EPDM Rubber/EVA Blend system

Parvane Masoudi[a], Ali Asghar Katbab[a*], Hossein Nazockdast[a]

[a]*Polymer Engineering Department, Amirkabir University of Techollogy, Tehran, Iran*
[*)] *Corresponding Auther:katbab@aut.ac.ir*

Abstract. In the present work, attempts have been made to prepare blends of EPDM rubber as an elastic and flexible matrix and ethylene vinyl acetate (EVA) with 19% VA for the prepare of acoustic absorber foam. As the morphology and extent of molecular interactions and viscoelastic characteristics play crucial roles, the influence of EVA content upon developed morphology and melt dynamic viscoelastic behaviour was studied. Blend with EPDM/EVA ratio of 80/20 exhibited much higher melt elasticity and viscosity than the pure EPDM, indicating good thermodynamic compatibility. These results implied a good phase compatibility at low EVA level. However, the blend exhibited more heterogeneous in nature at higher EVA content. This was evidenced by decreaseing both storage modulus and complex viscosity with increasing the percent of EVA as a result of increaseing the number of polar side groups of VA and hence decreaseing interfacial compatibility. SEM and AFM analysis revealed matrix-dispersed type of morphology for the blends comprising 20-30 wt.% of EVA, whereas at 60/40 ratio to co-continuous type of morphology was observed.

Keywords: Rheology, Morphology, EPDM, EVA, Compatiblizer.

INTRODUCTION

The technique of blending of two or more polymers (rubber–plastic) to produce an engineering combination of properties for the specific end-use requirements is a very active area in the field of polymer processing [1]. The prperties of polymer components (rubber–plastic) and their ratios play an important role in the desired properties of the blend. Ethylene propylene diene monomer (EPDM) elastomer has high crosslinking ability, excellent heat resistance, ozone resistance, impact strength, and flexibility. Therefore, this polymer is widely used in many applications such as wire and cable insulation, automotive, O-ring, gasket, etc. Ethylene vinyl acetate (EVA) is also an excellent thermoplastic polymer with polar side groups and has good engineering properties which make this polymer a good condidate for the preparation of blends with both rubbery and thermoplastic characteristics. Depending on the VA content, this polymer can be used for different applications such as electrical insulation purpose, cable jacketing, water proofing, corrosion protection, and packaging component [2]. EVA has a good barrier property, excellent low-temperature toughness, stress-crack resistance, waterproof properties, ultraviolet radiation, weather resistance, and excellent mechanical properties [3]. Blends of EPDM and EVA have a large number of applications. Several papers concerning EVA and EPDM blends have been published [4]. It is well known that morphology and therefor the blend ratio of polymer blends plays a significant role in determining their mechanical properties. Obviously, the effect of EVA on the EPDM/EVA blend morphology, interface, and driven properties are very important, which are left unstudied [5]. Most of the rubber/thermoplastic blends are incompatible and immiscible but EPDM/EVA blend shows partial miscibility in the amorphous region of EVA with EPDM [4]. Different techniques can be used to increase the compatibility. With the aid of a third component ("interfacial agent" or "compatibilizer") interfacial energy between the phases can be reduced, which allowes a finer dispersion at some stage in blending and improves the interfacial interaction [6]. It has been reported thatThe using EPDM-g-MA can improve the interfacial adhesion between the blend components for EPDM/EVA blends. In this study, we have prepared EPDM/EVA blends with various EVA content with and without compatiblizer. The effect of EVA content and compatiblizer on the morphology and rheological properties of the blends have been studied.

EXPERIMENT

Materials

EPDM (kep330 grade with moony viscosity, ML_{1+4} at 125°C= 28 gr/10 min) with ethylene content of 57% and ENB content of 7.9% and density of 0.86 g cm3, was manufactured by Kumho Polychem Co. Ltd, Korea.
EVA copolymer (VS430 grade, 19% VAc) in granular form was the product of Lotte International Co. Ltd, Korea. The melt flow index (MFI) and densitiy of the EVA are 2.5 g/10 min (ASTM D1238) and 0.939 g/cm^3, respectively. maleic anhydride Grafted EPDM (NG8002 grade) with graft content of 1% in granular form was provided by Shanghai Jianqiao Plastic Co. Ltd, China. The melt flow index (MFI) and densitiy of the EPDM-g-MA are 0.2 g/10 min (ASTM D1238) and 0.87 g/cm^3, respectively.

Blend Preparation

EPDM/ENA blend samples varying in blend ratio 80/20, 70/30, 60/40 and 50/50 were considered. All the samples were prepared by melt compounding using a 60 cm^3 internal mixer (Brabender Plasticorder W50) at 150 °C and rotor speed of 60 rpm for 20 min. Prior to the blending, EVA and EPDM-g-MA were dried at 70 C for 12h.

Characterization

Melt viscoelastic behaviour of the blends was studied using a rotational and oscillatory rheometer (Paar Physica USD 200) Was used to investigate the rheological responses of the blends. During the measurement, the melt temperature of sample was set at 190 C and the measurement was conducted within the linear viscoelastic strain range under the measuring frequency range of 0.01-1000 Hz.

SEM was used to characterize sample morphology. To characterize the sample morphology, the samples was first cryogenically fractured in liquid nitrogen, and then to enhance the conductivity fractured surfaces are gold sputtered prior to examining surface morphology.

Phase image of the blend films was studied on an AFM (USPM model, non-contant mode and silicon probe) with an osillation frequency of 2 Hz. The speed of the probe scan on the sample surface is 40 μm/s. Images are recorded in different sizes.

RESULTS AND DISCUSSIONS

Rheological Properties

Figure 1 shows the results of storage modulus (G') and complex viscosity (μ^*) as a function of frequency for neat EPDM, EVA, their blends with different blend ratio (80/20, 70/30, 60/40 and 50/50) and EPDM80/EVA20 sample with compatiblizer measured at 190 °C. As can be observed, the EVA shows a terminal behavior with slop almost equal 2 for G' versus frequency at a low frequency range (Rosian type behavior), whereas EPDM exhibits a pronounced positive deviation with a slop of 1 at the same frequency range. Also, the EPDM shows higher melt elasticity than EVA through frequency range.
In order to understand the rheological behavior of the blends better, storage modulus of samples at constatnt frequency of 0.106 s^{-1} was plotted versus EVA concentration and compared with logarithmic form of mixture law in Figure 2. All the EPDM/EVA blend samples showed a positive deviation from the mixture law, which is result of the strong interfacial interaction between two blend components. Also, strong molecular interactions can lead to partial miscibility in the amorphous region of EVA with EPDM. As it was expected, the G' shows greater positive deviation for the blends in which EVA phase is in the form of droplet dispersed in the EPDM matrix (EPDM80/EVA20 and EPDM70/EVA30 blends) which is due to elastic response associated with droplet deformation and/or interface, e.g., the maximum positive deviation is exhibited by EPDM70/EVA30 blend. This positive deviation of G' from mixture law is expected

to decrease for the blends with co-continuous morphology in which the droplet deformation-induced elastic contribution diminishes (EPDM60/EVA40 and EPDM50/EVA50 blends). Similar observations have been reported by others [7].

As shown in Figure 1, incorporating EPDM-g-MA (5 wt.%) into the blend composition resulted in enhancement of interfacial molecular interactions and compatibility between EPDM and EVA. This is evident in increase of G' for the compatiblized blend compared with uncompatiblized counterpart sample. This is due to the interaction between EVA and Maleic groups, which results in improved morphology and viscoelastic properties through addition of EPDM-g-MA. These polar components increase melt elasticity through dipole-dipole interactions. In fact, the inertia of the segments increases the elastic modulus of the polymer matrix.

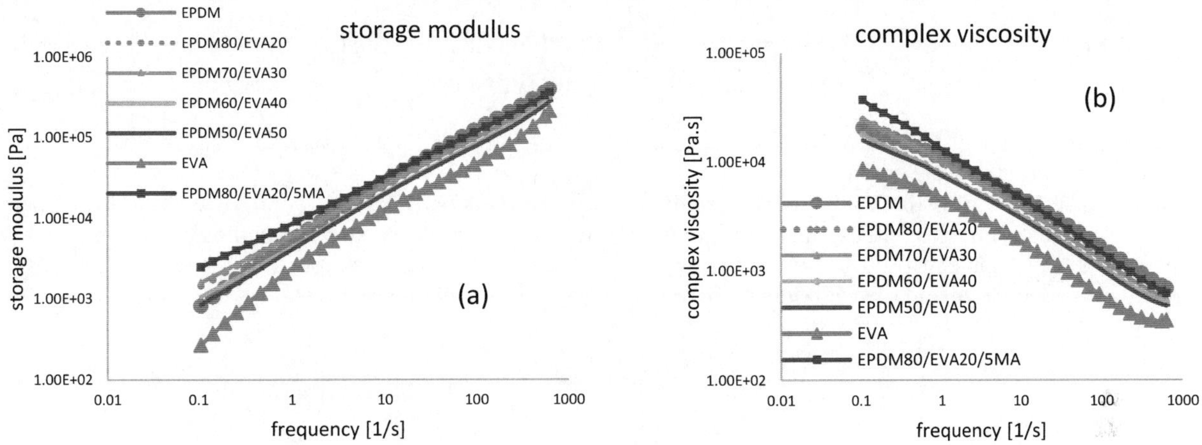

FIGURE 1. Rheological properties of all samples: (a) storage modulus, (b) complex viscosity

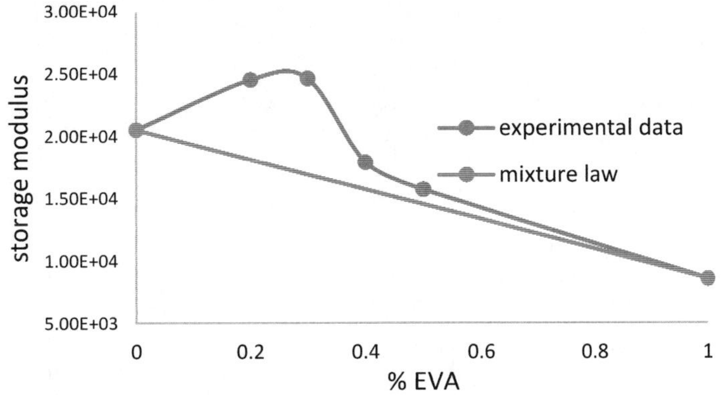

FIGURE 2. variation of storage modulus with EVA concentration of the blends measured at low frequency (0.106 s^{-1})

Morphology

Figure 3 shows the SEM micrographs of samples with 80/20 and 50/50 compositions as can be seen in Figure 3(a) EVA droplets dispersed randomly in EPDM80/EVA20 blend. On the other hand, the phase morphology was changed by increasing the EVA content to 50% (Figure 3(b)).

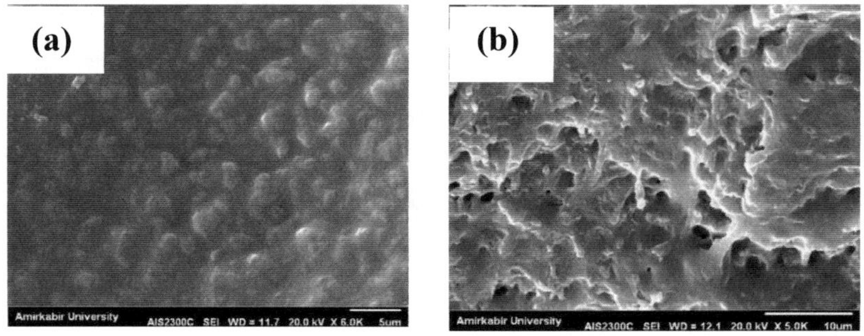

FIGURE 3. SEM images showing the morphologies of (a) EPDM80/EVA20, (b) EPDM50/EVA50 blends.

in order to reach better insight into structure and morphology of samples, AFM images of both compatibilized and uncompatibilized blends were studied.

In the 80/20 and 70/30 EPDM/EVA blend (Figure 4b and 4c), the EVA particles are found to be randomly dispersed as domains in the continuous EPDM matrix. Figures 4d shows the change in phase morphology of the blends upon increasing the EVA content to 40%. By increasing the EVA content in blend composition, the domain size of the dispersed phase decreased and then a more continuous morphology was obtained.

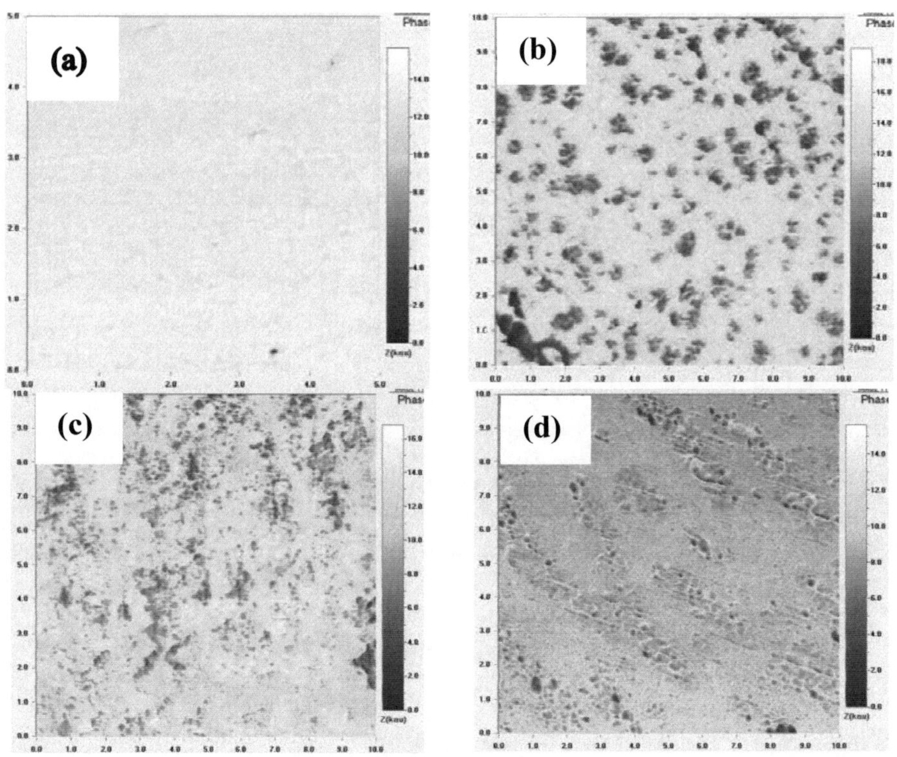

FIGURE 4. AFM images of (a) EPDM, (b) EPDM80/EVA20 blend, (c) EPDM70/EVA30, (c) EPDM60/EVA40 and (d) EPDM50/EVA50 blends.

The effect of the addition of 5% EPDM-g-MA on the morphology of EPDM80/EVA20 blend is presented in Figure 5. It is clear that upon introducing compatiblizer a fine morphology was achieved for the EPDM80/EVA20 sample (Figure 5(b)). The decrease in domain size indicated the compatibility between EVA and EPDM compared to those without compatibilizer resulting from enhanced interfacial adhesion between phases (Figure 5a). In other words, the addition of EPDM-g-MA compatibilizer leads to improvement in the affinity between EPDM and EVA.

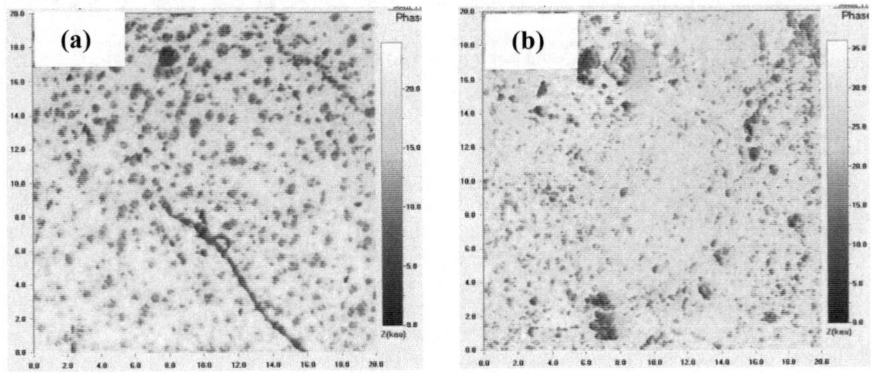

FIGURE 5. AFM images of of (a) EPDM80/EVA20 blend (b) EPDM80/EVA20 blend with compatiblizer

CONCLUSION

In this research microstructure development and melt viscoelastic properties of EPDM/EVA blends varying in composition were investigated. Frequency sweep test result revealed that all the EPDM/EVA blend samples showed a positive deviation from the mixture law, which was result of strong interfacial interaction between two blend components. Also, rheological measurments, SEM and AFM analysis revealed matrix-dispersed morphology for the blends comprising 20-30 wt.% of EVA, whereas at 60/40 ratio to co-continuous morphology was observed.

REFERENCES

1. M. K. Ghosh and C. K. Das, Effect of E/P ratio and VA content on flow behavior of EPDM and EVA blends, J. Appl. Polym. Sci., 1994, vol. 51, no. 4, pp. 643–650.
2. F. Khan, D. Kwek, E. Kronfli, and S. R. Ahmad, Photochemical crosslinking of ethylene-vinyl-acetate an ethylene-propylene-diene terpolymer blends, Polym. Degrad. Stab., 2008, vol. 93, no. 6, pp. 1238–1241.
3. T. Kuila, S. K. Srivastava, and A. K. Bhowmick, Ethylene vinyl acetate/ethylene propylene diene terpolymer-blend-layered double hydroxide nanocomposites, *Polym. Eng. Sci.*, 2009, vol. 49, no. 3, pp. 585–591.
4. I. Ray and D. Khastgir, EVA–EPDM blends as cable insulant, *J. Appl. Polym. Sci.*, 1994, vol. 53, no. 3, pp. 297–307.
5. S. Ray Chowdhury, B. K. Sharma, P. Mahanwar, and K. S. Sarma, Vinyl acetate content and electron beam irradiation directed alteration of structure, morphology, and associated properties of EVA/EPDM blends, *J. Appl. Polym. Sci.*, 2016, vol. 133, no. 21.
6. A. I. Khalaf, A. A. Ward, and N. N. Rozik, Investigation of physical properties and morphology of compatibilized EPDM/EVA blends, *J. Thermoplast. Compos. Mater.*, 2018, vol. 31, no. 3, pp. 376–391.
7. S. Tarashi, H. Nazockdast, Z. Javidi, and M. Mehranpour, Long-chain branch-induced interfacial interaction and its effect on morphology development in polypropylene/ethylene octene copolymer blend, *J. Mater. Sci.*, 2019, vol. 54, no. 8, pp. 6742–6751.

Accurate Characterization of Fiber and Void Volume Fractions of Natural Fiber Composites by Pyrolysis in a Nitrogen Atmosphere

Mehrad Amirkhosravi[a*], Maya Pishvar[a], Youssef K. Hamidi[b] and M. Cengiz Altan[a]

[a]*School of Aerospace and Mechanical Engineering, University of Oklahoma, Norman, OK 73019*
[b]*Mechanical Engineering Program, College of Science and Engineering, University of Houston-Clear Lake, Houston, TX 77058*

[*]*Reference author: mehrad@ou.edu*

Abstract. This paper proposes a simple and accurate technique to quantify both fiber and void volume fractions of natural fiber composites. This technique involves the separate pyrolysis of natural fibers, polymer resin, and natural fiber/polymer composite in a pure nitrogen atmosphere. The remained fractional residuals of these materials are used for the determination of fiber and matrix weight fractions. Consequently, the void and fiber volume fractions of the natural fiber composite are calculated by having weight fractions, as well as the density of natural fiber, epoxy resin, and composite sample. To validate this new method, a number of silk/epoxy composites at various levels of void content were manufactured. Specific results are given for the random mat silk/epoxy laminates made by wet lay-up vacuum bagging (WLVB) process, as well as plain weave silk/epoxy laminates fabricated by the vacuum assisted resin transfer molding (VARTM) process. Using the proposed method, the void volume fractions of the random mat and plain weave silk composites are calculated to be 9.5% and 1.7%, respectively. The fiber volume fraction of random mat and plain weave laminates are also determined to be 18.7% and 46.3%, respectively. The representative scanning electron microscopy (SEM) images of those laminates are also presented to visually compare the void and fiber contents of the parts. The results demonstrate that the proposed method enables an accurate determination of the fiber and void contents in silk/epoxy laminates and can be extended to composites containing other types of natural fibers.

Keywords: Void volume fraction, Fiber volume fraction, Natural fiber, Polymer composites, Pyrolysis.
PACS: 81.05.Qk, 81.20.Hy, 81.70.Bt

INTRODUCTION

Natural fibers such as silk, cellulose, and flax have recently attracted considerable attention as a low-cost, recyclable, and eco-friendly alternative to glass reinforcements in polymer composites [1-5]. However, the high level of moisture absorption and poor resin wettability of the natural fibers may lead to various microstructural defects such as voids in the final composite part [6, 7]. It is well-established that voids are detrimental to the integrity and properties of the composites, including flexural, compressive, and interlaminar shear strengths [8-10]. Another factor that significantly affects the composite performance is its fiber content, such that a higher fiber content of a composite typically results in improved mechanical properties. Therefore, accurate characterization of process-induced voids and fiber content of the laminate provides valuable information for predicting the part performance, as well as improving the composite processing techniques.

The characterization of voids is possible using either SEM or confocal microscopy techniques followed by image analysis methods [11]. However, characterization of voids through image analysis, specifically, for the large parts is time consuming and may not represent the overall void content of the part. In addition, ultrasonic testing [12] and X-ray micro-CT [13] are other well-known techniques used for the characterization of voids. However, these techniques require specialized equipment and are more complicated than the conventional methods such as digestion or ignition of the matrix. In addition, quantifying the fiber volume fraction of the composite parts may not be feasible with the microscopy, ultrasonic, and X-ray micro-CT techniques. The commonly used technique for the calculation of void and fiber volume fractions in traditional polymer composites is either ignition of the matrix in a furnace as specified by ASTM D2584 or digestion of the matrix in high-concentrated acid as explained in ASTM D3171, depending on the reinforcement type. In these techniques, the matrix is removed from composite samples, and the weight fraction of remaining fibers is obtained for the calculation of void and fiber contents. However, natural fibers undergo a complete decomposition during both the burn-off in an air atmosphere and acid digestion, thus making these techniques impractical for natural fiber composites.

Void and fiber volume fractions of natural fiber composites are commonly determined according to the total weight of fiber and resin used in the fabrication of the entire composite part [1, 14, 15]. However, this approach is not accurate for composite parts with non-uniform thicknesses or complex geometries. Therefore, to accurately quantify the void and fiber volume fractions, it is essential to calculate the fiber and matrix weight fractions of each composite sample separately.

In this paper, a novel technique is introduced to accurately quantify the void and fiber volume fractions of natural fiber composites by performing pyrolysis in a pure nitrogen atmosphere. The application of this method is then demonstrated by calculating the void and fiber volume fractions of random mat and plain weave silk/epoxy composites made by WLVB and VARTM processes, respectively. For visual comparison, the representative SEM images of the composite parts are also presented.

EXPERIMENTAL DETAILS

Quantification of Void and Fiber Volume Fractions of Natural Fiber Composites: Pyrolysis in a Nitrogen Atmosphere

To determine the void and fiber volume fractions of fiber reinforced polymer composites, the weight fraction of fiber and matrix for each composite sample, as well as density of fiber, matrix, and samples are required (ASTM D3171). The fiber and matrix weight fractions of glass fiber polymer composites are commonly obtained by performing resin burn-off in the air atmosphere to remove the polymer matrix from the composite. However, the presence of oxygen leads to complete decomposition of natural fibers, making this technique inapplicable for natural fiber composites.

This study, for the first time, introduces a novel technique to quantify the void and fiber volume fractions of natural fiber composites. This technique relies on the separate pyrolysis of composite constituents in a pure nitrogen atmosphere. First, the remained fractional residuals of natural fibers (R_f), polymer matrix (R_m), and natural fiber/polymer composite (R_c) are determined. Then, fiber and matrix weight fractions are obtained using the following equations:

$$R_c = w_f . R_f + w_m . R_m \tag{1}$$

$$w_f = 1 - w_m \tag{2}$$

where w_f and w_m are fiber and matrix weight fractions, respectively.

By having fiber and matrix weight fractions, as well as the density of natural fiber (ρ_f), polymer matrix (ρ_m), and composite sample (ρ_c), the void (v_v) and fiber (v_f) volume fractions are calculated using the following equations:

$$v_f = \frac{\rho_c}{\rho_f}(w_f) \tag{3}$$

$$v_v = 1 - \left(v_f + \frac{\rho_c}{\rho_m}(w_m)\right) \tag{4}$$

Materials and Fabrication Process

In this study, silk fibers with two different architectures: (a) random mat and (b) plain weave are used as the reinforcement (see Figure 1). Both random mat (Paradise Fibers, Inc.) and plain weave (Aurora Silk, Inc.) silk fibers are made from degummed cultivated Bombyx mori mulberry silk. An epoxy resin system, INF 114-INF 211 from PRO-SET, is used for all laminates. This is a low viscosity resin system (i.e. ≈300 mPa s) which is suitable for both wet lay-up and vacuum infusion processes. The cure cycle for this resin system is 5 h at room temperature, followed by 8 h at 60°C.

Random mat silk/epoxy laminates are fabricated by wet lay-up vacuum bagging process. In this process, 10 layers of random mat silk caps are wetted by INF 114/INF 211 resin system. The impregnation of the layers is carried out by a roller. Afterwards, the lay-up is sealed in a vacuum bag and full vacuum is applied to remove air and excess resin. The recommended cure cycle is then applied. The plain weave silk/epoxy laminates are fabricated using VARTM process. For that, 12-ply plain weave silk fabrics are stacked on top of each other and covered with a release film.

FIGURE 1. SEM images of (a) random mat silk cap and (b) plain weave silk fabric

The lay-up is then connected to the resin reservoir and vacuum line, and resin is infused under the vacuum pressure. After the part is filled, the inlet is clamped, and the same cure cycle is applied. Fabrication details for random mat and plain weave silk/epoxy laminates are presented in TABLE 1.

TABLE (1). Fabrication details for silk fiber/INF 114-INF 211 epoxy laminates.

Fabrication scenario	Fabrication Process	Fiber	Number of plies
W-RM-10	Wet lay-up vacuum bagging (W)	Random mat silk cap (RM)	10
V-PW-12	Vacuum assisted resin transfer molding (V)	Plain weave silk fabric (PW)	12

RESULTS AND DISCUSSION

Void and Fiber Volume Fractions

The pyrolysis is performed in a pure nitrogen atmosphere by heating samples from room temperature to 800°C at a constant heating rate of 10°C/min and keeping them at 800°C for 30 min. It should be noted that in this technique the same heating rate should be applied for fiber, resin, and composite samples. Figure 2 presents the weight percentage of the random mat silk cap, plain weave silk fabric, INF 114/INF 211 epoxy resin, and their composite samples as a function of time. Since both random mat and plain weave silk fibers are made from the same type of silk (i.e. Bombyx mori mulberry silk), their thermographs follow almost the same trend as shown in Figure 2. The difference between random mat and plain weave silk/epoxy laminates graphs is, however, due to the difference in the volume fractions of their constituents. It is observed that the weight loss for all materials is minimal after the temperature reaches 800°C. The final points in the thermographs of silk fiber, epoxy, and composite sample correspond to the remained fractional residuals of each material which is used for the calculation of weight fractions.

To investigate the repeatability of the final remained fractional residuals for silk fiber and epoxy, three samples are considered for each material. In addition, five samples per each fabrication scenario are burned separately in a pure nitrogen atmosphere to obtain the void and fiber volume fractions of the composite parts. The remained fractional residuals of random mat silk cap, plain weave silk fabric, INF 114/INF 211 epoxy resin, as well as void and fiber volume fractions of random mat and plain weave silk/epoxy laminates are presented in Table 2. The residuals of the random mat and plain weave silk fibers are around 27%, and the residual for INF 114-INF 211 is around 7%. The results also show that the void volume fraction of the random mat laminates (W-RM-10) is around 9.5% which is significantly higher than 1.8% void content of the plain weave laminates (V-PW-12). This difference in the void content can be due to both the fabrication process and fabric architecture.

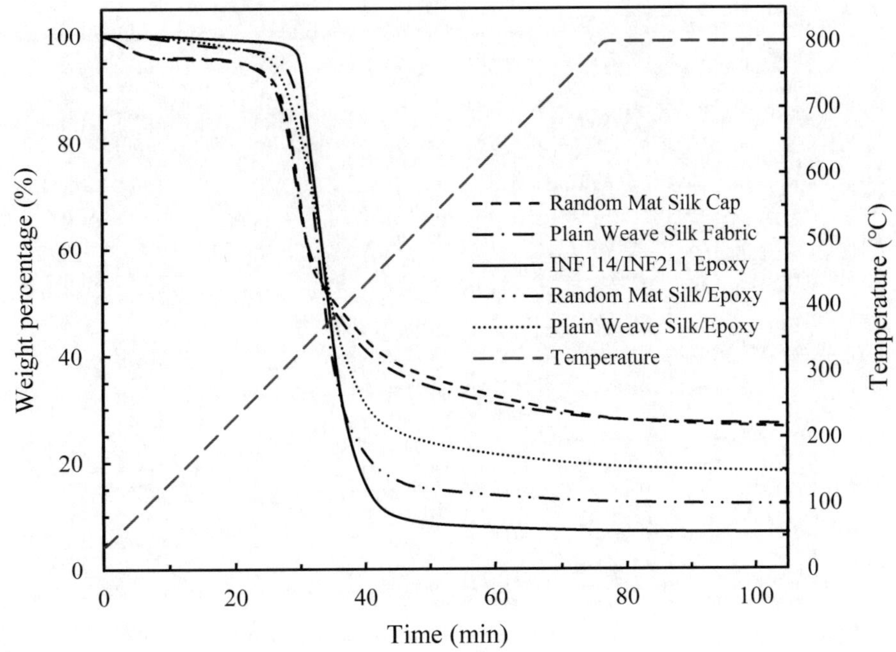

FIGURE 2. TGA thermographs of random mat silk cap, plain weave silk fabric, INF 114/INF 211 epoxy, as well as random mat silk cap/epoxy and plain weave silk fabric/epoxy composites under a nitrogen atmosphere. The temperature profile is also presented.

TABLE (2). The remained fractional residuals (R) of random mat silk, plain weave silk, INF 114/INF 211 epoxy, as well as void and fiber volume fractions of random mat and plain weave silk/epoxy laminates.

R_f for random mat silk cap (%)	R_f for plain weave silk fabric (%)	R_m for INF 114-INF 211 (%)
26.5 ± 0.9	27.1 ± 0.3	6.8 ± 0.7
Fabrication scenario	**Fiber volume fraction (%)**	**Void volume fraction (%)**
W-RM-10	18.7 ± 2.7	9.54 ± 1.44
V-PW-12	46.3 ± 3.4	1.75 ± 0.70

Microstructure of Silk/Epoxy Laminates

In order to visually compare the void and fiber volume fractions of random mat and plain weave laminates, SEM images of the through-the-thickness cross section of those laminates at 50x and 200x magnifications are given in Figure 3. In Figure 3(a), voids with different sizes and shapes, shown by arrows, contribute to the high void volume fraction (i.e. ≈9.5%) of the random mat silk laminate. In contrast, the size and number of voids in plain weave silk laminate shown in Figure 3(b) are considerably lower than those in random mat laminates, confirming the low void content of 1.7% in plain weave silk laminate.

As mentioned in Table 2, the fiber volume fraction of plain weave silk laminate (i.e. 46.3%) is much higher than that of random mat silk laminate (i.e. 18.7%). This can be confirmed by comparing the high-magnification SEM images of random mat and plain weave silk laminates in Figures 3(c) and (d), where: (i) more resin rich regions are observed in random mat laminate than those in plain weave laminate and (ii) more fibers (appear in light-grey in Figure 3) are recognized in the plain weave laminate compared to the random mat laminate.

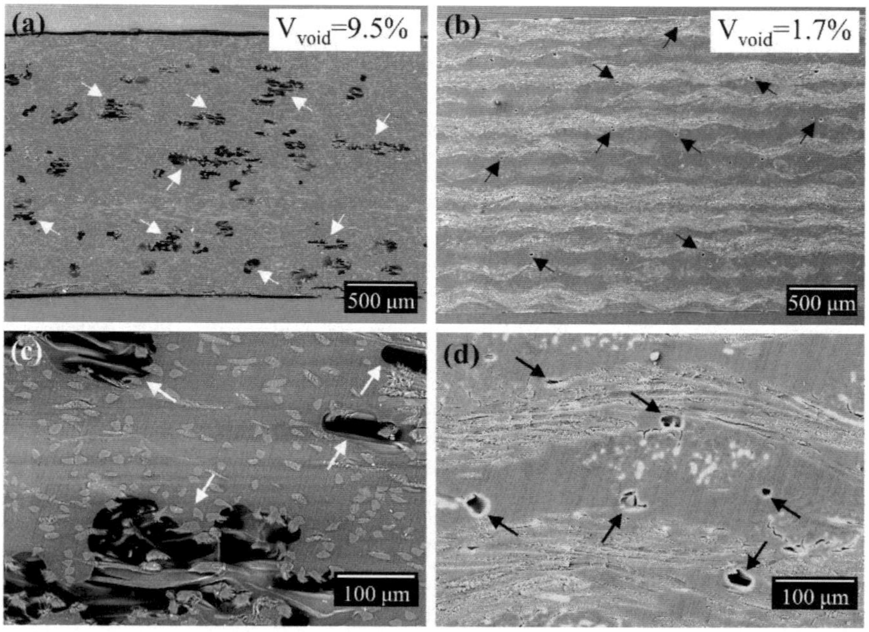

FIGURE 3. SEM images of the through-the-thickness surface of (a) random mat silk/epoxy and (b) plain weave silk/epoxy laminates (at 50x magnification). SEM images of (c) random mat and (d) plain weave silk/epoxy composites at a higher magnification (200x). Note: Arrows point to the voids ranging from a few microns to several hundred microns in the laminates.

CONCLUSION

This paper introduces an innovative technique for the quantification of void and fiber volume fractions of natural fiber composites. In this technique, the pyrolysis of natural fiber composite and its constituents are carried out in a pure nitrogen atmosphere to accurately obtain the fiber and resin weight fractions of composite samples. To investigate the suitability of the proposed method for natural fiber composites, random mat and plain weave silk/epoxy composite laminates are fabricated with wet lay-up and vacuum infusion processes, respectively. The materials and fabrication process are selected such that the resulting laminates have different levels of void and fiber contents. It is observed that the proposed method can accurately quantify different levels of void and fiber volume fractions of the composite parts. For example, the result shows that the random mat silk epoxy laminate contains 9.5% voids which is much higher than the 1.7% void content calculated for the plain weave laminate. Furthermore, the through-the-thickness microscopy images of the composites' microstructure confirm the range of void and fiber volume fractions obtained by the proposed technique.

REFERENCES

1. Y. K. Hamidi, M. A. Yalcinkaya, G. E. Guloglu, M. Pishvar, M. Amirkhosravi, and M. C. Altan, *Materials* **11**, 2135 (2018).
2. P. Wambua, J. Ivens, and I. Verpoest, *Compos. Sci. Technol.* **63**, 1259-1264 (2003).
3. S. Mahmoudi, A. Kervoelen, G. Robin, L. Duigou, E. Daya, and J. Cadou, *Compos. Struct.* **208**, 426-433 (2019).
4. C. Campana, R. Leger, R. Sonnier, L. Ferry, and P. Ienny, *Compos. Part A* **107**, 171-179 (2018).
5. F. Ansari, A. Sjöstedt, P.T. Larsson, L.A. Berglund, and L. Wågberg, *Compos. Part A* **74**, 60-68 (2015).
6. N. Agarwal, D. Hoagland, and R. Farris, *J. Appl. Polym. Sci.* **63**, 401-410 (1997).
7. C. Deo and S. Acharya, *J. Reinf. Plast. Compos.* **29**, 2513-2521 (2010).
8. M. Mehdikhani, L. Gorbatikh, I. Verpoest, and S.V. Lomov, *J. Compos. Mater.* (2018). DOI: 10.3390/ma11112135.
9. M. Amirkhosravi, M. Pishvar, and M.C. Altan, *J. Compos. Mater.* (2018). DOI: 10.1177/0021998318791311.
10. N. Niknafs Kermani, P. Simacek, M. Erdal, and S. G. Advani, Proceedings of the American Society for Composites Thirty-third Technical Conference, Seattle, WA, 2018.
11. Y. K. Hamidi, L. Aktas, and M. C. Altan, *Compos. Sci. Technol.* **65**, 1306-1320 (2005).
12. L. Liu, B. -M. Zhang, D. -F. Wang, and Z. -J. Wu, *Compos. Struct.* **73**, 303-309 (2006).
13. P. J. Schilling, B. R. Karedla, A. K. Tatiparthi, M. A. Verges, and P. D. Herrington, *Compos. Sci. Technol.* 65, 2071-2078. (2005).
14. C. Zarate, M. Aranguren, and M. Reboredo, *J. Appl. Polym. Sci.* **89**, 2714-2722 (2003).
15. R. Yahaya, S. Sapuan, M. Jawaid, Z. Leman, and E. Zainudin, *Polym. Compos.* **36**, 1469-1476 (2015).

"Development of Cost Effective In-Situ Microfibrillar Recycled PET/Carbon Fiber/Polypropylene Matrix Composites with High Mechanical Properties

Orkun Kaymakçı[a,b], Elif Büşra Ergin[c] and Nurseli Uyanık[d]

[a] Arçelik Central R&D, Tuzla - Istanbul - Turkey
[b] Istanbul Technical University, Polymer Science & Technology Prog. - Istanbul - Turkey
[c] TOBB Economy and Technology University, Materials Sci. and Nanotechnology Eng. - Ankara – Turkey
[d] Istanbul Technical University, Faculty of Science and Letter, Department of Chemistry - Istanbul - Turkey

Abstract. Carbon fibers (CF) are excellent reinforcers for thermoplastic polymer composites owing to their high thermal and mechanical properties. However, their applications are mostly limited to high-end applications such as in aerospace industries due to their relatively higher cost. In this study, it was aimed to improve the properties of CF – Polypropylene (PP) composites by supporting the matrix with microfibrillar PETs. Low cost recycled PET (rPET) flakes from used PET bottles were in-situ converted into PET fibers during the compounding of carbon fibers and PP matrix. Seven different composites were prepared with different rPET contents up to 15 phr, in order to understand the rPET's effect on the physical and mechanical properties of the composites. Maleic anhydride grafted PP (MA-g-PP) compatibilizer was used to improve interfacial adhesion between PP and both of CF and rPET. Additionally, the effect of the compatibilizer content on the mechanical properties was investigated by studying compounds of varying compatibilizer content (0wt%, 2.5wt% and 5wt%). The carbon fiber content was fixed to 10wt% in all composites. Scanning electron microscopy (SEM) and Energy-dispersive X-ray spectroscopy (EDX) studies showed the formation of microfibrillar PETs in CF-PP composite matrix. In addition, coupling efficiency of the compatibilizer was investigated in the morphology. Mechanical characterization results showed that the in situ formation of rPET microfibrils in CF-PP composites significantly improves the tensile, flexural and Izod impact properties. The orientation of the CFs and rPET microfibrils during injection molding process affect the tensile and flexural strength of the composites.

Keywords: Microfibrillar Recycled PET. Carbon Fiber, Polypropylene, Mechanical Properties
PACS: 81.05.Lg

INTRODUCTION

The use of carbon fiber(CF) reinforced thermoplastic composite materials is increasing day by day as the materials possess combination of properties such as high modulus, high strength, thermal stability, ease of processability and high wear resistance [1,2]. In addition to these superior properties, low weight of the composites makes them very attractive to automotive industry due to increasing global trends on carbon dioxide emission reduction and fuel efficiency [3]. However, relatively higher cost of the CF reinforced composites mostly limits the use and the applications of the materials to specific high-end industries such as aerospace and sporting goods.

Poly(ethylene terephthalate) (PET) is one of the most popular packaging materials and it is widely used to produce drink bottles. However, the massive use of the material creates a waste disposal problem. Recycling is the most prominent and widely used method for the disposal of PET wastes and it saves energy, water, CO2 emission and the consumption of limited petroleum based raw materials, resulting effective utilization of natural resources [4]. Rising environmental concerns such as global warming and environmental pollution increases the significance of recycling of PET based single-use plastics.

Formation of in-situ fibrils in polymer blends during compounding process is an effective method to produce high performance polymer blends of immiscible polymers with different melting temperatures. Among different microfibrillar reinforced composites (MFC) [5] such as PET/PA6 [6] pairs, PET/LDPE [7] pair, PET/PS [8] pair and PA6/PP [9] pair, PET/PP pair is studied more extensively than the other pairs, due to the optimum combination of the stiffness and toughness from the fibrillar dispersed phase and the matrix phase, relatively lower cost of the components and the abundance of the polymers in recycled forms. However, low immiscibility due to the differences of the polarity and the chemical nature of the different phases may result low mechanical properties. Therefore, several researchers compatibilized the PET phase and the PP matrix using different coupling agents. Freidrich et. al showed that ethylene glycidylmethacrylate (E-GMA) coupling agent significantly increases tensile modulus of the PET/PP microfibrillar composites [10]. Li et al. showed that Tg of PET in MA-g-PP compatibilized PET/PP microfibrillar composites shifts toward the Tg of PP indicating improved compatibility between PET and PP [11].

In this study, formation of in-situ microfibrillar Recycled PET in carbon fiber/polypropylene composites was investigated for the first time to develop cost-effective, high performance composites. The effect of rPET content and MA-g-PP compatibilizer content was investigated. The mechanical properties were studied by tensile, flexural and Izod impact tests. Formation of the microfibrillar PETs and the morphology of the composites were studied by Scanning electron microscopy (SEM) and Energy-dispersive X-ray spectroscopy (EDX).

EXPERIMENTAL

Materials

Borealis HE125MO homopolymer PP with 12 g/10min melt flow index (230°C/2,16 kg) was supplied by Borealis AG, Austria. Chopped Carbon fibers (CF) with silane sizing have supplied from Procotex SA, Belgium. Average monofilament diameter of the fibers was 7 μm and the mean length of the fibers was 6 mm. MA-g-PP coupling agent was kindly supplied by BYK Additives, Germany. BYK 6012 MA-g-PP is a solid phase grafted coupling agent with 1.8% chemically bound maleic anhydride content. Recycled PET flakes in the size of 1-4 mm sourced from used drink bottles were supplied by CevrePET Recycling Company, Turkey.

Composite Preparation

rPET flakes were dried in oven at 100°C for at least 8 hours to avoid hydrolysis during compounding process. In-situ microfibrillar rPET - CF - PP compounds were prepared using PRISM TSE24 IIC twin screw co-rotating compound extruder, with the screw diameter of 24mm and L/D ratio 28:1 at a screw speed of 200 rpm. Temperature profile of the compounder was set to 200°C – 220°C - 210°C – 240°C – 260°C - 235°C - 220°C. The extrudate was cold-streched using a self-designed drawing machine. Formulations that are produced in the compound extruder are listed in Table 1. Test specimens for tensile testing, Izod Impact testing and flexural testing were produced using a Arburg Allrounder 3200C injection machine.

TABLE 1: Sample Codes and Compositions of Microfibrillar rPET/CF/PP Composites

Sample	PP (%)	CF (%)	MA-g-PP (%)	rPET (phr)
PP/CF/2.5MA	87.5	10	2.5	0
PP/CF/2.5MA/2.5rPET	85	10	2.5	2.5
PP/CF/2.5MA/5rPET	82.5	10	2.5	5
PP/CF/2.5MA/7.5rPET	80	10	2.5	7.5
PP/CF/2.5MA/10rPET	77.5	10	2.5	10
PP/CF/2.5MA/12.5rPET	75	10	2.5	12.5
PP/CF/2.5MA/15rPET	72.5	10	2.5	15
PP/CF/10rPET	80	10	0	10
PP/CF/5MA/10rPET	75	10	5	10
PP/CF/2.5MA	87.5	10	2.5	0

Mechanical Characterization

Tensile tests were performed using Zwick Z020 universal tensile testing machine according to ISO 527 standards at room temperatures. Cross-head speed was kept constant at 50 mm/min. Flexural tests were performed in three-point bending mode using Instron 4505 universal testing machine according to ISO 78 standards at room temperature. Tests were performed at 5 mm/min cross-head speed rate. Izod Impact tests were performed using Zwick HIT Series Impact Tester according to ISO 180 standards at room temperature.

Scanning Electron Microscopy (SEM) / Energy Dispersive X-Ray (EDS) Analysis

SEM Analyses were performed by a Zeiss SEM coupled with EDS spectrometer at an acceleration voltage of 20kV. The specimens were fractured under liquid nitrogen and gold-sputtered before observation.

RESULTS AND DISCUSSION

In this study, rPET was compounded into PP matrix along with 10% carbon fibers to increase the stiffness of the composites and to reduce the overall cost of the compounds. The primary motivation of adding rPET flakes into the system was converting the flakes into PET microfibrils during the cold drawing process. Figure 1 illustrates the microfibrillar rPET formation in CF/PP composites.

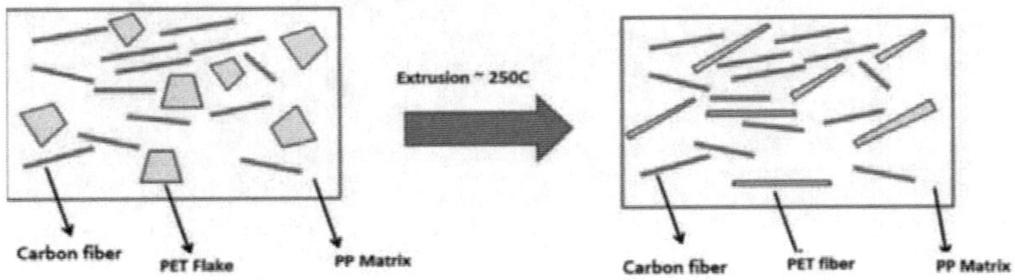

FIGURE 1: Transforming rPET flakes to PET microfibrills during compounding

Figure 2 illustrates the SEM micrographs and EDS mapping of the respective regions. It is clear that oxygen rich areas in the EDS maps correspond to PET phases in PP matrix. In addition, rPET microfibrills can be easily distinguished from CF with their uneven and flexible structure.

It shows that presence of the compatibilizer affects the morphology of the PET fibers. In the compatibilized specimen (Figure 2, (c-d)), PET microfibrills have regular, oriented shape, however, incompatibilized specimen (Figure 2, (a-b)) has PET domain that is irregular in terms of orientation and shape. In this case, MA-g-PP compatibilizer reduces the interfacial tension between the PP domain and the PET domain, resulting improved steric stabilization. Pang et al. Previously observed a similar effect of a compatibilizer on the morphology of the PP/PET blends [12].

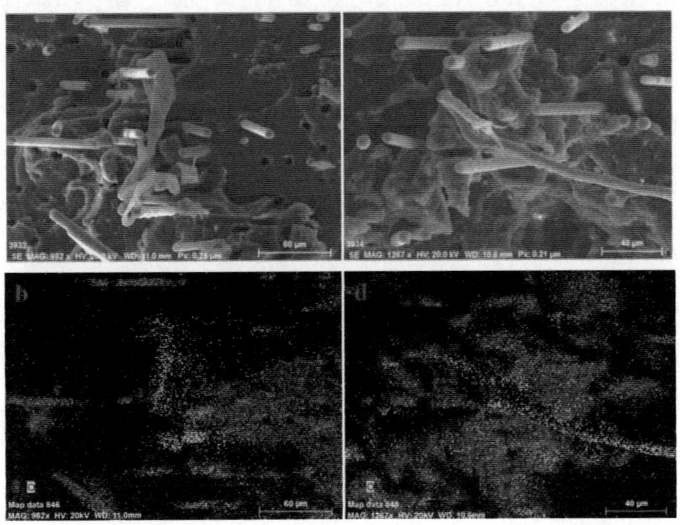

FIGURE 2: SEM and EDX images of PP/CF/10rPET (a-b) and PP/CF/MA/10rPET

It was also observed that MA-g-PP significantly improved the interfacial adhesion between PP matrix and the CFs. Figure 3, (a-b) shows the SEM micrographs of the incompatibilized specimen. In this specimen, CFs are not covered with PP matrix and CF fiber surfaces are very clear and smooth due to the poor interfacial interaction between the fibers and the polymer. In addition, fractured surfaces exhibit smooth fiber pull-out regions. Figure 3, (c-d) and (e-f) shows the SEM micrographs of 2.5% and 5% MA-g-PP compatibilizer added samples respectively. In the presence of the compatibilizer, fiber-matrix interaction was improved because of the hydrogen bonding interactions between the hydroxyl groups of the MA-g-PP and the hydroxyl groups of the silane sized CF surface. SEM micrographs show that surfaces of the specimen with 5% MA-g-PP exhibit better wetting efficiency with the PP matrix than the specimen with 2.5% MA-g-PP samples. In this case, adding more compatibilizer into the PP matrix improved the fiber-matrix interfacial adhesion.

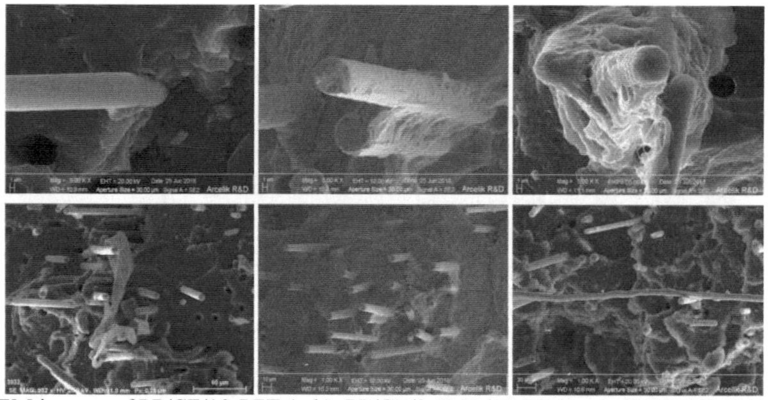

FIGURE 3: SEM images of PP/CF/10rPET (a-b), PP/CF/2.5MA/10rPET(c-d), PP/CF/5MA/10rPET(e-f)

Figure 4 shows the effect of rPET flakes on the tensile properties of the CF-PP composites in the presence of 2.5% MA-g-PP compatibilizer. It is shown that addition of rPET into CF-PP composites up to 5 phr significantly improved the elastic modulus and the tensile stress of the materials. The increase in the mechanical properties was due to the reinforcing effect of the rPET microfibrils that formed during compounding process. In addition, it is shown that both tensile stress and tensile strain increased with the addition of rPET into system. The improvements in the tensile stress and the tensile strain indicate good compatibility between the rPET microfibrils and PP matrix, possibly due to the presence of MA-g-PP coupling agent.

It is also shown that after 5 phr loading levels of rPET, the reinforcing effect of the rPET weakens. During the compounding process, only some portion of the rPET flakes were stretched into PET fibers. At higher loading values, the rPET flakes may still present in the system, disrupting the reinforcing effect of the fibers. In addition, as the rPET loading levels increase, the ratio of the carbon fibers in the total system decreases. Therefore, diluted carbon fiber levels resulted lower stiffness.

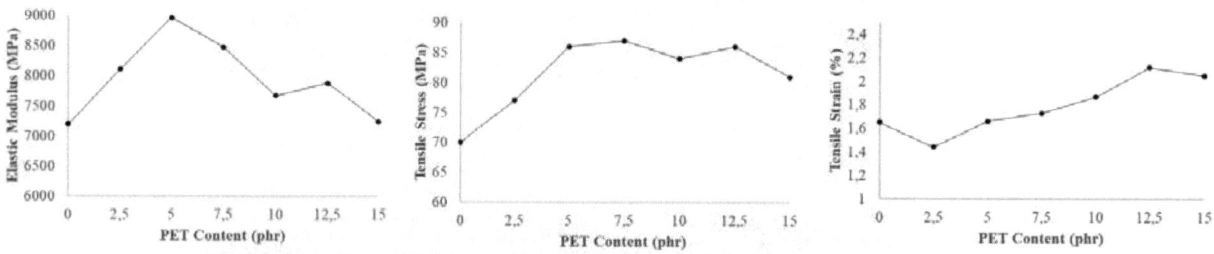

FIGURE 4: Effect of PET Content on the Tensile Properties of the Composites

Figure 5 shows the effect of rPET on the flexural mechanical and notched Izod impact properties of the composites. The effect of the rPET on the flexural properties was very similar to the effect on the tensile properties which shows that the rPET fibers have reinforced the network both in the flow direction and in the opposite of flow direction. In addition, similar to the tensile properties, the rPET have increased the flexural stiffness and the strength of the material up to 5 phr loading levels. Similar to the tensile and flexural mechanical properties, up to 5 phr loading levels, rPET fibers increased the Izod impact strength of the composites. The impact strength of the materials increased about 20% by the addition of 5 phr of rPET flakes.

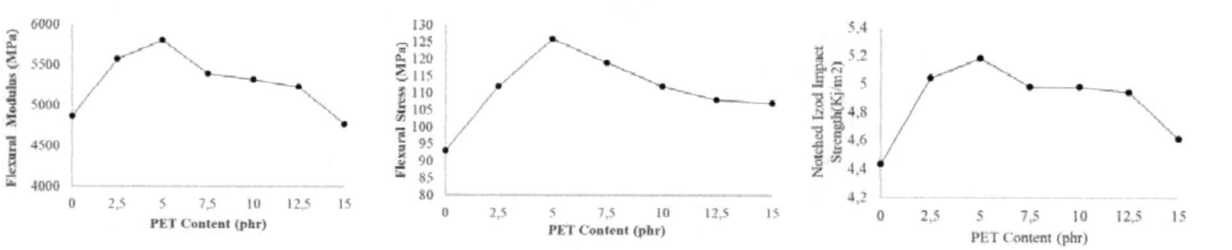

FIGURE 5: Effect of PET Content on the Flexural and Izod Impact Properties of the Composites

Table 2 summarizes the mechanical properties of rPET-CF-PP composites with different MA-g-PP contents. MA-g-PP compatibilizer was used to improve the interfacial adhesion between PP and both of CF and rPET. CF and rPET contents were set to 10% and 10 phr respectively. It is shown that MA-g-PP coupling agent is crucial for the interfacial adhesion between PP and CF – rPET. With 5% addition of MA-g-PP, both the tensile stress and tensile strain of the materials were almost doubled. The coupling agent affected the compatibility between both PP / CF and PP / rPET. SEM micrographs in Figure 3 supports the improvements in mechanical test results.

TABLE 2: Effect of MA-g-PP compatibilizer on the mechanical properties of the composites

Sample	MA-g-PP (%)	Elastic Modulus (Mpa)	Tensile Stress (Mpa)	Tensile Strain (%)	Flexural Modulus (Mpa)	Flexural Stress (Mpa)	Izod Impact Resistance (Notched) (kJ/m²)
PP/CF/0MA/10rPET	**0**	7720	61	1.09	4890	82	4.75
PP/CF/2.5MA/10rPET	**2.5**	7970	84	1.87	5320	112	4.98
PP/CF/5MA/10rPET	**5**	8520	102	2.30	6240	134	5.94

In summary, mechanical tests and SEM micrographs show that 5% compatibilizer results better adhesion to CF and rPET. Therefore, in the future studies, the compatibilizer loading levels will be increased to more then 5% for better understanding of the compatibilizer.

CONCLUSIONS

The effect of rPET and MA-g-PP compatibilizer on the morphological and mechanical properties of in-situ microfibrillar reinforced was investigated. rPET content was varied up to 15 phr in the presence of 2.5% compatibilizer and compatibilizer content was varied up to 5% in the presence of 10phr rPET content.

The formation of microfibrillar rPET during the compounding and cold drawing process was confirmed with SEM and EDS studies. It was found that morphology of the rPET microfibrilles depend on the presence of MA-g-PP compatibilizer due to steric stabilization effect of the compatibilizer. Tensile, flexural and Izod Impact results showed that addition of rPET up to 5 phr significantly improves the mechanical properties of the composites due to the reinforcing effect of the microfibrillar PETs. It was also shown that MA-g-PP compatibilizer significantly improves the mechanical properties of the in-situ microfibrillar recycled PET / CF reinforced PP composites due to the improved interfacial adhesion between PP and CF and PP and microfibrillar rPET.

REFERENCES

1. D. Kada, A. Koubaa, G. Tabak, S. Migneault, B. Garnier, and A. Boudenne, *Polymer Composites* **39**, (2016).
2. S.-S. Yao, F.-L. Jin, K.Y. Rhee, D. Hui, and S.-J. Park, *Composites Part B: Engineering* **142**, 241 (2018).
3. F. Rezaei, R. Yunus, N.A. Ibrahim, and E.S. Mahdi, *Polymer-Plastics Technology and Engineering* **47**, 351 (2008).
4. Oromiehie and A. Mamizadeh, *Polymer International* **53**, 728 (2004).
5. W. Li, A.K. Schlarb, and M. Evstatiev, *Journal of Applied Polymer Science* **113**, 1471 (2009).
6. M. Evstatiev, N. Nicolov, and S. Fakirov, *Polymer* **37**, 4455 (1996).
7. S. Fakirov, H. Kamo, M. Evstatiev and K. Friedrich, *Journal of Macromolecular Science, Part B 43*, 775 (2004).
8. S. Saikrasun, P. Limpisawasdi, and T. Amornsakchai, *Journal of Polymer Research* **16**, 443 (2008).
9. S. Naficy and H. Garmabi, *Composites Science and Technology* **67**, 3233 (2007).
10. K. Friedrich, M. Evstatiev, S. Fakirov, O. Evstatiev, M. Ishii, and M. Harrass, *Composites Science and Technology* **65**, 107 (2005).
11. W. Li, A.K. Schlarb, and M. Evstatiev, *Journal of Applied Polymer Science* **113**, 1471 (2009).
12. Y. Pang, D. Jia, H. Hu, D. Hourston, and M. Song, *Polymer* **41**, 357 (2000).

Chitosan Based Composite Membranes for Adsorbtion of Cu (II) and Pb (II) Ions in Aqueous Solution

Kurt Özge[a], Çelebi Hande[a*]

[a]*Eskisehir Technical University Department of Chemical Engineering - Eskisehir - Turkey*

Abstract. Shortage and contamination of global drinking and the removal of various pollutants such as heavy metals from wastewater has become a critical issue due to their adverse effects on human health and environment. Nanomaterials due to their good adsorption efficiency, higher surface area and greater active sites for interaction with pollutants have shown the effectiveness for cleaning industrial wastewater Chitosan is a biodegradable, biocompatible, non-toxic natural polymer produced from natural sources. In this study, chitosan composite membranes were prepared with nanocellulose (NC) and/or graphene oxide (GO) as additives by solution casting method. The obtained composites were further stabilized by cross-linking with gluteraldehyde. GO was synthesized from graphite by the modified Hummer's method. NC was prepared by acid hydrolysis of commercially available microcrystalline cellulose (MCC). A good dispersion of additives was obtained in chitosan solution because of the potential interaction between polymer chains and particles. The interactions between chitosan and additive materials were successfully observed with FTIR, XRD and SEM analyses. Heavy metal (Cu (II) and Pb (II)) removal from wastewater was investigated by ICP-OES (Inductively Coupled Plasma-Optical Emission Spectrometry) method. Chitosan based composite membranes can be used as replacement for conventional expensive active carbon–based adsorption technology as well as toxic magnetic material based process.

Keywords: Chitosan, Polymer Membrane, Heavy Metal Removal, Graphene Oxide, Nanocellulose.

INTRODUCTION

Impending depletion of fossil fuels and ecological concerns lead to the development of new biocomposites. Biocomposites are a new generation of composite materials used in materials science, life science and nanotechnology. Most biocomposites produced from renewable sources such as cellulose and starch are important for saving petroleum resources and protecting the environment [1]. Chitosan attracts attention among the polymer materials used as an alternative to petroleum products due to its unique properties such as biocompatibility, biodegradability, antibacterial properties and multiple functional groups [1]. Chitin, the raw material for chitosan, is the most naturally abundant national polymer on earth after cellulose. Chitosan is a natural polysaccharide obtained from alkaline deacetylation of chitin that is derived from shellfish [2]. Additives such as nanoclay, nanocellulose, carbon nanotube have been successful in improving the mechanical properties of chitosan. Graphene oxide has been successfully added by mixing to the the solution of polymers such as chitosan, and proved that is a new reinforcing filler. It has emerged that these composites may have promising new applications as biomaterials [3].

In this study, chitosan/graphene oxide/nanocellulose composite membranes were prepared by solution casting method. Thermal, mechanical and heavy metal removal properties of membranes are investigated for characterization.

EXPERIMENTAL

Materials

In experiments, graphite powder (Aldrich), sodium nitrate (Acros, 99 +%), sulfuric acid (Sigma Aldrich, 95-97%), acetic acid (Sigma Aldrich, 99,8%), potassium permanganate (Merck, ≥99%) , hydrogen peroxide (Merck, 35,7%), chitosan (Aldrich, middle molecular weight), microcrystalline cellulose (Merck), lead chloride (Roth, ≥98%), copper chloride (Aldrich, 97%), glutaraldehyde (Merck, 25%), sodium hydroxide (Merck), hydrochloric acid (Merck, 37%) were used.

Graphene Oxide Synthesis

Graphene oxide (GO) from graphite powder was synthesized by the modified Hummer's Method. 1 g of graphite powder, 0.5 g of sodium nitrate ($NaNO_3$) and 23 mL of sulfuric acid (H_2SO_4) were stirred in the magnetic stirrer for 1 hour. 3 g of potassium permanganate ($KMnO_4$) was slowly added to the ice bath solution to allow the oxidation of the graphite to ensure that the temperature remained below 10°C to prevent overheating and explosion. The mixture removed from the ice bath was brought to 35°C and stirred for 1 hour. In the second stage of the synthesis, 200 mL of deionized water was added to the mixture and brought to 95°C for 1 hour, followed by stirring for 2 hours. 50 mL of deionized water was added dropwise to the solution brought to room temperature. Hydrogen peroxide (H_2O_2, 35.7%) was added dropwise until the color of the mixture turned from brown to yellow and stirred for 1 hour. The solution washed with 5% hydrochloric acid, ethanol and distilled water was dried in a freeze dryer (VirTis Benchtop K2) to obtain graphene oxide.

Graphene Oxide Characterization

The surface morphology of the prepared graphene oxide samples was examined by Zeiss Evo 50 EP scanning electron microscope (SEM). Fourier Transformation of Infrared Spectroscopy (FTIR) analysis was performed to examine the functional groups of graphite and synthesized graphene oxide. The analyzes were performed using the Bruker Tensor 27 series FTIR-ATR in the range of 4000-600 cm^{-1} wavelength. X-ray diffraction (XRD) was used to investigate the change in crystalline graphite and graphene oxide layers. It was made with Rikagu Miniflex 600 diffractometer under 40kV and 15 mA with CuKα radiation with wavelength λ = 0,154 nm. The measurement was performed at a scanning speed of 2°/min in the range of 5-40°. The thermal degradability of graphene oxide was measured using the TGA Q500-TA Instruments thermal gravimetric analysis (TGA) instrument. The analysis was carried out in a nitrogen atmosphere, with a heating rate of 20°C/min from room temperature to 800°C. Malvern Mastersizer 2000 was used for the analysis of the particle size distribution of graphite and graphene oxide. UV measurements of graphite and graphene oxide were made with UV2600 SHIMADZU Spectrophotometer device in 200-700 nm wavelength range. Characterization results showed that graphene oxide was successfully synthesized.

Nanocellulose Synthesis

Acid hydrolysis method was used to obtain nanocellulose (CNW) from microcrystalline cellulose. For this, 10 g of microcrystalline cellulose was mixed in a magnetic stirrer until a homogeneous suspension was obtained in 100 mL of distilled water. The prepared 64% solution of 100 mL sulfuric acid was added dropwise to this suspension so that the temperature did not exceed 40°C. Then, the suspension was stirred for 2 h at 44°C in a magnetic stirrer. After the hydrolysis of the acid, centrifugation (Thermo Scientific SL 16 Centrifuge) (12000 rpm, 15 minutes) was applied to the suspension until the pH reached about 5 to remove the remaining sulfuric acid. After centrifugation, the suspension was filtered with Whatman no. 1 filter paper. After filtration, the suspension was dispersed by ultrasonic mixer (Sonics Vibra Cell), then dried in freeze dryer to obtain nanosellulose in powder form.

Nanocellulose Characterization

Nanocellulose characterization was performed with FTIR, XRD and Transmission Electron Microscopy (TEM, FEI, Tecnai G2 Spirit Biotwin). Characterization results showed that nanocellulose was successfully synthesized.

Composite Membrane Production

2 grams of chitosan were mixed in a 2% (v/v) acetic acid solution at 300 rpm for 24 hours in a magnetic stirrer. Centrifuge was performed at 1500 rpm for 15 minutes to remove impurities and insoluble material. 0.5%, 1% and 2% by weight of the amount of chitosan as graphene oxide and 5% by weight of nanocellulose were used (Table 1). Additive materials were homogenized in water for 10 minutes and added to the chitosan solution and stirred for 1 hour in a magnetic stirrer. The solution in the ice bath was mixed with the 20 min sonicator. Subsequently, the films were poured into teflon petri dishes and left to dry overnight under a fume hood. In order to neutralize the excess acid contained in the films, it was incubated in 4% NaOH (w/v) solution and then washed with distilled water.

Figure 1 shows the films that evaporate the solvent and remove from the petri dishes. Composite membranes were crosslinked in a glutaraldehyde solution and compared to non-crosslinked membranes by characterization.

TABLE 1. Mass Compositions of Membranes

Abbreviations	% Chitosan	% Graphene Oxide	% Nanocellulose	Glutaraldehyde
C	100	0	0	-
C0,5GO	99,5	0,5	0	-
C1GO	99	1	0	-
C2GO	98	2	0	-
C5CNW0,5GO	94,5	0,5	5	-
C5CNW1GO	94	1	5	-
C5CNW2GO	93	2	5	-
C cr	100	0	0	+
C0,5GO cr	99,5	0,5	0	+
C1GO cr	99	1	0	+
C2GO cr	98	2	0	+
C5CNW0,5GO cr	94,5	0,5	5	+
C5CNW1GO cr	94	1	5	+
C5CNW2GO cr	93	2	5	+

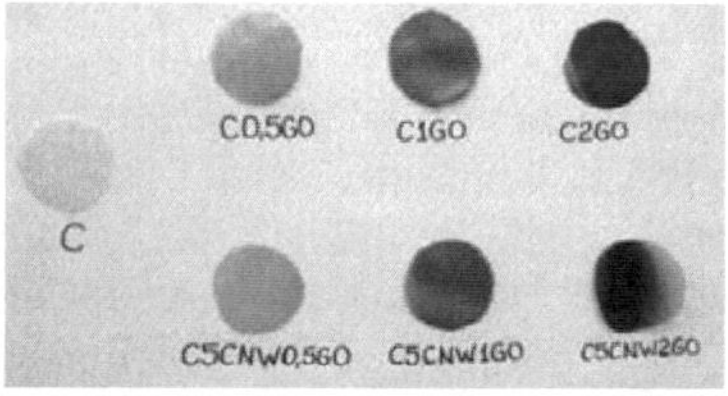

FIGURE 1. Membranes

Composite Membrane Characterization

The characterization of composite membranes was done by SEM, FTIR, XRD, TGA, tensile test (Instron 5944, 1 mm/min), contact angle measurement (Attension Tansiometer), UV-VIS and ICP-OES.

RESULTS

Graphene Oxide Characterization Results

SEM images of graphite and synthesized graphene oxide structures are given in Figure 2. The overlapping layers and at some areas distributed layers of graphene oxide are consistent with the literature [4].

a) b) c) d)

FIGURE 2. SEM Images of Graphite (a,b) and Graphene Oxide (c,d) (5 kX ,8 kX)

While no significant peak was observed in the FTIR spectrum of graphite, in the FTIR spectrum of graphene oxide C=O stretching which belongs to carboxyl groups was observed in 1723 cm^{-1}, and OH stretching belonging to carboxyl groups and hydroxyl groups were observed in 3391 cm^{-1} (Figure 3-a). In addition, the C-O stretching, which belongs to the epoxy groups which should be in the oxidized graphene structure, gave a wide peak at 1220 cm^{-1}, while the C-O stretching belonging to the alkoxy groups gave a wide peak at 1043 cm^{-1}. All these peaks reveal the formation of graphene oxide [4]. XRD analysis was used to investigate the change of graphite and graphene oxide layers and their crystal properties. Figure 3-b shows strong peaks of graphite (2θ=26.4°) and graphene oxide (2θ=10.2°). The peak at 10.2° is the characteristic peak of graphene oxide and disappearance of the peak at 26.4° after oxidation shows graphene oxide formation successfully [5].

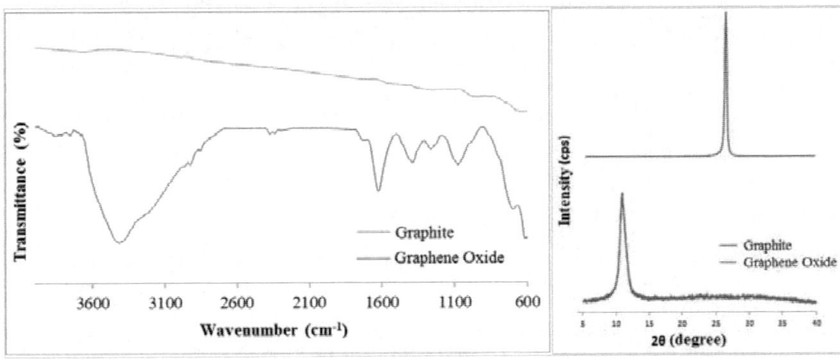

FIGURE 3. a) FTIR Spectrum b) XRD Spectrum of Graphite and Graphene Oxide

Composite Membrane Characterization

FTIR spectrum of chitosan and composite membranes in the range of 4000-600 cm^{-1} wave numbers were examined (Figure 4). The characteristic peaks of the chitosan belong to the primary amine group at 1644 cm^{-1} and the N-H bending at 1551 cm^{-1} to the secondary amine group. 1030 cm^{-1} shows C-O stretching, 1073 cm^{-1} O-H bending, and 1257 cm^{-1} C-N stretching. The wide peak between 3200-3450cm^{-1} can be explained by the overlap of the O-H and N-H stresses in the same region. The peaks in the range of 2960-2870cm^{-1} show symmetrical and asymmetric C-H vibrations [6]. The FTIR spectrum of chitosan-graphene oxide shows that most peaks overlap. The interaction between the amine group found in chitosan in 3400 cm^{-1} and the O-H bonds in the graphene oxide is observed [7]. The peak graphene formed at 1745 cm^{-1} belongs to the C=O bond from the carboxyl group [8]. In the FTIR spectra of the composites formed with chitosan, graphene oxide and nanocellulose, the wave number of the peak at 1575 cm^{-1} was shifted to 1551 cm^{-1}. This may be explained by the interaction between the ester groups on the surface of nanoscale and the amine group of chitosan [6]. FTIR spectra of glutaraldehyde-crosslinked membrane has the peak of CH_3 bending vibration at 1344 cm^{-1} in pure chitosan disappeared in cross-linked membranes. It shows the interaction between chitosan-glutaraldehyde formed in 1610 cm^{-1} [9].

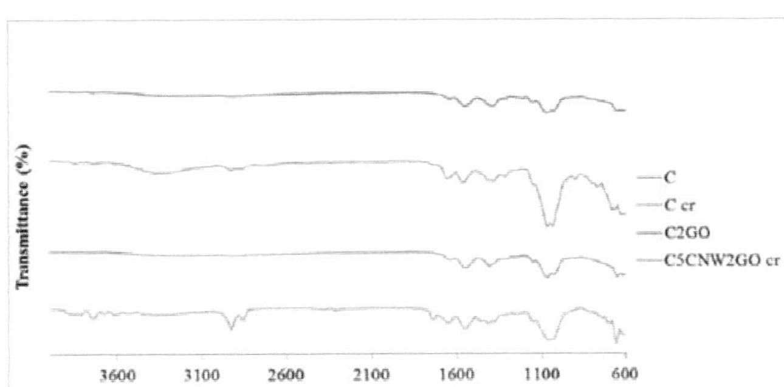

FIGURE 4. FTIR Spectrum of Membranes

The prepared membranes were placed in the membrane holder and 20 ppm Cu (II) and Pb (II) solutions were passed through the membranes. Heavy metal concentrations in the remaining solution were measured. According to the results of the inductively coupled plasma atomic emission spectroscopy (ICP-OES), the most successful membranes in Cu (II) ion treatment are C0,5GO and C2GO, and it is seen that the crosslinking process only has a positive effect on pure chitosan treatment. In Pb (II) treatment, crosslinking was found to be more effective (Table 2).

TABLE 2. ICP-OES Results of Membranes

Membranes	Cu(II) ppm	Pb(II) ppm
C	3,89	1,93
C cr	1,43	0,8
C0,5GO	<0,75	0,89
C0,5GO cr	1,8	1,1
C1GO	1,4	<0,75
C1GO cr	2,37	1,32
C2GO	<0,75	<0,75
C2GO cr	2,06	1,38
C5CNW0,5GO	1,01	<0,75
C5CNW0,5GO cr	2,97	<0,75

ACKNOWLEDGMENTS

The authors thank Eskisehir Technical University Scientific Research Projects (project 1707F462) for its financial support.

REFERENCES

1. Ashori, A., *Effects of graphene on the behavior of chitosan and starch nanocomposite films.* 2014. **54**(10): p. 2258-2263.
2. Vunain, E., A.K. Mishra, and B.B. Mamba, *1 - Fundamentals of chitosan for biomedical applications*, in *Chitosan Based Biomaterials Volume 1*, J.A. Jennings and J.D. Bumgardner, Editors. 2017, Woodhead Publishing. p. 3-30.
3. Cobos, M., et al., *Chitosan–graphene oxide nanocomposites: Effect of graphene oxide nanosheets and glycerol plasticizer on thermal and mechanical properties.* Journal of Applied Polymer Science, 2017. **134**(30): p. 45092.
4. YAZICI, M., et al., *Modifiye Hummers Yöntemiyle Grafen Oksit (Go) Sentezi ve Karakterizasyonu.*
5. Shahriary, L. and A.A. Athawale, *Graphene oxide synthesized by using modified hummers approach.* Int. J. Renew. Energy Environ. Eng, 2014. **2**(01): p. 58-63.
6. Celebi, H., & Kurt, A. (2015). Effects of processing on the properties of chitosan/cellulose nanocrystal films. *Carbohydrate polymers*, 133, 284-293.
7. Han, D., Yan, L., Chen, W., & Li, W. (2011). Preparation of chitosan/graphene oxide composite film with enhanced mechanical strength in the wet state. *Carbohydrate Polymers*, *83*(2), 653-658.
8. Guo, X., Qu, L., Tian, M., Zhu, S., Zhang, X., Tang, X., & Sun, K. (2016). Chitosan/graphene oxide composite as an effective adsorbent for reactive red dye removal. *Water Environment Research*, 88(7), 579-588.
9. Li, B., Shan, C. L., Zhou, Q., Fang, Y., Wang, Y. L., Xu, F., ... & Sun, G. C. (2013). Synthesis, characterization, and antibacterial activity of cross-linked chitosan-glutaraldehyde. *Marine drugs*, 11(5), 1534-1552.

Mechanical Properties of High Density Polyethylene (HDPE)/Uncarbonized and Carbonized Waste Corn Husk Polymer Composites

Ayberk AKA*, Gamze GULER, Ahmet Yigit TEKEREK, Munir TASDEMIR

Marmara University, Technology Faculty, Metallurgy and Materials Eng. Dep., Istanbul, 34722, Turkey

*akaayberk@gmail.com

Abstract: Lignocellulosic fibres are being utilized in various areas such as automotive, building and packaging applications to replace conventional man-made fibers. Various types of natural fibers have been the subject of intense research interest such as flax, sisal, jute, kenaf and hemp. Corn husk development as bioplastics is often used as a solution to maximize the potential of corn husks. This is because corn husks have high fiber contents of about 38%-50% and carbohydrate contents of about 38%-55%. In this study, high density polyethylene polymer matrix contributes to ground uncarbonized and carbonized waste corn husk powder is handled as 0, 10, 20, 30 and 40 wt% ratio will be mixed in the extruder. Mixture obtained from the extruder to be granulated and then the granules will be obtained as appropriate standard test sample of the injection molding machine. Mechanical and morphological tests will be applied such as elasticity modulus, tensile strength at break, % elongation, Izod impact strength, hardness, density. Also, SEM examination will be conducted to evaluate the microstructure of corn husk particles as well as material distribution in these experiments.

Key words: High density polyethylene, corn husk, mechanical properties, polymer composites.

1 INTRODUCTION

Corn is a popular crop commodity in Turkey. This commodity can also easily be found in every region in Turkey. Lignocellulosic fibres are being utilized in various areas such as automotive, building and packaging applications to replace conventional man-made fibres. Various types of natural fibres have been the subject of intense research interest such as flax, sisal, jute, kenaf and hemp [1-7]. This leaves another source of lignocellulosic fibres, such as agricultural byproducts, to be further investigated. These under-utilized agricultural waste materials include wheat, rice, corn straw, rice and corn husks, corn and okra stalks, pineapple leaf and banana bunch fibres [1,8-12]. Corn husks are composed of cellulose (~42%), lignin (~13%), ash (~4.2%), and other materials (~41%) [8]. Corn husk development as bioplastics is often used as a solution to maximize the potential of corn husks. This is because corn husks have high fiber contents of about 38%-50% and carbohydrate contents of about 38%-55% [13]. Bioplastic development itself is often conducted. One of its benefits is bioplastics ability to degrade more quickly than synthetic plastics. Another thing which supports bioplastic development is that synthetic plastics are made from petroleum byproducts, such as polyethylene. With petroleum as the basis, synthetic plastics have limited materials and cannot be renewed. Biodegradable plastics or bioplastics are one of the innovations created to reduce the amount of pollution caused by plastic waste. Bioplastics are made from a mix of synthetic polymer and natural materials such as starch or cellulose [14,15]. High-density polyethylene is one of the best polymers that are used as plastic composites. The nature of highly crystalline HDPE provides important thermal stability, excellent mechanical properties, and more chemical resistance [16]. HDPE was utilized as matrix material for natural composites, with an area of essential properties [17-19]. Many research works showed that the addition of natural fillers (corn flour, wood flour etc.) into HDPE matrix exhibited a useful approach, which helped to improve the properties of polymers [20].

In this study, HDPE matrix contributes to ground uncarbonized and carbonized corn husk powder is handled as 10, 20, 30 and 40 % ratio will be mixed in the extruder. Mixture obtained from the extruder to be granulated and then the granules will be obtained as appropriate standard test sample of the injection molding machine.

2 EXPERIMENTAL

2.1 Compositions and Materials: Five different polymer blends were prepared. Compositions of HDPE/corn husk powder (CHP) polymer composites that were formed are given in Table 1. High-density polyethylene (HDPE) (I-668 UV) was supplied by Petkim (Izmir-Turkey). Specific gravity is 0,970 g/cm^3. Melt flow rate is 5.2 g/10 min (190°C–2.16 kg). Yield strength is 28 MPa and notched Izod impact (23°C) is 12 kJ/m^2. Corn husks were collected from a local farmer's market in Tekirdağ Province of Turkey.

TABLE (1). Composition of the HDPE/CHP Polymer Composites Formulations

Groups	HDPE (wt %)	Corn Husk Powder (wt %)
1	100	-
2	90	10
3	80	20
4	70	30
5	60	40

2.2 Sample Preparation: Corn husk dried in a Yamato vacuum oven ADP-31 (Yamato/VWR Scientific Products, Japan) at 105 °C for 24 hours before being blended with HDPE. Mechanical premixing of solid compositions was done using a LB-5601 liquid-solids blender (The Patterson-Kelley Co., Inc. east Stroudsburg, PA-USA) brand batch blender for 20 min. Samples with various proportions of HDPE/CHP polymer composites were produced between 190-220 °C at 25-35 bar pressure, and a rotation rate of 27 rpm, with a Microsan co-rotating twin-screw extruder (Microsan Instrument Inc. Kocaeli - Turkey). HDPE/CHP polymer composites were also dried in vacuum oven at 105 °C for 24 hours after extrusion. Subsequently, test samples were molded in injection molding machine. Extrusion and injection conditions are given in Table 2.

TABLE (2). Extrusion and Injection Conditions of the HDPE/CHP Polymer Composites

Process	Extrusion	Injection
Temperature (° C)	190–220	190–220
Pressure (bar)	25-35	100–120
Waiting time in mold (s)	-	15
Screw speed (rpm)	27	27
Mould temperature (°C)	-	35-40

The corn husks were washed thoroughly with water to remove the adhered contaminants, and dried in an air oven at 105°C for 24 hours. Waste corn husk dry grinded with Siemens simatic C7-621 control system device to obtain unsegregated powders. The size corn husk particles varied between 1mm and 38µm. Powder preparations steps are given in Figure 1.

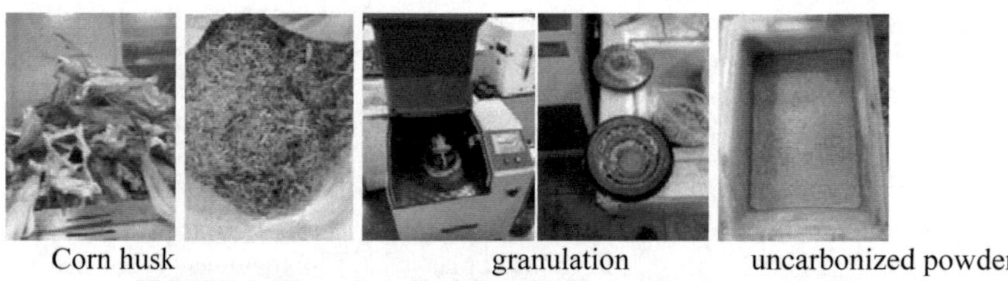

FIGURE 1. Waste Corn Husk Powder Preparation Steps

The powder was then packed in a graphite crucible and fired in electric resistance furnace at temperature of 1200°C (30 minute) to form carbonized corn husk particles. Corn husk carbonization steps and sieve analysis are given in Figure 2.

FIGURE 2. Corn Husk Powder Carbonization Steps and Sieve Analysis

2.3 Test Procedure: Tensile tests were performed according to ASTM D638 specification. They were carried out using a Zwick Z010 (Germany) testing machine with a load cell capacity of 10 kN at a cross-head speed of 50 mm/min. The elastic modulus, strength at break, max. tensile strength and percent elongation was determined from the stress–strain curves. Six samples were tested in each set and the average value was reported. The hardness test was done according to the ASTM D2240 method with Zwick hardness measurement equipment. To investigate fracture behavior, Izod impact test (notched) was done at room temperature according to the ASTM D256 method with Zwick B5113 impact test device. Determination of density was done according to ISO 2781 test standard. The fractured surfaces of the HDPE/CHP polymer blends were coated to thickness of 20 Å of a gold (Au) to prevent electrical charging by Polaron SC7640 (high resolution sputter coater) (United Kingdom). The surfaces of the prepared samples were observed by the FEI Sirion XL30 FEG (Nederland) scanning electron microscopy (SEM) at an acceleration voltage of 5-10 kV.

3. RESULT and DISCUSSION

3.1 Mechanical properties of HDPE/CHP polymer composites: Figure 3-A shows the effect of filler content on the elasticity modulus of HDPE corn husk composites. It can be seen that the elongation modulus of the HDPE corn husk composites increased with increase of filler content, showing that the ability of corn husk to impart greater stiffness to the HDPE composites. This can be understood from the fact that any filler which has higher stiffness than the matrix can increase the modulus of the composites which is reciprocal to elongation at break. With the increase in the filler content, the possibility of filler-matrix interaction increased which leads to an increase in the efficiency of stress transfer from the matrix to the filler. The increase in corn husk content eventually increases the modulus of elasticity of HDPE corn husk composites. The maximum elasticity modulus is observed at the 40 wt % carbonized corn husk powder concentration for HDPE. In comparison with the elasticity modulus of HDPE, the elasticity modulus increased by 44% for the composites with a 40 wt % carbonized corn husk powder concentration. As shown in Figure 2, the particle sizes of the carbonized groups are smaller than uncarbonized groups. Therefore, having small particles resulted in an increase in elasticity modulus. Carbonization process of corn husk increased elastic modulus as well. The relationship between the ratio percentage of the corn husk powder and max. tensile strength of HDPE is shown in the Figure 3-B. Max. tensile strength of composites shows an increment as the carbonized corn husk powder concentration increases from 0 to 40 wt %. On the other hand, max. tensile strength of composites shows a decrement as the uncarbonized corn husk powder concentration increases from 0 to 40 wt %. This is a common observation for any natural filler containing thermoplastic composites. The decrease in tensile strength with increase in filler content is due to low aspect ratio, and its inability to transfer stress from matrix. Another reason for the decrease in tensile strength is attributed to the poor wettability between hydrophilic corn husk and hydrophobic HDPE matrix. In comparison with the max. tensile strength of HDPE, the max. tensile strength increased by 7,6 % for the composites with a 40 wt % carbonized corn husk concentration. The strength at break of corn husk filled composites was measured, as shown in Figure 3-C. With increased loading, the strength at break of composites filled with corn husk is increased. The maximum strength at break is observed at the 40 wt % carbonized corn husk concentration for HDPE. In comparison with the strength at break of virgin HDPE, the strength at break increased by 283 % for the blends with a 40 wt % carbonized corn husk concentration. The relationship between the corn husk content and the % elongation of the polymer composites is shown in Figure 3-D. The value of % elongation decreased significantly from virgin HDPE to corn husk HDPE composites with increasing filler content, this may be possibly due to the friction between corn husk particles and HDPE matrix, generating a rigid interface, thereby inducing stiffness to the composite. The % elongation of the composites decreased (from 0 to 40wt %) with an increase weight percentage of uncarbonized and carbonized corn husk powder. The minimum % elongation is observed at the 40 wt% carbonized corn husk concentration for HDPE. In comparison with the % elongation of virgin HDPE, the elongation decreased by 98,6% for the composites at a 40 wt % carbonized corn husk concentration. Figure 3-E illustrates the effect of corn husk on the density of HDPE composites. The density increased as the carbonized corn husk particle concentration increased from 0 to 40 wt %. In comparison with the density of virgin HDPE, the density increased by 17,4% for the composites at a 40 wt % carbonized corn husk concentration. No significant change was observed in the uncarbonized group.

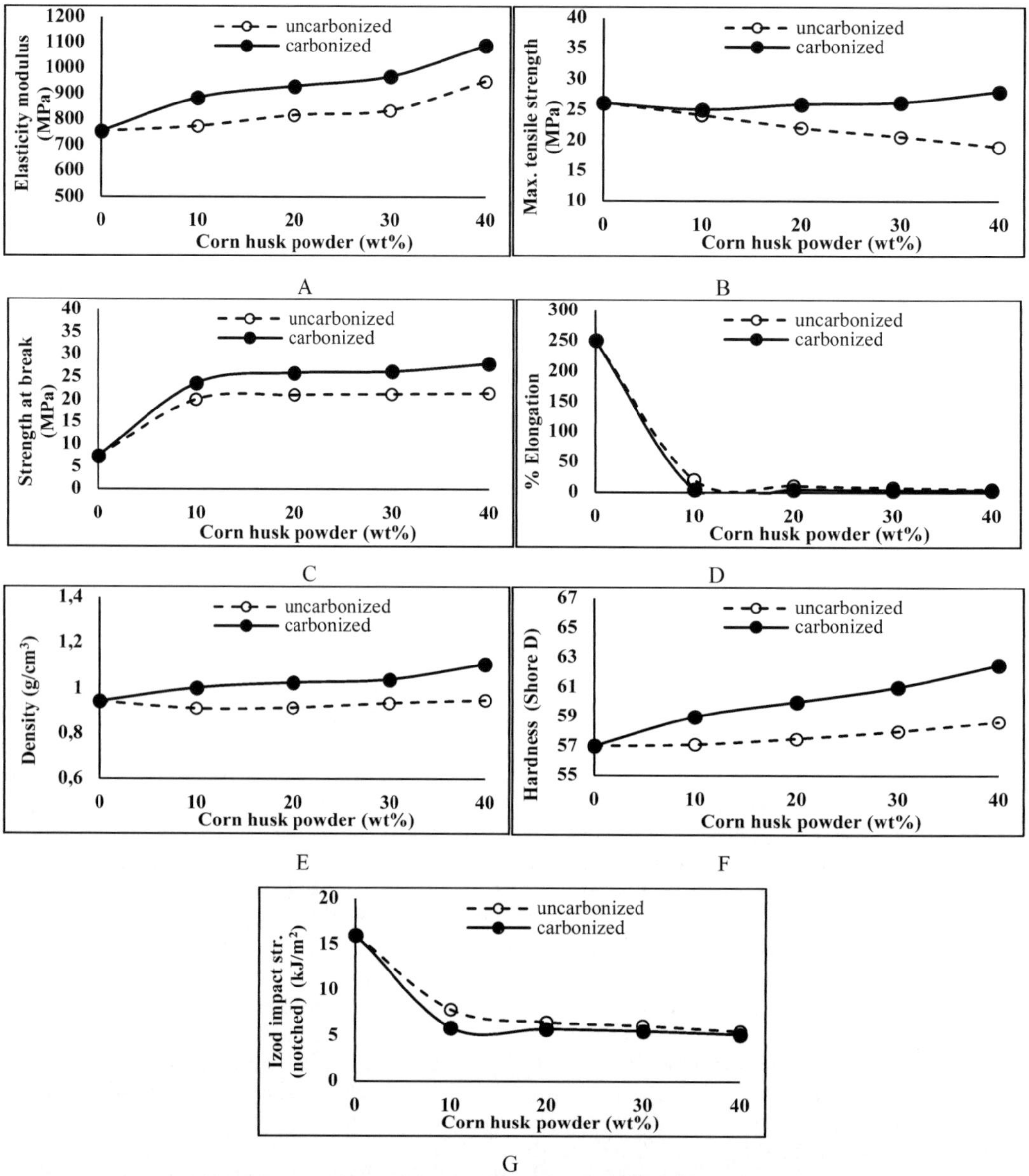

FIGURE 3. Mechanical Properties of the HDPE/CHP Polymer Composites

Figure 3-F illustrates the effect of corn husk on the hardness of HDPE composites. The hardness increased as the uncarbonized and carbonized corn husk particle concentration increased from 0 to 40 wt %. In comparison with the hardness of virgin HDPE, the hardness increased by 9,6% for the composites at a 40 wt % carbonized corn husk concentration. The relationship between the corn husk content and the Izod impact strength of the polymer composites is shown in Figure 3-G. The Izod impact strength of the composites decreased (from 0 to 40wt %) with an increase weight percentage of uncarbonized and carbonized corn husk powder. The minimum Izod impact strength is observed at the 40 wt% carbonized corn husk concentration for HDPE. In comparison with the Izod impact strength of virgin HDPE, the Izod impact strength decreased by 68% for the composites at a 40 wt % carbonized corn husk concentration. During the impact test, cracks travel through the polymer as well as along the weaker interfacial regions. The latter cannot resist crack propagation as effectively as the polymer region, hence reducing the impact strength. Incorporation of fillers is also a considerable reason to inhibit the mobility of polymer, thereby lowering the ability of the system to absorb energy during fracture propagation. Another reason for decrease in impact strength might be due to the stiffening of polymer chains which may be resulted due to the bonding between the corn husk and polymer matrix [21].

3.2 Morphological properties of HDPE/CHP polymer composites: The SEM study was carried out to study the dispersion of waste corn husk in the HDPE matrix. The boundaries and the contrast can be obviously seen between the corn husk and HDPE matrix on the fractured surfaces of polymer matrix (Figure 4). The micrographs indicate that the corn husk particulates are homogeneously dispersed on the fractured surfaces of HDPE polymer matrix. The SEM images show that the corn husk particles were poorly dispersed in the HDPE matrix leading to the poor interfacial adhesion between the HDPE matrix and the corn husk particles. The poor filler dispersion or poor interfacial bonds with increase in filler content, leading to poor stress transfer leading to the formation of cracks in the matrix material as well. These observations were also supported by filler agglomeration in the SEM analysis.

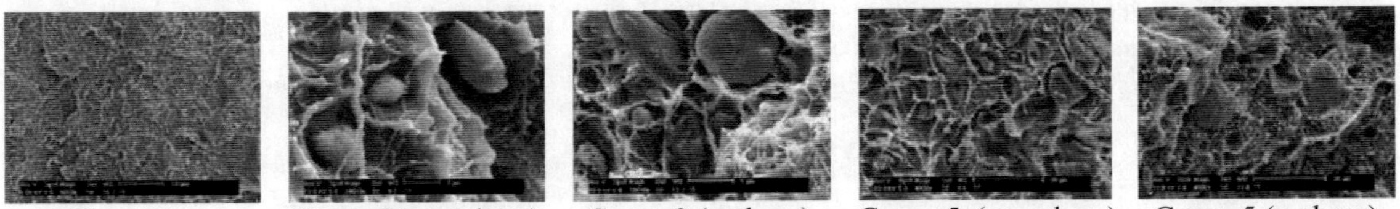

Group 1 Group 2(uncarbon.) Group 2 (carbon.) Group 5 (uncarbon.) Group 5 (carbon.)

FIGURE 4. SEM Micrographs of the HDPE/CHP Polymer Composites

4 CONCLUSION

This research was expected to serve as a study to initiate the role of corn husk as a bioplastic. The results of the study suggest the properties of corn husks that are expected to be a reference for the development of corn husks as a bioplastic. The biocomposite properties of corn husks are expected to be a reference for the direction of the application of bioplastics made from corn husks. The uncarbonized and carbonized corn husk powder as such was compounded with HDPE in various percentages (10wt%, 20wt%, 30wt% and 40wt %) and then molded and its mechanical properties were tested. The following results were obtained: Elastic modulus, strength at break, hardness and density increased gradually as the uncarbonized and carbonized corn husk powder content increased. % elongation and Izod impact strength decreased gradually as the uncarbonized and carbonized corn husk powder content increased. The micrographs indicate that the uncarbonized and carbonized corn husk particulates are homogeneously dispersed on the fractured surfaces of HDPE polymer matrix. Hence, this biocomposites can be used in low end applications, where strength is not considered to that extent such as in boards, baskets, household utensils, etc.

REFERENCES

1 S. L. Panthapulakkal, M. Sain, *J Appl Polym Sci,* 100 (2006) 3619.

2 H.U. Zaman, M.A. Khan, R.A. Khan, R.M. Arifur, L.R. Das, *Fiber Polym,* 11(2010)455-463.

3 J.Q. Zhang, J.C. Zhang, *Indian J Fibre Text Res,* 35(2010)115.

4 S. Debnath, M. Madhusoothanan, *Indian J Fibre Text Res,* 36(2010)122.

5 S. Nam, A.N. Netravali, *Fiber Polym,* 7(2006)372.

6 Q.T.H, Shubhra, A.K.M.M. Alam, M.A. Gafur, S.M. Shamsuddin, M.A. Khan, *Fiber Polym,* 11(2010)725.

7 N.D. Yilmaz, P.L. Banks, N.B. Powell, S. Michielsen, *J Appl Polym Sci,* 121(2011)2056.

8 S. Huda, Y. Yang, *Macromole Mater Eng,* 293(2008)235.

9 A. Ashori, A. Nourbakhsh, *J Appl Polym Sci*, 111 (2009)2616.

10 K.G.M. Arifuzzaman, M. Shaheruzzaman, M.H. Rahman, S.M. Abdur Razzaque, M. Sakinul Islam, M.A. Shamsul, *Fiber Polym*, 10(2009)65.

11 A.V.R. Parasad, R. K. Mohana, G. Nagasrinivasulu, *Indian J Fibre Text Res,* 34(2009)162.

12 B. Shyamal, N. Debasis, D. Sanjoy, *Indian J Fibre Text Res,* 36(2011)172.

13 A.A. Anis. *Physico Chemical and Mechanical Characteristics of Corn Husk as Packaging Material.* Departement of Agroindutrial Technology. Faculty of Agricultural Technology. Bogor Agicultural Univesity, 2006.

14 B. Raynasari, *Effect of Storage Temperature on the Physical and Mechanical Properties of Plastic Retail Packaging.* Departement of Agroindutrial Technology. Faculty of Agricultural Technology. Bogor Agicultural Univesity. 2012.

15 A.K.K. Deswita, G.I. Sudirman, *Material Science Indonesia*, 2008, 37.

16 S.A. Awad, C.M. Fellows, S.S. Mahini, Effects of accelerated weathering on the chemical, mechanical, thermal and morphological properties of an epoxy/multi-walled carbon nanotube composite. Polymer Testing 2018;66:70-77.

17 Q. Zhang, H. Cai, X. Ren, L. Kong, J. Liu, X. Jiang, The dynamic mechanical analysis of highly filled rice husk biochar/high-density polyethylene composites. Polymers, 2017; 9(11):628-637.

18 E. Khalaf, S. Awad, Improvement of mechanical and water absorbance properties of low-density polyethylene (LDPE) by using white kaolin powder (WK). Journal of Advanced Chemical Sciences, 2017; 3(1):426-427.

19 Y. Zou, Y. Feng, L. Wang, X. Liu, Processing and proper. of MWNT/HDPE composites. Carbon, 2004; 42(2):271-277.

20 S.A. Awad, E.M. Khalaf, Improvement of the chemical, thermal, mechanical and morphological properties of polyethylene terephthalate–graphene particle composites. Bulletin of Materials Science, 2018; 41(3):67-72.

21 D. Jagadeesh, P. Sudhakara, D.W. Lee, H.S. Kim, B.S. Kim, J. I. Song, Mechanical Properties of Corn Husk Flour/PP Bio-composites, Composites research, Vol. 26, No. 4, 213-217 (2013)

Machine Learning Applications in Polymer Composites

Youssef K. Hamidi [a,*], Abdelaziz Berrado [b], M. Cengiz Altan [c]

[a] *Mechanical Engineering Program, Engineering Department, University of Houston–Clear Lake, USA*
[b] *Ecole Mohammadia d'Ingénieurs, AMIPS, Mohamed V University in Rabat, Morocco*
[c] *School of Aerospace and Mechanical Engineering, University of Oklahoma, USA*

* *Email: hamidi@ou.edu*

Abstract. The primary interest in numerous research problems in both polymer composites and Machine Learning (ML) is to develop predictive models for one or more variables of interest using relevant independent variables, or inputs. However, these two fields have often adopted different approaches, where modeling of composite behavior is often based on physics-based models and phenomenological theories. These physical models are more precise and robust, but often suffer from restricted predictive capability since they are confined to a specific set of conditions. ML models, on the other hand, can be more efficient during the design phase as they allow managing massive and high dimensional data sets to extract the best fit or a predictive behavior for the application at hand.
In this context, material scientists would benefit from understanding and implementing some of the powerful ML methods, in order to predict or characterize a behavior of interest of a polymer composite. In this paper, we present a general methodology aimed at employing supervised machine learning models for predicting the properties of polymer composites, including thermo-mechanical properties, environmental effects such as moisture saturation level, durability, or other important behavior, based on the composite constituents, manufacturing processes, relevant process parameters, and expected life-span of the composite product.

Keywords: Polymer Composites, Behavior Prediction, machine learning.
PACS: 81.05.Qk, 83.60.-a

INTRODUCTION

Polymer composite materials enjoy a healthy combination of light weight and high performance, which led to a steady increase in popularity as load-bearing component in strategic industries such as aerospace, automotive, and energy [1]. Predicting and characterizing the behavior of polymer composites is a critical issue both in composites design and in composites research and development. Significant research has been conducted to characterize composite materials based on constituent materials and manufacturing process variables [2-4]. Physics-based models have been the standard way to predict key properties of composite parts, such as the elastic modulus. These thermomechanical models are often developed based on the properties of the reinforcing fibers and resin constituting the composite material. Nonetheless, these models are highly idealistic since they are limited to perfect microstructures and flawless interfacial conditions. Other constitutive models describing different physical phenomena of polymer composites were developed. However, these phenomenological models frequently utilize complex analytical expressions that are challenging to solve and/or computationally demanding. In addition, these models usually depended on several empirical parameters recovered through experimentation. This also limits the use of the phenomenological constitutive models since the recovered parameters are specific to the studied polymer composite [5]. From a design stand point, the final goal of these predictive models is to facilitate the design process of polymer composites. In other words, the predictive models are expected to help choose the most appropriate composite constituents (i.e. resin type, curing agent type, fiber type, fiber architecture, fiber content, fiber sizing, etc.), and the suitable process parameters (i.e. the process itself, and its parameters, along with resin curing cycle) so that the resulting composite part can carry the loads anticipated for specific applications.

The importance of characterizing and predicting the behavior of polymer composites has led to a remarkable number of published research on the subject. However, each of these investigations is often dedicated to a particular combination of constituent materials, manufacturing process and process parameters, or life-span conditions. Several studies focus on optimizing a specific manufacturing process by investigating process parameters effects on the composite behavior, while keeping the resin, fiber type, architecture and content constant. Others, on the other hand, explore the effect of manufacturing processes on the resulting composite [6]. Other complimentary studies focus on the effects of life-span conditions, such as loading and environmental history, on the behavior of the studied composites [7-8]. Each of these studies can definitely help understand and predict the behavior of a specific composite. However, they are not used beyond a particular application, thus limiting their broader utility. When designing a polymer composite, practitioners and researchers alike often refer to their own experience and rely heavily on predefined rules in materials classification, selection, and property predictions. Rarely would practicing engineers venture outside of their comfort zone to experiment with new constituents or processes, since introducing a new process would involve an intensive trial-and-error effort and a hefty cost before reaching process maturity and profitability. These costly trial-and-error experiments during the design of polymer composites can be avoided with the appropriate use of the collective knowledge in the polymer composite field. The large number of studies investigating the behavior of these materials can be exploited using machine learning (ML) in order to select the most appropriate constituent materials and manufacturing process for any specific composite application.

Unlike physical models, ML models can be more efficient during the design of polymer composites since they permit managing massive and high dimensional data sets to extract the best choice or a desired behavior [9-11]. ML techniques are currently used to build predictive models for material design. For instance, these techniques are applied successfully to predict certain properties of metals, such as microstructure [12-13] and plastic deformation [14]. They also are explored for atomistic simulations [15], electronic structure [16] and chemical similarity predictions [17]. In this context, material scientists would benefit from understanding and implementing some of the powerful ML methods, in order to predict or characterize a behavior of interest of a polymer composite. In a previous work [18], we successfully used ML to predict moisture saturation level for vinylester-based laminates. To the best of the authors' knowledge, no other attempt to use ML techniques has been made in designing composite laminates.

In this paper, we present a comprehensive methodology aimed at employing supervised machine learning models for predicting the behavior of polymer composites. This approach offers the potential of considerably enhancing the design of composite materials for a variety of applications. It would allow the appropriate design of composite constituents, manufacturing process, and process parameters in order to achieve a specific functionality for the resulting composite. Targeted properties include thermo-mechanical properties, environmental effects such as moisture saturation level, durability, or any other important behavior, based on the composite constituents, manufacturing processes, relevant process parameters, and expected service life of the composite product.

PROPOSED METHODOLOGY

The proposed methodology for using machine learning techniques to predict a targeted behavior of composite materials comprises three consecutive phases: (i) data preparation, (i) machine learning model building, and (iii) model evaluation. Each of the three phases is subdivided into different steps as described in the following sections. The essential ingredients of the proposed methodology are depicted schematically in Figure 1.

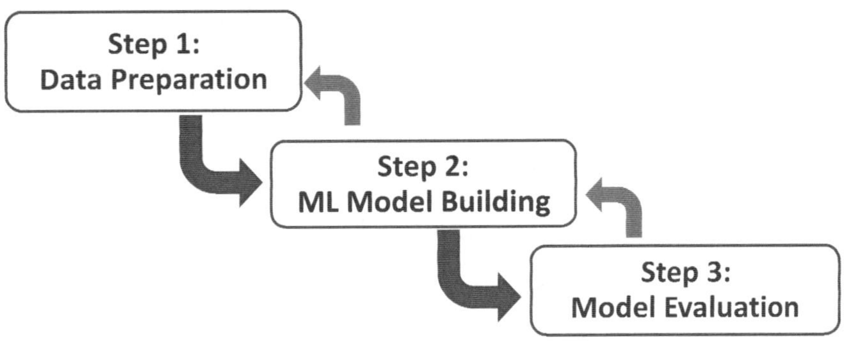

FIGURE 1. The Three Steps in Machine Learning Methodology for Predicting a Composite Material Behavior of Interest.

Phase 1: Data Preparation

The first phase involves collecting all available data relevant to the targeted polymer composite behavior. The goal is to compile a comprehensive database that can be exploited to construct a predictive ML model. This phase itself consists of three steps as presented in Figure 2: (a) data structuring, (b) data collection, and (c) data cleaning.

FIGURE 2. The Steps of Data Preparation Phase of the Machine Learning Methodology.

Data structuring

In this step, relevant material and process variables affecting the response at hand (i.e. the behavior to be modeled) need to be carefully determined. For example, all parameters documented or suspected to affect the moisture saturation level in polymer composites need to be included during this step to avoid duplicating the recovery process. The relative importance of these parameters, also referred to in ML context as explanatory variables, with respect to the behavior at hand or output can usually be measured once the ML model is built. Obviously, reinforcement or fiber-related parameters such as type, content, and architecture should be considered. Similarly, matrix type, matrix filler type and content, and curing cycle or processing temperature need to be considered. The manufacturing processes used to fabricate the polymer composites can also affect their behavior, and should be considered as well. In addition, parameters representing the life-span conditions or other important variables, i.e. void content and part thickness (or a characteristic dimension), need also to be considered since they are known to affect the behavior of these materials [19-20]. It should be noted that most data sources or published articles would not report all considered variables. Another attractive feature of several ML techniques is the possibility of constructing a model with incomplete, missing values in the compiled database.

In most cases, a classification of each explanatory variable is required. For instance, reinforcement/fiber architecture can be classified into unidirectional, multidirectional, random, and hybrid architectures. Matrix type can be categorized into the major known resins, and so forth. This step is critical as a careful selection of explanatory variables, and a well-defined classification for each of them is essential for a successful use of any learning algorithm.

Data collection

In this step, extensive composite material data relevant to the behavior to be predicted need to be collected from published research articles, technical reports, and available industrial data. Of all the steps involved in the proposed methodology, this step is the most tedious and time consuming. However, it is the building bloc upon which the model is constructed.

The accuracy of the ML predictive model will depend of the size and diversity of the compiled database. Therefore, a comprehensive, evolving database can be constructed over time. In the process, thousands of published materials reporting on the behavior of interest need to be reviewed before the data can be recovered. This step is preferably performed after data structuring in order to avoid duplicating data collection.

Data cleaning

In this step, the consistency of all recovered values needs to be assessed to ensure data quality. Incorrect data can tremendously hinder the accuracy of ML predictive models. These erroneous data may arise during both recovering data points from the literature and entering the data point into the database. All entered values need to be double checked to verify that no erroneous value was mistakenly included in the database. In case of uncertainty, publications of the same authors and/or research group need to be checked to verify the reported value.

Phase 2: Machine Learning Model Building

Phase 2 also comprises three steps as depicted in Figure 3. First, an appropriate machine learning technique needs to be selected. Then, a training data should be selected from the collected dataset in order to use the selected ML algorithm to build the model. The final step entails building a ML predictive model that can infer a relationship between a set of potential explanatory variables (or independent variables) and a specific material property (dependent variable, or response). The prediction accuracy of the model can be assessed to evaluate the ability of the input variables to explain the variation in the targeted composite behavior.

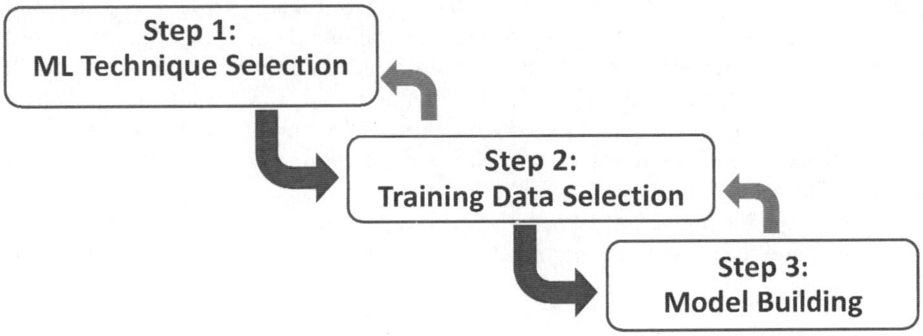

FIGURE 3. The Steps of Model Building Phase of the Machine Learning Methodology.

Machine learning technique selection

A multitude of advanced statistical learning algorithms are available, including artificial neural networks, Bayesian networks, and decision trees, among others [9-11]. This step involves a careful selection of the most appropriate supervised technique for modeling the behavior to be predicted. In case of missing values for instance, the Random Forests algorithm can be chosen since it copes well with missing values and mixed data types. Other aspects of the compiled database need to be considered such as robustness to outliers, and non-linearity handling.

Training data selection

In order to assess the predictive capability of the selected ML algorithm, the compiled dataset needs to be split into two subsets: (i) a training dataset comprising most of the collected data, and (ii) the remaining data, labeled unseen or validation data. The size of the training data should be a compromise between how well the model should be trained, and how extensively we want to evaluate its predictive capability.

The selected ML algorithm will be applied to the training data in order to build a predictive model that explains variations of the investigated composite behavior (referred to in ML community as dependent variable or response) based on a wide variety of material and process variables (independent or explanatory variables).

Model building

In this step, the selected algorithm is actually run on the training data. The resulting predictive model of can offer a relationship between the composite behavior and the explanatory variables with a quantifiable predictive accuracy. Most algorithms can assess the contribution of each of the independent variables in the explanation of the variability of the response in the resulting predictive model.

Phase 3: Model Evaluation

The goal of Phase 3 is to evaluate the predictive performance of the model. Therefore, the predictive performance of the ML model needs to be tested using the unseen data, also referred to as test data. The selected algorithm should be used to predict the response of the unseen dataset, and the predicted responses should be compared to the real responses. The obtained accuracy can be a good indicator of the predictive performance of the model for the behavior of composite laminates to be designed.

Another interesting feature of most ML techniques is their ability to distinguish the impact of explanatory variables on the predicted behavior. Hence, the explanatory variables can be classified according to their impact on the targeted composite behavior. This feature can be very helpful when expanding the database by redefining the number of explanatory variables to collect during the first phase.

EXAMPLE OF APPLICATION

Using the presented methodology, we successfully predicted moisture level saturation in vinylester-based composites [8] as it affects their durability, and thus of prime interest during the design of composite materials. A comprehensive database was compiled from more than 520 published research articles reporting moisture absorption in polymer composites. Using 90% of the compiled data as the training dataset, the Random Forests algorithm was used to build a predictive model for the moisture saturation level. Variables considered by the model comprised the manufacturing process, the fiber type and architecture, the fiber and void contents, the matrix filler type and content, and the conditioning environment and temperature. The model showed a predictive accuracy of $R^2_{training} \backsimeq 95\%$ on the training data, implying that the variables almost fully explain the variation in the moisture saturation levels. The model exhibited a predictive performance of $R^2_{test} \backsimeq 85\%$ on the unseen 10% of the compiled data. Additionally, the influence of the different variables on the moisture saturation level were weighed using the random forest model. The reinforcement type was found to be the most important factor influencing the moisture saturation level in vinylester composite laminates.

CONCLUSION

A general methodology is presented for predicting the behavior of polymer composites by employing supervised machine learning models. Targeted behavior of composite parts can vary from thermo-mechanical properties, to environmental effects such as moisture saturation level, durability, or other important behavior. The developed machine learning model would predict specific composite properties based on the composite constituents, manufacturing processes, relevant process parameters, and expected life-span of the composite product.

The proposed methodology, based on applying machine learning techniques to the collective knowledge in the composite materials field, involves 3 phases. First, a comprehensive database is compiled from published research articles, technical reports, and industrial data. Second, the most appropriate supervised advanced statistical learning is used to build a predictive model that explains the investigated material behavior based on a carefully selected material and process variables. Third, the predictive performance of the built model is assessed, and the importance of the used explanatory variables is assessed. This proposed approach offers the potential to significantly improve the design process for composite materials. It would alleviate the trial-and-error iterations by refining the initial selection of composite constituents and manufacturing process parameters for any specified application.

REFERENCES

(1) H. J. Barraza, K. Olivero, Y. Hamidi, E. A. O'Rear, M. C. Altan, *Compos. Interf.* **9**, 477-507 (2002).
(2) Y. K. Hamidi and M. C. Altan, *International Polymer Processing*, **32**, 1-18 (2017).
(3) H. J. Barraza, Y. K. Hamidi, L. Aktas, E. A. O'Rear, M. C. Altan, *Compos. Interf.* **24**, 125-148 (2017).
(4) K. A. Olivero, Y. K. Hamidi, L. Aktas, M. C. Altan, *J. Comp. Mat.* **38**, 937-957 (2004).
(5) G. E. Guloglu, Y. K. Hamidi, M. C. Altan, *Polym. Engr. & Sci.* **57**, 921-931 (2017).
(6) M. A. Yalcinkaya, E. M. Sozer, M. C. Altan, *Compos. A* **102**, 336-346 (2017).
(7) L. Aktas, Y. Hamidi, M. C. Altan, *J. Mat. Proc. & Manuf. Sci.* **10**, 255-267 (2002).
(8) Y. K. Hamidi, L. Aktas, M. C. Altan, PPS-32, 2016. *AIP Conference Proceedings* **1914**, *030012 (2017); doi.org/10.1063/1.5016699.*
(9) L. Breiman, J. Friedman, C. J. Stone, R. A. Olshen, *Classification and Regression Trees*, NY: Taylor & Francis, 1984, 1-386.
(10) L. Breiman, *Machine Learning*, **45**, 5-32 (2001).
(11) A. Berrado, G. C. Runger, *Data Mining & Knowledge Discovery* **14**, 409-431 (2007).
(12) H. Xu, R. Liu, A. Choudhary, W. Chen, J. *Mech. Design* **137**/051403, 1-10 (2015).
(13) M.A. Bessa, R. Bostanabad, Z. Liu, A. Hub, D. W. Apley, C. Brinson, W. Chen, W. K. Liu, *Comput. Meth. Appl. Mech. Engr.* **320**, 633-667 (2017).
(14) D. Versino, A. Tonda, C. A. Bronkhorst, *Comput. Meth. Appl. Mech. Engr.* **318**, 981-1004 (2017).
(15) J. Behler, *J. Chem. Phys.* **145**, 170901 (2016); *doi.org/10.1063/1.4966192*
(16) L. M. Ghiringhelli, J. Vybiral, S. V. Levchenko, C. Draxl, M. Scheffler, *Phys. Rev. Let.* **114**, 105503 (2015).
(17) G. Pilania, C. Wang, X. Jiang, S. Rajasekaran, R. Ramprasad, *Sci. Rep.* **3**, 2810 (2013); DOI:10.1038/srep02810.
(18) Y. K. Hamidi, L. Aktas, M. C. Altan, PPS-35, 2018. *AIP Conference Proceedings* **2065**, 030027 (2019); doi.org/10.1063/1.5088285.
(19) L. Aktas, Y. Hamidi, M. C. Altan, *J. Mat. Proc. & Manuf. Sci.* **10**, 239-254 (2002).
(20) H. J. Barraza, L. Aktas, Y. K. Hamidi, J. Long, E. A. O'Rear, M. C. Altan, *J. Adhesion Sci. & Techn.* **17**, 217-242 (2003).

Multiscale Simulation of Transverse Tensile Failure for Polymer Composites: From Atomistic Simulation to Structural Analysis

Takaya Kobayashia[1, a)], Kensuke Ogawa[1], Satoru Yamamoto[2], Riichi Kuwahara[2] and Ryosuke Matsuzaki[3]

[1] *Mechanical Design & Analysis Corporation (1-40-2 Fuda Chofu Tokyo, Japan)*
[2] *Dassault Systemes K.K. (2-1-1 Osaki Shinagawa Tokyo, Japan)*
[3] *Tokyo University of Science (2641 Yamazaki Noda, Chiba, Japan)*

[a)] Corresponding author: koba@mech-da.co.jp

Abstract. A multiscale modeling of a polymer composite, which is a combination of atomistic simulation and finite element structural analysis, was demonstrated. Heat curing and cross-linking reaction steps for thermosetting resins were calculated with activation energy and heat generation via molecular dynamics simulation. The microscale stress–strain curves suggesting fracture mechanisms of the molecular level were also estimated and were assigned to the finite element analysis. The crack propagation behavior, transiting from matrix-failure to interface-failure, was investigated.

MULTISCALE MODELING OF POLYMER COMPOSITES

Polymer composites formed by thermosetting resins and carbon fibers are promising materials in modern light weight structures. In thermosetting resins, combinations of epoxy and amine compounds are widely used as the base resin and curing agent, respectively. The mechanical behavior of the curing process can be characterized by thermo-rheological behaviors. From an experimental point of view, the expression of elasticity is evaluated by viscoelastic measurements, whereas the curing shrinkage is measured typically by volumetric dilatometry as shown in Figure 1. These measurements must be followed by an estimation of the curing (hardening) degree. For this purpose, differential scanning calorimeter (DSC) measurements provide the degree of curing, and the Kamal differential equation is a typical application to describe the progress of the curing reaction of the thermosetting resin.

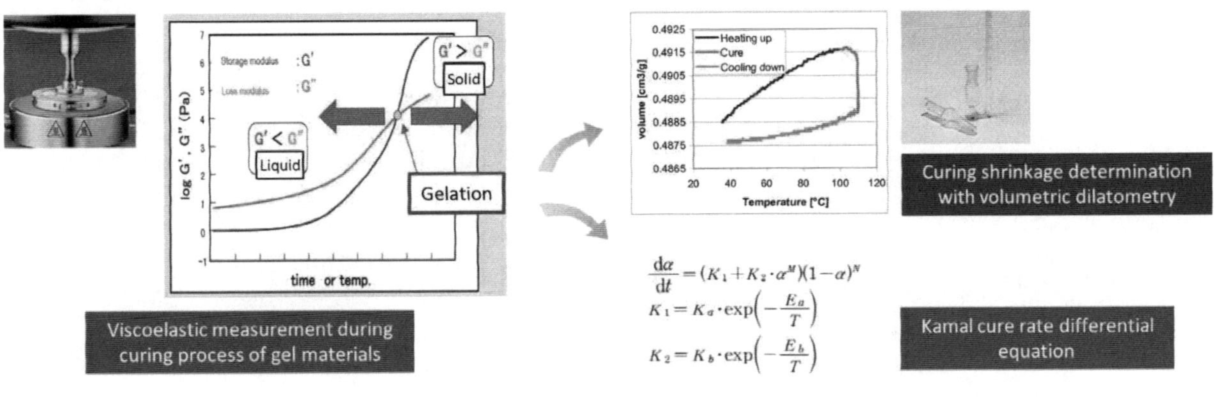

FIGURE 1. Experimental approaches to curing process of thermosetting resin.

ATOMISTIC SIMULATION

There are many varieties of epoxy and amine compounds, and their choice is a key strategy for material design. The conventional trial-and-error procedures are expensive and time-consuming, and therefore, computer modeling is desired. We propose a multiscale modeling procedure which consists of two parts: an atomistic simulation for curing reaction and estimation of the mechanical and thermal properties, and an FEM simulation for the molding process and structure. In the first part, a heat curing reaction is simulated by the molecular dynamics (MD) method in order to explore how combinations of the base resin and curing agent affect the cross-linked structure and material properties. Mechanical and thermal properties of the epoxy resin and the adhesive strength between the resin and fillers are also predicted via MD simulation.

Two epoxy compounds, DGEBA and TGDDM were considered in this study. The number of epoxy groups is different in DGEBA and TGDDM, i.e., 2 and 4. TGDDM is expected to provide firmer linking due to higher functionality. As the amine compound, 44DDS was chosen. The cross-linking (curing) simulation was performed using a molecular dynamics technique proposed by Okabe et al[1]. When the epoxy group approaches to the amino group within a certain reaction range (6Å), it is assumed that they are allowed to crosslink with each other. As shown in Figure 2, probability k and a random number P are used to determine the propriety of the reaction. If the reaction occurs, the reaction energy (H.O.F.) is added as a local kinetic energy of reacted molecules. Figure 3 shows the evolution of the cross-linked structures at 473 K. An isolated structure is indicated by a different color. It is clearly understood that the cross-linked structure increased with the conversion of the reaction. In case of DGEBA/44DDS, the percolated network emerged at 95% conversion. However, full percolation occurred at 65% conversion in the case of TGDDM/44DDS owing to the difference in the number of epoxy groups.

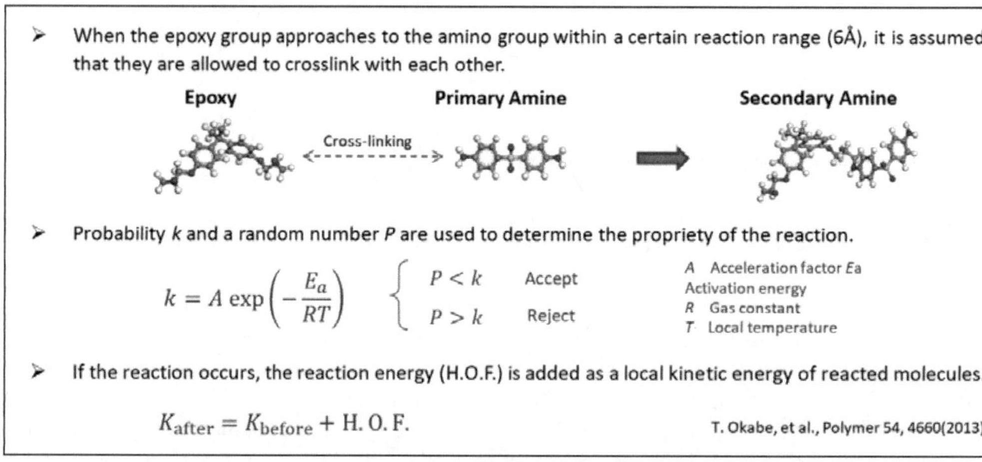

FIGURE 2. Cross-linking simulation method in molecular dynamics (MD).

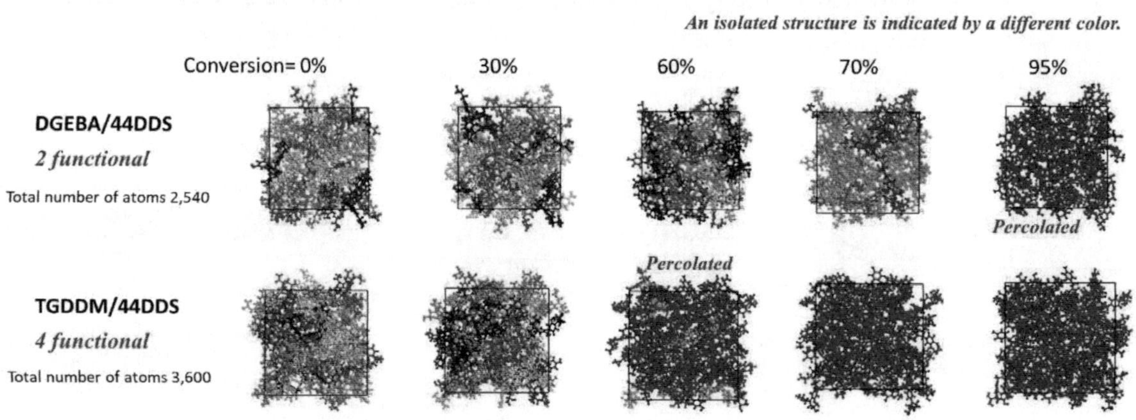

FIGURE 3. Evolution of cross-linked structures.

Next, the interfacial adhesion strength between a filler and epoxy resin was calculated. Here, we considered a copper substrate as a filler[2], because the surface of carbon fiber is not fully clarified at this time. As shown in Figure 4, a mixture of epoxy and amine compounds was placed over the copper surface, and the curing reaction was simulated to obtain initial structure. A displacement was applied in perpendicular and the stress was monitored via MD simulation. Two distinctive snapshots are shown for DGEBA and TGDDM. In DGEBA case, a void was initiated at strain 0.2 and the voids grew in size and number, and several chain segments became stretched. These are recognized as matrix-failure. On the other hand, In TGDDM case, the epoxy resin was peeled off at the interface. This may be caused by the high rigidity of TGDDM. This can be recognized as interface-failure. Interface-failure was observed in TGDDM at high conversion cases (>80%). The peak stress is clearly recognized owing to peel-off at the interface. Other cases showed matrix-failure drawing a gentle curvature.

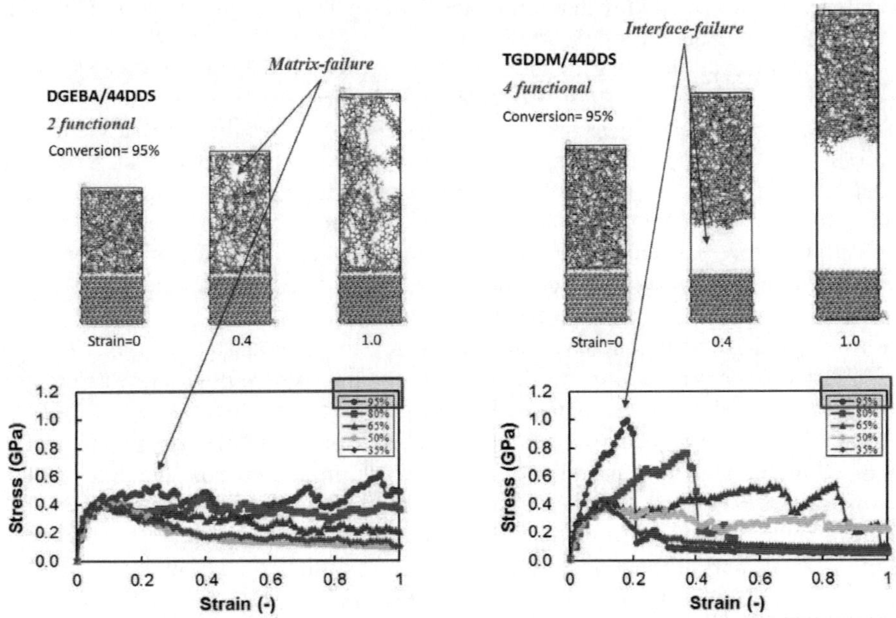

FIGURE 4. Peel-off simulation and obtained stress-strain curves.

STRUCTURAL ANALYSIS

Figure 5 shows the multiscale modeling of polymer composites in this study, which is a combination of Materials Studio and Abaqus. As shown in Figure 4, Materials Studio successfully provided the microscale stress–strain curves suggesting fracture mechanisms on the molecular level. Since fracture strength in the atomistic simulation can be recognized as a kind of ideal strength of materials, the difference from actual strength should be considered carefully. However, the analytical results of Materials Studio, which surely distinguish between a matrix-failure and an interface-failure, will be significantly useful in the selection of material in the polymer composite design. In the FEM part, Abaqus was used for its powerful nonlinear capability for fracture modeling[3]. Figure 5 also shows the finite element model, which is a periodic unit cell model containing randomly positioned 12 fibers in the matrix. A 2D plane strain condition was assumed, whereas the finite element mesh represented by the solid element C3D8 was pseudo three-dimensional in anticipation of practical needs for three-dimensional fracture analysis. The matrix and the fiber were assumed to be elastic, whose Young's moduli were 2,480 MPa and 298,000 MPa, respectively. The Poison's ratio was assumed to be 0.3 in common.

In this study, the characteristics of transverse failure in a unidirectional composite are simulated. A transverse crack, the so-called first-ply failure, causes considerable damage and can trigger fatal delamination. A numerical simulation of the progressive failure growth, consisting of initiation of transverse matrix cracking, crack propagation, and debonding of fiber–matrix interface, was demonstrated in this study. As shown in Figure 5, the eXtended finite element method (XFEM) was used to simulate the matrix-failure. As for the fiber–matrix interface, 8-node cohesive elements (COH3D8) were prescribed to simulate the delamination behavior with the cohesive zone model (CZM).

Figure 6 shows the multiscale modeling of the interface-failure in this study. During the interface crack growth process, two new surfaces are created. Before the crack is formed, these two surfaces can be assumed to be held together by traction to compose the interface. The CZM provides a traction-separation law for the gradual fracture formation took place in the local area (so-called cohesive zone, or fracture process zone) ahead of the crack-tip. As the surfaces separate, traction first increases until a maximum t_n^0 is reached, and then reduces to zero, which results in complete separation. The area under this curve is equal to the energy needed for separation G_c. Namely, The CZM assumes that the failure of the cohesive elements is characterized by progressive degradation of the material stiffness, which is driven by a damage process. As the maximum stress is limited to the cohesive strength of the material, this technique provides a stable solution for interface-failure problems, which tend to be singular by physical discontinuity due to debonding. As shown in Figure 6, if the "peel-off" cases in the microscale stress–strain curves obtained by Materials Studio are equated with an interface-failure, the maximum traction t_n^0 and the critical energy release rate G_c can be identified from the stress–strain curves.

Figure 7 shows the multiscale modeling of the matrix-failure using XFEM. The XFEM algorithm allows the presence of discontinuities such as cracks in an element by enriching the degrees of freedom with special displacement functions. As the crack tip changes its position, XFEM creates the necessary enrichment functions for the nodal points around the crack tip. Therefore, the technique is a very attractive way to simulate initiation and propagation of a discrete crack along an arbitrary, solution-dependent path without the requirement of remeshing in the materials. Similar to the above-mentioned idea, if the "void creation" cases in the microscale stress–strain curves obtained by Materials Studio are equated with a matrix-failure, the crack initiation and its growth can be identified from the stress–strain curves. The maximum principal stress criterion and the linear elastic fracture mechanics criterion were applied to the initiation and the propagation criteria, respectively.

Figure 8 shows a typical result of our analytical study for the relationship between nominal stress σ_n and nominal strain ε_n. The term "nominal" refers to the cross-sectional average of the stress and strain. Figure 8 also shows deformed shapes obtained from the analysis. It is helpful to review the process of failure by following the order of events. First, an initial crack occurs at the narrowest part of the matrix between two fibers ($\varepsilon_n = 0.0001$), where the maximum of the maximum principal stress is observed. In this portion, the deformation of the matrix is localized because this matrix area is held between two rigid fibers and given a forced displacement. The initial crack starts growing in accordance with the critical energy release rate G_c but the nominal stress increases as long as the failure is limited in the small part of the model ($\varepsilon_n < 0.0012$). Next, the crack propagates rapidly, and the nominal stress sharply decreases ($\varepsilon_n = 0.0015$). However, when the crack tip hits the outer surface of fiber, the propagation stops. This behavior is referred as a crack arrest, in which an unstable crack extension may start in a region of high stress, but that the surrounding material will have sufficient resistance to the crack extension to arrest the running crack. In fact, after the crack arrest, the nominal stress increases ($\varepsilon_n < 0.0031$), until an interface-failure occurs. As the fibers are positioned randomly in the matrix, other interface-failures take place successively ($\varepsilon_n = 0.0040$). Along with them, the load drops rapidly, yielding the loss of the load-bearing capacity ($\varepsilon_n = 0.0057$).

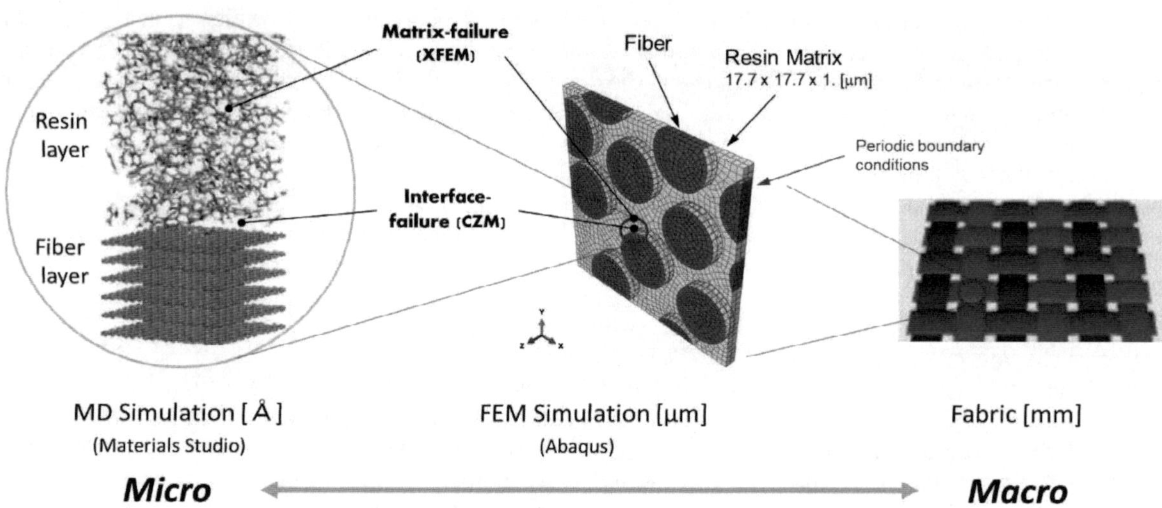

FIGURE 5. Multiscale modeling of polymer composites and periodic unit cell model.

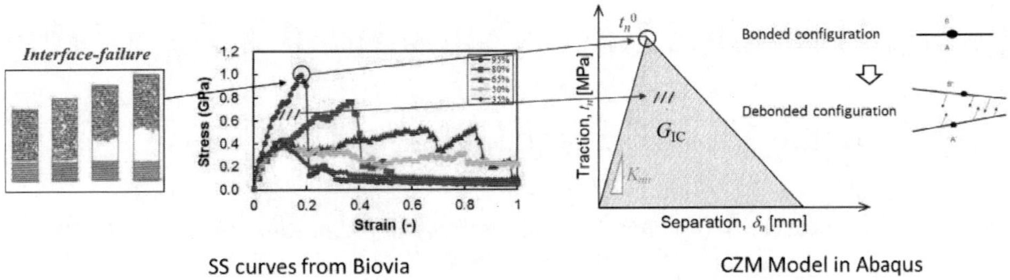

FIGURE 6. Multiscale modeling of interface-failure using CZM.

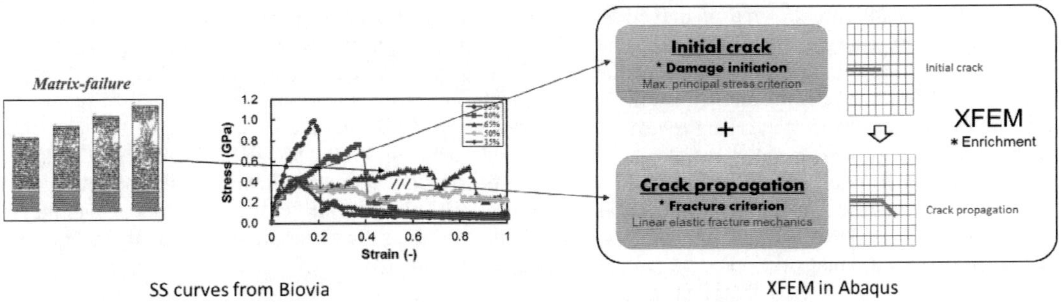

FIGURE 7. Multiscale modeling of matrix-failure using XFEM.

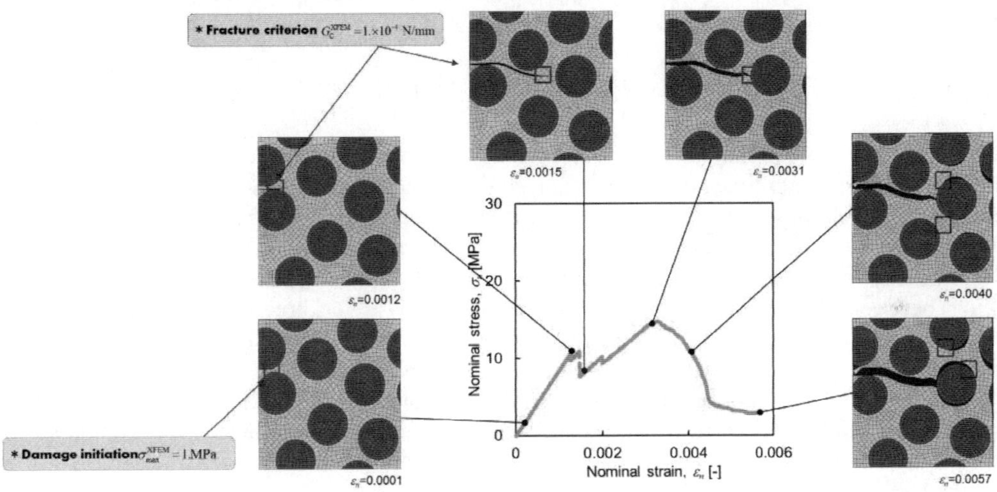

FIGURE 8. Failure behaviors transiting from matrix-failure to interface-failure.

CONCLUSIONS

A multiscale modeling of a polymer composite was demonstrated. The study showed the potential of the multiscale analysis to improve the quantitative aspect in design and material selection of the polymer composite, especially with respect to failure characteristics, which cannot be easily grasped by material testing alone.

REFERENCES

1. T. Okabe, Y. Oya, K. Tanabe, G. Kikugawa and K. Yoshioka, "Molecular Dynamics Simulation of Crosslinked Epoxy Resins: Curing and Mechanical Properties", European Polymer Journal, 80, pp.78-88, 2016.
2. A. Yang, F. Gao and J. Qu, "A study of highly crosslinked epoxy molding compound and its interface with copper substrate by molecular dynamics simulations" 2010 Electronic Components and Technology Conference, 128, 2010.
3. H. Fukui, A. Yoshimura and R. Matsuzaki, "Structural optimization for CFRP cryogenic tank based on energy release rate" Composite Structures, 152-15, pp. 883-890, 2016.

Magnetic Hydrogel Based Shoe Insoles for Diabetics

Rahul Patwa, Nabanita Saha* and Petr Sáha

Centre of Polymer Systems, University Institute, Tomas Bata University in Zlín, Třída Tomáše Bati 5678, 760 01 Zlín, Czech Republic

Abstract. According to WHO's Global report on diabetes 420 million adults are affected by diabetes worldwide. Foot infections are the most prevailing health complications in diabetics which arise due to neuropathy and peripheral vascular disease. The former concerns with nerve damage whereas the latter causes fatty depositions in blood vessels which reduce the blood flow causing pain, infection and slow healing wounds. One alternative therapy which is gaining popularity is the application of magnetic field to improve blood circulation by use of magnetic insoles. However, hydrogel based magnetic insoles have been scarcely reported. An efficient strategy is employed for the preparation of flexible magnetic hybrid hydrogels consisting of mixture of cobalt ferrite ($CoFe_2O_4$) and iron-oxide (Fe_3O_4) magnetic nanoparticles (MNPs) dispersed in PVP/CMC cross-linked hydrogels. Here the MNP ratio has been varied in the ratio of 1:0, 1:1, 1:2, 2:1 and 0:1, respectively. The method involves the formation of cross-linking between the two polymer chains under high pressure and temperature. The hybrid hydrogels were chemically and morphologically characterized using SEM, FTIR, VSM, swellability and mechanical testing. The morphological analysis showed the porous structure while using the VSM magnetic properties of the samples were evaluated. The rheological properties of the hydrogels were also investigated with the aim to verify their behaviour under an applied mechanical stress. The presence of MNPss, the hydrogel is capable of responding to an external magnetic field. Preliminary data show the possibility of use of the material for magnetic insoles for diabetics.

Keywords: magnetic; insoles; hydrogels; diabetics; nanoparticles

INTRODUCTION

Magnetic materials have been majorly used for various biomedical applications and devices such as magnetic resonance imaging (MRI), ferrofluids, hyperthermia, etc.[1] Magnetic hydrogels have found new applications in biomedical engineering including tissue engineering, enzyme immobilization, drug delivery and release, soft actuators and cancer therapy, etc. Recently magnetic hydrogels containing micron or nano-sized MNPs such as γ-Fe_2O_3, Fe_3O_4, $CoFe_2O_4$ have been reported which enhance their controllability.[2] Hydrogels are polymeric materials with high amount of cross-linking between adjacent polymer chains which give rise to three-dimensional structures which are hydrophilic in nature and non-soluble in water. Initially hydrogels were mainly used for treatment of open wounds as it was established that wet environment can enhance the wound healing process.[3] However, now hydrogel are readily used as soft materials for various bio-medical fields such as scaffolds for tissue engineering, super absorbents, drug delivery, contact lenses, control implants, substitutes for tendons, ligaments, cartilage and even bones, wound or burn dressing material and other medical device fabrication mainly due to the fact that most properties of hydrogels are comparable to the native extracellular matrix.[4]

Poly(N-vinyl pyrrolidone) (PVP) is a hydrophilic synthetic polymer with good biocompatibility discovered in 1939. It is mainly used in pharmaceuticals and personal care products (PPCPs) or food industries.[5] PVP however is known to be fragile and has a less swelling capacity. To obtain improved properties PVP is often blended with other polymers such as polysachharides. These days many polysaccharides are gaining popularity such as alginate, chitin/chitosan, cellulose/carboxymethyl cellulose (CMC), etc. mainly due to low-cost, abundance, non-toxicity and biodegradability

Cellulose, is the most abundant polymer which is a long chained polysaccharide joined by a B-1,4-glucosidic linkage. The carboxymethyl cellulose (CMC) is a hydrophilic, water soluble ether derivative in which H atoms in hydroxyl groups are replaced by carboxymethyl substituent, - CH_2COOH. It has been used for a variety of applications due to its low-cost, abundance, biocompatibility, good-water absorption behavior, etc.[6] Its properties can be improved by blending with other polymers such as gelatin, starch, PVP and glycerol etc.

Diabetes causes neuropathy and peripheral vascular disease which leads to nerve damage and fatty depositions in blood vessels, respectively.[7] The combined effect of two results in reduced blood flow, infection and slow healing wounds especially in the foot where in aggravated cases it might result in foot or leg amputation.[8] An alternative therapy is the use of magnetic insoles to improve blood circulation.[9]

In this work, we discuss the method of preparation of PVP/CMC based magnetic hydrogel its characterization (physical appearance, physicochemical structure, water uptake capacity, magnetic and rheological properties). A series of PVP/CMC blend hydrogels were prepared with different Co/Fe ratios to achieve effective content of composition. It is known that the viscoelastic properties of hydrogels correlate strongly with their microstructures, composition and the concentration of dispersed materials.

MATERIALS & METHODS

Poly (vinyl pyrrolidone) K-30 PVP, polyethylene glycol 3000 (PEG) and agar was purchased from Fluka, Switzerland; Carboxymethylcellulose (Na-CMC) was supplied from Sinopharm Chem. Reagent Co. Ltd. China. Glycerine was obtained from Lachema Ltd. Czech Republic. Cobalt iron oxide ($CoFeO_7$), Iron (II) chloride tetrahydrate 98% ($FeCl_2.4H_2O$), Iron (III) chloride hexahydrate 98% ($FeCl_3.6H_2O$) and ammonium hydroxide soln. (NH_4OH) was purchased from Sigma Aldrich, Czech Republic.

Preparation of PVP/CMC Blend Hydrogels

Iron-oxide nanoparticles are prepared by using standard Massart method.[10] Hydrogels were prepared by moist heat treatment method.[11, 12] Solutions of PVP (1wt%) and CMC (1wt%) were prepared separately by dissolving PVP and CMC in deionized water (40mL each) at 80 °C for about 2 h under continuous stirring. Then the two solutions were mixed to prepare homogeneous mixture into which agar (1wt%), PEG (3wt%), glycerol (3% w/v) were added to the homogeneous mixture. Finally, MNPs (1wt%) are added which are previously sonicated in 20mL of DI water. Then the mixtures were autoclaved for 20 minutes. After completion of the process, the hot polymer solution was cast in petriplates and allowed to cool at 37 °C in hot air oven. Finally, dried round flexible films were obtained and termed as PVP/CMC hydrogels. Some samples were also lyophilized.

Swelling Behavior

The lyophilized hydrogels without unbound water content were cut into small pieces (2×2 cm^2), weighed and immersed in physiological saline solution (0.9% NaCl) for a total time period of 330 mins. After every 30-minute time intervals the water on the swollen hydrogels was wiped off with wet tissue papers and the weight of the specimens was noted. The tests were conducted in triplicates and averaged values are reported along with standard deviation. Swollen piece of PVP/CMC blend hydrogel, with thickness of was put in a petridish and then heated in an oven. The temperature was set to 37 °C. the sample was weighed after an interval of 30 minutes. The Swelling degree and water retention percentage was calculated as per Eq. (2)[13]:

$$\text{(Swelling degree) \%} = [(W_1 - W_2)/ W_2] \times 100 \quad (1)$$
$$\text{(De-swelling degree)\%} = [(W_3 - W_4)/ W_4] \times 100 \quad (2)$$

where, W_1, W_2 are weights of swollen, dry and W_3, W_4 are weights of swollen and deswollen hydrogel, respectively.

Fourier transform infrared spectroscopy (FT-IR).

Pure PVP, pure CMC, and PVP/CMC gels after were analysed by FT-IR. The ATR-FTIR spectra were recorded using a NICOLET 320 FTIR over the range 4000-600 cm-1 at room temperature. A uniform resolution of 2 cm-1 was maintained in all cases.

X-ray diffraction

X-ray diffraction patterns of powdered samples of MNPs were obtained using Bruker D8-Discover X-ray diffractometer with Cu Kα radiation of wavelength 1.54056 Å.

Vibrating Sample magnetometer

Room temperature magnetic measurements were performed with a Lakeshore Model 7404 vibrating sample magnetometer (VSM).

Rheological properties

To evaluate the dynamic viscoelastic properties of swelled PVP-CMC hydrogels, a parallel plate Rheometer (Anton Paar, USA) testing machine with a software package was used. A 25 mm diameter parallel plate with a gap of 2-3 mm was used, and dynamic oscillatory flow measurements were carried out at 1% strain to maintain the measurements within the linear viscoelastic region. Frequency sweep tests were carried out at 25 °C to observe the storage modulus (G′), loss modulus (G′′), complex viscosity (η*), tensile stain and damping factor (tan δ) as a function of angular frequencies (0.1-100 rad/s).

Scanning electron microscope (SEM)

PVP/CMC blend hydrogels were lyophilized to get spumous samples. SEM image was made by scanning electron microscope. The SEM observation was carried out using Phenom world PRO instrument operating at high-vacuum/ secondary electron imaging mode at and accelerating voltage of 5-15 kV. The images were taken at a magnification of 100x to 10 kx magnification factor.

RESULTS AND DISCUSSION

The PVP/CMC blend hydrogels were successfully prepared using different ratios of MNPs. The hydrogels were prepared by simple air-drying (Fig. 1a) and freeze-drying (Fig. 1b). All the samples were highly flexible (Fig. 1c) and showed attraction towards magnet (Fig. 1d) indicating that the prepared hydrogels were magnetic in behavior.

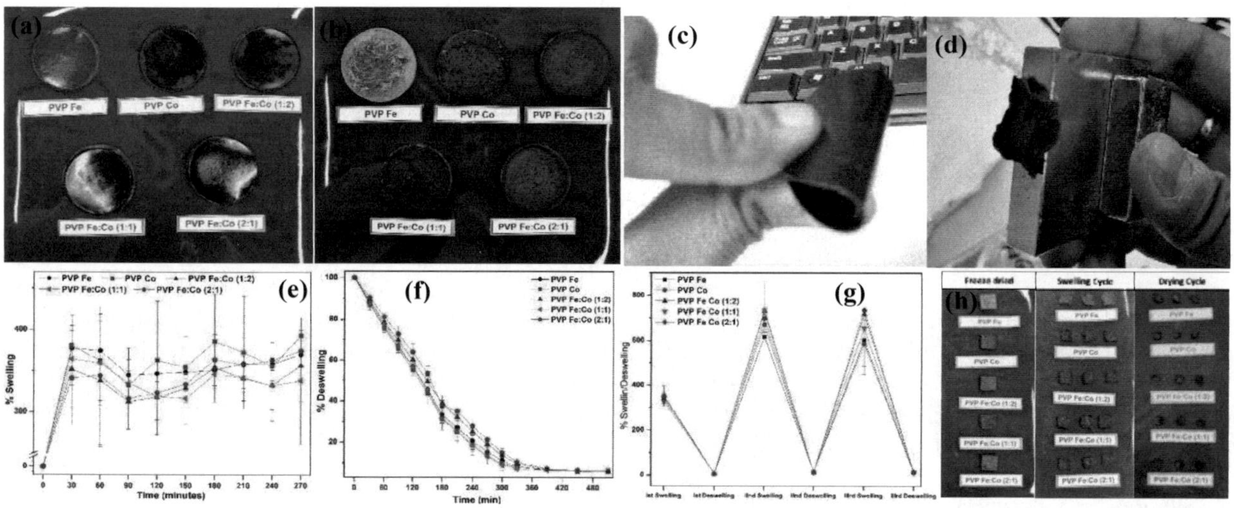

FIGURE 1. Pictorial representation of (a) air-dried, (b) freeze-dried, (c) flexible behavior of and (d) magnetic behavior of hydrogels, (e) Swelling degree, (f) Deswelling degree, (c) swelling-deswelling cycles and (d) pictorial comparison of original, swelled and dried hydrogels.

The swelling and de-swelling behaviour of the hydrogels is shown in Fig (1e and 1f). The swelling behavior of hydrogels is generally governed by amount of cross-linking and ionic strength of polymer. The maximum amount of water retention can be seen in the initial 30 minutes and reaches to more or less equilibrium state. It can be seen clearly in Fig 1e that hydrogels prepared using pure composition of MNPs such as PVP Co and PVP Fe have better swelling characteristics. As compared to swelling, the deswelling takes a longer time (~360 min) to take place due to the presence of hydrophilic groups (-CHO, -COO and –OH). Finally, it can be concluded that the water retention and water absorption characteristics of all the hydrogels is highly comparable. In Fig (1g and 1h), it can be seen that all the hydrogels do not undergo disintegration upon extended usage this was checked by subjecting the hydrogels to succesive swelling-de-swelling cycles.

FTIR spectra of pure PVP, CMC and fresh hydrogels of PVP-CMC are shown in Fig. (2a). The FTIR spectra of other components are not included to avoid confusion. PVP spectra contains the characteristic peaks at 1281 and 1658 cm-1 which are assigned to C-N and C=O stretching vibrations, respectively. The CMC shows absorption bands at 1588, 1428 and 1060 cm^{-1} which represents the carbonyl band, C=C stretching and β-1,4-glycosidic band, respectively. It is noteworthy to mention that PVP peak at 1658 cm^{-1} and CMC peak at 1588 cm^{-1} gets condensed and appear as one single peak around 1628 cm^{-1} indicating towards cross-linking or grafting between CMC and PVP chains in PVP/CMC blends.

X-ray diffraction patterns shown in Fig (2b) of the synthesized iron-oxide and purchased cobalt ferrite matched closely with the diffraction patterns of and Fe_3O_4 and $CoFe_2O_4$, respectively.[14, 15]

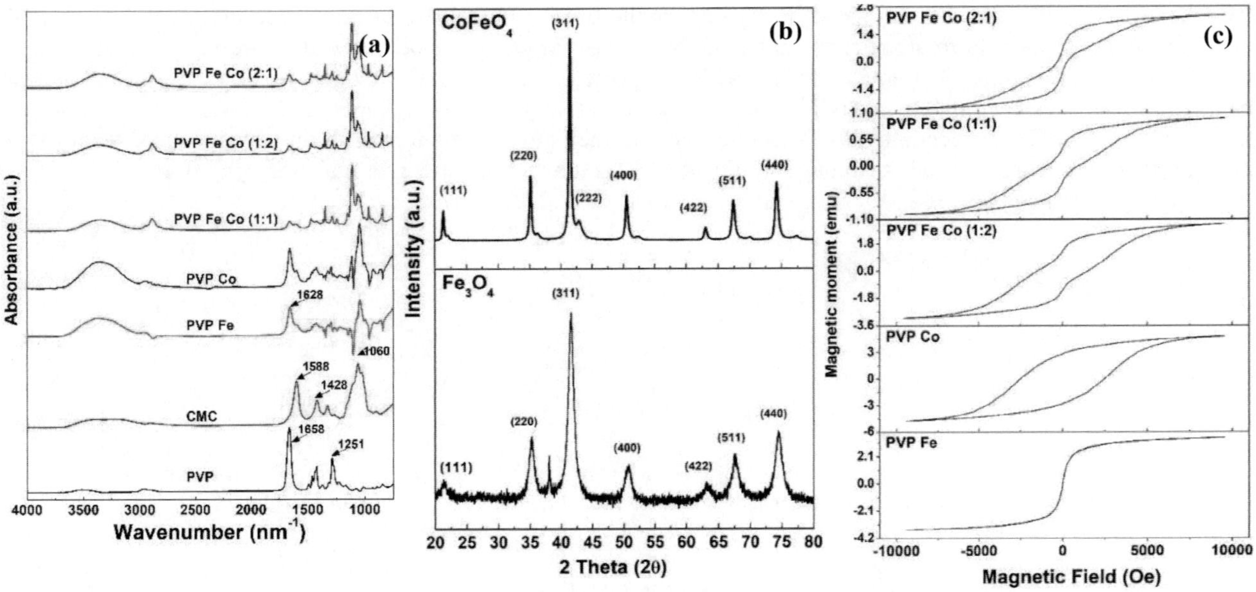

FIGURE 2. (a) FTIR spectra of prepared hydrogels; (b) x-ray diffraction patterns of MNPs; (c) VSM magnetic measurements of prepared hydrogels.

Magnetic properties measured by VSM revealed that cobalt ferrite MNPs have better Coercivity as compared to iron-oxide MNPs as shown in Fig (2c). As a result, the PVP Co showed superior magnetic properties. For the mixed MNPs formulations, PVP Fe Co (1:2) showed highest magnetic properties.

The SEM images of all lyophilized PVP/CMC based hydrogels are shown in Fig 3. The surface and cross-sectional images clearly show the presence of porous structures in all the hydrogels which arise from the polymer chain entaglements or physical interactions. The presence of pores allow these hydroels to absorb/retain water inspite of the fact that these hydrogels are prepared using water soluble polymers.

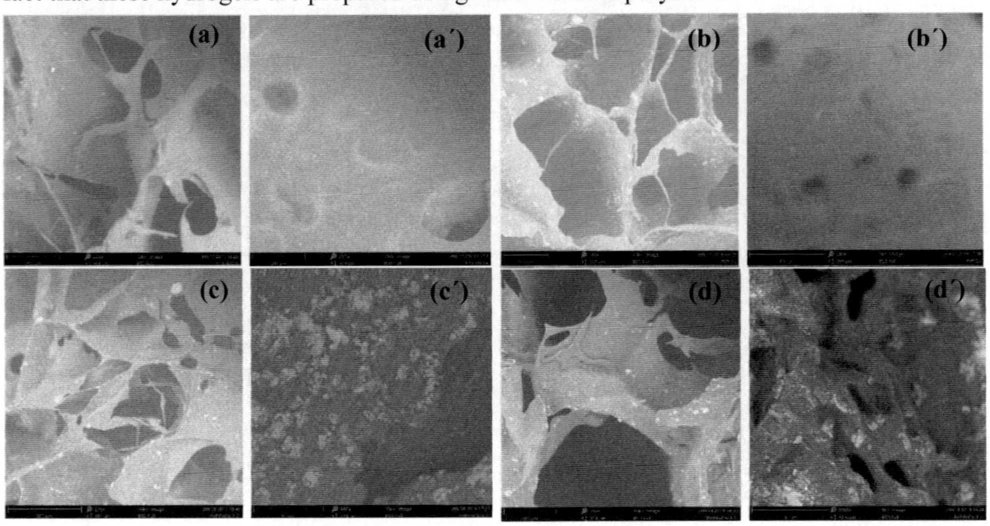

FIGURE 3. SEM micrographs of cross-sectional and the flat surfaces of (a, a′) PVP Fe, (b, b′) PVP Co, (c, c′) PVP Fe CO (1:2), (d, d′) PVP Fe CO (1:1) hydrogel.

It can be observed from the SEM micrographs that the PVP Fe Co (1:2) has the most dense network and numerous small pores which corroborates well with the rheological findings which shows that this hydrogel composition had the best rheological properties.

Hydrogels upon soaking in water depict viscoelastic behavior. In swollen form the water content is much larger in comparison to the polymer content which highly affects the rheological properties of the hydrogels. The incorporation of MNPs affect the magnetic properties of the hydrogels. Thus, the effect of MNPs on viscoelastic properties of

PVP/CMC hydrogels has been evaluated. Fig 4 depicts the storage modulus (elasticity) and loss modulus (viscous behaviour) with respect to frequency. It can be observed that the storage modulus values are higher than the loss modulus values throughout the frequency range which is typical for gels hence at low strain (~1%) they show rubbery behaviour. PVP Co shows lowest values of G´ and G´´ whereas PVP Fe shows higher G´ and G´´ values which indicate that Fe_3O_4 NPs act as lubricant and also it can be said that they induce better crosslinking as compared to CoFeO4 NPs. Furthermore it can be concluded that PVP Fe Co (1:2) is the most elastic hydrogel among all others.

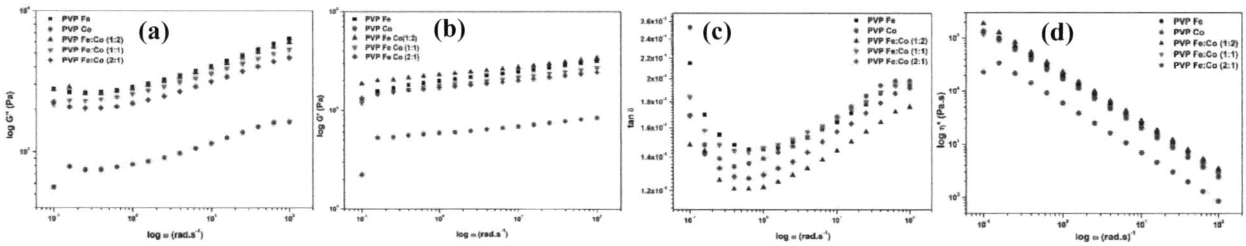

FIGURE 4. Effect of angular frequency (α) at 1% strain on (a) loss modulus (G´´, non-filled symbol), (b) storage modulus ((G´, filled symbol), (c) damping factor (tan δ), (d) tensile strain (%), (e) complex viscosity (η*) for PVP Fe, PVP Co, PVP Fe CO (1:2), PVP Fe CO (1:1), PVP Fe CO (2:1).

CONCLUSIONS

The main objective of this research work was to develop a hydrogel film as an alternative approach as diabetic shoe insoles. The materials used are all non-toxic and biodegradable. It was found that PVP/CMC hydrogel films containing MNPs could be a promising candidate as a shoe insole material for diabetic care. All the hydrogels are flexible in nature and shows good modulus which is required for durability. Moreover it was found that PVP Fe Co (1:2) hydrogel has the superior properties as compared with other prepared hydrogels in terms of swellability, magnetic, morphological and rheological properties. This hydrogel composition can be efficiently used as shoe insert by diabetics for increasing blood circulation. In addition, being a hydrogel the material can readily absorb moisture hence reduce bacterial growth thus preventing foul odor.

ACKNOWLEDGMENTS

This work was supported by Ministry of Education, Youth and Sports of the Czech Republic- Program NPU I (LO1504) and International Mobility of TBU researchers – CZ.02.2.69/0.0/0.0/16_027/0008464.

REFERENCES

1. A. L. Elrefai, T. Yoshida and K. Enpuku, J. Magn. Magn. Mater. **474**, 522-527 (2019).
2. E. Yan, M. Cao, X. Ren, J. Jiang, Q. An, Z. Zhang, J. Gao, X. Yang and D. Zhang, J. Phys. Chem. Solids **121**, 102-109 (2018).
3. A. Gupta, M. Kowalczuk, W. Heaselgrave, S. T. Britland, C. Martin and I. Radecka, Eur. Polym. J. **111**, 134-151 (2019).
4. M.-D. Veiga, R. Ruiz-Caro, F. Notario-Pérez, A. Martín-Illana and R. Cazorla-Luna, in *Design and Development of New Nanocarriers*, edited by A. M. Grumezescu (William Andrew Publishing, 2018), pp. 509-554.
5. N. Saha, R. Shah, P. Gupta, B. B. Mandal, R. Alexandrova, M. D. Sikiric and P. Saha, Materials Science and Engineering: C **95**, 440-449 (2019).
6. T. Heinze and A. Koschella, (unpublished).
7. S. Tesfaye, Medicine **47** (2), 92-99 (2019).
8. Y.-K. Jan, F. Liao, G. L. Y. Cheing, F. Pu, W. Ren and H. M. C. Choi, Microvasc. Res. **122**, 45-51 (2019).
9. R. Tao and K. Huang, Phys Rev E **84** (1), 011905 (2011).
10. Y.-k. Sun, M. Ma, Y. Zhang and N. Gu, Colloids Surf. Physicochem. Eng. Aspects **245** (1), 15-19 (2004).
11. N. Roy, N. Saha, T. Kitano and P. Saha, J. Appl. Polym. Sci. **117** (3), 1703-1710 (2010).
12. A. Gregorova, N. Saha, T. Kitano and P. Saha, Carbohydr. Polym. **117**, 559-568 (2015).
13. J. Liu, Q. Li, Y. Su, Q. Yue and B. Gao, Carbohydr. Polym. **107**, 232-240 (2014).
14. P. Karpagavinayagam and C. Vedhi, Vacuum **160**, 286-292 (2019).
15. O. Karaagac, B. B. Yildiz and H. Köçkar, J. Magn. Magn. Mater. **473**, 262-267 (2019).

INVESTIGATION OF SHAPE MEMORY BEHAVIOR AND PHYSICAL PROPERTIES OF CROSSLINKED LOW DENSITY POLYETHYLENE/OvPOSS/TAIC COMPOSITES

Ezgi BICER*, Gulsen KURT DEMIR, Mehmet KODAL, Guralp OZKOC

Kocaeli University, Chemical Engineering Department, Kocaeli, 41380, Türkiye

ezgibicer39@gmail.com, gulsenkurrt@gmail.com, mehmet.kodal@kocaeli.edu.tr, guralp.ozkoc@kocaeli.edu.tr

Abstract. The aim of this study is to investigate the shape memory behavior and physical properties of crosslinked low density polyethylene (LDPE)/Octavinyl functionalized polyhedral oligomeric silsesquioxanes (OvPOSS)/triallyl isocyanurate (TAIC) composites. Shape memory polymer can return from its temporary shape to the original (permanent) shape upon heating. One of the most widely used shape memory polymers is crosslinked polyethylene. Polyhedral oligomeric silsesquioxanes (POSSs) organic/inorganic hybrid materials are new the generation of nanofillers that can be used to increase the crosslinking efficiency of LDPE. OvPOSS nanoparticles having reactive functional vinyl groups and TAIC were used as coagents. On the other hand, 1 phr of di tert-butyl cumyl peroxide (BCUP) was used as a crosslinking agent in this study. LDPE/OvPOSS/TAIC/Peroxide composites were prepared in an Xplore model twin screw micro-compounder at a screw speed of 50 rpm and 135°C barrel temperature. The crosslinking was carried out using a hot-press operated at 175°C. The composites were characterized by performing tensile tests, differential scanning calorimetry (DSC), scanning electron microscopy (SEM), rheology and shape memory tests. Rheological test results showed that the degree of crosslinking, storage and loss modulus of LDPE/TAIC system increased by the addition of OvPOSS and TAIC. It was obtained from DSC analysis that LDPE/OvPOSS/TAIC/Peroxide composites exhibited lower melting enthalpy values, which can be attributed to the restricted chain mobility of LPDE in the presence of OvPOSS and PRX.

Keywords: Crosslinked polyethylene, Octavinyl POSS, Shape memory properties, Triallyl isocyanurate.
PACS: S08-239

INTRODUCTION

Polyethylene (PE) is the most widely used thermoplastic polymer in the world with a consumption of more than 80 million tons per year. Its main usage are packaging films, packaging materials (bottles, containers etc.), cable and pipe sector [1, 2]. Polyethylene has wide range of applications such as insulation of wire and cable because of its good mechanical properties, good resistance to chemicals, and good processability. However, it has some disadvantages such as low melting temperature, low thermal stability, high flammability, and poor compatibility with additives [3]. Moreover, the service area of PE is limited due to its low service temperature. For this reason, it is attempted to increase the service temperature of PE by crosslinking especially in industrial products such as pipes, cables and heat shrinkable films/tubes. Crosslinking of PE results in improving its mechanical properties; however, it decreased the molecular mobility [4]. Crosslinked polyethylene can be used in pipe and cable sectors due to its high thermo-mechanical properties, while it is preferred in the packaging and electrical/electronic sectors (applications such as heat shrinkable films and tubes) due to its shape memory feature.

Shape memory materials are capable to memorize a temporary shape unless they are exposed to an external stimulus that triggers them to recover their permanent shape [5]. Shape memory polymers allow molecular design of tailored stiffness and responsiveness to external stimuli such as heat, pH, electricity, and light [6].

Poly(hedral oligomeric silsesquioxane) (POSS) nanoparticles are structurally cage-like molecules and may contain different functional groups (vinyl, acrylate, methacrylate, epoxy, anhydride, hydroxyl, etc.), which are responsible for compatibility and reactivity between polymers and POSSs. POSS positively affects the thermal stability, thermo-mechanical strength, flow properties, and crystallization rates of polymers [7]. Coagents are used in order to improve the crosslinking efficiency by reducing undesirable side reactions such as chain breakage or disproportionation. They form bridges between polymer chains to be cross-linked.

The aim of this study is to investigate the synergistic effects between POSS and coagent on the physical and shape memory properties of crosslinked low density polyethylene composites.

EXPERIMENTAL

Materials

Polyethylene was purchased from DOW Chemical Company, Switzerland. According to the information supplied by the manufacturer, it has a density and MFI value of 0.980 g/cm^3 and 7.7 g/10 min (190°C/2.16 kg), respectively. Octavinyl POSS (OvPOSS) was obtained from Hybrid Plastics and triallyl isocyanurate (TAIC) was purchased from Sigma-Aldrich. Tert-butyl cumyl peroxide was supplied by AkzoNobel, Turkey.

Sample Preparation

LDPE/POSS/Peroxide, LDPE/Coagent/Peroxide and LDPE/POSS/Coagent/Peroxide composites were processed in an Xplore Instruments micro-compounder at a screw speed of 50 rpm and 135°C barrel temperature. The crosslinking was carried out by using a hot-press operated at 175°C.

Characterization

The LDPE based composites were characterized by performing tensile tests, differential scanning calorimetry (DSC), rheometer, scanning electron microscopy (SEM) and shape memory tests. Differential scanning calorimeter (DSC) analysis was performed using a Mettler Toledo DSC1 Star System between 25°C and 230°C at a heating rate of 10°C/min under the nitrogen atmosphere. Tensile tests were performed according to ISO 527 standard, using an Instron Universal Testing Machine. Morphological studies was carried out on tensile fractured samples by using scanning electron microscope (SEM, JEOL JSM 6040). The rheological properties of uncrosslinked polyethylene blends were determined at a crosslinking temperature of 175°C and 1 Hz oscillation. For this purpose, the molded samples were subjected to time scanning with the Antoon Paar MCR rheometer. The crosslinked samples were deformed at the melting point ($T_m = T_t = 100°C$) determined as transition temperature and fixed in the given transient atmospheric conditions for determination of the shape memory properties. Permanent shape return rates were also carried out in a hot oven.

RESULT AND DISCUSSION

Rheological analysis results are given in Figure 1. As can be seen in Figure 1 that the complex viscosity increased drastically due to crosslinking reactions. Storage modulus and the complex viscosity increased with the incorporation of Octavinyl POSS (OvPOSS) to LDPE. These increments can be attributed to the increased crosslinking density in the presence of POSS nanoparticles. Therefore, it can be concluded that POSS nanoparticles acted as a crosslinking agent for LDPE.

LDPE/OvPOSS/TAIC/PRX composites exhibited higher complex viscosity values as compared to both LDPE/TAIC/PRX and LDPE/OvPOSS/PRX system at a constant PRX concentration. It was obtained that degree of crosslinking and storage modulus of LDPE/Octavinyl POSS composites were higher than that LDPE/TAIC. On the other hand, the highest storage modulus and elastic properties were obtained from LDPE/OvPOSS/TAIC composites due to the synergistic effects between TAIC and OvPOSS. Moreover, crosslinking density of LDPE/PRX significantly increased in the presence of both TAIC and OvPOSS as can be seen from Figure 2.

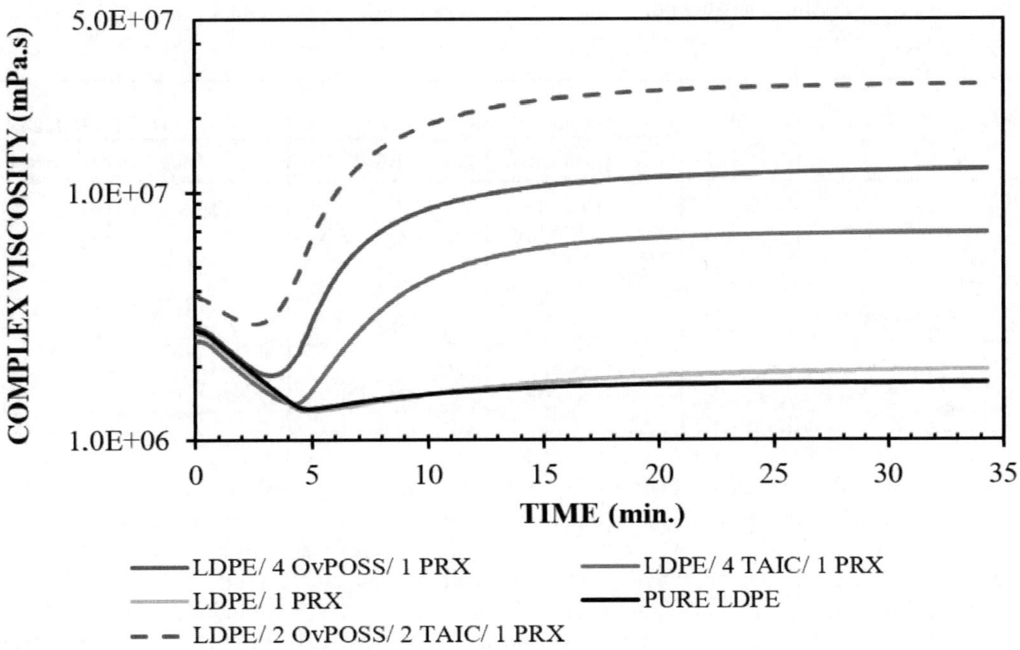

FIGURE 1. Complex viscosity of LDPE and its composites

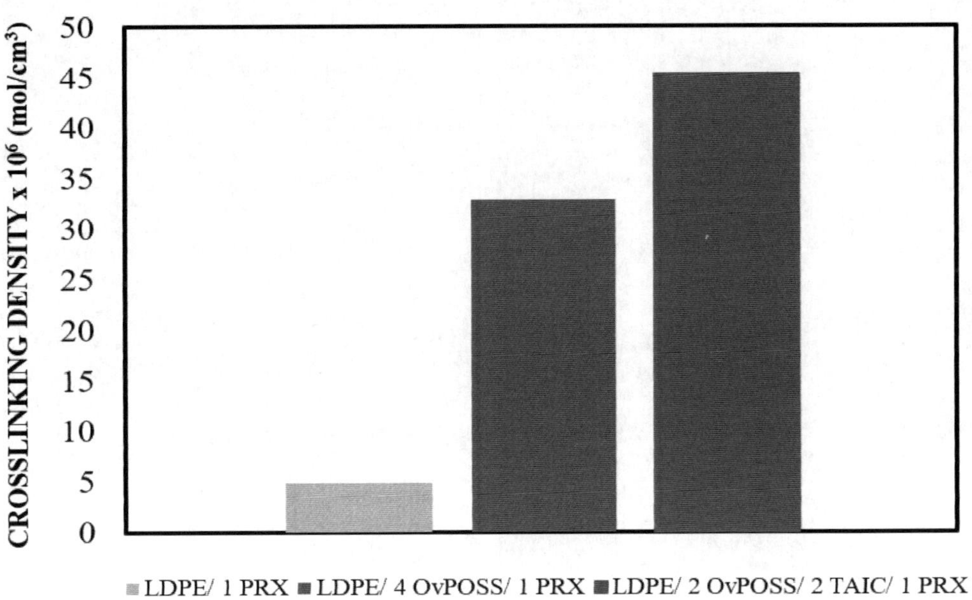

FIGURE 2. Crosslinking density of composites

Thermal properties of the pure LDPE and crosslinked LDPE samples were investigated by means of DSC and the results were summarized in Table 1. It was obtained that crystalline lamellar thickness (l_c) of low density polyethylene decreased in crosslinked samples, which could be attributed to the fact that crosslinking restricted the chain mobility of PE. Moreover, the lowest crystalline lamellar thickness value was obtained from LDPE/OvPOSS/TAIC/PRX system. Moreover, crosslinked samples exhibited lower percent crystallinity (X_c%) as compared to uncrosslinked ones.

TABLE 1. The changes of crystalline lamellar thickness and percent crystallinity of uncrosslinked and crosslinked composites of LDPE

COMPOSITES	UNCROSSLINKED		CROSSLINKED	
	X_c (%)	l_c (nm)	X_c (%)	l_c (nm)
Pure LDPE	19.5	3.0	16.6	2.9
LDPE/ 1 PRX	17.3	2.9	12.1	2.8
LDPE/ 4 OvPOSS	19.7	3.0	13.3	2.9
LDPE/ 4 OvPOSS/ 1 PRX	22.9	3.0	9.3	2.6
LDPE/ 4 TAIC/ 1 PRX	25.9	2.9	25.1	2.8
LDPE/ 2 OvPOSS/ 2 TAIC/ 1 PRX	28.3	3.1	27.6	2.5

The morphological analysis of LDPE/OvPOSS/TAIC/PRX composites were carried out by means of scanning electronic microscopy (SEM) on tensile fractured surfaces (Figure 3). SEM analysis were carried out to observe the size and distribution of OvPOSS and TAIC aggregates in LDPE before and after the crosslinking. The ductile fracture behavior of composites was obtained by SEM pictures. Before crosslinking, the crystals size of OvPOSS and TAIC ranging from hundreds of nanometers to several micrometers are randomly dispersed in the matrix (Figure 3a). However, OvPOSS and TAIC aggregates break to small particles with the size of tens of nanometers after crosslinking (Figure 3b). Moreover, it was obtained that POSS nanoparticles were dispersed homogeneously in LDPE matrix.

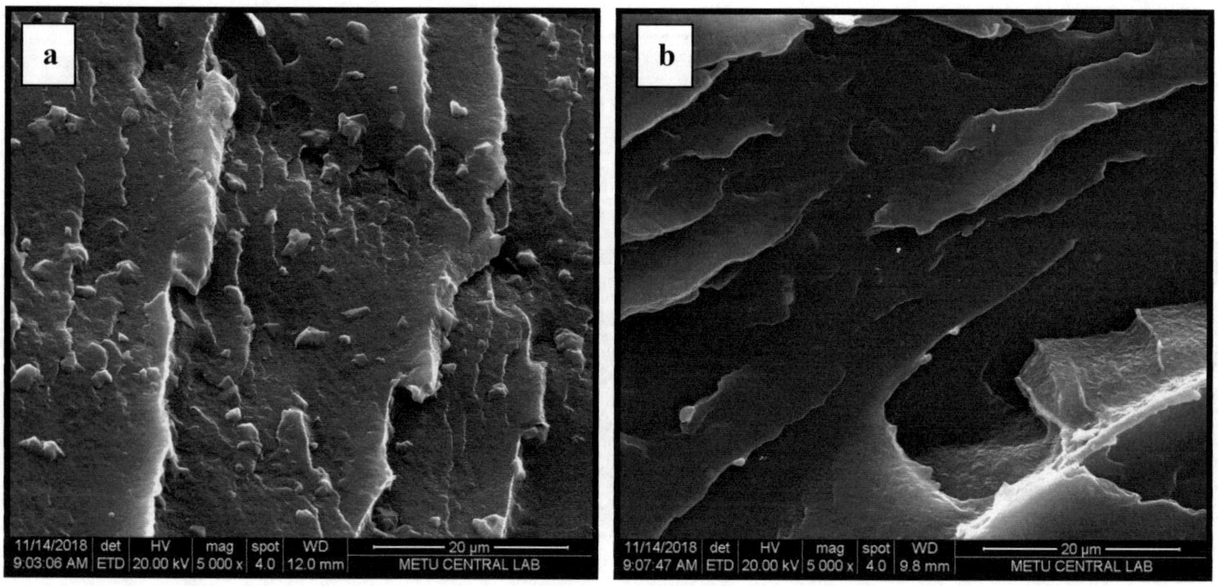

FIGURE 3. SEM micrographs of LDPE/OvPOSS/TAIC/PRX composites a) Uncrosslinked, b) Crosslinked

According to the shape memory test results, pure LDPE has a 25% shape recovery ratio. This ratio has increased by about 50% after crosslinking. The addition of OvPOSS and TAIC to the LDPE increased the shape recovery ratio. As the amount of crosslinking increases, there is also an increase in the rate of shape recovery.

CONCLUSIONS

Rheological analysis showed that the degree of crosslinking and storage modulus increased by the addition of OvPOSS and TAIC. A synergistic effect was observed between POSS and coagent. The composites with 2 phr OvPOSS, 2 phr TAIC and 1 phr PRX (LDPE/OvPOSS/TAIC/PRX) exhibited the superior rheological behaviors. The crosslinking density increased with the addition of coagent. Besides, higher crosslinking density also increased the rate of shape recovery.

ACKNOWLEDGEMENT

This study has been financially supported by TUBITAK (Grant Number: 116M981).

REFERENCES

1. A. J. Peacock, *Journal of Macromolecular Science-Polymer Reviews* **41**(4), 285-323 (2001).
2. I. Krupa and A. S. Luyt, *Polymer Degradation and Stability* **70**, 111-117 (2000).
3. A. A. Basfar, J. Mosnacek, T. M., Shukri, M. A. Bahattab, P. Noireaux, A. Courdreuse, *Journal of Applied Polymer Science* **107**, 642-649 (2008).
4. S-W.Kuo and F. C. Chang, *Progress in Polymer Science* **36**(12), 1649-1696 (2011).
5. R. Hoeher, T. Raidt, M. Rose, F. Katzenberg, J. C. Tiller, *Journal of Polymer Science, Part B: Polymer Physics* **51**, 1033-1040 (2013).
6. T. Chung, A. Romo-Uribe, P.T. Mather, *Macromolecules,* **41**, 184-192 (2008).
7. J. Miltz and M. Narkis, *Polymer* **9**, 173 (1968).

Cross-linked LLDPE composites in the presence of POSS nanoparticles and poly(ethylene glycol) dimethacrylate coagent: "Comparison of physical properties and shape memory behaviour"

Gulsen KURT DEMIR*, Ezgi BICER, Mehmet KODAL, Guralp OZKOC
Kocaeli University, Chemical Engineering Department, Kocaeli, 41380, Türkiye

gulsenkurrt@gmail.com,ezgibicer39@gmail.com, mehmetkodal32@gmail.com, guralp.ozkoc@kocaeli.edu.tr

Abstract. Shape memory polymers are smart materials having the ability to change their shape under external stimulus. One of the most widely used shape memory polymers is cross-linked polyethylene (PE). The cross-linked polyethylene can be triggered by temperature to transfer from its permanent shape to programmed shape. Polyhedral oligomeric silsesquioxanes (POSSs) are the new generation of hybrid nanoparticles that positively affect the thermal stability, thermo-mechanical strength, flow properties and crystallization rates of the polymers. POSSs can also be used as crosslinking agents. Besides, coagents are used in order to improve the crosslinking efficiency by reducing undesirable side reactions such as chain breakage or disproportionation. They form bridges between polymer chains to be cross-linked. The aim of this study is to investigate the synergistic effects between POSS and coagent on the physical and shape memory properties of cross-linked linear low-density polyethylene (LLDPE). Octavinyl POSS (OvPOSS) nanoparticles and poly(ethylene glycol) dimethacrylate (PEGDMA) were used as coagents in this study. The peroxide used in crosslinking was di tert-butyl cumyl peroxide (PRX). Its content was kept constant as 1 phr. LLDPE/OvPOSS/PRX, LLDPE/ PEGDMA/PRX and LLDPE/OvPOSS/PEGDMA/PRX composites were prepared in an Xplore model micro-compounder at a screw speed of 50 rpm and 135°C barrel temperature. The crosslinking was carried out using a hot-press operated at 175°C. The samples were characterized by performing differential scanning calorimetry (DSC), scanning electron microscopy (SEM), rheology, tensile tests and shape memory tests. Rheological analysis showed that the degree of crosslinking, storage modulus and complex viscosity increased in the presence of OvPOSS and PEGDMA. A synergistic effect was observed between OvPOSS and PEGDMA coagents.

Keywords: Linear low density polyethylene, Octavinyl POSS, Peroxide Crosslinking, Poly(ethylene glycol) dimethacrylate, Shape memory.
PACS: S08-234

INTRODUCTION

Polyethylene is preferred for packaging films, packaging materials, cable and pipe sector due to their properties such as cheapness, flexibility, resistance to low temperatures and easy processability [1]. However, critical physical properties of polyethylene decreased at high temperatures, which limits its versatile applications.

Crosslinking keeps desirable properties at high temperatures. For this reason, it is attempted to increase the service temperature by crosslinking polyethylene especially products such as pipes, cables and heat shrinkable films/tubes. [2]. Cross-linked polyethylene can be used in pipe and cable sectors due to its high thermo-mechanical properties, while it is preferred in the packaging and electrical/electronic sectors (applications such as heat shrinkable films and tubes) due to its shape memory feature.

Shape memory polymers are smart materials that can temporarily recall their original shape, which can be transiently transformed into permanent shapes controlled by an external stimulus externally applied and triggered by shape change [3, 4, 5]. Shape recovery is a temporary response under appropriate stimuli such as ambient temperature, force, light, sound, electricity, pH change, moisture change, ultraviolet light, electronic field, magnetic field, sonic field, organic solvent, specific ions and enzymes, mechanical stresses.

POSSs structurally cage-like molecules can contain different functional groups (vinyl, acrylate, methacrylate, epoxy, anhydride, hydroxyl, etc.) that determine both the compatibility and reactivity of POSSs. Thermal stability, thermo-mechanical strength, flow properties and crystallization rates of polymers can be increased with the incorporation of POSS nanoparticles [6].

Coagents uses in order to improve the efficiency of crosslinking by reducing undesirable side reactions, such as chain breakage or disproportion, which increases unsaturation. Coagents are compounds containing two or more unsaturated carbon bonds. They form coagulated bridges between the molecules to be crosslinked [7].

In the current study, it was aimed to determine the physical properties and shape memory behavior of cross-linked LLDPE composites in the presence of POSS nanoparticles and poly(ethylene glycol) dimethacrylate coagent.

EXPERIMENTAL

Materials

Polyethylene was purchased from DOW Chemical Company, Switzerland. According to the information supplied by the manufacturer, it has a density and MFI value of 0.980 g/cm^3 and 7 g/10 min, respectively. Octavinyl POSS (OvPOSS) was obtained from Hybrid Plastics. Poly(ethylene glycol) dimethacrylate (PEGDMA) as a coagent was obtained from Sigma Aldrich. Di tert-butyl cumyl peroxide (PRX) was supplied by AkzoNobel, Turkey.

Sample Preparation

LLDPE/OvPOSS/PEGDMA/PRX composites were prepared in an Xplore Instruments micro-compounder at a screw speed of 50 rpm and 135°C barrel temperature. PRX and OvPOSS-PEGDMA content was kept constant as 1 phr and 4 phr, respectively. OvPOSS/PEGDMA content was kept constant as 2/2 phr. The crosslinking was carried out using a hot-press operated at 175°C.

Characterization

LLDPE/OvPOSS/PEGDMA/PRX composites were characterized by performing tensile tests, differential scanning calorimetry (DSC), rheometer, scanning electron microscopy (SEM) and shape memory tests. Differential scanning calorimeter (DSC) analysis was performed by using a Mettler Toledo DSC1 Star System between 25°C and 230°C at a heating rate of 10°C/min under the nitrogen atmosphere. Tensile tests were performed according to ISO 527 standard. Morphological studies were carried out on tensile fractured samples by using scanning electron microscope (SEM, JEOL JSM 6040). The rheological properties of uncross-linked polyethylene blends were determined at a crosslinking temperature of 175°C and 1 Hz oscillation. For this purpose, the molded samples were subjected to time scanning with the Anton Paar MCR rheometer. The cross-linked samples were deformed at the melting point ($T_m = T_g = 110$ °C) determined as transition temperature and fixed in the given transient atmospheric conditions. Shape recovery ratios were also carried out in a hot oven.

RESULT AND DISCUSSION

Rheological analysis results were given in Figure 1. According to these results, when PRX is added to LLDPE, the complex viscosity, storage and loss modulus values increased in the presence of PRX, which resulted from the crosslinking reactions. The higher values compared to pure LLDPE are proof that the molecular structure is complex, that is, cross-linking occur. It can be seen from Figure 1 that that the complex viscosity, storage and loss modulus of LLDPE/4 OvPOSS/1 PRX, LLDPE/4 PEGDMA/1 PRX and LLDPE/2 OvPOSS/2 PEGDMA/1 PRX composites exhibited higher values as compared to of LLDPE/1 PRX. Moreover, it was obtained that, complex viscosity, storage and loss module values increased with the addition of OvPOSS and it was found a synergistic effect between PEGDMA and OvPOSS.

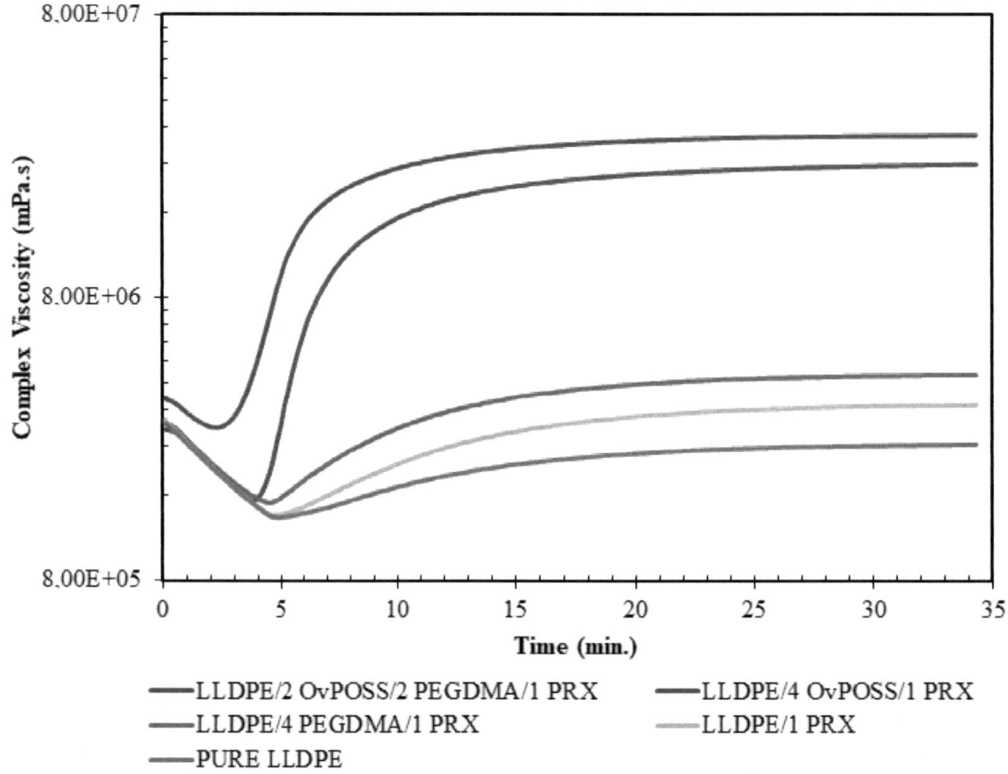

FIGURE 1. Complex viscosity of LLDPE composites

Figure 2 shows the crosslinking density of pure LLDPE and its composites. It was obtained that the crosslinking density of LLDPE increased in the presence of OvPOSS and PEGDMA. The crosslinking efficiency of OvPOSS for LLDPE was found higher than PEGDMA coagent. Moreover, the highest crosslinking density procured in LLDPE/OvPOSS/PEGDMA/PRX composites, which revealed the synergistic effect between OvPOSS and PEGDMA.

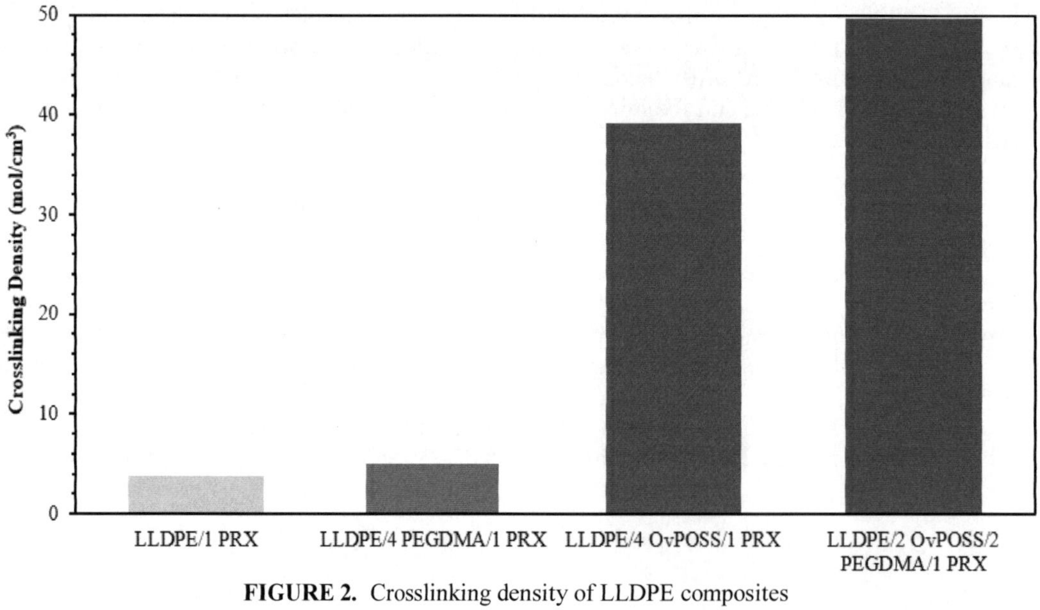

FIGURE 2. Crosslinking density of LLDPE composites

Thermal properties of the uncross-linked LLDPE and crosslinked LLDPE samples were investigated by means of DSC and the results were summarized in Table 1. As can be seen from Table 1 that the degree of crystallinity decreased with the crosslinking. This was due to the fact that the cross-linked polyethylene restricts the chain mobility and cannot easily fold to form crystals. In addition, it was found that the crystal thickness values decreased after crosslinking reactions.

TABLE 1. The changes of crystalline lamellar thickness and percent crystallinity of uncross-linked and cross-linked LLDPE composites

	UNCROSS-LINKED		**CROSS-LINKED**	
SAMPLE NAME	**X_c (%)**	**l_c (nm)**	**X_c (%)**	**l_c (nm)**
PURE LLDPE	17.4	5.80	13.3	5.27
LLDPE/1 PRX	19.6	5.20	11.6	4.60
LLDPE/4 OvPOSS/1 PRX	20.7	5.13	13.8	3.83
LLDPE/4 PEGDMA/1 PRX	31.2	5.32	28.9	4.68
LLDPE/2 OvPOSS/2 PEGDMA/1 PRX	28.7	6.24	26.6	3.64

The morphological analysis of LLDPE/OvPOSS/PEGDMA/PRX composites were carried out by means of scanning electronic microscopy (SEM) on tensile fractured surfaces (Figure 3). SEM analysis were carried out to observe the size and dispersion level of OvPOSS in LLDPE composites. The ductile fracture behavior of composites was obtained by SEM pictures. In a general conclusion, POSS nanoparticles were finely dispersed in LLDPE matrix.

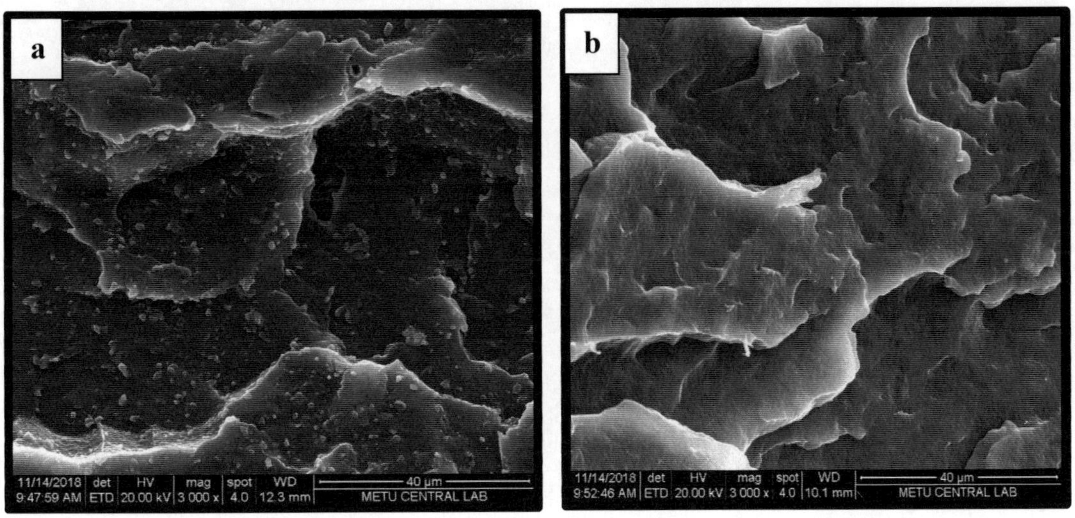

FIGURE 3. SEM micrographs of a) Uncross-linked LLDPE/2 OvPOSS/2 PEGDMA/1 PRX, b) Cross-linked LLDPE/2 OvPOSS/2 PEGDMA/1 PRX composites

According to shape memory tests, the pure LLDPE has 30% shape recovery ratio. When cross-linked with PRX, this ratio has increased by about 48%. The incorporation of both OvPOSS and PEGDMA to the LLDPE resulted in increasing the shape recovery ratio, which was due to the fact that they increased the crosslinking density of LLDPE.

CONCLUSIONS

The aim of this work was to investigate the physical properties and shape memory behaviour of cross-linked LLDPE composites in the presence of Octavinyl POSS (OvPOSS) nanoparticles and poly(ethylene glycol) dimethacrylate (PEGDMA) coagent. The addition of both OvPOSS and PEGDMA coagent with together to the LLDPE provided LLDPE with more crosslinking density. A synergistic effect was observed between OvPOSS and PEGDMA. It was obtained that shape recovery ratios increased as the crosslinking density increased.

ACKNOWLEDGMENTS

This study was financially supported by The Scientific and Technological Research Council of Turkey (TUBITAK) (Project No: 116M981).

REFERENCES

1. A. M., Almasri, *Materials Research* **18(5)**, 1109-1114 (2015).
2. S-W., Kuo, F. C., *Progress in Polymer Science* **36(12)**, 1649-1696 (2011).
3. A., Lendlein, S., Kelch, *Chem. Int. Ed.* **41**, 2034–2057 (2002).
4. C., Liu, H., Qin, P.T., *J. Mater. Chem.* **17**, 1543–1558 (2007).
5. D., Ratna, J., Karger-Kocsis, *J. Mater. Sci.* **43**, 254–269 (2008).
6. J., Miltz, M., Narkis, *Polymer* **9**, 173 (1968).
7. H.G., Dikland, T., Ruardy, A., Bantjes, *Rubber Chemistry and Technology* **66**, 693 (1993).

The synergistic effect between Phenoxy thermoplastic and Core-shell rubber (CSRs) particles on toughness properties of highly cross-linked epoxy resin

[1] *Institute of Condensed Matter and Nanosciences - Bio & Soft Matter (IMCN/BSMA), Université catholique de Louvain, 1 Place Croix du Sud, box: L7.04.02, 1348 Louvain-la-Neuve, Belgium*[1]
[2] *Institute of Mechanics, Materials and Civil Engineering, Materials and process engineering, Université catholique de Louvain, Place de l'Université 1, 1348 Louvain-La-Neuve, Belgium*[2]

Wael Ballout, [1,*] *Pascal Van Velthem,* [1] *Thomas Pardoen,* [2] *and Christian Bailly,* [1]
* *E-mail:* wael.ballout@uclouvain.be

Abstract. Two types of additives, phenoxy thermoplastic (PKHH) and core-shell rubber (CSRs) particles were studied at first separately to improve the toughness of highly cross-linked epoxy resin. The influence of the added amount as well as the mixing time have been investigated. 10 wt% of additive are selected as a compromise in order to obtain a low viscosity blend adapted to resin transfer moulding process, while keeping the mechanical and thermomechanical properties. The combination of both materials revealed a synergistic phenomenon. The results show that the critical energy release rate (G_{IC}) of the hybrid (5 wt% PKHH and 5wt% CSRs) modified epoxy system increased by more than ~210 % and 100% to the unmodified and single (10 wt% PKHH or 10 wt% CSR) modified epoxy system respectively. The resulting morphology of the fracture surfaces shows the presence of propagation cracks deviated/arrested by the CSRs particles surrounding the phenoxy nodules dispersed in epoxy matrix. These toughening mechanisms are underlying of G_{IC} improvement.

Keywords: Epoxy resin, Thermoplastics, Core-Shell rubbers particles, Interlaminar Fracture Toughness, Morphology.
PACS: Scanning electron microscopy, 68.37.Hk, High resolution transmission electron microscopy (HRTEM), 68.37.Og, Mechanical testing, 81.70.Bt, microstructure, 61.72.-y

INTRODUCTION

Highly cross-linked epoxy resins typically used in high performance automotive and aerospace applications lack molecular relaxation mechanisms making them inherently brittle. This lack of ductility and the resulting low fracture toughness limit applications in areas requiring high impact resistance and tensile strength. Therefore, many efforts to toughen epoxies have been reported in the past three decades. The most common strategy however consists in dispersing a second phase in the epoxy matrix, in order to delay crack initiation and to slow down crack propagation through various toughening mechanisms [1-4].The purpose of this study is to examine the effect of the compliant toughening agents on the mechanical, fracture mechanics of epoxy bulk. Preformed nano-sized core-shell rubber particles and phenoxy thermoplastic were used. The study examined the effect of each filler type solely and in combination thereof to uncover synergies.

EXPERIMENTAL

The epoxy resin used in this study was a N, N, N', N'-tetraglycidyl-4, 4'-diaminodiphenylmethane (TGDDM), Araldite MY721 crosslinked with a 4, 4'-diamino-diphenylsulfone (DDS), Aradure 9664-1 both from Huntsman Advanced Materials. Moreover, phenoxy thermoplactic, PKHP-200 (Tg= 92°C; M_w= 52 kg/mol) provided by Gabriel Performance Products was also selected. CSR particles, Kane Ace A MX416 is provided by Kaneka. Kane Ace A is a mixture containing 25 wt.% CSR pre-dispersed in TGDDM epoxy resin. Unmodified TGDDM/DDS blends and modified with various concentrations of phenoxy or CSR (5, 10, 15, 20 and 30 wt%) as well as 10 wt.% phenoxy/CSR (0/10, 5/5, 7.5/2.5, 9/1 and 10/0 wt.%) and 20 wt.% phenoxy/CSR (0/20, 2/18, 5/15, 10/10, 15/5, 18/2 and 20/0 wt.%) were prepared using a melt-mixing method. The Kane Ace A mixture has been diluted by unfilled TGDDM to get the right CSR concentration. In the case of phenoxy, the toughener in powder form was first added with TGDDM at 115°C and mixed using a propeller stirrer until homogenous mixtures were obtained. DDS was then incorporated in the mixture under continuous stirring at the same temperature. The mixtures were degassed in a vacuum oven at 120°C for 20 min. The curing cycle recommended by Huntsman and applied in this work consisted of three isothermal steps at 80°C for 2 hours, 100°C for 1 hour and 150°C for 4 hours. Finally, all cured blends have been post-cured at 200°C for 7 hours.

RESULTS AND DISCUSSION

The critical strain energy release rate (G_{Ic}) was measured through single edge notch bending tests (SENB) according to the ASTM D-5045 standard. Specimens were cut to 88 x 20 x 10 mm, including a sharp notch by machining. Subsequently, a natural crack by inserting a fresh razor blade was obtained by tapping. Samples were loaded in flexion using a Zwick Universal testing machine equipped with a 50 kN load cell, at 1 mm/min crosshead displacement speed. The measurements of load and displacement were recorded. The toughness properties of 5 specimens for each composition were calculated using the corrected modified beam theory. Moreover, the failure surfaces of the specimens were examined after SENB tests by scanning electron microscopy (SEM) using a Jeol 7600F SEM. Sample surfaces were coated with a thin layer (8 nm) of chromium by sputtering in a Cressington 280HR chamber.

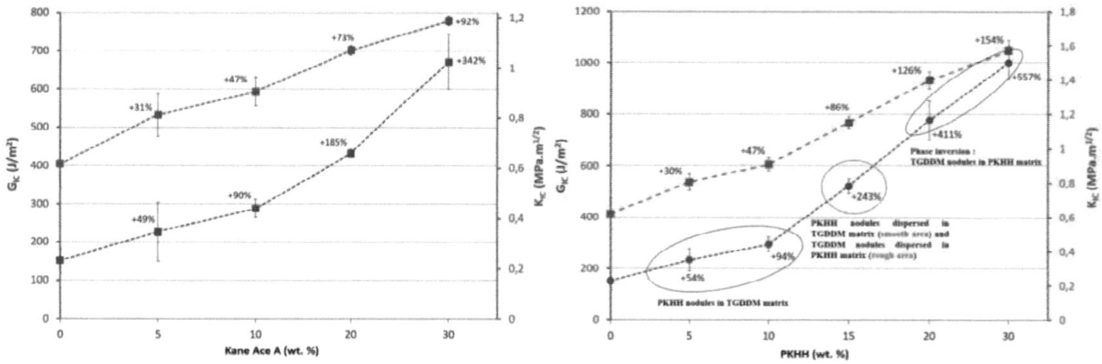

Fig. 1: Critical energy release rate required for mode I crack propagation (G_{IC}) of the Kane ace A and phenoxy modified epoxy resin.

The toughening performance of hybridized core-shell rubber particles and phenoxy thermoplastic on bulk highly cross-linked epoxy resin was investigated. The results show that the critical energy release rate (G_{IC}) of the modified epoxy system increased by more than ~100% (Figure 2). The resulting morphology of the fracture surfaces shows the presence of propagation cracks deviated/arrested by the CSRs particles surrounding the phenoxy nodules dispersed in epoxy matrix. These toughening mechanisms are underlying of G_{IC} improvement.

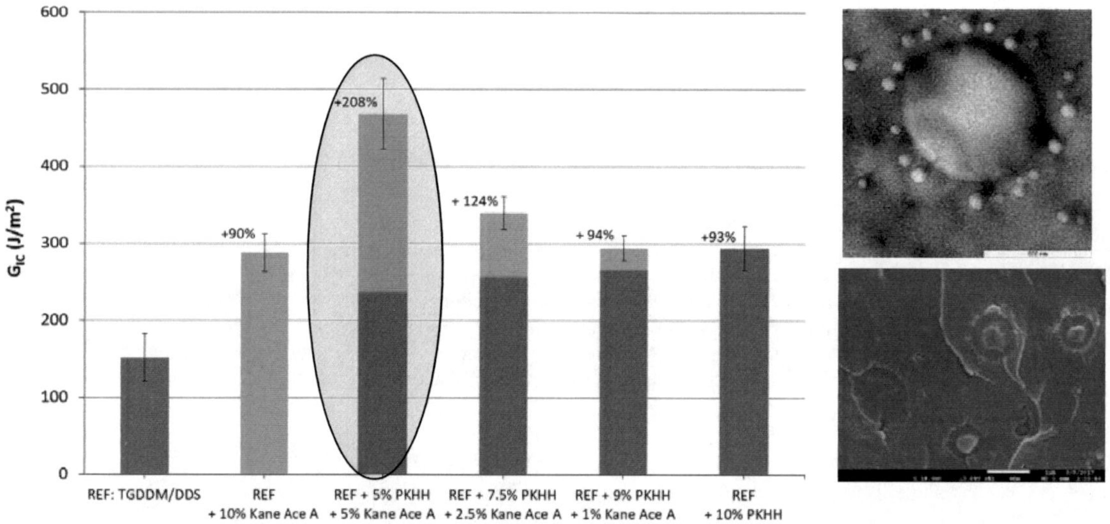

Fig. 2: Critical energy release rate required for mode I crack propagation (G_{IC}) of hybridized Kane ace A and phenoxy thermoplastic on bulk epoxy resin.

ACKNOWLEDGEMENTS:

The authors acknowledge the Wallonia Region (SKYWIN excellence cluster: ECOM project) for financial support. They gratefully thank Pascale Lipnik and Colette Douchamps from technical staff for experimental support.

REFERENCES:

1. Yee AF, Pearson RA. Toughening mechanisms in elastomer-modified epoxies. J Mater Sci 1986; 21(7):2462-74.
2. Pearson RA, Yee AF. Toughening mechanisms in elastomer-modified epoxies: part 3. The effect of cross-link density. J Mater Sci 1989; 24(7):2571-80.
3. Pearson RA, Yee AF. Toughening mechanisms in thermoplastic-modified epoxies: 1. Modification using poly(phenylene oxide). Polymer 1993; 34(17): 3658-70.
4. Huang Y, Hunston D, Kinloch AJ, Riew CK. Mechanisms of toughening thermoset resins. Adv Chem 1933; 233:1-35.

High Density Polyethylene/Uncarbonized and Carbonized Waste Corn Husk Polymer Composites: Physical and Wear Behaviors

Ayberk AKA*, Ahmet Yigit TEKEREK, Gamze GULER, Munir TASDEMIR

Marmara University, Technology Faculty, Metallurgy and Materials Eng. Dep., Istanbul, 34722, Turkey

*akaayberk@gmail.com

Abstract: Renewable resources such as natural fillers in the field of particles reinforced materials with their new range of applications represent an important basis in order to fulfill the ecological objective of creating environmental friendly materials. Nowadays, natural fibers reinforced composites exhibit the superior mechanical properties than synthetic fiber reinforced polymer composites due to its inherent properties. Natural fiber reinforced composites are renewable, biodegradable, environment friendly, light weight material when compared to the synthetic fiber reinforced composites. In this investigation, composites high density polyethylene and uncarbonized and carbonized waste corn husk powder were prepared. The effects of uncarbonized and carbonized waste corn husk powder ratio on the physical and wear properties of the polymer composites are presented. Corn husk powder, in four different concentrations (10, 20, 30 and 40 wt %), were added to HDPE to produce composites as well. The physical properties, including the Vicat softening point, heat deflection temperature, melt flow index, moisture content and limit oxygen index of the composites were investigated. On the other hand, the wear properties of the composites were investigated. The structure of the composites were investigated by scanning electron microscopy and compared to physical and wear properties as a function of uncarbonized and carbonized waste corn husk powder content.

Key words: High density polyethylene, corn husk, physical properties, wear, polymer composites.

1. INTRODUCTION

A large amount of agro waste materials are produced every year globally, and consequently crop residues can serve as a source material for obtaining bio-energy and developing bio-based polymer composites [1]. Plant biomass flour or fibers [2-4] are being increasingly used as organic reinforcing materials in thermoplastic composite materials. Wood–plastic composites (WPCs) are a new class of composites. They are used as engineering materials for housewares, automotive industries, and various construction applications. WPCs combine the favorable performance and low cost attributes of wood and plastics [5]. Various types of natural fibres have been the subject of intense research interest such as flax, sisal, jute, kenaf and hemp [6-12]. This leaves another source of lignocellulosic fibres, such as agricultural byproducts, to be further investigated.
High-density polyethylene is one of the best polymers that are used as plastic composites. The nature of highly crystalline HDPE provides important thermal stability, excellent mechanical properties, and more chemical resistance [13]. HDPE was utilized as matrix material for natural composites, with an area of essential properties [14-16]. Many research works showed that the addition of natural fillers (corn flour, wood flour etc.) into HDPE matrix exhibited a useful approach, which helped to improve the properties of polymers [17].
In this study, HDPE matrix contributes to ground uncarbonized and carbonized corn husk powder is handled as 10, 20, 30 and 40 % ratio will be mixed in the extruder. Mixture obtained from the extruder to be granulated and then the granules will be obtained as appropriate standard test sample of the injection molding machine.

2 EXPERIMENTAL

2.1 Compositions and Materials: Five different polymer blends were prepared. Compositions of HDPE/corn husk polymer composites that were formed are given in Table 1. High-density polyethylene (HDPE) (I-668 UV) was supplied by Petkim (Izmir-Turkey). Specific gravity is 0,970 g/cm^3. Melt flow rate is 5.2 g/10 min (190°C-2.16 kg). Yield strength is 28 MPa and notched Izod impact (23°C) is 12 kJ/m^2. Corn husks were collected from a local farmer's market in Tekirdağ Province of Turkey.

TABLE (1). Composition of the HDPE/CHP Polymer Composites Formulations

Groups	HDPE (wt %)	Corn Husk Powder (wt %)
1	100	-
2	90	10
3	80	20
4	70	30
5	60	40

2.2 Sample Preparation: Corn husk dried in a Yamato vacuum oven ADP-31 (Yamato/VWR Scientific Products, Japan) at 105 °C for 24 hours before being blended with HDPE. Mechanical premixing of solid compositions was done using a LB-5601 liquid-solids blender (The Patterson-Kelley Co., Inc. east Stroudsburg, PA-USA) brand batch blender for 20 min. Samples with various proportions of HDPE/CHP polymer composites were produced between 190-220 °C at 25-35 bar pressure, and a rotation rate of 27 rpm, with a Microsan co-rotating twin-screw extruder (Microsan Instrument Inc. Kocaeli - Turkey). HDPE/CHP polymer composites were also dried in vacuum oven at 105 °C for 24 hours after extrusion. Subsequently, test samples were molded in injection molding machine. Extrusion and injection conditions are given in Table 2.

TABLE (2). Extrusion and Injection Conditions of the HDPE/CHP Polymer Composites

Process	Extrusion	Injection
Temperature (° C)	190–220	190–220
Pressure (bar)	25-35	100–120
Waiting time in mold (s)	-	15
Screw speed (rpm)	27	27
Mould temperature (°C)	-	35-40

The corn husks were washed thoroughly with water to remove the adhered contaminants, and dried in an air oven at 105°C for 24 hours. Waste corn husk dry grinded with Siemens simatic C7-621 control system device to obtain unsegregated powders. The size of corn husks particles varied between 1mm and 38μm. Powder preparations steps are given in Figure 1.

FIGURE 1. Waste Corn Husk Powder Preparation Steps

The powder was then packed in a graphite crucible and fired in electric resistance furnace at temperature of 1200°C (30 minute) to form carbonized corn husk particles. Corn husk carbonization steps and sieve analysis are given in Figure 2.

FIGURE 2. Corn Husk Powder Carbonization Steps and Sieve Analysis

2.3 Test Procedure: Heat deflection temperature (HDT) and Vicat softening point tests were done according to ISO 75 and ISO 307 standard with determined by CEAST 6521 (Ceast SpA, Italy) HDT-Vicat test equipment. Flow behavior testing of all the mixtures was done according to ISO 1133 standard with Zwick 4100 MFI equipment. Moisture testing of all the mixtures was done according to ASTM D 6980 standard with Kern DBS 60-3 equipment. Limit oxygen index (LOI) testing of all the mixtures was done according to ISO 4589 standard with Devotrans LOI equipment. The wear tests were done according to the DIN 53 516 method with Devotrans DA5 (Devotrans quality control test equipment Istanbul-Turkey) abrasion test equipment. The thickness of the test specimens was 7.0 mm and diameter was 15.5 mm. Cylinder rotational speed was selected as 40 rpm and normal load (F_N) of 10N was used. Total sliding distances (L) were 20, 40, 60 and 80 m. The mass loss of the samples (Δm) was measured after the wear process, and the specific wear rates (Ws) were calculated using the following equation:

$$Ws=(\Delta m) / \rho.F_N.L \ (mm^3/Nm) \qquad (1)$$

Where Δm is the specimen's mass loss, ρ is the density of specimen, F_N is the normal load applied, and L is the total sliding distance. The fractured surfaces of the HDPE/CHP polymer composites were coated to thickness of 20 Å of a gold (Au) to prevent electrical charging by Polaron SC7640 (high resolution sputter coater) (United Kingdom). The surfaces of the prepared samples were observed by the FEI Sirion XL30 FEG (Nederland) scanning electron microscopy (SEM) at an acceleration voltage of 5 and 10 kV.

3. RESULT and DISCUSSION

Figure 3-A shows the effect of filler content on the heat deflection temperature (HDT) of HDPE corn husk composites. It can be seen that the HDT of the HDPE corn husk composites increased with increase of filler content. The maximum HDT is observed at the 40 wt % carbonized corn husk powder concentration for HDPE. In comparison with the HDT of HDPE, the HDT increased by 44% for the composites with a 40 wt % carbonized corn husk powder concentration. As shown in Figure 2, the particle sizes of the carbonized groups are smaller than uncarbonized groups. Therefore, having small particles resulted in an increase in HDT. Carbonization process of corn husk increased HDT as well. The relationship between the ratio percentage of the corn husk powder and limit oxygen index (LOI) of HDPE is shown in the Figure 3-B. LOI of composites shows an increment as the carbonized corn husk powder concentration increases from 0 to 40 wt %. On the other hand, LOI of composites shows a decrement as the uncarbonized corn husk powder concentration increases from 0 to 40 wt %. The melt flow index (MFI) of corn husk filled composites was measured, as shown in Figure 3-C. With increased loading, the MFI of composites filled with corn husk is decreased. The minimum MFI is observed at the 40 wt % carbonized corn husk concentration for HDPE. In comparison with the MFI of virgin HDPE, the MFI decreased by 72 % for the blends with a 40 wt % carbonized corn husk concentration. The relationship between the corn husk content and the moisture absorption of the polymer composites is shown in Figure 3-D. The moisture absorption of the composites increased (from 0 to 40wt %) with an increase weight percentage of uncarbonized and carbonized corn husk powder. The maximum moisture absorption is observed at the 40 wt% carbonized corn husk concentration for HDPE. In comparison with the moisture absorption of virgin HDPE, the moisture absorption increased by 53% for the composites at a 40 wt % carbonized corn husk concentration. Figure 3-E illustrates the effect of corn husk on the vicat softening point of HDPE composites. The vicat softening point increased as the carbonized corn husk particle concentration increased from 0 to 40 wt %. In comparison with the vicat softening point of virgin HDPE, the vicat softening point increased by 4,4% for the composites at a 40 wt % carbonized corn husk concentration. The effects of corn husk content on the tribological behaviors of HDPE composites were examined. The wear loss for various specimens sliding distance against the sand paper (#60) under 10N load and 0,32 m/s abrasion speed. The values of wear rate is shown in Figure 4-A and B. The wear rate of HDPE/CHP composites increases as the carbonated and uncarbonized corn husk concentration increases from 0 to 40 wt %.

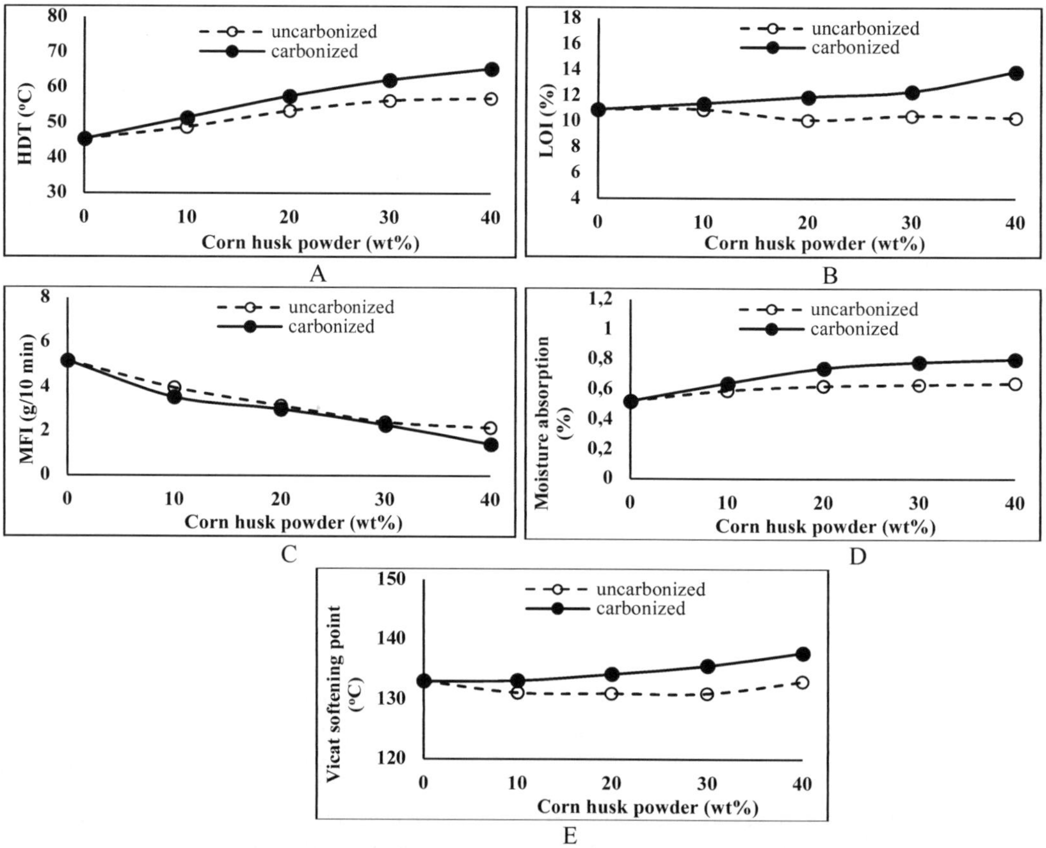

FIGURE 3. Physical properties of HDPE/CHP polymer composites

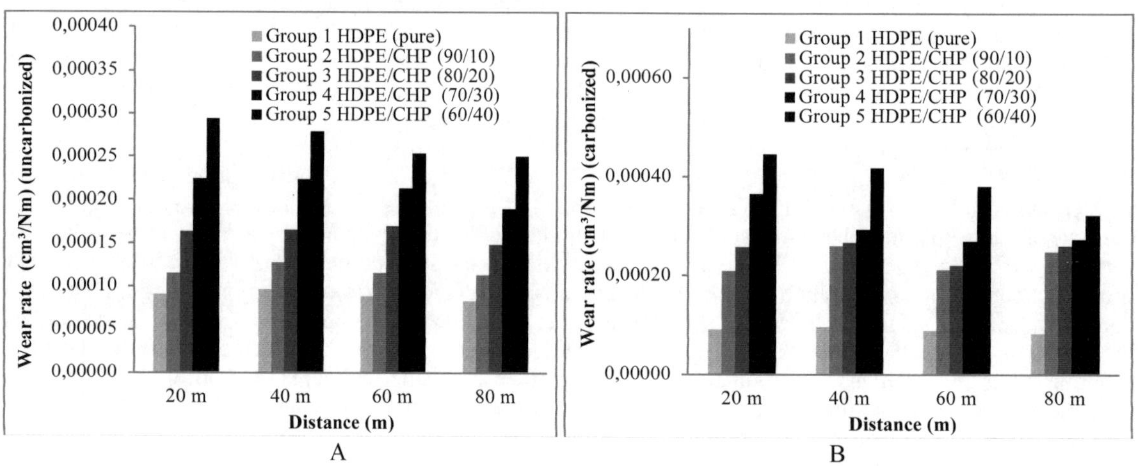

FIGURE 4. Wear properties of the HDPE/CHP polymer composites

The SEM study was carried out to study the dispersion of waste corn husk in the HDPE matrix. The boundaries and the contrast can be obviously seen between the corn husk and HDPE matrix on the fractured surfaces of polymer matrix (Figure 5). The micrographs indicate that the corn husk particulates are homogeneously dispersed on the fractured surfaces of HDPE polymer matrix. The SEM images show that the corn husk particles were poorly dispersed in the HDPE matrix leading to the poor interfacial adhesion between the HDPE matrix and the corn husk particles. The poor filler dispersion or poor interfacial bonds with increase in filler content, leading to poor stress transfer leading to the formation of cracks in the matrix material as well. These observations were also supported by filler agglomeration in the SEM analysis.

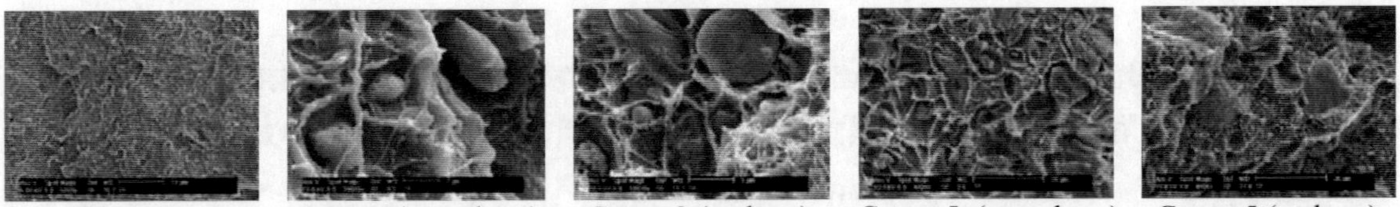

Group 1 Group 2(uncarbon.) Group 2 (carbon.) Group 5 (uncarbon.) Group 5 (carbon.)

FIGURE 5. SEM Micrographs of the HDPE/CHP Polymer Composites

4. CONCLUSIONS

The uncarbonized and carbonized corn husk powder as such was compounded with HDPE in various percentages (10wt%, 20wt%, 30wt% and 40wt %) and then molded and its physical properties such as heat deflection temperature, limit oxygen index, melt flow index, moisture content, vicat softening point and wear were tested. It has marginal effects on some properties of the composites. The following results were obtained: Heat deflection temperature, vicat softening point, limit oxygen index, moisture content and wear rate values of HDPE composites increased as the carbonized and uncarbonized corn husk concentration increases. On the other hand, MFI values decreased as the corn husk powder concentration increases. The micrographs indicate that the corn husk particulates are homogeneously dispersed on the fractured surfaces of HDPE matrix.

REFERENCES

1. R. Lou, S. Wu, G. Lv, Pyrolytic products from rice straw and enzymatic/mild acidolysis lignin (EMAL). BioResources 2010; 5: 2184–2194.
2. H. Kim, Y.G. Eom, Thermogravimetric analysis of rice husk flour for a new raw material of lignocellulosic fiberthermoplastic composites. Mokchae Konghak 2001; 29: 59–67.
3. K. Ross, G. Mazza, Characteristics of lignin from flax shives as affected by extraction conditions. Int J Mol Sci 2010; 11: 4035–4050.
4. F. Mengeloglu, K. Karaku, Thermal degradation, mechanical properties and morphology of wheat straw flour filled recycled thermoplastic composites. Sensors 2008; 8: 500–519.
5.S. Samarasinghe, A.J. Easteal, N.R. Edmonds, Biodegradable plastic composites from corn gluten meal. J Polym Int 2008; 57: 359–364.
6. S. L. Panthapulakkal, M. Sain, *J Appl Polym Sci,* 100 (2006) 3619.
7. H.U. Zaman, M.A. Khan, R.A. Khan, R.M. Arifur, L.R. Das*, Fiber Polym,* 11(2010)455-463.
8. J.Q. Zhang, J.C. Zhang, *Indian J Fibre Text Res,* 35(2010)115.
9. S. Debnath, M. Madhusoothanan, *Indian J Fibre Text Res,* 36(2010)122.
10. S. Nam, A.N. Netravali, *Fiber Polym,* 7(2006)372.
11. Q.T.H, Shubhra, A.K.M.M. Alam, M.A. Gafur, S.M. Shamsuddin, M.A. Khan*, Fiber Polym,* 11(2010)725.
12. N.D. Yilmaz, P.L. Banks, N.B. Powell, S. Michielsen, *J Appl Polym Sci,* 121(2011)2056
13. S.A. Awad, C.M. Fellows, S.S. Mahini, Effects of accelerated weathering on the chemical, mechanical, thermal and morphological properties of an epoxy/multi-walled carbon nanotube composite. Polymer Testing 2018;66:70-77.
14. Q. Zhang, H. Cai, X. Ren, L. Kong, J. Liu, X. Jiang, The dynamic mechanical analysis of highly filled rice husk biochar/high-density polyethylene composites. Polymers, 2017; 9(11):628-637.
15. E. Khalaf, S. Awad, Improvement of mechanical and water absorbance properties of low-density polyethylene (LDPE) by using white kaolin powder (WK). Journal of Advanced Chemical Sciences, 2017; 3(1):426-427.
16. Y. Zou, Y. Feng, L. Wang, X. Liu, Processing and proper. of MWNT/HDPE composites. Carbon, 2004; 42(2):271-277.
17. S.A. Awad, E.M. Khalaf, Improvement of the chemical, thermal, mechanical and morphological properties of polyethylene terephthalate–graphene particle composites. Bulletin of Materials Science, 2018; 41(3):67-72.

A New Anisotropic Fiber Orientation Model for Long Fiber Composites in Injection Molding Simulation

Huan-Chang Tseng[a)]

CoreTech System (Moldex3D) Co., Ltd., ChuPei City, Hsinchu County, 30265, Taiwan.

[a)] Corresponding Author: ivortseng@moldex3d.com

Abstract. For injection molded long fiber composites, the Anisotropic Rotary Diffusion (ARD) model developed by Phelps and Tucker is significant in its ability to predict fiber orientation. The five-parameter ARD tensor was assumed to be a polynomial tensor-valued function depending on both the orientation tensor and the rate-of-deformation tensor. However, most of the fitted parameters are not convenient for wide usage. We, now, propose a new ARD tensor in which the probability ratio of the fibers lie along eigenvector coordinates obtained from the orientation tensor. This tensor is independent of the rate-of-deformation tensor and is indirectly to the orientation tensor. For validation, using the proposed ARD tensor to predict the shell-core structure for long fiber composites in injection molding is in good agreement with the related experimental data.

Keywords: Fiber-reinforced thermoplastic composites; Fiber orientation prediction; Injection molding simulation.

1. Introduction

Previously, the famous Folgar-Tucker theoretical equation of fiber orientation tensor developed in relation to concentrated fiber suspension rheology became a standard model for predicting the flow-induced fiber orientation patterns of the shell-core structure in computer-aided-engineering (CAE) tools of injection molding simulations, especially for short fiber thermoplastic composites. However, such a model results in inaccurate predictions of the thinner or narrow core region due to the faster transient orientation rate [1]. In regard to this issue, the reduced strain closure (RSC) model of Wang *et al.* [2, 3] was primarily considered in the rheological objectivity of material frame indifference to slow down the orientation rate and widen the core thickness. Long fibers are usually longer than 1mm. Based on a previous study by Phan-Thien *et al.* [4] on the Folgar–Tucker constant for fiber suspension in a Newtonian fluid , Phelps and Tucker [5] adopted the two-dimensional diffusion tensor in surface spherical coordinates to derive the anisotropic rotary diffusion (ARD) model. They straightforwardly defined the five-parameter ARD tensor as a second-order polynomial tensor-valued function consisting of the identity tensor, the orientation tensor and the rate-of-deformation tensor. However, determining how to define the ARD tensor is a critical requirement. Following the development of the aforementioned fiber orientation models, Tseng *et al.* [6, 7] proposed an improved ARD (iARD) model along with a new Retarding Principal Rate (RPR) model. This coupled model, denoted as the iARD-RPR model, has been incorporated into another commercial injection molding simulation software, Moldex3D. With this model, the iARD tensor depends on the square of the objective rate-of-deformation tensor. A recent PNNL project [8, 9] (Predictive Engineering Tools for Injection-Molded Long-Carbon-Fiber Thermoplastic Composites), generated numerous experimental orientation data for long carbon fiber composites in injection molding. Thus, Tseng *et al.* [10] used the iARD-RPR model to derive accurate predictions of fiber orientation, which are in good agreement with the PNNL data.

The five-parameter ARD tensor depends on the orientation tensor **A** and the rate-of-deformation tensor, while the two-parameter iARD tensor is related to the tensor **D**. However, the rotary diffusion tensor only depending on tensor **A** has not yet been investigated. Based on the principal vectors or the rotation matrix of the orientation tensor, we aimed to propose a new "principal form of the ARD tensor", namely, pARD, which is independent of the tensor **D**. As with the iARD tensor, the fiber orientation predictions derived by using the pARD model attached with two parameters are compared with related experimental data.

2. Fiber Orientation Models

A single fiber is regarded axisymmetric and rigid. The bond's unit vector **p** along its axis direction can describe fiber orientation. Advani and Tucker [11] defined the second order orientation tensor as:

$$\mathbf{A} = \oint \psi(\mathbf{p})\mathbf{pp}d\mathbf{p} \tag{1}$$

where $\psi(\mathbf{p})$ is the probability density distribution function over orientation space. Various time-evolution equations of the second-rank orientation tensor **A** for determining the dynamic fiber orientation states are fixed on the material derivative, denoted as $\dot{\mathbf{A}}$. For completeness, below we summarize representative theoretical models of fiber orientation developed in the suspension rheological field.

2.1 Folgar-Tucker model

Starting from the Jeffery hydrodynamics (HD) and considering the Folgar-Tucker isotropic rotary diffusion or fiber-fiber interactions, Advani and Tucker replaced orientation vector **p** and the probability density function $\psi(\mathbf{p})$ with orientation tensor **A** to modeled the time-evolution of orientation tensor [5]:

$$\dot{\mathbf{A}} = \dot{\mathbf{A}}^{\mathrm{HD}} + \dot{\mathbf{A}}^{\mathrm{IRD}} \tag{2}$$

$$\dot{\mathbf{A}}^{\mathrm{HD}} = (\mathbf{W}\cdot\mathbf{A} - \mathbf{A}\cdot\mathbf{W}) + \xi(\mathbf{D}\cdot\mathbf{A} + \mathbf{A}\cdot\mathbf{D} - 2\mathbf{A}_4 : \mathbf{D}) \tag{3}$$

$$\dot{\mathbf{A}}^{\mathrm{IRD}} = 2\dot{\gamma}C_I(\mathbf{I} - 3\mathbf{A}) \tag{4}$$

where $\mathbf{L} = \nabla\mathbf{v} = \mathbf{W} + \mathbf{D}$ is the velocity gradient tensor with its component of $L_{ij} = \nabla_j v_i$, wherein v_i is the component of the velocity in the x_i direction; $\mathbf{W} = \frac{1}{2}(\mathbf{L} - \mathbf{L}^{\mathrm{T}})$ is the vorticity tensor; $\mathbf{D} = \frac{1}{2}(\mathbf{L} + \mathbf{L}^{\mathrm{T}})$ is the rate-of-strain tensor and $\dot{\gamma}$ is the scalar magnitude of the rate-of-strain tensor, $\dot{\gamma} = \sqrt{2\mathbf{D}:\mathbf{D}}$; L_f and d_f are fiber length and fiber diameter, respectively; $a_r = L_f / d_f$ is the fiber aspect ratio; the particle shape factor is $\xi = (a_r^2 - 1)/(a_r^2 + 1)$; since $a_r >> 1$ for a rod fiber, ξ is assumed to signify unity.

2.2 ARD and RSC models

Following previous anisotropic rotary diffusion studies of Phan-Thien *et al.* [4], Phelps and Tucker [5] further considered the 2-D surface rotary diffusion tensor $\mathbf{D}_r$ as the projection of a 3-D space matrix **B** onto the local surface coordinates of the unit sphere with **t** is a 3-by-2 matrix, namely, $\mathbf{D}_r = \mathbf{t}^T\mathbf{B}\mathbf{t}$. The space tensor **B** could presumably change with orientation vector **p**, as a function of a symmetric 3-D global spatial tensor **C**. Thus, they derived the ARD model coupled with the Jeffery hydrodynamics model $\dot{\mathbf{A}}^{\mathrm{HD}}$ as

$$\dot{\mathbf{A}} = \dot{\mathbf{A}}^{\mathrm{HD}} + \dot{\mathbf{A}}^{\mathrm{ARD}} \tag{5}$$

$$\dot{\mathbf{A}}^{\mathrm{ARD}} = \dot{\gamma}[2\mathbf{C} - 2\mathrm{tr}(\mathbf{C})\mathbf{A} - 5\mathbf{C}\cdot\mathbf{A} - 5\mathbf{A}\cdot\mathbf{C} + 10\mathbf{A}_4 : \mathbf{C}] \tag{6}$$

For simplicity, they further assumed that the spatial tensor **C** depending on both the orientation tensor **A** and the rate-of-strain tensor **D**, $\mathbf{C} = f(\mathbf{D}, \mathbf{A})$, is a general second-order polynomial tensor-valued function form with fiver parameters (b_1-b_5), as below:

$$\mathbf{C} = b_1\mathbf{I} + b_2\mathbf{A} + b_3\mathbf{A}^2 + \frac{b_4}{\dot{\gamma}}\mathbf{D} + \frac{b_5}{\dot{\gamma}^2}\mathbf{D}^2 \tag{7}$$

On the other hand, Tucker and Huynh [1] have early remarked the limitations of the Folgar–Tucker model with predicting a fast response rate of fiber orientation than experimental observation. Thereby, Wang *et al.* [3] developed the RSC model to slow down a response rate of fiber orientation kinetic:

$$\begin{aligned}\dot{\mathbf{A}}^{\mathrm{RSC}} = {} & \mathbf{W}\cdot\mathbf{A} - \mathbf{A}\cdot\mathbf{W} + \xi\{\mathbf{D}\cdot\mathbf{A} + \mathbf{A}\cdot\mathbf{D} - 2[\mathbf{A}_4 + (1-\kappa)(\mathbf{L}_4 - \mathbf{M}_4 : \mathbf{A}_4)] : \mathbf{D}\} \\ & + 2\kappa\dot{\gamma}C_I(\mathbf{I} - 3\mathbf{A}),\end{aligned} \tag{8}$$

$$\mathbf{L}_4 = \sum_{i=1}^{3} \lambda_i \mathbf{e}_i\mathbf{e}_i\mathbf{e}_i\mathbf{e}_i\,, \tag{9}$$

$$\mathbf{M}_4 = \sum_{i=1}^{3} \mathbf{e}_i\mathbf{e}_i\mathbf{e}_i\mathbf{e}_i\,, \tag{10}$$

where the slow-down factor κ is constrained between 0 and 1. Note that when $\kappa = 1$ the RSC equation can recover to the standard Folgar-Tucker equation. In addition, λ_i and $\mathbf{e}_i$ are eigenvalues and eigenvectors of the orientation tensor **A**, respectively; and both tensors $\mathbf{L}_4$ and $\mathbf{M}_4$ are the fourth-order orientation tensors depending on λ_i and $\mathbf{e}_i$.

For the ARD model, its the RSC form can be derived as

$$\begin{aligned}\dot{\mathbf{A}}^{\text{ARD-RSC}} &= \mathbf{W}\cdot\mathbf{A}-\mathbf{A}\cdot\mathbf{W}+\xi\{\mathbf{D}\cdot\mathbf{A}+\mathbf{A}\cdot\mathbf{D}-2[\mathbf{A}_4+(1-\kappa)(\mathbf{L}_4-\mathbf{M}_4:\mathbf{A}_4)]:\mathbf{D}\}\\ &+\dot{\gamma}\{2[\mathbf{C}-(1-\kappa)\mathbf{M}_4:\mathbf{C}]-2\kappa\text{tr}(\mathbf{C})\mathbf{A}-5(\mathbf{C}\cdot\mathbf{A}+\mathbf{A}\cdot\mathbf{C})\\ &+10[\mathbf{A}_4+(1-\kappa)(\mathbf{L}_4-\mathbf{M}_4:\mathbf{A}_4)]:\mathbf{C}\}\end{aligned} \tag{11}$$

This equation above is also called the ARD-RSC model for long fiber composites.

2.3 iARD and RPR models

Following the previous studies of fiber suspensions [4, 5], Tseng *et al.* [7] propose an new 3-D spatial tensor **C** related to velocity gradient tensor $\mathbf{L}$ with two parameters,

$$\mathbf{C}=C_I(\mathbf{I}-C_M\frac{\mathbf{D}^2}{\left\|\mathbf{D}^2\right\|}) \tag{12}$$

where the tensor contains the Folgar-Tucker fiber-fiber interaction parameter C_I and the fiber-matrix interaction parameter C_M; The scalar $\left\|\mathbf{D}^2\right\|=\sqrt{\frac{1}{2}\mathbf{D}^2:\mathbf{D}^2}$ is the norm of tensor $\mathbf{D}^2$. Such a spatial tensor $\mathbf{C}$ given to $\dot{\mathbf{A}}^{\text{ARD}}$ is called the "iARD" (improved ARD), namely, $\dot{\mathbf{A}}^{\text{iARD}}$. When $C_M=0$ is given, $\dot{\mathbf{A}}^{\text{iARD}}$ becomes the Folgar-Tucker IRD term. However, the tensor $\dot{\mathbf{A}}^{\text{iARD}}$ does not pass the classic rheological objective rule. It is hard to ignore the non-objective effect due to the fact that different coordinate systems can yield different answers.

As the RSC model, Tseng *et al.* [6] also developed the RPR (Retarding Principal Rate) model to slow down a response rate of fiber orientation kinetic:

$$\dot{\mathbf{A}}^{\text{RPR}}=-\mathbf{R}\cdot\dot{\mathbf{\Lambda}}^{\text{IOK}}\cdot\mathbf{R}^{\text{T}} \tag{13}$$

$$\dot{\Lambda}_{ii}^{\text{IOK}}=\alpha\dot{\lambda}_i\,,\quad i,j,k=1,2,3 \tag{14}$$

where $\dot{\mathbf{\Lambda}}^{\text{IOK}}$ is the material derivative of a particular diagonal tensor and its superscript indicates the intrinsic orientation kinetics (IOK) assumption [12]; $\mathbf{R}$ is the rotation matrix and $\mathbf{R}^{\text{T}}$; λ_i is the eigenvalues of $\mathbf{A}$, $\lambda_1\geq\lambda_2\geq\lambda_3$; and the rotation matrix $\mathbf{R}=[\mathbf{e}_1,\mathbf{e}_2,\mathbf{e}_3]$ is defined by eigenvector columns of $\mathbf{A}$. The parameter α is in dimensionless units; and the subscripts *i*, *j* and *k* are indices of permutation.

Tseng *et al.* [7] developed a new fiber orientation model to couple with Jeffery's hydrodynamic (HD) model, namely, the iARD-RPR model (known as Improved Anisotropic Rotary Diffusion model combined with Retarding Principal Rate model),

$$\dot{\mathbf{A}}=\dot{\mathbf{A}}_{\text{HD}}+\dot{\mathbf{A}}_{\text{iARD}}(C_I,C_M)+\dot{\mathbf{A}}_{\text{RPR}}(\alpha) \tag{15}$$

where $\dot{\mathbf{A}}$ represents the material derivative of **A**. C_I and C_M describe the fiber-fiber interaction and fiber-matrix interaction, respectively; α can slow down a response of fiber orientation. Details of the RPR and iARD models are described in reference. According to these criteria, we can fine-tune three parameters to obtain an accurate orientation prediction, namely, determining an optimal parameter set. The shell-orientation layer depends on the iARD parameters (C_I and C_M), while the core region and skin layer are controlled by the RPR parameter (α). For the range of these three parameters, the best practices are: $0 < C_I < 0.1$; $0 < C_M < 1$; $0 < \alpha < 1$. When C_M and α all set to 0, iARD-RPR model goes back to Folgar-Tucker model [7].

2.4 A new pARD model

We proposed a new principal form of the spatial tensor *coaxial* with the orientation tensor to dominate anisotropic changes in the orientation tensor as:

$$\mathbf{C}=C_I\mathbf{R}\cdot\hat{\mathbf{C}}\cdot\mathbf{R}^T=C_I\mathbf{R}\cdot\begin{bmatrix}\hat{C}_1 & 0 & 0\\ 0 & \hat{C}_2 & 0\\ 0 & 0 & \hat{C}_3\end{bmatrix}\cdot\mathbf{R}^T \tag{16}$$

where the principal spatial tensor $\hat{\mathbf{C}}$ is the spatial tensor's principal matrix with three diagonal components limited between 0 and 1, $0\leq\hat{C}_3\leq\hat{C}_2\leq\hat{C}_1\leq1$; the rotation matrix **R** is defined by the orientation tensor $\mathbf{A}$'s eigenvectors ($\mathbf{e}_1$, $\mathbf{e}_2$, $\mathbf{e}_3$).

The ARD equation using such a new form of the spatial tensor **C** above is called the *principal* anisotropic rotary diffusion (*p*ARD) model. Physically, the principal spatial components, $\hat{C}_1$, $\hat{C}_2$, and $\hat{C}_3$, indicate the probability ratio of the fibers lying along the principal vectors of the fiber orientation tensor for a group of fibers. When "$\hat{C}_3 = \hat{C}_2 = \hat{C}_1 = 1$" is given (i.e. $\hat{\mathbf{C}} = \mathbf{I}$), the anisotropic tensor **C** returns to the isotropic. In particular, we simplified the tensor $\hat{\mathbf{C}}$ below through one parameter Ω and the limitation $0 \le \hat{C}_3 \le \hat{C}_2 \le \hat{C}_1 \le 1$

$$\hat{\mathbf{C}} = \begin{bmatrix} 1 & 0 & 0 \\ 0 & \Omega & 0 \\ 0 & 0 & 1-\Omega \end{bmatrix} \tag{17}$$

where $\hat{C}_1 = 1$, $\hat{C}_2 = \Omega$ and $\hat{C}_3 = 1-\Omega$; $0.5 \le \Omega \le 1$.

3. Application to injection molding simulation

In this work, the existed iARD-RPR and the new pARD model have been implemented in Moldex3D software. One case study that is a center-gated disk mold filling is shown in **Figure 1**. The disk is 90 mm in radius with a thickness of $2h$ = 3 mm. The 3-D model was first developed by one of the CAD (Computer Aided Design) programs and then meshed by creating hexagonal elements with 20 layers. The 40wt% LGF/PP composite is used. A volume flow rate is 118.32cm^3/sec with a filling time of 0.65 sec. All of details are available elsewhere [13, 14]. The Folgar-Tucker, iARD-RPR, pARD-RPR models have been integrated into a commercial injection molding simulation software, Moldex3D, to enable fiber orientation predictions. Here, we adopted Moldex3D to perform the 3D injection molding simulation with creating hexagonal (HEXA) elements with 20 layers, as shown in **Figure 2**. All the parameters of the fiber orientation models are addressed in **Table II**.

Figure 3 presents that the orientation distribution exhibits the shell-core structure. As shown in the orientation component A11 distribution, the Folgar-Tucker model was examined to predict the long fiber orientation: obtaining good agreement between the prediction and the PNNL experiment for the shell is significant, while there is a discrepancy in the thinner core, namely, over-prediction. Thus, we demonstrated that the Folgar-Tucker model is unable to determine the broader core for long fiber composites. As a whole, the iARD-RPR and pARD-RPR predictions for the core region and shell layers agree fairly well with the PNNL-ORNL results in the experimental orientation data. In comparison, the pARD-RPR model is *somewhat* superior to the iARD-RPR model. Therefore, we demonstrated that a newly developed model of the pARD tensor is available to predict transient, anisotropic flow-induced orientation.

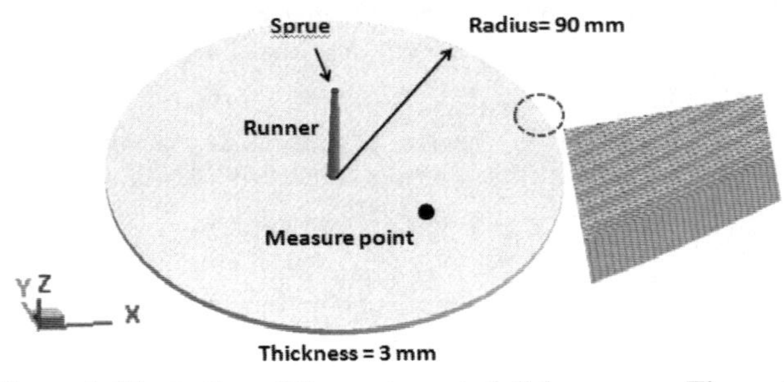

Figure 1. Illustration of the center-gated disk.

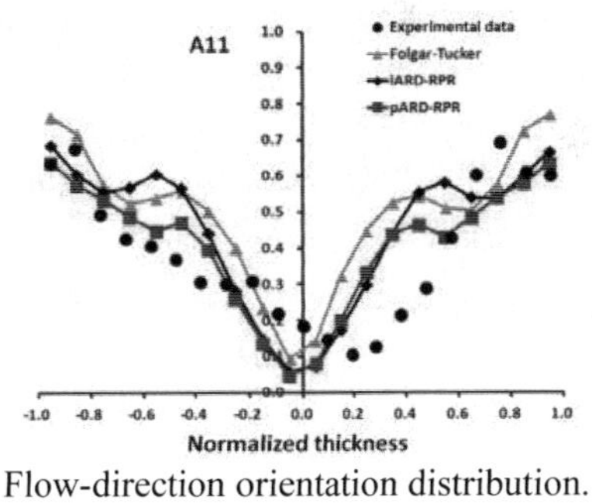

Figure 2. Flow-direction orientation distribution.

Table II. Parameters of the fiber orientation models for the 40wt% LGF/PP composite used in the 3D injection molding simulation of the **PNNL-ORNL center-gated disk**.

Folgar-Tucker		**iARD-RPR**		**pARD-RPR**	
Parameter	Value	Parameter	Value	Parameter	Value
C_I	0.02	C_I	0.01	C_I	0.01
		C_M	1	Ω	1
		α	0.1	α	0.1

4. Conclusion

In this work, for the Phelps-Tucker ARD model, we have proposed that a new principal form of the spatial tensor coaxial with the orientation tensor's principal directions is to control the orientation tensor, and its physical nature indicates the probability ratio of the fibers lying along the rotation matrix of the fiber orientation tensor. Our proposed principal spatial tensor used to the ARD model is useful for predicting fiber orientation in injection molding simulations of long fiber composites. The *p*ARD-RPR model results in the flow-induced fiber orientation patterns of the shell-core structure being in good agreement with the experimental data. In this article, we have presented a reasonably thorough review of various tensor-based orientation prediction methodologies. As a whole, the present two-parameter *p*ARD model is wieldy to derive, and superior to the prior five-parameter ARD model. In particular, we have pointed out that the proposed spatial tensor would change discontinuously for small changes in fiber orientation near the isotropic state. This issue is going to be deep investigated in future work.

References

[1] Tucker III CL, Huynh HM. Fiber orientation in short flow length parts: Limitations of current predictions. In: Proceedings of the 17th Annual Meeting of the Polymer Processing Society. Montreal, Quebec, Canada, 21-24 May, 2001.

[2] Tucker III CL, Wang J, O'Gara JF. Method and article of manufacture for determining a rate of change of orientation of a plurality of fibers disposed in a fluid. U.S. Patent No. 7,266,469 B1; 2007.

[3] Wang J, O'Gara JF, Tucker III CL. An objective model for slow orientation kinetics in concentrated fiber suspensions: Theory and rheological evidence. J Rheol 2008; 52(5) 1179-1200.

[4] Phan-Thien N, Fan X-J, Tanner RI, Zheng R. Folgar–tucker constant for a fibre suspension in a newtonian fluid. J Non-Newtonian Fluid Mech 2002; 103 251-260.

[5] Phelps JH, Tucker III CL. An anisotropic rotary diffusion model for fiber orientation in short- and long-fiber thermoplastics. J Non-Newtonian Fluid Mech 2009; 156(3) 165-176.

[6] Tseng H-C, Chang R-Y, Hsu C-H. Phenomenological improvements to predictive models of fiber orientation in concentrated suspensions. J Rheol 2013; 57(6) 1597-1631.

[7] Tseng H-C, Chang R-Y, Hsu C-H. An objective tensor to predict anisotropic fiber orientation in concentrated suspensions. J Rheol 2016; 60(2) 215-224.

[8] Nguyen BN, Fifield LS, Mathur RN, Kijewski SA, Sangid MD, Wang J, Jin X, Costa F, Gandhi UN, Mori S, Tucker III CL. Predictive engineering tools for injection-molded long-carbon-fiber thermoplastic composites - fy 2014 fourth quarterly report. the US Department of Energy, Pacific Northwest National Laboratory, PNNL Report under Contract DE-AC05-76RL01830 2014; PNNL-23842.

[9] Nguyen BN, Fifield L, Kijewski S, Sangid M, Wang J, Jin X, Costa F, Tucker III CL, Gandhi UN, Mori S. Predictive engineering tools for injection-molded long-carbon-fiber thermoplastic composites - fy 2015 first quarterly report. the US Department of Energy, Pacific Northwest National Laboratory, PNNL Report under Contract DE-AC05-76RL01830 2015; PNNL-24031.

[10] Tseng H-C, Chang R-Y, Hsu C-H. Numerical prediction of fiber orientation and mechanical performance for short/long glass and carbon fiber-reinforced composites. Compos Sci and Technol 2017; 144 51-56.

[11] Advani SG, Tucker III CL. The use of tensors to describe and predict fiber orientation in short fiber composites. J Rheol 1987; 31(8) 751-784.

[12] Tseng H-C, Chang R-Y, Hsu C-H. Method and computer readable media for determining orientation of fibers in a fluid. US Patent No. 8,571,828; 2013.

[13] Nguyen N, Jin X, Wang J, Phelps JH, Tucker III CL, Kunc V, Bapanapalli SK, Smith MT. Implementation of new process models for tailored polymer composite structures into processing software packages. the US Department of Energy, Pacific Northwest National Laboratory, PNNL Report under Contract DE-AC05-76RL01830 2010; PNNL-19185.

[14] Tseng H-C, Chang R-Y, Hsu C-H. An integration of microstructure predictions and structural analysis in long-fiber-reinforced composite with experimental validation. Int Polym Process 2017; 32 455-466.

Thermal conductivity and mechanical properties of synthetic graphite loaded polyphenylene sulfide composites

Gözde Seviğ TANTUĞ[a], Yoldas Seki[*b], Mehmet Sarikanat[c], Lutfiye Altay[c], Sena Gülen[c] and Akin İsbilir[a]

[a] *İzmir Eğitim Sağlık San. Yatırım A.Ş. IMS Polymers, Manisa Turkey*
[*b] *Dokuz Eylul University, Faculty of Science, Department of Chemistry, İzmir Turkey*
[c] *Ege University, Faculty of Engineering, Department of Mechanical Engineering, İzmir, Turkey*

Abstract. It is known that polyphenylene sulfide (PPS), which is a semi-crystalline aromatic polymer, exhibits high thermal stability, outstanding chemical resistance, and excellent flame retardancy properties. However PPS has a poor electrical and thermal conductivity, which limits their uses in some applications where thermal conductivity is required for heat dissipation or storing efficiently. In order to improve thermal conductivity of Polyphenylene sulfide (PPS), PPS- Synthetic graphite (SG) composites were produced by using twin-screw extrusion. SG at different weight fractions (10, 20, 30%) were loaded into PPS and thermal conductivity values of composites were measured in plane and through plane directions. The effect of loading SG into PPS on the mechanical properties of PPS was determined by using three-point bending tests. Charpy tests were conducted to determine the impact properties of SG loaded PPS composites. The effect of SG loading on density of PPS was also determined. Differential scanning calorimetry analyses of SG loaded PPS composites were carried out to determine Cp values. When 50 wt% SG was loaded into PPS, in plane and through plane thermal conductivity of PPS were obtained to be 6.54 and 1.5 W/mK, respectively. Considering ISO 180 standard, charpy-notched impact strengths of PPS-10SG, PPS-30SG, and PPS-50SG are determined to be 9.69, 8.75, and 7.9 kJ/m^2 respectively. From conducted result it was concluded that in spite of loading of SG, the mechanical properties of PPS has not affected greatly.

Keywords: Polyphenylene sulfide (PPS), thermal conductivity, synthetic graphite, mechanical properties

1. INTRODUCTION

Polyphenylene sulphide (PPS) shows outstanding chemical resistance, good stability at elevated temperature, and inherent flame retardancy property[1]. Besides PPS has excellent mechanical properties, high melting point (285–295∘C), and high degradation temperature (>450∘C) [2]. Due to its excellent performance, PPS has been commonly used in many areas, such as in heat sensor filters, high-temperature filters, automotive field, and aerospace industry [3]. Since PPS is rather brittle, with a lower elongation at break, the application of pure PPS is limited. Therefore PPS is modified by different materials [3]. However PPS has a poor electrical and thermal conductivity. This limits the usage of PPS in some applications where thermal conductivity is required for heat dissipation or storing efficiently. Various conductive fillers such as metal powder, carbon black, carbon nanotubes (CNTs), graphene, and graphite are used to improve the thermal conductivity of polymers [4,5]. Amongst these conductive fillers, graphite has attracted much attention due to its unique properties such as low cost, abundant availability, and easy functionalization [6]. In order to improve thermal conductivity of PPS, synthetic graphite (SG) loaded PPS composites were fabricated by using twin-screw extrusion. Thermal conductivity values of SG loaded PPS composites at different weight fractions (10, 20, 30%) were measured in plane and through plane directions. The effect of loading SG into PPS on the flexural and impact properties was obtained.

MATERIAL AND METHODS

PPS (M 2588) and SG (KS 44) were purchased from Toray and IMERYS, respectively. SG loaded PPS composites were produced by using a twin screw extruder (Leistritz Extruder Corporation, Model ZSE 27). SG loaded PPS composites were injection molded into test specimens by using injection molding machine (Bole model BL90EK). Thermal diffusivity and thermal conductivity values of SG loaded PPS composites were conducted by using Flash Diffusivity Analyzers (Discovery Xenon Flash DXF 200) with respect to ASTM E 1461 standard. Three-point bending tests were made to determine flexural strength and flexural modulus values of SG loaded PPS composites according to ASTM D790 standard. Charpy and Izod impact strength values of SG loaded PPS composites were determined by using a pendulum-type tester at 23 °C with respect to ISO 179 and 180 standards. Density values of SG loaded PPS composites were determined according to ASTM D792 standard.The obtained density values for PPS-10SG, PPS-30SG, and PPS-50SG are 1.40, 1.51, and 1.64 g/cm^3, respectively. Cp values of SG loaded PPS composites were obtained by using DSC analyzer (TA Instruments Inc., DSCQ20)

RESULTS AND DISCUSSION

The variation of Flexural strength and flexural modulus with the loading of SG is presented in Figure 1. With increasing the weight fraction of SG within PPS, flexural strength values decreased. However SG loading led to increases in flexural modulus values. It is known that the increase in flexural modulus of composites is a function of filler volume fraction and stiffness [7]. Since SG is stiffer material, flexural modulus values of SG loaded PPS composites increases with increasing the weight fraction of SG.

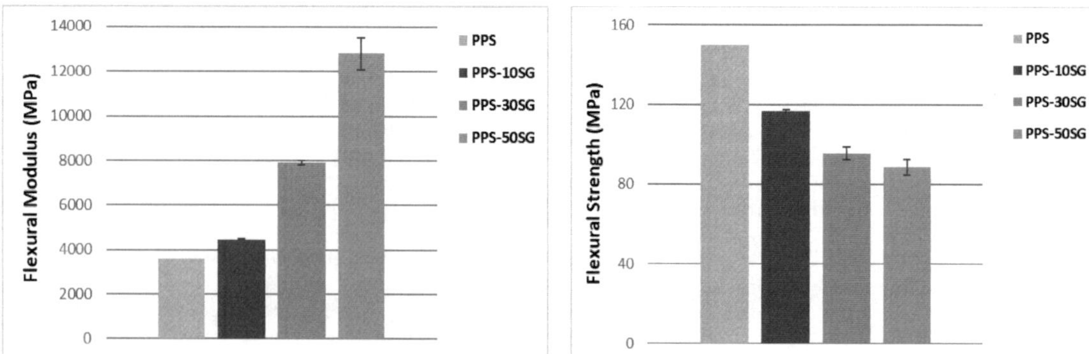

FIGURE 1. Flexural properties of SG loaded PPS composites

Charpy notched impact strength (CNIS) and charpy un-notched impact strength (CUIS) values are given in Figure 2. CNIS and CUIS values of PPS-10SG, PPS-20SG, PPS-30SG were obtained to be 9.7 and 39.6, 8.8 and 18.8, and 7.9 and 14.8 kJ/m^2, respectively. Similar phenomenon in izod impact, decrease in un-notched impact strength is more pronounced, was observed in charpy impact. The decreases in impact strength of the composites with increasing the weight fraction of SG are owing to poor compatibility between SG and PPS. Besides, loading of SG into PPS may lead to the formation of agglomerates of SG in the PPS matrix, which act as internal notches [8].

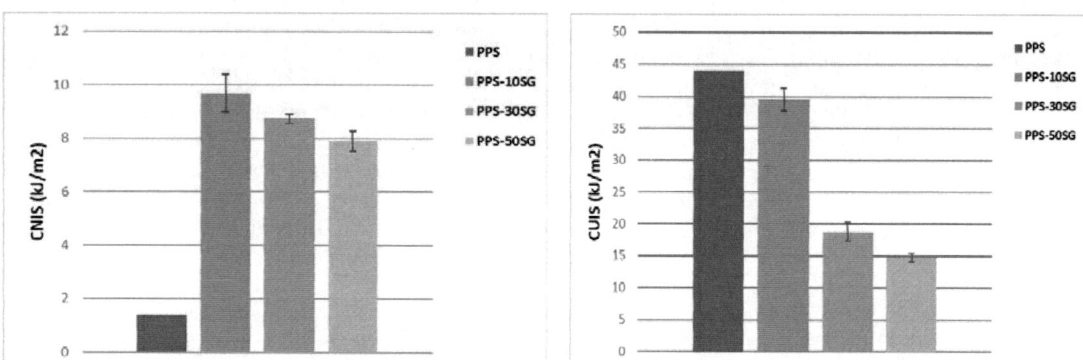

FIGURE 2. Charpy impact strength values of SG loaded PPS composites

The variation of thermal diffusivity values (In-plane and through-plane) are presented in Figure 4. Thermal diffusivity values (in-plane direction) of PPS-10SG, PPS-30SG, and PPS-50SG were determined to be 0.018, 0.022, and 0.035 cm^2/s respectively. Thermal diffusivity values (through-plane direction) were obtained as

0.003, 0.005, and 0.009 cm^2/s, respectively. An increasing trend both thermal diffusivity values in-plane and through-plane directions is noticeable.

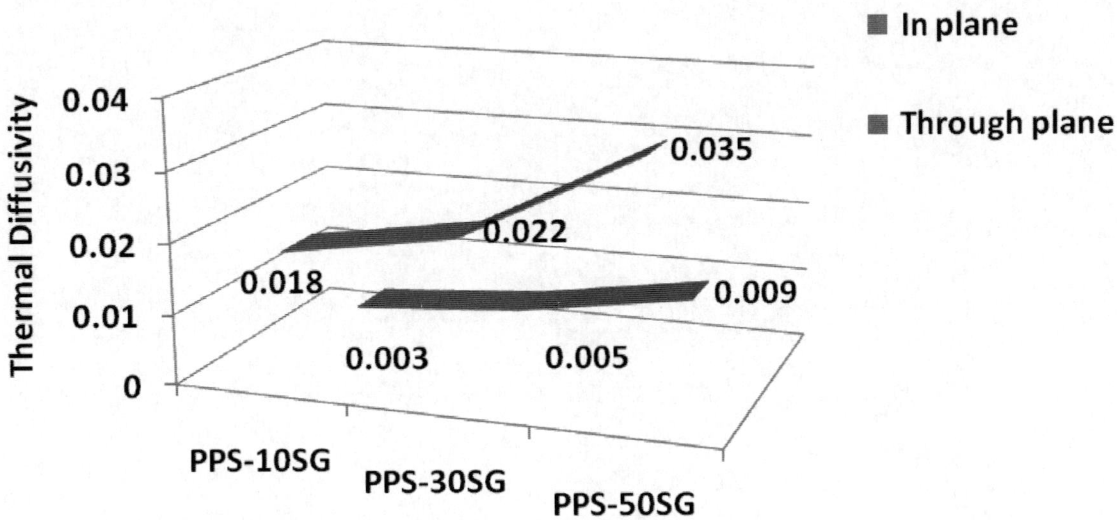

FIGURE 3. Thermal Diffusivity values (cm^2/s) of SG loaded PPS composites

In-plane thermal conductivity values of PPS-10SG, PPS-30SG, and PPS-50SG were obtained to be 3.89, 4.74 and 6.54 W/mK, respectively, which indicates about 19-, 23-, and 32-fold larger than that of PPS. Through-plane thermal conductivity values of PPS-10SG, PPS-30SG, and PPS-50SG are 0.62, 1.02, and 1,65 W/mK, respectively. This corresponds to 3-, 5-, and 8-fold increases compared to PPS. It is known that graphite merits special interest because of its high thermal conductivity (25–470 W m^{-1} K^{-1}) [9]. Since SG has a layered structure, heat pathways forms within the matrix and thus thermal conductivity increases [10].

SEM images of PPS and SG loaded PPS composites are given in Figure 4a-d. SG particles can be clearly seen in the micrographs of PPS-30SG and PPS-50SG. The micrograph of PPS-50SG shows that the graphite particles remained in a stacked form. As shown in Figure 4d, a closer contact between the graphite sheets was observed. Formation of a tightly conducted thermal paths explained higher thermal conductivity of PPS-50SG.

CONCLUSION

SG was loaded into PPS at different weight fractions to improve the thermal conductivity of PPS. When 50 wt% SG was loaded into PPS, in-plane and through- plane thermal conductivity of PPS increased by 32- and 8-fold, respectively. Additionally, the increases in through-plane conductivity are less than those in-plane conductivity. However SG loading into PPS has led to drops in mechanical properties of PPS composites. Besides, CNIS and CUIS values decreased with increasing the weight fraction SG.

ACKNOWLEDGMENTS

The authors would like to thank TUBİTAK "The Scientific and Technical Research Council of Turkey" for financial support under Project Number 117M088.

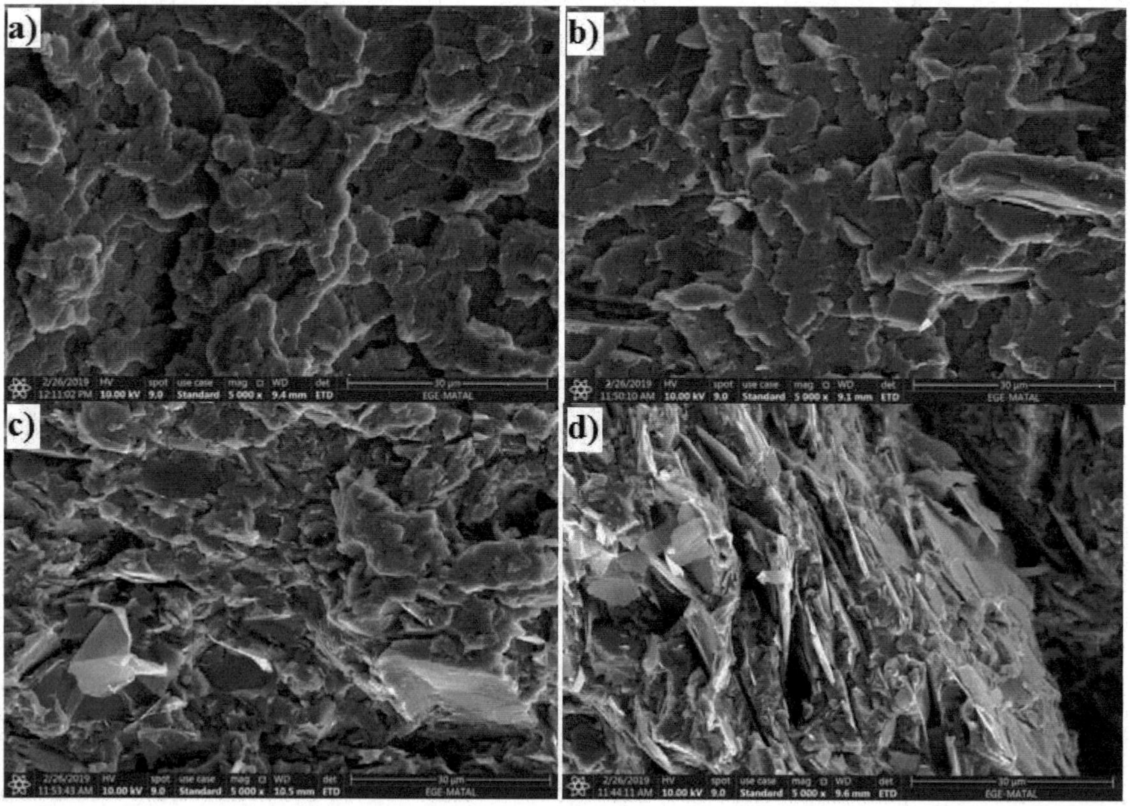

FIGURE 4. SEM images of PPS and SG loaded PPS composites. a) PPS, b) PPS-10SG, c) PPS-30SG, d) PPS-50SG.

REFERENCES

1. Z. Jiang, P. Hornsby, R. McCool, R., A. Murphy), *J. App. Poly. Sci* **123**, 2676–2683 (2012).
2. R.K. Goyal, K.R. Kambalea, S.S. Nenea, B.S. Selukar, S. Arbuj, U.P. Mulik, *Materials Chemistry and Physics* **128** 114–120 (2011).
3. J. Xing, B. Deng, Q. Liu, *High Performance Polymers*, **30**, 519-526 (2018).
4. J.R. Potts, D.R. Dreyer, C.W. Bielawski, R.S. Ruoff R.S., *Polymer* **52**, 5-25 (2011).
5. G. Chen, W. Weng, D. Wu, C. Wu., *European Polymer Journal* **39**, 2329-2335 (2003).
6. I. Krupa and I. Chodák, *European Polymer Journal* **37**, 2159-2168 (2001).
7. A.A: Ibrahim, *Ibn AL-Haitham Journal For Pure and Applied Science,* **24**, (2011)
8. S. Zhou, Q. Zhang, C. Wua, J. Huang, *Materials and Design* **44**, 493– 499 (2013)
9. C.M. Ye, B.Q. Shentu, Z.X. Weng ZX., Journal of Applied Polymer Science **101**, 3806-3810 (2006).
10. Y.X. Fu, Z.X. He, D.C. Mo, S.S. Lu, *Applied Thermal Engineering*. **66**, 493-498 (2014).

Calcium Filled Bacterial Cellulose Based Composite Hydrogel for Bone tissue engineering: Optical Microscopy and Bone Marker Analysis

Probal Basu [a] (Presenting Author), Nabanita Saha* [a] (Reference author), Radostina Alexandrova[b] and Petr Saha [a]

[a] *Centre of Polymer Systems, University Institute, Tomas Bata University in Zlin, Trida Tomase Bati 5678, 760 01 Zlín, Czech Republic*

[b] *Institute of Experimental Morphology, Pathology and Anthropology with Museum, Bulgarian Academy of Sciences, Sofia, Bulgaria*

***Corresponding/reference author**: nabanita@utb.cz*

Abstract

The primary focus of the work is the biological characterization (i.e. cell growth study through optical microscopy and bone marker analysis/alkaline phosphate expression study) of the novel inorganic calcium filled [in the form of calcium phosphate or/and calcium carbonate ($CaCO_3$)] bacterial cellulose (BC) based hydrogel composite scaffolds in the context of their utilization in bone tissue engineering application. The BC based scaffolds also contain the synthetic polymers, polyvinylpyrrolidone (PVP) and carboxymethyl cellulose (CMC). Calcium phosphate (in the form of hydroxyapatite [HA] and β–tri-calcium phosphate [β-TCP]) was incorporated in the BC based scaffolds (i.e. BC-PVP and BC-CMC) to prepare inorganic calcium phosphate filled BC based scaffolds (i.e. BC-PVP-β-TCP/HA and BC-CMC-β-TCP/HA) and then these were in *vitro* biomineralized ($CaCO_3$) to obtain biomineralized inorganic calcium (Calcium phosphate + $CaCO_3$) filled BC based hydrogel scaffolds (i.e. BC-PVP- β-TCP /HA-$CaCO_3$ and BC-CMC- β-TCP /HA-$CaCO_3$). The cell viability was found significant for BC-PVP-β-TCP/HA scaffold with both cell lines; human fibroblast cells, Lep-3 and mouse bone explant cells (BEC) after 5 days of incubation. The ALP expression in cell culture of BC-PVP-β-TCP/HA is also notable. In addition, the ALP expression of BEC for BC-CMC- β-TCP /HA-$CaCO_3$ hydrogel scaffold is significant. The cell growth study through optical microscopy and bone marker study (i.e. alkaline phosphatase expression) demonstrates the cell biological efficiency of inorganic calcium filled BC based scaffold; especially BC-PVP-β-TCP/HA composite scaffold, which indicates its possibility for the soft bone tissue engineering application.

Keywords: Bacterial cellulose, polymeric hydrogel composite, calcium phosphate, calcium carbonate, bone tissue engineering

INTRODUCTION

Structural framework of human body is majorly maintained by Bone [1]. Research showed that bone related diseases like osteoporosis causes more than 8.9 million fractures worldwide annually [2]. However, the polymeric hydrogel based bioactive composite scaffold could become a notable approach in bone tissue engineering due to its osteoconduction and osteoinduction property, notable mechanical property and cost-effective production attributes [3]. The polymeric hydrogel based composite scaffolds has the necessary attributes to become an efficient extra cellular matrix (ECM) [4, 5].

Study indicated that different polymer and polymer composite and bacterial cellulose (BC) based hydrogel scaffold can be a potential biomaterial for tissue regeneration [6, 7]. BC is biocompatible biopolymer [8]. It has high crystallinity, ultra-fine network structure and high water absorption capability. On the other hand, bioactive inorganic fillers like calcium phosphate e.g., β-tri-calcium phosphate (β-TCP), and hydroxyapatite (HA) could enhance the osteoconduction and osteoinduction property of the biomaterial [9]. The hydrogel scaffolds prepared from biopolymer, BC and synthetic polymers polyvinylpyrrolidone (PVP), carboxymethyl cellulose (CMC) were also reported in our earlier work [10]. However, in order to develop a more specific bone tissue engineering scaffold, inorganic calcium (in the form of calcium phosphate or/and $CaCO_3$) filled BC based composite hydrogel scaffolds are developed.

In the present work, the focus has been given on the biological characterization (i.e. cell growth study after 5 days of incubation with sample through optical microscopy and bone marker analysis) of the novel inorganic calcium phosphate or/and calcium carbonate ($CaCO_3$) filled bacterial cellulose (BC) based hydrogel composite scaffolds in the context of their application in bone tissue engineering application.

EXPERIMENTATION

Preparation of Inorganic Calcium phosphate and Biomineralized Calcium carbonate ($CaCO_3$) filled BC based Hydrogel Scaffold

BC (holding 99% H_2O) was synthesized in presence of basal synthetic Hestrin-Schramm (HS) nutritive medium (pH 7.0) using *Gluconacetobacter xylinus CCM 3611^T* (syn. *Acetobacter xylinum*) at 30°C for 15 days. 100 mL bacteriological culture bottles were inoculated with 5 mL of H.S. medium containing 96 × 10^8 cells/mL bacteria [bacteria counted at 550 nm wavelength with Grant-Bio McFarland Densitometer DEN-1B, Grant Instruments Ltd, UK]. The freshly prepared BC pellicle was then treated with 0.5 N NaOH solution and then heated at 80°C for 1 hours to remove the possible contaminations from the BC pellicle. Thereafter, a homogenous suspension of BC (particle size: 351.03 nm) from the obtained BC mat was prepared by grinding (10-12 minutes) the BC mat in distilled water. BC based hydrogel scaffolds were first developed by applying the BC (holding 99% water) with CMC and PVP. Polyethylene glycol (PEG) was also used in all the solutions to reduce the risk of tissue damage and other significant cytotoxic effects. Agar used as gelling agent and glycerin used as humectant. Additionally, β-TCP and HA were applied in the ratio of 20:80 [7] in the hydrogel scaffolds to produce the inorganic calcium phosphate filled BC based composite hydrogel scaffolds and henceforth will be termed as, "BC-PVP-β-TCP/HA" and "BC-CMC-β-TCP/HA". BC-PVP and BC-CMC hydrogel scaffolds were used as base scaffolds. All materials were used are reagent grade and majorly purchased from Sigma Aldrich.

The BC based hydrogel composite were prepared following the solvent casting method, applying moist heat and pressure. 100 mL polymer solutions were prepared in 250 mL sealed glass bottles under 15 lbs (107 KPa) pressure and 120°C temperature for 20 minutes [6]. Two sets of polymer solutions were prepared; where one set was without inorganic calcium phosphate, and another set was with inorganic calcium phosphate (β-TCP/HA). 25 mL polymer solution from each sealed glass bottles was poured into 75 mm diameter petri-dishes and allowed to cool at room temperature (22-25°C). Finally, smooth, round shaped, off white color BC based scaffold without inorganic calcium phosphate (henceforth termed as "BC-PVP", "BC-CMC"; Diameter: 75 mm; Thickness: 6.0-6.2 mm) and with inorganic calcium phosphate filled (henceforth termed as "BC-PVP-β-TCP/HA" and "BC-CMC-β-TCP/HA"; Diameter: 75 mm; Thickness: 5.5-6 mm) hydrogel scaffolds were attained.

The calcium phosphate filled BC based hydrogel scaffolds were further in *vitro* bio-mineralized for further incorporation of additional $CaCO_3$. The in *vitro* bio-mineralization process was performed following the simple liquid diffusion technique. Two different ionic solutions (i.e., Na_2CO_3 and $CaCl_2$) were used at fixed concentration ratios (5.25/100 mL and 7.35/100 mL) to prepare inorganic calcium phosphate and $CaCO_3$ filled BC based hydrogel composite scaffolds. The bio-mineralization of hydrogel matrix was carried out for 60 min by keeping the test samples in each ionic solution. The calcium phosphate filled BC based hydrogel scaffolds were first immersed in 100 mL solution of $CaCl_2$. H_2O for 30 min and then transferred into 100 mL of Na_2CO_3 solution and kept them for 30 min [11]. In this procedure, calcium phosphate and $CaCO_3$ filled BC based in *vitro* bio-mineralized hydrogel scaffold were achieved (Diameter: 75 mm; Thickness: 4.5-4.9 mm), which are finally termed as "BC-PVP-β-TCP/HA-$CaCO_3$" and "BC-CMC- β-TCP/HA-$CaCO_3$" respectively.

Cell Culture

Human diploid fibroblast; Lep-3 cells from lung of 3-month old human embryo and cell culture established from bone explants of ICR mice (Bone explant cells, BEC) [12] were used as model systems in our study. The cell cultures were grown in Dulbecco's Modified Eagle's medium (DMEM) supplemented with 10% fetal bovine serum (FBS), 100 U/mL penicillin and 100 µg/mL streptomycin. The cultures were kept in a humidified incubator (Thermo Scientific, HEPA Class 100) at 37°C under 5% CO_2 in air. For routine passages, the cells were detached using a mixture of 0.05% trypsin and 0.02 % Ethylenediaminetetraacetic acid (EDTA). The cell cultures were passaged 2-3 times per week (1:2 to 1:3 split) [12].

Cell Growth Study through Optical Microscopy

The cells (5 x 10^4 cells/well) were seeded directly on the material sample placed on the bottom of a 48-well cell culture plate and incubated for 5 days (120 h) in CO_2 incubator at 37°C. Cells were grown in wells without materials, served as controls. Cell morphology during the period was recorded by Inverted Light Microscope (Olympus, CK40-SLP,

Japan; Camera: CAM2800/XP2.0). 3-(4, 5-dimethylthiazol-2-yl)-2, 5-diphenyl tetrazolium bromide (MTT) assay was also performed later [12].

Alkaline phosphatase assay

The activity of alkaline phosphate was assessed in the modified culture medium with the BEC cell supernatant. The cells were seeded in 6-well flat-bottomed microplates at a concentration of 1 x 10^5 cells/well. After 24 hour, the culture medium was removed and changed with 3 days modified media of the compounds tested. After 6 days of incubation, the cells were first re-suspended in 100 – 200μL of assay buffer. Then they are centrifuged at 4°C at high speed for 15 minutes. Then the culture supernatants were removed. Thereafter ALP activity in the cell supernatant is assessed by manufacturer's protocol at 405 nm absorbance. Alkaline phosphatase activity was evaluated by standard colorimetric kit as recommended by the manufacturer protocol.

Statistical Analysis

The data are presented as mean ± standard error of the mean.

RESULT AND DISCUSSION

The novel inorganic calcium filled composite scaffolds "BC-PVP-β-TCP/HA", "BC-CMC-β-TCP/HA", "BC-PVP-β-TCP/HA-$CaCO_3$", and "BC-CMC-β-TCP/HA-$CaCO_3$" were developed. The "BC-PVP" and "BC-CMC" scaffolds were considered as control set of scaffolds. The cell growth/viability after 5 days of incubation by optical microscopy and bone marker analysis (i.e. alkaline phosphate expression study) were performed and discussed below.

The cell viability of two cell lines Lep-3 and BEC after 5 days of incubation with the hydrogel samples is shown in Fig. 1A. For this study, human fibroblast Lep-3 and mouse bone explant cell (BEC) was used. The Lep-3 cell viability in the direct contact with BC-PVP and BC-CMC scaffolds was found significant after 5 days of incubation. BEC viability was found significant with BC-PVP, BC-CMC, BC-PVP-β-TCP/HA after 5 days of incubation. The BEC and Lep-3 initially acclimatizes with the chemical and physical environment that provided by the scaffold sections and they show cell growth and viability after 5 days of incubation. (Fig. 1A). Interestingly, the growth of both the cells for BC-CMC-β-TCP/HA-$CaCO_3$ hydrogel composite scaffold was found less compared to BC-PVP-β-TCP/HA-$CaCO_3$. This might resulted from the combinatorial effect of polymers like BC, CMC and inorganic calcium (calcium phosphate and calcium carbonate).

The alkaline phosphate (ALP) expression by the BEC, in presence of the extract of six BC based hydrogel samples are depicted in Fig. 1B. The BEC cell supernatant were used to analyze the expression of alkaline phosphate. It can be seen from the Fig. 1B that, the ALP expression of BEC in cell culture containing the extract of BC-PVP-β-TCP/HA is notable among the all inorganic calcium phosphate filled hydrogel scaffolds. In addition, the ALP expression of BEC in the cell culture containing BC-CMC-β-TCP/HA-$CaCO_3$ hydrogel scaffold extract is also significant compared to the other inorganic calcium filled hydrogel scaffolds. Alkaline phosphatase expression has considered a significant marker for bone cell growth and development. The notable expression of ALP for BC-PVP-β-TCP/HA indicates its suitability for the bone cell growth.

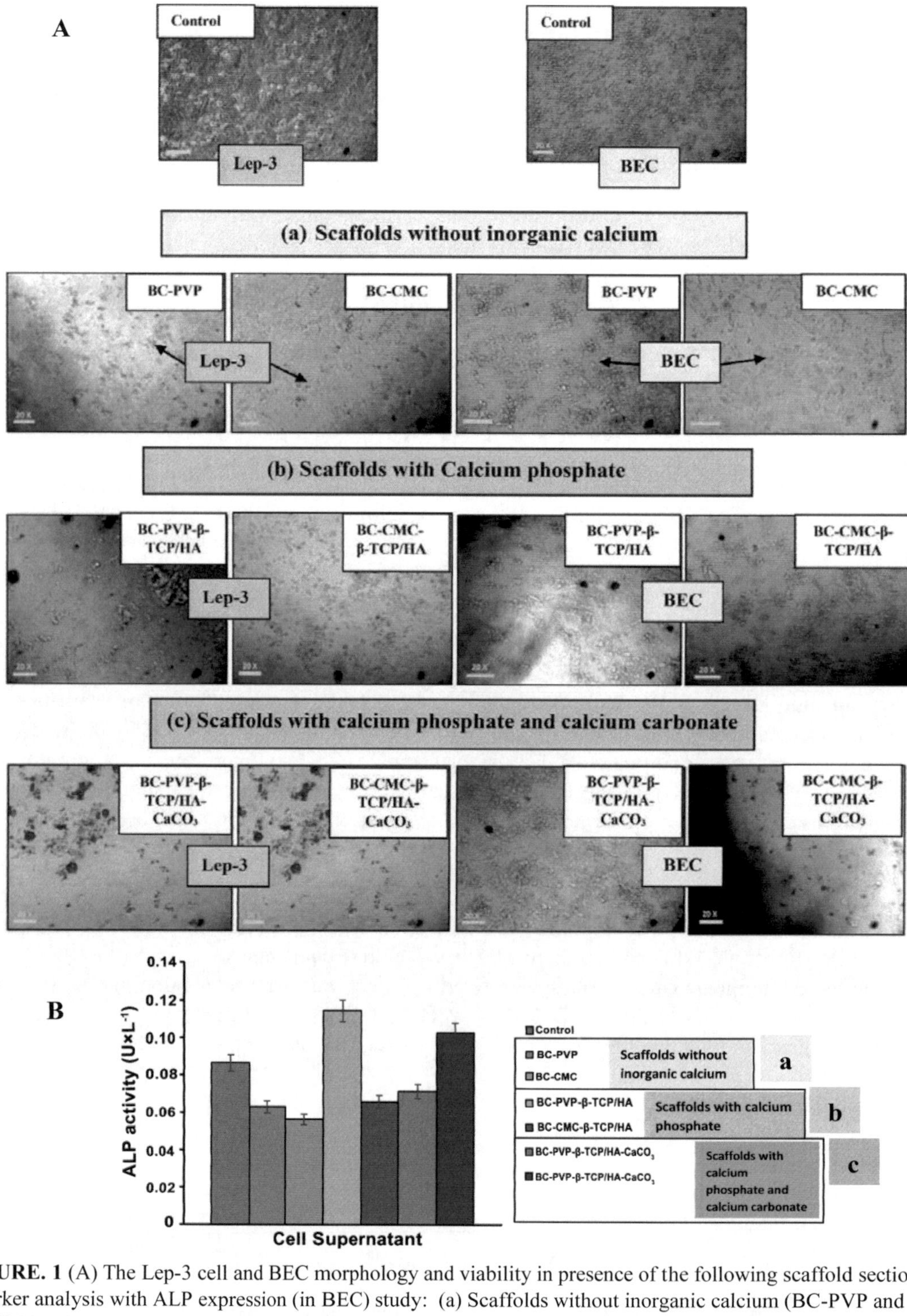

FIGURE. 1 (A) The Lep-3 cell and BEC morphology and viability in presence of the following scaffold sections: and B) Bone marker analysis with ALP expression (in BEC) study: (a) Scaffolds without inorganic calcium (BC-PVP and BC-CMC), (b) Scaffolds with calcium phosphate (BC-PVP-β-TCP/HA and BC-CMC- β-TCP/HA), (c) Scaffolds with calcium phosphate and $CaCO_3$ (BC-PVP-β-TCP/HA- $CaCO_3$ and BC-CMC-β-TCP/HA- $CaCO_3$).

CONCLUSION

The cell growth/viability study through optical microscopy and bone marker study (i.e. alkaline phosphatase expression) demonstrates the cell biological efficiency of the polymeric composite scaffolds. BEC viability was found significant with BC-PVP-β-TCP/HA after 5 days of incubation. The notable expression of ALP in BEC for BC-PVP-β-TCP/HA indicates its suitability for the bone cell growth. Thus, the calcium filled BC based hydrogel composite scaffold especially, BC-PVP-β-TCP/HA has the necessary attributes for the soft bone tissue engineering.

ACKNOWLEDGEMENTS

This work is supported by the Internal Grant Agency (Project No. **IGA/CPS/2019/003 and IGA/CPS/2017/003**), Tomas Bata University in Zlin, Czech Republic and Ministry of Education, Youth and Sports of The Czech Republic – **NPU Program I (LO1504)**. Moreover, this work was performed within the framework of COST Action MP1301"New Generation Biomimetic and Customized Implants for Bone Engineering" (www.cost.eu).

REFERENCES

1. P. R. Buenzli, and N. A. Sims, *Bone* **75**,144-150 (2015)
2. International Osteoporosis Foundation (IOF), 2017, accessed on 30th August, 2017, Available from: https://www.iofbonehealth.org/facts-statistics
3. K. Deligkaris, T.S. Tadele., W. Olthuis, *Sens Actuators B* **147**, 765–774 (2010).
4. N. Roy, N. Saha, T. Kitano, P. Saha., *Carbohydr Polym* **89**, 346–353 (2012).
5. K. J. L. Burg., S. Porter, J.F. Kellam, *Biomaterials* **21**, 2347-2359 (2000).
6. S. Stratton, N.B. Shelke, *Bioactive Materials* **1**, 93-108 (2016).
7. P. Basu,N. Saha,S. Bandyopadhyay,P. Saha., *AIP Conf. Proc* **1843**, 050008 (2017).
8. S. Keshk, *J Bioprocess Biotech* **4**, 1000150 (2014).
9. T. Muthukumar, A. Aravinthan, J. Sharmila, *Carbohydr.Polym* **152**, 566-574 (2016).
10. R. Shah.et al, *AIP Conf. Proc* **1662** (040007), 1-7 (2015).
11. P. Basu,N. Saha, P. Saha., *Int J Polym Mater* 1-11. (2018). DOI:10.1080/00914037.2018.1525733
12. P. Basu, N. Saha, R. Alexandrova, B. Andonova-Lilova, M. Georgieva, G. Miloshev, P. Saha, *Int. J. Mol. Sci* ***19***, 3980 (2018).

Influence of Screw Configuration in Twin Screw Extruder on Dynamic Viscoelastic Properties of Hemp Fiber Reinforced Plant-Derived PA1010 Biomass Composites in Molten State

Yosuke Nishitani[1, a)] and Maiko Morino[1)]

[1]*Department of Mechanical Engineering, Faculty of Engineering, Kogakuin University 2665-1 Nakano, Hachioji, Tokyo 1920015, Japan.*

[a)]Corresponding author: at13152@ns.kogakuin.ac.jp

Abstract. To develop new polymeric engineering materials made of only inedible plant-derived materials, we investigated the influence of the screw configuration in a twin screw extruder on the dynamic viscoelastic properties of molten state hemp fiber (HF) reinforced plant-derived polyamide 1010 (PA1010) biomass composites. PA1010 was made from sebacic acid and decamethylenediamine, which are obtained from plant-derived castor oil. HF, which is a type of natural fibers, was used as a reinforcement fiber. HF was surface-treated by two types of surface treatments: (a) alkaline treatment by sodium chloride ($NaClO_2$) solution and (b) surface treatment by ureidosilane coupling agents (A-1160). These biomass composites were extruded using a twin screw extruder, and compression-molded. In this study, two types of screw configurations were used in the twin screw extruder: original screw (conventional type, sc1) and new screw (mixing elements added type, sc2). The dynamic viscoelastic properties of molten state HF/PA1010 biomass composites were evaluated by oscillatory flow testing using a parallel plate-type rotational rheometer. It was found that the storage and loss moduli of HF/PA1010 biomass composites prepared by sc2 are remarkably lower than those of sc1 in the whole range of angular frequency. Thus the modification of the screw configuration has stronger influence on the dynamic viscoelastic properties of the biomass composites in the molten state. This may be attributed to the change in the internal microstructure formation such as fiber shape, fiber network formation and fiber dispersion in the composites.

INTRODUCTION

In recent years, natural fiber reinforced polymer biomass composites are attracting extensive interest as a solution for growing environmental threats [1-5]. To further enhance the eco-friendliness of materials, there is a strong need to use not only reinforcement fibers such as natural fiber but also biopolymers obtained from plant-derived materials as the matrix polymer of these biomass composites. In previous studies, we investigated the thermal, rheological, mechanical and tribological properties of hemp fiber (HF) reinforced plant-derived polyamide 1010 (PA1010) biomass composites to develop new engineering materials made of only inedible plant-derived materials [3-8]. In particular, we investigated the effects of fiber content, surface treatment of fiber, and the addition of other polymers, etc. on the various physical properties of HF/PA1010 biomass composites. We found that various physical properties improved with the surface treatment of fiber, and the addition of other polymers, etc. These may be attributed to the change in the internal microstructure of the composites such as fiber network formation, fiber dispersion, and the interfacial interaction between HF and PA1010. However, to date, little interest has been paid to the influence of polymer processing technology such as twin screw extrusion conditions on various physical properties of these biomass composites although the internal microstructure of these biomass composites was found to correlate closely with the polymer processing technology. To develop new polymeric engineering materials made of only inedible plant-derived materials, this study aimed to experimentally investigate the influence of the screw configuration in the twin screw extruder on the dynamic viscoelastic properties of molten state HF/PA1010 biomass composites.

EXPERIMENTAL

The materials used in this study were hemp fiber-reinforced plant-derived polyamide 1010 biomass composites surface-treated using two types of surface treatments: alkaline treatment and silane coupling agent. Plant-derived polyamide 1010 (PA1010, Vestamid Terra DS16, Daicel Evonic Ltd., Japan) was used as the matrix polymer. Hemp fiber (HF, ϕ50-100μm, Hemp Levo Inc., Japan) was used as the reinforcement fiber. The HF was precut into 5 mm pieces and surfaced-treated by two types of surface treatments: (a) alkaline treatment by sodium chlorite ($NaClO_2$) solution and (b) surface treatment by ureidosilane coupling agents (3-ureidopropyltrimethoxy silane, A-1160, Momentive Performance Materials Inc., Japan) [3,8]. These surface treatment methods for HF are described in detail in the previous conference paper [8]. The volume fraction of fiber V_f was fixed at 20 vol % in this study.

HF/PA1010 biomass composites were extruded using a twin screw extruder and compression-molded [3]. All the components were dried for 12 h at 80 °C in a vacuum oven beforehand until the moisture level was below 0.2%, then dry blended in a small bottle, and subsequently the melt was mixed at 85 rpm and 220 °C using a twin screw extruder (TEX-30HSS, Japan Steel Works, Ltd., Tokyo, Japan). In this study, two types of screw configurations were used in the twin screw extruder: original screw (conventional type, sc1) and new screw (mixing elements added type, sc2). Figure 1 shows the schematic diagrams of the screw configuration in the twin screw extruder used in this study. After mixing, the extruded strands of HF/PA1010 biomass composites prepared by different screw configurations in the twin screw extruder were cut into 5 mm pieces using a pelletizer and dried again at 24 h at 80 °C in a vacuum oven. 2mm-thick sheets were compression-molded at the conditions of 220 °C, 5 MPa, and 3 min, and cut into ϕ25mm disk shapes for measuring dynamic viscoelastic properties. To maintain the samples in a dry condition in all the measurements, they were kept in accordance with JIS K 6920-2 for at least 24 h at 23 °C in desiccators after molding.

The dynamic viscoelastic properties in the molten state were evaluated by oscillatory flow testing using a parallel plate-type rotational rheometer (ARES-G2, TA Instruments Japan, Japan). The diameter of the parallel plate was ϕ25mm, and the gap between the two plates was fixed at 2 mm. Under such a gap condition, the test specimens were slightly compressed in the molten state. The angular frequency was varied from 10^{-1} to 10^{2} rad/s, and the strain amplitude was set as 1%. These measurements were carried out at 220 °C. The dynamic viscoelastic properties of the molten state HF/PA1010 biomass composites were evaluated by storage modulus G', loss modulus G'', phase angle δ (=arctan G''/G'), complex modulus $|G^*|$ and complex viscosity $|\eta^*|$. The polished surface of these biomass composites was observed using a scanning electron microscope (SEM, JSM-6360LA, JEOL Ltd., Japan) coated with osmium (Os) sputter.

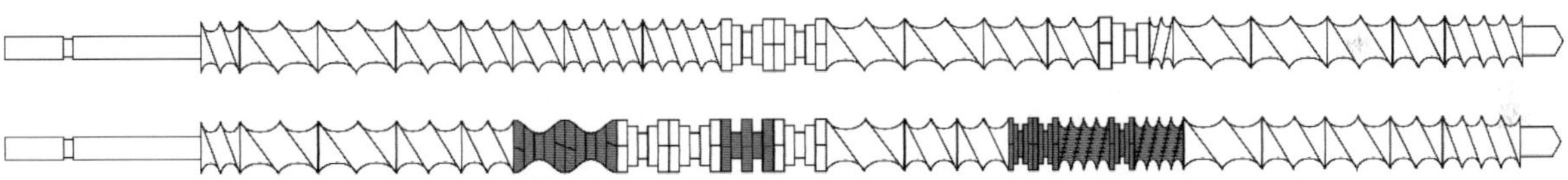

FIGURE 1. Schematic diagram of screw configuration in twin screw extruder: (a) Original screw (conventional type, sc1, top) and (b) New screw (mixing element added type, sc2, bottom).

RESULTS AND DISCUSSION

The dynamic viscoelastic properties of molten state HF/PA1010 biomass composites prepared by using different screw configurations in the twin screw extruder were evaluated by oscillatory behavior, which are considered to be strongly dependent on the internal microstructure of these biomass composites. The angular frequency dependence, which is a basic variable in dynamic viscoelastic properties, is discussed in the following. The dynamic viscoelastic properties of these biomass composites are plotted as a function of angular frequency ω at temperature 220 °C and strain 1% in Figure 2(a) (storage modulus G'), Figure 2(b) (loss modulus G''), and Figure 2(c) (complex viscosity $|\eta^*|$) respectively. The G'-ω curves of sc1 and sc2 show the typical G' of highly filled systems [9, 10]. The magnitude of G' of sc2 is lower than that of sc1 in the whole range of ω. The slope of G' against ω in the low ω region (ω=0.1 to 1 rad/s) became small, indicating a "second rubbery plateau", i.e., long-scale relaxation time [11, 12]. This second rubbery plateau behavior suggests the presence of an apparent yield stress, which can be attributed to the fiber network formed. However, the slope of G' against ω of sc2 is slightly higher than that of sc1. This behavior indicates the change in the internal structure of the composites such as fiber network formation, fiber dispersion and interfacial interaction between HF and PA1010 according to the different screw configurations in the twin screw extruder. On the other hand, the magnitude of *G'* of sc2 in the high ω region (ω=10 to 100 rad/s) is lower than that of sc1. In general, *G'* in the high ω region is dominated by the matrix polymer [13], suggested by the change in the molecular structure such as decrease in the molecular weight matrix polymer according to the difference in the screw configurations in the twin extruder. On the contrary, the loss modulus *G"* in Figure 2(b) and the complex viscosity $|\eta^*|$ have the same tendencies as *G'*. These behaviors can be explained by the mechanisms mentioned earlier.

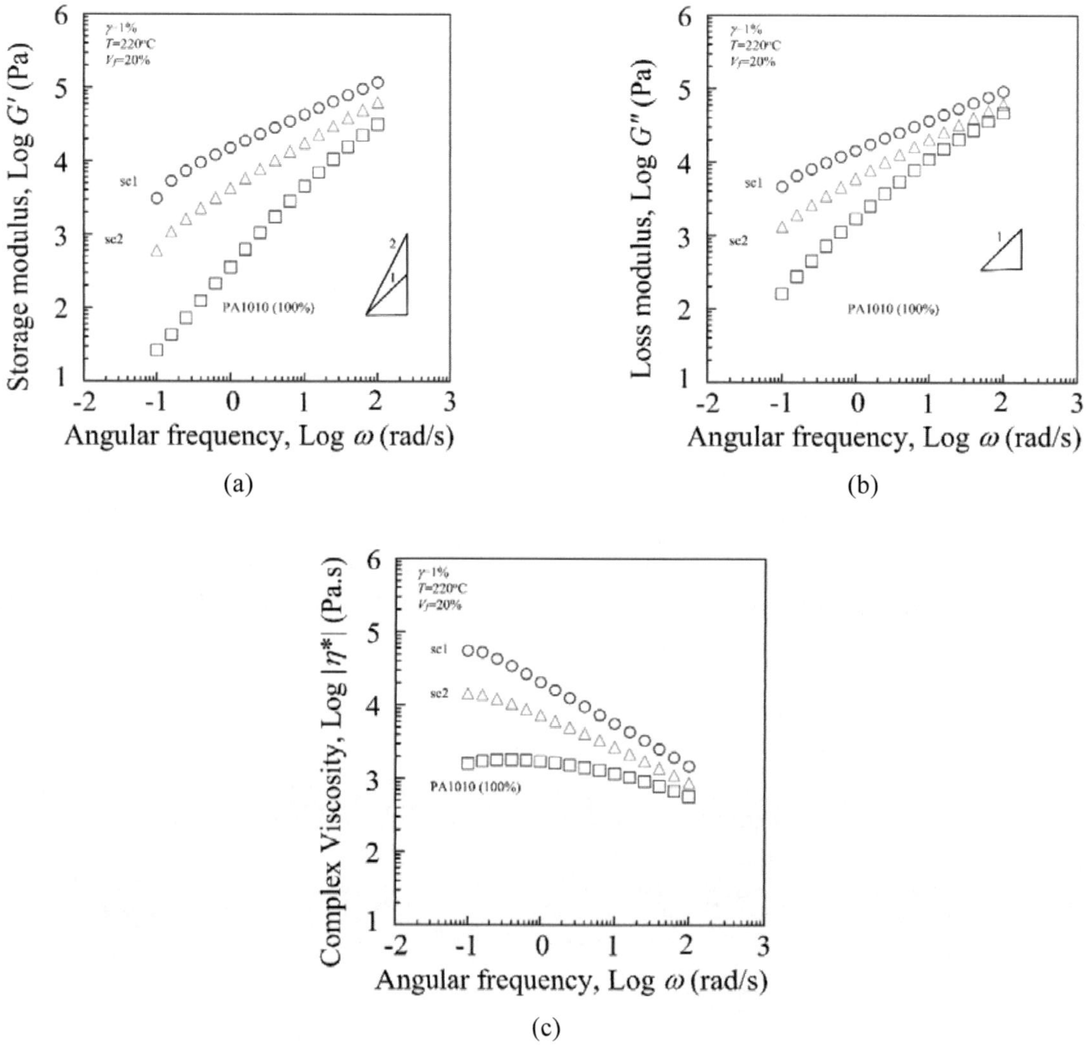

FIGURE 2. Dynamic viscoelastic properties as function of angular frequency for HF/PA1010 biomass composites prepared by different screw configurations in twin screw extruder: (a) Storage modulus, (b) Loss modulus and (c) Complex viscosity.

The influence of complex modulus on the viscoelastic properties of HF/PA1010 biomass composites prepared by different screw configurations in a twin screw extruder is discussed in order to understand the rheological behavior. The dynamic viscoelastic properties of these biomass composites are plotted as a function of complex modulus $|G^*|$ in Figure 3(a) (complex viscosity $|\eta^*|$), and Figure 3(b) (phase angle δ= arctan *G"/G'*), respectively. The former plot ($|\eta^*|$-$|G^*|$ curves) indicates the presence of an apparent yield stress in the low ω region. The apparent yield stress value of $|G^*|$ is a critical value which determines whether material can flow or not. In general, the $|\eta^*|$-$|G^*|$ curves show the following behaviors [3,14]. In Newtonian fluids, although $|\eta^*|$ slightly increases with decreasing $|G^*|$, $|\eta^*|$ has a constant value behavior. On the contrary, in non-Newtonian fluids, $|\eta^*|$ does not have a constant value behavior, and

$|\eta^*|$ dramatically increases with a slight decrease in $|G^*|$. This dramatic behavior may be due to the remarkable change in the structure of reinforcement fiber in the composites. These behaviors of the $|\eta^*|$-$|G^*|$ curves suggest that there is apparent yield stress. On the other hand, from the results of Figure 3(a), $|\eta^*|$ of neat PA1010 gradually increases with decreasing $|G^*|$, indicating that there is no apparent yield stress. On the contrary, the $|\eta^*|$ of sc1 and sc2 increases with a slight decrease in $|G^*|$, indicating that there is apparent yield stress. However, the yield stress value of sc2 is clearly lower than that of sc1. On the other hand, the latter plot (δ-$|G^*|$ curves) is called as van Gurp - Palmen plot [15]. This plot is considered to be a sensitive indicator for time-temperature superposition, presence of the long chain branch of polymer, gelation behavior, rheological percolation of polymer nanocomposites, change in the internal structure of the composites such as fiber network formation and fiber dispersion, etc. [3, 14, 16]. This is because the changes in the rheological properties, from which it is hard to understand the angular frequency dependence (G'-ω, G''- ω, tanδ-ω curves, etc.), are emphasized in this δ-$|G^*|$ plot. The δ-$|G^*|$ curve of neat PA1010 approaches 90° at the low $|G^*|$, indicating the flow behavior of viscous fluid. On the contrary, the δ-$|G^*|$ curves of sc1 and sc2 have tendencies that differ from that of PA1010. In the high $|G^*|$ region, the δ of sc1 and sc2 decreases gradually with decreasing $|G^*|$. However, in the low $|G^*|$ region, the δ of sc1 and sc2 decreases abruptly with a slight decrease in $|G^*|$. Moreover, the transition point of sc2 in the δ-$|G^*|$ curve is lower than that of sc1. These behaviors may be attributed to the change in the microstructures of composites.

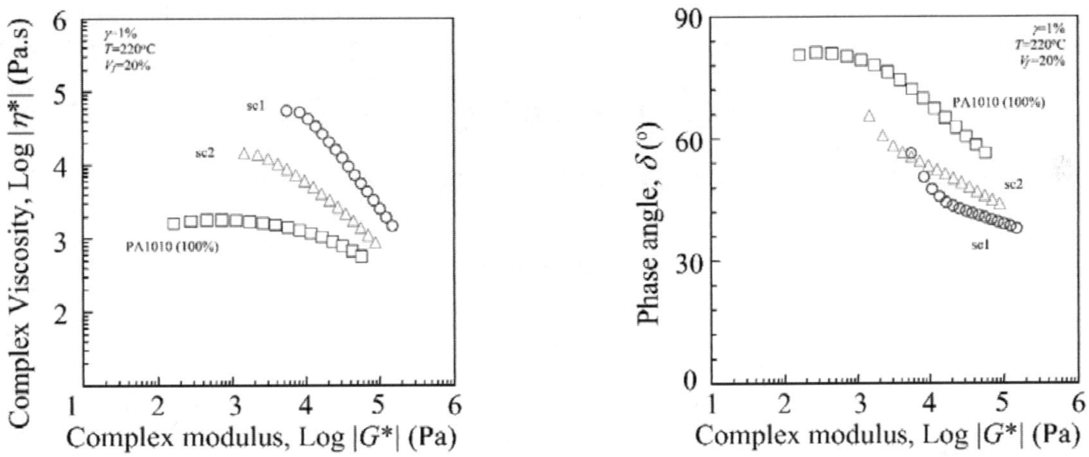

(a) (b)

FIGURE 3. Dynamic viscoelastic properties as function of complex modulus for HF/PA1010 biomass composites prepared by different screw configurations in twin screw extruder: (a) Complex viscosity, and (b) Phase angle.

(a) (b)

FIGURE 4. SEM photographs of polished surface of HF/PA1010 biomass composites prepared by different screw configuration in twin screw extruder: (a) original screw (conventional type, sc1), and (b) new screw (mixing elements added type, sc2).

To clarify the relationship between the dynamic viscoelastic properties and internal structure of HF/PA1010 biomass composites, we reviewed the morphologies of these biomass composites, such as internal microstructure like fiber network formation, fiber dispersion and interfacial interaction between HF and PA1010. Figure 4 shows the SEM photographs of the polished surface of these biomass composites prepared by different screw configurations in the twin screw extruder. The fiber dispersion of sc2 (Figure 4(b)) is better than that of sc1 (Figure 4(a)). In particular, no fiber agglomeration is seen in sc2, although the fiber shape, such as diameter and length in the composites of sc2, is smaller than that of sc1. These results demonstrate that the screw configuration in the twin screw extruder has strong influence on the dynamic viscoelastic properties of HF/PA1010 biomass composites.

CONCLUSION

To develop new engineering materials made of only inedible plant-derived materials, this study aimed to experimentally investigate the influence of the screw configuration in the twin screw extruder on the dynamic viscoelastic properties of molten state hemp fibers (HF) reinforced plant-derived polyamide 1010 (PA1010) biomass composites. It was found that the storage and loss moduli of HF/PA1010 biomass composites prepared by the new screw (mixing element added type, sc2) are remarkably lower than those of the original screw (conventional type, sc1) in the whole range of angular frequency. Thus the modification of screw configuration in the twin screw extruder has stronger influence on the dynamic viscoelastic properties of molten state biomass composites. These may be attributed to the change in the internal microstructure formation such as fiber shape, fiber network formation, and fiber dispersion in the composites.

ACKNOWLEDGEMENTS

This work was supported by JSPS KAKENHI Grant Number 16K06750. We would like to thank the Functional Microstructured Surfaces Research Center (FMS, MEXT, Japan) of Kogakuin University for funding this study.

REFERENCES

1. O. Faruk, A. K. Bledzki, H.-P. Fink and M. Sain, Prog. Polym. Sci., **37**, 1552–1596 (2012).
2. L. Yu., K. Dean and L. Li, Prog. Polym. Sci., **31**, 576–602 (2006).
3. Y. Nishitani, T. Yamanaka, T. Kajiyama and T. Kitano, "Thermal Properties of Hemp Fiber Reinforced Plant-Derived Polyamide Biomass Composites and their Dynamic Viscoelastic Properties in Molten State", in *Viscoelastic and Viscoplastic Materials*, edited by M. F. El-Amin (Intech, Rijeka, 2016), p.53–79.
4. Y. Nishitani, M. Hasumi and T. Kitano, AIP Conference Proceedings, **1664**, 060007 (2015).
5. Y. Nishitani, T. Kajiyama and T. Yamanaka, Materials, **10**, 1040 (2017).
6. Y. Nishitani, J. Mukaida, Y. Yamanka, T. Kajiyama and T. Kitano, AIP Conference Proceedings, **1713**, 120007 (2016).
7. Y. Nishitani, J. Mukaida, Y. Yamanka T. Kajiyama and T. Kitano, AIP Conference Proceedings, **1779**, 060004 (2016).
8. Y. Nishitani, Y. Ito and T. Kajiyama, The proceedings of 34th International Conference of the Polymer Processing Society (PPS-34), G06-260, Taipei, Taiwan (2018).
9. Y. Nishitani, I. Sekiguchi, B. Hausnerova, Y. Nagatsuka and T. Kitano, Polym. Polym. Compos., **9**, 199 (2001)
10. T. Kitano and T. Kataoka, Rheol. Acta., 19, 753 (1980)
11. J. D. Ferry, Viscoelastic Properties of Polymers, John Wiley & Sons, Incorporated, 1980, New York, USA.
12. Y. Nishitani, C. Ishii and T. Kitano, "Rheological Properties of Surface Treated Glass Fiber Reinforced Polypropylenes in Molten State", in *Polypropylene*, edited by Dr. Fatih Dogan (Intech, Rijeka, 2012), p.287–316
13. A. Durmus, A. Kasgoz, C. W. Macosko, Polymer, **48**, 4492–4502 (2007)
14. Y. Nishitani and T. Kitano, "Rheological Properties of Carbon Nanofiber-Filled Polyamide Composites and Blend of these Composites and TPE", in *Viscoelastic and Viscoplastic Materials*, edited by M. F. El-Amin (Intech, Rijeka, 2016), p.103–140.
15. M. Van Gurp. J. Palmen, Rheol. Bull., **67**, 5–8 (1998)
16. P. Potschke, M. Abdel-Goad, I. Alig, S. Dudkin, D. Lellinger, Polymer, **45**, 8863-8870 (2004)

Thermosetting polyurethane foams by physical blowing agents: chasing the synthesis reaction with the pressure

Cosimo Brondi[1], Maria Rosaria Di Caprio[1], Giuseppe Scherillo[1], Ernesto Di Maio[1,a)], Thomas Mosciatti[2], Sara Cavalca[2], Vanni Parenti[2], Maurizio Corti[3] and Salvatore Iannace[4]

[1]*Dipartimento di Ingegneria Chimica, dei Materiali e della Produzione Industriale, University of Naples Federico II, P.le Tecchio 80, 80125 Naples, Italy*
[2]*Dow Italia s.r.l, Polyurethanes R&D, Via Carpi 29, 42015 Correggio, Italy*
[3]*Cannon Afros S.p.A., Via Ferraris Galileo, 33, 21042 Caronno Pertusella, VA, Italy*
[4]*Institute for Macromolecular Studies, National Research Council, Via Edoardo Bassini, 15, 20133 Milan, Italy*

[a)]Corresponding author: edimaio@unina.it

Abstract. The aim of the present study was the introduction of a novel method to produce microcellular thermosetting polyurethane foaming by the gas-foaming technique, using high-pressure physical blowing agents. In particular, to tackle the inherent difficulties of imposing a rapid pressure quench O(10^{-2} s) to a material whose synthesis timing is much larger O(10^{2} s), we utilized a two-stage foaming. In a first stage, a rapid pressure quench O(10^{-2} s) was imposed to nucleate a large amount of bubbles, from the saturation pressure to a low, non-zero pressure; in the second stage, the growth of the nucleated bubble is then controlled by slowly O(10^{2} s) decreasing the pressure to ambient (zero) pressure. In this way, by separating the nucleation from the growth rate and by chasing the synthesis reaction with the pressure to avoid excessive stresses to the curing polymer, we achieved microcellular (size diameter below 10 um), medium-to-low density (below 200 g/L) thermosetting polyurethane foams.

INTRODUCTION

Thermosetting polyurethane foams (PUs) were introduced at the beginning of the last century as a synthesis product of a polyol and an isocyanate (*synthesis reaction*). Concurrent to the synthesis reaction, a *blowing reaction* of isocyanate with water provides the blowing agent, CO_2, responsible for foaming the curing matter. In this process, by recalling the classification of Physical Blowing Agent (PBA) or Chemical Blowing Agent (CBA) utilized in the foaming of thermoplastics [1], water is a CBA, while CO_2 gradually forms and inflates bubbles at relatively low pressures, not solubilizing in the polyol and/or the isocyanate phases prior to foaming, and cannot be considered a PBA, here [2]. PUs have been also produced by using PBAs, like chlorofluorocarbons (CFCs) (in the late 1950s) and hydro-chlorofluorocarbons (HCFCs) (at the beginning of 1990s), then replaced by hydrocarbons (HCs) (e.g. pentane) because of their negative environmental impact [3]. PBAs are solubilized at relatively low temperatures and pressures (typically, ambient temperature and pressure) in the polyol and/or isocyanate phases and then allowed to evolve in the gaseous state by the temperature rise following the exothermic synthesis reaction. With the temperature increase, PBA solubility decreases, and the solution undergoes phase separation (bubbles nucleation and growth). HCs use was limited by their inherent flammability, despite being inexpensive and with zero ozone depletion potential (ODP) [3]. In this context, CO_2 is considered more sustainable and safer, with zero ODP and the lowest global warming potential (GWP = 1) among known blowing agents [3]. As a drawback, CO_2 solubility is much lower than HCs and much larger pressures have be utilized in order to reach concentrations appropriate for low-density foams, pushing technology and know-how development.

High-pressure CO_2 foaming proved very effective with thermoplastics. The pressure drop rate (PDR) is the key to achieve microcellular or nanocellular foams, as introduced in the '80s by the MIT group [4] and, also at ultra-high PDR O(10^3 MPa/s), the number of bubbles nucleated per unit volume exponentially increase with the PDR [5]. In fact, with this method, utmost performances in terms of cell number densities have been reached with numerous thermoplastic polymers: microcellular and, more recently, nanocellular foams have been produced, characterized by improved thermal insulating and mechanical properties as compared to standard cell-sized foams [6]. So far, thermosetting foams stay orders of magnitude behind, in terms of cell size.

To the best of our knowledge, no papers addressed the use of CO_2 as a PBA in any thermosetting polymers, in which CO_2 solubilization is conducted before reactants mixing. Because of the environmental concerns and of the encouraging performance in producing microcellular and nanocellular foams with thermoplastic polymers, high-pressure CO_2 foaming of PUs has attracted a growing industrial interest [7-13]. To ground the foam processing on a scientifically sound base, a preliminary study included: i) the measurement and modeling of the physical properties of the polyol/CO_2 [14] and the isocyanate/CO_2 solutions [15]; ii) the measurement and modeling of the retarding effect of the CO_2 concentration on the PUs synthesis kinetics by FT-NIR spectroscopy [16]; iii) the development of a new experimental apparatus for processing and characterization of the foaming [17]. As a completion of this explorative study, we herein report the development of a novel method that uses CO_2/polyol and CO_2/isocyanate solutions (formed at high pressure) to produce microcellular (size diameter below 10 um), medium-to-low density (below 200 g/L) thermosetting polyurethane foams. The method, essentially, copes with the two very different time scales of the gas foaming O(10^{-2} s) and the gelling/curing reaction O(10^2 s), separating the nucleation and growth stages in two different processing stages.

MATERIALS AND METHODS

Materials

A polyether polyol and polymeric methylene diphenyl diisocyanate (PMDI) were supplied by DOW Chemical Italy S.r.l. (Correggio, RE, Italy) within the LIFE13-ENV/IT/001238 project [7] and used "as received". High purity grade CO_2 (99.95% pure) and N_2 (99.99% pure) were supplied by SOL (Naples, Italy).

Methods

To perform the foaming attempt, we utilized a novel pressure vessel designed to processing and studying the foamability of PUs by CO_2 as a PBA. The equipment was extensively described in [17]. The main novel features of the equipment were: i) the use of a rubbery impeller for the reactant sealing (and successive mixing), ii) a high-pressure-tight sapphire window mounted beneath an IR-transparent sample holder for remote NIR monitoring. Figure 1 reports 3D renderings of the proposed pressure vessel and images of the equipment in two configurations. With respect to the original configuration (see Fig. 1b), to apply a partial pressure quench, we added a gas tank to the evacuation system (see Fig. 1c).

(a) (b) (c)

FIGURE 1. Equipment utilized to perform the foaming tests. (a) 3D scheme showing a cut of the pressure vessel with the sample holder and the mixing shaft. (b) Picture of the equipment in the original configuration reported in [19]: (1) actuator for ball-valve

(2); (3) pressure vessel. (c) Upgraded configuration with a gas tank (4), in addition, to evacuate the pressure vessel to a non-zero pressure.

In a typical foaming attempt, the two components of the PU formulation are loaded in the cylindrical sample holder, kept separate by the rubbery impeller, and subjected, at 35°C, to a CO_2 pressure ranging from 2.0 to 10.0 MPa. After a defined sorption time, needed to achieve the desired average CO_2 mass fraction in the total formulation, the impeller is put in rotation for the curing reaction to start. Then, at a certain degree of curing is achieved, the actuated ball valve is opened for pressure release, at ambient pressure (configuration in Fig. 1b) or to a non-zero pressure (configuration in Fig. 1c). In the latter case, by using the valve on top of the gas tank the pressure is eventually brought to ambient after an additional curing period. The foamed samples are finally extracted from the sample holder for the characterization.

RESULTS

Figure 2 reports the effect of one-stage and two-stage foaming on the PU cell morphology. The reactants were kept under CO_2 sorption for 3 hours at 35°C and 4 MPa and then (after mixing) cured for 12 minutes. In the case of one-stage foaming, after the curing stage, the pressure was reduced to ambient pressure by slow pressure release (SPR) at 0.1 MPa/s. A bi-modal cell size distribution was generated with bigger cells around 70 μm and a density of 400 g/L. In the other case, two-stage foaming was successfully applied to achieve the foam reported in Fig. 2b. After the curing stage, the pressure was reduced to an intermediate pressure of 0.35 MPa by a fast pressure release (FPR) at 40 MPa/s. The PU sample was kept under the same pressure for 3 minutes and the second pressure reduction was conducted by SPR. The two-stage foaming allowed to reach a uniform cell distribution with an average cell size of 10 μm and a density of 150 g/L.

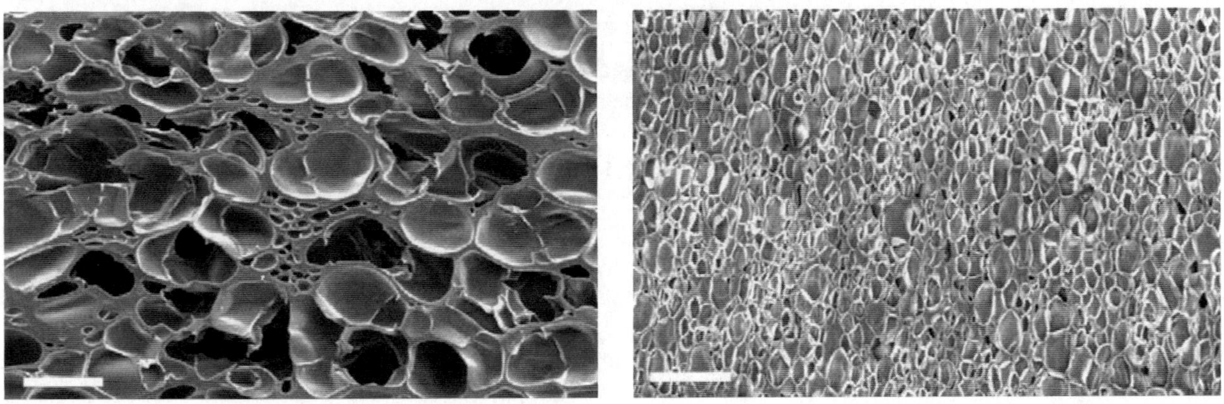

(a) (b)

FIGURE 2. SEM micrographs of PU samples kept under CO_2 sorption for 3 hours at 35°C and 4 MPa and then cured for 12 minutes under pressure. (a) One-stage foaming attempt by SPR. (b) Two-stage foaming performed by FPR, the subsequent holding stage and then the controlled SPR. Scale bars are 100 μm.

The main issue with one-stage foaming was, to our point of view, the very different time scales between reaction time, $O(10^2 s)$, and pressure quench, $O(10^{-2} s)$, and, more generally, the idea to overturn the long-term development of PUs foaming. The impossibility to achieve the desired goal by one-stage foaming induced us to derive a new method.

The main idea under the present two-stage method is to separate bubble nucleation from bubble growth and have the two phenomena following the two timescales, $O(10^{-2} s)$ and $O(10^2 s)$. The two-step foaming stages and their pressure history are shown in Figs. 3a and b respectively. In particular, in the first holding stage the pressure (p_{sat}) is kept constant (0-1) allowing the cure of the polyurethane mixture, in the first foaming stage (1-2) the bubble nucleation is promoted by fast pressure (FPR) quenching (as fast as possible by the specific equipment) the reacting mixture, the pressure (p_0) is kept constant during the second holding stage and, at last, the pressure is slowly reduced (SPR) to ambient pressure (3-4). In this stage, the gas contained within the bubbles will expand.

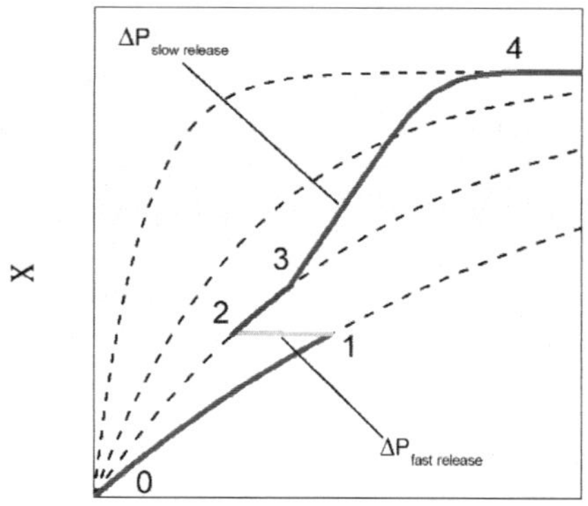

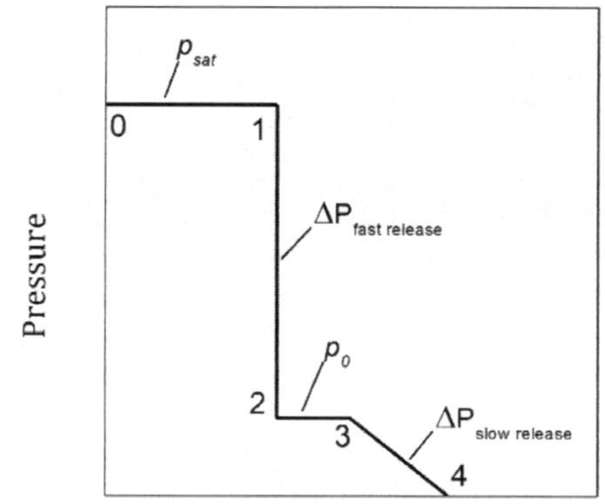

FIGURE 3. Two-step foaming stages and their pressure history: (0-1) first holding stage, (1-2) fast pressure release, (2-3) second holding stage, (3-4) slow pressure release.

It is possible to design the pressure drop history for the SPR in order to optimize foaming. As a proof of concept, and as it is also simpler to apply to an equipment, we decided to work at a constant pressure drop rate, $\dot{p}$, and derived a model to attain a desired value of final curing degree (X_f) when $p = 0$ (ambient pressure), starting from certain initial values of X_0 and p_0:

$$\dot{p} = \frac{A}{BH} \frac{\left(1-e^{-BHp_0}\right)}{ln\left(\frac{1-X_0}{1-X_f}\right)} \tag{1}$$

Where H is the Henry constant for CO_2 in the averaged polymeric matter. Equation 1 describes the effect of the reaction rate parameters (A, B), the Henry constant, H, the initial conditions (p_0, X_0) and the desired X_f on $\dot{p}$.

CONCLUSIONS

The first attempts to foam the reacting PU formulation were based on the idea that, within the PU curing reaction time O(10^2s), it was possible to guess the curing degree and, hence, a polymer viscosity suitable to tolerate the pressure quench, which has to be fast enough to generate a high number of bubbles O(10^{-2}s). However, the morphology results (average cell size of 70 μm and a density of 400 g/L) proved the impossibility to achieve such a goal.

The novel two-stage foaming method allowed to obtain a PU foam with uniform cell size distribution around 10 μm and a density of 150 g/L. In the first stage, a rapid pressure quench O(10^{-2} s) was imposed to nucleate a large amount of bubbles, from the saturation pressure to a low, non-zero pressure; in the second stage, the growth of the nucleated bubble is controlled by slowly O(10^2 s) decreasing the pressure to ambient (zero) pressure. In this way, by separating the nucleation from the growth stage, we may describe this method as "chasing the synthesis reaction with the pressure" (*pondus mixturam progredientem persequitur*), intending the use of external gas pressure to control expansion of the gas contained in the bubbles, while the curing reaction is inexorably running

ACKNOWLEDGMENTS

The European union's financial support (LIFE13-ENV/IT/001238 project [7]) is gratefully acknowledged.

REFERENCES

1. J. Štěpek and H. Daoust, in Additives for Plastics, Polymers-Properties and Applications (New York, NY, Springer, 1983)
2. D. Randall and S. Lee, in The polyurethanes book (New Jersey, John Wiley & Sons, Inc., 2003), p. 113.
3. D. Randall and S. Lee, in The Polyurethanes book (New Jersey, John Wiley & Sons, Inc., 2003), pp. 127.
4. C. B. Park, . D. F. Baldwin and N. P. Suh, Polym. Eng. Sci., 35, 432 (1995).

5. D. Tammaro, A. Astarita, E. Di Maio and S. Iannace, Ind. Eng. Chem. Res., 55, 5696, (2016).
6. S. Costeux, J. Appl. Polym. Sci., 131, (2014).
7. PU Disruptive technology to dramatically improve Energy Efficiency of Household Appliances, available at http://ec.europa.eu/environment/life/project/Projects/index.cfm?fuseaction=search.dspPage&n_proj_id=5077.
8. B. Barron and J. Dunlap , Air frothed polyurethane foams, U.S. Patent US3821130A (1972).
9. C. Fiorentini and A. C. M. Griffiths , Froth process for continuous manufacture of polyurethane foam slab-stocks, U.S. Patent US5665287A (1993).
10. E. Di Maio and E. Kiran, J. Supercrit. Fluids, 134, 157, (2018).
11. R. Eiben, W. Ebeling, H.M. Sulzbach and R. Raffel , Method and device for the production of foam using carbon dioxide dissolved under pressure, U.S. Patent US5883143A (1998).
12. S. Lindner, W. Friederichs, R. Strey, T. Sottmann, E. Khazova, L. Kramer, V. Dahl and A. Chalbi , Process for producing a polyurethane foam by means of supercritical or near-critical blowing agent, U.S. Patent US9139683B2 (2009).
13. M. Fricke, M. SCHÜTTE, T. M. STAUDT, C. Holtze, S. Koch and F. Bartels, Micro-emulsions, WIPO (PCT) Patent WO2013026813A1 (2011).
14. M. R. Di Caprio, G. Dal Poggetto, M. G. Pastore Carbone, E. Di Maio, S. Cavalca , V. Parenti and S. Iannace, Fluid Phase Equilib., 425, 342, (2016).
15. M. R. Di Caprio, B. Immirzi, E. Di Maio, S. Cavalca, V. Parenti, S. Iannace and G. Mensitieri, Fluid Phase Equilib., 456, 116, (2018).
16. M. R. Di Caprio, C. Brondi, E. Di Maio, T. Mosciatti, S. Cavalca, V. Parenti, S. Iannace, G. Mensitieri and P. Musto, Polyurethane synthesis under high-pressure CO_2, a FT-NIR study, Eur. Polym. J. (Submitted).
17. M. R. Di Caprio, D. Tammaro, E. Di Maio, S. Cavalca, V. Parenti, A. Fangareggi and S. Iannace, Polym. Test., 62, 137, (2017).

Expanded Polyamide 12 Bead Foams (ePA) — Thermo-Mechanical Properties of Molded Parts

Dominik Dörr[a], Daniel Raps[a], Dharan Kirupananthamb[b], Chris Holmes[b] and Volker Altstädt[a,c,*]

[a] *University Bayreuth, Department of Polymer Engineering, Universitätsstraße 30, 95447 Bayreuth, Germany*
[b] *adidas AG, Future Team, Adi-Dassler-Straße 1, 91074 Herzogenaurach, Germany*
[c] *Bavarian Polymer Institute, Universitätsstraße 30, 95447 Bayreuth, Germany*
[*] *Corresponding Author*

Abstract. The demand for bead foam products, typically made from expanded polystyrene (ePS) or expanded polypropylene (ePP), has risen sharply in the past. Bead foam products made from engineering thermoplastics with superior thermomechanical properties compared to existing products, such as a polyamide could open up completely new fields of application. Like other engineering thermoplastics, polyamide shows challenging properties with regard to foaming, such as a low melt strength and a narrow processing window. The present work shows the feasibility to produce expanded bead foams from polyamide 12 (ePA) via foam extrusion with underwater pelletizing. The beads thus obtained, with a bulk density of 66 g/l and an average cell size of 30 µm, were successfully molded into parts with a density less than 100 g/l. Finally, mechanical tests on the molded foam parts show the thermomechanical properties of the ePA bead foams to be significantly better than those of ePP.

Keywords: Cellular plastic, bead foam, particle foam, engineering thermoplastic, polyamide 12

INTRODUCTION

A life without any lightweight materials would be hardly imaginable and impossible to realize today. Lightweight materials such as polymer foams can be found almost everywhere. The main applications of polymer foams are the packaging and the construction industry. In most cases, polymeric foams are not directly visible, e.g. as dampers or seat cushions in cars [1–4]. In total it can be said, that at least one polymer foam can be found within a radius of 5 m. This wide range of applications can be traced back to the beneficial properties of such polymer foams. On the one hand, they exhibit useful acoustical and thermal insulation properties as well as energy absorption. On the other hand, they offer a huge weight reduction compared to compact materials [3,4].

Bead foams are a special type of polymer foams. Based on a two-step manufacturing process, which describes the process of bead production and the subsequent molding process to final parts, it is possible to obtain products with complex geometries and low densities [4]. In contrast, in other foam manufacturing processes, like foam injection molding and foam extrusion, only one of these advantages can be exploited [5,6]. Consequently, the interest in bead foam products is growing continuously. Currently only four commercially available polymers are well-established. These are the amorphous expandable and expanded polystyrene (ePS), the semi-crystalline expanded polypropylene (ePP), the expanded polyethylene (ePE) and expanded thermoplastic polyurethane (eTPU), known as BASF "Infinergy" or adidas "boost" [4].

The main disadvantage of the established bead foams are the low mechanical properties at high temperatures, which is 80 °C for ePS and 110 °C for ePP [7]. Recent developments include the production of expanded bead foams made of the engineering polymer poly(butylene terephthalate) (ePBT), which is only moldable after chemical modification and at high pressures [8,9], and of expandable polyethersulfone (ePESU), with a glass transition temperature of 225 °C [10]. Publications about engineering thermoplastic bead foams, which can be used at higher temperatures compared to standard bead foams, still lacking. In addition, various industries, like the automotive sector, are always searching for new or advanced materials either for weight reduction, lower production costs or the associated economic aspects or better material recycling. New automotive regulations on the recyclability states that at least 95 wt.-% of a car has to be reusable or recoverable [11]. This regulation could be efficiently met by building more parts from the same raw material that can be easily recycled without expensive separation processes.

This article introduces a new type of bead foam made from the engineering polymer polyamide 12 (PA). This expanded bead foam is a high temperature resistant, low density foam, which offers the opportunity to replace compact PA parts in the engine compartment.

MATERIALS AND METHODS

Materials

This study uses non-commercially available polyamide 12 grades allocated from EVONIK Industries AG *(Essen, Germany)* with different molecular weights. The polymers are designated as PA_{High}, and PA_{Medium} as defined by their melt-flow index. A lab-extrusion-line by Dr. Collin GmbH *(Ebersberg, Germany)* with an attached underwater pelletizer from Gala Kunststoff- und Kautschukmaschinen GmbH *(Xanten, Germany)* is used to produce the polyamide bead foam. During the reactive extrusion with a multifunctional polymeric epoxy oligomer (chain extender (CE)), the polymer is charged with supercritical CO_2 and ethanol as blowing agents. The steam chest-molding was conducted in an EPP-molding machine. For comparison purposes, a commercially available ePP (BASF, Neopolen P) with a density of 100 g/l is tested.

Methods

Morphology and Density

The produced beads and molded parts were investigated regarding bead surface, foam structure and fractured surface with the scanning electron microscope (SEM) *Jeol JSM 6510 (Freising, Germany)* and an acceleration voltage of 10 kV. The foamed beads were cut across the direction of extrusion and coated in the *Cressington Sputter Coater 108 auto (Watford, UK)* with a 1.3 nm platinum layer. The cell morphology of the bead foams were analyzed using SEM images and evaluated with at least 100 cells per sample with ImageJ [12].

Mechanical Testing

Three point bending experiments were carried out according to ISO 1209 with an universal testing machine 1475 from Zwick Roell *(Ulm, Germany)*. The specimen size was reduced to 90x25x15 mm, the span to 75 mm and the testing speed set to 10 mm/min. Heat deflection temperature was measured according to ISO 2796 at 150 °C for 20 h. Furthermore, dynamic mechanical temperature analysis (DMTA) was performed on a Gabo Eplexor 500N *(Ahlden, Germany)* in compression mode on a sample of the molded parts (diameter 25 mm, height 20 mm) with a load of 10 N, a frequency of 1 Hz and heating from 20 to 170 °C at a heating rate of 1 K/min.

RESULTS AND DISCUSSION

Round foam beads were obtained for the two materials used, PA_{High} and PA_{Medium}. Both grades had to be modified during extrusion with a CE. As already shown for various polyesters, such as polylactic acid or polybutylene terephthalate, polyamides are not foamable without a chemical modification due to their low melt strength [9,13]. As can be seen in FIGURE 1, the single beads show very comparable and homogenous foam morphologies at densities below 70 g/l. The average cell sizes are below 50 µm in both produced bead foams.

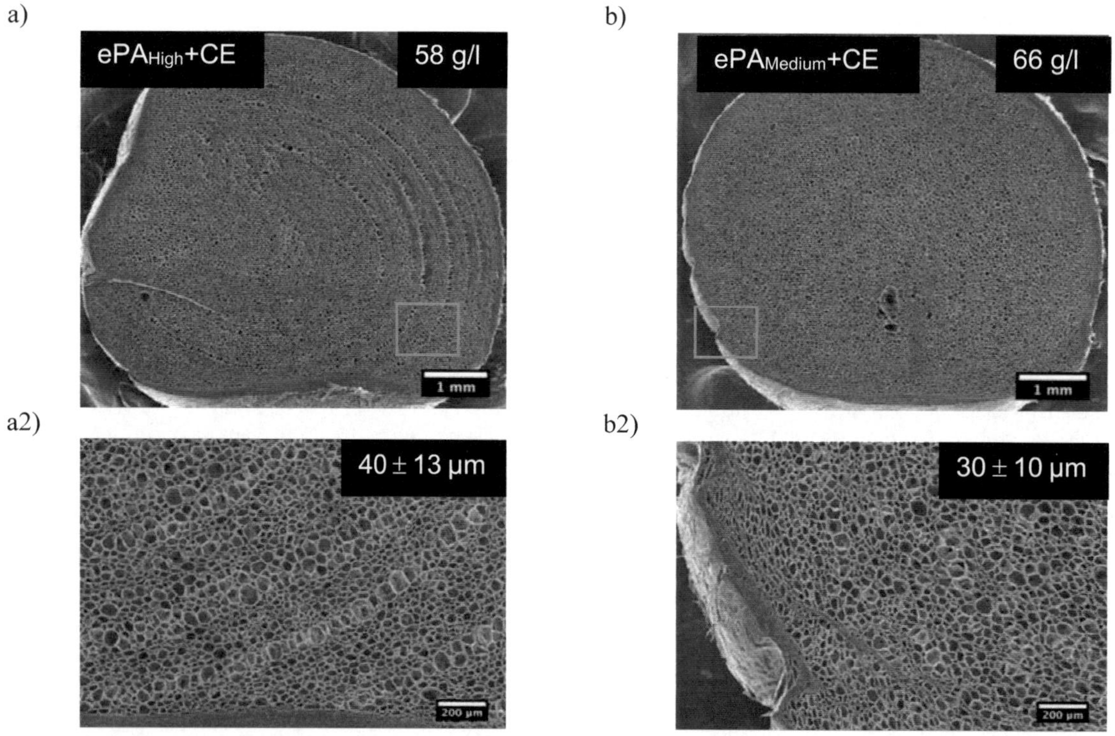

FIGURE 1. Foam morphologies of ePA beads. **a)** ePA_{High} beads and **b)** ePA_{Medium} beads. Densities and cell sizes are noted.

The beads were steam chest molded into parts and specimen were cut out for the mechanical analysis. The results of the three-point bending test can be found in FIGURE 2a.

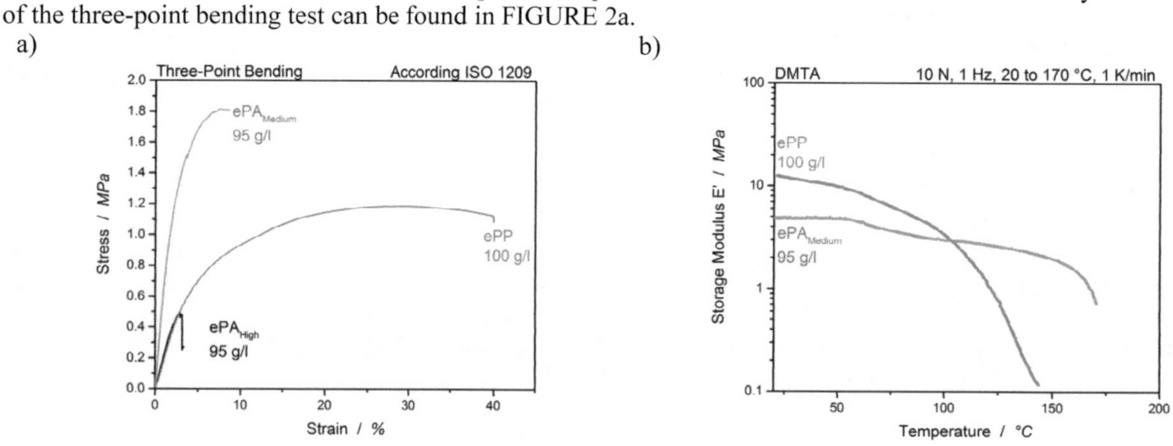

FIGURE 2. a) Three-point bending and **b)** DMTA measurement of the molded ePA beads in comparison with ePP with the same density.

The fracture behavior of the molded ePA parts are completely different. This can be explained by SEM pictures of the fracture surface shown in FIGURE 3. The degree of fusion, which is the ratio of trans- and interbead failure, represents the molding quality of the parts [14]. The fracture surface only shows interbead failures in the ePA_{High} part. Here only very low forces are required to break the plate and the weld itself is the weakest point of the part. In comparison, the ePA_{Medium} part shows a mixture of trans- and interbead failures and requires greater forces. The mechanics can be even more improved with a better steam-chest molding process. The deviations in molding quality could be explained by differences in the molecular weight. During the steam-chest molding process the bead surfaces are melted by the hot steam and the polymer chains start to move, which is generally described by reptation [15,16]. The reptation time is directly correlated to the molecular weight, so a lower molecular weight leads to lower reptation times and finally better fusion. Furthermore, the chain extension leads to a long chain branched structure, that can influence the crystallization behavior and finally the moldability [17,18]. Both effects on the moldability have to be further investigated in the future.

a)

b)

FIGURE 3. Fracture surface of molded plates (**a**) ePA$_{High}$ and **b)** ePA$_{Medium}$) after the three-point bending test. Scalebar 500 μm.

The differences between the ePA and ePP can be attributed to the difference in the molecular structure. The three-point-bending tests are performed at room temperature, about 20 °C above the glass-transition temperature of PP. This leads to a very elastic behavior of ePP compared to the rigid ePA, with a glass-transition temperature of about 55 °C. In addition, PA build hydrogen bonds, which lead to high stiffness and a resulting high Young's modulus and fracture toughness. In the following, only the well molded plate ePA$_{Medium}$ will be tested further.

FIGURE 2b shows a representative curve of ePA$_{Medium}$ and ePP as reference. Between room temperature and 100°C the values for ePP are higher. After the glass transition temperature of PA12, the storage modulus decreases only very slightly. Even at temperatures above 100 °C the modulus is very stable, whereby the ePP fails. This clearly shows that ePA has a good thermal stability over a wide temperature range. From 160 °C, however, the storage modulus decreases significantly. The thermal resistance of ePP is near 100 °C, while that of ePA it is around 150 °C. The heat deformation resistance, tested according ISO 2796, confirms this observation. The volume change after 20 h at 150 °C is <4 % of ePA compared to about 60 % of ePP (cf. FIGURE 4).

FIGURE 4. ePA and ePP samples after thermal resistance test at 150 °C for 20 h.

CONCLUSION AND OUTLOOK

Bead foams made of engineering thermoplastics, such as a PA12, offer superior thermomechanical properties as compared to existing ePP. This is a requirement for their usage in the engine compartment of current cars in order to dissipate the heat more efficient or as insulation material for electrically operated cars in order to withstand the elevated temperatures during charging. Therefore, two PA12 materials with different molecular weights were investigated regarding their suitability for bead foam manufacturing. It is possible to produce expanded PA bead foams via foam extrusion with underwater pelletizing. Beads with a low bulk density down to 66 g/l with small cells (35 μm average cell diameter) were obtained. Furthermore, the produced beads are moldable to parts with a density of 95 g/l, which is comparable with semi-crystalline ePP. The thermal stability of ePA is much better at temperatures above

100°C compared to ePP which exhibit a thermal resistance temperature of about 150 °C. Furthermore, ePA shows nearly unchanging dynamic mechanical properties in the tested range of 20 °C to 150 °C, which is very advantageous for temperature changing applications, like the engine compartment. Due to the high mechanical strength, ePA can displace commonly used compact polyamide parts as well as ePP parts in the car. Having more parts made from the same material would make it much easier to meet recycling regulations.

This article shows that it is possible to produce a bead foam out of the engineering polymer polyamide 12 and suggests the possibility of developing bead foams made from other polyamides, like PA6.6, with an even higher thermal resistance.

ACKNOWLEDGMENTS

We acknowledge Annika Pfaffenberger and Jacqueline Uhm for preparing the SEM pictures, the BPI KeyLab Electron and Optical Microscopy, Ute Kuhn for the analysis - and Sebastian Gröschel as well as Andreas Mainz for the experimental support. Further thanks to Jasmin Walter for the helpful discussions.

REFERENCES

[1] D. Eaves, Handbook of Polymer Foams, Rapra Technology, **2004**, ISBN:9781859573884.

[2] P. Viot, E. Plougonven, D. Bernard, Microtomography on polypropylene foam under dynamic loading: 3D analysis of bead morphology evolution, *Compos. Part A Appl. Sci. Manuf.* 39 (**2008**) 1266–1281. doi:10.1016/j.compositesa.2007.11.014.

[3] J.C. Venkata Reddy Nallagundla, Rakesh Lingam, Handbook of Manufacturing Engineering and Technology, Springer London, London, **2015**, ISBN:978-1-4471-4669-8. doi:10.1007/978-1-4471-4670-4.

[4] D. Raps, N. Hossieny, C.B. Park, et al., Past and present developments in polymer bead foams and bead foaming technology, *Polymer (Guildf).* 56 (**2015**) 5–19. doi:10.1016/j.polymer.2014.10.078.

[5] W. Zhai, Y.-W. Kim, C.B. Park, Steam-Chest Molding of Expanded Polypropylene Foams. 1. DSC Simulation of Bead Foam Processing, *Ind. Eng. Chem. Res.* 49 (**2010**) 9822–9829. doi:10.1021/ie101085s.

[6] S.-T. Lee, C.B. Park, Foam Extrusion, CRC Press, **2014**, ISBN:978-1-4398-9859-8. doi:10.1201/b16784.

[7] BASF SE, Neopolen P -Designed for New Ideas - Product brochure, *Prod. Broch.* (**2017**).

[8] T. Köppl, D. Raps, V. Altstädt, E-PBT—Bead foaming of poly(butylene terephthalate) by underwater pelletizing, *J. Cell. Plast.* 50 (**2014**) 475–487. doi:10.1177/0021955X14528524.

[9] T. Standau, B. Hädelt, P. Schreier, et al., Development of a Bead Foam from an Engineering Polymer with Addition of Chain Extender : Expanded Polybutylene Terephthalate, *Ind. Eng. Chem. Res.* 57 (**2018**) 17170–17176. doi:10.1021/acs.iecr.8b04799.

[10] I. Fonseca, J. Bräuer, G. Graessel, et al., Advances in high performance thermoplastic foams Results and discussion, *SPE Foam. 2018.* (**2018**).

[11] European Parliament and of the Council, Directive 2000/53/EC of the European Parliament and of the Council of 18 September 2000 on end-of life vehicles, (**2000**). https://eur-lex.europa.eu/legal-content/EN/TXT/?uri=CELEX:32000L0053 (accessed October 16, 2018).

[12] C.A. Schneider, W.S. Rasband, K.W. Eliceiri, NIH Image to ImageJ: 25 years of image analysis., *Nat. Methods.* 9 (**2012**) 671–5.

[13] S. Göttermann, S. Weinmann, C. Bonten, et al., Modifiziertes Polylactid für die Schaumextrusion, *(24. Stuttgarter Kunststoffkolloquium 25.-26. Febr. 2015).* (**2015**).

[14] J. Rossacci, S. Shivkumar, Influence of EPS bead fusion on pattern degradation and casting formation in the lost foam process, *J. Mater. Sci.* 38 (**2003**) 2321–2330. doi:10.1023/A:1023924211663.

[15] R.P. Wool, Self-healing materials: a review, *Soft Matter.* 4 (**2008**) 18. doi:10.1039/b711716g.

[16] M. Bousmina, H. Qiu, M. Grmela, et al., Diffusion at polymer/polymer interfaces probed by rheological tools, *Macromolecules.* 31 (**1998**) 8273–8280. doi:10.1021/ma980562r.

[17] M. Nofar, W. Zhu, C.B. Park, et al., Crystallization Kinetics of Linear and Long-Chain-Branched Polylactide, *Ind. Eng. Chem. Res.* 50 (**2011**) 13789–13798. doi:10.1021/ie2011966.

[18] Y.-M. Corre, A. Maazouz, J. Reignier, et al., Influence of the chain extension on the crystallization behavior of polylactide, *Polym. Eng. Sci.* 54 (**2014**) 616–625. doi:10.1002/pen.23595.

Foam Injection Molding of Phenolic Resins

Martin Bayer[a], Torsten Maenz[a], Gerrit Hülder[a], Pierre Eckold[a] and Michael Gehde[b]

[a]*Corporate Sector Research and Advance Engineering, Robert Bosch GmbH, Renningen, Germany*
[b]*University of Technology Chemnitz, Chemnitz, Germany*

Abstract. The process of foam injection molding is state of the art for thermoplastic materials, whereas very little research has been carried out applying this technique for thermosetting materials. This publication demonstrates the feasibility of a foam injection molding process with phenolic resin as a thermosetting material. Test series were carried out applying different foaming agents, which differ in their decomposition process. One is exothermic, another one is endothermic and the third creates foam due to thermal expansion. The maximum weight reduction achievable for each foaming agent has been determined by gradually decreasing the shot volume until incomplete filling occurs. The foam structures are analyzed by microsections and evaluated with a digital image processing method. Additionally, the effect on mechanical properties is discussed by evaluating the bending stress. The exothermic foaming agent shows the best foam patterns achieving a maximum weight reduction of approximately 50 %, whereas the endothermic and the thermal foaming agent can reduce the weight by approximately 25 % and 20 % respectively. The bending stress decreases significantly with an increasing proportion of foam in all samples.

Keywords: foam injection molding; chemical blowing agent; foam; thermosets; phenolic resin; lightweight construction; insulation

INTRODUCTION

Thermosetting materials have many advantages. Due to their chemical structure, thermosetting materials exhibit a good inherent fire resistance and can withstand solvent or heat attacks [1, 2, 3]. These excellent thermomechanical properties at high temperatures, as well as the good media resistance, meets the demand of the automotive industry for high performance plastics [1]. Additionally thermosets like PF are more cost effective, than most high performance thermoplastics for those applications, like PPS or PPA.

On the other hand thermosets typically have a higher density compared to most thermoplastic materials [4]. Since lightweight design has become increasingly important for the automotive industry, research has to be done to decrease their specific weight. A commonly used technique for this is foam injection molding. Structural plastic foams show, besides weight reduction, enhanced insulation properties and mechanical damping [5]. Due to shorter cycle times and materials savings, foam injection molding is also very economic [6]. Whereas the process of foam injection molding is state of the art for thermoplastics, very little research has been done to apply this technique for thermosetting materials so far. To combine the advantages of foamed plastics and thermosetting materials, a stable process has to be developed. Currently only one research group at IKV Aachen is dealing with this specific topic. In their studies they achived a weight reduction of approx. 10% with a combination of an epoxy molding compound and hydrocerol BIF as a foaming agent [7].

This paper focuses on a phenolic resin as the thermosetting material. Three different chemical blowing agents were tested and compared. The maximum weight reduction for each blowing agent is evaluated by gradually decreasing the shot volume until incomplete filling occurs. Additionally the foam structure, as well as the mechanical properties, are analyzed and discussed.

BACKGROUND & PROCESS DESCRIPTION

For foam injection molding a foaming agent is required. In general, there a two options to introduce the blowing agent in the system: One is to use a physical blowing agent like N_2 or CO_2, which is metered into the polymer melt under high pressure in liquid form and dissolves in the melt under supercritical conditions [8]. When the melt is injected into the cavity, a sudden pressure drop occurs leading to a transition of the blowing agent into the gaseous state and foaming of the cavity [8]. The second option is to use a chemical blowing agent, which is added to the plastic pellets in powder form. At elevated temperatures, the blowing agent dissociates gas due to a chemical reaction [3, 5]. For thermoplastic foam injection molding this reaction takes place in the plasticizing cylinder. Just like in the physical foaming process, the polymer melt has to be kept under high pressure forcing the gas to dissolve in supercritical conditions. To hold up the pressure in the plasticizing unit, a complex engineering system including nonreturn valves, back pressure control, and shut off nozzle is required [8, 9]. When dealing with thermosetting materials such complex systems are not applicable due to the risk of creating dead volume spots, where curing of the thermosetting material can occur within the plasticizing unit. The process of foam injection molding of thermosetting materials can therefore not be adopted from the commonly known process of thermoplastics.

A distinct difference between thermosets and thermoplastics is the polymer melt's temperature during the injection molding cycle. For thermoset injection molding, the temperature in the plasticizing unit is comparatively low (about 90 °C) whereas the cavity is heated. Therefore, the decomposition temperature of the chemical blowing agent can be chosen to be above the temperature in the plasticizing cylinder. Accordingly, no gas is evolving in the cylinder but the blowing agent, distributed in the polymer melt, is injected into the cavity still in a solid state. As the cavity is heated up to 150 – 200 °C for thermoset injection molding processes, the decomposition of the blowing agent is introduced leading to a gas formation simultaneous to the start of the resin's curing reaction. Thus, a pressure-sealed design of the plasticizing unit with non-return valves and shut off nozzles is redundant which only makes the process possible.

The crucial step for the foam formation in thermoset injection molding is the ratio of the resin's curing and the blowing agent's reaction rate. If the curing of the thermosetting material is much faster than the decomposition of the blowing agent, the polymer matrix is in a solid state too soon. The evolving gas no longer foams the already rigid polymer but builds up internal pressure instead, possibly leading to mechanical destruction. If the curing of the thermosetting material is much slower on the other hand, the gas is evolving in a too early stage of the process. The gas bubbles are not solidified in the polymer matrix but escape the system. Since the foam structure is influenced significantly by the speed of the chemical reactions, test series with three different blowing agents were carried out. Focus of research were the foam structure, the weight reduction and the bending stress.

MATERIALS

The test series were carried out with phenolic novolac resins as a molding compounds. The hardener is in all cases hexamethylentetramine, which hydrolysis between 110°C and 120 °C, starts the curing of the resin. For reinforcement the compounds are filled with anorganic fillers. Additionally, blowing agents are added during the compounding process. The used compounds are Bakelite® 2117, Bakelite® LPF 2135 and Bakelite® LPF 2129, all provided by Hexion GmbH, Iserlohn, Germany. Bakelite® LPF 2117 contains an exothermic blowing agent which decomposes between 150°C and 190 °C releasing nitrogen, carbon monoxide, carbon dioxide and ammonia gas [6]. The blowing agent in Bakelite® LPF 2135 forms carbon dioxide and water in an endothermic reaction in the temperature range of 130 to 170 °C [6]. Bakelite® LPF 2129 is filled with thermoplastic microspheres, which consist of a PMMA shell filled with pentane gas. The microspheres do not decompose in a chemical reaction but create a foam structure with increasing temperature due to their thermal expansion.

EXPERIMENTAL

The specimen were produced with a ARBURG Allrounder 370 S 700-100 injection molding machine with an electrically heated molding tool. The thermoset aggregate is tempered with water. A compressionless, 25 diameter screw without backflow valve and shut off nozzle is used. The test specimen are plates with 4 mm thickness and 80 mm edge length. The processing parameters are listed in table 1.

TABLE 1. Processing parameters

Parameter	Value	Unit
Barrel temperature	65	°C
Die temperature	90	°C
Mold temperature	160	°C
Die diameter	4	mm
Injection speed	6	cm^3/s
Curing time	80	s
Holding pressure	-	bar

To evaluate the bending stress, test specimen are cut in the center of the plates perpendicularly to the flow direction and measured following DIN EN ISO 178 on a ZwickRoell ProLine Z005 TN from Zwick GmbH & Co. KG, Ulm, Germany. At the same position, rectangular test specimen with 10 mm edge length are cut out for density measurements based on Archimedes' principle as well as weighing with a high precision scale YDK03 from Sartorius AG, Göttingen, Germany. To visualize the foam structure, microsections are prepared and photographed in the area of interest with Axio Scope.A1 microscope of Carl Zeiss AG, Oberkochen, Germany. The evaluation of the foam structure images is done using the software ImageJ.

RESULTS AND DISCUSSION

To evaluate the maximum weight reduction, the shot volume is gradually decreased until incomplete filling occurs. Figure 1 shows the local density in the center of the plate at different degrees of shot volumes. 100 % shot volume is defined as the maximum shot volume, which can be injected into the cavity at complete filling conditions including 250 bar holding pressure. The lowest data point for each blowing agent marks the minimum shot volume at which complete filing of the cavity by the blowing agent's foam is still possible.

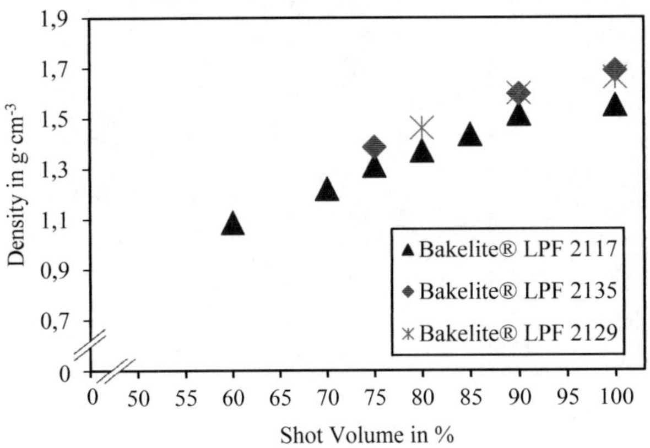

FIGURE 1: Density vs. parentage of shot volume

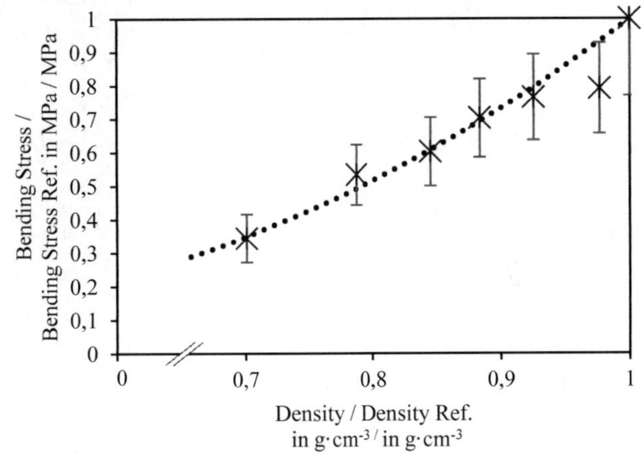

FIGURE 2: Mechanical properties of Bakelite® LPF 2117: bending stress vs. relative density

It is noticeable that, the density decreases in a linear manner with decreasing shot volume for all samples. The Compound Bakelite® LPF 2117 with exothermic foaming agent shows the highest foamability. Weight reductions up to 50 % can be achieved. Bakelite® LPF 2135 (endothermic) and Bakelite® LPF 2129 (thermoplastic microspheres) can reduce the weight by 25 % and 20 % respectively. These different foamabilities can also be visualized by microscopic investigations of microsections. Figure 3 illustrates the the center of the plate at 75 % shot volume for Bakelite® LPF 2117 and Bakelite® LPF 2135 compounds and 80 % shot volume for Bakelite® LPF 2129 respectively. Foam bubbles appear black in the reflected-light microscope pictures, whereas the synthetic material phase is white. The black portion of the microsection pictures quantifies the degree of foam (Figure 4). It can be seen that the compound Bakelite® LPF 2117 produces a dense foam structure with many small bubbles. Towards the top and bottom of the plate, the bubble size increases slightly. A decrease in the foaming degree can be observed over the flow path. The compound Bakelite® LPF 2135 forms fewer, but larger foam-bubbles. The effect of the growing bubbles towards the plate edge is more pronounced. A foam gradient over the flow path is clearly visible: While the degree of foam is 46 % directly behind the gate, it reduces to 26 % in the middle and 8 % at the end of the plate. The microspheres show foaming in the area of the gate at the beginning of the flow path. In the middle and at the end of the plate almost no foam formation can be observed.

The mould technology and the blowing agent's different types of reaction schemes can explain the results of the different pore size distributions. All blowing agents require heat from the outside to initiate foam formation. In thermoset injection moulding, heat is introduced by the heated mould cavity. This results in a heat flow from the plate's top and bottom to its center. As a result, a stronger growth of bubbles in the edge areas of the plates can be observed. Since the mould cavity is tempered with only one heating cartridge, installed near the gate, a mould temperature gradient of approx. 15 K over the flow path exists. The end of the plate is therefore less foamed than the beginning of the plate. Another influence is the different reaction mechanisms of the blowing agents. Bakelite® LPF 2117 contains an exothermic blowing agent. As soon as its activation energy is exceeded, it continues to react with additional heat formation and thus accelerates itself in an autocatalytic reaction. Therefore, the exothermic foaming agent can decompose in the middle and at the end of the plate at comparatively low temperatures as well. The result is a rather homogeneous, completely foamed plate. The endothermic blowing agent, on the other hand, requires a constant heat supply for decomposition. Consequently, a good foam result can be achieved in the hot zones at the edge of the plate and in the area of the gate. However, in the areas with lower temperatures the reaction is too slow to foam the plastic. The principle of microspheres is that their volume increases with thermal expansion. As a result, a round pore structure should appear. However, the irregular shaped pores, which occur exclusively in the gate area, suggest that the PMMA shell is largely destroyed by the high shear forces during injection and that the pentane gas escapes. Therefore, this process variant appears to be unsuitable for thermoset foam injection molding.

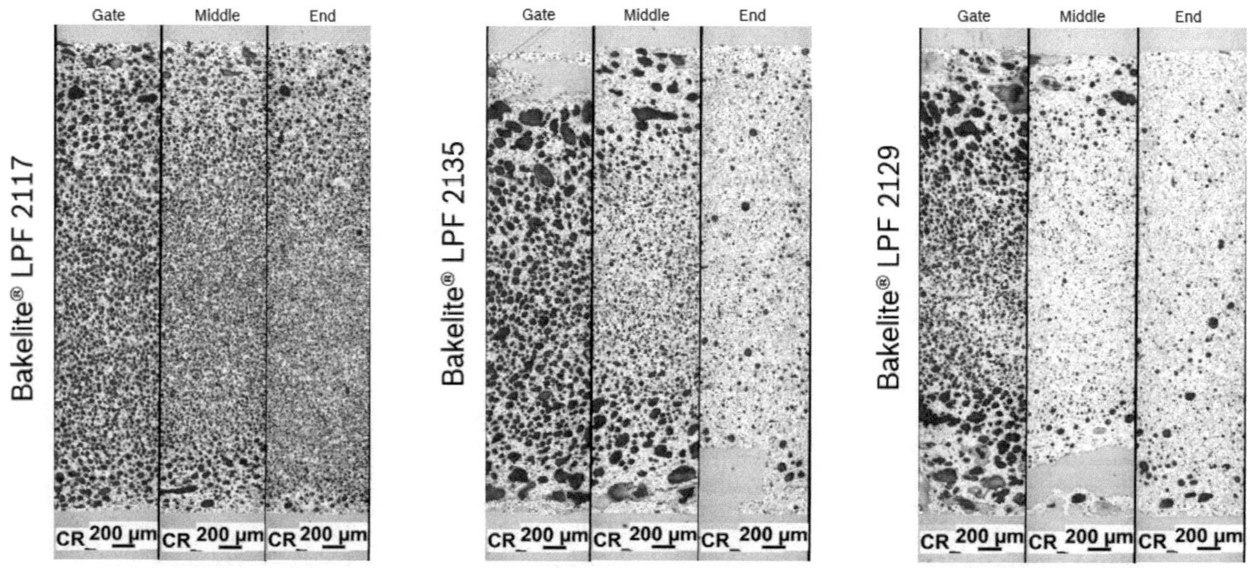

a) Exothermic blowing agent b) Endothermic blowing agent c) Thermoplastic Microspheres

FIGURE 3: Microsections of different foaming agents

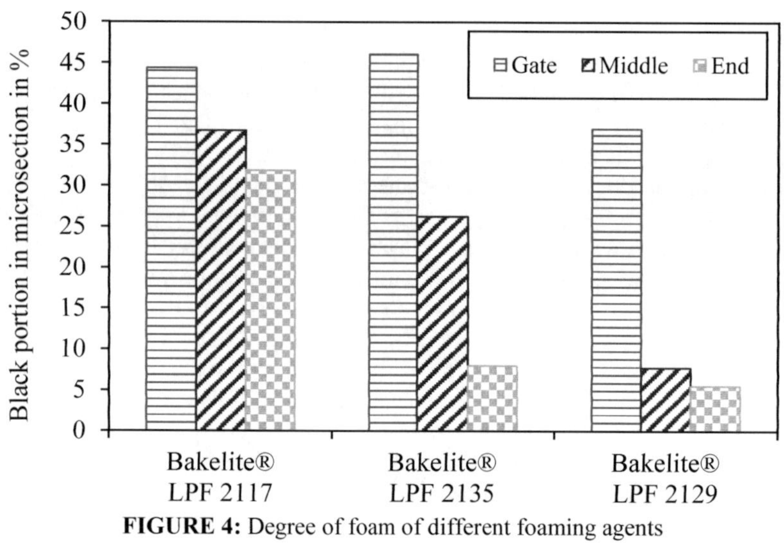

FIGURE 4: Degree of foam of different foaming agents

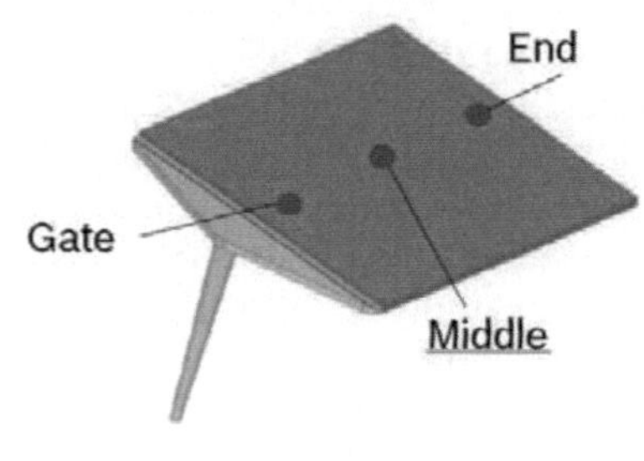

FIGURE 5: Measuring points

The mechanical properties are discussed using the bending stress of the samples foamed of compound Bakelite® LPF 2117. In analogy to thermoplastic foam injection molding, foamed thermosets also show a disproportionately strong decrease in mechanical properties with decreasing density. The decrease in bending stress can be described according to a power law (equation 1) with an exponent n between 2 and 2.5 for thermoplastic integral foams [6, 10, 11]:

$$\frac{X_F}{X_0} = \left(\frac{\rho_F}{\rho_0}\right)^n \quad (1)$$

X describes the mechanical property of the foam and ρ is the density. The index F refers to a foamed sample, whereas the index 0 refers to the compact sample at complete filling conditions without foam. Figure 2 illustrates the correlation of the relative bending stress and the relative density of the samples of compound Bakelite® LPF 2117. They can also be described according to equation 1, but the exponent n is 2.92. Thus, the reduction of the bending stress is more pronounced than with thermoplastic integral foams. The reason is the foam structure of the plates, which currently have no skin-core structure with foamed core and compact boundary layer, like integral foams. Therefore, the notch effect of the pores in the edge area lead to a strong decrease in the mechanical properties. The area of future research is therefore to produce a thermoset integral foam. For this purpose, test series are planned in which the foam structure is not produced by low-pressure foam injection molding (volumetric underfilling) but by an expanding mold technique.

CONCLUSION

In order to demonstrate the potential of different blowing agents for thermoset foam injection molding, test series were carried out with compounds, containing an exothermic blowing agent, an endothermic blowing agent and thermoplastic microspheres filled with pentane gas. It is shown, that with the exothermic blowing agent the component weight can be reduced up to 50 % before incomplete filling occurs. With the endothermic blowing agent and the thermoplastic microspheres, weight reductions of up to 25 % and 20 % respectively can be achieved. The foam structures of the different blowing agents differ significantly. The decomposition behavior of the exothermic blowing agent leads to a homogeneous pore distribution, whereas the endothermic blowing agent and thermoplastic microspheres show gradients in pore distribution and size.

The foam structure's impact on mechanical properties is shown exemplarily by the bending stress. Similar to the thermoplastic foams, the decrease of the bending stress with decreasing density can be described by a power law. Yet, the decrease is more pronounced compared to thermoplastic integral foams. Therefore, future field of research will be the process development for a thermoset integral foam with enhanced mechanical properties.

ACKNOWLEDGEMENTS

The authors acknowledge the financial support by the Federal Ministry for Economic Affairs of Germany in the project FoamSet (project number 03ET1424C) supervised by the project executing organization Juelich. The responsibility with regard to contents lies with the authors.

REFERENCES

1. H. Domininghaus, *Kunststoffe*. Berlin, Heidelberg: Springer Berlin Heidelberg, 2012. (033)
2. H. Domininghaus, P. Eyerer, P. Elsner, und T. Hirth, *Die Kunststoffe und ihre Eigenschaften*, 6., Neu bearb. und erw. Aufl. Berlin: Springer, 2005. (P05)
3. W. Woebcken, W. Adam, G. W. Becker, und D. Braun, Hrsg., *Duroplaste: mit 803 Bildern und 256 Tabellen*, 2., Völlig neubearb. Aufl. München: Hanser, 1988. (P04)
4. M. Bonnet, *Kunststofftechnik: Grundlagen, Verarbeitung, Werkstoffauswahl* und Fallbeispiele, 2., Überarb. und erw. Aufl. Wiesbaden: Springer Vieweg, 2014. (048)
5. Andreas Praller, „Physical Blowing Agents with Inert Gases", *Kunststoffe*, Bd. 2005, Nr. 06, S. 96–99, 2005. (008)
6. V. Altstädt und A. Mantey, *Thermoplast-Schaumspritzgießen*. München: Hanser, 2011. (TIB 001) (TIB001)
7. Christian Hopmann and Stefan Haase, „Investigation of chemical foaming agents application for thermoset injection moldng", in *ANTEC® 2017: proceedings of the technical conference & exhibition Anaheim*, Anaheim, USA, 2017, S. 667–671. (066)
8. M. Rohleder und F. Jakob, „Foam Injection Molding", in *Specialized Injection Molding Techniques*, Elsevier, 2016, S. 53–106. (046)
9. T. A. Osswald, L.-S. Turng, und P. J. Gramann, Hrsg., *Injection molding handbook, 2nd ed.,* Updated 2nd ed. Munich : Cincinnati: Carl Hanser Publishers ; Hanser Gardner Publications, 2008.
10. D. V. Rosato, D. V. Rosato, und M. G. Rosato, Hrsg., *Injection molding handbook*, Softcover reprint of the hardcover 3. ed. New York: Springer, 2000.
11. S. Semerdjiev, *Introduction to Structural Foam*. Connecticut: Society of Plastics Engineers, 1984.

Properties Of Bead Foams With Increased Heat Stability Made From The Engineering Polymer Polybutylene Terephthalate (E-PBT)

Tobias Standau[a], Peter Schreier[b], Kathrin Hilgert[a] and Volker Altstädt[a,b,c*]

[a] *University of Bayreuth, Dept. of Polymer Engineering, Universitätsstraße 30, 95447 Bayreuth*
[b] *Neue Materialien Bayreuth GmbH, Gottlieb-Keim-Straße 60, 95448 Bayreuth*
[c] *Bavarian Polymer Institute, Universitätsstraße 30, 95447 Bayreuth*

Abstract. Bead foams from the standard polymers polystyrene (EPS) and polypropylene (EPP) are widely established in many applications but possess a rather low heat stability. Currently other polymers such as engineering and high temperature polymers get more and more in the focus of research. For instance, bead foams made from polybutylene terephthalate (E-PBT) are able to withstand a higher thermal load. In the current study the mechanical behavior of this new material is described. Tensile tests showed, that E-PBT and EPP, have a similar Young's modulus. Judged on the fracture surface a good bead fusion can be achieved for E-PBT as predominantly trans-bead fracture occurs. Within this work it could be shown that E-PBT possesses a higher compressive strength compared to EPP, especially at elevated temperatures up to 150 °C. Furthermore, the burning behavior was evaluated revealing that E-PBT ignites later and in case of fire it releases less heat compared to EPP.

Keywords: Bead Foaming, Polybutylene Terephthalate, compressive behavior, burning behavior

INTRODUCTION

Using bead foams, it is possible to manufacture parts with low density and complex shapes and good insulating properties. Because of this superior combination of advantages the consumption is steadily growing. However, mainly bead foams from commodity polymers are established in the market, such as (i) expandable polystyrene (EPS) which is often applied as building insulation or as packaging material and (ii) expanded polypropylene (EPP) which can be typically found in automotive exterior and interior applications (e.g., bumper cores and sun visors). However, for some processes or applications these polymers reach their limit. For instance, if it is desired to have a bead foam that is biodegradable or has a high energy return or possess a higher heat resistance. Therefore, in the past decades research and development focused on applying bead foam technology on other polymers such as polylactide (EPLA) [1, 2] to create a biodegradable alternative for EPS packaging, or thermoplastic polyurethane (E-TPU) [3] which can be found meanwhile in many shoe soles. Several attempts to make bead foams with a higher thermal resistance were also made, including PMI (from Evonik AG) [4], PET (from Armacell S.A.) [5], PESU (BASF SE) [6] or PBT (from our own previous works) [7, 8]. This is necessary as the above-mentioned classical bead foams only can be used up to temperatures of 80°C (EPS) and 110°C (EPP). However, bead foaming of engineering and high temperature polymers is not trivial. Several approaches to produce bead foams are known [9]. Coupling a foam extrusion line with an underwater granulator is a convenient method for continuous bead foam production. In the case of polyesters, such as PBT the low melt strength makes it necessary to add chemical modifiers and also the fusion of the beads requires specialized processes as pointed out in our former work [8].

Even though some commercial products are available meanwhile, less reports on the final properties can be found in the scientific literature. Therefore, the temperature dependent compression behavior and the bead fusion quality is shown within the current work for E-PBT. Also, the burning behavior was quantified for the first time.

EXPERIMENTAL

Materials

PBT Pocan B1300 was purchased from Lanxess AG. It was modified during the foam extrusion process with 1 wt% of the multifunctional epoxy-based chain extender Joncryl 4468 ADR from BASF SE. Foam extrusion was carried out on a Dr. Collin Tandem line with attached under water granulator (UWG) from Gala Industries using 2 wt% CO_2 as blowing agent as described in our previous work [8]. The obtained bead foams were subsequently welded by steam chest molding process on a modified Teubert TVZ 162/10099 with a high pressure steam generator HaeCo II from the company UniBell. A more detailed description of the processing can be found in our former study[8]. As reference Neopolen P from BASF SE with a final part density of approx. 190 kg/m^3 was used.

Methods

Mechanical Analysis

Tensile testing was done on a universal testing machine Z020 from the company Zwick Roell according to standard DIN 53420.A pre-load of 5 N was applied. Test was started with a low testing speed of 1 mm/min allowing to determine the Youngs modulus and then testing speed was increased to a final value of 25 mm/min. Series of 6 samples were measured.

Shear testing was done according to ASTM C273 on a universal testing machine Z020 from Zwick Roell. Strain was detected optically with extensometer videoXtens from the same company. Samples with the dimension of 140 mm x 50 mm x 11 mm were glued on metal plates using scotch-weld DP490. For both, EPP and E-PBT four samples were tested using a pre-load of 20 N and a testing speed of 0.5 mm/min

Compression tests were carried out according to standard DIN 53412 with cuboids (50 mm x 50 mm x 30 mm) on a Zwick Roell Z1475 with a heated chamber with a pre-load of 10 N and a test speed of 3 mm/min. Measurements were carried out at 25, 110 and 150 °C. Each series consisted out of 15 samples.

Prior to all mechanical measurements the density was determined for each specimen using the Archimedes principle to ensure comparability of the samples in and in between the series.

Visual Analysis

Subsequent to tensile tests the fracture surface were sputter-coated with gold and investigated by using a scanning electron microscope JEOL JSM -6510 at 10 kV. Using ImageJ software the surface for inter- and trans-bead fracture were determined for 3 samples to evaluate the bead fusion.

Burning Behavior

To evaluate the burning behavior, combustion experiments on the basis of ISO 5660-1-2002 were carried out on a cone calorimeter iCone from Fire Testing Technologies (East Grinstead, United Kingdom) with servomex Xentra gas analysers (Crowborough, United Kingdom). Samples with dimensions of 100 mm x 100 mm x 30 mm were cut out of the welded plates and placed 25 mm under the ignition coil. An incident heat flux of 35 kW/m^2 was set for all experiments. Heat release rate (HRR), total heat release (THR), total smoke production (TSP) and maximum average of heat release emission (MAHRE) were measured during the experiments. Data point collection was stopped after the flame extinguished and HRR values underwent 30 kW/m^2. For each material 3 samples were tested.

RESULTS AND DISCUSSION

Mechanical Testing

Tensile Tests

For the compact materials it is known that unfilled PP possess a higher elongation at break of up to several hundred percent [10] compared to PBT (approx. 8%) [11]. The Young´s modulus of PBT is around 2700 MPa [11] and for PP between 1000 and 2000 MPa [12]. The results of the tensile tests of EPP and E-PBT are shown in Figure 1 . For both foams a similar modulus of approx. 95 MPa was measured. However, E-PBT fractures at lower stress and lower deformation. EPP exhibits a much higher deformation of approx. 12%. Furthermore, two representative SEM images of the fracture surfaces of E-PBT and EPP, respectively are shown on the right side of Figure 1 after tensile testing.

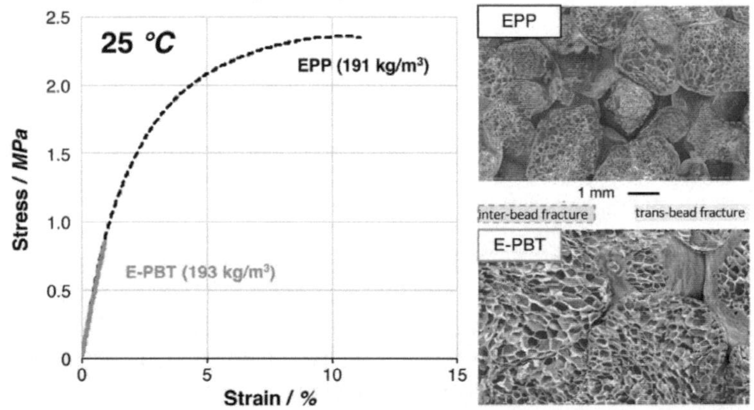

Figure 1: Left: Tensile tests for E-PBT and EPP. Right: SEM images of the fracture surfaces. Inter-bead fraction is marked orange.

Clearly the homogeneous foam structure can be seen for both materials. More interestingly here is that in both cases the fracture appears to be predominantly within the bead, i.e., so called trans-bead fracture revealing a good fusion of the individual beads as stated in the works of Rossacci et al. [13, 14]. The average of three samples is 68,3 +/- 5,5 % for EPP and 92.2 +/- 4 % for E-PBT. Thereby, the steam chest molding process can be judged as sufficient.

Shear Tests

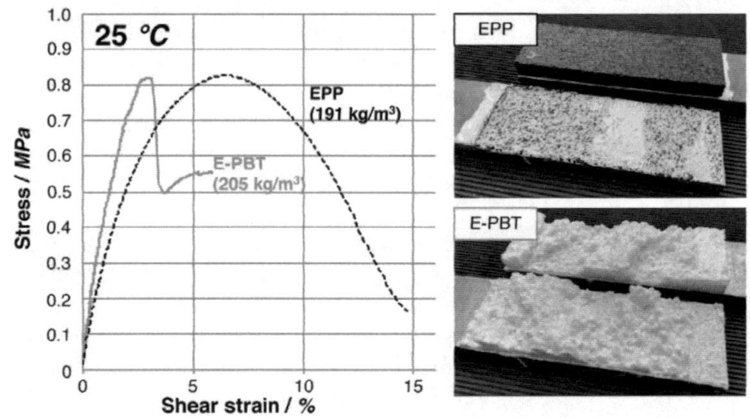

Figure 2: Shear testing of EPP and E-PBT.

Shear tests were conducted for samples of the same density range. The stress strain curves are shown in Figure 2. The shear modulus for E-PBT (35.7 MPa) is higher than for EPP (27.5 MPa). EPP shows a higher deformation. The shape of the curve with a strain maximum at 7% can be explained from the images of the fractured samples. It can be seen, that in case of EPP the sample holder was ripped of the sample as the failure happened within the glue. For E-PBT the failure happens within the sample. Hence, no reliable value for the shear strength can be determined for EPP; but it is anticipated to be higher than shown by the curve.

Compression Tests

In Figure 3 the stress-strain curves for EPP and E-PBT at different temperatures are shown. For E-PBT the compressive strength at 10% strain reduces from 1.82 MPa at room temperature down to 0.59 MPa at 110 °C. For the same strain, the compressive strength of EPP is 1.29 MPa at room temperature and 0.24 MPa at 110 °C. Obviously both materials have reduced compressive strength at this elevated temperature, because the testing temperature is carried out above the T_g. However the compressive strength of E-PBT is still almost twice as high. Because of its higher melting point, E-PBT has a nine times higher compressive strength of 0.45 MPa at 150°C. At this temperature the EPP starts to melt and cannot withstand the applied force. It can be summarized, that at all temperatures the compressive strength and modules are higher for E-PBT.

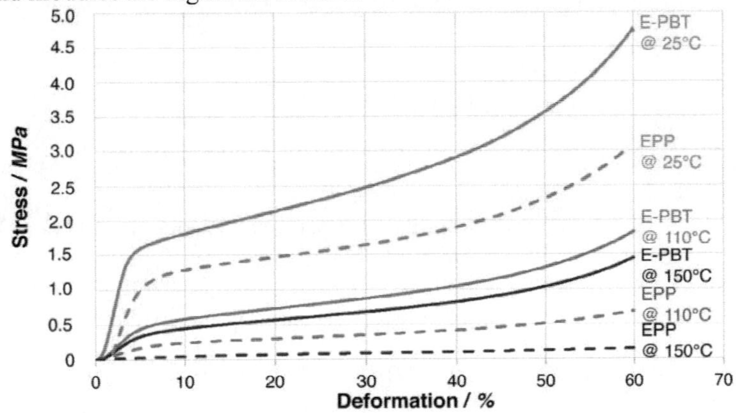

Figure 3: Compression tests of EPP (190 kg/m³) and E-PBT (220 kg/m³) at different temperatures.

Burning Behavior

The results from the cone calorimeter trials are shown in Figure 4. Neither EPP nor E-PBT leave residues after burning. For both materials similar total smoke production (TSP) between 25 and 27 m² was measured. The time to ignition is 16 seconds for EPP and 47 seconds for E-PBT. In the beginning EPP burns with a higher heat release rate (HRR). After approx. 3 minutes the HRR of E-PBT exceeds EPP and is culminating with a peak heat release rate (PHRR) of 1154 kW/m². E-PBT extinguishes earlier than EPP. It has to be mentioned, even though the density of E-PBT is higher (i.e., more mass) the total heat release (THR) and the maximum average of heat release emission is lower than for EPP as revealed by the total area under the shown curves.

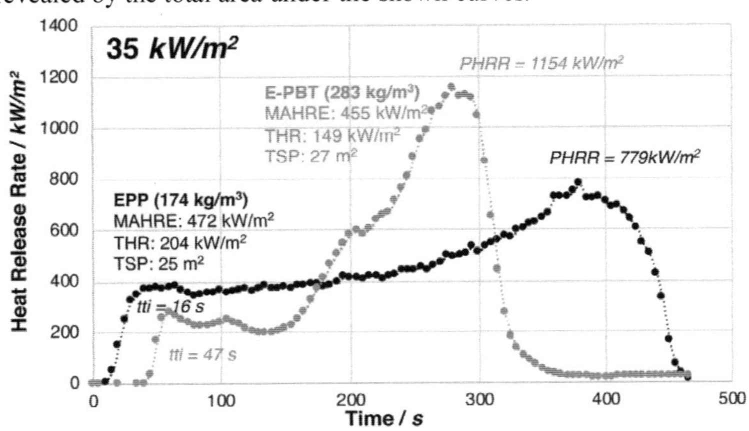

Figure 4: Results of the Cone measurements for E-PBT and EPP at 35 kW/m².

CONCLUSIONS

In this brief study on the mechanical behavior of bead foams made from the engineering thermoplastic polybutylene terephthalate the tensile-, shear and compression behavior has been investigated. It was shown that the tensile modulus of E-PBT is similar to EPP and that good bead fusion was achieved as trans-bead fracture dominates. Furthermore it could be shown, that EPP can deform more under shear stress. However, compressive strength is much higher for E-PBT even at elevated temperatures. The burning tests reveal that E-PBT burns with a higher peak heat release rate but with lower total heat release, a similar total smoke production and has a longer time to ignition. However, the incorporation of flame retardants seem to be a good approach to further reduce the flammability.

ACKNOWLEDGMENTS

We would like to express our acknowledgment for the financial support by German Research Foundation (grant number AL474/28-1). Also, we thank Sebastian Gröschel (University of Bayreuth) for carrying out the foam extrusion and Christian Brunner (Neue Materialien Bayreuth GmbH) for the steam chest molding trials.

REFERENCES

1. Britton RN, Hendrikus FA, Van Doormalen C, et al (2012) Coated Particulate Expandable Polylactic Acid, US Patent, US 8268901 B2. 2:1–7
2. Lohmann J, Sampath BDS, Gutmann P, et al (2013) Process for Producing expandable pelletized Material which comprises Polylactic Acid, US 2013/0345327 A1
3. Prissok F, Braun F (2007) FOAMS BASED ON THERMOPLASTIC POLYURETHANES. 25
4. Richter T, Schwarz-Barac S, Bernhard K, et al (2015) Bead Polymer for Producing PMI Foams US20150361236 A1
5. Strasser JP (2011) Process for Producing PET Pellets, and PET Pellets, WO2011063806A1
6. Fonseca I, Bräuer J, Graessel G, et al (2018) Advances in high performance thermoplastic foams Results and discussion. In: SPE Foams 2018
7. Köppl T, Raps D, Altstädt V (2014) E-PBT - Bead foaming of poly(butylene terephthalate) by underwater pelletizing. J Cell Plast 50:475–487 . doi: 10.1177/0021955X14528524
8. Standau T, Hädelt B, Schreier P, Altstädt V (2018) Development of a Bead Foam from an Engineering Polymer with Addition of Chain Extender : Expanded Polybutylene Terephthalate. Ind Eng Chem Res 57:17170–17176 . doi: 10.1021/acs.iecr.8b04799
9. Standau T, Zhao C, Castellón SM, Bonten C (2019) Chemical Modification and Foam Processing of Polylactide (PLA). 1–38 . doi: 10.3390/polym11020306
10. Tjong C, Shen JS, Polytechnic C, Kong H Mechanical Behavior of Injection Molded beta-Crystalline Phase Polypropylene
11. Lanxess Technical Data Sheet PBT Pocan B1300
12. Pukanszky B, Mudra I, Staniek P (1996) Relation of crystalline structure and mechanical properties of nucleated polypropylene. J Eng Appl Sci 2:2317–2321 . doi: 10.1002/vnl.10165
13. Rossacci J, Shivkumar S (2003) Bead fusion in polystyrene foams. J Mater Sci 38:201–206 . doi: 10.1023/A:1021180608531
14. Rossacci J, Shivkumar S (2003) Influence of EPS bead fusion on pattern degradation and casting formation in the lost foam process. J Mater Sci 38:2321–2330 . doi: 10.1023/A:1023924211663

Effect of TPU soft segment molecular weight on TPU's bead foaming behavior

Bati, B.[1], Kucuk, E.B.[1], Nofar, M.*[1,2]

1: Polymer Science and Technology Program, Istanbul Technical University, Maslak, Istanbul, 34469, Turkey
2: Metallurgical & Materials Engineering Department, Faculty of Chemical and Metallurgical Engineering, Istanbul Technical University, Maslak, Istanbul, 34469, Turkey
** Corresponding Author: nofar@itu.edu.tr*

The bead foaming behavior of three different thermoplastic polyurethane (TPU) samples with different polyol soft-segment molecular weights was investigated through a lab-scale bead foaming reactor. These TPU samples possessed similar hard-segment content and soft-segment molecular weights of 1000, 2000, and 3500 g/mol. The evolution of the hard-segment crystallization during the isothermal saturation as well as the increased solubility of supercritical CO_2 within the soft-segment region are, respectively, determinative parameters on improving the heterogeneous cell nucleation and expansion of the expanded TPU (ETPU) bead foams. The effect of saturation temperature, pressure, and time on the hard-segment crystallization and hence on the foaming behavior of TPU samples were explored. The results showed that the increase in soft-segment molecular weight increased the expansion ratio of the bead foams while the processing parameters differently influenced the hard-segment crystallization and CO_2 solubility and hence the foaming behavior of ETPU foams.

Keywords: TPU, bead foaming, soft segment, molecular weight, hard-segment, crystallization

INTRODUCTION

Plastic foaming has been revealed to be a solution to the continuous rise in the cost of plastic resins because less material is required. On the other hand, cheaper and light weight products could be developed, with improvements in mechanical property, thermal insulation, acoustical insulation, optical property, or a combination of these [1-2].

Apart from other foaming technologies, the bead foaming technology is considered as a highly promising alternative which can produce three-dimensionally shaped foam products with ultra-low densities. This manufacturing process involves molding and sintering of tiny foamed plastic beads into plastic foam components through steam chest molding. Despite the huge potential of the bead foaming technology in various applications, only a few polymers are currently being used as bead foam products in a wide range of applications. These bead foams could be listed as expanded polystyrene (EPS), expanded polypropylene (EPP), and expanded polyethylene (EPE) [1-5].

Recently, the development of expanded TPU (ETPU) bead foams has become of a great attention for sport and flexible packaging applications [5]. TPU is an elastomeric multi-block copolymer, whose molecular structure consists of hard and soft segments and these segments could be phase separated due to their polarity differences [6]. The phase separation could influence the molecular mobility and hence the crystallization of each domain; therefore, both soft and hard segments could form crystalline regions. This crystallization is generally more possible for the hard segment blocks which indeed provides the rigidity to the TPU and the remaining soft segments contributes to the flexibility and elastomeric behavior of TPU.

Few patents have been published on ETPU bead foam products [7-8], however, research is still ongoing to introduce ETPU bead foams with improved foaming behavior such as high expansion and high cell density. These foam properties could eventually influence the final properties of the bead foams [9]. In this study, TPUs that are different in soft segment length are foamed by depressurization through a small scale bead foaming reactor. For a better understanding on how the process parameters affect the overall cell morphology and final foam behavior, TPU samples were foamed under various saturation times, temperatures, and pressures. The differential scanning calorimetry (DSC) thermograms of samples were also explored to understand how the hard segment crystallization vary with the change in soft segment molecular weight.

EXPERIMENTAL

Materials and Foam Preparations

TPU samples were supplied directly from Huntsman Polyurethanes, Belgium. The TPUs possessed a similar hard segment content of around 36-37% and based on 4,4' MDI and 1,4 BDO. The soft segment of the TPU samples was based on a polyester polyol of adipic acid and 1,4 BDO with molecular weights of 1000, 2000, and 3500 g/mol, which are referred to as SS-1000, SS-2000, and SS-3500, respectively. The DSC first heating and cooling thermograms of the noted TPUs (using a Perkin Elmer DSC400) are disclosed in Figure 1. As seen, during the heating cycle the increase in soft segment molecular weight, increased the hard segment crystallinity while broadening the crystal melting peak indicating the existence of various sizes and various ranges of hard segment closed-pack crystals. This is because the longer soft segments provide a better molecular mobility for the hard segments to crystallize with various sizes and/or perfection ranges. According to the cooling graph, an increase in soft segment molecular weight accelerated the hard blocks crystallization rate due to the facilitated molecular motion.

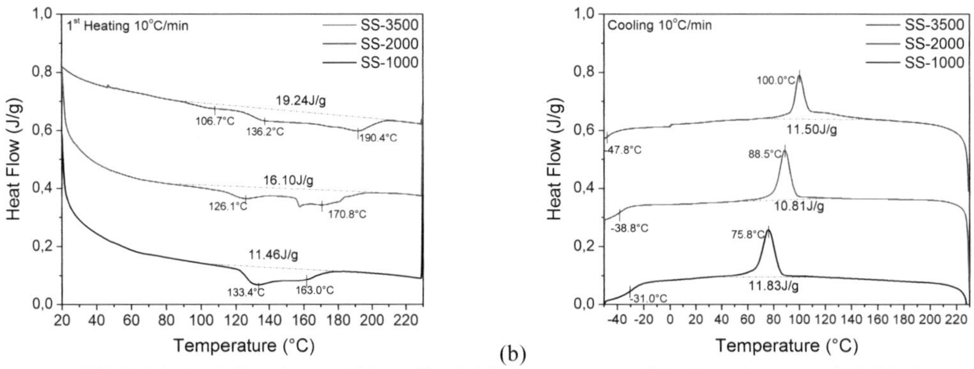

(a) (b)

FIGURE 1: (a) 1st heating and (b) cooling DSC thermograms of TPUs at the rates of 10°C/min.

TPU bead foams were prepared by firstly drying the samples at 50°C overnight in a vacuum oven. The saturation temperatures were determined according to the hard segment crystal melting peaks and the processing window was fixed by running series of experiments at different temperature ranges. Few pellets were loaded into the high pressure chamber. The chamber was then pressurized with CO_2 using a syringe pump at various saturation pressures of 1000, 1500, and 2000 psi at the desired temperatures for 30 min. The saturation time effect was also studied by saturating the samples at 2000 psi for 15, 30, and 60 min at the given temperatures. After the saturation, the foaming occurred through a fast pressure release via rapidly opening the ball valve connected to the chamber followed by quenching the chamber within a cold water bath.

Foam Characterization

Density measurement of the foamed samples obtained from experiments were done using a Mettler A1201 density measurement device. Archimedes method was used to calculate the density of the foamed samples. Then, the expansion ratio of each foamed samples was calculated using equation 1, where $\rho_{polymer}$ and ρ_{foam} are density of neat polymer and the foamed sample.

$$\emptyset = E.R = \frac{\rho_{polymer}}{\rho_{foam}} \tag{1}$$

Scanning electron microscope (SEM) (JEOL S410) was used to observe the cellular morphology of the foams. The cell density of each foamed sample was calculated using equation 2 where N is the number of cells, A (cm^2) is the area.

$$\text{Cell density} = \left(\frac{N}{A}\right)^{3/2} . \frac{\rho_{polymer}}{\rho_{foam}} \tag{2}$$

RESULTS AND DISCUSSIONS

Saturation temperature is one of the most important process parameters that control the cell nucleation and expansion ratio. Figures 2 and 3, respectively, reveal the SEM images of the foamed samples and the expansion ratio and cell density values after foaming at various saturation temperatures. As seen in all TPU foamed samples, the expansion ratio increases with the saturation temperature due to the reduced stiffness where with further increase of temperature the shrinkage of the expanded samples is more likely due to the reduced hard segment crystal content. All TPU bead foams showed morphologies with cell sizes in the range of 10-40 µm, and cell densities between 10^8 cell/cm^3 and 10^9cell/cm^3. The increase in saturation temperature could indeed further melt the hard segment crystals and hence fewer crystals could contribute to the heterogeneous cell nucleation and higher expansion ratios could be achieved. This is while the increase in temperature reduces the solubility of CO_2 in polymer/gas domains. The increase in soft segment molecular weight the processing window increases to a wider saturation temperature range due to the wider hard segment crystals melting peak existed in cases with longer soft segments (illustrated in Figure 1). Furthermore, the higher molecular weight polyol causes a higher solubility of CO_2 in TPU due to the increase in the free volume of soft segment chain with more flexibility [10].

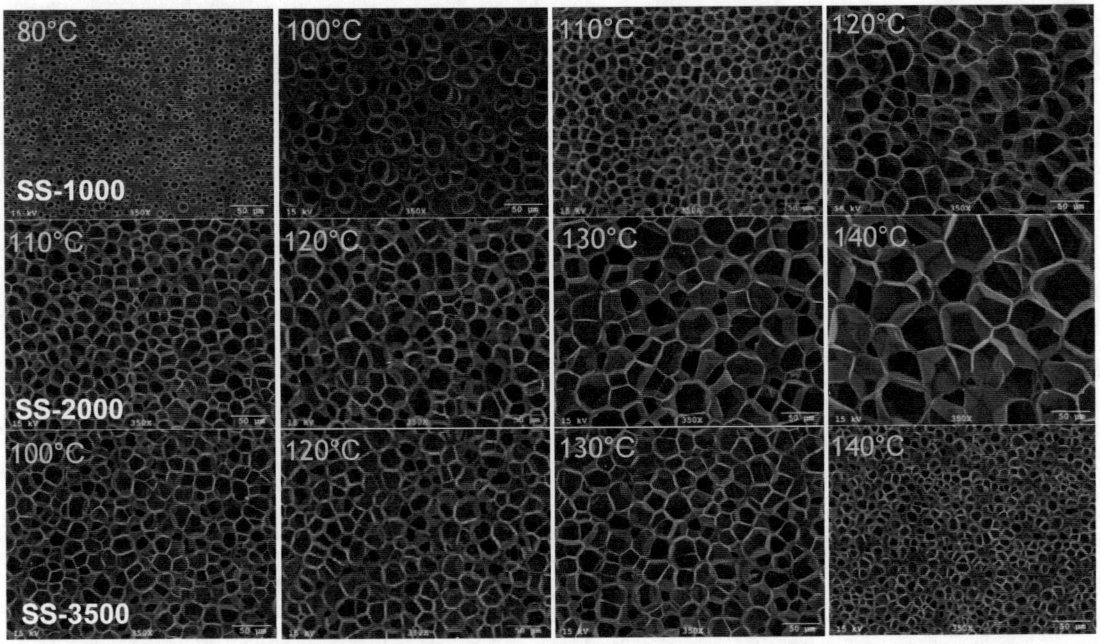

FIGURE 2: SEM micrographs of the TPU foamed samples after being saturated at various temperatures

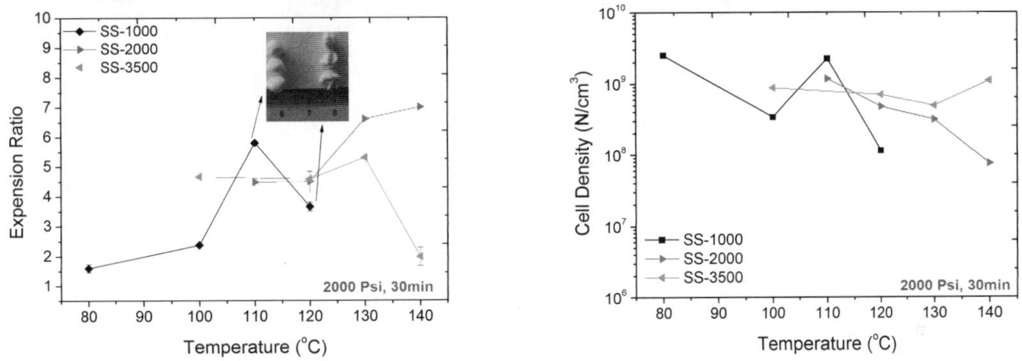

FIGURE 3: Expansion ratio (a) and cell density (b) of the TPU foamed samples at various saturation temperatures

To understand the effect of saturation pressure on foaming behavior of the noted three different TPUs, experiments were performed at the saturation pressures of 2000, 1500 and 1000 psi at the selected saturation temperatures of each sample for a given saturation time of 30 min. Figures 4 and 5, respectively, reveal the SEM images of the foamed samples and the expansion ratio and cell density values after foaming at various saturation pressures. Except the foaming at 1000 psi, other TPU foams showed microcellular morphologies with cell sizes in the range of 10-70 μm, and cell densities between 10^8 cell/cm^3 and over 10^9cell/cm^3 as shown in Figure 5. According to Figures 4 and 5, it can be observed that increasing the saturation pressure leads to an increase in foam expansion, cell density and decrease in average cell size. On the other hand, foam expansion was observed under 1500 psi, and even more under 2000 psi. According to the Cell Nucleation Theory, the higher saturation pressure not only improves the CO_2 solubility in TPU but also decreases the free energy barrier for bubble nucleation [11]. More nuclei are hence formed, causing a decrease in cell size and an increase in cell density, during the depressurization step. This is while the influence of existing hard segment crystals during the saturation on the heterogenous cell nucleation should also be taken into account.

The effect of saturation time on the foaming behavior of TPU samples was also studied by saturating samples for 15, 30 and 60 min under 2000 psi and at a given saturation temperatures. The SEM together with the expansion and cell density results are shown in Figures 6-7, respectively. All TPUs showed microcellular morphologies with cell sizes in the range of 10-40 μm, and cell densities between 10^8 cell/cm^3 and over 10^9cell/cm^3. SS-1000 and SS-2000 have higher expansion ratio and largest cell size at 15min comparing to SS-3500, nonetheless rapid growth caused shrinkage since CO_2 dissolution could not be completed at such a short time. On the other hand, expansion ratio of SS-3500 increased with longer saturation time. The reason of the difference in expansion behavior between the low and high molecular weight polyols versus saturation time is that SS-3500 has more entanglement because of its longer soft segment chains and more gas can dissolve with longer time. Moreover, longer saturation time results in obtaining more closely packed crystals [9].

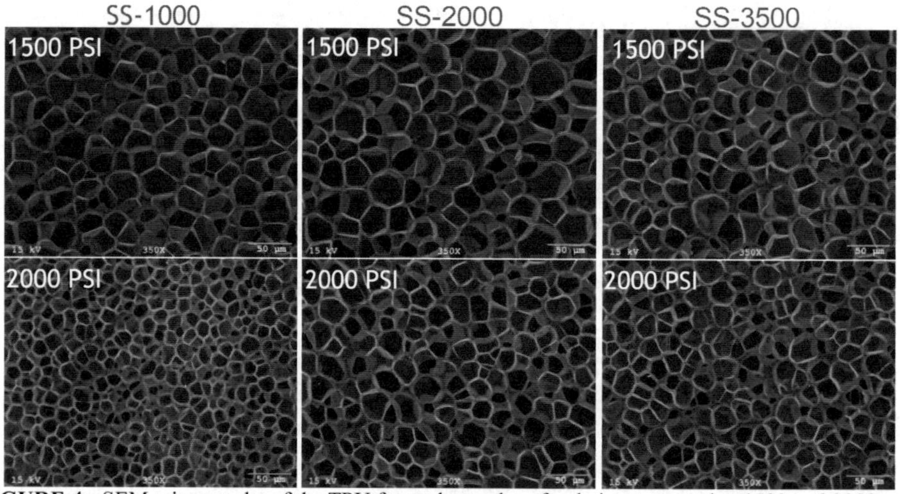

FIGURE 4: SEM micrographs of the TPU foamed samples after being saturated at 1500 and 2000 psi

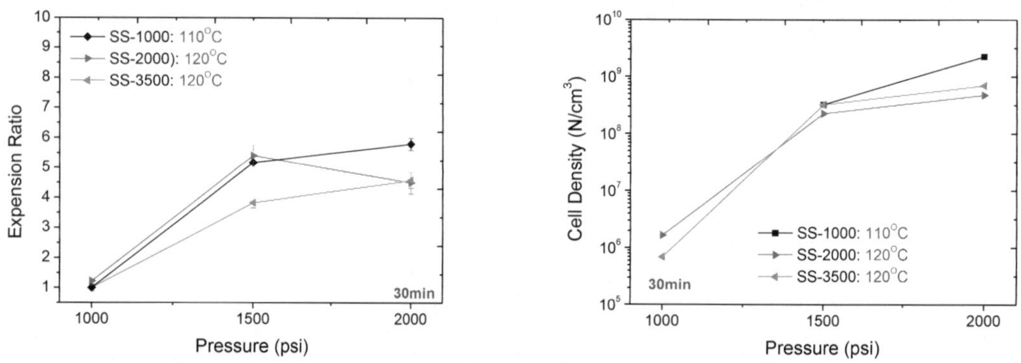

FIGURE 5: Expansion ratio (a) and cell density (b) of the TPU samples foamed at various saturation pressures

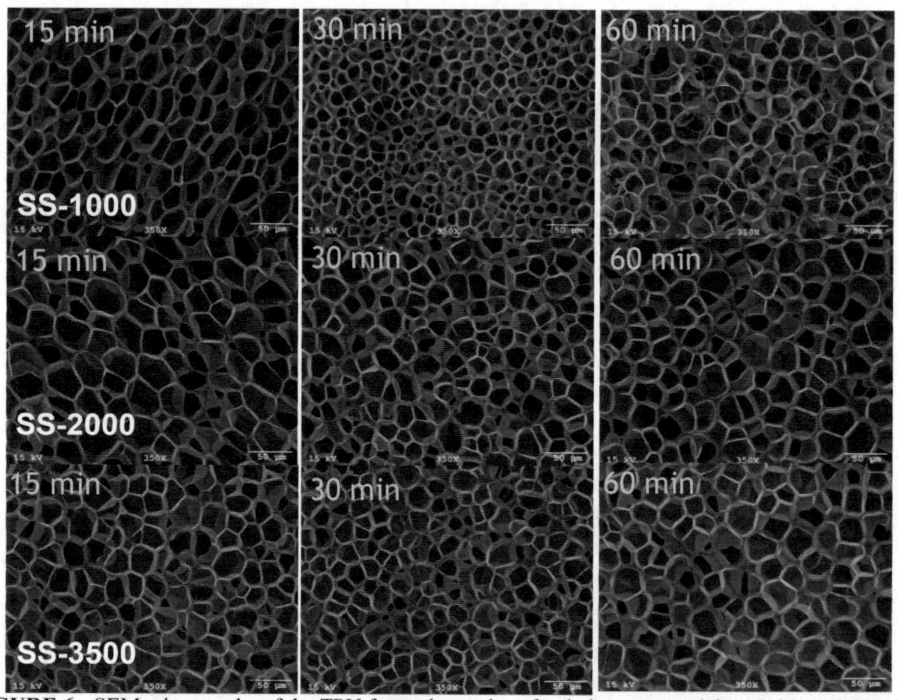

FIGURE 6: SEM micrographs of the TPU foamed samples after being saturated for 15, 30, and 60 min

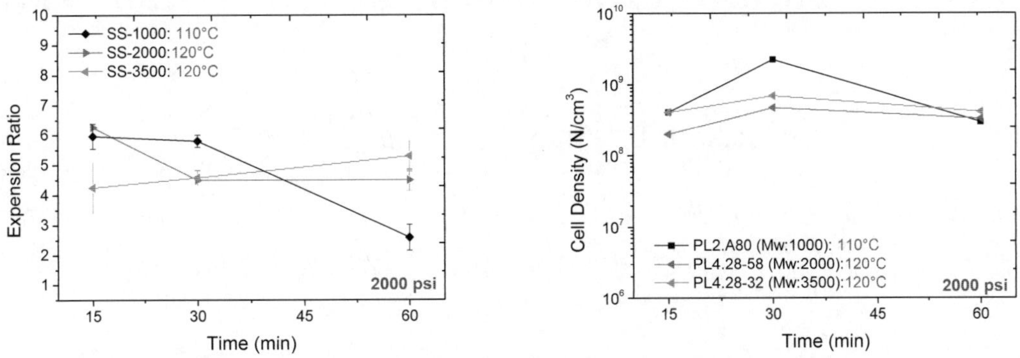

FIGURE 7: Expansion ratio (a) and cell density (b) of the TPU samples foamed at various saturation times

CONCLUSION

In this study, ETPU bead foams with different length of polyols were prepared in a lab-scale autoclave bead foaming setup. The effect of process parameters on the foaming behavior of TPUs was investigated. The saturation temperature had a significant effect on the foaming behavior of TPUs. The expansion ratio and cell density was improved with increasing temperatures for TPUs. Increasing the saturation pressure increased the CO_2 solubility causing a better cell nucleation and expansion. The increase in soft segment molecular weight increased the hard segment crystals range and hence the foam processing window increased in TPU with longer soft segment. On the other hand, the longer soft segment could have caused higher expansions although the foam shrinkage must also be considered.

ACKNOWLEDGEMENT

The authors like to thank Huntsman Polyurethanes, Belgium for suppling the materials and sponsoring the project.

REFERENCES

1. Nofar M, Park C.B, Polylactide Foams, Fundamentals, Manufacturing, and Applications, Elsevier, ISBN: 9780128139912 (2017)
2. Nofar M, Park C.B, *Prog. Polym. Sci*, **39**, 1721-1741 (2014).
3. Nofar M, Ameli A, Park C.B, *Polymer*, **69**, 83-94 (2015).
4. Nofar M, Guo Y, Park C.B, *Ind. Eng. Chem. Res.*, **52**, 2297-2303 (2013).
5. Raps, D., Hossieny, N., Park, C. B., & Altstädt, V., *Polymer*, **56**, 5-19 (2015).
6. Qi HJ, Boyce MC, *Mechanics of materials*, **37**, 817-839 (2005).
7. Prissok F, Braun F, *U.S. Patent Application No. 12/161,354* (2010).
8. Reinhardt SD, Wood DM, Wardlaw A, Robinson TK, Whiteman J, *U.S. Patent No. 9,788,606* (2017).
9. Hossieny NJ, Barzegari MR, Nofar M, Mahmood SH, Park CB, *Polymer*, **55**, 651-662 (2014).
10. Ito S, Matsunaga K, Tajima M, Yoshida Y, *J. Appl. Polym. Sci.*, **106**, 3581-3586 (2007).
11. Nofar M, Park CB, Chapter 5: *Heterogeneous cell nucleation mechanisms in polylactide foaming*. CRC Press, ISBN: 139781466561809 (2015).

Effect of TPU hard segment content on TPU's bead foaming behavior

Kucuk, E.B.[1], Batı, B.[1], Nofar, M.*[1,2]

1: Polymer Science and Technology Program, Istanbul Technical University, Maslak, Istanbul, 34469, Turkey
2: Metallurgical & Materials Engineering Department, Faculty of Chemical and Metallurgical Engineering, Istanbul Technical University, Maslak, Istanbul, 34469, Turkey
** Corresponding Author: nofar@itu.edu.tr*

A lab-scale bead foaming reactor was used to investigate the foaming behavior of thermoplastic polyurethane (TPU) samples with different hard segment contents of around 39 %, 49 % and 57%. The saturation temperature, pressure, and time differently influenced the formation of hard segment crystals with different close-packed structures and sizes. Consequently, with the increase in hard segment crystallization during the saturation the heterogeneous cell nucleation was improved and the foam expansion was controlled by the stiffness of the TPU/CO_2 mixture, solubility of CO_2, and final shrinkage of the foamed samples. As the saturation temperature increased, the expansion ratio tended to increase with a tendency to shrink more due to the reduced crystallites content. This is while the cell density decreased due to the reduced hard segment crystallites generated during the saturation. Increase in pressure also improved the CO_2 solubility and hence the expansion ratio was increased more significantly. The saturation time also influenced the amount and perfection degree of hard segment crystals. The interrelation between the crystallization behavior of the TPU bead foams and process conditions was reported using differential scanning calorimetry (DSC) and scanning electron microscope (SEM) analysis.

Keywords: TPU, bead foaming, hard segment, crystallization, supercritical CO_2

INTRODUCTION

Thermoplastic polyurethane (TPU) is an elastomeric polymer which is formed by combination of diisocyanates, long chain diols and short chain diols as chain extender. Diisocyanates with chain extenders create rigid hard segments, while long chain polyols create flexible soft segments. Polarity differences between these segments result in phase separation, which enhances crystallization of hard segments [1].

Plastic foaming has been developed to have final products with reduced cost and weight, while some mechanical and physical properties could also be improved [2-3]. Among the foaming technologies, bead foaming has been offered to be a solution to produce products with three dimensional complex geometries and high expansion which is not feasible in foam extrusion or foam injection molding [4]. Among the bead foam products, the generation of a double crystal melting peak during the bead foam saturation of polypropylene (PP) [5-6] and polylactide (PLA) [7-8] has been introduced to be a promising method to generate high cell density foam products while the intrabead sintering could be improved during their molding [5-8]. In this context, the presence of a wide hard segment crystal melting peak in TPU samples could be considered as an inherent benefit to have a better foaming behavior and intrabead sintering among the foamed beads. This is because the unmelted crystals during the saturation could contribute to the heterogeneous cell nucleation and improve the expansion with less shrinkage, while during the sintering in steam chest molding the sintering could occur by partially melting the hard segment crystals [9]. The bead foaming behavior of TPU has been explored in few studies and they have shown that the presence of hard segment crystals plays an important role on the final foam morphology and foam expansion [10-11].

In this work, expanded TPU (ETPU) bead foams were prepared using different TPUs that possessed different hard segment contents. The effect of process parameters (i.e, saturation temperature, pressure and time) on the foam morpholgy and expansion behavior was investigated. Foamed samples were studied using scanning electron microscope (SEM) to observe their cellular morphology and their cell densities.

EXPERIMENTAL

Materials and Foam Preparations

TPU samples were supplied directly from Huntsman Polyurethanes, Belgium. The TPUs possessed same hard segments and soft segments based on 4,4' MDI and 1,4 BDO, and a polyester polyol of adipic acid and 1,4 BDO with molecular weights of 1000. The content of hard segment in these TPU samples was around 39%, 49% and 57%, which are referred to as L-HS (low hard segment TPU), M-HS (medium hard segment TPU), and H-HS (high hard segment TPU). The thermograms of the neat TPUs from DSC (Perkin Elmer DSC400) were investigated and their thermal properties are disclosed in Figure 1. Higher and broader endothermic peaks are observed with increasing hard segment content. Because of different morphology of hard segment domains, there are multiple and broader melting peaks. At low temperature, crystals are related to short range order hard segments and at high temperature the melting crystals are relevant to the hard segment crystallites with long range order and closed pack structure. In the cooling step, the TPUs with high hard segment content crystallized faster and earlier which resulted in the formation of a more closed packed structure. Moreover, the increase in hard segment content increased the glass transition temperature of the TPU samples.

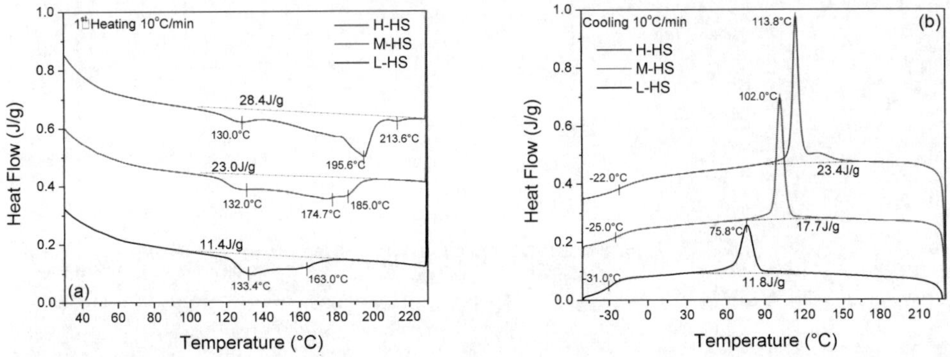

FIGURE 1. DSC first heating (a) and cooling (b) thermograms of TPUs with different hard segments

Before foam processing, all TPUs samples were dried at 50°C for overnight in a vacuum oven to remove moisture. Melt temperatures were examined from DSC results to specify the process temperatures. Few samples were placed inside the chamber. The system was heated to various temperatures to find the processing temperature window. During this study, pressure and time were fixed at 2000 psi and 30 min, respectively. After determining the most reliable structure of the bead foams, the saturation temperature was selected. Then, to understand the effect of the saturation pressure and time on the foaming behavior, various conditions were also studied. After finding the temperature at which high expansion foams without shrinkage were observed, pressure was adjusted to respectively 1000, 1500 and 2000 psi, and samples were foamed after 30 min. Finally, to understand the saturation time effect, at a given temperature and pressure (2000 psi), the samples were kept for 15, 30 and 60 min. At the end of the experiments, the pressure was released rapidly (in nearly less than 1 sec) by opening a ball valve, and the chamber was cooled in a water bath.

Foam Characterization

Density measurement of the foamed samples obtained from experiments were done using a Mettler A1201 density measurement device. Archimedes method was used to calculate the density of the foamed samples. Then, expansion ratio of each foamed samples was calculated using equation 1, where $\rho_{polymer}$ and ρ_{foam} are density of neat polymer and the foamed sample.

$$\emptyset = E.R = \frac{\rho_{polymer}}{\rho_{foam}} \tag{1}$$

Scanning electron microscope (SEM) (JEOL S410) was used to observe the cellular morphology of the foams. The cell density of each foamed sample was calculated using equation 2 where N is number of cells, A (cm^2) is the area.

$$\text{Cell density} = \left(\frac{N}{A}\right)^{3/2} \cdot \frac{\rho_{polymer}}{\rho_{foam}} \tag{2}$$

RESULTS AND DISCUSSIONS

Temperature is one of the critical parameters in bead foaming, especially for cell nucleation phenomena and expansion since it affects the flexibility of the matrix as well as the solubility of blowing agent in the polymer. CO_2 was used as a blowing agent which has a low critical temperature. Although, solubility of CO_2 decreases with increasing temperature, microcellular polymers cannot achieve high expansion ratio at lower temperatures due to the increased stiffness of TPU with more rigid hard segment regions. Therefore, finding the appropriate temperature for saturation has a significant effect during the process.

SEM micrographs, expansion ratio, and cell density of ETPU samples are given in Figures 2 and 3. Highest expansion was observed for the M-HS and H-HS TPUs at 130°C, while it was achieved at 110°C for L-HS TPU. The reason is that more energy was needed for higher HS content TPUs to melt and rearrange its HS crystalline parts; moreover, to become more flexible for cell growth. In general, lower expansion ratios were obtained for H-HS TPU comparing to M-HS and more obviously to L-HS TPU. This was due to the hindering effect of high amount of crystals for gas solubility and expansion. As can be seen the processing window shifts to higher temperatures with the increase in hard segment content and the expansion tends to reduce due to the increased content and perfection of hard segment crystals. A significant expansion could not be even observed at higher saturation temperatures for H-HS TPU. As can be seen tailored bead foams could be obtained by considering the hard segment content and the saturation temperature. According to the cell density values, although the cell sizes changes significantly in TPUs with various hard segment contents and after saturation at various saturation temperatures, the cell densities remained in the similar ranges for almost all cases which sometimes revealed to be over 10^9 cells/cm^3 which represents the formation of microcellular structure.

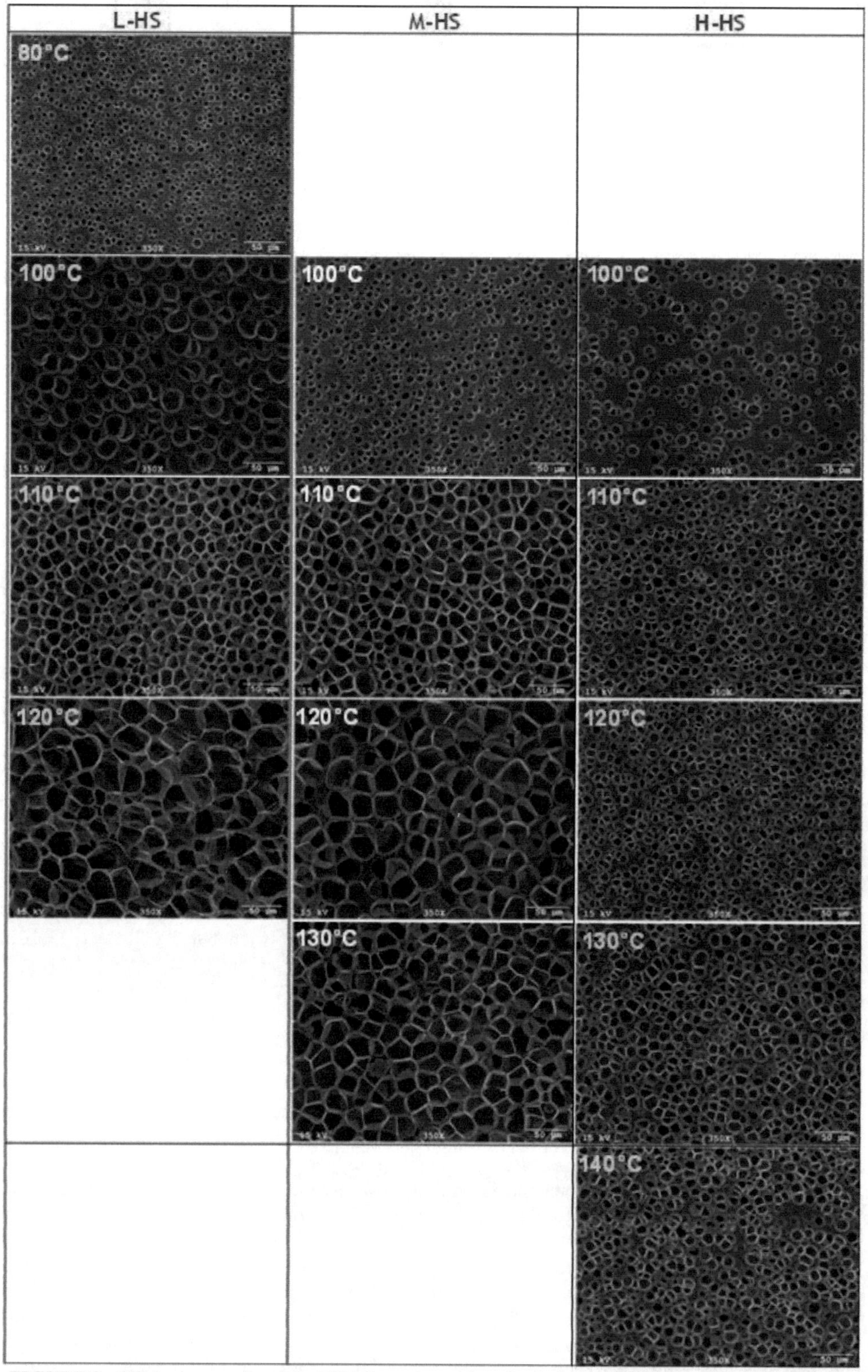

FIGURE 2: SEM micrographs of the TPU foamed samples after being saturated at various temperatures

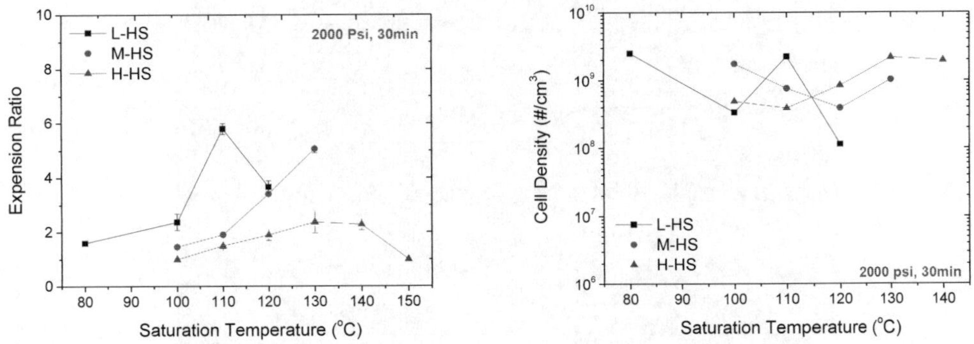

FIGURE 3: Expansion ratio (a) and cell density (b) of the TPU foamed samples at various saturation temperatures

To understand the effect of saturation pressure on the morphology of ETPU bead foams, polymer pellets were foamed under 2000, 1500 and 1000 psi at a defined foaming temperature for 30 min. As the saturation pressure increases the free energy barrier for cell nucleation decreases and the solubility of gas increases. Further, absorbed CO_2 acts as a plasticizer, and reduces the melt viscosity. Therefore, cell nucleation rate increases with dissolved CO_2 amount, and more absorbed CO_2 in polymer hence results in higher cell density. As can be seen in Figures 4 and 5 higher expansions and cell densities were obtained in TPU samples with increasing saturation pressure.

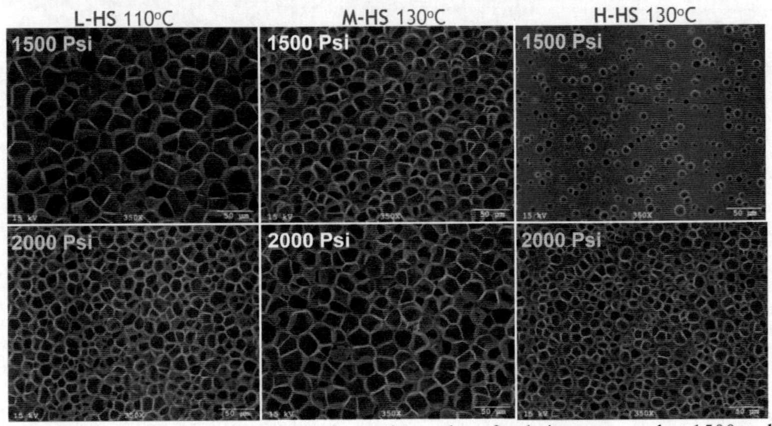

FIGURE 4: SEM micrographs of the TPU foamed samples after being saturated at 1500 and 2000 psi

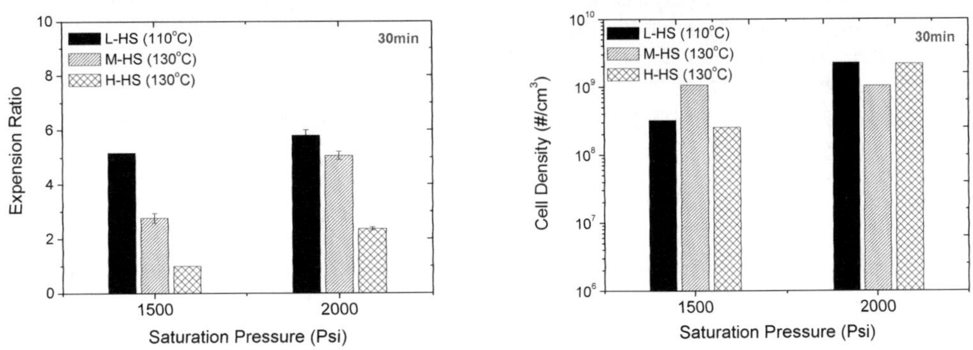

FIGURE 5: Expansion ratio (a) and cell density (b) of the TPU samples foamed at various saturation pressures

Determining the saturation time is crucial for a foaming process since it affects the crystallization as well as the foam morphology. Figures 6 and 7 reveal the SEM images and expansion and cell density values of the TPU foamed beads, respectively. It was observed that the change in saturation time could not influence the cell density much however, the expansion ratio values were significantly affected. This was due to the influence of the formation of more hard segment crystallites during a longer saturation which could hinder the expansion, although the shrinkage in some cases should also be taken into account. When comparing L-HS, M-HS and H-HS TPUs, low hard segment content provides high expansion and increasing time leads the high perfect degree crystalline domains. At 15 min saturation time, high expansion was obtained for L-HS; however, shrinkage was observed later. Expansion was very low for TPUs having higher HS amount, which can be explained by the fact that CO_2 could not find enough time to be dissolved in TPU with high hard segment crystals.

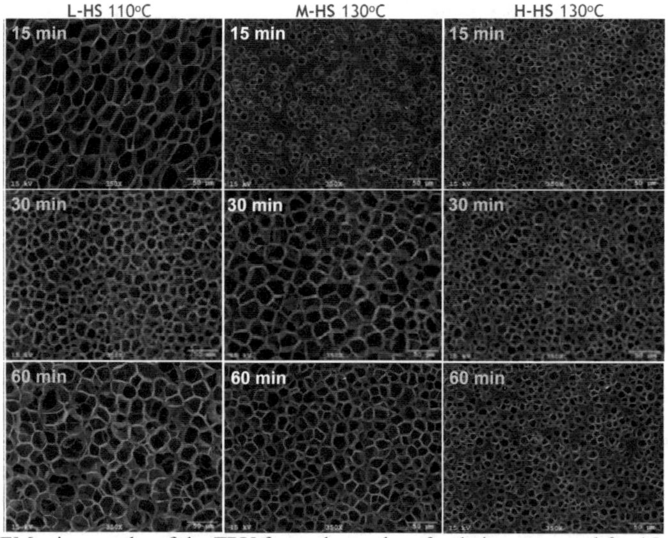

FIGURE 6: SEM micrographs of the TPU foamed samples after being saturated for 15, 30, and 60 min

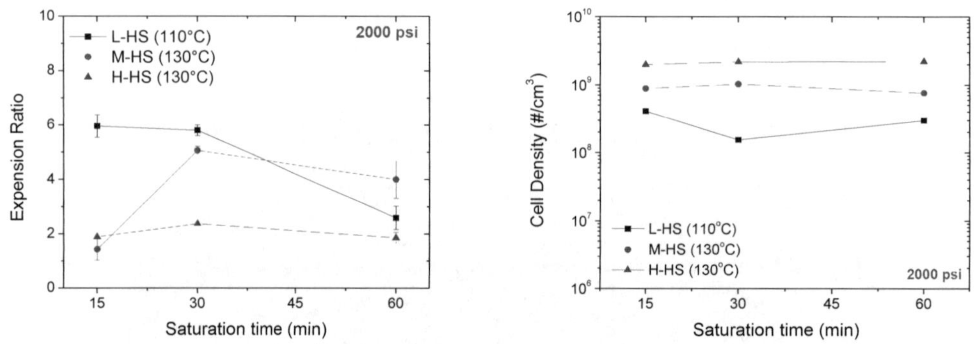

FIGURE 7: Expansion ratio (a) and cell density (b) of the TPU samples foamed at various saturation times

CONCLUSION

In this study, ETPU bead foams were prepared with TPUs having various hard segment contents. The results showed that the processing parameters such as saturation temperature, pressure, and time could influence the hard segment crystallization during the saturation, the CO_2 solubility, and the flexibility of the polymer/gas mixture. Therefore, the foam morphology, expansion ratio,and the shrinakage could be significanlt dependent on these parameters. The icnreasse in hard segmnet content also increased the processing window to higher temperatures, minmized the shrinake due to the existance of large content of crystals, as well as reduced the final expansion values of the foamed beads.

ACKNOWLEDGEMENT

The authors like to thank Huntsman Polyurethanes, Belgium for suppling the materials and sponsoring the project.

REFERENCES

1. Nitta KH, Kuriyagawa M, *Handbook of Thermoplastics*, **41**, 387 (2016).
2. Nofar M, Park C.B, Polylactide Foams, Fundamentals, Manufacturing, and Applications, Elsevier, ISBN: 9780128139912 (2017)
3. Nofar M, Park C.B, *Prog. Polym. Sci*, **39**, 1721-1741 (2014).
4. Raps, D., Hossieny, N., Park, C. B., & Altstädt, V., *Polymer*, **56**, 5-19 (2015).
5. Nofar M, Guo Y, Park C.B, *Ind. Eng. Chem. Res.*, **52**, 2297-2303 (2013).
6. Guo Y, Hossieny NJ, Chu RKM, Park C.B, Zhou N, *Chem. Eng. J.*, **214**, 180-188 (2013)
7. Nofar M, Ameli A, Park C.B, *Polymer*, **69**, 83-94 (2015).
8. Nofar M, Ameli A, Park C.B, *Materials & Design*, **83**, 413-421 (2015).
9. Nofar M, Park CB, Chapter 5: *Heterogeneous cell nucleation mechanisms in polylactide foaming*. CRC Press, ISBN: 139781466561809 (2015).
10. Hossieny NJ, Barzegari MR, Nofar M, Mahmood SH, Park CB, *Polymer*, **55**, 651-662 (2014).
11. Hossieny NJ, Shaayegan V, Ameli A, Saniei M, Park CB, *Polymer*, **112**, 208-218 (2017).

Thermo-Expandable Microcapsule as a Blowing Agent for Producing Thermoplastic Elastomer Vulcanized Syntactic Foam

M. Riou[a,b*], G. Ausias[a], T. Gaudry[b], J-M. Veillé[b], J. Férec[a], A. Primel[b] and Y. Grohens[a]

[a] *Univ. Bretagne Sud, UMR CNRS 6027, IRDL, F-56100 Lorient, France*
[b] *Cooper Standard, Route des eaux, 35503 Vitré, France*

Abstract. Thermoplastic Elastomer Vulcanized (TPE-V) syntactic foams made by Thermo-Expandable Microcapsules (TEMs) are becoming of a large interest for automotive industry. TPE-Vs combine advantages of class rubber compound properties with an easier manufacture process and are recyclable. Therefore, TPE-Vs are more and more used to manufacture automotive sealing system by an extrusion process, which is the focus of this research. Due to the two-phase structure of TPE-V, classical extrusion foaming technologies, such as chemical and physical foaming, are complex to be controlled. Thus, TEMs show an outstanding ability to be effective and produce repeatable microstructure with standard extrusion equipment. TEMs consist in a mixture of liquid hydrocarbons, encapsulated by a gas-proof polymeric shell. Exposed to elevated temperatures, the internal pressure drives the dilation of the TEMs, as a result a TPE-V syntactic foam is produced. The aim of this study is to understand where the expansion occurs and how can it be affected by extrusion process parameters. Temperature is a well-known key parameter to control final expansion of TEMs. Hence, measurements at high pressure and temperature were done to find out the location of expansion during extrusion process. Additionally, we have designed three groups of three dies to apply different pressure drops, pressure drop rates and residence time, inside the die. Under these process conditions, density, cell microstructure and viscosity have been investigated.

Keywords: Syntactic Foam – Thermo-Expandable Microcapsule – Thermoplastic Elastomer – Foam Extrusion
PACS: 83.50.Uv

INTRODUCTION

Since the middle of the twentieth century, researchers have developed several processes to lightweight and improve foam characteristics. Polymer foams have received attention from industries due to their good mechanical performance versus weight. Lightweight material is also an effective way to reduce the product cost. Foaming innovation is also driven by regulation evolutions. In fact, the understanding of the global warming and chemical effects on human health or the environment, led and may lead in the future to blowing agents restrictions such as: Chlorofluorocarbon (CFC) and more recently Azodicarbonamide (ADCA). ADCA was appointed as a Substance of Very High Concern by the European CHemicals Agency (ECHA) in 2012. However, ADCA is a high effective gas releasing chemical blowing agent. It still be widely used especially in automotive weather-strip industry [1]. An eco-friendlier alternative of ADCA must be found.

Historically, Ethylene-Propylene-Diene-Monomer (EPDM) rubber compounds remain the most used material to produce automotive door seals system. For fifteen years, ThermoPlastic Elastomer (TPE) has gained interest to substitute rubber material in glass run channel sealing systems due to its recyclability and processability [2]. Two kinds of TPE are found on weather-strip on current vehicles: Styrenic block copolymers TPE (TPE-S) and Vulcanized TPE (TPE-V). TPE-Vs usually selected for making weather-strip are a blend of polypropylene and small rubber particles. Rubber nodules are dynamically vulcanized during the compounding, as firstly mentioned by Gessler [3].

First TPE-V foaming investigation was done by Dutta and Cakmak in 1992 [4]. Santoprene TPE-V made by mixing PP and EPDM particles was foamed by ADCA blowing agent. They concluded that foaming occurs in PP plastic phase, consequently the maximum expansion ratio was rigorously dependent of the TPE-V composition. For softer grades, porosity needs to be introduced inside EPDM phase to achieve lower densities. In 1998, Kropp et al. [5] investigated physical foaming of TPEs with carbon dioxide and Hydrocerol® (sodium bicarbonate) as nucleating agent. More recently, others gas molecules have been tested such as water, hydrocarbons and nitrogen. [6, 7]. Solubility and diffusion ability of the gas are directly related to nucleation and cell growth phenomenon, and by the end, its foaming performance. CO_2 has gained in interest because finer cell size and uniform distribution, which enhance mechanical properties, can be obtained [8, 9]. Some studies have been done to investigate the effect of pressure drop rate on expansion ratio [10-13]. They showed that die design can significantly affect foam microstructure. They observed that a higher pressure drop rate is increasing cell nucleation, which is favorable to promote the expansion ratio. Nucleation and diffusion inside the polymer is also related to the polymer nature and temperature. Even if CO_2 can produce good foaming result, it still be complicated to control microstructure for various polymers and product designs at an industrial scale.

That is why, Thermo-Expandable Microcapsules (TEMs) have recently appeared as a new alternative to foam polymer. They consist in a mixture of hydrocarbons encapsulated within a rigid polymer shell. Exposed to elevated temperature, the shell becomes softer and the vapor pressure of hydrocarbons drives the microcapsule expansion. TEMs was firstly introduce by Dow Chemicals [14]. Originally used in wallpaper, paint and textile for textural properties, now TEMs are economically interesting to be used as a blowing agent in shoe soles and automotive industry. TEMs preserve its structure even after foaming, as a result, a closed cell microstructure is created. This kind of foam is called "syntactic foam". Only few literature papers are dealing with TPE syntactic foam produced with TEMs. [15, 16]. They achieved 25 to 40 % weight reduction. More recently, Guo et al. [17] worked on TPU foaming with TEMs during extrusion process. They showed at low shear rate (20 to 110 s^{-1}), that higher pressure drops along a capillary die increase the expansion ratio. They suggested for high pressure in the die, the main part of expansion occurs after die exit, whereas for low pressure in the die, the expansion can occur inside the die.

In this study, we intend to better understand foaming process of TEMs inside a low hardness TPE-V elastomer. In a first approach, we aim to find the location where the microcapsules are foaming during extrusion process. Then, inspired by the study done by Xu et al. in 2008 [12], we have designed a set of 9 dies to apply different pressure drop rate and time residence. The objective was to investigate in the range of thermoplastic elastomer sealing system extrusion, if die design could influence the expansion level of the produced TPE-V syntactic foams.

EXPERIMENTAL

Materials

A low hardness TPE-V. This material was chosen because it can exhibit good elasticity and compressibility properties on sealing geometry. TEMs has been used as blowing agent in this study. TEM grade was chosen because it is consistent with the temperature range used for TPE-V extrusion process. Main characteristics of materials are presented in Tables (1) and (2). A custom-made Masterbatch (MBTEM) was designed with 40% of TEMs and and Ethylene-Vinyl-Acetate (EVA), supplied by Repsol. The use of a low melting point polymer is necessary to avoid any expansion during the masterbatch compounding. In this work, the EVA inside MBTEM plays two distinct roles: firstly, as a carrier to dose TEMs during extrusion and secondly as a matrix during expansion pressure measurements. For the last, the idea was to roughly represent external rheology seen by TEMs during extrusion process.

TABLE (1). Polymer Material characteristics

Material	Density (ISO 1183) [kg/m³]	Melting point (Internal) [°C]	Hardness (ISO 868) [Shore A]
TPE-V	925	155	38
EVA	937	85	83

TABLE (2). TEMs characteristics

Material	Initial density (ISO 1183) [kg/m³]	Tstart (Internal) [°C]	Tmax (Internal) [°C]	Initial diameter (Internal) [µm]	Maximal diameter (internal) [µm]
TEMs	1100	170	208	16.8	58.8

To figure out how the expansion occurs during the extrusion process requires to investigate TPE-V and TPE-V foam rheology. For this study we performed rheological measurements with a home-made in-situ rheometric die. The section of the die is a ribbon strip of 2x40 mm. Rheological behaviors of TPE-V and foam TPE-V are presented in Figure 1 (a). Auto-heating during measurements are corrected by Arrhenius law from previous capillary rheology measurements. TPE-V and TPE-V foam are following power law, in agreement with literature [6, 18, 19]. As TPE-V foam does not highly affect rheology, the same Arrhenius law is applied to correct auto-heating. To neglect the influence of the EVA rheology on the overall mixture behavior, a small amount of masterbatch (1,62%) was introduced in the samples. Hence, TEMs effective amount is 0,65%. Due to the small amount of TEMs, TPE-V foam shows a slightly lower viscosity. This result was also expected following microcellular influence on foam [17, 20].

We measured EVA rheology with a plate and plate geometry rheometer and with a capillary rheometer at 120°C and 190°C. Cox-merz rule [21] and thermodependency, using Arrhenius law, were done to determine EVA mastercurve at 200°C (Figure 1 (a)). 4 parameters Carreau-Yasuda model (Figure 1 (b)) was used to fit experimental data and rheological parameter are presented in Table 3. As explained above, Figure 1 shows the viscosity at 200°C of EVA at low shear rate (η_{EVA} = 10 Pa.s) which is not so far from the viscosity of TPE-V (η_{TPV} = 30 Pa.s) when it flows into the slit die with a shear rate in the range of 10^3 s^{-1}.

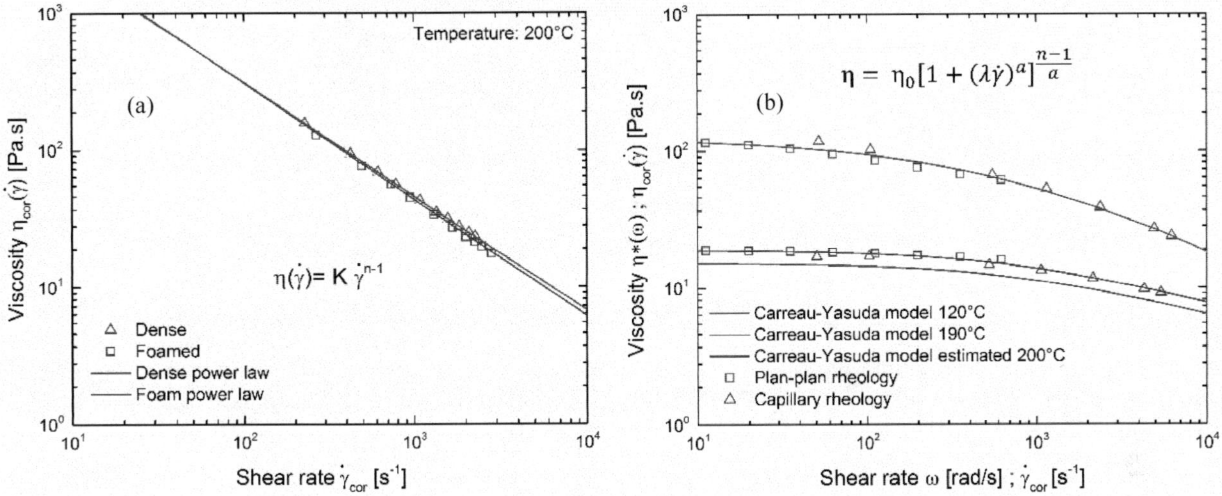

FIGURE 1. (a) Master curve at 200°C of viscosity determined by rheometric slit die and power law fitting model. (b) Experimental data from plan-plan rheology and capillary rheology with Carreau-Yasuda 4 parameters fitting model.

TABLE (3). Material rheology behavior at 200°C

Material	K Consistency [Pa.s^n]	η_0 viscosity at zero shear rate [Pa.s]	λ relaxation time [s]	n Pseudo-plastic index [-]	a [-]	Ea Activation energy [kJ.mol^{-1}]
TPE-V	14538	/	/	0.167	/	11.3
TPE-V + 0.65% TEM	15030	/	/	0.155	/	11.3
EVA	/	15.30	0.0013	0.678	0.868	41.1

Method and Results

Pressure Drop Threshold Measurements

This experiment was inspired by Kawaguchi et al. [23]. We subjected TEMs inside a masterbatch to high pressure at 205°C. Expansion does not occur after 10 minutes at 60 bar. Then, experiments were carried out with PDE 1700 MD apparatus equipped with pressure transducer and thermocouple developed by Eurotechnica GmbH. Acquisitions were done by a CCD camera and LabVIEW software. Samples were placed in a 10x10x20 mm cuvette, inside the pressurized cell. The pressure was controlled by N_2 flow. The area taken by MBTEM samples is measured by image processing (i.e., imageJ software). The expansion ratio is normalized by the highest level of area taken by the sample during the experiment. The pressure drop is controlled manually around 0.1 bar/s. Figure 2 shows clearly a pressure threshold around 10 bar below which the expansion is increasing widely. In case of TPE-V extrusion, the expansion occurs when the pressure is around 10 bar. Inside the extrusion unit, a sufficient elevated temperature and low pressure is only achieved at the end of the die. Experiments do not show any variation in dimension of the sample after the die. It suggests expansion may occur mostly before the die exit.

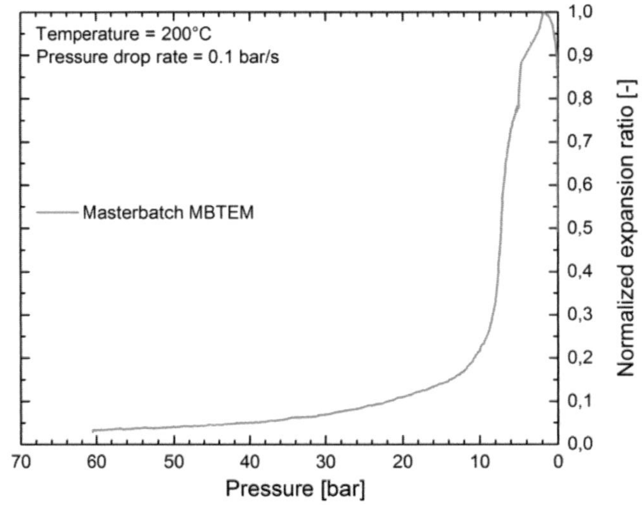

FIGURE 2. Expansion of MBTEM at 200°C in pressurized cell.

Pressure Drop Rate Study

9 slit dies were calculated numbered from 11 to 33: 3 groups of 3 dies to apply 3 different pressure drop rate (32 bar/s, 76 bar/s and 110 bar/s, respectively) and different time residence (1.4 to 0.2 s). Every die has a width versus thickness ratio higher than 10 to neglect wall shear rate and stress effects [24].

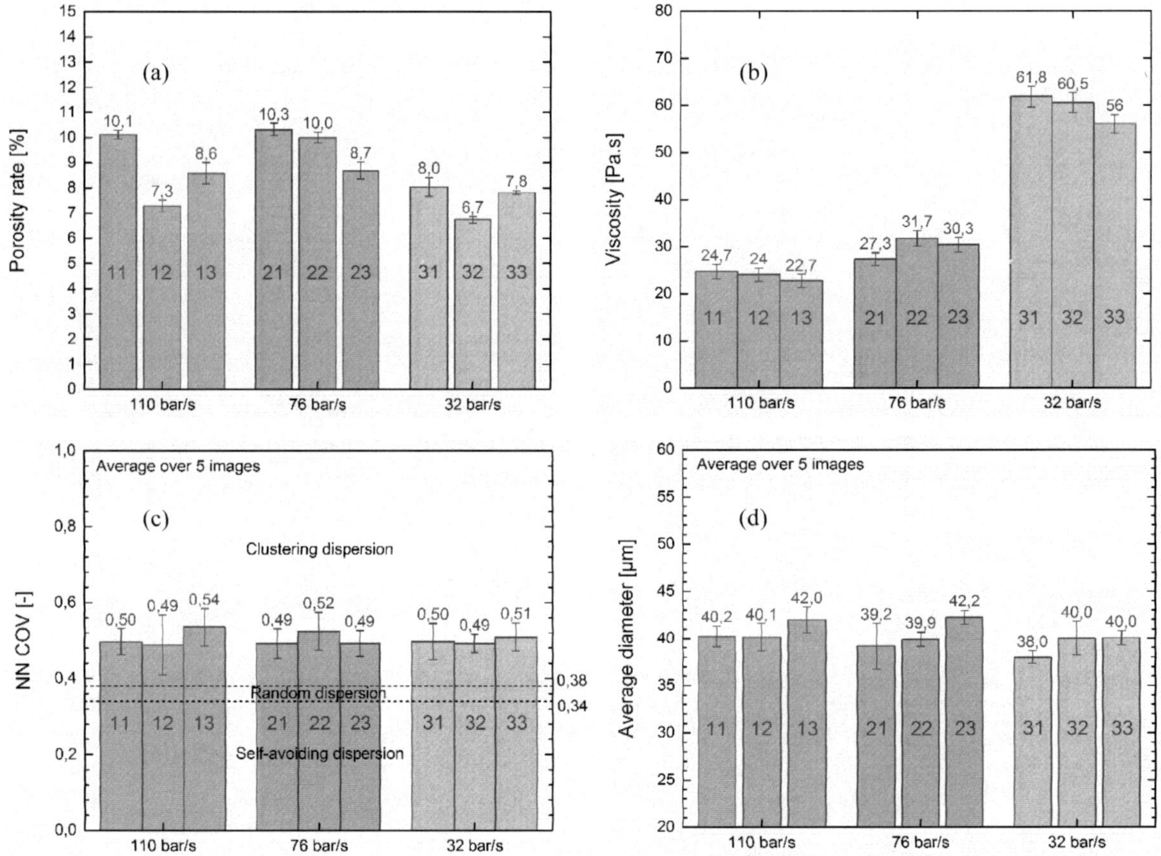

FIGURE 3. (a) Porosity rate calculated from density measurements. (b) Corrected viscosity of foam TPE-V, calculated from Rabinowitsch and thermo-dependency corrections. (c) 2D porosity dispersion parameter calculated as presented by Yang [25]. (d) Average diameter is calculated from apparent area of TEMs.

Temperatures were fixed at 205°C ± 2°C. The flow rate was fixed at 13500 ± 200 mm^3/s. Rheological models presented above are applied as reference to calculate and to correct rheological values. TPE-V melt density was determined by PVT measurements. Assumption is made to calculate the melt density in case of foaming: the porosity proportionally decreases the melt density. Corrected viscosity was evaluated from pressure measurement inside the slit die and by considering Rabinowitsch effect as described by Macosko [26]. Figure 3 presents the 4 main parameters variations in function of pressure drop rates. Average porosity achieved with 0.65% of TEMs is 8.5%. Every pressure drop rate corresponds to a material viscosity level. The dispersion is not fully homogeneous. The average size of porosities is 40 μm. The 4 parameters analyzed do not described any trend due to pressure drop rate or time residence. In the range analyzed, die design does not affect the TEM expansion during extrusion process.

CONCLUSION

In this work, TEMs expansion phenomena localization during extrusion process was investigated. A masterbatch was created with a EVA matrix which roughly exhibits similar rheology of TPE-V during extrusion. At low pressure drop we measured a pressure threshold at around 10 bar under which the expansion can occur. In classic automotive sealing system extrusion, conditions that enable expansion can only be achieved inside the die.

That is why, we have designed multiple die corresponding to three pressure drop rates and various residence times. Each die section applies a shear rate on melt which corresponds to a viscosity. The dispersion is not homogeneous, microcapsule tends to form cluster. It could be attributed to the original masterbatch pellets or a too short extrusion unit. Under these conditions, TPE-V syntactic foam microstructures are roughly the same. Die design does not affect the extrusion foaming process. Rheology of microcapsule shell needs to be investigated to compare with the rheology of the surrounded TPE-V polymer matrix.

ACKNOWLEDGMENTS

The authors acknowledge Eurotechnica GmbH for their experimental approach work.

REFERENCES

1. J. Stehr, *Int. Polymer Sci. Tech.* **43** (2016).
2. C. Harper, Handbook of plastics, elastomers & composites, McGraw-Hill Professional (2002)
3. A. M. Gessler and J. W. H. Haslett, U.S. Patent No. 3,037,954, (5 June 1962).
4. A. Dutta and M. Cakmak, *Rubber Chem. Technol.* **65**, pp. 932-955 (1992).
5. D. Kropp, W. Michaeli, T. Herrmann, and O. Schröder, *Journal of Cellular Plastics* **34**, pp. 304-311 (1998).
6. A. Sahnoune, *Journal of Cellular Plastics* **37**, pp. 149-159 (2001).
7. S. G. Kim, C. B. Park, and M. Sain, *Journal of Cellular Plastics* **44**, pp.53-67 (2008).
8. M. Shimbo, D. F. Baldwin, and N. P. Suh, *Polym. Eng. Sci.* **35**, pp.1387-1393 (1995).
9. L. M. Matuana, C. B. Park, and J. J. Balatinecz, *Polym. Eng. Sci.* **38**, pp. 1862-1872 (1998).
10. C. B. Park, D. F. Baldwin, and N. P. Suh, *Polym. Eng. Sci.* **35**, pp. 432-440 (1995).
11. X. Xu, C. B. Park, D. Xu, and R. Pop-Iliev, *Polym. Eng. Sci.* **43**, pp. 1378-1390 (2003).
12. X. Xu and C. B. Park, *J. Appl. Polym. Sci.* **109**, pp. 3329-3336 (2008).
13. Q. Guo, J. Wang, C. B. Park, and M. Ohshima, *Ind. Eng. Chem. Res.* **45**, pp. 6153-6161 (2006).
14. D. S. Morehouse Jr and R. J. Tetreault, U.S. Patent No. 3,615,972 (26 October 1971).
15. J. Paukkunen, "Foaming of thermoplastic elastomers", Master Thesis, Tampere University of Technology (2012).
16. S. Ries, A. Spoerrer, and V. Altstaedt, AIP Conference Proceeding, **1593**, pp. 401-410 (2014).
17. Q. Guo, S. A. Andrew Kenny, and G. Rizvi, *SPE ANTEC*, (2017).
18. L. A. Goettler, J. R. Richwine, and F. J. Wille, *Rubber Chem. Technol.* **55**, pp. 1448-1463 (1982).
19. P. K. Han and J. L. White, *Rubber Chem. Technol.* **68**, pp. 728-738 (1995).
20. D. F. Baldwin, C. B. Park, and N. P. Suh, *Polym. Eng. Sci.* **38**, pp. 674-688 (1998).
21. W. P. Cox and E. H. Merz, *J. Polym. Sci.* **28**, pp. 619-621 (1958).
22. K. Yasuda, R. C. Armstrong, and R. E. Cohen, *Rheol. Acta* **20**, pp. 163-178 (1981).
23. Y. Kawaguchi et al., *Polym. Eng. Sci.* **50**, pp. 835-842 (2009).
24. J. L. S. Wales, J. L. den Otter, and H. Janeschitz-Kriegl, *Rheol. Acta* **4**, pp. 146-152 (1965).
25. N. Yang, J. Boselli, and I. Sinclair, *J. Microsc.* **201**, pp. 189-200 (2001).
26. C. W. Macosko, Rheology: Principles, Measurements, and Applications, WILEY-VCH (1994).

Influence of Process Management on the Increase of Dynamic Solubility Limits in Injection Foam Molding

Clemens Kastner[a], Georg Steinbichler[b]

Johannes Kepler University Linz
Institute of Polymer Injection Molding and Process Automation
Altenbergerstrasse 69
4040 Linz, Austria
[a]clemens.kastner@jku.at
[b]georg.steinbichler@jku.at

Abstract. The foam injection molding process provides a multitude of advantages both during production and throughout the life cycle of foamed components. Yet still there are many uncertainties concerning the conditions inside the cylinder during gas injection. For a better understanding of these processes, the solubility of gases in polymers is crucial. In current literature, almost exclusively static solubility is investigated. It is questionable whether results from static solubility measurements are relevant for highly dynamic processes like injection molding. Additionally, it is unknown whether the limits detected in static measurements also count for dynamic systems. In this work, a novel measurement method based on the bulk modulus of gas containing polymer melts, which provides in-line determination of dynamic solubility limits, was applied. Using a talc filled polypropylene in a MuCell®-process the question was whether the solvable amount of gas can be influenced and increased by a larger gas injection time to dosing time ratio ($R_t=t_i/t_d$). Starting at a gas concentration that was initially beyond the solubility limit the ratio R_t was increased stepwise, while process pressure and gas content were kept constant. The bulk modulus method in combination with transmissive ultrasonic measurements proves that the dynamic solubility can be increased above 1.6 wt% nitrogen at 140 bar. In addition, compression work and injection work were monitored and compared to the respective states of solution. These findings show that the influence of material on solubility limits might only be relevant at long dosing and gas injection times as an equilibrium cannot develop in short times typical for injection molding. This approach is useful when large amounts of gas should be dissolved in the melt in order to achieve high density reductions. It opens new perspectives for injection foam molding.

Keywords: Polymer, Injection foam molding, Solubility, Ultrasonic measurement, Bulk modulus.

INTRODUCTION

New developments in injection molding should allow for a more economical production, enable the realization of components with new properties and structures or help save cost during the product life cycle (e.g. lightweight design). Injection foam molding is one of the oldest yet most important and current representative out of more than 100 special injection molding techniques that help reaching these targets [1][2]. Gases that are dissolved in the melt lead to a cellular structure as soon as the polymer-gas-solution experiences a pressure drop during injection into a cavity. Starting in the 1950s constant developments, especially at the Massachusetts Institute of Technology (MIT) in the early 1980s, and the introduction of supercritical fluids lead to achievable cell densities up to 10^{15} cells/cm^3 [3][4].

Besides some drawbacks like higher demands on process control or surface defects [5] reduced weight, increased specific mechanical properties, better dimensional stability etc. are among the advantages of this technology [6][7][8]. To make use of the numerous advantages and surmount difficulties a clear understanding of the underlying polymer physical processes is necessary. Above all, solubility and diffusivity need to be described in order to predict material behavior during processing. Current literature offers many studies on static measurements based on adsorption, absorption and diffusion. Using magnetic suspension balances [9], micro/electro balances [10] or the pressure decay method [11] measurement times in the range of hours are necessary. The relevance of these measurements for highly dynamic systems (i.e. under the presence of shear) like injection molding is therefore questionable. Many questions regarding the processes inside the cylinder remain unanswered.

Very little literature can be found on dynamic solubility measurements. Lee and Ramesh for instance do not attach any importance to the effect of shear processes on solubility [12]. Chen et al. [13] draw a different conclusion. They use a special heated pressure chamber that includes a rotor for shear application. Under a shear rate of 30 s^{-1} the solubility of carbon dioxide in high density polyethylene increases by roughly 25 %. This indicates that dynamic solubility is more a process-oriented property rather than a material constant.

In this work, the influence of process management on dynamic solubility limits was investigated using a novel measurement method based on the compressibility of polymer-gas-solutions. Beside a qualification of the method, this study should also offer new insights on the dependence of different process parameters on large gas contents in injection foam molding.

EXPERIMENTAL

Material

The measurements were conducted using a heterophasic polypropylene copolymer intended for injection molding supplied by Borealis Polyolefine GmbH (Linz, Austria). The neat material has a density of 895 kg/m^3 and provides a melt flow rate of 18 g/10min (230 °C/2.16kg). The material was compounded with 25 % high aspect ratio talc. Nitrogen was used as a blowing agent.

Experimental

The experiments were conducted on an injection molding machine duo 700 by Engel Austria GmbH (Schwertberg, Austria), equipped with an SII MuCell®-unit (Trexel, Inc. Wilmington, USA) and a 70 mm MuCell®-screw. In this work solubility is evaluated using the bulk modulus of the material. Figure 1 depicts the necessary measurement setup.

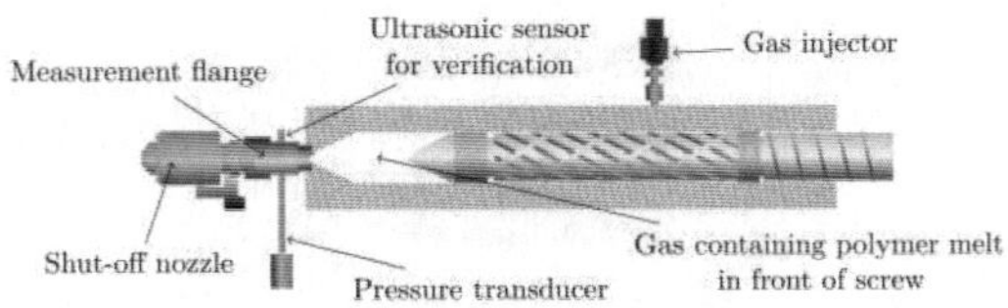

FIGURE 1. Measurement setup of the determination of bulk modulus

The bulk modulus method requires the shut-off nozzle to remain closed during the first phase of injection. An initially pressure-controlled forward motion of the screw leads to a pressure increase, Δp, to about 800 bar. The pressure increase as well as the corresponding volumetric change, ΔV, are monitored. After the compression stage, the nozzle is opened and the injection process is continued as usual. Based on Δp and ΔV the bulk modulus of the polymer-gas-mixture can be calculated according to equation 1, where V_0 is the melt volume prior to compression. The bulk modulus, C, allows for a process related determination of dynamic solubility. The compressibility, β, ($\beta=1/C$) increases sharply as soon as the solubility limit is exceeded. The reason for this is that the second discrete gas phase is more compressible than the polymer-gas-solution. Vice versa, a sudden increase of the bulk modulus indicates that the amount of gas falls below the solubility limit.

$$C = -V_0 \frac{\Delta p}{\Delta V} \tag{1}$$

Prior to the study the dynamic solubility limit at a MuCell®-Process-Pressure (MPP, i.e. back pressure) of 140 bar was determined. The gas content was increased stepwise until a drop of the bulk modulus curve was observed, as shown in figure 2. The limit was found to be between 1.2 and 1.4 wt% nitrogen.

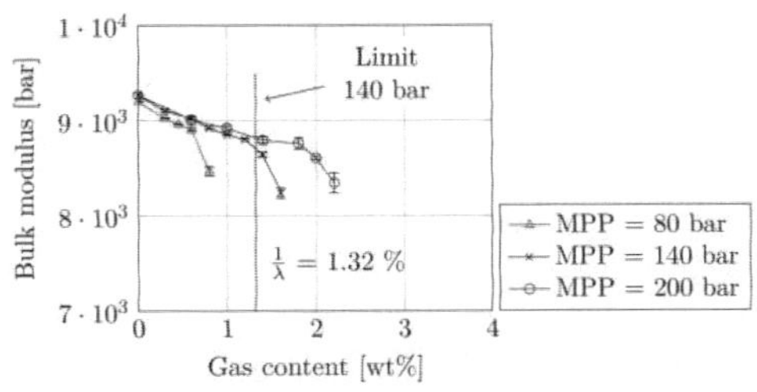

FIGURE 2. Determination of dynamic solubility limits based on the bulk modulus (melt temperature: 230 °C, injection/compression speed: 92 mm/s)

Based on these results, a design of experiments was created (table 1). The question was whether it is possible to raise the limit via appropriate process management. The MPP was fixed at 140 bar while the gas content was set to 1.6 wt%, which is beyond the limit determined in figure 2. The ratio, R_t, between dosing time, t_d, and gas injection time, t_i, is steadily increased. This should provide better dispersion and ad- and absorption behavior. Vice versa, a prolonged dosing time without adequate adaption of gas injection time leads to a decrease of R_t. However, this can offer even better conditions for dissolution of the gas due to the longer time available. Therefore, the second parameter, Γ_{sol}, according to equation 2 also takes the total time available for solution per cycle, t_{sol}, into account.

$$\Gamma_{sol} = t_{sol} * R_t = \left(\frac{s_{tot}-s_o}{s_{tot}} t_d - \frac{t_i}{2}\right) R_t \quad (2)$$

Herein, s_o and s_{tot} stand for the screw position at which the gas injector opens during dosing and the total feeding stroke, respectively. Analogously, t_o describes the time until the gas injector opens, t_{tot} represents the total dosing time. The concepts of R_t and Γ_{sol} are represented in figure 3. The bulk modulus was expected to increase at a certain R_t or Γ_{sol}, respectively. For verification, ultrasonic measurements were conducted during injection of the melt. A transmissive way of measurement allows for the detection of a second discrete gas phase that is present as long as the gas content exceeds the dynamic solubility limit. This phase is visible as interruptions on the measured signal.

Apart from a general investigation of dynamic solubility limits, the compression and injection work as well as maximum injection pressures were examined. This should allow for a comprehensive understanding of the whole process in these high gas content regions.

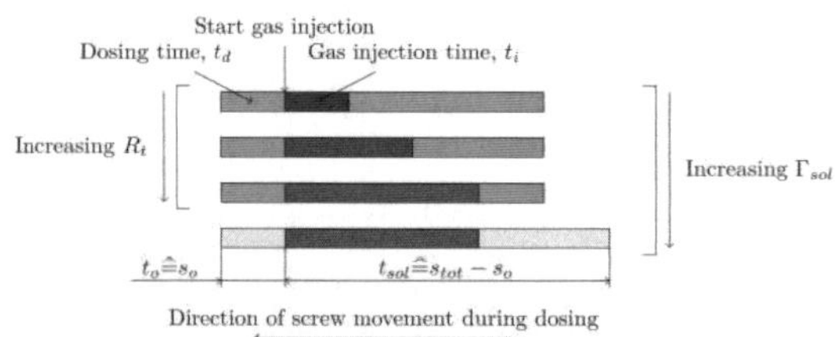

FIGURE 3. Schematic representation of different ratios between t_d and t_i

TABLE (1). Experimental design

No	MPP (bar)	Gas content (wt%)	Screw speed (rpm)	t_d (s)	t_i (s)	R_t (-)	Γ_{sol} (s)
1	140	1.6	75.6	32.79	6.8	0.21	4.73
2			75.6	32.99	8.8	0.27	5.87
3			75.6	32.85	10.8	0.33	6.86
4			75.6	32.06	12.8	0.40	7.68
5			75.6	33.07	14.8	0.45	8.53
6			75.6	32.99	16.8	0.51	10.00
7			72	34.38	19.8	0.58	11.52
8			54	45.23	19.8	0.44	13.49

RESULTS AND DISCUSSION

Bulk modulus measurements

Figure 3 (a) includes the results from bulk modulus measurements. Each data point is the average of 10 individual measurements. Standard deviations are represented by error bars and lie between 0.65 and 1.31 % of the mean. The bulk modulus decreases initially as the gas injection times are prolonged from 6.8 s to 10.8 s. A decrease in bulk modulus beyond the solubility limit (verified by ultrasonic measurements) corresponds to a larger undissolved gas phase. Although the prolonged injection time should lead to a better homogeneity, the possibility of larger amounts of injected gas due to longer injection time cannot be excluded. Starting from t_i = 10.8 s the bulk modulus rises again, which can be a first sign of an improved solubility at a larger time ratio R_t. Between 16.8 and 19.8 s an abrupt rise of the bulk modulus was observed. The first rise was achieved by increasing R_t from 0.51 to 0.58. This rise even increases when further increasing Γ_{sol}. The resulting "inverse knee" in the bulk modulus curve is in accordance with the expectation. The solubility limit could successfully be increased to over 1.6 wt%.

The results from ultrasonic measurements for verification purposes are shown in Figure 3 (b). Encircled interruptions on the signals stem from undissolved gas phases that pass the sensors during injection of the melt. At an injection time of 16.8 s these interruptions are clearly visible. At t_i = 19.8 s and the largest ratio, R_t, of 0.58 the material is just at the transition between a homogeneous polymer-gas-solution and a heterogeneous mixture (i.e. vanishingly small interruptions). Compared to measurements at gas injection times under 16.8 s the gas-polymer-mixtures at 19.8 s can be considered homogeneous. Thus, a first conclusion is that the solubility limit can be increased by increasing R_t. Yet still the homogeneity of the polymer-gas-solution further improves when raising Γ_{sol}. The bulk modulus increases to 8866 bar, which is in perfect agreement with the measurement from figure 2. An increased solubility limit would cause the bulk modulus to continue dropping linearly, which would lead to similar results at 1.6 wt% nitrogen.

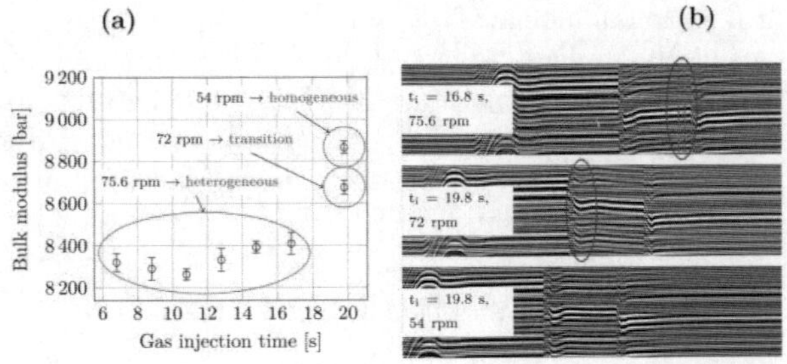

FIGURE 3. Results from bulk modulus (left) and ultrasonic (right) measurements

These findings show that the influence of material on solubility limits might only be relevant at long dosing and gas injection times as an equilibrium cannot develop in short times typical for injection molding. This approach can be useful when large amounts of gas should be dissolved in the melt in order to achieve large density reductions.

Compression and injection work

In addition to this study on the solubility limit also the work during compression and injection of the polymer melt should be referred to the state of solution. This work, W, can be calculated via equation 2 with cross sectional area of the hydraulic cylinder, A_H, pressure in the hydraulic cylinder, $p_H(s)$, and axial screw position, s (s_1 and s_2 represent the screw position prior to and after compression/injection, respectively) [2]. Without taking the cross sectional area of the hydraulic cylinder into account the calculated results are the specific compression work, $W_{comp,spec}$, specific injection work, $W_{inj,spec}$, and the specific total work, $W_{tot,spec} = W_{comp,spec} + W_{inj,spec}$, respectively.

$$W = A_H \int_{S_1}^{S_2} p_H(s) ds \quad (3)$$

As can be seen in figure 4, the behavior of the specific compression work strongly correlates to the bulk moduli in figure 3. A low bulk modulus is equivalent to a higher compressibility of the melt, which increases the distance the screw has to be moved forward in order to build up the predefined pressure and increases the area under the pressure-position-curve. Vice versa, the sudden increase in stiffness at long gas injection times causes the necessary work to drop. This parameter can be used as an alternative to the bulk modulus.

Interestingly, the specific injection work shows opposite behavior. At low gas injection times $W_{inj,spec}$ remains almost constant at 128 N/mm. When the solubility limit was increased beyond the injected amount of gas the work required for melt injection increased to almost 160 N/mm, which corresponds to an increase of 25 %. As Lee et al. show, the viscosity of polymer melts decreases with an increasing amount of dissolved gas [14][15]. From these experimental results, however, it can be concluded that an undissolved gas phase has an even larger viscosity reducing effect than the same amount of gas completely dissolved in the melt.

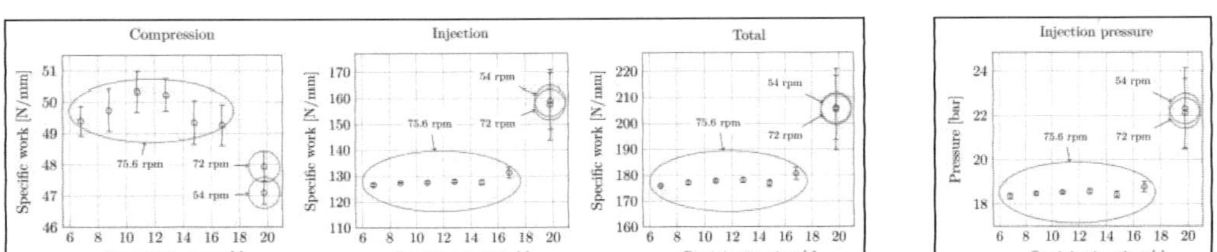

FIGURE 4. Influence of gas injection time on specific compression, injection and total work (left) and on maximum injection pressure (right)

These results are emphasized by the development of the maximum hydraulic injection pressure, $p_{H,max}$, as also shown in figure 4. Between measurements at gas injection times up to 16.8 s and those at 19.8 s an increase of 20 %

can be observed. One possible explanation could be that these separate gas phases are stretched during injection and form something similar to a lubricating film. The larger necessary work during injection even outnumbers the reduction of compression work due to the stiffer melt (see Total specific work). Furthermore the difference between the two settings at t_i = 19.8 s disappears. Under the assumption that the average bubble size of undissolved gas decreases with increasing Γ_{sol}, there seems to be a threshold size of these bubbles under which the flow behavior is similar to that of a homogeneous polymer-gas-solution and above which the viscosity of the mixture is reduced to a constant value independent of changes in bubble size.

The results from these analyses imply applications like thin-wall injection molding (e.g. in the packaging sector) where viscosity reduction should be achieved without foaming. This implicates that as little gas as possible should be dissolved in the melt. From a combination of bulk modulus measurements and analyses of injection work, it can be concluded that both low ratios, R_t, as well as low Γ_{sol} (i.e. short dosing times) prevent a complete dissolution of the injected gas while simultaneously promoting flowability.

SUMMARY AND CONCLUSION

In this work, the influence of process management on the solvable amount of gas and the state of solution was investigated using a bulk modulus measurement method. The possibility of increasing the solubility limit by appropriate time management was evaluated. Initially a gas content of 1.6 wt.% (i.e. beyond a priory determined limit of 1.32 wt.%) was selected. An increase of the ratio between dosing time and gas injection time, as well as an increase of the introduced parameter, Γ_{sol}, that considers the time available for solution, was performed. A combination of bulk modulus and ultrasonic measurements allowed for the monitoring of the state of solution. The results show that an increase of the dynamic solubility limit beyond 1.6 wt.% nitrogen was successful.

An analysis of compression and injection work as well as the maximum injection pressure was conducted simultaneously. As expected, beyond the solubility limit the melt became stiffer, reducing the compression work. The injection work on the other hand showed unexpected behavior. A state of complete dissolution caused the work to increase drastically. It is assumed that undissolved gas phases form something similar to a lubricating film. A comparison with the respective states of solution derived from bulk modulus measurements lead to the hypothesis of a threshold value for the bubble size. Below this value, the promotion of flowability is less effective.

These results show that the ultimate amount of dissolvable gas is subject to much more than just the material. Inappropriate process measurement can cause the solubility limit to decrease.

ACKNOWLEDGMENTS

Financial support and support in conducting all experimental work by Engel Austria GmbH and Borealis Polyolefine GmbH is gratefully acknowledged.

REFERENCES

1. E. Wintermantel and S.-W. Ha, *Medizintechnik: Life Science Engineering,* Berlin Heidelberg: Springer Verlag, 2009, p. 613.
2. F. Johannaber and W. Michaeli, *Handbuch Spritzgießen,* Munich: Carl Hanser Verlag, 2004, p. 397.
3. V. Altstädt and A. Mantey, „Historie des Schaumspritzgießen“ in *Thermoplast-Schaumspritzgießen,* Munich: Carl Hanser Verlag, 2011, pp. 7-13.
4. J. Xu, *Microcellular Injection Molding*, New Jersey: John Wiley & Sons, Inc., 2010, p. 2-3.
5. S.-T. Lee, "History and Trends of Polymeric Foams: From Process/Product to Performance/Regulation" in *Polymeric Foams*, edited by S.-T. Lee and D. Scholz, Boca Raton: CRC Press, 2009, pp. 1-40.
6. A. Kramschuster, R. Cavitt, D. Ermer, Z. Chen and L.-S. Turng, Polymer Engineering and Science **45**, pp. 1408-1418 (2005).
7. S. Wong, J. W. S. Lee, H. E. Naguib and C. B. Park, Macromolecular Materials and Engineering **293**, pp. 605–613 (2008).
8. K. Hikita, JSAE Review **23**, pp. 239-244 (2002).
9. Y. Sato, T. Takikawa, S. Takishima and H. Masuoka, Journal of Supercritical Fluids **19**, pp. 187-198 (2001).
10. B. Wong, Z. Zhang and Y. P. Handa, Journal of Polymer Science: Part B: Polymer Physics **36**, pp. 2025-2032 (1998).
11. S. Takishima, K. Nakamura, M. Sasaki and H. Masuoka, Journal of The Japan Petroleum Institute **33**, pp. 332-336 (1990).
12. S.T. Lee, „Introduction: Polymeric Foams, Mechanisms, and Materials“ in *Polymeric Foams: Mechanisms and Materials*, edited by S.T. Lee and N.S. Ramesh, Boca Raton: CRC Press, 2004.
13. L. Chen, H. Sheth and R. Kim, Polymer Engineering and Science **41**, pp. 990–997 (2001).
14. M. Lee, C. Tzoganakis and C.B. Park, Advances in Polymer Technology **19**, pp. 300-311 (2000).
14. M. Lee, C. Tzoganakis and C.B. Park, Polymer Engineering and Science **39**, pp. 99-109 (1999).

The Influence of Design and Process Parameters on Microcellular Foams for Automotive Applications

Burcu Girginer Özünlü[1, 2 a)] and F. Seniha Güner [3, b)]

[1] Farplas Automotive R&D Departmant, Çayırova, 41420 Gebze/Kocaeli, Turkey
[2] *Material Science and Engineering Department, Graduate School of Science Engineering and Technology, Istanbul Technical University, 34467 Ayazağa /Istanbul*
[3]. *Department of Chemical Engineering, Istanbul Techical University, Maslak 34467 Istanbul*

[a)]Corresponding author: b.girginer@farplas.com
[b)]guners@itu.edu.tr

Abstract. *Recently, there has been an increasing interest in lightweighting design and production technologies in automotive industry owing to energy efficient vehicle requirements. Microcellular physical foaming using supercritical fluid inert gases, known as Mucell, is an important approach to lightweighting technology for a wide range of scientific and industrial applications. Synergistic effect of design and foam structures enable ultimate weight reduction ratios. Therefore, understanding of relationship between design and foaming processes is important for optimization lightweight design and production of the plastic parts. Studies over the past two decades have provided significant information on microcellular foaming through batch process. In this study, the relationship between design parameters, process parameters and foam morphology were investigated through continuous process on industrial scale. The influence of key microcellular foaming parameters such as supercritical fluid gas amount, temperature of polymer melt, injection speed on variable design structures used in automotive applications such as rib, boss was evaluated. The weight reduction ratio and dimensional accuracy of microcellular molded samples were determined. The flexural and impact behavior of the microcellular molded samples were measured and microscopic investigation of microcellular foams was analyzed.*

Keywords: microcellular injection molding, foaming, automotive application, design, process optimization

INTRODUCTION

The microcellular physical foaming technology, commercially known as Mucell technology, is a revolutionized technology that enables design flexibility, dimensional accuracy, cost and weight saving for thermoplastic molded parts. Regarding the EU 2020 regulations limiting $CO_{2(g)}$ emissions, automotive industry shows great interest to microcellular physical foamings. Microcellular foaming technology provides advantages such as utilizing smaller injection molding machine due to lower clamping force requirements, faster cycle time and an alternative eco-friendly production methods [1,2]. However, the microcellular injection molding technology has not become prevalent as it has been expected due to lack of knowledge about designing automotive structures with predictable and safe failure modes and reliable, sustainable measurement of mechanical properties of microcellular molded via standard test specimens. On the other hand; polypropylene is widely used in automotive industry due to relatively low cost, good processability, higher environmental and chemical durability. The most common commercial grade of polypropylene materials is such as impact modified polypropylene, polypropylene copolymer. Unfilled polypropylene has weak melt strength and melt elasticity at high process temperatures. And also, the gas absorption and diffusivity in unfilled polypropylene is relatively low even the utilization of nitrogen as blowing agent. The optimum weight reduction for neat polypropylene material may reach ultimate 10% since the drastic decrease of mechanical properties. Mechanical properties of these materials can reduce up to 40% based on process conditions [3, 4].

Microcellular foamed part design has significant design approaches and rules compared to conventional solid (unfoamed) part design. Owing to high viscosity and balanced pressure of the polymer melt during foaming enables molding of thin wall sections of design elements like wall thicknesses, ribs, bosses [5, 6]. So, microcellular foaming design rules enables light weight design of molded parts. In other words, synergistic effect of design and process conditions is considered to utilize full advantages of microcellular molding technology and ultimate weight reduction ratios.

In this study, the influence of key microcellular foaming parameters such as supercritical fluid gas amount, temperature of polymer melt, injection speed on variable design structures used in automotive applications such as rib, boss was evaluated. In order to figure out influence of these parameters on mechanical properties, microcellular molded samples punched from industrial scaled component. The weight reduction ratio and dimensional accuracy of microcellular molded samples were determined. The flexural and impact behavior of the microcellular molded samples were measured and microscopic investigation of microcellular foams was analyzed.

MATERIALS AND PROCESSING METHODS

Materials

A commercial grade of nucleated heterophasic polypropylene copolymer, commonly used for automotive exterior trim applications Moplen EP548P, in pellet form (Melt Flow Rate (230 °C/2.16 kg) = 16g/10 min) was provided by the Lyondell Basell Technology Company. Industrial grade nitrogene gase was used as blowing agent and supplied from Linde Gas, Turkey.

Microcellular Injection Molding

A single cavity mold consisting automotive design structure given in Figure 1 was employed to produce samples into a 1600 ton-f injection molding machine (KM-MX1600 Model Krauss Maffei, Germany) equipped with Trexel SCF unit was used in experimental part. The mold has homogeneously distributed five direct hot nozzle- sprueless gate. The wall thickness of the part is 2.5 mm which is preferred principally in automotive applications.

The wall thickness measurements were accomplished via the same point from each region with different process parameters. And also, the cross overed ribs with 0,5 mm, 1 mm and 2 mm thickness samples were taken from the region 2 as indicated in Figure 1- a to measure thickness of each samples. The observed sink marks due to rib design among were evaluated via imaging.

The mechanical test samples (bars) were punched from the A1, A2, A3 region as indicated in Figure 1-b with equal distance from each other. Flexural modulus measurements were accomplished using Zwick Roell Mechanical testing system with 20kN capacity and unnotched impact strength measurements were accomplished using Zwick Roell HIT 5 J impact test machine. The cell morphology was investigated via images generated from Nikon SMZ 745 stereomicroscope. The microscope analysis samples were produced by following these steps: The test bars indicated A1, A2 and A3 notched via razor blade, immersed in to liquid nitrogen for 3 minutes and break into 2 pieces.

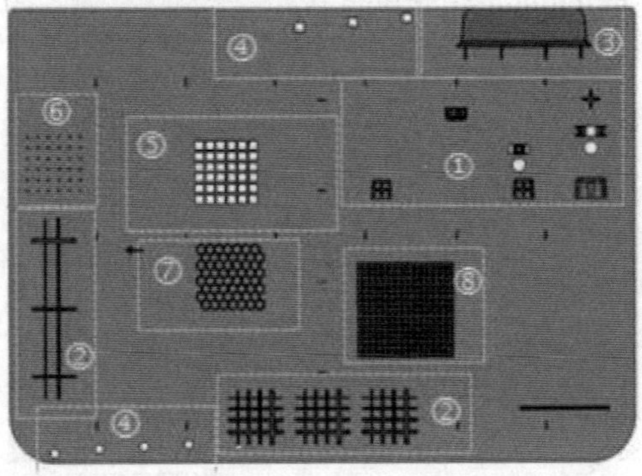

a)

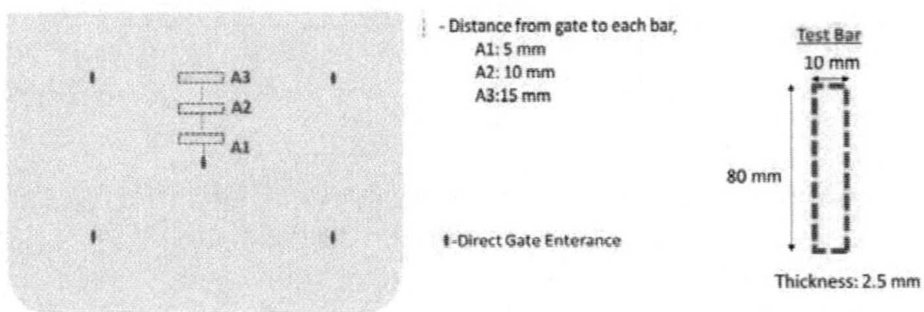

b)

FIGURE 1: a) The mold design has various design elements on it 1. Assembly structures (boss, screw hole), 2. Ribs, 3. Flap, 4. Assembly structures (holes)-, 5. Grill, 6. Dots for vibration welding, 7. Honeycomb ribs, 8. Pyramid structures for sound absorption,
b) Position of gates and mechanical test specimen with 80mm x 10 mmx 2.5 mm dimensions

The mold temperature was kept at 15 °C during experiments. Solid molded parts were produced at 220 °C with1600 ton-f clamping force. All microcellular molded parts produced by processing with 1200 ton-f clamping force. The variable values of key parameters were given in Table 1.

TABLE (1) Process parameters used in experiments

Parameter	***Low***	***High***
Melt Temperature, °C	180	220
SCF Amount, %	0.6	1
Injection Speed, mm/sec	80	100

RESULTS AND DISCUSSION

The void fraction, average cell size and wall thickness of microcellular parts produced with various process parameters were given in Table-2. According to the results, at lowered process temperatures; SCF gases solubility is increased and in the case of lowered injection speed; gas droplets collapsed and post-blowing and flash problems has occurred. The wall thickness measurement and average cell size were observed in wide tolerance range due to non-homogeneous foaming. Fluctuation of average wall thickness was strongly depended on the fluctuation of average cell size and post-blowing formation. The shrinkage behavior was obtained as expected. The higher melt temperature resulted with higher shrinkage ratio as well as lower injection speed.

TABLE (2): Characteristics of microcellular foamed parts produced with various process parameters

Proses Parameters	**Injection Speed, mm/sec: 80** **SCF Amount, %: 1** **Melt Temperature, °C: 220**	**Injection Speed, mm/sec: 120** **SCF Amount, %: 0,6** **Melt Temperature, °C: 220**	**Injection Speed, mm/sec: 80** **SCF Amount, %: 1** **Melt Temperature, °C: 180**
Void Fraction	0.044 *(±0.010)*	0.034 *(±0.015)*	0.066 *(±0.025)*
Average cell size, μm	180*(±50)*	120*(±50)*	200 *(±100)*
Average wall thickness, mm	2.50 *(±0.03)*	2.50 *(±0.10)*	2.50 *(±0.20)*
Shrinkage, % *(parallel)*	3.46	2.72	2.25
Shrinkage, % *(transverse)*	2.93	2.63	2.40
Observed Defect	Post-blowing of thick-walled sections; short-shot in thin-walled sections	Post-blowing of thick-walled sections; short-shot in thin-walled sections	Post-blowing of thick-walled sections; flashes

Mechanical properties of microcellular foamed parts which was punched out from different flow distance from gate resulted as shown in Figure 2. The results showed that, optimum cell formation and crystallization behavior occurred in the middle of flow path, A2 region. At the end point of filling path, A3 region, the foaming was unstable and resulted with lower flexural and impact strength. The non- homogenous foaming and collapsed cells may occurred due to cooling of polymer melts.

According to the results, surprisingly, microcellular parts with processing parameters at low level SCF, injection speed and melt temperature give the highest of flexural modulus, however the lowest impact strength. Because of high void fraction and post-blowing in local areas, the reduction in impact strength is expected. It is known that cells play role as notch in microcellular structures. On the other hand, for two polymer melt temperatures, the microcellular molded samples produced by low injection speed and high SCF amount were resulted with higher flexural modulus and impact strength values.

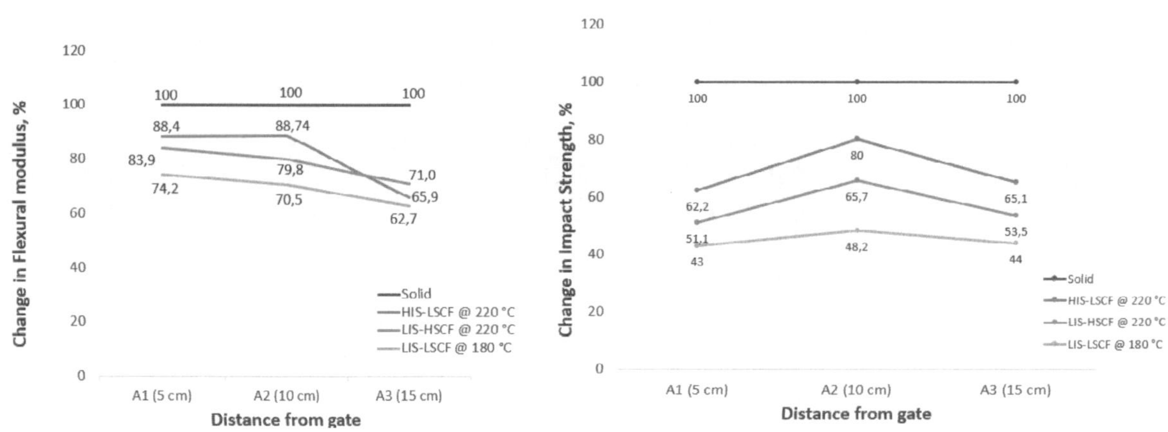

FIGURE 2: Mechanical properties changes of A1, A2, A3 region with high injection speed/low SCF amount and low injection speed/high SCF amount compared to solid (unfoamed) at 220 °C and low injection speed/low SCF amount at 180 °C.

It is seen in Table 3 that, the wall thickness to rib ratio, 2.5:1.25 was not applicable in traverse and 2.5:4 in transverse. All the wall thickness to rib ratio, 2.5:1,5 seems to be stable in transverse. Their fluctuation occurred in a tight zone. All other the wall thickness to rib ratios were convenient to application in reinforcing design, even the ratio 1:1 is restricted in standard injection molding. Although, according to results microcellular foaming utilized elimination of sink marks. **TABLE (3)**: Wall Thicknesses of microcellular foamed rib structure produced with various process parameters

Proses Parameters	**Rib to wall thickness: 1.25:2.5**		**Rib to wall thickness: 2.5:2.5**		**Rib to wall thickness: 4:2.5**	
	Parallel rib thickness,	Transverse rib thickness,	Parallel rib thickness	Transverse rib thickness	Parallel rib thickness	Transverse rib thickness
Injection Speed, mm/sec: 80 **SCF Amount, %: 1** **Melt Temperature, °C: 220**	Short shot, (not dully filled)	1.25 mm *(±0,010)* Sink marks on the B surfaces	2.5mm *(±0,3)* Sink marks on the B surfaces	2.5mm *(±0,03)* Sink marks on the B surfaces	Post blowing, sink marks on the B surfaces	4 mm*(±0,5)*
Injection Speed, mm/sec: 120 **SCF Amount, %: 0.6** **Melt Temperature, °C: 220**	Short shot, (not dully	1.25 mm *(±0,010)* no sink mark on the B surfaces	2.5mm *(±0,2)* no sink mark on the B surfaces	2.5 mm *(±0,02)* no sink mark on the B surfaces	Post blowing,	4 mm *(±0,1)*
Injection Speed, mm/sec: 80 **SCF Amount, %: 1** **Melt Temperature, °C: 180**	Short shot, (not dully filled)	1.25 mm *(±0,010)* no sink mark on the B surfaces	2.5mm *(±0,3)* no sink mark on the B surfaces	2.5 mm *(±0,05)* no sink mark on the B surfaces	Post blowing	4 mm *(±0,3)*

CONCLUSION

In this study, the relationship between design variables such as wall thickness, different wall-to-rib thickness, shrinkage ratio (parallel/transverse) and key proses parameters; melt temperature, SCF amount% and injection speed were investigated. The optimum process parameters for un-filled commercial grade of polypropylene copolymer is melt temperature, 220 °C; injection speed, 80 mm/s and SCF amount, %0.6. For this material adjustment of SCF amount and temperature is crucial. Even high melt temperature reduces the viscosity; the flash (burr) defects were obtained in samples produced by high SCF amount- low melt temperature parameters shows that amount SCF gases dissolved in polymer melt has significant effect on increasing the viscosity. According to the result, even using polypropylene copolymer with high crystallinity, to obtain homogenous and medium dense foam structure; addition of micro and nano reinforcements or fillers is suggested. As shown in literature and our study design approach for microcellular molding enables different wall thickness to rib thickness. Even 1.5 mm rib design can be microcellular molded in transverse direction. And also, our study supported literature in the point of the fact that microcellular foaming eliminates sink marks.

REFERENCES

1. Regulation (EC) 443/2009 of the European Parliament and of the Council setting emission performance standards for new passenger cars as part of the Community's integrated approach to reduce CO2 emissions from light-duty vehicles; *Journal of the European Union*, **L 140**, 05 June 2009
2. Harris, A., Lee, E., Peraro, W., Nunes, S. et al., "Automotive Interior Injection Molded Parts Using Microcellular Foaming Technology," *SAE Technical Paper*, **2014-36-0172**, 2014.
3. Hyde, L. J. et. al., How Microcellular Foam Molding Changes the Cost Structure of Injection Molded Automotive Components: A Review of the Process and Automotive Applications, *SAE Technical Paper*, **2002-01-0717**, 2012.
4. Gunkel, F. et. al., Understanding Melt Rheology & Foamibility of PP-Based TPO Blends, *Journal of Cellular Plastics,* 44, 4, pp. 307-325(2008).
5. J. Xu, "Design of Microcellular Injection Molding" in *Microcellular Injection Molding*, (John Wiley & Sons, Inc., Hoboken New Jersey, 2010), pp. 165–227.
6. Kishbaugh, L, Kolshorn U., Boss Design and Optimization for Microcellular Foam Injection Molded Parts, SPE ANTEC, 2011 Plastics Annual Technical Conference Proceedings, Vol. 1-3, pp. 1507-1511

EMI Shielding Performance of Natural Rubber-Based Nanocomposite Foams with Segregated CNT

Luigi Sorrentino[a,*], Maria Oliviero[a], Yanhu Zhan[b], Jian Wang[c], Andrea Sorrentino[a], Giovanna Buonocore[a], Marino Lavorgna[a], Hesheng Xia[c] and Salvatore Iannace[d]

[a]Institute of Polymers, Composites and Biomaterials, National Research Council, P.le Fermi, 1 - 80055 Portici, NA, Italy
[b]School of Materials Science and Engineering, Liaocheng University, Liaocheng 252000, China
[c]State Key Laboratory of Polymer Materials Engineering, Polymer Research Institute, Sichuan University, Chengdu 610065, China
[d]Institute for Macromolecular Studies (ISMAC - CNR) National Research Council, Milano 20133, Italy
[*]Corresponding author: L. Sorrentino (Email: luigi.sorrentino@cnr.it)

Abstract. Natural rubber/carbon nanotubes composite foams (F-NR/CNTs) with high electrical conductivity and excellent electromagnetic interference (EMI) performance were developed through a multi-step process. CNTs were assembled on natural rubber particles and then a pre-crosslinking treatment was applied. Such composite structures were foamed with supercritical carbon dioxide with a batch process and finally foamed composite samples were post-crosslinked. Closed-cell porous structure and segregated CNT network were induced in the resulting foams. The expansion ratio ranged between 0.3 and 2.7, the cell size was between 4 and 23 μm and the cell density was in the order of 10^8 cells/cm^3. Thanks to the segregated network morphology, composite foams exhibited good mechanical properties, and the electrical conductivity was proportional to the CNT content. Owing to the multiple radiation reflections and scattering between the cell-matrix interfaces, the composite foams showed a specific shielding effectiveness (SSE) up to 312.69 dB cm^2/g for composite foams containing 6.4wt% of CNTs. This performance is significantly higher than those already published for rubber composites containing comparable nanocomposite filler content. More importantly, the shielding effect is mostly due to absorption, implying that most of incident electromagnetic radiation is dissipated. This work provides the fundamentals for the design of innovative lightweight and efficient EMI shielding foams, characterized by a three-dimensional segregated CNTs network with valuable potentials in the electronics and aerospace industries.

Keywords: EMI Shielding, Natural Rubber, Nanocomposite Foams, Segregated CNT

INTRODUCTION

Electromagnetic interference (EMI) has become a relevant societal concern since it affects the functionality of electronic devices and may have negative effects on people health [1]. In order to mitigate this technological and safety issues, high-performance EMI shielding polymer-based composites have recently attracted high attention of both academic institutions and industrial companies [1-3] The inclusion in polymers of high content of conductive filler, such as carbon black (CB), carbon nanotubes (CNTs) and graphene nanoplatelets and its derivatives (GE), brings about outstanding improvement of both electrical conductivity and EMI SE of the resulting composite materials [4]. The high density of such systems tends to limit their use, therefore the development of porous polymer composites exhibiting excellent EMI shielding properties represents a promising solution for promising technological applications [5,6].

In nanocomposites the proper distribution of the conductive filler in a three-dimensional network fashion has proved to be an effective approach in improving electrical conductivity and, consequently, the EMI shielding effect (SE) of materials [7-13]. Segregated structures are obtained by assembling the conductive filler at the interface of polymer particles generally trough a solvent-wrapping or an ultrasonic assisted latex-based approach. In segregated structures the conductive filler is moved in the volume between polymeric particles. Such approach allows for a significant reduction of the percolation threshold to trigger the formation of a conductive network with respect to the case where the filler is randomly dispersed within the polymer [11].

An additional contribution to EMI SE is detected in conductive lightweight materials [14-16], where the presence of bubbles, induced by using carbon dioxide, reduces the density of the system and significantly increase the absorption of electromagnetic waves through the multiple reflections effect of the incident waves [15]. EMI shielding properties of polymeric foams can be dependent on the degree of foaming because, when density reduces, the effective conductive paths decrease, bringing about a sharp reduction of electrical conductivity and EMI shielding properties. For polymer composites with segregated filler network, their electrical conductivity and EMI shielding properties decrease after foaming because of the formation of some discontinuity during the foaming process. A new approach to prepare composites foams based on natural rubber (NR) and carbon nanotubes (CNTs) with a segregated network is proposed. It allows the precise control of the two concurrent processes occurring in such systems, namely rubber crosslinking and supercritical CO_2 foaming.

EXPERIMENTAL

Natural rubber (NR) latex (LCS, solid content: 67.5wt%) was supplied by Synthomer (UK). Dicumyl peroxide (DCP) was provided from Acros Organics (Thermo Fisher Scientific, Belgium). Carbon nantube (NC 7000, diameter: 10 nm, length: 1.5 μm, density: 1.75 g/cm^3) were purchased from Nanocyl S.A., Belgium. Cetyltrimethylammonium bromide (CTAB), as a surfactant, was obtained from Sigma Aldrich Chemicals (USA). All agents are used without further purification.

Nanocomposites with segregated network are coded as N-NR/CNTx, where x is the CNT concentration in weight percentage with respect to NR (Table 1). Composite foams (F-NR/CNTs) with the segregated morphology were prepared from N-NR/CNT samples. A pre-crosslinking step was performed and then samples were solubilized with CO_2 in a high pressure vessel set at 40 bar and 35 °C for 60 minutes. After CO_2 sorption, the pressure was quickly released to foam the samples. As a final step, foamed specimens were quickly placed into an oven at 170 °C for 90 s to completely crosslink the natural rubber. This procedure allowed to preserve the segregated morphology originated by the tailored assembling of CNTs around the NR latex particles.

TABLE 1. Compositions of N-NR/CNTs composites

Sample	NR latex (g)	CNT (g)	CTAB(g)	DCP(g)
NR	8.3	0	0	0.25
N-NR/$CNT_{0.82}$	8.3	0.05	0.2	0.25
N-NR/$CNT_{1.68}$	8.3	0.1	0.4	0.25
N-NR/$CNT_{3.31}$	8.3	0.2	0.8	0.25
N-NR/$CNT_{4.87}$	8.3	0.3	1.2	0.25
N-NR/$CNT_{6.40}$	8.3	0.4	1.6	0.25

RESULTS AND DISCUSSION

The assembling process of CNTs allowed to uniformly coat NR particles. One end of CNTs was attached to the polymer surface, while the other was free to move in the solvent (Fig. 1A). A representative morphology of the produced foams is shown in Fig. 1B. The pristine NR based material expands to a higher extent with respect to CNTs-based composite systems but this condition is not stable and the morphology significantly collapses during the thermal crosslinking treatment. This behavior has been related to the fact that the strained rubber macromolecules tend to recover their initial conformations state due to the increment of the entropic contribution that reduces the total free energy of the system. The presence of CNTs promotes strong interface forces and entanglements between CNTs and rubber chains which, acting as physical crosslink point, hinder the macromolecular chains retraction of foamed rubber during the post-crosslinking treatment. As the CNTs content increases, samples possess a higher number of physical crosslinks due to π-π carbon nanotubes interactions, which therefore contrast the bubble collapsing after the foaming process and quench the foamed structure. At 1.68 wt% of CNT, the maximum expansion ratio was achieved. At CNT contents higher than 1.68 wt%, the expansion ratio is lower because the higher CNT availability hindered the formation of large porosity (Fig. 2). The segregated morphology of highly CNT containing systems was responsible for hindering the cell growth in N-NR/CNT composites at high CNTs content.

The electrical conductivity of F-NR/CNTs composites is generally lower than that of N-NR/CNTs composites. The extent of reduction depends on the expansion ratio, the higher is the expansion ratio of the composite foams the bigger is the difference in the electrical conductivity between foamed and unfoamed samples. It is well known that the electromagnetic interference (EMI) shielding effect (SE) of materials depends on the reflection and absorption of the incident electromagnetic waves. The reflection of EM wave occurs as consequence of the impedance mismatch between air and absorber material. The absorption (SEA) originates from ohmic loss (i.e. attenuation of energy through the electrical current flow in phase with the applied EM wave), polarization loss (i.e. dissipation of energy by overcoming the momentum needed to reorient the dipoles in the alternated EM wave) and magnetic loss (i.e. attenuation of energy coming from magnetic particles in EM field) [2]. Since rubber and CNT do not possess magnetic properties, the EMI SE performance of segregated F-NR/CNTs originate from the ohmic loss and the polarization loss. The polarization relates to the materials' functional groups, defects, and interfaces [2]. In segregated F-NR/CNTs, the polarization loss is due to a) the rubber-gas interface on the surface of cells and b) the CNT-rubber interface. This combination enhances the interfacial polarization loss and induces a high absorption of EMI waves. The conductive three-dimensional network consequently dissipates electromagnetic energy by ohmic losses. The segregated network in F-NR/CNT foams behaves as black hole because the electromagnetic radiation cannot escape from the material without a large absorption and conversion of energy in form of heat. In order to show the role of the segregated network on the EMI SE performance, a foamed system (R-NR/CNT 3.31) was prepared with randomly dispersed CNTs and its performance was compared with the analogous provided with the segregated network. The EM wave was not absorbed by the R-NR/CNT foam (Fig. 2A), and its behavior was similar to that of the foamed NR (F-NR). This demonstrated that the presence of the segregated CNT network is the crucial factor for improving EMI SE of F-NR/CNTs foams. It can be concluded that. in general, the presence of a segregated conductive network gives electrical conductivity alongside with high number of interfaces, and it is responsible for the excellent EMI shielding performance of the foams (Fig. 3).

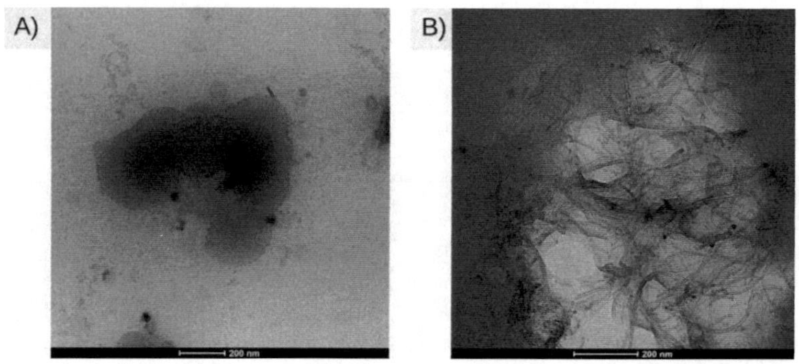

FIGURE 1. TEM images of composite with 3.31 wt% of CNTs: A) before foaming and B) after foaming.

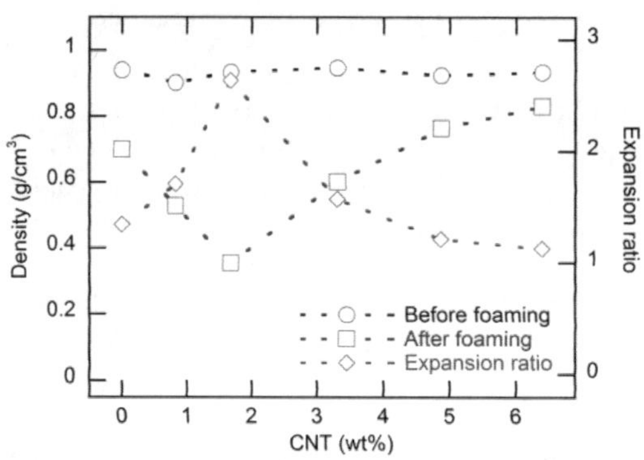

FIGURE 2. Density and expansion ratio of F-NR/CNTs composite foams

Moreover, since in lightweight shielding applications the specific shielding effectiveness (SSE) is often used to compare the effectiveness of shielding material, SSE of all samples was calculated as the ratio between SE and the product of mass density and sample thickness. SSE of F-NR/CNT3.31 and F-NR/CNT6.40 composites is equal to 269.77 and 312.69 dB cm^2/g, respectively, outperforming the results of about 181 dB cm^2/g exhibited by pioneering rubber composites in the 8.2-12.4 GHz frequency range already published by some of the authors (see Table 2). Although the EMI SE of F-NR/CNT6.40 composites is lower than that of N-NR/CNT6.40 composites with the same

CNT content (shown in Table 3), their SSE is similar with that of N-NR/CNT6.40 composites because of the lower density.

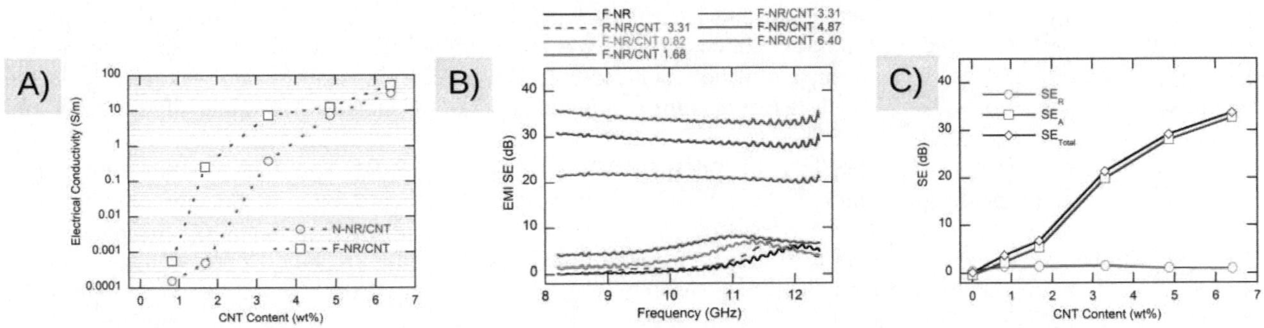

FIGURE 3. A) Electrical conductivity of N-NR/CNTs and F-NR/CNTs composites with segregated network; B) EMI SE as a function of frequency for segregated F-NR/CNTs composites and R-NR/$CNT_{3.31}$ foam; C) Shielding by reflection, absorption, and total shielding of segregated F-NR/CNTs at the frequency of 10.3 GHz

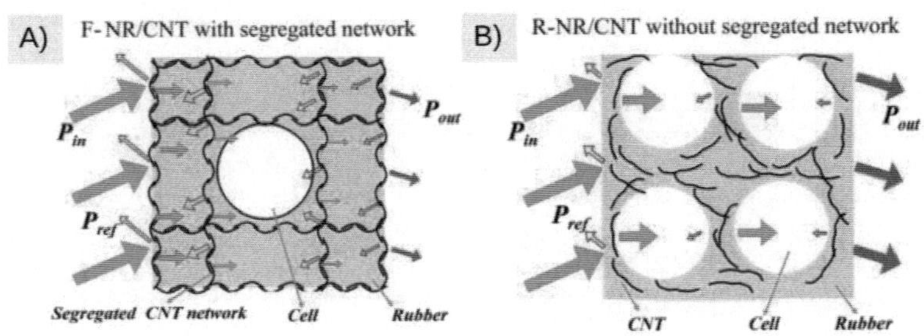

FIGURE 3. Schematics of the EMI shielding mechanisms of foams with (A) and without (B) a segregated network. (P_a=P_{in}-P_{ref}-P_{out}, where, P_{in}, P_{ref}, P_a and P_{out} are the incoming, reflecting, absorbing and outgoing power of electromagnetic radiation, respectively).

TABLE 2. EMI SE and SSE in 8.2-12.4 GHz frequency range for NR/CNT composites.

Sample	Thickness (mm)	CNT content (wt%)	SE (dB)	SSE (dB cm^2/g)	Ref.
F-NR	1.3	0	1.06	11.69	This work
F-NR/CNT0.82	1.3	0.82	3.66	53.23	This work
F-NR/CNT1.68	1.3	1.68	6.05	131.23	This work
F-NR/CNT3.31	1.3	3.31	21.11	269.77	This work
F-NR/CNT4.87	1.3	4.87	29.08	293.07	This work
F-NR/CNT6.40	1.3	6.40	33.74	312.69	This work
N-NR/CNT0.82	1.3	0.82	10.18	87.01	This work
N-NR/CNT1.68	1.3	1.68	23.14	190.78	This work
N-NR/CNT6.40	1.3	6.40	37.33	308.10	This work
CNT/NR	2.6	5.0	33.3	137.69	[12]
SBR/CNT	5	9.09	35.06	71.56	[17]
BR/CNT	1	7.41	13	14.13	[18]
NR/ENR/CB	4	23.08	23.58	63.38	[19]
TPNR/Fe3O4	9	12	25.52	30.48	[20]
NR/CB/silica	2.83	41.18	16.06	52.54	[21]
tire rubber/CNT	2.6	2 wt% CNT +13.1 wt% CB	36.8	152.19	[22]
NR/Fe3O4@GE	1.6	4.50 wt% GE +20.48 wt% Fe3O4	32.9	181.31	[13]

CONCLUSIONS

Composite foams consisting of segregated CNT network foamed particles were prepared via a latex mixing procedure able to promote CNT wrapping around latex particles in combination with CO_2 foaming. The formation of the cellular morphology and the CNT segregated network was assessed by TEM analyses. The expansion ratio of F-NR/CNTs wasn't in direct relationship with the CNT content. The highest expansion ratio was reached in the composite containing 1.68wt% of CNT as a result of the combined effect of physical-crosslinking between CNT and rubber.

The addition of CNT enhanced the electrical conductivity and EMI SE of F-NR/CNTs. The absorption efficiency was identified as the major contributor to the EMI shielding performance, hence most of the EM radiation was dissipated in form of heat. Specific SE was even more pronounced and reached 312.69 dB cm^2/g in the F-NR/CNT6.40 foam. It is very important to note that foamed systems showed SSE values comparable with solid systems but at a fraction of their conductivity, meaning that the shielding effect can be decoupled from the conductivity, and that low conducting materials can be used as shielding systems in EMI sensitive applications.

ACKNOWLEDGMENTS

This paper is based upon work supported by the Air Force Office of Scientific Research under award number FA9550-17-C-0116.

REFERENCES

1. F. Meng, H. Wang, F. Huang, Y. Guo, Z. Wang, D. Hui and Z. Zhou, *Compos. Part B*, 2018, 137, 260-277.
2. F. Sharif, M. Arjmand, A.A. Moud, U. Sundararaj and E.P.L. Roberts, *ACS Appl. Mater. Interfaces* 2017, 9 14171-14179.
3. R. Sun, H. B. Zhang, J. Liu, X. Xie, R. Yang, Y. Li, S. Hong and Z. Z. Yu, *Adv. Funct. Mater.* 2017, 27, 1702807.
4. P. Cataldi, I.S. Bayer, F. Bonaccorso, V. Pellegrini, A. Athanassiou and R. Cingolani, *Adv. Electron. Mater.*, 2015, 1, 1500224.
5. Z. Zeng, H. Jin, M. Chen, W. Li, L. Zhou and Z. Zhang, *Adv. Funct. Mater.*, 2016, 26, 303-310.
6. H. Zhang, G. Zhang, J. Li, X. Fan, Z. Jing, J. Li and X. Shi, *Compos. A Appl. Sci. Manuf.*, 2017, 100, 128-138.
7. H. Pang, L. Xu, D.X. Yan and Z.M. Li, *Prog. Polym. Sci.*, 2014, 39, 1908-1933.
8. Y. Zhan, M. Lavorgna, G. Buonocore and H. Xia, *J. Mater. Chem.*, 2012, 22, 10464-10468.
9. Y. Zhan, Y. Meng and Y. Li , *Mater. Lett.*, 2017, 192, 115-118.
10. N. Yan, G. Buonocore, M. Lavorgna, S. Kaciulis, S.K. Balijepalli, Y. Zhan, H. Xia and L. Ambrosio, *Compos. Sci. Technol.*, 2014, 102, 74-81.
11. L.C. Jia, D.X. Yan, C.H. Cui, X. Jiang, X. Ji and Z.M. Li, *J. Mater. Chem. C*, 2015, 3, 9369-9378.
12. L.C. Jia, D.X. Yan, Y. Yang, D. Zhou, C.H. Cui, E. Bianco, J. Lou, R. Vajtai, B. Li, P. M. Ajayan and Z. M. L, *Adv. Mater. Technol.*, 2017, 2, 1700078
13. Y. Zhan, J. Wang, K. Zhang, Y. Li, Y. Meng, N. Yan, W. Wei, F. Peng and H. Xia, *Chem. Eng. J.*, 2018, 344, 184-193.
14. L. Guadagno, M. Raimondo, V. Vittoria, L. Vertuccio, C. Naddeo, S. Russo, B. D. Vivo, P. Lamberti, G. Spinelli and V. Tucci, *RSC Adv.* 2014, 4, 15474-15488.
15. B. Shen, Y. Li, D. Yi, W. Zhai, X. Wei and W. Zheng, *Carbon*, 2016, 102, 154-160.
16. C. Liu, C. Wu, C. Hao, P. Liu, X. Guo, Y. Zhang and Y. Huang, *J. Appl. Polym. Sci.*, 2018, 135, 45996.
17. J. Ma, M. Zhan and K. Wang, ACS Appl. Mater. Interfaces, 2015, 7, 563-576.
18. J. Abraham, M. Arif P, P. Xavier, S. Bose, S. C. George, N. Kalarikkal and S. Thomas, Polymer, 2017, 112, 102-115.
19. N. Joseph, C. Janardhanan and M.T. Sebastian, Compos. Sci. Technol., 2014, 101, 139-144.
20. P. Zhao, L. Li, Y. Luo, Z. Lv, K. Xu, S. Li, J. Zhong, Z. Wang and Z. Peng, Compos. Part B, 2016, 99, 216-223.
21. I. Kong, S.H. Ahmad, M.H. Abdullah, D. Hui, A.N. Yusoff, D. Puryanti, J. Magn. Magn. Mater., 2010, 322, 3401-9.
22. A.A. Al-Ghamdi, O.A. Al-Hartomy, F. R. Al-Solamy, N. Dishovsky, M. Mihaylov, N. Atanasov, G. Atanasova and D. Nihtianova, J. Polym. Res., 2016, 23, 180.

Multi-Graded foams by gas foaming using periodic mass transport boundary conditions

M. Trofa, E. Di Maio[a)], P.L. Maffettone

Dipartimento di Ingegneria Chimica, dei Materiali e della Produzione Industriale, University of Naples Federico II, P.le Tecchio 80, 80125 Naples, Italy

[a)]Corresponding author: edimaio@unina.it

Abstract. Polymeric foams with a gradient in density and morphology exhibit improved performances in terms of energy and sound absorption as compared to the uniformly foamed ones. In this work, we report an improvement of the gas-foaming method to obtain, for the first time, well-defined multi-graded foams. At variance with the conventional procedure which involve constant conditions, we impose a time varying partial pressure of the foaming agent. This allows to form a complex blowing agent concentration profile within the polymer, in turn giving a structured foam (tuneable for specific applications) when supersaturation is attained by pressure drop. We use a unidimensional diffusion model to correlate the desired concentration profile to the pressure history imposed in the vessel. Following this new procedure, we used a periodic triangular pressure history of the blowing agent (CO^2) to obtain a 3-layers and a 5-layers polystyrene foam. The produced foam layers were characterized by different void fractions, ranging between 0 and 90%, and pore sizes, from 50 to 500 microns.

INTRODUCTION

Thermoplastic foams are ubiquitous in our life, as well as in the industry and technology, due to their versatility, low cost and performances. Not to mention the variety of foamable polymers, versatility derives from the possibility to tune foam density and morphology to the application. Reduced costs, with respect to unfoamed, dense polymers, arise essentially from the possibility to reduce material usage and parts weight. These advantages, of course, are obtained without sacrificing performances. From the other point of view, hence, foams are often the optimal material solution to boost performances/weight ratios. Nature, which has improved material usage through hundreds of millions of years of evolution, is in fact crowded with porous materials, for the functional and the structural properties as well: wood, bones, shellfish shells, the Nature structural materials, all are porous materials.

At a closer look, natural high performance porous materials have a complex structure, non-uniform in both density and morphology, e.g. to cope with the graded stress fields arising in the practical loading conditions or with the directional transport of nutrients. Scientists are trying to mimic these structures, and several strategies are reported to achieve foams with graded morphologies and/or densities. Some of the reported strategies, however, look a bit factitious, and difficult to scale and/or industrialize. Graded microballoon positioning [1], multiple layers gluing [2] are some examples. These papers, together with numerous theoretical studies, proved the utility of graded structures, such as in improving impact resistance or compressive strength. One of the approaches we consider more readily industrializable has been introduced by Zhou and colleagues and considers the use of "non-equilibrium gas concentration profiles"[3], in which, the material to be foamed by gas-release has not been spatially saturated by the blowing agent. If the mass transport transient has not fully developed, the inner portion of the sample to be foamed contains a minor concentration of the blowing agent as compared to the outer ones, in which the gas concentration readily reaches the equilibrium conditions with the external gas pressure. As a result, the inner portion of the sample will be less foamed or unfoamed, while the outer portions will be fully foamed. Time plays a role here, and the graded morphology will be achieved if sorption time, τ_s, is less than but comparable with the characteristic diffusion time, τ_d, in this diffusion-controlled problem, $\tau_d = \frac{L^2}{D}$ (eq. 1), where L is the sample thickness and D is the gas diffusivity in the polymer matrix. Despite the ease of achievement of a graded structure, this method is quite limited, as it does not allow anything different from a single-graded structure, less foamed in the inner of the sample.

With this approach, gradient symmetries depend on sample shape and gas access to the sample. One-side foamed samples come out, in particular, when the blowing agent penetrates solely from the top surface of a disk-shaped sample, inducing the expansion of the top layers, with minimal to null expansion of the bottom ones.

In this paper, we push forward this concept of using non-equilibrium mass transport conditions to achieve multiple graded (layered) foamed structures.

MATERIALS AND METHODS

Materials

The PS (N2380) was supplied by Versalis S.p.A., Mantova, Italy, and used as received. The density and melt flow index, are 1.05 g/cm^3 and 2.0 g/10 min at 200°C and 10 kg. The average molecular weight of the PS used is 3 kDa as supplied by Versalis S.p.A.

Methods

The foaming equipment utilized in this study is a custom-made pressure vessel with accurate control of the foaming processing variables [4]. The control of the temperature was achieved by means of a PID controller (model X1, Ascon-New England Temperature Solutions, Attleboro, MA, USA) and a Pt100 sensor. A pressure transducer (model P943, Schaevitz-Measurements Specialties, Hampton, VA, USA,) was used to measure pressure and the pressure history was registered by using a data acquisition system (DAQ PCI6036E, National Instruments, Austin, TX, USA). A syringe pump 500D (Teledyne Isco, Lincoln, NE, USA) was used to define the pressure history, controlled by a personal computer via a RS232 connection. The pressure release system consists of a discharge valve (model 10-80 NFH, High Pressure Equipment Company, Erie, PA, USA) and a pneumatic electro-valve. In a typical experiment, a cylindrical polymeric sample, 2 mm in thickness and 12 mm in diameter, or few spherical samples, 1 mm in diameter were placed in the pressure vessel. After sample loading, the vessel was heated to the saturation temperature (Tsat=100°C) and pressurization started according to the defined program. At the end of the pressure program, pressure was released in 0.05 s, still at 100°C. Samples were then removed from the vessel in 30 s. Foams were characterized to determine the number of bubbles per unit volume, N_b, and the average cell size, Φ, from Scanning Electron Microscope (SEM) images. The samples were first sectioned with a razor blade in liquid nitrogen and then coated with gold using a sputter coater. N_b was calculated as Nb =$(n/A)^{3/2}$, where n is the number of the cells in the area A of the SEM micrograph. Φ was measured according to ASTM D3576, and using ImageJ software. Details of the pressure program for the different experiments follow, starting from neat (without blowing agent) samples. All samples were allowed to expand freely.

RESULTS

When a periodic boundary condition (e.g. a triangular external CO_2 pressure wave) is applied to the top and bottom surfaces (Σ and Σ' in Figure 1A) of a cylindrical polystyrene (PS) slab with null initial concentration and uniform thickness, with impermeable (no-flux) lateral surface, Ω, the CO2 concentration profile presents, at a certain time instant, two concentration peaks (Figure 1B) and one concentration well at the center. Applying said pressure program is nowadays possible with pumping systems connected to gas foaming apparata. In our case, we utilized the pressure vessel reported in our previous study [4] and a programmable Isco Inc. piston pump (mod. 500D). When subjecting the PS sample to said pressure program with CO_2 as the blowing agent, and releasing the pressure when the gas concentration inside the sample matches the one shown in Figure 1B, the corresponding foam results in a three layer structure (not to mention the usual unfoamed skin layers), unfoamed at the center and well-foamed close to the external surfaces. Figure 1C reports an SEM micrograph of the overall foam morphology, evidencing the three-layers graded foam structure.

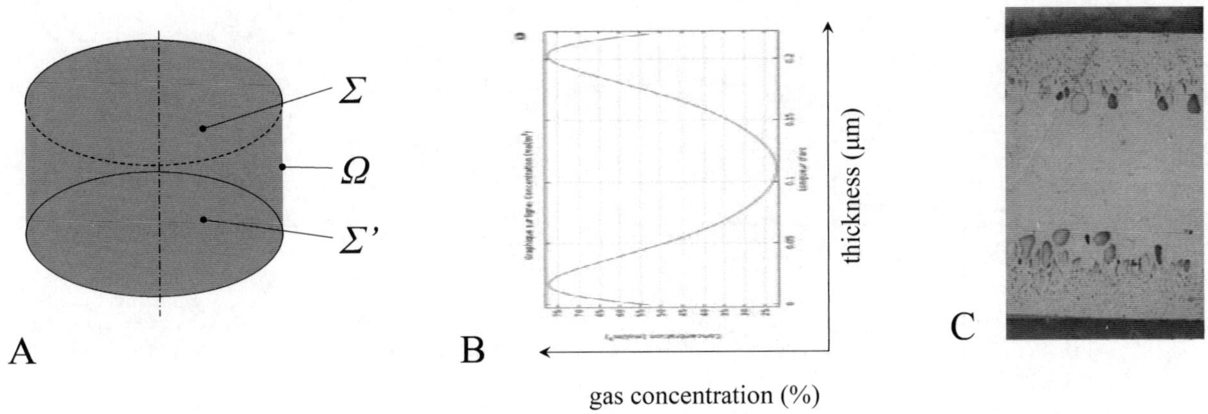

FIGURE 1. (A) Scheme of adopted samples. (B) Applying a time-varying external gas pressure – gas pressure history – to a polymeric sample results in a 3D gas concentration profile When the external pressure is rapidly quenched to atmospheric pressure, the super-saturated concentration profile induces bubble formation to an extent that depends locally on the instantaneous gas concentration. (C) SEM micrographs of foamed samples from (B).

The above problem of diffusion in the slab of thickness L, made impermeable on the lateral surface Ω and accessible to the blowing agent through the surfaces Σ and Σ' (see Fig. 1(A)), is well described by a one dimensional Fickian model with time varying boundary conditions:

$$\frac{\partial C}{\partial t} = D\frac{\partial^2 C}{\partial x^2}, \qquad 0 < x < L$$

$$b.c. \quad \begin{matrix} C(0,t) = \phi_1(t) \\ C(L,t) = \phi_2(t) \end{matrix}$$

$$i.c. \quad C(x,0) = f(x) = c_0$$

where we have assumed a constant gas diffusivity D and a uniform initial gas concentration c_0. Furthermore, the thickness of the polymer L is assumed to be constant during the sorption. This problem has a general analytical solution (see, e.g., [5]). If we consider, on both sides of the sample, a periodic boundary condition of period T, and we make the equations dimensionless by using T as the characteristic time and L as the characteristic length, the following dimensionless parameter arises $\Pi = DT/L^2$. This number can be seen as the ratio between the characteristic time of the boundary condition T and the characteristic time of diffusion L^2/D, or the ratio between the penetration depth (corresponding to one period) $\sqrt{DT}$ and the sample size L. The other parameter which has to be taken into account is the total process time $t_f^* = t_f/T$, i.e., the total number of periods. The time evolution of the dimensionless concentration profile inside an initially empty sample, with a sinusoidal boundary condition $\phi(t) = 0.5(1+\sin(2\pi t^* - \pi/2))$ on both sides, and for different values of the parameter Π is reported in Figure 2. Only half of the sample is considered for symmetry. When $\Pi \sim 1$ the boundary condition varies so slowly that the concentration inside the sample has the time to reach equilibrium and the profile is almost uniform (quasi-steady-state). In this condition, there is no difference at increasing the number of periods considered. When reducing Π, e.g., by using a faster boundary condition, the penetration depth decreases and concentration gradients arise. The core of the sample is initially unaffected by the concentration variation at the boundary, however, as the number of periods increases also the concentration inside the core starts to change, finally approaching the mean value of $\phi = 0.5$.

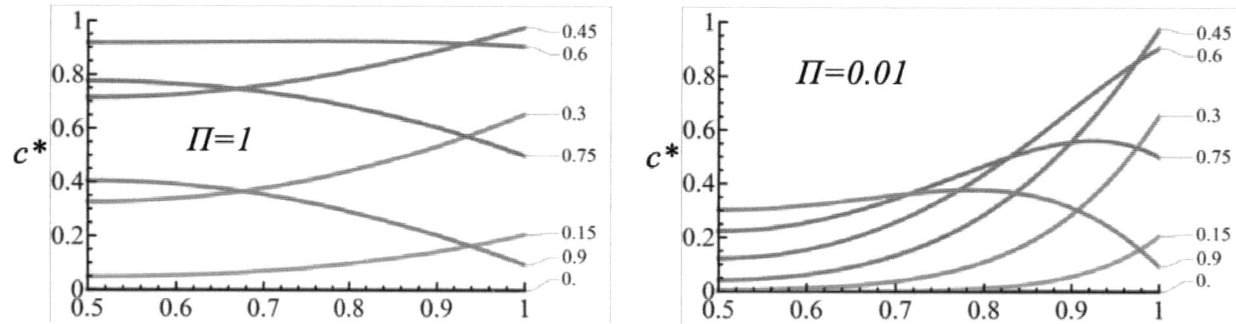

FIGURE 2. Time evolution of the concentration profiles for different Π. Only half of the sample is considered for symmetry. The concentration is normalized as respect to saturation.

Once the relevant variables and dimensionless number at play have been analyzed and various possible concentration profiles drawn, we performed the foaming tests to verify the efficacy of the proposed method to achieve layered foams. Results will be reported, for the different cases, in Fig. 3, illustrating the tight correspondence between the concentration profile (as a consequence of a specific saturation pressure history) and the morphology of the foam.

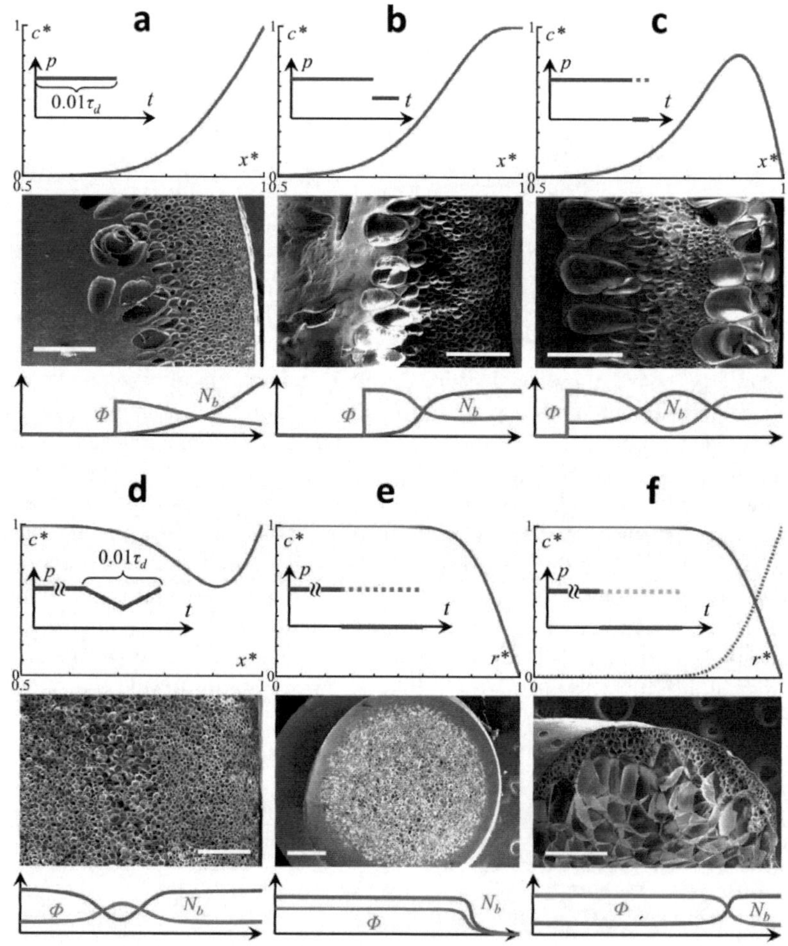

FIGURE 2. Graded foams as a result of different concentration profiles. top: normalized concentration profile (c* vs. x*) after a specific sorption pressure history (p vs. t in the inset); middle: corresponding SEM micrograph (scale bar is 500 μm); bottom: Nb and Φ qualitative profiles of the morphologies shown in the SEM image.

The morphology is reported in terms of the Scanning Electron Microscopy (SEM) image and of the qualitative analysis of the SEM image in terms of the bubble number density, Nb, and the average bubble size, Φ, profiles. A simple graded foam was obtained by shortcutting the sorption stage, at constant pressure on an initially empty polymer. The resulting gas concentration profile is described by the erf function. At the pressure release, a Nb profile, faithfully following the concentration profile, results (the bubble nucleation rate monotonically increases with the gas concentration). Φ has the opposite trend, as the larger the Nb, the less gas is available to inflate each bubble. In particular, numerous, tiny bubbles are observed close to the sample surface, gradually changing, towards the inner of the sample, to few large bubbles and, beyond, to a bulk, unfoamed structure (Fig. 3a). As a development of known prior art, we now apply time-varying gas pressures to achieve foams with a desired, advanced, morphology (Fig 3b-e). For example, if a more uniform external layer in the outer shell is desired with respect to the previous case, after a first sorption step at constant pressure, pressure has to be step-changed to a minor pressure in order to partially empty the external layers. If a less foamed-to unfoamed external layer is desired, then a gas switch is required, maintaining a constant p. In particular, we utilized a slow-diffusing gas, because of a higher molecular weight - namely R-134a - to substitute CO_2 at constant pressure. This results in a CO_2 emptying of the outer layer, without premature foaming, which is eventually achieved at pressure quench after the required empting period elapsed (Fig 3c). The following examples regard samples initially at a (high) gas concentration (they were saturated at a constant, high pressure for a period > τ_d). The depth of the layer where this morphology change is attained can be to a certain extent designed, by restoring the initial pressure and by adjusting timing and shape of the pressure history profile. If a triangular pressure profile is applied at the end of the sorption stage, a 5-layers morphology is achieved (Fig. 3d); if the external layer has to be stiff or impermeable, it is possible to render it bulk by substituting CO_2 with R-134a at constant pressure (Fig. 3e); finally if the external layer has to have a different morphology than the inner of the sample, then it is sufficient to substitute the CO_2 with a different, yet effective foaming agent, N_2, capable of giving finer morphologies (Fig 3e).

CONCLUSIONS

Many layered foamed structures are possible, keeping in mind that imagination is upper bounded by the physics of mass transport, as to have concentration gradients one has to hamper diffusion. A possible limit of the technique can be foreseen based on the extent of the foaming processing window available with polymers. These novel foams can be exploited, for instance, to produce new safety helmets, damping soles, and automotive bumper beams with improved energy absorption performance and reduced weight. As a final remark, we believe that the idea of using time varying boundary conditions can be applied to any process involving mass as well as energy transport (e.g., drying and quenching) to develop new layered, advanced materials.

ACKNOWLEDGMENTS

Materias S.r.l. financial support is gratefully acknowledged.

REFERENCES

1. Gupta, N. & Ricci, W. *Mater. Sci. Eng. A Struct. Mater.* **427** 331–342 (2006).
2. Wang, E., Gardner, N. & Shukla, A. Int. J. Solids Struct. 46 3492–3502 (2009).
3. Zhou, C., Wang, P. & Li W.. Compos. B Eng. 42 318–325 (2011).
4. Tammaro, D., Contaldi, V., Pastore Carbone, M. G., Di Maio, E. & Iannace, S. J. Cell. Plast. 52 533–543 (2016).
5. Carslaw, H. S. & Jaeger J. C. Conduction of Heat in Solids (Oxford Science Publications, Oxford, 1959).

Combining Supercritical Foaming and Polymer Leaching to Obtain Fibrillated and Highly Interconnected Porous Polymer Scaffolds

Zihui Li[a], Haiyang Guo[a], Jing Jiang[a,b*], Jianhua Hou[a], Xiaofeng Wang[a], Qian Li[a]

[a] *School of Mechanics and Engineering Science, Zhengzhou University, National Center for International Joint research of Micro-Nano Molding Technology,, Zhengzhou,450001, Henan, China*
[b]*School of Mechanical Engineering, Zhengzhou University, Zhengzhou, 450001,Henan, China*

Abstract. Tissue engineering scaffolds are applied to temporary frameworks to support the regeneration of tissues. Highly porous and interconnected 3D structures are crucial elements for scaffolds since they can support for the mass transport of cell nutrients and waste. Supercritical foaming technology is an environmentally-friendly and solvent-free way of manufacturing porous scaffolds. In this research, highly porous interconnected poly(ε-caprolactone) (PCL) scaffolds combined with supercritical carbon dioxide (CO_2) foaming and a polymer leaching process were fabricated by blending PCL with water-soluble poly(ethylene oxide) (PEO) as a sacrificial material. The effects of phase morphology of PCL/PEO blend on foaming behavior and pore morphology were firstly investigated. The incorporation of PEO not only facilitated the foaming of PCL by increasing its viscosity, but it also improved the porosity and interconnectivity of the post-leached PCL scaffolds. The open pore content up to 91% and fibrillated porous scaffolds were obtained after leaching process because of two different cell-opening mechanisms. Cell-opening on surface of scaffolds is one difficult point to prepare porous materials. In the end, a novel method for improving surface porosity was described as well to set up the so-called outer and inner porous PCL scaffolds. The information gathered in this study may provide a theoretical basis for research into porous tissue engineering scaffolds.

Keywords: Poly(ε-caprolactone), supercritical foaming, leaching, porous scaffolds
PACS: Replace this text with PACS numbers; choose from this list: http://www.aip.org/pacs/index.html

INTRODUCTION

In order to fulfill the requirements of tissue engineering, an ideal tissue scaffold should contain suitable pore sizes, homogeneous pore distribution and highly interconnected porous structures. These structures can not only facilitate cell adhesion, migration and growth into the interior of scaffold, but they can also provide mechanical support and transportation of cell nutrients and waste [1,2]. Various techniques have been developed to fabricate porous scaffolds [3,4]. Particle/polymer leaching (PL) is a cost-effective and attractive method for generating fully interconnected porous scaffolds. But the scaffolds fabricated by PL alone have the disadvantages of uncontrollable pore sizes and poor mechanical properties [5]. Supercritical CO_2 foaming is an environmentally-friendly and solvent-free way of manufacturing porous materials. During this process, CO_2 is widely used as physical blowing agent because of inexpensive, non-toxic, recoverable, reusable and high dissolution in polymer [6]. However, this process is limited to pore size variety, incontrollable pore morphology and low interconnectivity in porous scaffold. Combining polymer leaching and supercritical CO_2 foaming can yield finer pore size and homogeneous distribution, which can guarantee higher porosity and open cell content of fabricated scaffolds.

Solid-state S_cCO_2 batching foaming process and polymer leaching method was combined in this paper. The effects of blend composition and phase morphology on foaming behavior and final pore morphology, as well as polymer leaching efficiency were comparatively investigated. Pore size, pore density and open pore content were measured quantificationally. Furthermore, PCL scaffolds with desired pore size and highly interconnected porous structures were fabricated successfully. Finally, a novel method for improving surface porosity was described as well to set up the so-called outer and inner porous PCL scaffolds.

EXPERIMENTAL

Materials

PCL (Capa 6500 from Perstorp U.K. Ltd.), a medical grade semi-crystalline polymer, was used as the matrix material in this study. PEO (WSR N-10, DOW) was chosen as a sacrificial polymer due to its water solubility and similar melting temperature to that of PCL.

Sample preparation

The compounding of PCL and PEO was carried out with a twin-screw extruder (TSE-20). Different polymer blends with a variety of mass formulas-Solid PCL(Neat PCL), 80%PCL/20%PEO(PCL80), 70%PCL/30%PEO(PCL70), 60%PCL/40%PEO(PCL60), and 50%PCL/50%PEO(PCL50) were prepared. The PCL/PEO blend pellets were placed into a press to form a disk sample with a thickness of 2 mm. Each disk sample was cut into a rectangular plate with dimensions 15mm×15mm×2mm. The rectangular samples were pre-loaded into a house-made cylindrical mold, which was then placed in a lab-synthesized high-pressure vessel illustrated in Fig. 1. After batch foaming process, foamed PCL/PEO scaffolds were immersed into deionized water to leach out PET phase.

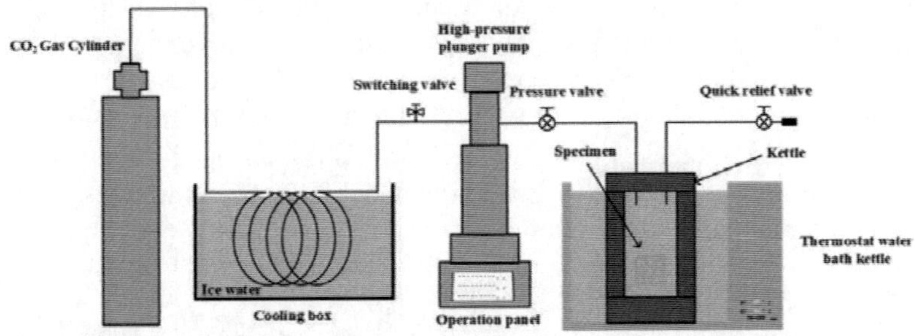

FIGURE 1. Schematic diagram of our supercritical foaming equipment

A novel method for improving surface porosity was illustrated in fig.2. Sacrifice material coating method was used before batch foaming process.

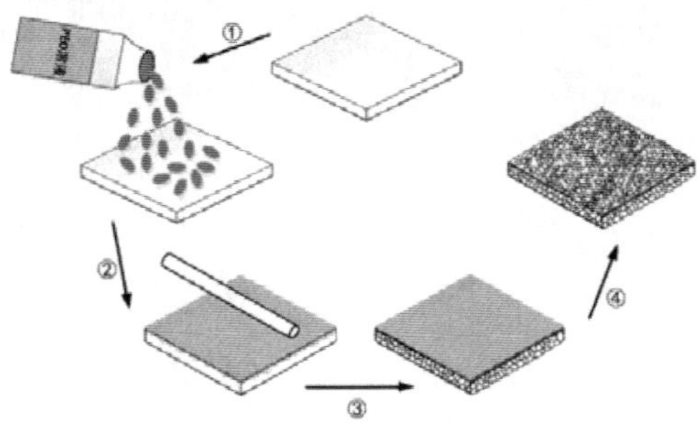

FIGURE 2 Schematic illustration of cell-opening on surface method

Characterizations

Scanning electron microscope (SEM)

The phase morphology of solid samples and the microstructure of batch foamed samples before and after leaching PEO were evaluated using a scanning electron microscope (FEI Quanta 200) with an accelerating voltage of 10kV. All specimens were frozen in liquid nitrogen and broken to expose the cross sections. SEM observations were made after sputtering the samples with a thin film of gold for 80s with 2mA.

The pore size and pore density were measured from the SEM micrographs utilizing the Image-J software. The average cell size (D) and cell density (N_f) of all of the pores in the SEM micrograph were calculated with Eqs. (1 and 2) :

$$D = \frac{\sum_{i=1}^{n} d_i}{n} \tag{1}$$

$$N_f = \frac{\left(nM^2\right)^{\frac{3}{2}}}{A} \tag{2}$$

Where, n represents the number of cells in the micrograph, d_i represents the equivalent diameter for each cell, N_f represents the number of cells in the micrograph, M is the magnification factor, and A represents the area of the micrograph (cm^2).

Open cell content and porosity

The open cell content was calculated by Eq.3:

$$Open\,cell\ content = \frac{V_{\text{cell}}}{V_{polymer}} \times 100\% \tag{3}$$

where V_{cell} represents the open cell volume, which was measured with a nitrogen pycnometer, and $V_{polymer}$ represents the total volume of the polymer sample. The porosity of the prepared scaffold was determined by Eq.4:

$$Porosity = (1 - R_v) \times 100\% \tag{4}$$

Where R_v is the expansion ratio of the foamed samples.

RESULTS AND DISCUSSION

Morphologies of PCL/PEO blends

Fig.3 gives the SEM images of fracture surface for Neat PCL and PCL/PEO blends with various blend ratios after leaching. All samples exhibit typical binary structures, which further prove the immiscibility of PCL and PEO. It can be seenthat neat PCL exhibited a smooth fracture surface. Dispersed droplets of the minor phase in the matrix are observed with the addition of 20% and 30% PEO(Fig.3b, Fig.3c), and the fractured surface became a little bit rough. As the content of PEO phase increased to 40%, the morphology changed significantly. The dispersed phase showed an elongated or fibrous structure clearly (Fig.3d) and partially co-continuous morphology was emerged. Numbers of void increased and interconnected each other as the amount of PEO added up to 50%(Fig.3e). Channels on the voids wall were visible, suggesting that the formation of fully co-continuous phase.

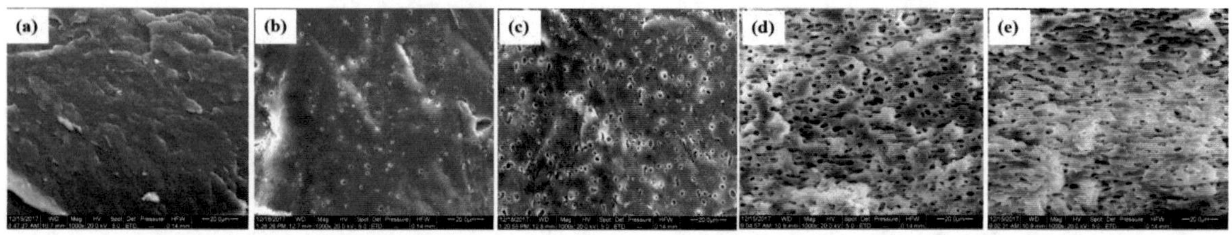

FIGURE 3. SEM micrographs after leaching PEO phase of the fractured cross section of neat PCL and PCL/PEO blends: (a) Neat PCL, (b) PCL80, (c) PCL70, (d) PCL60, (e) PCL50

Morphology of scaffolds before leaching PEO

It was found from fig.4 that neat PCL exhibited huge circular pore structure, and most of the pores yield closed pore structure. Some of pores were collapsed because of weak melt strength during foaming. As evidenced in the fig.4(b)-(e) from SEM micrographs, all pores were characterized by a sharp decreased size and an open-pore morphology. The shape of pores turned from regular polyhedral faces (fig.4a) to fully random polygon(fig.4d,4e). By increasing the concentration of PEO from 20wt.-% to 50wt.-%, the thickness of pores were increased and numbers of well-interconnected pores were decreased. Even no pores were observed in some rigions especially when PEO was increased by 50wt.-%. Another feature from morphologies of PCL/PEO foamed samples is that non-homogeneous structures were detected with narrow pore size distribution because of various foaming behavior of PCL and PEO.

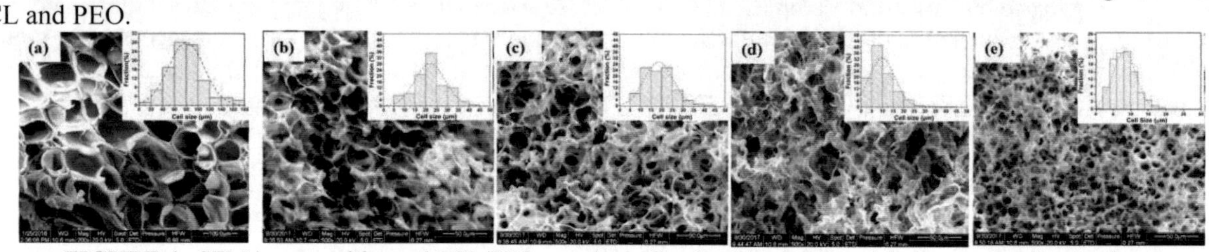

FIGURE 4. SEM images of foamed PCL/PEO samples before leaching PEO and their pore size distributions: (a) Neat PCL, (b) PCL80, (c) PCL70, (d) PCL60, (e) PCL50

It can be seen from fig.5 that the average pore size became smaller and decreased from 79.5μm to 8.2μm, while the average pore density increased from 1.5×10^5 to 8.9×10^8 cell per cm^3, as the PEO content increased. These results could be explained that the heterogeneous nucleation effect was strengthened after adding PEO, which increased numbers of nucleating points and restricted the expansion of cells[7].

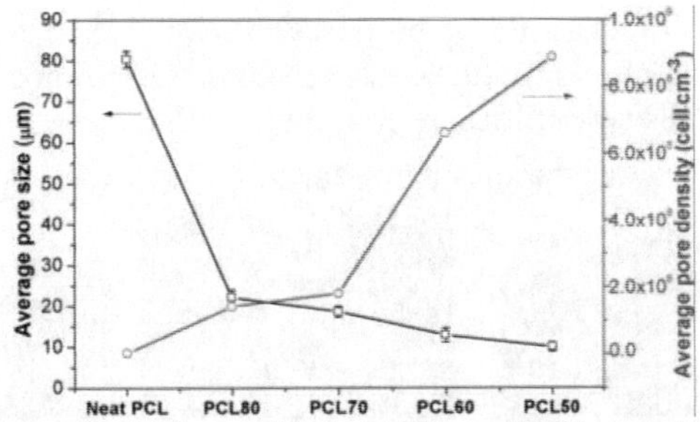

FIGURE 5. Average pore size and pore density of foamed PCL samples before leaching PEO

Morphology of scaffolds after leaching PEO

All samples shown highly interconnected pore structure through holes fractured in the pore walls. Fully random polygon shapes of pore were aslo found. Another feature that was noteworthy was that the pores were interconnected with many fibrillated structures with nano-scale for PCL80 and PCL70 foamed samples. The pore walls of PCL was stretched to a certain extent during cell expansion. The thickness of pore walls were decreased and resisted by dispersed PEO with higher melt strength under function of CO_2-induced plasticization effect[8]. The matrix PCL became too weak to sustain cell expansion, then the pores tend to be broken, which resulted in open pores and fibrillated structures among pore walls. It is clearly seen that many continuous channels appeared after leaching PEO. Multi morphologies porous structures were appeared, which were mixed with highly interconnectivity PCL pores come from foaming process and continuous channels duo to PEO leaching.

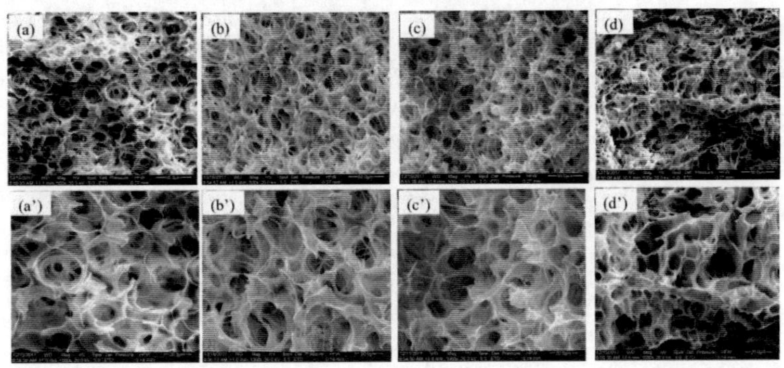

FIGURE 6. SEM images of foamed PCL/PEO samples after leaching PEO and their pore size distributions: (a) PCL80, (b) PCL70, (c) PCL60, (d) PCL50, scale bar=50μm; (a')~(d') are enlarged images, scale bar=20μm

As expected, increasing the concentration of porogen (PEO) induced in an increase in overall porosity. The porosity increased from 67% to 86% with increasing the concentration of PEO from 0% to 30%. Different from variation tendency of porosity, it was found that the open pore content increased first and then decreased as the amount of PEO was increased. Before leaching PEO, a maximum open pore content of 78% was obtained for PCL70. After leaching process, open pore content was improved obviously for all foamed samples, and the maximum porosity of 83% was detected for PCL70. Additionally, average increment of open pore content is about 5% after leaching for PCL80 and PCL70. However, average increment of open pore content is increased by 10% after the concentration of PEO is added from 30% to 50%.

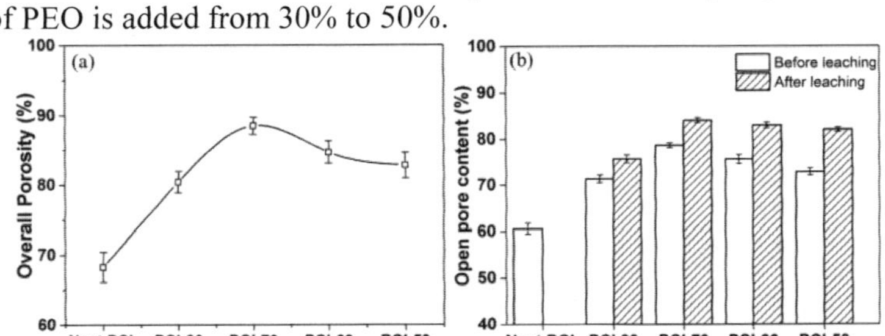

FIGURE 7 Compared statistical results for the porosity of foamed PCL/PEO scaffolds.

Surface porosity and interconnection with inner porosity

Cell-opening on surface of scaffolds is one difficult point to prepare porous materials [9]. From the results shown in Fig.8, it can be observed that compared with the surface porosity before treatment, More open homogenous pores were found after treatment on surface of scaffolds. The pores formed at the surface are clearly interconnected with those inside the sample. Furthermore, pore structures inside samples were not destroyed and Pore size on surface was increased to match endothelial cells by optimizing coating process.

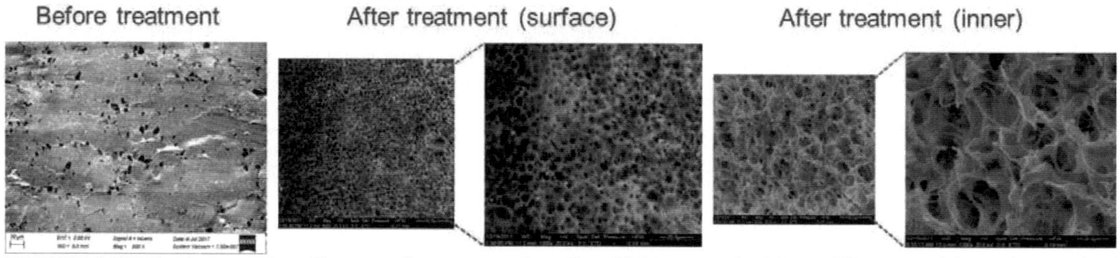

FIGURE 8 SEM micrographs of inner and outer porosity of scaffolds treated with sacrifice material coating method

CONCLUSIONS

In this work, an environmentally-friendly and cost-effective approach combining batch foaming and polymer leaching techniques was applied to preparing highly interconnected porous PCL scaffolds with open-cellular pore structures. The effects of phase morphology of PCL/PEO blend on foaming behavior and pore morphology were investigated. The incorporation of PEO not only facilitated the foaming of PCL by increasing its viscosity, but it also improved the porosity and interconnectivity of the post-leached PCL scaffolds. Because of their different melt strengths, PCL50 porous samples fabricated in this experiment had a bigger porosity than PCL70 scaffolds. Average open cell content of greater than 75%, even reached 85% for all porous scaffolds can be easily obtained, which can meet the physical demands of porous tissue engineering scaffolds. The pore statistical results demonstrated that the method of blending the immiscible polymers had a big effect on the interconnectivity of pores for PCL70. The leaching process made a greater contribution to the open pore content of porous scaffolds for PCL50 blends. A novel sacrifice material coating method for improving surface porosity was described as well to set up the so-called outer and inner porous PCL scaffolds in the end.

ACKNOWLEDGMENTS

The authors acknowledge China Postdoctoral Science Foundation (No.2017M612415), the International Technological Cooperation Project (No.2015DFA30550), and Key scientific research project plan of Henan high education institutions (No.18A430030) for their financial support of this project. The project was also sponsored by Scientific and technological research project of Henan Province (No.182102210188).

REFERENCES

1. Hutmacher DW, Scaffolds in tissue engineering bone and cartilage. *Biomaterials* **21 (24)**, 2529-2543(2000).
2. Zhang C,Turng LS et al.,The surface grafting of graphene oxide with poly(ethylene glycol) as a reinforcement for poly(lactic acid) nanocomposite scaffolds for potential tissue engineering applications. *Journal of the Mechanical Behavior of Biomedical Materials* **53**,403-413(2016).
3. Gao Q, Gu H, Zhao P et al., Fabrication of electrospun nanofibrous scaffolds with 3D controllable geometric shapes. *Materials & Design* **157**,159-169(2018).
4. Zhao H, Zhao P et al., Airflow-Assisted 3D Bioprinting of Human Heterogeneous Microspheroidal Organoids with Microfluidic Nozzle. *Small* **14 (39)**, 1802630 (2018).
5. Huaguo L, Yingjun W et al., Preparation of Poly(ε-caprolactone) Tissue Engineering Scaffold by Freeze-drying/Particle-leaching Method. *Materials Review* **2**,032(2007).
6. Kuang T, Peng X et al., Facile Preparation of Open-cellular Porous Poly (L-lactic acid) Scaffold by Supercritical Carbon Dioxide Foaming for Potential Tissue Engineering Applications. *Chemical Engineering Journal* **307**,1017-1025(2017).
7. Zeltinger J, Griffith L et al., Effect of pore size and void fraction on cellular adhesion, proliferation, and matrix deposition. *Tissue Engineering Part A* **7 (5)**, 557-572 (2001).
8. Sabapathy S, "Studies on co_2-induced crystallization and foaming behaviour of poly (lactic acid)/clay nanocomposites" ,Ph.D.Thesis, Indian Institute of Technology Delhi, 2018.
9. Castaño M, Pintado-Sierra M et al., Combining Breath Figures and Supercritical Fluids To Obtain Porous Polymer Scaffolds. *ACS omega*, **3(10)**, 12593-12599 (2018).

Characterization of Mechanical and Thermal Properties of PLA/Blowing Agents Composites

Tomasz Garbacz[a], Aneta Tor-Świątek[a], Vladimir Sedlaric[b]

[a]*Lublin University of Technology, Department of Technology and Polymer Processing, ul. Nadbystrzycka 36, 20-618 Lublin, Poland, t.garbacz@pollub.pl,*

[b]*Centre of Polymer Systems, Tomas Bata University in Zlin, tr. T. Bati 5678, 760 01 Zlin, Czech Republic*

Abstract. Biodegradable plastics find more and more applications. In addition to biodegradable packaging, the number and type of products also being used in medical applications is increasing. The goal of the presented research is development technique for preparation of microcellular foams based on biodegradable polylactic acid (PLA) and its co-polymers for biomedical applications. The goal is to understand the influence of microblowing agents on the course and efficiency of microcellular process of injection molding biodegradable polymer compounds as well as on the properties of the obtained products.

Polylactide (PLA) under the trade name NatureWorks Ingeo 2002D modified with chemical blowing agents having an exothermic and endothermic character of the distribution was used during the tests. The blowing agents used in the injection molding process were inserted into polymers processed in the following amounts: 0.5%, 1.5% and 3.0% wt. The article presents the characteristics of the process of injection of biodegradable plastics and the study of selected physical properties of the materials produced. The research on the structure of manufactured materials, density and the degree of porosity, mechanical strength, impact resistance and hardness are presented.

It was found that as a result of the increased temperature in the plasticizing system, the blowing agents decompose and the resulting numerous microcells, due to pressure and surface development, immediately mix with the plastic. The resulting microcells are filled with both air and gases, increasing the volume of PLA. There was a change in the examined physical properties of the tested materials, mainly mechanical strength of the tested PLA compositions.

Keywords: polylactide, blowing agent, injection molding, physical properties, mechanical properties

INTRODUCTION

Cellular of polymeric materials, including biodegradable materials, is aimed at giving new products, among others, specific physical and mechanical properties. In the case of materials difficult to recycle, the cellular process may lead to easier processing. Modification of the properties of materials during the cellular process is related to the modification of the technology of the production process, the construction of tools used in production, such as extrusion head, injection forms. The modification also concerns the composition of the material composition by adding poring and other factors such as plasticizers, fillers, and nucleating agents. The porous structure is obtained by introducing a blowing agent in the form of a gas or a solid into the material.

If the blowing agent is in a liquid or solid state, under certain conditions of the technological process, most often during plasticizing, it goes into gas state. The process of portraying the reason that the product obtained changes its structure from solid to porous. Properly selected technological conditions of the process enable the production of products from biodegradable materials with modified physical and mechanical properties [1, 2, 4, 6, 7, 8]. Porous materials have, among others, reduced mass, increased damping and insulating properties, and the possibility of using them after use. To obtain a cellular structure, the properties of the products are modified using the appropriate type of polymer or by incorporating blowing agents into the polymer [3, 4, 7, 10, 11]. The poring means are selected so as to correspond to the type of polymer used in such a way that the temperature of their decomposition is contained at the processing temperature of a given material. The desired structure of the porous material is retained when the final product is immediately cooled and solidified [2, 3, 4, 7, 9].

The aim of the research, presented in the article is to present the possibility of chemical cellular of biodegradable plastics in the injection molding process. The aim is also to analyze selected physical and mechanical properties and structure of processed plastics. The scope of work includes testing of density, adsorption of liquid substances and testing of mechanical properties, such as: tensile strength, stresses, elongation and impact resistance.

EXPERIMENTAL

During the tests, biodegradable plastic, polylactide (PLA) NatureWorks Ingeo 2002D was used. According to the manufacturer's data, PLA is characterized by a transparent color. It has the following properties: MFR (210°C, 2.16 kg): 5-7 g/10 min; tensile strength: 7.7 MPa; yield point: 8.7 MPa; Young's modulus: 500 MPa; elongation: 6.0%. Chemical chemical blowing agents in the form of granules were used in the conducted research, having an exothermic and endothermic character of decomposition during the process. Chemical blowing agents are modifiers that produce gases as a result of chemical reactions at elevated temperatures and the breakdown of organic or inorganic bonds, resulting in a cell structure due to affinity to the polymer. These are the following blowing agents: Hydrocerol 530, Hydrocerol ITP 810, (Clariant Masterbatches GmbH), Expancel 950 MB (Akzo Nobel) and LyCell-F017 (Ly-TeC GmbH). The chemical blowing agents decompose mainly on CO 2, a small amount of H_2O and N_2 nitrogen during the process of blowing. Selected properties of discussed blowing agents are summarized in Table 1.

TABLE (1). Selected properties of blowing agents used in PLA cellular process

Characteristic quantity	Distribution	Decomposition initiation, °C	Processing temperature, °C	Dosage, %
Ly-Cell F017	Endothermic	155	170 - 210	0.5 - 2.0
Hydrocerol 530	Exothermic	170	170 - 190	0.5 - 2.0
Hydrocerol ITP810	Endothermic	160	160 - 180	1.0 - 5.0
Expancel 950 MB	Microspheres	130	130 - 190	0.5 - 2.5

The blowing agents were introduced into the tested PLA in a mechanical mixing process prior to the processing. During the injection molding process, blowing agents are subject to the same steps as plastic processing, compression, homogenization, transport, even before decomposition into gas products. The blowing agents used in the injection molding process were introduced into PLAs processed in the following amounts: 0.5%, 1.5% and 3.0% by mass.

In the conducted tests, Arburg injecting molding machine, type All Rounder 320, was used, equipped with an injection mold, as shown in Fig. 2. The machine has a plasticizing system with a screw diameter of 30 mm. The injection mold used has two sockets, with the shape and dimensions of the samples for mechanical strength tests, for static tests. In order to obtain a high process efficiency for the above blowing agents, the temperature range of the plasticization of the PLA compositions tested in the injection molding machine plasticizing system was from 180 to 220°C. The following injection molding process parameters were adopted: injection and crimping time = 2s; mold cooling time = 35s, thermostated mold temperature = 50 ± 1°C.

FIGURE 1. Test stand used in the experiments, injection molding machine Arburg AllRounder 320C and porous PLA injection molded parts with an ingot and molded part, PLA +0.5 % blowing agents

RESULTS AND DISCUSSION

The appearance of the porous structure of the tested PLA compositions is shown in Figures 2. Materials having a porous core and a solid top layer were produced. Pictures of structures made using SEM microscopy (Tescam Vega /LMU) confirm the structural changes resulting from the cellular process. Density determining the degree of sporulation of PLA composition was examined. Figure 3 shows the results of density measurements of PLA compositions using blowing agents having exo and endothermic character of the distribution. Density, which is a property defining the volume of the gas phase in a cellular product, simultaneously determines the dispersal of a given material. The dependence of density on the content of the blowing agent shows that both the content of the blowing agent and its distribution characteristics are very important here. For example, the use of an exothermic blowing agent (Hydrocerol 530) resulted in a product with a larger number of pores and increased porosity compared to the product obtained using an endothermic agent (Ly-Cell F017).

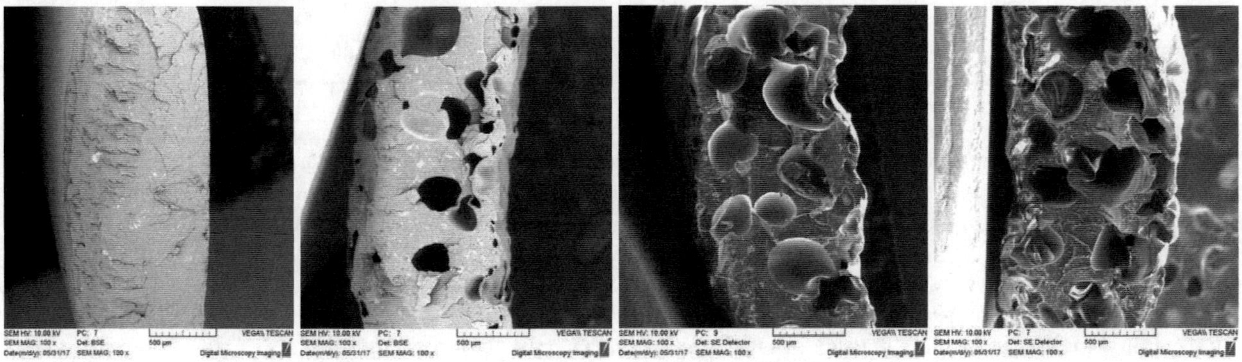

FIGURE 2. Cross section of porous structure of injection molded parts made from the examined polymers: a) solid PLA, b) PLA + 0.5% blowing agent Hydrocerol 530 (H530), c) PLA + 1.5% H530, d) PLA + 3.0% H530, magnified x 100

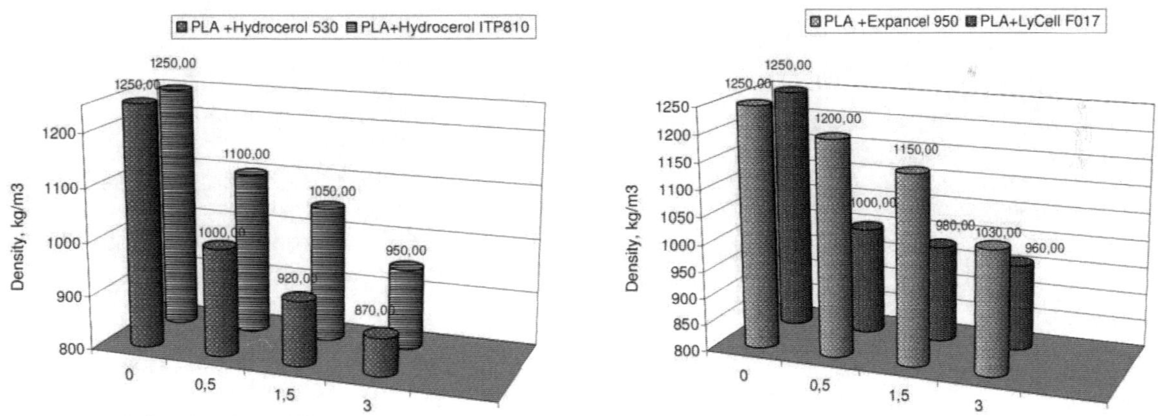

FIGURE 3. Dependence of normal density of the PLA injection molded parts on the content of blowing agents

Examples of water and oil absorption results are summarized in Table 2. Cellular composites made using an exothermic blowing agent showed a water absorption in the range of 1.5% (at 0.5% blowing agent content) to 2.8% (at 3.0% blowing agent content). after 48 hours and from 2.65 to 3.60%, respectively, after 7 days. Similar values of water absorption were characteristic of composites produced using an endothermic blowing agent. The porous materials from PLA were characterized by oil absorption in the range from 0.30 to 1.70%, with blowing agents content in the range of 0% to 3.0%, respectively. In some practical applications this may be too large to accept.

TABLE (2). Results of water and oil absorption of cellular composites

No.	Blowing agent type	Blowing agent content %	Water absorption, %		Oil absorption, %	
			after 48 h	after 168 h	after 48 h	after 168 h
1	Exothermic blowing agent	0	1.50	2.60	0.25	0.70
2		0.5	1.80	2.65	0.35	1.00
3		1.5	2.00	3.25	0.50	1.30
4		3.0	2.35	3.60	0.70	1.50
5	Endothermic blowing agent	0.5	1.50	2.55	0.25	0.70
6		1.5	2.30	3.50	0.60	1.10
7		3.0	2.80	4.10	1.05	1.70

In order to determine selected mechanical properties of porous PLA materials, tensile strength tests, yield strength and elongation at break were carried out in accordance with the relevant standards (ISO 868 (2003) and ISO 527-1 (2010). The thickness of the sample corresponded to the thickness of the injection molded part and was measured and recorded each time along with the width of the measurement length just before the tests. The strength properties of the injection molded parts under static stress were determined using the Zwick/Roel Z010 testing machine. Measurements were made at a tensile speed of 50 mm/min) and under a measuring load in the range of 0÷500 N. The results of determination of selected mechanical properties of injection molded parts, obtained at different blowing agent contents in the processed polymers, are presented in Figure 4.

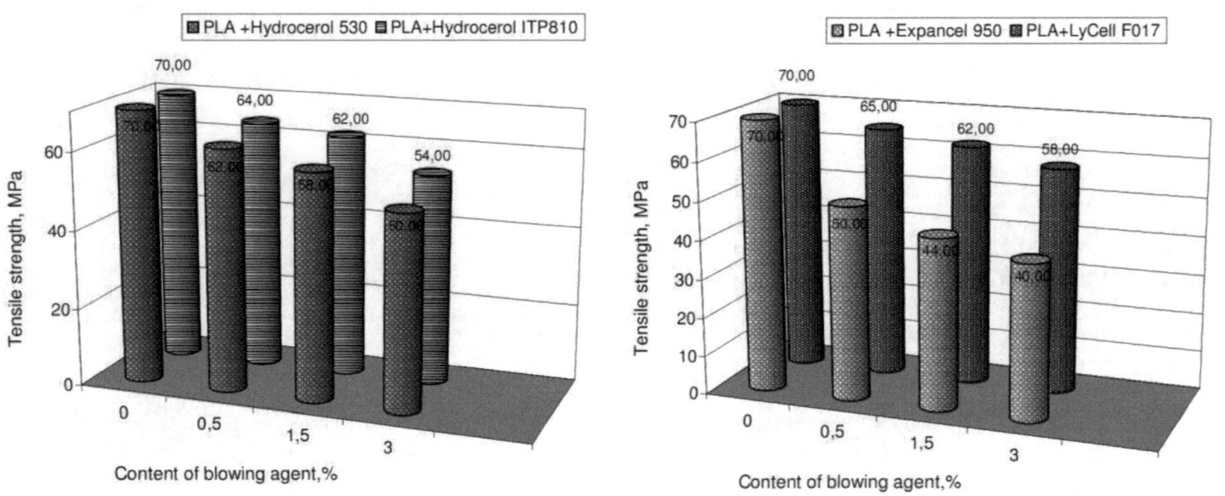

FIGURE 4. Dependence of tensile strength of the PLA injection molded parts on the content of blowing agents

As the content of the blowing agent in PLA increases, the yield point decreases. Elongation at break decreases monotonically, non-linearly, in the range of growth of the blowing agent content. This relationship is similar for each of the materials and blowing agents, regardless of the thermal decomposition characteristics of the substances used. For example, the Hydrocerol 530 blowing agent with endothermic decomposition characteristics, dosed in the range of 0 to 3.0% results in a reduction in tensile strength by approximately 30% in the PLA tested compositions. In the case of PLA with 3% content, LyCell F017 tensile strength reduces an average of 16%. Nevertheless, it was observed that the tensile strength depends on the type of blowing agents. The highest tensile strengths were obtained with blowing agents containing sodium bicarbonate and citric acid. Strength of PLA modified with 3% wt. Hydrocerol 530 is significantly lower than that obtained for other blowing agents, amounting to approximately 50 MPa. This means that it is 30% less than the tensile strength of pure PLA. PLA compositions modified with Expancel microspheres have the lowest tensile strength for all tested blowing agents. Tensile strength of about 48, MPa, or 34% less than for unmodified PLA. This may be related to the nature of pore formation as the volume of the hydrocarbon filled EVA microspheres increases. The microspheres do not break during pore formation and may later become the area of fracture initiation in tensile tests.

CONCLUSIONS

On the basis of the conducted research, the analysis of the obtained results showed a significant influence of the process of the poring, on the obtained selected physical and strength properties of the tested PLA compositions. Poring, using chemical blowing agents, used in the injection moldings process allows to make products with new, modified properties without much interference in the process technology. Changing these properties offers new possibilities for using these types of products while reducing the cost of producing biodegradable materials. A mixture of chemical compounds having endothermic decomposition causes a decrease in the density of the biodegradable PLA product. This density is determined by the ratio of the amount of solids to the number of gaseous pores. The blowing agent, containing sodium bicarbonate and citric acid, decomposes in the plasticizing system of the machine.

At the appropriate temperature of the plasticized material, numerous gaseous bubbles are formed, which due to pressure and surface development, immediately dissolve into the surrounding PLA. The resulting bubbles are filled with both air and gases, increasing the volume of PLA. The effect is to reduce the density of the obtained porous material. The results of resistance to water and oil absorption showed an increased absorption of the porous materials, depending on the time spent in the aggressive environment and the type of blowing agent used. However, PLA compositions with a blowing agent content of up to 1.5% meet the production requirements for pure PLA. According to the standards, acceptable water and oil absorption is 3-4% for various types of PLA. The mechanical properties of porous, biodegradable materials, including their hardness, impact strength, and damage resistance depend on the structure of the material, shape and pore orientation.

Increasing the content of blowing agent increases the volume of biodegradable material, the number and size of pores obtained during the blowing process. The smaller number of cross-links in the porous material causes a deterioration in the strength properties of such material. It was found that the characteristics of the poring agent distribution, whether it is an exothermic or endothermic agent, affect the properties of the biodegradable materials studied. The agent added to 3%, regardless of its performance, contributes to a significant deterioration of the strength properties of the tested biodegradable materials based on PLA.

ACKNOWLEDGMENTS

Article prepared as part of the project no. PPN/BIL/2018/1/00048. The project is co-financed by the Polish National Agency for Academic Exchange.

REFERENCES

1. K. T. Okamoto, *Microcellular processing,* Munich, Germany, Hanser Publishers, 2003.
2. S. Martial, F. Jacques, C. Audrey, N. Clémence, E. Rodier, *Progress in Polymer Science* **36**, 749-766 (2011)
3. T. Garbacz, L., Dulebova, *Chemistry and Chemical Technology* **7**, 1, 113-118 (2013)
4. T. Garbacz,L. Dulebova, V. Krasinsky, *Advances in Science and Technology* **8**, 18, 74-80 (2013)
5. L. Urbanczyk, M. Alexandre, Ch. Detrembleur, *Macromolecural Material Engenering*, **295**, 915–922 (2010)
6. M.C. Guo, M.C. Heuzey, P.J. Carreau, *Polymer Engineering and Science* **47**, 1070–1081 (2007)
7. A. Tor-Swiatek, *Maintenance and Reliability* **15**, 3, 225-229 (2013)
8. P. Palutkiewicz, P. Postawa, *Journal of Cellular Plastics* **52**, 4, 399-418 (2016)
9. E. Bociąga, P. Palutkiewicz, *Polymer Engineering and Science* **53**, 4, 679-704 (2013)
10. T. Jachowicz, I. Gajdoš, Chemical Industry (Przemysł Chemiczny) **93**, 11, 1983-1885 (2014)
11. P. Stloukal, P. Kucharczyk, *Polymer Degradation And Stability* **142**, 217-225 (2017)

Study on effect of dihydroxymethylphenol derivatives on conductivity enhancement of poly(3,4-ethylenedioxythiophene) doped with poly(styrene sulfonic acid) (PEDOT:PSS)

Shigeji Konagaya[1,a)], Yoshiichi Tawara[2] and Kazuaki Sanada[3]

[1] *Department of Research, Nagoya Industrial Science Research Institute, 1-13 Yotsuya-doori, Chikusa-ku, Nagoya-shi, Aichi464-0819 Japan*
[2]*Aichi Factory, Asahi Yukizai Corporation, 26-4 Fuso town, Niwa County, Aichi 480-0105, Japan*
[3]*Graduate School of Engineering, Toyama Prefectural University, 5180 Kurokawa, Imizu-shi, Toyama 939-0398, Japan*

[a)] *Corresponding author: skonagaya118@yahoo.co.jp*

Abstract. The study was made of the addition effect of DMPs on the conductivity of PEDOT:PSS thin films prepared from the aqueous coating mixture. There was a large difference in conductivity enhancement effect among DMPs, and DMPs with the solubility parameter (SP) ranging from 14 to 16 showed a good conductivity enhancement ability. The addition effect of DMPs was attributable to its good solubility to PEDOT:PSS molecules. UV-Vis spectroscopic study suggested some electronic interaction between 26DMPC and PEDOT:PSS in the slurry and in the solid state of 26DMPC/PEDOT:PSS, and the size and ζ-potential of PEDOT:PSS nano-particle changed with the increase of 26DMPC, which possibly were brought about by the dopant exchange between PSS and 26DMPC and the combination of PEDOT:PSS. Our assumption for the action mechanism of DMPs is as follows. PEDOT:PSS nano-particles are solvated with DMPs, followed by the removal of PSS from PEDOT:PSS. At that time the size and overlapping of PEDOT/PSS particle increase, which brings about the conductivity enhancement of PEDOT:PSS. Such assumption was supported by the computer simulation.

INTRODUCTION

Indium tin oxide (ITO) used as a transparent conductive material involves some problems such as poor mechanical flexibility, indium scarcity and high production cost. Many researchers have been studying new transparent conductive materials such as poly(3,4-ethylenedioxythiophene) (PEDOT:PSS), silver (Ag) nano-particles/wires, and nano-carbons such as carbon nanotube (CNT) and graphene as a substituent for ITO. PEDOT:PSS has been an attracting considerable attention, since its thin film exhibits a high conductivity and transparency, but it has no conductivity and transparency enough to be replaced for ITO. Therefore, lots of additives to enhance its conductivity, that is a secondary dopant for PEDOT:PSS, such as ethylene glycol (EG) and dimethyl sulfoxide (DMSO), dimethyl formamide (DMF) have been proposed and actually applied to transparent conductive materials[1-4)]. The authors have been studying the second dopant effect of diol compounds (glycols) like EG on PEDOT:PSS and made clear that there was a good relationship between their conductivity enhancement ability and their solubility parameter (SP), and that glycols with SP values ranging from 14 to 15 exhibited higher conductivity enhancement ability[5)]. Recently it was reported that phenol with phenolic OH group had a high conductivity enhancement ability to PEDOT:PSS[6)]. Therefore the authors have been interested in the conductivity enhancement ability of dihydroxymethylphenol compounds (DMPs) which have not only one phenolic OH group and two alcoholic OH groups.

In this paper the addition effect of DMPs on the conductivity enhancement of PEDOT:PSS and its probable action mechanism will be discussed.

EXPERIMENTAL

Materials and reagents

Aqueous solution of PEDOT:PSS was purchased from Sigma Aldrich company (now Merck company). Its concentration was 1.3% by weight, and the weight ratios of PEDOT and PSS in PEDOT:PSS were 0.5 wt.% and 0.8 wt.%, respectively. DMPs shown in Tab.1 were supplied by Asahi Yukizai co. ltd. Commercial grade of iso-propanol (IPA) and pure water were used without further purification. The solubility (g/100 ml) of DMPs was measured in the mixture of IPA and water (50 wt%). The solubility parameters (SP) of DMPs were calculated according to the reference[7,8].

TABLE 1. Names of DMPs used in this study and their abbreviation

DMPCs	X group	Name of DMPC compound	Abbreviation
4-X-2,6-dihydroxymethylphenol	CH_3	4-Methyl-2,6-dihydroxymethylphenol	26DMPC
	C_2H_5	4-Ethyl-2,6-dihydroxymethylphenol	26DMPEP
	$CH(CH_3)_2$	4-Isopropyl-2,6-dihydroxymethylphenol	26DMPIPP
	$C(CH_3)_3$	4-tert-Butyl-2,6-dihydroxymethylphenol	26DMPTBP
	OCH_3	4-Methoxy-2,6-dihydroxymethylphenol	26DMPMeHQ
	SCH_3	4-Methylthio-2,6-dihydroxymethylphenol	26DMPSMe
	F	4-Fluoro-2,6-dihydroxymethylphenol	26DMPFP
2-X-4,6-dihydroxymethylphenol	CH_3	2-Methyl-4,6-dihydroxymethylphenol	46DMOC
	C_2H_5	2-Ethyl-4,6-dihydroxymethylphenol	46DMOEP

Preparation of PEDOT:PSS solid films treated with DMPs

The thin films of PEDOT:PSS were prepared by coating the mixture of PEDOT:PSS aqueous solution (1.3wt%) and IPA/water solution of DMPs (1.0wt%) with the certain weight ratio of DMPs to PEDOT:PSS (0.10, 0.25, 0.50, 1.0, 2.0, 4.0) on the commercial glass substrate (2.0cm×5.0cm×1.0mm(thickness)) for an optical microscope followed by the heat-treatment of the coating layer on the hot plate at 70°C for 0.5 hr and in the drier at 120°C for 0.5hr. The thickness of the PEDOT:PSS/DMPs solid film was controlled to be almost the same. The thickness(t) of the solid film was calculated from the weights of the solid film coated glass and glass substrate.

Measurement of volume resistivity (ρ_v) of PEDOT:PSS/DMPs solid films

The surface resistivity (Rs, Ω/□) of the PEDOT:PSS/DMPs thin films formed on the glass plate was measured by the four-point probe device Loresta-GP (MCP-T610, DIA INSTUMENTs co. ltd.) or Model152 (Treck Japan co. ltd.). The volume resistivity (ρ_v, Ω·cm) was calculated from the Rs value and the thickness (t, cm) according to the formula, $\rho_v = Rs \times t$.

Evaluation of interactions between PEDOT:PSS and 26DMPC in slurries and solids

UV-Vis spectra of PEDOT:PSS/DMPs slurries and solids with the different weight ratios of DMPs/PEDOT:PSS were measured with UV-Vis spectrometry (JASCO UV-650). The particle size and zeta (ζ)-potential of PEDOT:PSS in the IPA/water slurries of PEDOT:PSS/DMPs were measured with the zeta-potential & size analyzer with a titrator, ELS-Z (Photal Otsuka electronics Co. Ltd in Japan).

Computer simulation for conductivity enhancement of PEDOT:PSS by 26DMPC

The finite element analyses were performed with a two-dimensional representative volume element (RVE) model of the PEDOT:PSS core-shell composites by using Digimat-FE (MSC software) in order to support our assumption.

RESULTS AND DISCUSSION

Effect of DMPs on the conductivity of PEDOT:PSS/DMPs solid films

Tab.1 shows the addition effect of DMPs on the ρ_v values of PEDOT:PSS/DMPs films, which were obtained at the equi-molar ratio of DMP/PEDOT:PSS. There was a large difference in the ρ_v value among the species of DMPs. 46DMOC showed the smallest ρ_v value, that is, the highest conductivity among them. The author made it clear that diols with the SP values ranging from 12 to 14 exhibited higher conductivity enhancement effect from the study on the influence of monools and diols on the conductivity of PEDOT:PSS[5]. Fig.1 shows the relationships between SP values of DMPs and ρv values of PEDOT:PSS/DMPs solids at the equi-molar ratio. It can be seen that DMPs with SP values ranging from 14 to 16 exhibit a higher conductivity enhancement effect to PEDOT:PSS, which means that the conductivity enhancement ability of second dopant is possibly related to its solubility to PEDOT:PSS.

TABLE 2. Effect of DMPs on the ρv values of PEDOT:PSS/DMPs solids

DMPCs	ρv (Ω/□)	Solubility in IPA/H_2O mixturre	SP calculated
26DMPMeHQ	7.82E-02	11.1	15.56
26DMPSMe	7.82E-02	11.1	15.17
26DMPFP	7.86E-02	11.1	16.04
26DMPEP	1.25E-01	11.1	14.38
26DMPC	2.35E-01	5.3	14.99
26DMPTBP	5.99E-01	11.2	13.27
26DMPIPP	5.65E+00	0	13.76
46DMOC	4.49E-02	11.1	14.99
46DMOEP	8.30E-02	8.7	14.38

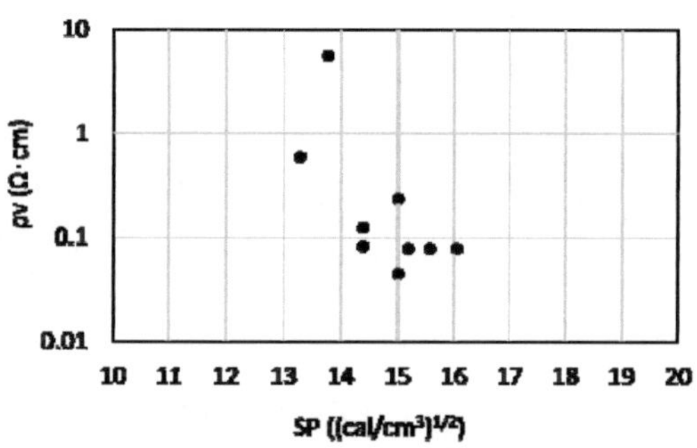

FIGURE 1. Relationship between SP values of DMPs and ρv values of PEDOT:PSS/DMPs solids

Study on interactions between PEDOT:PSS and 26DMPC in slurries and solids

UV-Vis spectroscopic study on the interaction between PEDOT:PSS and 26DMPC in slurries and solids

Fig.2 shows the UV-Vis spectra of PEDOT:PSS/26DMPC dispersions in IPA/H_2O mixture and their solids. In the spectra of the slurry, not only a peak at 291nm but also a shoulder peak ranging from 295nm to 300nm appeared with the increase of the ratio of 26DMPC/PEDOT:PSS. The former was attributable to 26DMPC, and the latter to 26DMPC phenolate anion[9]. It was found that 26DMPC was mixed with PEDOT:PSS to produce its phenolate anion, which suggested dopant species around PEDOT changed partially from PSS anion to 26DMPC anion. In addition, UV-Vis spectra of PEDOT:PSS/26DMPC solids showed a new shoulder peak around 350nm besides a peak at 293nm attributable to 26DMPC with the increase of 26DMPC ratio. Appearance of such a shoulder peak suggests an electronic interaction between PEDOT:PSS and 26DMPC resulting in the blue shift, which possibly leads to the conductivity enhancement of PEDOT:PSS.

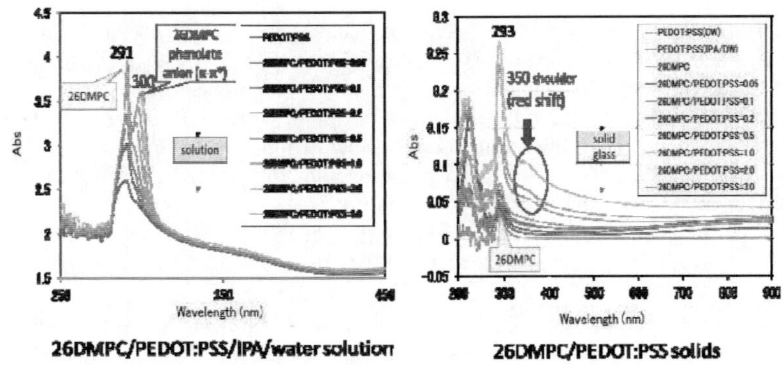

FIGURE 3. UV-Vis spectra of 26DMPC/PEDOT:PSS/IPA/water solution 26DMPC/PEDOT:PSS solids with its ratio of 0 to 3.0 (prepared from PEDOT:PSS conc. = 0.20 wt%)

Study on the structural change of PEDOT:PSS nano-particle by 26DMPC

PEDOT:PSS is dispersed as nano-particles in the slurry. So, the surface charge and particle size of PEDOT:PSS possibly change, if there is the above electronic interaction between PEDOT:PSS and 26DMPC. Then the authors checked the changes of the ζ-potential and particle size of PEDOT:PSS with the increase of 26DMPC in IPA/water slurry (PEDOT:PSS conc. = 0.20 wt%). Its ζ-potential changed from -45mV to -38mV and its particle size got larger than initial by the addition of 26DMPC, which suggested the change of the shell structure of PEDOT:PSS. Such changes mean the change of the structure of PEDOT:PSS nano-particles, possibly resulting from the substitution of PSS by 26DMPC (dopants change).

Computer simulation for action mechanism of conductivity enhancement of PEDOT:PSS by 26DMPC

The finite element analyses were performed with a two-dimensional representative volume element (RVE) model of the PEDOT:PSS core-shell composites by using Digimat-FE (MSC software). The form of the RVE model is based on the assumption of a square array of the PEDOT:PSS core-shell composite in an infinite free PSS. The authors assume that the composite of PEDOT:PSS/26DMPC is comprised of PEDOT:PSS core, PEDOT:PSS shell and free PSS removed from PEDOT:PSS by 26DMPC as shown in Fig.5, and then the thickness of PEDOT:PSS shell increases with the square of PEDOT:PSS core size. The effect of overlapping degree of the PEDOT:PSS shell on the conductivity of PEDOT:PSS was examined. Overlapping degree of PEDOT:PSS shell increased with its thickness and Increase of core size (overlapping degree) and shell conductivity enhance the conductivity of the composite as shown in Fig.4.

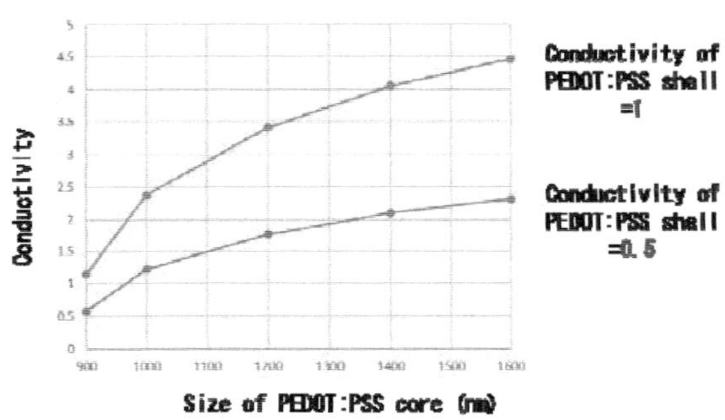

FIGURE 4. Computer simulation results for the effects of size and shell conductivity of PEDOT:PSS on its conductivity

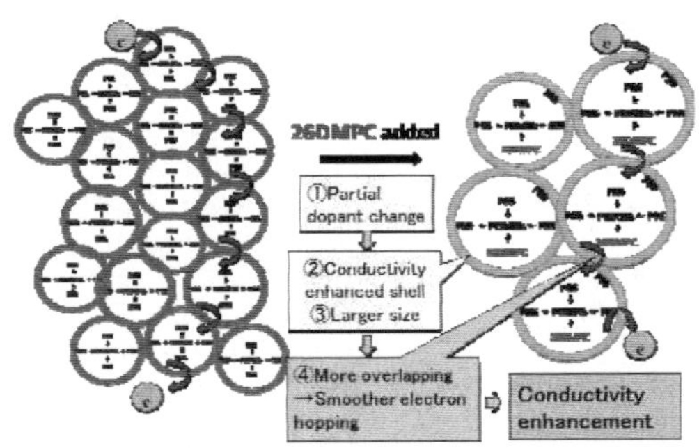

FIGURE 5. Probable action mechanism of 26DMPC on the conductivity enhancement of PEDOT:PSS

The SP values of PEDOT and PSS are 10.4 and 8.4 respectively[10,11]. The PS value of PEDOT:PSS (PEDOT doped with PSS) is expected to be larger than those of PEDOT and PSS, since PEDOT:PSS is a kind of salt and more ionic. DMPs with the SP values ranging from 14 to 16 exhibited higher conductivity enhancement effect as mentioned above. Such SP values are expected to be close to that of PEDOT:PSS. So PEDOT:PSS molecules are easily solvated with DMPs, which brings about the chemical interaction between PEDOT and PSS, followed by the removal of PSS from PEDOT:PSS. At that time the size and overlapping of PEDOT/PSS particle increase, resulting in the conductivity enhancement of PEDOT:PSS as shown in Fig.5.

CONCLUSION

1) There was a large difference in conductivity enhancement effect on PEDOT:PSS among DMPs, and DMPs with the solubility parameter (SP) ranging from 14 to 16 showed a good conductivity enhancement ability, which means the addition effect of DMPs was attributable to its good solubility to PEDOT:PSS molecules.
2) UV-Vis spectroscopy suggested some electronic interaction between 26DMPC and PEDOT:PSS in the slurry and in the solid state of 26DMPC/PEDOT:PSS, and the size and ζ-potential of PEDOT:PSS nano-particle changed with the increase of 26DMPC, which possibly were brought about by the dopant exchange between PSS and 26DMPC and the combination of PEDOT:PSS.
3) Our assumption for the action mechanism of DMPs is as follows. PEDOT:PSS nano-particles are solvated with DMPs, followed by the removal of PSS from PEDOT:PSS. At that time the size and overlapping of PEDOT/PSS particle increase, which brings about the conductivity enhancement of PEDOT:PSS. Such assumption was supported by the computer simulation.

ACKNOWLEDGMENTS

This work was partly supported by JSPS KAKENHI (Grant Number 25420727, 16K06762) and Nanotechnology Platform Program (Molecule and Material Synthesis) of the Ministry of Education, Culture, Sports, Science and Technology (MEXT), Japan.

REFERENCES

1. Hidenori OKUZAKI, "Plastic Electronics Based on Patterning of Conducting Polymers", *Shikizai*, 76〔7〕, 260-264(2003)
2. S. Ashizawa, R. Horikawa, H. Okuzaki, "Effects of solvent on carrier transport in poly(3,4-ethylenedioxythiophene)/poly(4-styrenesulfonate) ", *Synthetic Metals* 153 (2005) 5–8.
3. Hidenori OKUZAKI, "Conductivity Improvement of PEDOT/PSS and Its Application to Electric Devices", *Hyoumen Kagaku*, Vol. 32, No. 10, 653-658, 2011.
4. O.P. Dimitrieva, D.A. Grinkoa, Yu.V. Noskov, N.A. Ogurtsov, A.A. Pudb, "PEDOT:PSS films-Effect of organic solvent additives and annealing on the film conductivity", *Synthetic Metals*, Volume 159, Issues pp. 21–22, November 2009, Pages 2237-2239.
5. S. Konagaya, M. Mutou, M. Terada, "Addition effect of aliphatic monools and diols on the conductivity enhancement of poly(3,4-ethylenedioxythiophene) doped with poly(styrene sulfonic acid) (PEDOT:PSS)", *Proceedings of PPS34* Taipei (Taiwan) (2018)
6. Jaber Saghaei, Ali Fallahzadeh, Tayebeh Saghaei, "ITO-free organic solar cells using highly conductive phenol-treated PEDOT:PSS anodes", *Organic Electronics*, 24 (2015) 188-194.
7. Toshikazu MITOMO, Takuma TESHIROGI, "The simple method for calculation of solubility parameters (1)", *Kagaku Koyukuk*, vol.27 No.6, pp449-453 (1979).
8. Toshikazu MITOMO, Takuma TESHIROGI, "The simple method for calculation of solubility parameters (2)", *Kagaku Kyou-iku*, vol.28 No.1, pp70-74 (1980).
9. Lucila P.Méndez De LeoHugo L.Bianchi Roberto J.Fernăndez-Prini, "Thermodynamics of 4-methylphenol dissociation in water at high temperature using UV–Vis spectroscopy", *The Journal of Chemical Thermodynamics* Volume 34, Issue 9, September,2002,Pages1467-1479, https://doi-org.ejgw.nul.nagoya-u.ac.jp/10.1016/S0021-9614(02)00203-3
10. Wang, Gang; Jiang, Cheng-gun; Gao, Cai-yan; et al., "Dispersion of Electrically Conductive Polymer PEDOT in Organic Solvents", *ACTA POLYMERICA SINICA* Issue:11 Pages: 1532-1538 (NOV 20 2014).
11. Y. N. Sundhakar, D. Krishna Bhat, M. Selvakumar, "Miscibility studies of Starch and Poly(4-styrene sulfonic acid) blend system", *International Journal of ChemTech Research*, 6(7), pp3589-3601(2014).

Evaluation of Electro-spun Polymeric Nanofibers for Sound Absorption Applications

Mohamed Ali Elkasaby[a], Utkarsh[a], Nabeel Ahmed Syed[a], Remon Pop-Iliev[a], Ghaus Rizvi [a*] and Atef Mohany[a]

[a]*University of Ontario Institute of Technology, 2000 Simcoe Street North, Oshawa, Ontario, L1G0C5, Canada*

Abstract. The electrospinning process was used to manufacture nanofiber mats of polymeric materials that have low thickness and improved sound absorption characteristics. The processed nanofiber composites consisted of a number of formulated combinations of polymeric materials, i.e., Polyvinyl Alcohol (PVA), Polystyrene (PS), and Polyvinyl Chloride (PVC). These materials were further modified by addition of fillers like Carbon nanotubes (CNTs), Graphene (GN), Wollastonite (WS) and Fiberglass (FG) for the purpose of betterment of their soundproofing characteristics. This paper focuses on an experimental study for evaluating the sound absorption coefficient of these composites by implementing the standard test method of sound absorption using an impedance tube. The effects of back cavity, air gaps, layering and graded thickness of different combination of the produced nanofiber mats were also investigated in this study. The experimental results revealed that these nanofiber mats demonstrate good sound absorption potential in low and medium frequency ranges (Up to 2000Hz). These cost-effective nanofiber mats have the potential of being used as sound barriers, to reduce noise transmission and keep it contained and/or as sound-dampening and sound-deadening materials in applications where space and volume savings are critical, especially in the electronic and aerospace industries.

Keywords: Electrospinning, Sound absorption, Noise control, Nanofiber composite.

INTRODUCTION

Noise pollution is increasing and is now the 3rd biggest contamination issue which our world is witnessing today [1-3]. Therefore, it is imperative to develop strategies and acoustic materials for its mitigation. Nanofibers derived from polymers have higher surface area to volume ratio, lesser weight and lower thickness that offers better sound absorption in low frequency zone compared to conventional materials. [4, 5].

Nanofiber composites exhibits excellent properties such as large specific surface area along with high porosity and better aspect ratio, which make them attractive for sound absorption applications. Nanofibers can be produced by methods such as electrospinning, phase partition, and self-gathering out of which electrospinning is significantly consolidated [6]. Electrospinning can be used to fabricate a wide range of structures dependent on shape and porosity [7-9].

A wide variety of materials which can be dissolved or dispersed in a solution can be used in electrospinning. PVA is a biodegradable polymer that can be dissolved in water at certain elevated temperatures. It is an inexpensive and non-toxic polymer used in variety of fields [10]. PVC is an inexpensive polymer with excellent chemical stability and mechanical properties [11]. PS possess good dimensional stability and dielectric properties and is widely accepted in electrical and food packaging industries [12].

Nano-filler materials can contribute to the enhancement of functional characteristics of nanofibers and improve the mechanical and thermal characteristics of the composites. This study examines the sound absorbing characteristics of GN, CNT's, FG and WS as micro/nano-fillers by dispersing them in the polymeric solutions of PVA, PVC & PS prepared for electrospinning. Multi walled grade CNT's and GN, which are allotropes of carbon were used because of the ability to uplift the utility of composites based on mechanical, electrical and thermal properties in addition to the high specific surface area [13, 14]. Presence of graphene in the composites also improves the overall stability of the composites [14]. WS is a compound of calcium that possesses high resistance to moisture and oil absorption, and is generally used as filler to improve the flexural and tensile strength of the polymeric composites [15]. FG exhibits not only good sound absorption characteristics but also enhances the thermal and mechanical properties of the fibers [16].

The sound absorption capabilities of electro-spun mats made from the above discussed polymers and filler materials have not been fully investigated in literature. The composites of these polymers were prepared by adding different types of fillers in the solution to fabricate nanofiber tangles for sound absorption characterization. For similar intent, these membranes were layered up in different combinations by varying air gap (between layers), polymer material (multilayering) and back cavities.

EXPERIMENTAL

Materials

Polymeric materials like PVA (Mw = 130,000 g/mol), PVC (90,000 g/mol) and PS (Mw = 190,000 g/mol) were purchased from Sigma Aldrich. The solvents such as N, N-Dimethylformamide (DMF) or distilled water were provided by Sigma Aldrich. FG particles (size: 2μm) was purchased from Owens corning, multi-wall CNT (length 10-20 μm) with outer diameter (30-50nm) and GN (Size 5μm & surface area 120-150m^2/g) were provided by Dr. Simon Park (University of Calgary). WS (520U) was supplied by Fibertec, Inc. USA.

Sample Preparation

The sample preparation was carried out in two steps: First step involves filler dispersion in the solvent for 30 minutes by ultrasonic mixer. The second step involves mixing of the polymers in the sonicated solution by continuous stirring. For PVC composites, PVC 12wt% was added in the sonicated solution of DMF and filler (CNT/FG) and is continuously stirred for 4-5 hours at 70°C. For PVA composites, PVA (9wt% or 12wt %) was dissolved in the sonicated solution of distilled water and filler (CNT/GN/WS) by stirring for 3 hours at 70°C. For all materials the filler dispersion was improved by using Sodium dodecyl sulfate (SDS) as a surfactant in the solution. The electrospinning process was carried out with as prepared solutions at a flow rate of 1ml/hr for PVC composites and 0.3ml/hr for PVA composites. The remaining electrospinning parameters were set to: 15KV voltage at 1000rpm speed of collecting drum while maintaining 10cm needle-collector distance.

Testing and Characterisation

The morphology of the experimentally obtained nanofiber composite mats were analyzed using FE-SEM: JSM 6060. The impedance tube test set up conformed to ASTM1050 and ISO10534-2 [17, 18] for normal incidence sound absorption measurement. The test set up consists of several components: a) A 117cm long tube of 10cm diameter (Length > 3 times tube diameter) to ensure propagation of plane sound wave; b) An insulated box with a loudspeaker connected to a signal generator at one end of the tube that produces white noise; c) 1/4" size microphones are assembled in a tube at specific locations to provide decomposition of plane sound waves in a frequency range from 99 Hz to 2000 Hz; d) A stainless steel sample holder that holds the nanofiber sample at the other end which receives the transmitted signal from speaker. The back cavity of the sample can be adjusted by a movable piston cylinder (aluminum) arrangement. The samples were cut in circular shape of 10cm diameter and pasted on a ring for acoustic measurement. The principle of sound absorptive fibers is based on the energy interaction where the sound energy is converted into heat energy. The sound absorption coefficient values were measured in full frequency range and were analyzed by FFT and 1/3 octave plots. The test for acoustic characterization were performed on nanofiber membranes of: a) PVC nanocomposites, b) PVA nanocomposites, c) PVA/PVC – CNT nanocomposites (2 layers) with different air gaps, d) PVA/PVC-PS-PVA/PVC nanocomposite (3 layers) and e) multilayer nanocomposite with different back cavity.

RESULTS AND DISCUSSION

Sound Absorption Characteristics of PVC Nanocomposites:

As shown in Fig. 1(a) PVC nanofiber composites showed maximum sound absorption coefficient of 0.3 without filler. The correspondent fiber distribution for PVC 12wt% is shown in Fig. 1(b). The addition of 5wt % CNT in PVC composite increased the sound absorption by 62% in comparison with composite without filler and reached the maximum absorption coefficient value of 0.4 at 800Hz. It may be due to proper dispersion of the filler in the polymer solution which participates in energy interaction with sound waves [19-20]. The morphology of PVC composites with CNT is shown in Fig. 1(d). The nanofiber mat with FG as a filler did not show a significant enhancement in sound absorption. This may be due to the non-homogenous dispersion of fillers which reduces the air mobility which is a contributing factor for better sound absorption [21]. The fiber geometry and filler dispersion for this composite is shown in Fig. 1 (c).

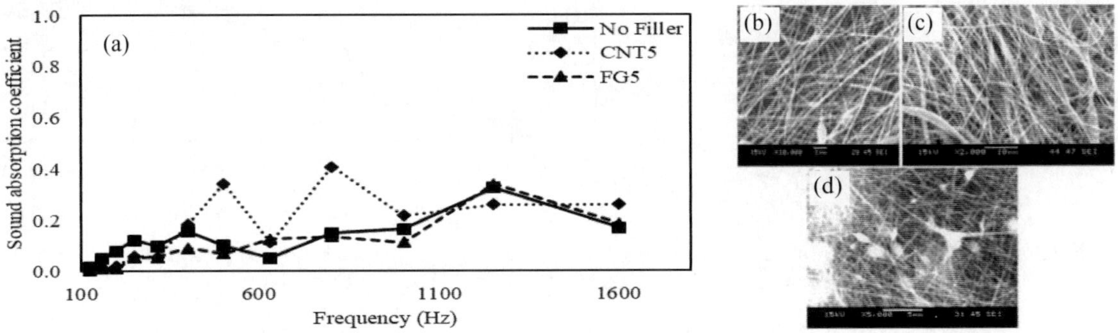

FIGURE 1. (a) Sound absorption performance of PVC 12wt % nanofiber membrane with or without fillers (Back cavity 3cm); SEM images of nanofiber composites of (b) PVC 12wt %; (c) PVC 12wt %-FG 5wt %; (d) PVC 12wt %-CNT 5wt %.

Sound Absorption Characteristics of PVA Nanocomposites:

The acoustic performance of PVA 12wt % nanocomposites with different fillers (5wt %) were investigated and a slight improvement in sound absorption was noticed in the frequency range of 1000-1600Hz [Fig. 2(a)]. The maximum absorption coefficient of pristine PVA 12 nanofiber mat was observed to be 0.24. The addition of CNT and GN by 5 wt% didn't show any noteworthy improvement in sound absorption coefficient value. This may be due to the dispersion non-uniformity of CNT/GN nanoparticles along with reduced surface area volume ratio that helps vibrations to propagate [22]. The micrographic representation of the PVA 12wt% with or without fillers is shown in Fig. 2(b-d). The addition of WS (Nano-clay) in polymer matrix increased the absorption by 50% compared to PVA composite without filler at 1600Hz. The improvement could be attributed to the increased energy absorption due to visco-thermal losses [23].

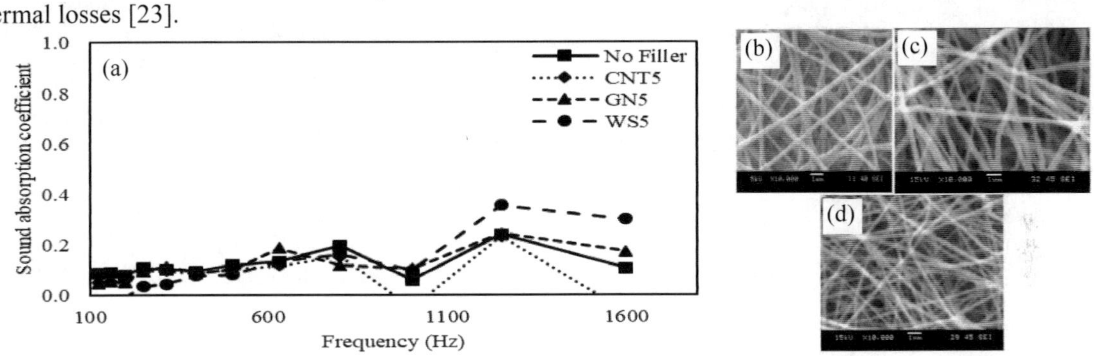

FIGURE 2. (a) Sound absorption performance of PVA 12wt % nanofiber membrane with or without fillers (Back cavity 3cm); SEM images of nanofiber composites of (b) PVA 12wt %; (c) PVA 12wt %-CNT 5wt %; (d) PVA 12wt %-GN 5wt %.

Sound Absorption Characteristics of Layered Polymer Nanocomposites:

Two nanofiber composite mats are combined together with different air gaps in between for sound absorption evaluation [Fig. 3(a)]. The absorption coefficient value of PVC12-CNT5 layered composite (with 5mm air gap) showed an increase of 46% & 10% at 500 & 1000Hz respectively when compared to composite without air gap. The increase is probably due to the gradient structure of composite where the increase in air gap increases the effective thickness of sample. Furthermore, a considerable increase of 50% & 60% were observed at 630 & 1000Hz, respectively, for PVA9-CNT5 composite with increase in air gap (from 0mm to 5mm). Overall, the sound absorption performance of PVC composites was better than PVA composites with maximum sound absorption coefficient value of 0.56 at 1000Hz.

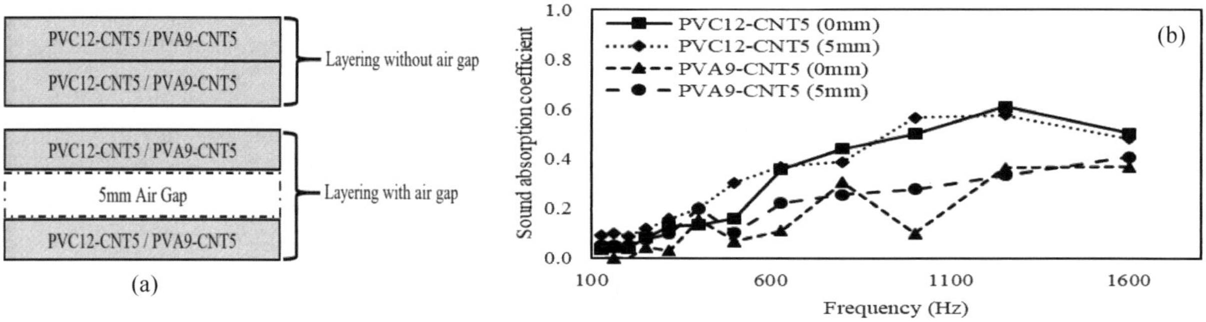

FIGURE 3. (a) Nanofiber mat assembly with or without air gap; (b) Sound absorption performance of PVC & PVA nanofiber composite with CNT as filler (Double layers with 3cm back cavity).

Sound Absorption Characteristics of Sandwiched Polymer Nanocomposites:

Three nanofiber mats are combined together in a series where nanofiber mat of PS (10 wt%/ or 20 wt %) is sandwiched between two nanofiber mats of either PVC12 or of PVA9. For PVC composites, the combined layers with PS20 showed an increase of almost 60% in absorption coefficient value at 630, 1000 & 1250Hz when compared to the combined layers with PS10 [Fig. 4(a)]. This may be due to the formation of the layered gradient structure by PVC composite where both PVC and PS layers are loaded with uniform fibers which provide adequate resistance to sound wave propagation. The cross-sectional image of the PVC composite shows the variation in fiber distribution at each layer in Fig 4(b) & (c). Similarly, for PVA composites, an increase of 47% in absorption value was observed at 1000 Hz with increase in PS concentration. The maximum absorption value of 0.7 was observed with PVA9wt% composite where PS 20wt% was sandwiched in between. This may be due to the more uniform fiber distribution in PVA and PS layer which is evident in micrographic images in Fig.4 (d) & (e).

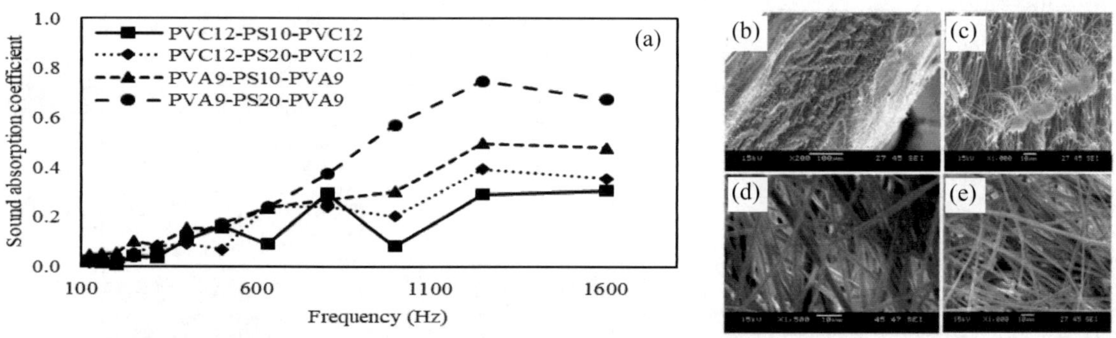

FIGURE 4. (a) Sound absorption performance of PVC/PVA nanofiber composite sandwiched with PS.; SEM images of nanofiber composites of (b) PVC12-PS20-PVC12 (Cross-sectional view); (c) PVC12-PS20-PVC12 (PS20 layer in focus); (d) PVA9-PS20-PVA9 (Top view); (e) PS20 membrane.

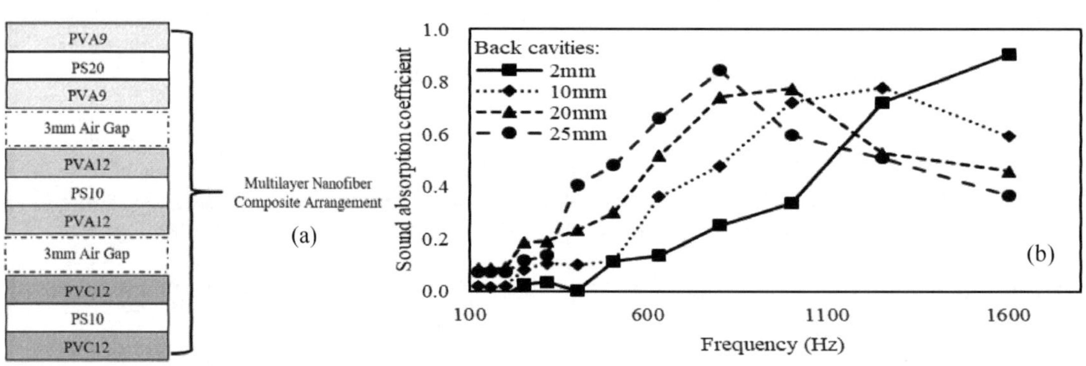

FIGURE 5. (a) Multilayer nanofiber mat arrangement; (b) Sound absorption performance of PVA9PS20PVA9–3mm (air gap)-PVA12PS10PVA12-3mm (air gap)-PVC12PS10PVC12 with different back cavities.

Sound Absorption Characteristics of Polymer Nanocomposite with different Back Cavity:

The effect of back cavity variation was investigated on the multilayer nanocomposite structured in a pattern as shown in Fig. 5(a). For such composite, it was noticed that the sound absorption coefficient curve is shifted to low frequency range with an increase in back cavity [Fig. 5 (b)]. The frequency shift took place from 1600Hz to 800Hz with back cavity variation from 2mm to 25mm. Also the maximum sound absorption coefficient value reaches to 0.85 (back cavity – 25mm) from 0.25 (back cavity – 2mm) which is a considerable improvement of 70% at 800 Hz. The frequency shift is probably due to the increase in effective thickness of the nanofiber sample with increase in back cavity where the longer wavelength of low frequency sound waves makes it easier for the nanofiber composite to vibrate and dissipate energy [24].

CONCLUSIONS

The study presents an evaluation of sound absorption coefficient of polymeric nanofiber composites. The presence of fillers in polymers matrix plays an important role for acoustic dampening. The experimental observations for nanofiber composites with or without filler showed up to 63% improvement in sound absorption with CNT as filler in PVC and the absorption coefficient reaches a maximum value of 0.4. Multiple layers of nanofiber/composite mats were also evaluated for their sound absorbing performance. The combination of PS sandwiched in between PVA membranes reaches absorption coefficient values of 0.5 and 0.7 in frequency ranges of 1000 and 1250 HZ. The highest sound absorption coefficient value of 0.85 was achieved with increase in back cavity to 25mm for multilayer polymer nanofiber composite. Based on these observations it was concluded that nanofiber membranes are the potential candidates for sound absorption applications.

ACKNOWLEDGEMENT

The authors are grateful to NSERC's Discovery Development Grant and NSERC's Chairs in Design Engineering Program for the financial support of this research. The authors would also like to acknowledge Dr. Simon Park from university of Calgary for his kind contribution.

REFERENCES

1. W. Yang and Y. Li, Sci. *China Technol. Sci.* **55**, 2278-2284 (2012).
2. U. Berardi and G. Iannace, *Building and Environment* **94**, 840-852 (2015).
3. M. Alziadeh and A. Mohany, *Experimental Thermal and Fluid Sci.* **102**, 38-51 (2019).
4. R. R. Asmatulu, W. W. Khan and M. B. Yildrim, *ASME* **15**, 223-227 (2009).
5. J. Wang, H. E. Naguib, and A. Bazylak, *Proc* ***SPIE*** *Int Soc Opt Eng* **8342**, 1-13 (2012).
6. R. Vasita and D. S. Katti, *Int J Nanomedicine* **1**, 15-30 (2006).
7. R. Dersch, T. Liu, A. K. Schaper, A. Greiner, and J. H. Wendorff, *J. Polym. Sci. A* **41**, 545-553 (2003).
8. B. Ding and J. Yu, *Nanostructure Science and Technology,* (2014).
9. M. Elkasaby, H. A. Hegab, A. Mohany, G. M. Rizvi, *Advances in Polymer Technology* **37**, 2114-22 (2018).
10. P. Supaphol and S. Chuangchote, *Journal of Applied Polymer Science* **108**, 969-987 (2008).
11. M. Es-Saheb, A. A. Elzatahry, E. M. Sherif, A. S. Alkaraki and E. R. Kenawy, *International Journal of Electrochemical Science* **7**, 5962–5976 (2012).
12. T. Uyar and F. Besenbacher, *Polymer* **49**, 5336-5343 (2008).
13. L. Y. Yeo, J. R. Friend, *Journal of Experimental Nanoscience* **1:2**, 177–209 (2006).
14. M. Wang, C. Yan and L. Ma, *Graphene Nanocomposites* **10,** 5772-50840 (2012).
15. A. S. Luyt, M. D. Dramićanin, Ž. Antić, and V. Djoković, *Polymer Testing Journal* **28:3**, 348–356 (2009).
16. P. W. Gibson, D. Ph, C. Lee, F. Ko, and D. Reneker, *Journal of Engineered Fibers and Fabrics* **2:2**, 32-40 (2007).
17. I. T. Method and D. Filters, ASTM E-1050, 1–12 (2014).
18. G. C. C. da Silva, M. A. de Araújo Nunes, R. V. Lopes and A. B. A. Júnior, *COBEM,* 105–115 (2013).
19. W. Khan, R. Sharma and P. Saini, *INTECH Open* Ch. 1, 1-45 (2016).
20. H. Xiang, S. Tan, X. Yu, Y. Long, X. Zhang, N. Zhao and J. Xu, *Chin J Polym Sci* **29,** 650-657 (2011).
21. J. Zhang, Y. Shen, B. Jiang and Y. Li, *Aerospace* **5**, 75 (2018).
22. N. M. S. Hasan, H. R. Sobuz, M. S. Sayed and M. S. Islam, *J Appl Sci Res* **12,** 831-839 (2012).
23. P. Soltani and M. Zarrebini, *Journal of the Textile Institute* ***104*****(9)**, 1011-1016 (2013).
24. C. Fan, Y. Tian, Z. Q. Wang, J. K. Nie, G. K. Wang, and X. S. Liu, *IOP Conference Series: Materials Science and Engineering* **213**, 012028 (2017).

Properties of Poly(L-Lactic Acid) Blends Based Nanocomposite Films Prepared with Metal Organic Frameworks

Selin Arıcan[1, b)], Sennur Deniz[1, a)] and Ceren Alpaslan Güler[1, 2, c)]

[1]*Yildiz Technical University, Department of Chemical Engineering, Davutpasa Cad. Nr: 127, 34210 Esenler, Istanbul-TURKEY*
[2]*Elif Ambalaj San. ve Tic. AS, Ikitelli OSB, Hurriyet Bulvari, Nr: 1/28, Basaksehir, Istanbul-TURKEY*

[a)]deniz@yildiz.edu.tr
[b)]selinarican93@gmail.com
[c)]cerenalpaslan@elif.com

Abstract. High petroleum prices, corporate environmental awareness and favorable legislations have provided an increase demand for bio-based and biodegradable polymers. Poly(lactic acid) the main commercially available bio-based and compostable polyester produced by the monomeric synthesis of L-lactic acid derived from renewable resources such as corn, cassava, or sugar beet has successfully gained market acceptance for consumer good applications. PLA is an aliphatic polyester with a low melting temperature, biodegradable and renewable properties, which in recent times gained attention in the packaging industry and works well on it. Metal organic frameworks (MOFs) are a class of porous crystalline compounds manufactured from metal ions bridged together by organic ligands forming multi-dimensional structures. Incorporating MOFs in polymers is one the various solutions to contain the cost of films and achieve the desired mechanical properties. In this study, poly(L-lactic acid) (PLA) resin was blended with thermoplastic starch and polycaprolactone (PCL) in different ratios. The PLA/PCL/starch blends and their nanocomposites with copper based MOF (CuMOF) are prepared by solution casting method at room temperature. CuMOF nanocrystals synthesized with solvothermal method were dispersed in polymer solutions. The mixtures were cast by using a casting knife to obtain a thin film. The effects of CuMOF contents on the mechanical and antibacterial properties of PLA blend films were investigated and compared with unfilled films. XRD, BET and SEM analyses were performed to characterize the crystal structure of CuMOF nanocrystals to be added into the polymer matrix. The antibacterial activity of films were evaluated using antibacterial tests. The durability and elasticity properties of the films were determined by mechanical tests.

INTRODUCTION

Nowadays, poly(lactic acid) (PLA) became a polymer of large interest not only in the traditional areas of biomedical and pharmaceutical applications but also in the commodity area, i.e. mainly packaging applications. Indeed PLA has biodegradability, annually renewable availability, and processability following standard melt processes such as solvent casting, injection moulding, film blowing or melt spinning [1]. In addition, its properties are comparable to, or even better than, those of petroleum-based polymers, such as polyethylene (PE), polypropylene (PP), polystyrene (PS), polycarbonate (PC), and polyethylene terephthalate (PET) [2]. PLA has high modulus, acceptable strength, perfect flavor and aroma barrier capability, ideal heat sealability and can be readily manufactured, thereby making it one of the most hopeful biopolymers for several applications [3].

Poly(ε-caprolactone) (PCL) is one of the most hopeful synthetic biodegradable polymers for the reason that it is biodegradable in an aqueous medium and biocompatible in biological applications. Thus, PCL has been used in medical devices frequently and is noted to be a convenient material for tissue engineering [4].

Starch is a potentially useful material. Because, it has the advantages of being found naturally and with low cost in production processes. However, the starch-based materials have some disadvantages such as poor mechanical properties and processability for some application. In order to solve some of these problems, it is contemplated to mix the starch-based plastics with synthetic polymers [5].

Nano-porous materials have attracted the attention of both academia and industry in many applications such as separation, gas storage, shape/size selective catalysis. Metal organic frames (MOFs) can be regarded as a relatively new family of crystalline nanoparticles with thousands of different structures. MOFs usually have low density (0.2-1 g/cm^3), very high surface area (500–4500 m^2/g), high porosity, and acceptable thermal and mechanical stability. The most important advantage of MOFs over other notable nanoporous materials such as zeolites and carbon nanotubes is the capability to set to the structure and functionality of MOFs during synthesis immediately [6].

In this study, we investigated the influence of three type starches in different ratio on the mechanical and antibacterial properties of PLA/starch and PCL/starch blends and their nanocomposites with copper based metal organic framework (CuMOF) synthesized in this study.

EXPERIMENTAL

Materials

Poly (L-lactic acid) (PLLA) biodegradable polymer containing 98% L-lactide was purchased from NatureWorks LLC (Blair, NE, USA). Polycaprolactone (PCL) CAPA 6500 polymer was obtained from Perstorp. Chloroform ($CHCl_3$) was used as the solvent and the LAB-SCAN brand was preferred. Dimethylformamide (DMF), ethyl alcohol (EtOH), dichloromethane (DCM) and triethylamine (TEA) were purchased from Acros Organics as analytical reagent grade. 1,3,5-benzenetricarboxylic acid (BTC, Acros) was used as received without further purification.

Synthesis of CuMOF

A crystalline porous copper-based metal organic framework, CuMOF, was synthesized from the reaction of BTC and copper(II) acetate monohydrate by a solvo-thermal method [7]. BTC (4.7 mmol) was added into 24 ml of a 1:1:1 mixture of DMF/EtOH/H_2O and $Cu(CH_3COO)_2.H_2O$ (8.07 mmol) was added into 24 ml of DMF/EtOH/H_2O solvent mixture. Two solutions were combined with stirring. Triethylamine (6.9 mmol) was added to the reaction mixture, which was stirred for 23 h at room temperature. The blue powder product was separated by centrifugation and washed with DMF for two times and with DCM for three times. The product then dried at 100°C for 6 hours, yielding CuMOF nanocrystals.

Preparation of PLA/PCL/Starch films with CuMOF

The PLA/Starch, PCL/Starch, PLA/PCL/Starch and their nanocomposites with the different amounts of CuMOF were prepared through a solution-blending method. The nominal CuMOFs content was set at 0, 0.5, 1, 1.5 and 2 wt%. Polymers were dried in an oven at 60°C for 24 h using a vacuum oven at 0.1 MPa. PLA, PCL or PLA/PCL with starch were dissolved in chloroform (8 wt.% based on polymer) at room temperature to form a transparent and homogeneous solution. To prepare PLA and PCL nanocomposites, CuMOF nanocrystals were dispersed in chloroform to form a uniform dispersion which was then blended with the prepared PLA or PCL solution. The mixtures were stirred for 4 h at room temperature and then cast on to a glass plate to obtain the thin film using a casting knife resulting film thickness about 50 μm. After casting, the films were dried at room temperature under the hood for 24 hours, and dried at 60°C for 48 hours in a vacuum oven to remove the remaining solvent.

Measurement and characterization

Tensile strength tests were performed in Devotrans universal testing machine. The structure and morphology of CuMOF nanocyrstals were analyzed using Rigaku D/max-2200 Ultimo X-ray diffractometer with CuKα radiation at a scan rate of 2θ=0.01o sec^{-1}. Scanning electron microscopy (SEM) was performed with a JEOL JSM-5910LV instrument (20 kV; Au coating). FTIR analysis was carried out using solid FTIR Avatar 360 ESP spectrometer instrument in the range of 650-4000 cm^{-1} with ATR accessory with Nicolet's OMNIC software. Nitrogen adsorption isotherms at -196 °C were obtained using the Micrometritics accelerated surface area and porousimetry (ASAP) 2020 analyzer. Prior to the measurement of adsorption isotherms, samples were prepared at 200°C under vacuum for 8 h. The Brunauer-Emmett-Teller (BET) method was applied to determine specific surface area of materials.

RESULTS and DISCUSSION

Characterization of CuMOF

The morphology of the CuMOF nanocrystals can be studied in more depth by taking SEM images. The SEM micrograph of CuMOF is shown in Fig. 1-a. The gross morphology of the nanocrystals indicates the presence of polyhedral structures which are dominated by agglomeration of crystals [8].

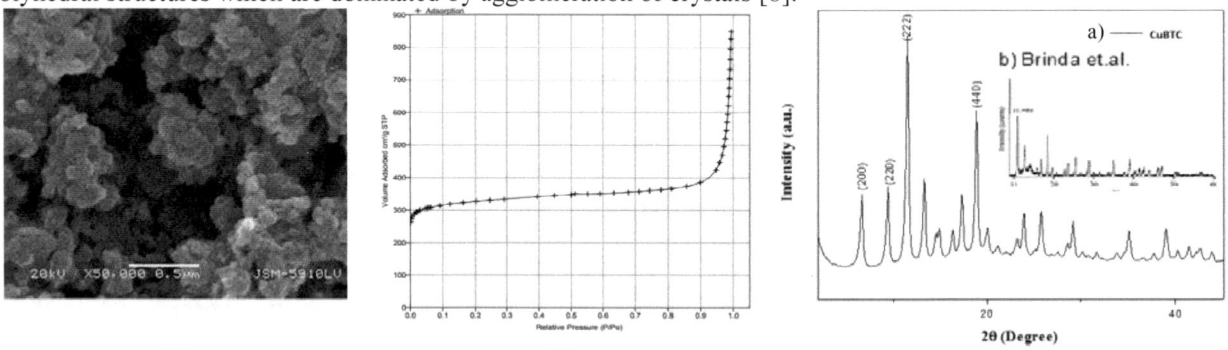

(a) (b) (c)

FIGURE 1. (a) SEM image, (b) The N_2 adsorption isotherm, and (c) XRD pattern of CuMOF nanocrystals

Nitrogen (N_2) gas was used as the adsorbant in the BET analysis (Fig. 1-b). As a result of the analysis, BET surface area was found as 1031 m^2/g. BJH Adsorption cumulative surface area of pores between 17 and 3000 A diameter was 237 m^2/g. Micropore surface area was determined by Dubinin-Astakhov equation and found as 1718 m^2/g.

Powder XRD pattern of CuMOF is presented in Fig. 1-c which clearly shows the highly crystalline reflections in accordance with the literature. The XRD reflections of MOF derived CuO exhibit phase-pure nature without any trace impurity phases like Cu, Cu_2O and $Cu(OH)_2$. The observed reflections are indexed according to the monoclinic structure with C_2/cspace group. Lattice parameter values were calculated during refinement and found to be a=4.682(7), b=3.427(6), c=5.132(2) and β=99.322 which are consistent with literature values [8].

FTIR Analysis

The FTIR spectrums of PLA, PCL and PLA/PCL films and nanocomposites prepared with CuMOF were shown in Fig.4. Infrared spectroscopy data of PLA was presented in the literature for semicrystalline and amorphous PLA [9]. The peak at 755 cm^{-1} can be assigned to the crystalline phase, while the peak at 869 cm^{-1} can be assigned to the amorphous phase. The bands at 1645 and 1590 cm^{-1} and at 1450 and 1370 cm^{-1} corresponds to the asymmetric and symmetric stretching vibrations of the carboxylate groups in CuMOF.

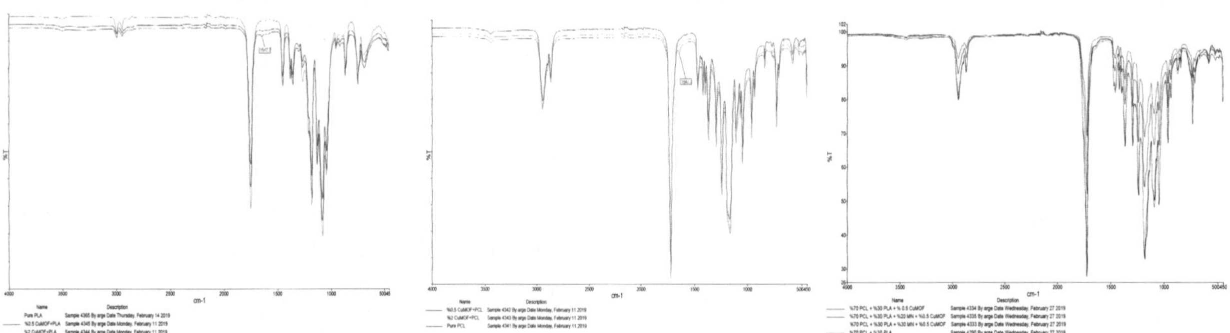

(a) (b) (c)

FIGURE 4. FTIR spectrums of PLA, PCL and PLA/PCL films prepared with CuMOF nanocrystals

Mechanical Tests of PLA and PCL Films

Mechanical properties such as tensile strength, elongation at break and E-modulus of PLA and PCL films blending with the different types of starch were determined and the results are shown in Table 1. Tensile properties of PLA/CuMOF, PCL/CuMOF and PLA/PCL/CuMOF nanocomposites are summarized in Table 2.

TABLE 1. Mechanical test results of PLA/starch and PCL/starch films.

	Tensile strength	Tensile elongation	Tensile strength at break	Elongation at break	E-Modulus
	N	%	N	%	N
PLA	42.00	2.00	42.00	2.00	2081.00
PLA/B-20	41.10	1.95	41.10	1.95	2624.51
PLA/B-40	22.80	1.08	22.80	1.08	2386.10
PLA/M-20	42.37	1.75	41.71	1.86	2923.86
PLA/M-40	34.34	2.87	34.08	2.98	2326.12
PLA/S-20	35.14	1.60	35.14	1.60	2469.72
PLA/S-40	36.46	2.53	36.46	2.53	2387.04
PCL	31.84	726.55	31.78	732.81	433.15
PCL/M-20	13.18	259.21	12.86	268.24	725.70
PCL/M-40	11.34	148.42	11.12	155.38	866.34

PLA films were prepared with three types of starch such as corn, wheat and soluble starch that are represented M, B and S, respectively. When the results were evaluated, corn starch gave better results and showed good mechanical properties in homogeneous dispersion. Therefore, PCL films were prepared with corn starch. The greater E-modulus, the stiffer the material, or the smaller the elastic strain that results from the application of a given stress. As shown in Fig. 5, it was observed from the mechanical test results that the elastic modulus and tensile strength values decreased as starch ratio increased in both PLA and PCL.

TABLE 2. Mechanical test results of PLA, PCL, PLA/PCL and PLA/PCL/starch films prepared with CuMOF

	Tensile strength	Tensile elongation	Tensile strength at break	Elongation at break	E-Modulus
	N	%	N	%	N
PLA-50/PCL-50	9.63	0.99	9.50	1.01	915.68
PLA-30/PCL-70	9.71	2.64	8.56	9.17	515.46
PLA/CuMOF-0.5	31.31	1.62	31.31	1.62	2095.75
PLA/CuMOF-1.0	31.03	1.70	31.03	1.70	2028.95
PLA/CuMOF-1.5	41.98	2.03	41.69	2.19	2178.29
PLA/CuMOF-2.0	38.13	2.04	37.72	2.13	2045.54
PCL/CuMOF-0.5	14.44	7.3	11.61	407.75	372.57
PCL/CuMOF-1.0	16.67	9.95	8.37	193.31	390.21
PCL/CuMOF-1.5	13.97	9.74	10.60	303.04	325.08
PCL/CuMOF-2.0	13.44	7.94	9.40	290.03	319.85
PLA-50/PCL-50/CuMOF-0.5	15.37	2.14	15.37	2.14	872.90
PLA-70/PCL-30/CuMOF-0.5	33.75	3.03	33.68	3.28	1672.01
PLA-30/PCL-70/CuMOF-0.5	13.02	5.58	12.79	5.74	481.11
PLA-30/PCL-70/M20/CuMOF-0.5	10.00	2.40	10.00	2.40	522.90
PLA-30/PCL-70/M30/CuMOF-0.5	9.75	2.98	9.75	2.98	471.94
PLA-30/PCL-70/M40/CuMOF-0.5	4.66	3.36	4.67	3.36	237.07

Antibacterial tests of PLA and PCL Films

In this study, the antibacterial activity was determined to PLA/CuMOF and PCL/CuMOF films. Antibacterial test was performed according to ISO 22196 standard method. The antibacterial activity of the films were determined on S. Aureus. The test was terminated to 24 hours and the number of bacteria grown was analyzed.

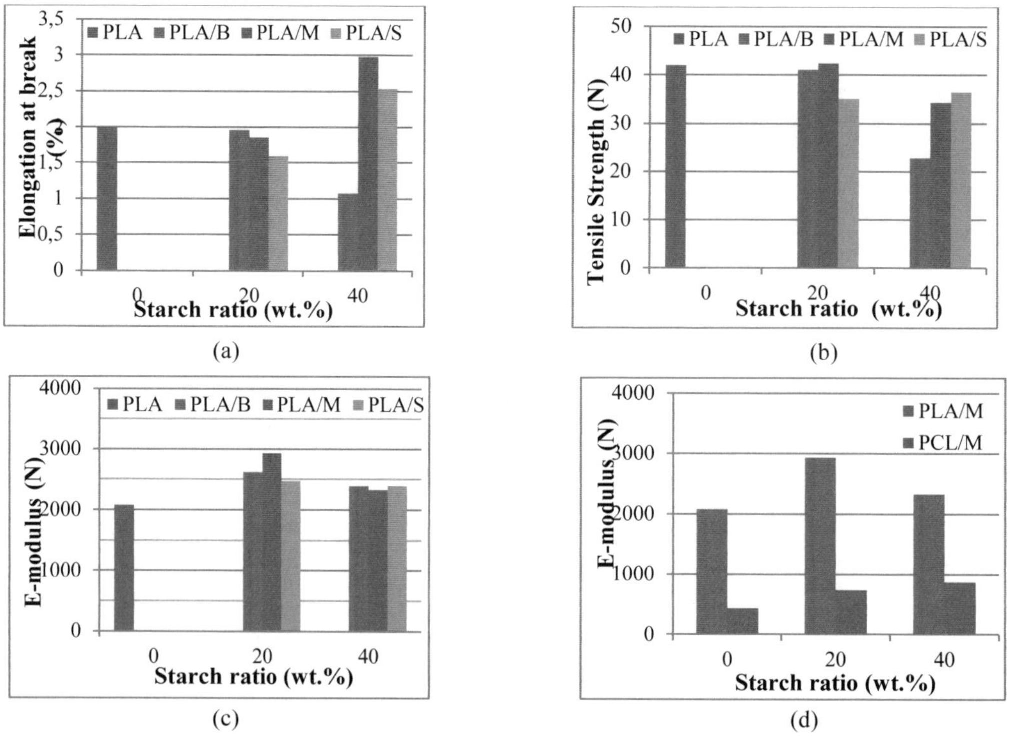

FIGURE 5. Mechanical test results: (a) Elongation at break, (b) Tensile strength, (c) E-modulus of PLA/starch films, and (d) E-modulus of PLA/starch and PCL/starch films

According to the results of the antibacterial tests as shown in Fig. 6, the amount of bacteria in the pure PLA and PCL film samples were found $9.75x10^3$ and $5.88x10^4$ cfu/ml, respectively. The amounts of bacteria in the PLA/CuMOF-0.5 and PCL/CuMOF-0.5 films were counted 6.25 cfu/ml. These values indicate that the adding 0.5 wt.% of CuMOF in to PLA and PCL films was sufficient to give the antibacterial activity and in a good agreement with literature [10].

CONCLUSION

PLA/starch, PCL/starch, and PLA/PCL/starch films were prepared by solution casting method. To determine the antibacterial activity, synthesized CuMOF nanocrystals were added to the PLA and PCL films. It was concluded that PLA biodegradable films had a harder structure than PCL films. The mechanical properties of PLA films were changed by adding PCL and starch. The films obtained with the addition of CuMOF were also found to have an antibacterial property of 99.97 %.

REFERENCES

1. F. Signori, M.B. Coltelli and S. Bronco, Polymer Degradation and Stability **94**, 74–82 (2008).
2. K. Hamad, M. Kaseem, M. Ayyoob, J. Joo and F. Deri, Progress in Polymer Science **85**, 83–127 (2018).
3. Q. Fang and M. A. Hanna, Industrial Crops&Products **10**, 47-53 (1999).
4. H. Mi, X. Jing, J. Peng, M. R. Salick, X. Peng and L. Turng, Cellulose **21**, 2727-2741 (2014).
5. J. Liu, H. Jiang and L. Chen, J Polym. Environment **20**, 810–816 (2012).
6. S. Keskin and S. Kızılel, Ind. Eng. Chem. Research **50**, 1799-1812 (2010).
7. D.J. Tranchemontagne, J.R. Hunt and O.M. Yaghi, Tetrahedron **64**, 8553-8557 (2008).
8. L. Brinda, K.S. Rajan and J. B. B. Rayappan, Journal of Applied Sciences **12**, 1778-1780 (2012).
9. D. Garlotta, Journal of Polymers and the Environment **9**, 63-84 (2001).
10. N.M. Mahmoodi and J. Abdi, Microchemical Journal **144**, 436-442 (2019).

Rheologically Determined Effect of Block Copolymer Morphology on Dispersion of CNT in SEBS/CNT Nanocomposites

Kosar Alavi[1, a)], Parvin Alaee[1, b)] and Hossein Nazockdast[1, c)]

[1] *Polymer Engineering Department, Amirkabir University of Technology, Tehran, Iran*

[a)] k.alavi@aut.ac.ir
[b)] parvinalaie@gmail.com
[c)] Corresponding author: nazdast@aut.ac.ir

Abstract. In the present work, an attempt was made to explore the effect of polystyrene-b-poly (ethylene-butylene)-b-polystyrene (SEBS) block copolymer morphology on carbon nanotube (CNT) dispersion in SEBS with cylindrical and spherical morphology which were melt mixed with different amounts of CNT. Microstructure of nanocomposite samples were studied by using both melt linear and nonlinear viscoelastic measurements. The frequency sweep test results showed a pronounced enhancement in storage modulus in low frequency region with greater extent for the SEBS with spherical morphology. This result could be explained in terms of lower viscosity and/or melt elasticity of SEBS with spherical morphology compared to that of cylindrical one. Lower viscosity and/or melt elasticity lead to greater diffusion of polymer chains into nanoparticle agglomerates and better dispersion. The startup flow test results evidenced the existence of a 3D network due to block copolymer phase separated microdomains in cylindrical SEBS and its nanocomposites which intensified through addition of CNT.

INTRODUCTION

Nowadays, block copolymers (BCPs) based nanocomposites are receiving attention in both industrial and academic fields due to their unique molecular architecture and the ability to form tailored nanostructures. Depending on their block compositions, BCPs exhibit different morphologies such as spherical, cylindrical, gyroid and lamellar [1-2]. Carbon nanoparticles such as CNT remarkably enhance mechanical, thermal and electrical properties of nanocomposites owing to their high surface area and aspect ratio. Of course the extent of enhancement is mainly dependent on dispersion of CNT which is still a great challenge. Employing BCPs as polymer matrices in nanocomposites helps reaching better dispersion state due to higher interaction of nanoparticles with one of the blocks comparing to particle-particle interaction [3]. Rheology is a powerful tool to study melt viscoelastic properties of polymer nanocomposites which is highly sensitive to structural development in such materials. Therefore, it is employed as the main analyzing method in present study. In this paper, using SEBS with two different nanostructures, effect of block copolymer morphology on dispersion state of CNT is studied via rheological measurements.

SAMPLE PREPARATION AND CHARACTERIZATIONS

SEBS with 13% of polystyrene block (Globalprene SEBS 9557) and 30% (Globalprene SEBS 9552) of polystyrene block was purchased from LCY chemical corporation, China. CNT (nc 7000) was supplied by Nanocyl, Belgium.

Prior to mixing, CNT was dried at 70 °C for 12 h in oven. Nanocomposites were compounded using a 60 cc internal mixer (Brabender Plasticorder, Germany) at 220 °C and 60 rpm for 15 min. The prepared samples are coded as SSCx and CSCx for spherical and cylindrical SEBS containing nanocomposites respectively in which x denotes CNT amount.

Rheological measurements were performed with Paar Physica UDS 200 rheometer. The instrument was equipped with parallel disks with diameter of 25 mm and 1 mm gap size. Strain sweep test was performed at angular frequency of $\omega = 1$ rad/s in strain rage of $\gamma = 0.1$-300%. Frequency sweep test was conducted within 0.1-1000 rad/s at 220 °C in linear viscoelastic region (γ = 1%). In order to study structure breakup of samples, startup test was employed at $\dot{\gamma} = 1/s$ for 500 s.

RESULTS AND DISCUSSION

Figure 1 depicts frequency sweep test results of samples based on spherical and cylindrical SEBS at 220 °C. As can be seen, the SS sample showed terminal behavior in low frequency region ($G' \propto \omega^2$) similar to that of pure homopolymer melts while deviation from such behavior was observed for the CS sample which contained higher PS content ($G' \propto \omega^{0.25}$). Also, an upturn appeared in complex viscosity versus frequency curve of the CS sample. This indicates that although the test was conducted at a temperature higher than order-disorder temperature of samples, the whole microdomain 3D network of matrix was not destroyed [4]. With addition of different amounts of CNT, storage modulus increased gradually in the whole frequency range in both groups of nanocomposites with higher extent for samples containing spherical SEBS. Generally, the extent of enhancement in rheological properties of nanocomposites strongly depends on dispersion state of nanoparticles obtained through diffusion of polymer chains into nanoparticle agglomerates.

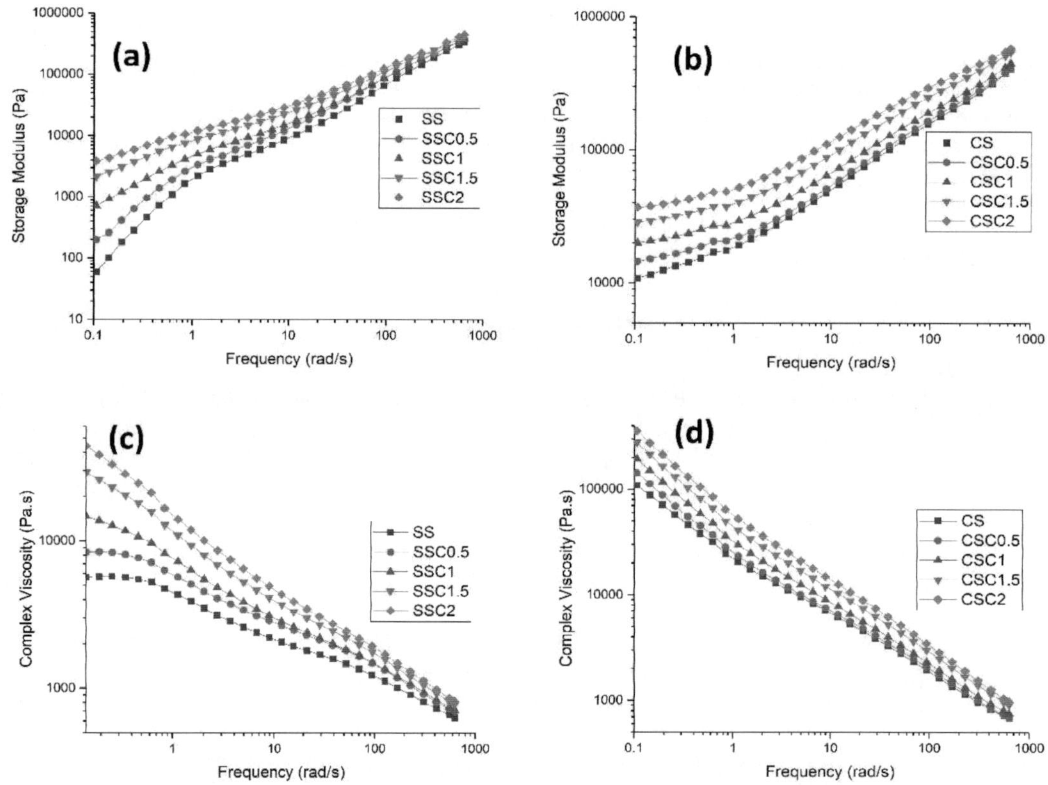

FIGURE 1. Frequency sweep results of samples at 220 °C: (a) and (c) storage modulus and complex viscosity of spherical SEBS containing samples; (b), (d) storage modulus and complex viscosity of cylindrical SEBS containing samples

With increasing hard block content in SEBS and changing morphology from spherical to cylindrical, a stronger microdomain 3D network is expected to be formed. Therefore, block copolymer chains stuck in cylindrical microdomains from both ends have slower dynamics hence less diffusion in CNT bundles [5]. Accordingly, nanocomposites with spherical SEBS showed more pronounced enhancement both in storage modulus and complex viscosity with increasing CNT amount.

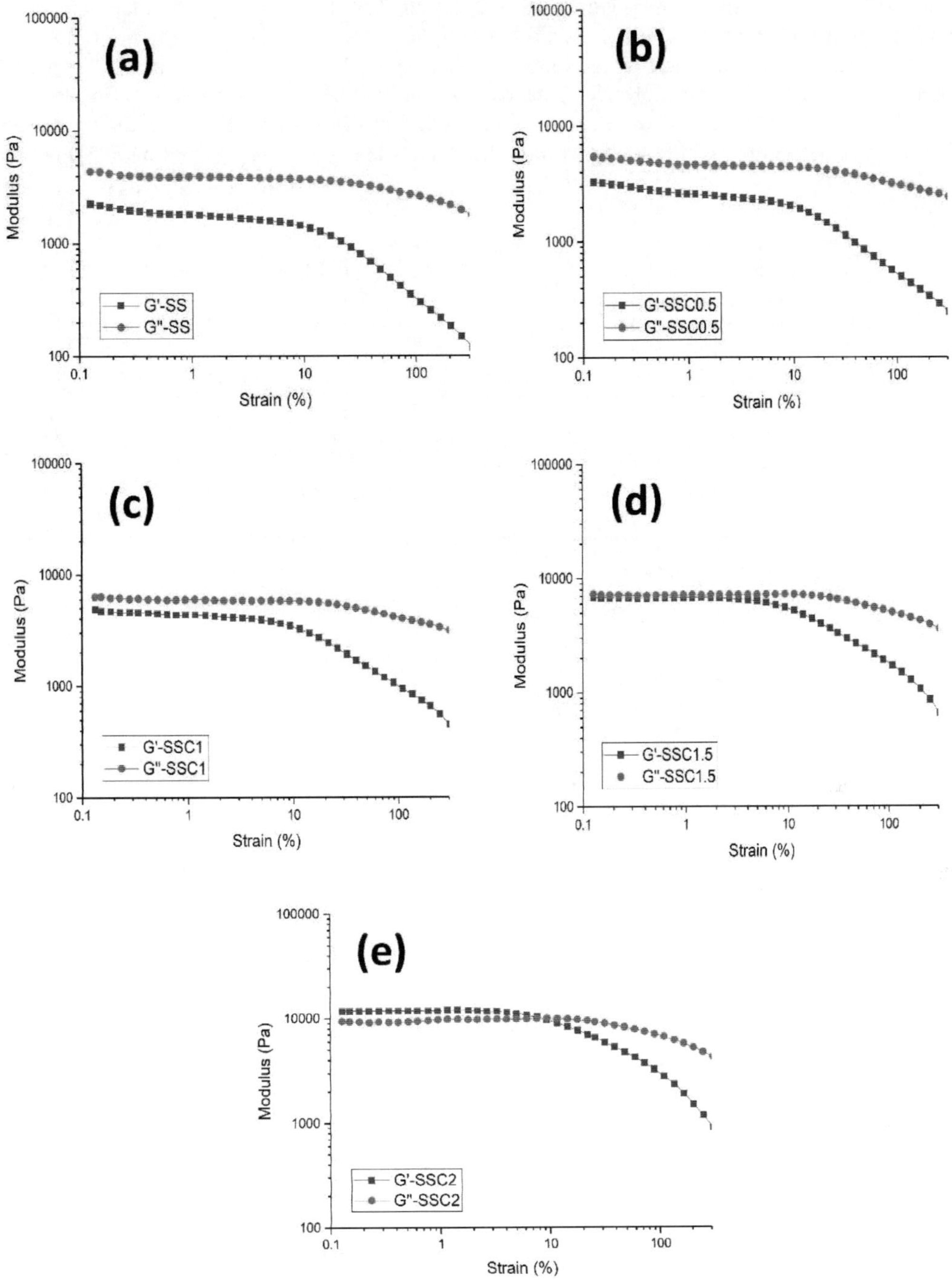

FIGURE 2. Strain amplitude test results of spherical SEBS and its nanocomposites

Figure 2 shows storage modulus (G') and loss modulus (G") versus strain amplitude of spherical SEBS and its nanocomposites measured at 220 °C. As it is evident, for the SS sample, G" was higher than G' in the whole strain range indicating liquid-like behavior of this sample in testing temperature. With addition of CNT and increasing its content, the difference between G' and G" started to decrease so that in the SSC1.5 sample, G' approached and exceeded G". Further increase in CNT amount leaded to dominance of G' over G" in lower strains which was a sign of increase in the volume fraction of nanoparticle-nanoparticle and/or nanoparticle-polymer 3D network [6] although a complete network was not achieved and G' of the SSC2 sample was still quit frequency dependent (figure 1).

According to figure 3, in cylindrical SEBS containing samples however, solid-like behavior (G' > G") was observed in low strains followed by yielding and fluidization in higher strains. Although increasing CNT content increased the difference between G' and G", it did not affect the yielding point and linear viscoelastic region of these samples significantly. This can be attributed to poorer dispersion of CNT in cylindrical SEBS containing samples comparing to spherical ones which was mostly originated from higher matrix elasticity and microdomain 3D network.

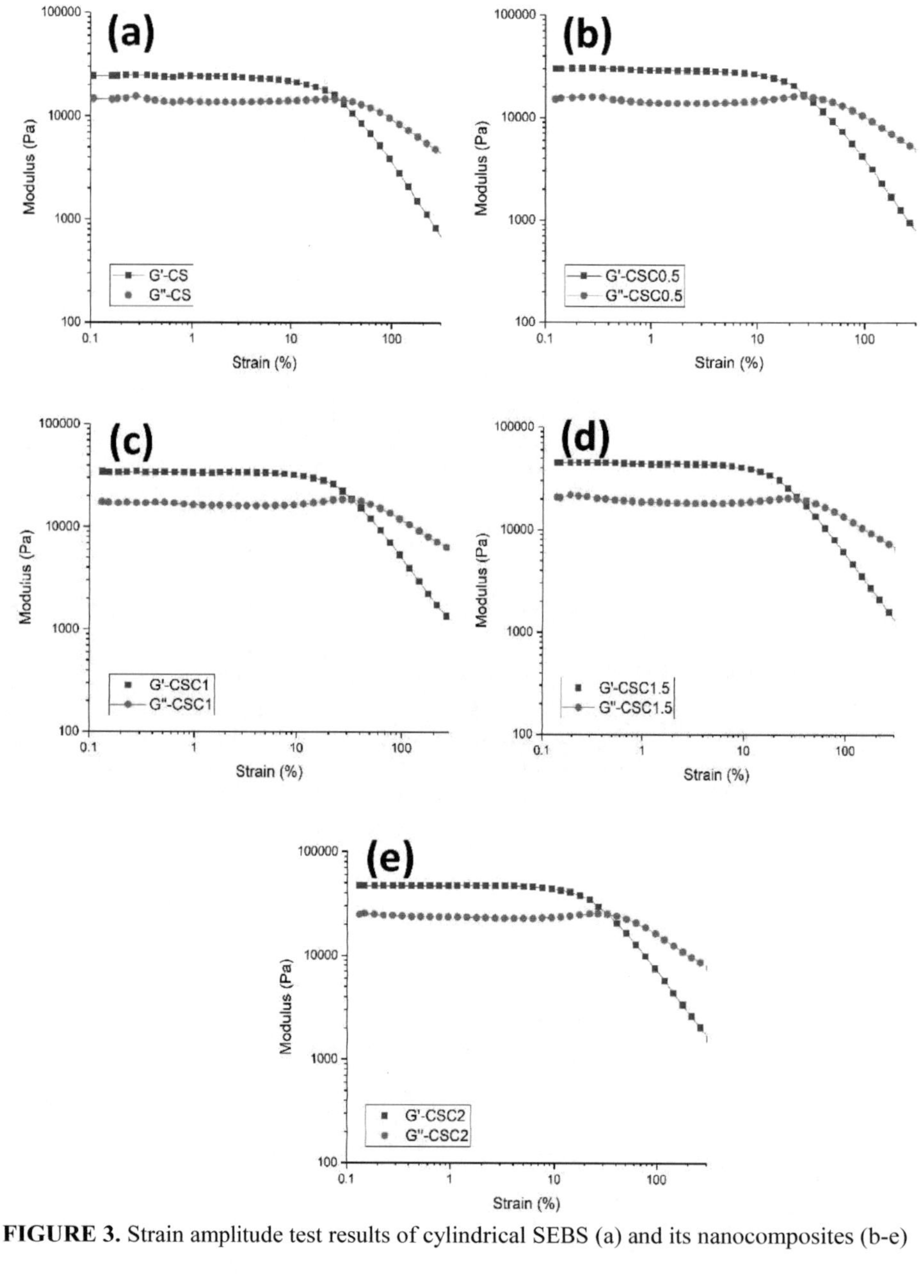

FIGURE 3. Strain amplitude test results of cylindrical SEBS (a) and its nanocomposites (b-e)

Figure 4 represents transient shear stress versus time in startup test. As can be seen, the SS sample exhibited a negligible stress overshoot which was mainly due to chain slippage and orientation in rotational flow field. In the CS sample curve a more considerable stress overshoot was observed comparing to that of the SS. Beside chain slippage, this was chiefly rooted in microdomain 3D network persistence against shear flow [5] that verifies frequency sweep test results. With addition of CNT and increasing its content, stress overshoot intensified in both group of nanocomposites.

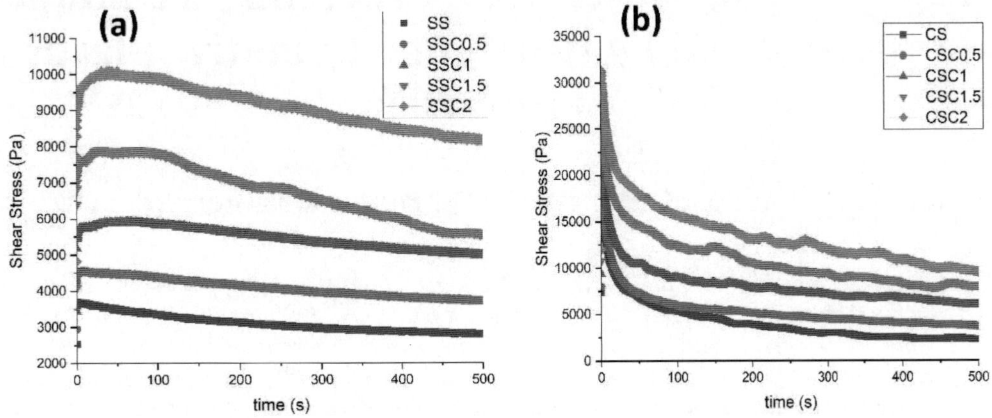

FIGURE 4. Startup test results of spherical SEBS and its nanocomposites (a) and cylindrical SEBS and its nanocomposites (b)

CONCLUSION

In this study, effect of SEBS matrix morphology on CNT dispersion was explored utilizing rheological measurements. Frequency sweep test results revealed that CNT dispersion in spherical SEBS containing samples was better than in cylindrical containing ones due to existence of microdomain 3D network in cylindrical SEBS. The same results were obtained in amplitude sweep test. Also, appearance of stress overshoot in startup flow results of cylindrical SEBS and its nanocomposites confirmed microdomain 3D network and its strengthening with increasing CNT content.

REFERENCES

1. G.H. Bates, F. Fredrickson, *Phys. Today*, **52**, 32-38 (1999).
2. A.E.Y. Mai, *Chem. Soc. Rev.*, **41**, 5969-5985 (2012).
3. I. Szleifer, R. Yerushalmi-rozen, *Polymer*, **46**, 7803-7818 (2005).
4. D. J. Carastan, N. R. Demaraquette, A. Vermogen, K. Masenelli-Varlot, *Rheol. Acta.*,**47**, 521-536 (2008).
5. N. Hasanabadi, H. Nazockdast, S. Balog, M. Lattuada, *Compos. Sci. Technol.*, **149**, 192-206 (2017).
6. S. Sadeghi, M. Arjmand, I. O. Navas, A. Zehtab Yazdi, U. Sundararaj, Macromolecules, **50**, 3954-3967 (2017).

Graphene from waste tire by recycling technique for cost-effective and light-weight automotive plastic part production

Burcu Saner Okan [a,b*], Yusuf Menceloğlu[a,b], Burcu Girginer[c,d], Yavuz Emre Yagci[c]

[a]Sabanci University Integrated Manufacturing Technologies Research and Application Center & Composite Technologies Center of Excellence, Teknopark İstanbul, 34906 Pendik, İstanbul, Turkey
[b]Nanografen Nanotechnological Products Limited Company, Istanbul Teknopark, 34906 Pendik, Istanbul, Turkey,
[c]Farplas Otomotiv A.S., Taysad Organize Sanayi Bölgesi, 41420 Çayırova/Kocaeli, Turkey
[d]Material Science and Engineering Department, Graduate School of Science Engineering and Technology, Istanbul Technical University, 34467 Ayazağa /Istanbul

Abstract. *Light weight vehicle design is a principal keystone to improving fuel efficiency and vehicle performance whilst reducing adverse environmental impacts. At this point, graphene – a solution with a high potential to further reduce vehicle weight and improve mechanical and thermal properties. With a unique combination of mechanical, thermal and electrical properties, graphene shows great potential for exploitation in the automotive industry. In the present work, we developed a technology for the production of graphene nanoplatelets (GNP) from recycled carbon black obtained from the pyrolysis of waste tire by using recycling and upcycling technology. This graphene is comparably cheaper than available ones in the market. This graphene as a reinforcing and nucleating agent was used to reduce glass fiber amount for the production of cost-effective and light-weight automotive plastic composite part by extrusion and injection techniques. With this newly developed compound, polyamide 66 (PA66) based automotive part was produced by the integration of 1 wt% GNP and providing 10% weight reduction. Consequently, this multidisciplinary work will favor the integration of new knowledge and will ensure significant innovation potential in the field of new thermoplastic based composites for the automotive industry.*

Keywords: Graphene, recycling, compounding, masterbatch, composites

INTRODUCTION

The automotive sector is responsible for nearly one third of global energy demand, being the major source of pollution and greenhouse gas emissions in urban areas [1]. This has led to many governments adopting stringent environmental sustainability measures, with the EU for example, setting a CO_2 emission target of 75 g CO_2/km by 2025 for vehicles. The most promising way of achieving this is by reducing vehicle weight which is a major determinant of fuel consumption rates and has been a constant design concern since the dawn of automobile engineering. Studies show that cutting a vehicle's weight by a mere 10% improves the fuel economy by 6-8% and reduces CO_2 emissions by 15-20 g/km travelled [2]. Adoption of advanced composites reinforced by glass fiber leads to a drastic weight reduction in vehicle components. Current glass fiber composites have been highly recommended by the automotive industry due to their advantages: 50-60% weight reduction vs steel, enable one-piece manufacture, design flexibility, reduced tooling costs, better mechanical and corrosion resistance. However, the automotive sector still considers them heavy, giving rise to opportunities for further weight reduction due to their density. Herein, graphene shows great potential for exploitation in the automotive industry with its unique characteristic mechanical and thermal properties [3, 4]. Among engineering plastics, polyamide (PA), referred as nylon, has distinguished characteristics such as toughness, elasticity, excellent resistance to chemical agents, good flame resistance, and increased tolerance at high temperature. In the literature, there are several attempts for the fabrication of graphene reinforced PA6, PA66 and PA12 composites but most of work have focused on the dispersion of graphene oxide during *in-situ* polymerization or solution coagulation [5, 6]. These mentioned techniques can provide solutions for the agglomeration of graphene sheets in bulk systems but these techniques require much more solvent and thus waste management become a serious problem. At this point, melt-compounding is the feasible and widely used technique and also provide advantages in high efficiency and low cost. However, the high cost of graphene is one of the major obstacles to its widespread adoption for commercial applications. Another huge hurdle for commercializing graphene is the lack of an industrially applicable process for large-scale production of high quality graphene at a low cost and in a reproducible manner. In addition, usage of toxic chemicals in the conventional methods used for the synthesis of graphene which results in the generation of hazardous waste and poisonous gases [7].

In the present work, it is aimed to produce new PA-based compounds having low glass fiber and graphene by melt compounding technology to fulfil automotive standards and formulate the recipe for new polyamide compound using hybrid approach (glass fibres and graphene). The developed process provided homogeneous dispersion of graphene nanoplatelets produced from recycled carbon in PA matrice at low ratios and thus resulting in the fabrication of graphene reinforced plastics for low weight automotive parts.

MATERIALS AND METHODS

Instead of graphite as a starting material for the production of graphene, NANOGRAFEN produces near prime graphene-like nanoplatelets (GNP) form recycled carbon obtained from the pyrolysis process of waste tire by applying chemical treatments. Nanographene can produce graphene as about 500 kg/month with a minimum of 10-ton waste tyres being pyrolyzed per day and 40% of this process producing carbon materials. Figure 1 represents the schematic representation of graphene manufacturing process from recycled carbon produced from the pyrolysis of waste tire. After the investigation of its characteristic properties by spectroscopic and microscopic techniques, compounding process was performed by thermokinetic mixer and twin screw extruder for the development of recipe for manufacturing graphene masterbatch and graphene/glass fiber polyamide compound. Specimens for flexural and tensile tests were prepared by injection moulding. Mechanical properties of specimens were investigated by Universal Testing Machine. High resolution transmission electron microscopy (TEM) and scanning electron microscopy (SEM) were used to observe the morphology and the dispersion behavior of graphene sheets in cross-sections of composites.

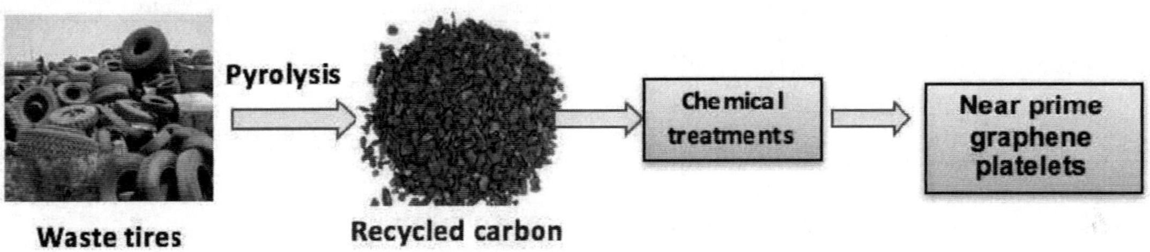

FIGURE 1. Graphene manufacturing process from recycled carbon produced from the pyrolysis of waste tire

RESULTS AND DISCUSSIONS

Before compounding process, the detailed characterization was carried out to examine the structural and morphological properties of GNP. Figure 1a represents TEM image of GNP having an average size of 50 nm plates resulting in better dispersion properties since conventional graphene platelets have a big particle size (5 microns=5,000nm). In addition, GNP has a specific surface area of 85 m^2/g with a chemical composition of 9.1 at% oxygen functional groups. Figure 1b exhibits Raman spectrum of GNP having the prominent characteristic D and G bands and a flat 2D region like graphene oxide produced by Hummers' method [7, 8]. A relatively high ratio of D/G intensity in the Raman spectrum of GNP verified an amorphous structure and the presence of single platelets.

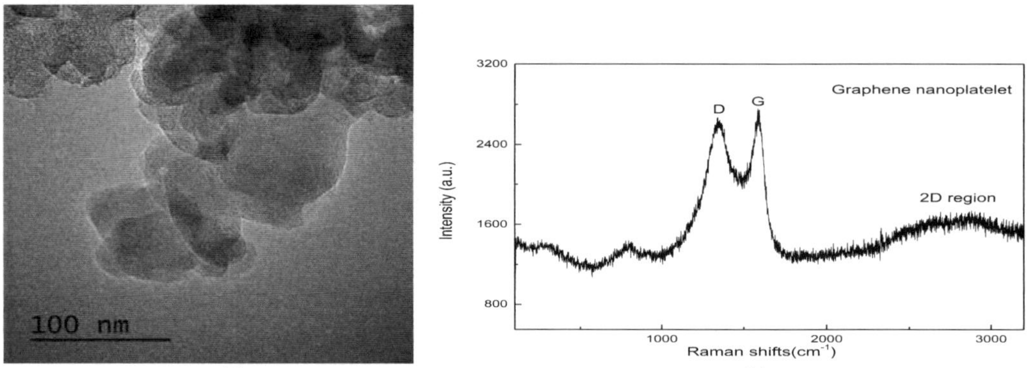

(a) (b)

FIGURE 1. (a) TEM image and (b) Raman spectrum of GNP

The dispersion of graphene sheets during melt compounding is the challenging issue since graphene sheets can restack leading to the stress concentration sites [9]. At this point, process parameters and mixing technique have a great influence on the dispersion state of graphene in the matrix. In the present work, GNP was dispersed at high loading ratio of 10wt% by thermokinetic mixer under high shear rates at melt-phase in order to produce masterbatch. Table 1 summarizes tensile and flexural test results of neat PA66 and 10wt% GNP reinforced PA66 masterbatch produced by thermokinetic mixing. Tensile modulus of 10wt% GNP/PA66 was increased by 43% but there is a decrease in tensile strength as about 8%. This stems from an increase in brittleness by the addition of graphene especially at high strain and thus a low tensile strength is observed. On the other hand, flexural modulus and strength values were enhanced by 23.5% and 9.6%, respectively. The results indicated that applied high shear rates at melt phase increased the diffusion of polymer chains through graphene platelets and thus provided exfoliated structure.

TABLE (1). Summary of tensile and flexural tests of neat PA66 and 10wt% GNP reinforced PA66 masterbatch by thermokinetic mixing

Specimen	Tensile modulus	Tensile strength	Flexural modulus	Flexural strength
Neat PA66	2916	71	2672	107
10wt% GNP/PA66	4160	66	3301	118

In order to understand the dispersion state of GNP, a microtomy analysis was carried out by TEM. Figure 2 exhibits cross-sectional analysis results of 0.5 wt% reinforced PA66 nanocomposite specimen. The dispersion of layers was clearly seen in the matrix in Figure 2a and separated GNP was observed with the average plate size of 50 nm at higher magnification as seen in Figure 2b which is complementary with Figure 1a.

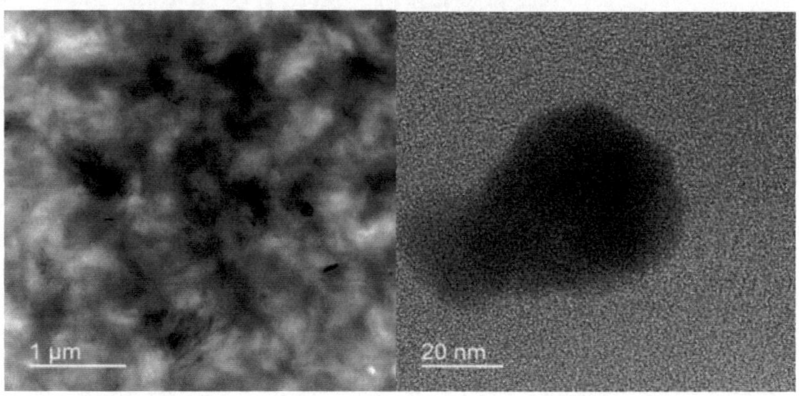

(a) (b)

FIGURE 2. (a) and (b) TEM images of 0.5 wt% reinforced PA66 nanocomposite specimen at different magnifications

Table 2 gives mechanical properties of glass fiber reinforced PA66 and GNP/Glass fiber reinforced PA66 composites produced by twin-screw extrusion. In order to monitor the effect of GNP in glass fiber reinforced PA66 matrix, just 1wt% glass fiber was reduced and replaced with 1 wt% GNP. The integration of GNP led to an increase in tensile modulus, flexural modulus and flexural strength. Also, elasticity modulus and impact strength values were enhanced by adding 1wt% GNP. These findings confirmed that GNP showed proper wettability due to the presence of surface oxygen groups in its structure resulting in exfoliated structure. In addition, enhancement of strength values is a good indication for uniform dispersion of GNP. Furthermore, polymer matrix was chosen as PA66 since it is a semi-crystalline thermoplastic polymer composed of a rigid crystalline phase and an amorphous phase. Graphene can act as both reinforcing and nucleating agent that can strengthen the voids and improve the interfacial interactions. Regarding DSC results, glass transition temperature of glass fiber reinforced PA66 increased from 42.1 to 50.6°C by the integration of 1wt% GNP and thus this alteration supported the nucleation effect of GNP in the matrix.

TABLE (2). Summary of mechanical properties of glass fiber reinforced PA66 and GNP/Glass fiber reinforced PA66 composites by twin-screw extrusion

Mechanical parameters	30wt% glass fiber/PA66	1wt% GNP/29wt% glass fiber/PA66	Improvement (%)
Tensile modulus	123	146	19
Elasticity modulus	7430	8300	12
IZOD impact strength	5.69	6.86	21
Flexural modulus	5970	7220	21
Flexural strength	206	235	14

In the proof of concept study, clutch pedal was chosen as a reference part in the utilization of graphene in automotive by injection process. In comparison of conventional part, the weight reduction was provided as about 10% by reducing glass fiber amount and using graphene as a reinforcement. Also, cost was decreased down to 15% by decreasing production cycle time.

CONCLUSIONS

In the present work, automotive plastic part was successfully produced by injection process in cost-effective manner by providing light-weighting as 10% by using graphene obtained from recycled carbon. A key challenge in the wide spread use of graphene is its high cost because of the source of the graphene (graphite) and the current production methods. Our starting material is recycled carbon produced from waste tires and we use a simple and robust production process, which lowers the graphene price drastically. Ease dispersion of graphene was achieved in PA66 matrix the due to the nature of graphene having surface oxygen functional groups that can make covalent interactions by improving interfacial interactions. Therefore, GNP was compounded with glass fiber and PA66 by twin screw extrusion effectively. These new hybrid compounds have a great potential to be used in engine components (e.g. oil pan), plastic structural parts (e.g. FEM, pedals), metal plastic hybrid structural parts (e.g. traverses, brackets) and underhood components (e.g. engine under cover) in automotive industry. Conclusions, graphene reinforced thermoplastic nanocomposites are promising materials for next-generation automobiles with the reduction in weight, CO_2 emission and fuel consumption.

ACKNOWLEDGEMENTS

The authors acknowledge Dr. Ilse Letofsky-Papst from Graz University of Technology for TEM analysis.

REFERENCES

1. A. Elmarakbi and W. Azoti, "State of the Art on Graphene Lightweighting Nanocomposites for Automotive Applications", in *Experimental Characterization, Predictive Mechanical and Thermal Modeling of Nanostructures and their Polymer Composites*, edited by Francesco Marotti de Sciarra and Pietro Russo, Elsevier, 2018, pp. 1-23.
2. W. J. Joost, *The Journal of the Minerals, Metals & Materials Society* **64**, 9 (2012).
3. J. S. M. Zanjani, B. Saner Okan and Y. Z. Menceloglu, *Mater. Chem. Phys.* **176**, 58–67 (2016).
4. B. Saner Okan, *Turk. J. Chem.* **41**, 381-390 (2017).
5. A. O'Neill, D. Bakirtzis and D. Dixon, *Eur. Polym. J.* **59**, 353-362 (2014).
6. Z. Cai, X. Meng, Y. Han, H. Ye, L. Cui and Q. Zhou, *Composites: Part A* **69**, 115–123 (2015).
7. D. C. Marcano, D. V. Kosynkin, J. M. Berlin, A. Sinitskii, Z. Sun, A. Slesarev, et al., *ACS Nano* **4**, 4806–4814 (2010).
8. A. Kaniyoor and S. Ramaprabhu, *AIP Advances* **2**, 032183 (2012).
9. X. Wang and M. Song, *Nanomaterials and Energy* **2**, 265 – 278 (2013).

Investigation on the Influence of Multi-step Processing on the Mechanical and Thermal Properties of Cellulose Reinforced EVOH Composites

Georg Graninger, Sandeep Kumar and Brian G. Falzon*

Advanced Composites Research Group, School of Mechanical and Aerospace Engineering, Queen's University Belfast, Ashby Building, Belfast BT9 5AH, United Kingdom
Email: b.falzon@qub.ac.uk, Web Page: https://www.qub.ac.uk/

Abstract. The industrial fabrication of thermoplastic composites involves the consecutive application of different melt processing techniques such as extrusion and compression moulding. In the course of processing, the material properties can undergo significant changes that determine the final properties of the product. Thermoplastic composites created by implementing cellulose - a relatively low cost, bio-renewable and bio-degradable polymer with high mechanical properties and low density – as a filler show promising results. In order to establish a cost-efficient, scalable and sustainable production in industry it is important to understand how melt processing affects the mechanical and thermal properties of cellulose thermoplastic composites. Microcrystalline cellulose (MCC) and cellulose nanocrystals (CNC) were combined, in turn, with ethylene vinyl alcohol (EVOH) using a high shear screw configuration for extrusion to create composites with improved dispersion. Melt processing steps included compounding in the extruder followed by compression moulding. No thermal degradation was found in the course of processing. Results revealed an elevated crystallization temperature for composites due to a nucleating effect of cellulose. When the compression moulding step was repeated, a processing-induced orientation led to an increase in crystallinity which is higher in pristine material as strong filler-matrix interaction inhibits molecular movement in composites.

Keywords: processing, cellulose, thermoplastic composites, crystallization, nucleating effect.
PACS: 81.05.Qk, 81.07.Bc, 81.20.Hy

INTRODUCTION

In recent years, an increasing number of 'green' composites were developed in academic and industrial research using renewable fillers like cellulose, chitin and starch [1, 2]. Restricted compatibility with non-polar matrices due to their polar nature, thermal stability regarding the processing with engineering thermoplastics and dispersion difficulties, caused by agglomeration, are major concerns in melt processing these fillers [2, 3, 4, 5]. In addition, fabrication of cellulose thermoplastic composites can involve multiple processing steps (extrusion, compression moulding, injection moulding) each possibly affecting the final product's properties [1, 4, 6].
Being extracted from plants, trees and animals, cellulose is a bio-polymer of β-1,4-glucosidic linked glucose units [4]. Strong intra- and intermolecular hydrogen bonds between hydroxyl groups of the molecule chains lead to the formation of a rigid semicrystalline structure [4, 5]. Using acid hydrolysis to remove amorphous parts, different types of cellulose fillers can be obtained from the fibrils that are suitable for mechanical reinforcement [1, 5]. Microcrystalline cellulose (MCC) consists of spherical particles with at least one micrometer-sized dimension. MCC is commercially available and has a thermal degradation onset of 300 °C [3, 7, 8]. Extensive acid hydrolysis produces needle-shaped cellulose nanocrystals (CNC) with diameters between 20-30 nm and up to 2000 nm in length. The resulting high aspect ratio (up to 80) and an elastic modulus of 143 GPa lead to an increased reinforcing effect in composites compared to MCC [3]. Sulfate groups that remain after acid hydrolysis reduce the thermal stability of CNC of around 200 °C [5, 9].

Ethylene-vinyl alcohol (EVOH), a commercially available thermoplastic polymer with high barrier performance used in industry for food packaging, pharmaceutical and biomedical purposes [5, 10]. EVOH possesses a large amount of hydroxyl (OH-) side groups due to its vinyl alcohol component. These side groups cause strong intra- and intermolecular hydrogen bond interactions responsible for the high barrier properties [10]. The ethylene content impedes the formation of hydrogen bonds causing reduced barrier properties and a lower melt temperature compared to poly(vinyl alcohol) (PVOH). An ethylene content between 28 mol% and 44 mol% gives a sufficient compromise between processability and barrier properties [11]. It is suggested that the presence of vinyl alcohol in EVOH can promote the compatibility with cellulose fillers [5, 10]. Sessini et al. reported a significant increase in elastic modulus (100 %) and thermal stability (4 %) of starch nanocrystals (5 wt.%)-reinforced ethylene-co-vinyl acetate (EVA) fabricated by extrusion. Improvements were attributed to strong hydrogen bonding between the OH-groups of filler and matrix [1].

Many studies investigate properties of composites that were fabricated using various consecutive processes. Sapkota's research dealt with nanocomposites made from CNCs and poly(vinyl acetate). These composites were first solution-cast, then extruded and finally compression moulded [3]. A similar processing strategy was chosen by Orr and Shofner to produce EVOH-CNC composites [5]. An internal mixer and microextruder followed by compression moulding were used for EVOH-lignin and EVA-starch nanocrystal composites, respectively [1, 6]. To our knowledge, no studies have investigated how mechanical and thermal properties change between different consecutive processing steps. In addition, while the influence of reprocessing on EVOH properties has been reported [10], its effect on EVOH-cellulose composites is not covered in literature.

The aim of this study is to investigate the effect of consecutive melt processing steps (extrusion and compression moulding) and of melt reprocessing (compression moulding) on the mechanical and thermal properties of EVOH-cellulose thermoplastic composites. The derived information is intended to provide a better insight into processing-structure properties of thermally sensible cellulose and to possibly support scalable melt processing for food packaging applications.

MATERIALS AND METHODS

EVOH copolymer (F101B) with an ethylene content of 32 mol% was used as matrix material (Kuraray Co. Ltd., Japan). The material had a glass transition temperature of 69 °C and a melt temperature of 183 °C. MCC (powder, particle size: 51 µm) (Sigma-Aldrich Company Ltd., United Kingdom [12]) and CNCs (powder, average length: 150 nm) (CelluForce, Canada [13]) were chosen as fillers.

EVOH-MCC and EVOH-CNC composites with 5 wt.% filler content, respectively, were manufactured in the course of five processing steps. Pristine EVOH was prepared accordingly to compare samples with the same processing history. In the first processing step, a cryogenic grinder (6850 Freezer/Mill, Rondol Technology Ltd., United Kingdom) was used to grind EVOH pellets into powder. Matrix and cellulose powder were dry-mixed using a blender (Rondol Technology Ltd., United Kingdom). Melt-blending was performed using a co-rotating 16 mm twin-screw extruder (L/D = 25, screw speed: 125 rpm) (HAAKE Rheomex OS PTW16, ThermoFisher SCIENTIFIC, United Kingdom). A high-shear screw configuration (kneading blocks 90° alternated in two sections) was chosen to improve the dispersion of cellulose [14]. The compounds were then compression molded two times in a platen press (P 200 P, COLLIN Lab & Pilot Solutions GmbH, Germany). Table 1 gives an overview of the various processing steps and parameters.

TABLE 1. Processing history of EVOH-cellulose samples.

Processing step	Temperature (°C)	Pressure (bar)	Residence time (min)
Grinding of EVOH	-196	/	57
Dry-blending of powders	25	/	4
Extrusion (EX)	210	75	1
Compression moulding 1 (CM1)	210	100	21
Compression moulding 2 (CM2)	210	100	21

Samples were taken after each melt processing step (EX, CM1, CM2) and compared. A PerkinElmer Pyris 6 calorimeter (PerkinElmer Inc., USA) was used to perform differential scanning calorimetry (DSC). Samples were subjected to two heating and one cooling cycle (temperature range: 30 °C to 250 °C) with a heating rate of 10 K/min and sample weights of 7.3+/-1 mg. The degree of crystallinity was evaluated from the melting peak in the second heating cycle using 128.1 J/g as the heat of fusion for 100 % crystalline EVOH. Dynamic mechanical thermal analysis (DMTA) was performed using a Tritec 2000 DMA (Triton Technology Ltd., United Kingdom). Measurements were conducted in single-cantilever bending mode with a heating rate of 2 K/min (temperature range: 30 °C to 150 °C), a frequency of 1 Hz and a displacement of 0.05 mm. The surfaces of cryo-fractured samples were observed with a scanning electron microscope (SEM), FlexSEM 1000 (Hitachi Ltd., Japan). Thermogravimetric analysis (TGA) was performed using a TGA/SDTA851e (Mettler Toledo, USA) (temperature range: 30 °C to 600 °C, purge gas: O_2). Material was dried for 8 hours at 80 °C prior to melt processing and DMA measurements.

RESULTS AND DISCUSSION

In order to evaluate the influence of multi-step processing on the material properties of the composites, both cooling and second heating scans from the DSC measurements were evaluated. The according thermal properties are summarized in Table 2.

TABLE 2. DSC measurement results of pristine EVOH and EVOH-cellulose composites for cooling and second heating scans.

Processing step	Crystallization Temperature (bar)	Melt Temperature (°C)	Crystallinity Degree (%)
EVOH-Pristine EX	155.07 (+/-0.11)	183.24 (+/-0.27)	54.63 (+/-5.68)
EVOH-Pristine CM1	154.31 (+/-0.10)	183.87 (+/-0.18)	52.68 (+/-8.29)
EVOH-Pristine CM2	155.49 (+/-0.19)	182.22 (+/-0.06)	66.34 (+/-3.48)
EVOH-MCC EX	158.28 (+/-0.02)	183.08 (+/-0.08)	57.41 (+/-3.19)
EVOH-MCC CM1	158.19 (+/-0.11)	182.91 (+/-0.09)	55.50 (+/-1.52)
EVOH-MCC CM2	158.04 (+/-0.08)	182.15 (+/-0.17)	58.26 (+/-1.97)
EVOH-CNC EX	158.32 (+/-0.25)	183.69 (+/-0.16)	54.22 (+/-1.01)
EVOH-CNC CM1	158.35 (+/-0.08)	183.57 (+/-0.08)	56.71 (+/-1.40)
EVOH-CNC CM2	158.22 (+/-0.09)	182.90 (+/-0.08)	59.68 (+/-1.16)

The crystallization temperature remains constant for all samples throughout all observed processing steps. EVOH-cellulose composites exhibit a higher crystallization temperature (EVOH-MCC: 158.17 +/-0.10 °C, EVOH-CNC: 158.30 +/-0.06 °C) than pristine material (EVOH-Pristine: 154.96 +/-0.49 °C) in all cases. Kim and Choi reported a similar increase in crystallization temperature (156.6 °C to 161.0 °C), when they incorporated chemically-functionalized graphene oxide sheets into EVOH (ethylene content of 32 mol%) [15]. Corrêa et al. found that nanocomposites from cellulose whiskers (length: 205 +/- 22 nm) coated with poly(amide) 6 (PA6) in PA6 also led to crystal formation at higher temperatures [4]. Both studies attribute the increased crystallization temperature to the filler particles acting as effective nucleating agents [4, 15]. As both cellulose types (MCC, CNC) lead to a very similar increase, it can be argued that their effectiveness as nucleating seeds is the same. As shown in our previous work, the application of a high-shear screw configuration leads to a delamination and decrease in size of MCC particles in the composite, thus providing similarly high degree of interaction as CNCs [14]. SEM images show that dispersion of fillers is further promoted by the continued processing in the platen press, see Figure 1.

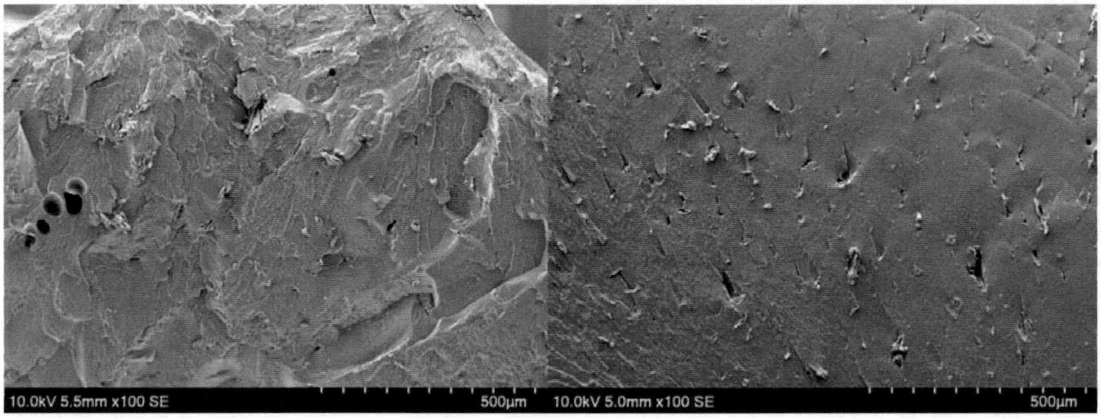

FIGURE 1. SEM images of the cryo-fractured surface of EVOH-MCC composites. Left: Composite surface after extrusion (EX): Dispersed MCC particles in EVOH matrix. Right: Composite surface after the first compression moulding cycle (CM1): Increased dispersion of MCC particles.

The melting peak temperatures of the second heating scan do not shift, but remain stable at 183.07 +/-0.28 °C. This indicates that different lattice modifications that might form due to processing and composition do not affect the melting behaviour [16].

In addition to a rise in crystallization temperature, an improved crystallinity can be observed from the DSC data. The degree of crystallinity increases for all compositions between processing via extrusion and the second compression moulding cycle. EVOH-cellulose composites show a higher crystallinity degree than pristine samples after extrusion and first compression moulding. After the second compression moulding cycle, pristine EVOH exhibits the highest crystallinity of 66.34 (+/-3.48) %. This can be attributed to an orientation of molecules [10], which increases with each processing step. While cellulose particles act as nucleating agents leading to an earlier crystallization, it can be argued that crystals growing perpendicularly to the cellulose surface and strong interaction between filler and matrix inhibit the crystallization of oriented EVOH molecules [1, 10, 17]. This leads to a lower crystallinity degree of composites after the second compression moulding. This crystallization behaviour is indicated by the second heating scans, shown in Figure 2.

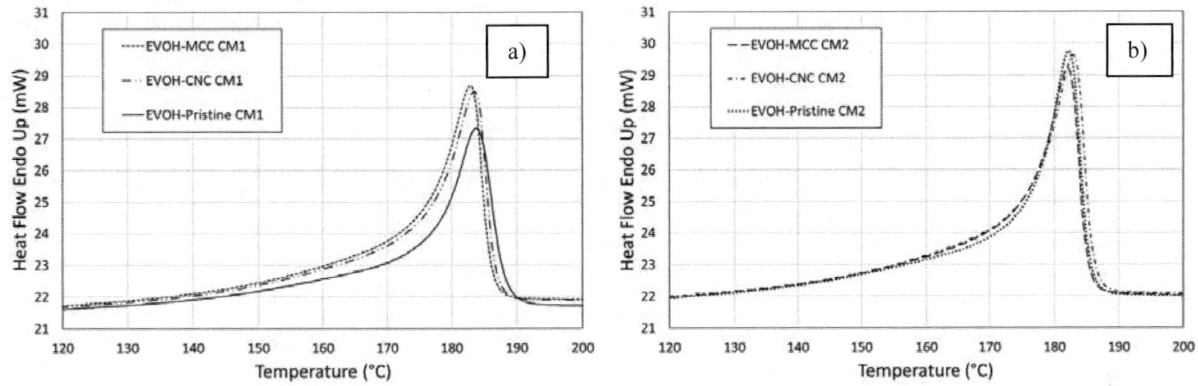

FIGURE 2. DSC second heating-thermograms: a) After compression moulding 1: Broader melt peaks of EVOH-cellulose composites with higher intensity than pristine EVOH indicate higher melt enthalpy and degree of crystallinity. b) After compression moulding 2: Narrow melt peak with highest intensity of pristine EVOH shows increased crystallinity due to uninterrupted crystallization of oriented molecule chains.

According to Sessini et al., the increased crystallinity can also be a consequence of the nucleation effect of the cellulose particles. This also leads to an increase in storage modulus due to interfacial interactions between cellulose and the matrix [1]. The storage moduli of cellulose composites in this study reflect this trend after the first compression moulding cycle. With an additional compression moulding processing step (CM1 → CM2), the storage modulus increases by more than 5,000 MPa (Figure 3). Franco-Urquiza et al. found an increase in storage modulus between non-processed EVOH film and film that was processed three times in a twin-screw extruder. This increasing trend came with a decrease in crystallinity with additional reprocessing [10]. It was suggested that oxidative cross-linking reactions led to confinement of molecule chains hindering crystallization [10, 15]. As TGA results show no decrease in thermal stability, it is proposed that the increased storage modulus is determined by the filler- and processing-dependent crystallization behaviour.

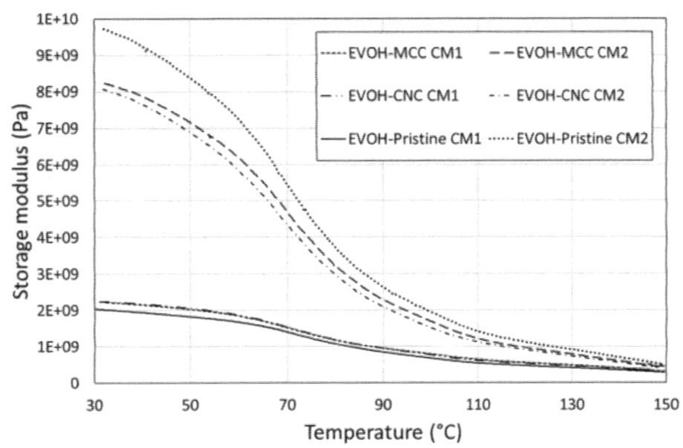

FIGURE 3. Storage modulus of pristine EVOH and EVOH-cellulose composites after the first and second compression moulding cycle, gained via DMTA.

CONCLUSION

EVOH thermoplastic polymer was combined with MCC and CNC to investigate how the mechanical and thermal properties of cellulose composites are affected by a series of consecutive processes, i.e. extrusion and two compression moulding cycles. Results showed that cellulose particles acted as nucleating agents exhibiting strong interactions with the matrix, thus causing an increase in crystallization temperature of the composites. A higher degree of crystallinity and an increased storage modulus in comparison to pristine material was observed after extrusion and first compression moulding. Processing-induced orientation of EVOH molecules led to higher crystallinity and storage modulus values after second compression moulding. SEM investigation suggested additional delamination of cellulose increasing the nucleating effect. TGA results showed no effect of processing on the thermal stability of EVOH-cellulose composites. The described processing give insight into scalable manufacturing of EVOH composites. Nano-sized cellulose fillers can thus be used to tailor crystallization in the course of processing.

ACKNOWLEDGMENTS

This research is funded by the Department for the Economy (DfE) (Northern Ireland, United Kingdom). The authors also want to express their gratitude to the members of the Polymer Processing Research Centre (PPRC) at Queen's University Belfast (QUB) for their continued support and advice.

REFERENCES

1. V. Sessini, J.-M. Raquez, J. M. Kenny, P. Dubois and L. Peponi, *Carbohydrate Polymers* **208**, 382-390 (2019).
2. F. Ansari and L. A. Berglund, *Biomacromolecules* **19**, 2341-2350 (2018).
3. J. Sapkota, "Processing, Structure, and Properties of Cellulose Based Nanocomposites", Ph.D. Thesis, University of Fribourg (Switzerland), 2016.
4. A. C. Corrêa, E. de M. Teixeira, V. B. Carmona, K. B. R. Teodoro, C. Ribeiro, L. H. C. Mattoso and J. M. Marconcini, *Cellulose* **21**, 311-322 (2014).
5. M. P. Orr and M. L. Shofner, *Polymer* **126**, 211-223 (2017).
6. B. Podolyák, D. Kun, K. Renner and B. Pukánszky, *International Journal of Biological Macromolecules* **107**, 1203-1211 (2018).
7. D. Trache, M. H. Hussin, C. T. H. Chuin, S. Sabar, M. R. N. Fazita, O. F. A. Taiwo, T.M. Hassan and M. K. M. Haafiz, *International Journal of Biological Macromolecules* **93**, 789-804 (2016).
8. A. Kiziltas, D. J. Gardner, Y. Han and H.-S. Yang, Thermochimica Acta 519, 38-43 (2011).
9. M. Mariano, C. Chirat, N. El Kissi and A. Dufresne, *Journal of Polymer Science, Part B: Polymer Physics* **54**, 2284-2297 (2016).
10. E. Franco-Urquiza, O. O. Santana, J. Gámez-Pérez, A. B. Martinéz and M. Ll. Maspoch, *eXPRESS Polymer Letters* **4**, 153-160 (2010).
11. M. Krasteva, M. L. Cerrada, R. Benavente and E. Pérez, *Polymer* **46**, 9831-9839 (2005).
12. Sigma-Aldrich, *Cellulose 435236*, https://www.sigmaaldrich.com/catalog/product/aldrich/435236?lang=en&region=GB (Accessed 11 February 2019).
13. Celluforce, *NanoCrystalline Cellulose*, https://www.celluforce.com/en/products/core-properties/ (Accessed 11 February 2019).
14. G. Graninger, S. Kumar and B. G. Falzon, *ECCM Conference Proceedings* **18**, 1-4 (2018).
15. S. W. Kim and H. M. Choi, *High Performance Polymers* **27**, 694-704 (2015).
16. M. L. Cerrada, E. Pérez, J. M. Pereña and R. Benavente, *Macromolecules* **31**, 2559-2564 (1998).
17. S. Ouchiar, G. Stoclet, C. Cabaret and V. Gloaguen, *Macromolecules* **49**, 2782-2790 (2016).

Elastomeric Bio-based Epoxy / Clay Nanocomposites

Andres Rigail-Cedeño (presenting author)[1,2] *Daniel F. Schmidt* (reference author)[4,5]
Grace Vera[1], *Monica Chavez*[1], *Yesenia Pacheco*[3], *and Mercedes Alava*[3]

[1]*Faculty of Mechanical Engineering and Production Science,*
[2]*Center of Research and Development in Nanotechnology,*
[3]*Faculty of Natural Science and Mathematics,*
ESPOL Polytechnic University Escuela Superior Politécnica del Litoral,
Campus Gustavo Galindo Km 30.5 vía Perimetral, Guayaquil, Ecuador.
[4]*Department of Plastics Engineering,*
University of Massachusetts Lowell
Lowell, MA, USA
[5]*Department of Materials Research & Technology,*
Luxembourg Institute of Science & Technology
Hautcharage, Luxembourg
arigail@espol.edu.ec, Daniel_Schmidt@uml.edu / daniel.schmidt@list.lu

Abstract. The emphasis of this research was to develop a fully bio-based elastomeric epoxy and its clay nanocomposites. Blends of sorbitol glycidyl ether (SGE) and epoxidized castor oil (ECO) were cross-linked with a dimer diamine (DDA) as curing agent in a solvent-free procedure. Blends of SGE and ECO (0/100, 30/70, 50/50, 70/30 and 90/10, w/w) were cured with stoichiometric amounts of DDA alone or following the addition of 5 wt% of one of two organically modified montmorillonite (OMMT) nanoclays, Cloisite 20A and 30B. Pure SGE was immiscible with DDA at room temperature, but SGE/ECO blends were compatible with DDA. All networks were cured at 60°C for 6 hours, with crosslinking kinetics monitored via near-infrared (NIR) spectroscopy. Epoxy conversion increased with increasing ECO content due to improved compatibility with DDA. The greatest concentration of unreacted primary and secondary amine groups was observed in the 90/10 SGE/ECO blend. Nanoclay dispersion and affinity for the matrix was assessed via x-ray diffraction (XRD), and an intercalated structure was confirmed regardless of nanoclay type or epoxy formulation. Mechanical properties were also investigated. A broad range of Young's Moduli (~0.7-4.8 MPa) and hardness values (30-81 Shore A) were obtained, with the highest stiffness and hardness observed in the 90/10 SGE/ECO blend reinforced with Cloisite 30B. Indeed, these results support the existence of a synergistic effect on mechanical properties due to nanoclay addition, thanks both to increased chemical crosslink density and nanofiller reinforcement. Based on these studies, SGE/ECO/DDA clay nanocomposites show promise as versatile elastomeric networks whose hydrophobicity and mechanical properties may be adjusted for a range of industrial applications.

Keywords: bio-based, elastomeric, clay, nanocomposite.
PACS: 81.05.Qk

INTRODUCTION

Thermosetting elastomers are mostly based on polyurethanes, silicones and vulcanized diene rubbers [1,2]. Polyurethanes can give a broad variety of mechanical properties; nevertheless, they have low thermal and hydrolytic stability compared to two-part silicones. However, silicones are usually softer, weaker materials than polyurethanes. Vulcanized diene rubbers are versatile materials that provide good mechanical performance and hydrolytic stability, but they are complex to formulate and study and generally require high processing temperatures. On the other hand, epoxy resins are among the most heavily used thermosets, and find use as stiff, rigid, glassy materials in a range of high-performance applications due to their excellent chemical and heat resistance, mechanical and electrical properties [3]. Most developments have been performed using elastomeric aromatic epoxy materials using a variety of different molecular weights of amine curing agents [2,4,5].

Additionally, epoxies have been of interest for many years as the basis for clay nanocomposites with unique properties. Researchers have studied several variables to enhance the ability of organically modified montmorillonite clays to disperse in epoxy networks [6]. In tandem, due to health concerns, replacements have been sought for aromatic epoxies based on bisphenol A (BPA) due to the recognition of BPA as an endocrine disruptor [7]. More generally, there is increasing demand for more sustainable polymeric materials based on renewable resources. In keeping with these trends, several bio-based epoxy resins have been crosslinked with petroleum-based amines in the presence of organically modified montmorillonite clays, both for reinforcement and to catalyze epoxy polymerization in the interlayer galleries [8,9].

The current report adds to this body of knowledge. Here, blends of sorbitol glycidyl ether (SGE) and epoxidized castor oil (ECO) resin have been cured with a dimer diamine (DDA). These fully bio-based networks were reinforced with two commercial organoclays. These bio-based nanocomposite networks show promise as replacements for traditional elastomeric thermosets, especially given the possibility to realize solvent-free processing at low temperatures. The combination of bio-based resins and mild processing conditions requiring minimal energy input promises enhanced sustainability as well. To better understand the potential of this family of bio-based epoxy nanocomposites, we have studied a range of formulations, and report on their processing and performance.

EXPERIMENTAL

Materials

Epoxidized castor oil (ECO), Heloxy Modifier 505, was provided by Hexion. ECO has an epoxy equivalent weight (EEW) in the range of 500 to 650 g/eq. Sorbitol glycidyl ether (SGE), Erisys GE-60, was supplied by Emerald Performance Materials. SGE has an EEW in the range of 160 to 195 g/eq. A fatty acid based 36 carbon diamine dimer (DDA), Priamine 1071, has an amine hydrogen equivalent weights (AHEWs) of 140 g/eq and was supplied by Croda Corporation. Figure 1 shows the chemical structures of SGE (a), DDA (b) and ECO (c). Two organoclays, bis(hydrogenated tallow alkyl)dimethyl quaternary ammonium montmorillonite, Cloisite 20A (C20A) and bis(hydroxyethyl)methyl tallow alkyl quaternary ammonium montmorillonite, Cloisite 30B (C30B), were purchased from Southern Clay Products. C20A and C30B give x-ray diffraction peaks associated with their interlayer spacings at ~25 Å and ~19 Å, respectively.

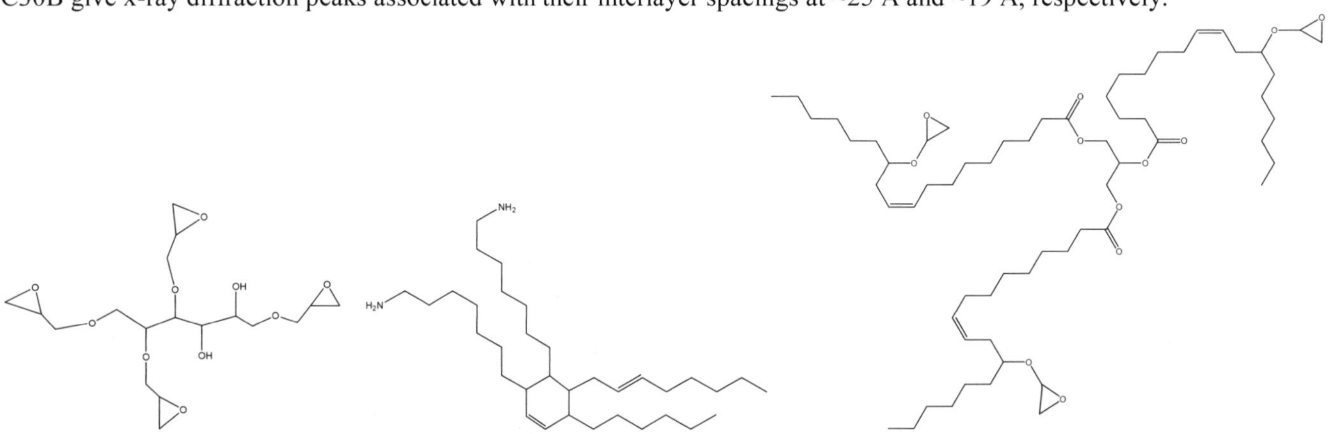

(a) (b) (c)

FIGURE 1. Chemical structure of SGE (a), DDA(b) and ECO (c)

Processing

Nanocomposites were prepared by dispersing 5 wt% of either C20A or C30B in various SGE/ECO blends (0/100, 30/70, 50/50, 70/30 and 90/10, w/w) using a Flacktek DAC 400.1 FVZ speed mixer operated at 2,500 RPM for 3 minutes. A stoichiometric amount of DDA was then added, after which mixing was repeated at 2,500 RPM for 10 seconds for both filled and unfilled systems. The final mixture was poured into customized silicone molds to obtain samples of an appropriate size and shape for further characterization and evaluation. For comparison purposes, all bio-epoxy/bio-amine formulations were prepared with and without organoclays. All mixing procedures were performed at room temperature, keeping the relative humidity below 50% with portable de-humidifiers. All systems were cured at 60 °C for ~5 hours.

Characterization and Evaluation

X-ray diffraction patterns were obtained using a Phillips PANalytical XPert-Pro model diffractometer equipped with a tube emitting Cu K_{α} (λ=1.54060 Å) radiation and energized at 40 mA and 45 kV with a ½" incident slit and ¼" detector slit, scanning from 1° to 9° (2θ) with a 0.05° step size and a 20 second count time per step. Curing analysis was carried out by gathering near-IR (NIR) absorption spectra of samples in the range of 7,500-4,000 cm^{-1}, using a Perkin Elmer Spectrum 100 spectrometer. 1" square pieces of glass microscope slides were separated by a 0.044" thick, flat, U-shaped silicone rubber gasket and held together with small binder clips to make disposable sample cells with a known path length. All mixtures were poured into these sample cells and allowed to cure, with the curing process monitored for 420 minutes at 60 °C. All NIR spectra were analyzed with the Thermo Nicolet EZ OMNIC software package (Version 7.3). Tensile samples measuring 4.5" × 0.75" × 0.055" were prepared by curing the mixture in a silicone mold. Tensile tests were performed at room temperature at a testing rate of 0.04 in/min using a Shimadzu AG-1S 10 kN load frame controlled by the Trapezium2

Software package (Version 2.33). For this test, the ASTM D638 standard (Type IV) was followed as the experimental procedure. Specimens for hardness testing measuring 0.39" cubed were also manufactured by pouring mixtures into an appropriate shape silicone mold. Shore A Hardness testing was performed according to the ASTM D2240 standard, using a Qualitest HPE Digital Durometer.

RESULTS AND DISCUSSION

Figure 2 shows XRD patterns of the bio-epoxy and bio-anime components in the presence of 5 wt% Closite 20A (a) and Closite 30B (b). Data from the neat bio-epoxy resins has been omitted due to a lack of diffraction peaks in these materials as previously reported [8,10], but diffraction data is shown for the neat DDA, as it exhibits diffraction peaks at ~47 Å and ~28 Å, presumably due to crystallization of the long alkyl substituents in the DDA structure. As far as the epoxy/nanoclay blends are concerned, there is clear evidence of clay intercalation by both bio-epoxy resins, with a shift in the interlayer spacing to ~35-45 Å except in the case of the SGE/C30B blend, where the clay peaks become so weak that they are difficult to resolve. While alone this does not prove exfoliation of the clay layers, it is at least consistent with this outcome. In the case of the DDA/nanoclay mixtures, broad diffractions in the same range as neat DDA were observed in both cases, making interpretation of the data less straightforward given that intercalated clay peaks should show up in the same region. However, as it has been reported that diamine curatives generally intercalates in organoclays [11]. In addition the diffraction peaks are observed to be significantly stronger following clay addition, and a new peak is observed at ~22 Å in DDA/C30B as well (with hints of a weaker peak observed in the same region in the DDA/C20A case). All of this points to intercalation of the DDA into both nanoclays as well.

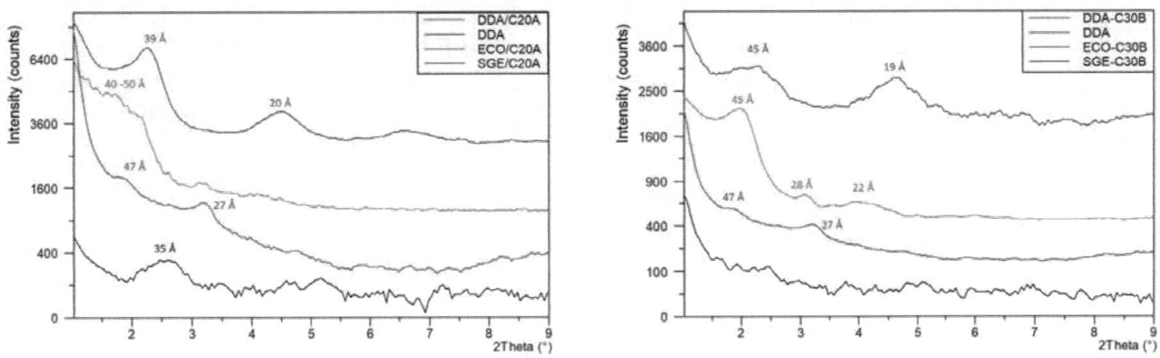

(a) (b)

FIGURE 2. X-ray diffraction patterns of neat DDA, as well as all components mixed with 5 wt% C20A (a) or C30B (b).

Figure 3 shows the XRD patterns of the bottom surfaces of the cured nanocomposites (NCs); diffractograms from the top surfaces are omitted given their similar appearance. Clay peaks at ~37-40 Å are observed for all cured C20A nanocomposites. However, no diffractions peaks are observed in any of the C30B nanocomposites. While this is consistent with exfoliation, the lack of a harmonic series of peaks has also been associated with disordered systems and related to the co-existence of multiple clay phases with different interlayer spacings [8,12,13].

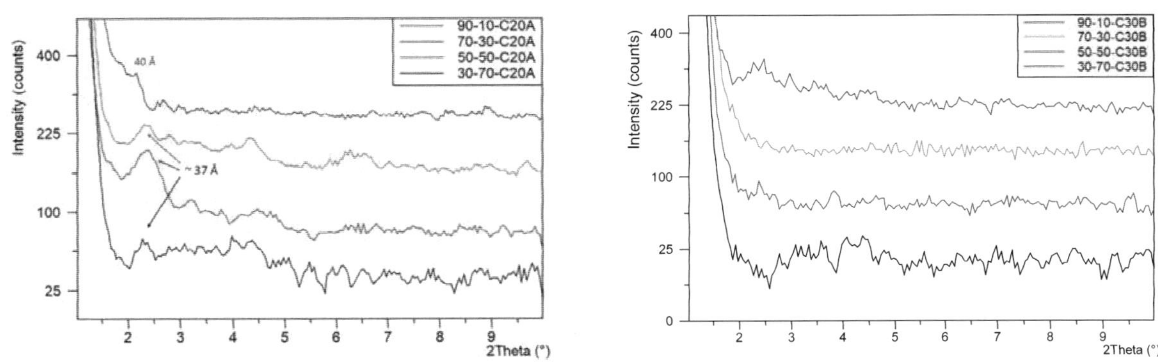

(a) (b)

FIGURE 3. X-ray diffraction of SGE/ECO/DDA network nanocomposites intercalated with C20A (a) and C30B (b).

Table (1) shows the final extent of epoxy conversion following curing for all bio-based networks studied, both neat and in the presence of the organoclays. The extent of epoxy conversion was calculated by quantifying the decrease in the area of the epoxy peak, as observed at ~4,530 cm^{-1} in the FT-NIR spectrum [14]. It was normalized to the C-H stretching overtone peak at ~5,990 cm^{-1}, which was assumed to be effectively constant regardless of the formulation being studied; the epoxy conversion (α_{NIR-Ep}) is calculated according to equation 1.

$$\alpha_{NIR-Ep} = 1 - \frac{[(A_{4530,t})/(A_{5990,t})]}{[(A_{4530,0})/(A_{5990,0})]} \quad (1)$$

The epoxy conversion results account for the conversion of epoxy groups from both bio-based epoxy resins. However, this does not mean that all epoxy groups in the system will be equally reactive; additional factors could affect the epoxy conversion of each bio-based epoxy component of the blends being studied. First, SGE which is more hydrophilic is expected to be less compatible with DDA than the more hydrophobic ECO. Also, ECO (with a functionality of three) has an EEW ten times higher than that of SGE (with a functionality of four). As a result of the local differences in epoxy group concentration this implies, SGE could be expected to react faster than ECO. Of these two factors, the former appears to be the more important, given the observation that epoxy conversion decreases as SGE content increases in the epoxy blend. For the nanocomposite systems, both nanoclays might be expected to decrease the chances of an encounter between an epoxy and an amine group via physical partitioning depending on the level of dispersion. As C20A is more hydrophobic than C30B, C30B is expected to be more compatible with the resins, giving rise to higher levels of dispersion. Indeed, the x-ray diffraction results presented previously are consistent with this conclusion, given the observation of reduced diffracted intensities in all cases. The x-ray diffraction data further imply that the SGE/C30B combination gives rise to a greater degree of disorder/exfoliation than any other combination of resin component and nanoclay. This, combined with the presence of hydroxyl groups in the modifier of the C30B, might preferentially enhance the reaction rate of the SGE given its proximity to the clay surface, leading to localized gelation, entrapment of residual amine hydrogens and suppression of their ability to react with the ECO in the system. These arguments may at least partially explain why the epoxy conversion values change little on addition of C20A but decrease on addition of C30B.

TABLE (1). Degree of epoxy conversion by Near-IR of SGE/ECO ratio cured with DDA

Epoxy Conversion (%)	30/70	50/50	70/30	90/10
Neat	100	92	85	78
Cloisite 20A	99	92	84	76
Cloisite 30B	98	80	71	70

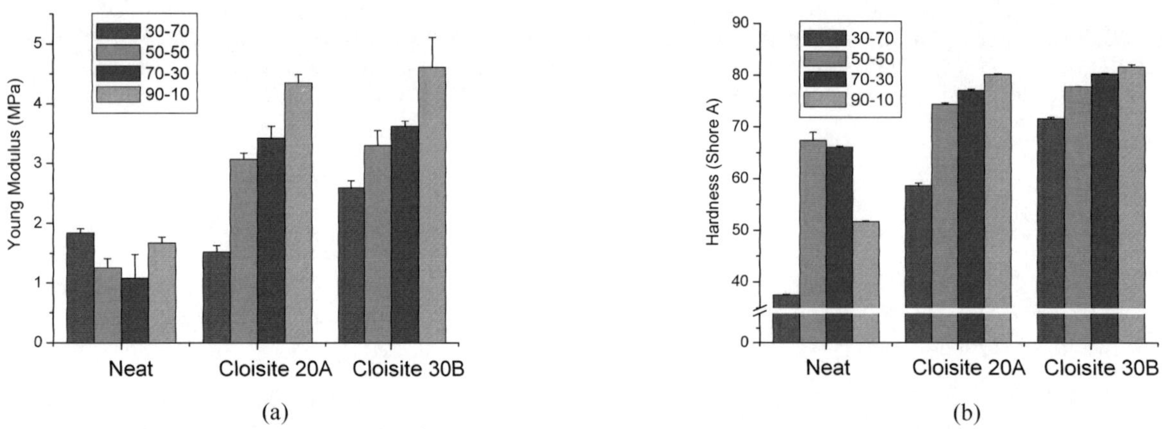

FIGURE 4. Young Modulus (a) and Hardeners (b) of neat bio-based epoxy networks and their nanocomposites containing Cloisite 20A and Cloisite 30B; the SGE/ECO ratio is given in the legend.

Figure 4 shows Young's Modulus and Shore A hardness data, respectively. The Young's Modulus is affected by the crosslink density in the elastic region, with the structure and flexibility of the chains between the crosslinks playing a role as well [15,16]. Likewise, it has been shown that hardness correlates well with modulus when the deformation during hardness testing is elastic [17]. The addition of SGE is expected to increase the crosslink density of the system [14], resulting in higher values of Young's Modulus and hardness; the same would be expected in both filled and unfilled systems in the rubbery state [18]. Moreover, the ability of clays to increase hardness in rubber has been reported as well [18]. On the other hand, DDA and ECO have greater affinity than DDA and SGE, resulting in higher epoxy conversion and more complete network formation when the ECO content is high. As a result of these competing factors, the highest Young's modulus for a neat network is observed when the SGE/ECO ratio is 30/70. Interestingly, however, this system also shows the lowest hardness, and more generally the hardness values do not track well with the Young's modulus values in the neat materials. One possible explanation for this could be fracture of the stiffer materials during hardness testing. This may be due to the phase behavior of the two bio-based epoxy components when blended with each other, with the resulting inhomogeneities reducing the strength of the materials. On nanoclay addition, on the other hand, well-correlated increases in Young's modulus and hardness are observed in all cases. This is consistent both with the reinforcing effects we would expect on nanocomposite formation and also the capacity of nanofillers to act as emulsifiers and to stabilize otherwise incompatible blends [19]

CONCLUSIONS

ECO/SGE/DDA networks and their nanocomposites display promise as easily processed elastomeric thermosets with enhanced sustainability vs. petroleum-derived materials. XRD provides information concerning intercalation of the individual network components into the nanoclays as well as the nanostructure of the cured systems, and gives evidence of interactions with both nanoclays regardless of the ratio of bio-epoxy components. Near-IR spectroscopy allowed us to assess the extent of epoxy conversion in these networks, and highlighted the differences between the nanoclays as far as their effect on network formation was concerned – results that may be rationalized in terms of changes in dispersion state. Superior mechanical properties vs. the unfilled networks were observed upon nanocomposite formation in all cases, with higher values of Young's modulus and hardness observed with increasing SGE content in spite of reductions in epoxy conversion. The mechanical properties reported here are comparable to petroleum-based elastomeric aromatic epoxy nanocomposites [20], and emphasize the need to more comprehensively evaluate the potential of these newly developed elastomeric bio-based epoxy systems as a replacement for current elastomeric thermosets in industrial applications.

ACKNOWLEDGMENTS

The authors express gratitude to the technical team of the Laboratory of Testing Materials (LEMAT) and Center of Nanotechnology Research and Development (CIDNA) at ESPOL. This research has been supporting by the Office of the Dean of Research at ESPOL. The authors would also like to thank Heloxy and Croda for kindly supplying materials for this research.

REFERENCES

1. J.E. Mark, *Thermoset Elastomers*. Applied Plastics Engineering Handbook 109-110, 2017
2. N. Wilmot, Duggal R, Medina J.C. Shah H, Schrock A. US Patent 9,006,385,385 B2
3. S.H. Goodman, *Epoxy resin Handbook of Thermoset Plastics*. William Andrew, 193-268, 1999
4. P. Krebaum, I. Lisle. USD Patent 5,952,446. Sep. 14, 1999
5. L. Corley, C.Ash. US Patent 9,840,588 B2 , Dec. 12, 2017
6. O. Zabihi, M. Ahmadi, S. Nikafshar, K. Chamdrakuma Preyesway, M. Naebe. *Composites Part B: Engineering* **134**, 1-24 (2018)
7. C. Beausoleil, E. Pasquier, C. Rousselle, C. Michel. *Molecular and Cellular Endocrinology*, **475**, 1-3 (2018)
8. A. Rigail-Cedeño, D.F. Schmidt. *AIP Conference Proceeding* 1914, 030023 ,2017
9. H. Salam, Y. Dong, I. Davies. *Woodhead Publishing Series in Composites Science and Engineering*, 101-132 (2015)
10. A. Rigail-Cedeño. Daniel F Schmidt, M. Chavez, G. Vera. *Abstract of paper of The Americana Chemical Society* **251** (2016)
11. H.C Greenwell, M. J Harvey, P. Boulet, A. A. Bowden, P. V. Covenery, A. Whitting, *Macromolecules* **38**, 6189-6200 (2005)
12. B. Soares, A. Silva, A.Solllymossy, R. Braga *Journal of Applied Clay Science* **83-84**, 244-252 (2013)
13. A. Rigail-Cedeño, A. Diaz-Barrios, J. Caceres, J. Pilataxi, M. Vera, and D. Vera, AIP Conference Proceeding to be published.
14. A. Rigail-Cedeño, Daniel F. Schmidt. *SPE- ANTEC* 2061-2865 (2015)
15. L. Shan, K. Verghese, C Robertson, K. Reinfsnider *Journal of Polymer Since: Part B, Polymer Physics* **37**, 2815-2819 (1999)
16. G. Kwak, S Parkm J. Lee. *Journal of Applied Polymer Science* **78**, 290-297 (2000)
17. A. Mix, J. Giacomin. *Polymer-Plastics Technology and Engineering* **50**, 288-291 (2011)
18. T. Ngo, N. Ton-That, S. Hoa, K. Cole Journal of Applied Polymer Science **107**, 1154-1162 (2008)
19 R. Iyer, S.Suin,, N.Shrivastava, S. Maiti, B. Khatua. *Polymer-Plastics Technology and Engineering* **52** 514-534 (2013)
20. M. Lipinska, J. Hutchinson. *Composite Science and Technology* **72**, 640-646 (2012)

Characterization of Nanoparticle Dispersion and Optimal Mixing Conditions for Polymer Nanocomposites

Jayadurga Iyer Ganapathi[a], Frank Fisher[b] and Dilhan Kalyan[c,d*]

[a]Intel, Hillsboro, Oregon 97124

[b]Department of Mechanical Engineering, Stevens Institute of Technology, Hoboken, New Jersey 07030

[c]Department of Chemical Engineering and Materials Science, Stevens Institute of Technology, Hoboken, New Jersey 07030

[d]Department of Biomedical Engineering, Stevens Institute of Technology, Hoboken, New Jersey 07030

Abstract. Different mixing processes and conditions give rise to significant differences in the agglomeration and deagglomeration of nanoparticles and how they are distributed and dispersed within a polymeric matrix. Homogeneous spatial distribution and effective nanoparticle dispersion are necessary to achieve optimal polymer nanocomposite properties. The focus of the current work is the identification of optimal mixing conditions for dispersion of CNTs and the impact of ultrasonication conditions on the dispersion and distribution of CNTs in the polymer nanocomposite. Here polymer nanocomposite samples of poly(caprolactone) (PCL) were compounded with carbon nanotubes (CNTs) using a solution processing technique where the samples are subjected to different durations of CNT/solvent (Stage 1) sonication. In addition to more standard characterization techniques, an analysis of the mixing indices and the shear-induced crystallization behavior of the polymer nanocomposites were used to study the effects of sonication time on nanoparticle dispersion within the samples. In particular, shear-induced crystallization has been shown to be highly sensitive to nanoparticle dispersion as crystal nucleation is highly sensitive to the amount of surface area provided by the embedded nanoparticles. Overall, various characterization results suggest that effective properties initially increase and then reach an optimum state as a function of processing time, after which further sonication leads to a decrease in the corresponding properties. The results suggest that mechanisms of demixing and damage of the CNTs co-exist and are responsible for the reduction in effective properties for excessive sonication times.

Keywords: polymer nanocomposites, ultrasonication, dispersion, shear-induced crystallization

PACS: Replace this text with PACS numbers; choose from this list: http://www.aip.org/pacs/index.html

1. INTRODUCTION

Incorporation of nanoinclusions into polymeric binders is a technological challenge. This is especially true for carbon nanotubes (CNTs) due to their very high aspect ratios (typically in the range of 500 to 1000) as well as their flexibility and ability to entangle with each other.[1-3] The success of incorporation of CNTs into a polymeric binder to achieve targeted functional end-use (ultimate) properties of the nanocomposite depends on the dispersion (agglomerate sizes) and distribution (spatial homogeneity) states of the CNTs within the nanocomposite. There are several processing techniques that can be used to incorporate CNTs into a polymeric binder. Such processing methods include ultrasonication, internal batch mixers, extruders, or in-situ polymerization methods.[1-2] Commonly, ultrasonication is employed for dispersion and stabilization of CNTs in the solvent and polymer.[4] While most studies of CNT nanocomposites found that the CNTs agglomerates can be broken down and the nanotubes separated from each other during the earlier stages of mixing, some reports suggest CNTs can be either damaged or demixed into various agglomerate shapes and sizes during subsequent deformation and processing.[4-5] For this study, multiwall carbon nanotubes (MWCNTs) were dispersed in a polycaprolactone (PCL) polymer matrix using a solution processing technique. The CNTs were incorporated into dichloromethane using a probe sonicator in conjunction with different durations of sonication (15, 30, 60 and 90 minutes). Using a variety of techniques, the dispersion and distribution states of CNTs were characterized while documenting the effects on the ultimate properties of the polymer nanocomposite.

2. EXPERIMENTAL

2.1 Materials

Polycaprolactone (PCL) was purchased from Scientific Polymers (CAT 1070 with molecular weight of 70,000 g/mol and solid density of 1145 kg/m3). Dichloromethane (DCM) (Pharmaco-Aaper, reagent grade) was used as the solvent for PCL. Multi-walled carbon nanotubes (CNTs) were obtained from Cheap Tubes Inc. of Boston, MA (density of 2100 kg/m^3, with reported elemental composition of 98.4% C, 0.45% Cl, 0.26% Fe, and 0.94% of Ni). The nanotubes typically exhibited outside diameters of 20-30 nm and lengths of 10 to 30 μm, resulting in an aspect ratio range of approximately 330-1500.

2.2. Solution Mixing of CNTs with PCL & Sample Preparation

Ultrasonication is a mixing technique where mechanical vibrations are transmitted through a horn tip immersed in a liquid solution. These vibrations generate ultrasonic cavitations which help to overcome the binding energy between nanotube surfaces which leads to deaggregation of the CNTs. The sonication apparatus used here was a Misonix XL2020 ultrasonic unit. Here a two-stage sonication procedure was applied. During the first stage in the mixing procedure (referred to as "Stage 1" sonication) the CNTs were sonicated and dispersed in DCM (0.05 g CNT with 50 ml of DCM) for 15, 30, 60, or 90 minutes as noted.[6-7] Separately, the dissolution of 9.95g of PCL in 100 ml of DCM was carried-out using a magnetic stirrer for 50 minutes under ambient temperature conditions. The Stage 1 CNT/DCM mixture was then directly added to the PCL/DCM solution, with the combined mixture (PCL/CNT/DCM) further sonicated for an additional 10 minutes (referred to as "Stage 2" sonication). For all sonication operations the power input to the sonicator was estimated to be in the range of 60–70W. It was verified using a thermal imaging camera (Inframetrics PM290) that under these power conditions the temperature of the suspension did not rise significantly. The resulting sonicated mixtures (i.e., after Stage 2) were vacuum dried at 50°C for 4 days until all solvent was evaporated. After complete drying (as verified with successive weight measurements) the nanocomposite samples were stored in low-density polyethylene bags in a desiccator to be used for molding and characterization as described below.

2.3 Microscopy

Microscopic methods in conjunction with image analysis are effective methods for the determination of agglomerate sizes and their distributions.[8-15] Optical characterization was carried out using a Nikon Eclipse E1000 microscope. For optical microscopy, 2μl drops of PCL/CNT/DCM obtained following Stage 2 sonication were placed on glass slides that were covered with glass cover slips. Optical images were obtained at 20X magnification and recorded using a digital camera at room temperature. For the analysis of microscopy images ImageJ software was used. The images were digitized to 8 bit and then subjected to binarization.[8-10] Binarization was carried out using the Triangle Thresholding algorithm of ImageJ.[10] Using this technique, the number and diameters of CNT aggregates and their area fractions were determined.

2.4 Rheological Characterization

It is known that the shear-induced crystallization behavior of polymer nanocomposites can be quite sensitive to the state of dispersion of CNTs within a semicrystalline polymer.[6-7] During shear-induced crystallization the nanoparticles can act as heterogeneous nuclei and alter the development of crystallinity for nanocomposites of CNTs dispersed into semi-crystalline binders.[16-18] The shear-induced crystallization tests were conducted using an ARES rheometer. The rheometer was equipped with a force rebalance transducer (2K-FRTN1) and an environmental chamber which maintains a constant temperature within ±0.1°C. For the shear-induced crystallization behavior characterization, the Stage 2 samples were first melted at 80°C for 5 minutes in the rheometer. A gap of 1 mm was then set between two parallel discs (25 mm diameter) and the excess material was trimmed. The environmental chamber was then closed again and the temperature allowed to re-equilibrate at 80°C post-trimming for at least 5 minutes. The crystallization temperatures of the pure PCL and the 0.5 wt% PCL/CNT nanocomposite are 28°C and 38°C, respectively.[7] After waiting approximately 2 minutes following thermal equilibration at 80°C, the shear-induced crystallization experiments were started by cooling the samples to the target temperature of 55°C at a cooling rate of 15°C/min. The dynamic properties were then collected as a function of time at 1% strain amplitude and 1 rad/s. The significant increases in the dynamic properties as a function of time are associated with the shear-induced crystallization behavior of the PCL/CNT nanocomposite. For each condition, three samples were tested.

2.5 Mechanical Properties

To evaluate the effects of the CNT dispersion and distribution states on mechanical properties, tensile properties of the PCL and 0.5wt% PCL/CNT nanocomposite samples were characterized. The storage moduli at room temperature were characterized using an INSTRU-MET floor model Instron with Testworks material software, with tests performed according to the ASTM D882. Rectangular samples with a thickness of 0.87 mm, width of 21.8 mm, and a gage length of approximately 25 mm were used for the tensile tests. The specimens were tested in displacement control mode at a crosshead speed of 0.25 mm/min. For each mixing condition, three samples were tested.

3. RESULTS

3.1 Morphological Characterization

Optical images have also been taken and used with image analysis to characterize the diameters and the area fraction occupied by the CNT agglomerates after 10 minutes of Stage 2 sonication. Figure 1(a-d) shows representative optical images of PCL/CNT samples subjected to different durations of Stage 1 sonication. It can be observed that as the Stage 1 sonication time increases the size of the CNT agglomerates decreases, with changes in the sizes of the agglomerates less discernible in samples subjected to longer sonication duration. For each Stage 1 sonication time, 24 images similar to those shown in Figure 2 were taken to analyze the distribution of the areas of the CNT agglomerates. Each of the images was analyzed using ImageJ software, with thresholding using the Triangle algorithm available within the software. Figure 1 (e) shows the distributions of the CNT agglomerate surface areas for different Stage 1 processed samples. Assuming that the surface areas of the agglomerates would be proportional to the volumes and hence the sizes of the agglomerates, Fig. 1 (e) suggests that Stage 1 sonication reduces the size of the agglomerates. It is observed that the number of agglomerates with surface areas less than 1-10 μm^2 increases as the Stage 1 sonication time is increased. Thus, the initial Stage 1 sonication plays an important role in the disentanglement of the CNTs and ultimately enables better distribution within the PCL matrix polymer.

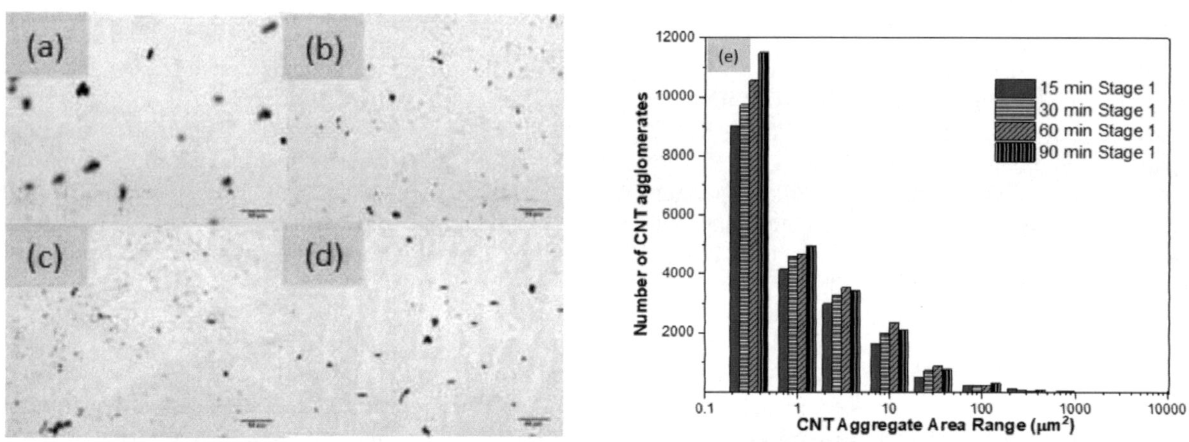

FIGURE 1. (Left) Optical images, respectively, of CNTs subjected to different Stage 1 sonication times: (a): 15 minutes; (b): 30 minutes; (c): 60 minutes; and (d): 90 minutes. (Right) (e) Distribution of CNT aggregate area from optical images of PCL/CNT samples subjected to different Stage 1 sonication times followed by 10 min of Stage 2 sonication. Optical Images taken at 20X and scale bars in the optical images represent 50 μm.

3.2 Shear-induced Crystallization Studies

Melts of semi-crystalline polymers undergo accelerated crystallization under shear and elongational deformations. CNTs can act as heterogeneous nucleating agents and accelerate the rate of crystallization under flow, and possibly also modify the morphology of the crystallites which form in comparison to those crystallized under quiescent conditions.[17-18] This is consistent with the overall finding in our previous work that the heterogeneous nucleation facilitated by the presence of the CNTs reduces the energy barrier required for polymer crystallization.[6-7] Thus the dispersion states as well as the sizes of the agglomerates of CNTs, and the associated changes in the surface areas of CNTs available for crystallite nucleation, can significantly impact the development of shear induced crystallization behavior.[6-7]

Figure 2 shows the typical flow-induced (small-amplitude oscillatory shear) crystallization behavior of PCL/CNT nanocomposite samples with 0.5 wt% loading of CNTs. The experiments were carried out via time scans of the linear viscoelastic material functions at 55°C with a constant strain amplitude of 1% and a constant frequency of 1 rad/s. It is observed in Figure 2 that for times less than 1000s (prior to the induction time where the storage modulus begins to increase), the 30 minute Stage 1 PCL/CNT samples exhibit the largest storage modulus values of the samples tested. From Figure 2, it is hypothesized that as the CNT aggregates are being broken down to a greater extent with increasing sonication time (Fig. 1e), greater CNT surface areas become available for nucleation, with the faster crystallization rate reflected in the corresponding rate of increases of the dynamic properties. The variation of the breadths of the confidence intervals of the dynamic properties with increasing sonication time is also noted.

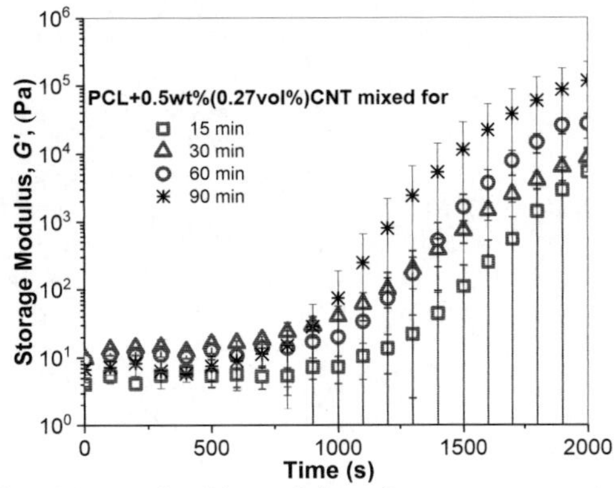

FIGURE 2. Average storage modulus G' as a function of time for nanocomposite samples undergoing shear-induced crystallization. Three samples for each condition were tested at 1% strain amplitude and 1 rad/s at 55°C. All samples were subjected to 10 minutes of Stage 2 sonication.

3.3 Mechanical Properties

Figure 3 shows the tensile modulus values of the PCL/CNT nanocomposites as a function of the Stage 1 sonication time. The tensile modulus of PCL/CNT samples subjected to 30 minutes of Stage 1 sonication is greater than those of PCL/CNT samples subjected to 15, 60, or 90 minutes of Stage 1 sonication. This trend confirms that the ultimate properties of the PCL/CNT nanocomposite initially increase as the dispersion and distribution of CNTs improves in the polymer, ultimately reaching an optimal value beyond which the elastic modulus and the tensile modulus decrease with increasing sonication time.

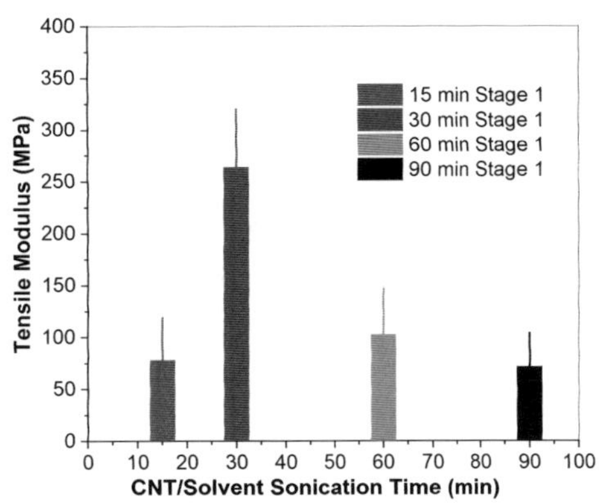

FIGURE 3. Tensile modulus of PCL/CNT samples with 0.5wt% CNT as a function of Stage 1 sonication times.

4. CONCLUSIONS

In this work, a systematic study to determine the optimal conditions for mixing/sonication of Stage 1 CNT/solvent and the resulting distributions (spatial homogeneity) and dispersions (agglomerate sizes) of CNTs within the polymer nanocomposite is described. Results from optical analysis and shear-induced crystallization suggest that the CNT agglomerate sizes monotonically decrease as the Stage 1 sonication time continues to increase. These optimal mixing conditions are also reflected in the ultimate tensile properties of the CNT/PCL nanocomposites, with these properties reaching maximum values at the optimal Stage 1 sonication time of 30 minutes. The documented demixing and reagglomeration that occurs at longer sonication times in turn leads to the deterioration of the mechanical properties.

ACKNOWLEDGMENTS

Microscopy resources within the Laboratory for Multiscale Imaging at Stevens were used in this work.

REFERENCES

1. Ma, P.-C.; Siddiqui, N. A.; Marom, G.; Kim, J.-K., *Composites Part A, 41* (10), 1345-1367 (2010).
2. Grady, B. P., *Macromol. Rapid Commun*, 31 (3), 247-257 (2010).
3. Pegel, S.; Pötschke, P.; Petzold, G.; Alig, I.; Dudkin, S. M.; Lellinger, D, *Polymer, 49* (4), 974-984 (2008).
4. Cheng, Q.; Debnath, S.; Gregan, E.; Byrne, H. J., *J. Phys. Chem. C, 114* (19), 8821-8827 (2010).
5. Yu, H.; Hermann, S.; Schulz, S. E.; Gessner, T.; Dong, Z.; Li, W. *Chem. Phys., 408*, 11-16 (2012).
6. Iyer Ganapathi, J.; Fisher, F. T.; Kalyon, D. M., *J. Polym. Sci., Part B: Polym. Phys. 54* (21), 2254-2268 (2016).
7. Iyer Ganapathi, J.; Kalyon Dilhan, M.; Fisher Frank, T., *J. Appl. Polym. Sci., 134* (14), 44681 (2017).
8. Alig, I.; Pötschke, P.; Lellinger, D.; Skipa, T.; Pegel, S.; Kasaliwal, G. R.; Villmow, T., *Polymer, 53* (1), 4-28 (2012).
9. Hui, L.; Smith, R.; Wang, X.; Nelson, J.; Schadler, L. In 2008 Annual Report Conference on Electrical Insulation and Dielectric Phenomena, IEEE: pp 317-320 (2008).
10. Kim, D.; Lee, J. S.; Barry, C. M.; Mead, J. L., *Microsc. Res. Tech., 70* (6), 539-546 (2007).
11. Lively, B.; Smith, P.; Wood, W.; Maguire, R.; Zhong, W.-H, *Composites Part A, 43* (6), 847-855 (2012).
12. Luo, Z.; Koo, J., *Journal of Microscopy, 225* (2), 118-125 (2007).
13. Bakshi, S. R.; Batista, R. G.; Agarwal, A., *Composites Part A, 40* (8), 1311-1318 (2009).
14. Khare, H.; Burris, D., *Polymer, 51* (3), 719-729 (2010).
15. Yazdanbakhsh, A.; Grasley, Z.; Tyson, B.; Al-Rub, R. K. A., *Composites Part A, 42* (1), 75-83 (2011).
16. Wurm, A.; Lellinger, D.; Minakov, A. A.; Skipa, T.; Pötschke, P.; Nicula, R.; Alig, I.; Schick, C., *Polymer, 55* (9), 2220-2232 (2014).
17. Chen, Y.-H.; Zhong, G.-J.; Lei, J.; Li, Z.-M.; Hsiao, B. S., *Macromolecules, 44* (20), 8080-8092 (2011).
18. Mago, G.; Fisher, F. T.; Kalyon, D. M., *Macromolecules, 41* (21), 8103-8113 (2008).

Release of Oregano Essential Oil from PHBV Active Films Containing Nanoclays

Renata Cerruti da Costa[a], Tales da Silva Daitx[b], Raquel Santos Mauler[b], Janaina da Silva Crespo[c], Marília Miotto[d] and Larissa Nardini Carli[e*]

[a] *Programa de Pós-Graduação em Ciência e Engenharia de Materiais, Universidade Federal de Santa Catarina – Santa Catarina, Brazil*

[b] *Instituto de Química, Universidade Federal do Rio Grande do Sul – Rio Grande do Sul, Brazil*

[c] *Área do Conhecimento de Ciência Exatas e Engenharias, Universidade de Caxias do Sul – Rio Grande do Sul, Brazil*

[d] *Departamento de Ciência e Tecnologia de Alimentos, Universidade Federal de Santa Catarina – Santa Catarina, Brazil*

[e] *Campus Blumenau, Universidade Federal de Santa Catarina – Santa Catarina, Brazil*

**larissa.carli@ufsc.br*

Abstract. The aim of the present work was the development of polymer films loaded with oregano essential oil (OEO)/clay nanoparticles for the active packaging of foods. The incorporation of OEO into poly(hydroxybutyrate-co-hydroxyvalerate) - PHBV films by melt blending is presented. In addition, three different clays (halloysite (Hal), kaolinite (Kaol), and sepiolite (Sep)) were added to the films to improve barrier properties of the polymer. The antimicrobial activity of systems was determined by turbidity method. Thermal properties (TGA and DSC) and oxygen permeability (OP) were also determined. Films with OEO effectively inhibited the growth of microorganisms and demonstrated improved barrier properties. Thermal analysis demonstrated that the addition of clays and OEO affected the thermal stability and increased crystallinity of the films. In conclusion, PHBV-based nanocomposite films containing OEO might be suitable to apply as an active packaging material to maintain the food quality and extend its shelf life.

Keywords: Active packaging, Essential oil, Poly(hydroxybutyrate-co-hydroxyvalerate), Clays, Nanocomposite, Controlled release.

INTRODUCTION

Recently, much research has been directed to the production of active food packaging by using materials that present antimicrobial and antioxidant characteristics. The use of polymeric films as carriers of such agents allows a progressive release of the compound on the surface of the product, thereby improving antimicrobial efficacy and enhancing food safety and quality throughout the storage [1]. Natural antimicrobial agents, such as essential oils, include extracts from various plants such as cinnamon oil, cloves, thyme, rosemary and oregano. Many studies have reported that oregano essential oil (OEO) has a strong potential to be used in food packaging as an antimicrobial agent [2].

The incorporation of essential oil into polymer matrices, however, can affect the physical properties of the materials depending on the polymer-oil interaction. To increase the effectiveness of functional packaging and improve the mechanical properties of PHBV films, one can also incorporate clay nanoparticles. The nanoparticles present a series of advantages such as atoxicity, improvement of the mechanical properties and formation of obstacles in the polymer matrix, capable of acting as a barrier to the diffusion of water vapor and oxygen, thus avoiding that the food loses its characteristics under the influence of the external environment [3].

Thus, this study aims to develop a new strategy for the production of active packaging with improved functional properties by the incorporation of an essential oil in PHBV films containing clay nanoparticles.

EXPERIMENTAL

Preparation of Active Nanocomposite Films

PHBV nanocomposite films were prepared by a direct melt processing using a Roller-Rotors R600, Rheomix 6002C mixer. Three different clays (halloysite (Hal), kaolinite (Kaol), and sepiolite (Sep)) and oregano essential oil (OEO) were added to the films in order to study the synergy between the additives on the improvement of barrier and antimicrobial properties of the systems. The total mass for each formulation was 50.0 g, with 3 wt% of clay and 8 wt% of OEO. The processing was performed at 100 rpm and 165 °C for 6 min. After blending, the obtained compounds were milled and compression molded at 190 °C for 2 min in a SL-11 Solab Equipamentos.

Characterization of Nanocomposites

The thermogravimetric analysis (TGA) allows visualization of the thermal stability of the materials as well as the maximum degradation temperature (T_d). The analyzes were conducted in a Perkin Elmer TGA 8000 equipment. The samples were heated from 30 °C to 700 °C with a heating rate of 20 °C/min under argon atmosphere (flow rate of 20 mL/min). In addition, with the TGA results it was possible to estimate the amount of oregano essential oil in the films after processing. The retained amount of OEO was calculated by the loss of mass associated with volatilization of the oil up to 270 °C. The incorporation efficiency was determined from Eq. 1.

$$\textit{Incorporation efficiency (\%)} = \frac{\textit{Percentage of oil measured}}{\textit{Theoretical percentage of oil incorporated}} \times 100 \qquad (1)$$

The crystallinity and thermal properties of the films were determined by differential scanning calorimetry (DSC) using a TA Instruments Q20 equipment. The materials were analyzed from 40 °C to 200 °C at a heating and cooling rate of 10 °C/min under inert atmosphere (50 mL/min N_2). The degree of crystallinity was determined using ΔH^0_m = 146 J/g for PHBV [4]. Oxygen permeability tests were performed at 23 °C with an oxygen flow of 100 cm^3/min, according to ASTM F1927-07 in a L100-5000 Systech Illinois equipment.

The antimicrobial activity of films was determined by qualitative analysis of turbidity in a medium susceptible to the growth of microorganisms. The formulations tested were the samples containing sepiolite (PHBV/Sep and PHBV/Sep/OEO), since these films presented higher release rate of OEO in the dissolution tests. Three tubes containing 9 mL of liquid culture medium (BHI broth) were prepared. One tube received the sample without oil (PHBV/Sep), and in another tube it was added the PHBV/Sep/OEO sample. A third tube was kept contaminated with 100 μL of *E. Coli* 10^4 UFC/mL for reference, and a forth tube was prepared containing only the BHI broth as a negative control. The tubes were shaken for 24 h at 25 °C and the inhibition was visually analyzed, recognizing the microbial growth by the turbidity of the medium [5].

RESULTS AND DISCUSSION

The results of the TGA analysis for the PHBV film are shown in Figure 1(a). For the PHBV, only one mass loss is observed in the range of 285-319 °C, associated to degradation of the polymer matrix [4]. Figure 1(b) shows the thermogram of OEO. The thermal decomposition of the oil starts at ~80 ºC, with a maximum loss of mass at ~198 ºC. At 270 ºC it already shows total volatilization of the oil. This information corresponds to that established by the literature for the boiling temperature of the oil at 246ºC [2].

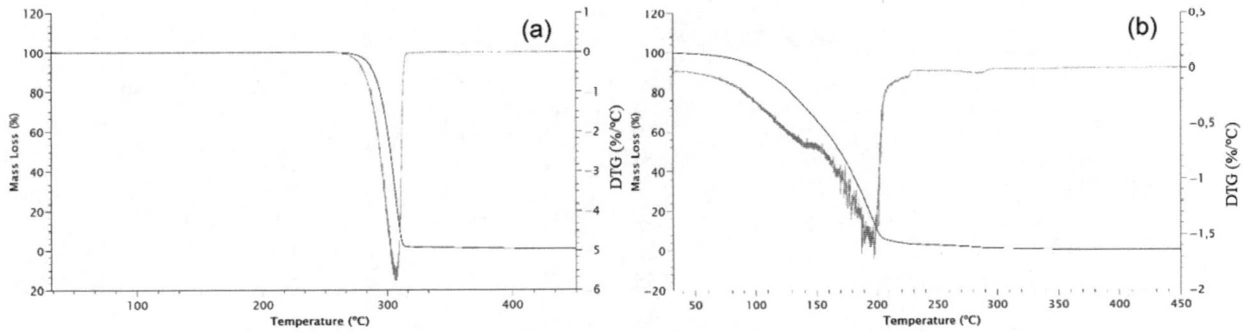

FIGURE 1. Thermogram of pure PHBV film (a) and oregano essential oil (b).

Figure 2 shows the thermograms of the nanocomposites containing OEO. It is possible to observe the mass loss related to the percentage of oil in the initial stages of the analysis, up to 270 °C. Above this temperature, the degradation behavior follows that expected for PHBV films: a mass loss starting at approximately 287 ° C.

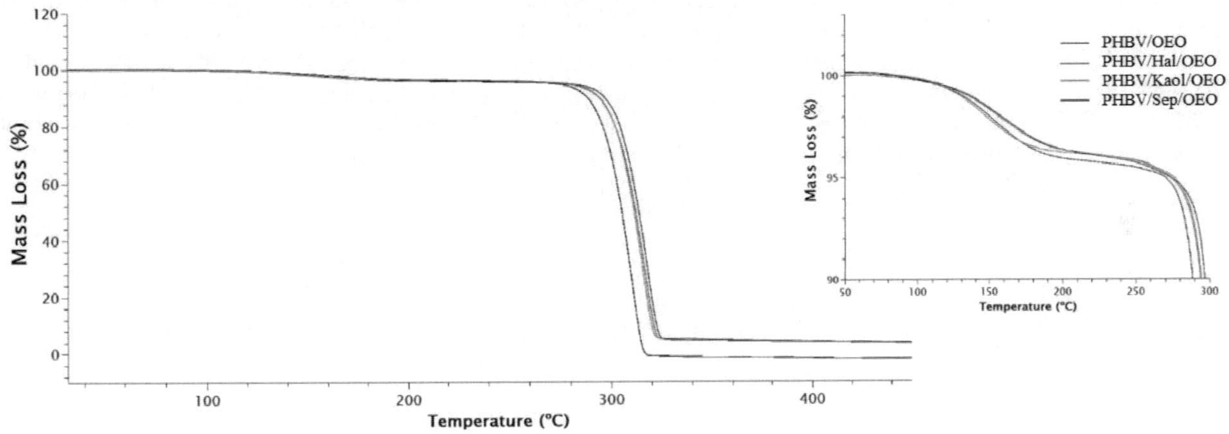

FIGURE 2. Thermogram of samples with oregano essential oil.

The oil retained in the films after processing corresponds to the difference between the mass percentage at the beginning of the analysis and the percentage observed at 270 ° C. Thus, it was possible to estimate through TGA analysis the incorporation efficiency of the oregano essential oil in the films. The results (Table 1) demonstrate good oil incorporation efficiency in the prepared materials, reaching 61.9% for PHBV/OEO and 61.1% for PHBV/Sep/OEO. Considering the temperature during the melt processing (165 °C) and the compression molding of the films (190 °C), a partial volatilization of the essential oil of oregano was expected.

TABLE (1). Determination of incorporation efficiency of oregano essential oil in the films after processing.

PHBV films	%wt at 270°C	Incorporated oregano (%)	Efficiency of incorporation (%)
PHBV/OEO	94.98	4.95	61.9
PHBV/Hal/OEO	95.17	4.69	58.6
PHBV/Kaol/OEO	95.25	4.74	59.2
PHBV/Sep/OEO	95.11	4.89	61.1

Table 2 presents the DSC results for PHBV samples in the presence and absence of oregano oil. It is possible to verify an increase of T_c in all the films where the nanoparticles were incorporated. This increase may indicate a nucleation effect of the clays on the crystallization of PHBV. In addition, it is also possible to observe that the crystallinity values of the films increased significantly, reaching values of 80% for the PHBV/SEP compared to the degree of crystallinity of 66% for the pure polymer.

TABLE (2). Thermal data obtained by DSC for the samples

Films	T_c (°C)	T_{m1} (°C)	T_{m2} (°C)	ΔH_m (J/g)	Xc (%)
PHBV	71	166	174	97	66
PHBV/OEO	70	164	171	112	84
PHBV/Hal	79	167	174	109	77
PHBV/Hal/OEO	78	164	171	103	79
PHBV/Kaol	79	167	174	110	77
PHBV/Kaol/OEO	74	164	172	100	77
PHBV/Sep	89	167	174	113	80
PHBV/Sep/OEO	85	165	172	103	79

The characterization of the PHBV/Clay/Oil films was also performed for the oxygen barrier property. Oxygen can cause oxidation, which is an early stage of various kinds of food spoilage. A film that acts as an adequate barrier to oxygen can help to improve its quality and extend its useful life [1]. Figure 3 shows the O_2 permeability result of the different samples. The pure PHBV film shows a permeability value of 392.39 mLmm/m^{2}d. After the incorporation of the different clays in the PHBV film, the values were significantly reduced, reaching 54.27 mLmm/m^{2}d for the nanocomposite PHBV/Sep. This is due to the barrier characteristic of nanoparticles and is consistent with most studies in the area [1, 6, 7]. When OEO was incorporated in these nanocomposites, a remarkable reduction of permeability was also observed and may be due to the increase of the crystallinity of the films compared to the films without oil.

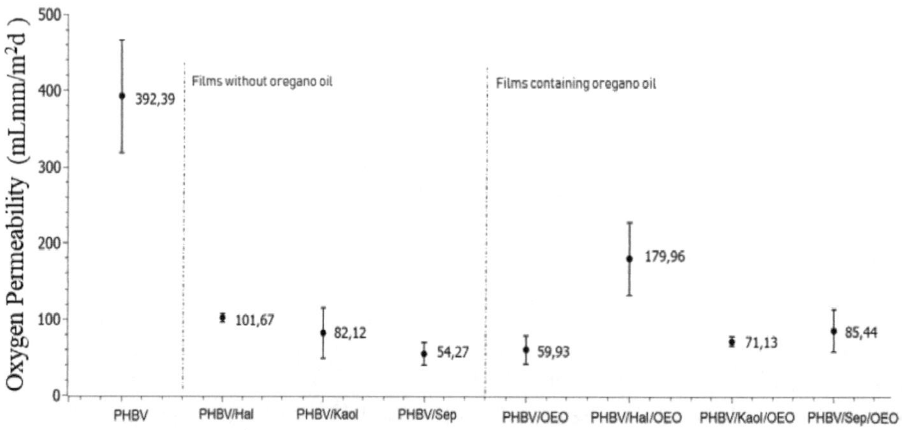

FIGURE 3. Results of oxygen permeability of the samples.

The antibacterial activity of nanocomposite films with and without OEO are shown in Figure 4. All tubes contain the BHI medium which serves as a nutrient for the growth of bacteria. After 24 h, it was possible to observe that in the tube containing PHBV/Sep/OEO there was no microorganism growth since there was no change in solution turbidity. However, for PHBV/Sep (without OEO), it was observed the growth of microorganisms and an increase in the turbidity of the solution. With this result it was possible to confirm that the films incorporated with essential oil inhibited the growth of bacteria, thus controlling the contamination of the foods.

FIGURE 4. Images of the microbial activity test in BHI broth for the PHBV/Sep and PHBV/Sep/OEO.

CONCLUSION

PHBV films with different types of clays were obtained by melt processing and compression molding. Films containing oregano essential oil (OEO) were also prepared. The properties of the PHBV films were affected by the addition of clay nanoparticles. Nanocomposite films showed higher crystallinity and thermal stability, compared with the control film (PHBV). The addition of OEO in PHBV significantly affected their antibacterial activity.

These results suggest that the physical, thermal and antibacterial properties of the PHBV films can be modified by the type of clay and the combination with essential oil, being possible to obtain polymer film with antimicrobial properties for use in active packaging.

ACKNOWLEDGMENTS

The authors are grateful to Conselho Nacional de Desenvolvimento Científico e Tecnológico (CNPq) and Coordenação de Aperfeiçoamento de Pessoal de Nível Superior/Fundação de Amparo à Pesquisa e Inovação de Santa Catarina (CAPES/FAPESC).

REFERENCES

1. P. Klangmuang and R. Sothornvit. *Food Hydrocolloids* **61**, 609-616 (2016).
2. R. Requena et al., *Polymer International* **65**, 883–891 (2016).
3. M.J. Khalaj et al., *Trends in Food Science & Technology* **51**, 41-48 (2016).
4. L. N. Carli et al., *Polymer International* **64**, 235-241 (2015).
5. A. Mickel; T. Nguyen and S. Chogle. *Journal of Endodontics* **29**, 257–258 (2003).
6. M. Farmahini-Farahani et al., *Applied Clay Science* **135**, 27-34 (2017).
7. S. Malmir et al., *Composites Part A: Applied Science and Manufacturing* **93**, 41–48 (2017).

Effect of Non-covalent Interactions on Strain Sensing Performance of 3D Printed Conductive Polymer Composites

Dong Xiang [a*], Xuezhong Zhang [a], Yuntao Li [a]

[a]*School of Materials Science and Engineering, Southwest Petroleum University, Chengdu 610500, China*

**Corresponding author: dxiang01@hotmail.com*

Abstract: The use of conductive polymer composites as strain sensors has been widely investigated due to their excellent elasticity and sensitivity. Such sensors may be manufactured using additive manufacturing techniques but there are some challenges to overcome in terms of performance if this technique is to be used. In this work. highly elastic strain sensors of carbon nanotube thermoplastic polyurethane (CNT/TPU) nanocomposites were fabricated using fused deposition modeling (FDM), and 1-pyrenecarboxylic acid (PCA) was introduced to non-covalently modify the CNTs and improve the polymer-nanofiller interactions. The results show that the electrical and sensing properties of the modified nanocomposites are significantly Increased as a result of more uniform CNT dispersion. The 3D printed sensors demonstrate outstanding performance with high sensitivity, large workable strain, excellent repeatability and regular signal responses within a wide measuring frequency range of 0.01-1 Hz. Also, the strain sensing ability of the sensor is greatly improved with the introduction of PCA.

Keywords: 3D printing; interfacial interactions; sensor; carbon nanotubes; polymer nanocomposites

INTRODUCTION

Recently, the use of conductive polymer composites (CPCs) as flexible strain sensors has been widely reported due to their excellent sensitivity, high elasticity and large workable range[1,2]. 3D printing has the advantage of low cost and mass production, which makes people develop polymer composite printing technology [3,4]. Although these studies have demonstrated the fabrication of flexible strain sensors by 3D printing technology[5-8], the interfacial properties of composites used for 3D printing are usually ignored. In this work, CNT/TPU nanocomposites were prepared by solution mixing and extruded into filaments using a single-screw extruder for fabricating flexible strain sensors by FDM. PCA was used to non-covalently modify the nanotubes in order to improve their dispersion and alter polymer-nanofiller interactions without destruction of the intrinsic structure of the carbon nanotubes [9]. The electrical and electromechanical properties of the 3D printed nanocomposites were also systematically investigated. It is observed that the strain sensors printed with non-covalent interactions exhibit better electrical conductivity and strain sensing performance

EXPERIMENTAL

Materials

Multi-walled carbon nanotubes (NC7000) were obtained from Nanocyl SA (Belgium). The CNT has a nominal average length of 1.5 μm and a nominal diameter of 9.5 nm. 1-pyrenecarboxylic acid was sourced from EA Corp with a molecular weight of 246.26 and a purity of 97%, and the density of the PCA is 1.85 g/cm^3. Thermoplastic polyurethane (TPU) in powder form (Bayer 2195) was supplied by Bayer Co. Ltd. with a melt flow index of 17.5g per 10 min (215 °C, at a pressure of 10 kg) and a density of 1.12 g/cm^3. Dimethylformamide (DMF) reagent was purchased from Chengdu Kelong Chemical Reagent Company (China).

Preparation of CNT/TPU Nanocomposites

Firstly, CNTs and PCA were dispersed in DMF via ultra-sonicating at 100 W and 40 kHz for 1 hour at room temperature. The weight ratio of CNTs and PCA is 4:1. TPU powder was dissolved in the resulting suspension by magnetic stirring for 2 hours (10 g TPU per 100 ml DMF). The mixed suspension was transferred into a glass container (140 mm×140 mm×40 mm) to dry for 24 h at 100 °C in an air-circulating oven in order to produce nanocomposite sheets. It should be noted that the nanocomposites without the addition of PCA were also prepared using the same method for comparison (Figure. 1a).

3D Printing of Nanocomposites

The nanocomposite sheets prepared were cut into small granules and fed into a benchtop single-screw extruder (Wellzoom Type C) to produce filaments with a diameter of 1.75 mm suitable for FDM 3d printing. The operating temperature of the extruder was 190 ℃, and the screw speed was 100 rev/min. CNT/TPU filaments were printed using an ET-K1 (ET Co. Ltd., China) desktop FDM 3D printer. A stacking mode was used with an interlayer angle of 90° (Figure 1.b).

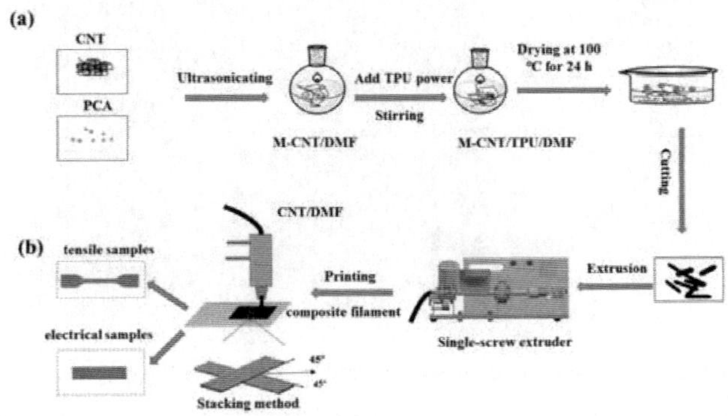

FIGURE 1. Schematic of (a) preparation process of CNT/TPU nanocomposites and (b) 3D printing of the nanocomposites.

Characterization

Field-emission Scanning Electron Microscopy (FE-SEM)

The morphologies of cast and printed nanocomposites were examined by FESEM (FEI Quanta 650 FEG) under an accelerating voltage of 20 kV. The specimens were cryogenically fractured perpendicular to the axial direction in liquid nitrogen. Then the fractured cross-section was gold sputtered before observation.

Electrical Resistivity of the 3D Printed Nanocomposites

The volume electrical resistivity (ρ) of the CNT/TPU nanocomposites was measured in cross-layer and through-layer directions using a two-point method combined with a picoamp-meter (Keithley 6485) and DC digital source meter (Tektronix PWS4323) at 3 V. The material strips had a dimension of 50 mm × 10 mm, and the electrode distance was 15 mm. Silver paste was used to minimize contact resistance between the specimen and electrodes.

Strain Sensing Measurement

The electromechanical performance of the printed strain sensors was measured using a MTS CMT4104 universal tester combined with a picoamp-meter and a digital source meter at a fixed voltage of 3 V. In order to precisely characterize the relationship of strain and relative resistance change ($\Delta R/R_0$), the dumbbell specimens for tensile testing were used, and the electrode distance was 20 mm.

RESULTS AND DISCUSSION

Morphologies of 3D Printed CNT/TPU Nanocomposites

The morphologies of cast and 3D printed samples were investigated by FESEM at different magnifications, as shown in Figure. 2. A clear multi-layer structure with a layer thickness of about 0.1 mm can be seen in the printed nanocomposites (Figure. 2c and e) compared with the cast nanocomposite (Figure. 2a). In addition, there are more voids in the printed samples in spite of good interlayer adhesion. It is observed in Figure. 2b and Figure. 2d that many small nanotube agglomerates are uniformly distributed in the cast and printed CNT/TPU nanocomposites, while there are very few agglomerates in the M-CNT/TPU nanocomposite (Figure. 2f). This demonstrates that the dispersion of CNTs in the TPU matrix is improved with the addition of PCA.

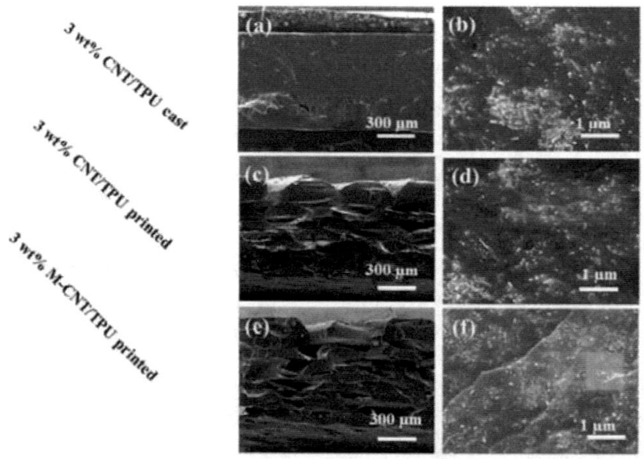

FIGURE 2. SEM micrographs of fractured cross-sections for the cast and 3D printed samples with 3 wt% MWCNTs: (a, b) cast CNT/TPU nanocomposite, (c, d) printed CNT/TPU nanocomposite, (e, f) printed M-CNT/TPU nanocomposite. It should be noted that the red circles indicate CNT agglomerates.

Electrical Conductivity of 3D Printed CNT/TPU Nanocomposites

The electrical resistivity, in different directions, of cast and printed nanocomposites is shown in Table 1. It can be seen that the forming method has a great influence on the resistivity of material with resistivity increasing in both directions in the printed material [10]. This can be attributed to the orientation of the CNTs during extrusion and 3D printing which results in the reorganization of the conductive network structure with preferential alignment in the extrusion/print direction. The addition of PCA results in a 37 fold drop in resistivity of the printed material compared with the unmodified 3 wt% CNT/TPU material and the modified material resistivity is also less sensitive to direction due to the enhanced dispersion of CNTs in the modified material. Comparing the modified material containing 3 wt% CNTs with that containing 1.5 wt% CNT we observe an increase in resistivity of approximately one order of magnitude which is to be expected as there are fewer CNTs in the network.

Table 1. The resistivity of nanocomposite materials in different directions.

Sample	Resistivity (Ω·m)	
	Through-layer	Cross-layer
3 wt% CNT/TPU cast	2.253×10^{2}	2.247×10^{2}
3 wt% CNT/TPU printed	1.516×10^{4}	4.285×10^{3}
3 wt% M-CNT/TPU printed	5.730×10^{2}	1.151×10^{2}
1.5 wt% M-CNT/TPU printed	1.059×10^{4}	1.765×10^{3}

Electromechanical Performance of 3D Printed CNT/TPU Nanocomposites

From Figure. 3, the sensor exhibited a high sensitivity (GF = 117213 at a strain of 250% for the printed sensor of 1.5wt% M-CNT/TPU) and a wide workable strain range [11,12]. Also, this sensor exhibited regular frequency responses within the test frequency range (0.01~1 Hz) indicating the potential for monitoring frequency changes. In cyclic stretching/release tests, shoulder peaks were observed in the ΔR/R0 results which were enhanced with an increase in the applied strain. In addition, this sensor exhibits excellent stability and repeatability during the cyclic strain testing up to 1000 cycles. The strain sensing properties of the sensor are greatly improved by the strong non-covalent interactions due to the introduction of PCA.

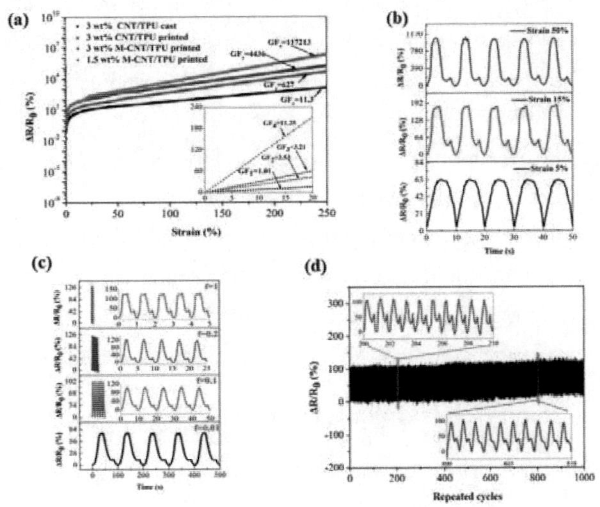

FIGURE 3. Relative resistance change of the strain sensors with different CNT content under monotonic loading; (b). different strain; (c). a cyclic strain of 10% and a frequency of 0.01~1 Hz; (d). during 1000 cycles at a strain of 10% and a frequency of 0.1 Hz.

Applications of the 3D Printed Sensors

3D printed sensors demonstrate the ability to monitor human motion, including finger movements (see Figure. 4a), joint motion (see Figure. 4b), respiratory rate (see Figure. 4c), and speech recognition (see Figure. 4d). This work provides a guideline for producing 3D printed high-performance flexible strain sensors with potential applications in intelligent robots, prosthetics and wearable devices where complex design and customizability are demanded.

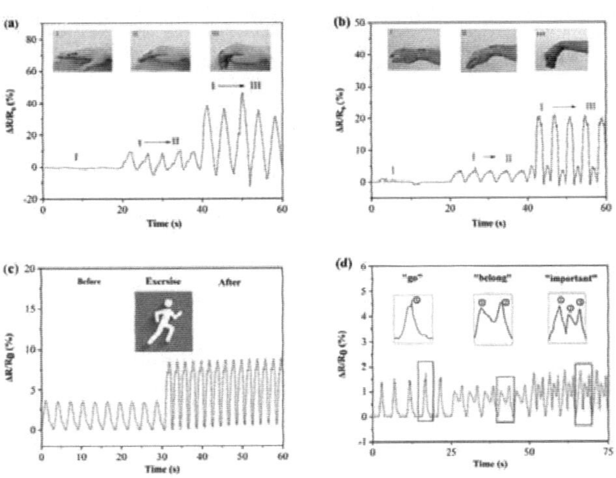

FIGURE 4. Electromechanical responses of the printed sensor to repeated (a) bending and unbending of the index finger at 30°and 90, (b) wrist, (c) breathing and (d) speaking "go", "belong", and "important".

CONCLUSIONS

In this work, highly elastic strain sensors of CNT/TPU nanocomposites were fabricated by FDM, and 1-pyrenecarboxylic acid was introduced to non-covalently modify the CNTs and enhance the interfacial interactions. The dispersibility of the CNTs in the TPU matrix is improved with the introduction of PCA. Compared with the printed 3 wt% CNT/TPU nanocomposite, the resistivity of M-CNT/TPU decreased by 37 times in the cross-layer direction at the same CNT loading. The sensor exhibited a high sensitivity (GF = 117213 at a strain of 250% for the printed sensor of 1.5wt% M-CNT/TPU) and a wide workable strain range. Also, this sensor exhibited regular frequency responses within the test frequency range (0.01~1 Hz) indicating the potential for monitoring frequency changes. In addition, this sensor exhibits excellent stability and repeatability during the cyclic strain testing up to 1000 cycles. The strain sensing properties of the sensor are greatly improved by the strong non-covalent interactions due to the introduction of PCA. We also demonstrated the sensor's capability in monitoring human motions, including finger movements, joint movements, respiratory frequency, and speech recognition. This work provides a guideline for producing 3D printed high-performance flexible strain sensors with potential applications in intelligent robots, prosthetics and wearable devices where complex design and customizability are demanded.

ACKNOWLEDGEMENTS

The authors are grateful for the financial support from the Department of Science and Technology of Sichuan Province (2017HH0086, 2017JY0152).

REFERENCES

[1] Chen Q, Cao P-F, Advincula RC. Advanced Functional Materials. 2018.

[2] Guo Z, Liu H, Li Y, Dai K, Zheng G, Liu C, Shen C, Yan X, Guo J. Journal of Materials Chemistry C. **4**, 157–166 (2015).

[3] Singh P, Campidelli S, Giordani S, Bonifazi D, Bianco A, Prato M. Chem Soc Rev. **38**, 2214–30 (2009).

[4] Sarlak N, Taherifar A, Salehi F. Journal of Agricultural & Food Chemistry. **62**, 4833–8 (2014).

[5] Wei H, Li K, Liu WG, Meng H, Zhang PX, Yan CY. Advanced Engineering Materials. 1700341 (2017).

[6] Giffney T, Bejanin E, Kurian AS, Travas-Sejdic J, Aw K. Sensors & Actuators A Physical. **259** (2017).

[7] Guo SZ, Qiu K, Meng F, Park SH, Mcalpine MC. Advanced Materials. **29**, 201701218 (2017).

[8] Muth JT, Vogt DM, Truby RL, Mengüç Y, Kolesky DB, Wood RJ, Lewis JA. Advanced Materials. **26**, 6307–6312 (2014).

[9] Simmons TJ, Bult J, Hashim DP, Linhardt RJ, Ajayan PM. Acs Nano. **3**, 865–70 (2009).

[10] Christ JF, Aliheidari N, Ameli A, Pötschke P. Materials & Design. **131** (2017).

[11] Cravanzola S, Cesano F, Muscuso L, Scarano D, Zecchina. Springer International Publishing Switzerland, P 51 **156**:51–73 (2015).

[12] Zhang R, Deng H, Valenca R, Jin J, Fu Q, Bilotti E, Peijs T. Composites Science & Technology. **74**, 1–5 (2013).

Effect of Halloysite Nanotube on Mechanical and Biological Properties of Polylactic Acid Scaffolds Fabricated by Foam Injection Molding

Meltem Eryildiz[a] and Mirigul Altan[a]

[a]*Yildiz Technical University, Istanbul Beşiktaş, Turkey*

Polylactic acid (PLA) scaffolds have been widely used in tissue engineering due to its biocompatibility and biodegradability. However, low melt strength and cell coalescence of PLA during foaming limit its processability in scaffold fabrication. In this study, halloysite nanotube (HNT) was used as nano filler in generation of PLA nanocomposite in order to improve mechanical and biological properties of the material. PLA/HNT nanocomposites were prepared by melt mixing method within HNT concentrations of 1-3-5%wt. Injection molding with chemical foaming agent, azodicarbonamide, was used to produce tensile test samples of PLA/HNT scaffolds. Universal tensile test was applied for determining mechanical properties and in vitro cell testing was used for observing cell viability and proliferation. Morphology of the polymer nanocomposite foams were investigated by scanning electron microscopy Results have shown that 3 wt% of HNT loading exhibited significantly improved and stable cellular structure in comparison with other foamed samples, leading to increased values of the tensile strength about 124% and elongation about 79%. Presence of HNT reduced the cell size, increased the cell density and improved cell morphology of the PLA. This brought enhancement in cytocompatibility of PLA based scaffolds apparently.

Keywords: Poly(lactic acid), halloysite nanotube, scaffold, injection molding, chemical foaming agent, microcellular foaming.
PACS: 42.70.Jk, 61.25.hk, 61.41.+e, 61.43.Gt, 62.23.Pq, 72.80.Tm.

INTRODUCTION

Poly(lactic acid) or polylactide (PLA) is a biobased, biodegradable and biocompatible aliphatic polyester and have found many applications due to their unique set of properties and has the potential to reduce the dependence on petroleum [1-2]. However, PLA was found to exhibit poor foamability because it suffers from low melt strength and elasticity, slow crystallization and narrow processing window [2, 3]. Many attempts have been made to tailor the properties of PLA by combining it with different dispersed phases such as micro- and nano-fillers, flame retardants and thermal stabilizers for many specific applications [4-5]. The incorporation of clay into the PLA matrix has attracted much attention in both the academic and industrial fields due to their enhanced physical and mechanical properties compared with pure PLA [6]. Halloysite clay nanotubes (HNT) are a tubular clay nanomaterial that has received considerable attention lately due to its environmental friendliness, biocompatibility, and natural abundance which make it an important nanomaterial for biomedical applications [7-8]. Therefore, using naturally available, cheap, high strength, and biocompatible halloysite nanotubes as fillers appears to be a promising alternative route for the preparation of PLA-based materials [5, 9]. Biodegradable polymeric scaffolds possessing three-dimensional architectures provide necessary transportation of nutrition, oxygen and growth factors for living cell proliferation until the tissues are totally restored or regenerated [10-11]. Various fabrication techniques have been attempted to produce a 3D porous scaffold structure such as 3D printing, porogen leaching, freeze drying, electrospinning and foam injection molding.

Foam injection molding has advantages such as a faster production cycle and easy processability [12, 13]. The literature studies on foam injection molding of PLA are just a few and quite recent, especially on PLA/HNT composite foaming by using chemical foaming agent. Some researchers have tried foaming of PLA by using supercritical nitrogen as physical blowing agent [12, 14]. But this technology is very expensive and when using chemical blowing agents, standard processing equipment can be used. PLA/HNT nanocomposite foaming is quite few in the literature. Wu et al. [6] studied Polylactide (PLA)/halloysite nanotube (HNT) nanocomposites with different HNT contents by using compression molding and batch foaming technology. Nizar et al. [13] developed PLA/HNTs nanocomposite scaffold by means of freeze extraction method. In this study, PLA/HNT porous scaffolds via injection molding with chemical foaming agent were prepared. Mechanical and biological properties of the samples were investigated.

MATERIALS AND METHODS

Commercial Poly(lactic acid) (PLA 3001D, Natureworks supplied by Resinex Inc. Turkey) was used in the experiments. The foaming agent was azodicarbonamide (Sigma-Aldrich). Halloysite nanotubes (HNT) were purchased from Sigma Aldrich (Turkey) with a diameter range of 30-70 nm and length range of 1-3 µm.

Before compounding, the PLA and HNT were dried at 80 °C for 24h. PLA/HNT composites were compounded on a twin screw extruder (Rondol, England, L/D=10) with a speed of the screw of 60 rpm and die temperature of 170 °C. PLA nanocomposites were produced within the loadings of 1, 3, 5 wt% HNTs. The extruded compound was pelletized and dried at 50°C overnight. Subsequently, the dried compounds were injection molded with 1 wt% chemical foaming agent in order to obtain PLA foams for the tests. The prepared mixture was injected into the standard tensile test (ISO 527) mold cavity at a melting temperature of 190 °C via the injection foaming machine.

The tensile properties of nanocomposite foams were measured with a universal tensile test machine (Instron 5982) with 1 mm/min of test speed at room temperature. The reported data were the average of three tests.

The scaffold samples were cultured for 24 hours, after which cytocompatibility tests were conducted using the methyl-thiazol-tetrazolium (MTT) technique. 10% DMSO solution was used as positive control, the culture medium as blank control was used as negative control. In the experiments, L929 cells were cultured at a density of 10^5 cells / ml.

In order to investigate the cell morphology of microcellular foamed samples, gold coated fracture surfaces of the foamed samples were observed on scanning electron microscope (XL 30 ESEM-FEG).

RESULTS AND DISCUSSIONS

Figure 1 and Table 1 show the evolution of both tensile strength, elongation and elastic modulus of the PLA and PLA/HNT based nanocomposite foams prepared with different HNT loadings, respectively. The incorporation of HNT improves the deformation behavior of nanocomposites (Figure 1) since it delays the cell coalescence and following crack formation thereby increasing the toughness. As the amount of HNT increased up to 5 wt %, the tensile strength decreased drastically for the PLA/HNT composite scaffolds in comparison to composites containing 1 and 3 wt % HNT. It can be said that the optimum clay content is 3 wt %. Increasing HNT concentration to from 0 to 3 wt % improved tensile strength (124%) and elongation (79%). Modulus increased almost linearly with the increase of HNT content, basically due to very efficient stiffening effect of rigid HNTs. When the amount of HNT reached 5 wt%, then elastic modulus of neat PLA increased about 156%. By comparison to neat PLA, both properties were gradually improved in nanocomposites by increasing the percentage of nanofiller up to 3%. However, higher loading (5% wt. HNT), revealed a dramatic decrease in tensile strength. Such an observation is generally attributed to inevitable aggregation of the nanofiller at higher loading [15-16]. The neat PLA is a brittle polymer with poor ductility. Upon HNT addition into PLA, ductility of the material increases dramatically. The reinforcing effect of HNT on PLA, frequently observed at low loadings, is generally ascribed to the rod-like and high aspect ratio structure, whereas the interfacial properties (nanotubes-polyester matrix) can play a key role as well. This behavior is quite remarkable and is commonly ascribed to the intrinsic stiffness of the nanotubes- resulting from their tubular structure, their good dispersion, as well as to the good interfacial properties between filler and matrix [15,17, 18]. As demonstrated by SEM in Figure 2, the HNT has a good dispersion in the PLA matrix. This increase in the strength and modulus is attributed to the reinforcement introduced by the HNT that enhances the stiffness of the nanocomposite, allows a greater degree of stress transfer at the interface. The good dispersion of HNT in the matrix plays the main role in improving the mechanical properties of polymers at the low HNT loading [6]. But higher loading (5% wt. HNT) a decrease in properties is noticed, this could be due to overloading of HNTs that may induce formation of aggregates. This reduction attributed to the presence of large voids in the microcellular specimens. These large voids may become regions of stress concentration and thus could adversely affect the mechanical properties [19]. It was assumed that the improvement in stiffness and ductility of the nanocomposites is due to the good dispersion of HNT inside the PLA as well as the strong interaction between them [20].

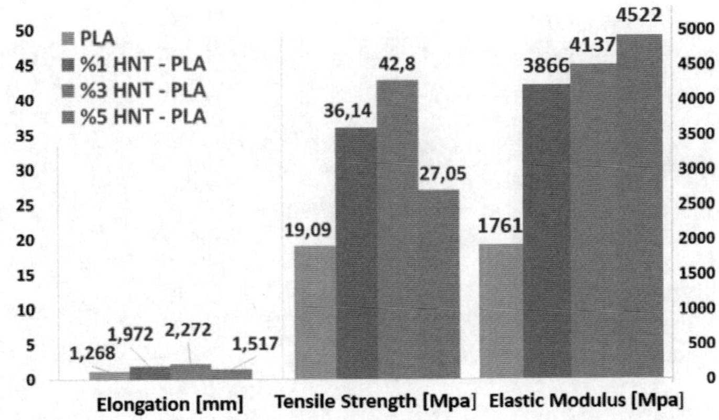

FIGURE 1. Tensile tests results pf PLA foams

TABLE (1). Tensile properties of PLA based composites

	Tensile Strength [MPa]	Elongation [mm]	Elastic Modulus [MPa]
PLA Foam	19,09	1,268	1761
%1 HNT	36,14	1,972	3866
%3 HNT	42,8	2,272	4137
%5 HNT	27,05	1,517	4522

As a general pattern, all foamed samples show a frozen layer structure (skin) and core structure. This is the usual behavior for injection-molded foams. Figure 3 represents SEM images of the fractured surfaces of foamed samples. It can be seen that the foamed structure with an HNT content of 3% in Figure 3(c) was the most uniform. Bigger cell sizes are and non-uniform foam structure can be seen in Figure 3(a) with no HNT addition, large gas pockets are formed in the core region of the cellular structure. While, in Figure 3(b), (c) and (d), presence of nanoparticles increased cell formation and made smaller cell size. Figure 3(b) and 3(c) shows that there is no significant distribution problem in the specimens with low HNT contents. With an increase in HNT content, the size of the cells increased. It has been seen that when amount of HNT is 5 wt%, then certain level of agglomeration may take place, cells coalesced and increased in size (Figure 3(d)). HNT behaved as a nucleating agent, acting as heterogeneous sites, promoting a fine distribution of cell nucleation, could enhance foaming homogeneity, mechanical and physical properties [6,11,12]. Compared to the result of the elongation shown in Table 1 and the foamed structure shown in Figure 3, it has been seen that there was a close relationship between the elongation and the foamed structure that the more uniformly distributed the foamed structure induced higher ductility.

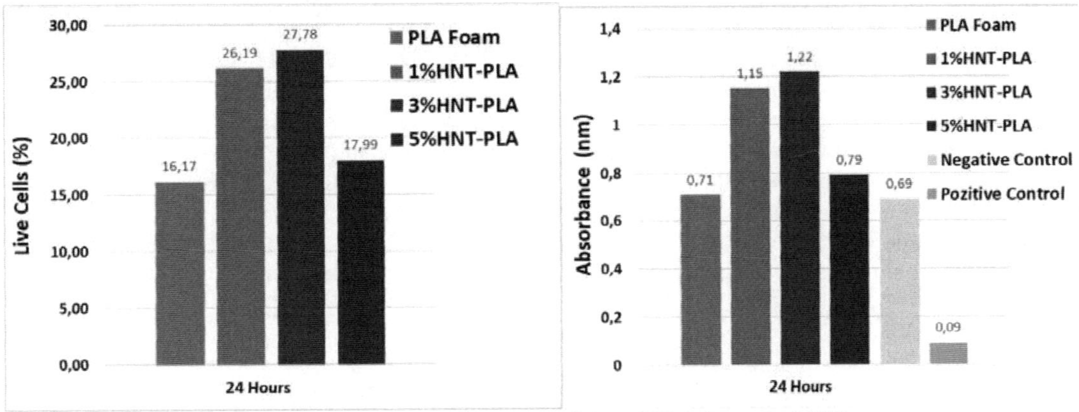

FIGURE 2. In-vitro cytotoxicity of PLA foams by MTT assay

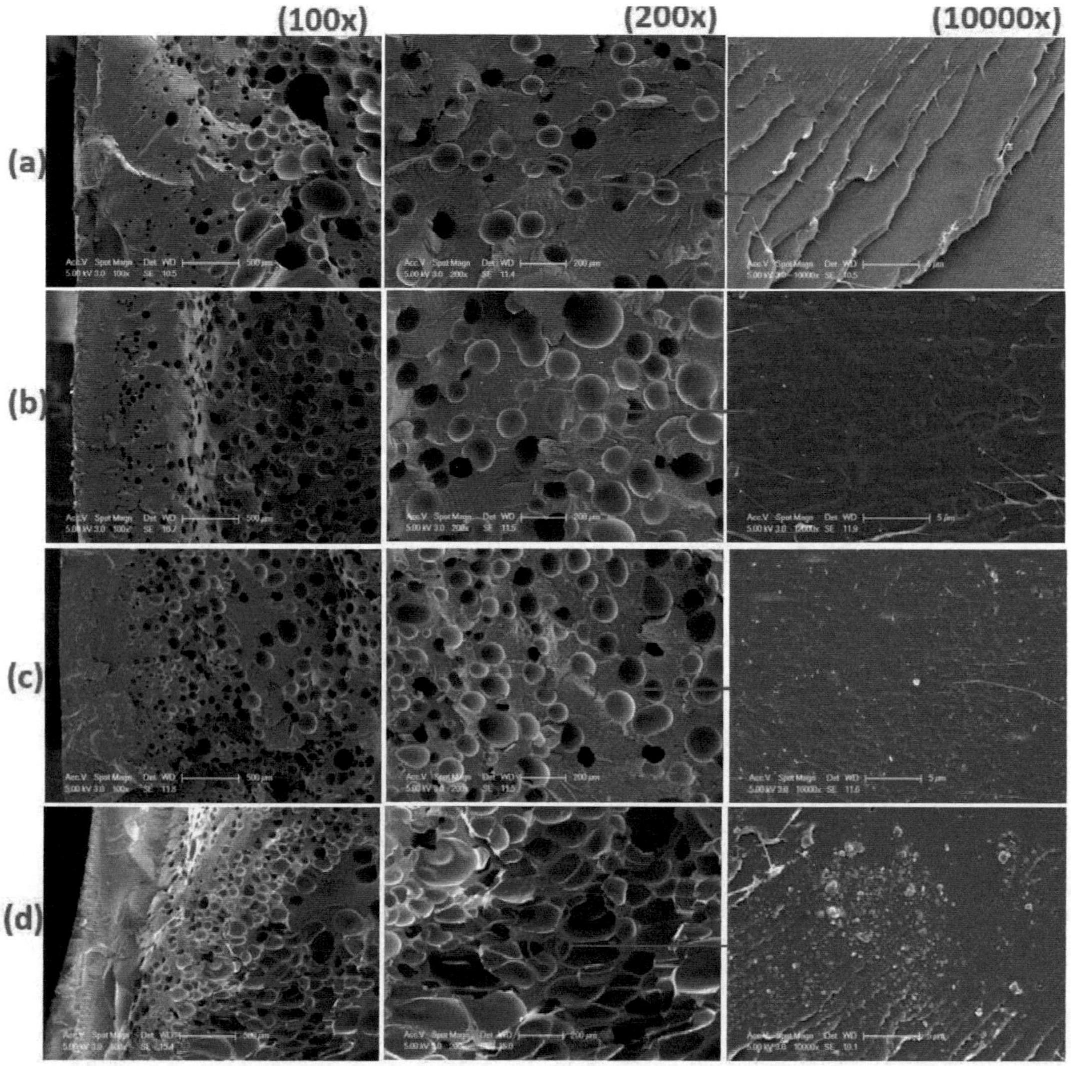

FIGURE 3. SEM images of foams a) neat PLA, b)1% HNT-PLA, c) 3%HNT-PLA, d)5% HNT-PLA

The biocompatibility of the scaffolds was investigated by evaluating cell proliferation for the 24 hours. The results are reported in Figure 2. The values represent the percentage of live cells relative to the condition where cells were cultured in the culture medium with solvent alone (control). With an increase in HNT content, the cell viability increased when compared with neat PLA foam. However, the presence of 5wt% of HNT gave reduction in cell viability due to the generation of larger pores while 3wt % of HNT showed the opposite. The reason of this behavior was due to the smaller and denser cell morphology of 3%HNT-PLA. As it known from previous studies [21-22], the small size pores in the scaffolds increase the surface area and depending on this cell proliferation increase. Also, high concentration of HNT may cause DNA breakage and result latent toxicity effects [23].

PLA foams were riskless to culture cells, cell proliferation occurs for all of the samples. Thus, it was shown that the scaffolds fabricated by foam injection molding could support cell growth satisfactorily.

CONCLUSIONS

A non-toxic, biocompatible nano filler, halloysite nanotube, was used to improve mechanical, biological and morphological properties of PLA based scaffolds fabricated by foam injection molding. It has been seen that 3 wt% of HNT loading exhibited significantly improved cellular structure in comparison with other foamed samples, leading to increased values of tensile strength about 124% and elongation about 79%. Also, presence of HNT up to 3wt% of concentraion brought enhancement in cytocompatibility of PLA based scaffolds. Over 3wt% of HNT, reductions in all properties were observed due to the possible agglomerations in the matrix.

ACKNOWLEDGMENTS

This work was supported by Research Fund of the Yildiz Technical University. Project Number: FDK-2018-3362.

REFERENCES

1. F.L. Jin, R.R. Hu and S.J. Park, *Comp. Part B* **164,** 287-296 (2019).
2. A. Ameli, M. Nofar, D. Jahani, G. Rizvi and C.B. Park, *Chem. Eng. J.* **262**, 78-87 (2015).
3. N. Najafi, M.C. Heuzey, P.J. Carreau, D. Therriault, C.B. Park, *European Polym. J.* **73,** 455-465 (2015).
4. J. Guo, J. Qiao and X. Zhang, *J. Appl. Polym. Sci.* **133,** 1-9 (2016).
5. X. Liao, A. V. Nawaby and H.E. Naguib, *J. Appl. Polym. Sci.* **124,** 585-594 (2012).
6. W. Wu, X. Cao, Y. Zhang and G. He, *J. Appl. Polym. Sci.* **130,** 443-452 (2013).
7. K. Feng, G.Y. Hung, J. Liu, M. Li, C. Zhou and M. Liu, *Chem. Eng. J.* **331,** 744-754 (2018).
8. A.C. Santos, C. Ferreira, F. Veiga, A.J. Ribeiro, A. Panchal, Y. Lvov, A. Agarwal, *Adv Colloid Interfac*. **257,** 58-70 (2018).
9. K. Prashantha, B. Lecouvet, M. Sclavons, M.F. Lacrampe, P. Krawczak, *J. Appl. Polym. Sci.* **128,** 1895-1903 (2013).
10. L. Sartore, S. Pandini, F. Bignotti and F. Chiellini, *AIP Conf. Proc*. **1981,** 1-4 (2018).
11. J. Chen, M. Yu, B. Guo, P.X. Ma and Z. Yin, J. *Colloid Interfac. Sci.* **514,** 517-527 (2018).
12. R. Pantini, A. Sorrentino, V.Volpe and G. Titomanlio, *AIP Conf. Proc.* **1593,** 397-400 (2014).
13. M.M. Nizar, M.S.A. Hamzah, S.I.A. Razak and N.H.M. Nayan, *IOP Conf. Series: Materials Science and Engineering* **318**, 1-8 (2018).
14. A. Larsen and C. Neldin, *Poly. Eng. Sci.* **53**, 941-949 (2013).
15. Murariu, M., Dechief, A.L., Paint, Y., Peeterbroeck, S., Bonnaud, L. and Dubois, P. *J. Polym. Environ.,* **20(4),** 932-943 (2012).
16. R.T. Silva, M. Soheilmoghaddam, K.L. Goh, M.U. Wahit, S.B.A. Hamid, S.P.C. and P. Pasbakhsh, *Polym. Comp*. **37**, 861-869 (2016).
17. B.J. Rashmi, K. Prashanta, M.F. Lacrampe and P. Krawczak, *eXPRESS Poly. Letters.***9(8), 721-735** (2015).
18. S. Therias, M. Murariu and P. Dubois, *Poly. Degradation and Stability*. **145,** 60-69 (2017).
19. H. Zhao, Z. Cui, X. Wang, L. Turng, and X. Peng, *Composites: Part B,* **51,** 79-91 (2013).
20. A. Thankappan, N. Kalarikkal, S. Thomas, A. Padinjakkara, *Eco-friendly Polymer Nanocomposites Processing and Properties,* Washington: Springer, 2015, pp. 308
21. J. D. Ciurana, L. Serenó, E. Vallès, *Procedia CIRP* **5**, 152-157 (2013).
22. J. Knychala, N. Bouropoulos, C.J. Catt, O.L. Katsamenis, C.P. Please, B.G. Sengers, *Ann Biomed Eng.* **41(5),** 917–930 (2013).
23. Y. Chen, A. Murphy, D. Scholz, L.M. Geever, J.G. Lyons, D.Mapp, *J. Appl. Polym. Sci.* **135(30),** 1-13 (2018)

Phase Morphology and Fracture Behaviour of Multiphase Epoxy Nanocomposite Matrices

Hong Ma[a], M. Ali Aravand[a, b] and Brian G. Falzon[a, b*]

[a]Advanced Composites Research Group, School of Mechanical and Aerospace Engineering, Queen's University Belfast, Ashby Building, Belfast BT9 5AH, United Kingdom
[b]Northern Ireland Advanced Composites and Engineering (NIACE) Centre, Airport Road, Belfast, UK
Email: bfalzon@qub.ac.uk, Web Page: https://www.qub.ac.uk/sites/acrg/

Abstract. CNT modified thermoplastic toughened epoxy nanocomposites used as potential matrix materials for fibre reinforced composites were prepared and compared. During the preparation step, CNTs were added, in two different ways to investigate the consequences of changing the order of addition in the final properties of the nanocomposites. It was shown that the presence of CNTs has a great impact on different aspects of the phase separation process of the thermoplastic particles and the resulting phase morphologies. Nanotubes in ternary blends with CNTs initially added to the epoxy phase showed a better dispersion in comparison with the other type where CNTs were initially added to the thermoplastic phase. The difference in morphology and microstructure of the blends including nanotubes with those without CNTs were attributed to the increase in the local viscosity, limiting the mobility and hence the growth of the thermoplastic phase following phase separation. This resulted in the formation of smaller particles with a wider distribution in particle size. Mode-I fracture toughness of the ternary blends with CNTs initially added to the epoxy phase, showed a synergistic increase where the gain in fracture toughness was higher than the combined increases resulted from addition of the thermoplastic and nanotubes individually. A more uniform distribution of carbon nanotubes in this system made it possible for the thermoplastic microparticles to exhibit a plastic deformation during crack propagation, as evident from the micrographs of the fracture surfaces. This together with the formation of shear bands in the epoxy matrix and the pull-out of the nanotubes were identified as the most important energy dissipation mechanisms, largely contributing into the observed synergistic increase in the fracture toughness.

Keywords: processing method, phase morphology, fracture toughness, CNTs, thermoplastic.
PACS: 81.05.Qk, 81.07.De, 82.35.Np

INTRODUCTION

Epoxy resins are used extensively in the fabrication of fibre reinforced polymer (FRP) composites due to their relatively high stiffness, superior compatibility and adhesion, and chemical resistance. However the brittleness of cured epoxy resins results in low resistance against crack formation and propagation which limits the performance level of the resulting fibre reinforced composite laminates in structural applications. A practical approach to address the brittleness of epoxies is incorporating toughness modifiers in their structure, in order to enhance energy dissipation mechanisms. Various types of inclusions have been used for this purpose with varying degrees of success. Among these inclusions, engineering thermoplastics (TP) have offered an effective mean for increasing the fracture toughness of epoxy materials without any significant deterioration of other mechanical properties, which is commonly the case with rubber modification [1, 2]. Nanoscale inclusions, such as carbon nanomaterials, can potentially offer great advantages in improving strength and damage tolerance of epoxy resins [3]. More importantly, the simultaneous addition of both thermoplastic and carbon nanomaterials to the matrix, could provide a higher degree of freedom in formulating multi-phase thermosetting matrices with superior fracture toughness without adversely affecting other mechanical properties.

The thermoplastic and epoxy blend forms a homogeneous solution before the onset of curing reaction. During curing, phase separation occurs mainly as a consequence of a decrease in the free energy leading to the formation of two separate phases (i.e., a TP phase and epoxy phase). The final mechanical properties of thermoplastic modified epoxy are greatly affected by its final phase morphology in addition to the composition and material type of the blend [2, 4]. In blends containing carbon nanotubes, dispersion of CNTs is also of vital importance in determining the fracture properties of nanocomposite [5]. Therefore, it is necessary to study the interactions between the toughness modifiers involved in the system in order to exploit the advantages of all ingredients and investigating possibilities for a synergistic enhancement in fracture toughness. In the present study, two different approaches for introducing the CNTs to the ternary matrix system were investigated: adding CNTs to the neat epoxy (method I) and adding CNTs to the thermoplastic phase (method II), prior to mixing the rest component. Distinct features of each system in terms of the phase morphology and dispersion state of CNTs in each system are compared. Mode-I fracture toughness of the cured systems are then measured and the fracture surfaces are examined for further clarification of the energy dissipation mechanisms.

MATERIAL AND EXPERIMENT

The epoxy monomer used in present study is diglycidyl ether of bis-phenol A (DGEBA). It was cured by a high temperature hardener, 4, 4'-Methylenebis [3-chloro-2, 6-diethylaniline] (MCDEA). Dichloromethane (DCM) was chosen to dissolve PEI. COOH functionalised multi walled carbon nanotubes (CNTs) with an average length of 10-30 μm and outer diameter of 20-30 nm were employed.

As mentioned earlier, two different approaches for introducing CNTs to prepare the ternary nanocomposites were employed. In the first approach (method I), CNTs were initially added to the epoxy monomer with stirring the mixture at 120°C overnight to detangle the CNTs. The ultrasonication was further conducted for 1 hr to improve dispersion of nanotubes. In the meantime, PEI pellets were separately dissolved in DCM. CNT/epoxy solution was then mixed with this thermoplastic solution. DCM was removed by keeping the solution at 180°C for at least 3 hrs followed by a degassing step under vacuum. After the addition of stoichiometric amounts of the hardener (MCDEA), the mixture was poured into a preheated rectangular mould with a depth of 4 mm. The blend was then cured at 170°C for 1hr, at 200°C for 2hrs and at 220°C for another 2hrs. The resulting nanocomposites are referred to as CDP, hereafter. The second approach for fabricating nanocomposite is started by mixing CNTs with PEI/DCM solution (method II). Again after 1 hr ultrasonication, the mixture was stirred with the epoxy monomer. Then the dispersion was dealt in the same way as method I. The cured modified epoxy prepared based on the second approach is denoted as CPD. The thermoplastic and nanotube content in the nanocomposite fabricated based on these two approaches were 10 and 0.3 wt%, respectively. Binary systems, such as CNT/epoxy with a nanotube content of 0.3 wt% and PEI/epoxy blends with 10 wt% of thermoplastic, were also prepared in the similar fashion, being referred to hereafter as CD and PD, respectively.

The cured bulk plates were cut and polished into thin layers with a thickness less than 100 μm using standard metallographic procedure. These thin layers were then examined by a transmission optical microscope (TOM, Nikon) to investigate the dispersion quality of CNTs. The fracture surfaces were sputter-coated with gold to prevent the build-up of electrical charge before further investigation using a field emission scanning electron microscope (SEM, JEOL). Fracture toughness of the cured blends was measured according to ASTM D5045 using a single-edge-notch bending (SENB) geometry. A wedge-shaped notch was machined at the midpoint of each specimen. A natural pre-crack was then introduced in the notch by employing a home-made drop weight setup equipped with a fresh razor blade. The tests were conducted on a universal testing machine (Lloyd Instruments) at a crosshead speed of 1 mm/min^{-1}. Critical strain energy release rate (G_{IC}, fracture energy) were calculated based on the relationships provided in ASTM D5045 standard. The reported values for fracture toughness tests were the mean average of at least five valid specimens.

RESULTS AND DISCUSSION

Interactions of CNTs and thermoplastic phases

TOM images of the cured epoxy materials are displayed in Fig. 1. Nanotubes in DGEBA form an interconnected network and the bright spots indicate regions with significantly lower CNT concentration. This implies that the carbon nanotubes are non-uniformly distributed. This is further confirmed by the SEM micrographs of the fracture surfaces shown in Fig. 3b, where two areas are divided by the dashed line. After simultaneous addition of PEI and CNTs into DGEBA, the dispersions of nanotubes in the matrix show obvious differences with the corresponding binary nanocomposite. No network-like structure is observed in both CDP and CPD. The dark points corresponding to the CNT agglomerates seem to be denser in CPD than those in CDP from the macroscopic inspection, as shown in Fig. 1b and c, indicating that nanotubes in CDP have a better dispersion. SEM images (Fig. 2) also suggest quite different dispersion behaviour of CNTs in ternary nanocomposites, in which some nanotubes are pointed by the yellow arrows. In the CDP blends, individual nanotubes could be identified at the boundaries of the TP particles (micron sized cavities and particles in Fig. 2d) where a number of them are observed in the matrix. In CPD blends, CNTs mostly form a loose cluster, being predominantly localised at the interfaces between particles and the matrix. This could be attributed to the significant formation of hydrogen bonds between the thermoplastic phase and the functional groups of CNTs during the mixing step.

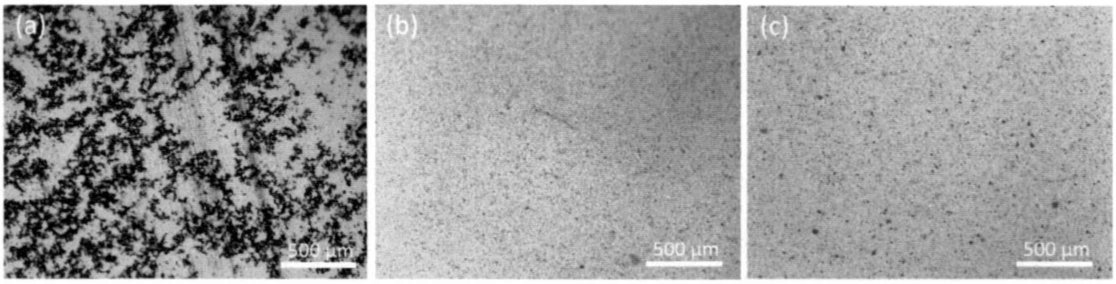

Figure. 1 TOM images of the modified epoxy resins: (a) CD, (b) CDP, and (c) CPD.

From the fracture surfaces, in the presence of CNTs, some TP particles represent a quite irregular morphology as marked by the dashed lines in Fig. 2b and d, where two spherical particles seem to have coalesced forming dumbbell shape morphology, with a number of coexisting smaller individual particles as pointed by pink arrows in Fig. 2a and c. This is an indication of a broader particle size distribution. An increased local viscosity at the vicinity of CNT agglomerates seems to be the main reason for such a different morphology as the phase separated particles easily get hooked by CNTs with a very limited mobility. This poses difficulties in particle growth following phase separation [6].

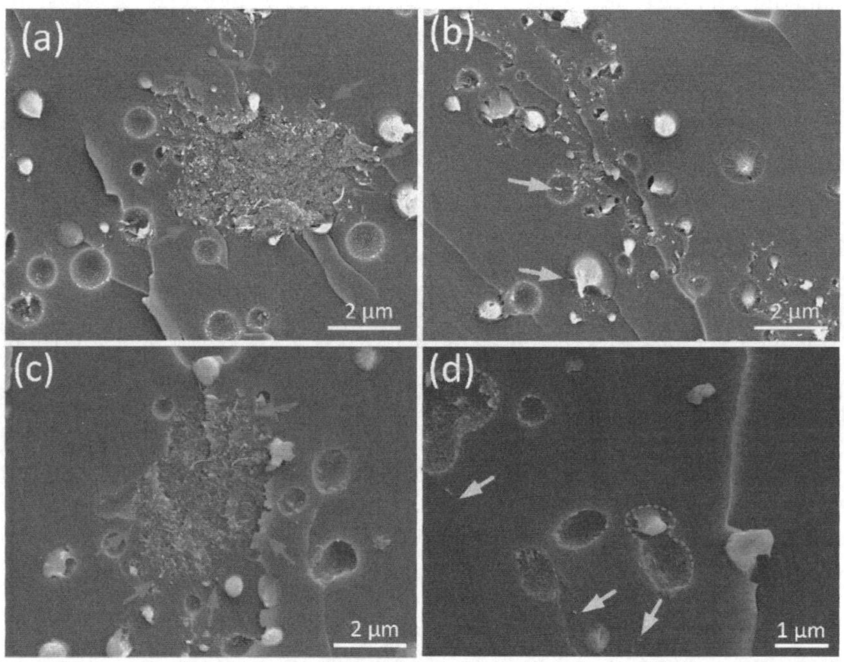

Figure. 2 Morphology of thermoplastic particles and distribution of nanotubes in modified epoxy resins: (a) and (b) CPD, (c) and (d) CDP.

Fracture behavior of modified epoxy resins

Table 1 summarises the mode-I fracture toughness results of the tested specimens. Both binary blends including either CNTs or PEI show increased fracture toughness in comparison with the baseline resin system (ND). Each system involves different types of energy dissipation mechanisms resulting in different gains in the corresponding values of the fracture toughness [2, 4]. The ternary nanocomposites show quite distinct fracture toughness when the two preparation routes are compared. The gain in fracture energy ($\Delta G_{IC} = G_{IC} - G_{IC,u}$, where G_{IC} and $G_{IC,u}$ represent fracture energy of the modified and unmodified epoxy respectively) is also listed in Table 1 to help the comparison. The gain in fracture toughness for CPD system is rather insignificant when it is compared with the CNT free binary TP/epoxy system (PD), while CDP shows a significantly higher fracture energy enhancement than the combined gains in both TP/epoxy and CNT/epoxy binary systems (PD and CD), indicating a synergistic increase in fracture toughness. Fractography of the modified epoxy resins shown in Fig. 3 provides further insights into prevailing mechanisms.

TABLE (1). Mode-I Fracture toughness of modified epoxy systems.

Materials	G_{IC} **(KJ·m^{-2})**	$G_{IC}/G_{IC,u}$[1]	ΔG_{IC} **(J·m^{-2})**[2]
ND	0.10±0.01	-	-
CD	0.11±0.01	1.14	13.3
PD	0.15±0.02	1.62	59.3
CPD	0.16±0.02	1.62	59.7
CDP	0.18±0.02	1.90	86.6

[1] $G_{IC,u}$ is the critical strain energy release rate of unmodified epoxy.
[2] $\Delta G_{IC} = G_{IC} - G_{IC,u}$

The PEI particles observed on the fracture surfaces of the binary PD blend are clearly stretched. This is an indication of the crack bridging mechanism where significant amount of the fracture energy is dissipated by the deforming particles. Additionally, the wrinkled texture of the fracture surface in the matrix is indicative of the so called shear banding. These seem to be the two major energy dissipation mechanisms for TP modified epoxy resin. The pull-out of nanotubes from the matrix in the binary nanocomposite can be clearly observed in Fig. 3b. It is also regarded that before pull-out of nanotubes, the debonding of nanotubes from matrix is likely to happen first [3]. So in the binary nanocomposite, pull-out of nanotubes followed by debonding is the main energy dissipation mechanism. For the ternary nanocomposite, the above-mentioned energy dissipation mechanisms can also be identified, however with some differences depending on the fabrication approach. In CPD systems, not all the thermoplastic particles are stretched; Instead, it seems that the crack just passes by these particles, leading to particle debonding as pointed by the arrows (Fig. 3c). As for the CDP blends, not only all the particles are stretched, the dispersion of nanotubes is also better, leading to higher levels of energy dissipation during the crack propagation where a synergistic increase in fracture toughness has been obtained.

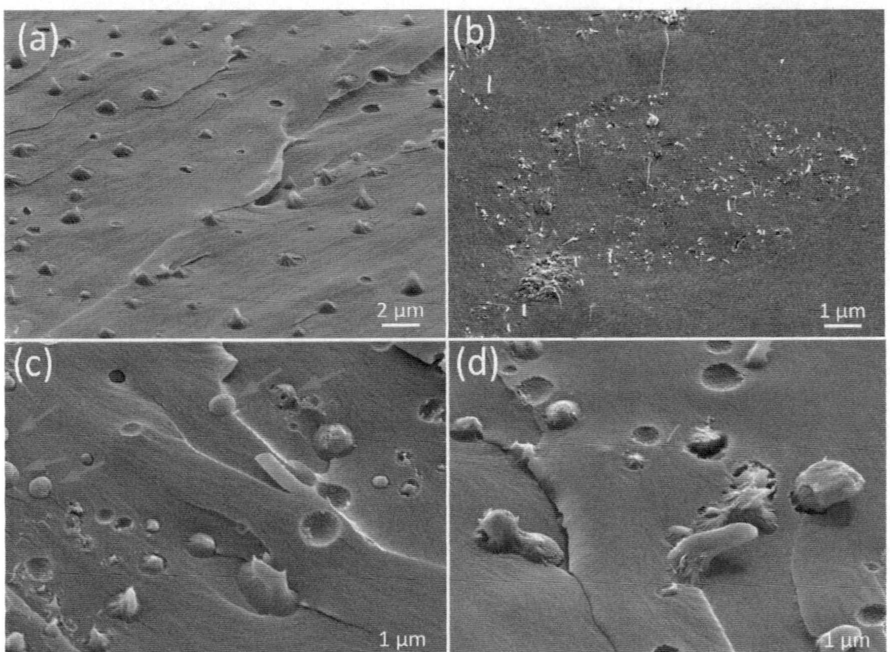

Figure. 3 Fracture surfaces of modified epoxy resins: (a) PD, (b) CD, (c) CPD and (d) CDP.

CONCLUSION

CNT/thermoplastic modified multiphase epoxy nanocomposites with nanotubes initially added to the epoxy phase showed significantly higher fracture toughness in comparison with other possible binary and ternary systems based on the same ingredients. The difference between the resulting fracture toughness values could be explained by the distinctive morphology and microstructure of the blends. CNTs could greatly affect the particles growth following the phase separation induced by the curing reaction, by increasing the local viscosity. Particle deformation and shear band formation in the continuous phase of the matrix together with the pull-out of CNTs during crack propagation were identified as the primary energy dissipation mechanisms. The synergistic increase in fracture toughness of the CDP ternary blends was attributed to a better dispersion of CNTs and a higher extent of thermoplastic deformation. This was not the case in the other ternary blend, CPD, where the nanotubes were initially added to the thermoplastic phase.

ACKNOWLEDGMENTS

The authors would like to gratefully acknowledge the funding from the Queen's University Belfast/China Scholarship Council (QUB/CSC) PhD Scholarship.

REFERENCES

1. C. K. Riew, A. J. Kinloch, *Toughened plastics I : science and engineering*, Washington, DC: American Chemical Society, 1993.
2. X. Cheng, Q. Wu, S. E. Morgan and J.S. Wiggins, *Journal of Applied Polymer Science*, **134,** 1–10 (2017).
3 F. Gojny and K. Schulte, *Composites Science and Technology*, **64**, 2303-2308 (2015).
4 J. Parameswaranpillai, N. Hameed, J. Pionteck and E.M. Woo, *Handbook of Epoxy Blends*, Springer International Publishing, Cham, 2017.
5 M. A. Aravand, S. V. Lomov, I. Verpoest and L. Gorbatikh, *Express Polymer Letters*. **8**, 596-608 (2014).
6 H. Ma, M. A. Aravand, S. C. Hawkins and B. G. Falzon, "Cure Kinetics and Phase Morphology of CNT Modified Multi-Phase Epoxy Nanocomposites" in *Materials Science: Nano Composites_3*, 18th European Conference on Composite Materials, Athens, 2018, pp, 1-5.

Effect of MWCNT on Rheological Behavior, Mechanical Performance, and Electrical Properties of SEBS/MWCNT Nanocomposites

Parvin Alaee[1, a)], Kosar Alavi[1, b)] and Hossein Nazockdast[1, c)]

[1] *Department of Polymer Engineering, Amirkabir University of Technology, Tehran, Iran*

[a)]Parvinalaie@aut.ac.ir
[b)]K.alavi@aut.ac.ir
[c)]Corresponding author: nazdast@aut.ac.ir

Abstract. In this study, the effect of multi-walled carbon nanotube (MWCNT) on microstructure development, mechanical performance, and electrical properties of MWCNT-filled Polystyrene-b-poly (ethylene-butylene)-b-polystyrene (SEBS) were studied. The nanocomposite samples with different amounts of MWCNT were prepared by melt compounding using an internal mixer. The result of frequency sweep experiment showed a pronounced low frequency nonterminal behavior which was considered as an indication of a 3D physical network formed between nanoparticle-nanoparticle and/or nanoparticle-matrix as a result of good dispersion of MWCNT. The MWCNT induced liquid-solid transition at low frequency range was explained in more details by Cole-Cole plot. The result of storage modulus as a function of MWCNT showed a rheological percolation threshold around 1.5%. The addition of MWCNT into the SEBS matrix was found to have a significant enhancing effect on modulus and tensile strength with little decrease in elongation at break. For example, the addition of 2% MWCNT enhanced tensile stress and modulus at 300% strain up to 125% and 143%, respectively. Electrical conductivity of the nanocomposites was dramatically increased between 1.5 wt. % and 2 wt. % of MWCNT, supporting the formation of conductive network within the polymeric matrix.

Keywords: Carbon Nanotube Nanocomposite, Microstructural Development, Mechanical Properties, Electrical Conductivity

INTRODUCTION

Carbon nanotubes have extremely high surface areas, large aspect ratios, remarkably high mechanical strength, and electrical conductivity [1]. Due to these unique properties, they have become good candidates for nanocomposite fabrication especially in those based on block copolymers like SEBS. SEBS/CNT nanocomposites are being utilized in many applications including temperature sensors, electromagnetic shieldings, and seals. The low percolation threshold of CNT containing polymer nanocomposites enhances their electrical conductivity while maintaining their other properties [2, 3, 4]. However, the final properties of nanocomposites depend on good dispersion of carbon nanotubes into the polymer which is still a challenging issue. In this study, an attempt was made to prepare and analyze electrically conductive and stretchable nanocomposites by uniform dispersion of MWCNT in matrix which is confirmed by using rheological, mechanical, and electrical measurements.

EXPERIMENTAL SECTION

Materials and Sample Preparation

SEBS (Globalprene 9557) with 13% polystyrene content and density of 0.902 gr. cm^{-3} supplied by Lee Chang Yung Chemical Industry Corporation. Multi-walled carbon nanotubes with the average surface area, diameter, and average length of 250-300 m^2, 9.5 nm, and 1.5 μm respectively, was supplied by Nanocyl SA, Belgium. The SEBS-MWCNT nanocomposite samples were prepared by melt compounding using an internal mixer (Brabender Plasticorder W50) at 240 °c with rotor speed of 60 rpm for 20 minutes. The MWCNT content was varied from 0.25 wt. % to 2 wt. % in the samples.

Properties Characterization

The melt rheological properties were determined using a rheometer (Paar Physica UDS 200). The test was performed by using parallel plate geometry with a diameter of 25 mm and a constant 1 mm gap under controlled shear deformation oscillatory mode. Dynamic frequency sweep test was conducted from 0.1 to 1000 rad/s range at 1% strain within the linear viscoelastic region of the samples.

For the mechanical and electrical measurements, all the melt compounded samples were hot pressed at 240 °c into sheets with the thickness of 1mm. For tensile testing, sheets were cut into dumbbell-shaped specimens and then tested by using a tensile testing machine with a crosshead speed of 500 mm/min. DC volume electrical conductivity of the samples was measured with electrometer (Keithley Instrument, 610C solid state) by 2 probe method according to ASTM D 257 standard.

RESULTS AND DISCUSSION

Rheological Properties

In order to determine MWCNT dispersion state and microstructure development of MWCNT containing nanocomposites, melt state rheological measurement was carried out. Figure 1 shows the frequency sweep results for neat SEBS and the nanocomposite samples with different MWCNT loading. As can be seen, neat SEBS showed a terminal behavior obeying power-law of $G' \alpha\ \omega^2$ as an indication of disordered state for SEBS at 240 °c [5]. Through addition of MWCNT, a pronounced low frequency nonterminal behavior was observed corresponding to 3D network between MWCNT/MWCNT and/or MWCNT/matrix as a result of good dispersion of nanofiller in matrix.

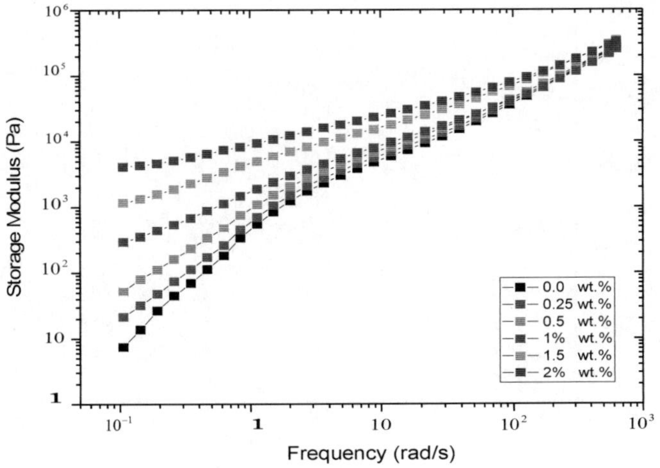

FIGURE 1. Storage modulus (G′) as a function of frequency (ω) for neat SEBS and its nanocomposites with indicated MWCNT loading.

The rheological percolation threshold of SEBS/CNT nanocomposites was determined by plotting G′ at 0.1 rad/s frequency as a function of MWCNT content as shown in Fig. 2. The sharp transition observed in G′ between 1.5 wt. % and 2 wt. % MWCNT can be considered an indication of rheological percolation threshold of nanocomposite samples.

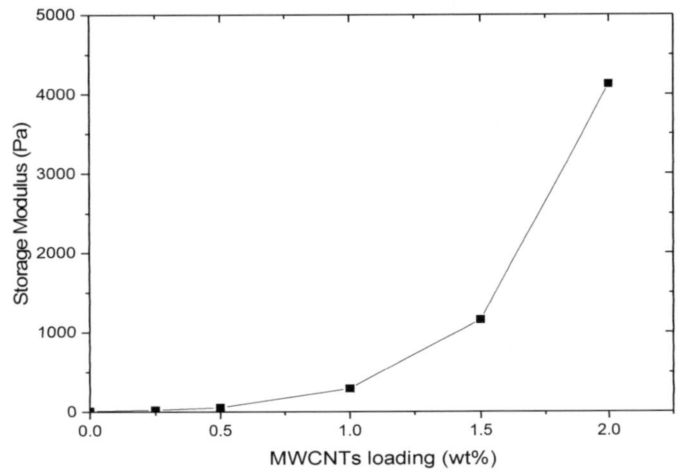

FIGURE 2. G′ as a function of MWCNT loading.

In order to provide more information about nanoscopic structure of the nanocomposites, Cole-Cole plots were obtained for the SEBS matrix and the nanocomposites. As shown in Fig. 3, the SEBS matrix and the nanocomposite samples containing 0.25 wt. % and 0.5 wt. % show a complete arc whereas the samples containing more than 0.5 wt. % exhibit a complete tail which implies a liquid like to solid like transition for these samples.

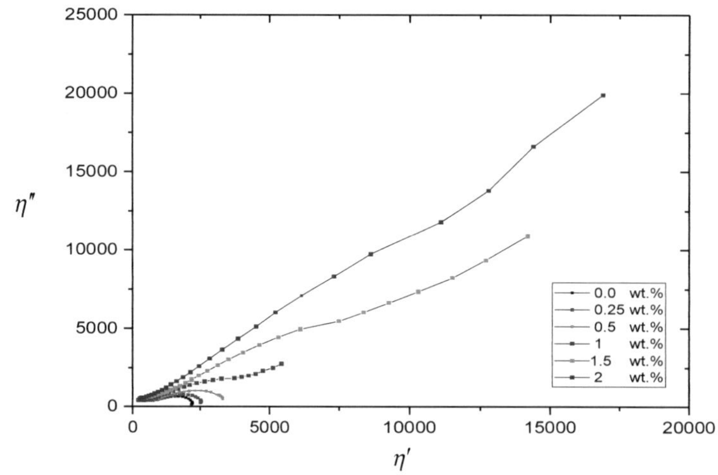

FIGURE 3. Cole-Cole plot of pure SEBS and its nanocomposites.

Mechanical Performance

The mechanical properties of SEBS nanocomposites reinforced with MWCNTs were studied using tensile test. Figure 4 shows the stress-strain curves for neat SEBS and its nanocomposites. Addition of MWCNT increased nanocomposites modulus with a slight decrease in elongation at break. In the sample containing 2 wt.% of MWCNT, tensile stress and modulus at 300% strain increased by 125% and 143% respectively while elongation at break decreased only by 17.5% compared to that of neat SEBS.

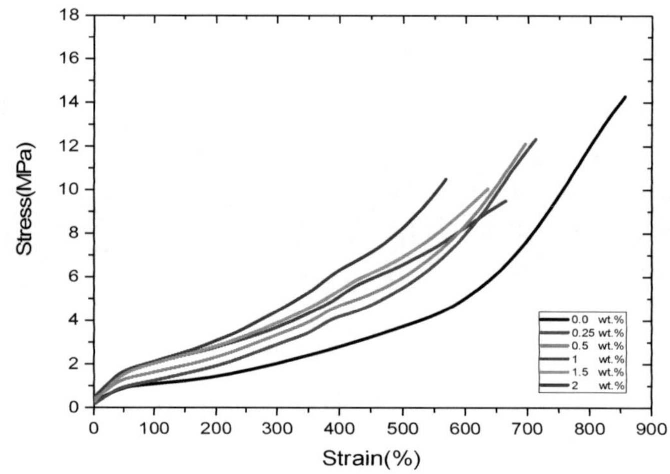

FIGURE 4. Stress-strain curves for neat SEBS and its nanocomposites with different MWCNT loading.

Electrical Properties

Figure 5 shows the volume electrical conductivity for SEBS and its nanocomposites with different amounts of MWCNT loading. The electrical conductivity of all nanocomposites increased with increasing MWCNT loading. The electrical conductivity of 2 wt. % MWCNT reinforced SEBS increased six orders of magnitude comparing to the neat SEBS sample. A sharp increase in electrical conductivity observed between 1.5wt. % and 2wt. % MWCNT loading suggests the formation of MWCNT network and conducting pathways facilitating charge transfer through samples. The rheological percolation threshold of the nanocomposite samples was found to be very close to the electrical percolation threshold. This is in contrast to many other polymer nanocomposite samples whose rheological percolation threshold has been reported to be lower than their electrical percolation thresholds [6, 7]. This can be attributed to high melt elasticity of the SEBS matrix used in this work.

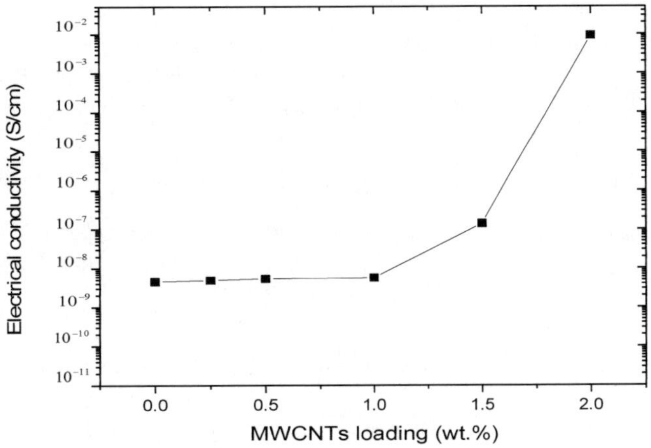

FIGURE 5. Electrical conductivity of SEBS/MWCNT nanocomposites as a function of MWCNT loading.

CONCLUSION

In this study, SEBS/MWCNT nanocomposites were prepared by melt compounding. A good dispersion of MWCNT and 3D network formed between MWCN-MWCNT and/or MWCNT-matrix was confirmed by frequency sweep test. The rheological and electrical percolation threshold was obtained between 1.5 wt. % and 2 wt. % MWCNT. SEBS filled with 2 wt. % MWCNT showed a DC electrical conductivity of 9.2×10^{-3} S/cm and an improved modulus with a slight decrease in elongation at break.

REFERENCES

1. Khalid Saeed and Ibrahim, Carbon Letters **14**,131-144 (2013).
2. N. Calisi, P. Salvo, B. Melai, C. Paoletti, A. Pucci, Materials Chemistry and Physics **186**, 456-461 (2017).
3. S. Kuester, G. Barra, J.C. Ferreira Jr., B. Soares, N. Demarquette, European Polymer Journal **77**, 43-53 (2016).
4. Y. Li and H. Shimizu, Macromolecules **42**, 2587-2593 (2009).
5. N. Hasanabadi, H. Nazockdast, S. Balog, M. Lattuada, Composites Science and Technology **149**, 192-206 (2017).
6. J. E. Jam and M. Ahangari, Polymer - Plastic Technology Engineering **51**, 1474–1482 (2012).
7. Q. Zhang, P. J. Lemstra, S. Rastogi, D. Chen, D. Lippits, Carbon **44**, 778–785 (2005).

Morphological and rheological characterization of carbon nanotubes/ poly (L-lactic) acid/ ethylene-co-vinyl acetate nanocomposites

Elaheh Abdollahzadeh Tabrizy Moghaddam[a], Hamid Garmabi[a*] and Hossein Nazockdast[a]

[a]Amirkabir university of technology, polymer engineering and color department, Tehran, Iran
**Corresponding Author: garmabi@aut.ac.ir*

Abstract. In this work, different contents of multi-wall carbon nanotubes (MWCNTs) were melt processed into an immiscible poly (L-lactic) acide/ ethylene-co-vinyl acetate (PLLA/EVA) blend. A constant PLLA/EVA 80/20 %wt. ratio that exhibited a sea-island structure was selected for preparation of PLLA/EVA/CNT nanocomposites. Rheological, and morphological properties of samples were comparatively investigated. The results demonstrated that the presence of CNT resulted in the formation of EVA particles with irregular shapes. A correlation between the observed morphology and the wetting coefficient calculated with experimental measurements of surface tensions allowed to predict CNT localization.

Keywords: Poly (L-lactic) acid, Ethylene vinyl acetate, Carbon Nanotubes, Rheology, Morphology.

INTRODUCTION

Polymeric materials derived from biomass have drawn considerable attention in recent years because of the environmental concerns and sustainability issues associated with petroleum-based polymers. However, there are serious limitations, notably brittleness and a low heat distortion temperature due to the low glass transition (Tg) of PLLA, and no back-up of the E-modulus above Tg due to a low rate of crystallization. In order to overcome the brittleness of PLLA, many approaches have been explored, blending with other polymers is the most effective and convenient way to modify PLLA [1]. PLLA was also modified with non-biodegradable polymers like ethylene co-polymers. Authors have achieved the impact strength 53.73 kJ/m2, by blending PLLA with EVA (VA contents 50 wt %), which is more than 25 times higher than neat PLLA and the same was termed super tough blend which is published somewhere else [2,3]. EVA is a co-polymer composed of two homo-polymer constituents polyethylene and poly(vinyl acetate) (PVAc) respectively, and possesses properties intermediate of these two polymers. Its properties are associated to its complex morphology which is composed of three phases: ethylene segments containing crystalline phase, an interfacial region composed of ethylene and vinyl acetate segments, and a complex amorphous phase of ethylene segments and non-crystallized VAc segments. PLLA was blended with EVA for performance improvement [3]. Good compatibility between PLLA and EVA was achieved without additional compatibilization [4].

Incorporating nanofillers into the polymer blends has attracted the interest of researchers. Nanofillers usually exhibit different interfacial affinities to the components of the immiscible polymer blends and consequently, nanofillers exhibit selective localization in one component and/or at the blend interface. Besides the thermodynamic factor relating to the interfacial affinity between components and nanofillers, the kinetic factors also greatly influence the migration and selective localization of nanofillers in the blend composites [5]. With the development of research on carbon nanotubes (CNT), it has been proved that CNTs exhibit greater ability in improving the conductivity of polymers due to their large aspect ratio which enables the formation of percolated network in a specific phase of the immiscible polymer blend at a much lower CNT content. The selective localization of nanofiller in a specific phase or at the interface in the immiscible polymer blend is governed by thermodynamic and kinetic factors. The former is related to the wetting coefficient, ω_a that can be calculated using the following equation:

$$\omega_a = \frac{\gamma_{nanofiller-polymer1} - \gamma_{nanofiller-polymer2}}{\gamma_{polymer1,2}} \qquad (1)$$

In this equation, γ are the interfacial energy between materials, respectively. If $\omega_a > 1$, nanofiller preferentially locates in polymer 2; If $\omega_a < -1$, nanofiller preferentially locates in polymer 1, and if $-1 < \omega_a < 1$, nanofiller locates at the interface between the two phases [6].

EXPERIMENTAL

Materials

PLLA or PLA (4032D) with D-isomer content of less than 2% was purchased from Nature Works, USA. EVA (Levapren®) with VA content ~50 wt% was kindly supplied by Lanxess Chemical Co., Ltd., Germany. Pristine CNTs (Nanocyl7000) were obtained from Nanosyl Company, Belgium.

Sample Preparation

Samples were prepared through melt mixing and compression molding processing. Before sample preparation, all the pristine materials were dried at 50 °C for 12 h to erase the effect of moisture on possible thermal degradation of PLLA and/or EVA. The blend of PLLA/EVA and PLLA/EVA -CNT nanocomposites were prepared by melt mixing at 190 °C and 60 rpm for 20 min using an internal mixer (Brabender Plasticorder W50). Here, the weight ratio between PLLA and EVA was maintained at 80/20 while the content of CNTs was varied from 1 to 5 wt%. The sample notation was defined as PLA80/EVA20/Cx, where x represented the content of CNTs. For making a comparison, the binary PLLA/EVA blend and binary PLLA/CNT and EVA/CNT nanocomposites were also prepared through the completely same processing procedures.

Characterization of Samples

The phase morphologies of the samples and the dispersion states of CNTs in the blend composites were investigated using a scanning electron microscope (SEM) with an accelerating voltage of 5.0 kV. To characterize the phase morphology and the dispersion states of CNTs, the sample was firstly cryogenically fractured in liquid nitrogen, and then the cryogenically fractured surface was immersed into cyclohexane at 50 °C for 2h to remove EVA phase.

The melt rheological properties of the samples were studied using a rheometric mechanical spectrometer (Paar Physica UDS 200). The oscillatory shear rheological measurements were conducted on the samples at a temperature of 190 °C using a parallel-plate geometry, with a diameter of 25mm and a gap of 1mm under controlled shear deformation mode. The linear viscoelastic properties of nanocomposites were measured within the angular frequency range of 0.07–1000 (1/s) with small strain amplitude of 1% to assure the nanocomposites remain in their linear elastic deformation regime.

Dynamic mechanical thermal analysis (DMTA) was carried out for the unfilled blend and the nanocomposites containing different MWCNTs contents using a Mettler Toledo DMA (Switzerland). The tests were performed in the bending mode between -50 °C and 120 °C at a frequency of 1 Hz with a heating rate of 3 °C min^{-1}.

RESULTS AND DISCUSSION

Dispersion of MWCNTs: Linear Viscoelasticity

Figure 1 exhibits the variations of storage modulus (G'), normalized storage modulus versus angular frequency for PLLA, EVA and their nanocomposites, Han plots of storage modulus versus loss modulus (G"). As it can be clearly seen in Figure 1 a, the PLLA/CNT with 1%wt. nanocomposite sample display a strong solid-like behavior in the low frequency range. This behavior is known to be an indication of 3D physical networks formed. Figure b presents the angular frequency dependency of normalized storage modulus ($G'_{Nanocomposite}/G'_{Matrix}$) of the PLA, and EVA nanocomposite samples containing 1 wt% of MWCNTs. At equal loading levels, this curve can provide insight about the affinity of the nanofiller toward each polymer. The stronger the solid-like response in the normalized storage modulus curve, the stronger the affinity toward that polymer. This concept was used to cast light on the affinity of the MWCNTs toward the blend components. It is evident in Figure 1 b that PLLA/CNT nanocomposite sample exhibited the MWCNTs induced solid-like response at low frequencies, but the response in the case of EVA was by far stronger. This suggests that the MWCNTs have a much greater affinity toward PLLA compared with that toward EVA.

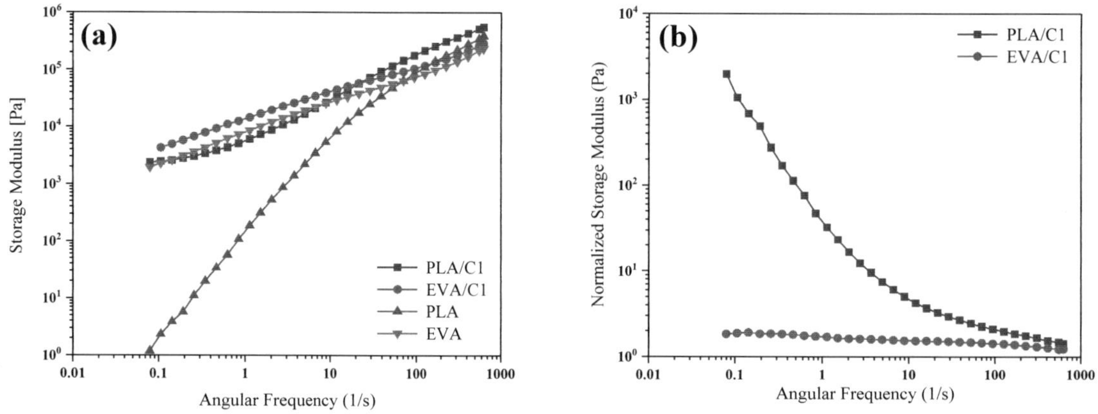

FIGURE 1. The effect of CNTs content on the storage modulus of neat PLA and EVA (a) measured at 190 °C, (b) normalized Storage modulus versus angular frequency for PLA and EVA nanocomposites samples containing 1 wt% of the CNTs.

Thermodynamic Evaluation of MWCNTs' Localization in Blend Nanocomposites

Surface tension (total, dispersive, and polar fractions) of the blend components and the MWCNTs were measured at 190 °C are shown in Tabe 1. For the PLLA/EVA blend nanocomposites, ω_a was found to be 0.09 respectively. Thus, at thermodynamic equilibrium, the MWCNTs should preferentially localize at the blend interface.

TABLE 1) The surface tension of the blend components and CNTs measured at 190 °C [8, 9, 10].

Component	γ^p (mNm^{-1})	γ^d (mNm^{-1})	γ (mNm^{-1})
PLA	11.3	4.9	16.2
EVA	2.3	33.5	35.8
CNT	10.2	17.6	27.8

Rheology of PLA/EVA/CNTs Nanocomposites

From Figure 1c one can see that although the PLLA80/EVA20/C1 sample exhibits very similar G' to that of the PLLA80/EVA20 specimen at all frequencies, further increasing CNT content leads to apparent change of G' at low frequencies. Specifically, for the PLLA80/EVA20/C2, PLA80/EVA20/C3 and PLLA80/EVA20/C5 samples, a plateau at low frequencies is observed. In this condition, the melt exhibits a solid-like response and nearly invariant G' is achieved. This indicates that CNTs form a physical network structure in the melt. The Han plots of G' versus G", which can be also used to demonstrate the microstructure of nanofillers in polymer melt, is shown in Figure 1d. Generally, when the Han plots exhibit a deviation from the linear relationship between G' and G", and the increase of G' is higher than that of G", the nanofiller forms the physical network structure which increases the elastic responses of the melt to the external shear stress. As shown in Figure 1d, the PLA80/EVA20/C1 exhibit very similar Han plots to that of the PLA80/EVA20 sample, and the two samples exhibit the linear relationship between G' and G". However, when a high content of CNTs (PLA80/EVA20/C2, PLA80/EVA20/C3 and PLA80/EVA20/C5) are present in the melt, the samples exhibit an apparent deviation from the linear relationship. This further indicates the formation of the physical CNT network structure in the melt of the blend composites.

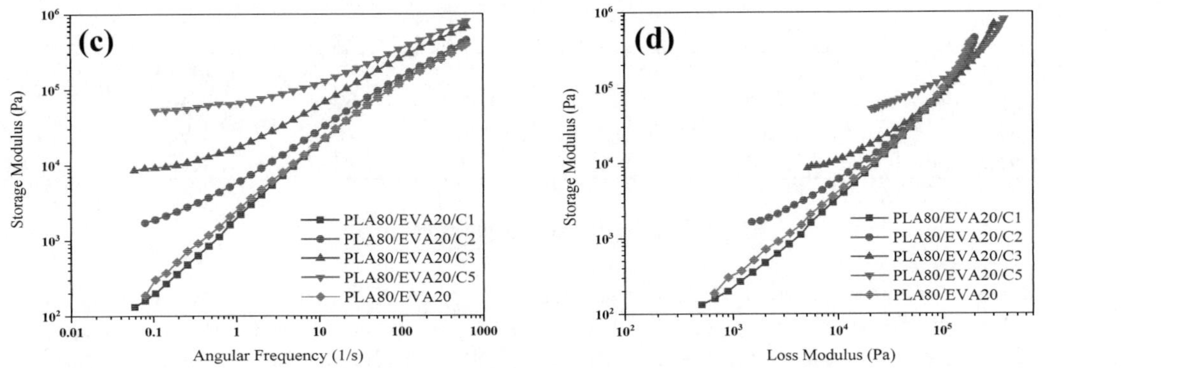

FIGURE 2. Rheological properties of the PLA/EVA/CNT blend composites. (a) Storage modulus, (b) Han plot.

Dynamic Mechanical Behavior of Samples

The formation of a physical network can limit the mobility of chain segments. To determine this, the relief behavior of the samples was investigated by mechanical dynamic measurement of DMTA. Figure 3 shows the temperature-loss-factor diagram for the DMTA test. In this figure, the EVA-related peaks are also larger. Also, in this diagram, the maximum discharge loss factor indicates the glass transition temperature of the material. Characteristics of Figure 2, such as glass transition temperature (Tg), maximum loss factor ($\tan\delta_{max}$), and integral below the surface of the graph (IR) for each of the peaks related to PLA and EVA are shown in Table 2. The lower the level below the peaks, it can be concluded that the chains are involved in a grid, resulting in less dynamics and less freedom to lose. The peak temperature of each peaks is also important. The higher the peak temperature, the lower the chains movement. It can be seen from the results of Table 2 that, by adding carbon nanotubes, the glass transition temperature of both PLA and EVA was transferred to higher temperatures. This means that the PLLA and EVA chains are involved with CNTs and their mobility is harder. Also, this chain engagement in PLLA has reduced the PLA loss factor. In the EVA phase, with increasing CNTs, the surface under the graph has increased. One possibility for EVA is that the physical network of carbon nanotubes has limited the EVA chain mobility. Another possibility can be attributed to the increased adhesion between the surface of the PLLA and the EVA.

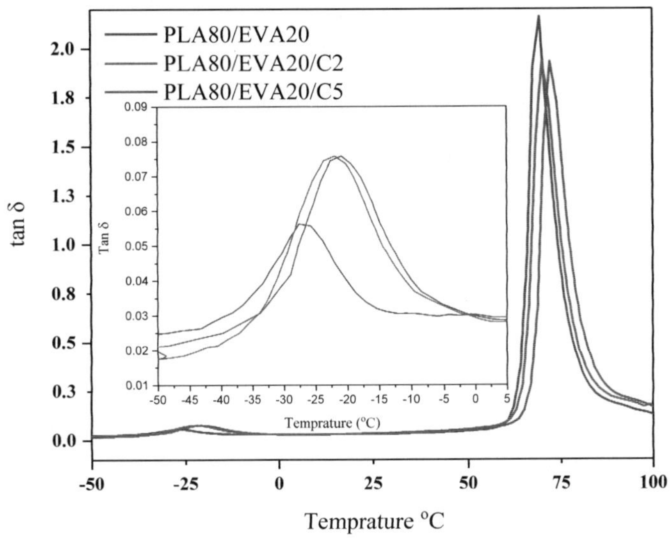

FIGURE 3. Loss factor of the binary PLA/EVA blend and the PLA/EVA/CNT blend composites obtained through DMA measurements.

TABLE 2) Relaxation behavior parameters obtained from DMA measurement of different PLLA/EVA/CNT specimens

	EVA			PLLA		
Samples	T_g (°C)	tan δ_{max}	I_R	T_g (°C)	tan δ_{max}	I_R
PLA80/EVA20	-27.61	0.06	0.35	69.39	2.16	19.92
PLA80/EVA20/C2	-21.89	0.08	0.90	70.07	1.93	18.19
PLA80/EVA20/C5	-20.86	0.08	0.71	72.19	1.93	15.58

Morphology Characterizations

Figure 4 shows FE-SEM micrographs of the cryofractured surface of the PLLA/EVA blend. The black holes represent the EVA particles which were previously removed by cyclohexane. From Fig. 4a one can see that the PLLA80/EVA20 specimen exhibits the typical sea-island morphology, namely, PLLA is the matrix while EVA is the dispersed particles. Furthermore, the average diameter of EVA particles is about 0.93 μm. Adding a small amount of CNTs leads to the apparent change of EVA morphology and especially, the more the CNTs in the blend composites, the more apparent the change of EVA morphology. In our work, the presence of CNTs induces the formation of irregular EVA particles with largely increased diameters. One possibility is that the migration of CNTs from PLLA to EVA and the selectively located CNTs in the EVA particles exhibit the role of volume filling.

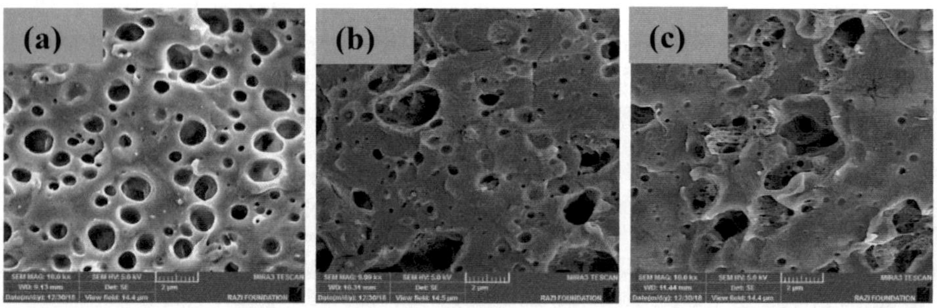

FIGURE 4. SEM micrograph showing the morphologies of PLLA/EVA/CNT blend composites with and without CNTs. Samples were immersed in cyclohexane at 50 °C for 2 h. (a) PLLA80/EVA20, (b) PLLA80/EVA20/C1 and (c) PLA80/EVA20/C2.

CONCLUSIONS

In summary, different contents of CNTs are introduced into the PLLA/EVA blend that exhibits sea-island morphology. Morphological characterizations demonstrate that CNTs induces the formation of irregular EVA particles with largely increased diameters. Rheological measurements demonstrate the formation of CNT physical network structure in the blend composites at relatively higher content of CNTs, which is believed to be the main reason for the morphological change of EVA particles. DMA measurements show that the glass transition of EVA particles and PLLA is influenced by CNTs.

REFERENCES

1. Ma P., Hristova-Bogaerds D.G., Goossens J.G.P., Spoelstra A.B., Zhang Y., Lemstra P.J., *Europ. Polym. J.*, 48, 146- 154, 2012.
2. Singla R. K., Maiti S. N., Ghosh A. K., *RSC Adv.*, 6, 14580-14588, 2016.
3. R. K. Singla, M. T. Zafar, S. N. Maiti, and A. K. Ghosh, *Polym. Test.*, vol. 63, pp. 398–406, Oct. 2017.
4. Q. Zeng, Y. Feng, R. Wang, and P. Ma, *E-Polymers*, vol. 18, no. 2, pp. 153–162, 2018.
5. Liu C. M., Ma F. F., Zhang Z. X., Yang J. H., Wang Y., Zhou Z. W., *Composites Part B*, 123, 1-9 ,2017.
6. Shi Y. Y., Yang J. H., Huang T., Zhang N., Chen C., Wang Y., *Composites: Part B*, 55, 463–469, 2013.
7. Li Y., Liu L., Shi Y., Xiang F., Huang T., Wang Y., Zhou Z., *J. of App. poly. Sci.*, 121, 2688-2698, 2011.
8. Shi Y. *et al.*, *J. Polym. Sci. Part B Polym. Phys.*, vol. 49, no. 4, pp. 267–276, 2011.
9. Shi Y. Y., Yang J. H., Huang T., Zhang N., Chen C., and Wang Y., *Compos. Part B Eng.*, vol. 55, pp. 463–469, 2013.
10. Li Y. *et al, Sensors Actuators, B Chem.*, vol. 255, pp. 2809–2819, 2018.

Properties of PS/Halloysite and PS/Sepiolite Nanocomposites Produced by Melt Blending

Nurseli Uyanık[a*], Refik Arat[b]

[a]*Istanbul Technical University, 34469 Maslak/Istanbul - Turkey*
[b]*Istanbul Kültür University, 34156 Bakirkoy/Istanbul - Turkey*

Abstract. In this study, the surfaces of naturally abundant inorganic nanoclays, namely halloysite nanotubes (HNT) and sepiolite (SEP), were modified with styrene-maleic anhydride copolymers (SMA) to improve compatibility between the nanoclays and the polymer matrix and to improve dispersibility of these nanoclays in polystyrene (PS) matrix. Four commercial SMA having different molecular weights and different styrene/maleic anhydride molar ratios were used for the surface modification. Modified nanoclays were initially characterized by FTIR and TGA. PS nanocomposites with 5wt.%, 10wt.% and 15wt.% nanoclays were prepared via a twin-screw micro extruder. Finally, structural and thermal properties of the nanocomposites were investigated. The results showed that modification of the nanoclays with SMA improved the dispersion and the compatibility of HNT and SEP into PS matrix.

Keywords: Polystyrene, Halloysite nanotubes, Sepiolite, Styrene-maleic anhydride copolymer, PS/clay nanocomposites.
PACS: 81.05.Lg

INTRODUCTION

Polymer composites containing nanoparticles are of great interest in the scientific and industrial field due to the significant effect of the nanoparticles on the mechanical and physical properties even at low nano-particle loading levels. [1–3]. The enhancement of properties depends on various factors, such as the size, shape and polarity of the nano-additives, the dispersion and orientation degree of the nano-additives in the polymer matrix and the interfacial interaction between the nano-additives and the polymer matrix [4–7]. HNT, a multi-walled inorganic nano-clay, has the molecular formula of $Al_2Si_2O_5(OH)_4.nH_2O$ and resembles carbon nanotubes (CNTs) in geometry [8]. Because of its relatively lower cost, good mechanical, thermal, flame retardant and biological properties, HNTs are widely used for preparation of nano-composites. Sepiolite is a kind of naturally occurring magnesium silicate clay, with a molecular formula of $(Si_{12}Mg_8O_{30})(OH)_4(H_2O)_4.8H_2O$. In particular, the sepiolite consists of an octahedral sheet of magnesia sandwiched between two tetrahedral silica sheets [9,10].

In this work, surfaces of HNT and SEP were modified with four different commercial SMA through esterification reaction as reported in our earlier study [11]. PS/HNT nanocomposites containing unmodified and modified clays were prepared by melt blending technique to investigate the effect of surface modification on the distribution of clays and on the physical properties of PS nano-composites. X-ray diffraction (XRD), differential scanning calorimetry (DSC), thermal gravimetric analysis (TGA) and scanning electron microscopy (SEM) were used to characterize the structural and thermal properties of the nanocomposites.

Materials and Method

PS was used as the polymer matrix to prepare the nanocomposites. Natural, inorganic, and hydrophilic nanoclays, HNT and SEP, were used as nanofillers. The commercial names, molecular weights, and styrene/maleic anhydride molar ratios of the four surface modifying agents are shown in Table 1. *p*-toluene sulfonic acid (PTSA) was used as catalyst in the surface modification reaction.

TABLE 1. Commercial names, molecular weights and styrene/maleic anhydride molar ratios of the four SMAs.

Commercial Name	Molecular Weight (g/mole)	Styrene/Maleic Anhydride Molar Ratio
SMA-1000	5500	1:1
SMA-EF40	10500	4:1
SMA XIRAN SZ26120	120000	2.67:1
SMA XIRAN SZ15170	170000	5.33:1

Surface Modification of Nanoclays

5 g of SMA was dissolved in THF solution via reflux and then 5 g of nanoclay was added to the solution (weight ratio of nanoclay/SMA was 1:1). A certain amount (about 4 wt% of the total weight of reactants) of PTSA catalyst was added to the reaction flask and the reaction was carried out in nitrogen atmosphere for 24 hours. The reaction was completed at the end of this period and the liquid mixture in the flask was centrifuged at 10000 rpm for 10 min to separate the modified nanoclay from the mixture. The modified nanoclay was washed again with THF to remove the unreacted SMA. Washing with THF was repeated three times to assure complete removal of unreacted SMA. Finally, the precipitate was transferred to a Petri dish and dried in oven at 60°C overnight.

PS/Clay Nanocomposites Preparation

PS composites containing 5wt%, 10 wt% and 15wt% unmodified and modified nanoclay were prepared by loading the nanoclay into PS matrix via a twin-screw micro extruder at 120 rpm rotational speed for 2 min at 200°C. Table 2 shows the details of PS/clay nanocomposites.

TABLE 2. PS/Clay nanocomposites content.

Clay	SMA Types	PS/Clay Nanocomposites
		PS/HNT-Control
		PS/HNT-1000
HNT		PS/HNT-EF40
	SMA-1000	PS/HNT-26120
	SMA-EF40	PS/HNT-15170
	SMA XIRAN SZ26120	PS/SEP-Control
	SMA XIRAN SZ15170	PS/SEP-1000
SEP		PS/SEP-EF40
		PS/SEP-26120
		PS/SEP-15170

Characterization

Characterization of Modified Clays

Figure 1(a) and Figure 1(b) show the FTIR and TGA characterization results of modified clay samples respectively. The characteristic absorption peaks of the pristine HNT, such as the O–H stretching of inner hydroxyl groups at 3625 cm^{-1}, O–H deformation of water at 1631 cm^{-1}, O–H bending of inner hydroxyl groups at 906 cm^{-1} and Si–O stretching at 1004 cm^{-1} can be seen in Figure 1(a). The new vibration bands of the modified HNT samples were observed in the FTIR spectra. The new bands at 1852 and 1776 cm^{-1} are assigned to the C=O stretching vibration. Furthermore, the FTIR bands at 3028–2919 cm^{-1} and 1495–1455 cm^{-1} are assigned to =C–H and C=C stretching of aromatic benzene ring, respectively. The FTIR spectrum of HNT-1000 displays the new absorption bands at 1707 and 1218 cm^{-1}, which can be attributed to the formation of ester bonding between anhydride and –OH group. The ester bonding formation shows that SMA-1000 copolymer was more efficiently grafted to HNT surfaces than the other copolymers due to its higher maleic anhydride content and lower molecular weight.

As shown in the thermal degradation analysis of Figure 1(b), the pristine HNT decomposed with two steps. The first mass loss in the range between 50 – 150 °C was due to the evaporation of physically adsorbed water. The major mass loss was in the range of 430 – 500 °C, corresponding 10wt% of a mass loss. This mass loss was attributed to the dehydroxylation of structural Al-OH groups of HNT. Char residue of pristine HNTs was 74wt% at 700°C. All modified HNT samples demonstrated two-step degradation process similar to the pristine HNT. For example, HNT-1000 started to decompose in the range of 80 – 200 °C due to evaporation of physically adsorbed water corresponding 10wt% of a mass loss. The second mass loss in the range of 350 – 430 °C was 41wt% and it resulted from the decomposition of the organic parts and dehydroxylation of the residual structural Al-OH groups of HNT. Finally, char residue was about 45wt% at 700 °C. The char residues of HNT-EF40, HNT-26120 and HNT-15170 at 700 °C were 63 wt%, 47 wt% and 49wt%, respectively. Thermal analysis results showed that hydroxyl groups of HNT cannot react efficiently with the anhydride of SMA compounds when the molecular weight is higher and anhydride ratio is lower.

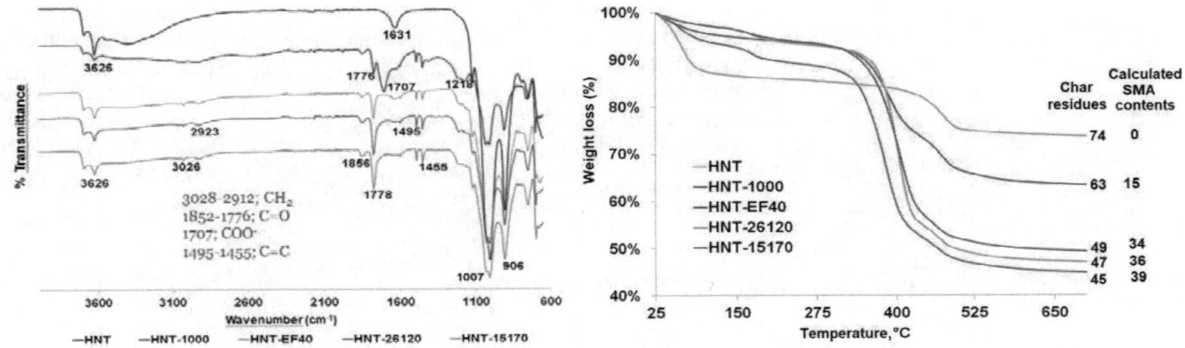

(a) (b)

FIGURE 1. (a) FTIR spectra, (b) TGA curves of modified HNT samples.

Figure 2(a) shows the details of the FTIR spectra of pristine and modified sepiolite samples. The broad peak at 3404 and 1464 cm^{-1} shows the O-H band stretching vibrations of zeolitic water in sepiolite. The band appeared at 1209 and 975 cm^{-1} shows Si-O bending and Si-O-Si stretching vibration, respectively. At the modified sepiolite spectra, the absorbance at 1853 and 1776 cm^{-1} shows the C=O stretching of anhydride. The absorbance at 1493 cm^{-1} and 1453 cm^{-1} exhibits C=C stretching of aromatic ring of benzene. The absorbance at 1194 cm^{-1} shows C–O–C stretching of anhydride. The other absorbances at 3028-2920 and 1600 cm^{-1} show -CH_2 bending and C=C stretching, respectively. All these observations demonstrated the modification of sepiolite surfaces with SMA compounds.

Figure 2(b) shows TGA curves of pristine and modified sepiolite samples. TGA curve of the pristine sepiolite shows that the weight loss for sepiolite during heating took place in 4 stages, at the temperature ranges of 25 – 150°C, 150 – 330°C, 550 – 700°C and 700 – 850°C, respectively. Sepiolite is a hydrated magnesium silicate, which contains zeolite water, coordinated water and hydroxyl group. The four weight loss may be attributed to the removal of surface-adsorbed water, zeolite water, coordinated water and hydroxyl group. Char residue of sepiolite was 76wt% at 900°C. Char residues of the modified sepiolite samples indicates that the SMA compounds with lower anhydride contents, such as SMA-EF40 and SMA-15170, were grafted onto the sepiolite surfaces.

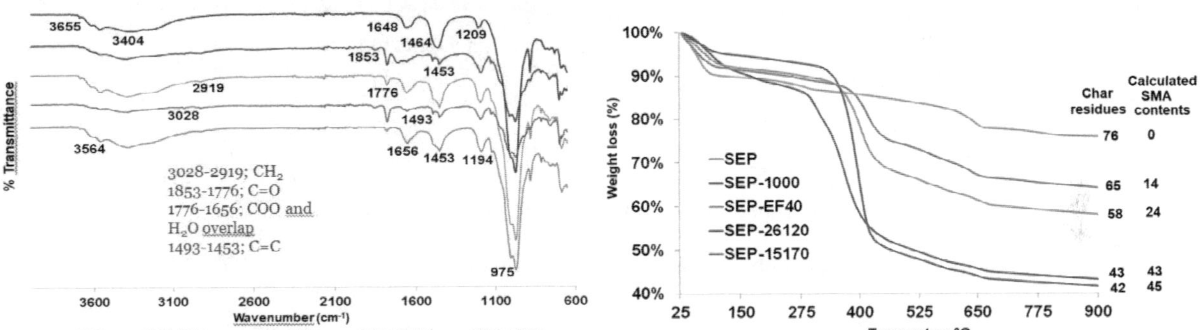

(a) (b)

FIGURE 2. (a) FTIR spectra, (b) TGA curves of modified SEP samples.

Characterization of PS/Clay Nanocomposites

The XRD patterns of the neat PS and the PS/HNT nanocomposites are shown in Figure 3(a). The characteristic (001) diffraction peak of HNT at 2θ = 12.0° is clearly visible in the figure. The diffraction peaks of the HNT are not apparent at 5 wt.% loaded nanocomposites, but become more visible, especially by the increased loading rate at 15wt%. The XRD patterns of PS nanocomposites based on SEP are shown in Figure 3(b). The characteristic (110) diffraction peak of SEP is located at 2θ=7.4°. As shown in the XRD patterns, the characteristic diffraction peak intensity of SEP increased with the increasing loading ratio of nano-fillers based on SEP.

The DSC endothermic curves for PS nanocomposites (PSNCs) are displayed in Figure 4. Obviously, pristine and modified HNT samples slightly increased the T_g of nanocomposites (Figure 4(a)). The increase may be attributed to the interaction between the PS matrix and the HNT, which limits the mobility of the polymer chains, increasing the glass transition temperatures of the nanocomposites. In addition, agglomeration of HNT nano-fillers may be the another reason for the increase of T_g due to inhibition of the molecular chain mobility of PS.

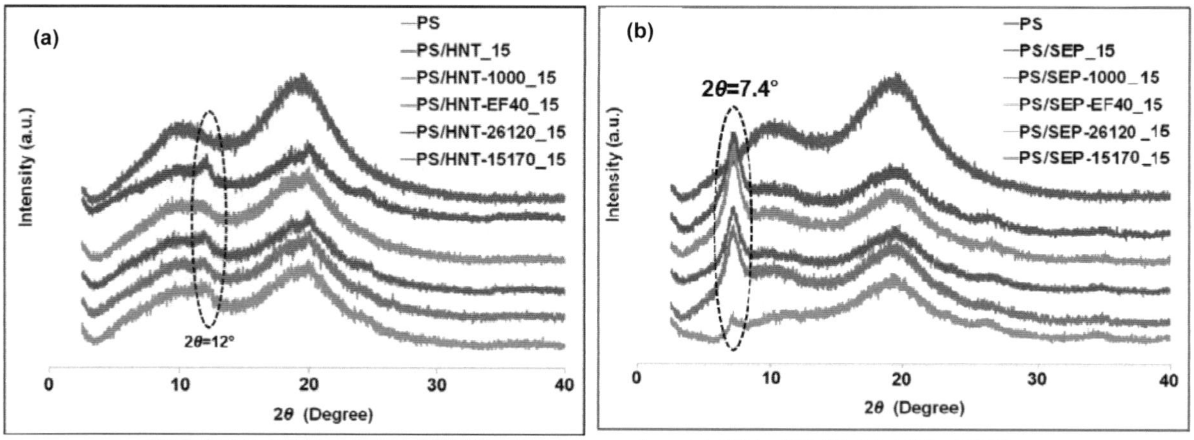

FIGURE 3. XRD patterns of PS/Clay nanocomposites; (a) 15% of HNT, (b) 15% of SEP.

Similar to the HNT samples, pristine SEP increased the T_g values of the PS nanocomposites (Figure 4(b)). However, the modified SEP samples increased the T_g values less than the pristine SEP. Herein, thanks to the surface modification with the SMA, modified SEP samples could be dispersed homogenously into PS matrices, compared to pristine SEP. Therefore, modified SEP nano-fillers slightly hindered the molecular chain mobility of PS less than pristine SEP.

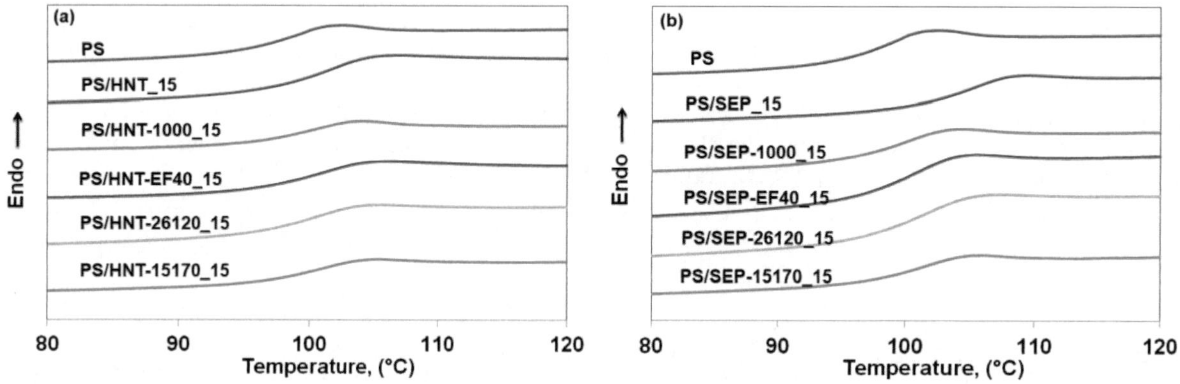

FIGURE 4. DSC curves of PS/Clay nanocomposites; (a) 15% of HNT, (b) 15% of SEP.

Figure 5 indicates the TGA curves of PS nanocomposites containing (a) 15wt% of HNT and (b) 15wt% of SEP nano-fillers. PS exhibited one-step degradation process. It showed a thermal stability up to 390 ºC and all the mass losses at 430 ºC were due to the thermal degradation of the organic chains. Moreover, it showed no residue at 600 ºC. The addition of 15 wt% of pristine HNT led to decrease the initial degradation temperatures (at 10 wt% weight loss, $T_{\%10}$) and the maximum degradation temperatures (T_{max}) of the nanocomposites compared to the degradation temperatures of the neat matrix. Herein, HNT displayed a catalytic effect on the PS pyrolysis due to the presence of hydroxyl groups (Al–OH) that acted as Brönsted active sites in catalysis. Therefore, HNT accelerated the escape of the volatile products in the degradation process. By the surface modification of HNT with SMA, the catalytic effect of nanotubes on thermal decomposition of PS can be avoided. The TGA results indicate that modified HNT samples, reduced the catalytic effect of nanotubes and improved the thermal stability of nanocomposites compared to the effect of pristine HNTs. Thanks to their tubular cavity, the modified HNT samples can capture the volatile products into their lumen, which enhances the thermal stability of nanocomposites (Figure 5(a)).

As seen from Figure 5(b), all kinds of SEP nano-fillers affected the degradation temperature of PS. Pristine sepiolite had the highest effect on the degaration temperature and increased it from 412 to 430 °C. PS nanocomposites containing 10wt% of sepiolite and 15wt% of SMA-15170 showed better thermal stability compared to the other PS nanocomposites. Thermal stability of PS nanocomposites increased as the SEP content increases. Therefore, sepiolite modified with SZ15170, namely SMA-15170, had the highest sepiolite ratio when compared to the others. Hence, SMA-15170 affected the thermal stability of PS the same as pristine sepiolite.

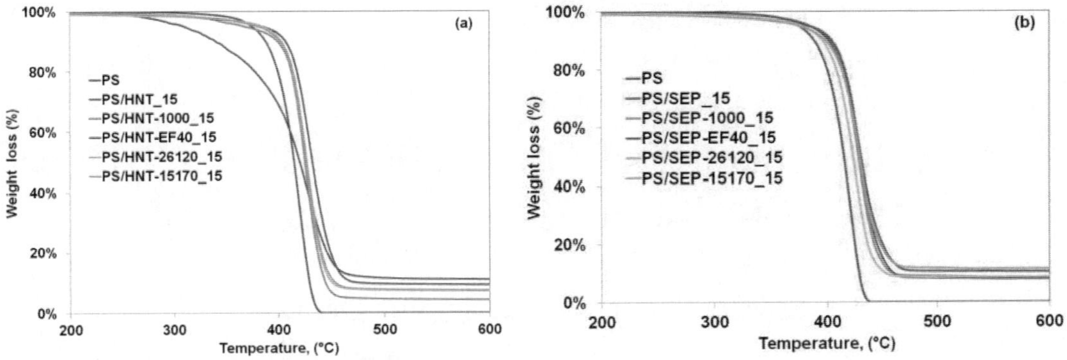

FIGURE 5. TGA curves of PS/Clay nanocomposites; (a) 15 wt% of HNT, (b) 15 wt% of SEP.

Figure 6 shows SEM images of PS/HNTs nanocomposites containing 5wt% and 10wt% of nanoclay. Pristine HNTs were agglomerated when added to the PS matrix in an amount of 5wt%. However, addition of 10 wt% HNTs resulted a decrease in the aggregate size. This was possibly due to the increased viscosity of the mixture at higher nanofiller loading levels. SMA-EF40 modified HNTs, namely HNT-EF40, showed the same behavior as pristine HNTs. HNT-EF40 has a lower organic content than HNT1000, it showed similar effect on the polymer matrix with the pristine HNTs. Conversely, HNT-1000 behaved differently than pristine HNT and HNTEF40. Since HNT-1000 has the highest organic content, it was uniformly distributed in the PS matrix even at 5wt% and 10wt% loading levels. As a result, surface modification with SMA-1000 increased the HNT distribution and prevented accumulation of nanotubes in the PS matrix.

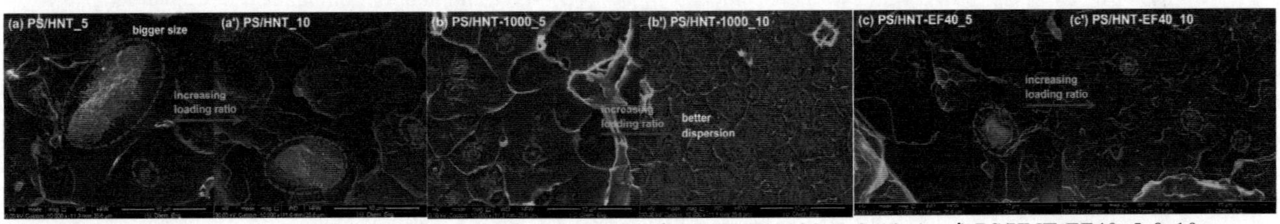

FIGURE 6. SEM images of PSNCs; (a–a') PS/HNT_5 & 10, (b–b') PS/HNT-1000_5 & 10, (c–c') PS/HNT-EF40_5 & 10.

Conclusion

In this work, PS/HNT and PS/SEP nanocomposites were successfully prepared via melt blending method to investigate the effect of the nanoclay dispersion on the structural and thermal properties of PS composites. In order to improve the dispersion of HNT and SEP into PS matrix and to improve compatibility between the nanoclays and the polymer matrix, modification with SMA-EF40 exhibited the best results.

ACKNOWLEDGMENTS

The authors would like to thank Eczacıbaşı ESAN, Assoc. Prof. Dr. Güralp Özkoç Group for all supports.

REFERENCES

1. M. Alexandre and P. Dubois, *Mater. Sci. Eng. Reports* **28,** 1–63 (2000).
2. S. Sinha Ray and M. Okamoto, *Prog. Polym. Sci.* **28,** 1539–1641 (2003).
3. R. Sengupta, S. Chakraborty, S. Bandyopadhyay, S. Dasgupta, R. Mukhopadhyay, K. Auddy, S. Deuri, *Engineering* **47,** 21–25 (2007).
4. N. Ning, Q. Yin, F. Luo, Q. Zhang, R. Du, Q. Fu, *Polymer* **48,** 7374–7384 (2007).
5. M. Du, B. Guo, M. Liu, D. Jia, *Polymer* **39,** 208–212 (2007).
6. H. Zou, S. Wu, J. Shen, *Chem. Rev.* **108,** 3893–3957 (2008).
7. B. Wang and H.X. Huang, *Polym. Degrad. Stab.* **98,** 1601–1608 (2013).
8. P. Yuan, P.D. Southon, Z. Liu, M.E.R. Green, J.M. Hook, S.J. Antill, C.J. Kepert, *J. Phys. Chem. C.* **112,** 15742–15751 (2008).
9. A. Nohales, R. Munoz-Espi, P. Felix, C.M. Gomez, *J. Appl. Polym. Sci.* **119(1),** 539–547 (2011).
10. D. Garcia-Lopez, J.F. Fernandez, J.C. Merino, J. Santaren, J.M. Pastor, *Compos. Sci. Technol.* **70(10),** 1429–1436 (2010).
11. R. Arat and N. Uyanık, *Nat. Resour.* **8,** 159–171 (2017).

Mechanical Properties of Polyurethane Films Synthesized from Bio-Based Hybrid Polyol Blends

Sevilay Nigar[1, b)] and Sennur Deniz[1, a)]

[1] *Yildiz Technical University, Department of Chemical Engineering, Davutpasa Cad. Nr: 127, 34210 Esenler, Istanbul-TURKEY*

[a)]deniz@yildiz.edu.tr
[b)]sevilay.nigar@gmail.com

Abstract. The increasing price of crude oil and eco-friendly approach has stimulated the development of polymers from renewable resources as an alternative. Several renewable materials like cellulose, protein, natural oil, lignin, starch and sugar have attracted the attention as an alternative renewable resource, among which, vegetable oils are the most advantageous because they are cost-effective, sustainable, eco-friendly and easily available. The vegetable oils, especially soybean oils, consisting of triglycerides with different chain lengths as well as number of double bonds are suitable for the several polymerization processes. The polymers with high molecular weight cannot be obtained from vegetable oils alone, unless their reactive groups are converted to more reactive hydroxyl, epoxy or carboxyl functionalities. Thus, efforts are being made to achieve a renewable approach via utilization of vegetable oils as a fundamental material for the synthesis of polymer. These efforts have brought vegetable oil-based polyurethanes (PUs) in limelight, especially due to their ease of synthesis and potential applications in various fields such as medical, footwear, packaging, automobile, paints and coatings. PUs are traditionally used as foams, elastomers, films, coatings, adhesives and varnishes, and in recent years they have also been used as biomaterials. Epoxidized soybean oil (ESBO) based polyols are a great alternative for PUs industry with their excellent properties. In this study, the ESBO-based polyols with a range of hydroxyl numbers were prepared using commercial epoxidized soybean oil by ring opening reactions in the presence of a catalyst. Flexible PU films were prepared via mixing poly(tetramethylene glycol) (PTMG1000), polycarbonate polyol (PC), polypropylene glycol (PPG425) and ESBO polyols by reacting with different diisocyanates (H12MDI, 4,4'-dicyclohexylmethane diisocyanate and IPDI, isophorone diisocyanate). Synthesized ESBO based polyols and PU films tensile strength and elongation were measured.

INTRODUCTION

PU materials attract much attention in commercally and scientific research due to their unique properties flexibility, durability, biocompatibility and biostability. PUs have a good abrasion resistance, excellent mechanical strength, good oil, grease and solvent resistant, good corrosion resistance and good heat resistance properties. Renewable raw materials for chemical industry are becoming more preferred than petrochemical-derived raw materials [1-3]. The demand for bio-renewable resources for polymer preparation that replace petroleum-based raw materials is increasing, due to escalating carbon footprints, depletion of nonrenewable resources, unsustainability of the petroleum supply and waste disposal, etc. [4]. Vegetable oil based polyols, which are relatively cheap, renewable and environmentally friendly, have been used to replace or partially replace conventional polyols derived from petroleum in many fields such as the polyurethane manufacture and fine chemicals [3].

PUs are produced by the exothermic reactions between polyols as a soft segment and isocyanates as a hard segment. Mainly petrochemical derived polyols are used for synthesis polyurethane. For polyols for PU, various kinds of vegetable oils, like castor, soybean, sunflower, corn oil, tung oil, castor oil and rapeseed oils have been used. Soybean oil is a good option for synthesis vegetable oil based polyols because of its ready availability, low cost and environmental sustainability [4]. The structure of the polyurethanes can be tailored by selecting appropriate raw materials at suitable ratios.

The aim of this study to investigate the synthesis of soy-based polyols and polyurethane films. In this study, epoxide soy bean oil has been used as alternatives to petrochemical-derived polyols. A comparison of soy-based polyols properties with their petroleum-based analogues were investigated. Also, the hybrid PU films were synthesized buy using soybean oil based polyols and petroleum based polyols are used as a soft segment.

EXPERIMENTAL

Materials

Dichloromethane (DCM, Lab-Scan), triethylamine (TEA, Acros), dimethylformamide (DMF, Lab-Scan), methyl ethyl ketone (MEK, JT. Baker), ethyl alcohol (EtOH, J.T. Baker), trimesic acid (1,3,5-benzenetricarboxylic acid, BTC, Acros), copper(II) acetate monohydrate (Acros) were used as received without further purification. Epoxidized soybean oil (ESBO) (f = 4.1) was supplied by Plastifay. The epoxy oxygen content (EOC) of ESBO is 0.385 mol/100 g. Isophorone diisocyanate (IPDI, 98 %), 4,4'-dicyclohexylmethane diisocyanate (H12MDI), dibutyl tin dilaurate (DBTDL), Amberlyst-15 and 1,4-butanediol (BuOH) were obtained from Sigma-Aldrich and used as received.

Synthesis of Vegetable Oil Based Polyol and Polyurethane

First step of this study is polyol synthesis. Under vigorous stirring, ESBO, methyl etyl ketone (MEK) and catalyst were added to a flask with a reflux condenser. The mixture was stirred at 100°C for 24 hours under nitrogen atmosphere. At the end of the reaction, ESBO based polyol was cooled down to room temperature.

Second step of this study is polyurethane synthesis. The synthesis of polyurethane includes the formation of prepolymer, followed by a chain extending reaction. Bio-based polyurethane are prepared by mixing different type of petroleum-based diisocyanates with different amount and soy-based polyol or/and petrochemical based polyols under nitrogen atmosphere. DBTDL as a catalyst, BuOH as a chain extender and DMF as the solvent added to the mixture step by step under vigorous stirring at 75°C for 5 h.

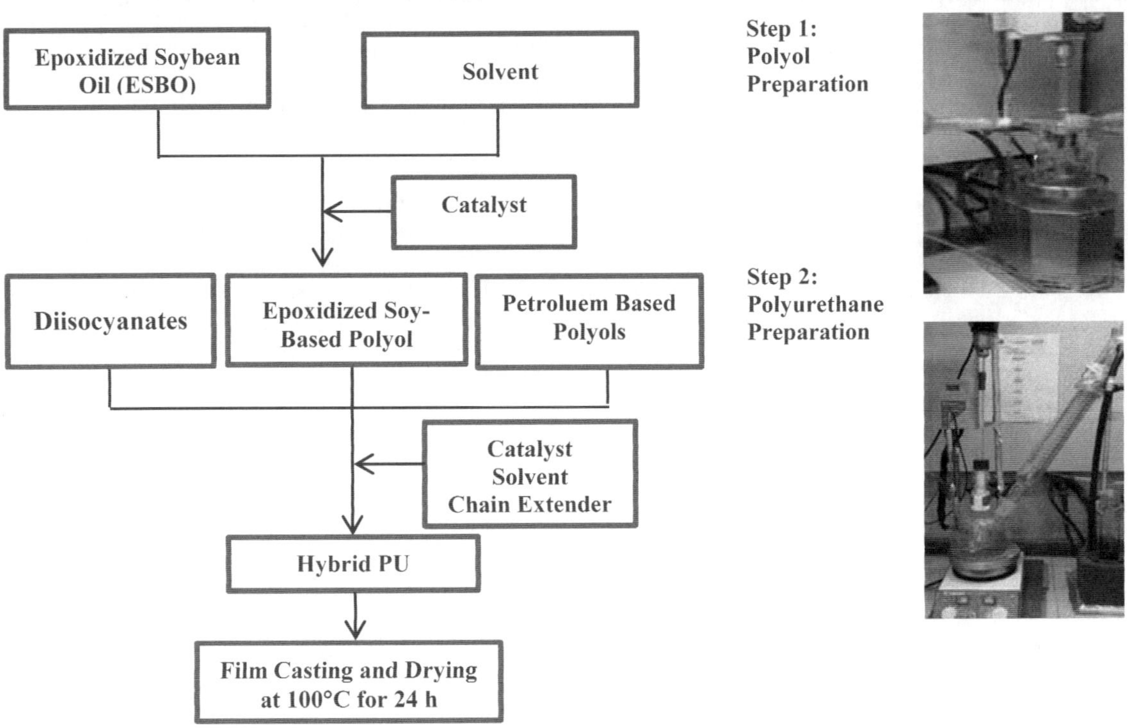

FIGURE 1. Schematic diagram and photos for the ESBO polyol and PU synthesis

As the PU reaction was completed, polyurethane mixtures were cast onto a transfer paper by using a casting knife. PU films were cured at 100°C for 24 h in oven and then cooled down to room temperature. Mechanical properties, such as tensile strength and elongation of the obtained films were determined. The preparation of polyols and polyurethanes were shown in Fig. 1. PU film preparation step of this work are shown in Figure 2.

(a) (b) (c)

FIGURE 2. (a) Synthesized PU film casting, (b) Drying of PU films and (c) Obtained PU films

The synthesis composition and properties of synthesized PU films were shown in Table 1.

TABLE 1. The combination of PU films preparation and film properties

Sample	ESBOBased Polyol	Diisocyanate	Petrochemical Based Polyols			Properties
			PC	PPG 425	PTMG 1000	
Sample1		IPDI	70		30	Flexible film formation High tensile strength and elongation
Sample2	50	H12MDI	50			Flexible film formation High tensile strength and elongation
Sample3	50	H12MDI		50		Flexible, weak film formation Low tensile strength and elongation
Sample4		IPDI	50	50		Flexible, weak film formation Low tensile strength and elongation

RESULTS AND DISCUSSION

Mechanical Properties of PU Films

Tensile strength testing and elongation at break of the polyurethanes were carried out using Devotrans universal testing machine with a load cell. Before and during the tensile strength and elongation test for PU film prepared with a bio-based hybrid polyol blend (Sample 2) were displayed in Fig. 3-a,b. The tensile strength and elongation at break test results for synthesized PU films were shown in Fig. 4 and listed in Table 2.

According to the results of mechanical tests, maximum tensile strength was obtained for Sample 2. The tensile properties of Sample 1 was observed the close to Sample 2. In this study, it was found that ESBO based polyols with proper petrochemical based polyols and isocyanates were a good alternative for the polyurethane manufacturing.

(a)

(b)

FIGURE 3. The photos of tensile strength test of PU film prepared with a bio-based hybrid polyol blend (Sample 2); (a) Before, and (b) During testing

TABLE 2. Mechanical test results of PU films

	Tensile strength	Elongation at break	Stress
	N	**mm**	**N/mm²**
● **Sample 1**	50.0	492.66	5.000
● **Sample 2**	63.0	357.01	6.300
● **Sample 3**	14.5	11.00	1.450
● **Sample 4**	9.5	12.44	0.950

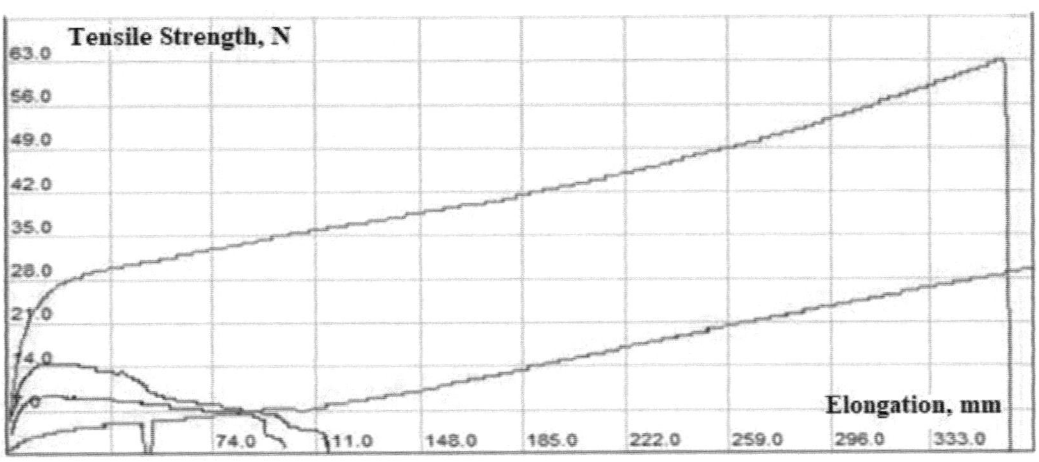

FIGURE 4. Tensile strength-strain curves of PU films

CONCLUSION

Polyurethane derived from soybean oil are alternative materials for polyurethane including the film applications. Soy-based polyols are important prepolymeric materials using the renewable resources that produce PUs films. They can compete in many aspects with polyols derived from petrochemicals. To obtain ESBO based polyols in this study, ESBO was polymerized by using ring opening reaction with a catalyst. For preparing of flexible PU films, PTMG1000, PC, PPG425 and ESBO polyols were used in different composition with diisocyanates. Maximum tensile strength and elongation were found by using PC (%50), ESBO (%50) and H12MDI. Also, the maximum tensile strength and elongation were found by using PC (%70), PTMG1000 (%30) and IPDI. According to our test results, it was found that ESBO can be used for preparing flexible polyurethane films.

ACKNOWLEDGMENTS

This work was supported by the Yıldız Technical University-Research Foundation (Project No: 2014-07-01-DOP05).

REFERENCES

1. G. Açık, M. Kamacı, Ç. Altınkök, H.R.F. Karabulut and M.A.Taşdelen, Progress in Organic Coatings **123**, 261-266 (2018).
2. J. R. Ashish, J. Smita, S. P. Vimal and Z. D. Bharatkumar, Bioresource Technology Reports **3**, 88-94 (2018).
3. P. Król, Ł. Urama, B. Król, K. Pielichowskab and M. Ł. Walczaka, Materials Science&Engineering C **93**, 483-494 (2018).
4. Y.S. Kim and B.G. Min, Fibers and Polymers **18**, 1841-1847 (2017).

Synthesis of Novel Terpolymer for High Temperature Drilling Applications

Hafiz Mudaser Ahmad[a], Muhammad Shahzad Kamal[b], Mamdouh Al-Harthi[c]

[a]*Department of Chemical Engineering, King Fahd University of Petroleum and Minerals, 31261 Dhahran, Saudi Arabia*
[b]*Center for Integrative Petroleum Research (CIPR), King Fahd University of Petroleum & Minerals, 31261 Dhahran, Saudi Arabia*
[c]*Department of Chemical Engineering, King Fahd University of Petroleum and Minerals, 31261 Dhahran, Saudi Arabia*

Abstract. Drilling fluids are considered as the blood of the drilling industry for the extraction of oil and gas. Many chemical additives are employed to formulate drilling fluids, but water-soluble polymers play a vital role to adjust rheological and filtration properties. Natural and conventionally used polymers in drilling fluids show degradation at high temperature drilling applications. For this purpose, a novel terpolymer based on 2-acrylamide-2-methyl propane sulfonic acid, acrylic acid, and acrylamide monomers was synthesized for high-temperature application of drilling fluids. The chemical structure and composition of terpolymer were analyzed using Fourier transform spectroscopy. The suitability of terpolymer for drilling fluid formulation was determined by performing rheological properties with different concentrations at a temperature ranging from 75°F to 150°F. The steady shear rheology experiments showed that increasing the concentration polymer nanocomposite in the base drilling fluid resulted in a significant increase in viscosity, shear stress and gel strength of modified drilling fluid. While increasing the temperature of modified drilling fluid up to 150°F with increasing terpolymer concentration, the rheological properties (shear stress and viscosity) shows no prominent change compared to the modified drilling fluid (B) at 150°F. The synthesized novel terpolymer with 2-acrylamide-2-methyl propane sulfonic acid, acrylic acid, and acrylamide monomers can be used to formulate drilling fluid for high temperature applications.

Keywords: Ent Polymer Synthesis, Rheology, Terpolymer, Drilling Fluids
PACS: Replace this text with PACS numbers; choose from this list: http://www.aip.org/pacs/index.html

INTRODUCTION

The water-soluble polymers have vast applications in cosmetics, paints, pharmaceutical and oil and gas industries. The water-soluble polymers are considered as primary additives in formulating drilling muds for various drilling applications [1]. These polymers have many functions in drilling fluids such as to control the rheology of drilling fluid, modify the filtration properties, as a lubricating agent, and shale inhibitors[2]. The synthesis of the water-soluble polymer has long been in the interest of chemists to enhance the drilling fluid properties. Based on the composition of drilling fluid additives, drilling fluids are classified as water-based drilling fluids and oil-based fluids. Oil-based drilling fluids have some advantages over water-based drilling fluids such as enhanced wellbore stability, superior filtration properties, less shale swelling, high lubricity, thermal stability, and better capacity to transport the drilled cuttings from bottom hole to the surface. However, there are some drawbacks of oil-based drilling fluids such as high cost, and strict environmental regulations. On the other hand, water-based drilling fluids are environmentally friendly and have less cost compared to the oil-based drilling fluids[3-5].

Recently, many synthetic and natural polymers have been employed to enhance the rheological and filtration properties of water-based drilling fluids. The natural polymers used in water-based drilling fluids include cellulose, xanthan gum, guar gum, polyanionic cellulose, and derivatives of cellulose. The synthetic polymers used in formulating drilling fluids include polymer acrylamide, partially hydrolyzed polyacrylamide, copolymers and terpolymers of acrylamide. Current research is mainly focused on the use of synthetic polymers with superior rheological, filtration and shale inhibition properties. In a recent study, a high molecular copolymer with varying concentration was used to enhance the rheological properties at 85°C. It was observed that yield stress, viscosity, gel strength, and filtration properties were enhanced at a higher temperature. The SEM analysis of filter cakes showed that homogeneous distribution of bentonite and polymer. The effect of the degree of hydrolysis of the different copolymer was assessed for water-based drilling fluid properties. The polymer molecules adsorb on the surface of bentonite platelets and resulted in an increase of rheological properties. The high molecular weight polymers are superior and less prone to thermal degradation at high temperature and pressure. Similarly, polymer nanocomposite and the addition of nanoparticles in the drilling fluids also increase the rheological properties[6-10].

In this article, the synthesis of terpolymer was discussed as one the additive for water-based drilling fluid. The terpolymer was synthesized by solution polymerization technique and characterized by FTIR spectroscopy. The terpolymer solution was mixed with the base drilling fluid to form modified drilling fluids. The steady shear rheological properties such as shear stress, viscosity, and gel strength were determined. The effect of time on viscosity and dynamic rheological properties were measured and compared with the base drilling fluid.

EXPERIMENTAL

Materials

Three different monomers 2-acrylamide-2-methyl propane sulfonic acid, acrylic acid and acrylamide were obtained from Sigma Aldrich. The initiator potassium persulfate and pH control agent sodium hydroxide were obtained with 99% purity from Sigma Aldrich. The bentonite was obtained from BDH Chemicals (UK) in powder form having pH ranges from 9.5–10.5 and a density of 2.5 g/cm^3 at 20 °C. Deionized water was used for making a solution of polymers and drilling fluid formulations.

Synthesis of Terpolymer

The terpolymer was synthesized using acrylic acid (AA), 2-acrylamide-2-methyl propane sulfonic acid (AMPS) and acrylamide (AM) monomers. The free radical solution polymerization technique was adopted to synthesize terpolymer. The stock solution was prepared using AMPS, AM, and AA with 6/5/4 molar ratio in 250 ml of deionized water. The monomer stock solution was stirred for 30 minutes in a three-neck round bottom flask. The monomer solution mixture was heated to 80°C in the oil bath and thoroughly mixed using magnetic stirrer. The whole apparatus was connected to nitrogen supply to carry the polymerization reaction under an inert atmosphere. After achieving 80°C temperature, the 10 ml freshly prepared 1 wt.% solution of potassium persulfate was injected dropwise in the reaction mixture. The polymerization reaction was allowed to take place for 4 hours. Terpolymer from the solution was separated by a mixture of acetone and ethanol solvent and dried in the vacuum oven at 50°C.

Preparation of Drilling Fluids

The base drilling fluid (A) was prepared by mixing 5% bentonite in deionized water and left in the beaker covered at room temperature for 24 hours. Before making modified drilling fluid, the solid form of terpolymer was dissolved in the deionized water to make to completely mixed terpolymer solution. The terpolymer solution was obtained by mixing 15 g of solid terpolymer in 100 ml of deionized water under magnetic stirring. The modified drilling fluid (B) was obtained by mixing 240 ml of base drilling fluid and 10 ml of terpolymer solution. The modified drilling fluid (C) was obtained by mixing 190 ml of base drilling fluid and 10 ml of terpolymer solution. The pH of all the drilling fluids was adjusted to 10 by adding an appropriate amount of sodium hydroxide. All the fluids were thoroughly mixed using mechanical stirrer and left undisturbed for 24 hours before performing the rheological experiment.

Characterization and Rheological Properties

The terpolymer was characterized using Fourier Transform Infrared spectroscopy to determine the chemical composition and molecular structure. The rheological properties of base drilling fluid and modified drilling fluid was determined at two different temperatures such as 75°F and 150°F. The steady shear rheological properties such as viscosity and shear stress were determined by increasing the shear the rate from 0.01 to 1000 (1/s). The gel strength of drilling fluid is an important rheological property that determined the cutting suspension capacity. Gel strength of base and modified drilling fluid was determined at 10 seconds and 10 minutes holding time according to the API standards. The viscosity of base and modified drilling fluids was determined at a constant shear rate at 75°F and 150°F over the time period of 30 minutes. The dynamic rheological properties such as storage and loss modulus were also measured at 75°F and 150°F of the base and modified drilling fluid. The effect of terpolymer concentration on rheological properties was measure and compared with the rheological properties of base drilling fluid.

RESULTS AND DISCUSSION

The FTIR spectroscopy was performed for terpolymer to determine the composition and chemical structure of terpolymer. The absorption peak of the terpolymer is shown in Figure 1. The absorption peak at 3410 cm-1 was attributed to the (H-N) stretching of the primary amide group in the acrylamide. The another (H-N) absorption peak was observed at 3126 which was due to the presence of secondary amide group in the AMPS. The (C-C) and (C-H) stretching peaks were observed at 2921 cm-1 and 2862 cm-1. The absorption peaks at 1047 and 1107 were attributed to the stretching of (S-O) and (S=O) bonds in the AMPS group.

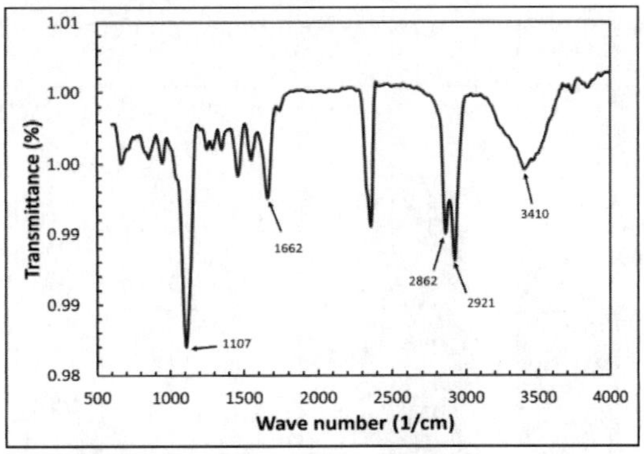

Figure 1. FTIR analysis of terpolymer

The viscosity profile of base and modified drilling fluids at different temperatures is shown in Figure 2. Figure 2 (a) shows that all the drilling fluids are shear thinning behavior which means that increasing the shear rate viscosity of fluid decreases. At the low shear rate, the incorporation of the polymer has significantly increased the viscosity. However, at high shear, the change in the viscosity was significant compared to the low shear rate viscosity. The incorporation of terpolymer to base fluid significantly increased the viscosity and this increase was attributed to the adsorption of polymer chains on bentonite. At higher temperature 150°F, the viscosity of base and modified polymer slightly increases compared to the viscosities at low temperature as shown in Figure 2(b). It was also noticed that increasing concentration of polymer at a higher temperature resulted in a slight decrease in the viscosity.

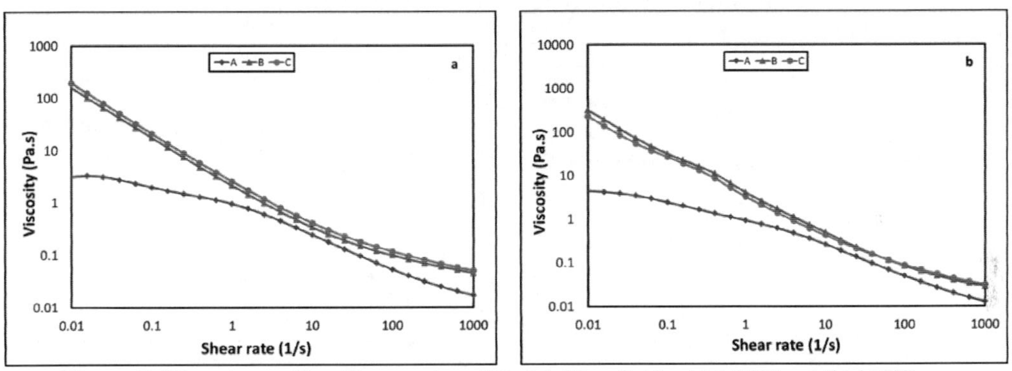

Figure 2. Viscosity profile of base and modified drilling fluids at (a) 75°F and (b) 150°F

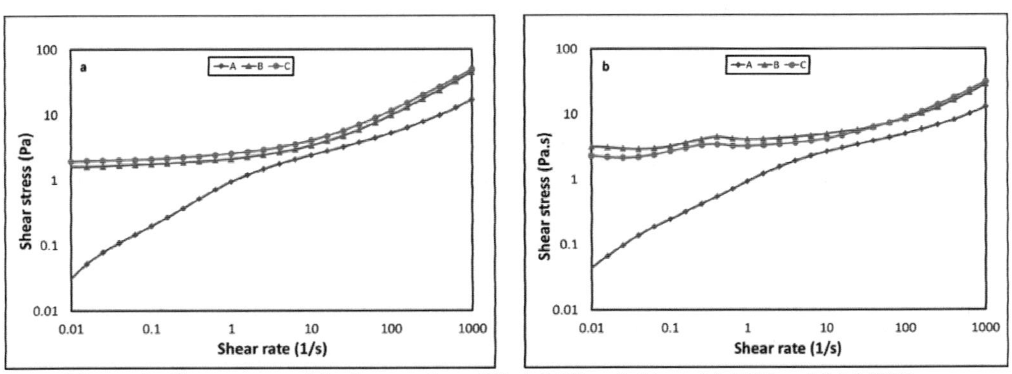

Figure 3. Flow curves of the base and modified drilling fluids at (a) 75°F and (b) 150°F

Figure 3 shows the flow curves of the base and modified drilling fluid at two different temperatures. Figure 3 (a) shows that incorporating terpolymer increased the shear stress significantly. At the low shear rate, the shear stress of modified fluid was very high compared to the base fluid. While at a high shear rate, the shear stress of base and modified fluid was close to each other. Figure 3 (b) shows the shear stress of the base and modified fluid at 150°F. The shear stress was slightly increased at higher temperature of all the fluids compared to the low temperature due to the enhance flocculation and adsorption of terpolymer with bentonite. The shear stress of modified fluid with highest terpolymer concentration shows slightly low shear stress at higher temperature.

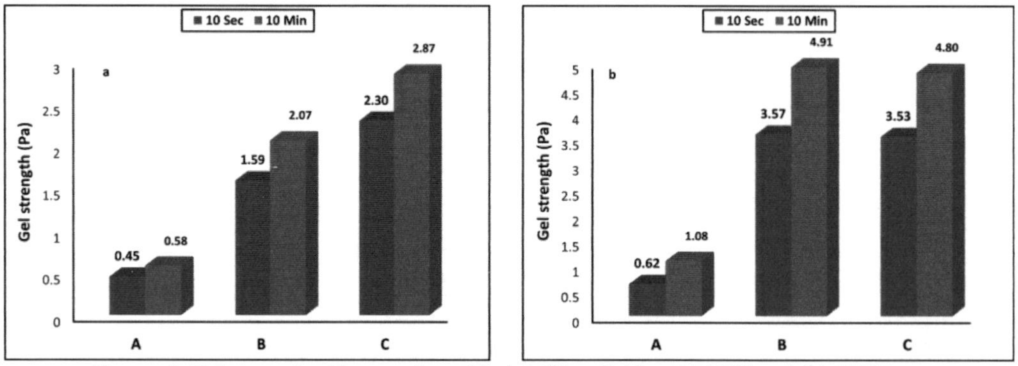

Figure 4. Gel strength of base and modified drilling fluid at (a) 75°F and (b) 150°F

Figure 4 shows the gel strength of the base and modified drilling fluids at two different temperatures. Figure 4 (a) shows that increasing the concentration of terpolymer at 75°F resulted in an increase of both 10-sec and 10-min gel strengths. While Figure 4 (b) shows the gel strength of the base and modified drilling fluid at 150°F. It showed that at 150°F, adding terpolymer to base fluid resulted in the increase of gel strength compared to the gel strength of base and modified fluid at 75°F. However, a further increase in the concentration of terpolymer resulted in a slight decrease in the gel strength compared to modified drilling fluid (B) at 150°F. The superior gel strength at higher temperature is beneficial to hold the drilled cutting for a specific time in the drilling operations.

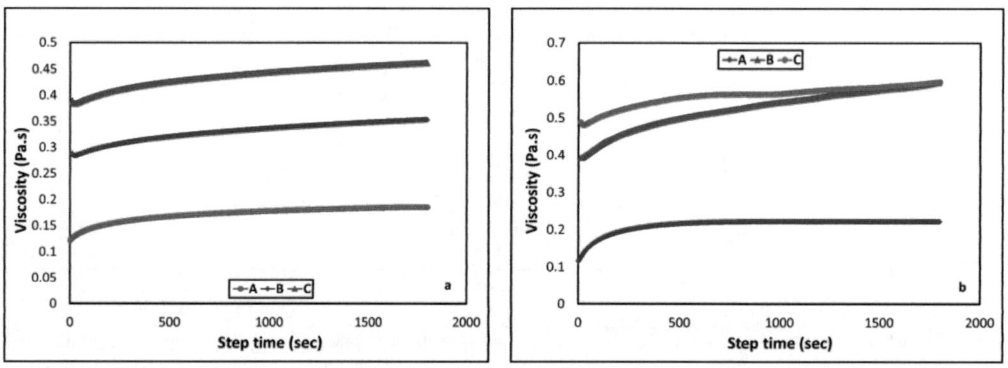

Figure 5. Viscosity vs time of base and modified drilling fluid at (a) 75°F and (b) 150°F

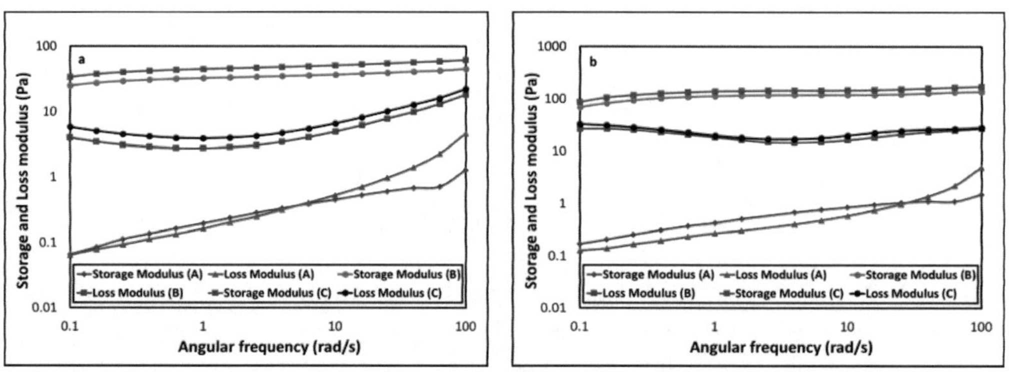

Figure 6. Storage and loss modulus of base and modified drilling fluid at (a) 75°F and (b) 150°F

Figure 5 shows the change in viscosity change of base and modified drilling fluid at two different temperatures over the time period of 30 minutes. Figure 5 (a) shows that the addition of terpolymer in the base fluid resulted in an increase in viscosity at 75°F. The viscosity of all the fluids slightly increased during the time period of 30 minutes at a constant shear rate. Figure 5 (b) shows the change of viscosity of base and modified drilling fluid at 150°F. The results showed that at higher temperature, the viscosity of all the fluids increased compared to the viscosities at low temperature. At 150°F temperature, the viscosity of drilling fluid (B) and (C) becomes the same after the time period of 30 minutes.

Figure 6 shows the storage and loss modulus of the base and modified drilling fluids at two different temperatures such as 75°F and 150°F. The storage loss modulus shows the elastic and viscous contributions of drilling fluids, respectively. Figure 6 (a) shows the storage and loss modulus of the base and modified drilling fluid at 75°F and it showed that incorporating terpolymer to base fluid resulted in the increase of storage and loss. The storage modulus is greater than the loss modulus of all the drilling fluids. Figure 6 (b) shows the storage and loss modulus of the base and modified drilling fluids art 150°F. It was observed that increasing the temperature resulted in the increase of storage and loss modulus compared to the low temperature storage and loss modulus.

CONCLUSION

The terpolymer was synthesized by free radical solution polymerization technique and characterized by FTIR analysis. The terpolymer was mixed with drilling fluid to form modified drilling fluids with different concentrations. The rheological properties of the base and modified drilling fluids were performed at 75°F and 150°F. The rheological properties of modified drilling fluid (B) at 150°F showed that addition of terpolymer in the base drilling fluid resulted in the enhanced rheological properties. Hence, the synthesized terpolymer offers enhanced rheological properties at higher temperature and can be used as one of the drilling fluid additives.

ACKNOWLEDGMENTS

The authors are thankful to the Department of Chemical Engineering and Center for Integrative Petroleum Research for providing research support.

REFERENCES

1. Vryzas, Z., et al., Effect of temperature on the rheological properties of neat aqueous Wyoming sodium bentonite dispersions. Applied Clay Science, 2017. **136**: p. 26-36.
2. Yue, Y., et al., Improving wellbore stability of shale by adjusting its wettability. Journal of Petroleum Science and Engineering, 2018. **161**: p. 692-702.
3. Seyedmohammadi, J., The effects of drilling fluids and environment protection from pollutants using some models. Modeling Earth Systems and Environment, 2017. **3**(1).
4. Noah, A.Z., et al., Enhancement of yield point at high pressure high temperature wells by using polymer nanocomposites based on ZnO & CaCO 3 nanoparticles. Egyptian Journal of Petroleum, 2017. **26**(1): p. 33-40.
5. Zhao, S., et al., Water-Based Drilling Fluid Technology for Extended Reach Wells in Liuhua Oilfield, South China Sea. Petroleum Science and Technology, 2009. **27**(16): p. 1854-1865.
6. Ahmad, H.M., M.S. Kamal, and M.A. Al-Harthi, High molecular weight copolymers as rheology modifier and fluid loss additive for water-based drilling fluids. Journal of Molecular Liquids, 2018. **252**: p. 133-143.
7. Ahmad, H.M., et al. Improving the Drilling Fluid Properties using Nanoparticles and Water-Soluble Polymers. in SPE Kingdom of Saudi Arabia Annual Technical Symposium and Exhibition. 2017. Society of Petroleum Engineers.
8. Ahmad, H.M., M.S. Kamal, and M.A. Al-Harthi, Effect of thermal aging and electrolyte on bentonite dispersions: Rheology and morphological properties. Journal of Molecular Liquids, 2018. **269**: p. 278-286.
9. Ahmad, H.M., M.S. Kamal, and M.A. Al-Harthi, Rheological and filtration properties of clay-polymer systems: Impact of polymer structure. Applied Clay Science, 2018.
10. Ahmad, H.M., et al., Synthesis and Experimental Investigation of Novel CNT-Polymer Nanocomposite to Enhance Borehole Stability at High Temperature Drilling Applications, in SPE Kingdom of Saudi Arabia Annual Technical Symposium and Exhibition. 2018, Society of Petroleum Engineers: Dammam, Saudi Arabia. p. 17.

Effect of the chain transfer agent on pressure-sensitive adhesive performance of butyl acrylate based latexes and determination of odor level caused by sulfur derivative compound

Ecem Temelkaya Bilgin[a, b], Murad Turna[b], Cansu Akarsu Dülgar[b]

[a] *Graduate School of Science, Engineering & Technology, Department of Polymer Science & Technology, Istanbul Technical University, Maslak 34469, Istanbul, Turkey*
[b] *Organik Kimya San. ve Tic. A.Ş., Eyüp 34075, Istanbul, Turkey*

Abstract. The purpose of the present study was to investigate the influence of bi-functional chain transfer agent, DMDO (1, 8-dimercapto-3, 6-dioxaoctane) at varying amounts on pressure-sensitive adhesive performance and determination of odor level according to VDA 270 test standard comparing to NDM species synthesized in the same manner. DMDO was incorporated at 0.15, 0.25 and 0.4% s/m ratio and NDM was incorporated at 0.25% s/m ratio into butyl acrylate and acrylic acid copolymer latexes via seeded semi-batch emulsion polymerization procedure. With the intended design, pressure-sensitive adhesive performance was investigated depending on the change in molecular weight. Since the increasing amount of DMDO obtained shorter polymer chains and consequently lower molecular weight, the peel adhesion and initial tack values were also increased enhancing better adhesive properties. NDM resulted in the highest initial tack value. However the peel adhesion value was between the peel adhesion values of DMDO-0.15% and DMDO-0.25%. According to the odor test results 0.15% and 0.25% DMDO content had an acceptable odor level whereas the one with 0.4% had very unpleasant and unacceptable odor like NDM at 0.25% for industrial scale uses. The present research is supported by Organik Kimya Research & Development Center, TURKEY.

Keywords: pressure-sensitive adhesive; adhesion; initial tack; odor determination.
PACS: 82.35.-x; 82.35.Gh

INTRODUCTION

Pressure-sensitive adhesives, PSAs form a bond when a light pressure is applied to bond the adhesive with the adherent[1]. As the name pressure-sensitive implies that the degree of bond and wet-out of adhesive on the substrate is related with the pressure applied. Pressure-sensitive adhesives can be categorized by type as acrylics, hot-melt and rubbers, by application as tape and label, and by end-use as permanent, removable and semi-removable. Pressure-sensitive adhesives are used in various areas including packaging, protective films, graphic arts, electronics, medical tapes, and specialty tapes etc[2]. Due to the environmental concerns and regulations water-based pressure-sensitive adhesives dominate the market passing through other types of PSAs[3]. The pressure-sensitive adhesives market was valued at USD 7.87 billion in 2015, and is projected to reach USD 15.76 billion by 2026 at a compound annual growth rate of 6.6% between 2016 and 2026[4]. The stringent requirements for some application areas complicate the design of the pressure-sensitive adhesive. Having both high peel adhesion along with high shear strength is challenging for manufacturers due to the fact that adhesion and cohesion are opposite characteristics to each other, in which the parameters that increase the adhesion decrease the cohesion, and vice versa. The balance between peel adhesion and shear strength may be regulated by the incorporation of chain transfer agents, which react with the free radical side of a growing polymer chain and interrupt chain-growth by reinitiating a polymer chain[5]. Thiols (mercaptans) are the most effective chain regulators among halogen compounds and some aromatic hydrocarbons[6]. However mercaptans have a pungent smell that can be easily detected by human's nose even in trace level concentration due to containing sulfur derivatives. Manufacturers, end-users and also raw material suppliers are in search of an alternating compound without or less odor to the most common type N-dodecyl mercaptan. 1, 8-dimercapto-3, 6-dioxaoctane, DMDO was introduced into the market as bi-functional and low odor mercaptan by its manufacturer.

The purpose of the present study was to investigate the influence of bi-functional chain transfer agent, DMDO at varying amounts on pressure-sensitive adhesive performance and determination of odor level according to VDA 270 test standard. DMDO was incorporated at 0.15, 0.25 and 0.4% s/m ratio and NDM was incorporated at 0.25% s/m ratio into butyl acrylate and acrylic acid copolymer latexes via seeded semi-batch emulsion polymerization procedure. Solid and coagulum content, pH and viscosity were determined and polymer latexes were successfully synthesized. Unreacted monomer content was measured by gas chromatography/mass spectroscopy to reveal the copolymerization. Particle size and particle size measurements were conducted, and molecular weights were determined by Fikentscher K-value equation. With the intended design, pressure-sensitive adhesive performance was investigated depending on the change in molecular weight. The present research is supported by Organik Kimya Research & Development Center, TURKEY.

MATERIALS and METHODS

Materials

Butyl acrylate (BA) and acrylic acid (AA) were supplied from Arkema and Mitsui & Co., LTD. NDM and DMDO were used as chain transfer agent and supplied from Adakem and ARKEMA respectively. The initiator was sodium persulfate (NaPS) and supplied from Siancity Xiamen Co., LTD. Sodium pyrophosphate was used as buffer to keep the pH up and justify the speed of reaction throughout the polymerization and supplied from VESER. The surfactant system consisted of a mixture of two different types of anionic surfactants. The surfactant system employed in the pre-emulsion was referred to as surfactant mixture. The seed polymer used for the polymerization was 50 nm of particle size. Surfactant system and seed composition cannot be revealed for proprietary reasons. For the post-polymerization processes tert-butyl hydroperoxide (t-BHP) and sodium formaldehyde sulfoxylate (SFS) were selected as redox couple and obtained from Gliss Chemicals and Bruggemann respectively. Ammonium solution (28%) was supplied from a local manufacturer. Distilled and deionized water were used throughout all experiments arranged here.

Preparation of Pre-Emulsion and Polymerization Procedure

Surfactant was dissolved in deionized water and placed in a vessel equipped with stirrer. BA, AA and DMDO or NDM were added into the water-surfactant mixture at specific amounts as indicated in Table 1 below. The mixing time required was 10 minutes for all the trials. The resulting monomer pre-emulsions were milky white in appearance and moderately viscous. The solid content for the trials was 50%.

TABLE (1). Amount of chain transfer agent, % s/m in latex formulations

Polymer	NDM-0.25	DMDO-0.15	DMDO-0.25	DMDO-0.40
n-dodecyl mercaptan, NDM	0.25	-	-	-
1, 8-dimercapto-3, 6-dioxaoctane, DMDO	-	0.15	0.25	0.4
Acrylic acid, AA	1	1	1	1
Butyl acrylate, BA	99	99	99	99

The polymerization was carried out by a seeded semi-continuous emulsion polymerization process in a 2 L three-neck round-bottom glass reactor equipped with a stainless steel stirrer, condenser and two feed ports. Deionized water and initiator solution (0.045% s/m) were charged into the reactor respectively and stirred at 100 rpm. The feeding times for the pre-emulsion and initiator solutions were 180 mins at 90 °C. For the first 20 min of feeding, the pre-emulsion and initiator solutions were fed slowly as ½ flow rate to have a controlled particle size and particle size distribution. After the feed was completed, the reaction was allowed to proceed for an additional 30 mins to increase monomer conversion at the reaction temperature. The latex was then cooled to room temperature and poured into a suitable container for further characterization and application tests. Specifications of synthesized copolymers are indicated in Table 2. Average particle sizes of synthesized latexes were measured as approximately 180 - 190 nm.

Sample Preparation for Adhesive-Performance Testing

Test specimens were prepared according to FINAT test procedures FTM1 and FTM9[7]. Coating weight was justified as 18±1 g/m^2 on 50 µm biaxially oriented polypropylene (BOPP) film by using 50 µm applicator and dried at 70 °C for 15 min. Then the specimens were smoothly cut in machine direction as pre-specified in test standards. The coated samples were conditioned in a controlled environment (22±2 °C and 50±5% relative humidity) for 30 min prior to testing.

Test specimens were prepared according to FINAT Test Method No: 1 (FTM1) for peel adhesion and FINAT Test Method No: 9 (FTM9) for initial tack measurement. In FINAT technical handbook, peel adhesion is defined as force required removing pressure-sensitive-coated material, which has been applied to a standard test plate under specified conditions from the plate at a specified angle and speed (FTM1) and initial tack value of PSAs is expressed as the force required separating, at a specified speed, a loop of material which has been brought into contact with a specified area of a standard surface (FTM9)[7].

In this study, a specimen of 25x400 mm was cut in the machine direction and laminated onto the clean stainless steel (SS) test plate for peel adhesion measurement. A specimen of 25x125 mm was cut in the machine direction and tested on SS test plate for initial tack measurements.

Characterization and Sample Preparation for PSA Performance Testing

Total solid content was measured by drying the polymer at 150 °C for 20 min after filtered through 60 micron according to ISO 124. Weight of polymer (W_1) and dried latex (W_2) were calculated by the following equation (1):

$$\text{Solid \%} = W_2 / W_1 \times 100 \quad (1)$$

pH of polymers was determined at 25 °C by a calibrated pH-meter according to ISO 976.

Viscosity was measured by Brookfield viscometer at 25 °C by LVT 3/60 according to ISO 3219.

Particle sizes were determined by dynamic light scattering using Malvern Zetasizer Nano S instrument.

Residual monomer content were measured by GC-MS/HS (Agilent Technologies 7890B, 5977B MSD, 7697A HS) and He was used as carrier gas.

Fikentscher K value calculation is a relative indication for molecular weight measurement. Fikentscher K value of polymers were measured by a Cannon-Fenske capillary flow viscometer comply with ISO/DIS 3105. The time Δt for the liquid level to move constant distance was measured and relative viscosity (η_{rel}) was calculated by using the time for polymer solution (t) and solvent (t_0) as illustrated in equation (2). THF was used as solvent in the preparation of solutions. Three drops of trifluoroacetic acid (TFAA) were added into each solutions to increase the solubility of polymers and the concentration of polymer solutions was adjusted to 0.5% (C).

$$\eta_{rel} = \eta/\eta_0 = t/t_0 \quad (2)$$

Fikentscher K value calculation was done according to the following equation by using the calculated ηrel as illustrated in equation (3).

$$K = 1000\,x\,\frac{1.5\,log\eta rel - 1 \pm \sqrt{1 + \left[\frac{2}{C} + 2 + 1.5\,log\eta rel\right]}\,x\,1.5\,log\eta rel}{150 + 300C} \quad (3)$$

The average force to peel the specimen from test plates was recorded at 180° at 300 mm/min speed by Adhesion Release Tester AR-1000 according to FTM1 standards. A loop of the specimen has been brought into contact with test plates on Initial tack Tester LT-1000 equipment according to FTM9 standards. All application tests were performed three times and the average values were shared on the graph. All samples were dwelled and all tests were performed at a controlled environment room.

The determination of odor level was conducted according to the VDA 270 test standard[8]. Same amount of sample was dried at room temperature on a Teflon coated cup for about 3 days until all the samples were totally dried. Then the test samples were removed from the cup and replaced in a pot with cap. The pots with caps closed placed into the oven at 80 °C for 2 hours. Then, the pots were conditioned at room temperatures for 10 min and the odor of test specimens were evaluated by at least 3 people and the level of odor was graded over 6 as "unbearable" to 1 as "no odor perceptible". The average value of grading gave the certain results for odor level.

RESULTS and DISCUSSION

Poly(BA-co-AA) latexes containing DMDO and NDM as chain transfer agent at varying amounts were synthesized via seeded semi-batch macro emulsion polymerization procedure to investigate the pressure-sensitive adhesive properties on SS plates. Solid content, pH, viscosity, particle size values and residual monomer contents of all polymers were indicated in TABLE (2). According to the results given on Table (2), polymers have been synthesized successfully via seeded semi-batch macro emulsion polymerization procedure.

TABLE (2). Specification of synthesized pressure sensitive adhesives

Polymer	NDM-0.25	DMDO-0.15	DMDO-0.25	DMDO-0.40
Solid content, %	50.2	50.3	50.7	50.2
pH, -	6.1	6.3	6.5	6.4
Viscosity, cps	100	90	110	120
Particle size, nm	188	183	189	189
Unreacted monomer content, ppm	77	12	111	159
Fikentscher K-value, -	37.5	60	45.3	34.5

The incorporation of DMDO at varying ratios did not change the particle size, however unreacted monomer content increased with the increasing amount of DMDO. It seems that DMDO inhibited the copolymerization when used at excess amount. Molecular weights decreased accordingly increasing DMDO content regarding the Fikentscher K-values as expected[9-11].

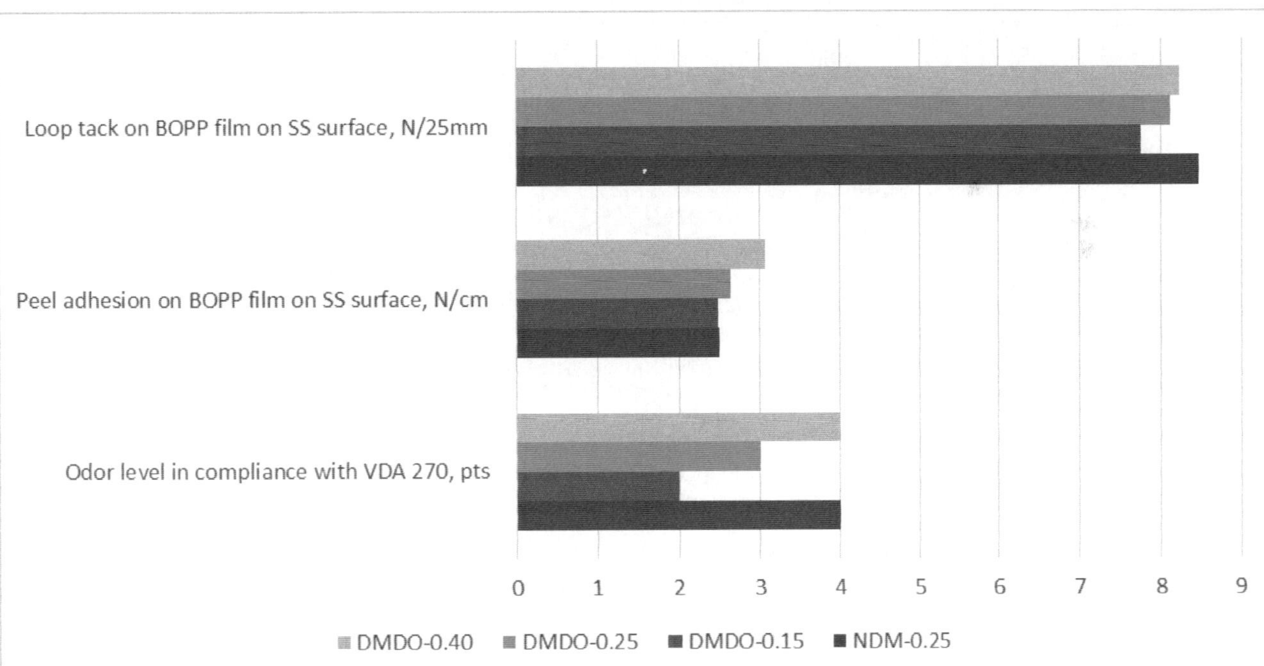

FIGURE 1. (a) The initial/loop tack (b) The peel adhesion and (c) Odor level determination

The addition of DMDO increased the adhesion values on SS as well as the initial tack values. Since the increasing amount of DMDO decreased the chain length and improved the mobility, the polymer chains were assumed to be more labile to interact with the surface and the substrate. The addition of NDM at 0.25% s/m resulted in the highest initial tack value. However the peel adhesion value was between the peel adhesion values of DMDO-0.15% and

DMDO-0.25%. The authors do not expect any correlation between the results of DMDO and NDM, since they have different chemical compositions leading to different copolymerization manner. These results conducted in this study fulfilled the authors' expectations. The primary concern of the present paper was to create an alternative to the odor generating sulfur compound due to the laborer's health and workplace safety regulations. DMDO was selected as chain transfer agent since it is double functionalized and low odor generating sulfur derivative compound. The odor level was determined in compliance with VDA 270 test standard. According to the average scaling results of 5 people who smelled the test specimens at the same time, 0.15% DMDO content had an acceptable odor level among the others. However when DMDO content increased up to 0.4%, the odor was very unpleasant and unacceptable for especially industrial scale uses. 0.25% NDM content resulted in an unpleasant odor such as DMDO with the highest amount[12-15].

CONCLUSION

The aim of the present study was to investigate the influence of bi-functional chain transfer agent, DMDO at varying amounts on pressure-sensitive adhesive performance and determination of odor level according to VDA 270 test standard. DMDO was incorporated at 0.15, 0.25 and 0.4% s/m ratio into butyl acrylate and acrylic acid copolymer latexes via seeded semi-batch emulsion polymerization procedure. NDM was also used at 0.25% s/m to compare the PSA performance and odor level with countertypes. With the intended design, pressure-sensitive adhesive performance was investigated depending on the change in molecular weight. Unreacted monomer content and particle size measurements revealed that copolymer latexes were successfully synthesized. Molecular weights were determined by Fikentscher K-value equation. With the increasing amount of DMDO, molecular weight was decreased because the chain transfer agent shortened the polymer chains. However unreacted monomer content resulted highest with the highest amount of DMDO. Peel adhesion and initial tack value were also measured for monitoring the adhesive performance. When the DMDO content was increased up to 0.4%, the peel adhesion and initial tack values were also increased enhancing better adhesive properties. NDM-0.25% resulted in the highest initial tack value and peel adhesion value was between the peel adhesion values of DMDO-0.15% and DMDO-0.25%. Regarding the odor test results 0.15% DMDO content had an acceptable odor level among the others. However when DMDO content increased up to 0.4%, the odor was very unpleasant and unacceptable such as 0.25% NDM content.

ACKNOWLEDGMENTS

The authors gratefully acknowledge the technical and financial support of Organik Kimya R&D Center.

REFERENCES

1. S. P. Athavale, *Handbook of Pressure Sensitive Adhesives and Coatings*, Chennai, Notion Press, 2018.
2. I. Benedek, *Pressure-Sensitive Adhesives and Applications*, New York, Marcel Dekker, 2004.
3. R. Jovanović and M. A. Dubé, J Macromol Sci Polymer Rev **44**, 1-51 (2007).
4. https://www.marketsandmarkets.com/Market-Reports/pressure-sensitive-adhesives-market-1135.html
5. M. D. Gower and R. A. Shanks, Macromol Chem Phys. **205**, 2139-2150 (2004).
6. C. Plessis, G. Arzamendi, J. R. Leiza, J. M. Alberdi, H. A. S. Schoonbrood, D. Charmot, and J. M. Asua, J. Polym. Sci., Part A: Polym. Chem. **39**, 1106-1119 (2001).
7. FINAT Technical Handbook 9th edn. Test Methods FTM 1, FTM 9 (2014).
8. Verband der Automobilindustrie Recommendations, VDA 270 Determination of the Odour Characteristics (01 June 2018).
9. Z. Kowalski and Z. Czech Int J Adhes Adhes **60**, 9-15 (2015).
10. A. Falsafi, M. Tirrel and A. V. Pocius, Langmuir **16**, 1816-1824 (2000).
11. K. Dhal, A. Deshpande and G. N. Babu, Polymer **23**, 937-939 (1982).
12. M. Nomura, Y. Minamino, K. Fujita and M. Harada J Polymer Sci Polymer Chem Ed **20**, 1261-1270 (1982).
13. J. Comyn, *Adhesion Science*, The Royal Society of Chemistry, 1997.
14. S. Sun, M. Li, A. Liu Int J Adhes Adhes **41**, 98-106 (2013).
15. Z. Czech Polym Int **52**, 347–357 (2003).

Obtaining black electrochromic materials by copolymerization of EDOT and star-shaped carbazole derivatives

Merve Guzel*, and Metin Ak

Chemistry Department, Faculty of Art and Science, Pamukkale University, Denizli 20070, Turkey

Abstract. Because of their electrochromic properties conductive polymers can be used in display technologies and smart windows. Despite the large number of electrochromic polymer synthesis, researchers have made great efforts to synthesize black-to-transparent electrochromic materials due to their superior optical contrasts. In this study, novel black-to-transmissive electrochromic materials based on copolymerization of EDOT and star-shaped carbazole derivatives (SC1 and SC2) have been designed and synthesized. The electrochemical and spectroelectrochemical properties of the copolymers prepared with different monomer feed ratios have been investigated and the optimum ratio has been determined for obtaining the black-to-transmissive electrochromic material.

Keywords: Thin films, Electrical properties of conducting polymers; Electrochromism; Electrochemical displays

PACS: 73.61.Ph; 78.20.Jq; 82.47.Tp

INTRODUCTION

Recent advances in the design and synthesis of conducting polymers (CPs) have been extensively studied owing to their practical applications in solar cells[1,2], smart windows[3–5] and sensors[6,7]. For conductive polymers, the electrochromism is defined as reversible color change during under an applied electric field. Carbazole has a the electron-rich unit, for this reason, it can be easily functionalized with 1,3,5-triazines. Carbozle-based compounds have properties such as high thermal stability, fast response time and high optical contrast[8,9]. When these compounds are in oxidized, their color alters from transparent to blue or green[10,11]. EDOT, which is widely used in copolymer studies, is transparent when it is oxidized[12,13]. Although numerous research focus on black-to-transmissive materials[14–16] such as carbazol, pyrrole and EDOT derivatives, it is very difficult to obtain these materials.

In this work, black electrochromic materials have been obtained by copolymerization of EDOT and star-shaped carbazole (SC1 and SC2) derivatives. Copolymers of EDOT and carbazole derivatives have been prepared using different monomer feed ratios by cyclic voltammetry. Properties of the obtained homopolymer and copolymer films have been comparatively investigated. The copolymers have been exhibited a black color with absorption the whole visible spectrum. As a result, the obtained black electrochromic materials make them applicable in smart window etc.

Synthesis of SC1 and SC2

2,4,6-tris((9H-carbazol-1-yl)oxy)-1,3,5-triazine (SC1) and 2,2'-((6-(quinolin-8-yloxy)-1,3,5-triazine-2,4-diyl)bis(oxy))bis(9Hcarbazole) (SC2) were synthesized according to our previous literatures[11,17].

Electropolymerization

The electrochemical polymerization and copolymerization of monomers were investigated by the technique of CV. In this technique, the electrochemical properties of the homopolymers and copolymers were performed by scanning potential from −0.5 to 1.5 V in 0.1 M $LiClO_4$/ACN electrolyte solution. (Figure 1). Onset oxidation potentials have been observed as 0.95 V for SC1, 0.78 V for SC2 and 1.1 V for EDOT. PSC1 showed broad oxidation peak at 0.55 and 1.03 V, while a reduction peak at 0.52 V showed. The polymerization of EDOT has been observed broad oxidation-reduction peaks. PSC2 demonstrated two oxidation peaks (E^{ox}=0.5 and 1.0 V) and a reduction peak at 0.51 V. With increasing the scan cycles in all monomers, the current has gradually increased and this is evidence of the formation of polymer films on the electrode surface.

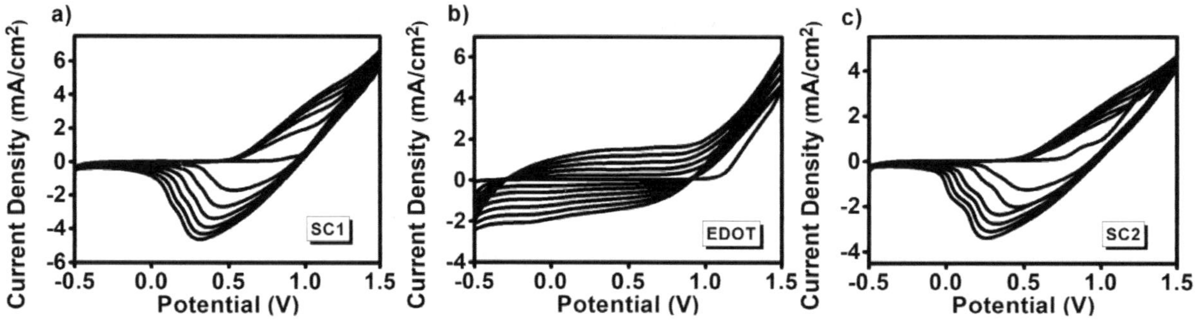

FIGURE 1. Cyclic voltammograms of electropolymerization of a)SC1 b) EDOT and c)SC2 in 0.1 M $LiClO_4$/ACN

To determine the best monomer feed ratio in copolymerization, copolymers were prepared with electropolymerization of different monomer feed ratios. This ratio has been selected based on the highest current value in cyclical voltammetry experiments and the highest optic contrast value in spectroelectrochemistry experiments. As can be seen from cyclic voltammograms in Fig. 1, SC1-co-EDOT and SC2-co-EDOT potential values and shapes are different from each other. The onset of oxidation is located at 0.85 and 0.75 V for SC1-co-EDOT and SC2-co-EDOT respectively. For comparison, CVs deposited in the different monomer feed ratios have been illustrated in Fig 2(a-f). The differences between the electrochemical properties of homopolymers and copolymers, indicate the formation of copolymerization.

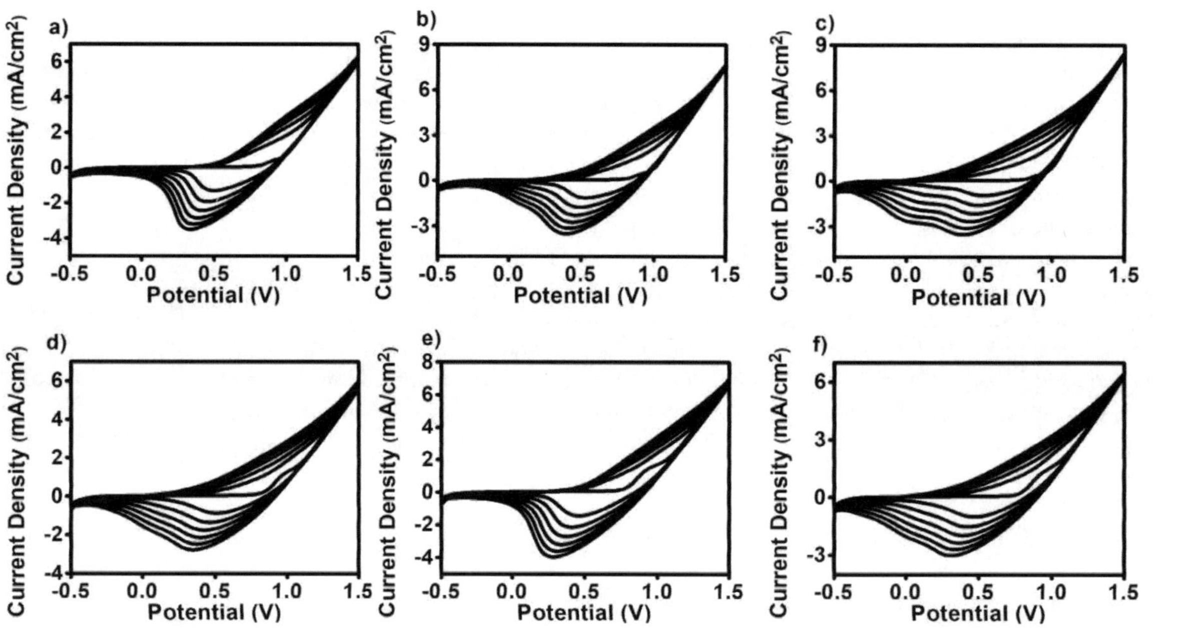

FIGURE 2. Cyclic voltammograms of SC1 and EDOT a) f_{SC1} = 0.75 b) f_{SC1} =0.85 c) f_{SC1} =0.95 and SC2 and EDOT d) f_{SC2}= 0.75 e) f_{SC2} =0.85 f) f_{SC2} =0.95 in monomer-free solution. f_{SC1} and f_{SC2} indicate mole fractions of SC1 and SC2 in the mixture with EDOT.

Spectroelectrochemistry

Spectroelectrochemistry is the best method for investigating the changes in the optical properties of homopolymer and copolymer films. Homopolymers and copolymers have been prepared as described above. After washing with ACN, homopolymer films were taken in a monomer-free solution. The absorption spectra of PSC1, PEDOT and PSC2 are presented in Fig. 3. At the neutral state, the PSC1 film has exhibited the absorption band at 327 nm due to the π-π* transitions. The maximum optical contrast of PSC1 has been calculated as 57% at 640 nm. PEDOT, in the neutral state, has shown a broad absorption at 605 nm, where the band gap is 1.65 eV (Fig. 3b). Besides, maximum optical contrast of PEDOT has been revealed as 52% (660 nm). PSC2 film has displayed the π-π* transition at 3.33 eV (328 nm) with the band gap at 1.73 eV. For PSC2 ,the maximum optical contrast has been calculated as 51% at 675 nm.

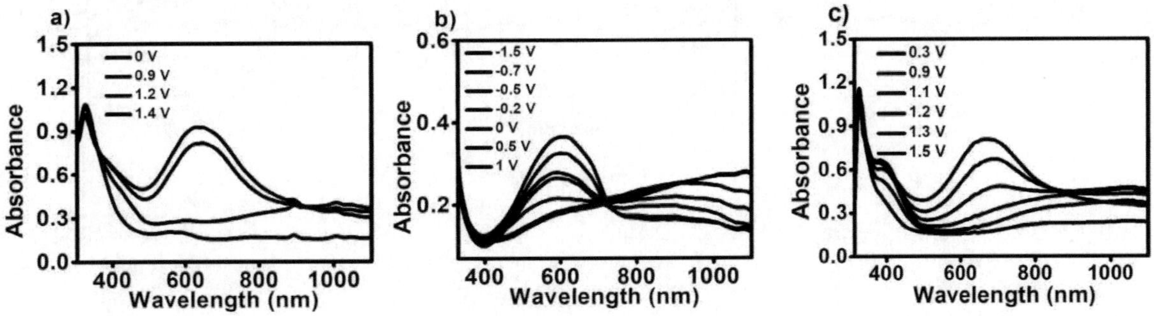

FIGURE 3. Optoelectrochemical spectras of a) PSC1 b) PEDOT c) PSC2 films in monomer-free solution.

The UV–vis spectra of the copolymer films was investigated 0.1 M $LiClO_4$/ACN solution solution. As shown in Figure 5, as the amount of EDOT decreased in the monomer mixture, the optical contrast values of the copolymers have increased. The maximum optical contrast has been observed at 665 nm for SC1-co-EDOT and at 450 nm for SC2-co-EDOT. In the light of the information obtained from cyclic voltammetry and spectroelectrochemical measurements, the optimum monomer feed ratio for both monomers (SC1 and SC2) has been specified as 0.95. In the neutral state, the P(SC1-co-EDOT) film has been exhibited the π-π* electron transition peak at about 328 nm and its band gap has been found as 1.73 eV (Fig. 4c). As the applied potentials increased, two bands at 422 and 660 nm arising from the formation of polaron and bipolaron have appeared. As shown in Fig. 4f, the π-π* transition has been showed 329 nm and the band gap has been found to be 1.8 eV for the P(SC2-co-EDOT). When spectroelectrochemistry graph of P(SC2-co-EDOT) have been examined, the polaron and bipolaron absorption bands have been shown at 405 and 665 nm respectively. The maximum optical contrast (%ΔTmax) of P(SC1-co-EDOT) and P(SC2-co-EDOT) films have been calculated to be 59% and 53%, respectively. When the results were compared, P (SC1-co-EDOT) was found to show higher optical contrast than P (SC2-co-EDOT). On the other hand, the colors of the copolymers changed from transparent to green and then to black.

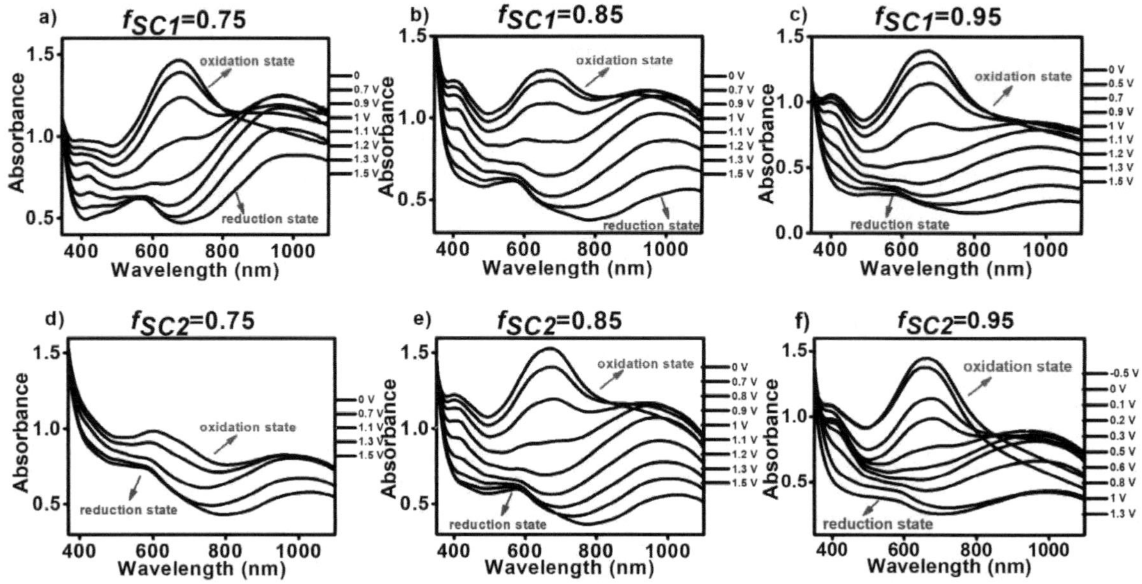

FIGURE 4. UV-visible spectra of P(SC1-co-EDOT) a)f_{SC1} = 0.75 b)f_{SC1} =0.85 c)f_{SC1} =0.95 and P(SC2-co-EDOT) d)f_{SC2}=0.75 e) f_{SC2} =0.85 f) f_{SC2} =0.95 in monomer-free solution. f_{SC1} and f_{SC2} indicate mole fractions of SC1 and SC2 in the mixture with EDOT.

In Table 1, the characteristic values for homopolymers and copolymers have been summarized. The energy levels (HOMO-LUMO) have been calculated using the onset of oxidation obtained from the CV as well as the band gap.

TABLE I. Optical and electrochromic characteristics of polymer films electrochemically synthesized on ITO

Polymer	E_{onset} (V)	Optical contrast (%)	HOMO (eV)	LUMO[a] (eV)	λ_{max}(nm)
PSC1	0.95	57	−4.09	−1.30	327
P(SC1-*co* EDOT)	0.85	59	-4.00	-2.27	422
PEDOT	1.13	52	-3.67	-2.02	605
PSC2	1.1	51	−4.79	−1.45	328
P(SC2-*co*-EDOT)	0.75	53	-4.04	-2.24	405

[a] LUMO energy levels calculated using optical band-gap values and HOMO energy levels.

Electrochromic properties of P(SC1-co-EDOT) and P(SC2-co-EDOT) films

The copolymer had exhibited multicolor electrochromism under various potentials. As the applied potential increases, the color of the copolymer films turned from transparent to black color (Figure 5).

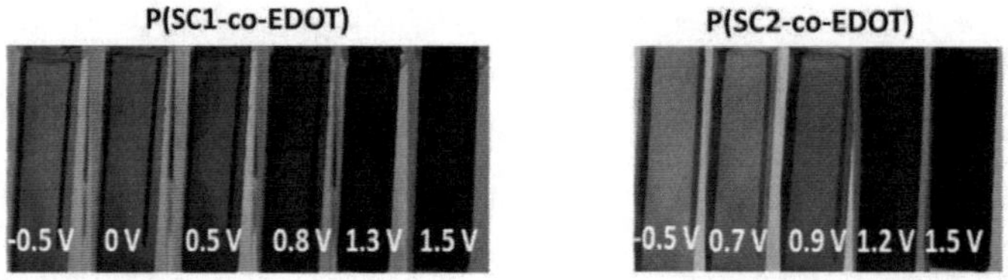

FIGURE 5. The images of P(SC1-co-EDOT) and P(SC2-co-EDOT) at different potentials

CONCLUSION

In conclusion, novel black-to-transmissive electrochromic materials have been designed from copolymerization of EDOT and star-shaped carbazole derivatives (SC1 and SC2). Electrochemical copolymerizations have been prepared using different monomer feed ratios in $LiClO_4$/ACN electrolyte solutions. The electrochemical and spectroelectrochemical characterizations of homopolymers and copolymers have been reported. The resulting copolymer absorbed all wavelength light in the entire visible region and this is evidence for the formation of black color.

ACKNOWLEDGMENTS

This work was supported by BAGEP Award of the Science Academy.

REFERENCES

[1] G. Dennler, M.C. Scharber, and C.J. Brabec, Adv. Mater. **21**, 1323 (2009).
[2] S. Günes, D. Baran, G. Günbas, A. Durmus, A. Fuchsbauer, N.S. Sariciftci, and L. Toppare, Polym. Chem. (2010).
[3] M. Guzel, T. Soganci, E. Karatas, and M. Ak, J. Electrochem. Soc. **165**, H316 (2018).
[4] M. Ak, V. Gancheva, L. Terlemezyan, C. Tanyeli, and L. Toppare, Eur. Polym. J. **44**, 2567 (2008).
[5] S. Soyleyici, M. Karakus, and M. Ak, J. Electrochem. Soc. **163**, H679 (2016).
[6] R. Ayranci, D.O. Demirkol, S. Timur, and M. Ak, Analyst **142**, 3407 (2017).
[7] R. Olgac, T. Soganci, Y. Baygu, Y. Gök, and M. Ak, Biosens. Bioelectron. **98**, 202 (2017).
[8] M. Guzel, T. Soganci, R. Ayranci, and M. Ak, Phys. Chem. Chem. Phys. **18**, 21659 (2016).
[9] M. Guzel, E. Karatas, and M. Ak, J. Electrochem. Soc. **165**, H437 (2018).
[10] M. Guzel, E. Karatas, and M. Ak, J. Electrochem. Soc. **164**, H49 (2017).
[11] M. Guzel, T. Soganci, M. Akgun, and M. Ak, J. Electrochem. Soc. **162**, H527 (2015).
[12] M. Akbayrak, E.G. Cansu-Ergun, and A.M. Önal, Des. Monomers Polym. **19**, 679 (2016).
[13] R. Ayranci, M. Ak, M. Karakus, and H. Cetisli, Des. Monomers Polym. **19**, 545 (2016).
[14] D. Weng, Y. Shi, J. Zheng, and C. Xu, Org. Electron. **34**, 139 (2016).
[15] S. Abraham, S. Mangalath, D. Sasikumar, and J. Joseph, Chem. Mater. **29**, 9877 (2017).
[16] J. He, L. You, and J. Mei, ACS Appl. Mater. Interfaces **9**, 34122 (2017).
[17] E. Karatas, M. Guzel, and M. Ak, J. Electrochem. Soc. **164**, H463 (2017).

Morphology Distribution within Injection Molded Parts, Interpretation on the Basis of Stretch and Work Evolutions

V. Speranza (reference author)[a], S. Liparoti[a], R. Pantani[a] and G. Titomanlio[a]

[a] *Department of Industrial Engineering, University of Salerno, Via Giovanni Paolo II, I84084 Fisciano (SA), Italy*
vsperanza@unisa.it, sliparoti@unisa.it, rpantani@unisa.it, gtitomanlio@unisa.it

Abstract. Isothermal experimentations carried out under steady shear clarified that, if the flow has adequate intensity, the polymer crystallization takes place into highly oriented morphological elements aligned along the flow direction. A criterion based on local critical values of both the shear rate applied, and the mechanical work spent was identified. However, rather than the shear rate, the molecular stretch is representative of the effect of flow on the molecules and it represents a more appropriate choice to adopt in a criterion.

The aim of this paper is to find a criterion based on the critical values of both a molecular stretch, and the mechanical work and to apply it to the complete processing of the material (from pellets to a final product through an industrial process). Since polymer processing operations are carried out under unsteady conditions at least according to the Lagrangian sense, the interest to a criterion formulation which considers unsteady conditions automatically came in. It is assumed that if a critical amount of mechanical work is carried out while the stretch is above its critical value, there a fibrillar morphology will form; however, if once the critical stretch is reached it relaxes below its critical value, the accounting of the mechanical work has to start again.

The criterion was applied to the analysis of a set of iPP injection molded samples obtained under a very wide range of thermal boundary conditions. The effects of temperature, pressure and flow on relaxation times, nucleation density, crystalline growth rate, as well as the interrelation among these quantities, were accounted for into a simulation software adopted to determine the evolutions of the molecular stretch and mechanical work. The results were combined into the criterion in order to obtain cross section morphology distribution into the final moldings. The dependence of the morphological layers widths upon the processing conditions was satisfactorily described.

Keywords: Thermoplastic polymers, Crystallization kinetics, Flow induced crystallization, morphology evolution, injection molding.
PACS: 47.85.md; 81.10.Aj; 47.85.M−

INTRODUCTION

The conditions required to achieve a shish kebab structure during the crystallization of thermoplastic polymers have been analyzed by several authors by experimentations carried out under steady shear and temperature conditions. It was found that there is a critical shear rate under which the crystallization gives rise to shish kebab structures, if the time required for carrying out all steps required for the morphology completion is available while the critical shear rate is active. the availability of such a time is assured by the achievement of a critical value of the mechanical work accumulated after the critical shear rate had reached its critical level. Recently it was recognized [1] that a parameter representing a critical molecular stretch is more appropriate than the shear rate to represent a critical effect of the velocity field on the polymer, and the critical mechanical work has to be accumulated after the critical stretch parameter reaches its critical level. Since all polymer processing operations take place under unsteady conditions, at least from the substantial point of view, the criterion has to be generalized to include also unsteady conditions.

To this purpose in this paper we will consider the morphology of isotactic polypropylene (iPP) injection molding samples (the injection molding process is unsteady also in the Eulerian sense and during the process the polymer undergoes even a relaxation immediately after the end of the filling step) obtained under very different thermal boundary conditions adopting thin heating devices layered on the cavity surface which allow a fast evolution of the cavity surface temperature (between the injection and the cooling channels temperature) during the process. With the aim of analyzing the morphology distribution along the cross sections of moldings previously obtained and reported in [2] detailed process simulations were carried out to obtain appropriate molecular stretch parameter and the mechanical work evolutions. The experimental information relevant to the simulation of those tests is summarized in the next section.

EXPERIMENTAL INFORMATION AND PROCESS SIMULATION

The polymer was injected into a cavity having thickness, width, and length of 1.50 mm, 12.7 mm, 110 mm, respectively. Thin heating devices were placed inside the mold, layered very close to the cavity surface (0.1 mm from it) in order to achieve a fast evolution of the cavity surface temperature during the injection cycle [3]. In order to avoid a cool contact of the polymer with the cavity, cavity surface heating was started a small time, t_a about 2 secs, in advance with respect to the first contact with the polymer; after few additional secs the cavity surface reached a steady value T_{cs} 80°C and 150°C were determined by the two heating powers adopted P_e, as reported in Table 1; also, a tests without activating the heating devices, Passive test, and a test where the heating devices were replaced by steel layers of the same geometry (Steel test) were carried out. All injection molding tests were performed with 2.3 cm^3/s average flow rate, 220 °C melt temperature, 25 C overall mold temperature, 720 bar holding pressure and about 11 s holding time.

Polarized optical micrographs (OM) of sample slices cut along the flow–thickness plane were obtained with crossed analyzer-polarizer and the flow direction of sample slices was rotated of 45° with respect to the analyzer. The slices were also analyzed with Atomic Force Microscopy (AFM).

TABLE 1 Injection molding conditions: P_e, surface power density of the heating devices; T_{cs} steady surface temperature value reached after several secs. heating; t_h, heating time; t_a, advance time the cavity heating starts with respect to the contact with the polymer

Test run	P_e (W/cm^2)	T_{cs} (°C)	t_h (s)	t_a (s)
Steel	0	25	0	0
Passive	0	25	0	0
80-07	4	80	0.7	2
80-1	4	80	1.3	2
80-6	4	80	6	2
150-07	9.5	150	0.7	2
150-1	9.5	150	1.3	2
150-6	9.5	150	6	2

The simulation of the injection molding process was performed describing the evolution of the temperature field along both flow and thickness directions [4]; accounting of the fast evolution of the cavity surface by effect of the heating by the thin heating elements. Momentum balance was solved within the lubrication approximation with a viscosity depending on shear rate, temperature, pressure and crystallinity. The viscoelastic nature of the polymer was considered through a relaxation time λ function of a molecular stretch parameter, Δ, evaluated as the difference between the largest and the smallest eigenvalues of the molecular conformation tensor [4] $\underline{\underline{A}} = 3\left(\langle \underline{R}\,\underline{R}\rangle - \langle \underline{R}\,\underline{R}\rangle_0\right)/\langle R_0^2\rangle$ whose evolution was described by a Maxwell-type contravariant equation [5,6]:

$$\frac{D}{Dt}\underline{\underline{A}} - \left(\underline{\nabla}\,\underline{v}\right)^T \times \underline{\underline{A}} - \underline{\underline{A}} \times \left(\underline{\nabla}\,\underline{v}\right) = -\frac{1}{\lambda}\underline{\underline{A}} + \left(\underline{\nabla}\,\underline{v}\right)^T + \left(\underline{\nabla}\,\underline{v}\right)$$

Mesomorphic and α crystallization processes, both competing for the same amorphous, were considered. The crystallization temperature was assumed function of both the pressure, and the molecular stretch parameter, Δ for the crystallization toward the α phase [6].

The temperature and pressure evolutions obtained by the simulations were satisfactorily compared with experimental results. Also, final crystallinity distributions inside the moldings obtained with the iPP T30G was satisfactorily described [1].

RESULTS

Several years ago, the criterion for the achievement of fibrillar morphology, as found in the shear layer of injection molded objects, was based on the achievement of a *critical* value of mechanical work accumulated while the shear rate is above a *critical* value adequate to stretch large molecules. The role of the mechanical work is to assure for the time needed for kinetic processes required for the creation of stable shish nuclei rows between which the polymer can crystallize as kebabs. The higher the stress, larger will be the shish density and shorter will be the time needed to achieve the critical mechanical work and thus faster are supposed to be the processes for the formation of a shish kebab structure.

Since the molecular stretch is more adequate than the shear rate to represent the effect of the flow on the macromolecules, the criterion was recently reformulated replacing the critical shear rate with a critical molecular stretch Δ_c [1]. Furthermore, since the injection molding process is highly unsteady, the interest to a criterion formulation which considers unsteady conditions automatically came in. It is therefore assumed that if, while the criterion as specified above is satisfied, by effect of some relaxation the stretch decreases below its critical value the accounting of the work has to start again, when and if the critical stretch is recovered.

The distributions of both the molecular stretch, Δ and the integral of the mechanical work, W, both at the end of the filling step, and at the end of the process, are reported in fig. 1 for some of the injection conditions considered in this paper. Also, the corresponding optical micrographs of the final molding cross section at about 15 mm downstream from the gate are reported on the top of each plot.

The poorly structured surface layer, appearing as a dark layer close the sample surface, when present, can certainly be identified in the optical micrographs. More difficult to be detected by optical microscopy is the shish–kebab morphology, especially if the oriented elements are very thin.

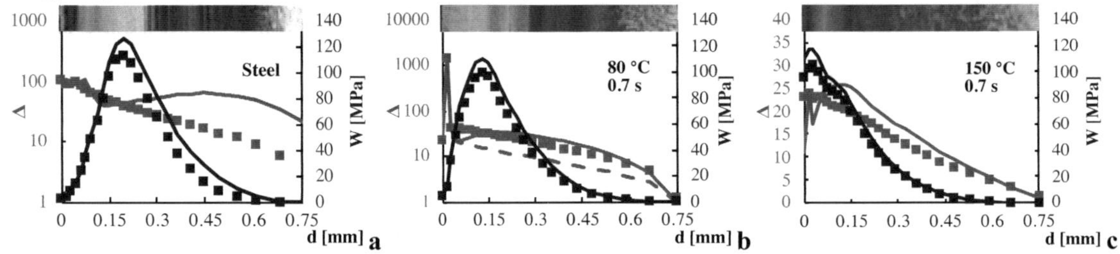

FIGURE 1. Calculated molecular stretch distribution, Δ in red, and integral of mechanical work, W in blue, 15 mm downstream from the gate for tests: Steel (a), 80-07 (b), 150-07 (c). Cross section distributions: at the end of the filling (symbols) and the process (full lines). The Δ distribution 2s after the end of filling (red dashed line) is also reported for the test with 80-07 (b).

Under some processing conditions, between the shish–kebab and the spherulitic morphology there is a transition region, where elongated elements, having thickness ranging from about 4 μm up to 6 μm, are aligned along the flow direction. Depending upon operating conditions, the thickness of the transition region can be very wide or, vice versa it can even disappear; sometimes the morphology of this region can (especially where the elements are thick) be detected by optical microscopy. What is difficult to identify by optical microscopy are both borders of this region, which, on the other hand, do not have a quantitative definition. A fifth curve (the dashed red one) reporting the distribution of the stretching parameter Δ, calculated two seconds after the end of the filling step, has been included in the plot related to the test 80-07. Considerable relaxation, with respect to the end of the filling step is clear. During such a relaxation, the packing flow starts and additional stretch builds up. The packing steps, and the eventual final relaxation determine the shape of the final stretch curve of each injection condition. Much smaller are the values of the final stretch distributions of the samples (150-07) obtained with higher cavity surface heating power; in this case, indeed, the packing flow is small and poorly effective for the stretch, whereas the relaxation is large.

As far as the integral of the mechanical work, close to the cavity surface the filling contribution appears dominant; the contribution of the packing step is minor on the whole cross section.

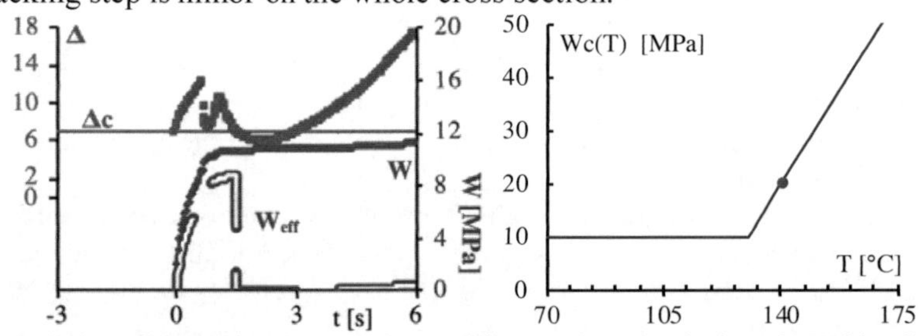

FIGURE 2. (a) Predicted evolution of the molecular stretch, Δ red line, and mechanical work, W blue lines and effective work W_e blue dashed line, at 0.44 mm from the cavity surface for the test 80-07. (b) Critical work as function of temperature adopted for the simulations reported in fig. 3; the red dot represents the experimental results at 140 °C obtained in for the iPP T30G

It has to be pointed out that the plots reported in fig. 1 are not sufficient to clarify if, according to the criterion illustrated above, at the end of the process, in a given position on the cross section, the morphology will be fibrillar or not. Indeed, the criterion for fibrillar morphology specified above clarifies that the fraction of mechanical work accumulated before the stretch reaches its critical value Δ_c has to be disregarded in the account and, furthermore, if by effect of a relaxation the stretch decreases below its critical value, the accounted mechanical work disappears and the accounting starts again if the critical stretch is reached again.

Let us, now combine the molecular stretch and the mechanical work evolutions for a case far from the surface, in particular at 15 mm downstream the gate and at 0.44 mm from the cavity surface for the test 80-07. For this case, the molecular stretch and the mechanical work evolutions are reported in Fig. 2a. The red line is the molecular stretch whereas the full blue line is the total work accumulated from the beginning of the process.

The morphologies of some moldings obtained without cavity surface heating with the same iPP adopted in this work analyzed by applying essentially the same criterion adopted in this work gave rise to good results with Δ_c and W_c equal to 7 and 10 MPa respectively [1] , these critical values will be the starting point of this analysis. According to the criterion adopted in this paper, the mechanical work accumulated before the molecular stretch has reached its critical value, has to be disregarded; one can denote as effective mechanical work, W_e, the residual work, which is also plotted in fig. 2a as dashed blue line. Since the molecular stretch, as consequence of the relaxation taking place after filling, drops below its critical value of 7, according to the criterion adopted in this paper, starting from that time, the effective work drops to zero, and, it starts to increase again at about 3 s when the molecular stretch reaches again its critical value. However, the work accumulated during the packing step is not sufficient to reach an effective critical work, of 10 MPa. Consequently, the condition for fibrillar morphology will not be achieved, although the stretch, and the integral of the overall mechanical work at the end of the process both reach their critical values.

On the other hand, experimental results reported by the Group of Sheffield University [8] indicated an increase of the critical work with temperature for some polymers and, furthermore, the results of an experimental analysis of the effect of shearing on the shish–kebab crystallization carried out at 140 °C on the same iPP adopted in this paper gave rise to a critical work of 20 MPa, rather than the 10 MPa identified in the previous paper [1]. Consistently, the predictions for thickness and location of oriented layers of the injection molded samples considered in this paper were calculated adopting critical stretch Δ_c equal to 7, as in previous analysis [1], and a critical work W_c function of the temperature as shown in fig. 2b, namely, equal to 10 MPa at low temperature, and linearly increasing for temperatures larger than 130 °C, considering that W_c was experimentally found 20 MPa at 140 °C (reported as a red dot in fig. 2b).

For each of the eight injection molded tests considered in this work, the predictions of the criterion (based on Δ_c=7 and W_c as shown in fig. 2b) for the morphology along the cross section 15 mm downstream the gate are compared in fig. 3 with a synthesis of the corresponding experimental morphology distribution as detected by the AFM analysis reported in [2]. Also, the corresponding polarized optical micrographs are reported in the figure. In particular, in the synthesis of the experimental morphology, layers of different colors, as indicated in the legend, correspond (after the surface layer which is black) to oriented morphological elements having thicknesses growing with the distance from the sample surface in the intervals 0.1 μm – 2 μm (blue) or 2 μm – 4 μm (orange) or, sometimes, a transitional layer toward a spherulitic morphology with elements of thickness ranging in the interval 4 μm – 6 μm (green); finally a spherulitic layer was always present up to the sample midplane. Only the layer 0.1 μm – 2 μm and the spherulitic layer were found in all injection molding conditions, details on this are given below. Within the interval 2 μm – 4 μm, and especially within the interval 4 μm – 6 μm, when present, the morphology characteristics were found to gradually change their order and homogeneity. In particular the order of the elements was found to decrease and the element thicknesses was found to increase with the distance from the sample surface including more and more cylindritic and sometimes small spherulitic structures.

In most of the cases, the evolution from oriented elements to the spherulitic morphology took place within a transitional layer of width of the order of only 70 μm and for the Steel test the evolution took place even and directly from the layer which includes aligned 0.1 – 2 μm thick structures to the spherulitic layer. Only the case with 150-07 showed a much more extended (nearly one third of the sample thickness) transitional layer; this has to be related to the intensity of the packing flow, which, on its turn, is determined by the amount, and rate of cooling and crystallization after filling. Indeed, in the case 150-07, during filling only a moderate cooling had taken place, and crystallization had not taken place yet (except for small percentages at the surface). To a smaller extent, similar observations can be made in relation to the 80-07 sample, which is second as far as width of the transitional layer thickness.

As far as the comparison, shown in fig. 3, between experimental results and the results of the criterion for oriented morphology distribution on the molding cross section, it appears that, with the choices described above for the critical stretch and for the critical work W(T), the surface layer widths for the Steel and Passive tests are correctly predicted, also the absence of the surface layer in all the other tests is properly described. Furthermore, simulation results reproduce, within a reasonable approximation, the additional widths from the surface layer (when present) up to the positions within which oriented and aligned morphological elements of average diameter up to 4 μm were identified [7]. The discrepancy is smaller than 15% (i.e. smaller than 40 μm) for all cases, except for the cases 80-1, 80-6 and 150-07, for which the discrepancies are 30% for the first one and 20% for the other two. It has to be considered that the approximations of the comparison reported in fig. 3 are the result not only of the approximations of the application of criterion for the formation of ordered morphological structures aligned along the flow direction but they include also all approximations within the model adopted for the calculation of the molecular stretch, and the mechanical work evolutions. For instance, it would be sufficient that one of the parameters (as, e.g. the ones determining the dependence of the viscosity or of the relaxation time upon pressure or temperature) is slightly either over or under estimated to determine a significant effect on the evolutions of both the molecular stretch and the mechanical work distributions, and, consequently on the final cross section morphology. Moreover, it has to be pointed out that the predictions of both the molecular stretch and the mechanical work evolutions and consequently the width of the oriented shear layer observed in the final cross section of the samples are sensitive to the distance from the cavity gate. A difference of some millimeters (less that few percent of the cavity length) in the position along the flow direction could significantly reduce the discrepancy between predictions and experimental observations.

The criterion, in its actual on/off formulation, does not include the existence of intermediate transitional layers between the highly oriented and the spherulitic layers. In most of the cases, these transitions took place within very thin zones; however, under special processing conditions, the transitional zone can be wide. The possibility of transitional layers could be included in the criterion by releasing its on/off approach; namely, by considering the cases when either the mechanical work or the molecular stretch do not satisfy completely their critical conditions.

Many authors point out that the intensity of the flow determines the shish density which, after lateral growth which occurs while the mechanical work is accomplished, in its turn gives rise to the thickness (and the density) of the morphological elements aligned along the flow direction. A more detailed analysis could have been carried out if the relationship among the stretch and the density of the morphological elements would have been quantitatively known. The identification of such a correlation could be attempted by comparing the experimental distributions (on the cross section) of aligned morphological elements density with the stretch distributions obtained applying the criterion adopted in this work; obviously very different experimental injection molding conditions should be simultaneously analyzed. In any case a decrease of the critical stretch in the criterion should determine a lower shish density and thus thicker elements.

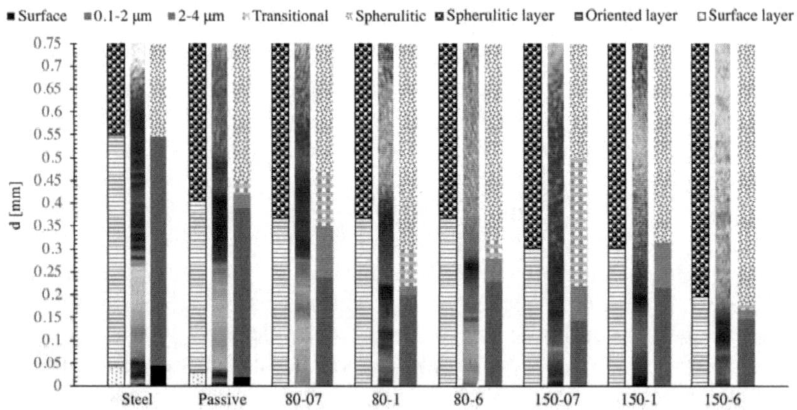

FIGURE 3. Comparison between experimental morphology distribution along the cross sections of the injection molded samples. The central bar is the optical micrograph; the right bar reports the morphological layers as detected by AFM [2]. The left bar reports the predictions for the thicknesses of the surface layer, of the oriented morphological layer and of the spherulitic layer. Each color of the bar on the right corresponds to the thickness of the morphological elements as specified in the legend.

CONCLUSIONS

The criterion (based on critical values of molecular stretch and of mechanical work) for the formation of ordered shish-kebab morphology has been adjusted to take account of unsteady conditions of both temperature and velocity fields. On the basis of the process simulation, the criterion has then been applied on the injection molding process (which is highly unsteady) and it has been tested with the experimental morphology distributions detected on moldings obtained under very different thermal boundary conditions. An overall satisfactory comparison of widths and positions of the shish-kebab shear layers was obtained adopting a mechanical work function of the temperature, complying the indications of previous experimentations performed at high temperature under steady conditions.

A more articulated criterion must be sought for a more detailed description of the morphology distributions, including the transition between the shish-kebab shear layers and the spherulitic layer around to the symmetry plane.

REFERENCES

1. R. Pantani, V. Speranza and G. Titomanlio, *Int. Polym. Process.* **33**, 355–362 (2018).
2. V. Speranza, S. Liparoti, R. Pantani and G. Titomanlio, *Materials* **12**, 424 (2019).
3. S. Liparoti, A. Sorrentino and G. Titomanlio, *RSC Adv.* **6**, 99274–99281 (2016).
4. R. Pantani, V. Speranza, and G. Titomanlio, *Int. Polym. Process.* **XXXI(5)**, 655–663 (2016).
5. G. Marrucci and G. Ianniruberto, *Chem. Eng. Sci.* **56**, 5539–5544 (2001).
6. R. Pantani, V. Speranza and G. Titomanlio, *Eur. Polym. J.* **97**, 220–229 (2017).
7. V. Speranza, S. Liparoti, R. Pantani and G. Titomanlio, *Materials* **12**, 505 (2019).
8. O.O. Mykhaylyk, P. Chambon, C. Impradice, J.P.A. Fairclough, N.J. Terrill, A.J. Ryan, *Macromolecules* **43**, 2389–2405 (2010).

Modelling and simulation of thermoset injection molding process

Ngoc Tu Tran (presenting and reference author)[1] and Michael Gehde[1]

[1] *Chemnitz University of Technology, Professorship of Plastics Engineering, D-09107 Chemnitz, Germany*

Abstract. Rheological and thermal equipment were employed to measure successfully material properties of thermoset injection molding compounds, which was then used as input data in writing a numerical method to model cure kinetics and reactive viscosity equations for thermosets injection molding simulation. All fitted parameters of the cure kinetics and reactive viscosity models were obtained via the developed numerical algorithm, which was directly imported into Moldex3D simulation tool to evaluate its application in prediction of mold filling behavior of thermoset injection molding compounds. In addition, the wall slip phenomenon of thermoset injection molding compounds in the presence of a wall surface which was investigated and visually detected during the experimental process was considered in the computation simulation process. A remarkable agreement between simulation and experimental results was found, indicating that the developed numerical method for generating thermoset material data for the reactive injection molding simulation process is reasonable and it is impossible to ignore the influence of wall slip during the thermoset injection molding simulation process.

Keywords: Reactive viscosity and cure kinetics model, Accumulation function, Thermoset material data, Thermoset injection molding, Wall slip, Injection molding simulation.
PACS: PPS-35

1. INTRODUCTION

To simulate the injection molding process, it is necessary to set material data. With thermoplastics injection molding simulation, a simulation software usually provides comprehensive material data bank to help users choose thermoplastics material data to simulate the entire phases of thermoplastics injection molding process. In addition, the mold filling behavior of thermoplastics injection molding compounds was successfully investigated. As a result, there have been some significant achievements in application of simulation packages to simulate the entire phases of thermoplastics injection molding process [1]. However, with thermosets injection molding simulation, the material data is found in limited sources and seldom available from data bank of simulation tools because of complication not only in rheological and thermal properties measurement but also in modeling rheological and cure kinetics mathematical models. Therefore, with thermoset injection molding compounds that its material data bank has not been found in data bank of simulation tools, before setting material data, it is necessary to create its own material data that simulation packages do not supply a tool. To solve this problem, it requires an extensive knowledge base in measurements of material properties as well as optimization algorithm. To create thermoset material data for the reactive injection molding simulation process, firstly, it is necessary to measure critical material properties including density, heat capacity, thermal conductivity, reactive viscosity, degree of cure and pvT. After that, based on the experimental data, optimization algorithm must be written to get all fitted parameters in rheological and thermal mathematical models that is used to describe material properties and being embedded in simulation tools. Rheological mathematical models are reactive viscosity models and thermal mathematical models are cure kinetics models. Viscosity of thermoset compounds is more complicated than the one of thermoplastics injection molding compounds. Because it not only depends on temperature and shear rate but also on degree of cure. At low degree of cure, the viscosity of thermoset resins will decrease due to the elevated temperature, which is the same for thermoplastic resins. As the temperature reaches a certain level, the rapid increase in the degree of cure and the cross-linking will cause a significant increase in viscosity. Consequently, the viscosity curve of thermoset materials is characterized by a parabolic curve. A complete mathematical model for characterizing viscosity of a reacting polymer must contain the effects of shear rate ($\dot{\gamma}$), temperature (T) and degree of cure (α). Castro Macosko model is the first reactive viscosity model developed by Castro Macosko [2], as shown in Eq. (1). The viscosity depends on temperature and degree-of-cure only and does not depend on shear rate. Therefore, a complete reactive model, namely, Cross Castro Macosko model was introduced and embedded in Moldex3D simulation tool. It is an extension to Castro Macosko model, with power-law type shear rate dependence, as shown in Eq. (2). This model is composed by two terms: the first is the Cross-WLF model and the second term is related to the gel model.

$$\eta(T,\dot{\gamma},\alpha) = \eta_0 \left(\frac{\alpha_g}{\alpha_g - \alpha} \right)^{(c_1 + c_2\alpha)} \quad with \;\; \eta_0(T) = B\exp(T_b / T) \tag{1}$$

$$\eta(T,\dot{\gamma},\alpha) = \frac{\eta_0(T)}{1 + \left(\frac{\eta_0(T).\dot{\gamma}}{\tau^*} \right)^{1-n}} \left(\frac{\alpha_g}{\alpha_g - \alpha} \right)^{(c_1 + c_2\alpha)} \quad with \;\; \eta_0(T) = B\exp(T_b / T) \tag{2}$$

In reactive viscosity models Eq. (1) and Eq. (2), η, T, $\dot{\gamma}$ are viscosity, temperature and shear rate, respectively. η_0 is the viscosity as shear rate gradually approaches zero. n is power-law index with value between 0 and 1. B is the pre-exponential and Tb represents the η_0 temperature sensitivity. τ^* is shear stress at transition from Newtonian to non-Newtonian flow. E_η is activation energy, R is gas constant and c_1, c_2 are experimental constants. α_g is cross-linking density (gel point) at which flow is no longer possible. α is degree of cure. The degree of cure (α) is measured via cure kinetics tests and described by some curing kinetics models, such as Kamal and Sourour Model, modified Kamal and Sourour Model and Deng-Isayev-Law model. Kamal and Sourour Model, as shown in Eq. (3), is the most used model to describe the cure kinetics of thermoset injection molding compounds [3]. In Kamal and Sourour Model, m and n are the reaction orders and k_1, k_2 are the Arrhenius overall constants. A, B is the fitted rate coefficient, E_a, E_b are the activation energy, R is the universal gas law constant, and T is the cure temperature.

$$\frac{d\alpha}{dt} = \left(k_1 + k_2\alpha^m\right)\left(1-\alpha\right)^n \; with \;\; k_i = a_i \exp\left(\frac{-E_i}{RT} \right) \tag{3}$$

Up to know, modelling of cure kinetics and reactive viscosity for the thermoset injection molding process simulation has not been adequately described. In another way, how to obtain the value of variables that is called as fitted parameters in cure kinetics and reactive viscosity models from experimental data has been found in limited knowledge only. There has not been any article that shows a complete way to create thermoset material data from measured experimental data (thermal data and rheological data) for the reactive injection molding simulation process. These problems are leading to the users 'dependency on the material data bank of simulation tools, leading to restriction in application of computer simulation in the thermoset injection molding simulation. For instance, in 2017, Kristin [4] investigated the flow behavior of phenolic injection molding compounds by doing injection molding experiments and simulation. In her thesis, phenolic 1110 of Hexion GmbH was used. A series of incomplete molded parts was molded and compared with the corresponding patterns predicted by a simulation tool. Unfortunately, due to lack of material data of phenolic 1110 simulation had to be done with material data of CRASTIN®HTI619. In addition, during doing experiments the existing problems [5] whether there is slip or no-slip of thermoset injection molding compounds during the filling phase was also not be answered. Therefore, simulation was done without taking slip boundary condition into account. As the result, a comparison between flow length generated from experiments and flow length predicted by simulation is not highly scientific.

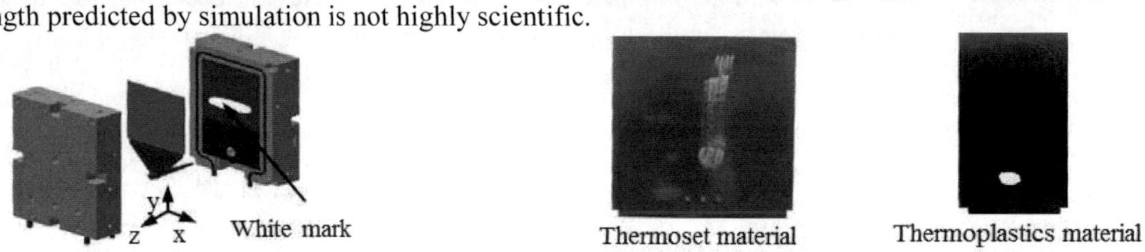

(a) The spotwise painting of the mold wall surface (b) Difference in the flow behavior between thermosets and thermoplastics

FIGURE 1. Developed method for testing slip phenomenon between polymer and wall surface

Recently, an exact answer to the existing question whether there is slip or no-slip of thermoset injection molding compounds during the filling phase was found by us [6]. A rather simple but effective and useful method, namely, the spotwise painting of the mold wall surface was developed to study the behavior of glass-fiber reinforced phenolic resin in the presence of the wall surface during injection molding under various processing parameters. In this study, a commercial phenolic molding compound, namely, Vyncolit X655 of Sumitomo Bakelite Co., Ltd was used for doing injection molding. Before doing a new shot of injection molding, the white marks were manually painted on the surfaces of the mold, as shown in Figure 1(a). The experimental results show clearly a strong slip of the phenolic melt on the cavity surface in the flow direction. The occurrence of slip begins immediately when the phenolic melt contacts the wall surface. The complete difference in the flow behavior of thermosets and thermoplastics polymer on the cavity surface in the process of filling a mold was found, as shown in Figure 1(b). There is a strong slip between the thermoset phenolic melt and wall surface, which is not found for the thermoplastics injection molding. As a result, velocity of the thermoset phenolic melt near to wall surface must be different from 0 and the mold filling behavior of these phenolic injection molding compound is plug-shear-flow instead of fountain flow like thermoplastics. Due to these differences assumptions of neglecting the influence of wall slip boundary condition on the mold filling characterization of thermoplastics injection compounds during simulation process is impossible to apply to the thermoset phenolic injection molding compounds.

In the present article, rheological and thermal properties of a thermoset injection molding compounds, namely, Vyncolit X655 were successfully measured. After that the numerical method was written to create cure kinetics data and reactive viscosity data of Vyncolit X655. The generated material data was directly imported into Moldex3D simulation tool to simulate the filling and curing phase of the reactive injection molding process with taking wall slip boundary condition into account. A comparison between results predicted by simulation and generated from experiments were conducted to estimate the accuracy of generated material data and investigate the applicability of slip model embedded in Moldex3D simulation tool.

2. EXPERIMENTAL

2.1. Selecting material

A commercial phenolic molding compound, namely, Vyncolit X655 of Sumitomo Bakelite Co., Ltd was used to do rheological, thermal tests and injection molding both experiment and simulation. It consists of 20 percent of resin, hardener, all other ingredients to improve molding process and 80 percent of short glass fiber and mineral.

2.2. Measuring rheological and thermal properties

The reaction heat of samples under non-isothermal conditions was measured via DSC Q2000 from TA instruments. The mass of the samples was about 10 mg of powder. Open aluminum pans were used to contain materials. The samples were heated from -60°C to 350°C by using multiple scanning rates of 5 K/min, 10 K/min and 20 K/min. AR 2000 from TA instruments was used to determine the flow curves of phenolic X655 at different heating rates of 1 K/min, 2 K/min, and 5 K/min. A test specimen of known dimensions (the diameter is 25 mm and the thickness is 1 mm) is held between parallel plates so that the material acts as elastic and dissipative element in a mechanically driven oscillatory system. The drive motor provides the mechanical deflection and the resulting torque transmitted through the test specimen is recorded by the torque sensor. The system is used in oscillation mode with a frequency of 1 Hz and a deformation of 0.1 %. Heat capacity of fully cured samples was measured by using DSC Q2000. DTC 300 from TA instruments was employed to measure thermal conductivity of fully cured samples.

2.3 Injection molding procedure

Two different study objects for the injection molding experiments were employed. The first part is a plate with a dimension of 150 mm x 150 mm x 4 mm and the film gate with a height of 2.5 mm. The second object is a spiral flow part with the flow length of 1385 mm. An injection machine CX 160-380 was employed to do injection molding experiments. With plate part, the injection speed was constantly kept at 32 cm^3/s and the mold temperature was 160°C, 175°C, and 190°C, respectively. With spiral flow injection molding process, the mold temperature was constantly kept at 175 °C and the injection speed varies according to ram position (8/16/32 cm^3/s).

3. SIMULATION

3.1. Creating material data

Material data of Vyncolit X655 is not currently found in material data bank of Moldex3D simulation tool. Therefore, before simulating it is necessary to create its material data. Based on measured material properties and the least-square estimation algorithm developed by Levender- Marquardt (LMA) embedded in Matlab program language, a cure kinetic and reactive fitting code was written. The fitting code was applied to model cure kinetic model (Kamal model) and reactive viscosity model (Cross Castro Macosko model), respectively. Consequently, all fitted coefficients of Kamal and Cross Castro Macosko model were obtained, which was employed to create material data sheet of Vyncolit X655 for its injection molding process simulation. A good comparison between fitted curves and experimental curves is shown in Figure 2 and 3, indicating the written fitting code is useful and reasonable. The fitted parameters of cure and reactive viscosity models is found and summarized in Table (1) and Table (2).

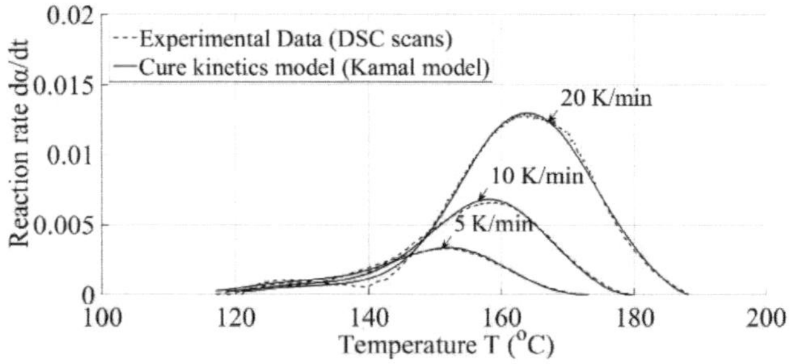

FIGURE 2. Fitting cure kinetics model

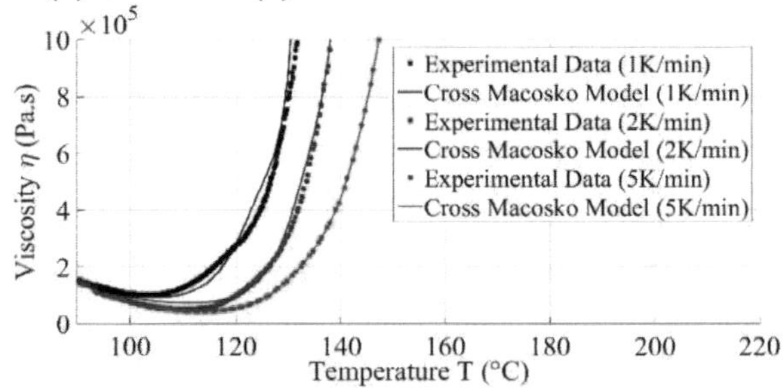

FIGURE 3. Fitting reactive viscosity model

TABLE (1). Fitted parameters of Kamal and Sourour Model

m	n	A (1/s)	B (1/s)	E_a (K)	E_b (K)
0.892	0.988	6.86e+08	5.06e+08	9.29e+04	8.52e+04

TABLE (2). Fitted parameters of Cross-Castro-Macosko model

α_g	c_1	c_2	B (Pa. sec)	T_b (K)	n	τ^* (Pa)
0.7	6.22	-5.05	1.7e+05	1.49e+04	0.432	2.8e-09

3.2. Simulation

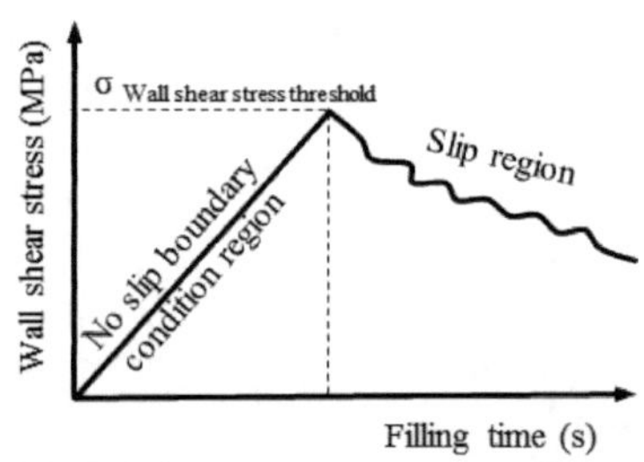

FIGURE 4. Slip boundary condition

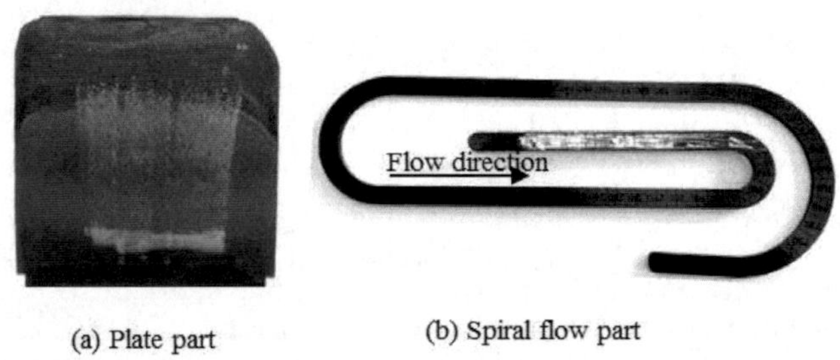

FIGURE 5. The movement of white stripe on the surface of molded parts.

TABLE (3). Parameters of wall slip boundary condition

Wall shear stress threshold (MPa)	Frictional coefficient
0; 0.001; 0.01; 0.09; 0.5; 1	0; 0.01, 0.05; 0.1; 0.5; 1

The generated material data of Vycolit X655 includes fitted parameters in Cross Castro Macosko model and Kamal - Sourour Model, thermal conductivity, heat capacity and density coefficient was directly imported into data bank of Moldex3D simulation tool to set as material data for simulating its filling and curing behavior during the injection molding process. The processing simulation conditions were the same as the processing conditions used in the injection molding experiments. In addition, slip boundary condition (Figure 4) was considered during simulation process. Friction coefficient is a dimensionless parameter and its value is 0 to 1. The wall shear stress threshold with unit of MPa represents the critical static friction from no-slip switch to slip. In this article, although a strong slip between the phenolic melt and wall surface that was found and shown in Figure 5, the value of frictional coefficient and shear stress threshold has not yet been measured. Therefore, simulation has been done by setting different values of frictional coefficient and shear stress threshold, as summarized in Table (3). After that the incomplete molded parts generated from experiments was used to compare with the corresponding filling patterns predicted by Moldex3D simulation tool. By using this reverse engineering technique, the value of frictional coefficient and shear stress threshold was estimated.

4. COMPARISON BETWEEN SIMULATION AND EXPERIMENT RESULTS

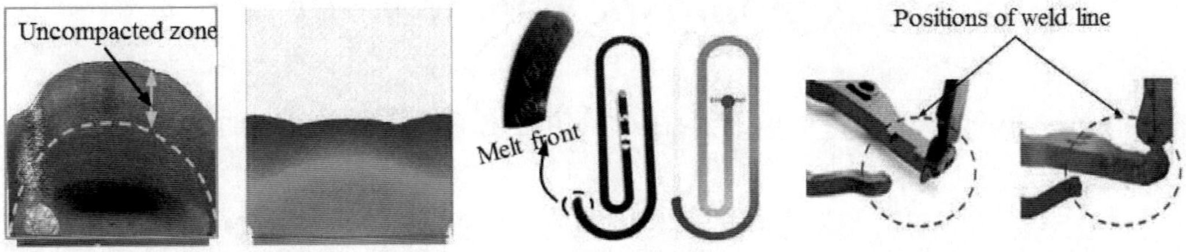

(a) Plate part (b) Spiral flow (c) Complex part

FIGURE 6. Comparison between simulation and experimental results

The filling patterns predicted from Moldex3D simulation tool were compared to the corresponding incomplete molded parts generated through interrupted filling experiments. Beginning with plate part, there is difference in the flow length between the incomplete injection molded part generated by experiments and the corresponding filling patterns predicted by Moldex3D. With all values of frictional coefficient and shear stress threshold used in simulation process, it was found that Moldex3D simulation tool has not yet been simulated exactly the uncompated zone that has an interrupted contact with the cavity surface. Therefore, with the same injected volume of 60 % the flow length of incomplete filling patterns predicted by simulation is shorter than the flow length of the corresponding incomplete molded parts generated by the injection molding experiment, as shown in Figure 6(a). However, difference in the melt front and flow length between simulation and experimental results depends on geometry of the cavity. For instance, if frictional coefficient of 0.5 and wall shear stress threshold of 0.09 MPa, with spiral flow part, the width and the thickness of part is only 8 mm and 4 mm respectively, the uncompacted zone seen to be disappeared and the melt front and flow length generated by simulation and experiment is found to be in good agreement, as shown in Figure 6(b). To prove this evaluation, molding filling behavior of a complex part from industry was simulated with frictional coefficient of 0.5 and wall shear stress threshold of 0.09 MPa. A comparison between the positions of weld line of incomplete injection-molded complex part generated from experiments and the corresponding weld line positions predicted by Moldex3D was conducted, as shown in Figure 6(c). The agreement appears remarkable, indicating that the generated thermoset material data plays an important role for well-found prediction. And frictional coefficient of 0.5 and wall shear stress threshold of 0.09 MPa could be used to predict the flow length of injection molded part during filling the cavity.

5. CONCLUSIONS

The results of this article are beneficial for understanding the slip phenomenon comprehensively, opening a new thinking about influence of slip on analysis, calculation and simulation of fiber orientation during the filling process of glass fiber reinforced thermoset injection molding compounds. Furthermore, the effect of wall slip on the cavity surface is not ignorable during thermoset injection molding process. Finally, with the developed numerical method, simulation guides become independent users of material data bank from simulation tools.

ACKNOWLEDGMENTS

The authors acknowledge the meaningful discussions with Prof. Dr. G. Mennig and Dr. Tham Nguyen-Chung on material properties of polymer, numerical method of creating material data and theory of injection molding simulation.

REFERENCES

1. T. Nguyen-Chung, G. Mennig, Rheol Acta 40, 67 – 73 (2001).
2. J.M. Castro, C.W. Macosko, Soc. Plast Eng. Tech. Papers 26,434-438 (1980).
3. M.R. Kamal and S. Sourour, Polymer Engineering and Science 13, 59-64 (1973).
4. R. Kristin, Grundlagenuntersuchungen zur Prozess- und Struktursimulation von Phenolharzformmassen mit Kurz- und Langglasfaserverstärkung, Ph.D Thesis, Technische Universität Chemnitz, 2017.
5. S. Englich, Strukturbildung bei der Verarbeitung von glasfasergefüllten Phenolformaldehydharzformmassen, Ph.D Thesis, Technische Universität Chemnitz, 2015.
6. Ngoc Tu Tran, M. Gehde, The International Journal of Advanced Manufacturing Technology 95, 4023-4029 (2018).

Non-standard warpage prediction for short fiber-reinforced thermoplastic components

Armin Kech (presenting, reference author)[a,*], Dr. Valérie Le Baube[a] and Vivien Le Baube[a]

[a]*Robert Bosch GmbH, CR/APP2, 71272 Renningen, Germany*

Abstract. The presented work will show how to use a combination of different software packages from MOLDFLOW® via DIGIMAT® to ABAQUS® to calculate warpage with material models that are user-defined.

The procedure to define an anisotropic thermo-viscoelastic material model on a commercial PBT-GF30 is presented. The measurement of the thermal expansion coefficients as a function of temperature and fiber orientation is needed as well as DMA measurements on the pure matrix material to determine the viscoelastic stiffness as a function of frequency and temperature. After performing DMA measurements, a mastering of the curves to one reference temperature is done with a Prony fit subsequently. Besides thermo-mechanical tests, information on the fiber orientation distribution is required. This can either be obtained by simulating the fiber orientation distribution with one of the advanced models implemented within MOLDFLOW or by measuring the data with a computer tomography and evaluating the scans.

The validation of the material model is done in two evolution steps. First by recalculating the DMA curves of the composite taken under different angles to the flow direction and second by applying it to stepwise tensile tests performed on milled tensile bars from an injection-molded plate with holding times on each step, done isothermal as well as non-isothermal in a temperature chamber.

Finally, the material model is applied to a real component made from the validated material. It can be shown that the prediction accuracy with this method is superior to the one from MOLDFLOW® alone.

Keywords: Simulation, visco-elasticity, warpage, injection molding, short fiber-reinforced thermoplastics,
PACS: 61.41.+e, 61.82.Pv, 62.20.−x, 62.20.Dc, 81.70.−q

INTRODUCTION

The current procedure for warpage prediction is based on micromechanical formulations for short fiber-reinforced materials. For glass fiber-reinforced polymers, the assumption is that the matrix (polymer) is behaving thermo-inelastic and the fibers linear thermo-elastic. By defining a material based on these two constituents, the composite behavior can be derived. The homogenization is done analytically based on Mori-Tanaka equations [1]. For the formation and relaxation of thermally initiated residual stresses it is assumed that the matrix is behaving thermo-visco-elastic.

The total strain is assumed to be the sum of the thermal strain ε_{th} and the visco-elastic strain ε_{ve} [2].

$$\varepsilon = \varepsilon_{th} + \varepsilon_{ve} \quad (1)$$

The temperature dependency was realized by applying the well-known WLF equation [3].

$$-\log(a_T) = \frac{C_1 \cdot (T - T_0)}{C_2 + T - T_0} \quad (2)$$

The parameters C_1 and C_2 are fitting parameters, T_0 is the reference temperature.
The linear visco-elasticity is used in the form of Prony coefficients based on a generalized Maxwell model for bulk and shear modulus, see ***FIGURE 1*** [4].

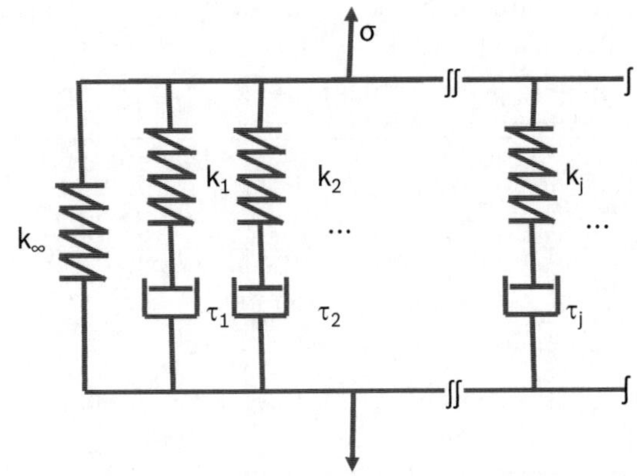

FIGURE 1. Scheme for a generalized Maxwell model

$$G_R(t,T) = G_0(T) \cdot \left[1 - \sum_{i=1}^{n} w_i \cdot (1 - \exp(-t/\tau_i)\right]; \; G_0(T) = G(t=0,T) \tag{3}$$

$$K_R(t,T) = K_0(T) \cdot \left[1 - \sum_{i=1}^{n} w_i \cdot (1 - \exp(-t/\tau_i)\right]; \; K_0(T) = K(t=0,T) \tag{4}$$

DEFINING MATERIAL PROPERTIES

To define the anisotropic thermo visco-elastic material model, it is necessary to perform some material characterizations for the pure matrix and for the composite. To define the thermo-visco-elasticity, DMA measurements on injection-molded specimens have to be performed. A temperature and frequency sweep was run in tensile mode to obtain the visco-elastic data. After determining the visco-elastic response for different frequencies from 0.01 Hz to 63 Hz and between 0 °C and 180 °C, a master curve is generated by a vertical shift with a defined reference temperature. The resulting curve for a PBT material provided by BASF SE can be seen in ***FIGURE 2***.

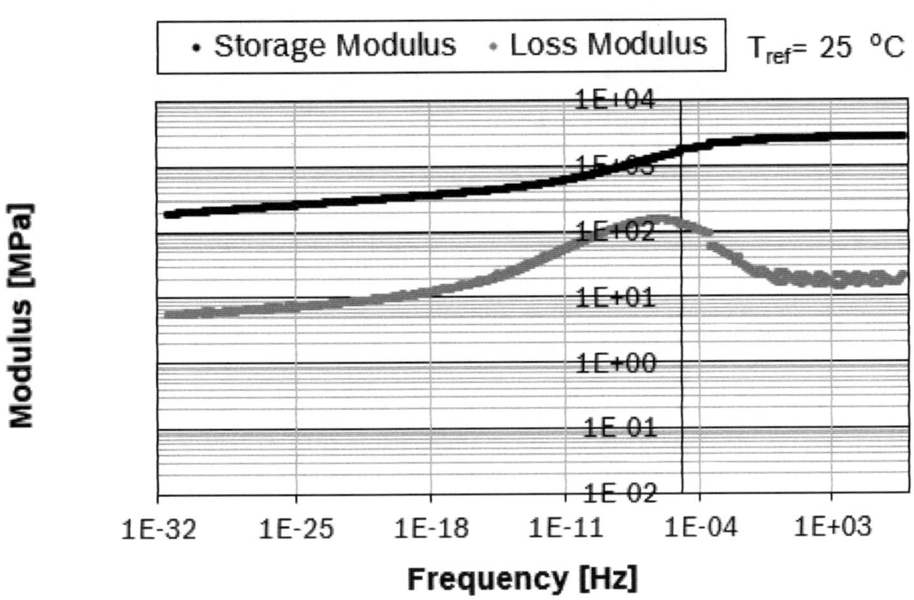

FIGURE 2. Master curve for PBT in tensile mode for loss and storage modulus; source: BASF SE.

The shift factors are fitted using the WLF approach and the parameters C_1 and C_2 are determined, see ***FIGURE 3***.

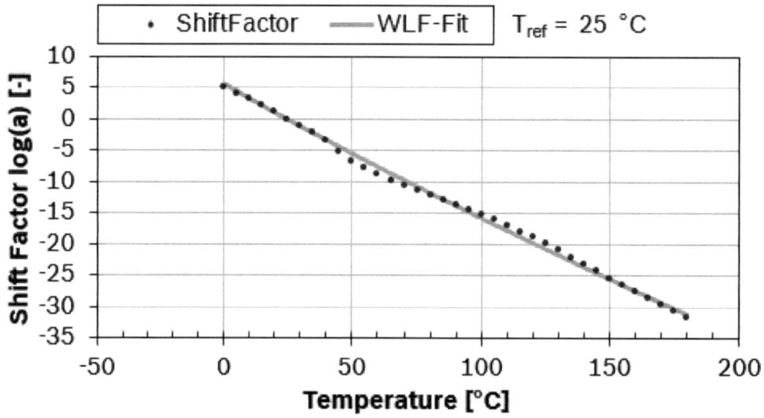

FIGURE 3. WLF fitting of shift factors for PBT and a reference temperature of 25 °C

For the definition of a composite material, also the anisotropic thermal expansion coefficients need to be known. The get these data, samples are cut from an injection-molded plate or tensile bar. With these samples, a TMA measurement for the main directions (in flow, transversal to flow and out-of-plane direction) needs to be performed. Furthermore, the orientation of the fibers is determined by doing some characterization using X-ray micro-computer tomography [5]. In ***FIGURE 4*** the distribution of the fiber orientation for a plate of 2 mm thickness made from PBT-GF30 can be seen. The typical skin-core layers can be observed with a highly oriented skin along the flow and a transversally oriented core.

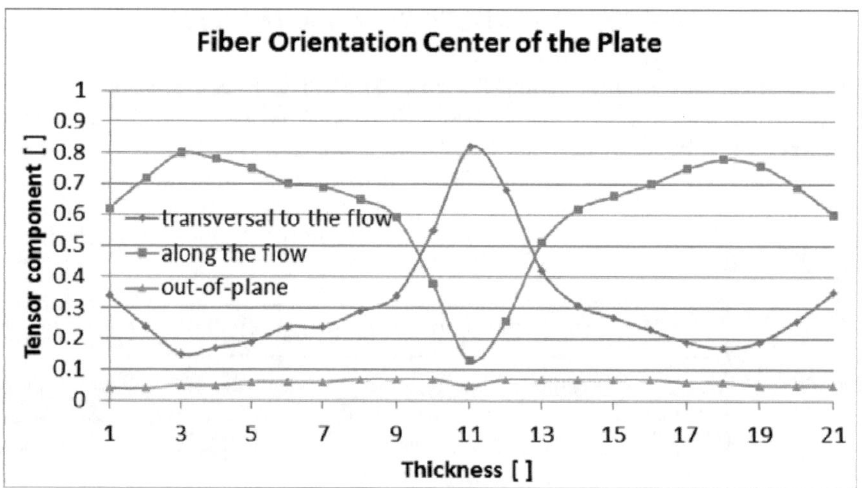

FIGURE 4. Fiber orientation determined for a plate of 2 mm made from PBT-GF30 in injection molding

For the homogenization the software DIGIMAT® was used.

VALIDATION OF THE MATERIAL MODEL

In order to validate the material model, two different mechanical tests were performed. First, at room temperature, a stress-controlled stepwise loading of a tensile bar cut from an injection-molded plate of PBT-GF30 was carried out. The testing data for a sample taken in flow direction can be seen in ***FIGURE 5***.

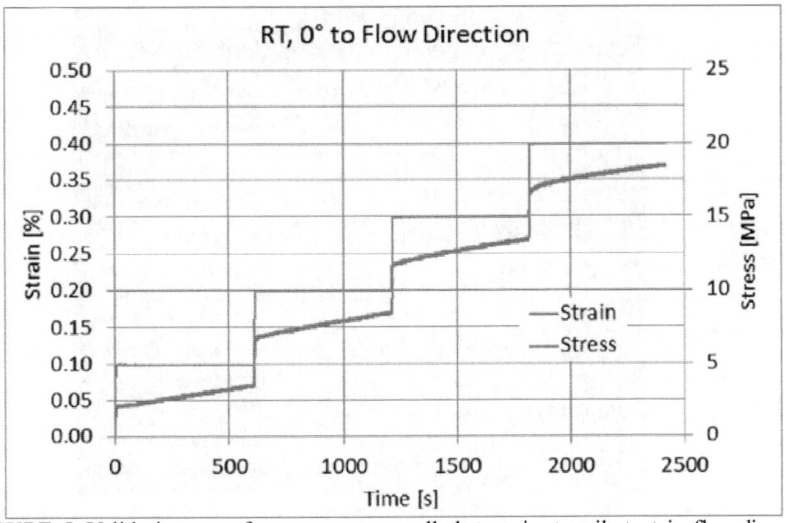

FIGURE 5. Validation case for a stress-controlled stepwise tensile test in flow direction

The second validation case was performed for a non isothermal tensile test at a fixed stress level in a temperature chamber. ***FIGURE 6*** shows the stress and strain propagation for a stepwise change of temperatures with defined holding steps.

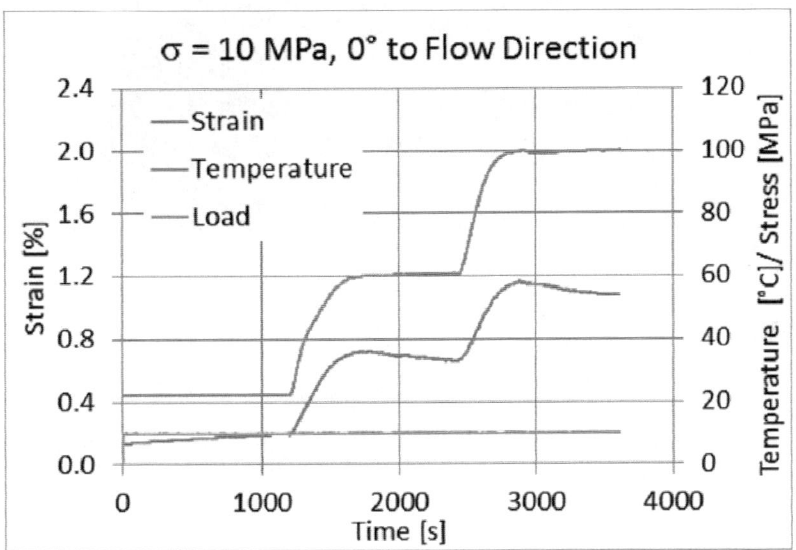

FIGURE 6. Validation case for a temperature-controlled stepwise load case at a fixed stress in flow direction

APPLICATION OF THE MATERIAL MODEL

The validated material model was applied to a consumer product from B/S/H/ home appliances, see ***FIGURE 7***. A leach container made from short fiber-reinforced thermoplastics was used as a first demonstrator project. For the application of the material defined several software codes were used.

1) AUTODESK MOLDFLOW AMI® for a calculation of the resulting fiber orientation distributions and a temperature field at the stage of volumetric filling
2) DIGIMAT MF® to define the material model and to generate a material card to be used in ABAQUS®
3) ABAQUS® to calculate thermally motivated residual stresses

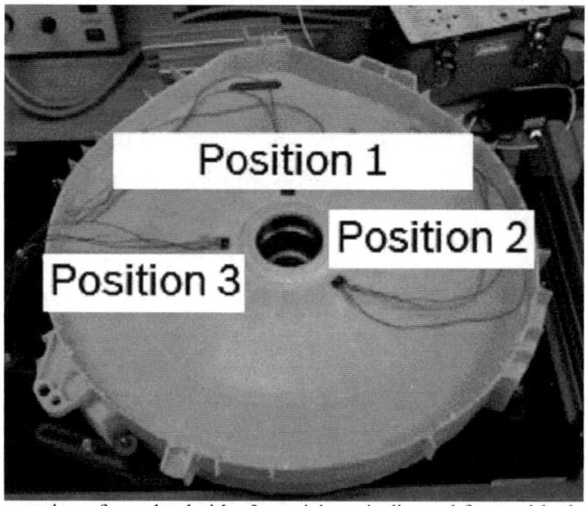

***FIGURE** 7*. Leach container from backside. 3 positions indicated for residual stress measurements

For the leach container, residual stress measurements based on the incremental hole drilling method were performed at the SKZ in Wuerzburg, Germany. The measurements were compared with simulations resulting directly from MOLDFLOW® and with the coupled simulations in ABAQUS®. A comparison of results can be seen in ***FIGURE 8***.

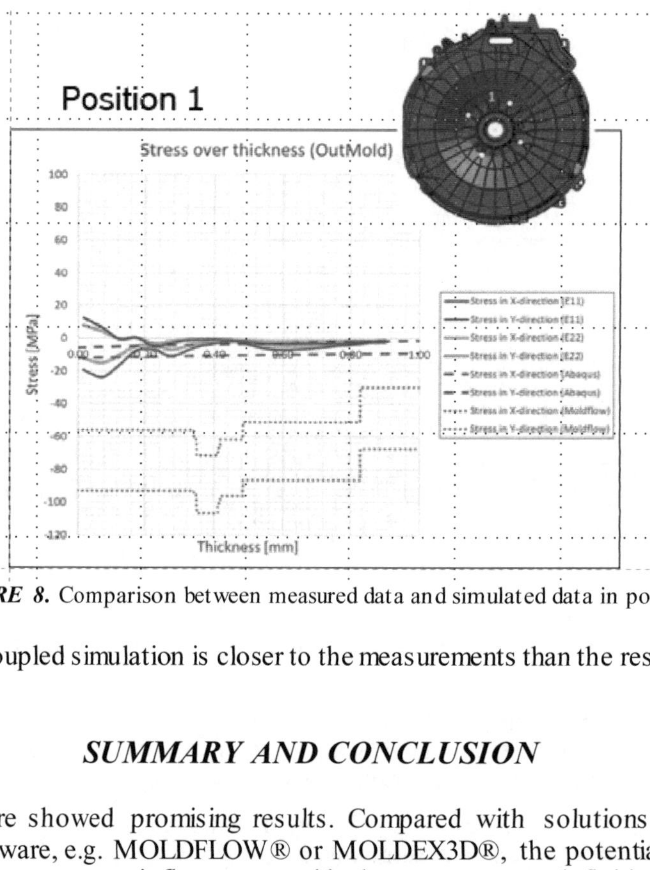

FIGURE 8. Comparison between measured data and simulated data in position 1

It can be seen that the coupled simulation is closer to the measurements than the results obtained directly from MOLDFLOW®.

SUMMARY AND CONCLUSION

The presented procedure showed promising results. Compared with solutions already implemented in commercially available software, e.g. MOLDFLOW® or MOLDEX3D®, the potential lies in the flexibility to introduce additional effects, e.g. pressure influence on residual stresses, contact definitions in the mold, etc. Further work needs to be done on the material characterization in DMA, as we could see scattering for some measurements that affects the final calculations. Finally, the calculation depends on the quality of the calculated data in MOLDFLOW®.

ACKNOWLEDGMENTS

This work was carried out within an internal project at Bosch Corporate Research, department CR/APP, with the help of several Bosch colleagues, also from department CR/ANA1. For the residual stress measurements, data was produced at SKZ, Wuerzburg, Germany.

REFERENCES

1. Mori, T., Tanaka, K. (1973). "Average Stress in the Matrix and Average Elastic Energy of Materials with Misfitting Inclusions". Acta Metall. 21: 571–574. doi:10.1016/0001-6160(73)90064-3
2. DIGIMAT User Manual
3. M. L. Williams, R. F. Landel, J. D. Ferry: The Temperature Dependence of Relaxation Mechanisms in Amorphous Polymers and Other Glass-forming Liquids, Journal of the American Chemical Society, 1955 (doi:10.1021/ja01619a008
4. F. R. Schwarzl, Polymermechanik, ISBN 978-3-642-61506-1
5. A. R. Clarke, C. N. Eberhardt, Microscopy Techniques For Materials Science, ISBN 1 85573 587 3

Efficient Parameter Identification for Macroscopic Fiber Orientation Models with Experimental Data and a Mechanistic Fiber Simulation

Susanne K. Kugler (presenting author) (reference author)[a*,b], Gregory M. Lambert[c], Camilo Cruz[a], Armin Kech[a], Tim A. Osswald[d] and Donald G. Baird[e]

[a] Corporate Sector Research and Advance Engineering, Robert Bosch GmbH, Renningen, Germany
[b] Institute for Polymer Technologie, Friedrich-Alexander-Universität, Erlangen-Nürnberg, Germany
[c] Department of Civil and Environmental Engineering, University of Tennessee, Knoxville, TN
[d] Department of Mechanical Engineering, University of Wisconsin Madison, WI
[e] Department of Chemical Engineering, Virginia Polytechnic Institute and State University, Blacksburg, VA

Abstract. Advanced macroscopic fiber orientation models depend on a variety of phenomenological parameters. The prediction quality is closely related to the choice of those parameters. Therefore, the aim of this research is to propose an efficient method for parameter identification. First, a macroscopic fiber orientation model for concentrated short fiber-reinforced polymers with a minimum number of parameters has to be identified. To define the macroscopic model a comparison with experimental data is used. A sliding-plate experiment with repeatable initial conditions is conducted for obtaining fiber orientation evolution under controlled shear and temperature conditions. Then the fiber orientation models are fitted to the experimental validation curve. Since the experimental curve generation for parameter fitting is time and cost consuming, a more efficient method is exploited: a mechanistic direct fiber simulation. The simulation can then be used to generate fiber orientation curves for varying physical descriptors (fiber length, fiber length distribution, volume fraction, viscosity, shear rate).

Keywords: Fiber Orientation, Sliding-Plate Rheometer, Mechanistic Model

INTRODUCTION

Short fiber-reinforced thermoplastics are a promising replacement for metal parts in terms of weight reduction, as they are easy to process and have high mechanical strength. Simulative prediction of failure and lifetime of reinforced polymer parts can reduce development costs significantly. To achieve this goal, a simulation chain is necessary. The first step is a simulation of the injection molding process to predict the flow-induced fiber orientation. This is used as an input to the structural simulation, since the mechanical properties of short fiber-reinforced thermoplastics are anisotropic and dependent on the fiber orientation. This work focuses on the fiber orientation models used in the process simulation.

Advanced fiber orientation models based on Jeffrey's equation [1] have been developed and implemented in commercial software to account for effects like isotropic diffusion (Folgar-Tucker model (FT) [2, 3]), strain reduction (SRF [4], RSC [5], RPR [6]), anisotropic diffusion (ARD [7], pARD [8], MRD [9]) and excluded volume effects (nematic model [10]). The prediction quality depends strongly on phenomenological parameters, which are difficult to determine.

The aim of this research is to define a model with a minimum number of phenomenological parameters capable of predicting the development of the measured fiber orientation correctly. This minimizes the effort for parameter adjustment. In order to identify the model, an experimental comparison is necessary. This research presents a sliding-plate experiment to measure fiber orientation evolution for short fiber-reinforced polybutylene terephthalate (PBT). The experiment ensures a repeatable initial state and is able to measure the steady state orientation at a high total strain. With this measurement, a validation of the main existing macroscopic fiber orientation model is performed and a model with reduced phenomenological parameters is tested.

In order to reduce the identification effort for future materials, a mechanistic single fiber simulation is presented and compared with the experimental data.

EXPERIMENTAL DETERMINATION OF FIBER ORIENTATION EVOLUTION

In-situ methods, for measuring the fiber orientation development for highly filled polymers do not exist. Alternatively, a homogeneous flow field can be stopped at different time steps to analyze the orientation. For this experiment, the sliding-plate method presented in [11] is used to provide a homogenous shear flow. The samples were prepared with great effort to ensure a repeatable initial condition. The pellets have to be aligned in one direction; a random orientation of the pellets results in a large standard deviation (Figure 1 a, b). To see a difference in the orientation evolution the pellets have been aligned perpendicular to the flow. The fiber orientation measurements were performed using the method of ellipses [12] . The 1-standard deviation is used in all plots. The process settings are stated in Table 1. For the experiment three repetitions at five different strains were measured (0, 25, 50, 75, 100 total strain). Steady state is reached at about 75 total strain (Figure 2).

a) 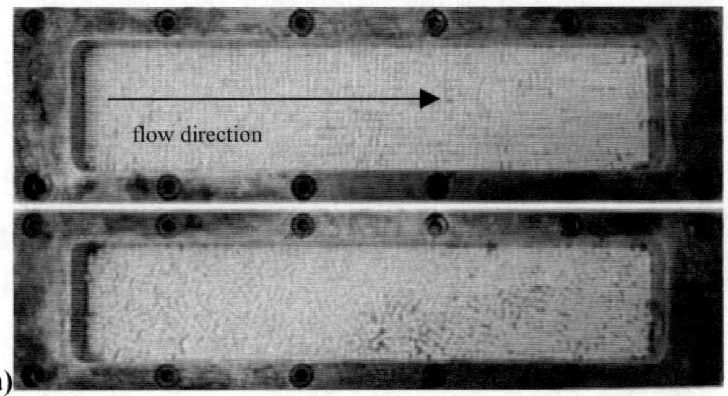

b) 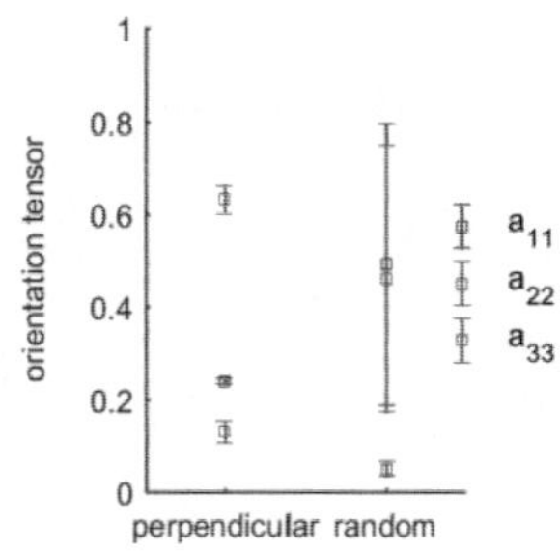

FIGURE 1. a) Sample preparation for the sliding-plate experiment: top perpendicular to the flow, bottom random b) Measured initial orientation for the initial plates with standard deviation

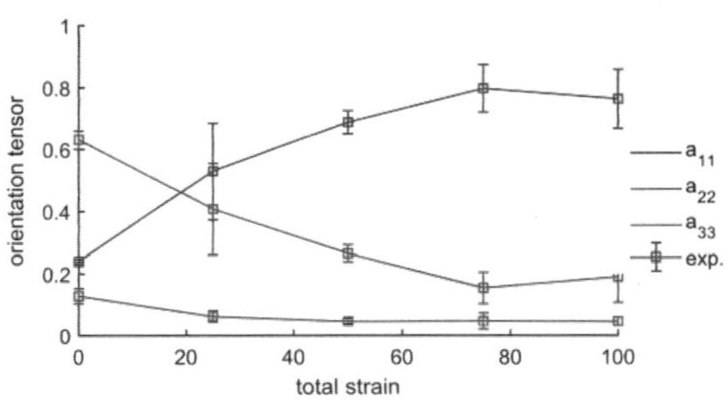

FIGURE 2. Measured orientation evolution with the sliding-plate rheometer for PBT GF30

TABLE (1) Process setting used in the sliding-plate experiment

Material	Sample thickness	Temperature	Shear rate	Max. shear strain
PBT GF30	1.5 mm	257 °C	$1\ s^{-1}$	100

PARAMETER FITTING OF MACROSCOPIC FIBER ORIENTATION MODELS

The macroscopic fiber orientation models depend on phenomenological parameters. The FT, nematic, RSC, ARD and ARD-RSC model are fit to the measured data with a generic algorithm (Table 2). Since the off-diagonal entries of the orientation tensor $\boldsymbol{A}$ show a significant standard deviation, the models are fit to the initial orientation of the plate without considering off-diagonal entries $\boldsymbol{A_0} = diag(0.24, 0.13, 0.63)$. For all needed closure approximations the OWE closure [13] is used. A retarding principal rate and an anisotropic diffusion are observed in the experimental curve (Figure 3). Therefore, the ARD-RSC model (eq. (1) and (3)), with six parameters, is necessary to fit the experimental data. To reduce the number of parameter, the MRD-RSC (eq. (2) and (4)) and the pARD-RSC (eq. (1)

and (4)) model are fit to the experimental data. Both models depend on five parameters. If the restrictions, stated in eq. (5), are applied, only three parameters have to be considered. For both models, a valid fit can be reached with the restriction. One additional advantage is the stability of the models for all parameter sets. There are only minor differences between the parameter fits and the evolutions for the MRD-RSC and pARD-RSC model (Figure 3). Therefore, it can be concluded that the reduced form of the ARD-model (eq. (2)) is sufficient for simple shear flows. The behaviour in more complex flow fields is not investigated at this point.

$$\dot{\mathbf{A}}^{ARD-RSC} = (\mathbf{W}\cdot\mathbf{A}-\mathbf{A}\cdot\mathbf{W}) + \xi\{\mathbf{D}\cdot\mathbf{A}+\mathbf{A}\cdot\mathbf{D} - 2[\mathbb{A}+(1-\kappa)(\mathbb{L}-\mathbb{M}:\mathbb{A})]:\mathbf{D}\} + \dot{\gamma}\{2[\mathbf{C}-(1-\kappa)\mathbb{M}:\mathbf{C}] -2\,\kappa(\mathrm{tr}\mathbf{C})\mathbf{A} - 5(\mathbf{C}\cdot\mathbf{A}+\mathbf{A}\cdot\mathbf{C}) + 10[\mathbb{A}+(1-\kappa)(\mathbb{L}-\mathbb{M}:\mathbb{A})]:\mathbf{C}\} \quad (1)$$

$$\dot{\mathbf{A}}^{MRD-RSC} = (\mathbf{W}\cdot\mathbf{A}-\mathbf{A}\cdot\mathbf{W}) + \xi\{\mathbf{D}\cdot\mathbf{A}+\mathbf{A}\cdot\mathbf{D} - 2[\mathbb{A}+(1-\kappa)(\mathbb{L}-\mathbb{M}:\mathbb{A})]:\mathbf{D}\} + \dot{\gamma}\{2\kappa[\mathbf{C}-tr(\mathbf{C})\mathbf{A}] \quad (2)$$

$$\mathbf{C}^{\mathbf{ARD}} = b_1\mathbf{I} + b_2\mathbf{A} + b_3\mathbf{A}^2 + \frac{b_4}{\dot{\gamma}}\mathbf{D} + \frac{b_5}{\dot{\gamma}^2}\mathbf{D}^2 \quad (3)$$

$$\mathbf{C}^{\mathbf{red}} = C_I\mathbf{R}_{\mathbf{A}}\,\mathrm{diag}(D_1, D_2, D_3)\mathbf{R}_{\mathbf{A}}^{\mathbf{T}} \quad (4)$$

$$D_1 = 1, D_2 + D_3 = 1, D_2 \geq 0.5 \Rightarrow D_1 \geq D_2 \geq D_3 \quad (5)$$

with $\mathbf{D}, \mathbf{W}$ symmetric, antisymmetric part of the velocity gradient, $\dot{\gamma}$ scalar magnitude of $\mathbf{D}$, $\xi = \frac{r_e^2-1}{r_e^2+1}$ with r_e aspect ratio, $\mathbb{L}, \mathbb{M}$ analytic functions of the eigenvalues and eigenvectors of $\mathbf{A}$, κ fitting parameters RSC model, b_i $i = 1, \dots, 5$ fitting parameters ARD model, $C_I, D_j, j = 1,2,3$ fitting parameters pARD/MRD model, $\mathbf{R}_{\mathbf{A}} = (\boldsymbol{e_1 e_2 e_3}), \boldsymbol{e_i}$ eigenvector of $\mathbf{A}$

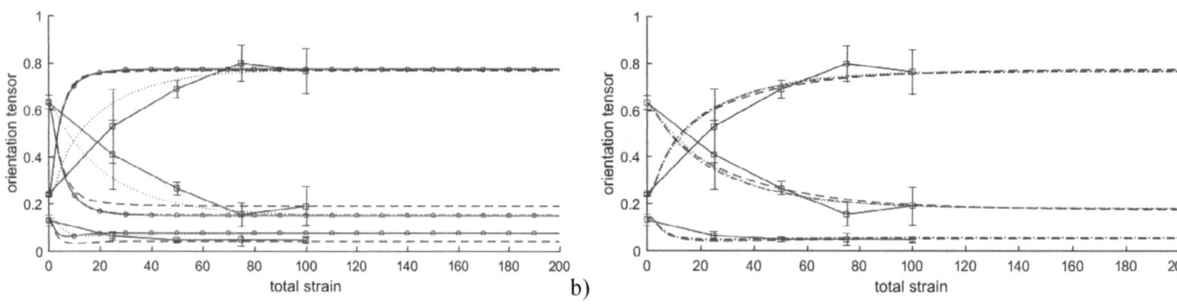

a) b)

FIGURE 3. Comparison of the macro models with the sliding-plate experiment with optimal parameters determined with a generic algorithm, — a_{11}, — a_{22}, — a_{33}, ┼ experiment

a) -·- FT, -o- nematic, ······· RSC, -- ARD

b) -·- ARD-RSC, ······· RSC-MRD, -- pARD-RSC

TABLE (2) Fitted parameters for the macroscopic models

Model	**Diffusion**		**Strain reduction**		**Nematic factor**	
	parameter	fitting	parameter	fitting	parameter	fitting
FT	C_I	0.01	-		-	
RSC	C_I	0.01	κ	0.3	-	
nematic	C_I	0.01	-		U_0	0.01
ARD	$\boldsymbol{b}$	(0.01 0.007 0.002 -0.001 -0.007)	-		-	
ARD-RSC	$\boldsymbol{b}$	(0.01 0.007 0.002 -0.001 -0.005)	κ	0.275	-	
MRD-RSC	$C_I, \boldsymbol{D}$	0.017 (1 0.66 0.34)	κ	0.275	-	
pARD-RSC	$C_I, \boldsymbol{D}$	0.012 (1 0.65 0.35)	κ	0.28	-	

A MECHANISTIC MODEL FOR PARAMETER FITTING

The use of a mechanistic single fiber simulation for parameter fitting instead of an experiment can considerably reduce cost and time. The simulation is based on the discrete element method (DEM). The overall structure of the simulation is described under [14]. Figure 4 shows the workflow of the simulation and the parameters, which have to be defined. The initial cells are created with the Fiber Generator based on [15]. Then a homogenous shear flow is applied to the RVE and the fiber movement in the flow is solved with the DEM. In a post-processing step the evolution of the orientation tensor is computed.

Each fiber is modeled as a chain of rigid cylindrical rods connected with ball and socket joins, similar to [16]. At each segment node, the position x_i, the velocity u_i and angular velocity ω_i are calculated. The fluid field is assumed as homogenous shear flow and a back coupling from fiber motion to fluid forces is not incorporated. The fibers are suspended in a viscous fluid and hydrodynamic flow forces are directly transmitted to the fibers. The forces are applied at each segment i: drag forces F_i^H from the fluid, interaction forces with segment j F_{ij}^C and intra-fiber forces from adjacent segments X_i. This leads to the translation equation of motion (eq. (6)). Equally, the rotational equation of motion (eq. (7)) is derived with the elastic recovery term $M^b$and the hydrodynamic torque T^H. In eq. (7) r_i describes the segment vector and d_{ij} the shortest distance vector between two segments. If fibers are divided in more than one segment, a constraint to enforce connectivity is necessary (eq. (8)).

$$0 = F_i^H + \sum_j F_{ij}^C + X_i - X_{i+1} \tag{6}$$

$$0 = T_i^H - r_i \times X_{i+1} + \sum_j d_{ij} \times F_{ij}^C + M_i^b - M_{i+1}^b \tag{7}$$

$$0 = u_i + \omega_i \times r_i - u_{i+1} \tag{8}$$

For the modeling of the hydrodynamic and elastic forces and the breaking of fibers, refer to [14]. The original fiber interaction equation is replaced by an advanced interaction model between fibers based on [16]. The interaction model is divided in three regimes: a lubrication one before contact F_{ij}^{lub}, a transition one F_{ij}^{trans} taking place at the scale of the fiber surface roughness δ_{sur} and mechanical contact F_{ij}^{mc} (eq. 9). Once mechanical contact is reached, Coulomb friction is added.

$$F_{ij}^C = \begin{cases} F_{ij}^{lub} & \delta_{sur} < d_{ij} < \delta_{lub} \\ F_{ij}^{trans} & |d_{ij}| < \delta_{sur} \\ F_{ij}^{mc} & d_{ij} < -\delta_{sur} \end{cases} \tag{9}$$

A comparison with the experiment shows good agreement with the steady state but a four times faster evolution speed (Figure 5). The simulation parameters are stated in Table 3. The reasons for this difference are still under investigation. One possible explanation is that each single fibers sees a smaller shear strain. The simulated curve with a reduced shear strain $\text{strain}_{reduced} = \frac{1}{4}\text{strain}$ shows good agreement with the experimental curve. At this point, it is not possible to conclude why we see the shear strain reduction. It is still not possible to conclude if the reduction is a global effect occurring in the experiment or a local effect in the simulation. Possible explanations are a non-linear shearing over the thickness in the experiment, due to thin lubrication layers or an overestimated shear rate applied in the simulation due to the neglected back coupling between the fiber motion and the flow.

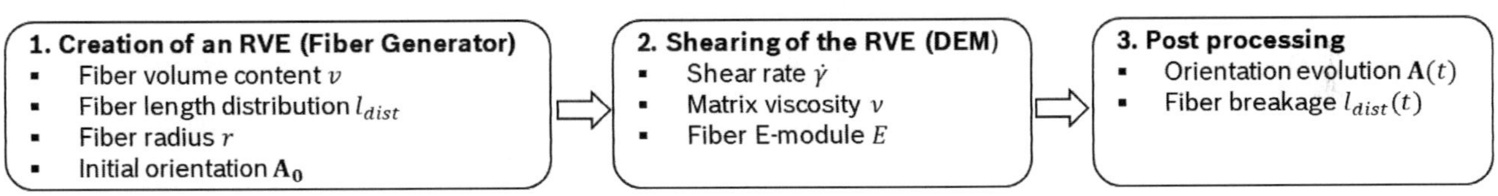

FIGURE 4. Flow chart of the mechanistic simulation

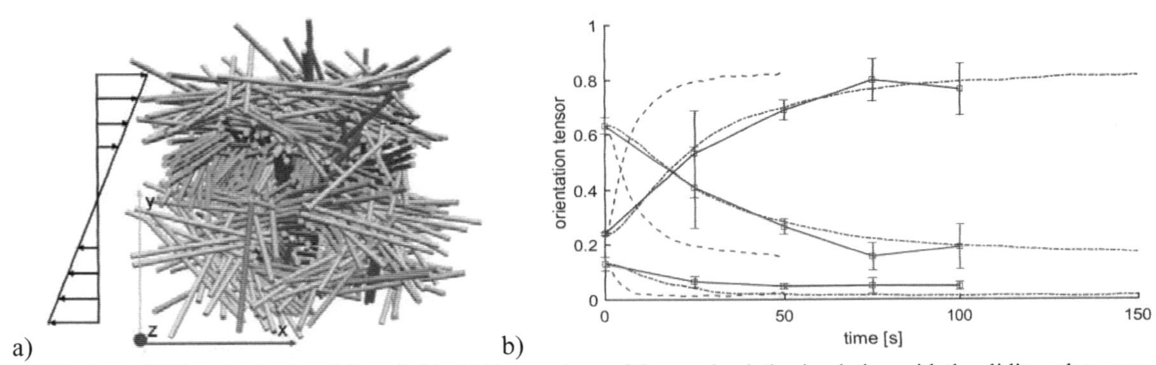

FIGURE 5. a) RVE with simulated flow field b) Comparison of the mechanistic simulation with the sliding-plate experiment — a_{11}, — a_{22}, — a_{33}, + experiment, --- micro model with process settings, -·-·- micro model retarded with ¼ of the strain

TABLE (3) Simulation parameter used in the mechanistic model

RVE size	Number of fibers	Fiber volume	Fiber radius	Fiber length	Shear rate	Viscosity
$(4e-4)^3\ m^3$	594	18.2 %	5e-6 m	2.5e-4 m	$1\ s^{-1}$	250 Pas

CONCLUSION

A fiber orientation evolution was successfully measured for 30% short fiber-reinforced PBT. To simulate the experiment, a model using strain reduction and anisotropic diffusion is necessary. The anisotropy can be modeled by an ARD approach, however, the original ARD model is complex with a high number of parameters. Moreover, not all parameter sets show stable behavior. Reduced parameter ARD models (pARD, MRD) yield equally good results and can be used instead of the complex model. In conclusion, only two parameters have to be fitted and stable behavior is ensured. The strain reduction can be incorporated with an RSC approach with one additional fitting parameter. Since the experiment is time and cost consuming, replacing it with a mechanistic simulation can be a great advantage. The presented simulation shows already good agreement with the steady state, but predicts a faster evolution. If the simulation is repeated with a reduced shear rate, a good agreement is reached. The investigation to explain the difference between simulation and experiment is still ongoing and more experimental validation is planned for the future.

REFERENCES

1. G. B. Jeffrey, "The motion of ellipsoidal particles immersed in a viscous," *Proceedings of the royal society of London A: Mathematical, physical and engineering sciences,* vol. 102, no. 172, pp. 161-179, 1922.
2. F. Folgar and C. L. Tucker, "Orientation Behavior of Fibers in Concentrated Suspensions," *Journal of Reinforced Plastics and Composites,* vol. 3, pp. 98-119, 4 1984.
3. S. G. Advani and C. L. Tucker, "The Use of Tensors to Describe and Predict Fiber Orientation in Short Fiber Composites," *Journal of Rheology,* vol. 31, pp. 751-784, 11 1987.
4. Huynh, HM, *Improved Fiber Orientation Prediction For Injection-Molded Composites,* MasterThesis University of Illinois Urbana-Champaign, 2001.
5. J. Wang, J. F. O'Gara and C. L. Tucker, "An objective model for slow orientation kinetics in concentrated fiber suspensions: Theory and rheological evidence," *Journal of Rheology,* vol. 52, pp. 1179-1200, 9 2008.
6. H.-C. Tseng, R.-Y. Chang and C.-H. Hsu, "Phenomenological improvements to predictive models of fiber orientation in concentrated suspensions," *Journal of Rheology,* vol. 57, pp. 1597-1631, 9 2013.
7. J. H. Phelps and C. L. Tucker, "An anisotropic rotary diffusion model for fiber orientation in short- and long-fiber thermoplastics," *Journal of Non-Newtonian Fluid Mechanics,* vol. 156, pp. 165-176, 1 2009.
8. H.-C. Tseng, R.-Y. Chang and C.-H. Hsu, "The use of principal spatial tensor to predict anisotropic fiber orientation in concentrated fiber suspensions," *Journal of Rheology,* vol. 62, no. 1, pp. 312-320, 2017.
9. A. Bakharev, H. Yu, R. Speight and J. Wang, "Using New Anisotropic Rotational Deffusion Model To Improve Prediction Of Short Fibers in Thermoplastic Injection Molding," in *ANTEC*, Orlando, 2018.
10. A. Latz, U. Strautins and D. Niedziela, "Comparative numerical study of two concentrated fiber suspension models," *Journal of Non-Newtonian Fluid Mechanics,* vol. 165, pp. 764-781, 7 2010.
11. K. Ortman, D. Baird, P. Wapperom and A. Aning, "Prediction of fiber orientation in the injection molding of long fiber suspensions," *Polymer Composites,* vol. 33, pp. 1360-1367, 2012.
12. M. Vélez-García, G. Wapperom and K. V., "Sample preparation and image acquisition using optical-reflective microscopy in the measurement of fiber orientation in thermoplastic composites: sample preparation and image acquisition," *Journal of Microscopy,* vol. 248, no. 1, pp. 23-33, 2012.
13. D. H. Chung and T. H. Kwon, "Improved model of orthotropic closure approximation for flow induced fiber orientation," *Polymer Composites,* vol. 22, pp. 636-649, 2001.
14. C. Pérez, "The Use of a Fiber Simulation to Predict Fiber Motion in Polymer Processing," in *PhD Thesis*, 2016.
15. M. Schneider, "The sequential addition and migration method to generate representative volume elements for the homogenization of short fiber reinforced plastics," *Computational Mechanics,* vol. 59, no. 2, pp. 247-263, 2017.
16. S. Linström and T. Uesaka, "Simulation of the motion of flexible fibers in viscous fluid flow," *Physics of Fluids,* vol. 19, no. 11, p. 113307, 2007.

New Analytical Approach to Calculate the Melt Flow within the Co-Rotating Twin Screw Extruder by Using the Performance Mapping Method

Frederik Sporkmann, Alexander Heide and Volker Schöppner

Kunststofftechnik Paderborn, Paderborn University, Warburger Str. 100, 33098 Paderborn, Germany
volker.schoeppner@ktp.upb.de; frederik.sporkmann@ktp.upb.de

Abstract. The analytical description of the melt flow within the co-rotating twin screw extruder has been issued in various publications and it is an integral part of calculation models, software solutions for simulating polymer compounding processes such as SIGMA. Nevertheless, the accuracy of an analytical model depends on the defined simplifications. The more simplifications within the analytical description, the more deviation between the technical system and the calculation model. Analytical models for twin screw processes are no exception and are subject to simplifications and adaptations due to the complex geometry and the transient flow field. Therefore, the use of the kinematical inversion principle, the exchange of the original screw channel geometry and the consideration of an evenly distributed pressure gradient are commonly used simplifications for calculations models. In this paper a new analytical approach to calculate the melt flow is presented. The model is based on numerical simulations and with the use of the performance mapping method an analytical calculation method is derived. The advantage of the performance mapping method is that the original channel geometry and kinematical behavior of the twin screws are considered and the transient flow field in the system is used to generate a wide range of performance maps. Within the scope of this work, the performance maps of the melt conveyance and the pressure gradient in the fully filled screw channel are quantitatively examined with the aid of specially defined evaluation routines. Characteristic operating points are identified by the conveying characteristics of the melt which are displayed in the performance maps. The validation of this new analytical approach by using the performance mapping method with numerical data demonstrated that the two-dimensional description of the melt flow within the screw channel flow has a deviation of less than 10 % for varying flow law exponents geometry and process parameters.

Keywords: Co-rotating twin screw extrusion, finite element method, throughput-pressure behavior, melt flow

INTRODUCTION

In order to design an optimum compounding process, the melt conveying behavior of the co-rotating twin screw extruder needs to be known. This has been investigated in numerous publications and mathematically described in [1-6]. Common analytical approaches are based on various simplifications which have influence on the accuracy of the calculation results. On the one hand, the material flow describing differential equations are simplified until they can be solved closed for simple geometries. On the other hand, specific simplifications are made for the co-rotating twin screw extruder, such as the assumption of kinematic reversal, the idealization of the channel geometry according to Erdmenger [3] to the rectangular channel and the assumption of a one dimensional channel flow [2][4][5]. The sum of all simplifying assumptions leads to deviations between the calculation model and the real co-rotating twin screw extruder but requires less calculation time than a numerical calculation approach. In order to increase the accuracy of the analytical calculation models, the performance mapping method is defined on the basis of numerical flow simulations with Extrud3D [7-8]. It is based on a two dimensional description of the complex behaviour of the polymer flow and considerates the real channel geometry according to Erdmenger [3]. This new calculation method is valid within the working range of a conveying Erdmenger-element.

PERFORMANCE MAPPING METHOD

Numerical flow simulations which are based on Extrud3D [7-8] are used to characterize the conveying behavior of the tightly intermeshing co-rotating twin screw extruder. It is evaluated appropriately with the open source and multi-platform tool ParaView. Within the numerical simulation the axial mass flow orthogonal to the cross section cut (FIGURE 1 f)) $\dot{m}_{ax}$ is given as essential condition for the simulations. Based on this global condition, the material transport within the Erdmenger-element is calculated. Further, the conveying characteristics will be analysed additionally with the use of individual channel cutplanes (FIGURE 1 d)) which follow the channel layout. The melt conveyance is induced by the rotation of the screws. In order to quantify the resulting flow conditions, several simulations were performed in the working range of a conveying Erdmenger-element. The resulting throughput and additional characteristic values such as the flow conditions in the conveyor channel, the axial pressure gradient, the share of the intermeshing area and the channels in the axial throughput $\dot{m}_{ax}$ located in the cross section of the twin screw extruder, are shown in FIGURE 1 a), c) and e).

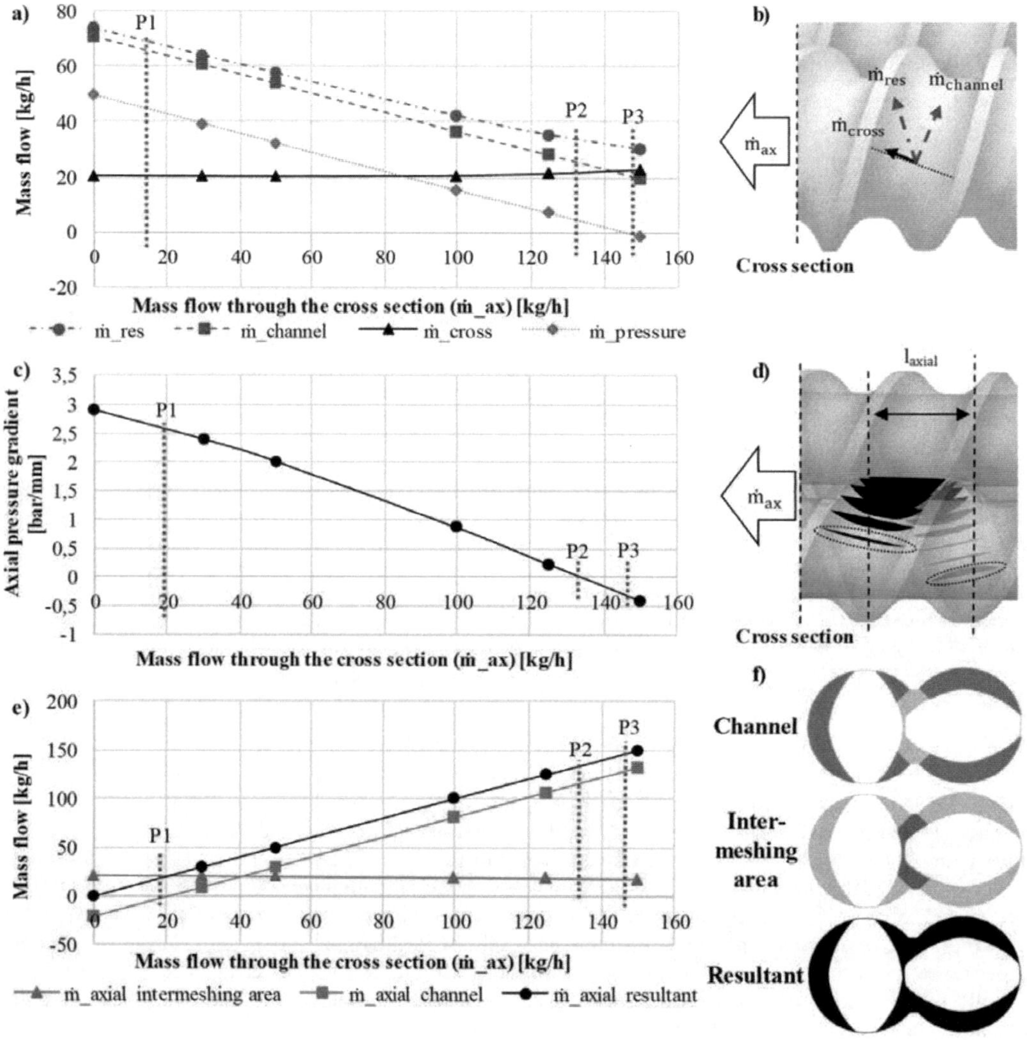

Figure 1. a) Flow Conditions in the Channel b) Pressure Gradient along the Conveying Elements c) Section Based Axial Mass Flow through the Cross Section

An orthogonal channel cross-section (FIGURE 1 b)) was generated to evaluate the flow conditions in the conveying channel. As shown in FIGURE 1 a), the resulting $\dot{m}_{res}$ and the mass flow in channel direction $\dot{m}_{channel}$ were determined directly from the flow simulations. The cross-flow $\dot{m}_{cross}$ results from equation (1).

$$\dot{m}_{cross} = \sqrt{\dot{m}_{res}^2 - \dot{m}_{channel}^2} \tag{1}$$

The pressure flow $\dot{m}_{pressure}$ is a part of the channel flow $\dot{m}_{channel}$. This can be calculated according to equation (2) with respect to the induced drag flow by the screw rotation $\dot{m}_{drag}$ according to [10].

$$\dot{m}_{pressure} = \dot{m}_{channel} - \dot{m}_{drag} \tag{2}$$

To determine the axial pressure gradient, in FIGURE 1 c), two orthogonal channel cross-sections were arranged around a screw at a 180° angle to each other (FIGURE 1 d)). The average pressure was determined for both channel sections, the difference was formed from both values and divided by the axial distance (l_{axial}). FIGURE 1 e) shows the axial conveying of the intermeshing area and the conveying channels (FIGURE 1 f)). In these evaluations, characteristic conveying points can be identified, which applies to all conveying elements phenomenologically. The first characteristic conveying point results from FIGURE 1 e) P1. At this point there is no axial conveying of the channels ($\dot{m}_{axial\ channel}$). The axial conveying below this point results exclusively from the intermeshing area. As a result, the maximum axial pressure gradient within the element (FIGURE 1 c) P1) is reached. The second characteristic conveying point is the reversal point of the pressure gradient and can be seen in FIGURE 1 c) P2. There is no axial pressure gradient at this point. Furthermore, FIGURE 1 a) P2 shows that, despite the missing pressure gradient a low pressure flow $\dot{m}_{pressure}$ is induced in the channel direction. This can be explained by the rotation of the screws and the resulting axial relative movement of the screw flights. The third conveying characteristic point is defined in FIGURE 1 a) P3. The pressure flow $\dot{m}_{pressure}$ turns to zero (2) and there is only induced drag flow remaining. Moreover, FIGURE 1 c) P3 underlines that the pressure gradient has reversed and compensates the flow component resulting from the screw rotation. FIGURE 1 e) displays that the axial conveying of the intermeshing area $\dot{m}_{axial\ intermeshing\ area}$ over the entire working range of a conveying Erdmenger-element can be assumed to be constant. The same applies to the cross-flow in the channel (FIGURE 1 a) $\dot{m}_{cross}$).

ANALYTICAL MODELING OF THE MELT FLOW

Since the characteristic conveying points are known, it is now possible to describe the melt conveying behavior of a conveying Erdmenger-element along the entire conveying range analytically. For this purpose, the characteristic conveying points are described by the channel cross-section (FIGURE 1 a)) of the conveying element in a vector field, as displayed in FIGURE 2. The resulting two-dimensional flow $\dot{m}_{res}$ in the conveying element is composed with each point of the transverse flow and the perpendicular channel flow (equation (3)).

$$\dot{m}_{res} = \sqrt{\dot{m}_{cross}^2 + \dot{m}_{channel}^2} \quad (3)$$

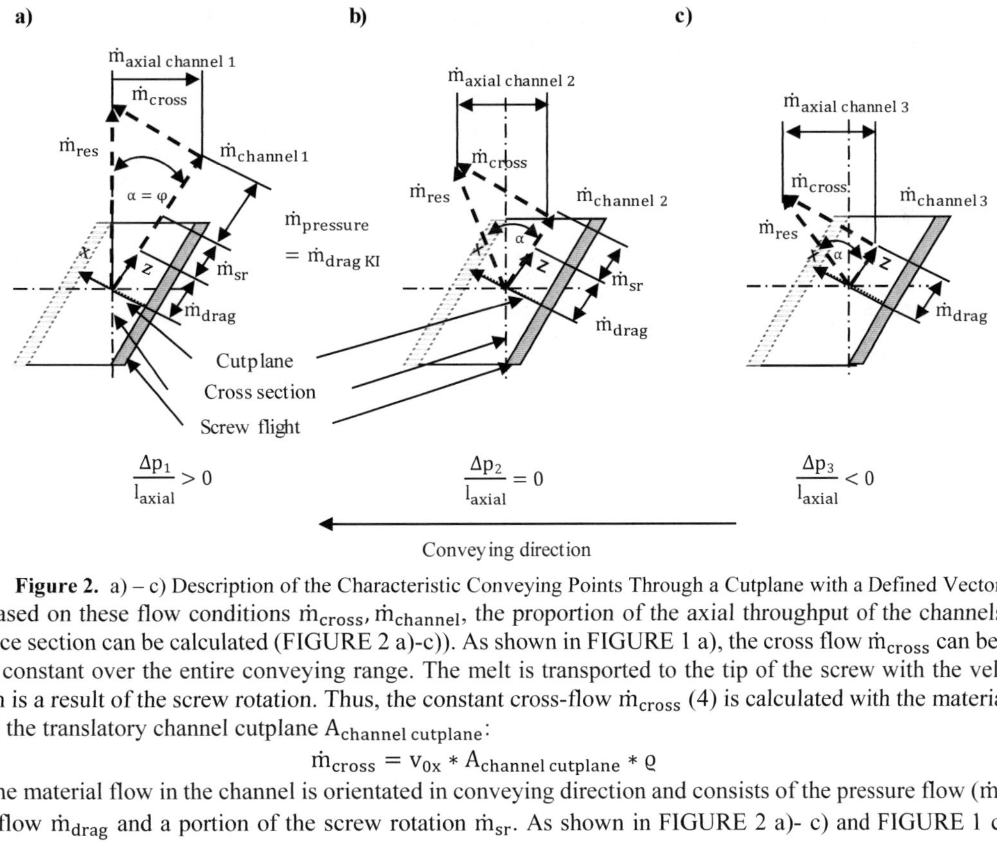

Figure 2. a) – c) Description of the Characteristic Conveying Points Through a Cutplane with a Defined Vector Field

Based on these flow conditions $\dot{m}_{cross}$, $\dot{m}_{channel}$, the proportion of the axial throughput of the channels through the face section can be calculated (FIGURE 2 a)-c)). As shown in FIGURE 1 a), the cross flow $\dot{m}_{cross}$ can be assumed to be constant over the entire conveying range. The melt is transported to the tip of the screw with the velocity v_{0x} which is a result of the screw rotation. Thus, the constant cross-flow $\dot{m}_{cross}$ (4) is calculated with the material density ϱ and the translatory channel cutplane $A_{channel\ cutplane}$:

$$\dot{m}_{cross} = v_{0x} * A_{channel\ cutplane} * \varrho \quad (4)$$

The material flow in the channel is orientated in conveying direction and consists of the pressure flow ($\dot{m}_{pressure}$), drag flow $\dot{m}_{drag}$ and a portion of the screw rotation $\dot{m}_{sr}$. As shown in FIGURE 2 a)- c) and FIGURE 1 c)- e), the channel flow increases with increasing pressure gradients, as a result the axial channel flow $\dot{m}_{\text{axial channel}}$ through the front section decreases. The consequence is, that there is no axial channel flow at the first characteristic point (FIGURE 2 a). The amount of the channel flow $\dot{m}_{\text{channel 1}}$ can be calculated according to equation (4), since the angle α corresponds to the pitch angle φ at this point.

$$\dot{m}_{\text{channel 1}} = \frac{\dot{m}_{\text{cross}}}{\tan\varphi} \quad (5)$$

The drag flow $\dot{m}_{\text{drag}}$ is calculated as described in [10] by dividing the channel geometry into individual rectangular segments and by assuming the viscous Couette-flow. The rotation of the screws is assumed. With the same approach, including the movement conditions for the kinematic reversal, the pressure flow component $\dot{m}_{\text{pressure}}$ is calculated. This is permissible because the hypothesis of the kinematic reversal induces a drag flow ($\dot{m}_{\text{drag KI}}$) in the conveyor channel, which is compensated by an opposite pressure flow. The amount of the drag flow corresponds to the kinematic reversal of the pressure flow $\dot{m}_{\text{pressure}}$ in FIGURE 2 a). The last unknown flow component resulting from the screw rotation ($\dot{m}_{\text{sr}}$) is determined according to equation (6).

$$\dot{m}_{\text{sr}} = \dot{m}_{\text{channel 1}} - \dot{m}_{\text{pressure}} - \dot{m}_{\text{drag}} \quad (6)$$

Based on the definition of the flow components, the two further conveying characteristic points in FIGURE 2 b) and c) can be calculated. The three characteristic conveying points represent the entire conveying range of a pair of conveying Erdmenger-elements. Since the three interpolation points are obtained in this way, a continuous function is set up by means of Newton's 2nd degree interpolation polynomial (equation (7)). It is possible to interpolate values between the points for each flow component.

$$f(x) = a_0 + a_1 * (x - x_0) + a_2 * (x - x_0) * (x - x_1) \quad (7)$$

Due to the three characteristic points and the quadratic interpolation approach, the prediction of the non-linear courses of the flow components can be analytically achieved by this method.

COMPARISON BETWEEN NUMERICAL AND ANALYTICAL RESULTS

To validate the model, isothermal simulations were carried out with a conveying Erdmenger-element (table 1 and table 2). The material data of Polyethylene (PE) was used for the simulations.

TABLE 1. Geometry data used for the calculations and simulations

Geometry Data	Unit	Erdmenger-Element			
Ratio D_o/D_i	-	1.45		1.89	
Outer diameter screw D_o	[mm]	25			
Barrel diameter D_b	[mm]	25			
Inner diameter screw Di	[mm]	17.2		13.2	
Axis distance of the screws	[mm]	21.1		19.1	
Element length l	[mm]	24	12	75	24
Pitch t	[mm]	24	12	75	24
Clearance between the barrel Ss	[mm]	0			

TABLE 2. Process parameters used for the calculations and simulations

Process Parameters	Unit	Erdmenger-Element			
RPM	[min^{-1}]	600			
Throughput $\dot{m}$	[kg/h]	0.1 - 150	0.1 - 100	0.1 - 350	0.1 - 150
Barrel temperature Tb	[°C]	215			
Melt temperature Tm	[°C]	215			
Power law index n	-	0.2	0.58	1	0.58

Within the scope of validation, the numerically and analytically determined material and channel flows are compared in FIGURE 3. By using the performance mapping method (PMM) the flow components and thus the flow conditions for varying material parameters, are analytically predictable with an accuracy of less than 10 % average (FIGURE 3 a)). Furthermore, FIGURE 3 b) displays that the flow conditions for varying conveying channel geometries close to the practicable limits are also recorded with approx. 10 % deviation from the numerically determined material flow.

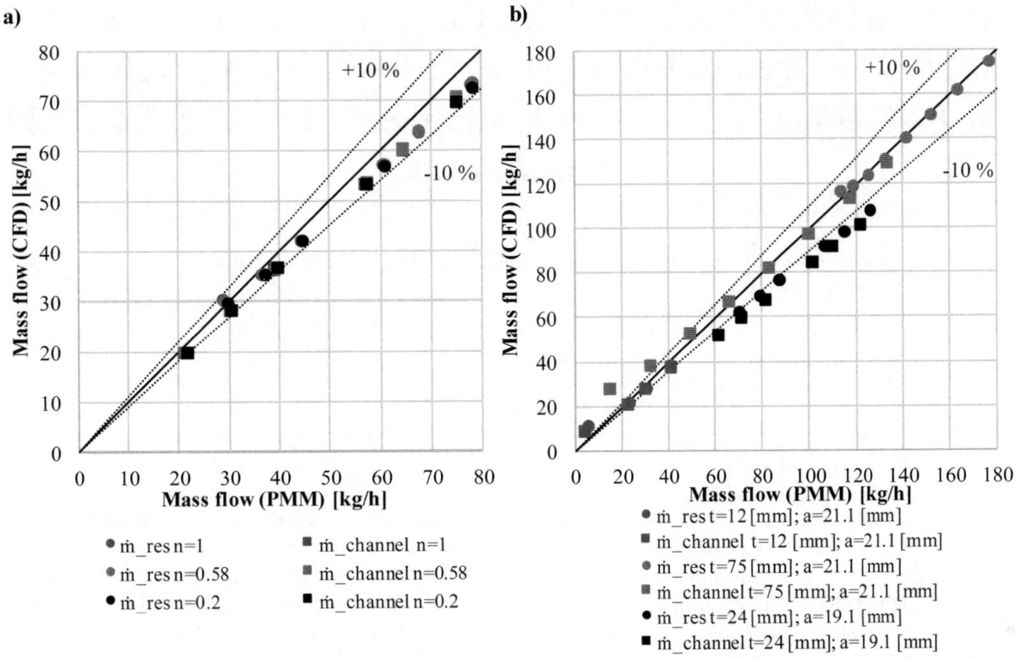

Figure 3. Comparison of the Analytical Results by Using the Performance Mapping Method with the Numerical Results of the Mass flow for Varying Power Law Exponents a) and Channel Geometries b)

SUMMARY AND OUTLOOK

In this paper it was proved, that it is possible to define analytical calculation method that describes the two dimensional material flow within the channel of an Erdmenger-element. The calculation considers the real channel geometry according to Erdmenger as well as the complex flow behavior of polymer melts. In this matter the performance mapping method (PMM) was used for the whole working range of a fully filled Erdmenger-element. The PMM extends the one dimensional models ([2][4][5]) by considering the two dimensional material flow regarding the calculation in twin screw processes. The validation with numerical data proved that the PMM shows an average deviation of less than 10 % for varying flow law exponents, geometry and process parameters. Future investigations will consider the extension of the PMM for different screw elements and screw designs.

REFERENCES

1. J. L. White and H. Potente, *Screw Extrusion*, Munich: Carl Hanser Publishers, 2002
2. K. Kohlgrüber, *Der gleichläufige Doppelschneckenextruder*, Munich, Carl Hanser Publishers, 2016
3. C. Rauwendaal, *Polymer Extrusion*, Munich: Carl Hanser Publishers, 2014
4. J. Ansahl, *Grundlagen für die Auslegung dichtkämmender Gleichdrall-Doppelschneckenextruder*, Ph.D. Thesis, University of Paderborn 1993
5. U. Melisch, *Grundlagen zur Simulation des Förder- und Plastifizierprozesses dichtkämmender Gleichdrall-Doppelschneckenextruder*, Ph.D. Thesis, University of Paderborn 1998
6. Z. Tadmor and C. G. Gogos, *Principles of Polymer Processing*, New Jersey, John Wiley & Sons, Inc., 2006
7. C. Becker, S. Turek, *FEATFLOW – Finite element software for the incompressible Navier-Stokes equation* – User Manuel Release 1.1. University of Heidelberg, Heidelberg, 1998
8. O. Mierka, T. Theis, T. Herken, V. Schöppner, F. Platte, *Mesh Deformation Based Finite Elmente - Fictitious Boundary Method (FEMFBM) for the Simulation of Twin-screw Extruders*; publication, IANUS Simulation GmbH, Dortmund, 2014
9. R. Erdmenger, German Patent No. 862668, 1953
10. F. Sporkmann and V. Schöppner, *New Method to Analyze the Throughput-Pressure Behavior with the Corrected Twin Screw Channel Geometry and the Induced Drag Flow of Screw Elements by Using 3D FEM Simulation;* publication Polymer Processing Society (PPS34), 2018

Analysis of the Characteristic Flow Areas for Erdmenger-Elements to Predict the Throughput-Pressure Behavior of Co-Rotating Twin Screw Extruders by Using 3D FEM Simulation

Frederik Sporkmann and Volker Schöppner

Kunststofftechnik Paderborn, Paderborn University, Warburger Str. 100, 33098 Paderborn, Germany
volker.schoeppner@ktp.upb.de; frederik.sporkmann@ktp.upb.de

Abstract: The analysis of the throughput-pressure behavior for Erdmenger-elements has been subject of a variety of publications and is still topic of controversial debates. However, the significance of an accurate prediction of the throughput-pressure behavior for developing and optimizing compounding processes for co-rotating twin screw extruders is undisputed. For this reason, there are different mathematical descriptions to predict the melt flow behavior within the conveying elements based on the Erdmenger geometry. Most of these models base on a two-dimensional simplification for the geometry and the kinematical inversion principle. Due to increasing demands regarding compound quality and the need of new complex screw geometries to reach higher compound quality, analytical approaches to predict the flow behavior are insufficient. The use of numerical simulations is mandatory. Within this paper a new method to analyze the throughput-pressure behavior based on 3D FEM simulations is discussed. In this method the characteristic flow areas channel, gap and intermeshing area of the co-rotating twin screw extruder are considered. Further the original kinematical behavior of the screws are respected. This new approach can be used to predict the melt flow more accurate and reveals potential for optimizing screw designs. Moreover, it is applicable for a variety of screw geometries. In this investigation the method will be displayed for different conveying Erdmenger-elements. The process and geometry parameters are varied to demonstrate the method to predict the throughput-pressure behavior based on the characteristic flow areas for co-rotating twin screw extruders.

Keywords: Co-rotating twin screw extrusion, finite element method, throughput-pressure behavior, melt flow.

INTRODUCTION

Due to constantly changing products and applications in the compounding sector which can be attributed to the increasing production volumes of the global polymer industry. The adaptation to higher production standards and improved efficiency in compounding is also necessary in the area of the twin-screw extrusion. Therefore, it is essential to analyse the characteristic conveying behaviour on co-rotating twin screw extruders to quantify it as a function of throughput and process parameters. In many approaches and models, the co-rotating twin screw extruder is reduced to the conveying channels, similar to a single screw extruder [1-3] and the actual physical behavior and complex geometry are not considered. As a result, there are barely any approaches for the description of characteristic flow areas. The approaches can be used as a starting point for new models regarding the prediction of material- and pressure throughput behaviour [3]. In this paper a new model is presented which considers and quantifies the characteristic flow areas and the real conveying conditions for fully filled conveying Erdmenger-elements [4] by using 3D FEM simulations.

DEFINITON OF THE CHARACTERISTIC FLOW AREAS FOR ERDMENGER-ELEMENTS IN CO-ROTATING TWIN SCREW EXTRUDERS

For the numerical investigations, the CFD-Solver Extrud3D was used. Extrud3D is an extension of the subroutine Featflow which was developed to calculate incompressible Navier-Stokes equations of fluid dynamics in the two and three dimensional area with the use of the FE-method [5-6]. It is developed in collaboration between the TU Dortmund, the IANUS Simulation GmbH and the Paderborn University. Within the FE simulations, the kinematical behavior is the same as in the real processes. According to FIGURE 1, the cross-sectional area of a complete Erdmenger element is divided into three characteristic flow areas (CFA).

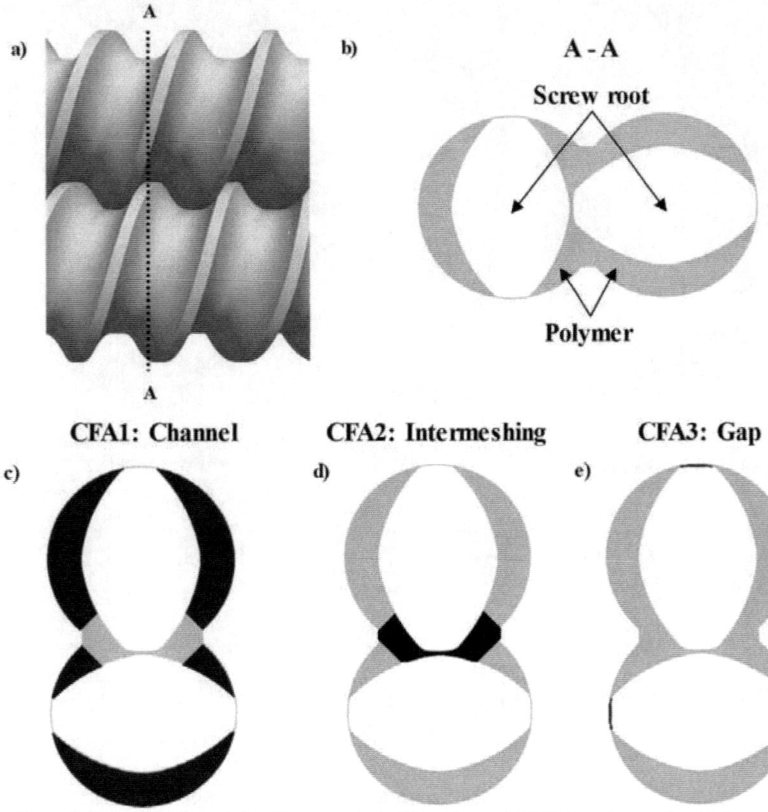

FIGURE 1. Overview of the Definition of the Characteristic Flow Areas (CFA) for Conveying Erdmenger-Elements a)-b), Channel c), Intermeshing d) and Gap e)

The aim of this division is to simplify the geometry of the Erdmenger element and to summarize the material flow for areas with similar flow characteristics. With the use of a geometry-dependent description for the CFAs, the reproducibly for all common pitches and diameter ratios could be achieved. Due to this parametric division of the cross-section, it is now possible to calculate the volume flows of CFA1 to CFA3. In the following, the individual volume flows for the individual CFAs are defined (1) - (4).

$$\dot{V}_{total} = \dot{V}_{CFA1} + \dot{V}_{CFA2} + \dot{V}_{CFA3} \tag{1}$$

$$\dot{V}_{CFA1} = \sum_{k=1}^{n=\text{Cells of CFA1}} v(z)_{cell,k} * A_{cell,k} \tag{2}$$

$$\dot{V}_{CFA2} = \sum_{k=1}^{n=\text{Cells of CFA2}} v(z)_{cell,k} * A_{cell,k} \tag{3}$$

$$\dot{V}_{CFA3} = \sum_{k=1}^{n=\text{Cells of CFA3}} v(z)_{cell,k} * A_{cell,k} \tag{4}$$

The sum of the individual volume flows (2) - (4) corresponds to the total volume flow (1) which is always conveyed through the cross-sectional area along the element. The procedure for determination of the individual volume flow is illustrated by the intermeshing area (CFA2) in FIGURE 2.

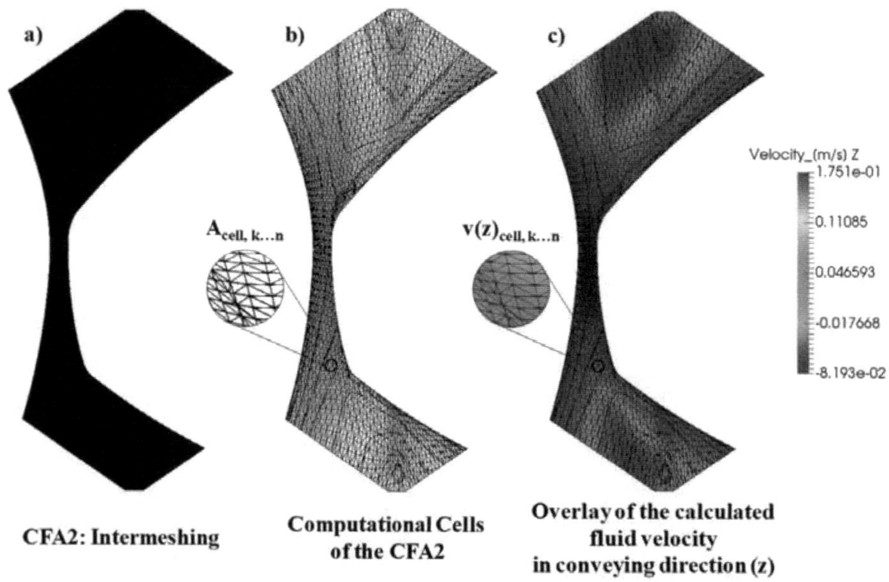

FIGURE 2. Procedure of the Numerical Calculation Regarding the Individual Throughput of the CFAs a)- c)

The calculation of the volume flows is based on the product of the individual area of the computational cells in the CFA and the results of the fluid velocity in conveying direction (z). These calculations are evaluated with the help of the ParaView software through parameterized geometry generation of the Erdmenger-elements and specially developed evaluation routines.

INVESTIGATION OF THE THROUGHPUT BEHAVIOR FOR THE CHARACTERISTIC FLOW AREAS

For the investigations, fully filled Erdmenger-elements with polymer melt were considered. Geometry (table 1) and process variables (table 2) were varied. All calculations and simulations were realized isothermally. The material data which was used is based on Polyethylene (PE).

TABLE 1. Geometry Data used for the Investigation

Geometry Data	Unit	Erdmenger-Element		
Diameter ratio D_o/D_i	-	1.45	1.55	1.64
Outer diameter screw D_o	[mm]		25	
Barrel diameter D_b	[mm]		25.3	
Inner diameter screw Di	[mm]	17.2	16.1	15.2
Axis distance of the screws	[mm]		21.1	
Clearance between screws Sŕ	[mm]	0	0.55	1
Element length l	[mm]		24	
Pitch t	[mm]		24	
Clearance between the barrel Ss	[mm]		0.15	

TABLE 2. Process Parameters used for the Investigation

Process Parameters	Unit	Erdmenger-Element
RPM	[min^{-1}]	300/600
Throughput $\dot{m}$	[kg/h]	5/10/30/50/75/100/150/200/225
Barrel temperature Tb	[°C]	215
Melt temperature Tm	[°C]	215
Power law index n	-	0.58

The results of the investigations are shown in the following graphes for 300 RPM. (FIGURE 3) and for 600 RPM (FIGURE 4). The results were plotted as a function of the pressure difference and the diameter ratio of the Erdmenger-element. For all CFAs the shares of the calculated volume flows regarding the total volume flow which is conveyed through the entire cross-section, were plotted.

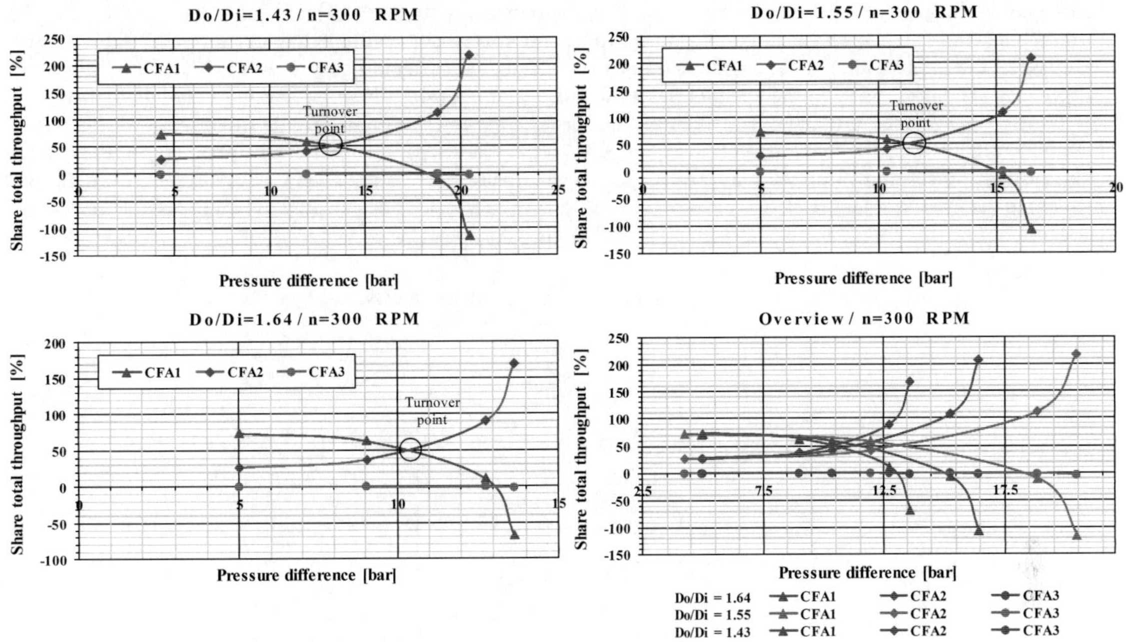

FIGURE 3. Investigation of the Throughput Behaviour of the Different CFAs for 300 RPM

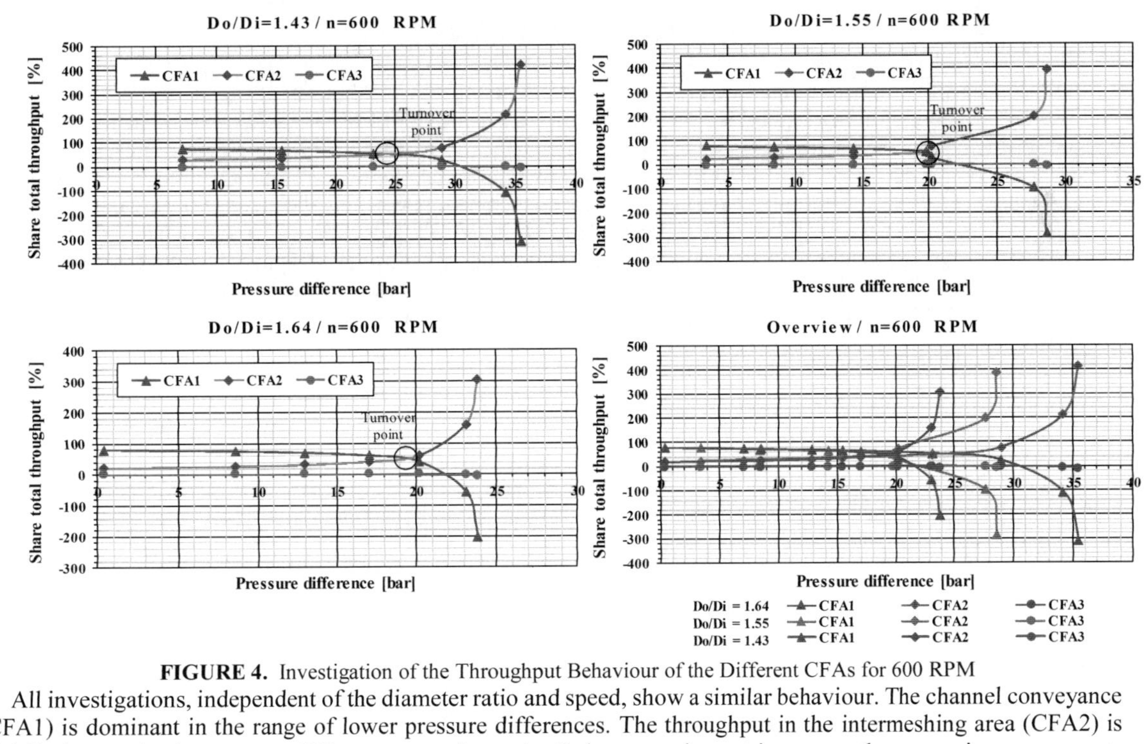

FIGURE 4. Investigation of the Throughput Behaviour of the Different CFAs for 600 RPM

All investigations, independent of the diameter ratio and speed, show a similar behaviour. The channel conveyance (CFA1) is dominant in the range of lower pressure differences. The throughput in the intermeshing area (CFA2) is initially lower. As the pressure difference over the entire Erdmenger-element increases, the conveying components are shifting in the same relation to each other. The channel conveyance in the CFA1 range decreases proportionally and the proportion of conveyance in the intermeshing area (CFA2) increases. The change in conveying from dominant channel conveying to dominant intermeshing conveying is referred to as turnover point. In all investigations, the conveying in the gap area (CFA3) is very low in the axial conveying direction (z) and has an average share of the total throughput on average of less than 0.5 %. Furthermore, the forced conveying behaviour in the CFA2 area, which is assigned to the intermeshing area, can be determined [7]. This is particularly noticeable in the area of high pressure differences. The conveying in the channel has changed in the opposite direction and corresponds to a reconveying behavior. This volume of material is then conveyed through the intermeshing in axial direction again. In addition to that, it should be noted that the turnover point depends on the diameter ratio and thus the geometry of the Erdmenger-element and the resulting free volume. The larger the diameter ratio is, the channel conveyance over the pressure difference over the Erdmenger element is pronounced. The turnover point has shifted to higher counter pressures. This shift of the turnover point can also be achieved by changing the speed, as shown in FIGURE 5 for example

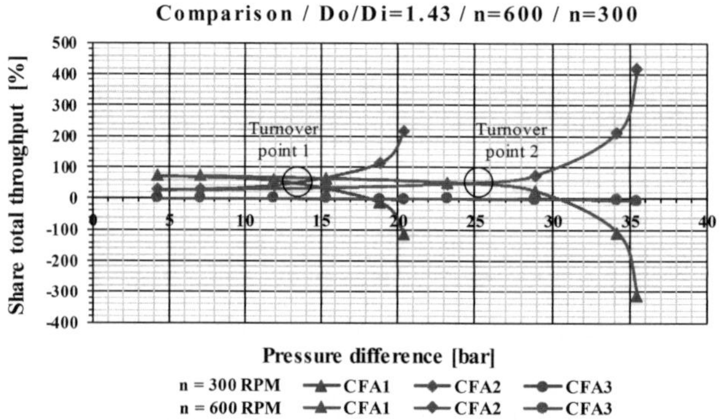

FIGURE 5. Comparison between the turnoverpoints of 300 and 600 RPM

SUMMARY AND OUTLOOK

In this paper a new method to analyze the throughput-pressure behavior of different characteristic flow areas for Erdmenger-Elements of a co-rotating twin screw extruder by using 3D FEM simulations is presented. With this method, it was possible to determine the characteristic flow behavior of different types of Erdmenger-elements. The forced conveying behavior of the intermeshing area was confirmed. Further, the turnover point regarding the dominance of the forced conveying behavior is dependent on the pressure gradient, the diameter ratio of the conveying Erdmenger-element and the rotational speed. For further investigations the prediction of the different shares on the total throughput for each characteristic flow area and the location of the turnover point is essential to develop new calculation models for the process- and material behavior of the co-rotating twin screw extruder. Conventional calculation models [7-8] can be extended and improved with the results of this investigation and screw designs can be optimized.

REFERENCES

1. J. L. White and H. Potente, *Screw Extrusion*, Munich: Carl Hanser Publishers, 2002
2. C. Rauwendaal, *Polymer Extrusion*, Munich: Carl Hanser Publishers, 2001
3. K. Kohlgrüber, *Der gleichläufige Doppelschneckenextruder*, Munich, Carl Hanser Publishers, 2016
4. R. Erdmenger, German Patent No. 862668 (8 July 1953)
5. O. Mierka, T. Theis, T. Herken, V. Schöppner, F. Platte, *Mesh Deformation Based Finite Elmente - Fictitious Boundary Method (FEMFBM) for the Simulation of Twin-screw Extruders*; publication, IANUS Simulation GmbH, Dortmund, 2014
6. C. Becker, S. Turek, *FEATFLOW – Finite element software for the incompressible Navier-Stokes equation – User Manual Release 1.1.*, University of Heidelberg, Heidelberg, 1998
7. W. Schuler, *Process Engineering Design of Co-Rotating Twin Screw Extruders*, Ph.D. Thesis, University of Wales Swansea, 1996
8. U. Melisch, *Grundlagen zur Simulation des Förder- und Plastifizierprozesses dichtkämmender Gleichdrall-Doppelschneckenextruder*, Ph.D. Thesis, University of Paderborn, 1998

The Use of Flow-Type Dependent Strain Reduction Parameter in Fiber Orientation Predictions for Injection-Molded Center-Gated Disk

Hongyu Chen[1], Peter Wapperom[2] and Donald G. Baird[1]

[1]*Department of Chemical Engineering, Virginia Tech, Blacksburg, VA 24061*
[2] *Department of Mathematics, Virginia Tech, Blacksburg, VA 24061*

ABSTRACT. Reliable predictions of the fiber orientation profile are important for injection molding processing. During the mold filling process, the flow types including extensional, simple shear, rotational flow and mixed flows, have a significant impact on fiber orientation kinetics. This study utilizes a flow-type parameter to locally describe the relative magnitudes of shearing, extension and pure rotation. As a result, a variable strain reduction factor is determined locally by the corresponding local flow-type and used in fiber orientation simulation. The application of the variable strain reduction factor in fiber orientation simulations for the injection molded center-gated disk allows the simulated fiber re-orientation rate to be dependent on local flow type. The predicted orientation results show improved agreement of the profile shape with the experimental data.

INTRODUCTION

There is an increased usage of injection molded discontinuous fiber-reinforced thermoplastics in the automotive industry [1] . The injection molding process will result in the generation of an anisotropic fiber orientation distribution (FOD) of the final injection molded parts [2] induced by the complex flow conditions in the mold [3,4]. The mechanical and other properties are highly dependent on the orientation distribution [5,6]. Therefore, it is important to obtain accurate fiber orientation predictions for the optimization of various performances of injection-molded parts. The well known Jeffery's [7]model describes the motion of fiber in flow with dilute fiber concentration in the absence of fiber-fiber interactions[7]. Folgar and Tucker [8] developed the isotropic rotary diffusion fiber orientation (FT) model for concentrated fiber suspensions. Huynh et al. [9] developed the strain reduction factor (SRF) model by straightforwardly scaling the right-hand side of the orientation equation by a factor α ($0<\alpha<1$) to better reflect slower orientation kinetics. Nevertheless, direct scaling of the FT model violates material frame indifference (Objectivity). To achieve an objective orientation equation, Wang [10] Is this a thesis? If so indicate it in the references developed the improved reduced strain closure (RSC) model.

The empirical parameters in fiber orientation equations need to be determined. Most recently, Lambert et al. [11] suggested that fiber orientation evolution rate in shearing flow deviates quite significantly from that in extensional flow. The fiber orientation evolution was evaluated in both pure shear (in a sliding plate) and pure extension (in lubricated squeeze flow), and they found that the fitted strain reduction factors from these two flows were significantly different. In continuation of obtaining further understanding toward fiber orientation kinematics according to flow-type, Lambert et al. [12] compared predictions of fiber orientation with experimental results during non-lubricated squeeze flow. It turned out that both the SRF an RSC models failed to accurately predict the fiber orientation profile through the entire sample thickness using a constant set of model parameters. In this study, we used a flow-type parameter proposed by Astarita et al. [13,14] check ref 14 in your list at end of paper!! to locally describe the relative magnitudes of shearing and extension in a flow. The local strain reduction factor was then interpolated between a typical value for simple shear flow, a value for extensional flow or a value specified for pure rotational flow using a function of this flow-type parameter. The application of a variable strain reduction factor according to flow conditions in fiber orientation simulations shows better qualitative agreement of the profile shape with the experimental data when compared with that using a constant value.

BACKGROUND

It is customary to use a unit vector **p** to specify the orientation of a rigid axially symmetric particle while a long and flexible fiber is represented as three beads connected by a pair of orientation vectors, **p,** and **q** as shown in **Error! Reference source not found.**[15]. To describe the orientation state of a suspension of rigid fibers, a tensor **A** defined as the second moment of the orientation vector **p** is given by[16]:

$$\mathbf{A} = \langle \mathbf{pp} \rangle = \int \mathbf{pp}\psi(\mathbf{p}, t)\, \mathrm{d}\mathbf{p} \tag{1}$$

where $\psi(\mathbf{p})$ is the probability distribution function statistically describing the orientation distribution the orientation state for a complete population of particles in a suspension.

To predict short (rigid) fiber orientation induced by flow, the reduced strain closure (RSC) model is widely used. This model has the advantage to capture slow orientation kinetics by only scaling the evolution of the eigenvalues of **A** as shown below:

$$\frac{\mathrm{D}\mathbf{A}}{\mathrm{Dt}} = \mathbf{W}\cdot\mathbf{A} - \mathbf{A}\cdot\mathbf{W} + \xi\{\mathbf{D}\cdot\mathbf{A} + \mathbf{A}\cdot\mathbf{D} - 2[\mathbf{A_4} + (1-\alpha)(\mathbf{L_4} - \mathbf{M_4}{:}\mathbf{A_4})]{:}\,\mathbf{D}\} \tag{2}$$

$$+2\mathrm{C_I}\dot{\gamma}(\mathbf{I} - 3\mathbf{A}),$$

$$\mathbf{L} = \sum_{i=1}^{3} \lambda_i \boldsymbol{e}_i\boldsymbol{e}_i\boldsymbol{e}_i\boldsymbol{e}_i, \tag{3}$$

$$\mathbf{M} = \sum_{i=1}^{3} \boldsymbol{e}_i\boldsymbol{e}_i\boldsymbol{e}_i\boldsymbol{e}_i, \tag{4}$$

where **L** and **M** are 4th order tensors in terms of the eigenvalues, λ_i, and eigenvectors, $\boldsymbol{e}_i$, of **A**. In Eq.(2), α is the strain reduction factor taking a value between 0 and 1.

During injection molding, all these types of flow conditions exist which have a great impact on fiber orientation kinetics accordingly. Specifically, the evolution of fiber orientation in extensional flow is much faster than that in a simple shear flow. In general, the flow condition or flow-type is often thought of as an intrinsic criterion to reflect the fundamental response of the fluid element in complex flow. Therefore, it is important to have a mathematical expression or classifier to classify the flow conditions or flow types. The criterion proposed by Astarita [13] was used in this study and is defined as:

$$\beta = \frac{\mathbf{D}{:}\,\mathbf{D} + \overline{\mathbf{W}}{:}\,\overline{\mathbf{W}}}{\mathbf{D}{:}\,\mathbf{D} - \overline{\mathbf{W}}{:}\,\overline{\mathbf{W}}} \tag{5}$$

$$\overline{\mathbf{W}} = \mathbf{W} - \boldsymbol{\Omega} \tag{6}$$

$$\boldsymbol{\Omega} = \frac{\mathrm{d}\mathbf{E}}{\mathrm{dt}}\cdot\mathbf{E}. \tag{7}$$

Continue paragraph not a new oneIn the definition, $\boldsymbol{\Omega}$ is the rate-of-rotation of the principle directions of **D** and **E** is the matrix of right(what do you mean by right?) eigenvectors of **D**. To make an objective flow type parameter β, $\boldsymbol{\Omega}$ is subtracted from the vorticity tensor **W** as in Eq.(6). When the eigenvectors of **D** do not rotate, $\boldsymbol{\Omega}$ vanishes and $\overline{\mathbf{W}} = \mathbf{W}$. In this study, we assumed $\overline{\mathbf{W}} = \mathbf{W}$ and the codes applying the objective version of β are under development.

The flow type can be identified by the non-dimensional scalar β. The value of β is 1 in purely extensional flows, 0 in shear flows, and -1 in a pure rotational flows. Values between 1 and 0 indicate a mixture of extension and shear, and values between 0 and -1 indicate mixed shear and rotation. According to the value of β, the local strain reduction factor of orientation model will be interpolated between a typical value of the strain reduction factor α_s for simple shear flow, a value α_e for extensional flow or a value specified for pure rotational flow. In this study, we set the parameter, α_s (0.2), to be the same as that obtained by Cieslinski et al. [2] by fitting the models to fiber orientation evolution data during startup of simple shear. Lambert and Baird [11] found that a value of the strain reduction factor close to 1 can reflect the fast fiber orientation rate during the startup of lubricated planar extension flow of a glass fiber reinforced polypropylene suspension. In a following paper, Lambert et al. [12] compared predictions of fiber orientation with experimental results during non-lubricated squeeze flow which includes features of pure shear , pure extensional and the combination of both flow conditions. Predictions using a constant value for the strain reduction factor were not capable of describing the fiber orientation state. Specifically, simulations using α_s only matched well with the experimental data near the wall where shear dominated, while applying a value of 1 for strain reduction factor only fitted the orientation near the center where extension dominated. When $\alpha_s < 0$, the flow has components of rigid body rotation. In rigid body rotation, the deformation is zero or so small it can be neglected [17,18]. That is to say, the distance between any two given points remains constant with time. The slow down effect for fiber orientation can be ignored for pure rigid body rotation and the original Folgar-Tucker model ($\alpha = 1$) can be used to describe fiber motions in such a case. Therefore, we set the value of α_e to be 1, and interpolate the local strain reduction factor from $\alpha_s$and α_e based on the value of β:

$$\alpha = |\beta|\alpha_e + (1 - |\beta|)\alpha_s \tag{8}$$

EXPERIMENTS

The composites under investigation were 30 wt% glass fibers with a polypropylene matrix. The material was supplied by SABIC Innovative Plastics as 12.5 mm long pellets created through a pultrusion process using 30 wt% formulations. The center-gated-disk as shown in Figure *1* was injection-molded from the as-received pellets. The temperature profile in the injection molder (BOY, 35E) was set to 200, 210, 230, 230, and 230 °C with a mold temperature of 90 °C. The mold filling time was about 1 s. The disks made have an internal radius $r_{in} = 4$ mm and an averaged outer radius $r_{out} = 56$ mm. The lengths of the glass fibers in the disks were determined from a burn-off procedure based on that used by Kunc et al [19]. The number average fiber length, L_n, was 1.14 mm, and the weight average fiber length L_w was 3.41 mm.

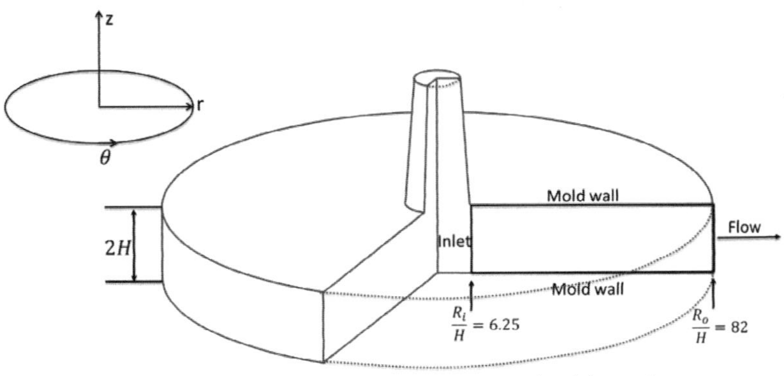

Figure 1. Center-gated disk geometry in this study.

RESULTS AND DISCUSSION

The simulation of the filling process of the injection molded center-gated disk started from the inlet of the disk. The experimental orientation was obtained at the inlet as shown in Figure 2 and specified as the inlet condition during simulation. The initial fiber orientations are asymmetric at the inlet and varies though the sample thickness. Near the walls, fiber orientation is flow-direction dominated, while the core region has preferential orientation that is transverse to the direction of flow.

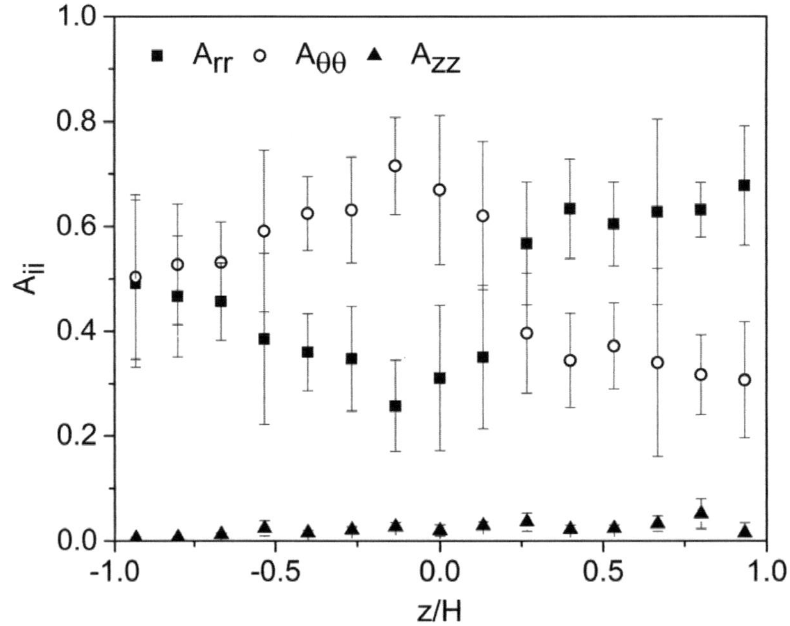

Figure 2. Initial fiber orientation state at the inlet of the disk. Here, Arr is the flow direction orientation component, $A_{\theta\theta}$ the neutral-direction orientation component, and Azz the thickness-direction orientation component.

The through the thickness the flowtype parameter β at 20% and 60% of the disk radius are shown in Figure 3. At the center extensional flow dominates, while near both the top and bottom walls shear flow dominates. The core region narrows as the radius increases. This is because the thickness of the region dominated by extension flow shrinks as the radius increases.

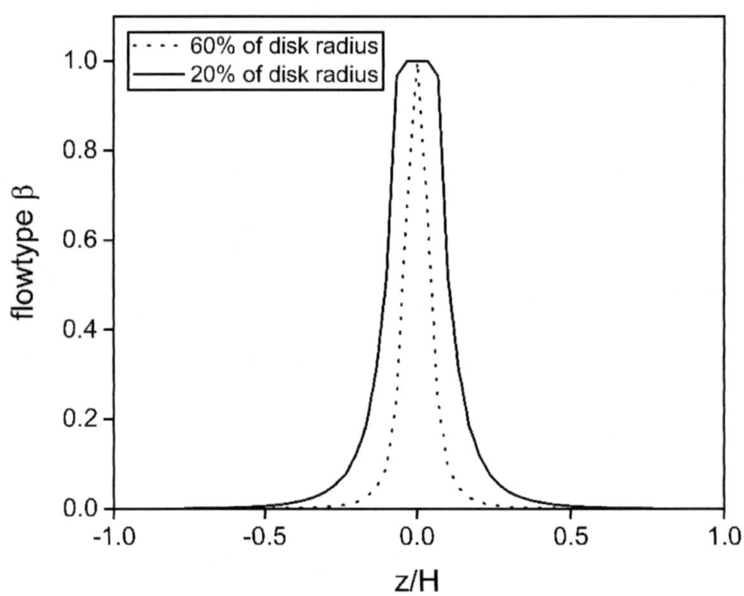

Figure 3. Calculated flowtype β through the thickness of the sample at 20% and 60% of the disk radius.

Through thickness A_{rr} profiles at 20% and 60% of the disk radius are shown in Figure 4. Scatter(???what does this mean?) The symbols are measured fiber orientation, which shows the typical shell-core-shell fiber orientation distribution through the thickness. At 20% of the disk radius, the prediction using α_e has the largest value from the transition region to the walls and gives the lowest value in the core region. From the transition region to the walls, the prediction using variable α overlapped with the prediction using α_s, this is because the shear flow dominates in this region. At the core region, especially close to the center where extensional flow dominates, the variable α prediction is very similar to that of α_e. At 60% of the disk radius, there is almost no difference among the predictions from the three types of strain reduction factors near the walls. This might be because that all the simulations have reached steady state at the shear flow dominated regions with high shear rate. In the transition regions ($z = -0.77H$ to $-0.1H$ and $z = 0.13H$ to $0.7H$), the model without the slow-down effect ($\alpha = \alpha_e$) predicts the highest flow-direction alignment. In the same regions, the results from using variable α and shear flow fitted κ_s are quite comparable, and there is only a slight difference around $z = 0.1H$ to $0.3H$ and $z = -0.2H$ to $-0.13H$. Around the center ($z = 0$), compared to the RSC model with the shear flow fitted α_s, the RSC model with variable α and the model without slow-down effect ($\alpha = \alpha_e$) predict considerably faster evolution of A_{rr}, where the flow is dominated by extension in the tangential direction. Notably, it seems that at both 20% and 60% of disk radius, using a constant strain reduction factor could not predict the orientation state through the entire thickness of the sample. This might be due to the great difference between the observed fiber orientation evolution kinetics during shear and extensional flow. The value of the variable strain reduction factor calculated is dependent on the flow-type. As a result, at the extensional flow dominated region, the predicted fiber orientation evolution can be much faster than that in the shear flow dominated region. Implicit in this method is that the variable strain reduction factor is empirical, and it is quite dependent on the parameters fitted in pure shear flow and pure extensional flow. However, this method is quite promising, since it allows the simulated orientation evolution rate to be dependent on the local flow conditions.

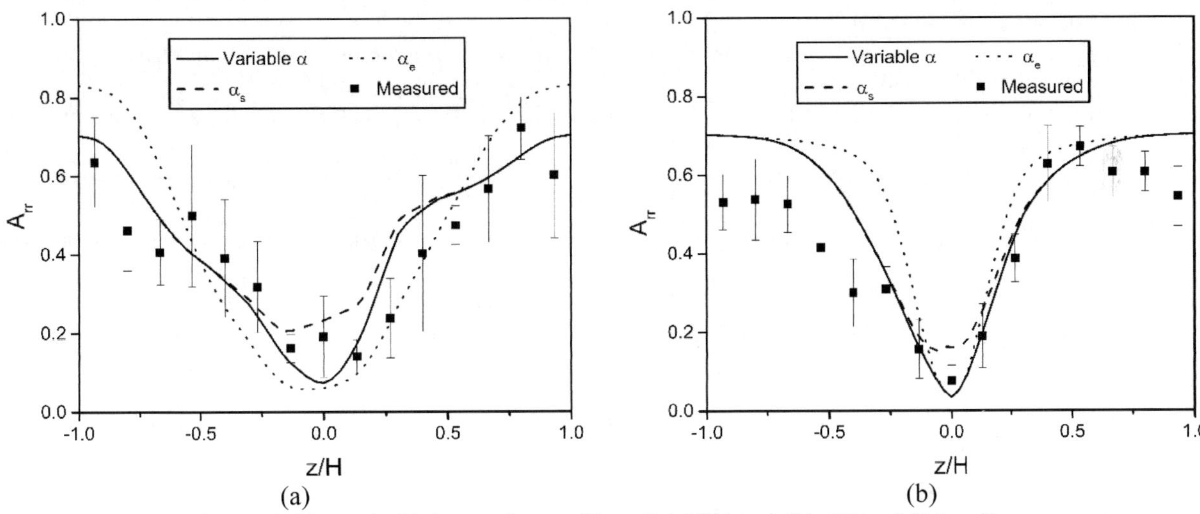

Figure 4. Through thickness A_{rr} profile at (a) 20% and (b) 60% of disk radius.

CONCLUSION

A variable strain reduction factor determined from the local flow type, was used to predict fiber orientation in the mold filling of injection-molded center-gated disks. The predictions in the shear dominated region were close to those using the shear-flow fitted strain reduction factor, while the predictions in extensional dominated region were similar to those using the model excluding the orientation slow-down effect (α = 1). There is a significant difference between the observed fiber orientation evolution rate during pure shear and pure extensional flows, and it seems that the fiber orientation model with a constant strain reduction factor was not capable of including all the fiber orientation states in a complex flow field. The variable strain reduction factor allows the simulated orientation evolution rate to be dependent on the local flow conditions. This empirical variable strain reduction factor helps to improve the fiber orientation predictions especially in complex flow.

REFERENCES

1. Henning, F.; Ernst, H.; Brüssel, R. Lfts for automotive applications. *Reinforced plastics* **49**, 24-33 (2005).
2. Cieslinski, M.J.; Wapperom, P.; Baird, D.G. Fiber orientation evolution in simple shear flow from a repeatable initial fiber orientation. *Journal of Non-Newtonian Fluid Mechanics* **237**, 65-75 (2016).
3. Mazahir, S.M.; Vélez-García, G.M.; Wapperom, P.; Baird, D. Evolution of fibre orientation in radial direction in a center-gated disk: Experiments and simulation. *Composites Part A* **51**, 108-117 (2013).
4. Mazahir, S.M.; Vélez-García, G.M.; Wapperom, P.; Baird, D. Fiber orientation in the frontal region of a center-gated disk: Experiments and simulation. *J. Non-Newton Fluid Mech.* **216**, 31-44 (2015).
5. Nguyen, B.N.; Bapanapalli, S.K.; Holbery, J.D.; Smith, M.T.; Kunc, V.; Frame, B.J.; Phelps, J.H.; Charles L. Tucker, I. Fiber length and orientation in long-fiber injection-molded thermoplastics — part i: Modeling of microstructure and elastic properties. *J. Compos. Mater.* **42**, 1003-1029 (2008).
6. Chen, H.; Baird, D. Prediction of young's modulus for injection molded long fiber reinforced thermoplastics. *J. Compos. Sci.* **2**, 47 (2018).
7. Jeffery, G.B. The motion of ellipsoidal particles immersed in a viscous fluid. *Proc. R. Soc. Lond. A* **102**, 161-179 (1922).
8. Folgar, F.; Tucker III, C.L. Orientation behavior of fibers in concentrated suspensions. *J. Reinf. Plast. Compos.* **3**, 98-119 (1984).
9. Huynh, H.M. Improved fiber orientation predictions for injection-molded composites. (University of Illinois at Urbana-Champaign, 2001).
10. Wang, J. *Improved fiber orientation predictions for injection molded composites*. (University of Illinois at Urbana-Champaign 2007).
11. Lambert, G.M.; Baird, D.G. Evaluating rigid and semiflexible fiber orientation evolution models in simple flows. *J Manuf Sci Eng* **139**, 031012 (2017).
12. Lambert, G.; Wapperom, P.; Baird, D. Obtaining short-fiber orientation model parameters using non-lubricated squeeze flow. *Phys. Fluids* **29**, 121608 (2017).
13. Astarita, G. Objective and generally applicable criteria for flow classification. *J. Non-Newton Fluid Mech.* **6**, 69-76 (1979).
14. Phelan, F.R.; Hudson, S.D.; Handler, M.D. Fluid dynamics analysis of channel flow geometries for materials characterization in microfluidic devices. *Rheologica Acta* **45**, 59-71 (2005).
15. Strautins, U. Flow-driven orientation dynamics in two classes of fibre suspensions. (Ph. D. thesis, Technical University Kaiserslautern, 2008).
16. Advani, S.G.; Tucker III, C.L. The use of tensors to describe and predict fiber orientation in short fiber composites. *Journal of Rheology (1978-present)* **31**, 751-784 (1987).
17. De Gennes, P. Coil - stretch transition of dilute flexible polymers under ultrahigh velocity gradients. *The Journal of Chemical Physics* **60**, 5030-5042 (1974).
18. Lumley, J.L. Drag reduction by additives. *Annual review of fluid mechanics* **1**, 367-384 (1969).
19. Kunc, V.; Frame, B.J.; Nguyen, B.N.; Tucker III, C.L.; Velez-Garcia, G. Fiber length distribution measurement for long glass and carbon fiber reinforced injection molded thermoplastics. *Research Gate*, (2007).

Polymer Sealing by Heat Transfer Modelling

Felice De Santis[a*] and Auli Nummila-Pakarinen[b]

[a]*Borealis Polyolefine GmbH - Linz - Austria*
[b]*Borealis Polymers Oy - Porvoo - Finland*

Abstract. Polymer films are used in the packaging of industrial and food products. Effective thermal bonding of the films requires that the interface between layers must be at least partially melted. There are limitations of the temperature of the heat seal, as too high a temperature could cause sticking of the films to the surface of the heat seal unit, and too low a temperature would affect throughput, ie sealing strength. To understand the problems and to be able to consider changes in operating conditions it is necessary to have a reliable model that predicts the temperature distribution of the film during heat sealing, and calculates the necessary dwell time to fuse the layers of the polymer. In developing heat transfer and energy balance models of the process it is necessary to realize the complexity of the problem, i.e., transient heat conduction in a composite solid material with multidimensional geometry. The situation is further complicated by the latent heat of fusion effects, incomplete/non-uniform contact of heat seal with the polymer, and the possible variation of properties upon heating. It is evident in other engineering applications that the control and accuracy of such models facilitate scale-up and improvement of existing systems as well as analysis of new applications. The development of an exact solution to the problem is possible, under some hypothesis.

Keywords: Sealing, Packaging, Heat transfer modeling
PACS: Replace this text with PACS numbers; choose from this list: http://www.aip.org/pacs/index.html

INTRODUCTION

Heat sealing is an important process in packaging, as many packages are closed with a heat seal. Heat seals are used primarily in flexible packaging, but they are also used in rigid and semi-rigid packaging. Heat sealing is a method of closing a package by applying energy to two pieces of thermoplastic material [2] (Selke et al., 2004). Thermoplastics return to a solid state upon cooling back down. The flow of melted thermoplastics and their cohesion/adhesion properties are exploited during heat sealing. Heat sealing is used to close single layer and multilayer packages and can fuse similar or dissimilar materials together (Morris & Darby, 2009 Morris, B. A., & Darby, D. O. (2009). In Yam K. L. (Ed.), The wiley encyclopedia of packaging technology (3rd ed.). Hoboken, New Jersey: John Wiley &). Its objective is to produce joints of sufficient strength to withstand stresses they may subsequently be subjected. In many applications the joint must additionally be recognized as "easy-to-open" or "peelable". Although there is a variety of heat seals and equipment, the basic sealing methodology still involves the bonding of two polymer surfaces by forcing them into intimate contact while they are in at least a partially molten state [3] (Theller1989). Obviously considering a multilayer film the external layers usually contribute to the sealing, while the inner core give (one or more layers) the main properties (as an example mechanical, barrier) to the whole film. For years, the technology of heat sealing has been practiced with more art than science, and still many approaches to the topic are practical. Furthermore, the authors of the scientific literature related to heat sealing are more linked to the industry and application than other polymer processes.

The view of the phenomena occurring during heat sealing were linked by some authors to the theory of diffusion of free and mobile ends of macromolecules on the surface of polymers, reported by Voyutskii in 1963 [4, 5]. The term "autohesion", or self-adhesion, refers to the ability of two surfaces of the same material in contact with each other to form a strong bond in some conditions. As the interface temperature of the two layers of material being heat sealed is increased from ambient, seals of significant strength begin to appear quite below the melting point of the heat seal layers. In the case of LDPE the onset of detectable heat sealing is 15 K – 20 K below the melting range, but in general this difference depends on the type of polymer, as shown in Figure 1 [6]. From the first appearance of measurable seal strength to a temperature almost to the melting point, bonds are formed that, when tested, fail at the seal interface-they peel apart. A peel seal is not necessarily a weak seal. Peel strengths range from just above zero up to the strength of a fused seal.

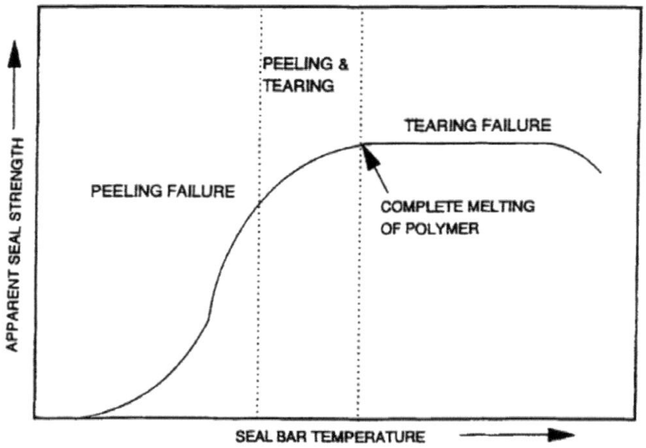

FIGURE 1. Schematic plot of a typical heat sealing curve [6].

The ultimate force that holds heat seals together is the same force that binds the chain molecules in the polymer structure itself: secondary valence forces. For these forces to come into play in the heat sealing process, enough movement and deformation of the surface polymer chains must take place for the molecules in the two surfaces to move into intimate contact with each other. The atomic groups on chains in the adjacent surfaces must move to within about 5 Angstrom units of each other-a primary prerequisite for secondary valence bonds to form. The bonds form instantaneously if the molecules have sufficient energy and the proximity condition is met. On a molecular scale the smooth surfaces of heat seal layers are of course not smooth, but very rough. Most polymers used in heat seal layers have a high amorphous content by design, and can be expected to have surfaces exhibiting clumps of tangled disordered chains and chain ends. The surface molecular order should nevertheless be somewhat greater than that of the bulk structure, due to the tendency of chains to lie parallel to the air interface. The first thing that must take place in the heat sealing process is to flatten the jaws to bring the two surfaces into molecular contact over as high a percentage of the total area as possible.

The main processing parameters

The process progresses through a series of phenomenological steps at the interface of the two films, comprising [7]:

- Heating
- Crystalline melting
- Surface wetting
- Chain interdiffusion and entanglement
- Recrystallization and cooling

In the paper by Theller in 1989 [3] a precision laboratory heat sealer was used on PE (LDPE, HDPE) films and the heat sealing and related parameter (i.e. temperature, time, and pressure) influencing the process were analyzed. As expected the temperature at the interface of heat seal layers is the most important factor affecting the final strength of the seam. On the other hand, the direct measurement of the interface temperature could be problematic, because it is a rapid thermal phenomena following the contact of the films with the hot jaws. Furthermore, the sensor placed between the surfaces should have a negligible heat capacity to register the actual interfacial temperature without a lag time. The simplest practical solution to this issue is to wait until the interface reaches the temperature of the jaws [3].

This aspect highlights the importance of the dwell (sealing) time, with the closed jaws, during which two events should take place:

- the seal interface must reach the temperature for diffusion of macromolecular chain ends;
- micro-Brownian movement of the macromolecules chain ends occur to achieve diffusion into the opposite surface.

Obviously, in the practical application of sealing, it is desired to minimize the total time required for both process to occur. Thus, as these diffusion transport phenomena are temperature dependent, the higher the temperature the shorter is the dwell time required.

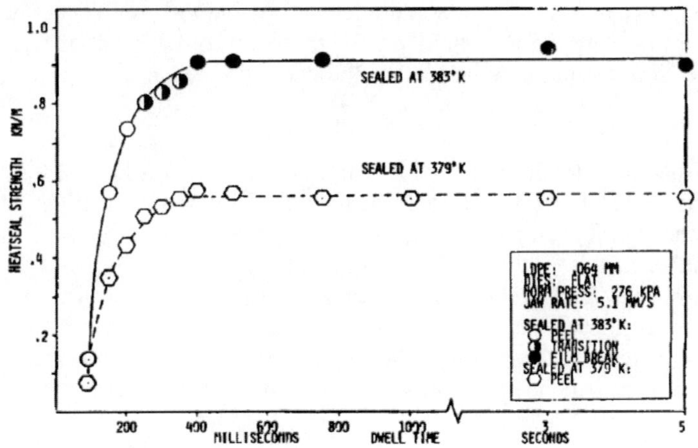

FIGURE 2. Effect of dwell time on heat seal strength [6].

In Figure 2, the heat seal strength of a LDPE film is reported as function of dwell time for two jaw temperatures [3]. It is clear that once the interface between layers is erased, longer dwell time has negligible effects on the heat seal strength. Furthermore, the temperature has a maximum value: the "burnthrough" temperature at which melting, burning, or distorting excessively in the seal are observed. To complete the picture, obviously, a minimum temperature have to be applied to realize a minimum sealing strength.

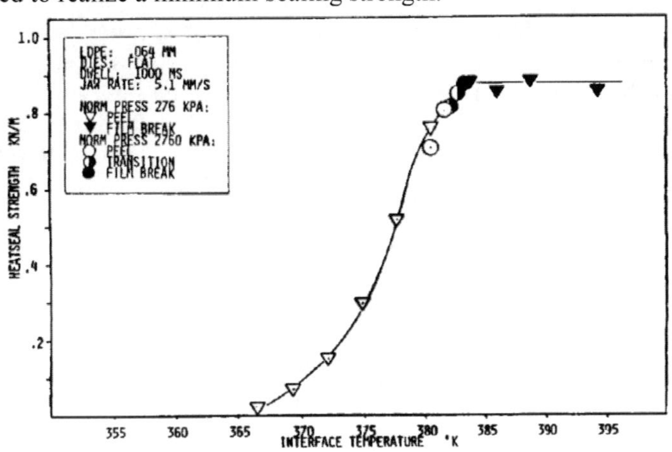

FIGURE 3. Effect of pressure on heat seal strength [3].

In Figure 3, the heat seal strength of a LDPE film is reported as function of pressure [3]. It is clear that once a certain level is reached, higher pressure has negligible effects on the heat seal strength. On other materials this behavior is confirmed with a very slight improvement in seal strength with pressure, after a certain level is reached. These results seem coherent with the diffusion phenomena, so that the pressure is needed just to create a sufficient contact between layers, flattening surface micro-roughness.

THE HEAT TRANSFER DURING THE SEALING

The transfer of thermal energy which occurs between the heat sealing plate and the multiple layers of packaging material is essential towards understanding how the seal process happens and the strength of the seal itself. The study of heat transfer and the application of various fundamental equations will allow for a model to be built which can relate the process parameters of material thickness, temperature and time to the material properties of the packaging materials.

This is a fundamental issue to understand in the usual testing with standard procedures of the sealing properties as during the processing. The transient heat transfer equation for this application involves various boundary conditions due to differing materials and therefore various thermal diffusivities. The transport phenomena could be quite complex to be analytically descripted, but sometimes is possible to simplify it considering and verifying some hypothesis. For example, evaluating the Biot number, Bi, that express the ratio between the conduction resistance within the body and convection resistance at the surface the body, defined as:

$$Bi = \frac{L/k}{1/h}$$

Where L is the layer thickness (or half thickness if the heat flow is on both sides of the layer), k the thermal conductivity of the layer and h the heat transfer coefficient. If the Biot number is high, i.e. Bi>10, the temperature of the surface of the layer reaches immediately the temperature of the external medium, as sketched in Figure 4a. On the other side if Biot number is low, i.e. Bi<0.1, there is a flat profile inside the layer, with no dependence with position the layer but just on the time. In this case, the problem could be solved with a lumped analysis, depending just on the time. In the most general case, without any assumption on Biot number, the temperature profile is sketched in Figure 4b, and the temperature at the surface is different from the temperature of the external medium.

To describe the temperature dependence with position and time a general energy balance equation (partial difference equation, PDE) should considered:

$$\frac{\partial}{\partial t}T = \alpha\frac{\partial^2}{\partial x^2}T \quad (1)$$

with the related initial and boundary condition

$$t \leq 0\ T = T_i \text{ Initial condition} \quad (2)$$

$$x = L \rightarrow -k\frac{\partial}{\partial x}T = h(T_s - T_\infty) \quad (3a)$$

$$x = 0 \rightarrow \frac{\partial}{\partial x}T = 0 \text{ “Symmetry”} \quad (3b)$$

where x is the position starting from the interface, t the time starting from the instant in which the layer is in contact with the external medium, $T(x,t)$ is the temperature, L is the thickness of each one of the two films (thus the layer thickness is $2L$), α is the thermal diffusivity of the layer, T_i is the initial temperature of the layer, T_s is the temperature of the layer at $x=L$ in contact with the external medium, T_∞ is the temperature of the external medium, h is the heat exchange coefficient in the external medium, as sketched in **Error! Reference source not found.**. The temperature evolution at different times ($t_1<t_2<t_3$) is also sketched: on the left side if $Bi>10$ so that $T_s=T_\infty$, and on the right side in the more general case in which $T_s(t)$ is time dependent.

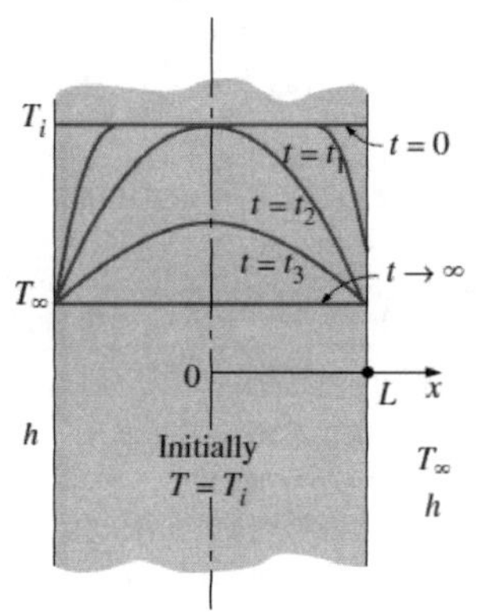

FIGURE 3. The temperature profile in a layer, if Bi>10 on the left side of the layer, and without assumption on Biot on the right side of the layer

Changing the value of the heat transfer coefficient h, the thermal contact with sealing jaw and subsequent cooling, for example in air or in another external medium, the temperature evolution in each position, and in particular at the interface, could be evaluated solving Eq. (1) and with related initial and boundary condition.

Some attempts to describe temperature evolution, solving the PDE in Eq. (1), during heat sealing can be found in literature. Zinsmeister et al. [8] simulated the heat sealing, evaluating the effect of the process parameters (only in-silico) with a numerical code and explicit finite difference technique.

Meka et al. [6] used a finite element analysis (FEA) modelling technique to predict the interfacial temperature (at the centre, neglecting any contact resistance in between the sealing layers) as a function of time during the sealing of semicrystalline polymer films. Consequently an experimental technique using micro-thermocouples to measure rapidly changing interfacial temperatures during sealing was developed. A good agreement between predicted interfacial temperature profiles and measured values for polyethylene films was stated, even if the heat transfer coefficient was used as a fitting parameter. Therefore a slight deviation in the temperature evolution was attributed to the melting and film-thickness changes occurring during sealing that are not taken into account in the calculations (Figure 6).

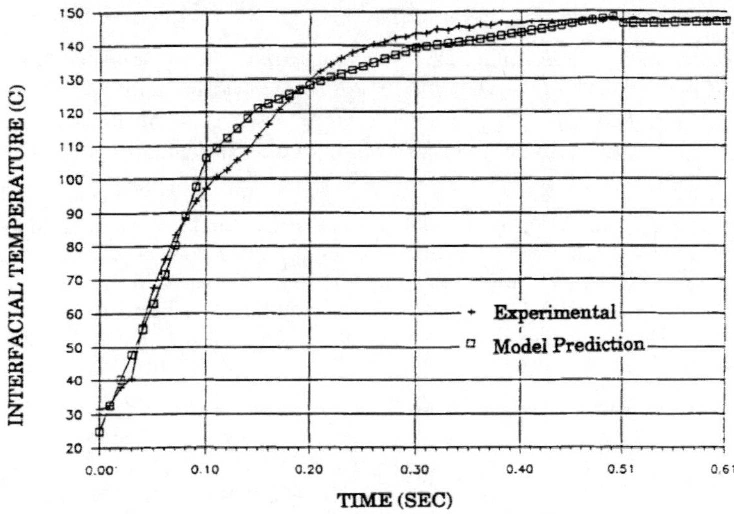

FIGURE 3. Interfacial temperature vs. dwell time as calculated by the FEA model in comparison with experimentally measured values for the condition of platen temperatures of 150°C on 50 μm-thick "Escorene 158BW" films [6]

CONCLUSIONS

Ut wisi enim ad minim veniam, quis nostrud exerci eliton ullamcorper suscipit lobortis nisl ut aliquip ex en commodo consequat. Duis te feugifacilisi per suscipit lobortis nisl ut aliquip ex en commodo consequat.Lorem ipsum dolor sit amet, consectetuer adipiscing elit, sed diem nonummy nibh euismod tincidunt ut lacreet dolore magna aliguam erat volutpat.

The reference section will follow the "Acknowledgment" section. References should be numbered using Arabic numerals followed by a period as shown below, and should follow the format as indicated in the below examples.

REFERENCES

1. M. P. Brown and K. Austin, *The New Physique*, Publisher City: Publisher Name, 2005, pp. 25-30.
2. M. P. Brown and K. Austin, *Appl. Phys. Letters* **85**, 2503-2504 (2004).
3. R. T. Wang, "Title of Chapter," in *Classic Physiques,* edited by R. B. Hamil, Publisher City: Publisher Name, 1999, pp. 212-213.
4. C. D. Smith and E. F. Jones, "Load-Cycling in Cubic Press" in *Shock Compression of Condensed Matter-2001*, edited by M. D. Furnish et al., AIP Conference Proceedings 620, American Institute of Physics, Melville, NY, 2002, pp. 651-654.
5. B. R. Jackson and T. Pitman, U.S. Patent No. 6,345,224 (8 July 2004)
6. D. L. Davids, "Recovery Effects in Binary Aluminum Alloys", Ph.D. Thesis, Harvard University, 1998.
7. R. C. Mikkelson (private communication).

Flow Analysis of Viscoelastic Fluid Stirred by Single Rotor

Takushi Saito[a], Yuki Noguchi[a], Kosuke Higashi[a, b], Kazuhisa Fukutani[b], Tastsuya Kawaguchi[a] and Isao Satoh[a]

[a]Department of Mechanical Engineering, Tokyo Institute of Technology
2-12-1, O-okayama, Meguro-ku, Tokyo, Japan
[b]Kobe Corporate Research Laboratories, Kobe Steel, Ltd.,
1-5-5 Takatsukadai, Nishi-ku, Kobe, Hyogo, Japan

Abstract. In this study, a simple kneading machine with a visualization window was prepared to experimentally analyze the behavior of the full filled viscoelastic fluid stirred by a single rotor. Several different types of viscoelastic non-Newtonian fluid were tested and particle image velocimetry (PIV) was used to quantitatively investigate the velocity distribution in the chamber. The obtained results showed that overall flow structure was dominated by the degree of non-Newtonian property. It was also shown that different flow patterns behind the trailing edge of the rotor tip depended on the degree of viscoelasticity.

Keywords: Mixing, PIV, Flow field analysis, Viscoelastic recovery
PACS: 47.15.G-, 47.80.-v, 47.80.Cb

INTRODUCTION

As represented by low fuel consumption tires, it is necessary to advance the kneading process in rubber production in order to improve the functionality of rubber products. For example, it is possible to balance tire fuel economy performance and wet grip at a high level by replacing conventional carbon black with silicate compound (1, 2). However, chemical treatment by adding a silane coupling agent is necessary in order to homogeneously disperse the inorganic material silica in rubber, which is an organic material. The additive material added to the rubber should be dispersed homogeneously. However, the behavior of molten rubber, which is a viscoelastic fluid, is complicated and it is difficult to fully predict and analyze by a numerical simulation.

Therefore, in this study, the flow phenomenon under stirring condition was experimentally analyzed using a non-Newtonian fluid having viscoelasticity and a single rotor mixer modeling an internal mixer. Specifically, a carboxyl methyl cellulose (CMC) aqueous solution and a xanthan gum (XTG) aqueous solution were adopted as the model fluid, and the velocity field was measured by the particle image velocimetry (PIV). In addition, the viscosity distribution was obtained from the measurement results, and the influence of the rheological characteristics of the fluid on the kneading process was discussed. The data obtained here is considered to be useful for verification of the results of computer simulation.

EXPERIMENTS

Mixing Chamber and Test Fluid

Figure 1 shows the structure of the kneader and its surroundings for measurement. In this apparatus, a 50 mm thick rotor having a triangular prism shape was inserted in the center of a transparent polycarbonate chamber (inner diameter of 100 mm) having a cylindrical hollow cross section so as to approximate a two-dimensional structure. In addition, the front side of the device was covered with a flat plate made of polycarbonate, and the flow field can be visualized and observed.

In the experiment, an aqueous solution of xanthan gum, an aqueous solution of carboxymethyl cellulose (CMC), glycerin were used as test fluids. The xanthan gum aqueous solution and the CMC aqueous solution are viscoelastic fluids. The former behaves relatively elastic, and the latter behaves relatively viscous. On the other hand, glycerin is almost Newtonian fluid, and shows constant viscosity irrespective of the shear rate. The viscosity characteristics and viscoelastic properties of the test liquids at room temperature are summarized in Figure 2. The concentrations of the CMC aqueous solution and the XTG aqueous solution in the graph were 1.0% by weight and 0.5% by weight, respectively.

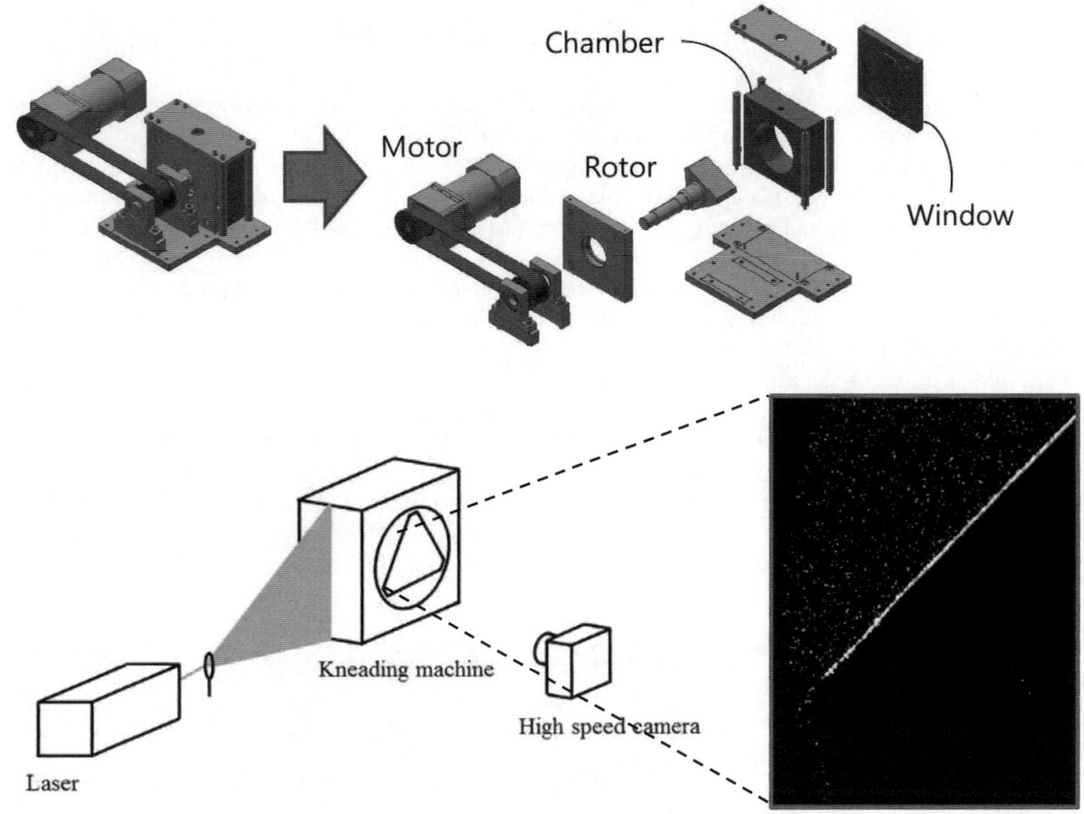

Figure 1 Experimental apparatus and PIV measurement

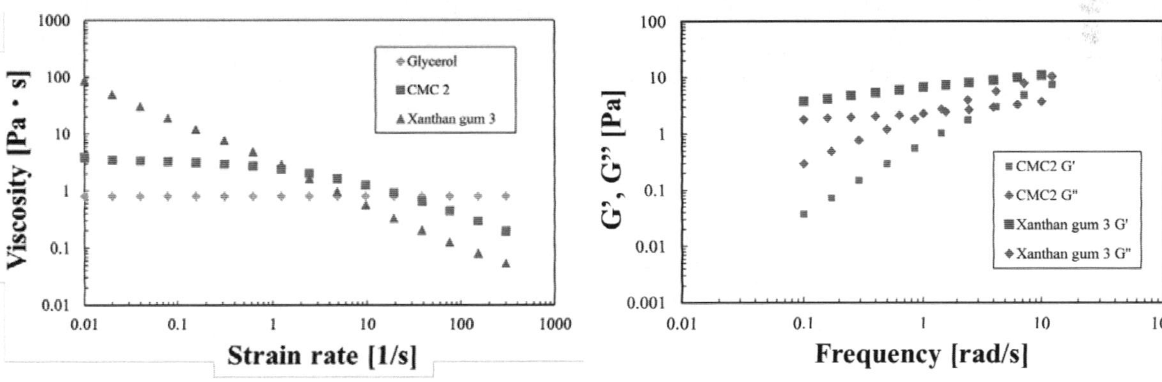

Figure 2 Measured viscosity, storage modulus (G') and loss modulus (G") of the test fluids

PIV Measurement technique

In this study, particle image velocity (PIV) measurement was used as the velocity distribution measurement method. In this method, tracer particles mixed in the flow field are illuminated with a laser sheet (Figure 1) and velocity vectors of particle groups are calculated from the time series images acquired with a high-speed camera. In the kneading phenomenon analyzed in this study, the relative motion of the fluid element to the rotor was important, so the rotational coordinate transformation was applied to the captured image and then analysis by PIV was performed. Therefore, the data obtained was the velocity distribution of the kneading field in the rotating coordinate system observed from the rotor.

RESULTS AND DISCUSSION

Figure 3 shows an example of measurement results at the rotation speed of 20 rpm and the clockwise rotation. In each figure, the rotor was fixed and the cylindrical wall surface moved to the counterclockwise direction.

In the case of glycerol, it was shown that a large vortex was generated in the region consisting of the walls of the rotor and the chamber. However, in case of the XTG aqueous solution, the flow occurred only in the vicinity of the wall surface of the chamber, and most of the other regions stayed. In order to explain this fact, the shear rate distribution and the viscosity distribution are shown in Figure 4. It should be noted that the XTG aqueous solution has a strong shear thinning viscosity. That is, at the vicinity of the wall surface, the shear rate was large and a sharp viscosity decrease occurred, and momentum transport from the vicinity of the wall surface to the rotational radial direction hardly occurred.

For the viscoelastic fluid, the elastic recovery behavior due to the velocity component in the radial direction of the rotor tip was observed (enlarged part in Figure 3 (b)). This result was presumed to be occurred because the elastic energy was not relaxed between the tip of the rotating rotor and the wall of the chamber and was released on the rear side of the rotor. In order to understand this behavior in detail, only the radial component was extracted from the velocity distribution behind the rotor tip and was shown as color mapping of scalar distribution (Figure 5). In the case of XTG aqueous solution where the elastic behavior predominates, stored elastic energy was rapidly released at the rear end of the chip. On the other hand, in case of CMC aqueous solution where viscous behavior is dominant, it was shown that the elastic energy gently released while descending.

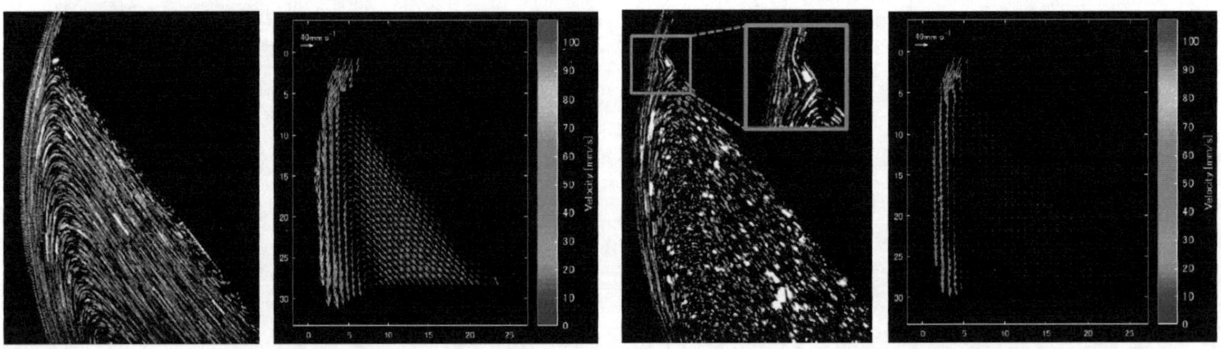

(a) Glycerol (b) Xanthan gum solution

Figure 3 Streamline and velocity vector map of (a) Glycerol and (b) XTG solution

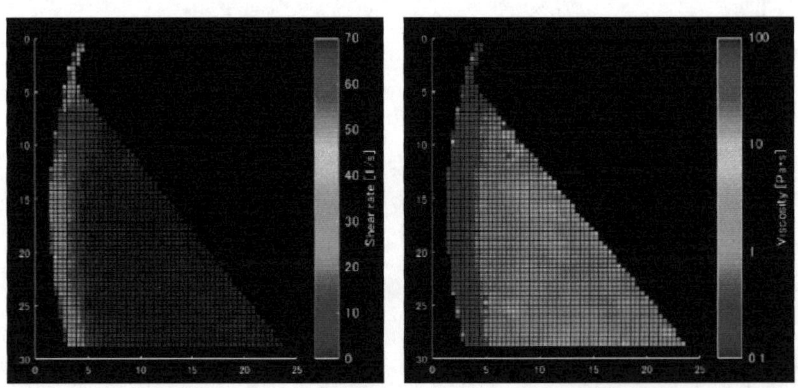

Figure 4 Calculated shear rate distribution (left) and viscosity distribution (right) of the XTG solution

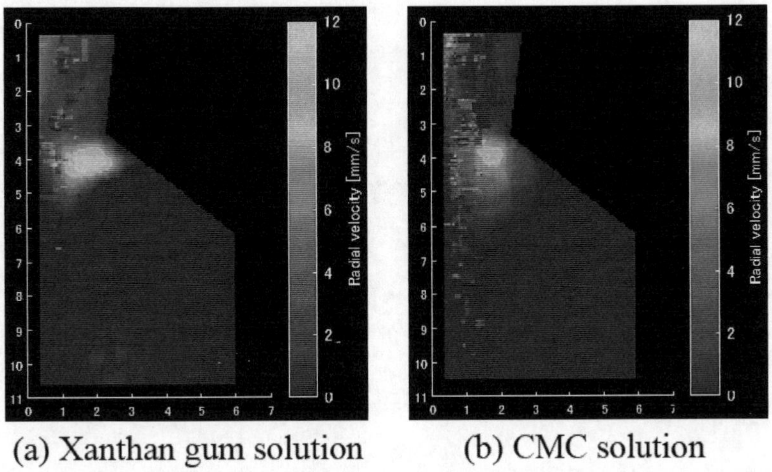

Figure 5 Contour map of the velocity component in the radial direction of the rotor

CONCLUSIONS

In this research, the influence of rheological characteristics of fluid on kneading behavior was experimentally analyzed by using visualization measurement method. As a result, it was shown that the global flow structure of the kneading field was dominated by the non-Newtonian nature of the fluid. Also, elastic recovery behavior was observed in which viscoelastic fluid jumped out toward the center of the chamber behind the rotor tip. This behavior was shown to be different depending on the viscoelastic properties of the test fluid.

REFERENCES

1. Pamela J. Martin, Paul Brown, Andrew V. Chapman, and Stuart Cook, *Rubber Chemistry and Technology*, Vol. 88, No. 3, pp. 390-411 (2015)
2. Luis González Hernández, Luis M. Ibarra Rueda, and Celia Chamorro Antón, *Rubber Chemistry and Technology*, Vol. 60, No. 4, pp. 606-617 (1987)
3. W. Dierkes, J. W. M. Noordermeer, *International Polymer Processing*, Vol. 22, No. 3, pp. 259-265 (2007)

Condition Monitoring of Non-Return Valves in Injection Molding Machines using available Process and Machine Data

Chen-Liang Zhao* and Reinhard Schiffers

University of Duisburg-Essen, Institute of Product Engineering (IPE), Engineering Design and Plastics Machinery, Lotharstraße 1, 47057 Duisburg, Germany

chen-liang.zhao@uni-due.de
reinhard.schiffers@uni-due.de

Abstract. This paper introduces a procedure to evaluate the wear state of non-return valves based on process and machine data of injection molding machines. In industry, non-return valves of injection molding machines are often replaced following specified maintenance intervals. Due to conservative lifetime estimations, the replacement of non-return valves is often carried out too early, since the current state of wear of the non-return valves is not known without elaborate disassembly and in case of failure other adjacent components can also be damaged. This paper shows an approach for an automated condition monitoring of non-return valves. During production, the corresponding wear state is evaluated without the use of additional sensors. By analyzing the curves of pressure, position signals etc. from process data with injection molding process monitoring tools, specific key figures were defined. These key figures allow wear-specific conclusions about the different signs of wear and thereby allow an estimation of the current wear state of the non-return valve. Based on the current state of wear, it is possible to predict the remaining service life of the non-return valve. The knowledge of the remaining service life leads to increased machine uptime while reducing storage and maintenance costs for plastic processing companies.

Keywords: injection molding, condition monitoring, non-return valve, wear state, remaining service life, maintenance strategy
PACS: 81.05.Lg, 81.20.Hy, 81.40.Np, 81.40.Pq

INTRODUCTION AND OBJECTIVES

The long-term vision of the plastic processing industry can be described as the fully autonomous injection molding machine [1]. The desired functionality of such an autonomous machine includes the abilities to automatically detect the state of all relevant machine components ("Condition Monitoring") and perform a residual service life prediction ("Predictive Maintenance") for each individual component to increase machine uptime while reducing inventory and maintenance costs.

Predictive maintenance of wear-prone components is highly desired as time and cost pressures are among the challenges faced by plastic processing companies. Major problems are the frequently used maintenance strategies. In the plastic processing industry, non-return valves of injection molding machines are often replaced following specified maintenance intervals. On the one hand, the component service life is not utilized to full capacity. Due to conservative lifetime estimations non-return valves are often replaced too early ("Preventive Maintenance"), since the current state of wear is unknown prior to the disassembly. On the other hand, in case of an unexpected failure ("Run-to-Failure Management") adjacent components can also be damaged. As a result, unplanned downtime can occur, which causes high costs [2]. Thus, an automated service life prediction of non-return valves is desired.

To estimate the current state of non-return valves, the machine operator can nowadays use a wide range of default parameters and tools to monitor the injection molding process [3]. By looking at the curves of pressure and position signals from sensors, it is possible to estimate the wear state of non-return valves by a manual test without the need for disassembly. However, this test only checks the leakage of the non-return valve and requires a production stop. In order to accomplish an automated condition monitoring of non-return valves, process and machine data can be used to detect the current state. Modern injection molding machines gather a variety of process and machine data in high quality [4], [5]. Various applications such as KraussMaffei APC (Plus) [6] are already using these data to compensate process fluctuations caused by the non-return valve. APC (Plus) helps reducing fluctuations in component quality but does not provide information about the current state of the non-return valve. Another approach to estimate the wear state of non-return valves is the evaluation of the torque signal profile. Different characteristics of the torque curve indicate non-uniform closing of the non-return valve and imply the wear of the non-return valve [7]. Unfortunately, only the axial wear of the check ring is taken into account, despite many other signs of wear occur at non-return valves.

This paper presents a procedure to evaluate the wear state of non-return valves during production based on process and machine data of injection molding machines without the use of additional sensors. All relevant signs of wear are taken into account regarding a holistic condition monitoring of the non-return valve.

CONDITION MONITORING OF NON-RETURN VALVES

In the plastic processing industry, the machine operator already has a number of proven methods at his disposal for estimating the wear state of non-return valves. The machine operator observes the residual melt cushion on the machine control system over several cycles or performs a manual test to check the leakage of the non-return valve. These established methods give an early indication of the condition of the non-return valve, but it is still necessary to dismount the non-return valve to check the actual condition. These methods do not always lead to the right assumption and an unnecessary disassembly of the plasticizing unit costs time and money. With a detailed look at the sensor signals, it is possible to determine the condition of the non-return valve more precisely. By using the standard sensor signals of the injection molding machine it is possible to monitor the condition during production and additional sensors are not required. Figure 1 shows the procedure for condition monitoring of non-return valves in injection molding machines using available process and machine data.

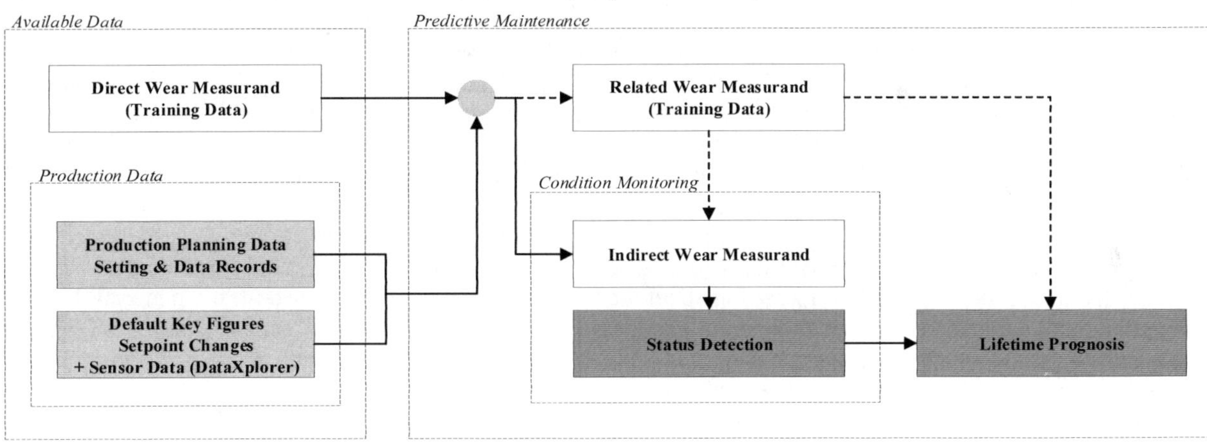

FIGURE 1. Structure of a component lifetime prediction model for non-return valves.

Available Data

Within an injection molding machine, the non-return valve is considered as a wear part and, at the same time, has a major influence on the injection molding process. In most applications, the standard three-part non-return valve is used, which is also the subject of this paper. Nowadays, the wear of non-return valves is intensified by the increasing use of filled and reinforced thermoplasts (glass fibres, silicates, metal powders etc.) for technical applications. In addition, the continuous increase in output, for example through higher screw speeds, increases wear. The main wear locations of a non-return valve are on the outer diameter (radial wear) and on the front contact surface of the check ring with the wings of the tip (axial wear). In this respect, wear can also take place on the wings of the tip (axial wear), cf. Figure 2 [8], [9].

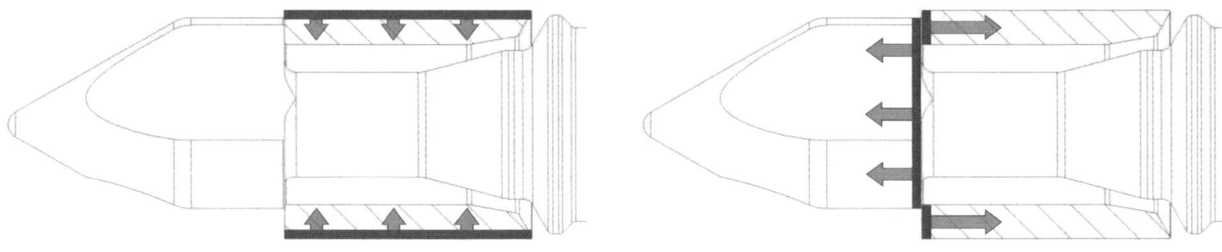

(a) (b)

FIGURE 2. The main wear locations of a three-part standard non-return valve with radial wear (a) and axial wear (b).

The wear on the outer diameter of the check ring is usually caused by the relative movement between ring and cylinder, which leads to friction during plasticizing and injection. When the check ring is in contact with the wing surface during plasticizing, the check ring begins to rotate due to static and sliding friction between the wing and check ring surfaces. The resulting relative movement between the check ring and cylinder leads to friction with the non-rotating inner wall of the cylinder during plasticizing. Additionally, the translational movement during injection causes wear of several tenths of a millimeter. Consequently, this impairs the closing behavior and the sealing of the non-return valve during the injection and holding pressure phase [8].

The different rotational speeds between wing and check ring surfaces during plasticizing can also cause material abrasion of several millimeters of the weaker body between the friction partners with increasing service life. The reason behind this wear is the melt film located between screw tip and check ring during plasticizing. In the case of polyethylene (PE) and polypropylene (PP), this film tears off due to high shear rates, resulting in solid friction [9].

The wear condition of non-return valves is mostly considered binary in the plastic processing industry. After disassembly, the machine operator visually inspects the non-return valve. Thereby, the non-return valve is only classified as functional or non-functional. Thus a quantitative classification of the condition does not take place. However, for the development of a model for condition monitoring, quantitative measurements of the condition are necessary. Such data are usually only available in the laboratory environment. With reference to radial wear, the outer diameter of the check ring can be determined as a direct wear measurand. Similarly, in the case of axial wear, the length of the check ring or the wings is considered. The direct wear measurands are used to evaluate the condition of the non-return valve and provides an indication of the accumulated damage.

In plastics processing production, a variety of different data types is available which can potentially be used to monitor the condition of machine components. The production data as well as the settings and data records are at the highest production level and provide information about the type and quantity of processed plastic material. Injection molding machines offer the option of selecting a limited set of default key figures ("actual values") and setpoint changes using a protocol function either manually via USB interface or automatically, e.g. via Euromap 63 interface to a production host computer. The machine operator can monitor the process by tracking these default key figures via the machine controller. However, as mentioned before, a holistic condition monitoring of the non-return valve is not possible by only using these default key figures. Detailed process data can be collected from KraussMaffei injection molding machines using the DataXplorer [5], which provides virtually continuous process signals (sampling rate 200 Hz) instead of single actual values. This allows the development of individual key figures indicating the wear state of non-return valves while omitting the need for additional sensors.

Condition Monitoring

As described before, the DataXplorer provides a detailed insight into the injection molding process. In order to perform condition monitoring of non-return valves solely utilizing available sensors, the data collected by DataXplorer must first be linked to the direct wear measurands during a training phase. Wear-specific anomalies in the data set are assigned to radial or axial wear. For the different types of wear, specific sensor signals are considered during corresponding phases. As an example, in the case of axial wear, a movement in the screw position signal during the holding pressure phase indicates a leakage of the non-return valve.

In order to link DataXplorer data and direct wear measurement variables in a systematic manner, worn non-return valves were examined. The outer diameter of two check rings was evenly reduced. In addition to a check ring in mint condition (Ø39.94 mm), three wear states of radial wear were available for an experimental test series: Ø39.94, Ø39.73 and Ø39.48 mm. Figure 3 shows the melt pressure and the screw position over time of a KraussMaffei 180-750 AX electric injection molding machine processing polystyrene (Styrolution GPPS PS165N) and high density polyethylene (Sabic HDPE M300056). The holding pressure level was set to 30 % of the maximum resulting injection pressure during the injection phase. In case of the processed PS (a), the screw path with the unworn check ring shows a holding pressure phase with low shrinkage compensation and a proper seal of the non-return valve. The corresponding melt pressure curve shows a deviating higher maximum value during the injection phase and, as a result, a higher set holding pressure height. Due to the correct sealing of the check ring, this deviating behavior was ignored. For smaller outer diameters of the check rings, the screw moves completely back to the 0 mm position during holding pressure phase. This observation can also be seen from the residual melt cushion on the machine control. However, only the position value at the end of the holding pressure phase is used on the machine control. With an even smaller outer diameter of the check ring, the position 0 mm is arrived much earlier: $t_{Ø39,48} < t_{Ø39,73}$. By this, the outer diameter of the check ring, for example, can be determined by creating such a wear-specific key figure. These conclusions can be expressed in indirect wear measurands by linking significant key figures and direct wear measurand. If the relevant process data on the injection molding machine are recorded and evaluated, according to the requirements, changes in the state of wear of the non-return valve can be detected online.

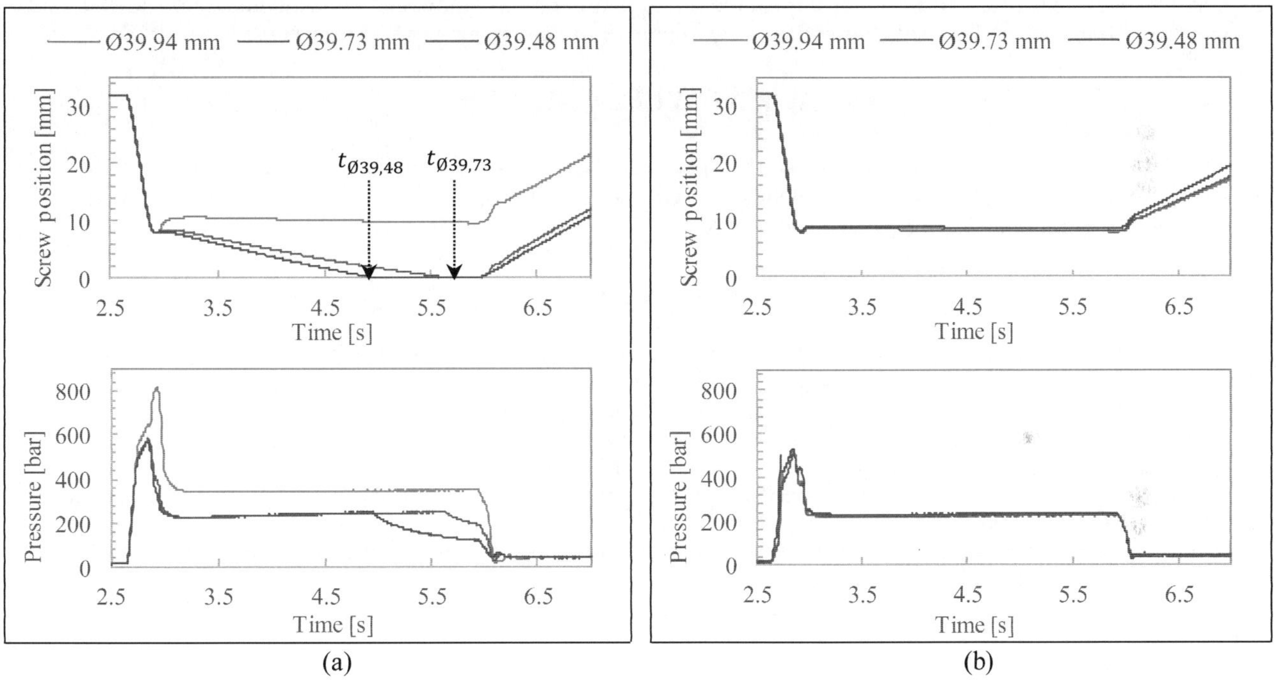

FIGURE 3. Curves of pressure of the translatory screw drive and the screw position for processing of polystyrene (Styrolution GPPS PS165N, a) and high density polyethylene (Sabic HDPE M300056, b) on a KraussMaffei 180-750 AX electric injection molding machine.

Key figures can largely vary due to material-, process- or machine-induced influences. The screw position signals during processing of higher viscosity HDPE (for injection molding applications) show no impairment due to the reduction of the outer diameter. Therefore, it is very important to identify and quantify regular influential factors by axial or radial wear and processed material. Wear-specific key figures can then be filtered and classified according to these factors. Hence, the filtered key figures can be visualized as a long-term trend and evaluated with statistical methods. Faults and irregular changes in the process can thus be detected and assigned to individual wear states. In addition to the wear-specific key figures, expert knowledge and best practice examples can be transformed into algorithms that can be used to monitor all injection molding machines.

Predictive Maintenance

Based on the current state of wear, the service life of a non-return valve can be predicted. In this respect, the condition-dependent rate of wear must be taken into account. The wear rate is determined by related wear measurand during production. Related wear measurand can be formed by analyzing the material quantity or produced parts with different states of wear. Depending on the state of wear, wear rates may show linear or non-linear behavior. Determining these parameters in the laboratory is challenging, as a large amount of continuous experimental data is required. However, the effort is well worth it, as it is a precondition to develop suitable models for the prediction of the service life of non-return valves (such as damage accumulation, regression models, etc.).

CONCLUSION

These days, plastic processing companies have to face requirements like adherence to schedules, cost reduction and resource optimization. Modern condition monitoring and predictive maintenance concepts could help solving these problems, but still do not find any use in the industrial production. With regards to predictive maintenance, the available sensor data from the injection molding machine contain valuable information. The paper shows a concept for condition monitoring of non-return valves through the intelligent use of available process and machine data for lifetime prediction. In this context, a reliable predictive maintenance system results in improved economic planning of maintenance measures, avoidance of unplanned shutdowns due to component failures, utilization of the entire component service life and, consequently, the increase in overall equipment effectiveness (OEE).

ACKNOWLEDGMENTS

Our special thanks go to the KraussMaffei Technologies GmbH (Munich, Germany), for supporting the experimental tests. Further, we apply our thanks to the iba AG (Fürth, Germany) for providing the continuous data acquisition and processing for the experimental tests.

REFERENCES

1. J. Wortberg, A. Schulze Struchtrup, C. Zhao, "Digitalisierung der Spritzgießproduktion - Intelligente Maschinen für effiziente Prozesse nutzen", in *VDI-Jahrestagung Spritzgießen*, Baden-Baden, Germany, 2017.
2. R. K. Mobley, "An Introduction to Predictive Maintenance", *Butterworth-Heinemann,* Oxford, UK, 2002.
3. F. A. Heinzler, S. Kruppa, R. Schiffers, J. Wortberg, "Useful Process Data from the Injection Molding Machine", *Journal Kunststoffe, Issue 02/2014,* München, Germany, 2014, pp. 42–46.
4. M. Lenz, "500 Werte für mehr Prozesssicherheit", URL: http://kmagazin.eu/index.php/2016/10/25/500-werte-fuer-mehr-prozesssicherheit/, accessed October 12, 2017.
5. S. Kruppa, R. Schiffers, M. Busl, U. Lettau, "Advanced Data Acquisition and Analysis for Injection Molders", *SPE-ANTEC Tech. Papers*, Anaheim, USA, 2017.
6. G. Holzinger, R. Schiffers, S. Moser, S. Kruppa, "An Adaptive Filling to Packing Switchover Method for Injection Molding", *SPE-ANTEC Tech. Papers*, Las Vegas, USA, 2014.
7. T. Uchiyama, "Abrasion amount estimation device and abrasion amount estimation method for check valve of injection molding machine", U.S. Patent US20170326771 A1, 2017.
8. G. Mennig, M. Lake, "Verschleißminimierung in der Kunststoffverarbeitung", *Carl Hanser Verlag*, München, Germany, 2007.
9. C. J. Gornik, "Neue Erkenntnisse zur Plastifiziereinheit basierend auf experimentellen Untersuchungen", Ph.D. thesis, Montanuniversität Leoben, Leoben, Austria, 2006.

Three Dimensional Numerical Simulations for Wet-RTM Process

Hsun Yang[a], Chih-Chung Hsu[a], and Chia-Hsiang Hsu[a]

[a]*CoreTech System (Moldex3D) Co., Ltd., Chupei City, Hsinchu Country, 302, Taiwan*

Abstract. Wet-RTM (Resin Transfer Molding) molding applies the resin onto the surface of the reinforcements and then the resin flows into the fabric with mold compression. This process is gaining attention thanks to car marker's use in recent years. Since both the fabric layup and charge location affect resin flow, it's not easy to know the resin flow behavior in complex mold cavity. In this study, we predict the resin flow in Wet-RTM process by using the novel three-dimensional simulation tool which can consider the Darcy's law and moving boundary problem in compression process. The simulation results demonstrate the effect of the anisotropic permeability and initial resin location variation by flow front evolution and pressure distribution during the compression stage. It could help user to determine the fabric layup and charge location before manufacturing processes to save try and error costs.

Keywords: Wet RTM, Darcy's law, Ply orientation, Moving boundary

INTRODUCTION

Resin transfer molding (RTM) process is widely used in composite industry, because it can make light and high strength products. There are 4 main stages in RTM process, preform preparation, mold closure, mold filling, resin cure. The advantages of the RTM process, including not only reduced solvent emissions but also manufactured dimensional accuracy and surface quality product [1]. However, there are some manufacturing problems in traditional RTM process, such as such as long filling time and poor wetting of fiber preform during making large dimensions or high fiber volume fraction product. In order to improve production efficiency and production acceleracy of the process. One is to increase inlet pressure, and the other is to use compression step. For example, HP-RTM uses the pressure-increasing method and Wet RTM uses the compression method. Wet RTM is a variant of the traditional RTM process. Wet RTM uses charge preparation and compression step instead of resin injection during the molding filling stage.

Y. J. Lee [2] used 2.5D(Shell) model to simulate the RTM process along the in-plane direction. They showed that filling behavior of simulation in the in-plane direction were in good agreement with the experiment. However, the filling behavior in thickness direction could not be predicted by 2.5D model. Hence, in this study the full 3D simulation were used to investigate the filling behavior in thickness direction. Shojaei et al.[3] used 3D simulation predict the flow behavior of resin in porous media. They showed that the simulation results have good agreement with analytical results. The objective of this study was to propose a full 3D numerical approach to model the resin compressed into the anisotropic permeability fabric during the

Wet RTM process. These predictions were performed by a 3D finite volume method (FVM) technique to solve the governing equations of resin flow and heat transfer. By full 3D simulation, the flow behavior insight the mold could be visualized to help improve the RTM process.

MODELING

In the study, there are two regions in system (Fig. 1): **Region 1** (without fabric region) and **Region 2** (fabric region).

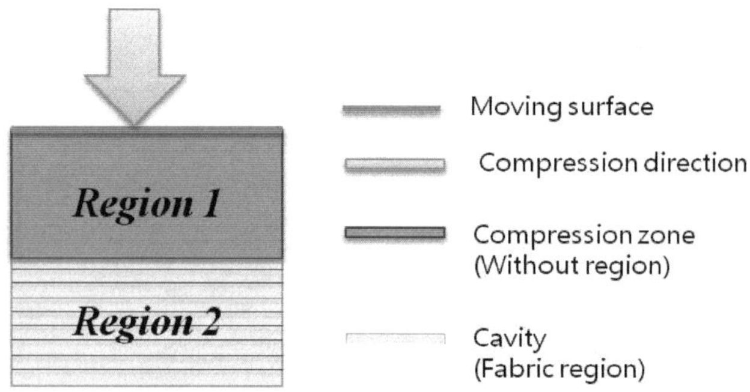

Fig. 1 Fabric region and without fabric region

It's necessary to choose the suitable governing equation for the simulation based on the regions.

In **Region 1**, Equation (1) is conservation of mass. The common approach to modelling resin flow is to use Generalized Newtonian Fluid (GNF) [4] as the momentum equation (2). The transport equation (3) describes the evolution of fluid over time.

$$\frac{\partial \rho}{\partial t}+\nabla\cdot(\rho\mathbf{u})=0 \tag{1}$$

$$\frac{\partial \rho}{\partial t}(\rho\mathbf{u})+\nabla\cdot(\rho\mathbf{u}\mathbf{u})-\nabla\cdot(-pI+\boldsymbol{\tau})=0 \tag{2}$$

$$\frac{\partial}{\partial t}(f)+\mathbf{u}\cdot\nabla f=0 \tag{3}$$

Where $\mathbf{u}$ is fluid velocity, T is temperature, t is time, p is pressure, $\boldsymbol{\tau}$ is total stress tensor, ρ is the density, f is the volume filled fraction, t is time, $\mathbf{u}$ is fluid velocity.

In **Region 2**, the assumption of resin is incompressible and steady flow, and the conservation of mass is written as equation (4). The common approach to modelling resin flow through anisotropic porous media is to use Darcy's law [5] and the transport equation should be modified in order to include porosity:

$$\nabla\cdot\mathbf{u}_s=0 \tag{4}$$

$$\mathbf{u}_s = -\frac{1}{\eta}\mathbf{K}\cdot\nabla P \tag{5}$$

$$\frac{\partial}{\partial t}(f\phi) + \mathbf{u}_s \cdot \nabla f = 0 \tag{6}$$

where $\mathbf{K}$ is permeability tensor, ∇P is pressure gradient, η is fluid viscosity, and $\mathbf{u}_s$ is superficial velocity, f is the volume filled fraction, t is time, ϕ is porosity. In equation (6), f=0 means pore region with no resin, f=1 means pore region filled with resin. The flow front is located within cells with 0<f<1.

$$\mathbf{K} = \begin{bmatrix} K_{xx} & K_{xy} & K_{xz} \\ K_{yx} & K_{yy} & K_{yz} \\ K_{zx} & K_{zy} & K_{zz} \end{bmatrix} = \begin{bmatrix} l_{11} & l_{12} & l_{13} \\ l_{21} & l_{22} & l_{23} \\ l_{31} & l_{32} & l_{33} \end{bmatrix} \begin{bmatrix} K_{11} & 0 & 0 \\ 0 & K_{22} & 0 \\ 0 & 0 & K_{33} \end{bmatrix} \begin{bmatrix} l_{11} & l_{21} & l_{31} \\ l_{12} & l_{22} & l_{32} \\ l_{13} & l_{23} & l_{33} \end{bmatrix} \tag{7}$$

where K_{ij} (i, j = x, y, or z) are components of the permeability tensor, K_{11}, K_{22}, and K_{33} are principal permeability in porous media, l_{ij} are the directional cosines of local coordinates. There is a more detailed description in Chen Kailing's master thesis [6] about permeability tensor.

The heat transfer can be mathematically described by the followings conservation laws for energy [7]:

$$\rho C_P\left(\frac{\partial T}{\partial t} + \mathbf{u}\cdot\nabla T\right) = \nabla(\mathbf{k}\nabla T) + \Phi \tag{8}$$

Where ρ is the density, C_P is heat capacity, T is temperature, t is time, p is pressure, $\boldsymbol{\tau}$ is total stress tensor, ρ is the density.

RESULTS AND DISCUSSIONS

In this section, the 3D simulation tool Moldex3D is used to predict the resin filling behavior during Wet RTM the compression stage. Two simulation example cases are presented to demonstrate the effect of the anisotropic permeability and the ply orientation. As shown in Fig. 2, geometry of the cavity geometry. The resin charge is set upon the fabric and compressed 1 mm/sec according to the compression direction. The charge dimension is 30×20×2 cm. The initial resin temperature is 25℃ and the mold temperature is 35.

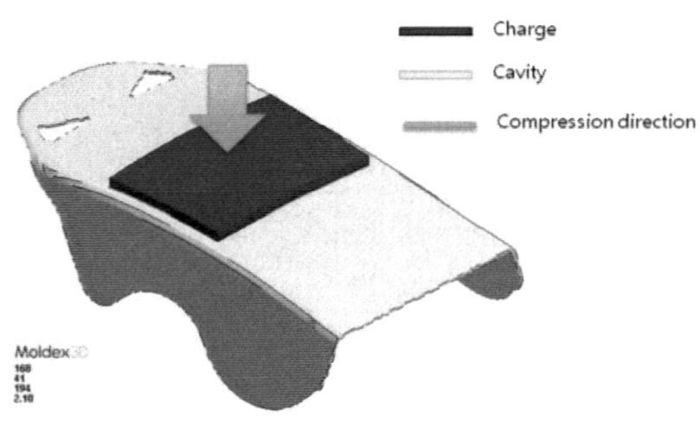

Fig. 2 Geometry and charge

Permeability is the reciprocal of the flow resistance, so resin flows faster the in high permeability direction than in the low permeability region. Case 1 is an isotropic permeability case the permeability ratio set as $K_{11} : K_{22} : K_{33} = 1:1:1$. Case 2 is an anisotropic permeability case the permeability ratio set as $K_{11} : K_{22} : K_{33} = 100:1:1$. L_1 and L_2 show in Fig. 3 are the permeability tensor axis direction. The main direction L_1 and L_2 of permeability tensor setting is based on the ply orientation.

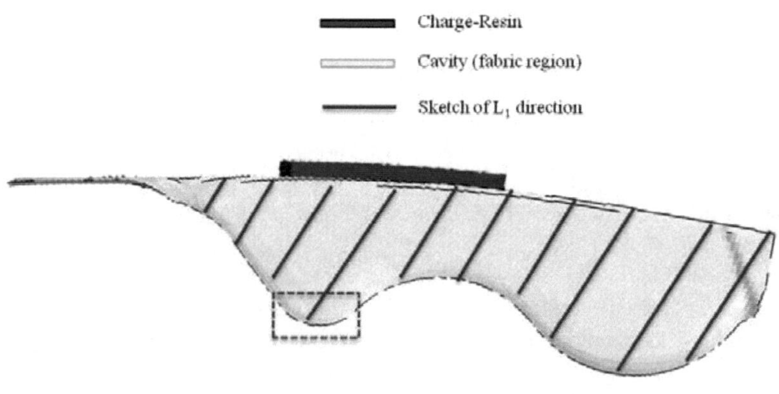

(a)

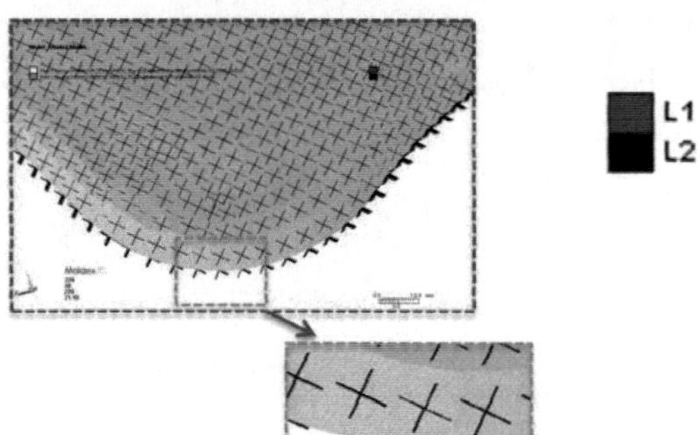

(b)

Fig. 3 Main direction the permeability tensor (a) sketch of the Main direction L_1 direction (b) L_1 and L_2 of each element

The flow front time distribution of simulation result the end of the compression stage could be seen from Fig. 4 and Fig. 5. In Case 1 main value of permeability tensor in L_1 direction equal to L_2, resulting in the symmetric flow pattern in Fig. 4.

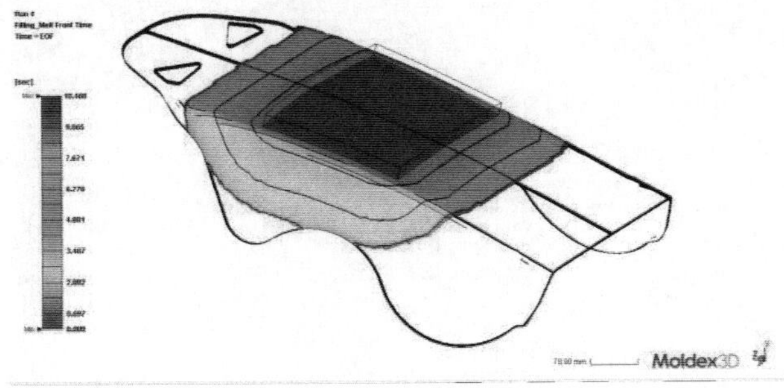

Fig. 4 Case 1 flow front time for isotropic permeability tensor case with $K_{11} : K_{22} : K_{33} = 1:1:1$

The asymmetric flow pattern shows in Fig. 5. The resin flows in L_1 direction much faster than in L_2 direction, because of the main value of permeability tensor in L_1 is 100 times as many as main value of permeability tensor. In Case 1 the resin flow pattern won't change with the ply direction, because of the isotropic properties. When using the fabric in Case 2, designer should be aware of design change of ply orientation.

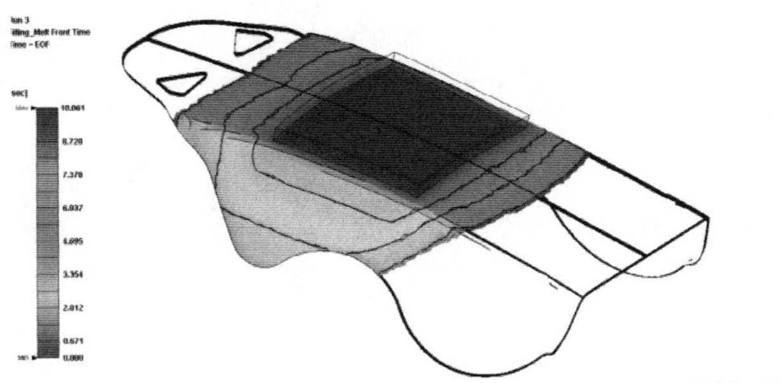

Fig. 5 Case 2 flow front time for anisotropic permeability tensor case with $K_{11} : K_{22} : K_{33} = 100:1:1$

The resin flow front evolution in anisotropic fabric demonstrates in Fig. 6 and in isotropic fabric demonstrates in Fig. 7. The result in Case 1 and Case 2 shows the influence of the fabric properties evidently. It's very difficult to observe the flow front evolution in Wet RTM process only when having the visualization system. It is helpful to use the simulation assist manufacturing execution, because of the visualized result.

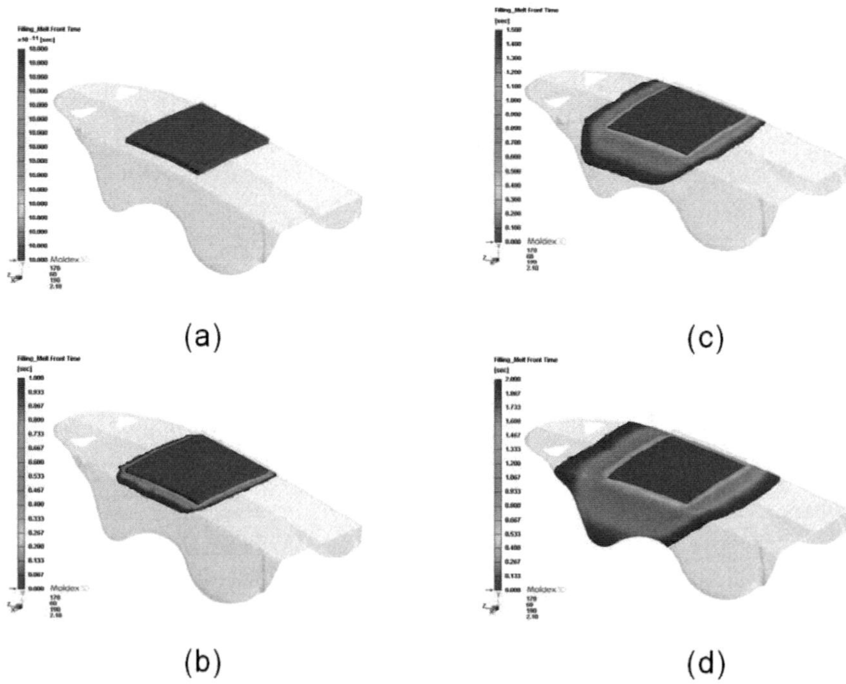

Fig. 6 Flow front evolution in Case 1 (a) t=0sec (b) t=1.0sec (c) t=1.5sec (d) t=2.0 sec

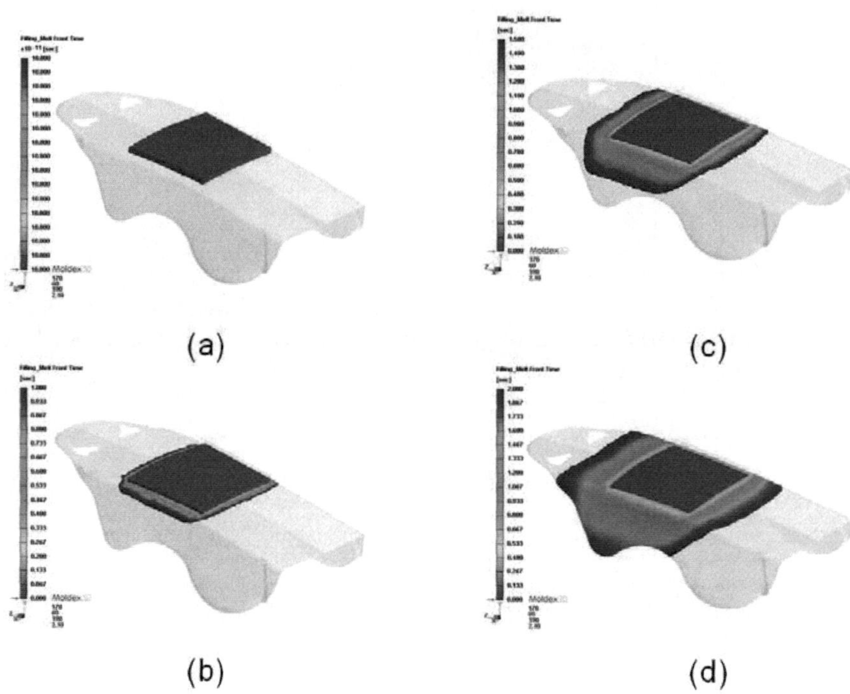

Fig. 7 Flow front evolution in Case 2 (a) t=0sec (b) t=1.0sec (c) t=1.5sec (d) t=2.0 sec

In experiment it's barely possible to get temperature distribution through the thickness direction. In simulation, it's easy to get local value by the use the slicing tool and sensor node.

Fig. 8 shows the slicing of the temperature distribution. Because the mold temperature is higher than initial resin temperature, the direction of the heat transfer is from the mold to the center region. The closer sensor node location is to the mold, the closer resin temperature is to the mold temperature. To get the accurate temperature value in local region, the proper way is use sensor node. The sensor node locations are in Fig. 9 and Fig. 10 is the through thickness direction temperature result recorded by the sensor node.

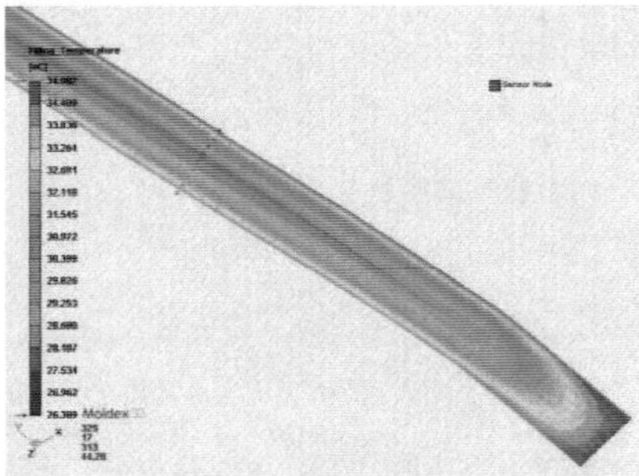

Fig. 8 Location of sensor node

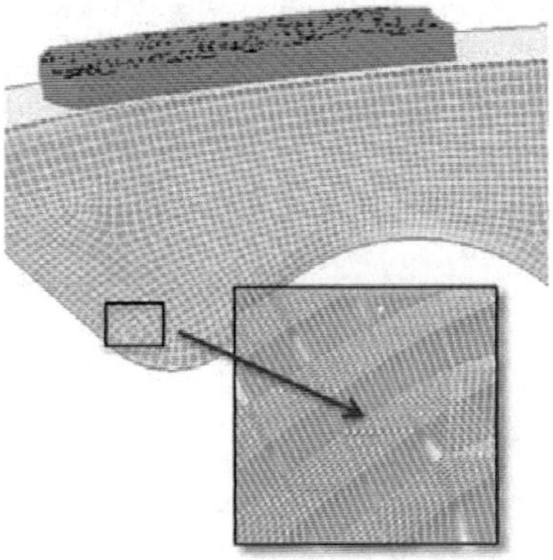

Fig. 9 Location of sensor node

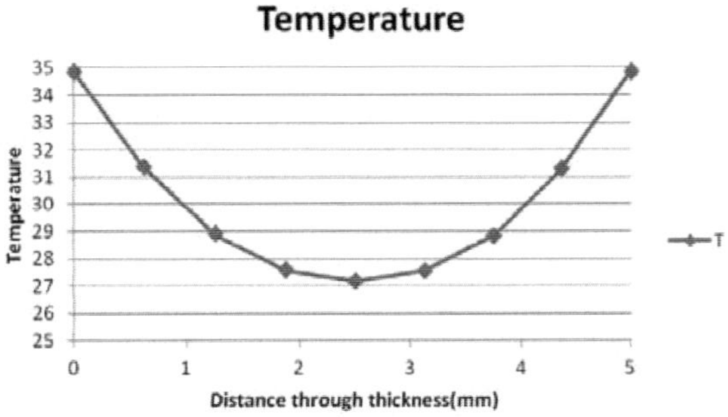

Fig. 10 Slicing of the temperature distribution

CONCLUSION

In this study, the true 3D simulation tool (Moldex3D) was used to simulate the flow of resin in the Wet RTM process during the compression stage. The material properties of the isotropic properties and the properties of the anisotropic materials were used to show the influence of the ply orientation on the resin flow behavior. The results show that the impact of the ply orientation on the flow results is significant. If designer has to change the ply orientation in the product design stage, the simulation results would be helpful for adjusting the process condition parameters.

Specifically, it's easier to control the filling behavior of the Wet RTM process through the visual analysis process by true 3D simulation tool. The simulation tool can effectively help manufacturers to control the actual process condition parameters, and reduces the expensive and time-consuming trial and error processes.

REFERENCES

1. YANG, B., JIN, T., LI, J. ET AL. APPL COMPOS MATER (2015) 22: 209.
2. K. L. CHEN AND Y. J. LEE," STUDY ON THE APPLICATION OF VARTM TECHNIQUE AND MOLD FLOW ANALYSIS TO THE LAMINATION OF SANDWICH PLATES", MASTER'S THESIS, NATION TAIWAN UNIVERSITY, (2005).
3. A. SHOJAEI, S.R. GHAFFARIAN, S.M.H. KARIMIAN, COMPOSITES SCIENCE AND TECHNOLOGY,62 ,861 (2002)
4. R.-Y. CHANG AND W.-H. YANG, INT. J. NUMER. METH. FLUIDS, 37, 125 (2001).
5. YANG, H & CHANG, W.-Y & HUANG, C.-Y & HSU, CHIH-CHUNG & CHANG, R.-Y. (2014). A FILLING BEHAVIOR STUDY IN RESIN TRANSFER MOLDING PROCESS THROUGH 3D SIMULATION AND EXPERIMENTAL VISUALIZATION. ANNUAL TECHNICAL CONFERENCE - ANTEC, CONFERENCE PROCEEDINGS. 2. 1557-1561.
6. K.-L. CHEN, "STUDY ON THE APPLICATION OF VARTM TECHNIQUE AND MOLD FLOW ANALYSIS TO THE LAMINATION OF SANDWICH PLATES," MASTER THESIS, DEPARTMENT OF ENGINEERING SCIENCE AND OCEAN ENGINEERING, NATIONAL TAIWAN UNIVERSITY, TAIWAN, 2005.
7. HSU, CHIH-CHUNG, ET AL. "THROUGH FIBER ORIENTATION INVESTIGATION TO VISUALIZE COMPRESSIONMOLDING." AIP CONFERENCE PROCEEDINGS. VOL. 1593. NO. 1. AIP, 2014.

Selective Modification to an Innovative New High ESCR HDPE

Farah Marcelo (a)*, Bueno Marcos Roberto (b), Mano Barbara Iria (c), Stocker Mariele Kaipers (b) Grassi Vinicius Galhard(a)

(a)Polymer Science, BRASKEM SA - RS - Brazil,
(b)Advance Characterization, BRASKEM SA - RS - Brazil,
(c) Renewables, BRASKEM SA - SP - Brazil

Abstract. In this document, we describe the emergence and advances of new promissory technology. It involves a selective reactive extrusion where this reaction brings a significant improvement in ESCR(Environmental Stress Cracking Resistance) without some usual technological movements like increasing molecular weight and reducing density. Its products have a much better HDPE properties balance. About the process, it is not usual in Polyolefins, based on free radical modification, where substantially modified PE's and PP's, it can selectively proceed the reactive extrusion, perfectly maintain the PP and PE original properties. To do that, first, a blend dispersion must be correctly prepared and then the disperse phase crosslink agent are fed to chemically modify the system. However, the correct sequence for feeding, controlling where the reaction must to take place and ideal composition were the challenges to be answer. Other important question is to remove the volatiles from the product. Therefore, a complex design of the process were done to solve and improve the technological challenges, becoming a suitable process to be scaled-up. The final knowledge were applied to develop a prototype to be tested in costumers. Several types of Crosslink agents were also used to define the most suitable system. The final HDPE are describe in terms of Mechanical properties, ESCR and processability in different stages of the blow molding process. The improvement in ESCR properties after optimization do not change the mechanical properties resulting in a much better resin balance and performance.

Keywords: reactive extrusion, HDPE, ESCR.

INTRODUCTION

Polyolefins such as polyethylene (PE) may be used to manufacture a varied range of articles, including films, molded products, foams, and the like. Polyolefins may have characteristics such as high processability, low production cost, flexibility, low density and recycling possibility. However, physical and chemical properties of polyolefin compositions may exhibit varied responses depending on a number of factors such as molecular weight, distribution of molecular weights, content and distribution of comonomer (or comonomers), method of processing, and the like.

Polyolefin materials may have a number of limitations, which can restrict application such as susceptibility to deformation and degradation in the presence of some chemical agents, suffering with the slow crack propagation in an aggressive environmental, as susceptibility to environmental stress cracking (ESC)[1,5] and Accelerated slow crack growth (SCG)[2,3,4], which may occur below the yield strength limit of the material when subjected to long-term mechanical stress. Property limitations may hinder the use of polyolefin materials in the production of articles requiring solvents/tensocatives resistance, such as packaging for food products, chemicals, agrochemicals, fuel tanks, water and gas pipes, and geomembranes, for example.

Conventionally, methods of altering the chemical nature of the polymer composition may include modifying the polymer synthesis technique or the inclusion of one or more comonomers. However, modifying the polyolefin may also result in undesirable side effects. By way of illustration, increasing the molecular weight of a polyolefin may produce changes in the SCG and ESC, but can also increase viscosity, which may limit the processability and moldability of the polymer composition.

Other strategies may include inclusion of a comonomer and/or blending polyolefins with other polymer classes and additives to confer various physical and chemical attributes. For example, polyolefins may be copolymerized with alpha-olefins having a lower elastic modulus, which results in a considerable increase in environmental stress cracking resistance (ESCR) and resistance to influence but adversely affects the stiffness of the polymer. The end result is the production of a polymer composition having reduced structural stiffness. It is also noted that, while advances have developed catalysts that increase the likelihood of displacing the incorporation of a comonomer to the highest molecular weight range, and that multiple reactors may be used to address these limitations, such modifications are expensive alternatives and not wholly effective in balancing resistance to impact and ESC without negatively affecting stiffness. The incorporation of a second phase into the matrix polymer, which generally has a different chemical nature, may increase the resistance to impact and ESC resistance in some cases. Other considerable problems is the effects of the blend morphological developments including effects of thermomechanical history and as dispersion and coalescence variations costumer to costumer.

This work describe an innovative process, when and polymeric polar Particle are added to the HDPE, interfering significantly in the SCG process and also altering the Solvent/tensoactive interaction, give to any HDPE an improved resistance do not described before. Many polar polymers could be used, but the more remarkable results come from EVOH and PVOH, but the technology also works to other polar polymers with suitable processing windows with HDPE. Unfortunately, the pure materials will lead to similar process of coalescence and worst, the PVOH is a soluble polymer. So a selective reactive extrusion process can transform a simple polymer blend dispersion in crosslinked particles dispersion, where no longer coalesces and processing dependence are present, but also transform soluble polymer is insoluble one.

PVOH reaction

The PVOH are easily reacted in presence of acids. In Spite of, the reaction is not acting over the HDPE that continues exactly as it is and this is the principle of the Selective Reactive extrusion. Some polyfuncionals(character polyacid) are suitable to generate the reaction, but the most cited is the maleic Anhydride[6,7].

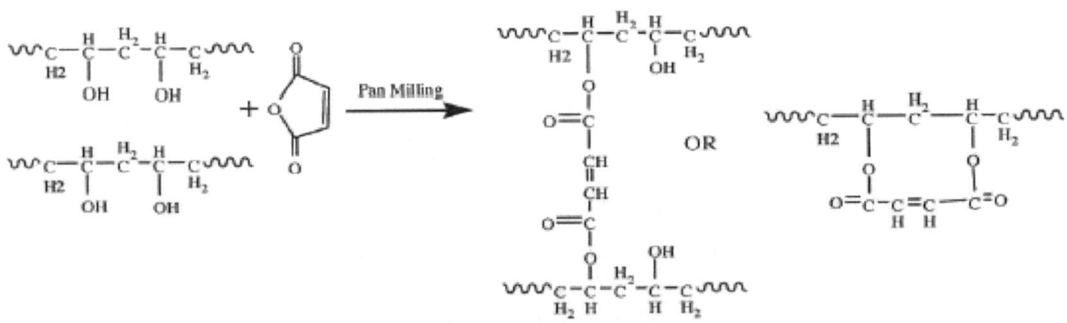

Figure 1 – Possible reactions of PVOH and maleic Anhydride[7].

It means that several previous approach as processing conditions (temperature, residence time, screw profile) and also material dispersion capability, CA(crosslink agent) and its concentration, compatibilizer and its concentration and also correct feeding sequence were studied before the master modification that could be another presented work.

MATERIALS AND METHODS

The materials used in this work are listed in on TABLE (1).

TABLE (1) - Materials used with selective modified masterbatch.

Materials	**MFI (2,16 kg)**	**MFI (5 kg)**	**Distribution of Molecular Weight**	**Density (g/cm³)**	**Remarks**
HDPE 1	--	0.30	Monomodal	0.956	--
HDPE 2	0.21	--	Bimodal	0.954	--
HDPE 3	0.34	--	Bimodal	0.960	--
HDPE-MAH	4.0	--	--	0.950	--
Competitor 1	--	0,41	Bimodal	0,953	--
Competitor 2	--	0,45	Bimodal	0,954	--
Polyvinyl alcohol (PVOH)	--	--	--	1,3	Degree of hydrolysis: 98%
EVOH	4,0	--	--	1,2	Usual barrier grade
Crosslinking agent(CA)	--	--	--	--	Maleic anhydride

The methods used are described in below in TABLE (2):

TABLE (2) – Characterization methods used in present work.

Methods	Sample Preparation	Standard	OBS
Environmental Stress Cracking Resistance (ESCR)	ASTM D-4703	ASTM D-1693	~2 mm thick plates
Flexural Modulus	ASTM D-4703	ASTM D-790	secant
IZOD Impact Resistance Test	ASTM D-4703	ASTM D-256	Room temperature
Scanning Electron Microscopy (SEM)	ASTM D-4703		
MFI		ASTM-D1238	
Density		ASTM D1505/D792	

Scanning Electron Microscopy (SEM)

Samples were examined using SEM after hot pressing the samples in accordance with ASTM D-4703 and polishing by cryo-ultramicrotomy. When indicated, some samples were submitted to the extraction process of the polar phase (PVOH) by contacting the sample with boiling water for 2 hours. Samples were dried and submitted to metallization with gold. The images were obtained by FESEM (Field Emission Scanning Electron Microscopy, Model Inspect F50, from FEI).

RESULTS

Morphological Characterization

Initially a description of the master development: as PVOH is a soluble polymer, the simple blend after treatment with give some "holes" by water extraction Figure 2(a). To a selective reactive extrusion, the particles are not removed by water Figure 2(b and c). The follow dispersion the particles will be the same observed in Figure 2(d).

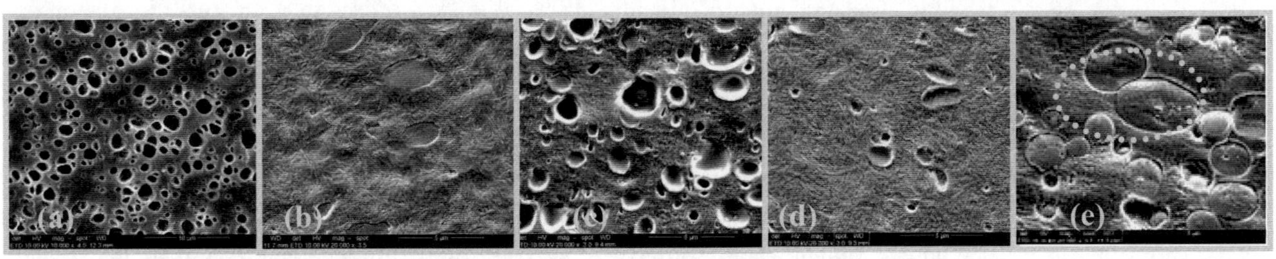

Figure 2 – (a) HDPE PVOH blend(mag. 10.000), (b) crosslinked particles(mag. 20.000x), (c) partially crosslinked particles (mag. 20.000x, (d)dispersed in HDPE particles(mag. 20.000x) and (e) particle collision and absence of coalescence .

Important observe, that some other crosslink agents also works, since generating a partially crosslinked morphology as shown in Figure 2(c). Other remarkable issue is that it is possible to see particles collision and absence of coalesce in crosslinked particles. It is remarkable to the technology, generating a blend stability, becomes a phase distribution changed to Particles distribution.

Mechanical and ESCR Characterization

The general properties are listed in

TABLE (3) below:

TABLE (3) – General new HDPE properties.

	Samples base resin	PVOH (%)	CA level	ESCR (h)	Flexural Modulus (MPa)	Izod Impact (J/m)
References	HDPE 1	0	0	180	1220	313
	HDPE 2	0	0	150	1100	168
	HDPE 3	0	0	14	1629	159
Competitor 1		0	0	610	1220	468
Competitor 2		0	0	860	1280	350
No CA Blend	HDPE 1	7	0	95,7	1282	356
Crosslink agent content	HDPE 1	7	I	>1000*	1337	303
		7	II	>1000*	1318	304
		7	III	>1000*	1272	315
PVOH content	HDPE 1	1	III	260	1210	225
		3	III	800	1204	247
		5	III	>1000*	1231	280
DPM	HDPE 2	7	III	>1000*	1112	163
Density	HDPE 3	7	III	66	1568	119
EVOH	HDPE 1	5	0	197	1250	225
EVOH mod	HDPE 1	5	III	490	1300	350

* standardized test must be stopped by Igepal Solution degradation.

First remarkable date is the effect of the simple blend of HDPE and the polar resin:. In this example, the effects of the PVOH and EVOH do not show significant improvement in ESCR properties and the effect to similar reacted blend as shown in Figure 3:

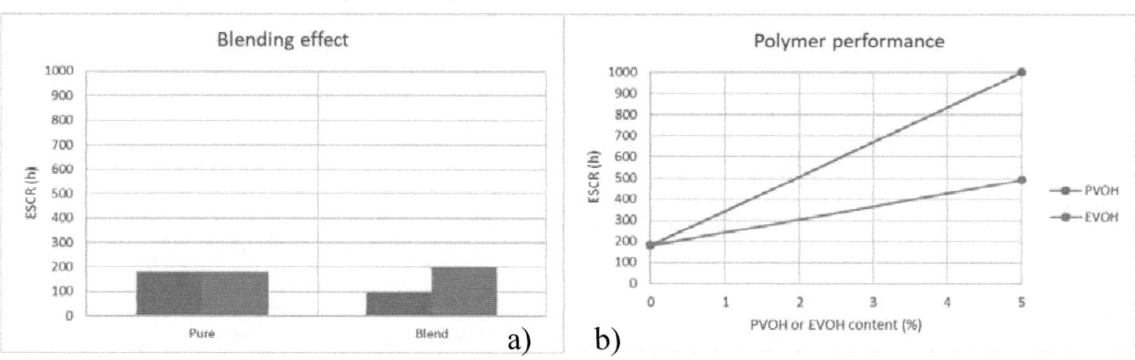

Figure 3 – Non-selective reacted blend with different polymers (a) and a selective reacted mixture (b).

Analyzing the mechanical properties, it is possible to conclude that the present selective reactive extrusion, do not interferes in the polymer mechanical properties, here exemplified by stiffness and Izop impact at room temperature. It means the improvement in chemical resistance could lead to an important additional tool in HDPE resin design.

In addition, it is possible to conclude that the pure blend is not something that could brought more chemical resistance (i.e. ESCR) to the system with PVOH neither EVOH. The reactive process not only become the PVOH insoluble but also increment the phase stability to a simple dispersion. Using a similar polymer, EVOH, an insoluble one, if the advantage EVOH processing can be applied the usual knowledge about blend dispersion, the results are inferior to PVOH as shown in Figure 3 (b).

The effects of PVOH loading in the HDPE is also observed in Figure 4. It increase quickly up to 3% and with 5% it active the test time limit. Comparing to usual commercial remarkable grades, it is possible to reach similar performance with a monomodal technology, an important industrial issue simplifying process and also redesign the HDPE density level.

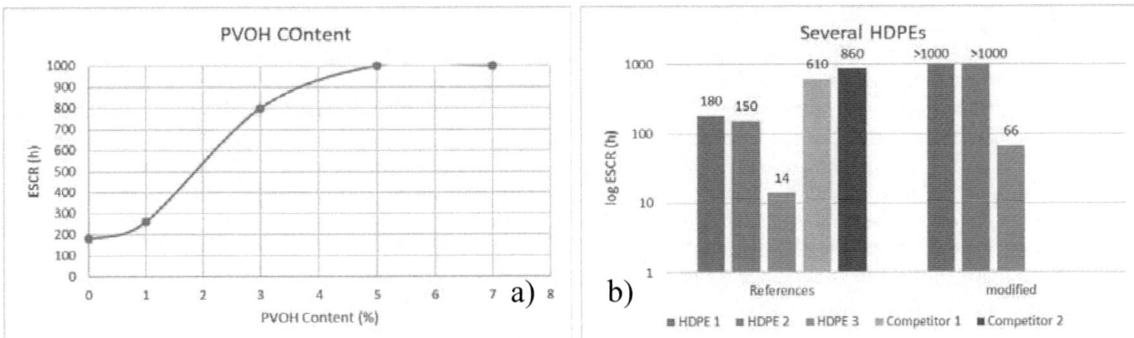

Figure 4 – Effects of the PVOH content in a HDPE 1 and the effect in seral other including competitor comparisons (log scale to be possible compare different resins design).

CONCLUSION

The present selective crosslinking technology as usual additive, with stabilized morphology, can bring a new HDPE design to several applications, where industrial assets, general packages weight or also performance security of dangerous products could be rethought, because it is a post reactor technology suitable to any reactor product.

ACKNOWLEDGEMENT

The authors would like to thanks Cristina PEREIRA, Rafael AMARAL, Alberto HABERT, Cristiano BORGES,; Jane FUJIYAMA-NOVAK and Maria GARCIA, also inventors of the patent application WO2017/054066 for the contributions and efforts given for the patent filing, which have been used as basis for some data/conclusions of this article.

REFERENCES

1. H. R. Brown, Polymer, **19**, 1186-1188 (1978).
2. X. Lu and H. R. Brown, Pol. Testing, **11,** 309-319 (1992).
3. L. Andena et al., Engineering Fracture Mechanics, **101**, 33-46(2013)
4. P. J. DesLauriers, M. J. Lamborn and J. S. Fodor, Polymer, **153** 422–429 (2018)
5. WO 2017/054066 A2, Polyolefin Composition with improved mechanical and Barrier Properties (2016).
6. W-Y Chiang, C-H Hu, J. *Appl. Pol. Sci.*,**30**, 3895-3910 (1985)
7. B. Zhao, C. H. Lu and M. Liang, *Chinese Chem. Letters*, **18**, 1353-1356 (2007)

Non-isothermal Crystallization Kinetics of Chain Extended Poly (l-lactic acid)

Elnaz Erfanian[a], Mohammad Moin Garmabi[a] and Hamid Garmabi[a], Omid Yousefzade[a]

[a]*Department of Polymer Engineering and Color Technology, Amirkabir University of Technology, Iran*

Abstract. Chain extended Poly (L-lactide acide) (PLLA) was prepared via reactive melt mixing. To investigate the effect of the chain extender content on non-isothermal crystallization kinetics, different crystallization rates of 2.5, 5, 7.5 and 10 ° C/min from the melt and different amount of an epoxied chain extender were used. The relative crystallinity and Avrami equations were used to study the kinetics of crystallization. Due to structural restriction on PLLA at high cooling rate (7.5 and 10 ° / min), the crystallization process for pure PLLA and PLLA with 0.25wt. % Joncryl did not occur, and the samples were completely amorphous. A higher crystallization temperature was indicated for the PLLA with 1wt. % Joncryl sample, which showed a faster crystallization kinetics. Using the 0.25 wt. % of Joncryl in chain extension reaction resulted in lower crystallization temperature due to the presence of branched chains. Results of Avrami n and k values showed that PLLA with 1wt. % Joncryl had higher crystallinity whereas this value for the PLLA with 0.25wt. % Joncryl was lowest among the other samples.

Keywords: Crystallization kinetic, Polylactide, Reactive mixing.
PACS: 87.15.Nn

INTRODUCTION

Studies on poly (L-lactide) (PLLA) are nowadays exponentially growing due to the performance of this polymer in various fields such as biomedical and packaging industries. PLLA can be found in both either completely amorphous or semi-crystalline forms, depending on its chemical structure (i.e., the enantiomeric content) and the processing conditions. In addition, physicochemical characteristics of PLLA deserve deeper studies according to a more fundamental/basic point of view. It is well known that PLLA slowly crystallizes from the melt state, which allows preparing materials with tuned degrees of crystallinity and different lamellar morphologies by changing the thermal history. Furthermore, cold crystallization becomes a fundamental process that allows modulating physical properties through subsequent thermal treatments. Different techniques, such as atomic force microscopy, transmission electronic microscopy, optical microscopy, small and wide-angle X-ray scattering/diffraction (SAXS and WAXD), and differential scanning calorimetry (DSC) were used to explore crystallization kinetics, structure, and morphologies of polymer crystals. The inherent difficulty of PLLA to crystallize as consequence among other factors of the high stiffness of molecular chains are the reasons to focus on crystallization process using two different additives: (1) incorporation of nucleating agents such as nanoparticles to enhance the crystallization process, (2) incorporation of additives with a plasticizing effect to decrease the glass transition temperature of the amorphous phase and the rigidity of the material. Recent research results have shown that applying the chain branching (i.e., connecting the terminal groups through a multifunctional chain extender) is an effective way to improve PLA processability as well as physical and mechanical properties. It was found that chain extension of PLLA influenced the kinetics of both isothermal and nonisothermal crystallization.

EXPERIMENTAL

Materials

Poly (L-lactide) (PLLA), grade 4032D, was purchased from Natureworks LLC, USA. It is a semi-crystalline material containing 2% of D-lactide units.Commercial multifunctional styrene–acrylic oligomers (BASF, JoncrylADR-4368) were used as a reactive agent. The sample is supplied in flake form and has the following physical characteristics according to the manufacturer: Mn < 3000, epoxy functions (fn) > 4, PDI > 3, epoxy equivalent weight (EEW) = 285 g/mol, and Tg = 55°C.

Sample Preparation

PLLA was prepared in the presence or absence of the selected chain modifier (i.e., Joncryl ADR) for reactive and nonreactive melt mixing. A lab-scale internal mixer (Brabender mixer, PL2200) operating at 190C and 60 r/min was used. Chain extender reactions were considered completed when a plateau in torque curves was achieved. This time was 15 min for the longest case and therefore was applied for all studied samples, including those prepared without Joncryl. The amount of chain modifier was 0.25, 0.5 and 1 wt. % for samples prepared via reactive melt. The studied samples are named by the abbreviation PLLAxR which "x" shows chain extender weight fraction in the samples.
All epoxy groups can be consumed during the chain extension reaction and the risk of having crosslinking reactions minimized. Thus, it can be hypothesized that the default epoxy groups of Joncryl can react in the melt state with the carboxylic groups of polymer chains, giving rise to a certain ratio of this hybrid structure.

Measurements

Molecular weights and polydispersity index (PDI) were measured by gel permeation chromatograpy (GPC) using a liquid chromatograph (Shimadzu, model LC-8 A) equipped with an Empower computer program (Waters). A PL HFIP gel column (Polymer Lab 300 mm × 7.5 mm) and a refractive index detector (Shimadzu RID-10 A) were employed. The polymer was dissolved and eluted in 1,1,1,3,3,3-hexafluoroisopropanol (HFIP) containing CF3COONa (0.05 M) at a flow rate of 0.5 mL/min (injected volume 100 mL, approximate sample concentration 1.5 mg/mL). The number and weight average molecular weights were calculated using poly (methyl methacrylate) standards.

Melting and crystallization characteristics of samples were examined using a METTLER TOLEDO DSC1 system. Samples with 9 mg weight were encapsulated in aluminum pans and treated at heating/cooling ramps in the range of 25C–200C with the heating rate of 10C/min and cooling rates of 2.5, 5, 7.5, and 10C/min. In addition, isothermal crystallization was performed for all samples at three different temperatures (110C, 120C, and 130C). Samples were heated to 200C and rapidly cooled to the selected isothermal crystallization temperature after erasing thermal history by keeping the sample in the melt state for 2 min. Crystallization process at the selected temperatures needed a maximum of 50 min to complete. The applied atmosphere was nitrogen with a flow rate of 50 mL/min.

RESULTS AND DISCUSSION

Nonisothermal crystallization

In order to investigate the effect of the chain extender on non-isothermal crystallization, different crystallization rates of 2.5, 5, 7.5 and 10 ° C/min were used for. Similar to isothermal crystallization, the relative crystallinity and Avrami equations were used to study the kinetics of crystallization.

In Figure 1, the DSC cooling curves with different cooling rates are shown for the four samples studied. In all samples, by increasing the cooling rate, the peak of the crystallization has been shifted to lower temperatures. Due to structural restriction of PLLA, the crystallization process for PLLA0R and PLLA0.25R samples did not occur at high cooling rate (7.5 and 10 ° / min), and the samples are completely amorphous. This phenomenon is due to the low mobility of the chains, the chains being arrested at low temperatures and was not placed in the lamella. This is related to the activation energy of crystallization for each sample. The values of the crystallization temperature (Tc), as shown in Table 1, indicate a higher crystallization temperature for the PLLA1R sample, which showed a faster crystallization kinetics in isothermal crystallization either. The PLLA0.25R sample also showed crystallization peak at lower temperature due to the presence of branched chains and low crystallization activation energy. To better understand, the relative crystallinity curves vs. temperature and relative crystallinity time were used. Similar to that observed in isothermal crystallization, the trend of relative crystallinity changes in Figure 2 shows that the sample PLLA1R has a faster crystallization kinetics and PLLA0.25R showed a lower kinetics compared to other samples. In order to determine the accuracy of the crystallization kinetics, Avrami model were used. For this purpose, in non-isothermal crystallization, the relative time of crystallization was calculated from the following equation:

$$t = \frac{T - T_0}{a} \quad (1)$$

In this case, T, T_0 and a are the sample temperature, crystallization start temperature and cooling rate, respectively.

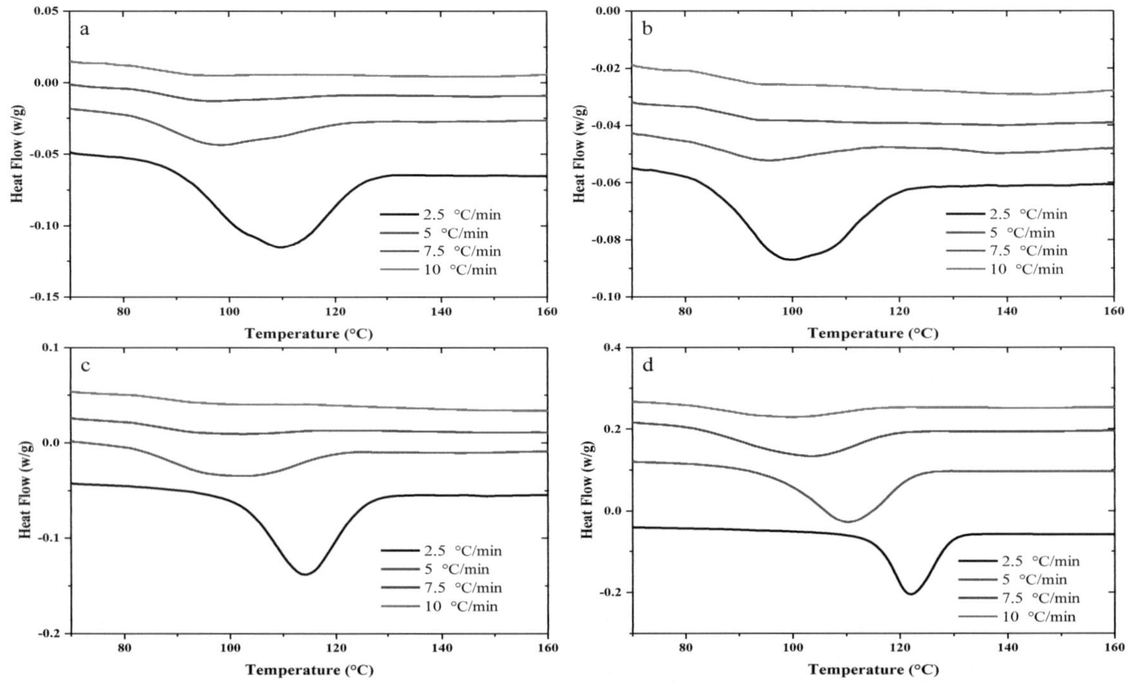

FIGURE 1. DSC curves for non-isothermal crystallization at different cooling rates of a) PLLA0R, b) PLLA1R, c) PLLA0.5R, d) PLLA0.25R

The general crystallization rate is also calculated from the following equation according to variability of the cooling rate:

$$\log Z_c = \frac{\log Z_t}{a} \quad (2)$$

In this case, Z_t is obtained by calculating the intercept of the Avrami curve. The results obtained from the Avrami equation are given in Table 1. As in the case of isothermal crystallization (data not shown here), the value of the total crystallinity for the PLLA1R sample was the highest value and for the PLLA0.25R sample showed the lowest value. However, unlike isothermal crystallization, different n values were observed in different samples. Although the Avrami equation does not have a physical concept for non-isothermal crystallization, the results can provide a very good comparison of material parameter changes. Data in Table 1 resulted in heterogeneous nucleation in three dimensions. The highest n value was obtained for the PLLA1R sample. An interesting phenomenon observed in the curve is the existence of two regions with very different variations. These two regions are referred to different stages of crystallization, which were observed in the PLLA1R sample. The results for two regions are reported in Table 1 which shows the different kinetic of crystallization in the two regions.

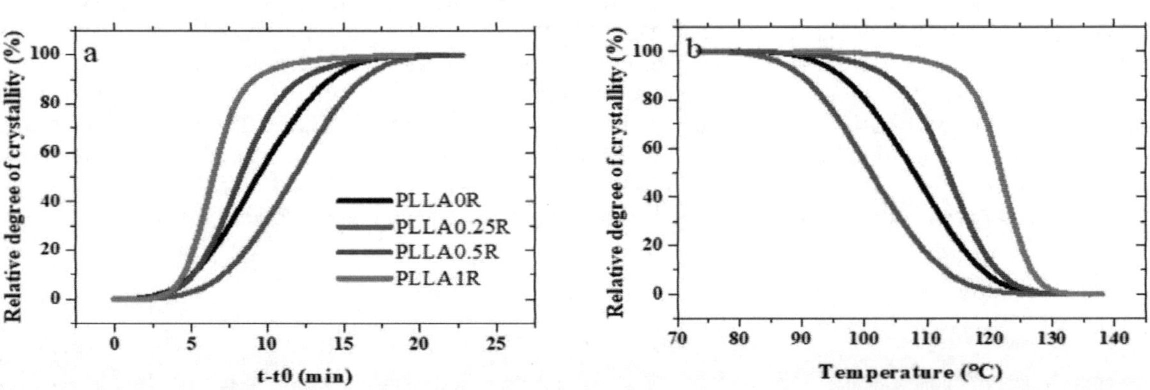

FIGURE 2. Relative crystallinity curves versus a) relative time & b) absolute temperature at cooling rate of 2.5 °C/min.

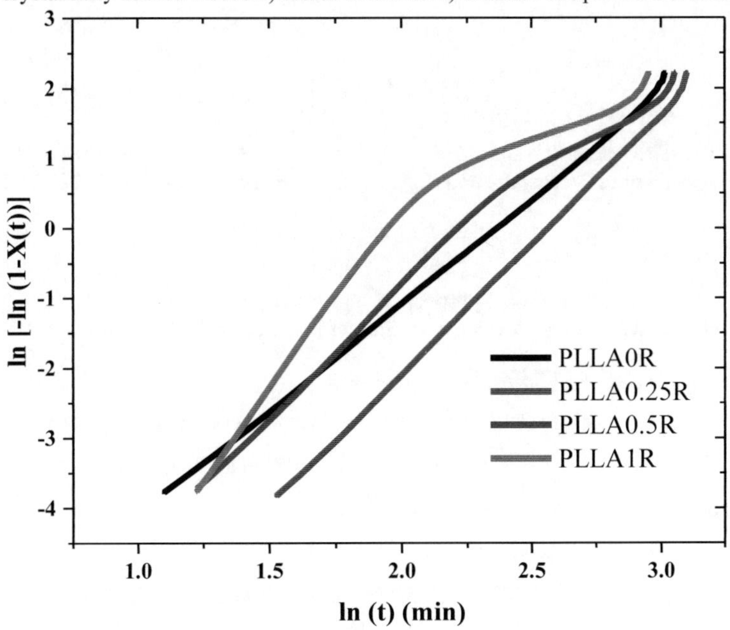

FIGURE 3. Non-isothermal Avrami plot at cooling rate of 2.5 °C/min.

TABLE 1 Non-isothermal crystallization parameters obtained and Avrami equation constant

Sample	Cooling rate (°C/min)	T_c (°C)	ΔH_c (J/g)	n_1	n_2	$k_1 \times 10^3$ (min^{-1})	$k_2 \times 10^3$ (min^{-1})
PLLA0R	2.5	110	31.1	3.0	-	388	-
	5	98	6.0	3.0	-	698	-
PLLA0.25R	2.5	101	16.0	3.7	-	358	-
	5	96	1.4	2.9	-	638	-
PLLA0.5R	2.5	114	32.5	3.5	-	410	-
	5	101	8.7	2.8	-	715	-
PLLA1R	2.5	122	36.1	4.7	1.5	449	500
	5	110	28.1	3.7	2.8	723	725

CONCLUSIONS

The current study investigated the influence of chain extender on non-isothermal crystallization of PLLA. The results indicated that higher crystallization temperature was calculated for the PLLA with using 1wt. % of Joncryl, which showed faster crystallization kinetics. Using the 0.25 wt. % of Joncryl in chain extension reaction resulted in lower crystallization temperature due to the presence of long chains. The relative degree of crystallinity for the PLLA1R sample was the highest, and the lowest was obtained for the PLLA0.25R sample. The results of Avrami showed that n value has been changed significantly in the range of 2.8 to 4.7. The overall crystallization rate (k) for all samples increased with increasing the cooling rate and in the PLLA0.25R sample had the lowest value. This parameter was the lowest value for the PLLA0.25R sample which is related to the mobility of the chains of the strings (the same as the chains formed in the body of the Joncryl molecule), as well as the role of entanglements in the nucleation process.

REFERENC9ES

1. O. Yousefzade, L. Franco, and M. Nami, Composite Materials (2019)0021998318822720.
2. O. Yousefzade, J. Puiggalí, H. Garmabi, and R. Macovez, Polymer Physics, **57.4**, 222-233 (2018).
3. Y. Wang, C. Fu, Y. Luo, C. Ruan, Y. Zhang, Y. Fu, Journal of Wuhan University of Technology-Mater. Sci. Ed., **25**, 774-779(2010).
4. R. Al-Itry, K. Lamnawar, A. Maazouz, European Polymer Journal, **58**, 90-102 (2014).
5. J.F. Mano, J.L. Gómez Ribelles, N.M. Alves, M. Salmerón Sanchez, Polymer, **46**, 8258-8265, (2005).
6. A. Magoń, M. Pyda, **50**, 3967-3973(2009).
7. S. Saeidlou, M.A. Huneault, H. Li, C.B. Park, Progress in Polymer Science, 37, 1657-1677(2012).
8. R.M. Michell, A.J. Müller, Progress in Polymer Science, **54-55**, 183-213(2016).
9. M.L. Di Lorenzo, C. Silvestre, **24**, 917-950(1999).
10. M. Avrami, Granulation, **9**, 177-184 (1941).
11. F. De Santis, R. Pantani, G. Titomanlio, Thermochimica Acta, **522** ,128-134(2011).
12. R. Legras, J.P. Mercier, E. Nield, Nature, **304**, 432-434(1983).
13. J.J. Kolstad, Journal of Applied Polymer Science, **62**, 1079-1091(1996).
14. X. Wang, H. Zhou, B. Liu, Z. Du, H. Li, Advances in Polymer Technology, **33** (2014).
15. R. Al-Itry, K. Lamnawar, A. Maazou, Polymer Degradation and Stability, **97**, 1898-1914(2012).
16. B.G. Risch, G.L. Wilkes, J.M. Warakomski, Polymer, **34**, 2330-2343(1993).
17. A. Jeziorny, Polymer, **19**, 1142-1144(1978).

Impact of Spacer on Rheological Properties of Polyoxyethylene Cationic Surfactant-Polymer Solution

Muhammad Shahzad Kamal, S. M. Shakil Hussain, Lionel T. Fogang, Ahmad Mahboob, and Sarmad Zafar

Center for Integrative Petroleum Research, King Fahd University of Petroleum & Minerals,

Abstract. Enhanced oil recovery (EOR) include all those methods in which an external source of energy are used to recover additional oil after water flooding. Thermal EOR, gas EOR and chemical EOR are three major tertiary recovery techniques. In chemicals EOR, different chemicals such as surfactants, polymers, and alkalis are injected to alter the fluid-fluid and rock-fluid interactions which improve the oil recovery. Polymers are injected to improve the viscosity of the injected water. Surfactants are used to reduce the interfacial tension between water and oil. In a few cases, surfactants and polymers are co-injected to achieve high viscosity and low interfacial tension, simultaneously. In this work, we synthesized three polymeric cationic gemini surfactant containing different spacer length and rigidity. The rheological properties of these surfactants with the commercially available acrylamide-based copolymer was investigated using Discovery hybrid rheometer (DHR-3) at various temperatures. The viscosity and storage modulus reduced by increasing the concentration of the surfactant. It was also observed that the spacer length and rigidity affect the rheological properties of the surfactant-polymer solution. The addition of surfactant with a rigid spacer causes more reduction in the viscosity of the polymer. However, at high shear rates, the effect of added surfactant was negligible. For field applications, the rigidity of the spacer becomes less important as for as rheological properties are concerned due to high shear rate.

Keywords: Rheology, Surfactant, Polymer, Enhanced Oil Recovery,

INTRODUCTION

Enhanced oil recovery (EOR) is a technique to recover additional oil from a reservoir after primary and secondary injection. Chemical EOR, thermal EOR, and gas EOR are most popular EOR methods [1-3]. In thermal EOR, hot water or steam is injected to reduce the viscosity of the oil that improves the oil recovery. In gas injection, certain gases such as nitrogen or CO_2 is injected to increase oil recovery. In chemical EOR, few chemicals such as surfactants, polymers and/or alkalis are injected to improve the oil/water mobility ratio and to enhance the recovery factor. Water-soluble polymers are used to increase the viscosity of the injection fluid. These water-soluble polymers improve the oil/water mobility ratio by increasing the viscosity of the water. It has been proved in some recent studies that these also microscopic efficiency due to elasticity in addition to improving the macroscopic displacement efficiency [4-6].

A number of the polymer have been tested in various fields for EOR [7]. Polyacrylamide is most common polymers among various water-soluble polymers. However, its application is limited to only certain reservoirs with less severe conditions. At high temperature, it can degrade and precipitate in the presence of salts. For high temperature, several modifications to polyacrylamide have been proposed by incorporating thermally stable monomers. Rheological properties are very important in screening the polymers for cEOR applications. Rheological properties can give an indication of polymer flow behavior in reservoir conditions.

Surfactants are injected to reduce the oil/water interfacial tension and/or to alter the rock wettability [8]. An ultra-low interfacial tension is required to improve the capillary number. A number of surfactants have been reported for EOR applications [9, 10]. Depending on the head group, the surfactant can be anionic, zwitterionic, nonionic and cationic. Most of the previous work has been conducted using anionic surfactants. However, for carbonate reservoirs, anionic surfactants are not a good choice due to the high adsorption rock surface. Cationic surfactants can be an attractive choice for carbonate reservoirs.

Surfactant and polymers can be injected together to get lower interfacial tension and better displacement efficiency simultaneously [11]. In this work, three cationic surfactants were synthesized with different spacer rigidity. The impact of surfactant concentration and spacer nature on the rheological properties of the polymer was evaluated.

EXPERIMENTAL

Water soluble cationic polymers manufactured by SNF FLOERGER were used and received as a powder. Three polyoxyethylene types cationic gemini surfactants were synthesized in the lab. These surfactants have a similar structure except the spacer. The surfactant 1 has saturated linear spacer, surfactant 2 has unsaturated spacer (double bond), and surfactant 3 has a ring in the spacer. All the rheological measurements (steady shear and dynamic) were conducted using DHR-3 instrument provided by TA Instruments. A concentric cylinder geometry with a vapor trap was used for all rheological measurements. The steady shear measurements were conducted using a shear rate range from 0.01/s to 1000/s. The storage modulus was determined over a frequency range from 0.1 rad/s to 100 rad/s.

RESULTS & DISCUSSION

Figure 1 shows the effect of surfactant 1 concentration on the steady shear viscosity of the polymer. The viscosity of the polymer decreases with an increase in the concentration of the surfactant. The reduction in the viscosity is associated with the charge screening effect. The polymer is a cationic polymer while all the surfactants used were also cationic in nature. In absence of surfactant, the higher viscosity of the polymer is associated with the charge repulsion on polymer backbone that keeps surfactant chains in stretched form. However, the addition of cationic surfactant also brings anionic counterions. The interactions of the surfactant counterions with the cationic polymer results in charge screening and chain coiling. This chain coiling reduces the viscosity of the polymer. The reduction in the viscosity on surfactant addition is only significant at the low shear rate. At the high shear rate, the viscosity of the polymer is not changing with the increase in surfactant addition. It is due to the fact that at a high shear rate, the effect of shear becomes more significant compared to the charge screening. The similar viscosity reduction was observed with the addition of surfactant 2 and surfactant 3 as shown in Figure 2 and Figure 3.

The comparison of three surfactants, when used at 0.1% concentration, is shown in Figure 4. It is clear from Figure 4 that the viscosity of the surfactant-polymer system with surfactant 1 is highest followed by surfactant 2 and surfactant 3. This shows that for these typical surfactant systems, the spacer has also impact on the rheology of the cationic polymers. At higher concentration (0.05 %), a similar observation was noted (Figure 5).

The elasticity of a material is also important as it contributes to improving the microscopic sweep efficiency. The storage modulus is a direct indication of elasticity of a material. The effect of surfactant 1 concentration on the storage modulus is shown in Figure 6. The addition of surfactant reduced the storage modulus of the polymer. The increase in surface concentration further reduced it. This reduction is also significant at a lower frequency. However, at high frequency the effect of added surfactant becomes negligible. At low frequencies, the reduction in storage modulus is associated with the charge screening. However, at high frequency the charge screening effect becomes negligible. The surfactant 2 and surfactant 3 also showed similar behavior (Figure 7 and Figure 8). The comparison of three surfactants suggested that surfactant 1 has the highest storage modulus followed by surfactant 2 and surfactant 3. In summary, the spacer of polyoxyethylene gemini surfactants affect the rheological properties of surfactant-polymer solutions.

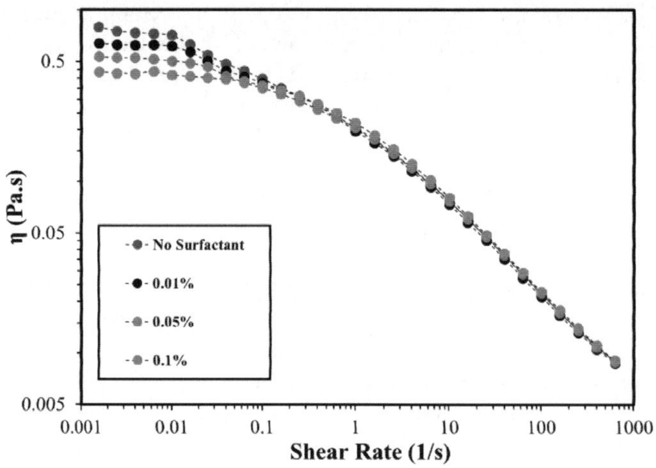

FIGURE 1. Effect of surfactant 1 concentration on polymer viscosity

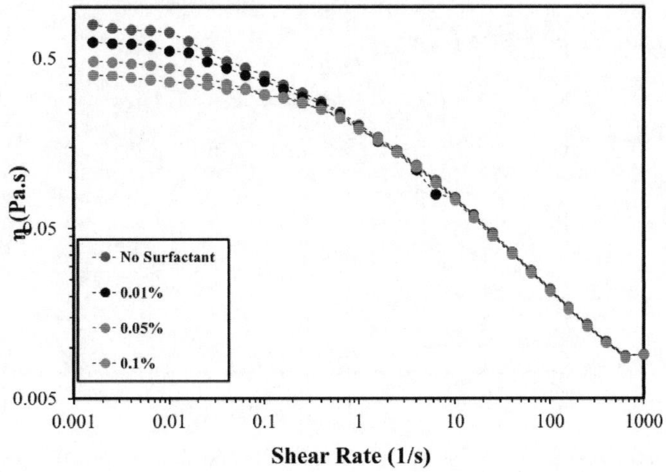

FIGURE 2. Effect of surfactant 2 concentration on polymer viscosity

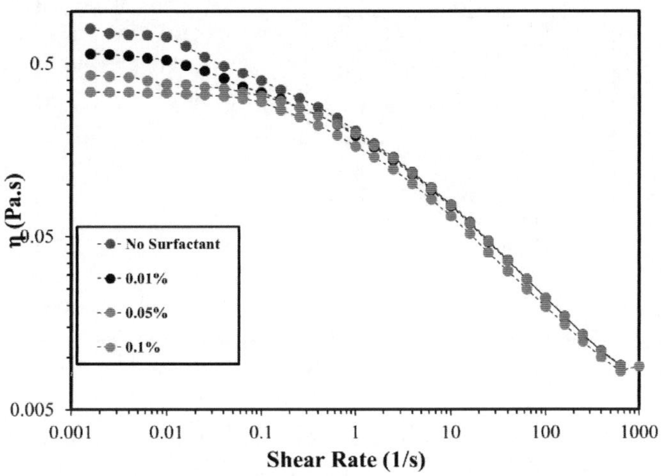

FIGURE 3. Effect of surfactant 3 concentration on polymer viscosity

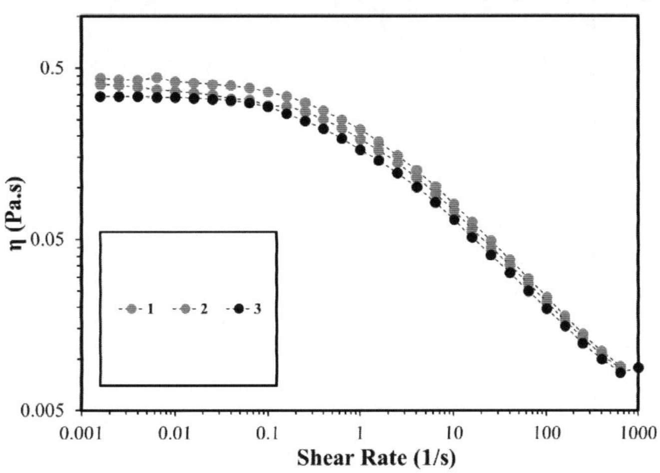

FIGURE 4. Surfactant comparison at a fixed concentration of 0.1%

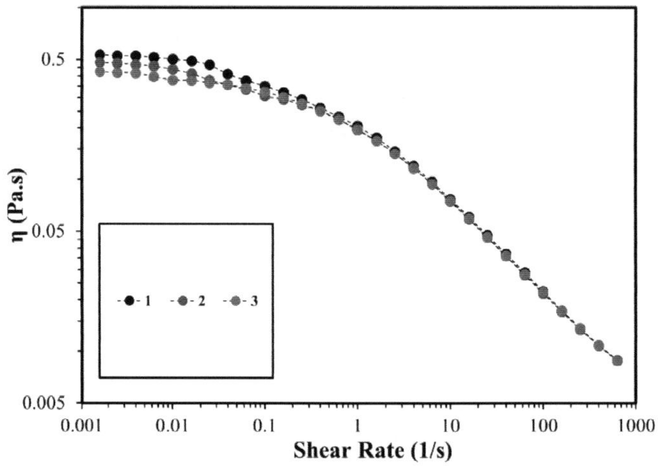

FIGURE 5. Surfactant comparison at a fixed concentration of 0.05%

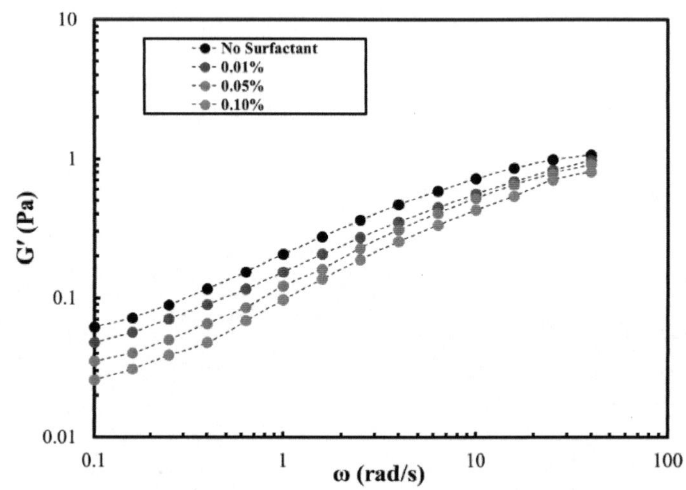

FIGURE 6. Effect of surfactant 1 concentration on the storage modulus of the polymer

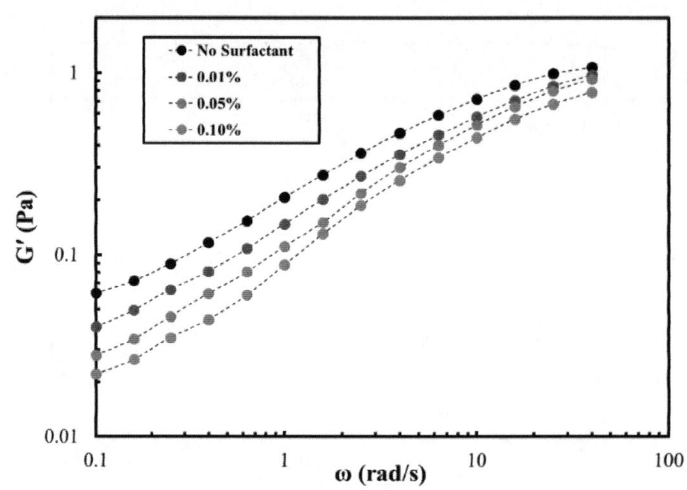

FIGURE 7. Effect of surfactant 2 concentration on the storage modulus of the polymer

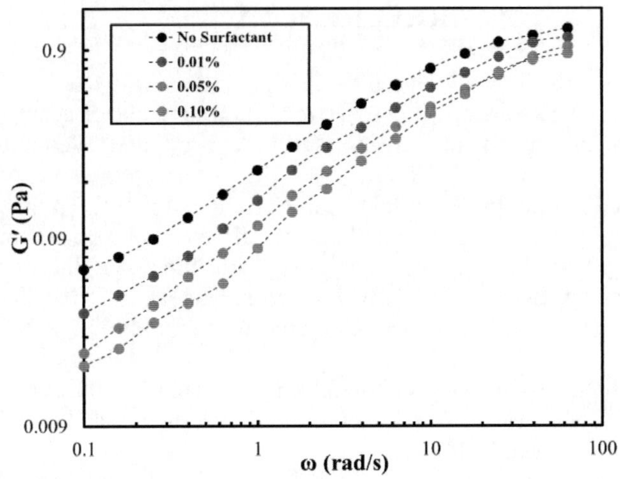

FIGURE 8. Effect of surfactant 3 concentration on the storage modulus of the polymer

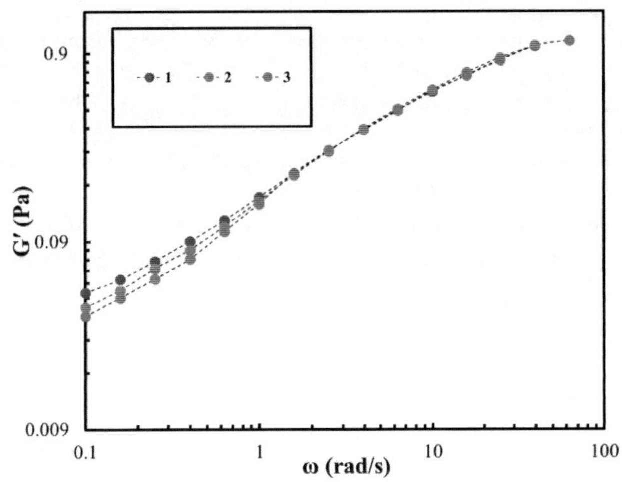

FIGURE 9. Frequency sweep comparison at 0.01%

CONCLUSIONS

In this work, we synthesized three polymeric cationic gemini surfactant containing different spacer length and rigidity. The rheological properties of these surfactants with the commercially available acrylamide-based copolymer was investigated using Discovery hybrid rheometer (DHR-3) at various temperatures. The viscosity and storage modulus reduced by increasing the concentration of the surfactant. It was also observed that the spacer length and rigidity affect the rheological properties of the surfactant-polymer solution. The addition of surfactant with a rigid spacer causes more reduction in the viscosity of the polymer. However, at high shear rates, the effect of added surfactant was negligible. For field applications, the rigidity of the spacer becomes less important as for as rheological properties are concerned due to high shear rate.

ACKNOWLEDGMENTS

Authors would like to thank College of Petroleum Engineering & Geoscience for providing financial support through a joint collaborative project with UT Austin.

REFERENCES

[1] M.S. Kamal, A.A. Adewunmi, A.S. Sultan, M.F. Al-Hamad, U. Mehmood, Recent Advances in Nanoparticles Enhanced Oil Recovery: Rheology, Interfacial Tension, Oil Recovery, and Wettability Alteration, Journal of Nanomaterials, 2017 (2017).

[2] S. Al-Mutairi, S. Kokal, EOR Potential in the Middle East: Current and Future Trends, in.

[3] A. Fuseni, M. Han, A. Al-Mobith, Phase Behavior and Interfacial Tension Properties of an Amphoteric Surfactant for EOR Application, in: SPE Saudi Arabia Section Technical Symposium and Exhibition, 2013.

[4] H. Xia, D. Wang, J. Wu, F. Kong, Elasticity of HPAM solutions increases displacement efficiency under mixed wettability conditions, in: SPE Asia Pacific Oil and Gas Conference and Exhibition, Soceity of Petroleum Engineers, Australia, 2004.

[5] H. Xia, D. Wang, W. Wu, H. Jiang, Effect of the visco-elasticity of displacing fluids on the relationship of capillary number and displacement efficiency in weak oil-wet cores, in: Asia Pacific Oil and Gas Conference and Exhibition, Society of Petroleum Engineers, Indonesia, 2007.

[6] Z. Zhang, J. Li, J. Zhou, Microscopic Roles of "Viscoelasticity" in HPMA polymer flooding for EOR, Transport in porous media, 86 (2011) 199-214.

[7] M.S. Kamal, A.S. Sultan, U.A. Al-Mubaiyedh, I.A. Hussein, Review on Polymer Flooding: Rheology, Adsorption, Stability, and Field Applications of Various Polymer Systems, Polymer Reviews, 55 (2015) 491-530.

[8] S.S. Hussain, M.S. Kamal, Effect of large spacer on surface activity, thermal, and rheological properties of novel amido-amine cationic gemini surfactants, Journal of Molecular Liquids, 242 (2017) 1131-1137.

[9] M.S. Kamal, A Review of Gemini Surfactants: Potential Application in Enhanced Oil Recovery, Journal of Surfactants and Detergents, 19 (2016) 223-236.

[10] M.S. Kamal, I.A. Hussein, A.S. Sultan, Review on Surfactant Flooding: Phase Behavior, Retention, IFT, and Field Applications, Energy & Fuels, 31 (2017) 7701-7720.

[11] I.A. Malik, U.A. Al-Mubaiyedh, A.S. Sultan, M.S. Kamal, I.A. Hussein, Rheological and thermal properties of novel surfactant-polymer systems for EOR applications, The Canadian Journal of Chemical Engineering, 94 (2016) 1693-1699.

Advanced robust rheological characterisation for rigid PVC extrusion

F. Desplentere[a], M. Vandaele[a], S. Deceur[a]

[a]*Materials Technology TC, KU Leuven Bruges Campus*
Spoorwegstraat 12, 8200 Bruges, Belgium
Corresponding author: frederik.desplentere@kuleuven.be

Abstract. As polymer processing and its applications are more and more demanding, the request for accurate numerical simulations is highly present. Accurate predictions using numerical simulation tools are inevitably linked to accurate material data. In the case of rigid PVC processing, one has to deal with two extra difficulties, namely the high sensitivity for slip at the wall during processing and the highly thermal sensitivity of the material. To cope with the high temperature sensitivity, the company ARTechnologies developed the so called Rheoart® using a rotating rod to lower the heating time for the polymer to be measured. Using this equipment using different settings correlating well to real processing conditions, allowed to define an appropriate measurement procedure for rigid PVC. Using this procedure allowed to generate relevant data necessary to extract shear viscosity as function of shear rate applicable for the extrusion process together with the slip at wall as function of the occurring shear stress at wall. Using the appropriate models for the shear viscosity and slip at wall, the results of three commercial software packages are compared which are able to cope with the slip at wall. The results are also compared with the experimental data obtained from the material characterization together with the theoretical result generated with an in house developed software tool for 2,5D modelling of extrusion profiles taking into account slip at wall.

Keywords: Rigid PVC, rheological characterization, numerical simulations for extrusion.

INTRODUCTION

Within extrusion of polymers, the products are more and more challenging in the sense that they contain more and more innovative features (resulting in different wall thicknesses) combined with challenging materials. Those materials are highly filled in some cases influencing the rheological behavior. On one hand, the shear viscosity behavior is influenced, on the other hand, due to the high filler content, the slip at wall behavior is also influenced. To support the development of the extrusion dies for plates and profiles, more often appropriate numerical simulation tools are used to predict the processing behavior. The quality of the numerical simulation results is highly depending on the input data for the material data. For the extrusion process, the shear viscosity behavior is crucial information to predict the processing behavior. If one concentrates on the extrusion of rigid PVC which is typically provided in powder form, one has to deal with high thermal sensitivity. Nowadays, also more and more materials are recycled. For this type of materials, the amount of thermal stabilization is also limited. From economical point of view, no extra additives will be applied if there is no strong request for it. During standard extrusion for these type of materials, they are only exposed to higher temperature for only short amount of time (some minutes). To provide accurate numerical simulation results, it is necessary to have representative rheological data. To cope with the large shear rate ranges valid within extrusion, typically a capillary rheometer is used with the main disadvantage of a larger measurement time together with the difficulty to handle powder materials. The RheoArt® equipment allows to cope with this difficulties. A picture and schematic drawing of this equipment is shown in Fig. 1. The equipment is promising but there are also some pitfalls in using this equipment. The main disadvantage is the uncertainty about the temperature of the melt to be measured. Some reference measurements are performed to compare the shear viscosity data with a standard capillary rheometer for a non thermal sensitive material resulting in promising default settings for a broad range of materials. Within this paper, first results on the rheological characterization for rigid PVC are illustrated together with the characterization for slip at wall. One has to be aware that the characterization for slip at wall is highly depending on the surface finish of the measurement geometry. The obtained data with this equipment is only indicating, to obtain the real slip behavior for a material in combination with a tool, similar surface finish is requested which is most probably different from the surface quality within a capillary which is used within a capillary rheometer [2].

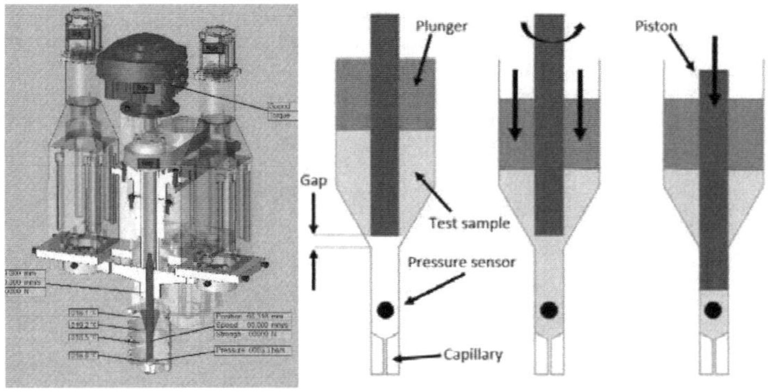

FIGURE 1. RheoArt® equipment and the different steps within a rheological measurement [3]

MEASUREMENT STRATEGY

Figure 1 illustrates the different steps to perform a measurement with the RheoArt®. The first schematic picture shows the starting position for the measurement. The operator has to load the test sample manually into the cavity below the plunger. During this loading, no compression is applied on to the material. The second schematic in Fig. 1 illustrates how the test sample is heated. The whole barrel is heated causing on one hand heating by thermal conductivity. On the other hand, the material is forced by the plunger through the gap indicated in Fig. 1. The speed of the plunger can be modified to adjust the applied shearing to the test sample. On top of the flow through a narrow gap, the piston is also rotating while the material is forced through the gap. Also the rotation speed of the piston can be modified. By this transportation through a narrow gap, there is also a controllable heating of the sample by shear. This combination of heating allows fast heating, which is in favor to avoid too long measurement times for thermal sensitive materials, however, the temperature of the test sample is influenced by the amount of shear which is applied. The image on the right on Fig. 1. illustrates the standard procedure for the measurement using a capillary of choice in the bottom of the equipment. During this measurement, the piston is moving down with a controlled speed. Within the RheoArt® software, different shear rates can be applied within one measurement. The shear viscosity data can be derived from the pressure sensor which is located at the entrance of the capillary. Applying the Bagley and Rabinowitsch corrections allows to define the shear viscosity based on different lengths for the same diameter of capillaries. If different diameters are used with the same L/d values, the slip at wall behavior can be defined using a Mooney analysis. In the labs of ProPoLiS in KU Leuven Bruges Campus, where this RheoArt® is installed, capillaries of 1mm diameter and a length of 0, 10 and 20mm are available. For 2mm, 0mm, 20mm and 40mm length are available. For 3mm, a length of 0mm and 30mm is available. The 3mm capillaries are mainly used for filled materials.

ROBUST MEASUREMENT AND CHARACTERISATION FOR SLIP AT WALL FOR RIGID PVC

Within the tool design for extrusion of rigid PVC, one has to deal with the high sensitivity for slip at wall for rigid PVC material. The slip behavior is of crucial importance to end up with a material distribution within the final profile as designed. At first, one has to define the temperatures which are relevant for the processing of the material of interest. Next, it is necessary to define the shear rate range which has to be characterized. Extrapolation of data in this field should be avoided, as the phenomena has some inherently stochastic nature. Within the extrusion process, one should also avoid shear stress levels beyond 300kPa. From this stress level on, the material is mechanically highly loaded and pressure levels are unacceptable within extrusion tools. The diameters of the capillaries used should also be appropriate for the material to be tested, in case of highly filled materials, a 1mm die is not recommended.

The data presented below is obtained for a standard rigid PVC used for window profiles. For each of the measurements, three measurements are performed, the repeatability is high due to the exclusion of the manual loading using the RheoArt® equipment. The coefficient of variation is typically below 3% which is acceptable within rheological measurements of this nature. All measurements are performed at a barrel temperature of 195°C which is a typical processing temperature for rigid PVC. The measurements were performed at shear rates 20, 50, 100, 200, 300, 400 and 500 1/s. For both capillary diameters (1 and 2mm), three L/D values are measured, namely, 0, 10 and

20. By performing Bagley plots for both capillary diameters, one can plot the shear stresses as function of shear rates (see Fig. 2 on the left). By applying the Mooney analysis, the results shown in Fig. 2 on the right are obtained. Figure 3 left shows the obtained shear viscosity data for the characterized rigid PVC data. On this graph, also the real shear viscosity is displayed if one neglect slip at wall velocities. As expected, to end up with similar pressure values if one neglect the slip at wall, the overall viscosity level is lower and the obtained slope is also indicating higher shear thinning behavior. On Fig. 3 on the right, one can see the slip velocity as function of the occurring shear stresses. A high power law exponent is noticed indicating a high sensitivity for high shear stresses.

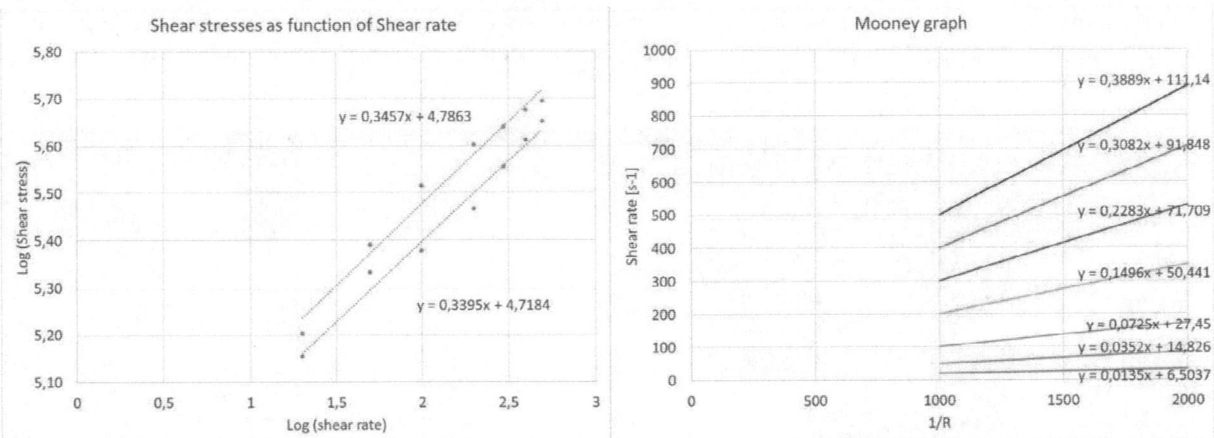

FIGURE 2. (left) Shear stresses as function of apparent shear rate (upper curve is for the 2mm capillary, the lower one is for the 1mm capillary. Results of the Mooney analysis for different shear stress levels results in the figure on the right.

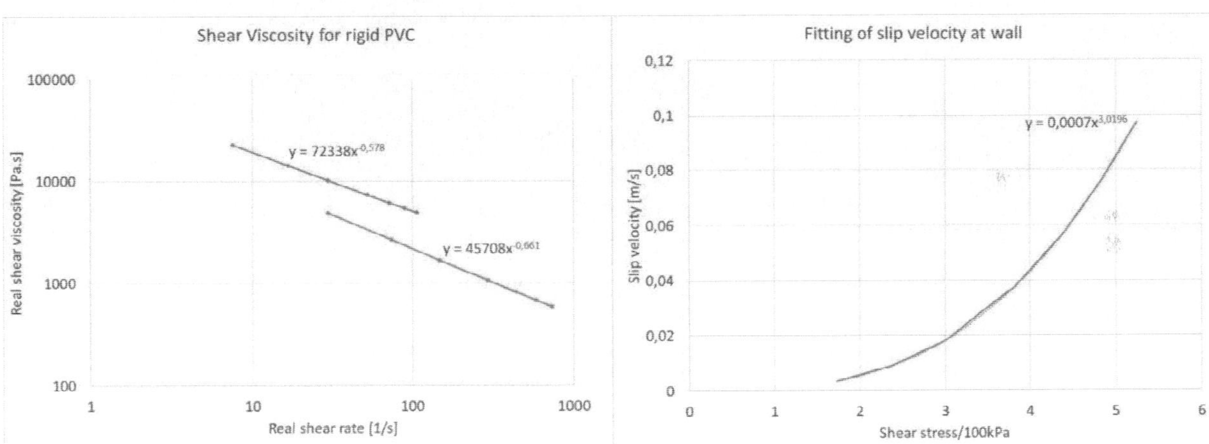

FIGURE 3. (Left) Real shear rate viscosity taking into account slip at wall (upper curve) and with neglecting slip at wall (lower curve) and Slip as function of shear stress level (right).

REVERSE ENGINEERING APPROACH TO FIND REAL SLIP BEHAVIOR

Within commercial numerical simulation tools, a lot of them already are capable to cope with slip at wall behavior [4,5], however, in some cases, they are not really well suited to cope with large slopes in the slip behavior causing non stable or non-converging behavior. To avoid this, a small tool named Profile Balancing has been developed within the ProPoLiS research group to solve for simple geometries taking into account the slip behavior neglecting temperature changes. For the shear viscosity description, a power law description is used combined with an Arhenius temperature dependency [6]. For the slip at wall behavior, a power law model is used based on the shear stress at the wall [6] . Those functions are also applied on Fig. 3. The FEM 3D tool of Virtual Extrusion laboratory 6.9 has been used to generate first numerical simulation results to calculate with the characterized data. The velocities occurring in the mold are shown in Fig 4. Within this software it is also possible to calculate the pressure drop taking into the slip at wall behavior. To perform the reverse engineering based on available experimental data (pressure as function of flow rate and temperature of the melt measured at the die entrance), the pressure drop on the entrance of the die have been estimated using this initial simulation. For this particular mold, the pressure drop is 22% of the total pressure value. In that way, the measured pressures can be reduced to be fed into the Profile Balancing tool to look for the best fit shear viscosity behavior and slip velocity description. Performing this analysis, results in data which only deviates 10% in the worst case based on slightly adapted shear viscosity data and slip behavior compared to the data acquired using the Rheoart® equipment. An overview of the data (valid for other PVC data due to confidentiality) is shown in Table 1.

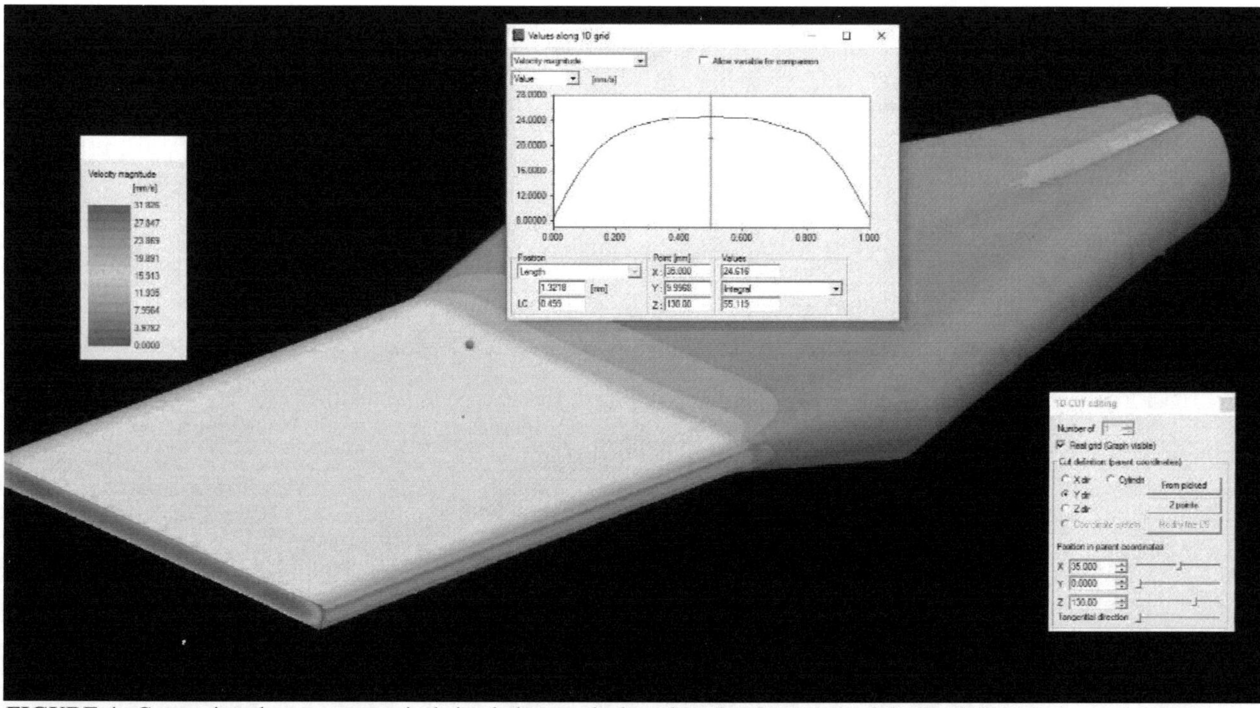

FIGURE 4. Comparison between numerical simulation results based on the characterized data and the experimental data

TABLE 1. Overview of numerical simulation results.

Situation	Flow rate [kg/hr]	Experimental pressure	Simulated pressure	Relative error [%]
2,65mm 177°C	6,7	16,9 MPa	17,3MPa	-2,2%
3,30mm 177°C	6,7	13,6 MPa	12,3 MPa	+9,5%
2,65mm 181°C	6,7	15,1 MPa	16,3MPa	-9,3%
3,3mm 181°C	6,7	12,0 MPa	11,6 MPa	3,3%
2,65mm 181°C	9,9	17,7 MPa	18,5 MPa	-4,5%
3,3mm 184°C	9,9	12,9 MPa	12,5 MPa	3,0%

CONCLUSIONS

This paper gives an scientific overview of the potential of the RheoArt® equipment together with numerical simulation tools. This combination allows to end up with acceptable fitting of experimental data within this highly sensitive domain within extrusion.

ACKNOWLEDGMENTS

The author also would like to thank the SIM funding system for providing funding to perform research on the (re) processing of PVC materials. This research has been performed within the Pocobuild project (SIM-VLAIO HBC.2017.0680)

REFERENCES

1. J.P. Villemaire, J.F. Agassant, *Apparent viscosity measurements using a capillary viscosimeter with preshearing*, Polymer Processing Engineering, 1, 221, (1983).
2. S. Apostol, J. Miethlinger, V. Putz, *Wall slippage of PLA and PVC,* PPS 2015 conference proceedings, Graz (2015)
3. http://www.rheoart.com/en/origine.html.
4. http://www.compuplast.com/product/vel-modules/profile-die.
5. http://www.compuplast.com/product/vel-modules/3d-fem.
6. http://www.compuplast.com/product/vel-modules/material-properties.

Rheological Behavior of Xanthan-based Hydrogels Crosslinked with bi-Functional/tetra-Functional Cross-linkers

Ghazaleh Azizi Saadatlou[a], Gholamreza Pircheraghi[a]

[a] *Polymeric Materials Research Group (PMRG), Materials Science & Engineering Department, Sharif University of Technology, Tehran, Iran*

Abstract. This work aims to study the rheological behavior of Xanthan-based hydrogels crosslinked using Citric Acid and Borax crosslinking agents at different temperatures to regulate the rheological properties and flowability of Xanthan-based hydrogels at various temperatures. The samples were synthesized via common film casting method, followed by heating at desired temperature to ensure the crosslinks formation. Rheological behavior of the samples have been studied in 25 (room temperature), 37 (human body temperature) and 60 (high temperature) degrees Celsius using oscillatory shear rheometer. The results have shown that the storage and loss moduli of the samples were increased with temperature, which has been attributed to conformational reformation of Xanthan chains at higher temperatures. The increase in storage and loss moduli for crosslinked hydrogels was lower, compared to the samples with no crosslinking agents. This behavior was attributed to the suppressed conformational changes of Xanthan chains due to the formation of crosslink bonds. The crosslinking was further approved by FTIR analysis and the results, alongside with the accepted mechanism for crosslinking of Xanthan with Citric Acid, were used to propose a mechanism for the crosslinking of Xanthan in the presence of Borax. Also, the results have shown that the blocking effect of Borax on Xanthan chains conformational change is superior to that of Xanthan – Citric Acid hydrogels, indicating stronger interactions of Borax with functional groups of Xanthan chains.

Keywords: Xanthan Hydrogel, Citric Acid, Borax, Rheology Behavior
PACS:

INTRODUCTION

Xanthan gum, a faint yellow powder, is a natural occurring, water soluble polysaccharide produced predominantly by Xanthomonas Campestris in aerobic conditions [1]. The monomer of the anionic heteropolysaccharide consists of glucose, mannose, and glucuronic acid units with a molecular ratio of 2:2:1. Trisaccharide side chains attached to alternate glucose units on C-3 positions, are composed of a D-Glucuronic acid located between two D-mannose units. It is approved that the structure of Xanthan in distilled water at low temperatures is a partially ordered broken helix which demonstrates the rheological behavior of a weakly structured material [2]. A disorder-order transition has been observed in presence of salts which improves the gel-like behavior of the solution. Also an order-disorder conformational change has been observed at high temperatures which turns the gel-like behavior to Newtonian behavior [2]. It is worth mentioning that Xanthan with ordered conformation demonstrates gel-like (non-Newtonian) behavior while disordered conformation facilitates Newtonian flow. The shear rate for Newtonian to gel-like behavior transition seems to be independent of the concentration [2].

The molecular weight of Xanthan gum has been reported between $4-12\times10^{6}\ g/mol$ in the literature. An accurate amount for M_w cannot be measured due to very high molecular weight, the stiffness of the molecule and the presence of aggregates [3].

Rheological behavior of Xanthan solutions depend on various factors including concentration of the biopolymer, pH and temperature [1]. At high concentrations (around 1 wt.%) Xanthan holds an ordered conformation while at lower concentrations a disordered conformation is induced [4].

Although dilute solutions of Xanthan have extensively been studied in the literature, to best of our knowledge, there are not comprehensive studies on concentrated solutions and thus, current research aims to regulate the rheological behavior of 4 wt.% Xanthan solutions.

Reddy et. Al. (1997) suggested a mechanism for crosslinking of Xanthan via Citric Acid. This mechanism has been illustrated in figure 1 [8] and has been the basis of our hypothesis on crosslinking Xanthan via Borax, as it will be discussed later.

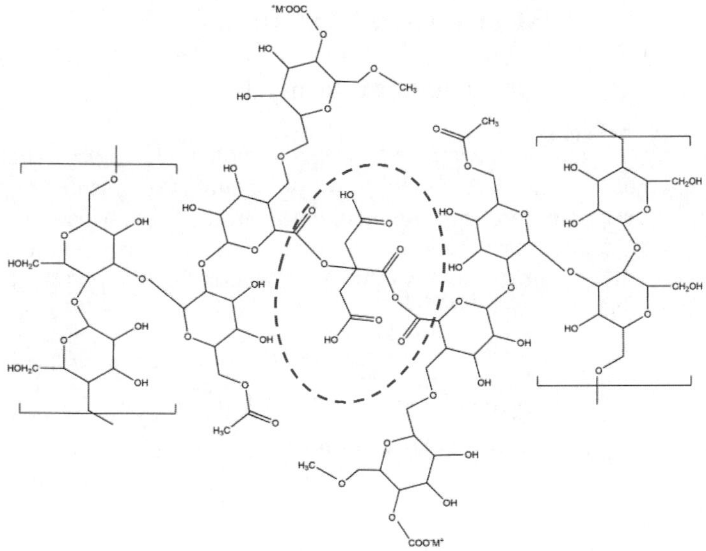

FIGURE 1. Schematic illustration of crosslinking Xanthan with Citric Acid; suggested by [8]

EXPERIMENTAL

Materials

Xanthan gum, citric acid (powder) and borax (powder) was purchased from Sigma Aldrich. All materials were analytical grade and used with no further process.

Hydrogel Formation

Three different xanthan films were synthesized via film casting method. The matrix consisted of 4wt% Xanthan. For films with Borax and Citric Acid the concentration of Xanthan remains the same while 2 wt% Borax/Citric Acid was added to the matrix. The solutions were mechanically stirred until the gel-like material is obtained. The time of stirring increased with the increased concentration of citric acid and borax due to acidic environment. The samples were heated at 165°C for 7-10 minutes prior to casting to ensure gel formation. The hydrogels were then cooled in room temperature in petri dishes and kept for further use. The samples were coded as mentioned in table 1.

Table 1. Sample codes

Sample Code	%wt. Xanthan	%wt. Citric Acid	%wt. Borax
Xan	4	-	-
Xan-C2	4	2	-
Xan-B2	4	-	2

Rheology Analysis

The rheological behavior of the samples were studied by dynamic shear rheometer, Physica MCR 301. The analysis were conducted in 25°C (room temperature), 37°C (human body temperature) and 60°C (for high temperature study).

FTIR Analysis

FTIR analysis was conducted to study the crosslinking of the samples and the formation of bonds.

RESULTS & DISCUSSION

Rheology Analysis

Figure 2 shows the rheological behavior of Xanthan hydrogel in room temperature in the absence of cross-linkers. The higher amount of storage modulus (G') in comparison to loss modulus (G'') is due to dominant elastic response of the hydrogel. In this temperature range Xanthan adopts a partially ordered conformation due to chain entanglements and hydrogen bonds [6, 7]. Additionally, the decrease in viscosity with increasing frequency is an indicator of shear thinning behavior in the system. Rheological analysis in higher temperatures showed the exact same trend and thus, is not worth mentioning in this context.

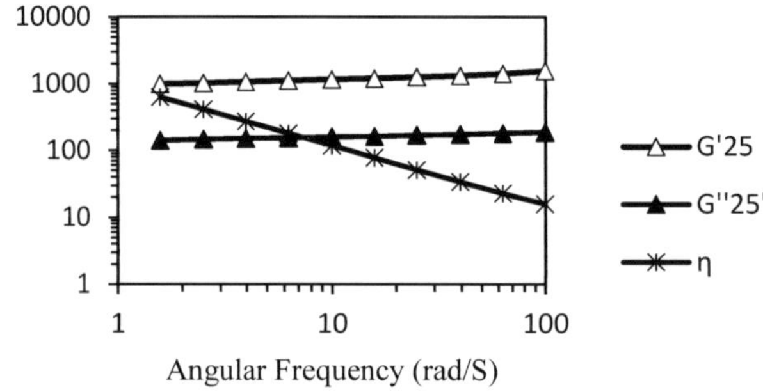

FIGURE 2. Rheological behavior of Xanthan hydrogel in 25°C.

As seen in figure 2 the amount of storage modulus for native Xanthan solution in 25°C is around 1000 Pa and almost independent of shear rate. It has been noted that the partially ordered conformation is dominant in high concentration Xanthan solutions which is due to hydrogen bonds and polymer entanglements. It is known that Borax hydrolysis occurs in aqueous solutions providing an alkaline environment which decreases the cross-links density. This could result in a reduction in G' amount in the presence of Borax.

Increasing temperature has a significant effect on the samples in the presence of Borax; It is known that Xanthan goes through an order - disorder conformational change at high temperatures. According to Morris [8], the ordered conformation consists of a single helix stabilized by ordered packing of side chains while the disordered conformation consists of a coaxial double helix which doubles the molecular weight and mass per unit length. The order – disorder transition temperature range could differ due to various factors one of which is the presence of crosslinks. Increasing temperature accelerates the hydrolysis of Borax and thus provides more alkaline condition which reduces the entanglements and crosslinks density of the polymer in the presence of Borax. This could facilitate the single – double helix transition and doubling of the molecular weight.

Figure 3 shows the storage modulus values for Xanthan samples with 2wt% Citric Acid/Borax. At 25°C the values for all samples are equal indicating that no transitions occur at this situation. This could be attributed to the higher cross-links density due to higher amounts of agents which prevents any chain mobility. At 37°C a significant loss of storage modulus values is observed for Xan-B2 sample. This observation could be explained assuming that hydrolysis effect of Borax is stronger than higher crosslinks density and thus, more crosslinks are degraded which results in higher mobility of the chain and consequent decrease in storage modulus. Increasing temperature from 37°C to 60°C favors the order-disorder transition and doubling effect of Xanthan disordered conformation which recovers the 1000 Pa initial value of the storage modulus.

It is shown that presence of Citric Acid does not affect the values of storage modulus significantly. In other words, Citric Acid increases the order – disorder transition temperature and thus no change is seen in the values of storage modulus below 60°C.

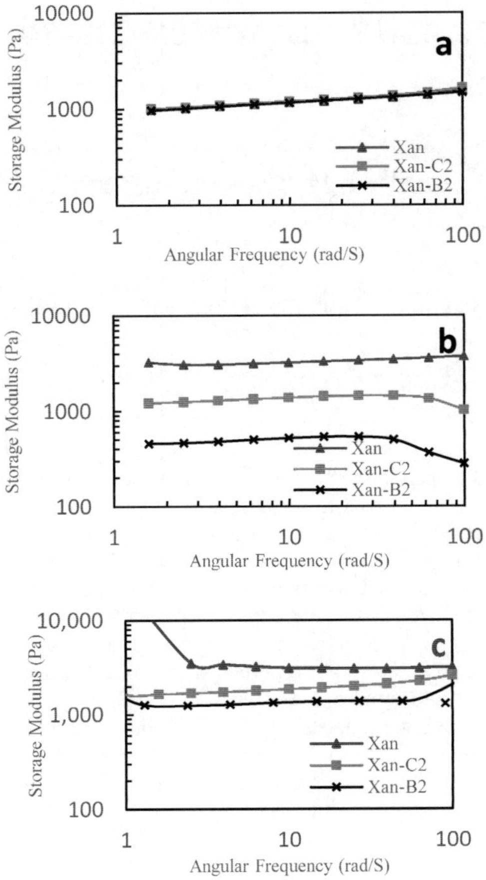

FIGURE 3. Comparing the storage modulus of Xan-C2 & Xan-2 in a) 25°C, b) 37°C and c)60°C

FTIR Analysis

Figures 4 demonstrate the FTIR analysis of the samples. For Xan-C2 sample, the 799 and 893 cm^{-1} peaks are thought to appear because of the formation of C-O-C bonds due to the formation of cross-links. Also, in FTIR analysis of the Xan-B2 sample, the 2261 and 2364 cm^{-1} peaks, represent the B-O-C bonds formed during crosslinking.

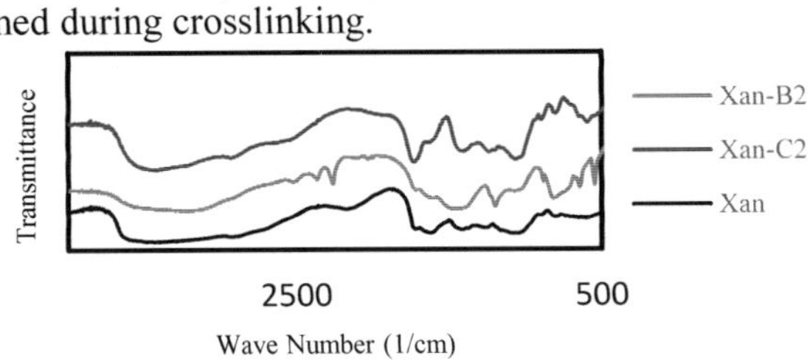

Figure 3- FTIR Analysis

CONCLUSION

It is known that the Xanthan chains go through conformational change in high temperatures. According to the rheological study, Xanthan hydrogels possess an increase in the storage and loss modulus with increasing temperature. This behavior has been attributed to the prevention of conformational change due to chain entanglements and crosslinks. It is shown that the prevention effect of Citric Acid on Xanthan chains is higher compared to Borax. Additionally, the FTIR results were used to suggest a mechanism for crosslinking of Xanthan in the presence of Borax. To better understand the rheological behavior of Xanthan hydrogels, it is suggested to consider more dilute solutions to eliminate the effect of physical entanglements.

REFERENCES

[1] V. S. J. C. E. G. s. F. Garcia-Ochoa, "Xanthan gum: Production, recovery, and properties," 2000.

[2] D. M.-A. M. J. L. P. F. M.-B. F. J. Reinoso, "Rheological characterisation of xanthan gum in brine solutions at high temperature," Carbohydrate Polymers, vol. 203, no. June 2018, pp. 103-109, 2019.

[3] D. V. L. D. P. B. Dr. Karin Born, "Xanthan," in Polysaccharides and Polyamides in the Food Industry, Wiley-VCH, 2005, pp. 259-291.

[4] L. M. M. R. J. C. Q. D. F. Aline F.Dário, "The effect of calcium salts on the viscosity and adsorption behavior of xanthan," Carbohydrate Polymers, vol. 84, no. 1, pp. 669-676, 2011.

[5] C. Q. W. X. &. K. I. Yang, "Ester cross-linking of cotton fabric by polymeric carboxylic acids and citric acid," Textile Research Journal, pp. 334-342, 1997.

[6] S. B. Ross-Murphy, "Structure–property relationships in food biopolymer gels and solutions," Journal of Rheology, vol. 39, no. 6, 1995.

[7] S. M. Willie E. Rochefort, "Rheology of Xanthan Gum: Salt, Temperature, and Strain Effects in Oscillatory and Steady Shear Experiments," Journal of Rheology, vol. 31, no. 4, 2000.

[8] E. R. Morris, "Ordered conformation of xanthan in solutions and "weak gels": Single helix, double helix e or both?," Food Hydrocolloids, vol. 86, no. June 2017, pp. 18-25, 2019.

[9] S. Dumitriu, Polysaccharides: Structural diversity and functional versatility, NewYork: Marcel Dekker, 2004.

[10] J. M. Bemmelen, Der Hydrogel und das kristallinische Hydrat des Kupferoxydes, Springer, 1907.

[11] R. B. H. C. F. S. P. Vania Blasques Bueno, "Synthesis and swelling behavior of xanthan-based hydrogels," Carbohydrate Polymers, pp. 1091-1099, 2012.

[12] M. P. M. H. Lulian Corneliu Alupei, "Superabsorbant hydrogels based on xanthan and PVA, The study of the swelling properties," European Polymer Journal, pp. 2313-2320, 2002.

[13] R. Z. W. X. Z. B. Y. Z. S. Z. Y. X. D. Y. Yongzhen Tao, "Rheological behavior and microstructure of release-controlled hydrogels based on xanthan gum crosslinked with Sodium trimetaphosphate," Food Hydrocolloids, pp. 923-933, 2015.

[14] M. P. L. P. a. D. L. C. Anca Bejenariu, "Effect of Concentraton, pH , and Temprature on Xanthan Conformation: A Preliminary Study Before Crosslinking," Revue Roumaine de Chimie, pp. 147-152, 2010.

Synthesis of novel polymer nanocomposite for water-based drilling fluids

Hafiz Mudaser Ahmad[a], Muhammad Shahzad Kamal[b], SM Shakil Hussain[b], Mamdouh Al-Harthi[a]

[a]*Department of Chemical Engineering, King Fahd University of Petroleum and Minerals, 31261 Dhahran, Saudi Arabia*

[b]*Center for Integrative Petroleum Research (CIPR), King Fahd University of Petroleum & Minerals, 31261 Dhahran, Saudi Arabia*

Abstract. Polymer nanocomposite has fascinated the strong interest due to its capitative properties and potential use for smart drilling fluids in the petroleum industry. Water-based drilling fluids are employed for drilling oil and gas wells to extract hydrocarbons from reservoirs. The formulation of drilling fluid includes various additives to control rheological and filtration properties. For this purpose, a novel associative hydrophobic polymer-based nanocomposite was synthesized using functionalized carbon nanotubes, acrylamide, maleic acid, and 2-Acrylamido-2-methylpropane sulfonic acid via solution polymerization technique. The chemical structure of novel polymer nanocomposite was characterized with Fourier transform spectroscopy to confirm the presence of covalent bond (ester linkage) between hydroxyl group from functionalized carbon nanotubes and a carboxylic group from malic acid. The rheological properties were evaluated by mixing an appropriate amount of polymer nanocomposite with base drilling fluid (modified drilling fluids) at 25°C. The steady shear rheology experiments showed that increasing the concentration polymer nanocomposite in the base drilling fluid resulted in a significant increase in viscosity, shear stress and gel strength of modified drilling fluid. The dynamic rheological experiments of polymer nanocomposite were also performed with two different concentration in the base drilling fluid to access the viscoelastic properties. The synthesized polymer nanocomposite can be used as one of the additives in water-based drilling fluids to modify rheological.

Keywords: Polymer Nanocomposite, Drilling Fluids, Rheology, Polymer Synthesis
PACS: Replace this text with PACS numbers; choose from this list: http://www.aip.org/pacs/index.html

INTRODUCTION

Drilling fluids have a variety of applications in the drilling industry for the exploration of oil and gas wells. The main uses of drilling fluids include well cleaning, prevent the dispersion of drilled cuttings, maintain the wellbore stability, prevent the bit balling and lubricating the drilling assembly during the drilling operations [1-3]. The water-based drilling fluids contain a variety of drilling fluid additives that maintain the rheological and filtration properties of drilling fluids. Conventionally available drilling fluids mainly contain bentonite as one of the main constituents of drilling fluid additives. Bentonite provides superior rheological and filtration properties upon swelling in the drilling fluid. But the use of a high quantity of bentonite in drilling fluids sometimes causes severe issues during drilling operations such drill bit sticking, high pumping power, and a low penetration rate of drilling bit. On the other hand, low quantity of bentonite resulted in low rheological properties that are insufficient to maintain the properties of the fluid in the wellbore. The better alternative of bentonite in the drilling fluid is to use water-soluble polymers to achieve optimum rheological properties. Various natural and synthetic polymers such as cellulose, carboxymethylcellulose (CMC) and hydroxypropyl starch and lignosulfonate hydrolyzed polyacrylamide[2, 4].

In a recent study, the impact of different polymer structures on drilling fluid rheological and filtration properties was discussed. The two different polymers such as a copolymer of acrylamide and 2-acrylamido-2-methylpropane sulfonic acid and a terpolymer of acrylamide, 2-acrylamido-2-methylpropane sulfonic acid, and N-Vinylpyrrolidone was used to study the rheological and filtration properties. The results showed that terpolymer showed superior properties at high temperature due to the presence of temperature bearing ability monomers in the polymer chain[5]. In another study, the effect of different monomers types in the polymer chain and degree of hydrolysis of polymer molecules was studied to two different temperatures to evaluate the rheological and filtration properties of drilling fluid. The polymer with a high degree of hydrolysis and high bulky groups in the polymer chain showed high rheological properties[6].

Nanotechnology is widely employed in the oil and gas industry specifically for modifying drilling fluids. Recently, different types of nanoparticles are used in formulating of drilling fluid having superior rheological and filtration properties. Iron oxide nanoparticles were used to reduce the filtration and prevent the invasion of drilling fluid in the pores of shale by making a thin impermeable filter cake. Carbon nanotube along with high molecular weight polymer was used to improve the filtration and rheological properties[7, 8]. The main issue of using nanoparticles in the drilling fluid formulation is through dispersion and it can be solved by making polymer nanocomposite. In this article, in situ solution polymerization of nanoparticles and three different monomers were studies and it resulted in the complete dispersion of nanoparticles in the drilling fluid.

In this article, the synthesis of polymer nanocomposite was discussed as one the additive for water-based drilling fluid. The polymer nanocomposite was synthesized by solution polymerization technique and characterized by FTIR spectroscopy. The polymer nanocomposite solution was mixed for base drilling fluid to form a modified drilling fluid. The steady shear rheological properties such as shear stress, viscosity, and gel strength were determined. The effect of time on viscosity and dynamic rheological properties were measured and compared with the base drilling fluid.

EXPERIMENTAL

Materials

The monomers acrylamide, acrylic acid, and 2-acrylamide-2-methyl propane sulfonic acid were obtained from Sigma Aldrich. The initiator potassium persulfate and sodium hydroxide were obtained with 99% purity from Sigma Aldrich. The bentonite was obtained from BDH Chemicals (UK) in powder form having pH ranges from 9.5–10.5 and a density of 2.5 g/cm^3 at 20 °C. The carbon nanotubes functionalized with (-COOH) groups was obtained from cheap tubes Inc. with 99.9% purity and more than 3% functionalized content.

Synthesis of Polymer Nanocomposite

The polymer nanocomposite was synthesized using acrylamide (AM), acrylic acid (AA) and 2-acrylamide-2-methyl propane sulfonic acid (AMPS) monomers. The free radical solution polymerization technique was adopted to synthesize polymer nanocomposite. The stock solution was prepared using AMPS, AM, and AA with 6/5/4 molar ratio in 250 ml of deionized water. The 1 wt.% CNT's dispersion was mixed with monomer stock solution and stirred for 30 minutes in a three-neck round bottom flask. The whole apparatus was connected to nitrogen supply to carry the polymerization reaction under an inert atmosphere. The whole solution mixture was heated to 80°C in the oil bath and thoroughly mixed using magnetic stirrer. After achieving 80°C temperature, the 10 ml freshly prepared 1 wt.% solution of potassium persulfate was injected in the reaction mixture. The polymerization reaction was allowed to take place for 4 hours.

Preparation of Drilling Fluids

The base drilling fluid (A) was prepared by mixing 5% bentonite in deionized water and left in the beaker covered at room temperature for 24 hours. Before making modified drilling fluid, the solid form of polymer nanocomposite was dissolved in the deionized water to make to completely disperse solution. The polymer nanocomposite solution was obtained by mixing 15 g of solid polymer nanocomposite in 100 ml of deionized water under magnetic stirring. The modified drilling fluid (B) was obtained by mixing 90 ml of base drilling fluid and 10 ml of polymer nanocomposite solution. The modified drilling fluid (C) was obtained by mixing 85 ml of base drilling fluid and 15 ml of polymer nanocomposite solution. The pH of all the drilling fluids was adjusted to 10 by adding an appropriate amount of sodium hydroxide. All the fluids were thoroughly mixed using mechanical stirrer and left undisturbed for 24 hours before performing the rheological experiment.

Characterization and Rheological Properties

The polymer nanocomposite was characterized using Fourier Transform Infrared spectroscopy to determine the chemical composition and molecular structure. The rheological properties of base drilling fluid and modified drilling fluid was determined at 25°C. The steady shear rheological properties such as viscosity and shear stress were determined by increasing the shear the rate from 0.01 to 1000 (1/s). The gel strength of drilling fluid is an important rheological property that determined the cutting suspension capacity. Gel strength of base and modified drilling fluid was determined at 10 seconds and 10 minutes holding time according to the API standards. The dynamic rheological properties such as storage and loss modulus were also measured at 25°C of the base and modified drilling fluid. The effect of polymer nanocomposite concentration on rheological properties was measured and compared with the properties of base drilling fluid.

RESULTS AND DISCUSSION

The FTIR spectroscopy was performed for polymer nanocomposite to determine the composition and chemical structure of polymer nanocomposite. The absorption peak of polymer nanocomposite is shown in Figure 1. The absorption peak at 3460 cm-1 was attributed to the (H-N) stretching of the primary amide group in the acrylamide. The another (H-N) absorption peak was observed at 3126 which was due to the presence of secondary amide group in the AMPS. The (C-C) and (C-H) stretching peaks were observed at 2921 and 2862. The absorption peaks at 1047 and 1107 were attributed to the stretching of (S-O) and (S=O) bonds in the AMPS group. The main peak at 2358 was associated (O-H) of polymer chain stretch which is strongly bonded with (-COOH) group of carbon nanotubes. This peak confirms the presence of ester linkage between the polymer chain and carbon nanotube.

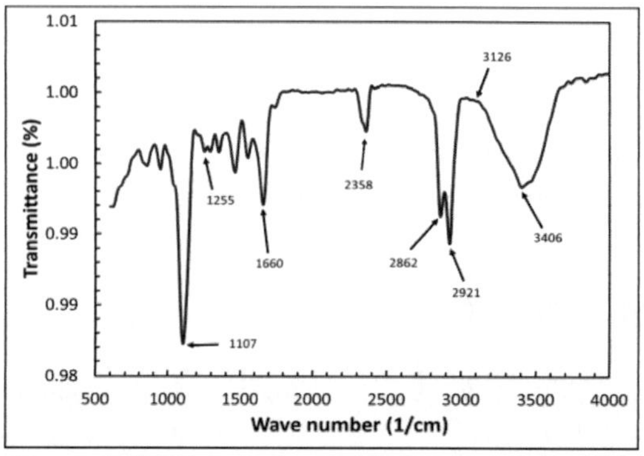

Figure 1. FTIR analysis of polymer nanocomposite

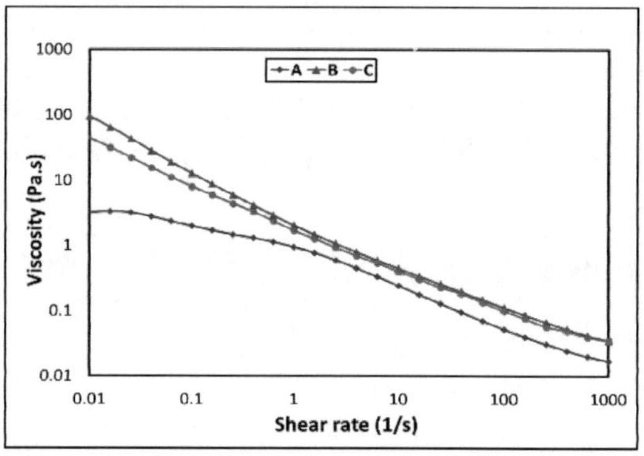

Figure 2. Viscosity vs shear rate of the base and modified drilling fluids with nanocomposite

The steady shear rheological properties of the base and modified drilling fluid were determined such as viscosity vs shear rate as shown in Figure 2. The viscosity of all the drilling fluids decreases with increasing shear rate depicting that all the fluid shows shear thinning behavior. At the low shear rate, the viscosity of base drilling fluid was very high due to the card house like structure at basic pH rang but the at a very high shear rate, the viscosity of base drilling was reduced to 0.001 (Pa.s) and ordered card house structure was destroyed. The addition of polymer nanocomposite resulted in an increase of viscosity compared to the base drilling fluid. The increase in viscosity was attributed to the adsorption of polymer nanocomposite on bentonite which gives three-dimensional structure and increases the viscosity of modified drilling fluid (B). In the case of modified drilling fluid (C), the viscosity was slightly reduced compared to the modified drilling fluid (B). This decrease in viscosity was due to the chain slipping effect with a high concentration of polymer nanocomposite.

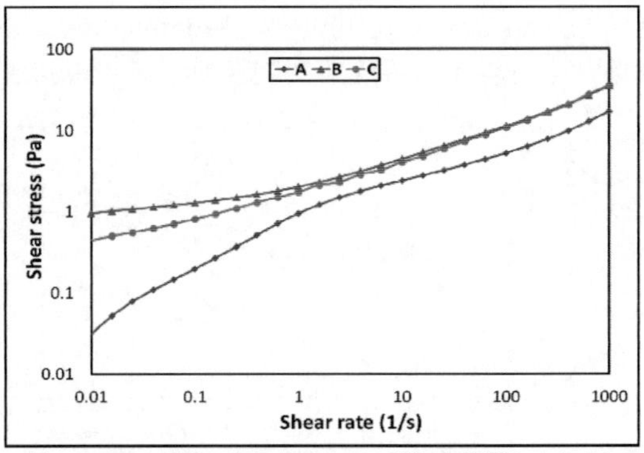

Figure 3. Shear stress vs shear rate of the base and modified drilling fluids with nanocomposite

The shear stress of base and modified drilling fluids is shown in Figure 3. At the low shear rate, the base drilling fluid showed a low value of shear stress while increasing the shear rate resulted in the increased shear stress. The addition of polymer nanocomposite resulted in an increase in the shear stress of modified drilling fluid (B). The change in shear stress was more prominent at a low shear rate compared to the change in shear stress at the high shear rate. The shear stress of modified drilling fluid (C) was slightly less at a low shear rate while at a high shear rate, the shear stress of modified drilling fluids (B) and (C) become similar.

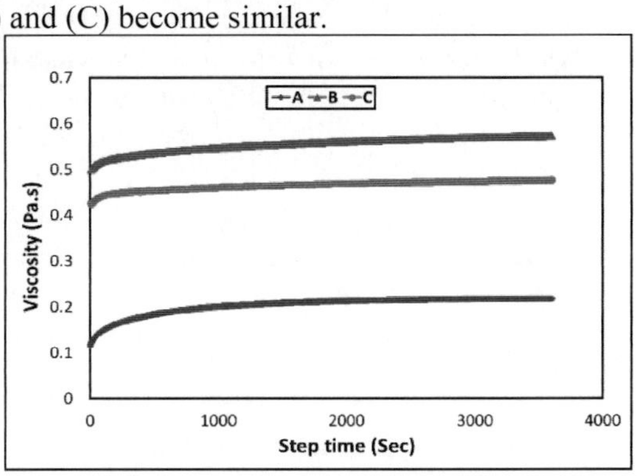

Figure 4. Viscosity vs time of the base and the modified drilling fluids with nanocomposite

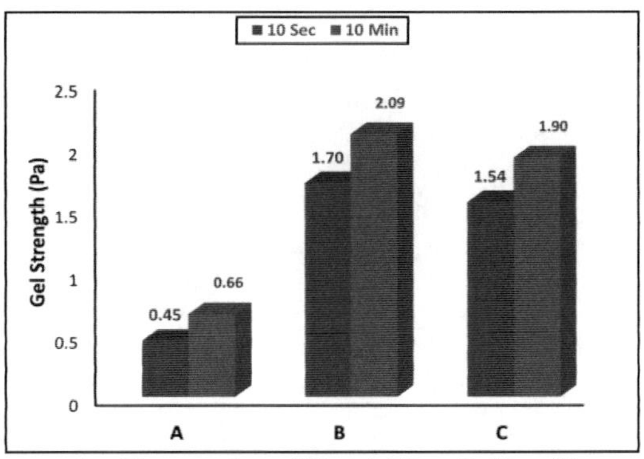

Figure 5. Gel strength of base and modified drilling fluids with nanocomposite

The effect of increasing time on the viscosity of drilling fluids is shown in Figure 4. The shear rate of all the experiments was kept constant at 10 (1/s) to study the effect of time on the viscosity of drilling fluids. The experiment was carried out for 60 minutes. It was observed that increasing time up to 500 seconds, increased the viscosity of base drilling fluid and after that, no prominent change in the viscosity was observed. Similarly, the viscosity of modified drilling fluids (B) and (C) was slightly increased in the starting time period of experiment and it remains constant throughout the end of viscosity vs time experiment. The viscosity vs time experiments confirmed the steady shear rheology experiment that the addition of polymer nanocomposite improves the rheological properties of modified drilling fluid. The gel strength experiment was performed at 25°C for the base and modified drilling fluids as shown in Figure 5. The 10-sec and 10- minutes gel strengths were determined for all the drilling fluid. The higher the gel strength of drilling fluid, the better the capacity of drilling fluid to hole the drilled cutting. The addition of polymer nanocomposite in the base fluid resulted in an increase in the gel strength compared to the base drilling fluid.

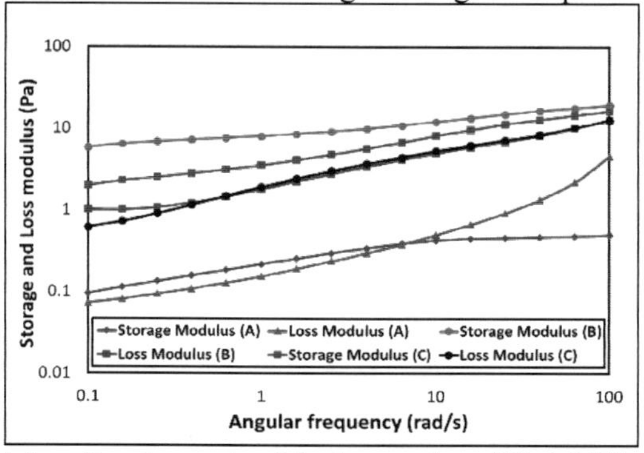

Figure 6. Storage and Loss Modulus of vs shear rate of the base and modified drilling fluids with nanocomposite

The storage and loss modulus of the base and modified drilling fluids is shown in Figure 6. The storage and loss modulus attribute the elastic and viscous contribution of drilling fluids. For base drilling fluid at a low angular frequency, the storage modulus was high than loss modulus. But at an angular frequency higher than 10, the loss modulus becomes higher than the storage modulus. The incorporation of polymer nanocomposite enhanced the storage and loss modulus of modified fluid (B). A higher angular frequency, the loss modulus of modified drilling fluid (B) and (C) are almost same whereas the storage modulus of modified fluid (B) was higher than modified fluid (C).

CONCLUSION

The polymer nanocomposite was synthesized by free radical solution polymerization technique and characterized by FTIR analysis. The polymer nanocomposite was mixed with drilling fluid to form modified drilling fluids with different concentrations. The rheological properties of the base and modified drilling fluids were performed at 25°C. The rheological properties of modified drilling fluid (B) showed that the addition of polymer nanocomposite in the base drilling fluid resulted in the enhanced rheological properties. Hence, the synthesized polymer nanocomposite offers enhanced rheological properties and can be used as one of the drilling fluid additives.

ACKNOWLEDGMENTS

The authors are thankful to the Department of Chemical Engineering and Center for Integrative Petroleum Research for providing research support.

REFERENCES

1. van Oort, E., *On the physical and chemical stability of shales.* Journal of Petroleum Science and Engineering, 2003. **38**(3-4): p. 213-235.
2. Li, M.-C., et al., *Cellulose Nanocrystals and Polyanionic Cellulose as Additives in Bentonite Water-Based Drilling Fluids: Rheological Modeling and Filtration Mechanisms.* Industrial & Engineering Chemistry Research, 2016. **55**(1): p. 133-143.
3. Ismail, A.R., et al., *The novel approach for the enhancement of rheological properties of water-based drilling fluids by using multi-walled carbon nanotube, nanosilica and glass beads.* Journal of Petroleum Science and Engineering, 2016. **139**: p. 264-275.
4. Ahmad, H.M., M.S. Kamal, and M.A. Al-Harthi, *Effect of thermal aging and electrolyte on bentonite dispersions: Rheology and morphological properties.* Journal of Molecular Liquids, 2018. **269**: p. 278-286.
5. Ahmad, H.M., M.S. Kamal, and M.A. Al-Harthi, *Rheological and filtration properties of clay-polymer systems: Impact of polymer structure.* Applied Clay Science, 2018.
6. Ahmad, H.M., M.S. Kamal, and M.A. Al-Harthi, *High molecular weight copolymers as rheology modifier and fluid loss additive for water-based drilling fluids.* Journal of Molecular Liquids, 2018. **252**: p. 133-143.
7. Ahmad, H.M., et al. *Improving the drilling fluid properties using nanoparticles and water-soluble polymers.* in *SPE Kingdom of Saudi Arabia Annual Technical Symposium and Exhibition.* 2017. Society of Petroleum Engineers.
8. Ahmad, H.M., et al., *Synthesis and Experimental Investigation of Novel CNT-Polymer Nanocomposite to Enhance Borehole Stability at High Temperature Drilling Applications*, in *SPE Kingdom of Saudi Arabia Annual Technical Symposium and Exhibition*. 2018, Society of Petroleum Engineers: Dammam, Saudi Arabia. p. 17.

Promoting the Strain-induced Crystallization of NR/HNT Composites through the Use of Modified Palm Stearin

Nabil Hayeemasae[a)], Nureeyah Jehsoh[b)], and Kannika Sahakaro[c)]

Department of Rubber Technology and Polymer Science, Faculty of Science and Technology, Prince of Songkla University, Pattani Campus, 94000, Pattani, Thailand

[a)]Corresponding author: anbil.h@psu.ac.th
[b)]nureeyah.psu56@gmail.com
[c)]kannika.sah@psu.ac.th

Abstract. In the palm oil industry, Palm stearin is derived from the refinery process of crude palm oil. The chemical constituents of palm stearin are very interesting. It can be used to synthesize a potential combatibilizer in the composite material. In this study, modified palm stearin was used as compatibiliser to promote the strain-induced crystallization for natural rubber (NR)/halloysite nanotubes (HNT) composites. It was varied in a range of 0.5 – 2 phr. The addition of modified palm stearin significantly facilitated strain-induced crystallization of the composites. This was clearly verified by the synchrotron wide angle X-ray scattering (WAXS). It proves that the modified palm stearin acts as a dispersant to improve the distribution of the HNT and also promotes the interaction between the NR and HNT through hydrogen bonding. Such corresponding interaction has greatly influenced on assisting the strain-induced crystallization of composites.

Keywords: Natural Rubber; Halloysite Nanotubes; Modified Palm Stearin; Wide Angle X-Ray Scattering.
PACS:

INTRODUCTION

Adding fillers to rubber matrix has attracted tremendous attention recently [1]. This is to improve several properties of rubber. However, it depends on many factors including filler aspect ratio, dispersibility, and the orientation of the filler particles. Among interesting fillers, Halloysite nanotubes or HNT is considered one of the most used nanofiller in many polymer matrices [2-4]. It has unique surface chemistry which is formed by surface weathering of aluminosilicate minerals. The compositions are mainly aluminum, silicon, hydrogen and oxygen. HNT consists of two interlayer surfaces composed of AL-OH groups which are located inside the tubes and siloxane groups which cover the outer surface of the HNT.

The drawback of HNT especially when using with non-polar rubber such as natural rubber (NR) is the compatibility. Many works have been made to solve this problem. For examples, the use of silane coupling agents [5], modified rubber as compatibilizers [6], and modifying the methods of preparation [7]. The use of a monomeric compatibilizer has been found to have an influence on the properties of the composites. Most compatibilizers are based on existing chemicals while less effort has been focused on the use of naturally occurring compatibilizers.

Palm stearin is one of the minor products derived from the fractionating process of crude palm oil. It consists of high fatty acid fraction of palm stearin and triglyceride [8], its value is lower than that of the palm olein which is the main product from the same process [9]. Their methyl esters or triglyceride can react with primary and secondary amines to produce the corresponding alkanolamides or modified palm stearin (MPS) [10, 11]. Such MPS gains special attention due to their application in surfactants, lubricants, fungicides and anti-blocking agents in the plastic processing industry [12, 13]. Based on the surface chemistry of MPS, it was expected to improve the compatibility between the silanol and/or siloxane groups and other related properties of the composites.

The compatibility of filler and rubber has been made relation with the strain-induced crystallization of rubber in many research works [14-16] .The incorporation of compatibilizer into filled rubberenables to induce crystallization because it acts as a hook to pull the surrounding molecular chains and speeds up the crystallization process. In this study, tensile properties and its strain induced crystallization of NR/HNT composites in the presence of MPS were focused. The crystallization characteristics of the composites were studied through synchrotron wide-angle X-ray scattering (WAXS) where extension apparatus was also equipped to capture the crystallization under deformation.

EXPERIMENTAL DESIGN

Materials

The NR used was STR 5L grade purchased from Chalong Latex Industry Co., Ltd., Thailand. HNT was mined and supplied by Imerys Tableware Asia Limited, New Zealand. The main compositions of the HNT are as follows (wt%); SiO2 (49.5%), Al2O3 (35.5%), Fe2O3 (0.29%), TiO2 (0.09%) with traces of CaO, MgO, K2O and Na2O. Stearic acid was purchased from Imperial Chemical Co., Ltd., Bangkok, Thailand. ZnO was obtained from Global Chemical Co., Ltd., Samut Prakan, Thailand. N-cyclohexyl-2-benzothiazole sulfenamide (CBS) was purchased from Flexsys America L.P., West Virginia, USA and sulfur was purchased from Siam Chemical Co., Ltd., Samut Prakan, Thailand. The palm stearin used was manufactured by Chumporn Palm Oil Industry PCL., Chumporn, Thailand. Other chemicals used to synthesize MPS such as sodium methoxide, diethanolamine, diethyl ether and saturated sodium chloride were supplied by Sigma-Aldrich (Thailand) Co., Ltd. Bangkok, Thailand.

Synthesis of MPS

The synthesis of MPS was done in a laboratory. First, methanol was added and mixed with sodium methoxide while stirring. The mixture of palm stearin and diethanolamine was later added to the solution and mildly stirred. Next, the mixture was heated and the reaction temperature was kept constant at 70°C for 5 h. The resultant mixture was extracted with diethyl ether, and washed with saturated sodium chloride solution. The crude product was finally purified with anhydrous sodium sulfate, and concentrated by a rotary evaporator. The MPS was stored in a desiccator prior to be used.

Preparation of the Composites

Table 1 lists the starting materials used for compounding the compatibilized NR/HNT composites, the entire amounts of NR, HNT, MPS and other additives were prepared using a Brabender (Plastograph® EC Plus, Mixer W50EHT 3Z). The compounds produced are hereafter designated as MPS0, MPS0.5, MPS0.7, MPS1 and MPS2 for the composites with 0, 0.5, 0.7, 1 and 2 phr of MPS, respectively. After dumping, the compounds were passed through a tight nip (~2 mm) two-roll mill to remove excess generated heat. The compounds were then compression molded into specific shapes by a hydraulic hot press, using the curing times determined by a moving-die rheometer (MDR) as described in the following section.

Table 1. Compound formulation of the NR/HNT composites with and without MPS.

Ingredients	Amount (phr)				
	MPS0	MPS0.5	MPS0.7	MPS1	MPS2
STR 5L	100	100	100	100	100
Stearic acid	1	1	1	1	1
ZnO	5	5	5	5	5
HNT	5	5	5	5	5
MPS	-	0.5	0.7	1	2
CBS	2	2	2	2	2
Sulfur	2	2	2	2	2

Measurement of Tensile Properties

Dumbbell-shaped samples were cut from the molded sheets according to ASTM D412, and tensile tests were performed at a cross-head speed of 500 mm/min. The tensile tests were carried out with a universal tensile machine (Tinius Olsen, H10KS).

Monitoring Strain-Induced Crystallization

The crystallization behavior of NR/HNT with and without MPS was analyzed using an in situ synchrotron wide angle X-ray scattering (WAXS). The WAXS was performed using a Beamline 1.3W at the Siam Photon Laboratory, Synchrotron Light Research Institute (SLRI), Nakhon Ratchasima, Thailand. The wavelength was 0.138 nm and the distance of sample-to-detector was 115.34 mm. 4-Bromobenzoic acid was used as a standard chemical to calibrate the scattering angle. The scattering pattern of the samples was captured by a CCD detector (Rayonix, SX165) with a diameter of 165 mm. The beam intensity before exposing and after passing through the samples was monitored by an ionization chamber installed in front of the sample holder and a photodiode mounted in front of the beam stop. The dumbbell-shaped specimens were mounted between the grips of a stretching machine. WAXS was recorded while stretching the samples with a crosshead speed of 50 mm/min. A 2D air image was subtracted from the 2D WAXS images, and the resulting images were converted to 1D WAXS profiles by averaging each 2D image in the azimuthal direction using the SAXSIT program developed by SLRI. This 1D profile was integrated to calculate the degree of crystallinity [14, 15].

RESULTS AND DISCUSSION

Tensile Properties

The stress-strain curves of the NR/HNT composites with and without MPS are shown in Figure 1. The typical strain-induced crystallization of NR/HNT composites with and without MPS can be observed from the stress-strain curves. Initially, stress increases slightly when the strain is applied, then increases steeply due to the strain-induced crystallization of the NR during tensile stretching. The stress and strain values appear to differ when the NR is with or without MPS. From the stress-strain curves, it is possible to predict the crystallization of the composites from the turning point of the stress and strain. It is obvious that the stress started to increase towards the MPS content. This is responsible to the formation of interaction in the system.

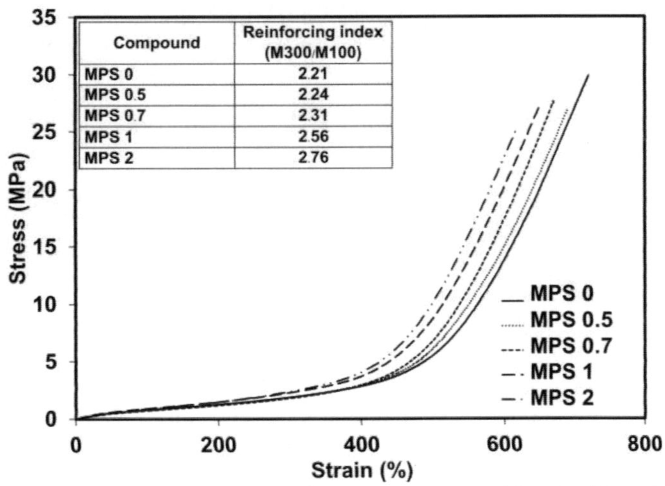

FIGURE 1. Stress-strain curves of NR/HNT composites with and without MPS.

Even the tensile strength dropped slightly upon the inclusion of MPS which was unexpected since the main reason for incorporating the MPS was to improve the compatibility of the NR and HNT. The reduction in tensile strength was simply due to the MPS itself, which had a strong effect on the interaction between the NR and HNT. As a result, the stress concentration point was observed at the interacting point, creating flaws in the rubber samples during the tensile testing. This phenomenon caused a similar effect to the elongation at break of the NR/HNT composites.

A better interaction between NR and HNT is formed by the interaction of the amide groups available in the MPS and the siloxane groups on the outer layer of the HNTand these reactions created an increase in the reinforcing efficiency of the NR/HNT composites. As can be seen from Table embedded in Figure 1, the reinforcing efficiency (M300/M100) of the composites increased upon the addition of MPS which not only acts as compatibilizer but also plays an important role in enhancing the dispersibility of HNT throughout the rubber matrix. As noted above, MPS is wax-like in nature and this helps to improve the dispersibility of the HNT.

Strain Induced Crystallization

The strain-induced crystallization was observed from WAXS analysis. The outputs from this technique is depicted in Figure 2. Considering Figure 2A, the peaks appeared at the 2θ degree of 12.5° and 18.0° which corresponding to planes 200 and 120 indicating the crystallization of the NR/HNT composites [17]. It was also found that the intensity slightly increased with increasing MPS, particularly at the 120 plane, suggesting that the crystallization taken place upon the addition of MPS. Apart from that, the degree of crystallinity (X_C, see Figure 2B) is also estimated from the integrated peak area of the intensities at 200 and 120 diffractions [18]. X_C increases with increasing the strain where the crystallinity starts to occur at the lower strain i.e., 250% for the composite without MPS, and 190% to 150% for MPS0.5 to MPS2.0 respectively. Here, it is obvious that MPS shows an important role in inducing the crystallization of the NR/HNT composites. This phenomenon is attributed to the better interfacial interaction of NR and HNT gained by the addition of MPS. Such interaction can promote an intense interaction point which then acts as a hook to pull the surrounding molecular chains during stretching. Subsequently, the molecular chain orientation and alignment are further enhanced.

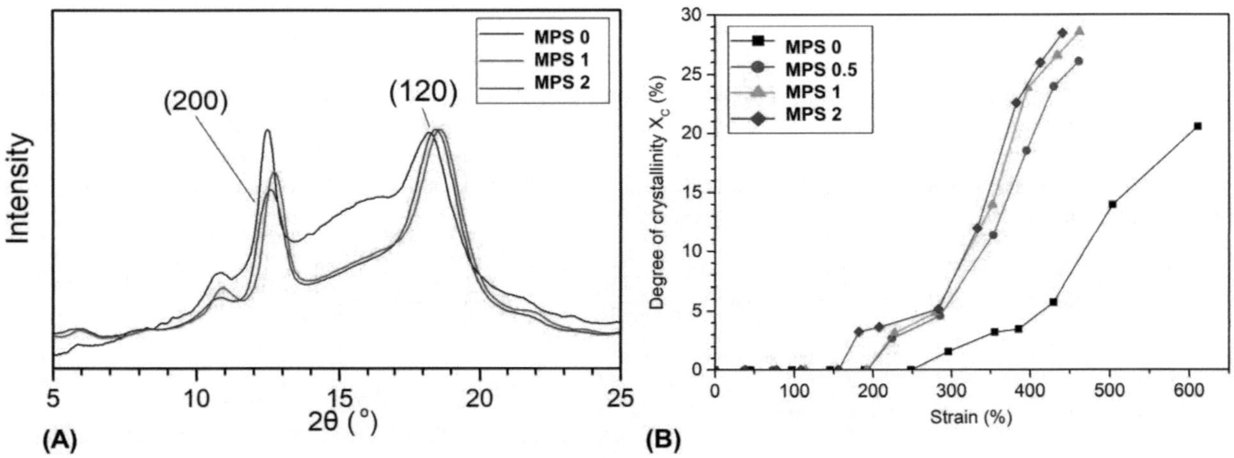

FIGURE 2. WAXS profiles as a function of scattering angle (A) and the degree of crystallinity determined from 200 and 120 reflection planes (B) of NR/HNT with and without MPS.

CONCLUSION

The overall properties of NR/HNT composites were clearly enhanced by the addition of MPS. The strong interfacial adhesion was basically due to the hydrogen bonding raised by the amines available in the MPS and the corresponding silanol and aluminol groups of the HNT. Another significant factor is the waxy nature of MPS, which can enhance the dispersion of HNT throughout the NR matrix. The degree of crystallinity of the NR/HNT composites increased with increases in the applied strain and the onset of strain started earlier after the addition of MPS. MPS acted as a hook to attract the surrounding molecular chains and thereby increase the crystallinity of the composites. It was also found that the orientation parameters tend towards parallel alignment, suggesting that the molecular chains were oriented and aligned easily in the presence of MPS. All the supporting evidences are in good relation to the results observed previously in the stress-strain behaviours and WAXS profiles.

ACKNOWLEDGMENTS

We gratefully acknowledge the financial support by Prince of Songkla University through a Research Grant for Fiscal Year 2017 (Grant No. SAT600101S). The second author wishes to thank the Faculty of Science and Technology, Prince of Songkla University, Pattani Campus for providing a personal scholarship during the period of her MSc. study.

REFERENCES

1. V. Arrighi, I.J. McEwen, H. Qian, M.B. Serrano Prieto, The glass transition and interfacial layer in styrene-butadiene rubber containing silica nanofiller, Polymer (Guildf). (2003). doi:10.1016/S0032-3861(03)00667-0.
2. H. Ismail, P. Pasbakhsh, M.N.A. Fauzi, A. Abu Bakar, Morphological, thermal and tensile properties of halloysite nanotubes filled ethylene propylene diene monomer (EPDM) nanocomposites, Polym. Test. (2008). doi:10.1016/j.polymertesting.2008.06.007.
3. Z. Jia, Y. Luo, B. Guo, B. Yang, M. Du, D. Jia, Reinforcing and flame-retardant effects of halloysite nanotubes on LLDPE, Polym. - Plast. Technol. Eng. (2009). doi:10.1080/03602550902824440.
4. V. Vahedi, P. Pasbakhsh, S.P. Chai, Toward high performance epoxy/halloysite nanocomposites: New insights based on rheological, curing, and impact properties, Mater. Des. (2015). doi:10.1016/j.matdes.2014.12.010.
5. S. Rooj, A. Das, V. Thakur, R.N. Mahaling, A.K. Bhowmick, G. Heinrich, Preparation and properties of natural nanocomposites based on natural rubber and naturally occurring halloysite nanotubes, Mater. Des. (2010). doi:10.1016/j.matdes.2009.11.009.
6. S.M.R. Paran, G. Naderi, M.H.R. Ghoreishy, XNBR-grafted halloysite nanotube core-shell as a potential compatibilizer for immiscible polymer systems, Appl. Surf. Sci. (2016). doi:10.1016/j.apsusc.2016.04.087.
7. M. Du, B. Guo, Y. Lei, M. Liu, D. Jia, Carboxylated butadiene-styrene rubber/halloysite nanotube nanocomposites: Interfacial interaction and performance, Polymer (Guildf). (2008). doi:10.1016/j.polymer.2008.08.042.
8. A.R. Norizzah, C.L. Chong, C.S. Cheow, O. Zaliha, Effects of chemical interesterification on physicochemical properties of palm stearin and palm kernel olein blends, Food Chem. (2004). doi:10.1016/j.foodchem.2003.09.030.
9. Y.B. Che Man, T. Haryati, H.M. Ghazali, B. a. Asbi, Composition and thermal profile of crude palm oil and its products, J. Am. Oil Chem. Soc. (1999). doi:10.1007/s11746-999-0224-y.
10. I. Surya, H. Ismail, A.R. Azura, Alkanolamide as an accelerator, filler-dispersant and a plasticizer in silica-filled natural rubber compounds, Polym. Test. 32 (2013) 1313–1321. doi:10.1016/j.polymertesting.2013.07.015.
11. H. Kolancilar, Preparation of laurel oil alkanolamide from laurel oil, JAOCS, J. Am. Oil Chem. Soc. (2004). doi:10.1007/s11746-006-0947-y.
12. U. Biermann, W. Friedt, S. Lang, W. Lühs, G. Machmüller, J. Metzger, Rüsch Gen Klaas M, H. Schäfer, M. Schneider, New Syntheses with Oils and Fats as Renewable Raw Materials for the Chemical Industry., Angew. Chem. Int. Ed. Engl. (2000). doi:10.1002/1521-3773(20000703)39:13<2206::AID-ANIE2206>3.0.CO;2-P.
13. H. Maag, Fatty acid derivatives: Important surfactants for household, cosmetic and industrial purposes, J. Am. Oil Chem. Soc. 61 (1984) 259–267. doi:10.1007/BF02678778.
14. J. Carretero-González, H. Retsos, R. Verdejo, S. Toki, B.S. Hsiao, E.P. Giannelis, M.A. López-Manchado, Effect of nanoclay on natural rubber microstructure, Macromolecules. (2008). doi:10.1021/ma800893x.
15. A. Masa, R. Saito, H. Saito, T. Sakai, A. Kaesaman, N. Lopattananon, Phenolic resin-crosslinked natural rubber/clay nanocomposites: Influence of clay loading and interfacial adhesion on strain-induced crystallization behavior, J. Appl. Polym. Sci. (2016). doi:10.1002/app.43214.
16. A. Masa, S. Iimori, R. Saito, H. Saito, T. Sakai, A. Kaesaman, N. Lopattananon, Strain-induced crystallization behavior of phenolic resin crosslinked natural rubber/clay nanocomposites, J. Appl. Polym. Sci. (2015). doi:10.1002/app.42580.
17. W. Kuang, Z. Yang, Z. Tang, B. Guo, Wrapping of polyrhodanine onto tubular clay and its prominent effects on the reinforcement of the clay for rubber, Compos. Part A Appl. Sci. Manuf. (2016). doi:10.1016/j.compositesa.2016.02.015.
18. M. Hernández, M.A. López-Manchado, A. Sanz, A. Nogales, T.A. Ezquerra, Effects of strain-induced crystallization on the segmental dynamics of vulcanized natural rubber, Macromolecules. (2011). doi:10.1021/ma201021q.

New Material for Highly Dispersed Silica Filled ESBR Compound; ESBR/silica Wet Masterbatch

Wonho Kim[a*], Muhammet Iz[a], Jong Woo Bae[b], Hee Seok Oh[c], Woo Jung Son[d]

[a]Department of Polymer Science and Chemical Engineering, Pusan National University, Busan 46241, Republic of Korea
[b]Rubber Research Division, Korea Institute of Footwear & Leather Technology, Busan 47154, Republic of Korea
[c]Young Polymer Co., Ltd., Busan 46020, Republic of Korea
[d]Material Research Team, DRB Holding Co., Ltd., Busan 46329, Republic of Korea

Abstract. Filler dispersion in the rubber compound is an important factor related to the processability and the compound performance. Especially, when silica is used as a filler, the agglomeration becomes severe due to the hydrogen bonding between silica (filler-filler interaction), which causes the processability and the mechanical properties are deteriorated. Therefore, many researches are being conducted on the improvement of the silica dispersion in the compound. In this study, ESBR/silica WMB using emulsion styrene-butadiene rubber (ESBR) was studied. In detail, the degree of hydrophobation of the bis [3-(triethoxysilyl)propyl]tetrasulfide (TESPT) modified silica was measured. Then, we had researched the effect of coagulant type on the processability, silica dispersion, wear properties, and mechanical properties. As a result, all of WMB compounds with different coagulants exhibited better processability than dry masterbatch compound (DMB, conventional mixing). Especially, WMB compound with calcium chloride as a coagulant had excellent silica dispersion, high filler-rubber interaction, and abrasion resistance.

Keywords: ESBR/silica wet masterbatch, silica dispersion, latex/silica co-coagulation
PACS: 81.05Lg

INTRODUCTION

Rolling resistance and wet traction are the important factors in tire performances related to fuel consumption and safety during the driving of the vehicle. These factors must be considered in the design of rubber compounds used for tire tread [1]. In general, it is known that the good wet traction performance of a tire (high tan δ value at 0 °C) is shown when there is high filler content, functionalized polymer with high filler-rubber affinity, and good filler dispersion [2]. The rolling resistance is also known to be lower (good fuel efficiency characteristics) when tan δ at 60 °C of the vulcanizates is lowered. Highly effective methods for low rolling resistance include reducing the filler content, improving filler dispersion, and using functionalized polymer with high filler affinity. However, when functionalized polymer is used, the raw material cost is higher than that of non-functionalized polymer, so improving filler dispersion are the priority.

Currently, silica is the dominant filler used in the tire industry. Hydrophilic silanol groups (Si–OH) are present on the silica surface, so the interaction with hydrophobic unfunctionalized polymer is low and it is difficult to obtain sufficient filler-rubber interaction (F–R interaction), even when silica-friendly functional polymer is used. Therefore, a silane coupling agent must be used to effectively disperse silica in silica filled rubber compounds by silanization reaction. However, despite the use of a silane coupling agent, the filler-filler interaction (F–F interaction) is still strong due to hydrogen bonding between the residual silanol groups when a high content of silica is added, which remains a disadvantage [3]. F–F interactions are particularly problematic when high silica content in tire tread compounds is used to improve wet traction performance. The result is that the rolling resistance is increased when silica content is increased. Therefore, in order to simultaneously improve both performances in the trade-off relationship, a new material is required, containing a high content of silica with a good dispersion. To prepare this material, silica wet masterbatch (silica WMB) technology has been developed [4].

Silica WMB technology refers to a technique of preparing composites of rubber/silica/silane coupling agent by mixing a rubber solution or rubber latex with surface-modified silica, followed by a stripping or coagulation process.

The advantages of this technology are the production of silica compounds with a high silica content and good dispersion, improved energy efficiency of the tire manufacturing process, reduced emission of ethanol, and lower hysteresis (lower rolling resistance)

In this study, we researched the effect of coagulants on the properties of ESBR/silica compounds. In detail, we determined the reaction mechanism of emulsion SBR silica WMB and the suitable manufacturing conditions. The study was carried out using three types of coagulants commonly used in manufacturing of ESBR/silica WMB. The effect of coagulants on the silica dispersion, mechanical properties, and dynamic viscoelastic properties of ESBR/silica WMB compounds was investigated.

MATERIALS AND EXPERIMENTAL METHODS

Materials

12 wt % TESPT modified silica was selected for use in ESBR/silica WMB experiments. The ESBR latex and processing oil used were SBR-1712 (styrene content: 23.5%, non-oil extended) and TDAE oil. For the coagulation process of ESBR/silica WMB, three types of coagulants were used; aqueous solutions of 1 M sulfuric acid, 25 wt % sodium chloride solution, and 2 wt % calcium chloride solution.

The DMB compound (T-1) was prepared, for comparison with the WMB compounds, using silica, TESPT, and SBR-1723 (styrene content: 23.5 wt %, 37.5 phr TDAE oil extend). Zinc oxide (ZnO) and stearic acid (St/A) as activators, and *N*-(1,3-dimethyl-butyl)-*N'*-phenyl-p-phenylenediamine (6PPD) as an antioxidant were added during the compounding process in the kneader. In the final masterbatch (FMB) step, sulfur, *n*-cyclohexyl-2-benzothiazole sulfonamide (CBS), and diphenyl guanidine (DPG) were added as crosslinking agents and cure accelerators.

The hydrophobation of TESPT modified silica was measured using dibutylamine, petroleum benzine, chloroform, crystal violet indicator, perchloric acid, and acetic anhydride. The crosslink density was determined using tetrahydrofuran, *n*-hexane, toluene, and acetone.

Experimental methods

Manufacturing of ESBR/silica WMBs

The 89.6 phr of 12 wt % TESPT-modified silica was added to 1L distilled water and the mixture was stirred at 60 °C for 15 min to prepare a modified silica slurry. The modified silica slurry was mixed with SBR-1712 latex containing 100 phr solid ESBR, heated to 60 °C and further stirred for 30 min. After that added 37.5phr TDAE oil emulsion into heated mixture, stirring for 30 min. WMBs were prepared using three types of coagulants: 1 M aqueous sulfuric acid solution 200 ml (WMB T-2), 1 M aqueous sulfuric acid solution 15 ml combined with a 25 wt % sodium chloride solution 300 ml (WMB T-3), and 2 wt % calcium chloride aqueous solution 300 ml (WMB T-4). Thereafter, ESBR/silica WMBs were washed 4 times using distilled water before drying in a convection oven at 60 °C for 12 hrs.

Compounding of DMB compound and WMB compounds

The rubber compound was prepared by applying a fill factor of 0.7 using a closed mixer. Compounding was carried out in a silica masterbatch (SMB) stage at a starting temperature of 120 °C for 12 min, dump temperature was 155 °C. The detailed formulations for T-1 to T-4 is shown in Tables (1).

TABLE (1). Formulation of wet masterbatch (WMB) compounds and dry masterbatch (DMB) compound according to the coagulant type.

Components (phr)	T-1	T-2	T-3	T-4
SBR-1723	137.5	-	-	-
WMB T-2 (H_2SO_4)	-	227.1	-	-
WMB T-3 (H_2SO_4 + NaCl)	-	-	227.1	-
WMB T-4 ($CaCl_2$)	-	-	-	227.1
Pure silica	80	-	-	-
TESPT	9.6	-	-	-
ZnO	3	3	3	3
St/A	2	2	2	2
6PPD	1	1	1	1
Sulfur	1.5	1.5	1.5	1.5
CBS	1.5	1.5	1.5	1.5
DPG	1.5	1.5	1.5	1.5

Analysis of WMB and DMB vulcanizates

The sheeted FMB compound prepared by a two-roll mill, which was tested to analyze the cure characteristics using a moving die rheometer for 30 min at 160 °C with vibration conditions of ± 1°. The vulcanization was carried out at 160 °C for the measured cure characteristics.

To measure the crosslinking density of the vulcanized compounds, they were cut into dimensions of 10 mm × 10 mm and stored in a vial containing tetrahydrofuran (50 mL) for 48 h. The sample was then treated with n-hexane (50 mL) for 48 h to remove organic chemicals. After drying for two days at room temperature, the mass of the sample was measured, and then the weight was measured by swelling in toluene for 48 h. The measured weight values were substituted into Equation (1) to calculate the crosslink density.

$$\upsilon = \frac{1}{2M_c} = -\frac{\ln(1-V_1) + V_1 + \chi V_1^2}{2\rho_r V_0 \left(V_1^{\frac{1}{3}} - \frac{V_1}{2}\right)} \qquad (1)$$

where v is the crosslink density (mol·g^{-1}), M_c is the average molecular weight between crosslink points (g·mol^{-1}), V_1 is the volume fraction of rubber in the swollen gel at equilibrium, V_0 is the molar volume of solvent (cm^3·mol^{-1}), ρ_r is the density of the rubber sample (g·cm^{-3}), and χ is the polymer-solvent interaction parameter.

To measure the mechanical properties of the vulcanizates, 3 specimens per each compound were prepared according to ASTM D 412 and evaluated using a universal testing machine. The abrasion resistance was evaluated with a DIN abrasion tester according to ASTM D 5963. Four specimens per each compound were manufactured in a cylindrical shape with a diameter of 16 mm and a thickness of 8 mm and the mass change of the specimens was measured by abrading the specimen with 40 m under a vertical load of 5 N using a cylindrical drum with a polishing cloth.

The dynamic viscoelastic properties of the vulcanizates were measured according to ASTM D 4065 using a dynamic mechanical thermal analyzer under the temperature range of −80 to 80 °C. The glass transition temperature (T_g), and tan δ at 60 °C (rolling resistance index) were measured.

RESULTS AND DISCUSSION

Analysis of uncured compounds

Table (2) show the result of the Mooney viscosity, Payne effect, and bound rubber analysis. The results of the Payne effect show that the ESBR/silica WMB compound exhibits lower $\Delta G'$ value than DMB compound. In particular, the lowest $\Delta G'$ value occurs when calcium chloride is used due to the increased hydrophobation of the silica via the alkali-silica reaction.

The T-4 compound with calcium chloride as a coagulant showed high bound rubber content and showed excellent results with respect to the F–R interaction. The F–R interaction is improved as the hydrophobation of the silica is increased by the alkali–silica reaction.

TABLE (2). Payne effect, Mooney viscosity, and bound rubber content of the T-1 to T-4 compounds.

Compounds	T-1	T-2	T-3	T-4
$\Delta G'$(KPa)	983.1	812.0	810.0	788.1
ML_{1+4}@100°C	114	107	104	103
Bound rubber (%)	31.3	29.5	28.5	36.6

Analysis of DMB and WMB vulcanizate

The mechanical properties are shown in Table (3). The T-1 compound with the highest crosslink density showed the highest 300% modulus. The T-3 and T-4 compounds showed similar crosslinking densities, but as the T-4 compound had excellent F–R interaction, the 300% modulus for the T-4 compound was better.

The DIN abrasion test showed that the T-4 compound had the best value to support the bound rubber test result and the reinforcing effect calculation result. The DMB compound, T-1, also showed good results.

TABLE (3). Mechanical properties and DIN abrasion of the T-1 to T-4 compounds.

Compounds	Unit	T-1	T-2	T-3	T-4
$M_{100\%}$	kgf·cm^{-2}	24.8	20.8	21.0	18.9
$M_{300\%}$	kgf·cm^{-2}	94.1	78.5	80.3	83.4
$M_{300\%}/M_{100\%}$	-	3.79	3.77	3.82	4.41
Elongation at break	%	600	670	650	590
Tensile strength	kgf·cm^{-2}	252	270	257	244
DIN abrasion	mg	115.1	120.5	118.2	114.0

The results of dynamic viscoelastic properties are shown in Table (4). Since only the type of coagulant was different and ESBR with the same microstructure was used in all cases, the glass transition temperature of the compound was measured to be almost the same for all samples. The value of tan δ at the glass transition temperature is higher when the dispersion of silica is better [31]. Therefore, the T-4 compound, which had the highest silica dispersion in Payne effect and Mooney viscosity measurements, showed the highest tan δ value at T_g.

The value of tan δ at 60 °C, which is a measure of rolling resistance during driving, generally shows a lower value as the crosslinking density is higher [39]. As a result, the DMB compound, T-1, showed the lowest value, and excellent results were obtained. The T-4 compound had a low crosslinking density compared with T-1 but had a similar rolling resistance because of the good silica dispersion and excellent F–R interaction. The results suggest that the use of calcium chloride in the WMB technology results in a lower total crosslink density than the DMB applied compound. However, as the degree of hydrophobation of the silica increases, the excellent silica dispersion and the improved filler-rubber interaction complement the low crosslinking density, showing dynamic viscoelastic properties similar to the DMB compound.

TABLE (4). Viscoelastic properties of the T-1 to T-4 compounds.

Compounds	**T-1**	**T-2**	**T-3**	**T-4**
tan δ at T_g	0.7772	0.8134	0.8122	0.8461
tan δ at 60 °C	0.1305	0.1365	0.1314	0.1291
Rolling resistance index (%)	100.0	95.6	99.3	101.8

CONCLUSION

From the results of the Payne effect measurements, it was confirmed that the ESBR/silica WMB applied compounds had a high silica dispersibility, in particular T-4 compound was improved by 25% than T-1 DMB compound when calcium chloride was used as a coagulant. Also, Mooney viscosity results showed lowest value (ML_{1+4}@100°C: 103) when calcium chloride was used, confirming the excellent silica dispersion. This occurred because the surfactant constituting the micelle in the rubber latex remained after the reaction with the coagulant, thereby acting as a processing aid and increasing the silica hydrophobation through the alkali-silica reaction. Bound rubber measurements showed more than 5% bound rubber content than other compound when the WMB compound using calcium chloride due to the better hydrophobation of silica.

Measuring various properties of the vulcanizate showed that the T-2 and T-3 compounds had a loss of accelerator activity due to an acid–base reaction between residual sulfuric acid and residual emulsifier, which decreased the crosslink density. This lower crosslink density was a disadvantage compared with the WMB compound made with calcium chloride with respect to the mechanical properties, dynamic viscoelastic properties, and abrasion resistance. The T-4 compound, a WMB compound with calcium chloride, showed excellent physical properties compared with other coagulants, but the 300% modulus was 11% lower than that of the DMB compound due to the lower crosslink density because of residual calcium carboxylate. However, the high F–R interaction measured in the bound rubber test, the Payne effect, and the excellent silica dispersion measured in the Mooney viscosity test complement the properties of the WMB compound, thus exhibiting abrasion resistance and rolling resistance characteristics (tan δ at 60 °C) similar to the DMB compound.

ACKNOWLEDGMENTS

This work was supported from Minister of Trade, Industry and Energy Grant funded by the Korean Government [Project Number 10067128] and BK21 PLUS Centre for Advanced Chemical Technology [21A20131800002], Republic of Korea.

REFERENCES

1. D.E. Hall, *Rubber Chem. Technol.* **74**, 525–539 (2001).
2. P.S. Pillai and G.S. Fielding-Russell, *Rubber Chem. Technol.* **65**, 444–452 (1992).
3. S.S. Choi, *Poly. Int.* **50**, 524–530 (2001).
4. Y. Wang, L. Liao and Z. Peng *J. Appl. Polym. Sci. 133 (2016).*

Effect of polymer relaxation in automated micro polishing

Soufian Ben Achour[a], Wule Zhu[b], Giuliano Bissacco[a], Anthony Beaucamp[b], Leonardo De Chiffre[a]

[a]Department of Mechanical Engineering, Technical University of Denmark, DK-2800 Kongens Lyngby
[b]Department of Micro-Engineering, Kyoto University, Kyoto, Japan

Abstract. Micro polishing is a process used when a fine surface quality is desired. In optics, it represents the key process to achieve low surface roughness and form error required for optics application. Automated micro polishing with polymer based tools requires a better understanding of the interaction between tool and workpiece than conventional polishing and an important phenomenon that has to be taken into account is polymer relaxation. Stress relaxation is due to the viscoelastic non-linear behavior of polymers. This relaxation is a time-dependent phenomenon that causes a decrease in stress although the strain is kept constant. In this work, the effect of polymer stress relaxation on polishing is studied. While this effect can be neglected in conventional polishing, it becomes an important and dominant factor in micro polishing. Characterization tests are conducted on a very common extruded polyurethane (LP-66) used in polishing. Subsequently, bonnet polishing using a spherical tool with a diameter of 1 mm is performed for 8 minutes on a nickel sample. Due to relaxation, the induced stress on the tool decreases during polishing, causing a reduction in the polishing pressure. This implies a reduction in the material removal rate in accordance with Preston's law. Experimental results are in accordance with the physical phenomenon. The polymer relaxation characteristics can be used to compensate the tool path to achieve uniform removal along the polished surface.

Keywords: Stress relaxation, Micro polishing, Polymer.

INTRODUCTION

Polishing is an essential manufacturing process used when a fine surface quality together with sub-micron form tolerance are needed [1]. Micro polishing tools offer several advantages when used for deterministic polishing. Among others, accessibility issues can be overcome and small features can be polished. Besides, as the tool contact area is smaller than in conventional polishing, a more uniform material removal can be achieved. On the other hand, the process requires careful control of the contact conditions. Preston's equation (eq.1) describes the factors affecting material removal in polishing. The equation states that the MRR is directly proportional to the applied pressure p and the relative velocity v between the workpiece and the polishing tool. A constant K (Preston coefficient) comprises all mechanical and chemical interactions not directly linked through pressure and velocity, e.g., grain size, lubrication, abrasive concentration and properties of the workpiece/tool [2].

$$MRR = \frac{dz}{dt} = K \cdot p(\vec{x}, t) \cdot v(\vec{x}, t) \qquad (1)$$

According to Preston's equation, when the pressure decreases so does the material removal. This implies that when the real contact pressure departs from its nominal value, form error on the polished surface may occur. It is, therefore, crucial to understand which factors can potentially change during polishing, and if needed compensate them. Since polishing tools are usually made from a polymeric material, a stress relaxation phenomenon is inevitable. This relaxation is due to the viscoelastic non-linear behavior of polymers. Stress relaxation is defined as a decrease in stress in response to a nominal constant strain [3,4]. This phenomenon is a time-dependent problem. Contact stress can be related to the contact pressure as shown in [5]. This implies that if the contact stress decreases while polishing then the pressure and consequently the material removal would have a similar trend. It is for these reasons that understanding and quantifying this effect can minimize form error in polishing and provide a more deterministic approach towards ultra-precision polishing. As a general trend in polymeric materials, the higher the strain, the higher the stress relaxation. In this work, the effect of polymer relaxation is characterized and its effect on the rate of polishing material removal is studied.

METHODOLOGY

The following methodology is adopted to prove that material removal rate is strongly affected by polymer stress relaxation, and provide quantification of this effect.
The tool used in all polishing experiments is made with a common extruded polyurethane (Universal Photonics LP-66) used for polishing applications, which was dressed to a spherical shape with radius 0.5 mm. the polished sample is a nickel disk of radius 2.5 mm. The experimental activities are divided into three stages: relaxation tests, tool wear, and temperature variation during polishing. Four hypotheses are stated and proven through the polishing experiments. The hypotheses are listed as follows:

- Stress relaxation is one of the main contributors to form error in polishing
- Wear of the polishing tool is negligible over the typical run time
- Temperature variations during polishing are negligible
- The temperature reached during polishing influences neither the relaxation behavior nor the elastic modulus

Polishing tool relaxation test

First, a relaxation test is performed to quantify the relaxation of the polymer under study. The experiment is conducted by pressing the tool against the sample for a fixed amount of time and a prescribed offset. Beneath the sample, a highly sensitive 3-axis dynamometer (Kistler 9027C) measures the contact forces (see figure 1a). The tool stays in contact with the sample with a tool offset of 100 μm. This tool offset is the same as used for the polishing experiments. Due to drift behavior of the charge amplifier of the dynamometer, only the ten first and last seconds of the experiments are sampled and acquired. This is done because acquiring data for long period of time generates thermal drift on the charge amplifier that cannot be adequately compensated. By doing so, the trend of the relaxation test cannot be observed but the absolute value between the start and the end of the experiment can give a useful quantification on the total relaxation as a function of time (see figure 1b).
To quantify and subsequently use the results of these relaxation tests, we define, in this work, the total relaxation Re at time t* as follows in equation 2

$$\mathrm{Re}(t^*) = \frac{F(T_{in})-F(t^*)}{F(T_{in})} \times 100 \qquad (2)$$

Where $F(T_{in})$ is the force at initial contact condition and $F(t^*)$ the force at the time the relaxation is calculated. This simple metric has the objective to quantify the effect of relaxation on material removal. Relaxation tests are repeated five times to assess measurement repeatability. At time t*=8 min the relaxation correspond to 43.2 ±1.81%
It has to be noted that a relaxation test is usually conducted whilst maintaining a constant real strain. In this experiment this cannot be assured. Nevertheless, the nominal strain is constant due to the nominal tool offset value. This makes these results applicable only to this particular configuration. The methodology is however of a more general nature.

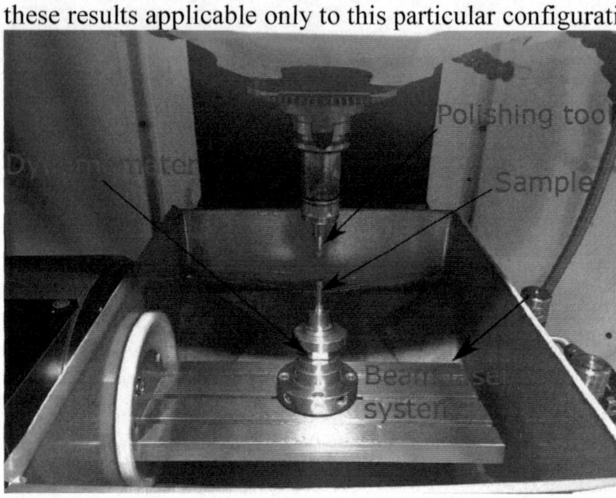

Figure 1a, Relaxation setup

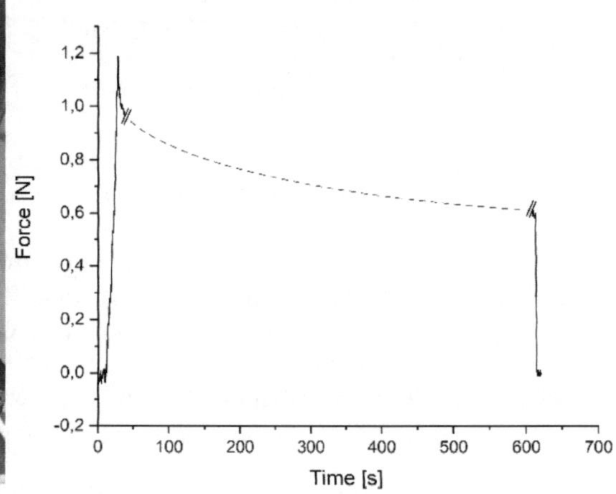

Figure 1b. Relaxation test behaviour

Tool wear in polishing

Tool wear hypothesis is tested by profiling the spherical tool with a beam laser interruption system before and after polishing. If the tool profile does not change, within a given tolerance after polishing, then it can be concluded that tool wear does not affect the material removal rate. Polishing tests were conducted using the parameters reported in table 1 to asses whether the tool wear can be neglected or not. Tests are repeated 5 times. The tool wear observed was 1 µm. Being that the tool offset is 100 µm, tool wear represents only 1% of the total offset. Moreover, from the relaxation test, the residual deformation immediately after interrupting the load is approximately 45 µm. The residual deformation is obtained as a subtraction between the measured profiles, before and after the relaxation test. The measurement is also performed with a beam laser interruption system. Based on these observations, it is safe to conclude that the tool wear does not affect the material removal and it can be neglected as a source of error.

TABLE 1. Polishing parameters.

Process parameters	Values	Units	Tool wear
Feed velocity	50	mm/min	
Spindle speed	2000	rpm	
Precess angle	30	deg	1±0,7 µm
Tool radius	500	µm	
Abrasives	Alumina	/	
Abrasives size	200	nm	

Temperature variation in polishing

Lastly, the temperatures reached during polishing were experimentally characterized. Temperature affects the Young modulus of the material. The Young modulus decreases with increasing temperature and ultimately the contact pressure follows the elastic modulus. A reduction in contact pressure results in a lower material removal rate according to Preston's law. Temperature affects also the relaxation of the polishing tool material, making it more severe. For these reasons if the temperature during polishing increases considerably, then the hypothesis of stress relaxation being the main contributor to form error cannot hold any longer.

To characterize the temperature during polishing both a thermal camera and thermocouple are used. Measuring the temperature is a challenging problem in micro polishing. Due to the small contact areas and accessibility issues, the required resolution of a thermal camera must be small enough to sense small variations. Thermocouples have to be placed as close as possible to the heat source and inevitable convective heat flow causes losses before reaching the thermocouple. In this paper, a method that overcomes these issues is presented that is suitable for polishing setups.

Polishing is performed directly on the thermocouple's junction (see Figure 2a). This is possible because of the small size of the tool and because of the small material removal rate in polishing. Some important assumptions are made in this experiment. First, the heat flow produced during polishing is considered to be nearly completely transferred to the thermocouple. This is equivalent to neglecting convective heat flow losses. Secondly, the cross-sectional dimension of the thermocouple is considerably larger than the material removed during polishing. The thermocouple's junction has a cross section of 0.4x0.4 mm^2 and the material removed is in the order of 100 nm, thus it is safe to accept this assumption. The thermocouple is a type K with a sensitivity of 41µV/°C, chosen to match the material of the sample to be polished. Type K thermocouples are made of two different alloys, Chromel and Alumel, both of which have a minimum content of nickel higher than 90%. With these choices and assumptions, a reasonable temperature during polishing can be inferred.

Spindle speed of 2000 rpm with a tool offset of 100 µm was tested. These parameters are chosen to be consistent with the ones used for the polishing experiments in the next section. The test is conducted with the same polishing tool as previously used, and a tool influence function (removal footprint) is generated with the tool being stationary (feed = 0) for 1 minute. A tool influence function represents the material removal produced by the polishing tool in a unit of time. Tests are performed with polishing slurry to replicate the real polishing conditions. Temperature measurements were conducted with a micro-bolometer array thermocamera (Optris PI 640), for characterizing the process temperature directly on the sample. In figure 2b, it is possible to see the results of the temperature characterization. The temperature during polishing increases by about 2 °C when polishing starts, stabilizes, and then gradually decreases when polishing ends. It has to be noted that the local maximum temperature could potentially be higher in some points. However, the temperature in the bulk material is the main factor affecting both the stress relaxation and Young modulus. Because of this observation, it can be concluded that, with the used parameters, the temperature variation due to polishing is not expected to influence the stress relaxation nor the Young modulus significantly.

2a) Setup for polishing on thermocouple

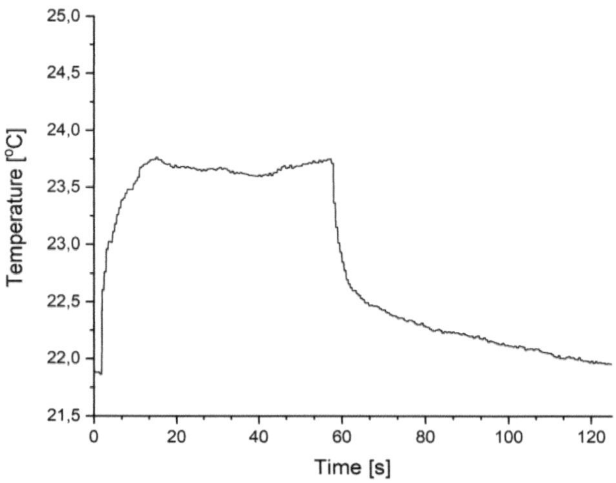

2b) temperature variation when performing polishing on thermocouple

Uniform polishing experiments

Having tested the above-listed hypotheses, polishing was conducted on a flat nickel disk of radius 2.5. Polishing experiments are performed on a CNC machine adapted for polishing purposes. The tool is a polyurethane sphere of radius 500 μm. The workpiece used for the polishing experiments is a diamond turned nickel disk presenting an initial surface roughness of 4.1 nm Sa. The used abrasives are alumina particles dispersed in a slurry with low viscosity to improve the slurry flow and promote a uniform distribution of abrasives in the polishing area. The slurry is then continuously pumped towards the polishing area assuring constant and uniform distribution of abrasives. The CAM program is generated using the ZephyrCam software developed by Zeeko Ltd. The parameters used for these tests are listed in table 2.

TABLE 2. Polishing parameters.

Process parameters	Column Header Goes Here	Units
Feed velocity	50	mm/min
Spindle speed	2000	rpm
Precess angle	30	deg
Tool radius	500	μm
Abrasives	Alumina	
Abrasives size	200	nm

Again, the polishing tool profile was measured with a beam laser interruption system before and after polishing to assess tool wear. The form of the sample is measured with a Fizeau laser interferometer from Veeco before and after polishing. The two surfaces are then subtracted to obtain the material removal distribution. This experiment is repeated 3 times to assess results repeatability and consistency.

RESULTS AND DISCUSSION

Figure 3 shows the material removed after 8 minutes of polishing and the raster path direction. It is possible to see how the material removal decreases in the cross feed direction. This is associated with the previously described relaxation behavior. It can be seen that the relaxation starts immediately and that at the end of the polishing it accounts for ΔRe=38.59% of the material removal. This demonstrates the importance of quantifying and consequentially compensate for this behavior. By extrapolating the relaxation behavior from the relaxation tests at t*= 8 minutes a total relaxation of 43.2 ±1.81% was obtained. This result is in accordance with the conducted polishing tests.

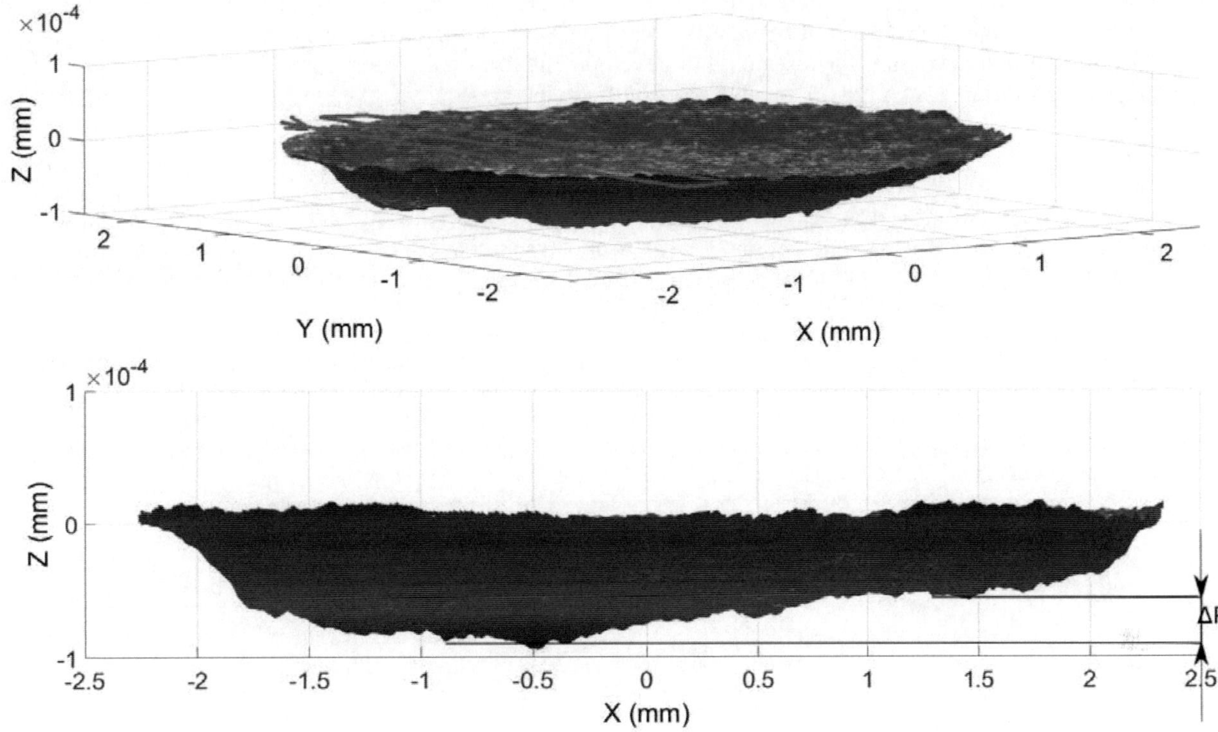

Figure 3. The material removal decreases along the direction of the polishing path. The trend and the final material removal are coherent and consistent with the relaxation test.

It has to be observed that conventional polishing operations are in general longer than 8 minutes and therefore the polishing removal rate becomes more stable in relation to the stress relaxation. Moreover, it is common to polish a surface multiple times, with different scan directions, in order to minimize the effect of polymer relaxation. In addition, when dealing with conventional size polishing tools this effect is more averaged on the surface area than in micro polishing, where the spot size and tool offset are considerably smaller than the micro-domain process. Moreover, due to the very low material removal rate in micro polishing, it becomes important to take the relaxation effect into account. The results are also biased by the stationary condition of the relaxation tests.

CONCLUSION

Polymer stress relaxation has been observed to be an important factor in high precision polishing. If this factor if not taken into account, form errors may occur that are not negligible any longer when tight tolerances are required. Experimental tests were conducted to isolate the effect of stress relaxation. The paper presents an innovative approach to quantify and measure polishing process temperatures by performing the polishing on the thermocouple itself. The hypothesis of negligible tool wear and temperature variation were proven to be solid and non-refutable. As a main result, the effect of temperature variation on the bulk material during polishing was seen to be insignificant on both the Young modulus and the relaxation behavior of the material. Tests were finally conducted to prove the main hypothesis retained in this paper, i.e. relaxation as main contributors to form error

ACKNOWLEDGMENTS

The work is undertaken in the context of MICROMAN project. MICROMAN is a European Training Network supported by Horizon 2020, the EU Framework Program for Research and Innovation (Project ID: 674801). This work was also supported by the Grant-in-Aid for Scientific Research No. 17K14571 from the Japan Society for Promotion of Science, and the grant program for research and development from the Mazak foundation. The authors finally acknowledge support from Zeeko Ltd. in loaning some of the polishing equipment and software.

REFERENCES

[1] E. Uhlmann, B. Mullany, D. Biermann, K. P. Rajurkar, T. Hausotte, and E. Brinksmeier, "Process chains for high-precision components with micro-scale features," *CIRP Ann. - Manuf. Technol.*, vol. 65, no. 2, pp. 549–572, 2016.

[2] D. W. Kim, W. H. Park, S.-W. Kim, and J. H. Burge, "Parametric modeling of edge effects for polishing tool influence functions.," *Opt. Express*, vol. 17, no. 7, pp. 5656–5665, 2009.

[3] R. F. Landel and L. E. Nielsen, *Mechanical properties of polymers and composites*. CRC press, 1993.

[4] T. D. Nguyen, H. Jerry Qi, F. Castro, and K. N. Long, "A thermoviscoelastic model for amorphous shape memory polymers: Incorporating structural and stress relaxation," *J. Mech. Phys. Solids*, vol. 56, no. 9, pp. 2792–2814, 2008.

[5] S. T. Patton, C. Chen, J. Hu, L. Grazulis, A. M. Schrand, and A. K. Roy, "Characterization of thermoplastic polyurethane (TPU) and Ag-carbon black TPU nanocomposite for potential application in additive manufacturing," *Polymers (Basel).*, vol. 9, no. 1, 2017.

The Assessment of Elastomers in Machine Direction Oriented Breathable Hygiene Films for Strengthening Machine Direction Tear Performance

Tolga Çeper, M. Fatih Ergüney

Hayat Kimya A.Ş. Ar-Ge Merkezi, 41230, Kocaeli/TURKIYE

Abstract. It is a well-known paradigm that the more the film is oriented, the worse it has tear resistance through one direction and the use of fillers such as calcium carbonate (CaCO3) has further negative effect. Machine direction orientation (MDO) process is vital for high breathability and downgauging of hygiene films while it sharply reduces tear resistance. Tear strength is crucial for sanitary napkin backsheet films from an application point of view. This paper outlines the effects of ethylene based elastomers on machine direction tear resistance in the case of MDO processed breathable polyethylene hygiene films. Additionally, an application based test method was developed for film characterization since such breathable films give almost no result when they are tested by conventional tear strength test equipment. Besides that, dart drop impact and stiffness tests have been performed to come up with some correlation to estimate the tear strength on the final product. Our results show that ethylene based elastomers can be used to provide better tear performance and the following parameters should be optimized; thickness, CaCO3 content, MDO parameters, polymer matrix, and dosing ratio of some functional additives.

Keywords: Breathable hygiene films, elastomer, polyethylene, MDO process, blown film.
PACS: 81.05.Lg

INTRODUCTION

Breathable films have drawn grate attention to cover disposable diapers and sanitary napkins in hygiene industry. Breathable covers keep the skin drier, prevent heat increase and more comfortable to wearer. [1] The breathability of the films are conventionally provided by billions of interconnected microspores in the film structure. These pores are smaller than a drop of water in dimension, while they are larger than water vapor molecule to allow the passage of steam through the film. [2] In general, high-breathability films use in disposable goods are made of polyethylene existing an inorganic porogen such as CaCO3, talc, or Zeolite. The tortuous path is necessary for air and moisture flow through in principle. When unfilled plastic films is stretched enough at high stress level, some stresses cracking occurs, however the number of pores is not reached to sufficient level to permit pore-to-pore connections. In literature, it has been proved that the mixture of polyolefin and inorganic fillers is able to stretch under normal ambient condition. Such stretching process creates the force on film and strain rate that is necessary to form microporous films. [3] Recent concept to make breathable film is to mix only polyolefins and inorganic filler in extrusion process although processing additives such as paraffin oil and glycerine monostearate were thought to use in the early years (1960's to 1980's). [4, 5] These polyolefins are typically polyethylene, polypropylene, and their copolymers, subsequently linear low density polyethylene is mostly used because it is tougher among polyethylene polymers and has high tear strength. There are three common techniques for stretching microporous film precursor. These are machine direction orientation (MDO), cross machine direction tentering and incremental stretching. MDO is a common technique by which the film passes through at least two pairs of rollers under controlled temperature. Second roller runs faster than first roller that running speed ratio of such two rollers determines the degree of stretching. [6] Orientation contributes to increase the gloss, tensile strengths, moduli, and barrier properties while machine direction tear strength and dart drop decrease, likewise, the relationship is seen in blown film process when comparing machine direction tear strength and draw-down ratio. [7, 8] In addition, machine orientation of film increase linearly with decreasing blow-up ratio. [9]

EXPERIMENTAL METHODS

Film Preparation

All breathable films were prepared by three layer blown film process using in-line MDO equipment. Film layer composition is 1/2/1. Ethylene based elastomer, which was supplied from Dow Chemical Company, was used in the middle layer of three layer film by blending with commercial LLDPE (1 MFI 190 ^{0}C/ 2.16 kg, 0.918 g/cm^3 density) and CaCO3 (approximately 2 micron in diameter). Skin layers of the film only contain LLDPE and CaCO3, and total CaCO3 ratio in film formulation was either 50% or 40%. All films were made on blown film extrusion line with 1:3.9 blow-up ratio, a melt temperature of 200-210 C, a 1.25 mm die gap and a 50 cm die diameter. The cooling conditions and winding tension were also kept constant before stretching by MDO unit that has eight rollers. After heating up the film with first two rollers, films were stretched at 90 to 95 C, then they were annealed at 95 to 100 C before cooling. Stretching ratio of the films was in the range from 4.6 to 3, and unit weight of the films were varying 30 to 20 gram per square meter. The temperature of fast and slow rollers were set to 80 ^{0}C in MDO process for all film samples. 70% filled CaCO3 prepared by compounding with LLDPE was used in film formulations. Table 1 shows the prepared film samples with different contents and conditions.

Film Characterization

Machine direction tear resistance of the films has great importance to evaluate the film properties whether it is suitable for sanitary napkin backsheet film with special wing design. In order to characterize tear performance of the films, an application based test method was developed, which is called wing tear test. It is well known that the films made by MDO process are weak in tear-propagation strength through machine direction so conventional methods such as Elmendorf tear strength test do not yield significant results for accurate comparison. The film samples were cut into shape of actual design of wing. Wing tear test were performed with following parameters: grip to grip seperation at the start position is 275 mm (samples are held on glass plates) and tensile test speed is 1000 mm/min. The results of wing tear test are given in N as a force to resist to tear in the form of special sample design.

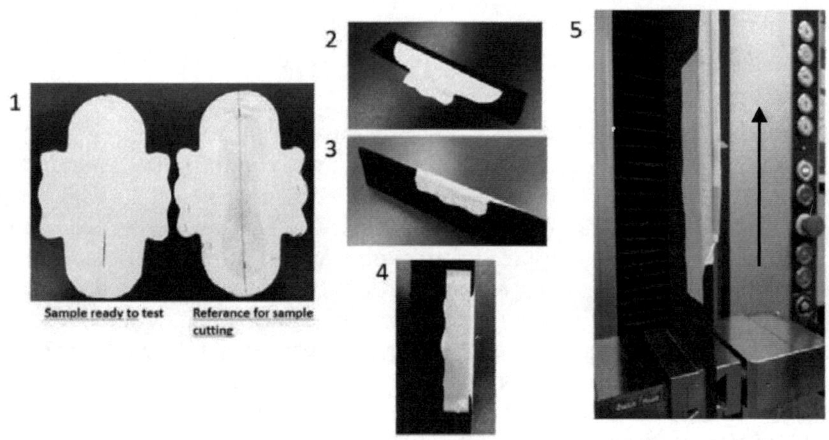

FIGURE 1. (1-4) Sample preparation for wing tear test. (5) Tensile test for the simulation of which sanitary napkin wing performs.

The film samples were also cut according to standard tensile test, ASTM D882 – 10 and standard unit weight test method WSP 130.1 (09). In tensile test, the speed of crosshead was set at 500 mm/min with 350 kN cell. Impact Resistance of Plastic Film by the Free-Falling Dart Method ASTM D1709 was performed on all film samples. For dart drop test, method specifies a dart with a 38 mm diameter dropped from 0.66 m. In order to check the CaCO3 weight of the film samples, ash content test method were carried out by which 10 g of film sample burns up at 600 ^{0}C for 15 min and is cooled to ambient temperature in desiccator and then residue is weighted. The breathability of the films were tested by Mocon water vapor transmission rate device at 38 C and 90% of humidity.

RESULTS AND DISSCUSSION

Breathable Films without Elastomer

The film mechanical and breathable properties, and process parameters are presented in Table 1. As we expected, the increase of the base weight of the films decrease breathability independent of CaCO3 content of the films. If we compare the breathability of film samples 2 and 7, 7 is less breathable than sample 2 because of the fact that 7 is less stretched than 2. This shows 15% of increase in MDO ratio from 4 to 4.6 give rise to increase 33% of breathability. Sample 4, 5 and 6 are the version containing higher CaCO3 content of 1, 2 and 3, respectively. The increase of CaCO3 amount from 50% to 57% by weight promotes the breathability of the films in varying range from 14% to 22%. The results show that the positive effect of CaCO3 amount in film on breathability is limited when the base weight of the film increase.

TABLE (1). Film compositions and properties, MDO stretching parameter.

Sample No.	Unit Weight (gsm)	MDO Ratio	Film Composition (wt. %)			Film Properties				
			Ethylene based elastomer	LLDPE	CaCO3	%2 Sec Mod MD (MPa)	%2 Sec Mod CD (MPa)	Dart Drop (g)	Wing Tear (N)	Breathability (g/m2/day)
1	25	4,6	-	50	50	550	205	110	3,4	3418
2	30	4,6	-	50	50	604	232	130	3,8	2923
3	35	4,6	-	50	50	643	264	150	6,5	2272
4	25	4,6	-	43	57	439	181	135	3,3	4177
5	30	4,6	-	43	57	536	207	150	3,4	3498
6	35	4,6	-	43	57	635	256	170	6,2	2609
7	30	4	-	50	50	480	230	145	4	2190
8	30	4,6	2,5	47,5	50	426	239	300	4,6	1987
9	30	4,6	5	45	50	428	240	355	5,2	900
10	25	4	10	50	40	329	245	255	4,5	621
11	25	4	15	45	40	285	191	295	5,5	317
12	25	3	15	45	40	227	190	310	5,6	242
13	20	4	15	45	40	230	121	265	4,2	260

Figure 2 shows the results of dart drop impact and wing tear force of breathable films without elastomer in respect to base weight. Wing tear force of the film can be enhanced by increasing base weight of the film. In parallel, dart drop impact test results show that the toughness of the film are improved. The increase of toughness could contribute to raise wing tear force of the material. When it comes to the effect of CaCO3 amount in the film, while calcite particles cause an increase of dart drop impact of the films, wing tear force slightly reduced. Calcite particles seem to help the film absorb more energy while there is an adverse effect on wing tear performance. In addition, the result in Table 1 suggests that the films with 57 wt. % CaCO3 has lower stiffness than the films with 50% CaCO3. This drop in stiffness could possibly influence on wing tear performance negatively. When 2 is compared with 7 in terms of mechanical properties, the less orientation through machine direction, the better wing tear resistance. MDO ratio is among the major factors that could manipulate wing tear performance because it is well known that orientation is fundamental to change the stiffness/toughness balance of the films. Wing tear is complex phenomenon resulting from number of interacting factors. This is the reason of which an application based test method is needed and conventional tear strength test method is not suitable to understand the behavior of such backsheet films in practice. A machine direction oriented film is extremely weak through MD that they cannot withstand the tear-propagation. On the other hand, initial tear resistance test (ASTM D1004) can be more appropriate to examine the efficiency on wing tear performance, however, wing tear performance is supposed to be combination of MD tear initial resistance, toughness/stiffness balance of the film.

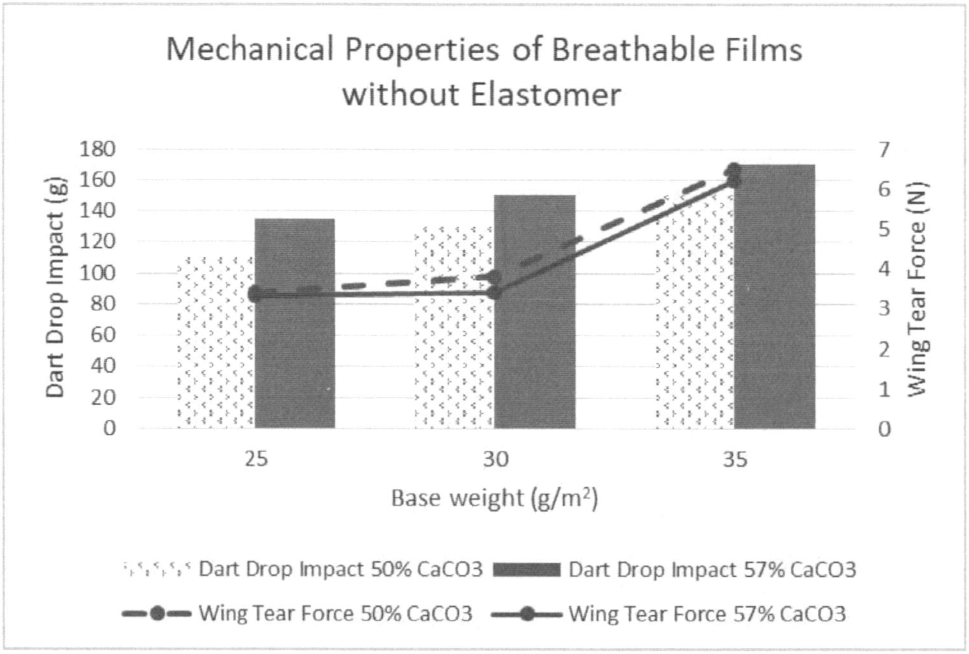

FIGURE 2. Dart drop impact and wing tear force of breathable films without elastomer in respect to base weight.

Breathable Films with Elastomer

The studies are basically conducted to evaluate the use of ethylene based elastomers and their effects on wing tear forces. As expected, the presence of elastomer in the film has negative effect on the breathability. Table 1 shows that the breathability of the films has dropped dramatically even when 2.5 wt. % of elastomer added to formulation. If we compared 2 and 8, breathability value dropped from 2923 g/m2/day to 1987 g/m2/day. This is because elastomer blocks the formation of interconnected pores through the film due to the fact that it can absorb the stress during the event of orientation.

Table 1 suggests that the addition of ethylene based elastomer reduces stiffness of the films while it increase the toughness of the films. If we compare 8 and 9, stiffness of the film in both MD and CD seems to remain stable with adding 2.5 wt. % of elastomer. However, dart drop impact value is significantly increased from 300 g to 350 g. Elastomer ratio in the film linearly raise the wing tear force if we consider the result of 2, 8 and 9 which are 3.8, 4.6 and 5.2 respectively. A similar increase trend can be seen when 10 compared with 11, which wing tear force increased from 4.5 to 5.5. Interestingly, the results of 11 and 12 shows the effect of MDO ratio on the films with elastomer is slight unlike the films that do not contain elastomer. When MDO ratio reduced from 4 to 3, wing tear force increased from 5.5 to 5.6. When consider the comparison of wing tear force between 11 and 13, which only base weight varies, it can be found out that the most effective factor on wing tear force is the base weight of the film in presence of elastomer in the film formulation. The use of elastomer in MDO based breathable film is promising to make thinner hygiene films while moderating the wing tear force.

Processability of Elastomers in In-line MDO Process

Elastomers were intentionally added into middle layer of three layers film. Elastomers cause blocking problem as long as they are present in the skin layers due to the sticky nature. Besides that, they cause film tearing during MDO process and create huge holes on the film. Tearing problem is thought to stem from adhesion of the film that can be hold by the MDO rollers. On the other hand, it is well known that the film separates during blown film production after nip rolls, so the film should have certain hardness to be able to cut without creeping. This issue restricts the use of elastomer in the films in blown film process. 15 wt. % of ethylene based elastomer could reasonably be used according to our trials.

CONCLUSION

Breathable films intended to hygiene business were investigated in terms of application based tearing performance. Base weight, MDO ratio and CaCO3 amount were found as factors in breathable films regarding wing tear performance. Wing tear performance could test with a method by which mimics the function of sanitary napkin wings. An ethylene based elastomer was used in breathable hygiene film formulation and their enhancing effect on wing tear performance is proved. Breathability was sharply reduced by the use of elastomer. This elastomer should be used in the middle layer of the film for better processability. Wing design of sanitary napkin limits base weight of the backsheet film to 35 gsm of MDO based breathable film because the wings could be teared in case lower base weight film is used. Ethylene based elastomer provides downgauging of breathable film even if the film is being used as a backsheet of sanitary napkin that has certain wing design that can be easy to tear in use.

ACKNOWLEDGMENTS

This study has been financially supported by HAYAT KİMYA A.Ş.

REFERENCES

1. F. Akin, M. Spraker, R. Aly, J. Leyden, W. Raynor and W. Landin, Pediatric Dermatology 18, (2001).
2. P. Wu, G. Jones, C. Shelle and B. Woelfli, Journal of Engineered Fibers and Fabrics 2, (2007).
3. B. Shi, M. Shlepr, J. Wang and G. Wideman, Journal of Applied Polymer Science 131, (2014).
4. Kao Corporation, J6-2280-234A, (1986).
5. Kao Corporation, J6-2280-235A, (1986).
6. L. Cancio, R. Jezzi, P.C. Wu, Nonwovens World, 65-77, (2004).
7. Equistar Chemicals, US6613841, (2003).
8. R. Patel, T. Butler, K. Walton and G. Knight, Polymer Engineering and Science 34, (1994).
9. L. Lee, R. Register and D. Dean, Journal of Polymer Science Part B: Polymer Physics 43, (2005).

Integration of Acceleration Sensors in Polymer Parts produced with Selective Laser Sintering

P. Stoll[a], M. Schmid[a,1], M. Schmock von Ohr[b] and K. Wegener[c]

[a]*Inspire, Innovation Center for Additive Manufacturing Switzerland (icams), 9014 St. Gallen, Lerchenfeldstrasse 3, Switzerland;* [1]*corresponding author: manfred.schmid@inspire.ethz.ch*

[b]*Robert Bosch GmbH, Postfach 13 42, 72703 Reutlingen, Germany*

[c]*Swiss Institute of Technology, ETH Zürich, Institute for Machine Tools and Manufacturing (IWF), 8092 Zürich, Leonhardstrasse 21, Switzerland*

Abstract. Selective Laser Sintering (SLS) is regarded today as the preferred choice of polymer Additive Manufacturing (AM) technologies able to produce functional parts for industrial applications. The high degree of part complexity along with sufficient mechanical properties designates SLS parts attractive for numerous technical solutions. The AM technologies for polymers as of today are typically performed in "closed" machine setups which do not facilitate manipulations during the manufacturing process. For the SLS-process a homogeneous and well controlled temperature distribution during several hours production time is particularly important since any irregularity or interruption of the complex process will damage the build job most likely. However, integration of functional parts like sensors or other active systems for further part optimization are considered as very beneficial in order to amplify application possibilities of SLS parts. This study describes one of the first approaches to overcome processing hurdles and incorporate external parts into SLS parts. A small-scale SLS machine was manipulated in a way that acceleration sensors could be embedded in polymer parts during their manufacturing. The single steps of the operating procedure as well as the difficulties encountered along the course of sensor integration are described. Two main challenges, powder removal from sensor cavity and fixation of the SLS part for manipulations in the powder bed, turned out to be the core problems to be solved. Finally, sensors could be embedded successfully in SLS-parts and their continuing activity was demonstrated based on results of a shaker test.

Keywords: Selective Laser Sintering, Function Integration, Hybrid Parts, Acceleration Sensor, Sensor Integration
PACS: 81

INTRODUCTION

Today the various processes for AM with polymers are established and widely used in industry for different purposes [1]. Recently conducted projections and surveys reveal that the integration of additional features or elements into AM parts anticipate the next step in development [2, 3]. Those incorporations are expected to considerably improve the possibilities of AM parts for various technical applications. Until now, the focus of the OEMs was to provide systems running stable and facilitate repeatable part manufacturing. Thus, the commercially available setups are closed machines with hardly any or no external manipulation possibility, e.g. regarding sensor integration. These closed manufacturing systems can be considered as one main reason that, despite the attractive combination of sensors and AM polymer parts, currently, there are only a few approaches presented with this regard. Castillo et al. [4] described the production of a helmet by Stereolithography (SLA) with an integrated acceleration sensor and Jentoft et al. [5] used Fused Deposition Modeling (FDM) to create AM parts with photo transistors. Particularly FDM is used attempting to integrate further active components like copper wires [6], copper grids as capacity sensor [7] and printed electronics in general [8]. Liu et al. [9] summarized 3D-printing applications with tactile sensors based on Material Jetting (MJ), FDM and SLA technique. Due to convincing part property profiles, powder bed based processes like SLS are considered today to be the AM-method of choice for industry needs in terms of different quality aspects and long-term behavior [10]. In this context, however, only one approach regarding sensor integration with SLS is described currently. Maier et al. embedded a fiber optical sensor in a Polyamide 12 (PA12) part during its manufacturing on a commercial SLS system (EOS P760) [11].

In the study presented in this paper, a further approach is pursued aiming at the integration of acceleration sensors into 3D-printed SLS parts. In a proof-of-concept the feasibility of sensor integration shall be investigated generally and finally validated by proofing the general functionality of the sensor after SLS processing.

MATERIALS AND METHODS

In order to overcome size issues regarding material consumption and machine handling the present investigation was performed on a commercially available, small scale SLS equipment, purchased from Sintratec (Brugg, Switzerland) (Fig. 1a)). One specific feature of this machine is the laser diode (λ = 445 nm) to selectively melt the powder. This diode requires the usage of black powder in order to achieve sufficiently large energy absorption for complete polymer melting. Thus, the powder used for the manufacturing of demonstrator parts was PA 2202 black from EOS (Krailling, Germany). Process parameters used to manufacture parts were:

- laser velocity (scan head controlled): 550 mm/s
- surface temperature of powder bed: 170 °C (for heat-up and build cycle)
- temperature in process chamber: 140 °C (for heat-up and build cycle)

The acceleration sensor used in this study was supplied from BOSCH (Reutlingen, Germany). Figure 1a) depicts the SLS equipment and the part which should be built during the SLS process can be seen in Figure 1b. The desired acceleration sensor, not depicted in detail for confidentiality reasons, will be integrated in the desired part 1b) during SLS processing.

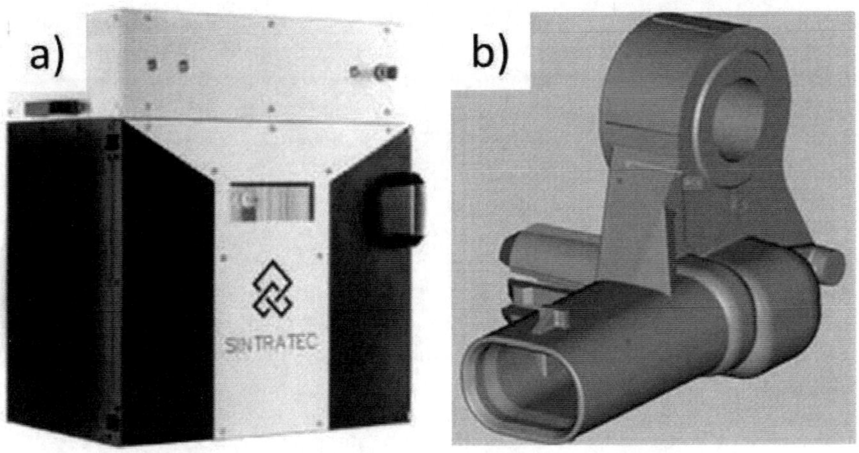

FIGURE 1. a) SLS lab equipment; b) SLS part to be built;

PROCESS MANIPULATION AND PART PRODUCTION

The general procedure for integrating parts during AM processes is schematically described in Figure 2. At a certain point the running process has to be interrupted for two decisive actions: cavity excavation (powder removal) and sensor integration (placement), before the process can be continued.

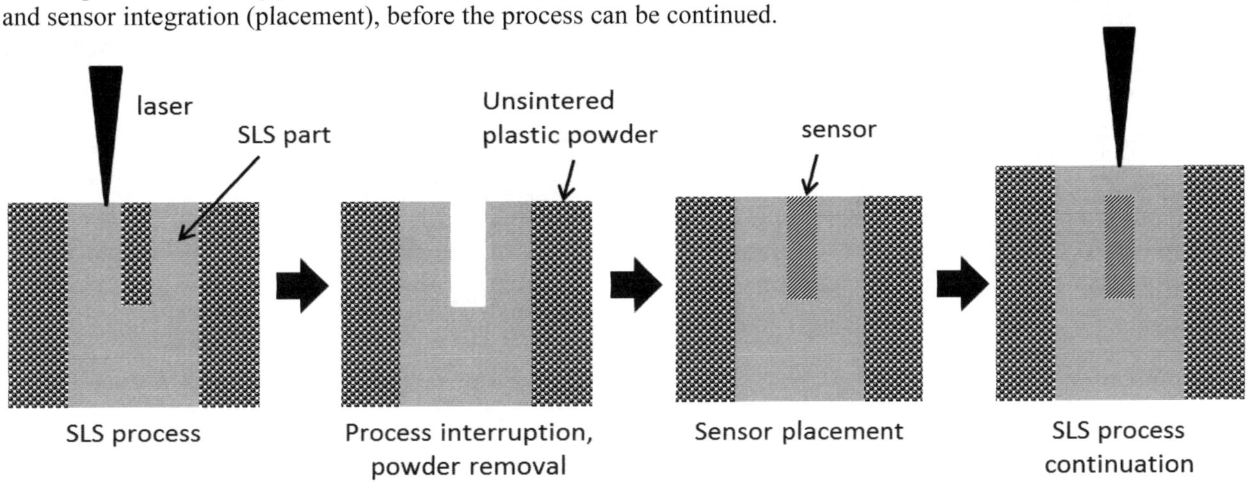

FIGURE 2. Principle process steps for integration of external parts into SLS parts; modified based on [12].

Machine adjustments

Following the procedure in Figure 2 it is obvious that machine adjustments are crucial to intervene in the running process to perform the required manipulation steps. Since the machine control software is not facilitating a designated process interruption, the machine was additionally equipped with manual switches. Hence the motors of the two chambers (powder supply and built chamber) and of the powder recoater (blade) could be controlled manually, enabling a stop of the motions at any desired time.

The goal of the study presented is the "proof-of-concept" for acceleration sensor integration in SLS parts. For that reason, an automation of this embedding procedure is not pursued, but instead the manipulation of the SLS part in form of cavity excavation and sensor integration is done from outside of the machine through a very small slot in the front door. In order to disturb the thermal management of the system as little as possible, this slot is closed with an insulating plug while the standard SLS process is running. It is merely removed for manipulation purposes.

Part modification and powder removal

Regarding the basic project idea, the desired part (Figure 1b)) must be cut into two halves to integrate the sensor during the process course. After building the lower part as a first step, the SLS-sequence has to be stopped and all residual powder has to be removed from the cavity. Given the fact that part manipulation during the running process is necessary, it turned out very quickly in pre-tests that fixation of the part within the build chamber are indispensable. Thus, the main problems to be solved are: part fixation and cavity excavation.

During the course of the project several different ideas were tried out to solve both problems. For powder removal the solution, which turned out best, was to build a sacrificial part (dummy) which fills the desired cavity as good as possible without contacting the part frame. The dummy contains a loop on the top (see Figure 3a)) in order to be grabbed and removed with the pincer, manipulated from outside through the slot in the front door.

In addition to the forces that can displace the component during powder (dummy) removal, further forces are exerted on the component during the insertion of the sensor. This is due to the requirement that the sensor should have no (or as little as possible) clearance to the surrounding SLS housing since it is supposed to move synchronously with the entire part and it must not have a relative motion to it. In order to absorb the forces during sensor insertion the lower part of the geometry must be fixed in the loose powder precisely. Also for this fixation several different ideas were tested, e.g. different frame concepts fitting into the building chamber. Finally, the best solution was a base plate made from a fiber-filled PA 12 (Duraform® HST) which was permanently mounted on the bottom of the building platform (Figure 3b)). The fixation of the SLS part on that base plate is achieved with a pedestal (Figure 3c)). With these auxiliary constructions, gap-filling dummy and parts connection to the base plate, several demonstrators with embedded sensors were manufactured.

FIGURE 3. a) "Dummy" part with loop to grab for easy removal; b) base plate mounted on the building platform of Sintratec machine; c) part with pedestal fixed to the base plate.

Figure 4 illustrates exemplarily the sequential process steps of the process continuation after positioning of the sensor. Figure 4a) and d) show very homogeneously and evenly spread powder layers from which a precise positioning and alignment of the sensor with the SLS part can be concluded.

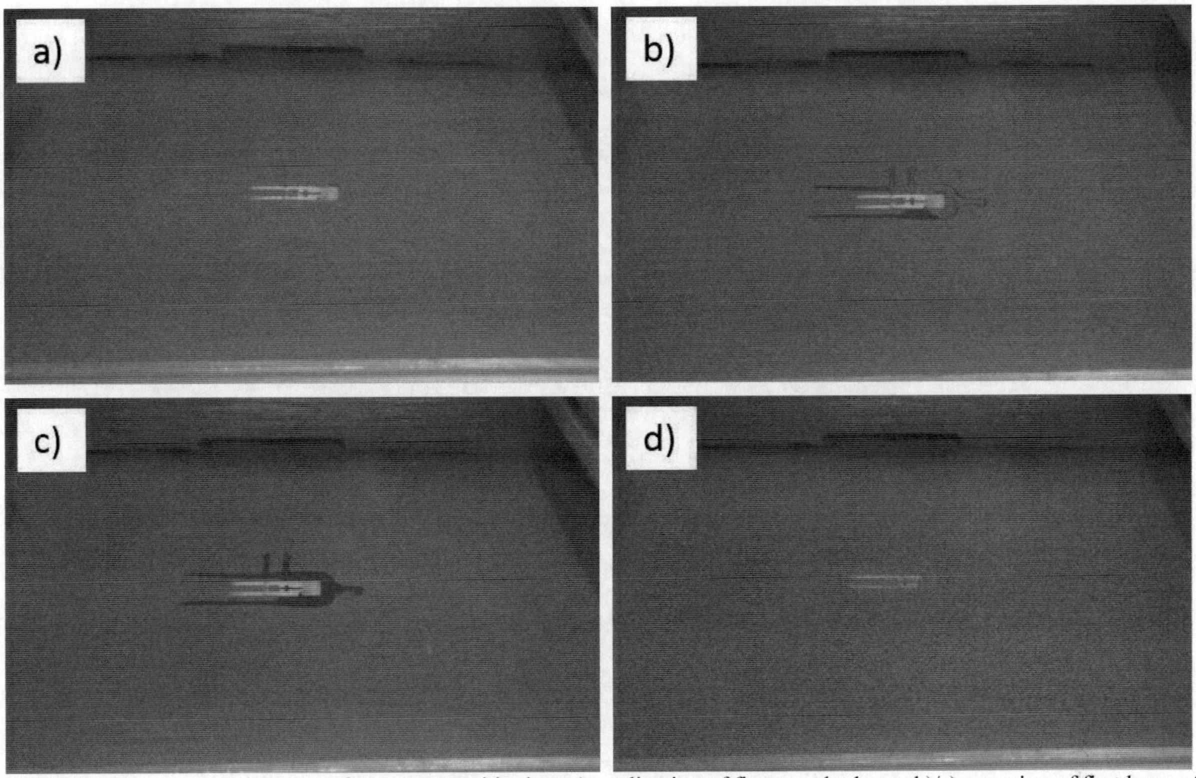

FIGURE 4. Process continuation after sensor positioning. a) application of first powder layer; b)/c) scanning of first layer of top part; d) application of second powder layer.

Figure 5a) depicts the part in the as built version, where the contactors can be seen clearly from front view. Likewise, the proper integration of the sensor was demonstrated by X-Ray analysis. Figure 5b) and 5c) present the respective X-ray images showing the side- and the top-view of the part. The sensor is represented by dark color whilst the polymer body is given by a light grey contour. Furthermore, the plug connecting the contactors to external data logging instruments is displayed in Figure 5b) and 5c).

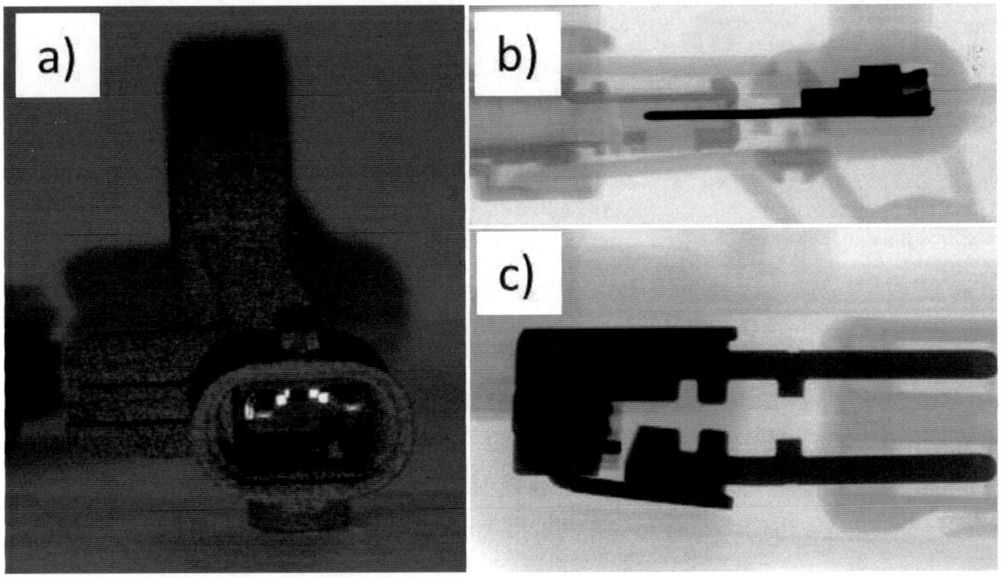

FIGURE 5. a) Finished part with built-in sensor; b) and c) X-Ray image of the SLS part from side and top; light grey: contour of SLS part; dark grey: integrated acceleration sensor.

Sensor activity

After the successful embedding of the acceleration sensor, its remaining sensing capability needs to be proven as well, since the harsh conditions of the SLS process, fine powder particles as well as temperatures of 140 °C in the process chamber, might negatively affect the sensor. The function of the sensors was checked on a shaker device (Figure 6a) and 6b)) with regard to the transfer function. The investigated sensors were functional and could be read out during the test run. The excitation of the sensor is done by the shaker which passes through a frequency profile. The behavior of the SLS part with embedded sensor is similar to the original one. The difference is caused by design modifications that lead to less weight of the SLS part. That means the sensor is fully functional and could be embedded without relevant side clearance.

FIGURE 6. a) Shaker test bench; b) SLS part mounted on test bench;

CONCLUSION AND OUTLOOK

This study described the successful integration and validation of acceleration sensors for various applications into SLS parts. However, a lot of manual operations have to be made and severe machine and part adjustments have to be implemented to achieve this goal. Nonetheless, smart solutions for the challenges of part fixation for manipulation during processing and of powder excavation from the cavity were developed.

The results are encouraging to transfer the project work to larger SLS machines closer to industrial purposes and to develop concepts to automatize process steps like powder removal from the cavity and positioning of the sensors, e.g. with robotic arms or similar apparatus.

REFERENCES

1. S.C. Ligon et al., Polymers for 3D Printing and Customized Additive Manufacturing, Chem. Rev. 2017, 117, 10212–10290.
2. D. Lehmhus et al., "Customized Smartness: A Survey on Links between Additive Manufacturing and Sensor Integration," *Procedia Technology,* vol. 26, pp. 284-301, 2016/01/01/ 2016.
3. R. Jiang, R. Kleer, and F. T. Piller, "Predicting the future of additive manufacturing: A Delphi study on economic and societal implications of 3D printing for 2030," *Technological Forecasting and Social Change,* vol. 117, pp. 84-97, 2017/04/01/ 2017
4. S. Castillo, D. Muse, et al. (2009), "Electronics Integration in Conformal Substrates Fabricated with Additive Layered Manufacturing," in 20th Annual International Solid Freeform Fabrication Symposium SFF- Austin, TX
5. L. P. Jentoft, et al. (2014), "Intrinsic Embedded Sensors for Polymeric Mechatronics: Flexure and Force Sensing," Sensors (Basel, Switzerland), 14(3). 3861-3870
6. D. Espalin, et al. (2014), "3D Printing multifunctionality: structures with electronics," Int J Adv Manu Tech, 72(5), 963-978
7. C. Shemelya et al. (2015), "Encapsulated Copper Wire and Copper Mesh Capacitive Sensing for 3-D Printing Applications," IEEE Sensors Journal, 15(2), 1280-1286
8. M. Matho, B. Sniderman (2017), 3D opportunity for electronics, A Deloitte series on additive manufacturing.
9. C. Liu et al. (2018), "3D Printing Technologies for Flexible Tactile Sensors toward Wearable Electronics and Electronic Skin," Polymers, 10(6), 629
10. M. Schmid, Laser Sintering with Plastics, 2018, Hanser Verlag, Munich, ISBN 9781569906835;
11. R. Maier et al. (2013), Embedded Fibre Sensors within Additive Layer Manufactured Components, IEEE Sens 13(3), 969
12. Lausch H, Töppel T, Petters R, Gronde B, Herrmann M, Funke K (2016). Multi-material Approach to Integrate Ceramic Boxed Temperature-sensitive Components in Laser Beam Melted Structures for Bio and Other Applications. Proceedings of 3rd Fraunhofer Direct Digital Manufacturing Conference (DDMC). ISBN (E-Book): 978-3-8396-1001-5.

Additive Manufacturing of Large Functional Assemblies through a Hybrid Approach

Christian Hopmann[a] and Lukas Pelzer[a*]

[a]*Institute for Plastics Processing (IKV) at RWTH Aachen University,*
Seffenter Weg 201
52074 Aachen

Abstract. Because of the layer-based process, plasticizing additive manufacturing technologies offer a lot of freedom in terms of geometrical design and functional integration and are therefore not only suitable for prototyping but also for final part production. Considering the latter, production time is a crucial factor regarding economic efficiency of the process. Furthermore, most parts have to be post-processed before being used.

To address these challenges, we developed a hybrid manufacturing cell, which is capable of producing large parts in a short amount of time. To achieve this, a screw-based extrusion was utilized which opens up the possibility of using standard pallet material. By using a standardized coupling system, we are able to incorporate subtractive manufacturing to ensure tight tolerances where needed and a pick and place system to insert additional components, such as bearings, bushings or electrical circuits. The machine is therefore capable of producing large functional assemblies autonomously.

In terms of mechanical properties, we are able to achieve tensile strengths which are as good or better when compared to injection moulded parts in x and y direction, because we are able to process highly filled materials. In z direction, layer bonding has to be improved to achieve similar results. To improve the mechanical properties of finished parts while reducing the amount of material being introduced, the use of infill structures is necessary. In its current use, infill is homogenously distributed throughout the component. By analysing the load case, we are able to redistribute the material inside the part, such that highly stressed regions receive more infill while regions with low strength requirements can be produced with less material, resulting in stronger yet lighter parts.

Keywords: additive manufacturing, hybrid manufacturing, functional integration, infill

INTRODUCTION AND MOTIVATION

In recent years, additive manufacturing (AM) has experienced a huge boom, both in personal as well as in industrial applications. This is represented by the amount of search requests for the term 'Additive Manufacturing', shown in figure 1 (a), and is caused by the potential that additive manufacturing offers in combination with the increasing accessibility of the machines and software. Especially for plasticizing processes, machines can be obtained for less than 100 $. Combined with the benefits that additive technologies offer, such as tool-less manufacturing, geometric freedom and complexity [1], functional integration [2] and a very limited occurrence of wasted material during production [2], this means a lot of business opportunities, especially in regard to small production runs and mass customization. Mass customization describes the process of being able to personalize a high-volume product to the needs of the customer. An example would be a computer mouse, which is manufactured based on a scan of the user's hand, therefore fitting its contour perfectly.

If one compares the cost of an injection moulding process to that of additive manufacturing, it can be seen that for smaller lot sizes, AM is cheaper than injection moulding. This is mainly due to the fact that there's no tool which has to be manufactured upfront. Furthermore, adding complexity increases the cost of an injection-moulded part while it is basically free in AM. There, however, increasing the lot size and increasing the part size will increase the cost, as illustrated by figure 1 (b). This is why small production runs and mass customization are typical applications for AM.

Other factors that limit the widespread adoption of AM in final part production are the speed of production [3], dimensional accuracy and surface quality of the parts [4], the integration of AM into existing production lines [5], the repeatability and quality control of the process [6] and finally the anisotropic properties, which are a result of the layer-based process [7].

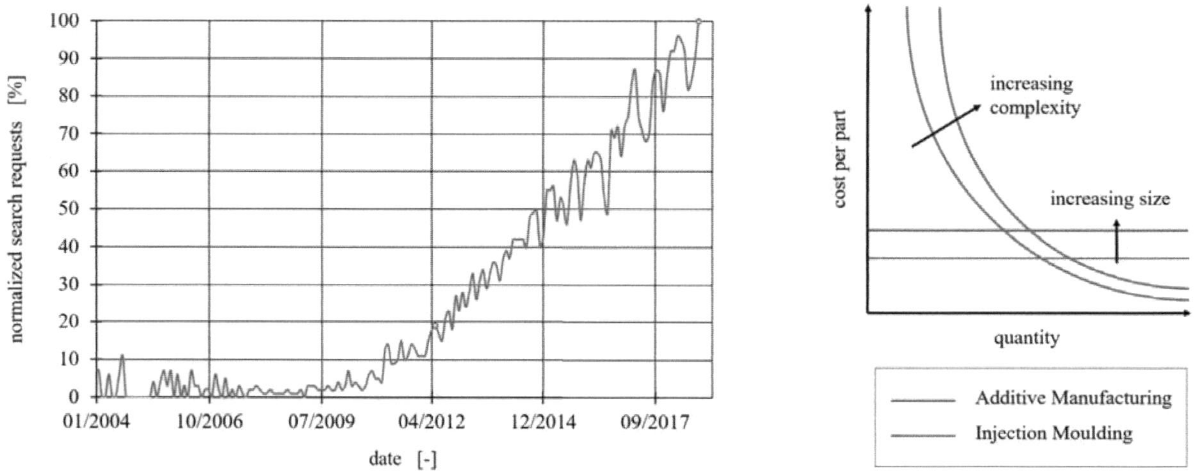

FIGURE 1. Google search results for the term 'Additive Manufacturing' (a) and comparison between costs per part in additive manufacturing and injection moulding (b)

To address those limitations while at the same time using as many of the advantages as possible, we developed a hybrid manufacturing cell at the IKV in Aachen.

PRODUCTION SPEED

Limitations in the speed of production in typical FDM machines are due to the time it takes to plasticize the plastic filament [7]. To address this topic, we developed a screw-based extruder, which is capable of processing pallet-based thermoplastics. This allows us to work with standard materials, circumventing the need for the intermediate step of producing filament, while multiplying the material throughput aby a factor of 20 as compared to a standard FDM process. While decreasing the production time, both factors also keep the production cost to a minimum. In regard to the material, filament is typically ten to 30 times more expensive than pallets, while decreasing the time it takes to produce one part allows for an increase in production volume, effectively multiplying the amount of parts which can be manufactured in an economical way.

The extruder is based on a screw with three distinct zones and mixing elements, which is 16 mm in diameter and 110 mm long. It is driven by a servomotor whose speed is reduced by a factor of 320 through the utilization of a planetary gearbox. The screw helps in homogenizing the material while the shearing forces start the process of plasticization. This is then further assisted by three 100 W heating elements, which are located right above the nozzle. We currently utilize nozzles with a diameter between 0.4 mm and 3 mm, allowing us to produce layer heights between 0.1 mm and 3 mm.

The material is supplied by a dosing unit and is provided to the extruder by a vacuum conveying system, which guarantees uniform feeding and prevents the build-up of pallet bridges in the delivery tubes. The liquid cooled intake zone of the extruder prevents premature plasticization.

Compared to a typical FDM process, the pressure inside the extruder is higher, allowing it to process highly filled materials. Possible filler materials include carbon fibres, glass fibres or graphite particles. While the former two increase the stiffness and strength of the produced part, graphite particles open up the possibility to create electrically conductive components or individual conductive tracks inside a non-conductive part. In our tests using a PA6, which was filled with 20 wt% graphite particles, we were able to achieve a resistivity of 49.67 Ω/m $\pm$ 10.29 Ω/m in the horizontal direction, meaning in the same layer. In the vertical direction, or across the layers, the resistivity was significantly higher at 88.25 Ω/m $\pm$ 18.98 Ω/m. This still shows that it is not only possible to create three-dimensional circuit boards, but that the conducting tracks can be implemented into structural parts, like housings, eliminating the need for separate components and assembly steps.

SURFACE QUALITY AND HYBRID APPROACH

To address challenges like dimensional accuracy and surface quality of the printed parts as well as the integration of AM into existing production lines, we decided to pursue a hybrid approach. This means that the previously described extruder for additive manufacturing is complemented by subtractive tools. This allows the production of a near-net shaped part using AM, which is finished by a mill, creating a very high surface quality and dimensional accuracy. Furthermore, this combination of tools allows the manufacturing of precise fits, which cannot typically be produced through additive means alone because of the technology's inconsistencies.

Because all tools should utilize the same kinematics – a six-axes robot arm – a standardized coupling system is necessary. For this system to work efficiently and autonomously, it is required to couple all available material, energy and information streams. In a previous attempt, a milling spindle was used as a motor for all tools. This meant that the torque of the spindle had to be coupled as well. This also meant that the spindle had to be sized to fit all tools, resulting in a large motor for the extruder and therefore a heavy tool. Because of that and the added complexity, the current version of the hybrid manufacturing cell utilizes different motors for different tools, decreasing complexity and weight for each tool.

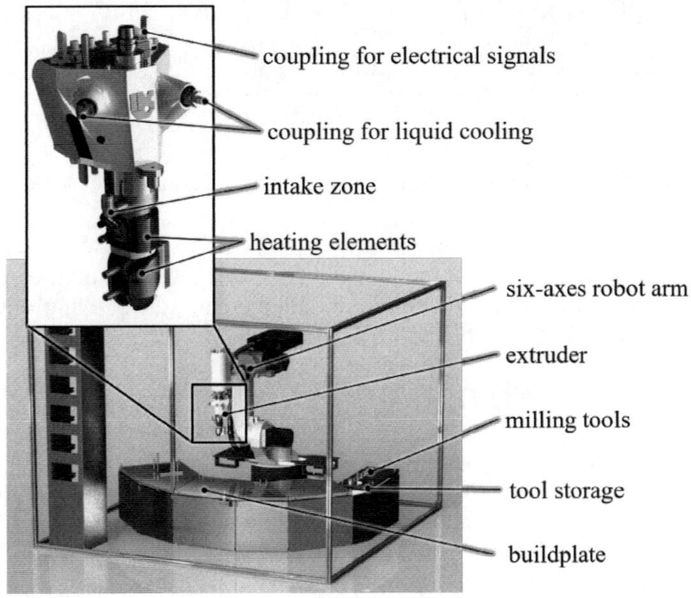

FIGURE 2. Digital mirror of the hybrid manufacturing cell and the screw-based extruder

Besides combining additive and subtractive manufacturing, there are more benefits to a standardized system for changing and coupling tools. For example, a pick-and-place tool is currently being developed, which allows the handling of finished parts. Furthermore, additional components, like bearings, bushings or computer chips, can be added without manual interference.

The process of adding material additively, creating precise surfaces and fits subtractively and inserting components through a pick-and-place system can be repeated multiple times. This further enables the autonomous production of finished parts.

REPRODUCIBILITY AND QUALITY CONTROL

In terms of reproducibility and quality control, additive manufacturing has a lot of catching up to do. While conventional processes are well understood and can therefore be controlled through a closed-loop system, additive machines are up until now mostly driven via an open-loop system. To guarantee a constant and repeatable quality for additively manufactured parts, feedback loops need to be implemented, allowing the machine to sense certain quality related parameters and adapt accordingly.

To be able to implement such a closed-loop control, our system is based on state-of-the-art control technology, allowing us to implement sensors and program all actuators to act based on the sensed input. One example would be a consistent wall thickness. Current FDM machines do not measure the material throughput, they rather calculate upfront how much material has to be extruded at any given time and then calculate the length of filament based on that. If a problem occurs, like the nozzle being blocked or the diameter of the filament changing slightly, the system cannot correct for that. Applying the same principle to the screw-based extruder meant creating a model for the extruded amount of material as a function of the screw's rotational speed. For this, the non-isothermal model of Potente [8] and Tadmor [9] was used. The accuracy of the result was improved by implementing the correction of White [10]. Figure 2 shows the comparison between the calculated throughput and the measured amount of material at 14.25 rpm and 22.5 rpm. The results show a significant discrepancy between the calculated and measured output. Because the measured output is always lower than the expected one, using this model would result in underfilled and weak parts.

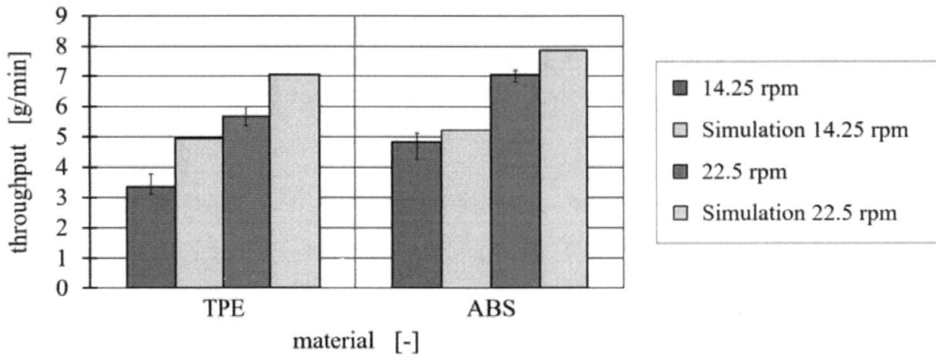

FIGURE 3. Comparison of simulated and measured throughput extruding TPE and ABS through a 0.4 mm nozzle

One solution to circumvent this problem, which can be applied to our hybrid manufacturing system, is the measurement of the current wall thickness. This value is fed back into the system, allowing it to vary the movement speed of the tool accordingly. If, for example, the applied walls were too thick, a slight increase in movement speed without increasing the material output would correct that inaccuracy.

ANISOTROPIC PROPERTIES

Currently, all plasticizing additive technologies are layer-based. This means that the manufactured part is made up of thin slices, which have a height according to the set layer height and which are, to some degree, dependent on the size of the nozzle. The resulting interface surfaces between those layers are a weak point of the part, because the adhesion between an already solidified layer and a new strand of molten material is typically weaker than the adhesion inside the same layer. Therefore, additively manufactured parts are typically weaker in the z- or build-direction, resulting in anisotropic mechanical properties or even delamination of layers.

At the same time, the layer-based process creates new opportunities, like the possibility to create complex internal structures easily. This benefit is often used in additive technologies in the form of infill. Infill structures give the necessary internal support without the need to fill the part completely. This safes production time and material, therefore reducing the cost. It also reduces the weight of the part.

To minimize these negative effects and use these benefits to their full potential, design guidelines have to be established. For other manufacturing processes, this is nothing new. In injection moulding for example, parts are always created with a constant wall thickness to achieve uniform cooling and therefore reduce warping. Up until now, a lot of those design principles have been applied to additive manufacturing. This was warranted when additive manufacturing was still called rapid prototyping and was used for exactly that – prototypes. In 2017 however, more than a third of additively manufactured parts were functional parts, while more than 60 % of the revenue was made in the production of functional parts [11]. This shows that additive technologies are not just a way to prototype for different manufacturing processes anymore, entailing the need for design guidelines which take the specific properties of AM into account.

Concentrating on infill structures, which are typically distributed homogenously throughout the part, we developed an algorithm which takes into account the applied load case and distributes infill accordingly. Because it is made up of specific unit cells, printability is guaranteed.

SUMMARY AND OUTLOOK

The discussed hybrid manufacturing system unites the advantages and minimizes the disadvantages of additive manufacturing. This is achieved through a combination of additive and subtractive processes, the latter of which are capable of circumventing the moderate surface quality and dimensional accuracy of additively manufactured parts. A standardized coupling system allows the usage of different tools via a six-axes robot arm. By utilizing the large working area of the robot arm in combination with the screw-based extruder, the machine is capable of producing large parts in a short amount of time. The integration of a pick-and-place system allows the manufacturing of complete assemblies. Reproducibility can be achieved by implementing closed-loop control systems. Those properties allow the manufacturing of final parts and assemblies in an economical fashion.

In future research, the anisotropic properties of produced parts will be examined further and a system will be implemented which minimizes the corresponding effects and therefore increases the mechanical properties of the products.

ACKNOWLEDGMENTS

We thank each partner who has supported this research project with physical and material donations.

REFERENCES

1. W. Gao et al., "The status, challenges, and future of additive manufacturing in engineering" in *Computer-Aided Design*, Amsterdam: Elsevier, 2015, pp. 65-89.
2. E. Hofmann and K. Oettmeier, "3D-Druck - Wie additive Fertigungsverfahren die Wirtschaft und deren Supply Chains revolutionieren" in *Zeitschrift Führung und Organisation*, Stuttgart: Schäffer-Poeschel Verlag, 2016, pp. 84-90.
3. P. Fastermann, *3D-Drucken – Wie die generative Fertigungstechnik funktioniert*, Berlin: Springer Vieweg, 2016.
4. S. O. Onuh and Y. Y. Yusuf, "Rapid prototyping technology: applications and benefits for rapid product development" in *Journal of Intelligent Manufacturing*, 1999, pp. 301-311.
5. C. Klahn et al, "Design Strategies for the Process of Additive Manufacturing" in *Procedia CIRP 36*, Amsterdam: Elsevier, 2015, pp. 230-235
6. N. Guo and M. C. Leu, "Additive manufacturing – Technology, applications and research needs" in *Frontiers of Mechanical Engineering*, 2013-8, pp. 215-243.
7. I. Gibson et al., *Additive Manufacturing Technologies – Rapid Prototyping to Direct Digital Manufacturing*, New York: Springer, 2010.
8. H. Potente, *Tagungsband zum Seminar Einschneckenplastifizierung*, Paderborn: Shaker Verlag, 2001.
9. Z. Tadmor and C. Gogos, *Principles of Polymer Processing*, New York: John Wiley & Sons, 1979.
10. J. L. White and H. Potente, *Screw Extrusion,* München: Carl Hanser Verlag, 2003.
11. T. Wohlers et al., *Wohlers Report 2017*, Fort Collins, Colorado: Wohlers Associates, 2017.

Mechanical Behavior of Scaffold-Like Structures: Research of Relationships between Properties and Geometry

Claudia Pagano[a*], Lara Rebaioli[a], Francesco Baldi[b] and Irene Fassi[a]

[a]Institute of Intelligent Industrial Technologies and Systems for Advanced Manufacturing, National Research Council, Via Alfonso Corti, 12, 20133 Milan, Italy
[b]Department of Mechanical and Industrial Engineering, University of Brescia, Via Branze 38, 25123 Brescia, Italy

Abstract. Scaffolds are structures used to support the growth of cells in order to replace damaged parts of tissues. They are typically simple structures with a porous arrangement, which allows cells adhesion, but also the flow of the biological fluid. The scaffold geometry and structure are, however, defined by the specific application and customized according to the pathology and the patient. Based on their functions, scaffolds are characteristically made of biocompatible and often biodegradable materials. A deep knowledge of the properties of the scaffold, and their peculiarities, is essential for a correct design. This work aims at analyzing whether the mechanical behavior of a typical scaffold structure could be described by referring to properties intrinsic to the system (independent on the geometry/size), or a geometry/size dependence should be taken into account. Only in the former case, the structure could be treated as an effective "3D material", so that scaffold design could be easily produced and its performance predicted. In order to explore this aspect, scaffold-like structures made of poly(lactic acid) have been manufactured using a Fused Deposition Modelling 3D printing machine. Several specimens with different sizes have been fabricated and their mechanical stiffness and strength measured. The results show that concerning the mechanical strength the structure behaves as a "3D material", whereas for the stiffness a size dependence is observed.

Keywords: Scaffold, Fused Deposition Modelling, Mechanical Properties.
PACS: 81.05.Lg, 81.70.Bt, 89.20.Bb.

INTRODUCTION

Tissue engineering deals with the procedures to rebuild critical size defects in human tissues using suitable artificial matrices (scaffolds) to grow cells and obtain new tissue. For instance, scaffolds are widely employed to treat large-scale bone defects, instead of using tissues taken from the patient himself (autograft reconstruction) or grafts extracted from cadaveric and living sources (allograft reconstruction) [1-3].

Scaffolds are three dimensional (3D), porous and interconnected structures that allow cell adhesion, but also the flow of biological fluid, to promote the tissue growth. Based on their functions, scaffolds are characteristically made of biocompatible and often biodegradable materials. Their geometry (shape and size) and structure (shape and interconnectivity of pores, etc.) are defined by the specific application and customized according to the pathology and the patients [1-3].

For the development of design strategies of parts with a complex 3D-structure, such as the scaffolds, it is essential to verify if the 3D-structured material can be given a profile of mechanical properties that could be considered intrinsic to the system (*i.e.* independent from component shape and size, at least on a length-scale much larger than that characteristic of its structure). In order to investigate this aspect, in this study, the relationships between the basic mechanical properties (stiffness and strength) and the size of the specimens have been researched for polymeric parts having a typical scaffold-like structure. More specifically, parallelepiped-shaped samples having the same base but different heights, made of poly(lactic acid), PLA, have been manufactured using the Fused Deposition Modelling (FDM) 3D printing process, and their mechanical response investigated under compression. Without any change in the filament deposition sequence the specimen height was controlled by the number of layers. In order to know the mechanical behaviour of the constituent material (PLA) in form of filament extruded through the 3D printer nozzle, some portions of the extrudate have been collected and tested under uniaxial tensile tests.

MATERIALS AND METHODS

PLA from NatureWorks, Ingeo™ Biopolymer 4043D, has been selected for this study. According to the manufacturer data sheet, its density and melt flow rate are 1.24 g/cm^3 and 76 g/10 min (evaluated at 210°C with an extruding mass of 2.16 kg, as in the ASTM D1238 [4]), respectively. In order to obtain filaments to feed the FDM machine, the PLA pellets have been processed using a LabTech co-rotating twin-screw extruder (screw diameter = 16 mm, screw length to diameter ratio L/D = 40); the screw rotation speed was set at 110 rpm, feed rate at 2.0 kg/h and barrel temperatures at 180°C except for the first zone, set at 140°C. Filaments with average diameter of 1.75 mm have been obtained.

A desktop FDM printer, Sharebot NG, has been used to manufacture the scaffold-like specimens. The CAD model has been sliced using the software Simplify3D in order to generate the instruction file for the FDM machine. After a preliminary experimental campaign, the printing parameters have been selected and set to: nozzle temperature 220°C, bed temperature 50°C, printing speed 400 mm/min, Cooling "ON".

Starting from the geometry proposed by the standard ASTM D695 [5], parallelepiped-shaped specimens with square base of 12.5×12.5 mm^2 and heights of 6, 12, 18 and 24 mm have been manufactured (10 replicates for each height). The structure selected for this study consists of a simple scaffold-like geometry with a porosity obtained with a 50% infill, which controls the distance between the beads extruded by the FDM machine, and layers deposited with a 90° orientation (FIGURE 1a). In FIGURE 1b a specimen for each height is shown. The four different heights correspond to different number of the layers (15, 30, 45 and 60 layers), whose thickness has been chosen equal to 0.4 mm. Some portions of PLA filament extruded through the 3D printer nozzle using the same processing conditions employed for the printing operations were also collected.

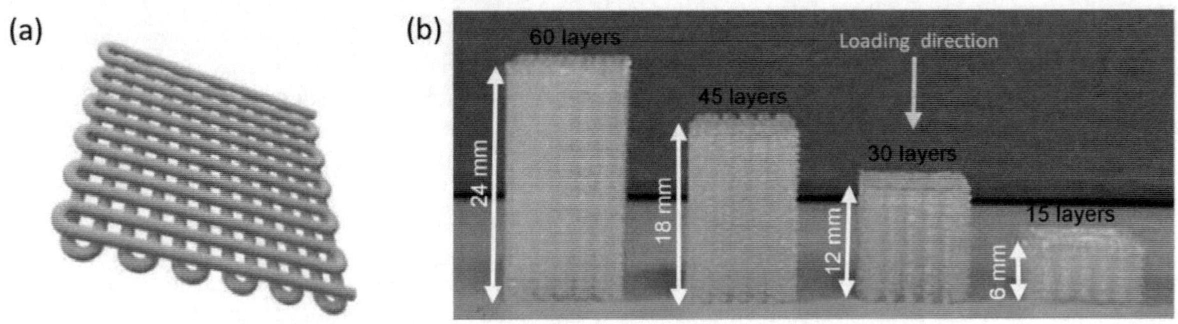

FIGURE 1. (a) Model of the structure geometry, (b) Specimen examples

For each 3D-structured specimen (3D specimen hereafter), the density and porosity degree have been evaluated from the specimen dimensions, measured with a micrometer (resolution of 1 μm), and the mass has been measured with an analytical balance. The porosity degree of the specimens has been calculated as the ratio of the difference between the total volume and the volume occupied by the filament, and the total volume. The volume occupied by the material has been estimated from the ratio of the mass of the scaffold and the density of the filaments. This density has been evaluated as mass over volume, from some filament samples extruded with the same process parameters of the compressive specimens.

The 3D specimens, together with the PLA extrudate filament portions, have been kept in an anhydrous environment until the execution of the mechanical tests, which have been performed with an Instron Universal Testing System (model 3366), at room temperature.

In the compression tests on the 3D specimens (FIGURE 2), the loading direction was parallel to the specimen height (h_0) that is its building direction, according to the layer-to-layer fabrication approach of the FDM machine.

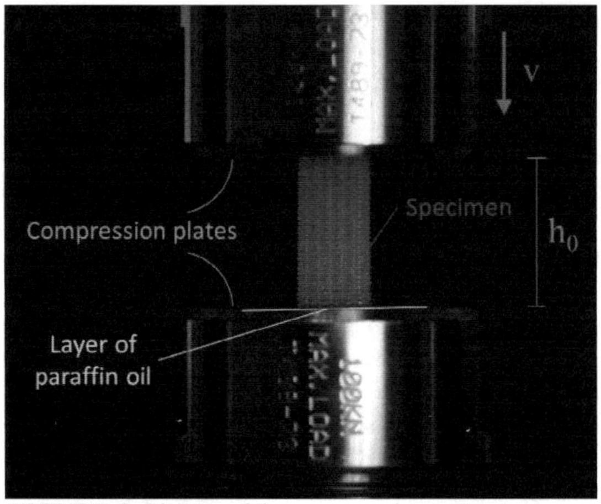

FIGURE 2. Picture of the compression test set-up.

Based on the standard ASTM D695 [5], the crosshead velocity (v) was set equal to 1.2 mm/min for the highest specimen (60 layers), which corresponds to an "apparent" compression strain rate of 0.05 min^{-1}. This "apparent" strain rate has been used also for the tests on the other specimens, by varying the crosshead velocity in dependence on the height. A thin layer of paraffin oil has been deposited on the plate surfaces in order to reduce the friction between the specimen and the compression plates.

The tensile tests on the PLA filament portions were carried out at a strain rate of 0.05 min^{-1}, using a free-to-deform length of 150 mm. Specimen handling and positioning between the grips was made easier by resorting to the use of an *ad-hoc* paper-mask. The results have showed that the PLA exhibits a ductile behavior. A Young's modulus (E_{PLA}) of 3130 (±100) MPa and a yield stress ($\sigma_{y,PLA}$) of 42 (±3) MPa have been obtained.

RESULTS AND DISCUSSION

A typical load (P) versus crosshead displacement (u) curve, obtained from a compression test on a 3D specimen is showed in FIGURE 3. Images of the specimen under testing are also reported on the loading curve in the corresponding testing instant. After the initial "foot" region, corresponding to the establishment of the contact between the moving plate and the specimen, the loading curve shows three typical regions: a first region (R1), where the load increases linearly; a second region (R2), characterized by an abrupt reduction of slope (knee) and a third region (R3), just beyond the knee, where the load increases up to test interruption (in the final region, a non-monotonic trend can even be observed). By analogy with the behavior of cellular materials [6], and by considering the compression direction and the specific 3D-structure (FIGURE 1), it is likely that: in the linear elastic region R1, the load is mostly borne by the material in the filament junctions of the adjacent layers; at the knee (R2), the plastic collapse of the structure occurs, based on localized yielding phenomena of the constituent polymer; in R3, the porous structure undergoes a progressive accumulation of plastic deformation and the filaments crush together, resulting in an evident distortion of the specimen.

For each specimen, the P vs u curve has been converted in an "apparent" stress (σ_{app}) vs "apparent" strain (ε_{app}) curve. σ_{app} and ε_{app} are engineering quantities, since they have been determined from the initial dimensions of the 3D specimen (the stress as P over the initial cross-section; the strain as u, properly corrected to remove the "foot" zone of the P *vs* u curve, divided by the initial height). Moreover, stress and strain have an "apparent" character, since they have been evaluated considering the cross-section despite of the porous structure. From the σ_{app} vs ε_{app} curve of each specimen, the material stiffness and strength have been evaluated. The stiffness, represented by the "apparent" elastic modulus ($E_{app,3D}$), has been evaluated from the slope in the initial linear region of the curve (region R1). The strength has been taken as the stress at the knee in the region R2, which is at specimen failure. This stress value ($\sigma_{app,f,3D}$) has been determined by resorting to a bi-linear representation of the curve around the knee.

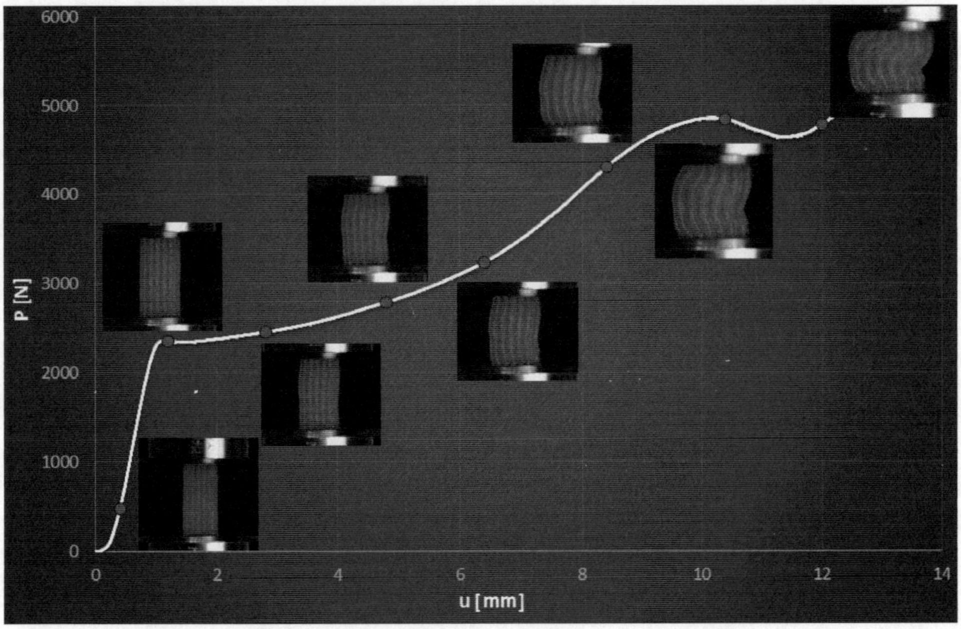

FIGURE 3. Load (P) vs crosshead displacement (u) curve from a compression test on a 3D specimen (60 layers); images of the specimen under testing are also reported with different magnification, only for the sake of qualitative comparisons.

FIGURE 4 reports the $E_{app,3D}$ and $\sigma_{app,f,3D}$ data for the various 3D specimens, represented as relative quantities according to the approach typically used with cellular materials. For each property examined (stiffness and strength), the ratio between the value of the property obtained from the compression test on the 3D specimen and the value of the corresponding property measured for the constituent solid polymer (PLA) with the tensile test, is plotted as a function of the 3D specimen relative density, $\rho_{rel,3D}$, that is the ratio between the effective density measured for the 3D specimen and the density of PLA. $\rho_{rel,3D}$ is the complement to 1 of the porosity degree, ϕ_p. This mode of representation allows highlighting directly the effect of the structure porosity on the mechanical response, and verifying also the possibility to establish a relationship between the relative property and $\rho_{rel,3D}$.

FIGURE 4 clearly shows that both stiffness and strength strongly depends on porosity of the 3D-structure: even a small variation of ϕ_p, due to repeatability errors in the manufacturing process, produces a measurable effect on the mechanical response of the sample. Moreover, it emerges that the strength of the 3D structure ($\sigma_{app,f,3D}$) is independent on the height of the specimen, whereas for the stiffness ($E_{app,3D}$) a dependence on specimen height is observed. With reference to this latter aspect, the higher the number of layers of the specimen tested, the stiffer the 3D material generally appears, for a given ϕ_p. In agreement with the cellular material analysis approach, each series of ($\rho_{rel,3D}$; $E_{app,3D}/E_{PLA}$) data has been fitted with a power-law function:

$$\frac{E_{app,3D}}{E_{PLA}} = C \cdot \rho_{rel,3D}^{n}$$

By imposing $C = 1$ for all the series, n-values between 2.8 (60 layers) and 3.7 (15 layers) have been obtained, with n increasing by reducing the number of layers. The difference between the value of $E_{app,3D}$ obtained for the 15 layer specimens and that of the 60 layer specimens appears rather pronounced (the latter is 1.64 times the former, by considering the averages of the data at the various $\rho_{rel,3D}$). However, since ε_{app} is determined directly from the crosshead displacement, u, a possible contribution of the machine compliance on the specimen height dependence of $E_{app,3D}$ cannot be completely ruled out. Therefore, further studies are currently in progress in order to confirm the specimen height dependence of $E_{app,3D}$ here observed.

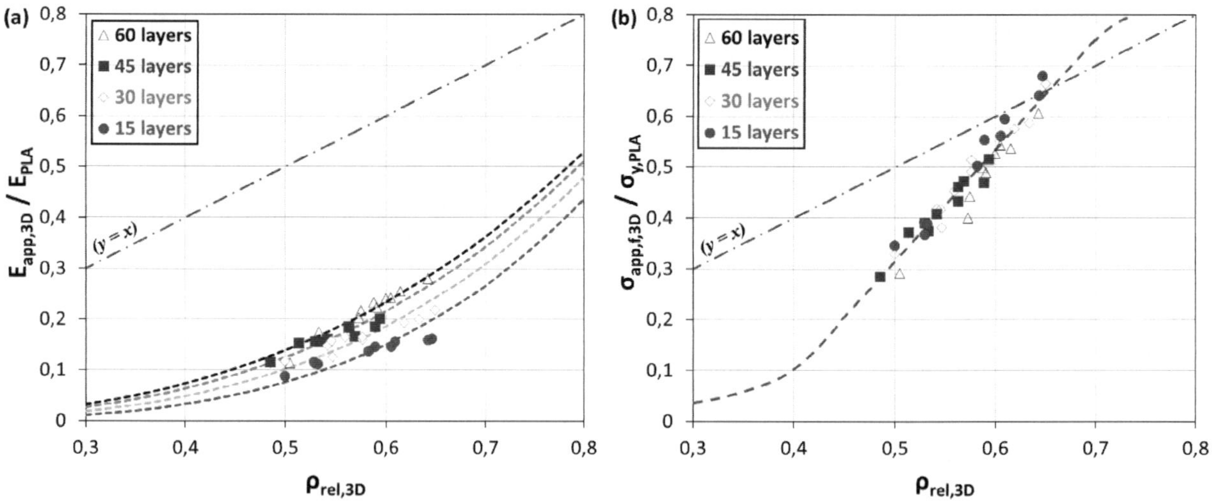

FIGURE 4. (a) Relative stiffness, $E_{app,3D}/E_{PLA}$; dashed lines indicate the fitting curves and (b) relative strength, $\sigma_{app,f,3D}/\sigma_{PLA}$, as a function of the relative density, $\rho_{rel,3D}$, of the 3D-structure; the dashed line is a guideline for the eye.

Interestingly, for the 3D-structure here examined, the contribution given by the constituent polymer (PLA) to the 3D material stiffness is different from that given to the 3D material strength. Indeed, considering the averages of all the data for the 60 layer specimens regardless the $\rho_{rel,3D}$, $E_{app,3D}/E_{PLA}$ is equal to 0.22 whereas $\sigma_{app,f,3D}/\sigma_{PLA}$ to 0.47. This greater influence of the polymer strength, with respect to the stiffness, should be considered for the selection/design of the polymeric material for the construction of a scaffold.

CONCLUSIONS

Since the scaffold geometry and structure are defined according to the pathology and the patient, a deep knowledge of the properties of the scaffold structure and their peculiarities is essential for a correct design. In case the mechanical behavior of a typical scaffold structure could be described by referring to properties intrinsic to the system (independent on the geometry/size) the structure could be treated as an effective "3D material", and the scaffold design could be easily produced and its performance predicted.

In order to explore this aspect, scaffold-like structures made of poly(lactic acid) have been manufactured using a Fused Deposition Modelling 3D printing machine. Several parallelepiped-shaped specimens with different sizes have been fabricated and their mechanical stiffness and strength measured by compression tests. The results showed that the porosity degree controls the stiffness and strength of the 3D-structure. Only the strength, taken as the stress at failure, is intrinsic to the examined structure (thus behaving as a "3D material" concerning the mechanical strength), whereas for the stiffness, a specimen size dependence has been observed. The polymer properties have a stronger influence on the 3D-structure strength rather than on its elastic response.

REFERENCES

1. J. Idaszek, E. Kijeńska, M. Łojkowski, and W. Swieszkowski, *Appl. Surf. Sci.* **388** 762-774 (2016).
2. G. Turnbull, J. Clarke, F. Picard, P. Riches, L. Jia, F. Han, B. Li, and W. Shu, *Bioact. Mater.* **3** 278-314 (2018).
3. S.Bose, S. Vahabzadeh, and A. Bandyopadhyay, *Mater. Today* **16**, 496-504 (2013)
4. ASTM D1238-13 Standard Test Method for Melt Flow Rates of Thermoplastics by Extrusion Plastometer
5. ASTM D695-15 Standard Test Method for Compressive Properties of Rigid Plastics
6. L.J. Gibson, M.F. Ashby, *Cellular solids: structure and properties,* Cambridge University Press, 1999

Carbon Fiber Reinforced Thermoplastics for Material Extrusion Additive Manufacturing

Joamin Gonzalez-Gutierrez [a], Manuel Treitler [a], Martin Spoerk [a], Florian Arbeiter[b], Stephan Schuschnigg [a], Herfried Lammer [c], Maximilian Lackner [d], Mohamed Aburaia [d], Guenther Poszvek [d], Zhang Haiguang [e], Janak Sapkota [a] and Clemens Holzer [a*]

[a] *Institute of Polymer Processing, Montanuniversitaet Leoben, Otto Gloeckel-Strasse 2, 8700 Leoben, Austria*
[b] *Institute of Material Science and Testing of Polymers, Montanuniversitaet Leoben, Otto Gloeckel-Strasse 2, 8700 Leoben, Austria*
[c] *Kompetenzzentrum Holz GmbH, Klagenfurter Strasse 87-89, 9300 St. Veit an der Glan, Austria*
[d] *Faculty of Industrial Engineering, University of Applied Sciences Technikum Wien, Hoechstaedtplatz 6, 1200 Vienna, Austria*
[e] *Shanghai University, Shangda Road, Bao Shan District, Shanghai, 200444, China*

Abstract. In an effort to broaden the engineering applications of material extrusion based additive manufacturing (MEAM), new materials are being developed. Adding carbon-fibers (CF) has been one strategy to increase the mechanical performance of different thermoplastics. One challenge is to determine the amount of CF needed to increase the mechanical performance without affecting the "printability" of the compounds. In this paper, different amounts (10, 15, and 20 vol.%) of CF were added to recycled polypropylene (rPP) and polyamide 12 (PA12). A compatibilizer was used for rPP, but not for PA12. Filaments for MEAM were extruded from the different compounds and the viscosity as well as the tensile properties were measured and compared to the processed polymeric matrices. It was observed that the viscosities at the angular frequencies relevant for MEAM (100 to 200 rad/s) were not significantly different for rPP+CF compounds, but it was higher for PA12+CF compounds. As expected, the elongation at break significantly decreased with the addition of CF for all compounds. For the composites with an rPP matrix, the Young's modulus and the ultimate tensile strength (UTS) continuously increased as the CF content increased to 20 vol.%. For PA12-based materials, the Young's modulus and the UTS increased with CF content, but adding more than 15 vol.% did not further improve these values. Therefore, it was concluded that for PA12 the maximum amount of CF that should be added was 15 vol.%. Using scanning electron microscopy, it was observed that the CF were homogeneously dispersed in the rPP matrix, but not so well in the PA12 matrix, with fibers being more concentrated towards the rim of the filament. Finally, filaments of rPP, rPP+20CF, PA12 and PA12+15CF were used to print complex geometries by means of MEAM, and it was observed that CF helped to reduce the warpage compared to the unfilled filaments. A potential application of this phenomenon could be the reduction of the bed temperature to develop a more energy efficient MEAM process for semi-crystalline polymers.

Keywords: Additive manufacturing, carbon fiber, material extrusion, recycled polypropylene, polyamide

INTRODUCTION

Material Extrusion based Additive Manufacturing (MEAM) [1], also known as Fused Filament Fabrication (FFF) or Fused Deposition Modeling (FDM™), is one of the most widely used additive manufacturing techniques. The main reasons of its popularity have been its reliable, safe and simple fabrication process, low cost, and the availability of a variety of materials [2]. Most of the applications of parts produced by MEAM are for hobby or prototyping applications in part due to the properties of the materials available; therefore in an effort to increase the engineering applications of parts produced by MEAM, new materials are being modified to increase the quality of the parts produced by MEAM. Of particular interest are semi-crystalline thermoplastics, such as polypropylene and polyamide due to their desirable properties in engineering applications.

One common way to modify the properties of semi-crystalline thermoplastic materials to improve the dimensional stability by MEAM is to add fillers. For example, volcanic ash particles and glass spheres have been added to a polypropylene random copolymer in order to reduce the warpage and shrinkage of the printed parts, which resulted in parts with improved geometrical accuracy [3, 4]. The mechanical properties of the same polypropylene were improved by adding short carbon fibers (CF), such fibers also improved the geometrical accuracy of the parts produced by MEAM [5]. Among the different fillers, CF seem to be the most suitable candidates for producing filaments for engineering applications, since they are easy to combine with different matrices and those compounds can easily be extruded to filaments for MEAM. Moreover, short fibers are also used to provide mechanical properties between continuous fiber-reinforced composites [6], primarily in structures in the aerospace and automobile industries, and the unreinforced polymers used in everyday applications [5].

One challenge is to determine the amount of short CF needed to increase the mechanical performance without affecting the processability of the compounds by MEAM. In this paper, different amounts (10, 15 and 20 vol.%) of CF were added to postindustrial waste recycled polypropylene (rPP) and polyamide 12 (PA12).

MATERIALS AND METHODS

Two polymeric matrices were selected in this study: recycled polypropylene (Kruplene-C PP multicolor MFR 0.4 - 1, Kruschitz GmbH, Austria) and polyamide 12 (Vestamid L1940 natural color, Evonik Resource Efficiency GmbH, Germany). Both materials were received as granules. rPP granules were a mixture of green and white pellets since they were recycled from postindustrial waste. PA12 pellets had natural color. The reinforcement materials for both matrices were short carbon fibers (CF.LS-MLD180, Apply Carbon S.A., France) with a diameter of 7 µm, a length of 180 µm and a density between 1.7 and 1.9 g/cm^3, as mentioned by the producer [7]. In order to improve the compatibility of the carbon fibers to the rPP matrix the compatibilizer Scona TSPP 102013 GB (BYK-Chemie GmbH, Germany) was used. No compatibilizer was used for the PA12 compounds. Due to its hygroscopic nature, PA12 granules were dried overnight in a vacuum oven at 80 °C (VD 23, Binder GmbH, Germany) before being compounded. rPP granules were not dried before compounding, since rPP is not hygroscopic. Three different amounts of filler addition were investigated for each of the matrices selected. The preliminary compounds were prepared in a Plasticorder PL2000 kneader (Brabender GmbH & Co. KG, Germany). The formulations and the processing conditions used in the kneader are shown in Table 1.

TABLE 1. Formulations and processing conditions used to produce the CF-reinforced compounds.

Name of compound	Vol.% Matrix	Vol.% CF	Vol.% compatibilizer	Kneader Temp. (°C)	Kneader speed (rpm)	Kneading time (min)
rPP kneaded	100.00	0.00	0.00	200	60	30
rPP+10CF	89.50	10.00	0.50			
rPP+15CF	84.25	15.00	0.75			
rPP+20CF	79.00	20.00	1.00			
PA12 kneaded	100.00	0.00	0.00	190		
PA12+10CF	90.00	10.00	0.00			
PA12+15CF	85.00	15.00	0.00			
PA12+20CF	80.00	20.00	0.00			

Filaments for preliminary trials were prepared in a capillary rheometer Rheograph 2002 (Gottfert Werkstoff-Prüfmaschinen GmbH, DE) with a cylinder of 290 mm in length and 12 mm in diameter, and a round die with a length of 30 mm and a diameter of 1 or 1.75 mm. The smaller die diameter was used for unfilled polymers, since extrudate expansion ("die swell") was observed. For filled polymers no die swell was observed, thus the target diameter for the filament was used (1.75 mm). The piston speed during extrusion was 1 mm/s. The extrudate of the capillary rheometer was pulled away from the die using a conveyor belt (Geppert-Band GmbH, Germany).

Rheological measurements of the compounds were performed in a MCR501 rotational rheometer (Anton Paar GmbH, Austria). The measuring geometry was a plate-plate with diameters of 25 mm and a measuring gap of 1 mm. Measurements were performed in the linear viscoelastic range in oscillatory mode at frequencies between 0.1 and 500 1/s. The specimens for rheometry were compression molded disks with a diameter of 20 mm and thickness of 2 mm. Compression molding was done in a vacuum press (P200 PV, Dr. Collin GmbH, Germany) at 200 °C for all materials.

The tensile properties of the preliminary filaments were measured directly on the produced filaments in a universal testing machine (Z001, Zwick GmbH & Co. KG, Germany) with a 1 kN load cell. The measurement length was 50 mm and the crosshead speed was 10 mm/min. All measurements were done at 23 °C and 50 % relative humidity.

The dispersion of the fibers in the filaments was investigated by means of scanning electron microscopy (SEM) in a Tescan Vega II (Tescan Brno s.r.o., Czechia). The filament samples were cryo-fractured using liquid nitrogen and after fracture they were fixed to the SEM platform and were gold-sputtered.

The reinforced compounds that showed the best tensile properties were scaled up (rPP+20CF and PA12+15CF) along with the virgin materials for comparative reasons. First, the compounds were pre-mixed in a kneader with a capacity of 300 cm^3 (Polylab Rheomix 3000, Thermo Scientific, Germany). Compounds with an rPP matrix were kneaded at 200 °C and compounds with PA12 at 190 °C. All of them were mixed for 10 min at a rotational speed of 60 rpm. Three batches per material were prepared. After the compounds were granulated, they were passed through a co-rotating twin-screw extruder (ZSE 18 HP 48D, Leistritz Extrusionstechnik GmbH, Germany) to ensure a proper dispersion of the CF. Extrusion temperatures between 25 and 180 °C (from feeding zone to extrusion die) were set in the compounder and the rotational speed was kept constant at 600 rpm. After compounding the extrudates were granulated. Filaments were produced from the granulated materials using a single-screw extruder (FT-E20T-MP-IS Dr. Collin GmbH, Germany). The heating zones of the extruder were set between 185 and 215 °C from the feeding zone to the die. The screw´s rotational speed was set to 70 rpm. The extrudate was cooled in a 3 m water bath and the filament was pulled using a self-developed haul off unit. The filaments were wound into spools in a self-developed winding unit. Before the filaments were spooled their diameter and their ovality were measured using a Sikora Laser 2010 T with Ecocontrol process-system (Sikora AG, Germany).

Gears were produced with filaments made out of rPP and rPP+20CF using a Duplicator i3 (Wanhao 3D printer, China) and PA12 and PA12+CF15 using a Hage3Dp-A2 (Hage Sondermaschinenbau GmbH & Co KG, Austria). The reason for using two machines was because the printing substrate was a PP plate for the rPP materials [8], which could be easily attached to the smaller printer. A standard glass mirror was used for printing the PA12 materials. Hair spray was used to enhance the adhesion of PA12 materials and for unfilled PA12 a brim was used in the first layer to prevent warpage and detachment from the printing bed. Set printing parameters are summarized in Table 2.

TABLE 2. Processing conditions used to produce the unreinforced and CF-reinforced compounds by MEAM

Parameter	rPP	rPP+20CF	PA12	PA12+15CF
Nozzle diameter (mm)	0.6	0.6	0.5	0.5
Printing speed (%)	50	60	150	65
Flow rate (%)	151	195	100	95
Extrusion temperature (°C)	230	230	240	250
Bed temperature (°C)	60	70	110	90
First layer brim (no. lines)	0	0	6	0
Hair spray on bed	No	No	Yes	Yes

RESULTS AND DISCUSSION

The rotational viscosity as a function of angular frequency at 190 °C of all the materials investigated is shown in Figure 1. It can be seen that the viscosities at the angular frequencies relevant for MEAM (100 to 200 rad/s) were not significantly different for any of the compounds with rPP. The compounds with PA12 had 9 to 69% higher viscosity values compared to the rPP compounds at the same filling grades and temperature. This indicates that the printing temperature has to be higher for PA12 compounds compared to rPP compounds in order to obtain good cohesion between the deposited layers. When fibers were added to the PA12 matrix the viscosity increased, but there is no significant difference between the compounds having 10 and 15 vol.% CF. A significantly higher viscosity was observed for the compound PA12+20CF, which indicates that this material might require higher processing temperature during MEAM to have a similar flow rate.

The tensile properties of the filaments from the different compounds are shown in Figure 2. As expected, the elongation at break significantly decreased with the addition of CF for all compounds (Figure 2a). For the compounds with an rPP matrix, the tensile modulus (Figure 2b) and the ultimate tensile strength (Figure 2c) continuously increased as the CF content increased to 20 vol.%, which hints at a decent compatibilisation of the fillers. For PA12 compounds, the Young´s modulus (Figure 2b) and the ultimate tensile strength (Figure 2c) increased with CF content, but adding more than 15 vol.% did not further improve these values. Therefore, it was concluded that for PA12 the maximum amount of CF that should be added was 15 vol.%. Thus the materials that were scaled-up were rPP+20CF and PA12+15CF as well as the pure materials.

Figure 3 shows scanning electron microscopy images for rPP+20CF and PA12+15CF. It can be observed that the CF were homogeneously dispersed in the rPP matrix, and a strong filler-matrix interface evolved. For the PA12 matrix, fibers were more concentrated toward the rim of the filament. Thus the compatibilizer added to rPP compounds was appropriate. Furthermore, the filaments of PA12+15CF have more pores than the rPP+20CF; this might explain the poor reinforcement of the CF in the PA12 matrix and indicates a bad compatibility between the fibers and the PA12 matrix.

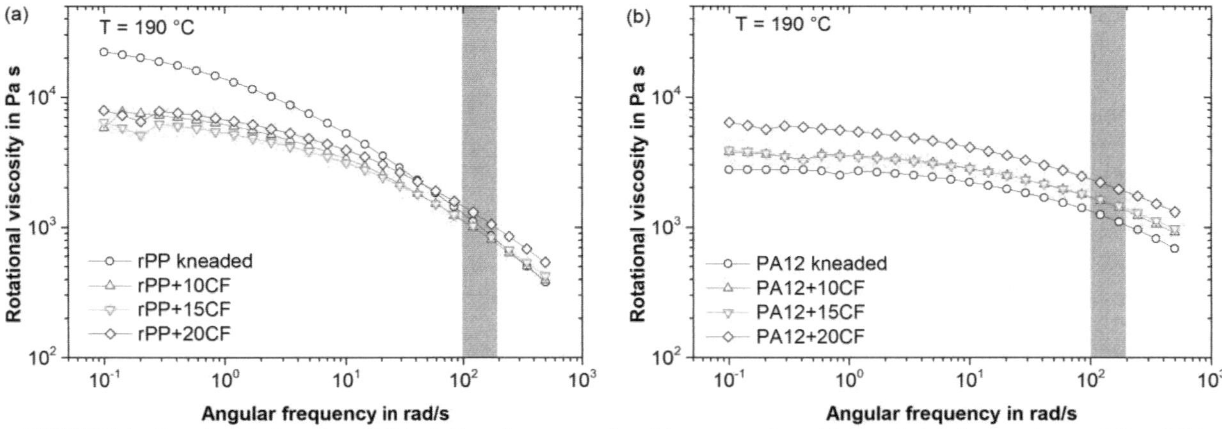

FIGURE 1. Rotational viscosity of CF-reinforced compounds with two matrices (a) recycled polypropylene - rPP and (b) polyamide 12 - PA12. Shaded areas are the oscillatory frequencies relevant for the MEAM process (100 to 200 rad/s).

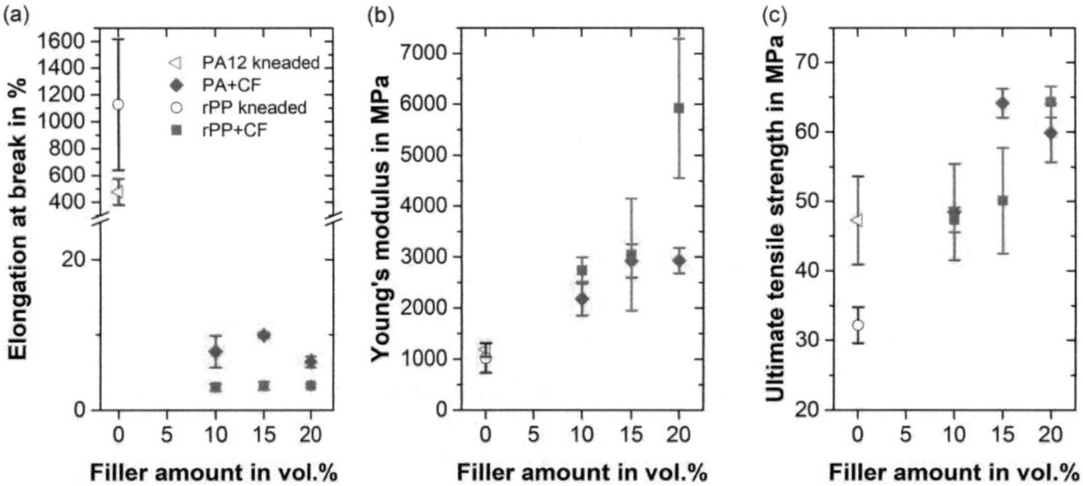

FIGURE 2. Filament tensile properties of CF-reinforced filaments as a function of filler amount: (a) elongation at break, (b) Young's modulus and (c) ultimate tensile strength for rPP and PA12 compounds.

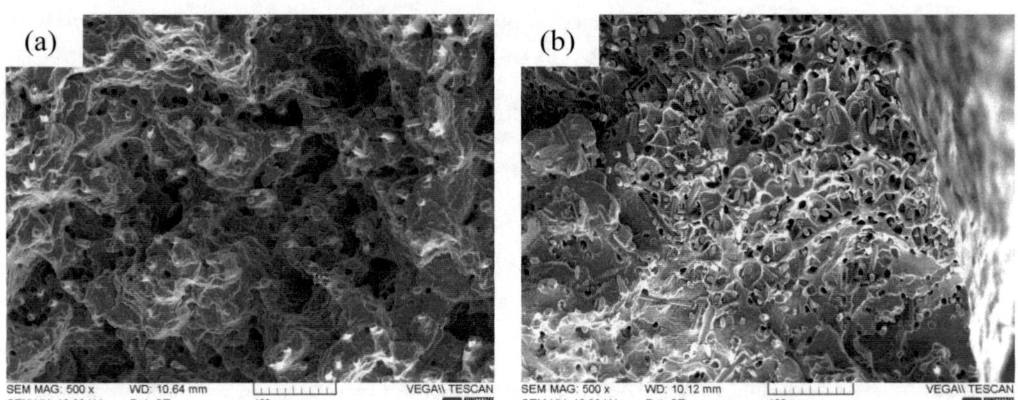

FIGURE 3. Scanning electron microscopy images of cryo-fractured filaments of (a) rPP+20CF and (b) PA12+15CF.

Using the filaments that were scaled-up (rPP kneaded, rrPP+20CF, PA12 kneaded, and PA12+15CF), gears were produced by MEAM. Figure 4a shows gears prepared from the four different materials. All the materials were printable, but unfilled materials had a tendency to warp, while filled materials remained flat against the printing bed. Thus it can be concluded that the addition of CF improves the geometrical stability of rPP and PA12. The results are in line with the results shown for a PP random copolymer [5]. For unfilled PA12, the only way to print the gears was by using a brim (Figure 4b). The brim prevented detachment from the printing bed. For PA12+15CF the brim was not necessary (Figure 4c) and the gears could be printed without detachment from the printing bed.

FIGURE 4. (a) Gears produced by MEAM with 4 materials, (b) printing of PA12 with a brim, and (c) printing of PA12+15CF without a brim.

CONCLUSIONS

Carbon fiber (CF) were added to recycled polypropylene (rPP) and polyamide 12 (PA12) at different concentrations (10, 15 and 20 vol.%). Filaments were produced and characterized in terms of viscosity and tensile properties. It was observed that the viscosity at the relevant shear rates for MEAM of CF-reinforced rPP did not increase with the addition of CF, but for PA12 there was an increase of 76 % as the content reached 20 vol.% compared to unfilled PA12. The tensile properties of rPP continously increased with the addition of CF, but for PA12 the properties reached a maximum at 15 vol.%; this suggests that CF were properly compatibilized for rPP, but not for PA12. The lack of compatibility was also observed in SEM images that show an accumulation of the CF towards the rim of the PA12-based filaments, but a homogeneous distribution for the rPP-based filaments. Similarly to other fillers and matrices, the addition of CF improved the geometrical stability of the printed parts. CF prevented warpage and detachment from the printing bed, which was particularly important for the PA12 materials. A potential application of this phenomenon could be the reduction of the bed temperature in an attempt to develop a more energy efficient MEAM process. Other compatibilizers will be investigated in the future to improve the mechanical performance of PA12 based materials.

ACKNOWLEDGEMENTS

Authors acknowledge the financial support of the Austrian Promotion Agency (FFG) and Shanghai University for the Austria-China project "Natural3D" under Project No. 860384. Authors acknowledge Kruschitz GmbH, Evonik Resource Efficiency GmbH and Apply Carbon S.A. for supplying the materials used in this study.

REFERENCES

1. ISO/ASTM DIS 52903:Standard specification for material extrusion based additive manufacturing of plastic materials (2016)
2. J. Gonzalez-Gutierrez, S. Cano, S. Schuschnigg, C. Kukla, J. Sapkota and C. Holzer. *Materials* **11**, 840 (2018)
3. M. Spoerk, J. Sapkota, G. Weingrill, T. Fishinger, F. Arbeiter and C. Holzer. *Macromol. Mater. Eng.* **302**, 1700143 (2017)
4. M. Spoerk, C. Savandaiah, F. Arbeiter, J. Sapkota and C. Holzer. *Polymer Composites* **40**, 638-651 (2017)
5. M. Spoerk, C. Savandaiah, F. Arbeiter, G. Traxler, L. Cardon, C. Holzer. and J. Sapkota. *Composites Part A* **113**, 95-104 (2018)
6. A.N. Dickson, J.N. Barry, K.A. McDonnell, and D.P. Dowling. *Addit. Manuf.* **16**, 146-152 (2017)
7. Applied Carbon S.A. CF.LS-MLD180 https://www.procotex.com/assets/downloads/CF.LS-MLD180-TDS-2015.pdf (2015)
8. M. Spoerk, J. Gonzalez-Gutierrez, C. Lichal, H. Cajner, G.R. Berger, S. Schuschnigg, L. Cardon, and C. Holzer. *Polymers* **10**, 490 (2018)

Authors do not want to publish a paper, since results have already been published elsewhere.

Development of Eco-Friendly Composites Based on Polypropylene and Cellulose for Additive Manufacturing (Fused Deposition Modeling)

Elena Mastalygina[a,b*], Petr Pantyukhov[a,b], Anatoly Popov[a,b], Alessio Massironi[c], Andrea Morelli[c], Dario Puppi[c], and Federica Chiellini[c]

[a]*Scientific laboratory "Advanced composite materials and technologies", Plekhanov Russian University of Economics,36 Stremyanny lane, Moscow, Russia (elena.mastalygina@gmail.com)*
[b]*Emanuel Institute of Biochemical Physics, Russian Academy of Sciences, Moscow, Russia*
[c]*Department of Chemistry and Industrial Chemistry, University of Pisa, UdR INSTM PISA, 13 Via Giuseppe Moruzzi, Pisa, Italy*

Abstract. The paper is devoted to the preparation of a composite material based on isotactic polypropylene (PP) and a microcrystalline cellulose (MCC) filler for additive manufacturing. MCC was treated by polypropylene-graft-maleic anhydride (PP-g-MA) in toluene solution. This method of modification had a compatibilizing effect on the composite structure. Filament samples of neat PP and PP composites loaded with either untreated or PP-g-MA-treated MCC were obtained by extrusion. Comparing technological properties of different filament types, it was stated that PP-g-MA modification conduced to lower shrinkage and uniform diameter of the filament. The developed filaments were employed to develop 3D polymeric devices with predefined geometry and porosity by fused deposition modelling with a prototyping machine.

Keywords: additive manufacturing, fused deposition modeling, biocomposites, polypropylene, cellulose.

INTRODUCTION

The production technology of additive manufacturing is characterized by high flexibility and allows developing polymeric products for different applications. The most common by number of users is a material extrusion technique called fused deposition modeling (FDM) [1]. There are a few types of polymers used for FDM. Usually PLA and ABS polymers are used for this purpose because of low values of shrinkage and high rigidity [2]. However, these materials are more expensive than polyolefins. Using polyolefins and composites on their basis is very promising idea. Polypropylene is used less often than ABS and PLA polymers in additive manufacturing. The reason for this is a large shrinkage, from 1.3 to 2.4% (for comparison: the shrinkage of ABS plastic is 0.4-0.7%). Such a defect leads to significant deformations of the models already after the printing is completed [3]. In addition, this material adheres to the cold print area, so for better effect it is recommended to use a heated plate.

To enhance technological and mechanical properties of thermoplastic polymers, inorganic and organic fillers are commonly used. For instance, the addition of cellulose to the polyolefin matrix allows creating biocomposites with a higher tensile strength and modulus of elasticity than pure polyolefins [4, 5]. Furthermore, the partial substitution of polyolefin with a biodegradable natural filler like cellulose is the way of creating more eco-friendly materials with a high ability to biodegradation. In a previous work [6], the effectiveness of cellulose as an additive stimulating polyolefin composites biofouling by mold fungi was shown.

This work is devoted to developing filled polypropylene composites for additive manufacturing. The main target of the work was to obtain uniform filaments based on polypropylene with small amount (5 wt.%) of microcrystalline cellulose and their processing by FDM into 3D porous structures. For this reason, different types of filler modification were used, including compatibilizer additive (polypropylene grafted maleic anhydride) and ultrasonic treatment. In future experiments, increasing of filler content is planned.

MATERIALS

Isotactic polypropylene (PP) (MOPLEN HP500L) with MFI of 12g/10min (230°C and 2.16 kg load) and melting point of 163±2°C provided by Lyondell Basell (Houston, TX, USA) was used as a matrix. Microcrystalline cellulose (MCC) (101 grade, Progress, Kemerovo, Russia) with particles size less than 100 μm was used as a filler for composites. Polypropylene-graft-maleic anhydride (PP-g-MA) with MA content of 8-10 wt. % provided by Sigma-Aldrich Co. was used as a compatibilizer. PP-g-MA has the following molecular mass characteristics: M_w = 9100 Da, M_n=3900 Da (GPC).

According to the optical microscopy analysis (Axio Imager Z2m Carl Zeiss, Germany), MCC consists of elongated particles (Figure 1a). The average length to diameter ratio of MCC particles was 2.5. Fraction analysis of MCC grade 101, obtained from the factory, showed that the majority of particles are bigger than 100 μm (Figure 1b). A significant amount of MCC particles aggregated to agglomerated structures. For the preparation of composites, fractions with size less than 100 μm was used. The FTIR analysis of MCC (Spectrum 100, Perkin Elmer) showed uniform composition of used cellulose without impurity inclusions [7]. To study the thermal stability of MCC, thermogravimetric analysis was performed by thermobalance NETZSCH TG 209 F1 Iris [8]. Cellulose is more stable than other polysaccharides: MCC was thermally stable up to 270±5°C, for comparison corn starch curve is also given at Figure 1 (c). The high degradation temperature of MCC allows obtaining composites based on different synthetic polymer matrices.

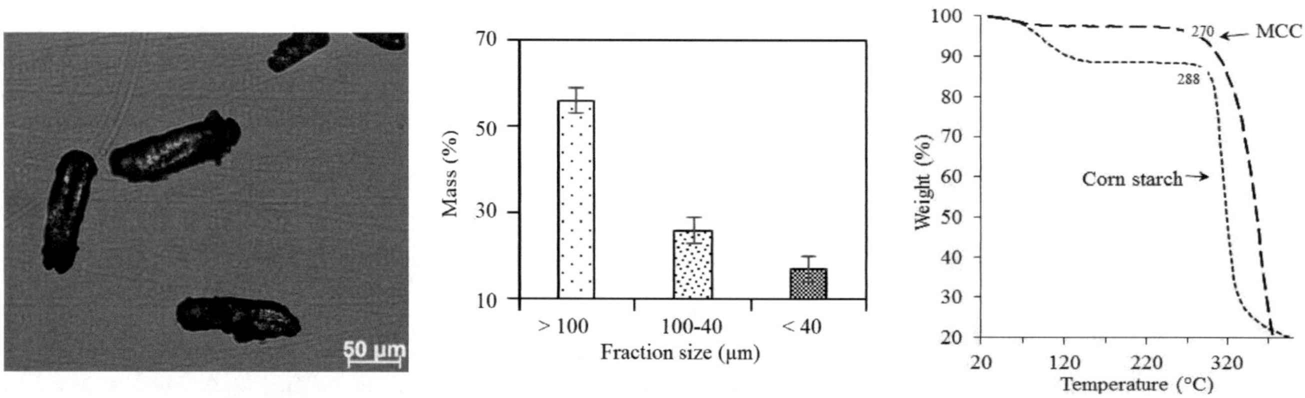

(a) (b) (c)

FIGURE 1. (a) Microphotograph of MCC, optical microscopy (reflection light, magnification of 200x); (b) Particle-size distribution of MCC grade 101; (c) Thermogravimetric curves for MCC and corn starch (reference sample).

DEVELOPED TECHNOLOGY

Before mixing, PP pellets were crushed by a cutting-grinding mill (MF 10, IKA) after freezing at a temperature of minus 20°C for 24 h. The milling process had two stages: pre-milling with a 8 mm sieve and repeated milling with a 2 mm sieve. The rotation speed was 4500 rpm and the mill was cooled down by means of a constant airflow. As a result, the PP powder was obtained.

Fractionating of MCC was performed using a sieve analyzer (A059-02KIT, Matest) with a brass sieve of 300×50 mm in size and sieve openings of 0.1 mm. Prior to treatment and compounding, cellulose was conditioned at a temperature of 105°C for 24 h.

To reduce the size of MCC particles and modify MCC surface by PP-g-MA, ultrasonic treatment in liquid medium (toluene, CP) was performed. Under the same conditions, MCC without PP-g-MA coupling agent was prepared.

0.33 g of coupling agent (PP-g-MA) was added to 16.7 mL of toluene and magnetically stirred (rotor speed of 1000 rev/min) for 1 h at 100°C. To avoid the evaporation of toluene a backflow condenser was used. After dissolution of PP-g-MA in toluene, 1.67 g of MCC was added. After that, MCC was dispersed in the obtained solution by a disperser IKA ULTRA-TURRAX T 50 at a frequency of 30 kHz and an amplitude of 100%. Dispersion was performed simultaneously using an ULTRA-TURRAX disperser and a magnetic stirrer without additional heating (Figure 2 (a)). Cellulose dispersed by this method of treatment formed a stable suspension with PP-g-MA in toluene. Cellulose without PP-g-MA treated by the same method in toluene, settled to the bottom of a vessel (Figure 2 (b), (c)). It was concluded that chosen compatibilizer is effective for this filler.

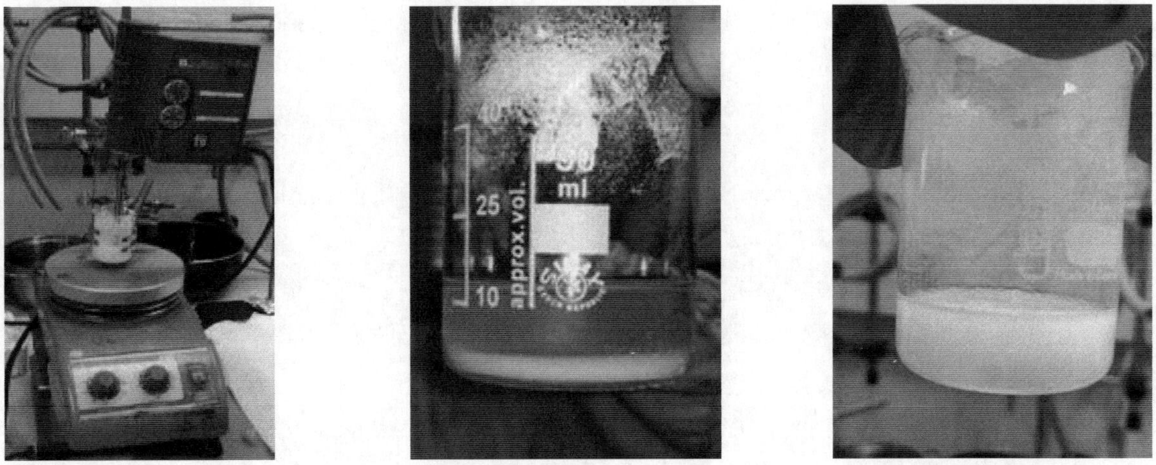

(a) (b) (c)

FIGURE 2. (a) Treatment of MCC in toluene solution with/without PP-g-MA by simultaneous action of a sonicator and a magnetic stirrer; (b) MCC in toluene after treatment; (c) MCC in toluene with PP-g-MA after treatment.

The particles size of MCC in the obtained suspension in toluene with PP-g-MA was measured by a particle analyzer (Delsa Nano C, Beckman Coulter). It was shown that the average particle size in suspension is about 15 μm (Table 1).

TABLE 1. Particle analysis of MCC suspension in toluene with PP-g-MA after treatment

№	Diameter (nm)	Polydispersity Index	D (10%) (nm)	D (50%) (nm)	D (90%) (nm)
1	19579	-13.77	8377	9322	10379
2	8774	-0.78	6623	7711	8982
3	17583	-3.79	9862	12222	15192
Average	15312	-6.11	8287	9752	11518

After the analysis, toluene was evaporated under vacuum using a rotary evaporator (Rotavapor R-114, Buchi) for 30 min at 50°C. The treated MCC was dried using a vacuum pump for 3 h at 50°C to remove any residual solvent.

Pre-milled PP was mechanically mixed with either treated or untreated cellulose powders. Composites were obtained by means of a single-screw extruder (Next 1.0, 3DEVO, Netherlands) at screw rotation rate of 2.9-3 rpm. Temperatures of the extruding zones were 215, 220, 225, 220°C. The diameter of the filament was 1.65-1.85 mm. The chemical composition of the obtained filaments is given in Table 2, photograph of filament extrusion is also provided.

TABLE 2. Chemical composition of the obtained filaments

№	PP, wt. %	MCC, wt. %	PP-g-MA, wt. %
1	100 (pellets)	-	-
2	100 (powder)	-	-
3	95 (powder)	5 (untreated)	-
4	95 (powder)	5 (treated)	-
5	94 (powder)	5 (treated)	1 (PP-g-MA)

All the composite filaments with MCC had yellow gradation compared to neat PP. The filaments made from non-modified MCC and MCC treated in toluene without PP-g-MA (№ 3, 4 in Table 2) had a lack of uniformity that

was responsible for numerous stops and filament breaking during FDM processing. In contrast, the blend filament from PP/MCC/PP-g-MA (№ 5 in Table 2) was characterized by uniform thickness and high flexibility. Figure 3 shows knots from neat PP and PP/MCC/PP-g-MA filaments.

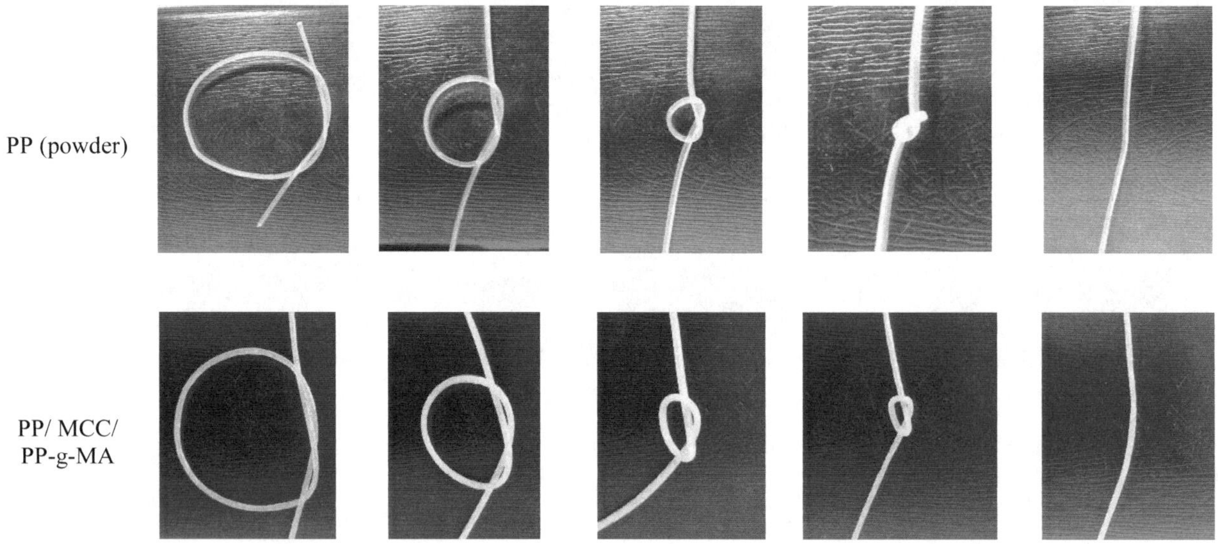

FIGURE 3. Recovery of filament original shape upon knot untying. Filament prepared with PP/MCC/PP-g-MA display higher flexibility.

The obtained filaments from pure PP and composites were used for FDM. The FDM technology was described in a previous paper [9]. The parameters of manufacturing were the following: velocity of deposition 4 mm/min; time of sample printing 3 min 56 sec; extruder multiplier 1.5 mm/min. Diameter of printed fiber resulted to be 0.4 mm. Environmental conditions were: air temperature (21±2)°C, relative humidity – (25±5)%. Extrusion temperature was (215±1)°C for pure PP and (250±1)°C for the composites, bed temperature – (90±1)°C. The theoretical dimensions of the samples were 12×12×2.8 mm (9 layers).

Figure 4 shows the obtained samples. Scaffold from untreated MCC was characterized by shrinkage with samples deformation. PP-g-MA modification conduced to lower scaffold shrinkage.

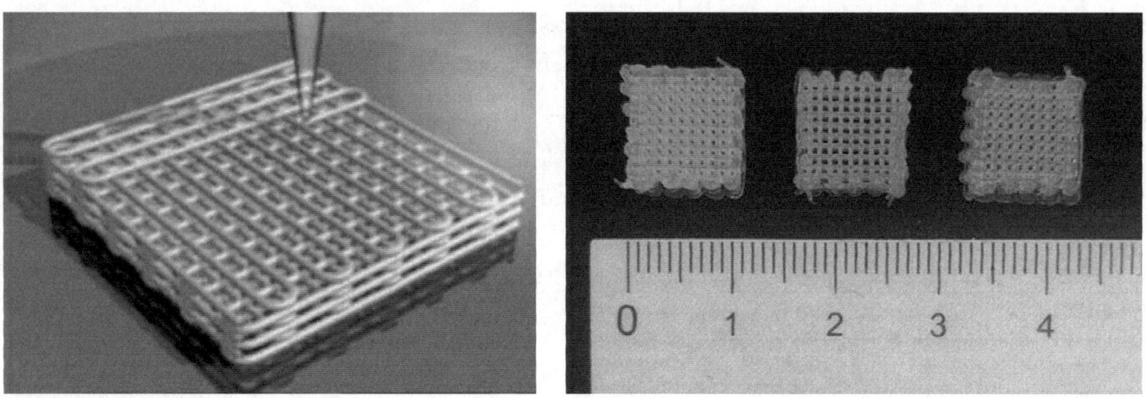

FIGURE 4. Schematic representation of the FDM process (left); representative picture of scaffolds by FDM made of PP (left), PP/MCC (untreated) (center), and PP/MCC/PP-g-MA (right).

CONCLUSION

Filaments for FDM made of PP or composites based on PP loaded with untreated or treated cellulose were successfully extruded and employed for fabricating 3D porous constructs by FDM. The developed method for cellulose treatment in a toluene solution containing PP-g-MA was confirmed to be effective in optimizing filament extrusion and FDM process. The obtained material appears as a promising eco-friendly FDM alternative to pure PP with comparable technological properties. Future research will be dedicated to mechanical, thermal and structural characterization of the obtained composite materials.

REFERENCES

1. I. Gibson, D. W. Rosen and B. Stucker, *Additive Manufacturing Technologies*, Springer, New York, 2015.
2. X. Wang, M. Jiang, Z. Zhou, J. Gou and D. Hui, *Compos B Eng*, **110**, 442-458 (2017).
3. O. Carneiro, A. Silva and R. Gomes, *Materials & Design*, **83 (1)**, 768-776 (2015).
4. E. E. Mastalygina and A. A. Popov, *Solid State Phenomena*, **265**, 221-226 (2017).
5. A. Zykova, P. Pantyukhov and A. Popov, *Polym Eng Sci*, **57 (7)**, 756-763 (2017).
6. E. E. Mastalygina, A. A. Popov and P. V. Pantyukhov, *IOP Conf Ser-Mat Sci*, **213 (1)**, 012011 (2017).
7. A. Y. Anpilova, E. E. Mastalygina, I. A.Mikhaylov, A. A. Popov and Z. S. Kartasheva, *AIP Conf Proc*, **1909**, 020008 (2017).
8. P. V. Pantyukhov, T. V. Monakhova, N. N. Kolesnikova, A. A. Popov and S. G. Nikolaeva, *Journal of the Balkan Tribological Association*, **19 (3)**, 467-475 (2013).
9. C. Mota, D. Puppi, F. Chiellini and E. Chiellini, *J Tissue Eng Regen Med*, **9 (3)**, 174-190 (2015).

Additive Manufacturing For Extrusion Tooling

Senne Van den Broeck[a], Tim Claeys[a], Wim Six[b] and Frederik Desplentere[b]

[a]*Vives University College, Doorniksesteenweg 145, 8500 Kortrijk, Belgium*
[b]*KULeuven Campus Brugge, Spoorwegstraat 12, 8200 Brugge, Belgium*

Abstract. Practical research on the implementation of metal additive manufacturing in extrusion tooling applications has been deducted by VIVES University College and the ProPolis research group from KULeuven in Belgium. In polymer extrusion lines, molten plastic is forced through a die to define the rough shape of plastic profiles. Calibration units are positioned right behind the mould; their function is to cool the plastic profiles in order to give them their final and exact shape. This cooling process needs to occur as uniform as possible to prevent geometrical distortion, and as fast as possible to maintain high production speeds. To achieve this, calibration units feature highly complex cooling channels and vacuum slots. These often intricate channels and slots are very challenging to create using conventional tooling such as CNC-milling and electrical discharge machining, resulting in sub-optimal cooling processes and more complex cooling block assemblies. Where the above subtractive technologies run into their limits, additive manufacturing opens up new possibilities for designers and engineers because of the technology's geometrical freedom: improved cooling, shorter lead times, faster production speeds and therefore a shorter time to market. These possibilities are explored in the aforementioned research through a number of practical case studies for Belgian extrusion companies. Through thermal simulations, calibration units modelled for conventional CNC machining/EDM are compared to new designs made specifically for additive manufacturing. These simulations provide insight in pressure drops and heat transfer from extrusion profile to cooling water, and therefore aid in the optimization of cooling channel layout. Once optimized, the new design is printed with selective laser melting technology and tested in production. Measurements on heat transfer are executed and compared to the conventional cooling units, showing great promise for additive manufacturing as an alternative production method for calibration units in polymer extrusion.

Keywords: 3D Printing, Additive Manufacturing (AM), Selective Laser Melting (SLM), Metal Powders, Polymer Extrusion, Calibration Unit, Conformal Cooling, Thermal Simulation, Heat Transfer

INTRODUCTION

Creating qualitative and reproductive polymer profiles using extrusion technology requires a precise combination of critical parameters. Arguably the most important part of the extrusion process (Fig. 1), is the calibration unit or cooling block found right behind the mould or die, where the molten polymer is solidified to its final shape. These calibration units should feature performant cooling combined with proper suction of the profile to the cooling block's wall using vacuum, as well as proper slip between said wall and the profile itself. If all settings are optimized, then the resulting profile will have the correct physical properties and geometrical accuracy.

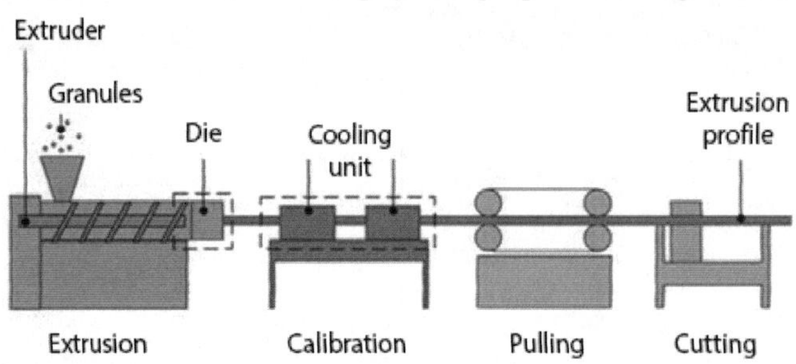

FIGURE 1. Schematic representation of a general polymer extrusion line, cooling unit displayed in purple

Currently, these cooling units are produced using a combination of conventional tooling methods such as CNC milling, drilling, electric discharge machining (EDM) and wire erosion. These technologies have become well established in the field of extrusion throughout the years, but are lacking when it comes to the creation of complex internal cooling channels and vacuum slots due to the geometrical restrictions they come with. This leads to engineers having no choice but to go for suboptimal cooling layouts in their designs.

METAL AM IN EXTRUSION APPLICATIONS

Additive manufacturing has steadily become a revelation in the manufacturing field, due to its completely different build method. The previously mentioned conventional techniques are subtractive and thus start from a solid block of material of which material is subtracted in order to create the final part. In 3D printing, layers of material are added on top of each other in order to build up the part. This creates an immense geometrical freedom for designers, allowing them to create very complex structures and internal channels with ease. With cooling being such a critical aspect in extrusion applications, it is evident that metal additive manufacturing could open up new and unexplored opportunities within this market, and help solve complex problems along the way.

Conformal Cooling

When it comes to the optimization of cooling capacity in applications such as extrusion and injection moulding, the term 'conformal cooling' may appear frequently. The term refers to the design of cooling passageways that follow the shape or profile of a mould core or cavity in order to perform rapid and uniform cooling along the wall of polymer parts. Conformal cooling has been explored extensively in injection moulding [1] and blow moulding applications, yet literature on extrusion applications is scarce. It is however realistic to assume that the advantages and disadvantages of conformal cooling [2] in different polymer processing techniques are interchangeable, and given in the table below:

TABLE 1. Advantages & disadvantages of conformal cooling

Advantages	Disadvantages
More geometrical design freedom	Complex cooling channel design
Cooling time reduction leading to higher productivity	Limitations of AM technology (not everything is printable)
Stable production quality through improved cooling	Higher surface roughness might cause plugging
Improved surface quality on end-use parts	Higher cost

CASE STUDIES

The case studies described in this research, are performed on existing extrusion profiles produced by a number of Belgian companies. Conventional tooling required to produce these profiles already exists and is being used, which allows the creation of a reference point. A total of five cases were examined, two will be discussed within this paper for the sake of conciseness.

Case Study I

As a base for this case study, a smaller part of a complete cooling unit is chosen. This part features a critical protruding detail (fig. 2a), which proves to be a problem area in the original design. Due to the very narrow passageway, it is very difficult to incorporate sufficient cooling through conventional techniques, whereas 3D printing allows more complex conformal cooling channels to be integrated in the design. The part itself needs to be 215mm long, but due to build volume restrictions it is printed in two halves as can be seen in fig. 2a below.

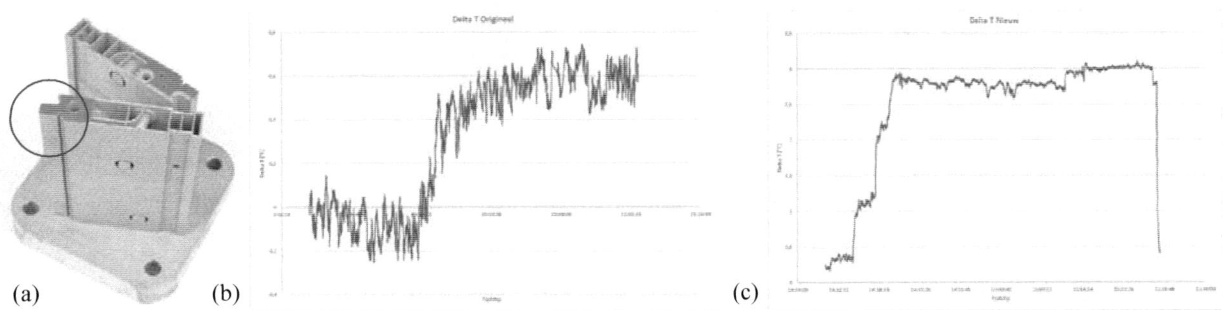

FIGURE 2. (a) printed halves with critical detail highlighted (b) temperature measurements on cooling water of original design (c) temperature measurements on cooling water of newly designed and printed part

First, a temperature measurement is executed on the original cooling unit design. The measurement is taken during the start and following 15 minutes of an extrusion run, showing the difference in temperature of the entering and exiting cooling water running through only the specific part of the cooling block assembly. Secondly, the original part is replaced by the newly designed and 3D printed part and the same measurement is repeated. From these graphs, it can easily be visualized that the delta T is much higher in the second design, meaning that more heat is transferred to and drained by the cooling water in the printed calibration unit even though the flow rate is about 33% slower. As can be seen from the table below, the energy drained by the cooling water in the redesigned and 3D printed caliber is between 60 and 100% higher than the original design (depending on the error of measurement).

TABLE 2. Resulting heat transfer from case study I

(average taken over a time of 15 minutes)	Original design	New design
Delta T [°C] over printed part	0.59	2.79
Flow rate [L/min]	7.41	2.67
Drained energy [J]	257-303	480-519

Case study II

A similar profile for a different manufacturer and application is chosen as a second case, yet it is showing entirely different practical results. The part in question also features a critical protruding detail, which proves to again be a problem area in the original design. Fig. 3a shows the original cooling channel layout: a combination of drilled holes leading to a large lateral channel in the critical detail. Said channel remains relatively far away from the surface of the profile, thus providing insufficient cooling. Fig. 3b shows the newly designed conformal cooling channel in the 3D printed design. Here, the cooling channel follows the contour of the profile as much as possible to provide uniform cooling.

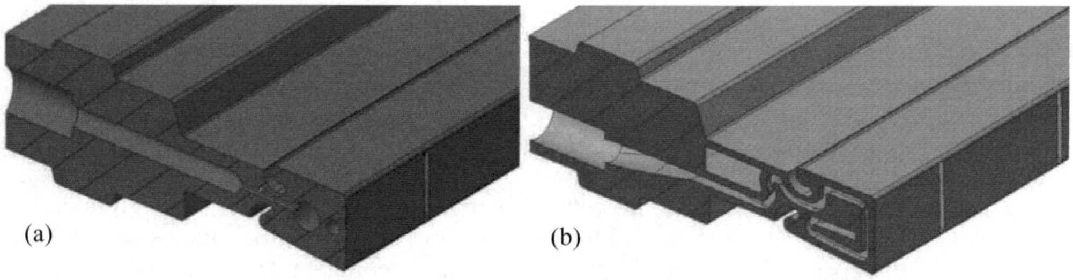

FIGURE 3. (a) Original cooling channel design (b) Conformal cooling channel design in 3D printed part

Simulations on flow and heat transfer are carried out in SimCenter NX 12 on both old and new design in order to compare the cooling performance before physically printing and testing the part. This allows for an iterative approach in the optimization of the design without having to print multiple design versions.

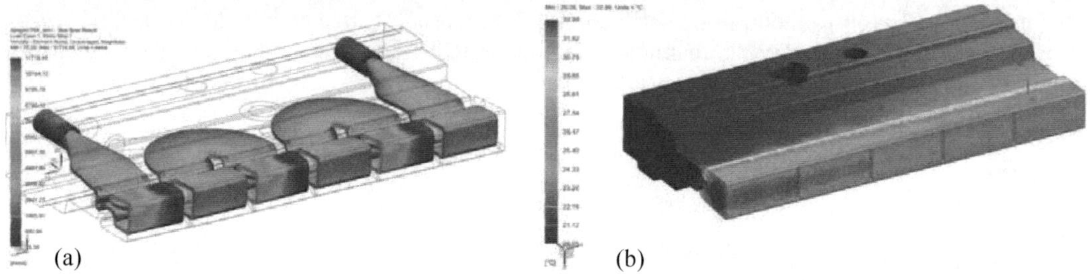

FIGURE 4. (a) Cooling water flow and temperature simulation in conformal cooling design (b) Wall temperature simulation of 3D printed part

Measurements identical to case study 1 are performed, but show no great difference in energy being drained by cooling water in the old versus new design: the delta T is pretty much identical. It should however be noted that the original design for this case study is manufactured using a material with a much higher heat conductivity (116W/mK) than the 3D printed material, being 316L Inox (50W/mK). If the newly designed print was made using a more suitable material for heat transfer applications, the results would be different and an improved cooling capacity would be expected.

Since the first step in this second case study focuses on only a small part of the cooling unit, the process is repeated with the complete calibration unit to verify whether this allow for more noticeable improvement. Again, a conformal cooling channel layout is added to the top and bottom parts of the block. Due to build volume restrictions on the printer used, the units have to be printed in separate halves once more as can be seen in figure 5b.

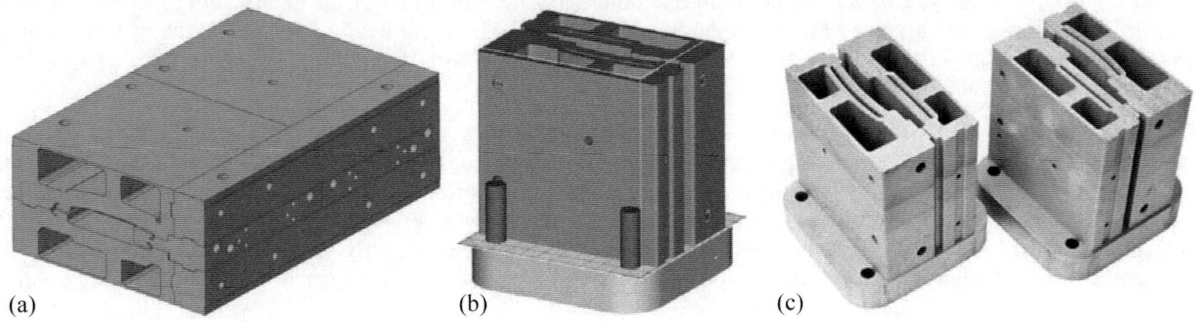

(a) (b) (c)

FIGURE 5. (a) Complete cooling unit assembly in 3D CAD (b) Two halves of cooling unit, prepared in print software (c) All four remaining halves (two top, two bottom), printed using SLM technology and ready for finishing

Identical measurements are executed on the top and bottom halves of the cooling unit, with Delta T being the difference in temperature of entering and exiting cooling water running through the cooling channels. The results are given in table 3. The Delta T shows an increase of 1.4 °C in the new design, but at a slower flow rate of the water due to smaller cooling channels and resulting pressure drop. When this difference is taken into account, the proportional improvement of the cooling capacity of the new versus conventional design is 8%.

TABLE 3. Resulting heat transfer from case study II

	Original design	**New design**
Delta T [°C]	1.5	3.9
Flow rate [L/min]	9.02	3.75
Heat transfer [W]	944	1020.33
Proportional improvement		**8%**

CONCLUSIONS

Practical Feasibility

In close cooperation with all stakeholders within the extrusion industry, this research project demonstrates that metal AM technology can certainly add value to the production of calibration units in extrusion. This however does not imply that metal printing in its current form will completely rule out the conventional technologies, because AM does bring along certain technical challenges of its own.

Added value

Cooling performance: the case studies prove that with metal AM, a higher cooling capacity can be achieved. This however strongly depends on the extrusion profile and accompanying calibration unit itself. Much is decided by the design and layout of the cooling channels, with space being the limiting factor. The more space a designer gets, the easier it will be to use this space advantageously when it comes to active cooling. This space is mostly defined by the shape of the extrusion profile itself, as well as the build volume of the printer and whether or not the redesign needs to fit into a pre-existing cooling unit.

Geometrical freedom: the enormous geometrical freedom of the metal AM technology clearly offers new possibilities towards the optimization of cooling channels, certainly in combination with simulation software as described within the case studies. Fluent transitions and conformal cooling make a huge difference in the active cooling of extrusion profiles, with most improvement being measured in smaller details where conventional techniques fail: critical details and small passageways are very challenging to run cooling channels through in conventional production methods, but have proven to be the ideal playground for designers to incorporate intricate sweeping channels that are made with ease using SLM technology.

Lead times: during the redesign phase of calibration units within the case studies, it frequently became apparent that the lead times of new iterations can be shortened dramatically. While conventional tooling mostly refers to lead times in months, additive manufacturing allows for the lead times to be shortened to weeks or even days. No matter how complex or intricate the design is, the printing time stays relatively identical. This is not the case for conventional techniques as CNC milling. Conventionally manufactured parts are hard to modify when certain problems arise. With 3D printing, extreme changes can be made in the 3D file one day, while still being able to print the next day and have them tested in production the next week. With CNC milling, this would require a lot more (manual) labor, and therefore a lot more time.

Optimization of Extrusion (process): all of the above factors point out that the extrusion process can be drastically optimized: when the cooling capacity of the calibration units increases, then this will manifest itself in higher quality end products with less geometrical distortion. This effect is most noticeable in critical details and confined areas, and therefore more evident in complex PVC extrusion than foam extrusion. Nevertheless, conformal cooling can optimize heat transfer in any extrusion line and therefore imply a higher production speed. In an uninterrupted manufacturing process such as polymer extrusion, this could be a critical factor for both company and customer.

Challenges

Limited build volume: one of the most limiting factors during the research project, was the build volume of the SLM printer. Measuring only 125x125x125mm, it was a constant challenge to create functional parts that would fit in such confined space. In certain cases, it was required to split certain designs up in multiple parts, which is in contrast with the possibilities of the technique itself: one of the biggest advantages of metal AM is the integration of multiple functions and parts within one print. Limiting the amount of different parts in the assembly of the cooling units would for instance increase the mounting speed when replacing the production line.

It should be noted that the printer used was one with a relatively small build volume. Higher capacity SLM printers already feature build volumes of over 500x300x300mm, which would drastically decrease these limitations. In certain cases with larger profiles, it might however still be a challenge.

Limited material choice: suitable metal powders for the SLM technology are still not as diverse as what is generally available for conventional techniques. Specific alloys with high heat transfer coefficient are not readily available yet, but it should be noted that more and more powder manufacturers are releasing different materials suitable for metal AM. The case studies described were all executed using standard 316L Inox and results were convincing, so with more suitable print materials the improvements could be even more drastic.

Complex finishing process: during the research project, it became apparent that that certain surfaces and features of metal prints require finishing in order to achieve correct tolerances, fitments and surface roughness. This finishing however requires very new strategies and ways of thinking, which occasionally led to accidentally opening cooling channels during CNC milling or wire erosion. It is crucial to stay in very close contact with manufacturers during the finishing process, with a thorough systematic guide of each individual move required.

Financial Feasibility

From these first few practical case studies for existing manufacturers, it has become clear that the technique is not too expensive to put into practice. From a company's financial point of view it would not be suitable to replace the complete calibration unit with metal printed parts as that would prove too costly, however replacing and optimizing certain critical details will certainly add value to the R&D department's workflow.

ACKNOWLEDGEMENTS

The authors wish to thank the Flemish Government for the funding in the TETRA HBC 2016.0094project. The results discussed in the present paper were obtained within this project.

REFERENCES

1. W. Six, F. Appermont et al. "Design and simulation of conformal cooling channels in SLM produced injection molds", Conference Proceedings of the International Conference of the Polymer Processing Society, pp. S13-250, 2016
2. K. Soete and F. Desplentere, "Simulation of the effects of using conformal cooling channels in SLM produced plastic extrusion calibrators", Conference Proceedings of the International Conference of the Polymer Processing Society, 2017

Fabrication and Characterization of 3D Printed PLA Scaffolds

Jassimran Kaur Arora[a], Pooja Bhati[a,b]

[a] *Indira Gandhi Delhi Technical University for Women (IGDTUW), Kashmere Gate, Delhi, India*
[b] *Indian Institute of Technology, Delhi*

Abstract. Three-dimensional printing (3DP) is one of the rapidly growing additive manufacturing (AM) technologies due to its ability to print biocompatible polymeric scaffolds directly. Fused Deposition Modelling (FDM), one of the many 3D printing methods, uses specialized printers in which the extruded material is deposited, layer after layer, in the desired shape. This method for scaffold fabrication is widely applied to various types of scaffolds due to the precise control it offers over the scaffold architecture at the micron-level. In the past, many 3D scaffold designs have been fabricated to explore the mechanical and biological viability of a 3D printed Polylactic acid (PLA) scaffold as a construct which could support cellular life.
However, all AM approaches require a computer-aided design (CAD) file, from which the complex shapes are read via a 3D scanner. Moreover, the commercial 3D printer is expensive as compared to the 3D pens. This study aims to compare the mechanical properties of the scaffolds fabricated by using a 3D printer and a 3D pen. A 3D pen can perform the same functions as that of a 3D printer, with the advantage being that no CAD files are required. 3D pens are cheaper than 3D printers and lie in the price range of 15 USD to 150 USD. They allow for higher flexibility and creativity, to suit the desired application and consume lesser time for fabrication. Compression testing was conducted on the 3D printed scaffolds which were printed using a 3D printer and 3D pen. Fourier Transform Infrared Spectroscopy (FTIR) was utilized to characterize both the bulk and surface chemistry of the samples. Scanning Electron Microscopy (SEM) and Differential Scanning Calorimetry (DSC) were used to describe morphology, crystallinity and the glass transition temperature following printing. Key differences between the samples in terms of compressive strength were found and reported.

Keywords: Fused deposition modelling (FDM), 3D printing, additive manufacturing, PLA, tissue engineering

INTRODUCTION

Bone is the second most transplanted tissue in the human body, owing to many factors like patient traumas, aging populations, osteoporosis, and bone tumour[1]. While bone is known for its self-healing capabilities, large-scale bone defects cannot always be treated successfully by the body. Among different treatment options lie autografts and allografts. However, these therapies have drawbacks associated with them - limited tissue availability, pain, donor site morbidity, immune responses in host and difficulty in producing anatomical shapes[2]. Engineered tissue implants have arisen as a promising alternative. They are developed with the aim to regenerate bone to restore, maintain or improve its functions *in vivo*. Bone scaffolds plays a very important role in bone tissue engineering. Scaffolds serve primarily as osteoconductive moieties on which newly formed bone is deposited through creeping substitution from an adjacent living bone. An ideal scaffold should be (i) biocompatible and porous to support cell proliferation and differentiation, (ii) biodegradable to be gradually replaced by the host tissue, (iii) osteoconductive and osteo inductive and (iv) manufactured in a specific shape to precisely match complex bone defects[2].
Conventional fabrication methods for porous scaffolds include gas foaming, solvent casting with particle leaching, freeze drying, and electrospinning[3]. While use of these methods allows control over macro architecture, the ability to control microarchitecture through these methods is limited to 3D shapes and geometries determined by molds and manual processing. Over the past two decades, 3DP has emerged as one of the vital tools to produce customized bone substitutes. Rapid Prototyping (RP), Computer Aided Design (CAD) and advanced imaging techniques, together allow greater and precise control over the scaffold micro and macro-architecture. Fused Deposition Modelling (FDM) is one of the most commonly used 3DP method in which the raw material used, is in the form of a filament. It is extruded and deposited to produce a three-dimensional model in a layer-by-layer fashion. It allows greater control over porosity and pore size. However, this method requires a patient-specific scaffold design in an STL file format, which is then entered into a computer connected to a 3D Printer. A commercially available 3D Printer lies in the cost range of 500 USD to 1500 USD. A 3D pen can essentially perform the same functions as that of a 3D Printer, with the added advantage of greater control over flexibility and creativity and at a lower cost. It requires lesser printing time, eliminates the need for a CAD file and is easy to customize according to the patient.
In this work, we compare the fabrication and characterization of 3D scaffolds printed from a common PLA spool, using a 3D Pen and 3D Printer.

Materials and Methods

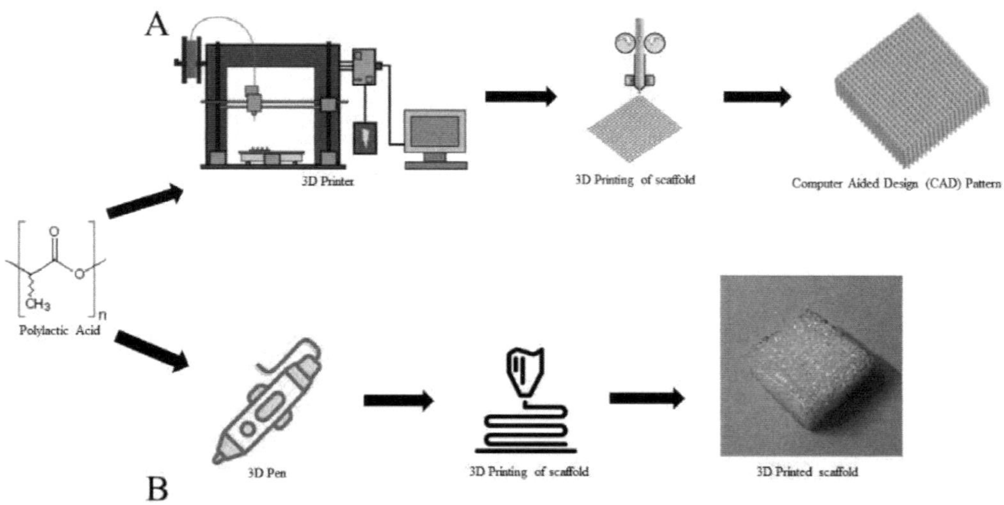

FIGURE 1. Schematic diagram of A) 3D Printer and B) 3D Pen

PLA scaffolds were printed using WOL3D Premium 1.75mm transparent PLA filament (RoHS approved).

FIGURE 2. A) REES52 3D Printing Pen and B) Raise3D N2 Plus 3D Printer

TABLE (1). Processing Parameters during scaffold fabrication

Parameters/Sample	3D Pen	3D Printer
Model	REES52 3D Printing Pen	Raise3D N2 Plus 3D Printer
Layer orientation	0-90°	0-90°
Printing Speed (cm s^{-1})	1.35	2.307
Nozzle diameter (mm)	0.4	0.4
Nozzle Temperature (°C)	190	210
Hotbed Temperature (°C)	NA	72

CAD and STL models of the scaffolds, having a strut diameter of 0.35 mm and strut spacing of 0.65-0.7mm, were generated in PTC Creo Parametric 5.0.0.0. Scaffolds of dimensions 20 x 20 x 10 mm were manufactured, as shown in Figures 3a and b. The filament and the scaffolds were stored in air-tight containers to prevent degradation by moisture.

Scanning Electron Microscopy (SEM): Morphological analysis of the 3-D structures was carried out by scanning electron microscopy (Hitachi-TM 1000) to visualize and evaluate the architecture, surface morphology and structural stability of the fabricated scaffolds.

Differential Scanning Calorimetry (DSC): Thermal transitions of the processed polymer were analyzed by differential scanning calorimetry (Perkin Elmer, MS-632) to observe the glass transition temperature (T_g), crystallization temperature (T_c) and malting temperature (T_m). Thermogravimetric Analysis (TGA) was conducted to determine the decomposition behavior.

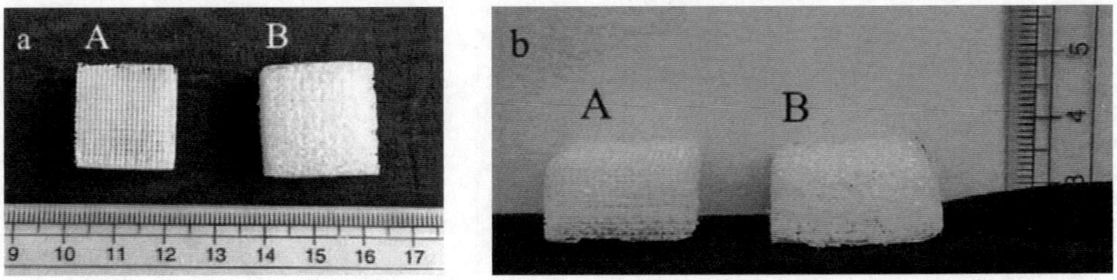

FIGURE 3. Image of A- 3D Printer and B- 3D Pen printed scaffolds before mechanical testing a) top view b) front view

Mechanical Properties of Scaffolds: The 3DP scaffolds were tested for axial compressive force on a universal testing machine (Instron 5582) with a 100 kN load cell. A preload of 0.01kN was applied on the samples and the rate of loading the samples during testing was kept at 1 mm min^{-1}. The samples were loaded till 60% strain was achieved.

Fourier Transform Infrared Spectroscopy (FTIR): FTIR was conducted on a Nicolet iS50 FTIR spectrometer to collect the FTIR spectra of the PLA spool used for the fabrication of the samples. The samples for FTIR were in the powdered form, obtained by crushing the PLA.

Porosity: The theoretical volume porosity percentage was calculated for each sample using the initially designed geometries based on a reference cube, whereby the strut diameter and spacing between the layers were equal, that is, no overlapping due to the fusion between struts from one layer to the adjacent one was assumed.

$$\text{Porosity} = \frac{(V_a - nV_c)}{V_a} \times 100$$

'V_a' is the Apparent Volume of reference sample (mm^3) = l x b x h

'V_c' is the True Volume (mm^3), that is the volume of a cylinder = $\pi r^2 h$ and 'n' is the number of fibers per sample

Surface Wettability: Water Contact Angle (WCA) measurements were performed to evaluate the sample's surface wettability. The sessile drop method was used to measure the contact angle by depositing a deionized water droplet (0.005 mL) on the surfaces of the polymer scaffolds, using a syringe, at room temperature (21°C) and 82% relative humidity. Three samples of each type, with dimensions 20 x 20 x 10 mm^3 were used for the study. ImageJ software was used to determine the water contact angle.

Results and Discussion

1. SEM: The architecture, surface morphology and structural stability of the deposited struts and layers are shown in Figure 4. The fiber diameter of 3D pen sample is observed to be higher than the 3D printed specimen. However, variation in fiber diameter is found in both the samples. The values of the strut diameter and spacing between the struts of both the samples i.e. 3D pen and 3D printer are shown in Figures 4A and B respectively. As evident from the SEM, 3D printed sample has more regular morphology as compared to 3D pen fabricated sample. The higher variation in the strut diameter during 3D pen fabrication is due to the unsteadiness of the human hand and thus, it is difficult to achieve precise control over the strut diameter using 3D pen.

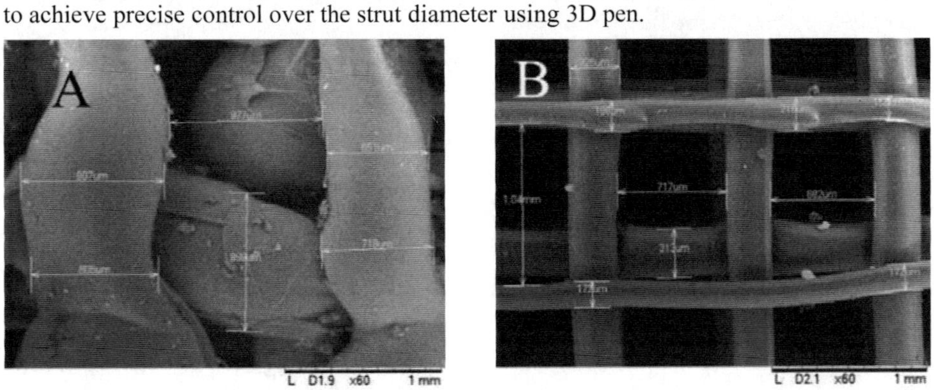

FIGURE 4. Scanning Electron Microscopy images of A) 3D Pen and B) 3D Printer scaffolds showing strut diameter and spacing

2. DSC: Figure 5 shows the thermal properties of the material, postprocessing. The values of T_g, T_c and T_m can be clearly seen from the thermograph. The values observed does not deviate from the usual values obtained by other researchers.

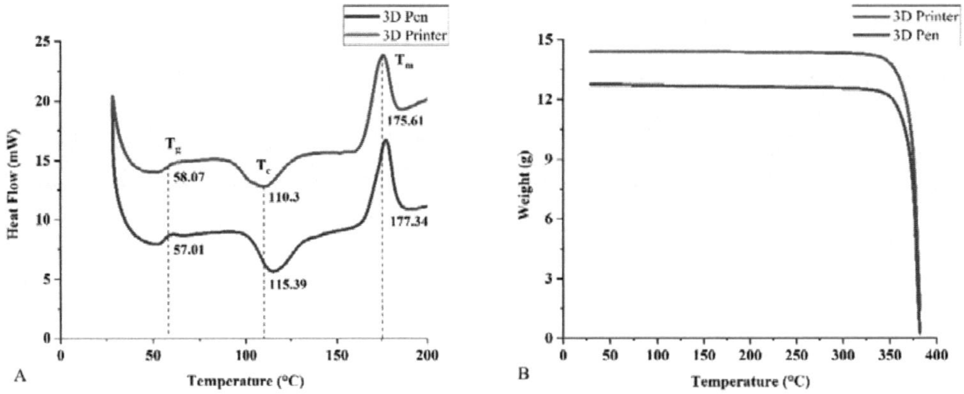

FIGURE 5. A) DSC and B) DSC TGA curves for PLA samples

3.Mechanical properties: Figure 6A shows the stress strain curve of the tested 3D fabricated samples. It is quite evident from the curve that 3D pen samples have a greater slope during testing and thus, a higher load bearing ability. Significant changes in maximum compressive strength are also observed between the samples fabricated by 3D pen and 3D printer. Higher maximum compressive strength (average of 3 samples) for the 3D pen samples is due to higher strut diameter of the 3D pen sample as compared to 3D printer sample.

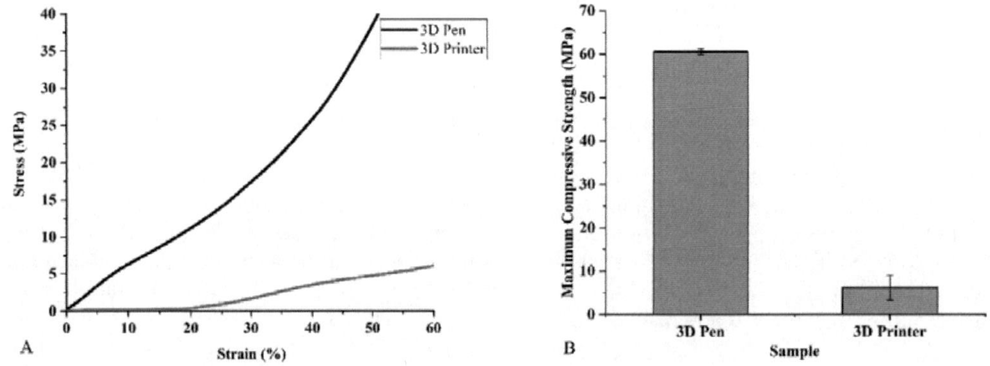

FIGURE 6. A) Stress vs. Strain curves and B) Maximum Compressive Stress for 3D Pen and 3D Printer scaffolds

4. FTIR: FTIR spectra of PLA spool is shown in Figure 7. The main characteristic bands of pure PLA are present in the FTIR spectra as given in Table 2. No new peak formation or any shift in the characteristic band is observed in the PLA spectra.

TABLE (2). Typical vibrational frequencies of functional groups[5]

Major Functional Groups	Frequency Region
C=O	1760
$=\text{-CH}_3$	1452
C-O-C	1215-1185
C-O	1100-1090

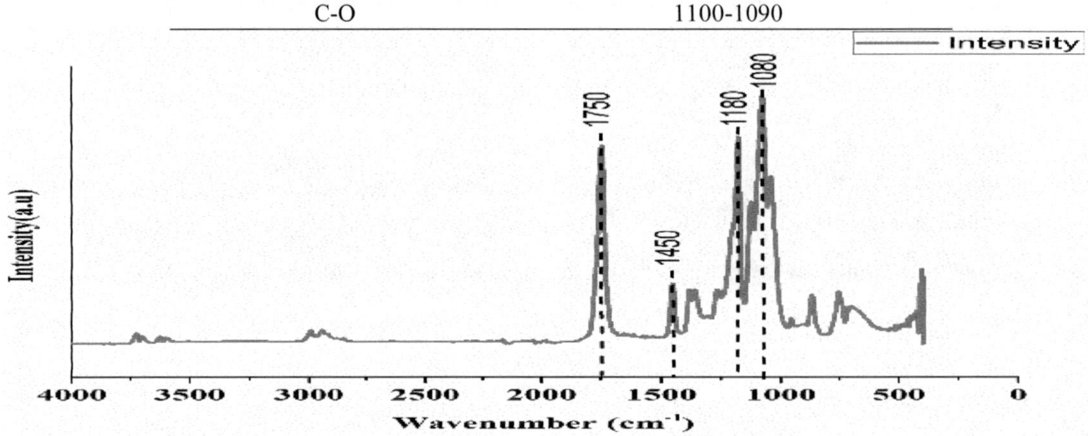

FIGURE 7. FTIR results of PLA sample

5. Porosity: Assuming strut diameter of 0.35 mm and a reference sample of dimensions 20 x 20 x 10 mm, the porosity of the scaffolds is shown in Table 3.

TABLE (3). Porosity of 3D Pen and 3D Printed Scaffolds

Sample	Number of fibers in one-layer	True Volume, V_c (mm^3)	Porosity (%)
3D Pen	18	1039.08177	74.02295
3D Printer	20	1154.5353	71.13

6. Surface wettability: The WCA of the 3D pen and printer printed PLA sample is shown in Figure 8. The WCA of the 3D printed sample is higher than that of the 3D pen. Both the samples have WCA, which is in agreement with the WCA obtained by other researchers. However, the WCA of 3D pen fabricated samples is observed to be 86.2^o which is 2.4% lower than the WCA of 3D printed. This difference in WCA of two samples could be due to different processing temperature used during the fabrication of scaffolds.

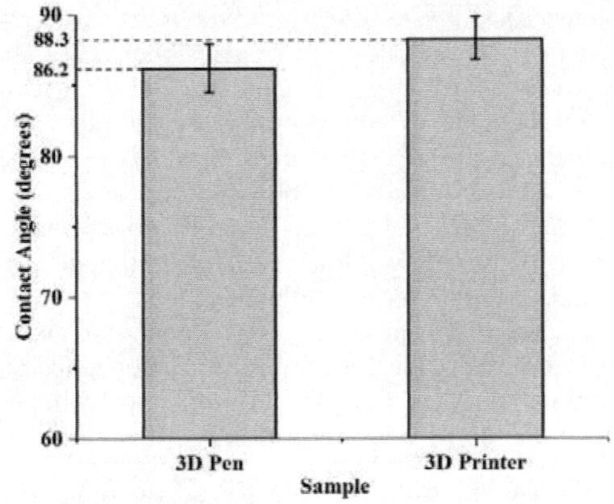

FIGURE 8. WCA of 3D Pen and 3D Printer samples

Conclusion

In this study, we have shown that bone scaffolds can be successfully printed by a 3D Pen with the same dimensions, using the same filament and with similar porosity as that of a 3D Printer. However, on the basis of compressive mechanical properties, we can conclude that scaffolds printed by a 3D Pen perform better as compared to those manufactured by 3D Printer. The results suggest that 3D Pens could be used for the fabrication of PLA scaffolds, by an experienced professional. It is a cheaper and time-saving alternative to 3D Printer for manufacturing scaffolds. However, further detailed studies are required to examine the biological viability of 3D Pen manufactured scaffolds.

ACKNOWLEDGMENTS

The authors would like to thank Dr. S.R.N. Reddy of CSE department at IGDTUW, Kashmere Gate, Delhi, India for assisting with the 3D Printing process. We would also like to thank Professor Naresh Bhatnagar at IIT Delhi, for allowing us to make use of facilities at PE lab.

REFERENCES

1. Lucas D. Albrecht, Stephen W. Sawyer, and Pranav Soman, *Developing 3D Scaffolds in the Field of Tissue Engineering to Treat Complex Bone Defects*
2. Agathe Grémare, Vera Guduric, Reine Bareille, Valérie Heroguez, Simon Latour, Nicolas L'Heureux, Jean-Christophe Fricain, Sylvain Catros, Damien Le Nihouannen, *Characterization of printed PLA scaffolds for bone tissue engineering*
3. Mian Wang, PelagieFavi, Xiaoqian Cheng, Negar H. Golshan, Katherine S. Ziemer, Michael Keidar, Thomas J. Webster, *Cold Atmospheric Plasma (CAP) Surface Nanomodified 3D Printed Polylactic Acid (PLA) Scaffolds for Bone Regeneration*
4. Susmita Bose, Sahar Vahabzadeh, Amit Bandyopadhyay, *Bone Tissue Engineering using 3D Printing*
5. G. Kister, G. Cassanas and M. Vert, *Effects of morphology, conformation and configuration on the IR and Raman spectra of various poly (lactic acid) s*

Design of 3D Scaffold Geometries for Optimal Biodegradation of Poly(lactic acid)-Based Bone Tissue

Pedram Karimipour-Fard[a*], Remon Pop-Iliev[a], Holly Jones-Taggart[a], Ghaus Rizvi[a]

[a]*University of Ontario Institute of Technology - Ontario – Canada*

Abstract. Additive manufacturing is utilized to produce different types of scaffolds to mimic the microcellular structure of bone tissue. Characterizing the biodegradation rate of engineered tissue scaffolds and their respective biodegradation patterns against the 3D structure design of the scaffold is of crucial importance. The research focus of this paper is to understand the relationship between the geometry of biomimetic 3D printed microcellular structures and their biodegradation properties through a comparative experimental assessment study. Polylactic acid (PLA) is one of the commonly used materials in tissue engineering and PLA filaments are widely available for additive manufacturing. To use PLA 3D printed structures as tissue scaffolds, a non-toxicity biological test is conducted using a mammalian cell line, to prove the biocompatibility. Then, multiple common as well as novel microcellular structures are manufactured to assess the effect of the geometry of biomimetic 3D printed scaffolds on the pattern and rate of biodegradation. Microcomputed tomography (Micro-CT) technique is used to monitor the morphology evolution in 3D structures before and after each step of biodegradation. Phosphate buffered saline (PBS) medium plus 5% Carbon Dioxide gas in a CO2 incubator is used to simulate an in vitro environment for performing biodegradation tests. The obtained experimental results are promising as they provide a more advanced understanding on how biodegradation and microcellular structure geometries influence each other; and are helpful for manufacture of optimized biomimetic product for specific needs.

Keywords: Additive manufacturing, 3D printing, Bone scaffold, Biodegradation, Poly(lactic acid).

INTRODUCTION

Additive manufacturing (AD) was developed for the first time in 1990s and in the last decade it became one of the prominent tools for tissue engineering [1, 2]. Fused deposition modeling (FDM) technique is one of the most common forms of AD that extrudes a filament onto a platform. The process starts with a computer aided design (CAD) generated file which is converted into a format suitable for FDM processing. The product is manufactured using a layer by layer deposition to achieve the whole geometry [2]. FDM is commonly used to manufacture microcellular structures for bone tissue engineering [3-5]. These structures provide a suitable environment to help and improve native bone cells growth for a short period during which the new native cellular structure replaces the polymeric tissue scaffold that undergoes biodegradation at a certain rate determined by its pattern [2].

Poly(lactic acid) (PLA) is a biocompatible and biodegradable material and is commercially available as FDM filaments [2, 3]. Most of the FDM 3D printers are optimized for PLA filaments, so they are good choices to assess the effect of geometry on the other properties of the final product.

The geometry of microcellular structures significantly affects the properties of scaffolds [6]. Numerous types of structures have been proposed to mimic the native tissue of bone scaffold including Primitive, Diamond and Gyroid structures [6, 7]. The Gyroid structure is one type of triply periodic minimal surfaces that contain no straight lines and no planar symmetries [6, 8].

Microcomputed tomography (Micro-CT) is a 3D characterization and visualization technique to assess morphology of a 3D structure [9-11]. The resolution of visualization, can be altered by X-rays current and voltage of energy [11]. This technique, because of its high resolution (up to 0.5 μm), can be used to assess internal structure of 3D scaffolds.

In this paper, a FDM printer is used to manufacture 3D scaffolds to assess the effect of structural properties on biodegradation rate and pattern of the scaffolds. In the literature, there are a very few studies to relate different structures of bone tissue scaffolds to biodegradation rate and pattern. Therefore, this paper aims to further understand the biodegradation rates based on the pattern of a scaffold and try to control biodegradation by geometric design.

To assess the effect of geometric properties on the biodegradation rate, six structures are designed. These scaffolds include one randomly distributed, one cubic, and four Gyroid structures. Siemens NX 12 and MATLAB are used to design and produce the final 3D design files and Utlimaker-Cura 3.6.0 is used as the slicer. An In-Vitro environment is simulated inside CO_2 incubator and MTT assay with human cells (HT29 cell line) in low passages is done to assess the toxicity. At the end, Micro-CT is employed to characterize the morphology of the samples before and after the biodegradation.

EXPERIMENTAL

Additive Manufacturing

The first step in additive manufacturing is to use CAD to generate geometry of the samples. MATLAB and NX 12 are used to generate a randomly shaped, a cubic and 4 Gyroid structures. Coding in MATLAB presents the opportunity of making the structures by using the precise mathematical descriptions. Figure 1 shows the six scaffolds that were studied. Sample 3 has high porosity and the wall thickness is same as that of sample 4. Sample 5 has the same structure as sample 4 but with 30% lower wall thickness; and sample 6 has high porosity and high wall thickness (twice as that of sample 4).The images are obtained from the Micro-CT scans of internal structure of the scaffolds.).

In the next step, Ultimaker-Cura 3.6.0 slicing software is used to slice the STL files provided by MATLAB and NX 12 to format them for the 3D printer. Scaffolds are manufactured by Ultimaker 3 printer with layer resolution up to 20 μm and the nozzle size of 250 μm.

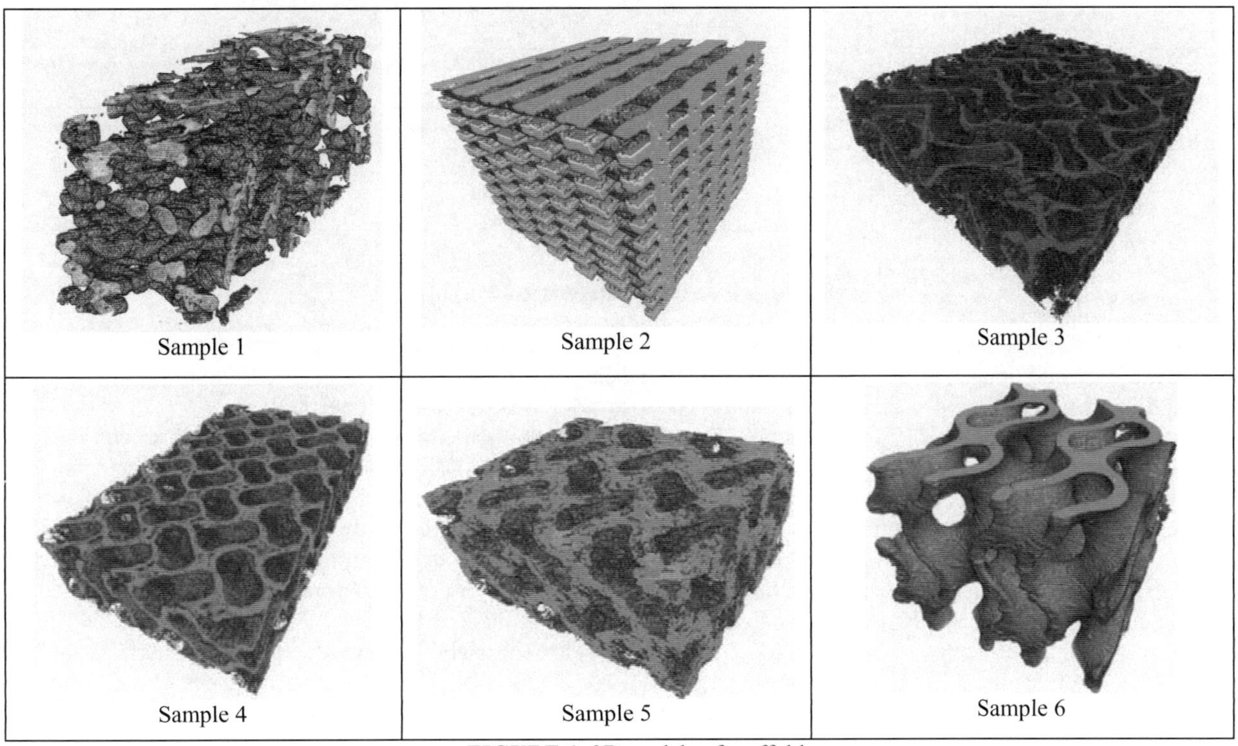

FIGURE 1. 3D models of scaffolds

Cell Metabolic Activity Test

To estimate toxicity of the scaffolds for human cells, MTT assay is used to assess cells metabolic activity on the PLA-based scaffolds. Human cells (HT29 cell line) in a low passage number in RPMI-1640 Medium (from Sigma-Aldrich) are cultured on the samples. The assay is done for half and one hour to assess acute effect of the scaffold on the cellular activity.

Microcomputed Tomography and Biodegradation Assessment

High resolution SkyScan 1172 (Bruker, Belgium) micro-CT is employed for 3D morphology characterization and internal structure visualization of the scaffolds. The scan is performed by 360° rotation of specimen in steps of 0.5°

and with 69 kV and 167 μA. The samples are scanned before and after the biodegradation procedure to evaluate the changes in morphology and visualize the biodegraded samples structure.

Phosphate Buffer Saline (PBS) solution inside a CO2 incubator is employed to assess biodegradation rate of the scaffolds. The samples are kept in the solution for 1 day and then they are air dried in 40 °C for 8 hours; in the next step, they are weighed and returned into the solution for 21 days and dried again with same procedure.

RESULTS AND DISCUSSION

Cell Metabolic Activity

Table 1 and 2 shows the results of MTT assay for acute cell metabolic activity of human cells for 3D printed PLA scaffolds. The cells activity is very high in both of the time slots, which proves biocompatibility and non-toxicity of the products to human cells.

TABLE (1). Cell viability in 30 min

30 min	Replicate 1	Replicate 2	Replicate 3	AVERAGE	%Viability
Medium Control	2.51	2.23	2.49	2.36	100
Scaffold	2.21	2.56	2.53	2.43	103.18

TABLE (2). Cell viability in 1 hour

60 min	Replicate 1	Replicate 2	Replicate 3	AVERAGE	%Viability
Medium Control	2.38	2.68	2.64	2.66	100
Scaffold	2.01	2.51	2.78	2.44	91.57

Biodegradation Rate and Pattern

Biodegradation rate and pattern is different in the scaffold based on their structure, original wall thickness, cell size distribution and contact surface with the media. In general, based on the weight change, sample 1 and 5 have the highest degradation rates and samples 4, 3, 2 and 6 had lower degradation rates, respectively.

Samples 1, 2 and 4 have different structure but similar original wall thickness and are therefore, compared to understand the impact of their general 3D design. Under biodegradation, the cubic sample 2 had lower degradation rate than samples 4, which had lower degradation rate than sample 1. Examination of degradation pattern shows that sample 1 degrades randomly and the scaffold structure deteriorates, and therefore the percentage of closed cell in the scaffold increases, which is an undesired phenomenon. Samples 2 and 4 degradation do not result in increment of closed cells or destruction of the scaffolds shape. Therefore, scaffolds with random porosity have higher degradation rate than cubic scaffolds, which in turn have higher rate than the gyroid scaffolds.

In figure 2, trabecular thickness distribution for samples 1-4 are shown before and after biodegradation. In samples 1, 2 and 3, the percentage of walls with lower thicknesses is reduced leading to an increase in the percentage of thicker walls, as a result of biodegradation of the scaffolds, In these samples, percentage of walls thinner than 200 μm is high and distribution of trabecular thickness is not perfectly arranged. In sample 4, the wall thicknesses are well distributed in the range of 13 μm to 351 μm due to its original design and there is an increase in percentage of thinner walls after degradation.

In figure 3, samples 2 and 4 are shown before and after degradation to compare the degradation pattern of a scaffold with gyroid based structure and a cubic based scaffold. In sample 2, walls' integrity is starting to breakdown by biodegradation as shown in figure 3 part A and B. The walls have regular FDM extruded filaments shape and after degradation in addition to small evidences of degradation on the surface, and hollow regions in the middle of the walls can be observed. These cracks on the walls may be the results of breakage of the layers' connections which shape the walls and this shows that the connections degrade on the surface at a faster pace than the surface of the main walls. On the other hand, in figure 3 part C and D, it is obvious that the well-connected surface of the pores are degraded and make new walls with lower thicknesses before the complete degradation of the layers. Thus, it can be concluded that higher contact surface of the scaffolds internal structure results in biodegradation on the surface area instead of the back bone of the structure.

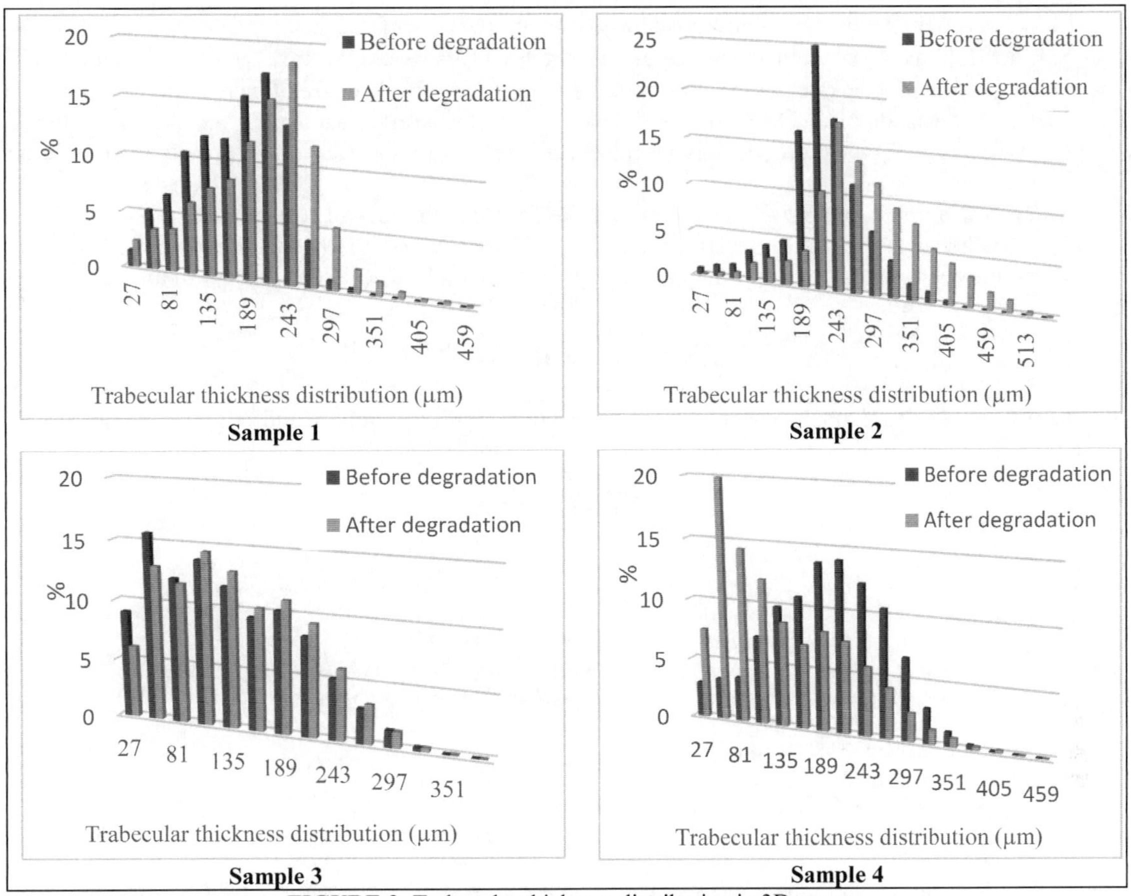

FIGURE 2. Trabecular thickness distribution in 3D space

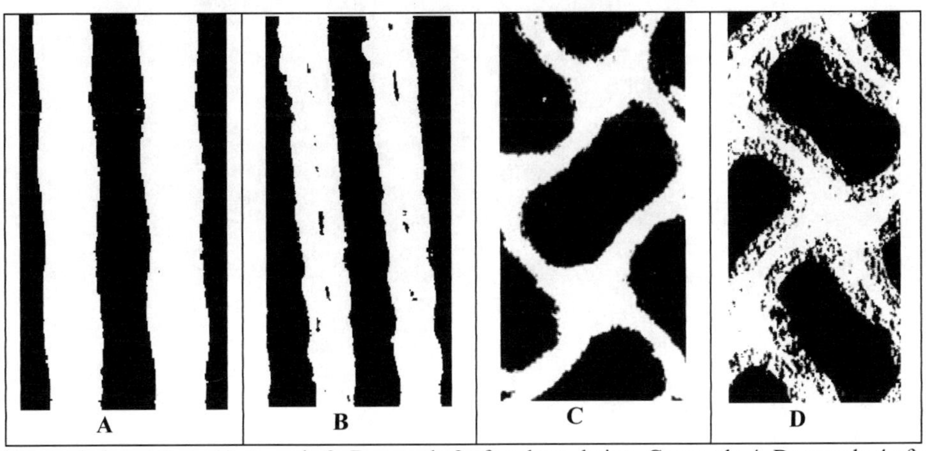

FIGURE 3. Biodegradation pattern (A: sample 2, B: sample 2 after degradation, C: sample 4, D: sample 4 after degradation)

CONCLUSION

In this paper, the main objective is to assess biodegradation rate of different 3D patterned scaffolds made of PLA made by additive manufacturing. An Ultimaker 3 FDM based 3D printer is employed to prepare the samples. The 3D scaffolds are designed with NX 12 and MATLAB with random, cubic and gyroid based structures. MTT assay is done to show acute response of human cells (HT29 cell line) to PLA based scaffolds and assess the cell metabolic activity to prove non-toxicity of the scaffolds to human cells.

The results of this research are summarized as follow:

- MTT assay shows biocompatibility and non-toxicity of the scaffolds to human cells.
- 3D scaffold with high contact surface and minimum edges would increase the biodegradation rate.
- High biodegradation rate in random structures are not desirable because of increase in percentage of close cells and breakage of the structure. Gyroid based scaffolds with the same wall thickness solve the problem by promoting degradation with saving the surface integrity in a pattern that does not lead to close cells formation.
- Biodegradation in cubic structure starts to make fractures in the wall structures by dissolving the weaker regions between the scaffold filaments. In gyroid based structures biodegradation pattern is focused on the well-connected surface areas and does not attack the joints between adjacent filaments.

ACKNOWLEDGMENTS

The authors would like to thank Michael Jeffrey for assistance with cell culture and cell based assays.

REFRENCES

1. E.M. Sachs, et al., *Three-dimensional printing techniques*, Google Patents, 1993.
2. Y. Deng and J. Kuiper, *Functional 3D tissue engineering scaffolds: materials, technologies, and applications*, Woodhead Publishing, 2017.
3. C. Esposito Corcione, et al., *Journal of Applied Polymer Science* **134**(13), 44656 (2016).
4. J. Korpela et al., *Journal of Biomedical Materials Research Part B: Applied Biomaterials* **101**(4), 610-619 (2013).
5. S. Dul, L. Fambri, and A. Pegoretti, *Composites Part A: Applied Science and Manufacturing* **85**, 181-191 (2016).
6. N. Yang et al., *Materials Letters* **173**, 136-140 (2016).
7. D.J. Yoo, *Biomaterials* **32**(31), 7741-7754 (2011).
8. B. Derby, *Science* **338**(6109), 921-926 (2012).
9. R. Pop-Iliev et al., *Radiation and applications* **3**(2), 143-146 (2018).
10. P. Karimipour-Fard, et al., "The Use of Microcomputed Tomography to Evaluate Integral-Skin Cellular Polyolefin Composites," in *Abstract Book; Unilever Res Dev Netherlands: Vlaardingen, The Netherlands*, 2018.
11. K. Holmes, M. Elkington, and P. Harris, *Clark's Essential Physics in Imaging for Radiographers*, CRC Press, 2013.

New hybrid nano additives for thermoplastic compounding: CVD grown carbon fiber on graphene

Elcin Cakal Sarac[a,b], Leila Haghighi Poudeh[c], Jamal Seyyed Monfared Zanjani[d], , Fevzi Çakmak Cebeci[e], Ismail Aydin[b], Yusuf Menceloglu[c,e], Burcu Saner Okan[c]

[a] *Kordsa Teknik Tekstil A.S., Teknopark Istanbul, 34906, Pendik, Istanbul, Turkey (elcin.sarac@kordsa.com)*
[b] *Istanbul University-Cerrahpasa, Faculty of Engineering, Chemical Engineering Department, Avcilar Campus, Avcilar 34320, Istanbul, Turkey*
[c] *Sabanci University Integrated Manufacturing Technologies Research and Application Center & Composite Technologies Center of Excellence, Teknopark Istanbul, 34906, Pendik, Istanbul, Turkey*
[d] *Faculty of Engineering Technology, University of Twente, 7500AE Enschede, The Netherlands*
[e] *Faculty of Engineering and Natural Sciences, Materials Science and Nano Engineering, Sabanci University, 34956, Tuzla, Istanbul, Turkey*

Abstract. Nano additives have unique characteristics widely used in high technology applications due to their ultrahigh mechanical and thermal properties. They are not preferred in price sensitive sectors especially in automotive applications because of their high cost. On the other hand, there is a growing interest to use graphene as a reinforcing agent in composite production. At this point, graphene platelet (GNP) produced from the recycle source was used as a template for carbon nanofiber production by using chemical vapor deposition (CVD) technique to overcome commercialization barrier. This bicomponent and novel structure is a good candidate to be used as a reinforcing agent in compound formulations. This produced hybrid additive was dispersed in thermoplastic resin by thermokinetic mixer to get homogeneous dispersion and provide strong interfacial interactions. In the current work, the outstanding properties of graphene with carbon fibers were combined into one type structure. With the further research, the number of graphene layer were adjusted in this hybrid structure to bring a new insight in graphene and its composite applications. After the fabrication of graphene and carbon fiber based reinforcements with different graphene sources, mechanically and thermally improved Polyamide 6.6 were developed at very low loadings by a thermokinetic high shear mixer. This developed technology will utilize an innovation to produce advanced thermoplastic prepregs including graphene and its hybrid additives with high mechanical properties and increased recycling degree by decreasing manufacturing costs.

Keywords: Graphene, Hybrid structure, Mechanical properties, Thermal properties

INTRODUCTION

There is an increasing trend to use thermoplastic materials for composite part production due to their advantages such as high fracture toughness, high service temperature, shorter production time, and reusing and recycling potential [1][2][3]. Matrices of fiber based thermoplastic composites such as polypropylene and polyamide have already been used notably in automotive, industrial, sports and leisure industries owing to their cost saving and light weighting properties as well as being easy to process at the low processing temperature and pressure [4]. Polyamide 6.6 (PA6.6) as an engineering thermoplastic polymer which has high stiffness, toughness, and resistance to dynamic fatigue was selected in the current study [4][5]. In accordance with this purpose, several micro and nanoparticles have been investigated for the reinforcement of PA in fundamental research and industrial applications [5]. Graphene which has 2-dimentional structure and high surface area is a promising reinforcing agent in composites because of its high mechanical and thermal properties [6][7][8][9]. However, homogeneous dispersion of graphene sheets in the polymeric matrix is still a challenging issue [9][10]. Recently, many studies have focused on the improvement of mechanical and thermal properties of nanocomposites by integrating nanofillers such as graphene, carbon nanotubes (CNT) and nanofibers (CNF) by overcoming the agglomeration problem of nanoparticles. In addition, graphene and CNT have been used together in polymeric matrices to get the synergetic effect to attain an ideal composite structure.

In the current study, carbon nanofibers (CNF) were grown on three different graphene templates which were graphene nanoplatelets (GNP), multi-layer thermally exfoliated graphene oxide (TEGO) and single-layer graphene oxide (GO) by applying CVD technique. These graphene materials have differences in terms of carbon/oxygen ratio, layer number and production technique. In order to understand the performance of these produced hybrid nanoadditives, melt-compounding process were applied for nanocomposite production. These new design additives were used as reinforcing agents in PA6.6 for to understand their 3-dimensional structure on morphological, crystalline, thermal and mechanical properties of the nanocomposites.

EXPERIMENTAL

Synthesis of Hybrid Nanomaterials

Single graphene nanoplatelet (GNP) and thermally exfoliated graphene oxide (TEGO) having 25 graphene layers were purchased from NANOGRAFEN Co. Single layer graphene oxide (GO) was synthesized from graphite flakes by improved Hummers' technique [11,12]. Iron was used as a catalyst in CVD process to grow CNF on the surface of graphene based templates [13]. As-received graphene (GNP) contains 0.5 at% iron in its structure obtained from X-ray photoelectron spectroscopy (XPS). Iron oxide impregnation process was applied for GO and TEGO in different solvent systems of water and DMF, respectively. During CVD process, Fe impregnated graphene samples were exposed to 750°C by purging gas mixtures of He and H_2 and then samples were kept about 10 min at this temperature for annealing. At the end of this stage, ethylene gas was purged to provide carbon source and grow CNF on graphene templates for 15 min at 750°C.

Morphologies of hybrid additives and the length of fiber size were investigated by JEOL2100 Lab6 High Resolution Transmission Electron Microscope (TEM). X-ray diffraction (XRD) and Raman measurements were carried out by using Bruker D2 PHASER Desktop with a CuKα radiation source and Renishaw inVia Reflex Raman Microscopy System, respectively to understand the structural changes. Thermo Scientific K-Alpha X-ray Photoelectron Spectrometer System (XPS) was used for quantitative elemental analysis of hybrid specimens.

Compounding by Thermokinetic Mixer

Hybrid reinforcements were compounded with PA6.6 matrix at 4700 rpm and 300 °C by using Thermokinetic Mixer. PA6.6 chip was supplied from Dupont. The content of hybrid additives were adjusted as 0.5 wt% to attain ideal dispersion and prevent agglomeration. For mechanical studies, test specimens were prepared by injection moulding. Crystallinity behaviour of PA6.6 based nanocomposites was examined by Differential Scanning Calorimeter (DSC) and the softening temperatures and glass transition temperatures (Tg) were examined by Thermomechanical Analysis (TMA) method using a 3 mm ball-point probe. The morphological structures of the nano additives were analyzed by JEOL2100 Lab6 High Resolution TEM.

RESULTS AND DICUSSIONS

The Characteristic Properties of Hybrid Nano additives

TEM images of hybrid structures in Figure 1 indicates the formation of CNF on different graphene structures and are supported the formation of 3-dimensional structure having two different components. Iron particles with the average size of 100 nm were seen clearly in TEGO and GO based additives as seen in Figure 1(b) and 1(c) whereas the length of CNF in GNP is comparably shorter than TEGO and GO based additives. Table 1 summarizes the average diameter and length of CNF, C/O ratio obtained from XPS and intensity ratio of D and G (I_D/I_G) received from Raman spectroscopy of the produced nano additives in this study. CNF-TEGO has the highest C/O ratio as about 6.9 but GNP based hybrid has more oxygen functional groups on its surface.

TABLE (1). Summary of the characteristic properties of hybrid nano additives

Type of graphene in hybrid structure	Production method of graphene	The length of CNF (μm)	The diameter of CNF (nm)	C (at%)	O (at%)	C/O ratio	I_D/I_G
Graphene platelet (GNP)	Recycling and upcycling	1.0	40	91.7	5	6.6	0.98
Multi-layer graphene oxide (TEGO)	Thermal exfoliation	2.0	28	98.1	1.9	6.9	1.5
Single-layer graphene oxide (GO)	Improved Hummers' method [11] [12]	1.6	34	96.8	3.2	5.5	1.5

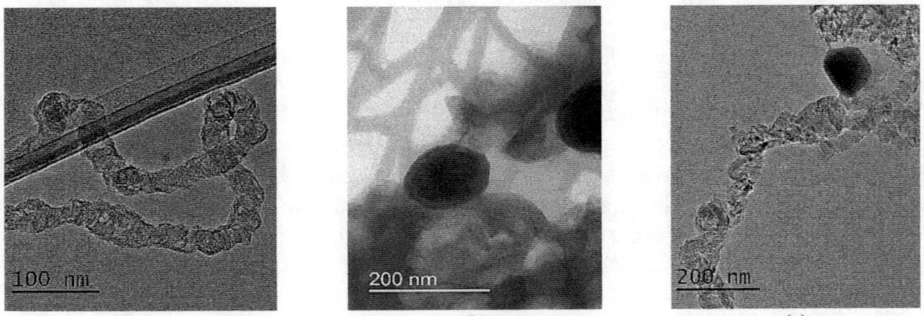

(a) (b) (c)

FIGURE 1. TEM images of **(a)** CNF-GNP, **(b)** CNF-TEGO and **(c)** CNF-GO

The structural changes of hybrid structures of GNP, GO and TEGO were monitored by XRD analysis is as seen in Figure 2(a). In all of the specimens, a diffraction peak at $2\Theta \approx 25°$ was corresponded to the (002) reflection of graphitic plane [14]. During CNF growth process, iron oxide catalyst was reduced to metallic Fe and Fe_3C on all the graphene types labelled in Figure 2(a) [15]. XRD verified CNF growing process and addition of carbon atoms on metal catalysts deposited on graphene surface. Among three types of additives, TEGO based hybrid has the highest crystallinity.

Raman characterization is an efficient technique to examine the changes in sp^2 and sp^3 degree of structure and the formation of additional carbon structure and defects. Graphene has three characteristic Raman peaks of D, G and 2D. In addition to these peaks, a second-order peak known as S3 at around 2929 cm^{-1} in the Raman spectra was appeared due to the interactions of intervalley and intravalley phonon scatterings between graphene and CNF [16]. As shown in Figure 2(b) and Table 1, a higher result in I_D/I_G of CNF-TEGO and CNF-GO compared to CNF-GNP showed the formation of more sp^3 bonds regarding the intensity of D band [17]. In conclusion, the differences in surface chemistry and crystalline structure of these fabricated nano hybrids can directly affect the performance of PA6.6 nanocomposites explained in the following sections.

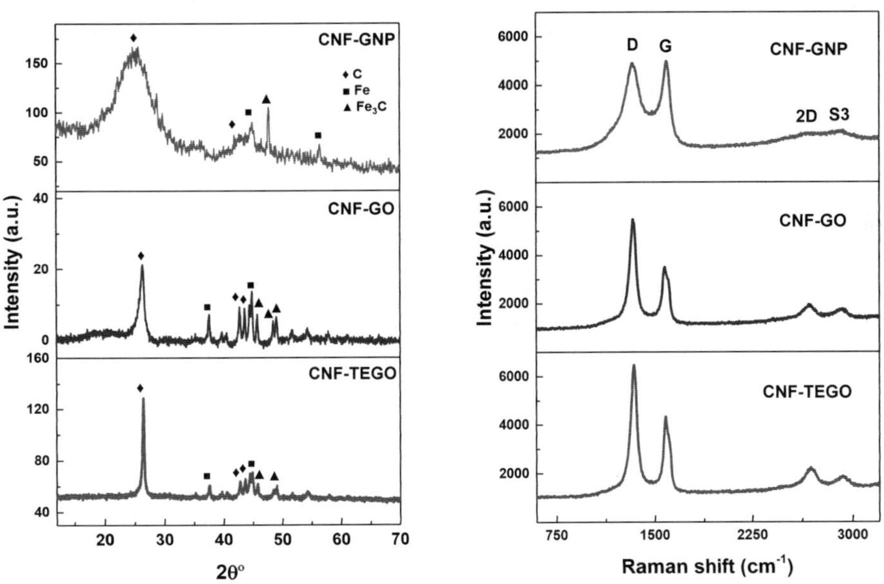

FIGURE 2. **(a)** XRD spectra and **(b)** Raman spectra of CNF-GNP, CNF-GO and CNF-TEGO

Crystalline and Thermal Behaviours of Nano Hybrid Reinforced Nanocomposites

The crystalline behaviour of nanocomposites were investigated by DSC technique. As seen in Table 2, there is no big difference in melting temperatures (T_m) of nanocomposites specimens in comparison of pristine PA6.6. The percent crystallinity degree is determined using *Eq. 1*:

$$X_c = (\Delta H_m / \Delta H_m^{100\%}) \times 100\% \quad (1)$$

where X_c is crystallinity degree, ΔH_m is melting enthalpy, and $\Delta Hm^{100\%}$ is purely crystalline PA6.6 enthalpy [18]. The significant increasing was observed in X_c values by the integration of each type of graphene hybrid additive which were acted as nucleating agents. In addition, an increase in X_c values of 0.5 wt% CNF-TEGO and CNF-GO-based PA66 nanocomposites is same compared to neat PA6.6 whereas X_c of CNF-GNP reinforced composites is slightly lower due to its production source which is recycled carbon and structural difference.

TABLE (2). DSC results of nano hybrid reinforced nanocomposites

Samples	T_m(°C)	ΔH_m (J/gr)	X_c (%)
Pristine PA6.6	262	48.2	25.6
PA6.6+0.5 wt% CNF-GNP	261	68.7	36.5
PA6.6+0.5 wt% CNF-TEGO	262	74.2	39.4
PA6.6+0.5 wt% CNF-GO	262	74.2	39.4

Table 3 summarized the Tg received from TMA analysis and softening temperatures for pristine PA6.6 and hybrid nano additive reinforced PA6.6 based nanocomposites. No big difference was observed in Tg values of each composite specimens by integration of each nano additive. On the other hand, softening temperatures of CNF-TEGO and CNF-GO loaded nanocomposites were slightly decreased when compared to pristine PA6.6. In case of CNF-GNP reinforced nanocomposite, softening temperature was increased about 6 °C and this indicates an increase in resistance to thermal deformation and thus CNF-GNP can block the movement of polymer chains and improve its process conditions.

TABLE (3). TMA results of hybrid nano fillers loaded PA6.6 based nanocomposites

Samples	Tg (°C)	Softening Temperature (°C)
Pristine PA6.6	42.7	227.0
PA6.6 + 0.5 wt% CNF-GNP	42.4	233.3
PA6.6 + 0.5 wt% CNF-TEGO	43.8	225.0
PA6.6 + 0.5 wt% CNF-GO	43.9	226.0

Mechanical Test Results

0.5 wt% hybrid reinforcements were compounded with PA6.6 polymer at high shear rates. Mechanical tests of the produced nanocomposites were performed with 2 mm/min speed by using Instron 5982 Static Test Machine for tensile (ISO 527-2) and flexural (ASTM D 790) tests. Table 4 summarizes mechanical characteristics of nanocomposite specimens. Tensile modulus calculated from the linear regions indicated that modulus of PA6.6 nanocomposite was enhanced up to 34% and 24% by the incorporation of 0.5 wt% CNF-GNP and multi-layer CNF-TEGO into the PA6.6, respectively. On the other hand, in the specimen reinforced by CNF-GO hybrid additive, the incorporation of 0.5 wt% reinforcement showed lower tensile modulus by 20%. This might come from the agglomeration of this type of hybrid additive and broken its fibers at high shear during compounding. This statement was also confirmed by microtomy study by TEM. Figure 3 represents TEM images of cross-sections of hybrid reinforced nanocomposites. CNF-GO based nanocomposite has agglomerated sheets in Figure 3(b) and in CNF-GNP based specimen, fibers are not aligned and aggregated as seen in Figure 3(a). In contrast, CNF-TEGO based nanocomposite has separated graphene, fibers and also iron metal catalyst. Tensile strength showed improvement in all cases of nanocomposites compared to neat PA6.6 specimen but exhibited different behaviour regarding the type of reinforcement. In case of only CNF-TEGO reinforced nanocomposites, the highest tensile strength was achieved at the loading of 0.5 wt%. The flexural modulus of all types of hybrid additives loaded PA6.6 nanocomposites are shown in Table 4. The best flexural property was achieved by using 0.5 wt% CNF-GNP providing 23% improvement in flexural modulus. On the other hand, CNF grown on TEGO at 0.5 wt% loading showed better performance in both tensile and flexural tests since the diameter of fibers on TEGO was smaller than the one grown on single-layer GO and this hybrid might reduce stress concentration of nanocomposites.

TABLE (4). Summary of mechanical properties of CNF-GNP, CNF-TEGO and CNF-GO loaded PA6.6 nanocomposites

Samples	Tensile Modulus (MPa)	Tensile Modulus Improvement (%)	Tensile Strength (MPa)	Tensile Strength Improvement (%)	Flexural Modulus (MPa)	Flexural Modulus Improvement (%)
Pristine PA6.6	2400	-	69.8	-	2570	-
PA6.6/0.5 wt% CNF-GNP	3225	34	73.4	8	3170	23
PA6.6/0.5 wt% CNF-TEGO	2971	24	80.6	15.5	2933	14
PA6.6/0.5 wt% CNF-GO	2868	20	71.7	2.7	2365	-8

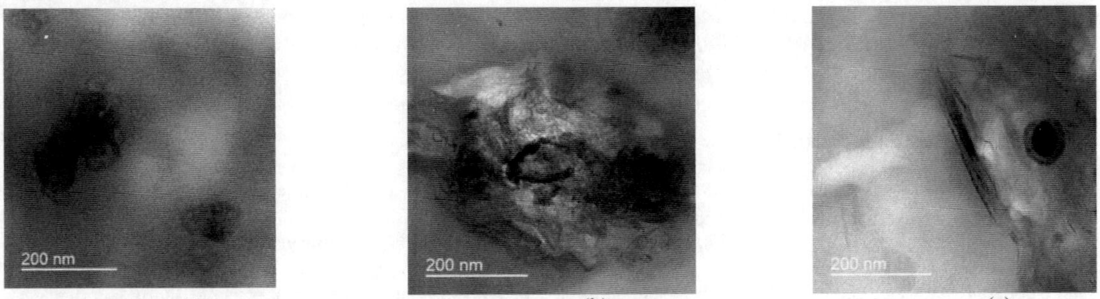

(a) (b) (c)

FIGURE 3. TEM images of **(a)** CNF-GNP, **(b)** CNF-GO and **(c)** CNF-TEGO reinforced nanocomposites at 0.5 wt% loadings

CONCLUSIONS

To conclude, CNF were successfully grown on Fe loaded three different graphene templates by CVD method and the structural formation of each sample was confirmed by spectroscopic and macroscopic techniques. CVD grown CNF on graphene acted as a reinforcing and nucleating agent in thermoplastic resins. Stronger interfacial interactions were achieved by CNF-TEGO nanoadditive and thus its nanocomposite showed better mechanical performance. Further improvement in graphene structure will influence the performance of structural composites especially used for load bearing applications in a positive manner.

ACKNOWLEDGMENTS

The authors would like to thank Kordsa Teknik Tekstil A.S. Company for providing financial support and Dr. Ilse Letofsky-Papst from Graz University of Technology for TEM analysis.

REFERENCES

1. I.Y. Chang and J. K. Lees. *Journal of Thermoplastic Composite Materials* **1**, 277-296 (1988).
2. J. Diaz and L. Rubio. *Journal of Materials Processing Technology* **143-144**, 342-346 (2003).
3. Z. Shen, S. Bateman, D. Y. Wu, P. McMahon, M. Dell'Olio, J. Gotama. *Composites Science and Technology* **69**, 239–244 (2009).
4. B. Mouhmid, A. Imad, N. Benseddiq, S. Benmedakhène, A. Maazouz. *Polymer Testing* **25**, 544–552 (2006).
5. A.M. Hadizadeh, A. Shojaei, R. Bagheri. *18th International Conference on Composite Materials* n.d.:1–5.
6. M. Bhattacharya. *Materials* **9**, 1-35 (2016).
7. Y. Zhu, L. Li, C. Zhang, G. Casillas, Z. Sun, Z. Yan, et al. *Nature Communications* **3: 1225**, 1-7 (2012).
8. J. Seyyed Monfared Zanjani, B. S.Okan, Y. Menceloglu. *Materials Chemistry and Physics* **176**, 58-67 (2016).
9. X. Sun, H. Sun, H. Li, H. Peng. *Advanced Materials* **25**, 153-176 (2013).
10. R. S. Edwards and K. S. Coleman. *Nanoscale* **5**, 38-51 (2013).
11. D. C. Marcano, D. V. Kosynkin, J. M. Berlin, A. Sinitskii, Z. Sun, A. Slesarev, et al. *ACS Nano* **4**, 4806–4814 (2010).
12. D. C. Marcano, D. V. Kosynkin, J.M. Berlin, A. Sinitskii, Z. Sun, A. S. Slesarev, et al. *ACS Nano* **12**, 2078–2078 (2018).
13. A. Gohier, C. P. Ewels, T. M. Minea, M. A. Djouadi. *Carbon* **6**, 2-9 (2008).
14. I. Stamatin, A. Morozan, A. Dumitru, V. Ciupina, G. Prodan, J. Niewolski, et al. *Physica E: Low-Dimensional Systems and Nanostructures* **37**, 44-48 (2007).
15. R. Kumar, B. Sahoo. *Nano-Structures and Nano-Objects* **14**, 131-137 (2018).
16. M. S. Dresselhaus, G. Dresselhaus, R. Saito, A. Jorio. *Physics Reports* **409**, 47-99 (2004).
17. L. Haghighi Poudeh, B. Saner Okan, J. Seyyed Monfared Zanjani, M. Yildiz, Y. Menceloglu. *RSC Adv* **5**, 91147–91157 (2015).
18. L. Lim, I. J. Britt, M. Tung. *Journal of Applied Polymer Science* **71**, 197–206 (1998).

Tex2Mat: Development of New Treatment Methods For Multi Material Textile Waste

Uta Jenull-Halver [a*], Clemens Holzer[a], Benjamin Piribauer[b], Felice Quartinello [c]

[a] *Montanuniversitaet Leoben, Chair for Polymer Processing, Otto Gloeckel-Strasse 2, 8700 Leoben, Austria*
[b] *TU Wien – Institute of Chemical, Environmental and Bioscience Engineering/E166*
Getreidemarkt 9/166, 1060 Wien, Austria
[c] *IFA Tulln, Institute of Environmental Biotechnology, Konrad-Lorenz-Strasse 20*
3430 Tulln an der Donau, Austria

Abstract. Polyethylene terephthalate (PET) and polyamide (PA) are high performance technical polymers for demanding applications. Even if up to date production methods work highly efficiently there will always be a certain amount of waste during production and furthermore when the products reach their end of life. In the presented study the waste streams come from two weaving companies (home textiles and industrial textiles) and one large scale laundry (home textile). The waste streams of the weaving companies come from the trimming of the edges at the loom. This means that the waste can be collected very easily on the spot, is relatively clean and can be separated according to the polymer type. In the case of the home textile company the waste is a yarn blend of PET and cotton. Here, a physical separation is not possible. In the scope of this study an enzymatic process was developed to dissolve the cotton and obtain clean and pure PET-fibres. In various tests and analyses, the properties of the new and the recycled material were tested and it was investigated, if the recycled PET-fibres were suitable again for spinning. The properties tested were intrinsic viscosity (IV), melt volume flow rate (MVR) and thermal properties using differential scanning calorimetry (DSC) method. Based on the results the spinning ability of the rPET was assessed and further necessary steps taken. For the PA and the PA-blends similar tests were performed with the objective to determine the suitability of the recycled fibre material for injection moulding. Based on the results injection moulding tests were performed and tensile test bars were produced.

Keywords: Polymer recycling, polymer textiles, cotton, spinning, injection moulding

INTRODUCTION

The EU's new circular economy directive presents a substantial challenge for the polymer processing industry, in the course of which this study has been initialized. The aim is to upcycle textile waste from various sources and to produce textiles again or technical parts from the recycled material with a high demand on mechanical and thermodynamic properties. In this study three textile waste streams were examined. There were two waste streams from home textiles (towels, bed sheets, etc) and one waste stream from industrial textiles (sieves and felts for paper and cement industry). The home textiles are a blend of ~50 wt.% polyethylene terephthalate (PET)-fibre and ~50 wt.% cotton fibre. One central task of this study was to develop an enzymatic process to break down the cotton fibres and to retrieve the pure PET-fibres. In the following steps the gained PET fibres will be spun again to fibres and woven into towels, as in the original application. Additionally, the textiles will be tested according to the oecotex standard. The industrial textiles waste stream - three different polyamide (PA) and PA blends - will be tested for selected products which have to comply with very high standards and are partially used in safety-critical applications. The objective here was to compare the recycled fibre material to standard injection moulding grades. For a material to be suitable for spinning it has to have a significantly higher molecular weight and also narrower molecular weight distribution compared to injection moulding grades. At the beginning of the study it was not clear if a material designed for spinning can also be used for injection moulding.

EXPERIMENTAL

Definition of Three Case Studies

At the beginning of the project three case studies were defined. The definition was made according to the origin of the waste stream and the later treatment as well as their processing and usage.

Case Study A: In this case study laundry (bed sheets, table cloths, etc) made from ~50 wt.% cotton and ~50 wt.% PET-fibre was examined. During its life time the laundry is being washed several dozen times at high temperatures and with detergents containing e.g. peroxyacetic acid for sterilization. Additionally there is the wear and tear during usage. Individual pieces get sorted out when the fabric is worn out or has non-removable stains. Summarised, the waste stream in this case study is clean, washed several times and has a certain age.

Case Study B. In this case study the textile waste comes from the edge trimming from the loom of a company producing towels, bath robes and similar. Again, a fibre composition of ~50 wt.% cotton and ~50 wt.% PET was chosen. This waste stream is not washed and is treated with various chemicals to make the processing steps of spinning and weaving more efficient. These chemicals are esters, oils and tensides.

Case Study C: This case study is very different from the first two. This waste stream comes from edge trimmings at the loom of the production of felts and sieves e.g. for the paper- or cement industry. Here the materials are various blends of PA6, PA6.6 and PA 6.10.

In all three case studies it is not possible to separate the individual components of the textiles. So for case study A and B an enzymatic process was developed. In this process the cotton is being broken down by specially selected enzymes. Before the enzymatic process a pre-treatment makes the cotton fibres more easily accessible for the enzymes. A cleaning process afterwards removes the by-products from the decomposed cotton. Cotton fibres have besides their chemical composition a very complex structure which makes them more stable than viscose. For these case studies the aim was to gain rPET from the enzymatic process, test its properties and - if necessary - to use additives or a recondensation process to re-establish the suitability for spinning. The aim is to produce several towels in sufficient quality again.

The polyamides from case study C are not separable. The PA-mixtures were examined and prepared for injection moulding trials.

Enzymatic Hydrolysis of PET Cotton Mixtures

To be able to regranulate the PET in the cotton/PET mixtures, the cotton has to be completely removed, while the PET should not be damaged by the used process. While there are possibilities to remove cotton chemically with the use of a strong acid such as sulfuric acid for example [1], these chemicals may damage the PET, too. Using enzymes however, damage to the PET can be avoided.

Enzymes are produced by living organisms, however, they themselves are not alive, this is making them easy to handle and ideal for a sustainable process. Enzymes are big molecules, consisting of different amino acids connected to each other by peptide bonds. These peptide chains fold in specific ways and create big three-dimensional structures, which are needed to correctly bind the reactants (2).

Each type of enzyme only catalyses very specific reactions, making them ideal to remove the cotton from the cotton/PET mixed textiles, without unwanted side reactions, which could damage the PET. The types of enzymes needed for this process are called cellulases, named so because they hydrolyse cellulose into its monomer glucose. The cellulase enzymes are however not one specific molecule, but a wide range of different ones that fulfil the same, or similar functions.

For the implementation of the process, it is important to generate a big surface that the enzymes can access. Therefore, the cotton/PET textiles were first mechanically broken down into short single fibres through a two-stage cutting and milling process. For even better hydrolysis results, an alkaline pretreatment opened the cotton structures further up, making it easier for the bulky enzymes to reach the cellulose chains.

After the preparation of the pre-treated substrate, a citric acid buffer solution was prepared to keep the pH value constant throughout the whole hydrolysis process. The solution was then heated up to the optimal operating point of the enzyme mixture. After reaching the temperature, the pre-treated textiles were added. After rechecking the pH value, finally, the enzyme mixture was added to the process. Depending on the pretreatment and other operating parameters, everything had to be rigorously stirred and kept at the operating temperature for one to seven days.

After the reaction is finished, the textile was washed extensively, since any residues of the enzyme mixture can start getting mouldy. The enzyme can afterwards be re-used after removing it from the solutions via a membrane process. The solution containing high amounts of glucose can be used as a platform chemical in different processes.

Analysis of the PET and PA material

A set of tests were selected to determine if the polymers were deteriorated during usage and the enzymatic process.

1) The following test methods were selected for the examination of the recycled materials:
2) DSC: Determination of melting points and differences of melting points of different rPEt- lots.
3) MVR: The MVR should give a first indication of a change in the molecular weight of the recyclates and also if the material designed for spinning process is suitable for the injection moulding process as well.
4) Intrinsic viscosity (IV) and relative viscosity (RV): Both parameters, IV (for PET) and RV (for PA) are relevant for the evaluation of the spinnability of materials.
5) Mechanical testing: Tensile test and impact test: These tests were done in order to assess if the material is suitable for the application for technical parts subject to defined mechanical stress.
6) Viscosity: High pressure capillary viscosimetry: This method was selected in order to compare the flow behaviour of the recycled material to that of the virgin material.

All these tests were performed on the rPET fibres and compared to the properties of virgin PET. It was tried to trace the value chain back to the producer of the PET raw material, but it was not possible to determine the actual producer of the polymer. Therefore, the PET grade F008 (by Invista) was chosen as a reference material, because it was used by one of the project partners and is a common material in the textile industry.

RESULTS

Due to limited space in this paper, the results shown here focus on DSC measurements and spinning trials.

DSC

Table (1) shows the DSC measurement results along with the sample names taken from the first batch of the PET gained in the enzymatic process. To that batch the name "PET sulphuric acid long" was given and it consists of a large quantity of small pieces of textile cloth and loose fibres from laundry. From the batch a cluster of fibres and several small individual pieces of textile were taken as samples and analysed (samples 3 to 8). As a reference virgin Invista F008 and a virgin fibre "Semi dull fibre" was taken. Figure 1 shows the curves of the second heating.

The samples were chosen due to their visual appearance since no other difference could be detected at first. These pieces of clothes varied strongly, e.g. thickness of thread, twisting and weaving technique. It was decided to take samples which differed the most. Sample 3 from table 1 is a cluster of fibres without any obvious connection to a certain piece of cloth. The samples 4-8 are from individual pieces of cloth.

The results showed no significant difference between the virgin material and the recycled material. In table 1 this can be seen in the melting temperatures and peak areas of the second heating cycle.

TABLE (1). Overview of the DSC measurement results peak temperature and peak area along with the description of the samples.

Sample no.	Material	Explanation	2nd peak temperature (°C)	2nd peak area (J/g)
1	Invista F008	Virgin material, not spun	255.4	35.7
2	Semi Dull fibre	New fibre material	254.4	33.8
3	Cluster of textile fibres from TU	Cluster of fibres from TU batch	254.4	33.9
4	Textile from TU batch	Sample from piece of textile	252.5	29.9
5	Textile from TU batch	Sample from piece of textile	254.1	32.1
6	Textile from TU batch	Sample from piece of textile	252.8	29.5
7	Textile from TU batch	Sample from piece of textile	254.9	34.8
8	Textile from TU batch	Sample from piece of textile	255.4	35.3

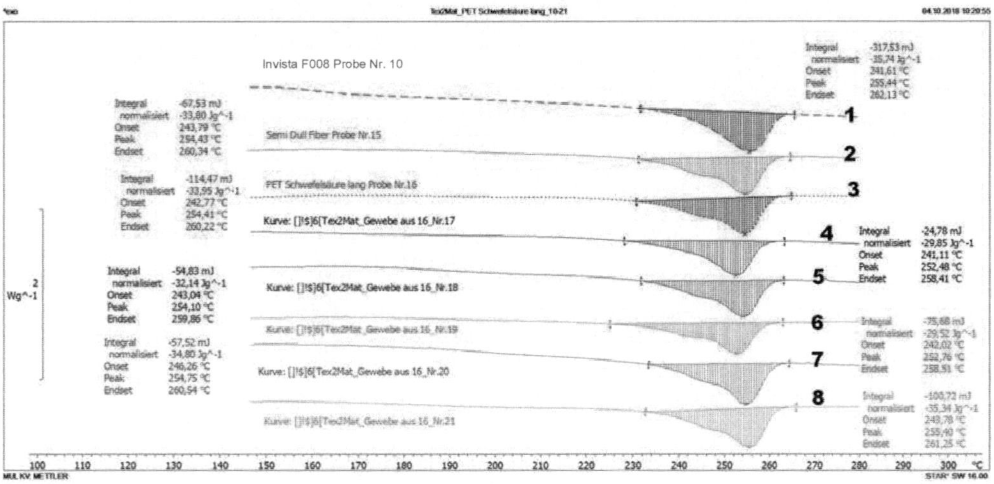

FIGURE 1. Curves from DSC measurements

Spinning trials

An essential part were the first spinning trials. These trials performed on a smaller quantity showed that the recycled PET was indeed suitable for spinning. The individual threads did not break when leaving the die nor was a spinning beard formed. These two phenomena would have been a sign for impurities and an insufficient IV. Also, the test showed that the material did not suffer from discolouration during spinning, again indicating that the recycled material was sufficiently free from contamination (residue of spinning and weaving finish, residue from the enzymatic process).

The results (processing parameters and mechanical properties of the fibres) will be the basis for the large-scale trials with a larger quantity of rPET which are planned as next steps.

SUMMARY AND OUTLOOK

rPET

The first tests, especially DSC and IV showed that in principle the enzymatic process is able to produce a rPET fibre that is suitable for spinning and weaving. This again was confirmed in the spinning trails.

The only necessary action to be taken was a recondensation process during regranulation.

For future trials and eventually for a large-scale production the enzymatic process has to be accelerated, though, as it takes several days at the moment, due to the high resilience of the cotton fibre. Currently, the process is being optimised in that direction.

rPA

The three types of PA recyclates (PA 6, PA6/6.6, PA6/6.10) had the properties required by the injection moulding companies. This applies to the requirements of the injection moulding process as well as their application properties.

The first tests showed that products made from the recycled material complied to various mandatory standards.

Nevertheless, further trials with additives are planned in order to broaden the field of application.

ACKNOWLEDGMENTS

The research work of this paper was performed within a framework of the Austrian Research Promotion Agency (FFG-Project no. 865383 91).

REFERENCES

1. Textiles-Quantitative chemical analysis-Part 11: Mixtures of cellulose and polyester fibres (method using sulfuric acid), ISO 1833-11:206
2. R. Singhania, M. Adsul, A. Pandey, A. Patel: Current Development in Biotechnology and Bioengineering. 2017

Properties of Lignocellulosic Composites of Coffee Husk Filled Polypropylene

Henrique de Almeida Leal, Alex Sandro Babetto, Baltus Cornelius Bonse*

Materials Engineering Department, Centro Universitário FEI. Av. Humberto de A.C.Branco 3972 São Bernardo do Campo – SP- Brazil

Abstract. Ground coffee husk has been incorporated into polypropylene (PP) at 20, 30 and 40 wt%, along with a PP maleated compatibilizer at 10 wt% relative to the coffee husk, by means of a co-rotating twin-screw extruder and subsequent injection molding into test specimens. In relation to neat PP, the 40 wt% composite showed an increase in flexural strength, flexural modulus, tensile modulus and heat distortion temperature of about 35, 90, 75 and 45%, respectively. However, strain at break, impact and tensile strength decreased by around 95, 40 and 20%, respectively. The presence of coffee husk practically did not affect PP melt temperature, but increased both composite's degree of crystallinity, measured by DSC, and temperature at maximum degradation rate, measured by TGA.

Keywords: Lignocellulosic Composites. Coffee Husk. Polypropylene
PACS: 72.80.Tm

INTRODUCTION

Brazil is currently the largest coffee producer and exporter in the world. According to the 2018 Coffee Research Consortium [1], in the 2017-2018 crop year, Brazil produced 51 million of 60-kg bags of coffee, which represented around 32% of the world production.

When coffee is processed, 30 to 50% of its mass becomes agricultural waste. Thus, in 2018, more than 1 million ton of waste was generated. Such residues are divided into coffee husks (CH) and pulp, from the peeling of coffee cherries during dry (husk) or wet (pulp) processing [2].

The reuse of coffee husks and pulp has not yet been fully explored, and the large amounts of this waste may result in environmental problems. A number of studies have attempted to find sustainable and cost-effective uses of these wastes, including animal feed, solid biofuel, replacement of wood in the production of particle boards, source of bioactive compounds and feedstock for the production of liquid and gaseous biofuels, low-cost adsorbents and fermentation products such as enzymes, food flavorings and aromas [2].

However, there is still need for research on alternative and profitable uses of coffee husk and pulp. An alternative in the search for techniques and products for the recovery, reuse and recycling of this type of waste would be its incorporation in thermoplastic matrix, such as polypropylene (PP), to produce thermoplastic lignocellulosic composites, adding hence value to this waste.

EXPERIMENTAL

PP pellets (H503) from Braskem with MFR (230°C/2.16kg) = 3.5 g/10min, density = 0.905g/cm³ were used as matrix. Compatibilizer used was PP grafted maleic anhydride, Polybond 3200 (MFR = 110 g/10min at 230°C/2.16kg), supplied by Crompton (Chemtura), with MA content of around 1 wt%. The agricultural residues containing mainly coffee husks with some twigs and stems were supplied by Coffair Brasil (Franca, SP, Brazil).

The residues were ground in a knife mill and sieved through a 2 mm screen steel sieve. Subsequently PP, 40 wt% ground coffee husks (CH) and compatibilizer, at 10wt% relative to CH, were tumble mixed and extruded in a Thermo Scientific HAAKE PolyLab OS RheoDrive co-rotating twin-screw extruder (L/D =25 mm; D=16 mm). From this 40wt% masterbatch, by mixing with neat PP, where required, test specimens of PP containing 20, 30 and 40 wt% CH were injection molded. A 30 wt% CH formulation without compatibilizer was also injection molded for comparison sake.

Tensile and three-point flexural tests were carried out according to ASTM D638 and D790 respectively, at speeds of 50 mm/min and 1.3 mm/min, respectively. For tensile modulus measurements a speed of 5 mm/min was used. Charpy notched impact test were performed in an Instron Ceast 9050 tester, according to ASTM D6110. Heat deflection tests were carried out in a CEAST HDT-Vicat 6 Stations tester in accordance with ASTM D 648. DSC and TGA runs were performed under nitrogen atmosphere in a TA Instruments Q20 Calorimeter and SDT 600 analyzer, respectively. DSC was performed at a heating rate of 10°C/min from 25 °C to 200 °C, whereas TGA at 20°C/min from 25 °C to 650 °C.

RESULTS AND DISCUSSION

The results of the tensile and flexural tests are shown in Fig.1, impact strength and HDT in Fig. 2, DSC and TGA results in Table 1.

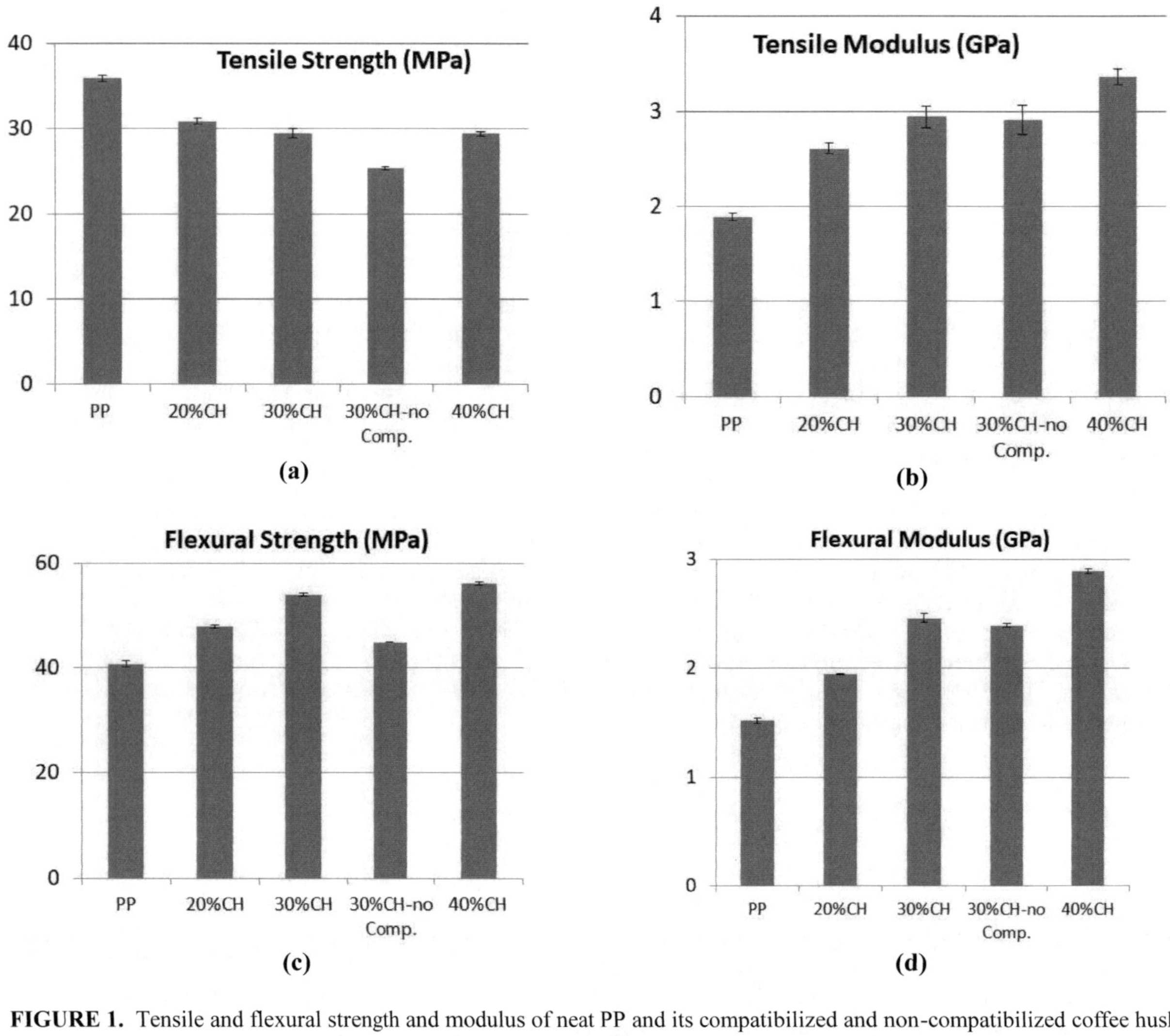

FIGURE 1. Tensile and flexural strength and modulus of neat PP and its compatibilized and non-compatibilized coffee husk composites

Figure 1(a) reveals a reduction of 15 to 20% in tensile strength when the ground coffee husk waste is added to PP, contrary to expectations, since the presence of stiff natural fibers with adequate amounts of compatibilizer tend to increase composite tensile strength [3-5]; the 30 and 40 wt% CH, showed practically equal TS, according to statistical analysis ANOVA ($p = 0.89$), which may indicate that the amount of compatibilizer was not adequate, that is, the fibers did not adhere completely to the polymer base. On comparing the 30wt% CH composite with and without compatibilizer, the absence of compatibilizer resulted in a reduction in composite TS exceeding 10%. However, tensile modulus is unaffected by the absence of compatibilizer as seen in Fig. 1b, on comparing the 30wt% CH composite. This can be explained by the fact that while modulus is measured at relatively small strains, TS is measured at relatively high strains, and only at sufficient high strains the lack of stress transfer from the matrix to the filler will be felt. A filler is considered a reinforcement when stress can be adequately transferred from the matrix to the filler, resulting in higher composite TS in relation to the matrix [6]. Hence, this residue did not perform as a reinforcement, despite the increase in flexural strength of up to 35% as seen in figure 1(c).

The presence of CH increased composite stiffness, as evidenced by the increase in both tensile and flexural moduli of up to 75 and 90% in relation to neat PP, Fig. 1(b) and (d), likely due to polymer chain mobility restrictions, brought about by the CH.

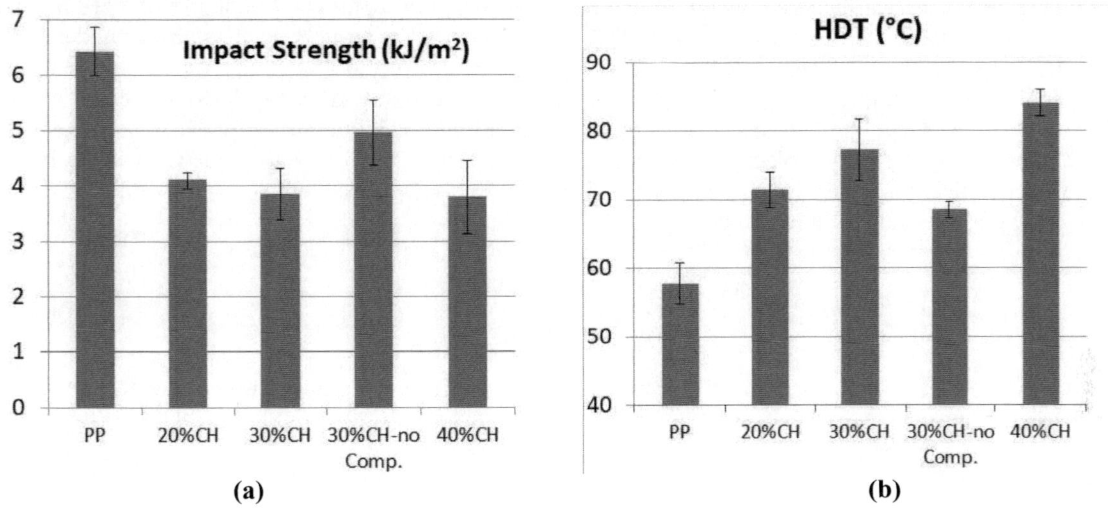

FIGURE 2. (a) Charpy notched impact strength and (b) heat distortion temperature of neat PP and its compatibilized and non-compatibilized coffee husk composites

Chain mobility restrictions also may have been the cause of the decrease in composite impact strength as shown in Fig. 2 (a), as well as lack of adequate compatibilization [6]. According to ANOVA (p=0,57) increasing CH content in the compatibilized composites has, statistically, no effect on impact strength.

Increased composite stiffness together with increased crystallinity (Xc) (Table 1), as CH content increases, accounted for increased composite HDT (Fig. 2 b), i.e., composite's shape stability under heat is increased with increasing CH content.

TABLE 1 – DSC and TGA results of neat PP and its compatibilized and non-compatibilized coffee husk composites

		DSC		TGA
wt% CH	Tm (°C)	ΔHm (J/g)	Xc (%)	Tmax
0	168	82.5	39.8	461
20	172	66.5	41.2	468
30	167	71.0	51.2	471
30 (no Comp.)	169	68.5	47.3	462
40	168	66.0	57.0	474

Table 1 show some DSC (melt temperature: Tm, heat of fusion: ΔHm, and degree of crystallinity: Xc) and TGA (temperature of maximum rate of degradation: Tmax) results. Degree of crystallinity was calculated from $Xc = [\Delta Hm/(w*\Delta Hm^{100\%})]*100$, where Xc: degree of crystallinity, w: mass fraction of PP in the composites, and $\Delta Hm^{100\%}$ the heat of fusion of 100 % crystalline PP, listed as 207 J/g [7].

As can be seen PP melt temperature does not undergo significant changes when CH is added. However, PP crystallinity increases steadily as CH content increases, likely because the filler adsorbed the polymer molecules and acted as crystal nuclei [6].

The TGA results show an increase of 13°C in Tmax when comparing the 40 wt% CH composite to neat PP, indicating that the presence of coffee husk increases the thermal stability of the composite.

CONCLUSIONS

Ground coffee husk has been successfully incorporated into polypropylene, acting rather as filler than as reinforcement, evidenced mainly by the decrease of up to 20% in composite tensile strength in the presence of the coffee husk, despite increase in flexural strength of up to 35%. Composite stiffness increased, evidenced by increase in tensile and flexural modulus of up to 75% and 90%, respectively. Composite's shape stability under heat also increased with increasing coffee husk content, evidenced by increase of up to 45% in heat distortion temperature. Ductility, as measured by strain at break, and impact strength were negatively affected, showing reductions of up to 95 and 40%, respectively, likely due to restrictions imposed on polymer chain mobility by the stiffer husks. Coffee husks also hindered polymer flow, evidenced by reduction in melt flow rate of up to 50%. Composite DSC melt temperature was practically unaffected by the presence of the husks, however composite's relative thermal stability increased as shown by thermogravimetric analysis.

ACKNOWLEDGMENTS

Thanks are due to Centro Universitário FEI for the undergraduate research scholarship granted to H.A. Leal, and to Coffair Brasil for donating the coffee husks.

REFERENCES

1. Consórcio pesquisa café. Retrieved Sep 1, 2018, from http://www.consorciopesquisacafe.com.br.
2. L.S. Oliveira; A.S. Franca An overview of the potential uses for coffee husks. In: V. Preedy.(Org.). Coffee in Health and Disease Prevention, chapter 31. 1ed.Amsterdam: Elsevier, 2014, v.1, pp. 283-292
3. S.H.P Bettini; B.C. Bonse; E.A. Melo; P. A. Muñoz, Effect of Sawdust Surface Treatment and Compatibilizer Addition on Mechanical Behavior, Mophology, and Moisture Uptake of Polypropylene/Sawdust Composites. Polymer Engineering and Science, v. 50, p. 978-985, 2010.
4. S.H.P. Bettini; A. B. L. C. Bicudo, I. S. Augusto; L. A.; Antunes, Morassi, P. L.; R.Condotta; B. C. Bonse Investigation on the Use of Coir Fiber as Alternative Reinforcement in Polypropylene. Journal of Applied Polymer Science, 2010.
5. B.C. Bonse; M. C. S. Mamede; R. A. da Costa; S. H. P. Bettini, Effect of Compatibilizer and Bamboo Fiber Content on the Mechanical Properties of PP-g-MA Compatibilized Polypropylene/Bamboo Fiber Composites. In: Proceedings of the Polymer Processing Society 26th Annual Meeting ~ PPS-26 ~ July 4-8, 2010 Banff
6. M. N. Subramanian, *Polymer Blends and Composites*, Beverly: Scrivener Publishing, 2017.
7. R. L. Blaine, Thermal applications note: Polymer heats of fusion. 2002, Retrieved Sep 1, 2018, from http://www.tainstruments.com/pdf/literature/TN048.pdf

Development of vanilline based flame retardants for PLA Composites

P. Zhao and A. Leuteritz*

*Leibniz-Institut fuer Polymerforschung Dresden e.V., Hohe Str. 6, 01069 Dresden, Germany
Leuteritz@ipfdd.de

Abstract. In this study, a halogen-free flame retarded for PLA based on vanillin was synthesized. The chemical structure of the bio-based and phosphor-containing modified vanillin (VP) was confirmed by ^{1}H-, ^{13}C- and ^{31}P NMR and FTIR. The flame retardant properties and the mechanical properties of PLA/VP composites were evaluated. PLA composite with 5 wt.-% VP presented notable enhancement in the LOI test and passed UL-94 V-0 classification. With addition of montmorillonite as passive flame retardant Loi could be further increased, but deteriorated mechanical behavior.

Keywords: Bio-based, Vanillin, Poly(lactic acid), Flame retardant, Mechanical properties

INTRODUCTION

The extensive use of polymer materials causes environmental environmental issues requiring more and more attention. The development and use of biopolymer has become growing focus. Among all the biopolymers, poly(lactic acid) (PLA) is a promising candidate regarding its excellent mechanical and optical properties. Still, due to the high flammability and brittleness PLA based materials are not in wide use. The present work addresses therefore the reduction of flammability and brittleness at the same time by compounding of poly(lactic acid) and modified vanillin to result in a bio-based polymeric composites. The aim of this work was to introduce flame retardant components into PLA and to study the burning behavior.

EXPERIMENTAL

Materials

Poly (lactic acid), Ingeo Biopolymer 6202D was derived from NatureWorks LLC. Vanillin (VA) was purchased from TCI Deutschland GmbH. Phenylphosphonic dichloride (PPDC, 90%), dichloromethane, triethylamine (TEA) and sodium sulfate anhydrous were purchased from Sigma-Aldrich Corporation. Modified montmorillonite (MMT) with the trade name Cloisite® 30B was supplied by BYK-Gardner GmbH. All the materials were used without further purification.

Preparation

Synthesis of bis(5-formyl-2-methoxyphenyl) phenylphosphonate (VP) was done by simple substitution reaction of 0.2 mol Vanillin (VA) and 0.1 mol of Phenylphosphonic dichloride with aid of 0.2 mol triethylamine (TEA) in an inert solvent at 0 °C. The product was identified by NMR and FT-IR spectroscopy. PLA composites were prepared by means of melt compounding. Prior to processing, all the materials were dried in a vacuum oven at 40 °C overnight. The mixtures were afterwards pelletized and injection molded into specimens for further measurements. PLA composites containing 5 wt.-% and 10 wt.-% of VP are denoted as, PLA5VP and PLA10VP. For comparison, pristine PLA and PLA composites containing 10 wt.-% of Vanillin, denoted as PLA10VA, were prepared under the same processing condition owing to have the same processing and heat history. MMT was introduced into PLA and PLA/VP system at 5 % and 10 % level, respectively. For instance, PLA/VP composite containing 10 wt.-% MMT was denoted as mPLA10MMT to differ from PLA/MMT composites. An oxygen index meter (FTT) was applied to determine the LOI value, and the UL-94 tests were carried out using a horizontal vertical flame chamber. The mechanical properties of the materials were studied by the means of tensile test.

RESULTS AND DISCUSSION

Flame retardancy

LOI and UL-94 tests are the most used small-scale measurements to evaluate the flammability of polymeric materials. In this work, these tests were applied according to ISO 4589 and ASTM D 3801. The results of LOI and UL-94 tests of PLA and PLA composites are summarized in **TABLE 1**. As an easily flammable polymeric material, PLA exhibited an LOI value as low as of 21.4 %. As observed in the UL-94 test, the sample of pristine PLA burned fiercely to the sample clamp with intensive flaming dripping behavior. With 10 wt.-% of VA, some of the samples presented self-extinguish effect. However, the underlaid cotton was ignited by the burning droplets, and the LOI value decreased to 19.5 %, indicating lowered flame resistance. By contrast, the samples incorporated VP showed a positive effect both in LOI and UL-94 tests. When the loading of VP were 5 wt.-% and 10 wt.-%, the LOI values of the composites were improved to 25.8% and 26.3%. Furthermore, both of the composites passed V-0 level in UL-94 tests. Consequently, the results of LOI and UL-94 tests suggested that VP can provide good flame resistance to PLA.

TABLE 1: Results of LOI and UL- 94 tests of PLA and PLA composites

Sample	LOI (%)	UL-94 Rating *	Ignition of cotton
PLA	21.4	No rating	Yes
PLA10VA	19.5	No rating	Yes
PLA5VP	25.8	V-0	No
PLA10VP	26.3	V-0	No
PLA5MMT	20.9	No rating	Yes
PLA10MMT	23.6	No rating	Yes
mPLA5MMT	26.7	V-0	No
mPLA10MMT	26.1	V-0	No

* Sample thickness of UL-94 tests was 3.2mm

MMT is widely used as a passive flame retardant for polymeric materials, especially as anti-dripping and char agent. Despite MMT could introduce protection char layer into PLA, it also prevented the taking away of heat by melting flow of polymer. Hence, at low loading, the sample PLA5MMT showed lower LOI value than pristine PLA, while at higher loading (10 wt.-%), more protection char layer was formed, and the LOI value increased. However, regardless of the anti-dripping effect, both PLA5MMT and PLA10MMT presented no rating in the UL-94 test and were burned out faster to the clamp compared to PLA. Large pieces of burning materials were observed, and the cotton beneath sample was ignited.

As observed during the test, the melting flowability of materials of mPLA/MMT composites, which was most responsible for the enhanced flame retardancy, was significantly improved by the introduction of VP. Higher MMT loading contributed to higher restriction of dripping. Thus, the LOI value decreased with rising MMT contentin samples mPLA5MMT and mPLA10MMT. Thus, both mPLA5MMT and mPLA10MMT passed V-0 in UL-94 test, indicating dramatic improvement on flame retardancy comparing to PLA/MMT composites.

Mechanical properties

The first selective criterion for a specific material in the application is the mechanical performance. The stress-strain plots of PLA and PLA composites are shown in **FIGURE 1**, and the results of Young's modulus, tensile strength and elongation at break are summarized in **TABLE 2**. As a stiff thermoplastic polymer, pristine PLA presented a high rigidity (300 MPa), a high tensile strength (61 MPa) and a low elongation at break (6 %).

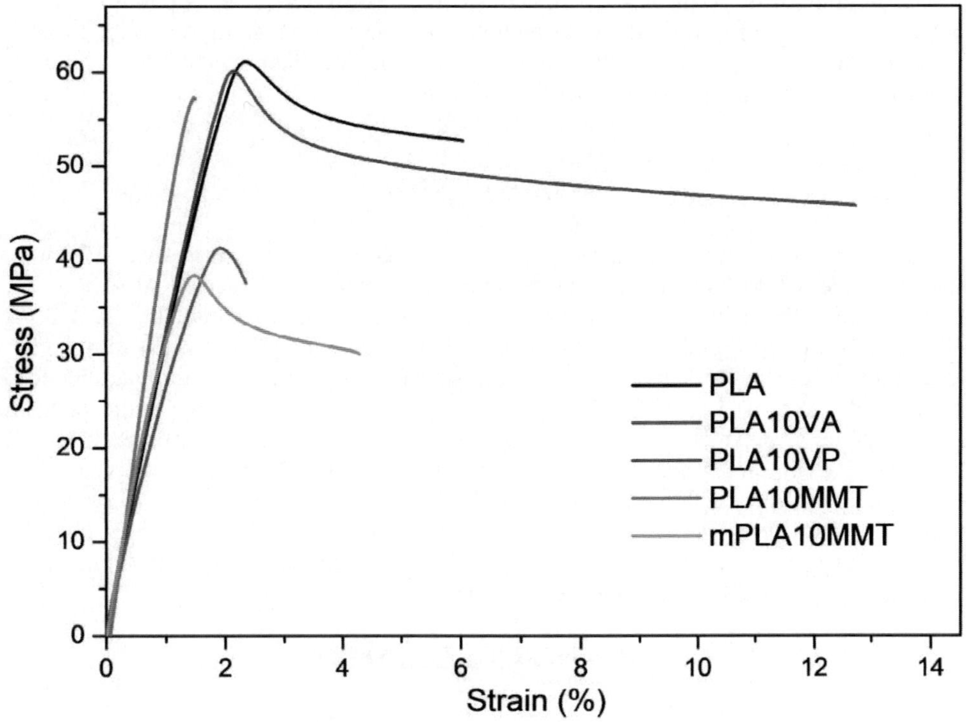

FIGURE 1: Tensile test plots of PLA and PLA composites

The introduction of VA led to overall deterioration of mechanical properties: the Young's modulus decreased to 2900 MPa while the tensile strength declined to 42 MPa. In case of PLA/VP composites, the elongation at break of PLA/VP composite was significantly improved (13 % for PLA10VP), while the changes of tensile strength and Young's modulus were negligible. This result can be explained by the lubricant effect of VP, which enhanced the chain mobility of PLA.

TABLE 2: Data obtained from tensile test of PLA and PLA composites

Sample	Young's modulus (MPa)	Tensile strength (MPa)	Elongation at break (%)
PLA	3300±400	61±1	6±3
PLA10VA	2900±100	42±1	4±2
PLA10VP	3400±100	61±2	13±3
PLA10MMT	4600±100	57±1	1±1
mPLA10MMT	3800±100	39±1	5±1

With the introduction of MMT, the tensile strength of PLA10MMT and mPLA10MMT were altered to 57 and 39 MPa, while their Young's modulus was 4600 and 3800 MPa, respectively. Various researches have focused on the mechanical properties of clay reinforced PLA composites. As reported, the incorporation of clay leads to obvious improvement of Young's modulus because the rigidity of inorganic fillers is normally much higher than that of polymeric materials. Furthermore, due to the high specific surface area of the nano-size particles and also the existence of organic modifier, the adhesion between clay and matrix can be kept at a relatively good level. Thus, for PLA/MMT composites, the tensile strength didn't decrease a lot. In the case of mPLA/MMT composites, antagonistic effects were found. Tensile strength and Young's modulus of mPLA were lowered due to the plasticization effect of VP. However, the addition of MMT elevated again the Young's moduli of the mPLA/MMT composites. On the other hand, an unavoidable fact is the incorporation of inorganic fillers apparently led to the deterioration of ductility of PLA. The elongation at break declined from 6 % of PLA to 1 % of PLA10MMT.

Interestingly, the elongation at break of mPLA10MMT was reduced to only 5 %. A possible explanation was that there was some interaction between MMT and PLA, which made the material so rigid that the lubricant effect of VP could not enhance the chain mobility of PLA that obvious anymore. Nevertheless, the improvement in elongation at break was still notable when comparing to PLA/MMT composites.

CONCLUSION

In this work, a bio-based flame retardant (VP) based on vanillin for PLA was successfully synthesized and characterized by NMR and FTIR. VP exhibited remarkable flame retardant efficiency. At very low loadings (5 and 10 wt.-%), PLA/VP composites achieved V-0 rating in UL-94 tests and improved notably the LOI value from 21.4 % for PLA to 25.8 % for PLA5VP and 26.3 % for PLA10VP. The most important is, the introduction of VP didn't cause any deterioration of the mechanical properties like many flame retardants, in contrast, the elongation at break of PLA was significantly improved from 6 % to 13 % (PLA10VP). This results may indicate VP as a promising flame retardant also for other polymer systems and offer new possibilities for bio-based multifunctional polymeric composites. However, the dripping behavior of PLA composites by burning is still unsolved. Therefore, VP was introduced into PLA composite with commercialized flame retardant MMT. The incorporation of VP into PLA/MMT composites improved the flame retardancy in small scale tests like LOI and UL-94. A remarkable improvement was observed in ductility and processability of the composites. The introduction of VP has notably changed the high rigidity of PLA/MMT composites.

ACKNOWLEDGMENT

The authors would thank the financial support from the China Scholarship Council (201308080080).

REFERENCES

1. R. P. Babu, K. O'Connor and R. Seeram, *Progress in Biomaterials*, **2**, 8 (2013).
2. M. Pluta, A. Galeski, M. Alexandre, M. A. Paul and P. Dubois, *Journal of Applied Polymer Science,* **86**, 1497-1506 (2002).
3. Y.-C. Li, J. Schulz, S. Mannen, C. Delhom, B. Condon, S. Chang, M. Zammarano and J. C. Grunlan, *ACS Nano*. **4,** 3325-3337 (2010).
4. G. Laufer, C. Kirkland, A. A. Cain and J. C. Grunlan, *ACS Applied Materials & Interfaces*, **4**, 1643-1649 (2012).

The effect of Olefin Block Copolymer and Organoclays in Recycled HDPE/PP nanocomposites

Andres Rigail-Cedeño (presenting and reference author)[1,2], *Antonio Diaz-Barrios*[4], *Bolivar Alban*[5], *Julio Caceres*[1,2], *José Pilataxi*[1], *Vanessa Molina*[3], *and Miriam Lazo*[3]

[1]*Faculty of Mechanical Engineering and Production Science,*
[2]*Center of Research and Development in Nanotechnology,*
[3]*Faculty of Natural Science and Mathematics,*
ESPOL Polytechnic University Escuela Superior Politécnica del Litoral,
Campus Gustavo Galindo Km 30.5 vía Perimetral, Guayaquil, Ecuador.
[4]*School of Chemistry and Engineering, Yachay Tech, Urcuqui, Ecuador*
[5]*Bactechnology, Guayaquil, Ecuador*
arigail@espol.edu.ec, adiaz@yachaytech.edu.ec, balban@bactechnology.com

Abstract. The circular economy promotes the recycling of plastics to develop new materials and products. The blends of immiscible plastics increase the availability of raw materials. This research aims to process and relate the morphology, rheological and thermo-mechanical properties of recycled high density polyethylene (rHDPE) and recycled polypropylene (rPP) modified by the addition of commercial organoclays (OMMT) and olefin block copolymer (OBC) compatibilizer. Given amounts of post-consumer blends of rHDPE coming from packing and rPP from lids were mixed in a torque rheometer. Mixes of virgin HDPE and PP in percentages of 0, 20, 40, 60, 80, 100 wt% were prepared to determine the concentration of these plastics in the blends. The corresponding DSC calibration curve of enthalpy of fusion versus weight percent was obtained. The resultant mixture of rHDPE/rPP (~83 wt%/~15 wt %) was melted with two organoclays (Cloisites20A and Cloisite 93A) (3 wt%) and with or without the presence of OBC (D5541) (5 wt%). Intercalated structures were observed in both clay nanocomposites, but Cloisite 20A showed greater basal spacing than those of Cloisite 93A. It was found that the organoclays and the OBC decreased the crystallinity on rHDPE and rPP. Rheological measurements using a plate-plate rheometer showed a viscosity increase of blends in the presence of both organoclays, at low shear rates (~ 10 s^{-1}), being the Cloisite 20A the one that showed the highest values. On the other side, the OBC decreased the viscosity in all composite blends, at higher shear rates (60-100 s-1). Both organoclays increased the stiffness of the rHDPE/rPP, but the OBC in conjunction with the organoclays declined the stiffness. The impact resistance was improved in the presence of the organoclays and the OBC. Based on these results, it can be concluded that the two organoclays used in this study can be used as compatibilizers for rHDPE/rPP blends.

Keywords: recycled plastics, circular economy, clay, nanocomposite.
PACS: 81.05.Qk

INTRODUCTION

The plastics industry has grown globally due to the increase in the worldwide population. Consequently, plastic waste has increased due to the growing of industrial markets, such as textile, automotive, electrical and packaging. A recent study concludes that 95% of plastic packing material is lost after the first use [1], and estimated that only 14% of the plastics used in this type of application are recycled. The World Bank estimates that 2.42 billion tons/year of plastic will be disposed by 2025[2]. The synthetic production of plastics is globally dominated by polyolefins such as polyethylene (PE) and polypropylene (PP) [3].

For the above mentioned reasons, The World Economic Forum (WEF) encourages to rethink the future of plastics and establish a circular economy to preserve the resources of the planet. In this sense, plastics recycling is one of the keys strategies to transform the global market of plastics packing industry, in line with circular economy. Furthermore, the New Plastics Commitment of the WEF encourages to increase the use of recycling plastics and increase their quality in order to make durable products by 2025 [4]. Thus, the worth of recycled plastics should be radically improved to make the plastics materials more economical, with better performance and environmentally sustainable.

According to the Ecuadorian government, 1.4 million PET bottles are produced in Ecuador, and 1.1 million are recycled [5]. Therefore, plastics solid waste (PSW) comprise mostly PE and PP, which have been traditionally used in the packing industry in Ecuador. Moreover, local regulations encourage to use recycled plastics in the products, so plastics recyclers have been dealing with unknown amounts and compositions of recycled plastics. For this reason, it is relevant to Ecuador to develop a methodology to determine the relative concentrations and characterization of PE/PP mixtures and decide the appropriate strategy to upgrade these mixtures. Mostly, for this type of upgrades, one should consider the incorporation of high-performance reinforcing agents such as organoclays [6]. We have successful processed recycled plastics nanocomposites using a single-screw extruder incorporating a dispersive and distributives mixers [7,8]. Therefore, this research aims is to evaluate an Olefin Block Copolymer OBC together with two commercial organoclays with the purpose of developing a high-performance material, with improved structural, thermal and mechanical properties, for industrial applications. Also, an objective of this research is to demonstrate the use of either OBC or organoclays as a potential substitute to other commercial compatibilizers frequently used for polymeric products made with recycled HDPE and PP.

EXPERIMENTAL

Materials

High density polyethylene (HDPE), Braskem HS5502XP, and polypropylene (PP), Braskem H 201, were donated by Nutec America. HDPE –HS5502XP– has an Izod impact strength of 150 J/m, a density of 0.955 g/cm^3, a melt flow index of 0,35 g/10 min and an extrusion temperature range from 190 ºC to 210 ºC. PP – resin H 201– has an Izod impact strength of 20 J/m, a density of 0.905 g/cm^3, and a melt flow index of 20 g/10 min. These virgin polymers were utilized to generate composition calibration curves using DSC. A recycled material unknown ratio mix of high density polyethylene (rHDPE) and polypropylene (rPP) was kindly donated by Bac-technology. Experimental D5541 Olefin block copolymer (OBC), was supplied by the Dow Chemical Company. This copolymer has a density of 0,863 g/cm^3, a melt flow index of 6,5 g/10 min and a melting point of 190 ºC (DSC). Organoclays, Cloisite 93A (C93A) and Cloisite 20A (C20A), were purchased from Southern Clay Products.

Table 1. Recycle plastics compositions

Blend	Description
1	rHDPE/rPP
2	rHDPE/rPP/3OBC
3	rHDPE/rPP/3C20A
4	rHDPE/rPP/3C93A
5	rHDPE/rPP/3C20A/3OBC
6	rHDPE/rPP/3C20A/3OBC

Processing

A torque rheometer, Brabender Plastograph EC, was used to prepare mixes of virgin HDPE and PP in percentages of 0, 20, 40, 60, 80, 100 wt%. Enthalpy of fusion for HDPE/PP blend was obtained by differential scanning calorimetry (DSC), and the corresponding calibration curve of enthalpy of fusion versus concentration was obtained following procedure found in the literature [9, 10]. The rHDPE/rPP mix was homogenized using a PAVAN rotatory drum dryer's under rotational function during 20 minutes at 25 ºC. Six different mixtures were prepared by dispersing 3 wt% of organoclay or OBC in 30 g of rHDPE/rPP mix, as illustrated in Table 1. The nanocomposites were processed following a temperature profile 220/220/230 ºC and a screw speed of 60 rpm for 10 minutes. The melted composites were transferred to a pneumatic heat press, compressed into required test specimens at 3000 psi, 160 ºC for 10 minutes, and cooled down to 130 ºC at a rate of 3 ºC/min.

Characterization

X-ray diffraction measurements were obtained using a Phillips PANalytical XPert-Pro model diffractometer equipped with Cu Kα (λ=1.54060 Å) radiation operated at 40 kV and 30 mA with a 1/8" divergent slit and 1/16" anti-scatter slit, scanning range 1° to 40° (1θ), 0.05° step size, 20 seconds per step. The heat flow curves were obtained using DSC 200 operating from 0ºC to 200ºC at a heating/cooling rate of 10ºC/min. Rheological properties were tested using a Malvern Kinexus Pro+ Rotational Rheometer at 180 ºC consisting of parallel plates with a 25 mm gap setting, angular frequencies varied between 0,01 and 30 s^{-1}. The rheological curves were smoothed using an Origin 6.0 software.

Evaluation

Tensile (ASTM D638) and impact (ASTM D256) samples were shaped using steel molds by compression molding, dimensions according to ASTM standards. Tensile tests were performed at room temperature using a Shimadzu AG-1S 10 kN, the initial testing rate of 0.5 mm/min shifting to 50 mm/min after a 3% stroke strain. Impact resistance was measured at room temperature using a Jin Jian pendulum impact machine.

RESULTS AND DISCUSSION

Quantification of rHDPE/rPP ratio was performed by DSC procedures suggested in the literature [9,10]. DSC thermograms of neat PP, neat HDPE and PP/HDPE blends helped to elaborate calibration curves for HDPE and PP (△ H vs. % HDPE and △ H vs% PP). As reported by other authors, results showed endothermic peaks from the melting temperature (T_m), HDPE and PP at about 130ºC and 162 ºC, respectively [12]. The corresponding area of the endothermic peak when were matched with the corresponding calibrations curves showed a composition of 82.6 ± 2 % for rHDPE and 15.0 ± 0.5 % for rPP and 2.5 ± 0.1 % for the unknown samples of other plastic wastes. Figure 1 shows a semiquantitative identification of elements by SEM-EDX. The results showed that ~ 94.97% of the material is composed of carbonaceous groups. Other authors, using DSC measurements, have been able to determine concentrations of plastic traces above 1 wt % [10]. These traces might be PET or PVC due to the presence of oxygen (O) and chlorine (Cl), in the SEM-EDC analysis.

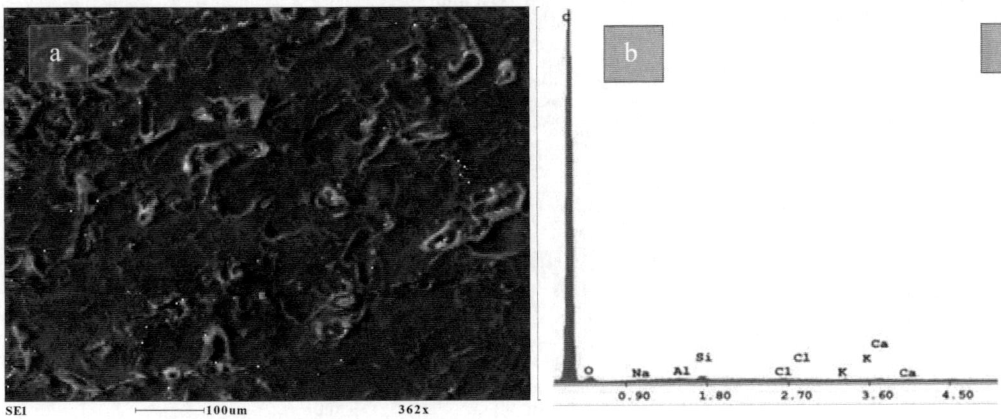

c

EDS Quantitative Results		
Element	**Wt%**	**At%**
C	94.97	96.63
O	3.73	2.85
Na	0.23	0.12
Al	0.16	0.07
Si	0.33	0.15
Cl	0.17	0.06
K	0.14	0.04
Ca	0.25	0.08

Figure 1. Micrograph of SEM (a) and corresponding EDX composition (b, c) of recycled plastics waste

Figure 3 (a) illustrates the effect of the organoclays and the compatibilizer in the structure of the recycle blends obtained by compression molding. The XRD patterns show the crystal structure of the rHDPE/rPP mix. rPP displayed diffraction peaks around 14° and 15°, whereas rHDPE exhibited the corresponding crystalline lattice around 22° and 24° [**¡Error! Marcador no definido.**]. All recycle blends that include organoclays, or the OBC showed a decrease in the crystal stability for both recycled plastics at the surface level of the sample. Figure 3 (b) illustrates the effect of the organoclays and of the compatibilizer in the structure of the recycle blends obtained by compressive molding. The organoclays have first reflections at about ~27 Å for Cloisite 93 A (C93A) and ~25 Å for Cloisite C20A (C20A) as previously reported [7,8,11]. The C20A in the recycle plastics blend showed a primary diffraction peak around 2.5 ° (~39 Å). The incorporation of OBC increases the primary diffraction peak of C20A at about ~ 41 Å slightly. The C93A and C93A/OBC displayed peaks at about ~36 Å and ~38 Å, respectively. Thus, it seems the C20A has a better affinity to the rHDPE/rPP blend compared to C93A. Nevertheless, the presented results tend to point out that both organoclays were able to intercalate in rHDPE/rPP blends.

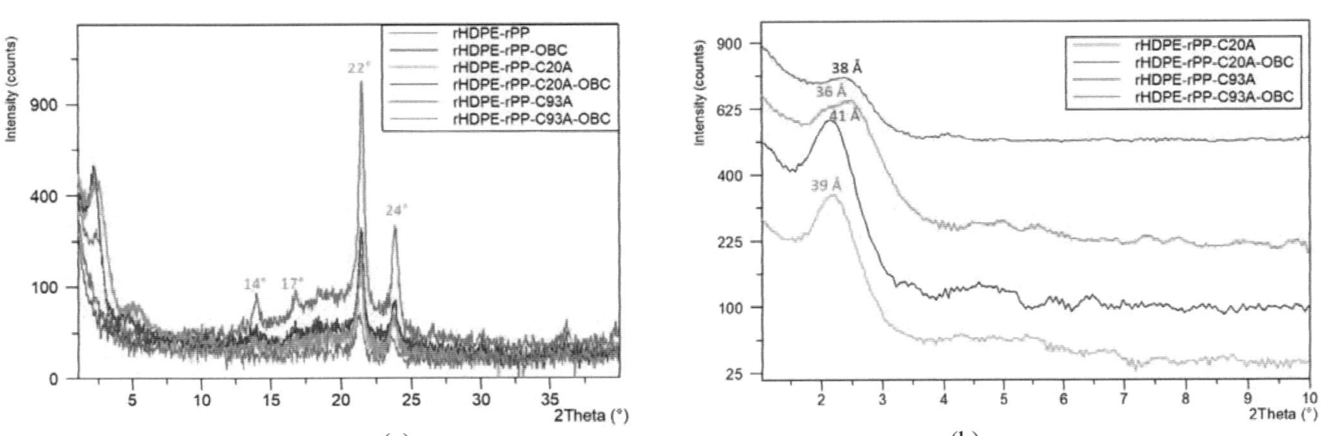

(a) (b)

FIGURE 3. XRD of recycled plastic blends with organoclays and olefin block copolymer.

Figure 4 shows the flow curves of recycling plastics that contain the organoclays and OBC. Thus, the viscosity of the mixtures at low shear rates increased with the incorporation of either organoclays or OBC. This increment in the Newtonian region appears to be related to the dispersion and interaction of the additives with the recycle plastics [12]. However, the highest viscosity is observed in all mixtures when OBC is added. The effect of the OBC on flow curves and Non-Newtonian behavior, at low shear rate, has also been reported by other authors [13]. At higher shear rates, the viscosity of the recycle plastics that content OBC and the C20A organoclay appears to be more shear sensitive than the observed for other systems. Moreover, the alignment and reorientation of the clays have an influence in the viscosity at a high shear rate [12,14]. Thus, the rheological behavior of these materials gives indications of the synergy between the organoclays and OBC. It can be said that organoclays and OBC potentially improve the processing behavior of rHDPE/rPP.

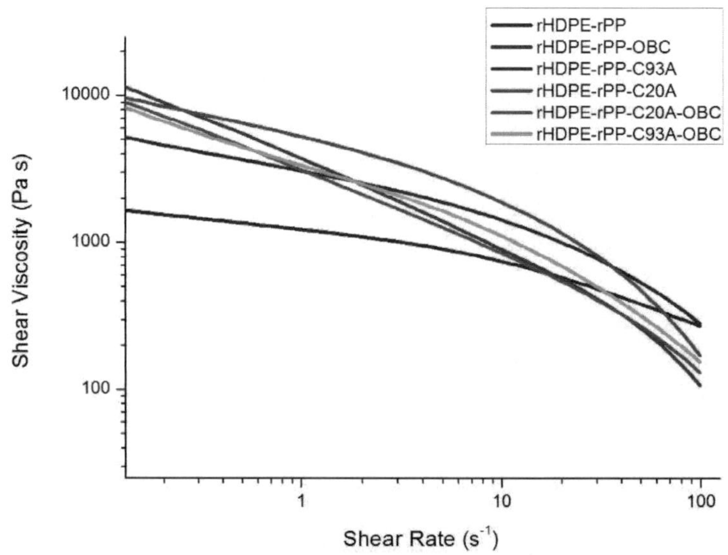

FIGURE 4. Viscosity as a function of shear rate in all recycled composite materials.

Table (2) illustrates the mechanical properties of these recycled plastics mixes. The addition of OBC did not increase the Young Modulus of these mixes, however, improves their corresponding impact resistance. Compatibilizer usually has this type of effect in mechanical properties [7]. In the case of nanocomposites, both organoclays increased the Young Modulus (>25%) compared to rHDPE/rPP. As illustrated previously in XRD results, good interaction and intercalation were observed with the recycle matrix materials. The Ultimate Strength was also improved by the addition OBC, organoclays and OBC/organoclays mixtures. As reported, stiffness and strength area usually enhanced due to a good interaction and dispersion of organoclays in the polymer matrix [15,16]. This is related to the ability of the organoclays to reinforce and stabilize immiscible blends of thermoplastics [8,17]. OBC favored the interaction between both recycled plastics however affected the organoclays and restricted the mobility of the chains [15]. It seems [15] that the synergistic effect among C20A, OBC and this recycled blend is related to an increase in stiffness, strength, and toughness. The organoclays increased the impact energy in both nanocomposites due to the good interaction of organoclays with both recycle plastics [7,18,19]. This type of OBC also increased the impact energy of rHDPE/rPP, but the impact resistance decreased when used in conjunction with organoclays. At high strain rates, the recycle blend is rich in PE and becomes less ductile in the presence of OBC/organoclays. This reduction in impact energy is also accompanied by a decrease in the Young Modulus of OBC/organoclays nanocomposites.

TABLE (2). Mechanical Properties of rHDPE/rPP systems

	Young's Modulus (MPa)	**Ultimate Tensile Strength (MPa)**	**Impact Energy (J/m)**
rHDPE/rPP	890.66 ± 17.78	21.22 ± 0.44	58.39 ± 1.55
rHDPE/PP/OBC	887.27 ± 14.99	22.38 ± 0.25	71.19 ± 1.68
rHDPE/PP/C20A	1145.09 ± 9.02	21.51 ± 0.08	72.38 ± 3.37
rHDPE/PP/OBC/C20A	1098.11 ± 39.70	23.74 ± 0.38	60.48 ± 1.68
rHDPE/PP/C93A	1010.01 ± 14.17	23.81 ± 0.38	71.79 ± 3.09
rHDPE/PP/OBC/C93A	974.53 ± 9.21	22.26 ± 0.24	61.67 ± 1.68

CONCLUSIONS

This research proves the feasibility of using organoclays or OBC to process and enhanced the properties of rHDPE/rPP mixes. The identification and quantification of a recycled blend of rHDPE/rPP were determined by DSC. The concentration in this blend was rHDPE/rPP was ~84/15 (wt/wt)%. rHDPE properties were significantly improved with the presence of rPP, organoclays and OBC. The XRD showed the intercalation of both organoclays in the recycled plastics. The OBC enhanced the organoclays intercalation and also improved the processing characteristic of the materials by decreasing the viscosity, in the extrusion processing range. Both organoclays increase stiffness, strength and toughness of composites. The structure-processing characteristics and mechanical properties of rHDPE/rPP composites illustrated some potential for applications for industrial profiles. Further research will be conducted in order to measure the energy consumption in recycled polymer-clay nanocomposites.

ACKNOWLEDGEMENTS

The authors acknowledge the technical workforce of the Laboratory of Testing Materials (LEMAT) and Center of Nanotechnology Research and Development (CIDNA) at ESPOL. The authors would also like to thank Bactechnology for kindly supporting the recycled plastics and OBC compatibilizer and also to Jorge Salgado of ENTEC Polymers for technical information about the OBC.

REFERENCES

1 The New Plastics Economy. World Economic Forum. http://www3.weforum.org/docs/ WEF_The_New_Plastics_Economy.pdf Last accessed: 21-01-2019
2. In South America, Plastic Trash Will Be as Good as Money. http://www.takepart.com/article/2014/04/28/south-america-picking-beachside-litter-comes-special-rewards Last accessed: 21-01-2019
3. The world's plastic problem in numbers. World Economic Forum. https://www.weforum.org/agenda/2018/08/the-world-of-plastics-in-numbers . Last accessed: 21-01-2019
4. The New Plastics Economy global commitment. Ellen MacArthur Foundation. https://www.ellenmacarthurfoundation.org/assets/downloads/Global-Commitment-Document-to-download-on-website-2.pdf Last accessed: 21-01-2019
5. Recycling: The Price is Right. International Development Bank (IADB) https://blogs.iadb.org/ciudades-sostenibles/en/recycling-the-price-is-right/ Last accessed : 01/10/2019
6. Zare, Y., *Waste Management,* **33** (3), 598-604 (2012)
7. A. Rigail-Cedeño, C, Perugachi, C. Paredes. L Baquerizo. 14th LACCEI International Multi-Conference for Engineering, Education, and Technology. San José, Costa Rica (2016)
8. A. Rigail-Cedeño, A. Diaz-Barrios, J. Gallardo-Bastidas1c, S. Ullaguari-Loor, N. Morales-Fuentes. To be presented in 9th International Conference on Key Engineering Materials (9th ICKEM 2019) and Published in Key Engineering Materials
9. A. Manivannan, M.S. Seehra.*Preprints Amer. Chem. Soc. Div. Fuel Chem.*, **4**, 1028-1032 (1997)
10. W. Camacho and S. Karlsson. *Polymer Engineering and Science*. **41** (9), 1626-1635 (2001)
11. A. Rigail-Cedeño, D. F. Schmidt. AIP Conference Proceeding 1914, 030023 (2017).
12. V. Pasanovic-Zujo, R. KumarGupta, S. Bhattacharya. *Rheology Acta*, **43**, 99-108 (2004)
13. S. Veervoort, J. Doelder, E. Tocha, J. Genoyer, K. Walton T. Hu, J. Munro, K. Jeltsch. *Polymer Engineering and Science*, **58**, 4, 460-465 (2018)
14. J. Caceres, J. Pilataxi, M. Vera, and D. Vera, A. Diaz-Barrios, A. Rigail-Cedeño. To publish in AIP Conference Proceeding (2019).
15. S. Tjong. *Material Science and Engineering* R: Reports, **53**, 73-197 (2006)
16. E. Garofalo , M. Claro, P. Scarfato, L.D. Maio and L. Incarnato. AIP Conference Proceeding, 1965, 020053 (2015)
17. R. Iyer, S. Suin, N. Shrivastava, S. Maiti, B. Khatua. *Polymer-Plastics Technology and Engineering*, **52**: 514-534, 213
18. D. Meerinska, H. Kubisova, A. Kalendova, P. Svoboda. *Journal of Thermoplastics Composites Materials*, **25**, 115-131 (2012)
19. N. Phuong, V. Gilbert, B. Chuong. *Journal of Reinforced Plastics and Composites*, **27** (18), 1983-2000 (2008)

A Highly Stretchable and Thermo-recoverable Polysaccharide based Double Network Hydrogel Containing Graphene Oxide

Sara Tarashi[a], Hossein Nazockdast[a*] and Gholamhossein Sodeifian[b]

[a]*Polymer Engineering Department, Amirkabir University of Technology, P.O.B. 15875-4413, Tehran, Iran*
[b]*Chemical Engineering Department, Faculty of Engineering, University of Kashan, P.O.B. 87317-53153, Kashan, Iran*

Abstract. Combining graphene oxide (GO) nanosheets with a carrageenan (Car)/polyacrylamide (PAm) double network, we have synthesized a Car/PAm polysaccharide based double network (DN) hydrogel containing GO nanosheets using a one-pot UV curing technique. The resulting nanocomposite double network (NCDN) hydrogel at the optimal condition revealed free-shapeable moldability, remarkable flexibility and stretchability, great toughness, and high-efficiency toughness recovery. The enhanced mechanical properties of the Car/PAm/GO NCDN hydrogel compared to the Car/PAm DN hydrogel without GO, originated from new fabricated energy dissipation mechanisms. As an exciting result and according to the rheological behavior, incorporation of GO nanosheets into the Car/PAm DN hydrogel had no adverse effect on the thermo-reversible property of carrageenan physical network. Therefore the present work makes several noteworthy contributions to invent new NCDN hydrogels with the combined advantages of excellent mechanical property and thermo-reversibility which can significantly extend the application of hydrogels, especially for the stimuli-responsive load-bearing tissues.

Keywords: Double Network Hydrogel, Nanocomposite, Mechanical Properties, Thermo-reversibility, Carrageenan.
PACS: A-10-1214-1.

INTRODUCTION

Hydrogels are hydrophilic polymeric networks which are mainly composed of polymer chains connected by physical and/or chemical cross-linking points.[1] Hydrogels are favorable candidate materials in various biomedical and industrial applications, such as tissue engineering, wound healing, drug delivery, water treatment, smart actuators and substitutes for soft tissues. For the application of hydrogels, excellent mechanical properties are required. Over the last decade, various methods have been suggested to improve mechanical properties. Among them, double network (DN) and nanocomposite (NC) hydrogels exhibit the most improving. Gong et al.[2] invented the first DN hydrogel with significantly enhanced mechanical properties using a free-radical polymerization method which was consisting of chemical cross-links in both networks. They proposed the sacrificial bond theory to explain the toughening mechanism of DN hydrogel. According to this theory, during deformation, the first network is fractured into small clusters and acts as the sliding cross-linkers of the second network to stop crack propagation. Nevertheless, the Gong's DN hydrogel was lack of recoverability. To address this problem, reversible physical bonds were introduced to the DN hydrogels and a number of DN hydrogels were developed by a combination of the first, physically and the second, chemically cross-linked networks to achieve both recoverable and tough property. Until now, several DN structures have been made to demonstrate the DN hydrogels, for instance, alginate/polyacrylamide (PAm),[3] agar/PAm,[4] carrageenan/PAm,[5] guar-gum/PAm,[6] gelatin/Pam[7] all of which enabled to introduce remarkable fracture energy and different degrees of mechanical recovery.

By introducing different nanoparticles to the hydrogels, a variety of NC hydrogels have been created. Nowadays, incorporating of nanoparticles into the suitable hydrogels is an effective approach to fabricate hydrogels with satisfactory mechanical performance. Graphene oxide (GO) is the most widespread nanoparticles for improvement in the mechanical property of hydrogels.

As mentioned above, constructing DN structure and NC methods were both considered to endow hydrogels with developed mechanical properties and tunable high toughness. To marry the advantages of two methods, a number of nanocomposite double networks (NCDN) hydrogels were developed. Zhu et al.[8] used GO to enhance the mechanical properties of agar/PAm DN hydrogel, whose fracture strain and strength reached to 4600% and 332 kPa, respectively and NCDN hydrogel shows a favorable fatigue resistance. Huang et al.[9] investigated a novel gelatin/PAm/GO NCDN hydrogel and found that high mechanical toughness (strength > 400 kPa and broken strain > 500%) was achieved by a combination of the double network and GO. Yan et al.[10] also noted that at the optimal condition, gelatin/PAm/GO NCDN hydrogel demonstrated an elastic modulus of 187.3 kPa, fractured stress of 0.324 MPa and fractured energy of 10.18 MJ/m3. Fan et al.[11] studied alginate/PAm/GO NCDN hydrogel and indicated superior mechanical strength and toughness. Despite the promising studies, few existing NCDN hydrogels have demonstrated recoverability and fatigue resistance properties.

Carrageenan is a family of linear sulfated polysaccharides derived from red seaweeds. It is made up of alternating α(1-3)-D-galactose-4-sulfated and β(1-4)-3,6-anhydro-D-galactose units. Among different types of carrageenan (kappa, iota, and lambda).[12] The carrageenan hydrogel is obtained through the coil-helix transition followed by aggregation of double helices. Moreover, the formation and melting of carrageenan hydrogels are thermally reversible.[13]

However, although a great success in synthesizing carrageenan-based DN hydrogel has been achieved, a suitable comprehension of the nanoparticles effect on the mechanical properties and thermo-reversibility of carrageenan-based DN hydrogel is missing. The present work makes several noteworthy contributions to invent new hydrogels with the combined advantages of excellent mechanical property, recoverability, stretchability, and anti-fatigue behaviors which can significantly extend the application of NCDN hydrogels, especially for the stimuli-responsive load-bearing tissues.

EXPERIMENTAL SECTION

Materials

Graphene oxide in powder form (GO) was purchased from Angstron Materials LCC (Dayton, Ohio USA) and is comprised of few-layer graphene oxide platelets with a thickness of 2-3 nm and a lateral dimension of approximately 7 um. Carrageenan powder (Car) was the product of Sigma-Aldrich (Germany). Potassium chloride (KCl), Acrylamide (Am), N, N'-Methylene bisacrylamide (MBA) and 2-hydroxy-4'-(2-hydroxyethoxy)-2-methylpropiophenone were purchased from Sigma-Aldrich (Germany) and used without further purification. Deionized water was obtained from a Millipore purification apparatus via deionization and filtration.

Synthesis of Nanocomposite Double Network Hydrogels

The physically-chemically cross-linked carrageenan/polyacrylamide/GO NCDN hydrogels were prepared by a one-pot solution polymerization method from the initial solutions including of GO, Car, KCl, Am, MBA, and a UV-initiator. The desired amount of dried GO was firstly dispersed and exfoliated in 10 mL deionized water by sonication for 1 hr and then vigorous stirring for 2 hr at room temperature. Thereafter Car and Am monomer were added into the GO suspension under stirring at room temperature. Subsequently, KCI, MBA, and UV-initiator were added into a reactor, and the reactor was sealed under N_2 gas protection after 30 min degassing. Then the mixture was gradually heated up to 90 °C in an oil bath to dissolve all the reactants in the aqueous solution and was stirred magnetically at this temperature for 4 hr to obtain a homogeneous solution. The resulting solution was injected into glass mold, and this was followed by cooling the solution at 4 °C for 30 min to form a physical Car first network. Subsequently, the molds were placed under UV light to carry out the polymerization reaction of Am at room temperature for 1 hr to fabricate the NCDN hydrogels. Car/PAm DN hydrogels was also prepared by a similar process, with the exception that GO was not added. The GO concentrations were fixed at 0.1, 0.3 and 0.5 wt% relative to the weight of Car. The NCDN hydrogel samples were named as DNGOx, where x is the weight percentage of GO-based on Car.

Characterization of Hydrogels

Mechanical properties of hydrogels were performed on an Instron machine (Model 5566, 50 N load cell), at room temperature. For uniaxial tensile tests, the gel was cut into a dumbbell-shaped with a gauge length of 30 mm, a width of 5 mm, and a thickness of 2.5 mm. The stretch rate of upper clamp was kept constant at 100 mm/min. The elongation ratio (λ) was defined as the deformed length (l) to the original one (l_0). The elastic modulus (E) was calculated in the range of linear relationship between stress and strain, and the fracture energy (W) was estimated by area below the stress-strain curve. For hysteresis measurement, gel specimens were first stretched to a pre-determined elongation ratio and then unloaded at the same stretch rate. The dissipated energy (U_{hys}) during deformation was determined from the area of the hysteresis loop. For the recovery tests at different temperature and rest time, the samples were firstly measured by a cycle of loading-unloading at room temperature. And then the samples were stored in a fixed-temperature of room temperature and/or 90 °C. The chemical characteristics of the hydrogels were investigated using a fourier transform infrared (FTIR, SRG 1100G, BOMEM, Canada) measurements with milled dried gels by the conventional KBr disk tablet method in the wave number range of 4000 - 400 cm^{-1}. Viscoelastic properties of hydrogels were performed using a stress-controlled rheometer (Paar-Physica, MCR301) with a parallel-plate system. The temperature was controlled by means of a Peltier plate.

RESULTS AND DISCUSSION

In order to quantitatively investigate the mechanical properties of the κ-Car/PAm/GO NCDN hydrogels with different concentration of GO, tensile tests were performed. As highlighted in Figure 1a and b, GO nanosheets can notably improve the tensile properties. After introducing GO into the DN hydrogel, NCDN hydrogels exhibit much improved mechanical properties compared to PAm and PAm/GO SN hydrogel and κ-Car/Pam DN hydrogel. Additionally, GO concentration indicates considerable effect on the tensile properties of NCDN hydrogels. As can be seen, the values of failure strain and stress and W show an optimal value at GO concentration of 0.3 wt%. However, the E values constantly enhance with increasing the GO content. Therefore, these results imply that our fabricated NCDN hydrogels have a significant mechanical property which can originate from the role of GO nanosheet as a reinforcing bridge between the two physical and chemical networks.

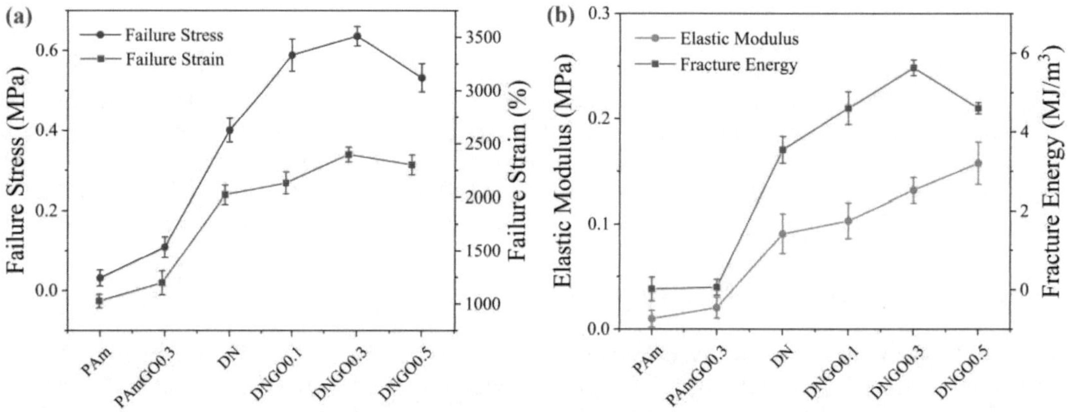

FIGURE 1. Comparison of (a) stress and strain at failure point and (b) elastic modulus and fracture energy between the hydrogels under the uniaxial tensile test.

FTIR spectra were employed to characterize the interaction between GO and the polymer chains of the NCDN hydrogels. In the spectrum of the PAm SN hydrogel (Figure 2), the apparent peaks at 3420 cm^{-1} and 1626 cm^{-1} are related to N-H stretching and C=O stretching, respectively. According to the spectrum of the DN hydrogel, the characteristic peaks of PAm are approximately kept unchanged in comparison with PAm SN hydrogel and the emerged peak at 1087 cm^{-1} is attributed to the glycosidic linkage of Car chains, indicating there are only hydrogen bonds between the Car and PAm chains. After the introduction of GO nanosheets, the characteristic peaks of PAm and Car shifted to the lower wavenumbers in the Car/PAm/GO NCDN hydrogels. Further, the shift of the peaks become more detectable with increasing the GO content, suggesting that the hydrogen bonding interactions formed between the polar functional groups of polymer chains (Car and PAm) and oxygen-containing groups of GO nanosheets.

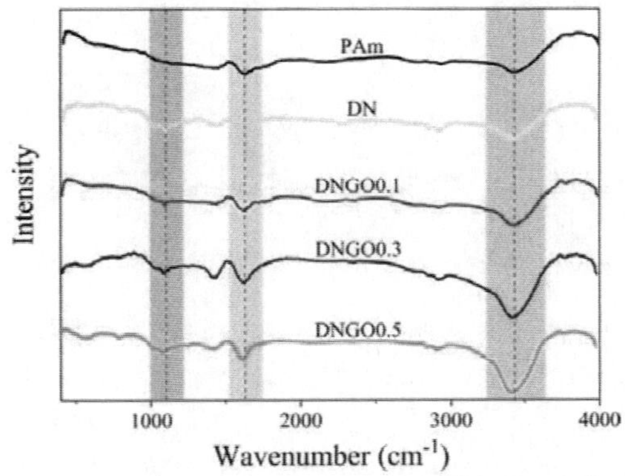

FIGURE 2. FTIR spectra of PAm SN hydrogel, DN and NCDN hydrogels with different GO loading.

It is common knowledge that the gelation and melting process of κ-carrageenan between sol and gel states in aqueous solution are thermally reversible. At the same time, the afore-mentioned interactions between the two networks and between the PAm and/or Car and GO nanosheets are reversible non-covalent. It is consequently expected that the Car/PAm/GO NCDN hydrogel can demonstrate recovery property under suitable conditions. In order to identify time-evolution recovery capacity of Car/PAm/GO NCDN hydrogel, cyclic loading tests were performed at two discrete temperatures below (room temperature) and above (90 °C) the melting point of the Car after the first loading cycle. For quantitative comparison, the toughness recovery percentage is defined by the ratio of dissipated energy at different resting times to that of the first cycle at $\lambda_{max} = 8$. The recovery percentages at room temperature increase progressively within 24 hr so that after 24 hr reach the values of 53.93 % and 67.07 % for the DN and DNGO0.3 hydrogels, respectively (Figure 3a). The reason for more evident recovery of DNGO0.3 hydrogel arises from the synergistic effect of new generated reversible interactions owing to the bridge of GO between the Car and PAm networks on the recovery process. Figure 3b indicates that the toughness of the DN and DNGO0.3 hydrogels can recover to 83.63 % and 97.23 % respectively after storing at 90 °C for 16 mins. This observation obviously points out that temperature responsive sol-gel-sol transition of Car is responsible for the recovery at elevated temperature. It is worthwhile noting that under higher temperature, polymer chains can diffuse more quickly therefore, not only the damaged Car aggregations can organize larger network bulk in shorter times but also more Car and PAm chains can reabsorb on the surface of GO.

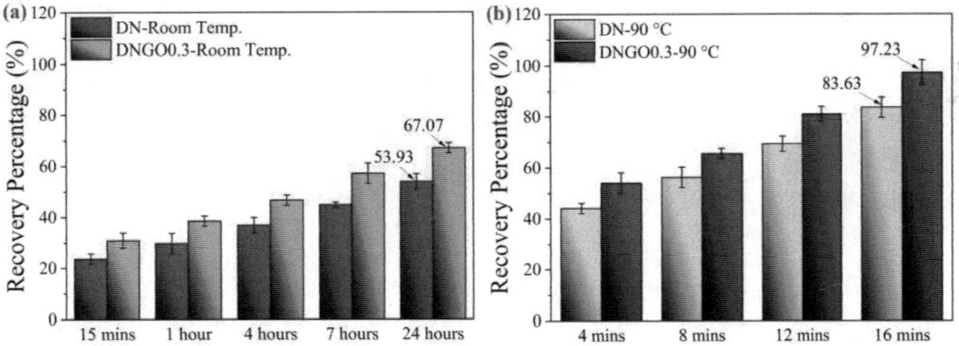

FIGURE 3. The recovery percentage of DN and NCDN hydrogels as a function of resting times at (a) room temperature and (b) 90 °C.

Rheological experiments were carried out to provide considerable insight into the temperature dependence of double network structure after the incorporation of GO nanosheets. Figure 4 pinpoints the rheological temperature sweep during a cooling-heating cycle (90 °C → 10 °C → 90 °C) at a rate of 2 °C/min with a fixed angular frequency of 1 rad/s and small strain amplitude of 1%. It is apparent that during the cooling at about 53 °C, contributions of elastic components increase suddenly, unlike in the heating process at about 64 °C, contributions of viscous components dominate, which are in accordance with the gel and sol points of the Car network, respectively. Moreover, tan δ values are always lower than the unit. This result also reveals that the thermo-reversibility of Car network remains unchanged after the addition of GO nanosheets in the DN hydrogel.

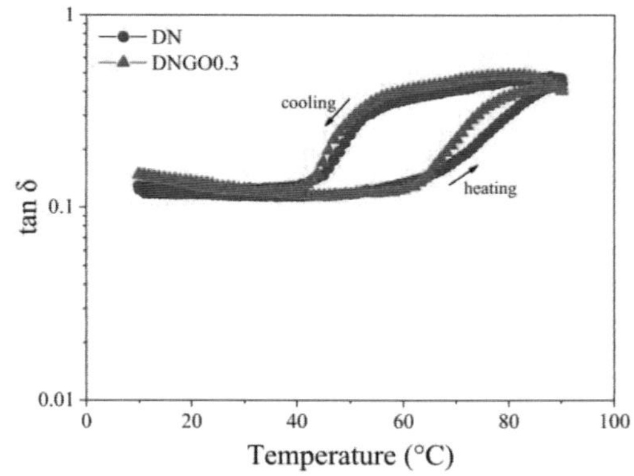

FIGURE 4. Dependence of tan δ = G" / G' on temperature during a rheological temperature sweep at a cooling and heating rate of 2 °C/min with a fixed angular frequency of 1 rad/s and a strain amplitude of 1% for the DN and DNGO0.3 NCDN hydrogels.

CONCLUSIONS

To sum up, we have applied GO nanosheets to fabricate a novel Car/PAm/GO NCDN hydrogel. As compared to the Car/PAm DN hydrogel without GO, the Car/PAm/GO NCDN hydrogels showed improved tensile properties. The enhanced mechanical properties of the Car/PAm/GO NCDN hydrogels not only arose from unzipping of the hydrogen bond and ionically cross-linked junctions between the double helix aggregations in the Car physical network and dissociation of hydrogen bond interactions between the Car and PAm networks, but also originated from the synergic effect of peeling off PAm and/or Car chains from the surface of GO nanosheets. This research deepened our understandings towards the effect of nanoparticles on the mechanical properties of DN hydrogels. Furthermore, introduction of GO endowed the Car/PAm/GO NCDN hydrogel with a remarkable thermo-responsive and toughness recovery properties which could significantly extend the application of Car/PAm/GO NCDN hydrogels in the engineering fields.

ACKNOWLEDGMENTS

This research was made possible by a grant from Amirkabir University of Technology.

REFERENCES

1. C. Creton, *Macromolecules* **50**, 8297-8316 (2017).
2. J. P. Gong, Y. Katsuyama, T. Kurokawa and Y. Osada, *Adv. Mater.* **15**, 1155-1158 (2003).
3. J.Y. Sun, X. Zhao, W. R. Illeperuma, O. Chaudhuri, K. H. Oh, D. J. Mooney, J. J. Vlassak and Z. Suo, *Nature* **489**, 133 (2012).
4. Q. Chen, L. Zhu, L. Huang, H. Chen, K. Xu, Y. Tan, P. Wang and J. Zheng, *Macromolecules* **47**, 2140-2148 (2014).
5. X. Lu, C. Y. Chan, K. I. Lee, P. F. Ng, B. Fei, J. H. Xin and J. Fu, *J. Mater. Chem. B* **2**, 7631-7638 (2014).
6. J. Zhu, S. Guan, Q. Hu, G. Gao, K. Xu and P. Wang, *Chem. Eng. J.* **306**, 953-960 (2016).
7. X. Yan, Q. Chen, L. Zhu, H. Chen, D. Wei, F. Chen, Z. Tang, J. Yang and J. Zheng, *J. Mater. Chem. B* **5**, 7683-7691 (2017).
8. P. Zhu, M. Hu, Y. Deng and C. Wang, *Adv. Eng. Mater.* **18**, 1799-1807 (2016).
9. D. L. Taylor and M. in het Panhuis, *Adv. Mater.* **28**, 9060-9093 (2016).
10. X. Yan, J. Yang, F. Chen, L. Zhu, Z. Tang, G. Qin, Q. Chen and G. Chen, *Compos. Sci. Technol.* **163**, 81-88 (2018).
11. J. Fan, Z. Shi, M. Lian, H. Li and J. Yin, *J. Mater. Chem.* A **1**, 7433-7443 (2013,).
12. B. T. Nguyen, T. Nicolai, L. Benyahia and C. Chassenieux, *Carbohydr. Polym.* **112**, 10-15 (2014).
13. S. Liu and L. Li, *ACS appl. Mater. interfaces* **8**, 29749-29758 (2016).

Comparative analysis of bacterial cellulose based polymeric films for food packaging

Smarak Bandyopadhyay[1], Nabanita Saha[1*], and Petr Saha[1]

[1] *Centre of Polymeric Systems, University Institute, Tomas Bata University in Zlin, Tř. T. Bati 5678, 760 01, Zlin, Czech Republic*

*Corresponding author: nabanita@utb.cz

Abstract. The excessive use of petroleum based packaging material has done enough harm to this planet. Largely to deal with this catastrophic cause, the scientific community has shifted towards the biopolymer based packaging resources. Bacterial cellulose (BC) is an edible packaging material but lacks the thermoplastic nature for wider application. In our present study we have integrated BC with carboxymethyl cellulose (CMC) and polyvinylpyrrolidone (PVP) to prepare polymeric films: PVP-BC and PVP-CMC-BC. The biopolymer, BC is hydrolyzed using an ultrasound treatment and used as an additive. Morphological, structural, rheological and water vapor permeability analysis were performed for the films. PVP-CMC-BC films has better thermal stability than PVP-BC films. Results from water vapor permeability analysis also suggest lesser penetration of oxidative degradation causing gasses like oxygen through PVP-CMC-BC films. Thus, it can be concluded that the polymeric film PVP-CMC-BC is appropriate for fresh fruits and vegetables for enhancing their shelf life. Further, PVP-CMC-BC can be recommended for its use as a food packaging biomaterial/film.

Key words: Bacterial cellulose, PVP-CMC, Hydrogel, Rheology, Food packaging
PACS: 01.10.Fv

INTRODUCTION:

The environmental concern for the soil and water pollution in the last few decades has raised interest of the scientific community for an alternative eco-friendly packaging material. Biopolymers has emerged as an important alternative material to the conventional plastic due to their abundancy in nature, low material cost, eco-friendly, biodegradability and bio compatibility. Packages composed of biopolymers also have an excellent selective barrier properties which prohibits the entry and growth of microbial contamination in the packaging environment. Apart from extending the food shelf life, biopolymers are also good integrating agent with different other active packaging copolymers and additives like antimicrobial agent, antioxidants, nutrients, color components, oxygen scavengers, lubricants, nucleating agents, light stabilizers, blowing agents, plasticizers, antifogs, slips and antistats (1).

Cellulose from microbial origin has already been established as a multifunctional biopolymer for its purity, water holding capacity, biocompatibility and biodegradability. Bacterial cellulose (BC) is secreted by the bacteria as an extracellular polysaccharide protective layer, which after treatment has also been utilized as an edible package. BC has also been applied in the production of paper and paper products, apart from using as wound dressing material. Since BC lacks the sealing ability, in our study we have to added thermoplasticity to BC filmswith the integration of other biodegradable and biocompatible copolymers PVP and CMC (Fig. 1). The evaluation is done on the basis of their morphological, rheological and barrier properties.

MATERIALS AND METHODS:

Materials

PVP was procured from Sigma Aldrich, USA and CMC was purchased from Sinopharm Chemical Reagent Co. Ltd. Polyethylene glycol (PEG) and Agar were provided by Fluka, CZ. Glycerine (Gly) was purchased from Lach-Ner, CZ. Sodium hydroxide (NaOH) were bought from VWR chemicals, CZ.

Methods

Production and cleaning of Bacterial Cellulose (BC)

The BC was produced as per the protocol followed in our previous work (2). To mention in brief, 100 mL H.S culture medium was inoculated with 5 mL H.S inoculum having 5 loops of bacterial culture, incubated for 3 days at 30°C. The 100 mL cultures were in turn incubated for 15 days at 30°C. The BC produced after 15 days were collected and treated with 0.5N NaOH, followed by repeated wash with distilled water until neutral pH.

Hydrolysis of BC

The BC was hydrolyzed using ultrasound treatment as per the method mentioned by Bandyopadhyay et al. (2). The hydrolyzed BC particles were measured in a Zetasizer of average size of 209 ± 18 nm.

Preparation of neat BC film and hydrogels (using hydrolysed BC)

The BC film was casted utilizing 100 g of hydrolyzed BC in a silicone tray for 2 days at 50°C. The PVP-BC hydrogels were prepared by mixing the hydrolyzed BC with PVP, PEG, Agar and Glycerin in the percent ratio 0.66: 0.66: 1: 1: 1 (dry weights) in 150 mL water. PVP-CMC-BC hydrogels were also prepared in a similar fashion by mixing PVP, CMC, BC, PEG, Agar and Glycerin in the percent ratio 0.66: 0.33: 0.33: 1: 1: 1 (dry weights) in 150 mL water. The physical cross linkage in the hydrogel (PVP-BC) is formed by exposing the mixture to 15 bar pressure and 121°C for 35 minutes. After the treatment the mixtures are poured in polyvinyl alcohol plates (with a dimension of 25 cm x 25 cm) and allowed to cool, then incubated for 48h at room temperature to dry.

SEM study

The morphological studies of the transverse-section of the samples were analyzed under a field emission scanning electron microscope (Nova NanoSEM™ by FEI™, CZ s.r.o, Brno). The samples were gold coated for 60 sec at 30 mA in SC7620 Mini Sputter Coater(Quorum Technologies Ltd, UK). The micrographs were taken under 1000x, magnification. The operating voltage was 5 kV.

IR analysis

Fourier transfer infrared spectra analysis was done using Nicolet iS5 (Thermo Scientific,USA) in ATR mode. The scans were performed between infrared regions of 2000 to 800 cm^{-1}, with an average from 64 scans.

DMA quantification

The DMA analysis was performed in dynamic oscillatory mode using DMA Q-800 (TA instruments, USA). The films were cut in a dimension of 30 mm length and 7 mm width. The thickness of BC film and PVP-BC hydrogel film were 0.02 mm and 0.12 mm respectively. The temperature range was between -25 °C to 100 °C with a heating rate of 3 °C/min. The frequency and the oscillatory amplitude for the experiments were maintained at 1 Hz and 10 um respectively.

Analysis of water vapour permeability

The water vapour permeability (WVP) was measured according to ASTM E96-00 method. The cup was filled with silica gel within 6 mm of the samples. The orifice of the cup or the test area of the samples was 9.42 cm^2. The set up was placed inside a climate chamber for 24 hrs with air velocity of 0.25 ms^{-1}, 20 °C temperature and 90 ± 2 % relative humidity.

RESULTS AND DISCUSSION:

Morphological analysis

In contrast to the PVP-BC hydrogels, PVP-CMC-BC film is more translucent (Fig. 1). This blurriness of the PVP-CMC-BC film has its advantages over the other two films. The lower transmittance of light will protect the food from photo oxidation. PVP-CMC-BC films are more see through than the pure BC films but more yellowish than the PVP-BC.

Figure 1 (d-f) depicts the detail surface view of the samples under scanning electron microscope. It is evident from Figure 2 that all the samples have different topologicalarrangement. Thus, it is also apparent that the addition of other co-polymers to BC has resulted in diverse morphological features. These structural differences may attribute to the varied mechanical properties of the films and hydrogels.

Structural analysis

Fig. 2d shows the FTIR spectra of BC films and BC based hydrogels. The characteristic peaks assigned to BC found in all the samples can be ascribed as follows: 1146-1160 cm^{-1} for asymmetric stretching of (C-O-C) bond or CH deformation, 1111 cm^{-1} for stretching of (C-C) ring in polysaccharides or cellulose and 1059 cm^{-1} for C-O bond stretching (2). Moreover the distinguishing peaks for PVP – CMC based hydrogels can be seen in spectrum for PVP-BC and PVP-CMC-BC: 1285 cm^{-1} for C-N stretching vibration, 1654 cm^{-1} for C=O vibration and 1062 cm^{-1} for 1,4-β-D-glucoside stretching vibration (2).

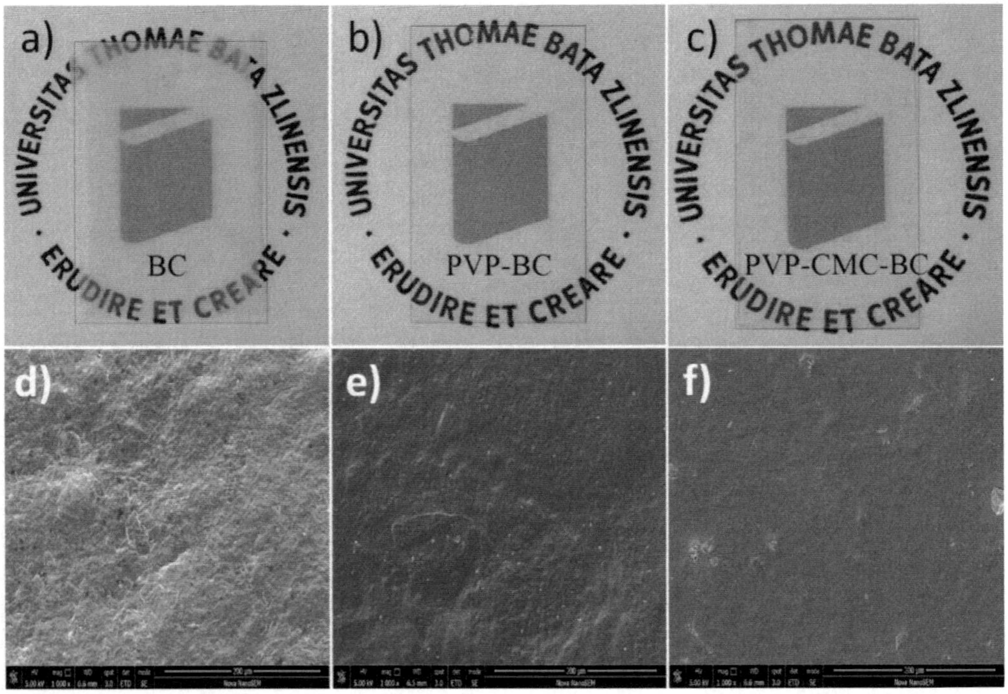

FIGURE 1. Physical appearance (a-c) and SEM (d-f) surface images of films a) BC b) PVP-BC c) PVP-CMC-BC d) BC films SEM e) PVP-BC SEM films and f) PVP-CMC-BC films SEM

Rheological properties

The viscoelasticity of the films were compared under varying temperatures and carried out by dynamic oscillatory experiments at 1 Hz. The temperature dependent curves for elasticity (storage modulus); viscosity (loss modulus) and damping properties (tan δ) are showed in Fig. 2 (a-c). It is evident from the graphs that PVP-BC has a wider deformation range than PVP-CMC-BC (Fig 2a). During the final use of the PVP-BC films the temperature region ranging from 15-35 °C (and higher) is not suitable for larger deformation, since tan delta raise lies in that region. The higher tan δ values for PVP-CMC-BC signifies it ability to dissipate more absorbed energy and giving more cushioning effect to the packed items (Fig 2c). The reason behind the loss of elasticity in the BC film with incorporation of the PVP and CMC may be due to the blocking of the intramolecular hydrogen bonds, the bonds which mainly attributes to the elastic nature of the BC (3).

Barrier properties

The water vapor permeability (WVP) is an indirect assessment of the porosity in the films. Figure 2e presents the water vapor permeability of the two films. Water vapor passes through the micro pores present in the structure of the films. The PVP-CMC-BC film has less micro pores than the PVP-BC film. The addition of additives to the BC has resulted in the formation of cross linking structure with bigger pores in surface than BC film. Since, the transmission of water vapor across the film occurs by the process of absorption and de-absorption of water molecules by the hydrophilic groups in the films, depending on a vapor pressure difference on either side of the film. Thus the presence of hydrophilic components in the hydrogel can also increase its WVP values in comparison to BC film. Based on the result PVP-CMC-BC film is more suitable for food packaging since less oxygen will pass through the pores thereby curbing the oxidative degradation and microbial spoilage of the vegetables packed inside.

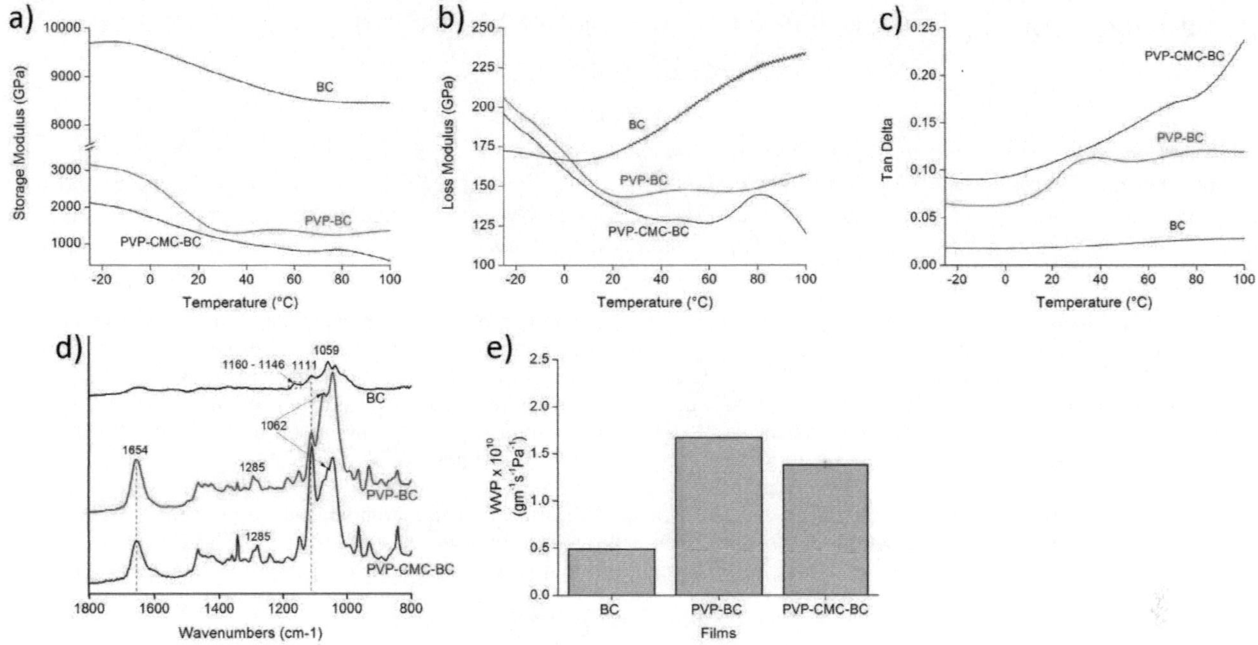

FIGURE 2. Effect of temperature (a-c) on storage modulus (a) loss modulus (b) and tan delta (c); FTIR spectra (d) and Water vapor permeability of the films (e)

Summary and Conclusion

The DMA and WVP analysis of the films suggest that PVP-CMC-BC film is best suited for food packaging application due to its better damping and barrier properties. It is also important to note that further analysis of color, chroma, lightness and resist to direct pull are important aspects of these packaging materials, to be done in future.

ACKNOWLEDGEMENT

The authors are grateful to the Internal Grant Agency IGA/CPS/2019, Tomas Bata University in Zlin, Czech Republic and to the Ministry of Education, Youth & Sports of the Czech Republic - NPU Program I (LO1504).

REFERENCES

1. Imran M, Revol-Junelles A-M, Martyn A, Tehrany EA, Jacquot M, Linder M, et al. *Critical Reviews in Food Science and Nutrition***9**, 799–821 (2010).
2. Bandyopadhyay S, Saha N, Brodnjak UV, Saha P. *Mater Res Express***5**, 115405 (2018).
3. Gea S, Reynolds CT, Roohpour N, Wirjosentono B, Soykeabkaew N, Bilotti E, et al. *Bioresour Technol* **19**, 9105–9110 (2011).

MFC Concept as a Possible Solution for Closed-loop Recycling of Food Packaging Trays

Maja Kuzmanović[a], Laurens Delva[a], Carla Isabel Martins[b], Ludwig Cardon[a] and Kim Ragaert[a]

[a] *Centre for Polymer and Material Technologies, Department of Materials, Textiles and Chemical Engineering, Faculty of Engineering and Architecture, Ghent University, Technologiepark 130, Zwijnaarde 9052, Belgium;*
[b] *Institute for Polymer and Composites/I3N, University of Minho, Campus de Azurém, 4800-058 Guimarães, Portugal*

Abstract. Flexible plastic films from food packaging trays make up one of the largest fractions of the plastic waste stream and recycling is one of the most important actions to deal with this fraction, reducing the impact of these plastics on the environment. However, the recyclability of multi-layered films is not straightforward and in most of the cases these fractions are landfilled or incinerated. Recycling of discarded mixed polymers without previous separation often results in low mechanical properties which have lead researchers to investigate novel solutions for recycling. In this research, the concept of microfibrillar composites (MFCs) was investigated aiming to upcycle mixed polymer waste streams. A blend based on polypropylene (PP) and poly(ethylene terephthalate) (PET) at a weight ratio of 80/20 PP/PET was studied. The final step of the MFC processing was conducted using a conical twin screw extrusion. The morphological results confirmed the presence of PET microfibrils in the composites, leading to an improvement in mechanical properties such as in the tensile yield strength and strain at break. Subsequently, the MFC samples were successfully moulded into trays via thermoforming.

Keywords: microfibrillar composites, morphology, mechanical properties, recycling

INTRODUCTION

Polymer products became an integral part of the modern life. With the increase in plastics consumption, the biggest challenge is how to solve the issue of plastic waste disposal, lowering its impact on the environment [1]. The success of polymers is mainly attributed to their low cost, low density and numerous mechanical and thermal properties that make them suitable for a wide range of applications. Packaging materials such as food packaging trays typically have very short lifetimes and a huge amount of plastic waste is generated from packaging applications. The most used polymers in food packaging applications are polyolefins and polyesters, and quite often these products are produced in complex multi-layered films, hindering the separation of the individual polymers in those plastic products. It is well known that many polymers are immiscible in the melt state, which causes phases separation and decreasing mechanical properties of the final mechanically recycled product [2]. Therefore, many researchers [3-5] investigated new methods to overcome the problem of immiscibility in polymer waste streams aiming to improve the processing of recycled blends and the final properties of the materials. The most common approach to mitigate the immiscibility is the addition of compatibilizers or different types of additives into polymer mixtures [2, 6]. Alternatively, production of microfibrillar composites (MFC) was presented as a worthy and easy way of making new composites with increased mechanical properties [7, 8]. The key factor in the development of MFCs lies in the immiscibility of two polymers, as this is one of the important requirements for their production, as well as the difference in melting temperatures (e.g. between polyolefins and polyesters).

Hence, considering the previous studies and proposals, this research focuses on developing the MFC concept as a possible solution for the recycling of food packaging trays. The samples were prepared in a similar way to the conventional MFC production, except for the last production step, where extrusion was used instead of injection moulding. Until now, there are no known studies where extrusion and thermoforming were used as final steps in the MFC processing. Therefore, virgin polymers were used in this study due to possible limitations that might be encountered during processing of recycled mixtures, such as degradation, viscosity issues, contaminations etc. [5, 9]. Our goal is to produce the MFC sample into a sheet via extrusion and convert it into a thermoformed tray. In our previous work, the influence of compatibilizers was pointed out, as well as the improvement of the mechanical properties such as tensile yield strength and strain at break obtained with rubber-based compatibilizers [4]. During the thermoforming process, the composite is exposed to extensive stretching; hence, the strain at break is an important characteristic. Therefore, a compatibilizer was also used in this study. The presence of the PET microfibrils in the composites may contribute to increased mechanical properties and possibly decrease permeability. Therefore, MFCs can be a possible solution for the multi-layered packaging waste, enabling a completely closed-loop recycling, however, at this point without taking into account the food contact legislation.

EXPERIMENTAL

Materials

The virgin materials used in this study were polypropylene (PP) purchased from Sabic (Sabic 575P, Bergen op Zoom, The Netherlands) with a melt flow rate (MFR) of 11 g/10 min (2.16 kg, 230 °C), and poly(ethylene terephthalate) (PET) (LIGHTER C93) from Equipolymers (Schkopau, Germany), which is a bottle-grade material with an intrinsic viscosity of 0.80 ± 0.02 dL/g. PET was dried in a vacuum oven for 15 h at 80 °C, and 2 h before processing at 120 °C, while PP was used as received. Polyolefin-based elastomer grafted with maleic anhydride (POE-g-MA), (Acti-Tech 16MA13) was used as a compatibilizer in this study, which is a Exxon performance polymer based, kindly donated by Nordic Grafting Company (NGC, Hellerup, Denmark). The grafting percentage of the maleic anhydride (MA) group onto the backbone of the compatibilizer was 1.3 wt%, according to the data sheet. Compatibilizer was dried at 60 °C for 15 h before processing.

Sample Preparation

The samples were prepared in a weight ratio of 80/20 PP/PET, where PP was used as a matrix and PET as a reinforcing component. To compatibilize the PP/PET blend, POE-g-MA was added in 6 wt %, while the same PP/PET ratio was maintained. The preparation of the blends (with and without compatibilizer) was conducted via twin screw extruder (Coperion ZSK18, Stuttgart, Germany) with two co-rotating screws of 18 mm diameter, *L*/*D* = 40 and a die opening of 19 mm × 2 mm. The screw speed was set at 120 rpm and the barrel temperatures were set between 205 and 260 °C. The extrudate was obtained as a sheet with dimensions of 25 mm × 1 mm, by passing through calander rolls, which were cooled down to ~15 °C. The received cooled extrudate was cold drawn by a pair of rolls passing through a hot oven (200 °C, 55.5 cm × 60 cm) above the glass transition temperature of PET. During drawing, the surface temperature of the extrudate was measured and amounted to approximately 95 °C. The speed of the rolls was adjusted to obtain a draw ratio of 8. Shredded pure blend and microfibrillar blends were extruded again via conical twin screw extruder (MAS24, Schulz GmbH, Hobelweg, Austria) into sheets with dimensions 150 mm x 1 mm. The conical twin screw extruder was set at temperatures of 170, 190, 200 and 210 °C and speed of 70 rpm. The obtained sheets were cooled down by passing between cast rolls at a temperature of 60 °C. The MFC sheets were thermoformed by a Formech508FS machine (Harpeden, United Kingdom) at 135 °C for 65-75 seconds into round trays with a diameter of 55 mm and 10 mm height.

Characterization

Small angle light scattering (SALS) was used to determine the spherulites size of the PP in the final sheets. The samples were microtomed into 15 μm thick layers and placed between two microscopic slides. SALS patterns were obtained with a 632.8 nm He-Ne laser (1 mm beam size) with the source of polarized monochromatic light. SALS H_v patterns were captured using a Hamamatsu digital camera (Hamamatsu Photonics K.K., Hamamatsu City, Japan) and analysed with Hipic 6.3.0 software (Hamamatsu Photonics K.K., Hamamatsu City, Japan). The equivalent radius (R_0) of the spherulites was estimated with the following Equation (1):

$$R_0 = \frac{1025}{\pi} \frac{\lambda}{\sin(\frac{\theta_{max}}{2})},$$

where λ is the wavelength of light in the medium. The distance from the centre of the H_v pattern to the intensity maximum in one lobe, in conjugation with the known sample-to-film distance, is a measure of the polar angle θ_{max}. (θ_{max} = tan (distance from the centre of the H_v pattern to the intensity maximum in one lobe/sample-to-film distance)) [4]. The structural characterization of the samples was performed by scanning electron microscopy (SEM) FEG SEM JEOL JSM-7600F 202 (Tokyo, Japan). The sheets were immersed in liquid nitrogen and subsequently fractured, and afterwards the surfaces were sputtered with gold by a Bal-Tec SCD005 sputter coater (Bal-Tec, Balzers, Liechtenstein). Micrographs were obtained with an accelerating voltage of 15 kV. Raman spectroscopy was used as an additional technique to confirm the dispersion of PET fibrils in the composites. Raman spectra were obtained on a Horiba LabRAM HR Evolution confocal microscope (Horiba Scientific, Longjumeau, France) using a laser excitation of 785 nm. The objective lens (×100) was used to focus the laser onto the sample. The scanned Raman red maps 40×35 μm were obtained with a step size of 1 μm and analysed using the LabSpec 6 software (also from Horiba Scientific). The tensile bars (ASTM D412T.C) were die cut from the sheets with dimensions 115 mm × 6 mm ×1 mm and gauge length of 33 mm. They were tested by an Instron 5565 tensile device (Norwood, MA, USA) according to standard ISO 527. During the tests, two test speeds were used: 1 mm/min as a first cross-head rate, and 5 mm/min after the Instron dynamic extensometer was removed (type catalogue 2620-603 with a gauge length of 12.5 mm). Analysis of the results was performed with the Bluehill software. The differences between the samples were calculated by t-independent test via the software package SPSS Statistics 24 (Armonk, NY, USA) with a probability value of 0.05.

RESULTS AND DISCUSION

Morphology study

It is well known that the structural development is of huge importance for good mechanical properties of composites [4], hence the extruded sheets were subjected to intensive microscopy study via SEM and Raman spectroscopy. In Figure 1, SEM micrographs of the blend, MFC and MFC_{POEgMA} sheets in longitudinal direction are presented. As can be seen, the blend shows a non-uniform dispersion and distribution of PET particles with a coalescence phenomenon present (black circles in Figure 1a) in the centre of the sheets, while at the edges of the sheet (Figure 1a'), the presence of stretched PET particles can be noticed. These stretched particles could be a consequence of the extrusion process as the final processing step and high viscosity of the melt, which has made the production of the sheets more difficult. We assumed that beside the high viscosity, another reason could be increased shear rate in die edges; hence, an additional stretching may have happened.

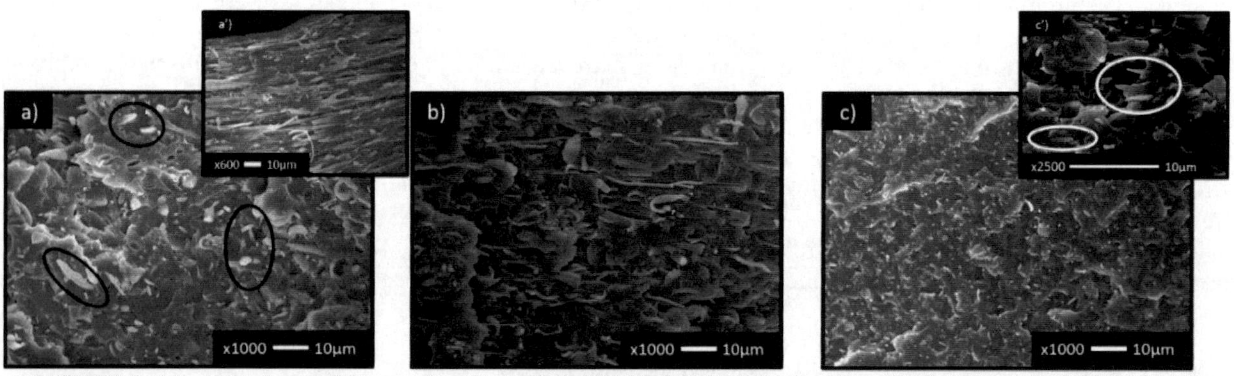

FIGURE 1. Microstructures of the samples obtained via SEM in longitudinal direction: (a), (a') blend, (b) MFC, (c), (c') MFC_{POEgMA}

Further on, in the MFC sample PET microfibrils were detected with high aspect ratios (Figure 1b), as well as a good level of orientation. Although, some fibril clusters were also found along the MFC sample. The reason for the clusters formation might be the relaxation of PET fibrils that takes place during the extrusion process. Observing the MFC containing POE-g-MA, it can be noticed that the fibrils are much shorter (Figure 1c') than in the non-compatibilized MFC, however, a relatively uniform dispersion and distribution can be noticed (Figure 1c). The presence of the compatibilizer may prevent coalescence of the fibrils by covering the PET particles during melt blending, which facilitates the dispersion of droplets and consequently reducing the final fibril lengths [4]. The red maps obtained via Raman spectroscopy, shown in Figure 2, confirm the dispersion of the second phase in the matrix. The maps show PP as a black phase, while the red one represents the PET phase. The red phase in Figures 2a and 2b show the PET clusters which have been detected in the composites while in the case of MFC_{POEgMA} it can not be seen. Similar findings have been reported in our previous study [4].

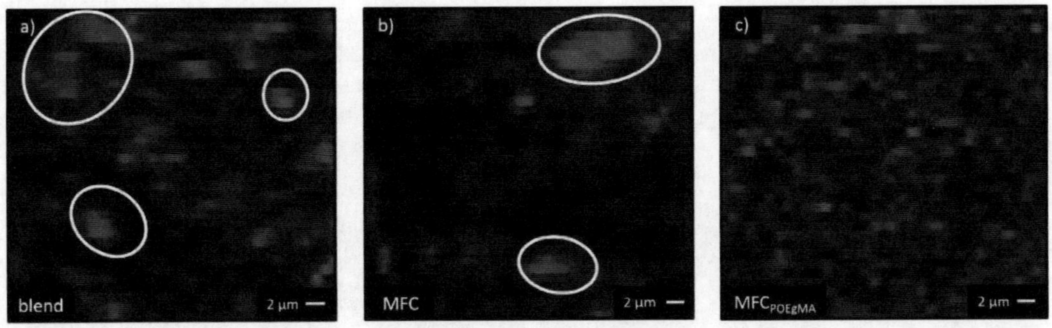

FIGURE 2. Red maps of the samples obtained via Raman spectroscopy: (a) blend, (b) MFC, (c) MFC_{POEgMA}. Matrix PP represents black phase, and PET is the red phase.

In addition to the structural development, the average spherulite size of PP was quantitatively measured by SALS. The average diameters of the PP spherulites in the composites are listed in Table 1. The MFC_{POEgMA} shows the lowest spherulite size when compared to the blend and the MFC. The reason for that lies in well dispersed and distributed PET fibrils which have a strong nucleating effect on the PP matrix, since they are covered by the compatibilizer [4]. In our previous work [4], it was explained that the nucleating effect may arise from isolated compatibilizer particles too, as there is always a certain amount of compatibilizer that does not react with the PET.

TABLE 1. Average diameter of PP spherulites in the composites

Material	Spherulite diameter, μm
Blend	10.4 ± 1.0
MFC	8.3 ± 0.9
MFC_{POEgMA}	3.6 ± 0.7

Tensile Properties

In order to discuss mechanical properties, the tensile modulus, tensile yield strength of the samples and strain at break were defined and the results are listed in Table 2. An increase in stiffness was observed for the MFC sample in comparison with the blend and the MFC_{POEgMA}, which can be explained by the presence of long PET fibrils which will contribute to increased stiffness. Due to the presence of the long fibrils in this sample, it is considered that sufficient interfacial contact between the microfibrils and the matrix phase is present, even without the presence of a compatibilizer [4]. MFC_{POEgMA} shows a lower stiffness, due to the shorter PET fibrils caused by the presence of the compatibilizer containing a rubber (soft) backbone [2, 4] and higher strength in comparison to other two samples. The reason for higher tensile yield strength for MFC_{POEgMA} lies in a better adhesion between the phases. Both PP and PET uniformly stretch during the tensile test. Furthermore, MFC_{POEgMA} shows a higher strain at break result, which confirm that stress transfer between the fibrils and the matrix is considerably improved.

On the other hand, the blend presented a higher stiffness, probably due to the presence of the PET fibrils at the edges of the sample (see Figure 1a'), but they did not affect the tensile yield strength. The stress transfer in the blend is not enough effective due to lower adhesion between phases, which causes for the lower yield strength value, as well as the low strain at break.

TABLE 2. Tensile properties of extruded MFC plates

Material	Tensile Modulus, GPa	Yield Strength at Zero Slope, MPa	Strain at break, %
Blend	1.91 ± 0.09	28.6 ± 0.5	15.9 ± 1.2
MFC	2.15 ± 0.13	30.3 ± 0.9	14.5 ± 3.8
MFC_{POEgMA}	1.59 ± 0.30	32.5 ± 5.5	250.0 ± 229.0

Preliminary Thermoform Experiments

As shown from the results above, there is an improvement in mechanical properties of the MFC samples, particularly in the yield strength and the strain at break for the sample MFC_{POEgMA}. In Figure 3, thermoformed parts are shown for the samples MFC and MFC_{POEgMA}. As can be seen from the images, MFC_{POEgMA} shows a smooth and uniform surface, while the MFC sample shows a rougher surface. From Figures 3a' and 3b', can be noticed that details in the MFC_{POEgMA} are more profound than in the MFC sample. Comparing the strains at break values, it is clear that MFC_{POEgMA} has a higher ability to stretch during the thermoforming process and thus obtained a better shape at the edges then the MFC sample. Moreover, during the thermoforming process, the MFC sample would break much easier due to its brittleness.

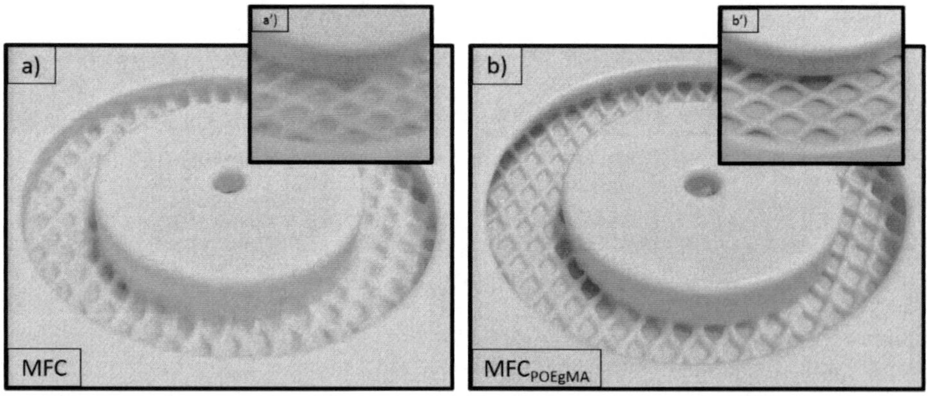

FIGURE 3. Thermoformed parts for the MFC samples: (a), (a') MFC, (b), (b') MFC_{POEgMA}.

CONCLUSION

This research presented the MFC concept as a possible solution for the recycling of food packaging trays aiming at a closed-loop recycling path.

Microstructural analysis has confirmed the presence of PET microfibrils in the MFC samples, as well as an improved dispersion and distribution in the MFC_{POEgMA} sample. SALS measurements showed small spherulite sizes in all samples, confirming the theory of heterogeneous nucleation of PP promoted by the PET fibrils.

The mechanical results showed a significant increase in the tensile yield strength and in the strain at break for the sample MFC_{POEgMA}, as the compatibilizer promoted a better interfacial adhesion between PP and PET phases. Based on these findings, the MFC samples were successfully moulded into trays via thermoforming.

In the future, the optimisation of the thermoforming process will be necessary, likewise taking into consideration the food contact regulations as the products would be prepared from recycled mixtures. Moreover, investigation of the permeability of the trays and its relation to deformation of PET fibrils during thermoforming would be of high importance.

ACKNOWLEDGMENTS

This research was financially supported by Ghent University, Belgium. We acknowledge the University of Minho, Portugal for collaboration on the project. M. K. would like to thank prof. Carla Isabel Martins for supervising her during the stay at the University of Minho and Paulo Lopes for the support with the Raman measurements.

REFERENCES

1. S. Al-Salem, P. Lettieri and J. Baeyens, *Waste Manage.* 2009. **29**(10): p. 2625-2643.
2. N. Mys, L. Delva, M. Kuzmanovic, T. Wieme and K. Ragaert, *Functional evaluation of compatibilization systems for recycled PP-PET blends.* in *International Conference on Polymers and Moulds Innovations-PMI 2018*. 2018. Institute of Polymers and Composites, University of Minho, Portugal.
3. W. K Pek, U. K. Ghosh, and S. Energy,Environ. *Prog. Sustainable Energy,* 2015. **34**(4): p. 1113-1119.
4. M. Kuzmanović, L. Delva, D. Mi, C. I. Martins, L. Cardon and K. Ragaert, *Polym.* 2018. **10**(3): p. 291.
5. K. Ragaert, L. Delva, and K. Van Geem, *Waste Manage.* 2017. **69**: p. 24-58.
6. H. T. Chiu and Y. K. Hsiao, *J. Polym. Res.* 2006. **13**(2): p. 153-160.
7. S. Fakirov, D. Bhattacharyya, R. J. Shields,*Colloids Surf., A*2008. **313**: p. 2-8.
8. R. Shields, D. Bhattacharyya, and S. Fakirov, *J. Mater. Sci.* 2008. **43**(20): p. 6758.
9. R. Stein, *Polymer recycling: opportunities and limitations.* Proceedings of the National Academy of Sciences, 1992. **89**(3): p. 835-838

Effect of the Chemical Modification on the Thermal and Rheological Properties of Different Polylactides for Foaming

Svenja Murillo Castellón[a*], Tobias Standau[b], Volker Altstädt[b] and Christian Bonten[a]

[a]*Institut für Kunststofftechnik, University of Stuttgart, Pfaffenwaldring 32, 70569 Stuttgart, Germany*
[*]*svenja.murillo-castellon@ikt.uni-stuttgart.de*

[b]*Lehrstuhl für Polymere Werkstoffe, University of Bayreuth, Universitätsstraße 30, 95447 Bayreuth, Germany*

Abstract. Polylactide (PLA) is one of the most important bioplastics on the market with a permanent growth in production capacity. Due to its good mechanical properties, it is comparable to polystyrene. But in terms of foaming the melt strength and extensibility of commercial polylactide are rather low. Therefore such commercial grades often need to be chemically modified to induce crosslinking, chain extension or grafting by means of reactive extrusion on a twin-screw extruder. In this work different polylactide grades were used, to better understand the influence of the molecular characteristics, such as molecular weight, acid value and D-content, not only on the modification mechanism, but also on the foam properties.

Keywords: Polylactide, bioplastic, foams, modification.
PACS: Replace this text with PACS numbers; choose from this list: 82.35.Pq, 87.15rp, 82.70.Rr

INTRODUCTION

Polylactide (PLA) is known as one of the most promising bioplastics on the market. It is derived from renewable feedstock and is biodegradable.

Due to increasing annual production rates of commercial PLA and falling prices it has become more and more important in scientific research and industrial use. PLA has comparable mechanical properties to the well-known fossil-based plastic polystyrene (PS), which is used in foams for packaging and insulation applications. But, PS causes some environmental issues, which lead to the demand for new materials. PLA is a promising plastic to substitute PS. However, the low melt strength and extensibility is one drawback for PLAs foaming. For foaming, PLA must be modified. One way is the chemical modification of PLA to induce crosslinking, chain extension or branching.

Yet, a wide variety of different chemical modifiers has been studied in the literature, as well as different PLA grades with different properties. Even though in the literature, the low melt strength and extensibility of PLA is often referred to the main challenge in regards of PLA´s foaming only a few publications quantify the melt strength [1-5]. Furthermore, the molecular influences of PLA e. g. molecular weight, number of functional groups or the d-content on the melt viscosity and also on the reaction mechanism during the chemical modification is not fully understood yet. Due to the chiral nature of lactic acid (the monomer to produce PLA), PLA occurs in two different enantiomeric forms (PLLA, PDLA and PDLLA). The D-lactide content strongly influences the thermal properties (melting temperature, glass transition temperature and the crystallinity are decreasing with increasing D-lactide content). Furthermore, a high D-lactide content leads to amorphous PLA. Mihai and its coworkers [6, 7] investigated already the foaming behavior of amorphous and semi-crystalline PLA in regards of foaming. Their work showed that crystalline PLA foams have a higher foam cell nucleation and higher foam expansion, which can be attributed to the crystalline nuclei that contributed to the foam cell nucleation and a better foaming performance.

In our previous work, detailed results of different modifiers and their effects on the rheological, thermal and foam properties of PLA were published [5, 8, 9]. The most effective results were achieved with an organic peroxide. With this modifier the molecular weight, the melt viscosity, the melt strength and the crystallization rate were improved significantly. The modifier was found to be more effective than a commercial available modifier (multifunctional epoxide).

The aim of this work is to investigate two different polylactide grades (with different molecular properties) in regards to foaming but also in regards to the chemical modification to increase PLA´s melt strength in order to enhance the latter foam properties.

EXPERIMENTAL

Materials

Two different commercially available PLA types were used in this study. PLA Ingeo™ 4060D (PLA_a) and PLA Ingeo™ 4044 D (PLA_{sc}), both from NatureWorks LLC. Both grades are used for the production of oriented films, card stock and graphic arts. PLA_{sc} is a reactive extrusion grade, while PLA_a is normally used for heat seals layer in coextruded oriented films. An overview of the materials´ properties is listed in Table i.

TABLE i). Overview of the PLA grades [10, 11]

Material	D-lactide content in %	T_g in °C	T_m in °C
PLA_{sc}	~ 4	55–60	155–160
PLA_a	~ 12 %	55–60	-

Reactive Extrusion

The modifications were carried out on a twin-screw extruder type ZSK26MC from Coperion (Stuttgart, Germany). Depending on the modification, the temperature profile of the twin screw extruder was varied between 180 °C and 190 °C. The screw speed (150 min^{-1}) was kept constant. The PLA granulate and the modifiers in solid powder form were added gravimetrically in zone 1. A double-strand die was used and the strands were first cooled in a water bath and then granulated. For the modification an organic peroxide with a concentration of 0.2 m.-% and a multifunctional epoxide with 0.7 m.-% were used.

Characterization

The linear viscoelastic shear deformation behavior of the materials was investigated with a parallel-plate rheometer Discovery HR-2 hybrid from TA Instruments Waters (New Castle, USA) (plate diameter 25 mm, gap 1 mm). The materials were vacuum dried for 12 hours in an oven at 40 °C. All measurements were done under nitrogen atmosphere at 180 °C.

The molecular weight was characterized by size exclusion chromatography (SEC). The average molecular weight was measured by using an Agilent 1260 (Santa Clara, US) equipped with a PSS SDV (10 µm) 8.0 mm x 50 mm pre-column and three linear PSS SDV (10 µm) 8.0 mm x 300 mm (10^3-10^6 Å). The samples were measured in chloroform at a temperature of 30 °C and a flow rate of 1 mL / min.

The thermal behavior of the materials was determined using a DSC type 2/400 from Mettler Toledo (Columbus, USA). Nitrogen was used as inert gas. Approximately 10 mg of sample were weighted into the sample crucibles. The measurements were carried out in a temperature range of -20 to 200 °C at a heating or cooling rate of 10 K / min.

Batch Foaming and Foam Characterization

The foaming was performed on rectangular melt pressed samples (18 mm x 11 mm x 0,8 mm) in a custom made high pressure autoclave initiated due to a sudden pressure drop which results in a thermodynamic disequilibrium. As a physical blowing agent CO_2 was used. The pressure was 180 bar and the saturation time 30 minutes. The foaming temperature was varied between 40 and 130 °C. To get further information on the foam properties, the density of the foams was determined using the Buoyancy Method using a Mettler Toledo (Columbus, USA) balance with density kit.

RESULTS AND DISCUSSION

The viscous and elastic properties of a polymer melt are essential for further processing e. g. foaming. A high melt viscosity at a low angular frequency is characteristic for materials which can be processed by foaming processes, where a sufficient high melt viscosity and melt strength is required to build a stable foam structure and to prevent cell collapse. Figure 1 shows the complex shear viscosity $|\eta^*|$ as a function of the angular frequency ω for the two investigated PLA. Both show a shear thinning behavior, which is typical for linear polymers. The rheological characteristics are similar, but PLA_{sc} shows a higher complex viscosity and also a higher zero shear viscosity.

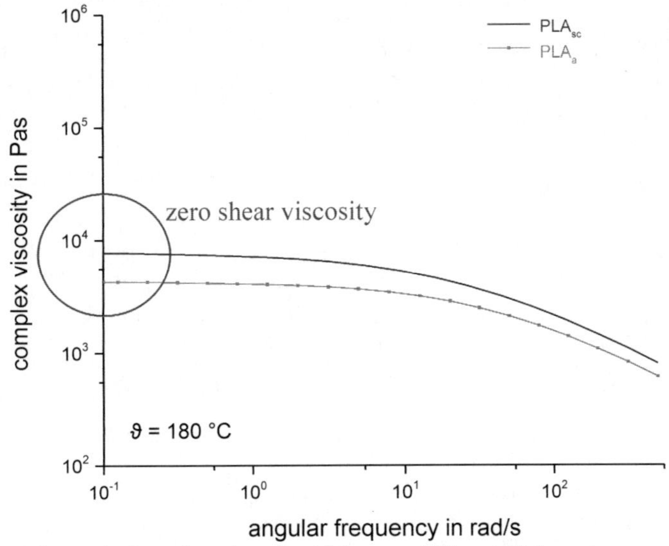

Figure 1: Complex viscosity of the two different PLA grades

Both are characterized by similar molar mass but different D-lactide content. The number average molar mass M_n is similar and the weight average molar mass M_w is slightly different, whereby PLA_{sc} shows a higher weight average molar mass as it can be seen in Table ii, which is also the reason for the higher zero shear viscosity in Figure 1.

TABLE ii). Overview of the PLA grades

Material	M_n in g / mol	M_w in g / mol	PDI
PLA_{sc}	117,000	230,000	1.97
PLA_a	118,000	217,000	1.84

Afterwards the two PLA were foamed and the effect of the saturation temperature was investigated. The foam density in dependence of the saturation temperature is shown in Figure 2.

Even though the two materials show a similar rheological behavior there is a high discrepancy in the foaming behavior. The achievable density reduction is similar but the foaming window is significantly shifted for both grades. PLA_a shows much higher expansions at lower temperatures close to its T_g. For this grade the lowest density is about 110 kg/m^3 when foamed at 65°C. At higher temperatures of above 80°C the foam densities are significantly increased. Here, less expansion is achieved, as the foams are collapsing. In the temperature range between 50 and 70 °C where PLA_a foams showed their lowest density PLA_{sc} can only be foamed with moderate density reduction. PLA_{sc} foams show a significant drop in foam density at around 80°C and the lowest foam density was achieved at 110°C (87 kg/m^3). At too high temperatures the viscosity decreases resulting in a lower melt strength. Hence, a foamed structure cannot be maintained and cells are coalescing and collapsing.

The reason for the differences in the foaming behavior can be mainly found in the D-content of the two polymers. The amount of D-lactide significantly influences the crystallinity of polylactides. With increasing D-lactide content, the crystallization rate becomes lower and the crystallinity is reduced. With a D-content of 12% PLA_a is fully amorphous [12, 13]. On the other hand PLA_{sc} can crystallize. As shown in the study of Mihai et al. [7] the crystallinity of PLA foams strongly depend on the D-content and the amount of CO_2. Therefore, in PLA_{sc} with a

D-content of about 4% crystals can form during the saturation time and / or while foaming leading to the observed temperature difference of the foaming temperature. The formed crystals could also act as nucleating point resulting in higher foam nucleation densities.

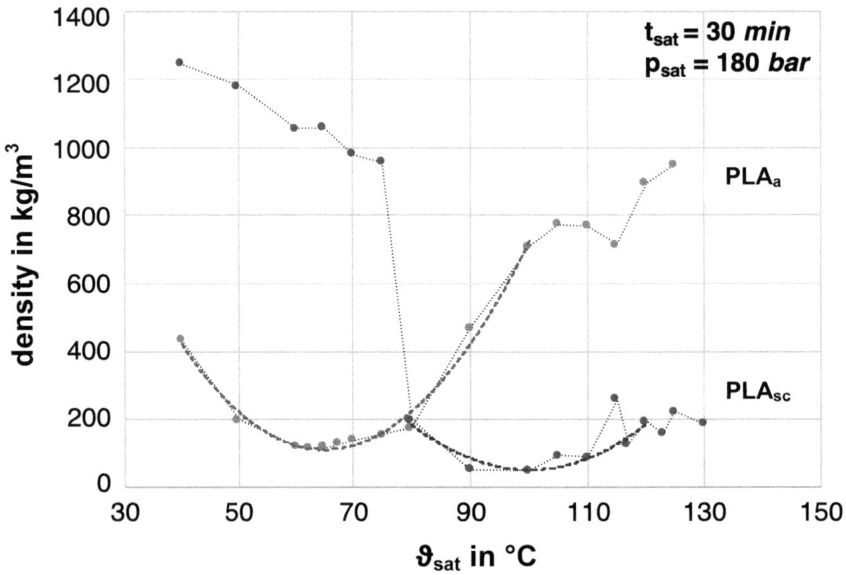

Figure 2: Resulting foam densities at different saturation temperatures

By varying the saturation temperature the melt viscosity can be influenced. The lower the temperature is, the higher the melt viscosity but also the melt strength of the materials, which is promoting the expansion and leading to lower foam densities, which can also be observed in Figure 2. For both grades a minimum in foam density can be seen. However a too low foaming temperature hinders the expansion as the resulting flow behavior is unfavorable.

Our previous works focused on the effect of different modifiers on the rheological and thermal properties of PLA. The results are reported in detail in [5, 8, 9]. The investigated organic peroxide was found to be more effective than a commercial available multifunctional epoxide (benchmark modifier). The molecular weight, the melt viscosity, the melt strength and the crystallization rate were improved significantly. In the next step PLA$_{sc}$ and PLA$_a$ will be chemically modified and afterwards foamed again, to investigate the effects of the modification on the foamability and foam properties. Figure 3 shows the effect of the modification on the rheological properties of PLA$_{sc}$.

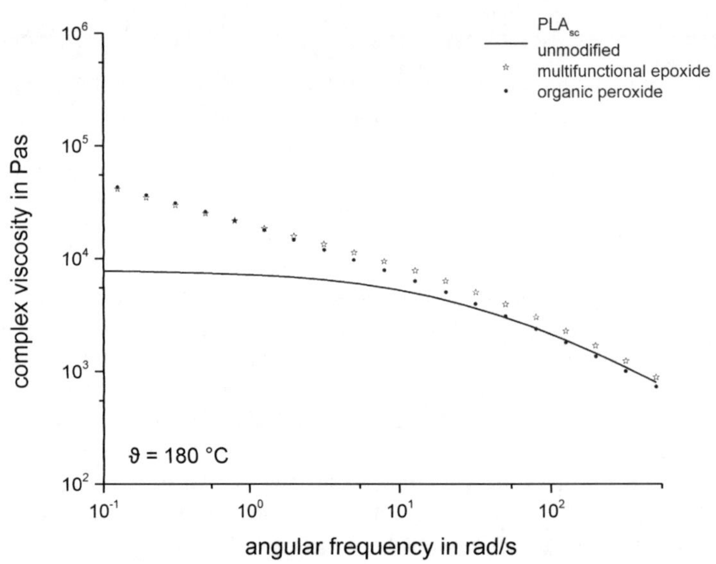

Figure 3: Complex viscosity of modified PLA$_{sc}$

It is visible that both modifiers, the organic peroxide and the multifunctional epoxide, increase the melt viscosity. By the modification we expect in case of PLA_{sc} either a further decrease of the foam density or a finer foam morphology. The effect of the D-lactide content on the modification has to be investigated.

CONCLUSION

Within this study it was shown that foaming conditions, i. e. the foaming temperature is strongly dependent on the D-content of the PLA. The molecular characteristic of PLA has an influence on the foaming behavior and the resulting foam density. The investigated grades had a nearly similar molecular weight and can be distinguished by their D-content. The lowest density was achieved at 65°C for the grade with the high D-content while the lower D-content grade showed its best expansion at 110°C. The shifted foaming windows can be explained by the different crystallization behavior resulting from the different D-content. By inducing crosslinks or branches into PLAs chains the melt viscosity and also the melt strength can be increased [14]. Further work will be targeted on the use of the modifiers for PLA_{a} in regards to further increase the melt strength. So far there is no publication known, in which the chemical modification of PLA with a high D-content with an organic peroxide or the multifunctional epoxide as well as the foaming behavior was investigated.

ACKNOWLEDGMENTS

The authors gratefully acknowledge the financial support by the Germany Research Foundation (DFG) for funding the project BO 1600/15-1 and AL 474/24-1.

REFERENCES

1. W. Liu, X. Wang, H. Li, D. Zhongjie and C. Zhang, *J. Cell. Plast.* **49**, 535–554 (2013).
2. J. Ludwiczak and M. Kozlowski, *J. Polym. Environ.* **23**, 137–142 (2015).
3. L. Yu, G. Toikka, K. Dean, S. Bateman, Q. Yuan, C. Filippou and T. Nguyen, *Polym. Int.* **62**, 759–765 (2013).
4. K. M. Dean, E. Petinakis, S. Meure, Y. Long and A. Chryss, *J. Polym. Environ.* **20**, 741–747 (2012).
5. S. Göttermann, T. Standau, S. Weinmann, V. Altstädt and C. Bonten, *Polym. Eng. Sci.* **57**, 1242-1251 (2017).
6. M. Mihai, M. A. Huneault and B. D. Favis, *Polym. Eng. Sci.* **50**, 629-642 (2010).
7. M. Mihai, M. A. Huneault and B. D. Favis, *J. Appl. Polym. Sci.* **119**, 2920-2932 (2009).
8. S. Göttermann, T. Standau, V. Altstädt and C. Bonten, SPE Conference Proceedings, SPE ANTEC, Anaheim, CA, 2017, pp. 452-456.
9. S. Göttermann, T. Standau, V. Altstädt and C. Bonten, AIP Conference Proceedings 2055, AIP Publishing, 2019, pp. 020003-1-020003-5.
10. G. Gorassi and R. Pantani, *J. Polym. Degrad. Stab.* **98**, 1006-1014 (2013).
11. R. Gendron and M. Mihai, "Extrusion Foaming of Polylactide", *in Polymeric Foams: Innovations in Processes, Technologies, and Products*, edited S. T. Lee, Boca Raton: CRC Press, 2016, pp. 135.
12. J. Reignier, R. Gendron and M. F. Champagne, *Cell. Polym.* **26**, 83-115 (2007).
13. L.-T. Lim, R. Auras and M. Rubino, *Prog. Polym. Sci.* **33**, 820-852 (2008).
14. S. Göttermann, „Modification, Preparation and Characterization of Polylactide Foams", SPE Foams Conference 2017, Bayreuth, Germany, 09.-12. May 2017). Bayreuth, 2017

Preparation thermoresponsive semi IPN magnetic nanogels based on poly(N vinylcaprolactam/carboxymethyl cellulose) for localized drug delivery

Hamed Mohammad Gholiha[1], Morteza Ehsani[2,3], Ardeshir Saeidi[1], Mojgan Bahrami[4], Sogand Safari[2]

1. **Department of Polymer engineering, Science and Research branch, Islamic Azad University, Tehran, Iran.**
2. **Plastic Department, Iran Polymer and Petrochemical Institute (IPPI), Tehran, Iran**
3. **Department of Polymer Engineering, Faculty of Engineering, South Tehran Branch, Islamic Azad University, Tehran, Iran.**
4. **Department of pharmacy, Tehran university of medical science, Tehran, Iran**

Abstract. Biomaterials are widely used as a drug delivery system. In recent years local drug delivery has been extensively studied to improve the pharmacokinetics of drugs. Thermoresponsive polymers are mostly used as a drug delivery system. Poly N-vinylcaprolactam (PNVCL) is a thermoresponsive polymer that exhibit LCST behavior in aqueous solution around 32 °C. Thermoresponsive polymers switch its behavior at the higher temperature of LCST. The hydrophilic polymer change to hydrophobic behavior due to disappeared the interaction between carbonyl groups of PNVCL and hydroxyl groups of Aquos environments. Control LCST near the human body is the most important parameter for successful protein delivery system. To increase phase transition temperature, carboxymethylcellulose is a hydrophilic biocompatible polymer that makes it appropriate for copolymerization with PNVCL. In this work, we prepared thermoresponsive PNVCL/ carboxymethylcellulose semi IPN nanogels(MTRSI-NGs). At first, PNVCL were synthesized via free radical polymerization, and then carboxymethylcellulose was crosslinked in Fe3O4 presence. FTIR spectra confirmed the formation of MTRSI-NGs. The phase transition temperature was determined by Rheological and turbidimetry analysis. The turbidimetry analysis was shown LCST of PNVCL and magnet-free nanogels around 32 and 37 °C respectively. Rheological analysis showed phase transition temperature around 38 °C that confirm turbidimetry results. On the other hand, the rheological analysis was confirmed semi IPN nanogels by the thermoresponsive behavior of magnetic nanogels.

Keywords: **CMC semi IPN, Thermoresponsive nanogels, drug delivery system, Poly(N-vinylcaprolactam)**

Introduction

OVER the past two decades, biopolymers have been studied extensively for medical applications [1]. A wide variety of materials usually derived from biological sources such as microorganisms, plants, or trees can be described using the term "biopolymer". Materials produced by synthetic chemistry from biological sources such as vegetable oils, sugars, fats, resins, proteins, amino acids, and so on can also be described as biopolymer[2]. Nanotechnology has been attracting considerable attention since the 1980s and has been adapted into many engineering fields such as electronics, mechanical, biomedical and space engineering. researchers have shown great interest in the application of nanoparticles as biomaterials for delivery of therapeutic molecules such as drugs and genes, and also for tissue engineering[3]. Macromolecules provide many ways of creating stimulus sensitive behavior. Synthetic methods for the creation of polymeric molecules, structures, and architectures that can be assembled in a variety of ways to perform useful functions are available. Polymers responsive to temperature changes are the most studied class of environmentally sensitive polymers as they have potential applications in the biomedical field. Polymers that respond to temperature changes and, more specifically, those undergo a phase transition in aqueous solution are currently gaining special attention due to their potential applications in the biomaterials field. Most widely studied type of thermo responsive polymers are polymers that undergo a solution liquid–liquid phase transition. Thermosensitive polymers are polymers for which their solutions are characterized by a lower critical solution temperature (LCST). We are reported the formulation of thermoresponsive semi-IPN nanogels based on poly (N-vinylcaprolactam). The LCST can control by hydrophilic and biocompatible polymers such as polysaccharides. Here, we prepared semi-IPN of chitosan-poly (N-vinylcaprolactam) nanogels as drug carrier and LCST of a system determined near the human body.

Synthesis of PVCL

PNVCL was synthesized by controlled free radical polymerization in IPA solution. First, a mixture of VCL, TG as a chain transfer agent, and the solvent was added to a sealed flask. The oxygen was removed using a vacuum. Then AIBN was added to the reaction mixture. The reaction was carried out under nitrogen atmosphere at 70 °C. After the desired time, the polymerization was stopped and the product was purified by adding antisolvent N- Hexan and filtered by Teflon filter under vacuum condition.

PREPARATION OF BLANK THERMORESPONSIVE CHITOSAN POLY (NVINYLCAPROLACTAM) SEMI IPN NANOPARTICLES (TRI-NPS)

The poly N-vinylcaprolactam (PNVCL)was prepared as described previously. then 1 g of PNVCL was dissolved in acetic acid (1% solution) and then 0.06 g of carboxymethyl cellulose (CMC) was added to solution, and the resulting solution was crosslinked by aluminium sulfate octadecahydrate (Merck) in 1:9 ratio. the solution was under high-speed magnetic stirrer mixing about 800 rpm.

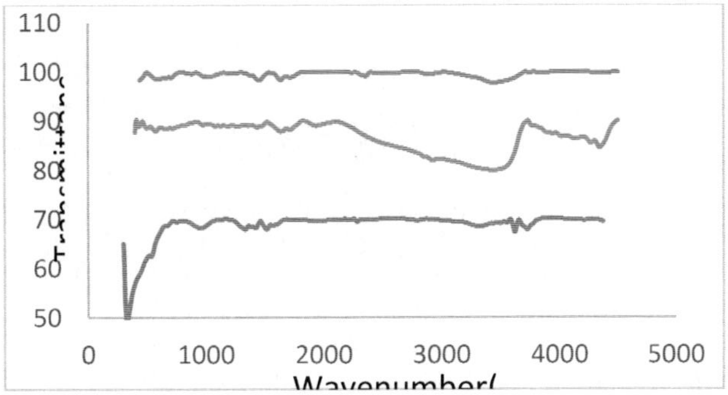

FIGURE 1. FTIR spectera of CMC, PNVCL, PNVCL/CMC semi Ipn

Second Level Heading (Heading 2) with Each Initial Letter Capitalized Capitalized (Note: prepositions and articles should be lowercase)

FTIR spectral studies:

FTIR spectroscopy has been applied to confirm the polymerization of NVCL and other prepared systems. The presence of the amide I band depends on the degree of the hydrogen bonding and physical state of the compound. FTIR spectra of polyvinycaprolactam and vinylcaprolactam monomer are shown in figure1 . a broadband at3441 cm is attributed to carboxyl groups in the synthesized polymer. The vibration at 2930 cm−1 and 1433 attributed to the aliphatic C-H and -ch2. stretching. the dip of 1717 and 1673 cm−1 are associated to C=O and amide I band vibrations. Fig. 2. demonstrated the FTIR spectra of crosslinked CMC, PNVCL and CMC-polyvinylcaprolactam semi IPN particles. The bands in the FTIR spectrum of CMC (Figure1) can be seen as follows: The amine peak at 1632 in crosslinked chitosan and 1637 in PNVCL were shifted to 1639 cm−1 in PNVCL-chi semi IPN due to the interaction of chitosan with PNVCL. these peaks are attributed to amide I band and carboxylic groups which are overlapped to each other[4]. In semi IPN sample two new peak appeared. The dip of 1082 is assigned to C–O–C that observed in PNVCL and PNVCL-CMC semi ipn sample. another new peak was 1320 corresponding to CH2 and OH bending vibrations that may due to interaction of ch2 group presence in PNVCL and OH group in CMC[5,6]. The stretching vibration bands at 3443 in CMC and 3441 in PNVCL spectra.

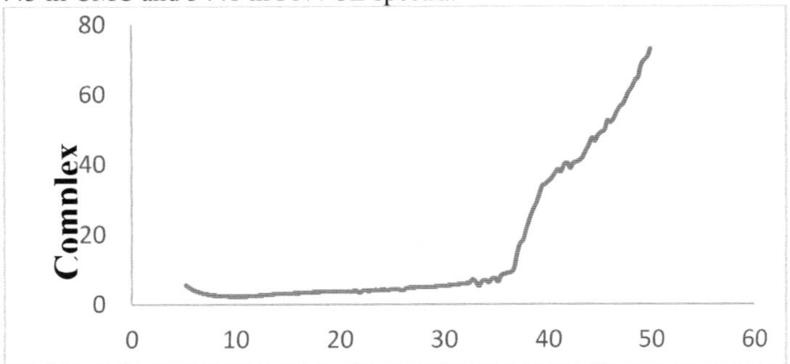

Rheological analysis

Rheology measurement was performed to evaluated the temperature-dependent viscoelastic properties of the thermoresponsive nanogels in a wide range of temperatures. Rheological behavior can help to explain LCST processes in nanogels by their performance in response to shear rate applied at various temperatures. Figure 2 display the results of a temperature sweep test. There is an evident increase in complex viscosity the storage at approximately at 37 ˚C for semi IPN of CMC-PNVCL nanogels.

Conclusion:

Poly N-vinyl caprolactam was synthesized with free radical polymerization. FTIR analysis was confirmed presence of vinyl caprolactam in CMC nanogels. Rheological analysis exhibited increasing in storage modulus around 37 c that represent LCST temprature of polymers. Semi IPN nanogels of chitosan caprolactam is suitable for smart local drug delivery system.

REFERENCES

1. M. P. Brown and K. Austin, *The New Physique*, Publisher City: Publisher Name, 2005, pp. 25-30.
2. M. P. Brown and K. Austin, *Appl. Phys. Letters* **85**, 2503-2504 (2004).
3. R. T. Wang, "Title of Chapter," in *Classic Physiques,* edited by R. B. Hamil, Publisher City: Publisher Name, 1999, pp. 212-213.
4. C. D. Smith and E. F. Jones, "Load-Cycling in Cubic Press" in *Shock Compression of Condensed Matter-2001*, edited by M. D. Furnish et al., AIP Conference Proceedings 620, American Institute of Physics, Melville, NY, 2002, pp. 651-654.
5. B. R. Jackson and T. Pitman, U.S. Patent No. 6,345,224 (8 July 2004)
6. D. L. Davids, "Recovery Effects in Binary Aluminum Alloys", Ph.D. Thesis, Harvard University, 1998.
7. R. C. Mikkelson (private communication).

1. Ottenbrite, Raphael M., and Sung Wan Kim, eds. *Polymeric drugs and drug delivery systems*. CRC Press, 2000.

2. Schmidt, A.M., 2007. Thermoresponsive magnetic colloids. *Colloid and Polymer Science*, *285*(9), pp.953-966.

3. Liu, R., Fraylich, M. and Saunders, B.R., 2009. Thermoresponsive copolymers: from fundamental studies to applications. *Colloid and Polymer Science*, *287*(6), pp.627-643.

4. Owens, D.E., Jian, Y., Fang, J.E., Slaughter, B.V., Chen, Y.H. and Peppas, N.A., 2007. Thermally responsive swelling properties of polyacrylamide/poly (acrylic acid) interpenetrating polymer network nanoparticles.*Macromolecules*, *40*(20), pp.7306-7310.

5. Patel, M.P., Patel, R.R. and Patel, J.K., 2010. Chitosan mediated targeted drug delivery system: a review. *Journal of Pharmacy & Pharmaceutical Sciences*, *13*(4), pp.536-557.

6. Kriwet, B., Walter, E. and Kissel, T., 1998. Synthesis of bioadhesive poly (acrylic acid) nano-and microparticles using an inverse emulsion polymerization method for the entrapment of hydrophilic drug candidates. *Journal of controlled release*, *56*(1), pp.149-158.

Impact of Electrosprayed PCL/Nano-Hydroxyapatite Particles on the Properties of Electrospun PCL Nanofibers

Pedram Karimipour-Fard[a]*, Ghaus Rizvi[a], Remon Pop-Iliev[a]

[a]*University of Ontario Institute of Technology - Ontario – Canada*

Abstract. Electrospun polycaprolactone (PCL) nanofibers are widely used in bioengineering applications. The main research objective of this paper is to conduct an experimental comparative study to assess the effect of electrosprayed PCL, nano-Hydroxyapatite (nHA), and PCL/nHA particles on the mechanical strength, morphology, and biodegradation-rate of electrospun PCL nanofibers. Electrospinning is utilized to process PCL nanofiber specimen scaffolds of various diameters out of three solutions of PCL with varying concentrations of 5%, 7.5% and 10%, respectively. To assess the effect of nanoparticle density and size on properties of the final samples, PCL particles are sprayed on nanofibers by using solutions of 1% to 5% concentrations and nHA is used in original particle sizes of less than 200 nm. Dimethylformamide (DMF) is used as a solvent for both the electrospinning and electrospraying processes. Phosphate buffered saline (PBS) medium together with 5% Carbon Dioxide gas in a CO2 incubator is used to simulate an in vitro environment for performing biodegradation tests. To assess the effects on mechanical strength, dynamic mechanical analysis (DMA) tensile test is performed and scanning electron microscope (SEM) is used to investigate the morphology evolution of the samples. The obtained experimental results indicate that it is feasible to manufacture an engineered electrospun tissue scaffold with controlled mechanical strength, morphology, and biodegradation-rate.

Keywords: Electrospinning, Electrospraying, Polycaprolactone, Hydroxyapatite, Nano fibers, Nano particles.

INTRODUNCTION

Through the recent decades electrospinning has attracted remarkable attention because of its ability to manufacture micro/nano fibers [1]. Electrospinning process involves a polymeric solution inside a syringe with metallic needle, a metallic collector and high voltage source. High voltage produces an electrostatic field that directs the polymers toward the collector. When high voltage is applied to flow of polymer solution, it changes the droplet on the head of the needle to a conical shape, called Taylor's cone which can be controlled to change electrospinning into electrospraying

Electrospinning and electrospraying have broad applications in manufacturing of polymeric tissue scaffolds because of their characteristics [1, 2]. They have the ability to control morphology of a scaffold consisting of micro/nano fibers and particles manufactured by using biocompatible and biodegradable polymers [4].

To characterize morphology of polymeric composites, there are many choices based on the size of prominent morphological components, including scanning electron microscopy (SEM) [5, 6], optical microscopy [5], microcomputed tomography [7, 8] and a few other techniques. Among these characterization methods, electrospun/electrosprayed composites in nano scale are characterized by SEM because of its precision up to lower than 20 nm [1, 5, 6].

Polycaprolactone (PCL) is a semi-crystalline aliphatic polymer with appropriate properties for tissue scaffolds including biocompatibility and great mechanical strength. PCL is compatible to electrospinning and electrospraying processes and due to its superior viscoelastic and rheological properties. There are numerous studies on [9, 10]using PCL micro/nano electrospun fibers or its composites with other materials as fibrous scaffolds [2, 5, 6]. Wu and Clark [3], studied the porous PCL particles microstructures manufactured by electrospraying to control the morphology of the product. Hydroxyapatite (HA) has been widely investigated in the literature as a biocompatible biomaterial to enhance mechanical properties of polymeric scaffolds and has proven its effectiveness [1, 4].

There are many studies on PCL electrospun fibers and on PCL electrospraying, but there is lack of research on impact of using electrospraying and electrospinning simultaneously on the final product. The main objective of this research is to use electrospinning and electrospraying simultaneously for multiple combinations of PCL solutions with different concentrations to assess the effect on morphology, mechanical properties and biodegradation rate of the samples. HA/PCL electrospraying is also used to demonstrate the effect of HA particles on properties of the electrospun PCL scaffolds.

EXPERIMENTAL

Materials

Polycaprolactone (PCL) (average Mn 80,000) pellets and Hydroxyapatite (HA) nanopowder (<200 nm particle size) from Sigma-Aldrich are used for electrospinning and electrospraying process. Dimethylformamide (DMF) is used as solvent for all of the solutions. An ultrasonic mixer is employed to prepare the solutions. Electrospinning solutions are stirred for 6 hours and electrospraying solutions are stirred for 4 hours in the mixer.

Electrospinning and Electrospraying

For the electrospinning and electrospraying processes a flat collector is usually used. The high voltage values applied for electrospinning and electrospraying processes are in the range of 8-10 kV. Three different concentrations of PCL are employed for each of the processes, 5%, 7.5% and 10% for electrospinning and 1%, 3%, 5% for electrospraying. 1% HA with 3% PCL is used to electrospray alongside electrospinning of 10% PCL solution for the last sample. All combinations of the experiments in this study are listed in table 1.

TABLE (1). Manufactured samples by electrospinning and electrospraying

Sample number	Electrospinning	Electrospraying
1	5% PCL	-----
2	7.5% PCL	-----
3	10% PCL	-----
4	10% PCL	1% PCL
5	10% PCL	3% PCL
6	10% PCL	5% PCL
7	10% PCL	3% PCL + 1% HA

Biodegradability

Biodegradation of the scaffolds are assessed by using Phosphate Buffer Saline (PBS) solution inside a CO_2 incubator. The samples are kept in the solution for 1 day and then they are washed and air dried in 40 °C for 4 hours; in the next step, they are weighted and placed back into the solution for 10 days and dried again using the same procedure. The difference in mass is reported as biodegradation rate.

Mechanical and Morphological Analysis

Dynamic mechanical analysis (DMA) is used for a comparative study of changes in tensile strength of the scaffolds by adding electrospraying process to electrospun samples. Film tension clamp (TA instrument) is used to mount the samples and carry out the stress-strain mechanical analysis of the samples (initial strain of 0.1%, 0.2% per min to 7% strain).

Morphology of the products are assessed by using scanning electron microscopy (SEM) to visualize and evaluate the changes in morphology when altering the manufacturing processes details. SEM is also employed to assess the impact of biodegradation on morphology of the samples.

RESULTS AND DISCUSSION

Mechanical Properties

Tensile test results of the fibrous samples 1-3 demonstrate that mechanical strength improves with increases in concentration of PCL, likely, because it leads to higher fiber diameters. By adding the electrospraying process for low PCL concentration, the sample 4, the mechanical strength is lower than sample 3 as young modulus is decreasing from 15.48 MPa to 10.19 MPa, probably because of the irregularities that small PCL particles make in the fibers thereby reducing its strength. For sample 5, the mechanical strength is almost the same as sample 3, likely due to larger and more numerous particles. For sample 7 the strength is significantly higher due to both larger size and numbers of particles and the reinforcement provided by HA, which has superior mechanical properties. Figure 1, shows the stress-strain carve results for samples 3, 4 and 7 mechanical properties.

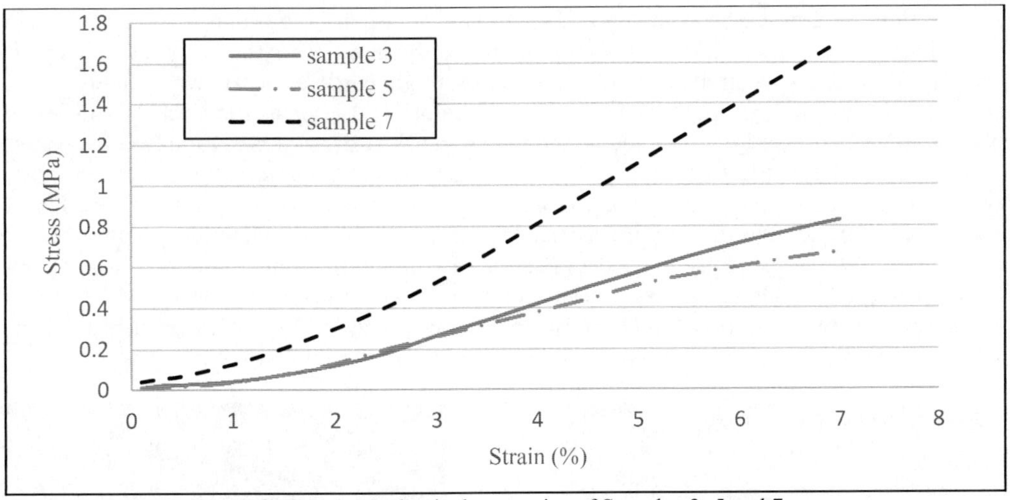

FIGURE 1. Mechanical properties of Samples 3, 5 and 7

Biodegradation Rate

Biodegradation rates of samples 1-7 are shown in figure 2 (the results are presented for two specimens per sample). For samples 1-3, biodegradation rate is decreased by increase in solutions PCL concentration (the resistance to biodegradation increases). The reason is increase in fibers diameter as PCL concentration increases in electrospinning solution that results in higher mechanical strength of the fibers. Sample 4 has highest degradation rate, which seems surprising. On closer inspection of the sample's morphology, it appears that nano particles in some areas divide the fibers, thus creating a weak connection between the fibers and particles. This can result in breakage of the connections due to biodegradation and expose the fibers to the active media. Comparison of samples 4-6, show a reduction in biodegradation rate on increasing the electrospraying solution concentration, possibly due to reduction in nano fibers percentage. Sample 7 demonstrates that addition of HA nano particles improve the resistance to biodegradation by 42% and it has the lowest biodegradation rate among the samples.

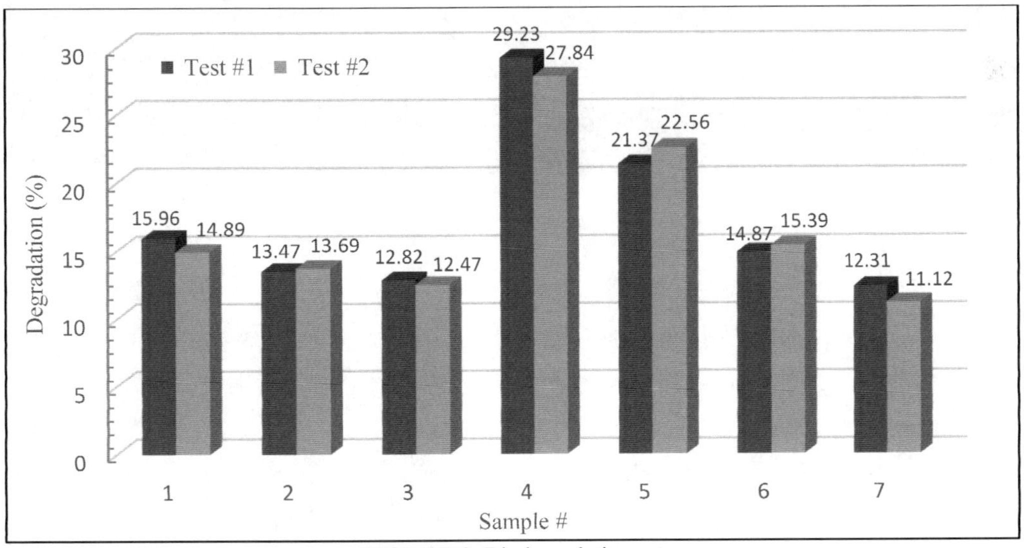

FIGURE 2. Biodegradation rate

Morphology

The specimen fabricated by electrospinning and electrospraying consist of nano fibers and particles. Due to the electrospraying process, the fiber diameters varied significantly throughout the samples and in places they are connected by particles that are accumulated at one point or by single particles, as shown in figure 3.

Figure 3 shows the morphology of samples 3, 5 and 7 before and after degradation in vitro environment. The structure of sample 3 is conserved under biodegradation and the fiber diameters are reduced by average of 30-50%, as shown in figure 3A and 3B. Electrospraying of PCL, increases biodegradation rate in low concentration solutions electrosprayed, but the resistance to biodegradation increases when the concentration increases. Figure 3C and 3D, show that the fiber connections to nano/micro particles are broken due to biodegradation and the fibers are more exposed to the media and consequently to biodegradation. HA particles enhance the resistance to biodegradability because of improvement in biodegradation properties of the fibers and improvement in mechanical properties that results in stronger bond between particles and nano fibers. Figure 3E and 3F show that the fibers are weakened because of biodegradation and their diameter is reduced, but their bond to the particles still exist.

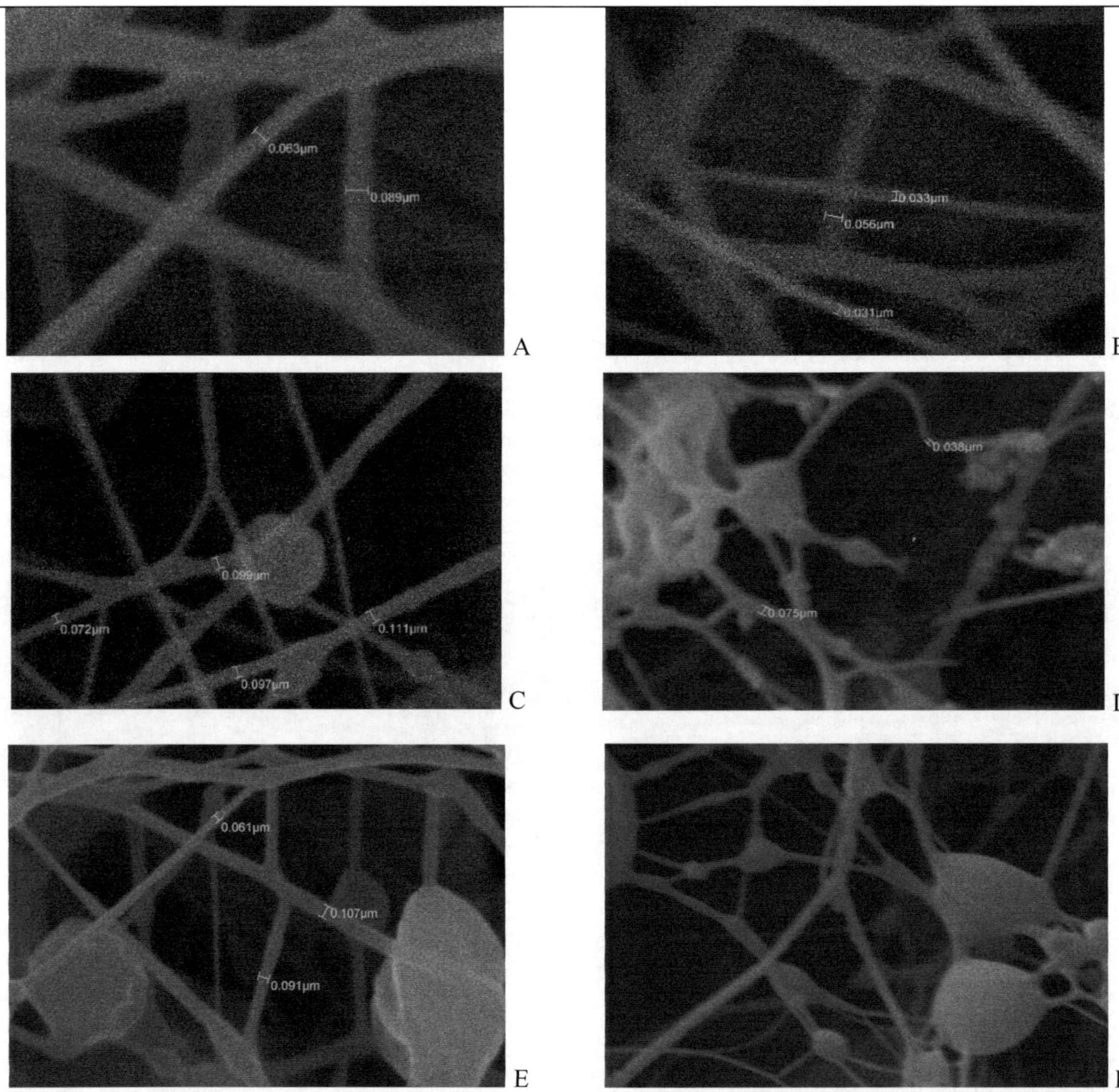

FIGURE 3. Morphology of samples 3, 5 and 7 (A: sample 3 before degradation, C: sample 5 before degradation, E: sample 7 before degradation, B: sample 3 after degradation, D: sample 5 after degradation, F: sample 7 after degradation.)

CONCLUSION

The aim of this study is to assess the effect of electrospraying PCL/HA blend simultaneously with electrospinning of PCL fibers, on the final products properties. To achieve this goal, multiple samples with different concentrations are manufactured. DMA, SEM and In-Vitro simulation environment are utilized to evaluate the impact of manufacturing processes on mechanical properties, morphology and biodegradation characteristics of the samples.

The results of this research are summarized as follow:

- PCL/PCL-HA sample shows superior mechanical properties due to addition of HA particles to PCL electrosprayed/electrospun samples.
- Biodegradation rate is increased by electrospraying process with the same material (PCL) that means the resistant to biodegradation was reduced. This result is important for controlling the biodegradation rate for specific applications.
- HA because of its inherent superior mechanical strength and low biodegradation rate property, increased the resistance to biodegradation.
- Electrospraying/electrospinning samples morphology are not stable under biodegradation for PCL/PCL samples and is more stable in PCL/PCL-HA sample.
- Fibers dimeter, percentage of electrosprayed particles and strength of the connection between particles and nano fibers, play the most important role in biodegradation and mechanical properties of the samples.

REFRENCES

1. Deng, Y. and J. Kuiper, *Functional 3D tissue engineering scaffolds: materials, technologies, and applications*, Woodhead Publishing, 2017.
2. Q. P. Pham, U. Sharma, and A.G. Mikos, *Tissue engineering* **12**(5), 1197-1211 (2006).
3. Y. Wu and R.L. Clark, *Journal of colloid and interface science* **310**(2), 529-535 (2007).
4. Y. F. Goh, I. Shakir, and R. Hussain, *Journal of Materials Science* **48**(8), 3027-3054 (2013).
5. Y. Zhang et al., *Journal of Biomedical Materials Research Part B: Applied Biomaterials* **72**(1), 156-165 (2005).
6. E. Chong et al., *Acta biomaterialia* **3**(3), 321-330 (2007).
7. P. Karimipour-Fard et al., "The Use of Microcomputed Tomography to Evaluate Integral-Skin Cellular Polyolefin Composites," in *Abstract Book; Unilever Res Dev Netherlands: Vlaardingen, The Netherlands,* 2018.
8. R. Pop-Iliev et al., *Radiation and applications* **3**(2), 143-146 (2018).
9. M. A. Woodruff and D. W. Hutmacher, *Progress in polymer science* **35**(10), 1217-1256 (2010).
10. R. Ghobeira et al., *Polymer* **157**, 19-31 (2018).

Evaluation of the Properties of Corn Cob Hemicellulose Based Biodegradable Films Obtained by Extrusion and Solvent Casting Methods

Büsra Akınalan Balık[a], Necati Özkan[b], Ufuk Bölükbaşı[c*]

[a]*Department of Food Engineering, Yeditepe University, 34755, Atasehir, Istanbul, Turkey*
[b]*Department of Polymer Science and Technology, Middle East Technical University, 06800 Çankaya, Ankara, Turkey*
[c]*Department of Chemical Engineering, Middle East Technical University, 06800 Çankaya, Ankara, Turkey*

Abstract. The remarkable consumption and environmental concerns related to the utilization of fossil based resources have led researchers to focus on renewable raw materials. Lignocellulosic biomass, the fibrous part of plants with a complex matrix comprising different polysaccharides, mainly cellulose, hemicellulose and lignin, is the most abundant, inexpensive, non-starch renewable source in the world. In the present study, hemicellulose, which was obtained from corn cobs by alkali extraction method (with a yield of 30%), was used to produce biodegradable film by extrusion followed by rolling process and solvent casting techniques. The morphological, mechanical and thermal properties were investigated within these two techniques while the thicknesses of films were comparable. The effect of extrusion temperature (75, 90°C) and screw speed (50, 100 rpm) on film properties were also investigated. Homogenous and crack free films were obtained by extrusion for all conditions. Discontinuous film formation was observed by solvent casted method even they were observed as homogenous and crack free via scanning electron microscopy. The onsets of thermal degradation temperature for solvent casted and extruded films were determined via thermal gravimetric analysis at 300°C. Increasing the extrusion temperature had a slightly positive effect on the mechanical properties of films, whereas screw speed did not have a noticeable effect ($p>0.05$). While the toughness of the extruded films was 13 MJ/m^3, that of the solvent casted films was 5 MJ/m3. The elongation at break values of the extruded and solvent casted films were 28% and 11%, respectively.

Keywords: Hemicellulose, biodegradable polymers, extrusion, mechanical properties
PACS: Replace this text with PACS numbers; choose from this list: http://www.aip.org/pacs/index.html

INTRODUCTION

Lignocellulosic biomass is the fibrous, non-edible and non-starch part of plants with a complex structure comprising mainly cellulose (40-50%), hemicellulose (25-30%) and lignin (15-20%).[1] Hemicellulose is a heterogeneous polysaccharide and composed of glucuronic acids and different sugar monomers including xylose, galactose, glucose and arabinose.[2] In the recent years, biodegradable film and coating production from hemicelluloses receive significant interest, particularly due to the good oxygen barrier properties of hemicellulose based polymeric materials, which makes them suitable for food packaging applications.[3,4]

Film production technology influences the film properties and quality. Extrusion is a more efficient, lower energy and time required film manufacturing method compared to solvent casting. Previously, the extrusion of hemicellulose was illustrated successfully in our laboratory.[5,6] The aim of this study is to investigate the effect of extrusion conditions in terms of temperature and rotational speed on the mechanical properties of corn cob based hemicellulose films produced by extrusion followed by rolling and to compare the results with the films obtained using traditional solvent casting method. Also, thermal and morphological properties of the films were analyzed.

MATERIALS & METHODS

Corn cobs were obtained from Hatay and were milled to pass a 10 mesh (2 mm) screen, then, the particles between sizes of 16 to 100 mesh (0.15 mm to 1.19 mm) were selected. Hemicellulose was isolated from the milled corn cobs according to the method given in the study of Zilliox and Debeire (1998).[7] Following the extraction, hemicellulose was desalted by washing three times with ethanol-water mixture (3:1).

For producing solvent casted films, 1 gram of extracted hemicellulose was dissolved in 15 ml of de-ionized water and left for stirring on a magnetic stirrer for 24 hours at room temperature. Following this step, the suspension was ultrasonificated to maintain better dispersion of solid particles. Then, the solution was poured into a petri dish and left to dry in a conditioning cabinet (Medcenter 111 Climacell, MMM Group, Munich, Germany) at 23°C and 50% relative humidity for 2 days.

A co-rotating, conical twin-screw (Diameter: 4-15 mm; Length: 109.4 mm) extruder (Thermo HAAKE Mini CTW) with two heating zones was used for the extrusion of corn cob hemicellulose. Temperature and rotational speed of screws were controlled digitally. Each time 5 grams of polymer were fed to produce nearly 1.5 meter of strip. Thickness of extruded strips was rendered similar to the solvent casted films by rolling through a roller (Yilmaz Machine Industry, Turkey). Before rolling, strips were held in 90% relative humidity containing desiccators for 1 day.

Mechanical properties of films were analyzed by using the tensile testing equipment. Prior to tensile testing, films were cut as dog bone shape by using a cutting press (ZCP 020, ZwickGmbH&Co., Ulm, Germany) and conditioned at 23°C and 50% relative humidity containing cabinet for 24 hours before testing. Thermal characterizations of the films were investigated using thermogravimetric analysis (TGA) equipment (Pyris 1, Perkin Elmer). Film samples were heated from room temperature to 950°C, at a heating rate of 10°C/min with nitrogen flushing to obtain thermogravimetric curves. A scanning electron microscopy (QUANTA 400F Field Emission SEM) was used to investigate cross sectional and surface images of the films.

RESULTS & DISCUSSION

Hemicellulose was extracted from corn cobs at a yield of 30% (w/w) based on dry weight. The hemicellulose based films were produced successfully by extrusion and followed by rolling steps. The effects of extrusion temperature and screw speed on mechanical properties of films were analyzed. Also, the solvent casted films were prepared for comparison of morphological, thermal and mechanical properties.

Effects of Extrusion Temperature and Speed on the Mechanical Properties of Corn Cob Hemicellulose Based Films

The effects of extrusion temperature and rotational speed on the mechanical properties of films were given in **TABLE 1**. The influences of the extrusion temperature or extrusion speed (rpm) on the Young's modulus (E) and ultimate tensile strength (UTS) of the films were insignificant ($p>0.05$). Approximate E and UTS values of these films were 1120 MPa and 52 MPa, respectively. A significant difference was observed between flexibilities and toughness of films that are extruded at different temperatures ($p<005$). During extrusion, the hemicellulose was shaped under heat and shear. When the extrusion temperature was increased, more heat was supplied to hemicellulose granules. Thus, the strips extruded at 90°C became more uniform than the strips extruded at 75°C. Also, defects formed on the strip surfaces decreased when the extrusion temperature is increased. Thus, the films with higher elongation at break values were produced using higher extrusion temperature for both the extrusion screw speeds of 50 and 100 rpm. As a result, the toughness of the films extruded at 90°C (approximately 13 MJ/m^3) was higher than the films which were extruded at 75°C (approximately 9.6 MJ/m^3) which means that extrusion of hemicellulose at 90°C is more preferable than 75°C.

TABLE 1. The effect of extrusion temperature and rotational speed on the mechanical properties of films.

Extrusion Parameters	Modulus (MPa)	Ultimate Tensile Strength (MPa)	Elongation at Break (%)	Toughness (MJ/m^3)
T=75°C, Speed=50 rpm	1095±51	50±3	20±2	8.9±1.3
T=75°C, Speed=100 rpm	1106±91	52±2	21±3	9.5±1.5
T=90°C, Speed=50 rpm	1085±94	52±3	24±2	10.9±1.3
T=90°C, Speed=100 rpm	1064±34	52±2	27±3	12.4±1.4

Comparison of the Extruded and Solvent Casted Films

Morphology

Morphological properties of the extruded films were investigated by SEM analysis as given in **FIGURE 1**. The cross sectional areas of the hemicellulose films exhibited a relatively homogenous, smooth and compact structure. The surfaces of all hemicellulose based films also appear homogenous. Parallel lines on the surfaces of the extruded films were observed which can be related with the rolling process because of the rough surface of the roller.

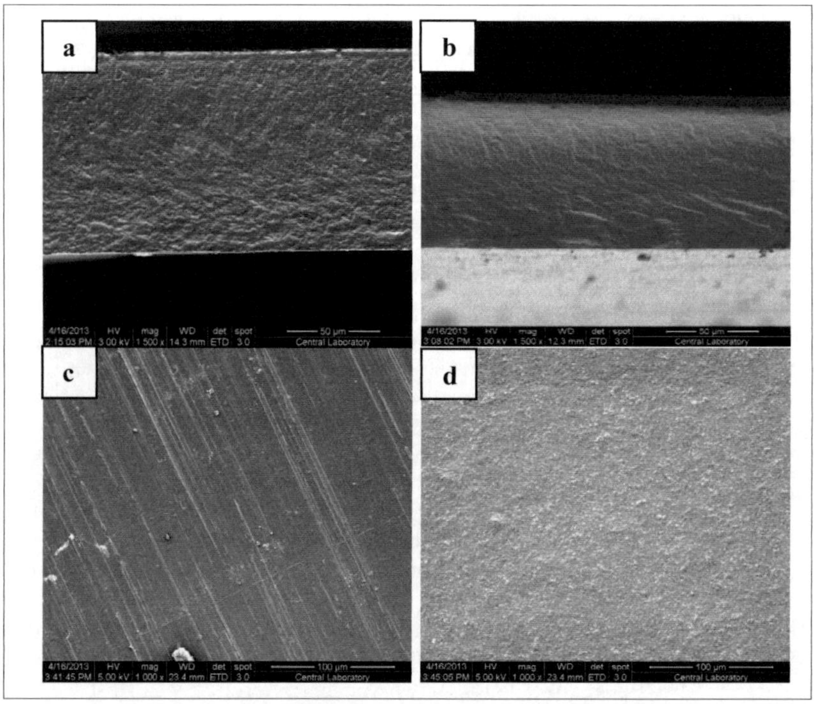

FIGURE 1. The cross sectional SEM images of (a) extruded, (b) solvent casted films and the surface SEM images of (c) extruded (d) solvent casted films.

Mechanical Properties

The mechanical properties of the extruded and solvent casted films were given in **TABLE 2**. Films were extruded at 90°C and 50 rpm. The elongation at break and toughness of the extruded films were more than two times higher than those of the solvent casted films. While the extruded films had a toughness of 13 MJ/m^3, the solvent casted films only had 5 MJ/m^3.

TABLE 2. The mechanical properties of solvent casted and extruded films.

Method	Modulus (MPa)	Ultimate Tensile Strength (MPa)	Elongation at Break (%)	Toughness (MJ/m^3)	Thickness (μm)
Solvent Casting	1306±300	60±9	11±2	5±0.5	107±18
Extrusion	1120±35	52±1	28±3	13±1.6	114±12

Thermal Properties

The TGA curves of extruded and solvent casted hemicellulose films were given in **FIGURE 2**. There was no difference on the degradation onset temperatures of the films produced by the two methods, which corresponds to approximately 300°C. Obviously the choice of film production method did not influence the thermal properties of films.

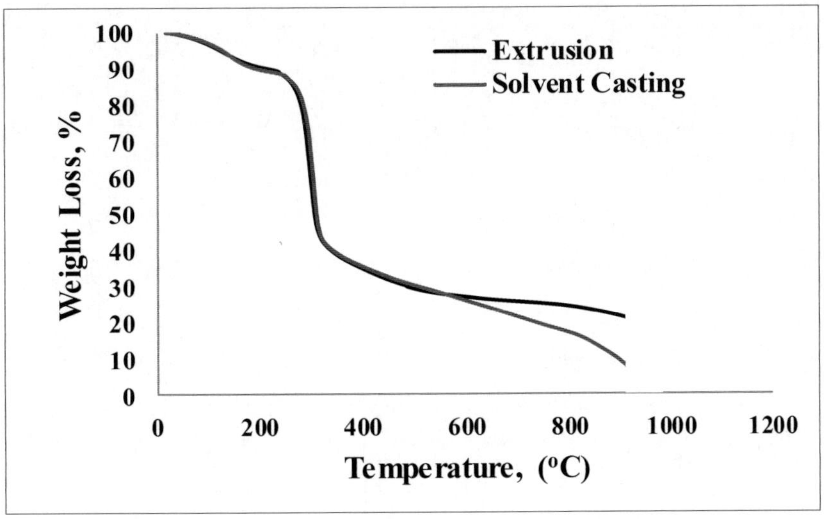

FIGURE 2. TGA curves of corn cob hemicellulose based films obtained via extrusion and solvent casting.

CONCLUSION

Hemicellulose was isolated from corn cob and hemicellulose based films were produced via extrusion and solvent casting with a thickness of approximately 100 μm. Considering the mechanical properties of the films, the extrusion temperature of 90°C was determined as a slightly better extrusion temperature, whereas no significant effect of the extrusion screw speed was observed. The extrusion technique was determined as a better production method than that of the solvent casting, since the films produced using the extrusion method had better mechanical properties especially the toughness of the extruded films (13 MJ/m^3). The onsets of thermal degradation temperatures for the solvent casted and extruded films were determined as 300°C using TGA.

For further studies, additional characterizations can be performed including biodegradability, oxygen permeability or water vapor transmission of films.

ACKNOWLEDGMENTS

The present study was supported by The Scientific and Technological Research Council of Turkey (TÜBİTAK) via grant number 112M353. The authors would like to thank to METU Central Laboratory for assistance during SEM analysis and mechanical testing.

REFERENCES

1. M. Knauf and M. Moniruzzaman, *International Sugar Journal* **106**, 147-150 (2004).
2. N. M. Hansen and D. Plackett, *Biomacromolecules* **9**, 1493-1505 (2008).
3. Z. Li and X. Pan, *Rev Environ Sci Biotechnol* **17**, 47-69 (2018).

4. S. Sousa et al., *Industrial Crops and Bioproducts* **94**, 682-689 (2016).
5. E. Bahcegul et al., *Bioresource technology* **149**, 582-585 (2013).
6. M. Akkus et al., *RSC Advances* **4(107)**, 62295-62300 (2014).
7. C. Zilliox and P. Debeire, *Enzyme and Microbial Technology* **22(1)**, 58-63 (1998).

Dual Responsive Nanogels Based on Chitosan-NIPAM Decorated with Au/Fe_3O_4 Nanoparticles as a Carrier for Controlled Drug Release

Mina Bagherifard[a], Mohadeseh Doroudian[a] and Ali Pourjavadi[a *]

[a]*Polymer laboratory of chemistry department, Sharif university of technology, Tehran, Iran*

Abstract. Drug delivery by smart carrier systems is one of the interesting issues for scientists to reduce side effects of drugs during chemotherapy for diseases like cancer and also drug delivery for the remote and difficult regions. External stimuli are more controllable and convenient compared to internal stimuli. In this study, light beam with high accuracy, lower response time and practical control with magnetic field are used as external stimuli. Light can provoke a drug carrier, which is sensitive to the light due to the presence of gold nanoparticles (GNPs). Light irradiation warms up the drug carrier through SPR effect of GNPs and subsequently drug releases. Various tests such as SEM, DLS, FTIR, and XRD were used to confirm successful synthesis of the Cs-NIPAM-GNPs-Fe_3O_4. Drug release experiments were monitored by UV-Vis spectroscopy. Results have shown that nanocarrier with average diameter of 100 nm has been synthesized. Drug release in different conditions was done, however the best result have been achieved in the presence of green light.

Keywords: Drug delivery, Chitosan, Gold nanoparticles, Smart carriers, Thermosensitive

PACS: S22-197

INTRODUCTION

One of the most important concerns in the world related to prevalence of cancer and the most effective way to remove abnormal cells from body is surgery which usually after it patients must do chemotherapy or radiotherapy. Both of these treatment destroy abnormal cells as well as normal one. Immune cells and Subsequently, body's defense against disease are reduced. In addition, several side effects could occur. [(1)]

Recently, for decreasing these disadvantages of chemotherapy, different kinds of nanocarrier have been used such as micelle, liposomes, nanogels, nanoparticles and etc. these nanocarriers are loaded with drugs and because of small size of them, they are able to go into cells.[(2)] For targeted delivery, these nanocarriers must response to special stimulus like pH, temperature, light, magnetic field or any other types which this behavior reduce side effects of chemotherapy about normal cells and improve ability of drugs for curing abnormal cells at these areas.[(3)] Among of these stimuli, external ones which prove nanocarriers to release drug by remote controlling, are more suitable, comfortable and controllable than others.[(4)] Because of using nanocarriers in the body, they must be biodegradable, biocompatible and have stability in the blood circulation system. For this purpose, polysaccharides are candidate to utilize as nanocarrier but they must be functionalized to response about special stimulus. Some of these polysaccharides like Chitosan could be functionalized easily by different ways. [(5)]

In this study, dual responsive nanogel by using gold nanoparticles (GNPs) and magnetic nanoparticles (MNPs) in thermosensitive cross-linked chitosan-g-NIPAM was fabricated. Both of these nanoparticles could response to external stimuli. The important feature of GNPs related to surface plasmon resonance (SPR) of them. According to shape and size of GNPs, they can absorb light with different wavelength. For this purpose, GNPs which were able to response to light with 532nm wavelength were synthesized by a green method. [(6)] Enhancement in SPR for GNPs could occur under light with 532 nm and create heat. NIPAM as thermosensitive monomer was grafted to Chitosan and when green light was used, due to existence of GNPs which were able to create heat, nanogel shrunk and drug released. MNPs could improve drug release into targeted area.

EXPERIMENTAL

Materials

Chitosan (medium molecular weight, CAS number: 9012-76-4); N-isopropyl acryl amide, 97%, from Sigma Aldrich; HAuCl4 (99.5%), acetic acid (96%), $FeCl_2.4H_2O$, $FeCl_3.6H_2O$, Ammonium hydroxide solution (25% W/W), Polyethylene glycol-600 (PEG), N, N-methylene bisacrylamide, ammonium persulfate, acetone and ethanol all of them from Merck; Sodium metabisulfite, food grade, chemical company in Iran; deionized water; Drug: azithromycin.

Preparation of Chitosan-GNPs

GNPs were synthesized by method of Pestova et al. (2015); 0.30 g of chitosan was dissolved in 100 mL of acetic acid solution (4%) at 70 °C for 1 hour. Then, 6 mL of Au(III) solution (1000 ppm) was added and stirred for 6 hours.

Preparation of MNPs (Fe_3O_4)

1.99 g $FeCl_2.4H_2O$, 5.41 g $FeCl_3.6H_2O$ and 1 g PEG-600 were added into 50 Ml distilled water and Ammonium hydroxide solution (25% W/W) was added drop by drop until pH=11 under stirrer at 50 °C. after 1-hour sediment was gathered by magnetic field and washed by distilled water two times.

Synthesis of Nanogels

Fe_3O_4 (2% weight ratio of nanogel) was dispersed by ultra sound bath into Chitosan-GNPs solution. 0.30 g NIPAM and 0.123 g MBA were dissolved into it and degassed by N_2 for 30 minutes. The mixture was placed in water bath at 60 °C and 4mL APS and 4mL SBS fresh solution were added and reaction was kept for 12 hours. Product was centrifuged and washed with DI-water two times.

Nanogels Characterization

Fourier-transform infrared (FTIR) spectra were measured by ABB Bomem (MB-100 series). X-ray diffraction (XRD) analysis was achieved by X'Pert PRO MPD (PANalytical Company). The hydrodynamic diameter of nanoparticles in water was determined by Dynamic Light Scattering method (DLS) using a Nano S (red badg-632.8 nm, Malvern, Inc.) analyzer at room temperature. UV–vis spectra were recorded by Perkin-Elmer (model Lamba 2S) at ambient temperature. The morphology of nanogel was investigated by Field Emission Scanning Electron Microscope (FESEM: TeScan – Mira III).

Drug loading, Drug content and Drug release conditions

Nanocarrier was stirred in 15 mL of saturated solution of drug for 24 hours and then drug loaded nanogel were centrifuged and rinsed with phosphate buffer solution (pH=7.4) three times.

0.2 g of nanogel was stirred in 10 mL of ethanol for 24 hours, then was centrifuged and the amount of loaded drug was determined by using UV-Vis spectroscopy.

0.2 g of the drug-loaded nanogel was dissolved in 5 mL of buffer solution and poured in dialyzing bag against of 10 mL buffer solution into water bath (37°C - 45 °C- under green light). For drug release under green light, temperature was kept at 37 °C. At the determined times, the outer buffer solution was analyzed by UV-Vis spectroscopy.

RESULTS AND DISCUSSION

The Composition Identification and Morphology of Nanogels

FTIR was applied to confirm successful synthesis of nanogels. Spectra of NIPAM. Chitosan and nanogels were shown in figure 1.

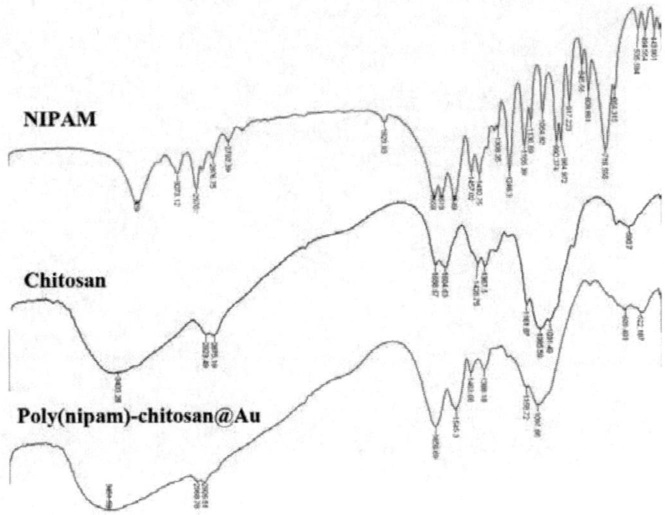

FIGURE 1. FTIR spectra of NIPAM, Chitosan and nanogel

To confirm successful synthesis of GNPs and MNPs, XRD test was used. The XRD pattern (figure 2) clearly shows the existence of chitosan (at ~ 11.0° and 21.0°), PNIPAM (two broad diffraction peaks at 2θ=7.5° and 2θ=20°) and the crystalline nature of Au nanoparticles (from (111), (200) and (220) planes) and Fe_3O_4 nanoparticles (from (220), (311), (400), (422), (511) and (440)).

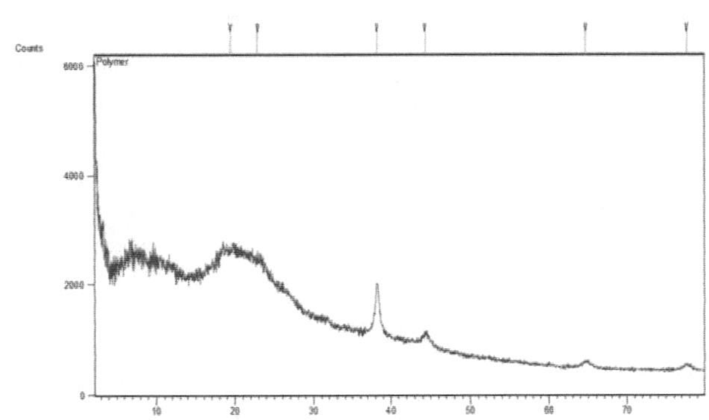

FIGURE 2. XRD pattern of nanogels with GNPs (left) and MNPs (right)

For investigating size and surface morphology of nanogels, SEM was applied. According to image in figure 3 the average size of nanogels is around 100 nm that is appropriate for cellular uptake drug delivery.

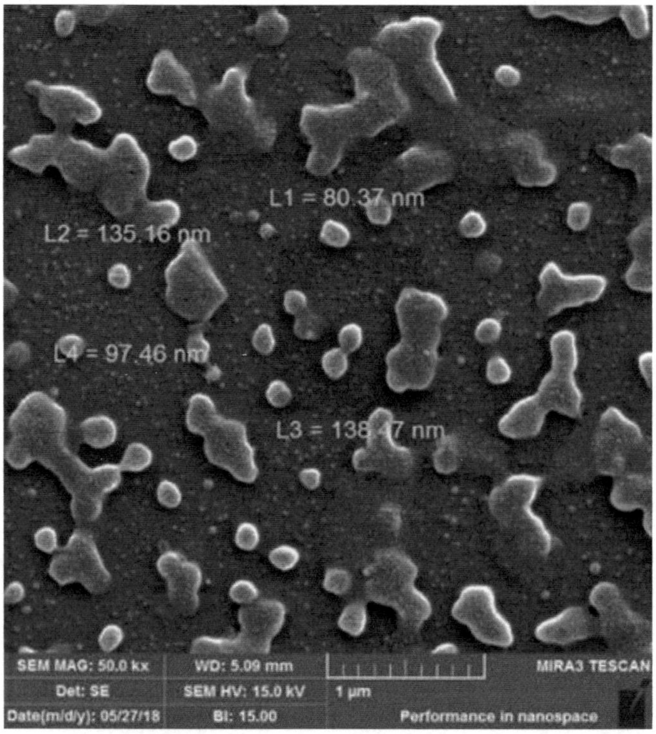

FIGURE 3. SEM images of nanogels.

The size distribution and hydrodynamic diameter of nanogels were determined by DLS. The diagram of size of Nanogel by percent number shown in figure 4. Two maximum peak in 115 and 338 nm related to distinct Nanogels and aggregation of some polymeric nanogels, respectively.

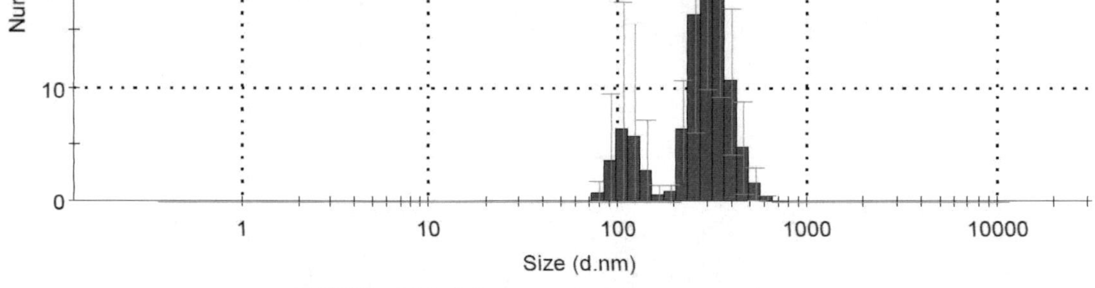

FIGURE 4. DLA diagram, size by percent number.

Drug Release Profile

Figure 5 shown drug release profile at the three conditions (37ºC - 45 ºC- under green light). The best result related to green light condition. Results of 37ºC related to drug releases under physiological condition in the body, results of 45 °C mimics minimum amount of heat that GNPs could create by SPR and the green light condition related to use of light (532nm) at the biological condition.

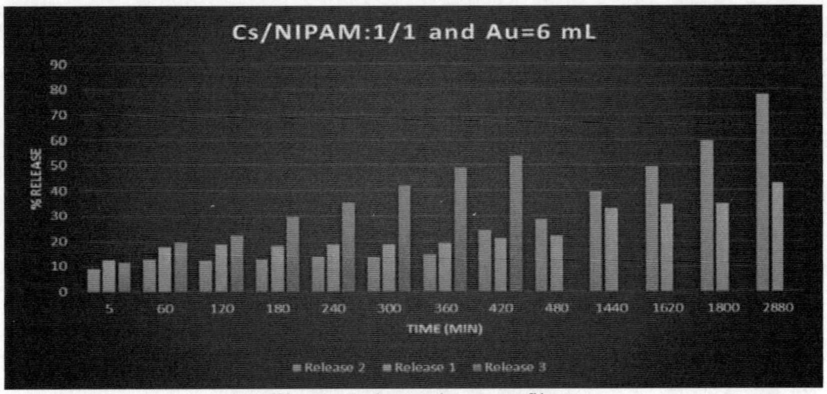

Figure 5. drug release profile.

CONCLUSIONS

In this article, by using Chitosan and NIPAM, dual responsive nanogels that decorated with Au/Fe_3O_4 nanoparticles, were fabricated. Nanogels could response to green light and magnetic field as external stimuli. Drug loaded and drug releases were investigated by using azithromycin as drug model. The results shown controlled drug release under green light.

REFERENCES

1. H. Chen and Y. Zhao, *ACS Appl. Mater. Interfaces* **10**, 21021–21034 (2018).
2. S. Jia et al., *Chem. Mater* **30**, 2873–2887 (2018).
3. S. Chen et al., *Polym. Chem* **8**, 6150–6157 (2017).
4. C. Hang et al., *Colloids and Surfaces B: Biointerfaces* **158**, 547–555 (2017)
5. A. Khan et al., *Iran Polym J* **24**, 317–328 (2015).
6. L. Sun et al., *Carbohydrate Polymers* **178**, 105-114 (2017).

Influence of Composition, Uni/Biaxial Stretching and Constrained Annealing on the Evolution of Structure and Dielectric Properties of PEI/PBT Blends

Zeynep Mutlu[1, 3, b)], Mukerrem Cakmak [1, 2, 3, a)]

[1]*School of Materials Engineering, Purdue University, West Lafayette, Indiana*
[2]*School of Mechanical Engineering, Purdue University, West Lafayette, ,Indiana*
[3]*Birck Nanotechnology Center, Purdue University, West Lafayette, Indiana*

[a)]Corresponding author: mcakmak@purdue.edu
[b)]zmutlu@purdue.edu

Abstract. Understanding of how processing of polymer materials affect the dielectric properties like dielectric loss and breakdown is of great importance for a wide range of applications for electronics. The aim of this work is to study structural changes that occur during stretching and heat setting by blending PEI and PBT. It has been found out that with the addition of the PEI, the blends become quenchable into glassy state from melt and cold crystallization peak of PBT moves to higher temperatures with increase of PEI content due to the dilution effect. Stretching of 70/30 sample material shows rubber like deformation behavior where birefringence increases linearly with the increase of stress. Heat setting of this material shows increase of crystallinity in crystalline domains.

INTRODUCTION

Improvements in energy storage systems is critical for upcoming different applications like electromagnetic rail guns, electric/hybrid vehicles electromagnetic armor and more. Capacitors are one of the popular areas in energy storage systems due to their high ability to transfer energy and short charge time[1]. Polymers like biaxially oriented polypropylene (BOPP), polyethylene terephthalate (PET) and polyether imide (PEI) are widely used in capacitor applications due to their excellent properties like low cost and high dielectric constant.

Conventional polymer capacitors are made from thin polymer films that are either melt cast or solution processed. During melt casting, tenter frame machines are used where polymer films are subjected to multi- stage process, such as uni/biaxial stretching followed by heat setting that would boost crystallinity while relaxing amorphous orientations ultimately making the material more thermally stable. Processing in each of these stages has a critical impact on evolution of structural hierarchy and final polymer performance. Due to the lack of lab scale machines, mechano-optical and structural behaviors of these films in each step have not yet been completely understood. Here we investigate the effect of PBT on the development of crystalline orientation of PEI/PBT blends. In this presentation , we primarily focus on the kinetics of structural evolution in real time birefringence measurements during constrained annealing of films that underwent uniaxial constrained stretching.

MATERIALS

Poly(ether imide) (Ultem 1000) was kindly provided by Sabic company in pellet form. Its molecular weight was about 61,000 (Mw), (Mn=32,000). Poly(butylene terephthalate) (Crastin FG6129 NC010) was purchased from

Polyone and used as received. Materials were dried in vacuum oven at 150⁰C for at least 12 hours to eliminate moisture before performing blending, thermal analysis and compression molding processes. Co-rotating Euro Lab twin screw extruder was used to melt blend PEI and PBT. Screw diameter was 15.6 mm and the length of the screw is 50cm. Screw speed kept constant at 150 rpm. Different concentrations of PEI/PBT blends were made (wt. %) : 80/20, 70/30, 60/40, 50/50 and 30/70.

Uniaxial free width samples were compression molded by using 6" x 8" x 0.86" mold by using hydraulic press and then cut into a dog-bone shape. Uniaxial constrained width samples were compression molded without mold by using hydraulic press and cut as a 14.5cmx14.5cm square samples.

EXPERIMENTAL

DSC experiments were performed using Netzsch Instruments, DSC Polyma, under nitrogen atmosphere. Samples were heated from room temperature to 280⁰C at the heating rate of 10⁰C/ min to understand their thermal behaviors. DSC heating curves of melt-quenched PEI/PBT films, prepared by compression molding are shown in Fig. 1. All the compositions have a single Tg, and it shifts to higher temperatures with the increase in PEI fraction, as quantitively

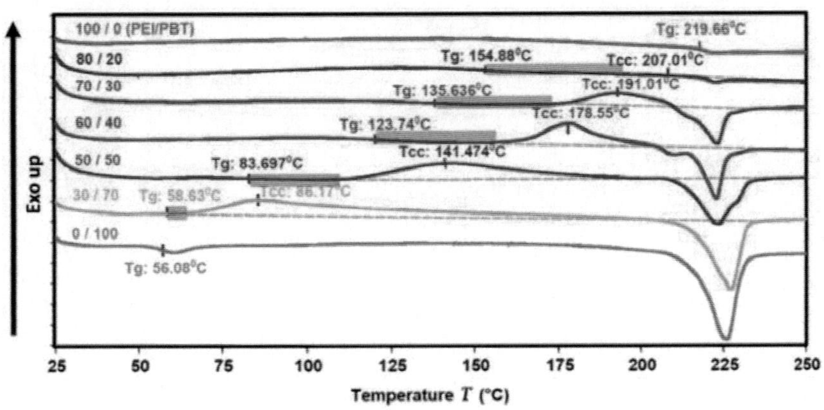

FIGURE 1. DSC thermograms of PEI/PBT blends

presented in Fig.1 . At temperatures above Tg, PBT starts to crystallize at 170⁰C and melting begins to occur at 200⁰C. With the addition of the PEI, cold crystallization peak of PBT moves to higher temperatures due to the dilution effect by the increase of amorphous PEI chain concentrations. This dilution will reduce the thermodynamic driving force of the PEI width and height ratio of the melting peak of the blends. This is an indication of a detrimental change in the crystalline domains. In the PEI/PBT blends, increasement in Tcc is higher than the increase in Tg with the increase of PEI concentration which leads to the widening of processing window (Tcc-Tg) as it shown in Fig. 1.

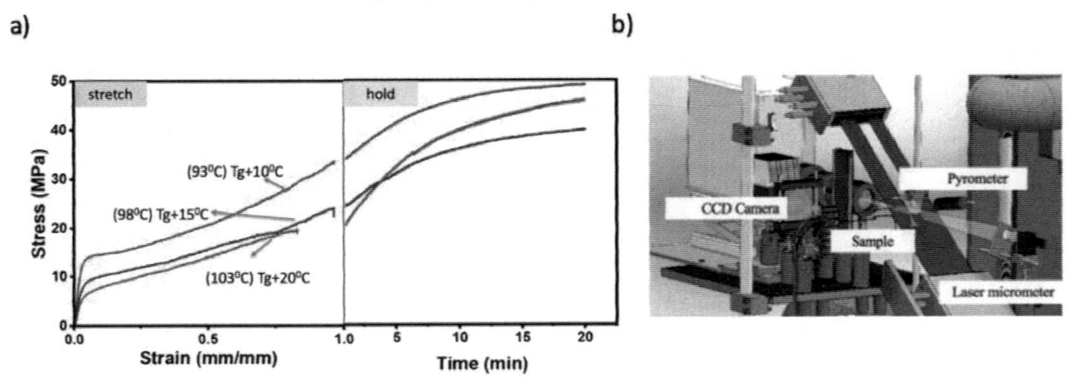

FIGURE 2. (a) Stress-strain curve of 50/50 (PEI/PBT) sample in different temperatures and stress time curve (20 minute holding) **(b)** custom built uniaxial mechano-optical stretching device

To investigate the effect of stretching temperature on the mechano-optical behavior of PEI/PBT blends, cast films were stretched above their respective glass transitions temperatures, where thermal crystallization was suppressed and material was still in rubbery state. Dog-bone shaped 70/30 blends were cut from the compression molded films and placed in between the clams for mechano-optical testing device which is designed by our group (Fig 2b). This system tracks simultaneously optical retardation Γ_t values and continuously records sample width W_t and force F_t during stretching. which allows real-time measurement of the cross sectional area during the process[2]. By assuming (a) simple uniaxial extension and (b) incompressibility real time thickness D_t is calculated by using;

$$D_t = \left(\frac{W_t}{W_0}\right) D_0 \tag{1}$$

where W_o is initial sample width, W_t is real time sample width and D_0 is the initial sample thickness. True stress, true strain and real time birefringence is calculated by using

$$\Delta n_t = \frac{\Gamma_t}{D_t} \tag{2}$$

$$\text{true strain} = \frac{\text{elongation}}{\text{initial length}} = \frac{L_t}{L_0} - 1 = \left(\frac{W_0}{W_t}\right)^2 - 1 \tag{3}$$

$$\text{true stress} = \frac{F_t}{W_t D_t} = \frac{F_t}{\left(\frac{W_t^2}{W_0}\right) D_0} \tag{4}$$

where Γ_t is real time optical retardation and D_t is sample thickness. L_t, L_o, F_t are sample length, initial sample length and real time force respectively. These samples were uniaxially stretched in their rubbery state at temperatures above T_g with a constant speed of 20 mm/min. As it is shown in Fig 2a. at the temperatures near Tg, strain hardening was observed at the end of the plastic region, indicated by a steep rise in true stress with true strain. When temperature is increased the onset of strain hardening suppressed and moves to higher strains. At temperatures close to the glass transition temperature the birefringence-stress curves indicate temperature dependency indicating that true stress optical behavior is not obeyed due to proximity of these temperatures to the glass transition temperature. (They are in between Tg and so-called Tll (liquid liquid transition). (Fig 3a).

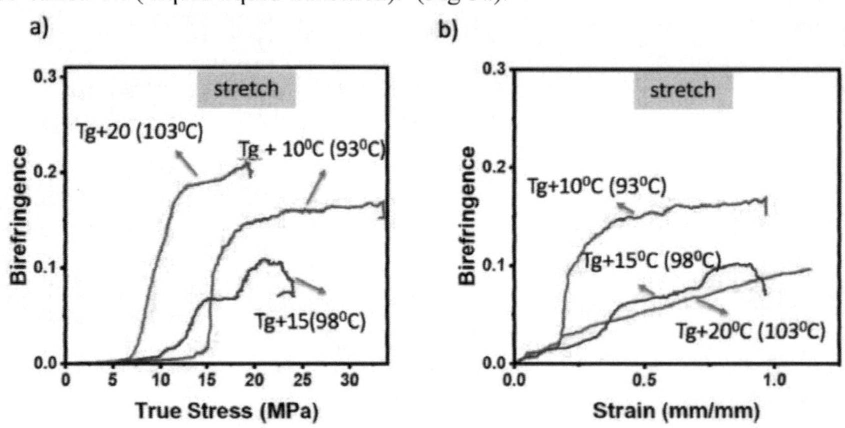

FIGURE 3. (a) Birefringence- True Stress **(b)** Birefringence-True Strain graphs of 50/50 (PEI/PBT) samples in different temperatures.

Fig 3b shows the strain optical behavior of PEI/PBT samples as a function of temperature. The behavior is non linear particularly at low strains and increase of temperature decreases the rate of increase (the slope) as the relaxation processes become increasingly dominant during stretching.

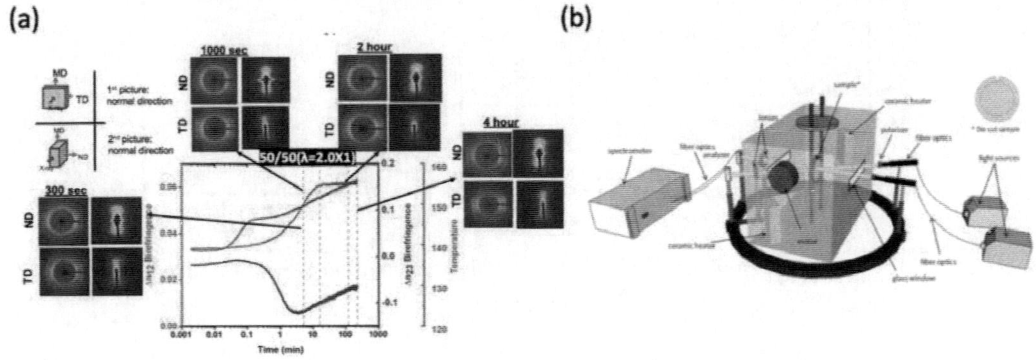

FIGURE 4. (a) time slice annealing study of 50/50 UCW sample 2x1 stretched film (stretching temperature 103 °C , annealed @ 155°C) **(b)** multistage heat setting set up

After stretching heat setting is applied to enhance the thermal stability of the stretched films. Multi angle custom made constrained annealing machine as it is shown in Fig 4b. used to detect the optical changes during annealing. Sample is mounted in a circular holder that constrains the film (to prevent shrinkage), and later inserted into a heated chamber to detect the changes. Loading and unloading of the sample was accomplished very quickly. In plane birefringence calculated by using Eq 2. and out of plane birefringence calculated by using;

$$\Delta n_{23}(t) = -\frac{1}{d(t)} \frac{\left[R_0(t) - R_{45}(t)\left(1 - \frac{Sin^2 45}{n^2}\right)^{0.5}\right]}{\frac{Sin^2 45}{n^2}} \qquad (5)$$

R_o zero degree and R_{45} degree retardation values and average refractive index n. Fig 4a. In plane birefringence increases first and subsequently indicate a relaxation followed by further increase in birefringence. Initial stages of this process is nonisotherm shows the WAXS diffraction patterns on samples captured at indicated times by rapid freezing at different time-slicing stages during the annealing of the 2.0 x 1 UCW sample. As it is shown in the figures highest birefringence changes were observed in the first 10 minutes. With the increase of annealing time crystalline intensification and sharpness increases. As it is shown in Fig 5a. polyimides have dielectric constants higher than BOPP due orientational polarizability of the imide functional groups[3] with the annealing dielectric constant is not changing much however annealing lowers down the dielectric loss (Fig 5b).

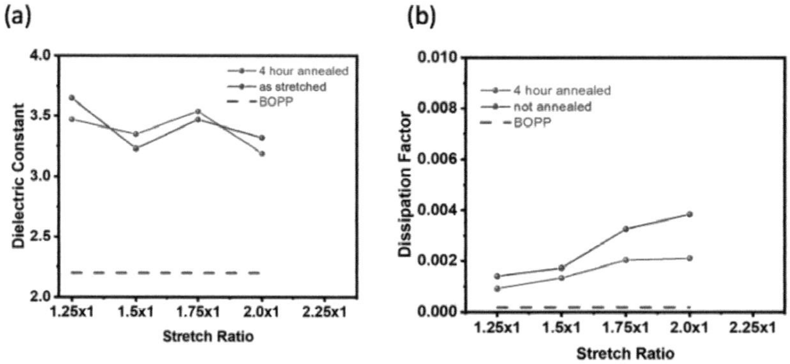

FIGURE 5. (a) Dielectric constant and **(b)** dielectric loss change with stretching and annealing (stretching temperature for all films: 103 °C , annealed @ 155°C)

CONCLUSION

Increase of PEI concentration lead to increase of glass transition temperature as these films are melt miscible. The increase of PEI concentration suppresses thermal crystallization of the films allowing them to be quenched into glassy state opening up a processing window between Tg and Tcc where they behave rubbery and where thermal crystallization rates are low. Rubbery state deformation lead to orientation development . Crystalline phase of PBT exhibit a great deal of crystal distortion. The heat setting process increases the crystallinity of the stretched films.

REFERENCES

1. A. Arsoy, Y. Liu, B.K. Johnson, P.F. Ribeiro, and M.L. Crow, Proc. IEEE **89**, 1744 (2002).
2. E. Unsal, B. Yalcin, I. Yilgor, E. Yilgor, and M. Cakmak, Polymer (Guildf). **50**, 4644 (2009).
3. R. Ma, A.F. Baldwin, C. Wang, I. Offenbach, M. Cakmak, R. Ramprasad, and G.A. Sotzing, ACS Appl. Mater. Interfaces **6**, 10445 (2014).
4. J. Choi, T. Serhatkulu, and M. Cakmak, Polym. (United Kingdom) **97**, 569 (2016).

AUTHOR INDEX

AUTHOR INDEX

AUTHOR INDEX

AUTHOR INDEX